# contents

# technical drawing

## Books by the Authors

Basic Technical Drawing, 2nd revised ed., by H. C. Spencer and J. T. Dygdon (Macmillan, Inc., 1968)

Basic Technical Drawing Problems, Series 1, by H. C. Spencer and J. T. Dygdon (Macmillan, Inc., 1972)

Descriptive Geometry, 4th ed., by E. G. Paré, R. O. Loving, and I. L. Hill (Macmillan, Inc., 1971)

Descriptive Geometry Worksheets, Series A, 3rd ed., by E. G. Paré, R. O. Loving, and I. L. Hill (Macmillan, Inc., 1968)

Descriptive Geometry Worksheets, Series B, 3rd ed., by E. G. Paré, R. O. Loving, and I. L. Hill (Macmillan, Inc., 1972)

Descriptive Geometry Worksheets, Series C, 2nd ed., by E. G. Paré, R. O. Loving, and I. L. Hill (Macmillan, Inc., 1967)

Engineering Graphics by F. E. Giesecke, A. Mitchell, H. C. Spencer, I. L. Hill, and R. O. Loving (Macmillan, Inc., 1969)

Engineering Graphics Problems, Series I, by H. C. Spencer, I. L. Hill, and R. O. Loving (Macmillan, Inc., 1969)

Technical Drawing, 6th ed., by F. E. Giesecke, A. Mitchell, H. C. Spencer, and I. L. Hill (Macmillan, Inc., 1974)

Technical Drawing Problems, Series 1, 4th ed., by F. E. Giesecke, A. Mitchell, H. C. Spencer, and I. L. Hill (Macmillan, Inc., 1967)

Technical Drawing Problems, Series 2, 3rd ed., by H. C. Spencer and I. L. Hill (Macmillan, Inc., 1974)

Technical Drawing Problems, Series 3, 2nd ed., by H. C. Spencer and I. L. Hill (Macmillan, Inc., 1974)

*Courtesy Oldsmobile Division, General Motors Corp.*

## Frederick E. Giesecke

M.E., B.S. in ARCH., C.E., Ph.D., Late Professor Emeritus of Drawing, Texas A&M University

## Alva Mitchell

B.C.E., Late Professor Emeritus of Engineering Drawing, Texas A&M University

## Henry Cecil Spencer

A.B., B.S. in ARCH., M.S., Late Professor Emeritus of Technical Drawing, Formerly Director of Department, Illinois Institute of Technology

## Ivan Leroy Hill

B.S., M.S., Professor of Engineering Graphics, Chairman of Department, Illinois Institute of Technology

## Sixth Edition

Revised by Henry Cecil Spencer and Ivan Leroy Hill

# technical drawing

---

**Macmillan Publishing Co., Inc.**
New York

**Collier Macmillan Publishers**
London

Printed in the United States of America



Macmillan Publishing Co., Inc.
866 Third Avenue, New York, New York 10022

Collier-Macmillan Canada, Ltd.

Library of Congress Cataloging in Publication Data

Giesecke, Frederick Ernest, (date)
Technical drawing.

Bibliography: p.
1. Mechanical drawing. I. Title.
T353.G52 1974 604'.2 72-84877
ISBN 0-02-342700-0

Printing: 2 3 4 5 6 7 8 Year: 4 5 6 7 8 9

# preface

This book is intended as a class text and reference book in technical drawing. It contains a great number of problems covering every phase of the subject, and it constitutes a complete teaching unit in itself. In addition to the problems in the text, three complete workbooks have been prepared especially for use with this text: *Technical Drawing Problems, Series 1,* by Giesecke, Mitchell, Spencer, and Hill; *Technical Drawing Problems, Series 2,* by Spencer and Hill; and *Technical Drawing Problems, Series 3,* by Spencer and Hill. Thus, there are available four alternate sources of problems, and problem assignments may be varied easily from year to year. In general, it is expected that the teacher who uses this text together with one of the three workbooks will supplement the workbook sheets by assignments from the text, to be drawn on blank paper. Many of the text problems are designed for $8\frac{1}{2}'' \times 11''$ sheets, the same size as the workbook sheets, for easier filing.

The extensive use of this text during the past forty years in college classes, technical schools, and industrial drafting rooms has encouraged the authors to continue with the original aim, which was and still is to prepare a book that *teaches the language of the engineer,* and to keep it in step with the latest developments in industry. The idea has been to illustrate and explain basic principles from the standpoint of the student—that is, to present each principle so clearly that the student is certain to understand it, and to make the text interesting enough to encourage the student to read and study on his own initiative. By this means we hope to free the teacher from the repetitive labor of teaching every student individually those things that the textbook can teach and to permit him to give his attention to students having real difficulties.

Our purpose in preparing this new edition is not simply to enlarge the book, although this has been done to some extent, but primarily to bring the book completely up to date with the latest trends in engineering education and with the newest developments in industry, especially with the various sections of ANSI Y14 *American National Standard Drafting Manual.* The increased educational emphasis on the design function of the engineer is reflected throughout the text and especially in Chapter 14, "Design and Working Drawings." Much new material has been added to this chapter to help the student better to understand the fundamentals of the design process.

Many of the problems and illustrations have been redrawn or revised to bring them completely up to date. With regard to dimensioning, a large number of drawings have been converted to the decimal-inch system, now that it has come into extensive use in industry. Many problems in Chapter 14 also present an opportunity for the student to convert the dimensions to the metric system.

An important objective has been to maintain and, where possible, to improve the quality of the drafting in the illustrations and problems. It is logical that in a drawing book the drawings are more important than anything else.

An outstanding feature is the emphasis on technical sketching throughout the text as well as in the complete chapter given early in the book. This chapter is unique in integrating the basic concepts of views with freehand rendering so that the subject of multiview drawing can be introduced through the medium of sketches.

The chapters on dimensioning and tolerancing have been revised to bring them into conformity with the latest American National Standard, *Dimensioning and Tolerancing for Engineering Drawings.*

The chapter on electronic diagrams has been revised by Professor R. O. Loving of the Illinois Institute of Technology. The chapters on structural drawing and topographic drawing and mapping have been revised by Professor Emeritus E. I. Fiesenheiser of the Illinois Institute of Technology. The chapters on graphs, alignment charts, empirical equations, and graphical mathematics have been revised by Mr. E. J. Mysiak, Mechanical Engineer, International Electro-Magnetics, Inc. The chapter on gearing and cams has been revised by Professor Emeritus B. L. Wellman of Worcester Polytechnic Institute; and the chapter on shop processes has been revised by Professors J. George H. Thompson and John Gilbert McGuire of Texas A&M University. We also are indebted to Professor H. E. Grant for certain problem material and valuable suggestions and to Mr. David Wong for his assistance with the revision of many of the illustrations.

Every effort has been made to bring the book completely abreast of the many technological developments that have occurred in the past few years. Through the cooperation of leading engineers and manufacturers, we have been able to include many commercial drawings of value in developing the subject. We wish to express our thanks to these persons and firms, and others too numerous to mention here, who have contributed to the production of this book.

Students, teachers, engineers, and draftsmen are invited to write concerning any questions that may arise. All comments or suggestions will be appreciated.

Henry Cecil Spencer

Ivan Leroy Hill
Illinois Institute
of Technology
Chicago, Illinois
60616

# contents

# 1

# the graphic language and design

**1.1 Evolution of Design.** The old saying "necessity is the mother of invention" continues to be true, and a new machine, structure, system, or device is the result of that need. If the new device, machine, system, or gadget is really needed, people will buy it, providing it does not cost too much. Then the questions arise: Is there a wide potential market? Can this device or system be made available at a price that people are willing to pay? If these questions can be answered satisfactorily, then the inventor, designer, or the officials of a company may elect to go ahead with developing production and marketing plans for the new device or system.

A new machine, structure, or system, or an improvement thereof, must exist in the mind of the engineer or designer before it can become a reality. This original conception is usually placed on paper and communicated to others by the way of the *graphic language* in the form of freehand *idea sketches,* Figs. 1.1 and 5.1. These idea sketches are then followed by other sketches, such as *computation sketches,* for developing the idea more fully.

**1.2 The Young Engineer.** The engineer or designer must be able to create idea sketches, calculate stresses, analyze motions, size the parts, specify materials and production methods, make design layouts, and supervise the preparation of drawings and specifications that will control the numerous details of production, assembly, and maintenance of the product. In order to perform or supervise these many tasks, the engineer makes liberal use of freehand drawings. He must be able to

Fig. 1.1 **Edison's Phonograph.***

record and communicate quickly his ideas to his associates and support personnel. Facility in freehand sketching, Chapter 5, or the ability to work with computer-controlled drawing techniques, §14.12, requires extensive training in drawing with instruments and a thorough knowledge of the graphic language.

A typical engineering and design department is shown in Fig. 1.2. Many of the men have considerable training and experience; others are recent graduates who are gaining experience. There is much to be learned on the job and it is necessary for the inexperienced man to start at a low level and advance to more responsibility as he gains experience. Very much to the point is the following statement by the chief engineer of a large corporation:†

"Many of the engineering students whom we interview have the impression that if they go to work at the drafting board, they will be only draftsmen doing routine work. This impression is completely erroneous, because all of our engineers work at the board at least occasionally. Actually, drawing is only one phase of responsibility which includes site evaluations, engineering calculations, cost estimates, preliminary layouts, engineering specifications, equipment selection, complete drawings (with the help of draftsmen), and follow-up on construction and installation.

"Our policy is to promote from within, and it is our normal practice to hire engineers at the time they finish school, and to give them the opportunity for growth and development by diversified experience. These newly hired engineers without experience are assigned to productive work at a level which their education and experience qualify them to handle successfully. The immediate requirement is for the young engineer to obtain practical engineering experience, and to learn our equipment and processes. In design work, these initial assignments are on engineering details in any one of several fields of engineering study (structural, mechanical, electrical, etc.). Our experience has shown that it is not wise to give a newly graduated engineer without experience a problem in advanced engineering, such as creative design, on the assumption that he can make quick sketches or layouts and then have them detailed by someone else. Rather than start a young engineer at an advanced responsibility level where he may fail or make costly mistakes, we assign him initially to work which requires him to make complete and accurate detail drawings, and his assignments become increasingly complex as he demonstrates the ability to do work of an advanced engineering caliber. If he demonstrates the capacity to assume design responsibility, he is given direction of other engineers with less experience who in turn do detailed engineering for him."

## 1.3 The Graphic Language.

Many of the troubles of the world are caused by the fact that the various peoples do not understand one another. The infinite number of languages and dialects that contributed to this condition resulted from a lack of intercommunication of peoples widely separated in various parts of the world. Even today when communication is so greatly improved, the progress toward a world language is painfully slow—so slow,

*Original sketch of Thomas A. Edison's first conception of the phonograph; reproduced by special permission of Mrs. Edison.

†C. G. R. Johnson, Kimberly-Clark Corp.

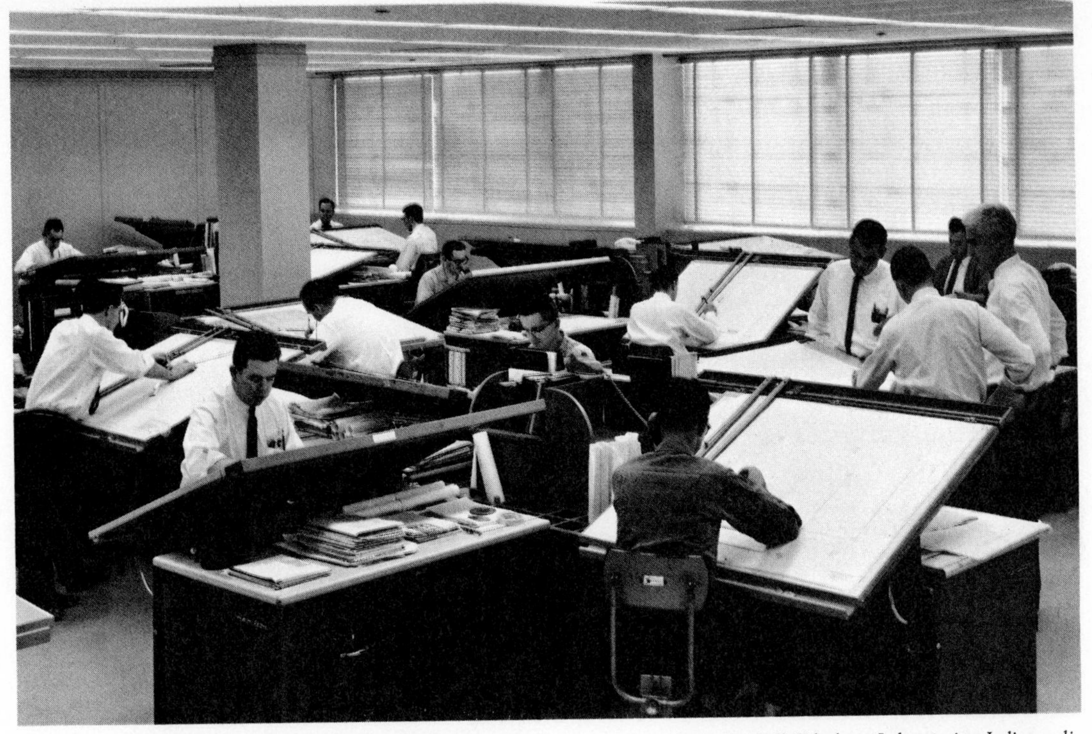

Fig. 1.2 **Section of an Engineering Department.** *Courtesy Bell Telephone Laboratories, Indianapolis*

indeed, that we cannot foresee the time when it will be a fact.

Although men have not been able to get together on a world language of words and sentences, there has actually been a universal language in use since the earliest times: the *graphic language.* The idea of communicating thoughts from one person to another by means of *pictures* occurred to even the earliest cave dwellers, and we have examples still in existence to prove it. These earliest men communicated orally, undoubtedly by grunts and other elementary sounds, and when they wished to *record* an idea, they made *pictures* upon skins, stones, walls of caves or whatever materials they could find. The earliest forms of writing were through picture forms, such as the Egyptian hieroglyphics, Fig. 1.3. Later these forms were simplified and became the abstract symbols used in our writing today. Thus, even the letter characters in present word languages have their basis in drawings. See §3.1.

A drawing is a *graphic representation* of a real thing, an idea, or a proposed design for construction later. Drawings may take many forms, but the graphic method of representation is a basic natural form of communication of ideas that is universal and timeless in character.

Fig. 1.3 **Egyptian Hieroglyphics.**

## 1.4 Two Types of Drawings.

Man has developed graphic representation along two distinct lines, according to his purpose: (1) Artistic and (2) Technical.

From the beginning of time, artists have used drawings to express aesthetic, philosophic, or other abstract ideas. In ancient times nearly everybody was illiterate. There was no printing, and hence no newspapers or books as we know them today. Books were hand lettered on papyrus or parchment and were not available to the general public. People

Fig. 1.4 **The Circus Maximus in Rome.** *The Bettmann Archive, Inc.*

learned by listening to their superiors and by looking at sculptures, pictures, or drawings in public places. Everybody could understand pictures and they were a principal source of information. In our museums and in ruins of antiquity are thousands of examples of story-telling or teaching by means of drawings. If someone wished to preserve his own image or a friend's, he had to have the job done in stone, in bronze, in oil on canvas, or in some other art medium—there were no photographs. The artist was not just an artist in the aesthetic sense—he was a teacher or philosopher, a means of expression and communication.

The other line along which drawing has developed has been the technical. From the beginning of recorded history, man has used drawings to *represent* his design of objects to be built or constructed. Of these earliest drawings no trace remains today, but we definitely know that drawings were used, for man could not have designed and built as he did without using fairly accurate drawings. In the Bible the statement is made that Solomon's Temple was "built of stone made ready before it was brought thither."* Each stone and timber was carved or hewn into shape, brought to the site, and fitted into place. It is evident that accurate drawings were used, showing the exact shapes and sizes of the component parts for the design of the temple.

Moreover, we can see today the ruins of fine old buildings, aqueducts, bridges, and other structures of antiquity that could not possibly have been erected without carefully made drawings to guide the builders. Many of these structures are still regarded as "Wonders of the World," such as the Temple of Amon at Karnak in ancient Egypt, completed in about 980 B.C., which took seven centuries to construct. In sheer mass of stone, this building exceeded any roofed structure ever built, so far as we know, being 1200 feet long and 350 feet wide at its greatest width. Likewise, the Circus Maximus in Rome was a large structure, Fig. 1.4; according to the historian Pliny, it seated a total of 250,000 spectators.

*1 Kings 6:6.

## 1.5 Earliest Technical Drawings.

Perhaps the earliest known technical drawing in existence is the plan view for a design of a fortress drawn by the Chaldean engineer Gudea and engraved upon a stone tablet, Fig. 1.5. It is remarkable how similar this plan is to those made by architects today, although "drawn" thousands of years before paper was invented.

The first written evidence of the use of technical drawings was in 30 B.C. when the Roman architect Vitruvius wrote a treatise on architecture in which he said, "The architect must be skillful with the pencil and have a knowledge of drawing so that he readily can make the drawings required to show the appearance of the work he proposes to construct." He went on to discuss the use of the rule and compasses in geometric constructions, in drawing the plan and elevation views of a building, and in drawing perspectives.

In the museums we can see actual specimens of early drawing instruments. Compasses were made of bronze and were about the same size as those used today. As shown in Fig. 1.6, the old compass resembled the dividers of today. Pens were cut from reeds.

The theory of projections of objects upon imaginary planes of projection (to obtain *views*, §6.1) apparently was not developed until the early part of the fifteenth century—by the Italian architects Alberti, Brunelleschi, and others. It is well known that Leonardo da Vinci

Fig. 1.5 **Plan of a Fortress.** This stone tablet is part of a statue now in the Louvre, in Paris, and is classified in the earliest period of Chaldean art, about 4000 B.C. *From Transactions ASCE, May 1891*

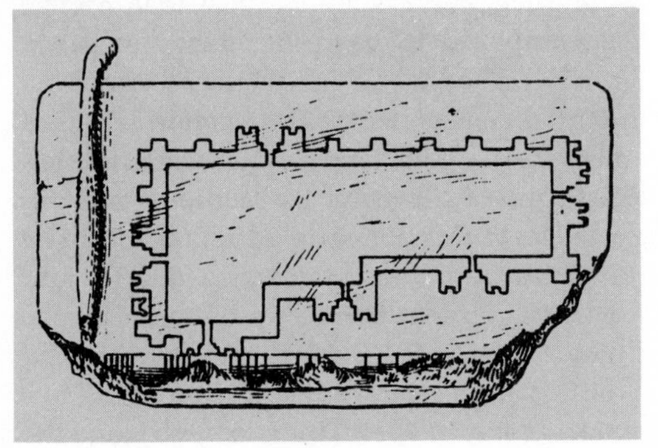

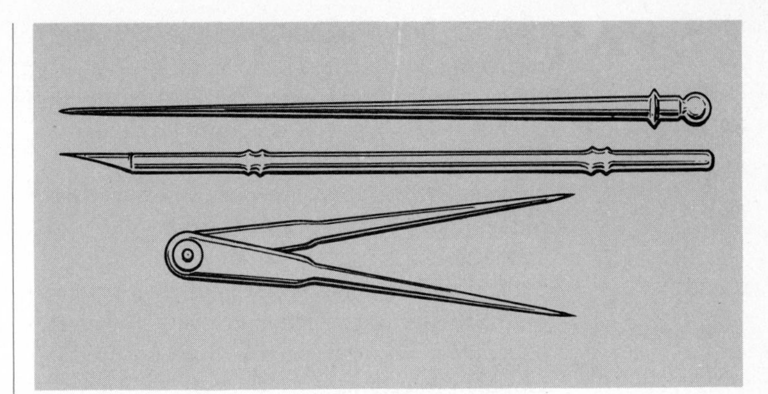

*From Historical Note on Drawing Instruments, published by V & E Manufacturing Co.*

Fig. 1.6 **Roman Stylus, Pen, and Compass.**

Fig. 1.7 **An Arsenal, by Leonardo da Vinci.** *The Bettmann Archive, Inc.*

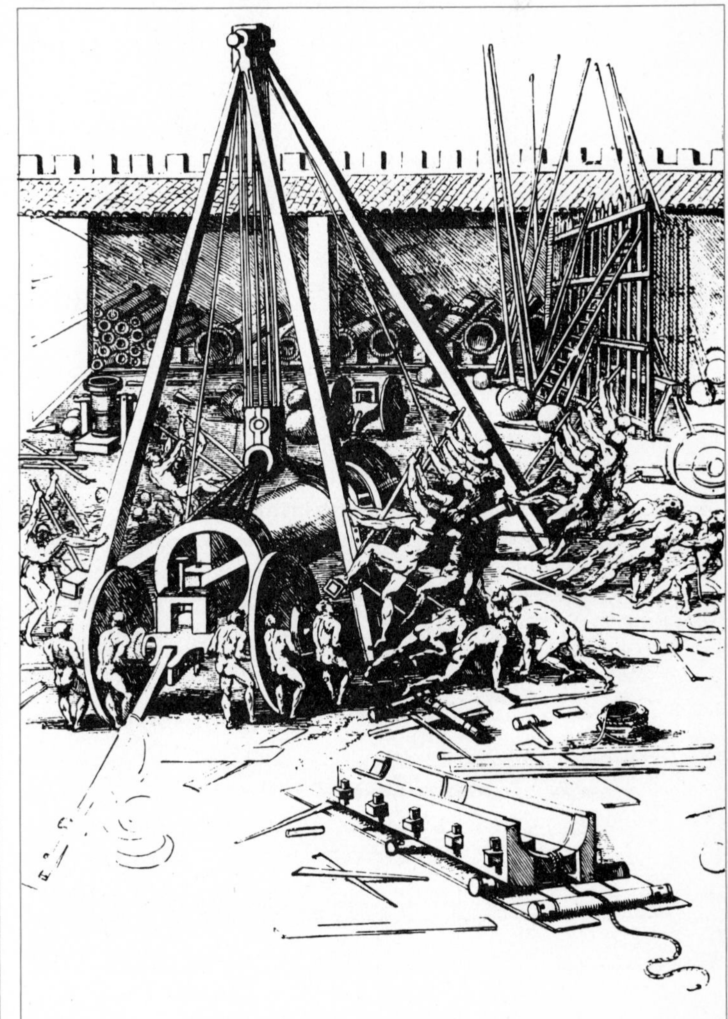

used drawings to record and transmit to others his ideas and designs for mechanical constructions, and many of these drawings are in existence today, Fig. 1.7. It is not clear whether Leonardo ever made mechanical drawings showing orthographic views as we know them today, but it is probable that he did. Leonardo's treatise on painting, published in 1651, is regarded as the first book ever printed on the theory of projection drawing; however, its subject was perspective and not orthographic projection.

The compass of the Romans remained very much the same during Leonardo's time, Fig. 1.8. Circles were still scratched with metal points, since graphite pencils were not invented until the eighteenth century, when the firm of Faber was established in Nuremburg, Germany. By the seventh century reed pens had been replaced by quills made from bird feathers, usually those of geese (hence: goose-quill pens).

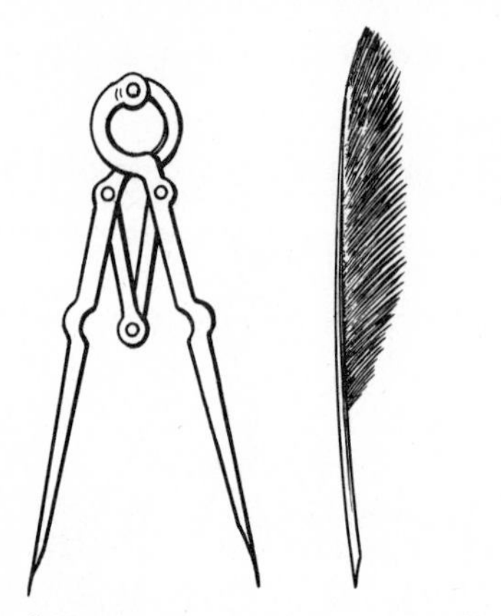

*From Historical Note on Drawing Instruments, published by V & E Manufacturing Co.*

Fig. 1.8 **Compass and Pen, Renaissance Period.** Compass after a drawing by Leonardo da Vinci.

Fig. 1.9 **George Washington's Drawing Instruments.**

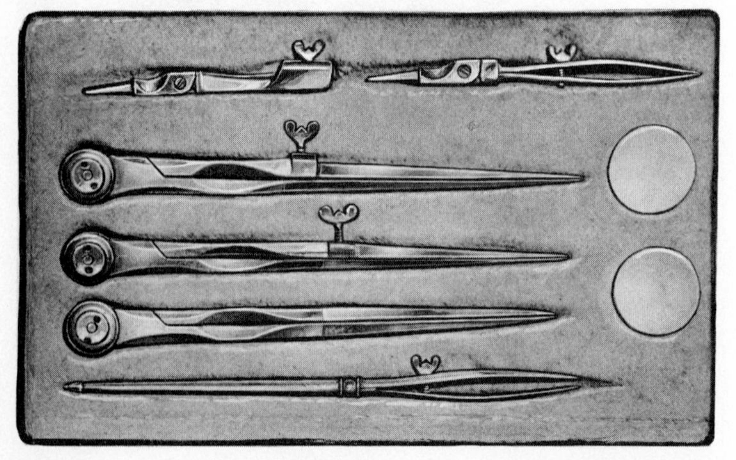

*From Historical Note on Drawing Instruments, published by V & E Manufacturing Co.*

The scriber-type compass gave way to the compass with a graphite lead shortly after graphite pencils were developed. At Mount Vernon we can see the drawing instruments used by the great civil engineer, George Washington, and bearing the date 1749. This set, Fig. 1.9, is very similar to the conventional drawing instruments used today, consisting of a divider and a compass with pencil and pen attachments plus a ruling pen with parallel blades similar to the modern pens.

**1.6 Early Descriptive Geometry.** The beginnings of descriptive geometry are associated with the problems encountered in designs for building construction and military fortifications of France in the eighteenth century. Gaspard Monge (1746–1818) is considered the "inventor" of descriptive geometry, although his efforts were preceded by publications on stereotomy, architecture, and perspective in which many of the principles were used. It was while he was a professor at the Polytechnic School in France near the close of the eighteenth century that Monge developed the principles of projection that are today the basis of our technical drawing. These principles of descriptive geometry were soon recognized to be of such military importance that Monge was compelled to keep his principles secret until 1795, following which they became an important part of technical education in France and Germany and later in the United States. His book, *La Géométrie Descriptive,* is still regarded as the first text to expound the basic principles of projection drawing.

Monge's principles were brought to the United States in 1816 from France by Claude Crozet, a professor at the United States Military Academy at West Point. He published the first text on the subject of descriptive geometry in the English language in 1821. In the years immediately following, these principles became a regular part of early engineering curricula at Rensselaer Polytechnic Institute, Harvard University, Yale University, and others. During the same period, the idea of manufacturing interchangeable parts in the early arms industries was being developed, and the principles of projection drawing were applied to these problems.

## 1.7 Modern Technical Drawing.

Perhaps the first text on technical drawing in this country was *Geometrical Drawing*, published in 1849 by William Minifie, a high school teacher in Baltimore. In 1850 the Alteneder family organized the first drawing instrument manufacturing company in this country (Theo. Alteneder & Sons, Philadelphia). In 1876 the blueprint process was introduced in this country at the Philadelphia Centennial Exposition. Up to this time the graphic language was more or less an art, characterized by fine-line drawings made to resemble copper-plate engraving, by the use of shade lines, and by the use of water color "washes." These techniques became unnecessary after the introduction of blueprinting, and drawings gradually were made plainer to obtain best results from this method of reproduction. This was the beginning of modern technical drawing. The graphic language now became a relatively exact method of representation, and the building of a working model as a regular preliminary to construction became unnecessary.

Up to about the turn of the nineteenth century throughout the world, drawings were generally made in what is called *first-angle projection*, §6.38, in which the top view was placed *under* the front view, the left-side view was placed at the *right* of the front view, etc. At this time in the United States, after a considerable period of argument pro and con, practice gradually settled on the present *third-angle projection* in which the views are situated in what we regard as their more logical or natural positions. Today, third-angle projection is standard in the United States, but first-angle projection is still used throughout much of the world.

During the early part of the twentieth century, many books on the subject were published in which the graphic language was analyzed and explained in connection with its rapidly changing engineering design and industrial applications. Many of these writers were not satisfied with the term "mechanical drawing" because they were aware of the fact that technical drawing was really a *graphic language*. Anthony's *An Introduction to the Graphic Language*, French's *Engineering Drawing*, and this text, *Technical Drawing*, were all written with this point of view.

## 1.8 Drafting Standards.

In all of the above books there has been a definite tendency to standardize the characters of the graphic language, to eliminate its provincialisms and dialects, and to give industry, engineering, and science a uniform, effective graphic language. Of prime importance in this movement has been the work of the American National Standards Institute (ANSI), with the American Society for Engineering Education, the Society of Automotive Engineers, and the American Society of Mechanical Engineers. As sponsors they have prepared the *American National Standard Drafting Manual—Y14*, which will contain twenty-seven or more separate sections when completed. These sections are published as approved standards as they are completed. See Appendix 1.

These sections outline the most important idioms and usages in a form that is acceptable to the majority, and are considered the most authoritative guide to uniform drafting practices in this country today. The Y14 Standard gives the *characters* of the graphic language, and it remains for the textbooks to explain the *grammar* and the *penmanship*.

## 1.9 Definitions.

After surveying briefly the historical development of the graphic language and before starting a serious study of theory and applications, the definitions of a few terms should be considered.

*Descriptive Geometry* is the grammar of the graphic language; it is the three-dimensional geometry forming the background for the practical applications of the language and through which many of its problems may be solved graphically.

*Instrumental* or *Mechanical Drawing* should be applied only to a drawing made with drawing instruments. Mechanical Drawing has been used to denote all industrial drawings, which is unfortunate not only because such drawings are not always mechanically drawn, but also because it tends to belittle the broad scope of the graphic language by naming it superficially for its principal mode of execution.

*Engineering Drawing* and *Engineering Drafting* are broad terms widely used to denote the graphic language. However, since the language is not used by engineers only, but also by a much larger group of people in diverse fields who are concerned with technical work or with industrial production, the term is still not broad enough.

*Technical Drawing* is a broad term that adequately suggests the scope of the graphic language. It is rightly applied to any drawing used to express technical ideas. This term has been used by various writers since Monge's time at least and is still widely used, mostly in Europe.

*Engineering Graphics* or *Engineering Design Graphics* is generally applied to drawings for technical use and has come to mean that part of technical drawing which is concerned with the graphical representation of designs and specifications for physical objects and data relationships as used in engineering and science.

*Technical sketching* is the freehand expression of the graphic language, while *mechanical drawing* is the instrumental expression of it. Technical sketching is a most valuable tool for the engineer and others engaged in technical work, because through it most technical ideas can be expressed quickly and effectively without the use of special equipment.

*Blueprint reading* is the term applied to the "reading" of the language from drawings made by others. Actually, the blueprint process is only one of many forms by which drawings are reproduced today (see Chapter 15), but the term "blueprint reading" has been accepted through usage to mean the interpretation of all ideas expressed on technical drawings, whether the drawings are blueprints or not.

*Computer Graphics* is the application of conventional computer techniques with the aid of one of many graphic data processing systems available to the analysis, modification, and the finalizing of a graphical solution.

**1.10 What Engineering and Science Students Should Know.** The development of technical knowledge from the dawn of history has been accompanied, and to a large extent made possible, by a corresponding graphic language. Today the intimate connection between engineering and science and the universal graphic language is more vital than ever before, and the engineer or scientist who is ignorant of or deficient in the principal mode of expression in his technical field is *professionally illiterate.* That this is true is shown by the fact that training in the application of technical drawing is required in virtually every engineering school in the world.

The old days of fine-line drawings and of shading and "washes" are gone forever; no artistic talent is necessary for the modern technical student to learn the fundamentals of the graphic language. For its mastery he needs precisely the aptitudes and abilities he will need to learn the science and engineering courses that he studies concurrently and later. The student who does poorly in the graphic language courses is like to do poorly in his other technical courses.

The well-trained engineer, scientist, or technician must be able to make correct graphical representations of engineering structures, designs, and data relationships. This means that he must understand the fundamental principles, or the *grammar* of the language, and must be able to execute the work with reasonable skill, which is *penmanship.*

Drawing students often try to excuse themselves for inferior results (usually caused by

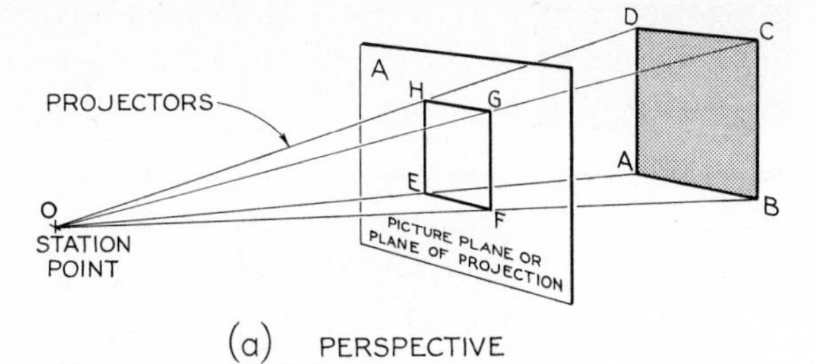


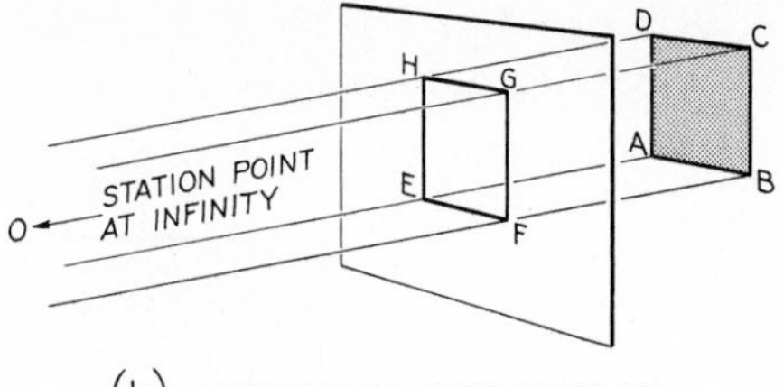


Fig. 1.10 **Projections.**

lack of application) by arguing that after graduation they do not expect to do any drafting at all; they expect to have others make any needed drawings under their direction. Such a student presumptuously pictures himself, immediately after graduation, as the accomplished engineer concerned with bigger things and forgets that his first assignment may well be "on the board" and that he will be the one who will make the drawings under the direction of a really experienced engineer. Though he may not realize it, entering the engineering profession via the drawing board is fortunate for him, since it affords an unexcelled opportunity to learn the ropes and prepare him to supervise others.

Even if the young engineer has not been too successful in developing a skillful penmanship in the graphic language, he still will have great use for its grammar, since the ability to *read* a drawing is of utmost importance, and he will need this ability throughout his professional life. See §14.1.

Furthermore, the young engineer is apt to overlook the fact that in practically all the subsequent technical courses he will take in college, he will encounter technical drawings in most of his textbooks, and he will be called upon by his instructors to supplement his calculations with mechanical drawings or sketches. Thus, a mastery of his course in technical drawing will aid him materially not only in professional practice after graduation but more immediately in his other technical courses, and it will have a definite bearing on his scholastic progress.

Besides the direct values to be obtained from a serious study of the graphic language, there are a number of very important training values which, though they may be considered by-products, are fully as essential as the language itself. Many a student learns for the first time in his drawing course the meaning of *neatness, speed,* and *accuracy*—basic habits that every successful engineer and scientist must have or acquire.

All authorities agree that the ability to *think in three dimensions* is one of the most important requisites of the successful scientist and engineer. This training to visualize objects in space, to use the constructive imagination, is one of the principal values to be obtained from a study of the graphic language. The ability to *visualize* is possessed in an outstanding degree by persons of extraordinary creative ability. It is difficult to think of Edison, De Forest, or Einstein as being deficient in constructive imagination.

With the increase in technological development and the consequent crowding of drawing courses by the other engineering and science courses in our colleges, it is doubly necessary for the engineering or science student to make the most of the limited time devoted to the language of his profession, to the end that he will not be professionally illiterate, but will possess an ability to express himself quickly and accurately through the correct use of the graphic language.

## 1.11 Projections.

Behind every drawing of an object is a space relationship involving four imaginary things: the *observer's eye* or *station point,* the *object,* the *plane* or *planes of projection,* and the *projectors.** For example, in Fig. 1.10 (a) the drawing EFGH is the projection on the plane of projection A of the square

*Also called *visual rays* and *lines of sight.*

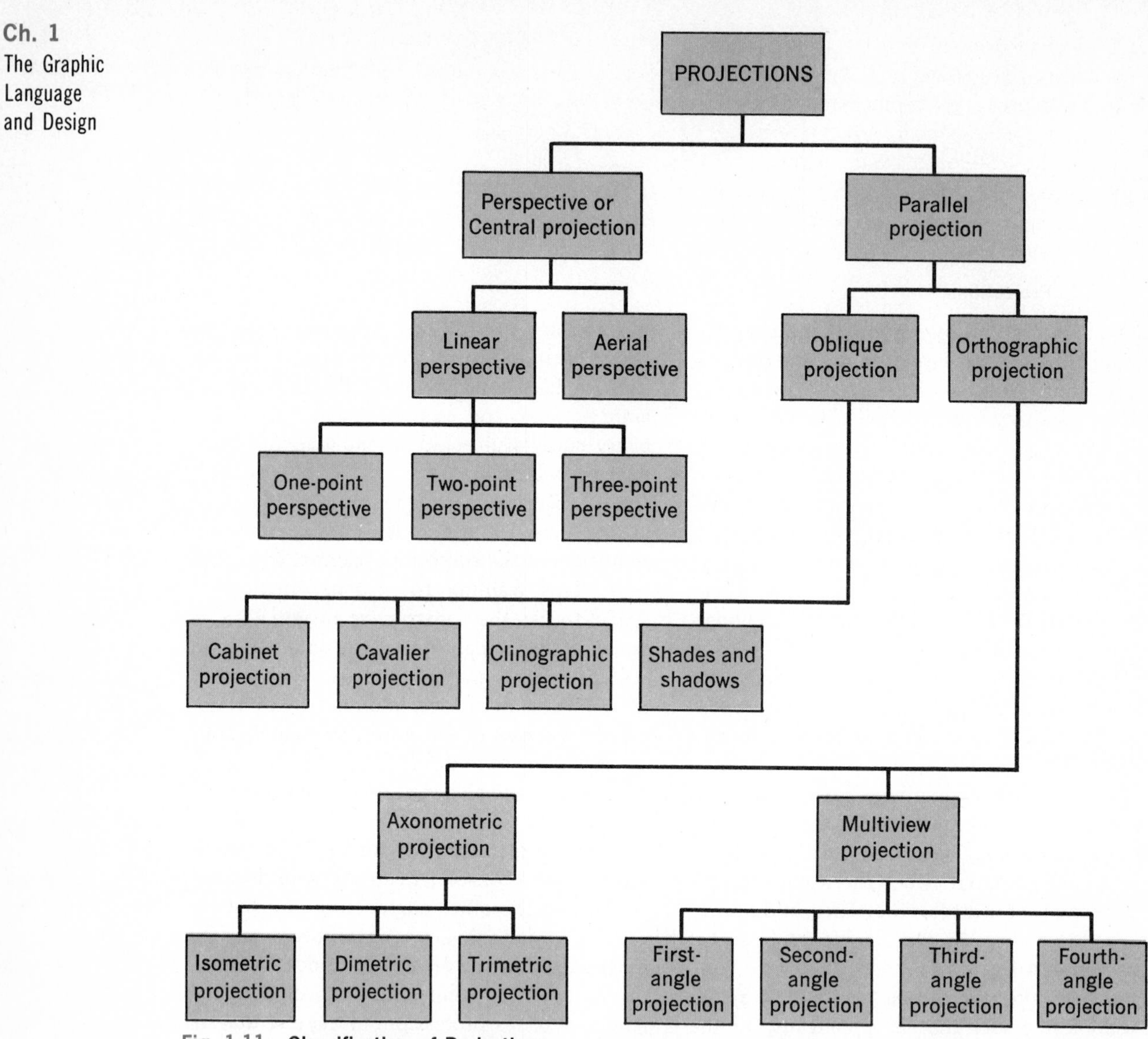


Fig. 1.11 **Classification of Projections.**

Fig. 1.12 **Classification by Projectors.**

| Classes of Projection | Distance from Observer to Plane of Projection | Direction of Projectors |
|---|---|---|
| Perspective | Finite | Radiating from station point |
| Parallel | Infinite | Parallel to each other |
| Oblique | Infinite | Parallel to each other and oblique to plane of projection |
| Orthographic | Infinite | Perpendicular to plane of projection |
| Axonometric | Infinite | Perpendicular to plane of projection |
| Multiview | Infinite | Perpendicular to planes of projection |

ABCD as viewed by an observer whose eye is at the point O. The projection or drawing upon the plane is produced by the piercing points of the projectors in the plane of projection. In this case, where the observer is relatively close to the object and the projectors form a "cone" of projectors, the resulting projection is known as a *perspective.*

If the observer's eye is imagined as infinitely distant from the object and the plane of projection, the projectors will be parallel, as shown in Fig. 1.10 (b); hence, this type of projection is known as a *parallel projection.* If the projectors, in addition to being parallel to each other, are perpendicular (normal) to the plane of projection, the result is an *orthographic** *projection.* If they are parallel to each other but oblique to the plane of projection, the result is an *oblique projection.*

These two main types of projection—perspective and parallel projection—are further broken down into many subtypes, as shown in Fig. 1.11, and will be treated at length in the various chapters that follow.

A classification of the main types of projection according to their projectors is shown in Fig. 1.12.

* *Orthographic* means written or drawn at right angles.

# 2

# instrumental drawing

*Details are trifles, but trifles make perfection, and perfection is no trifle.*

BEN FRANKLIN

**2.1 Typical Equipment.** The principal items of equipment needed by students in technical schools and by draftsmen and designers in professional practice are shown in Fig. 2.1. To secure the most satisfactory results, the drawing equipment should be of high grade. When drawing instruments (item 3) are to be purchased, the advice of an experienced draftsman or designer, or of a reliable dealer,* should be sought because it is difficult for beginners to distinguish high-grade instruments from those that are inferior.

A complete list of equipment, which should provide a satisfactory selection for students of technical drawing, follows. The numbers refer to Fig. 2.1:

1. Drawing board (approx. 20″ × 24″), drafting table, or desk.
2. T-square (24″, transparent edge), drafting machine, or parallel ruling edge, §§2.59 and 2.60.
3. Set of instruments, §§2.33 and 2.34.
4. 45° triangle (8″ sides).
5. 30° × 60° triangle (10″ long side).
6. Ames Lettering Guide or lettering triangle.
7. Triangular architects scale (or flat mechanical engineers scale), see also Fig. 2.34.
8. Triangular engineering scale (or flat decimal scales), see also Fig. 2.34.

*Keuffel & Esser Co., New York; Eugene Dietzgen Co., New York; Charles Bruning Co., Mt. Prospect, Ill.; Frederick Post Co., Chicago; V & E Manufacturing Co., Pasadena, Calif.; and the Gramercy Guild Group, Inc., Denver, are some of the larger distributors of this equipment; their products are available through local dealers.

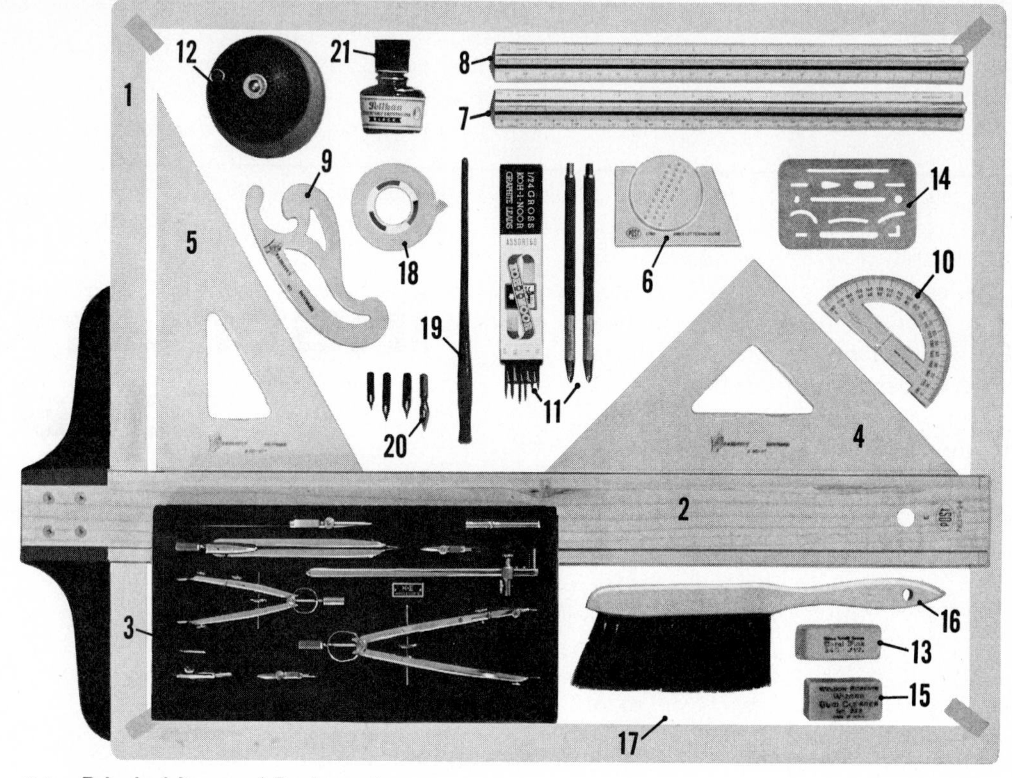


Fig. 2.1 **Principal Items of Equipment.**

9. Irregular curve.
10. Protractor.
11. Mechanical pencils and HB, F, 2H, and 4H to 6H leads or drawing pencils, see also Fig. 2.9.
12. Tru-Point pencil pointer, sandpaper pad, or file.
13. Pencil eraser.
14. Erasing shield.
15. Artgum or other cleaning eraser.
16. Dusting brush.
17. Drawing paper, tracing paper, tracing cloth, or films as required. Backing sheet (drawing paper—preferably white—to be used under drawings and tracings).
18. Drafting tape.

*Optional equipment.*

19. Pen staff.
20. Pen points (Gillott's 303, 404; Hunt's 512; Leonardt's ball-pointed 516F).
21. Drawing ink (black waterproof) and pen wiper.

*The following items (not shown in Fig 2.1) may also be included if necessary:*

Metric scale, Fig. 2.34 (e).
Drop pen, detail pen, proportional dividers, beam compass, and contour pen.
Arkansas oil stone (for sharpening ruling pens).
Pocket knife.
Slide rule.
Dust cloth or cleansing tissue.

**2.2 Objectives in Drafting.** On the following pages the correct methods to be used in instrumental drawing are explained. The student should learn and practice correct manipulation of the drawing instruments so that correct habits may be formed and maintained. Eventually he should draw correctly by habit so that his full attention may be given to the problems at hand. The instructor will insist upon absolutely correct form at all times, making exceptions only in cases of physical disability.

The following are the important objectives the student should strive to attain:

1. Accuracy. No drawing is of maximum usefulness if it is not accurate. The student must learn from the beginning that he cannot be successful in his college career or later in his professional employment if he does not acquire the habit of accuracy in his work.

2. Speed. "Time is money" in industry, and there is no demand for the slow draftsman or engineer. However, speed is not attained by hurrying; it is an unsought by-product of *intelligent and continuous work.* It comes with study and practice, and the fast worker is usually mentally alert.

3. Legibility. The draftsman or engineer should remember that his drawing is a means of communication to others, and that it must be clear and legible in order to serve its purpose well. Care should be given to details, especially to lettering, Chapter 3.

4. Neatness. If a drawing is to be accurate and legible, it must also be clean; therefore, the student should constantly strive to acquire the habit of neatness. Untidy drawings are the result of sloppy and careless methods, §2.13, and will not be accepted by the instructor.

**2.3 Drafting at Home or School.** In the school drafting room, as in the industrial drafting room, the student is expected to give thoughtful and continuous attention to the problems at hand. If he does this, he will not have time to annoy others. He must work in quiet surroundings without distractions. Technical drawing requires headwork. The efficient draftsman sees to it that he has the correct equipment and refrains from borrowing—which is a nuisance to everyone. While the student is drawing, the textbook—his chief source of information—should be available in a convenient position, Fig. 2.2.

When questions arise, first use the index of the text and endeavor to find the answer for yourself. Try to develop self-reliance and initiative. On the other hand, if you really need help, ask your instructor. The student who goes about his work intelligently, with a minimum waste of time, *first* studies the assignment carefully to be sure that he understands the principles involved; *second,* makes sure he has the correct equipment in proper condition (such as sharp pencils); and *third,* makes an effort to dig out answers for himself (the only true education).

One of the principal means of promoting efficiency in drafting is *orderliness.* All needed equipment and materials should be placed in



Fig. 2.2 **Orderliness Promotes Efficiency.**

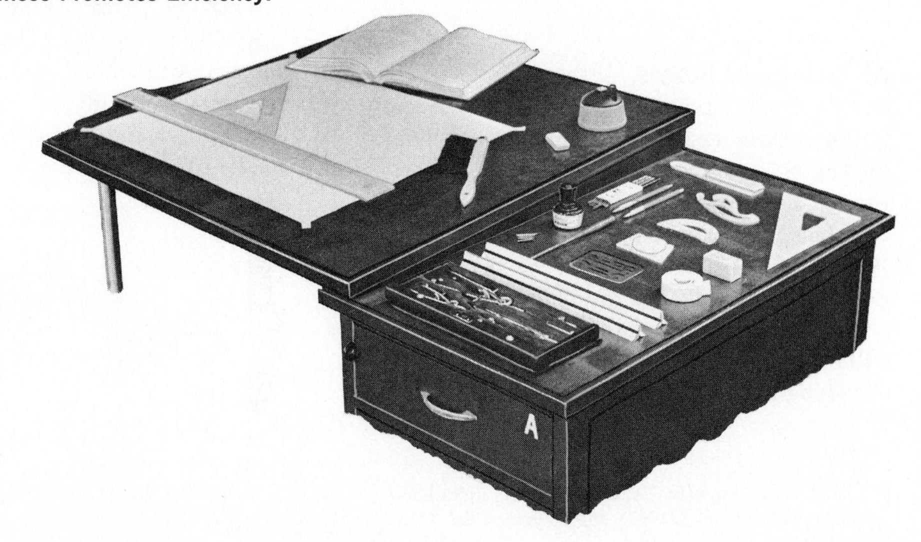

an orderly manner so that everything is in a convenient place and can readily be found when needed, Fig. 2.2. The drawing area should be kept clear of equipment not in direct use. Form the habit of placing each item in a regular place outside the drawing area when it is not being used.

When drawing at home, it is best, if possible, to work in a room by yourself. A book can be placed under the upper portion of the drawing board to give the board a convenient inclination, or the study-table drawer may be pulled out and used to support the drawing board at a slant.

It is best to work in natural north light coming from the left and slightly from the front. Never work on a drawing in direct sunlight or in dim light, as either may be injurious to the eyes. If artificial light is needed, the light source should be such that shadows are not cast where lines are being drawn, and such that there will be as little reflected glare from the paper as possible. Special draftsman's fluorescent lamps are available with adjustable arms so that the light source may be moved to any desired position.* Drafting will not hurt eyes that are in normal condition, but the exacting work will often disclose deficiencies not previously suspected.

LEFT-HANDERS. Place the head of the T-square on the right, and arrange the light source from the right and slightly from the front.

## 2.4 Drawing Boards.

If the left edge of the drafting table top has a true straight edge and if the surface is hard and smooth (such as masonite), a drawing board is unnecessary, provided drafting tape is used to fasten the drawings. It is recommended that a *backing sheet* of heavy drawing paper be placed between the drawing and the table top.

However, in most cases a drawing board will be needed. These vary from 9″ × 12″ (for sketching and field work) up to 48″ × 72″ or

*Dazor Mfg. Corp., St. Louis, Mo. 63110; Luxo Lamp Corp., Port Chester, N.Y. 10573, and Sausalito, Cal. 94965.

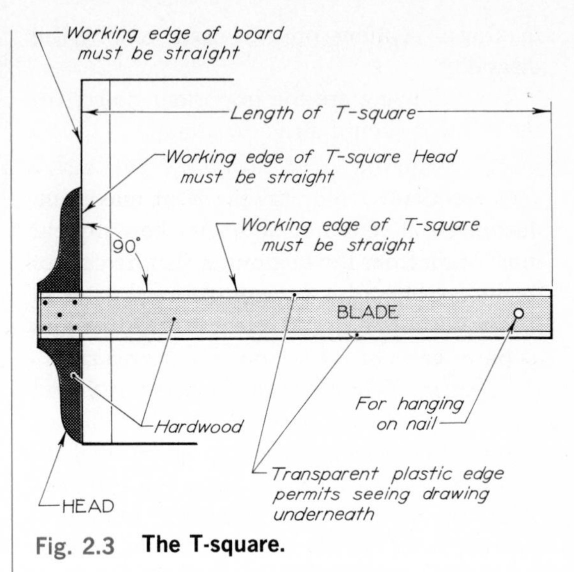


Fig. 2.3 **The T-square.**

larger. The recommended size for students is 20″ × 24″, Fig. 2.1, which will accommodate the largest sheet likely to be used, 17″ × 22″ (see inside back cover).

Drawing boards traditionally have been made of soft woods, such as white pine, so that thumbtacks can be easily pushed down. However, after considerable use the board is likely to be full of objectionable thumbtack holes. Many draftsmen now prefer to use drafting tape, which in turn permits surfaces such as hardwood, masonite, linoleum, or other materials to be used for drawing boards.

The left-hand edge of the board is called the *working edge*, because the T-square head slides against it, Fig. 2.3. This edge must be straight, and you should test the edge with a framing square or with a T-square blade that has been tested and found straight, Fig. 2.4. If the edge of the board is not true, it must

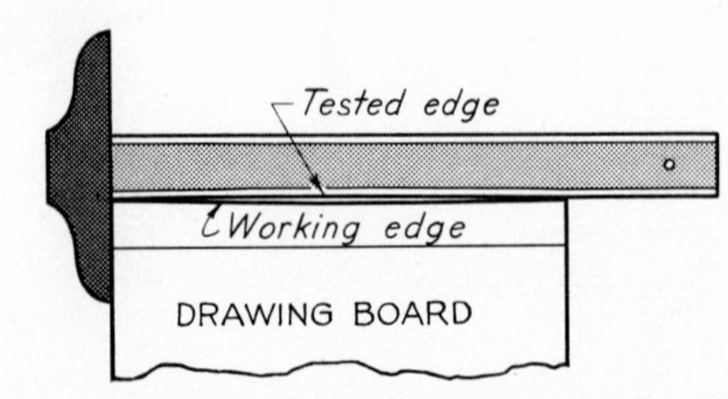


Fig. 2.4 **Testing the Working Edge of the Drawing Board.**

be run through a jointer or planed with a jack plane.

**2.5 T-square.** The T-square, Fig. 2.3, is composed of a long strip, called the *blade,* fastened rigidly at right angles to a shorter piece called the *head.* The upper edge of the blade and the inner edge of the head are *working edges* and must be straight. The working edge of the head must not be convex or the T-square will rock when the head is placed against the board. The blade should have transparent plastic edges and should be free of nicks along the working edge. Transparent edges are recommended, since they permit the draftsman to see the drawing in the vicinity of the lines being drawn.

*Do not use the T-square to drive tacks into the board or for any rough purpose. Never cut paper along its working edge, as the plastic is easily cut and even a slight nick will ruin the T-square.*

**2.6 Testing and Correcting the T-square.** To test the working edge of the head, see if the T-square rocks when the head is placed against a straight edge, such as a framing square or a drawing board working edge that has already been tested and found true. If the working edge of the head is convex, remove the head and run it through a jointer, or plane it by hand with a block plane until it tests straight. In replacing the blade on the head, use furniture glue in addition to the screws.

To test the working edge of the blade, Fig. 2.5, draw a sharp line very carefully with a hard pencil along the entire length of the working edge; then turn the T-square over and draw the line again along the same edge. If the edge is straight, the two lines will coincide; otherwise the space between the lines will be twice the error of the blade.

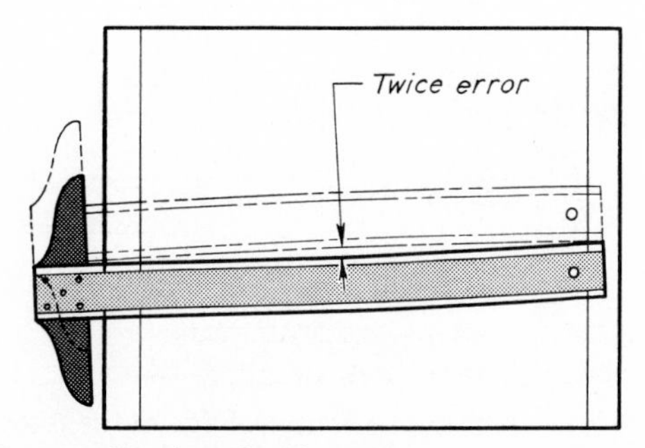


Fig. 2.5 **Testing the T-square.**

It is difficult to correct a crooked T-square blade, and if the error is considerable, it may be necessary to discard the T-square and obtain another. However, if care is taken, the blade can be made true by scraping the edge with a scraper or a sharp knife, as in truing a triangle, Fig. 2.25, and then sanding with #0 or #00 sandpaper wrapped around a block.

**2.7 Fastening Paper to the Board.** The drawing paper should be placed close enough to the working edge of the board to reduce to a minimum any error resulting from a slight "give," or bending of the blade of the T-square, and close enough to the upper edge of the board to permit space at the bottom of the sheet for using the T-square and supporting the arm while drawing, Fig. 2.6.

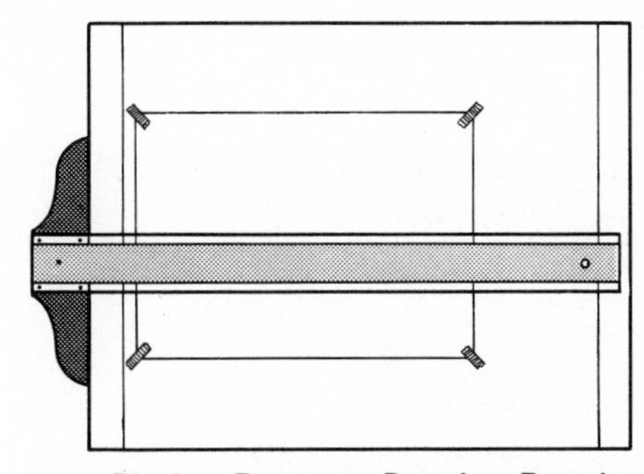

Fig. 2.6 **Placing Paper on Drawing Board.**

To fasten the paper in place, press the T-square head firmly against the working edge of the drawing board with the left hand, while the paper is adjusted with the right hand until the top edge coincides with the upper edge of the T-square. Then move the T-square to the position as shown and fasten the upper-left corner, and then the lower-right corner, and finally the remaining corners. Small sheets may require fastening only at the two upper

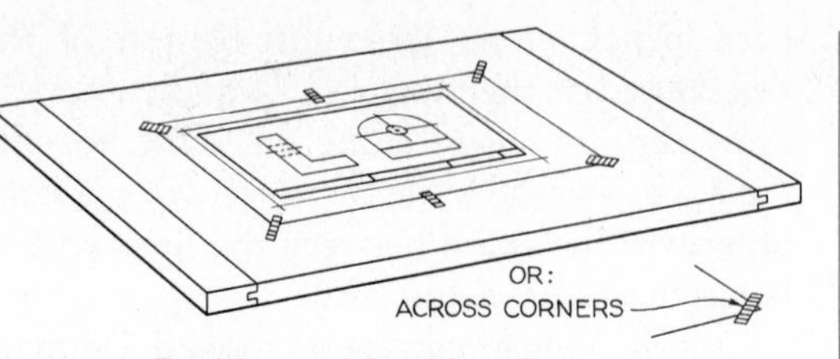


Fig. 2.7 **Positions of Drafting Tape.**

corners, while large sheets may require additional fastening.

Many draftsmen prefer drafting tape, Fig. 2.7, for it does not damage the board and it will not damage the paper if it is removed by *pulling it off slowly toward the edge of the paper.* Some draftsmen prefer to use cellophane tape, since it does not tend to roll up under the T-square, but it is more tedious to remove and tends to damage the paper if not handled carefully. Therefore, such tape should be placed outside the trim line of the drawing, if possible.

If thumbtacks are used, a type with a thin smooth head should be selected and the heads should be pushed down flush with the paper to minimize interference with the T-square.

Another method for fastening paper to the board is to use wire staples. A special draftsman's stapler is shown in Fig. 2.8.

Tracing paper should not be fastened directly to the board because small imperfections in the surface of the board will interfere with the line work. Always fasten a larger backing sheet of heavy white drawing paper on the board first, then fasten the tracing paper over this sheet.

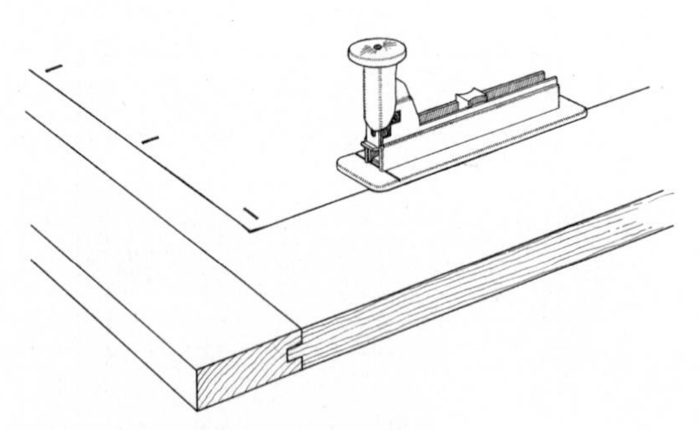

Fig. 2.8 **Draftsman's Stapler.**

## 2.8 Drawing Pencils.

High-quality *drawing pencils,* Fig. 2.9 (a), should be used in technical drawing—never ordinary writing pencils.

Many makes of mechanical pencils are available, Fig. 2.9 (b), together with refill drafting leads of conventional size in all grades. Choose the holder that feels well in the hand and one that grips the lead firmly without slipping. Mechanical pencils have the advantage of maintaining a constant length of lead (no more need to draw with a stub!), of permitting use of a lead practically to the end, of being easily refilled with new leads, of affording a ready source for compass leads, of having no wood to be sharpened, and of easy sharpening of the lead by various mechanical pencil pointers now available.

Also, thin-lead mechanical pencils are available with .3 mm or .5 mm drafting leads in several grades. These thin leads produce uniform width lines without sharpening.

Mechanical pencils are recommended for they are less expensive in the long run.

Fig. 2.9 **Drawing Pencils.**

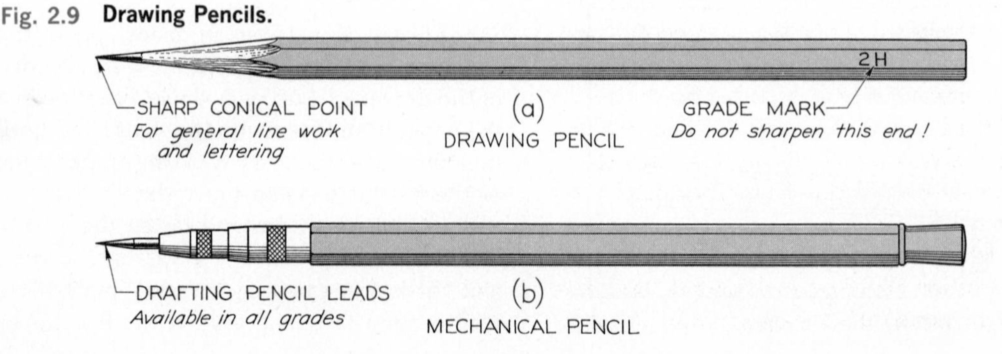

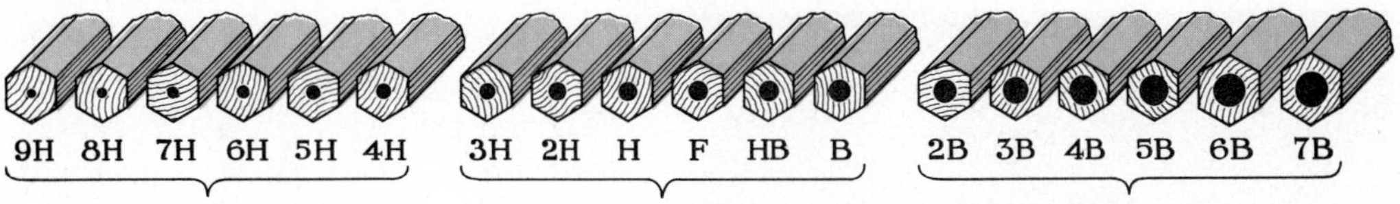

9H 8H 7H 6H 5H 4H | 3H 2H H F HB B | 2B 3B 4B 5B 6B 7B

**Hard**
The harder pencils in this group (left) are used where extreme accuracy is required, as on graphical computations and charts and diagrams. The softer pencils in this group (right) are used by some for line work on engineering drawings, but their use is restricted because the lines are apt to be too light.

**Medium**
These grades are for general purpose work in technical drawing. The softer grades (right) are used for technical sketching, for lettering, arrowheads, and other freehand work on mechanical drawings. The harder pencils (left) are used for line work on machine drawings and architectural drawings. The H and 2H pencils are widely used on pencil tracings for blueprinting.

**Soft**
These pencils are too soft to be useful in mechanical drafting. Their use for such work results in smudged, rough lines which are hard to erase, and the pencil must be sharpened continually. These grades are used for art work of various kinds, and for full-size details in architectural drawing.

Fig. 2.10 **Pencil Grade Chart.**

**2.9 Choices of Grade of Pencil.** Drawing pencil leads are made of graphite with kaolin (clay) added in varying amounts to make eighteen grades from 9H (the hardest) down to 7B (the softest), Fig. 2.10. The uses of these different grades are shown in the figure. Note that small-diameter leads are used for the harder grades, while large-diameter leads are used to give more strength to the softer grades. Hence, the degree of hardness in the wood pencil can be roughly judged by a comparison of the diameters. The leads for mechanical pencils tend to be more uniform in diameter.

Specially formulated plastic-base leads are available also in several grades for use on the polyester films now used quite extensively in industry. See §2.64.

Unfortunately, pencil grades are not sufficiently standardized; one can depend on the grade marks only in a general way. Thus an F lead in one brand may be about the same as 2H in another. Therefore, it is necessary to select a brand and then experiment with the various grades of that brand. The draftsman must first know the character of line required and be able to tell at once by inspection whether or not the line is made with the correct pencil.

To select the grade of pencil, first take into consideration the type of linework required. For light construction lines, guide lines for lettering, and for accurate geometrical constructions or work where accuracy is of prime importance, use a hard pencil, such as 4H to 6H.

For mechanical drawings on drawing paper or tracing paper, the lines should be *black*, particularly for drawings to be reproduced. The pencil chosen must be soft enough to produce jet-black lines, but hard enough not to smudge too easily or permit the point to crumble under normal pressure. This pencil will vary from F to 2H, roughly, depending upon the paper and weather conditions. The same comparatively soft pencil is preferred for lettering and arrowheads.

Another factor to consider is the texture of the paper. If the paper is hard and has a decided tooth, it will be necessary generally to use harder leads. For smoother surfaces, softer leads can be used. Hence, to obtain dense black lines, the paper should not have too much tooth.

A final factor to consider is the humidity. On humid days the paper absorbs moisture from the atmosphere and becomes soft. This can be recognized because the paper expands and becomes wrinkled. It is necessary to select softer leads to offset the softening of the paper. If you have been using an F lead, for example, change to an HB until the weather clears up.

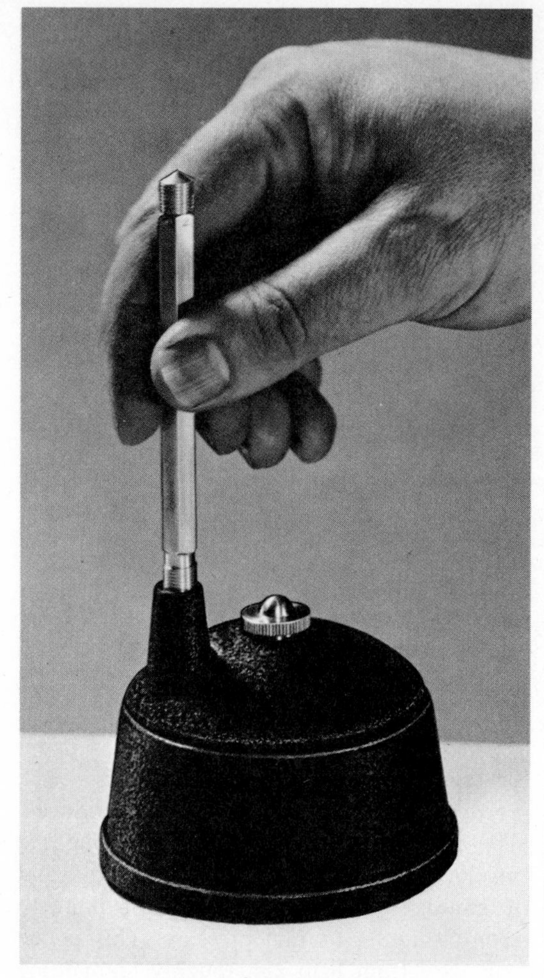

*Courtesy Elward Manufacturing Co.*

Fig. 2.11 **Tru-Point Pencil Lead Pointer.**

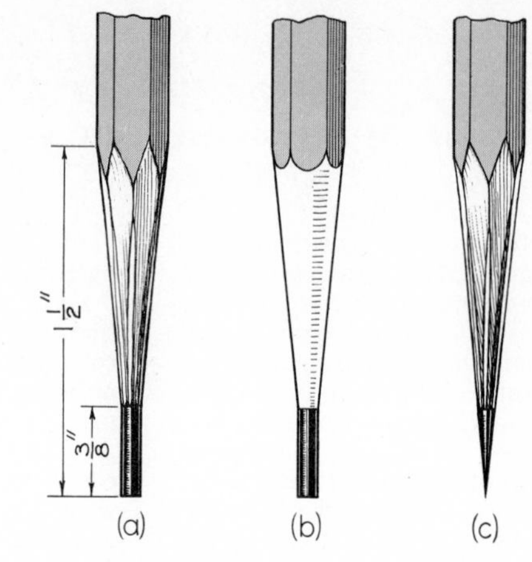


Fig. 2.12 **Pencil Points.**

## 2.10 Sharpening the Pencil.

*Keep your pencil sharp!* This is certainly the most frequent instruction needed by the beginning student. A dull pencil produces fuzzy, sloppy, indefinite lines, and is the mark of a dull and careless student. Only a *sharp pencil* is capable of producing clean-cut black lines that sparkle with clarity.

If a good mechanical pencil, Fig. 2.9 (b), is used, much time may be saved in sharpening, since the lead can be fed from the pencil as needed. An excellent lead pointer for mechanical pencils is shown in Fig. 2.11. It has the advantages of one-hand manipulation and of collecting the loose graphite particles inside where they cannot soil the hands, the drawing, or other equipment.

If a wood drawing pencil is used, Fig. 2.9 (a), sharpen the unlettered end in order to preserve the identifying grade mark. First, the wood is removed with a knife or a special draftsman's pencil sharpener for about $1\frac{1}{2}''$ from the end with about $\frac{3}{8}''$ of uncut lead exposed, Fig. 2.12 (a) and (b). Next the lead is shaped to a sharp conical point and the point wiped clean with cloth or paper tissue to remove loose particles of graphite.

The procedures for shaping the wood and the lead are illustrated in Fig. 2.13. Mechanical devices for removal of the wood are shown in Fig. 2.14 (a) and (b).

*Never sharpen your pencil over the drawing or any of your equipment.*

Many draftsmen burnish the point on a piece of hard paper to obtain a smoother, sharper point. However, for drawing visible lines the point should not be needle-sharp, but very slightly rounded. First sharpen the lead to a needle point, then stand the pencil vertically, and with a few rotary motions on the paper, wear the point down slightly to the desired shape.

*Keep the pencil pointer close by, as frequent pointing of the pencil will be necessary.*

When the sandpaper pad is not in use, it should be kept in a container, such as an envelope, to prevent the particles of graphite from falling upon the drawing board or drawing equipment, Fig. 2.14 (c).

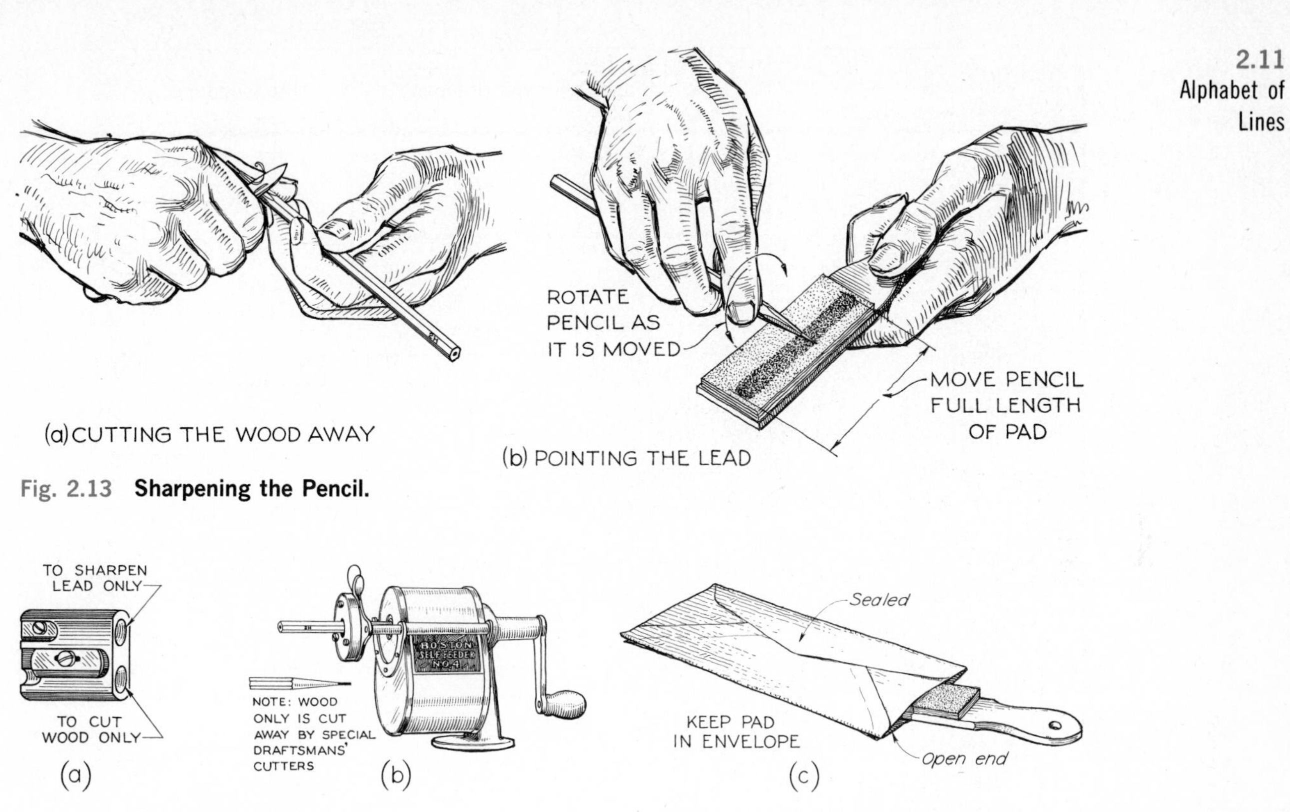


Fig. 2.13 **Sharpening the Pencil.**

Fig. 2.14 **Pencil Sharpeners.**

## 2.11 Alphabet of Lines.

Each line on a technical drawing has a definite meaning and is drawn in a certain way. The line conventions recommended by the American National Standards Institute (ANSI)* are used in Fig. 2.15, together with illustrations showing various applications.

Two widths of lines are recommended for use on drawings. All lines should be clean-cut, dark, uniform throughout the drawing, and properly spaced for legible reproduction by all commonly used methods. Spacing of .06″ or more between parallel lines is usually satisfactory for all reduction and/or reproduction processes. The size and style of the drawing and the smallest size to which it is to be reduced govern the actual width of each line. The contrast between the two widths of lines should be distinct. Pencil leads should be hard enough to prevent smudging, but soft enough to produce dense black lines so necessary for quality reproduction.

If photoreduction and blowback are not necessary, as is the case for most drafting laboratory assignments, three weights of lines may improve the appearance and legibility of the drawing. The "thin lines" may be made in two widths—regular thin lines for hidden lines and stitch lines and a somewhat thinner version for other secondary lines such as center lines, extension lines, dimension lines, leaders, section lines, phantom lines, and long-break lines.

For the "thick lines"—visible, cutting plane, and short break—use a relatively soft pencil such as F or H. All thin lines should be made with a sharp medium-grade pencil such as H or 2H. All lines (except construction lines) must be *sharp and dark.* Make construction lines with a sharp 4H or 6H pencil so thin that they barely can be seen at arm's length and need not be erased.

* ANSI Y14.2–1973.

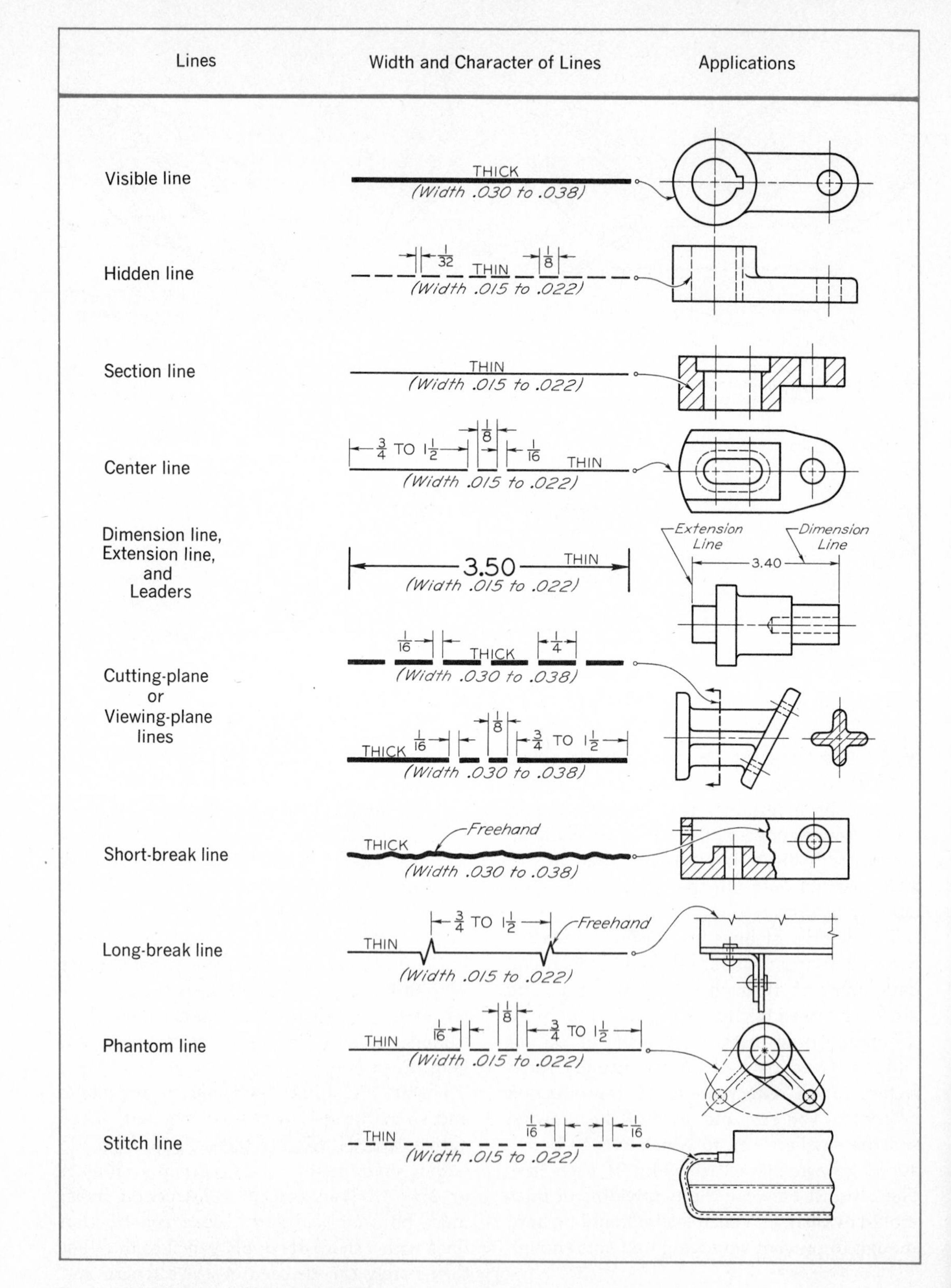


Fig. 2.15 **Alphabet of Lines (Full Size).**

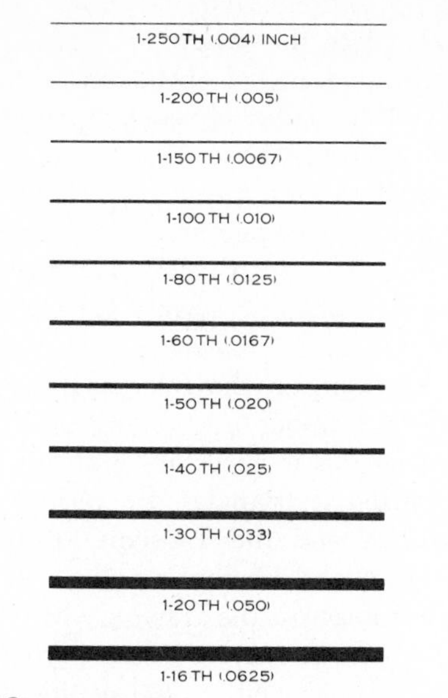

Fig. 2.16 **Line Gage.**

The high-quality photoreduction and reproduction processes used in the production of this book permitted the use of three weights of lines in many illustrations and drawings for increased legibility.

In Fig. 2.15, the ideal lengths of all dashes are indicated. It would be well to measure the first few hidden dashes and center-line dashes you make, and then thereafter to estimate the lengths carefully by eye.

For inking procedures, see §§2.48 to 2.55. The *line gage,* Fig. 2.16, is convenient when referring to lines of various widths.

## 2.12 Erasing.

Erasers are available in many degrees of hardness and abrasiveness. For general drafting, the Weldon Roberts India eraser or the Eberhard Faber Ruby is suggested, Fig. 2.17 (a). These erasers are used for erasing either pencil or ink. Avoid gritty erasers, even for erasing ink, as they invariably damage the paper. A soft pink pencil eraser is preferred by many draftsmen to erase light lines during the construction stage of a drawing. Best results are obtained if a hard surface, such as a triangle, is placed under the paper being erased. If the surface has become badly grooved by the lines, the surface can be improved by burnishing with a hard smooth object or with the back of the fingernail.

The *Artgum,* Fig. 2.17 (b), is recommended for general cleaning of the large areas of a drawing or for removing pencil lines from an inked drawing. The Artgum should not be

Fig. 2.17 **Erasers.**

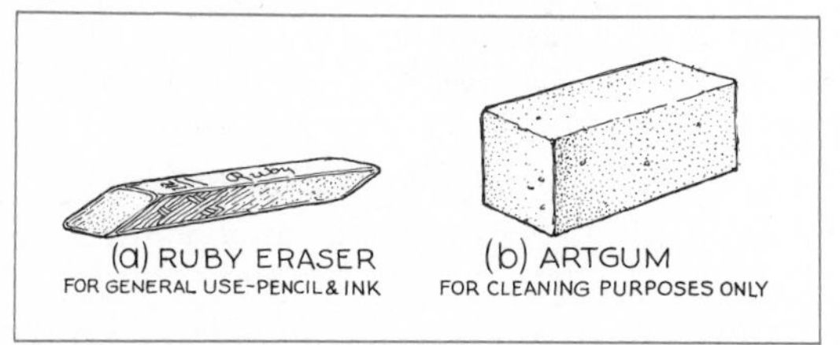


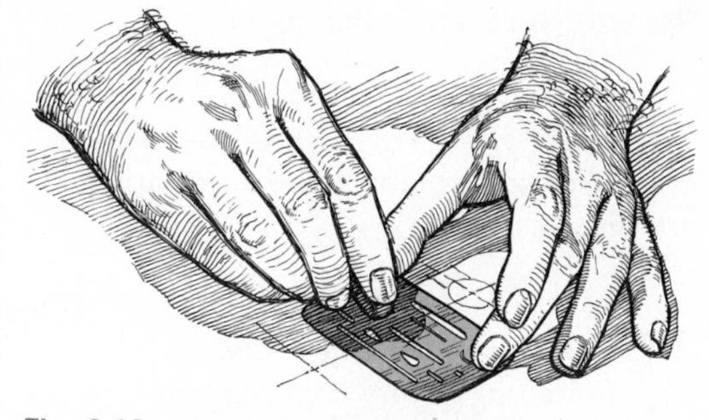

Fig. 2.18 **Using the Erasing Shield.**

Fig. 2.19 **Electric Erasing Machine.**
*Courtesy Pierce Corp.*

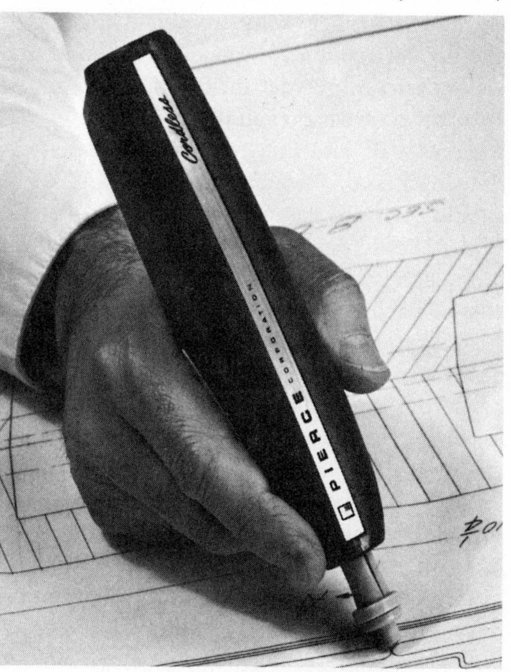

used as a substitute for the regular pencil eraser.

The *erasing shield,* Fig. 2.18, is used to protect the lines near those being erased.

The *electric erasing machine,* Fig. 2.19, saves time and is essential if much drafting is being done. (For design details of the eraser, see §14.9.)

Fig. 2.20 **Draftsman's Dusting Brush.**

A *dusting brush,* Fig. 2.20, is useful for removing eraser crumbs without smearing the drawing.

## 2.13 Keeping Drawings Clean.

Cleanliness in drafting is very important and should become a habit. Cleanliness does not just happen; it results only from a conscious effort to observe correct procedures.

*First,* the draftsman should keep his hands clean at all times. Oily or perspiring hands should be frequently washed with soap and water. Talcum powder on the hands tends to absorb excessive perspiration.

*Second,* all drafting equipment, such as drawing board, T-square, triangles, and scale, should be wiped frequently with a clean cloth. Water should be used sparingly and dried off immediately. Artgum or other soft erasers may also be used for cleaning drawing equipment.

*Third,* the largest contributing factor to dirty drawings is *not dirt, but graphite* from the pencil; hence the draftsman should observe the following precautions:

1. Never sharpen a pencil over the drawing or any equipment.
2. Always wipe the pencil point with a clean cloth or cleansing tissue, after sharpening or pointing, to remove small particles of loose graphite.
3. Never place the sandpaper pad or file in contact with any other drawing equipment unless it is completely enclosed in an envelope or similar cover, Fig. 2.14 (c).
4. Never work with the sleeves or hands resting upon a penciled area. Keep such parts covered with clean paper (not a cloth). In lettering a drawing, always place a piece of paper under the hand.
5. Avoid sliding anything across the drawing. A certain amount of sliding of T-square and triangles is necessary, but this can be minimized if triangles are picked up by their tips and the T-square blade tilted upward slightly before moving.
6. Never rub across the drawing with the palm of the hand to remove eraser particles; use a dust brush, Fig. 2.20, or flick—don't rub—the particles off with a clean cloth.

When the drawing is completed, it is not necessary to clean it if the above rules have been observed. The practice of making a pencil drawing, scrubbing it with Artgum, and then retracing the lines, is poor technique and a waste of time, and this habit should not be acquired.

At the end of the period or of the day's work, the drawing should be covered with paper or cloth to protect it from dust.

## 2.14 Horizontal Lines.

To draw a horizontal line, Fig. 2.21 (a), press the head of the T-square firmly against the working edge of the board with the left hand; then slide the left hand to the position shown, so as to press the blade tightly against the paper. Lean the pencil in the direction of the line at an angle of approximately 60° with the paper, (b), and draw the line from left to right. Keep the pencil in a vertical plane, (b) and (c); otherwise, the line may not be straight. While drawing the line, let the little finger of the hand holding the pencil glide lightly on the blade of the T-square, and rotate the pencil slowly between the thumb and forefinger so as to distribute the wear uniformly on the lead and maintain a symmetrical point.

When great accuracy is required, the pencil

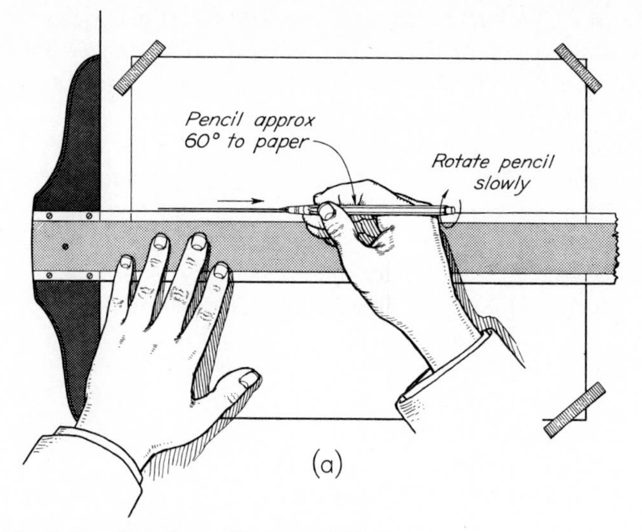


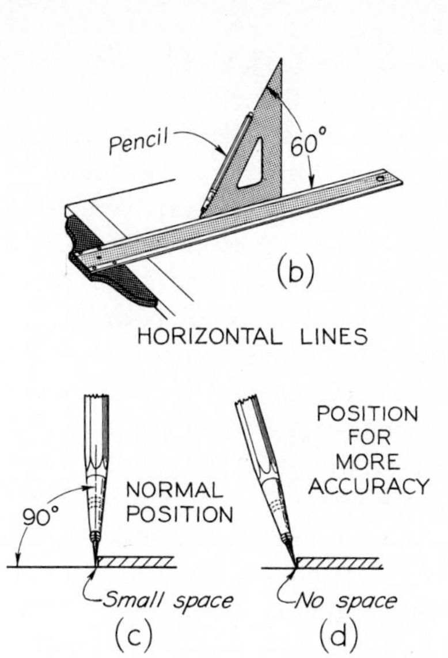


**Fig. 2.21 Drawing a Horizontal Line.**

may be "toed in" as shown at (d) to produce a perfectly straight line.

Left-Handers. In general, reverse the above procedure. Place the T-square head against the right edge of the board, and with the pencil in the left hand, draw the line from right to left.

Triangles and T-squares, especially when new, often have very sharp edges which tend to cut into the pencil lead and cause a trail of graphite on the drawing. To prevent smearing of these particles, blow them off at intervals. If the edges of the triangles or T-square are too sharp, they can be sanded *very lightly* with #00 sandpaper—just enough to remove the sharp edges.

**Fig. 2.22 Drawing a Vertical Line.**

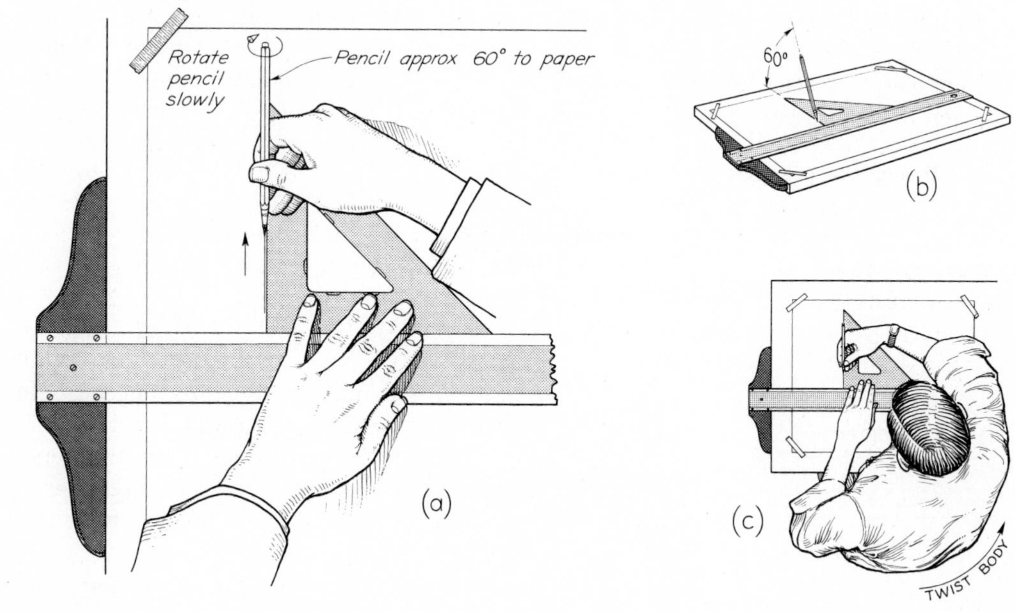

**2.15 Vertical Lines.** Use either the 45° triangle or the 30° × 60° triangle to draw vertical lines. Place the triangle on the T-square with the *vertical edge on the left* as shown in Fig. 2.22 (a). With the left hand, press the head of the T-square against the board, then slide the hand to the position shown where it holds both the T-square and the triangle firmly in position. Then *draw the line upward,* rotating the pencil slowly between the thumb and forefinger.

Lean the pencil in the direction of the line at an angle of approximately 60° with the paper and in a vertical plane, (b). Meanwhile, the upper part of the body should be twisted to the right, as shown at (c).

LEFT-HANDERS. In general, reverse the above procedure. Place the T-square head on the right and the vertical edge of the triangle on the right; then, with the right hand, hold the T-square and triangle firmly together, and with the left hand draw the line upward.

The only time it is permissible for right-handers to turn the triangle so that the vertical edge is on the right is when drawing a vertical line near the right end of the T-square. In this case, the line would be drawn downward.

**2.16 The Triangles.** Most inclined lines in mechanical drawing are drawn at standard angles with the *45° triangle* and the *30° × 60° triangle,* Fig. 2.23. The triangles are made of transparent plastic so that lines of the drawing can be seen through them. A good combination of triangles is the 30° × 60° triangle with a long side of 10″, and a 45° triangle with each side 8″ long.

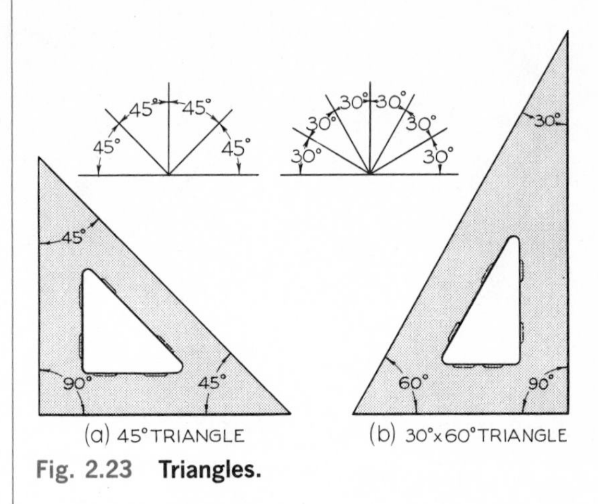


Fig. 2.23 **Triangles.**

**2.17 Testing and Correcting the Triangles.** Triangles are subject to warping, sometimes even before they are sold by the dealer. Therefore, the purchaser should test his new triangles immediately after purchase to determine if they are true. If they are not found to be correct, they should be returned at once to the dealer.

Test the sides of the triangles for straightness in the same manner as for the T-square blade, §2.6. To test the right angle of either of the triangles, Fig. 2.24 (a), place the triangle on the T-square and draw a vertical line; then turn the triangle over (like turning a page in

Fig. 2.24 **Testing the Triangles.**

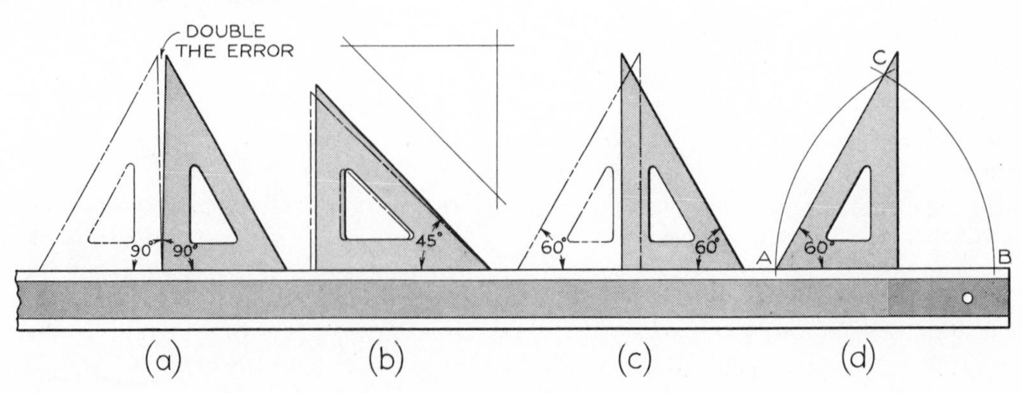

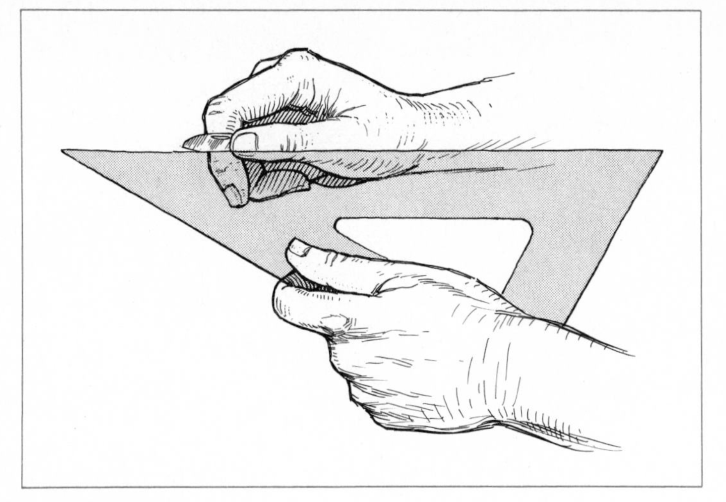

Fig. 2.25 **Scraping the Triangle.**

a book) and draw the line again along the same edge. If the two lines thus drawn do not coincide, the right angle is not 90° and the error is equal to half the angle between the two lines.

To test the 45° angle, place the triangle on the T-square, as at (b), and draw a line along the hypotenuse; then turn the triangle over, and using the other 45° angle of the triangle, draw a line along the hypotenuse. If the two lines do not coincide, there is an error in one or both 45° angles. A direct test of the 45° angle can be made by drawing a right triangle. The sides adjacent to the 90° angle will be equal if the two 45° angles are correct (assuming the 90° angle to be correct).

To test the 60° angle of the 30° × 60° triangle, draw an equilateral triangle, as shown at (c). If all three sides are not exactly equal in length, the 60° angle is incorrect. Another method of testing the 60° angle, (d), is to draw a horizontal line AB slightly shorter than the hypotenuse of the triangle, and to draw arcs with A and B as centers and AB as radius, intersecting at C. When the triangle is placed as shown, its hypotenuse should pass through C.

To true up the edge of a triangle, make a rough cut by scraping the edge with a sharp knife, Fig. 2.25. Or place the triangle in a vise and plane with a sharp block plane set for a very shallow cut. Then hold the triangle flat against the edge of a table top, with the edge of the triangle level with it, and sand the edge with #0 or #00 sandpaper wrapped around a block.

## 2.18 Inclined Lines.

The positions of the triangles for drawing lines at all of the possible angles are shown in Fig. 2.26. In the figure it is understood that the triangles in each case are resting upon the blade of the T-square. Thus, it is possible to divide 360° into twenty-four 15° sectors with the triangles used singly or in combination. Note carefully the directions for drawing the lines, as indicated by the arrows. Note that all lines in the left half are drawn *toward the center,* while those in the right half are drawn *away from the center.*

## 2.19 Protractors.

For measuring or setting off angles other than those obtainable with the triangles, the *protractor* is used. The best protractors are made of nickel silver and are capable of most accurate work, Fig. 2.27 (a). For ordinary work the plastic protractor is satisfactory and is much cheaper, (b). To set off angles with greater accuracy, use one of the methods presented in §4.21.

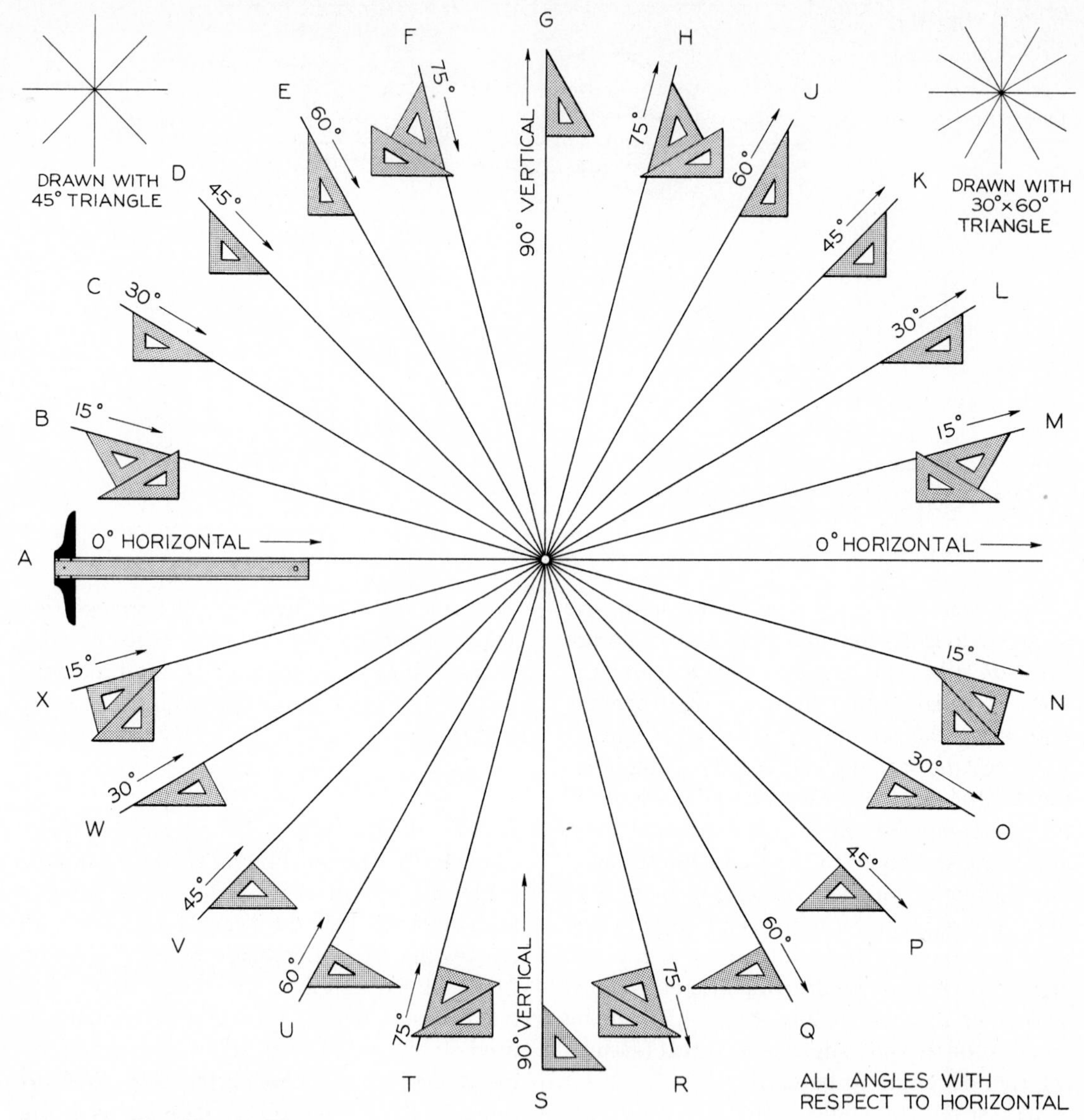


Fig. 2.26 **The Triangle Wheel.**

Fig. 2.27 **Protractors.**

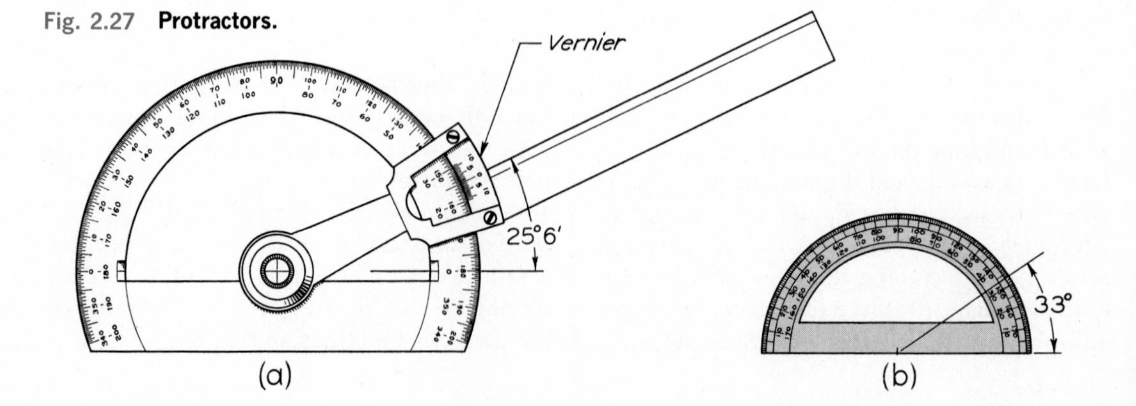

## 2.20 **Drafting Angles.**

A variety of devices combining the protractor with triangles to produce great versatility of use are available, one type of which is shown in Fig. 2.28.

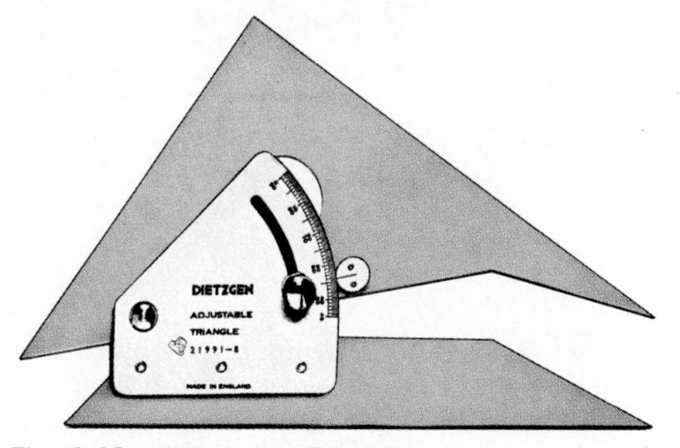


Fig. 2.28 **Adjustable Triangle.**

## 2.21 **To Draw a Line Through Two Points.**

To draw a line through two points, Fig. 2.29, place the pencil vertically at one of the points, and move the straightedge about the pencil point as a pivot until it lines up with the other point; then draw the line along the edge.

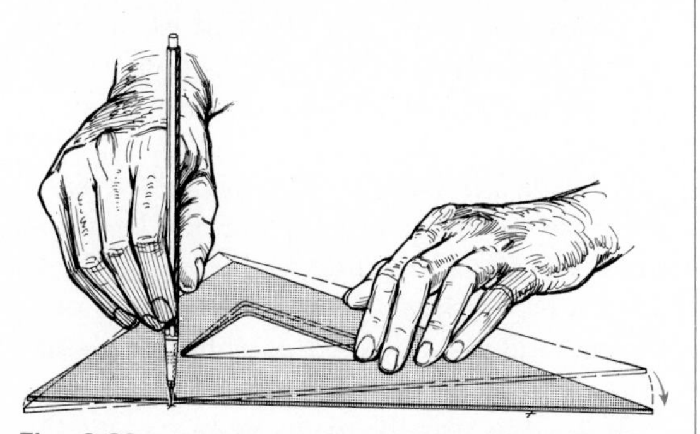

Fig. 2.29 **To Draw a Pencil Line Through Two Given Points.**

## 2.22 **Parallel Lines.**

To draw a line parallel to a given line. Fig. 2.30, move the triangle and T-square as a unit until the hypotenuse of the triangle lines up with the given line, (a); then, holding the T-square firmly in position, slide the triangle away from the line, (b), and draw the required line along the hypotenuse, (c).

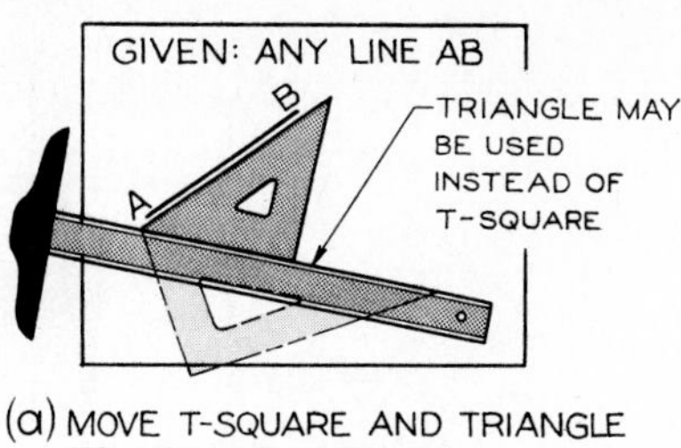


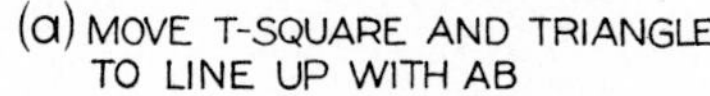

(a) MOVE T-SQUARE AND TRIANGLE TO LINE UP WITH AB

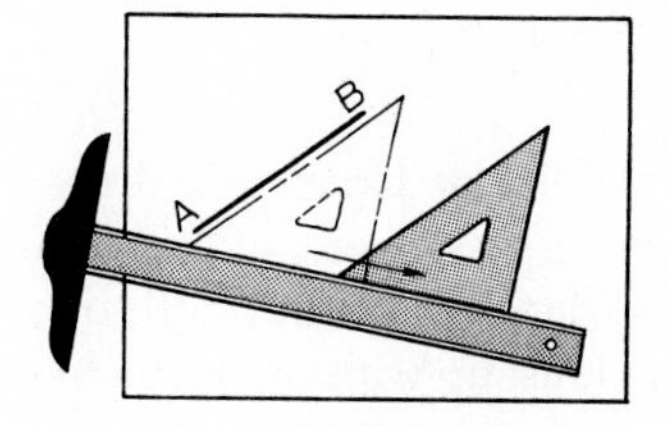


(b) SLIDE TRIANGLE ALONG T-SQUARE

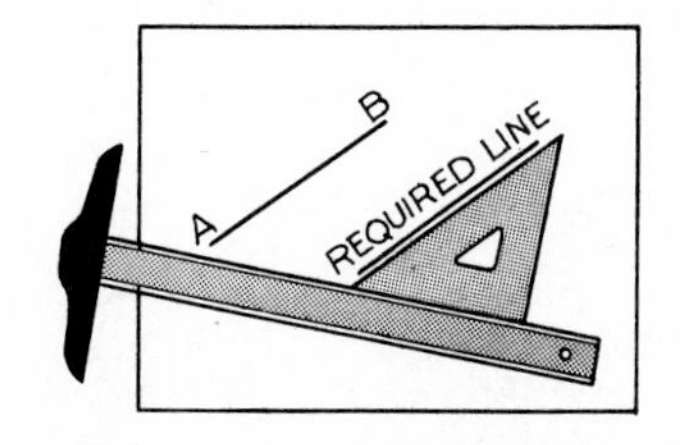


(c) DRAW REQUIRED LINE PARALLEL TO AB

Fig. 2.30 **To Draw a Line Parallel to a Given Line.**

Obviously any straightedge, such as one of the triangles, may be substituted for the T-square in this operation, as shown at (a).

To draw parallel lines at 15° with horizontal, arrange the triangles as shown in Fig. 2.31.

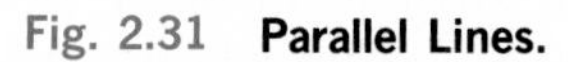

Fig. 2.31 **Parallel Lines.**

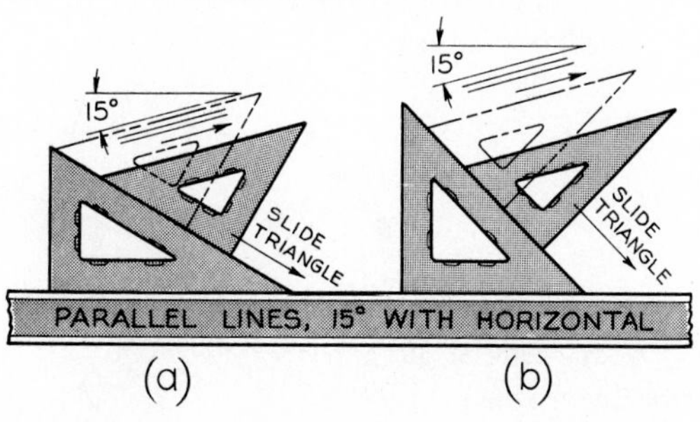

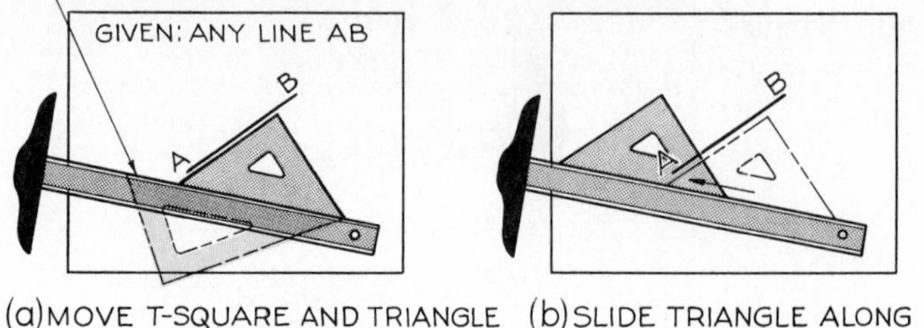


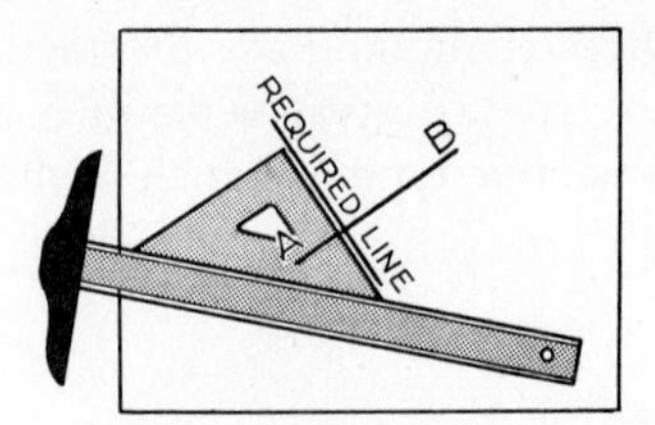

Fig. 2.32 **To Draw a Line Perpendicular to a Given Line.**

**2.23 Perpendicular Lines.** To draw a line perpendicular to a given line, move the T-square and triangle as a unit until one edge of the triangle lines up with the given line, Fig. 2.32 (a); then slide the triangle across the line, (b), and draw the required line, (c).

To draw perpendicular lines when one of the lines makes 15° with horizontal, arrange the triangles as shown in Fig. 2.33.

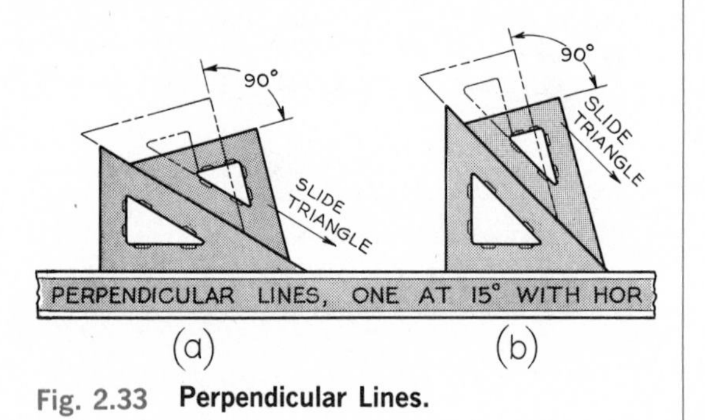


Fig. 2.33 **Perpendicular Lines.**

**2.24 Lines at 30°, 60°, or 45° with Given Line.** To draw a line making 30° with a given line, arrange the triangle as shown in Fig. 2.34. Angles of 60° and 45° may be drawn in a similar manner.

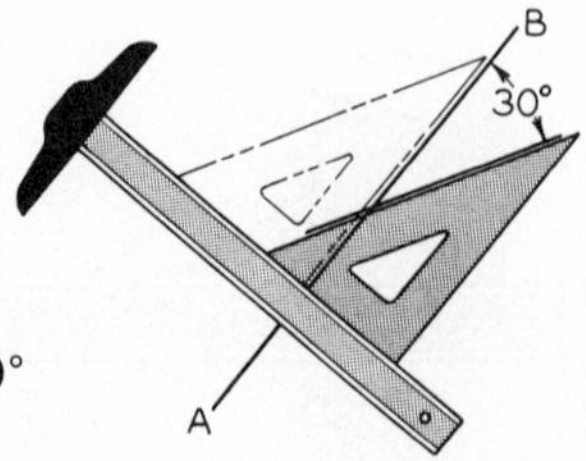


Fig. 2.34 **Line at 30° with Given Line.**

**2.25 Scales.** A drawing of an object may be the same size as the object (full size), or it may be larger or smaller than the object; in most cases, if not drawn full size, the drawing is made smaller than the object represented. The ratio of reduction depends upon the relative sizes of the object and of the sheet of paper upon which the drawing is to be made. For example, a machine part may be half size ($\frac{1}{2}'' = 1''$); a building may be drawn $\frac{1}{48}$ size ($\frac{1}{4}'' = 1'\text{-}0''$); a map may be drawn $\frac{1}{1200}$ size ($1'' = 100'\text{-}0''$); or a gear in a wrist watch may be ten-times size ($10'' = 1''$ or $\frac{10}{1}$).

Scales, Fig. 2.35, are classified as *architects scale* (a), *engineers scale* (b), *mechanical engineers scale* (c), the *decimal scale* (d), and the *metric scale* (e).

A *full-divided scale* is one in which the basic units are subdivided throughout the length of the scale, Fig. 2.35 (a) upper scale, and (b) to (e). An *open-divided scale* is one in which only the end unit is subdivided, as is the lower scale at (a).

Scales are usually made of boxwood, the better ones having white plastic edges. Scales are either triangular, Fig. 2.36 (a) and (b), or flat, (c) to (f). The triangular scales have the advantage of combining many scales on one stick, but the user will waste much time looking for the required scale if a *scale guard,* (g), is not used. The flat scale is almost universally used by professional draftsmen because of its convenience, but several flat scales are necessary to replace one triangular scale, and the total cost is greater. Since machine drawings are made full, half, quarter, and eighth size, these scales may be obtained on one or two

1656
DIETZGEN
U S STD

(a) Architects Scale

1667
DIETZGEN
U S STD

(b) Engineers Scale

HALF SIZE (6IN.=1FT.)
FULL SIZE - 32
DIETZGEN
EXCELLO
1582

(c) Mechanical Engineers Scale

50
FULL SIZE-50(.02)
DIETZGEN
EXCELLO
1593
ASA STYLE G

(d) Decimal Scale

METER
DIETZGEN
EXCELLO
1574C

(e) Metric Scale

**Fig. 2.35 Types of Scales.**

**Fig. 2.36 Sections of Scales and Scale Guard.**

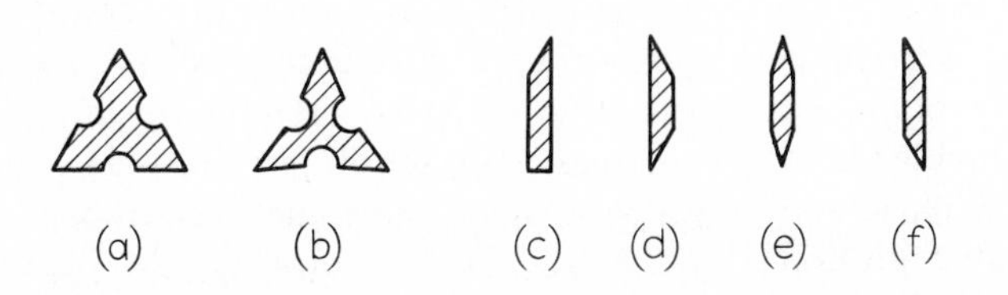


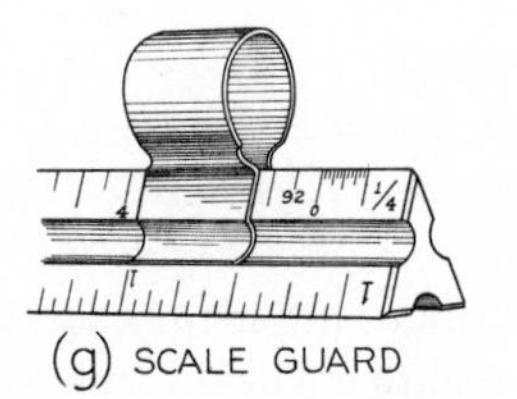

flat scales of the type shown in Fig. 2.35 (c), (d), and (e).

## 2.26 Architects Scale.

**Architects Scale.** Fig. 2.35 (a). The *architects scale* is intended primarily for drawings of buildings, piping systems, and other large structures which must be drawn to a reduced scale to fit on a sheet of paper. The full-size scale is also useful in drawing relatively small objects, and for that reason the architects scale has rather general usage.

The architects scale has one full-size scale and ten overlapping reduced-size scales. By means of these scales a drawing may be made to various sizes from full size to $\frac{1}{128}$ size. *Note particularly: In all of the reduced scales the major divisions represent feet, and their subdivisions represent inches and fractions thereof.* Thus, the scale marked $\frac{3}{4}$ means $\frac{3}{4}$ inch = 1 foot, not $\frac{3}{4}$ inch = 1 inch; that is, one-sixteenth size, not three-fourths size. Similarly, the scale marked $\frac{1}{2}$ means $\frac{1}{2}$ inch = 1 foot, not $\frac{1}{2}$ inch = 1 inch; that is, one-twenty-fourth size, not half-size.

All of the scales, from full size to $\frac{1}{128}$ size, are shown in Fig. 2.37. Some are upside down, just as they may occur in use. These scales are described as follows:

Full Size. Fig. 2.37 (a). Each division in the full-size scale is $\frac{1}{16}''$. Each inch is divided first into halves, then quarters, eighths, and finally sixteenths, the division lines diminishing in length with each division. To set off $\frac{1}{32}''$, estimate visually one half of $\frac{1}{16}''$; to set off $\frac{1}{64}''$, estimate one fourth of $\frac{1}{16}''$.

Half Size. Fig. 2.37 (a). Use the full-size scale, and divide every dimension mentally by two (do not use the $\frac{1}{2}''$ scale, which is intended for drawing to a scale of $\frac{1}{2}'' = 1'$, or one-twenty-fourth size). To set off 1″, measure $\frac{1}{2}''$; to set off 2″, measure 1″; to set off $3\frac{1}{4}''$, measure $1\frac{1}{2}''$ (half of 3″), then $\frac{1}{8}''$ (half of $\frac{1}{4}''$); to set off $2\frac{13}{16}''$ (see figure), measure 1″, then $\frac{13}{32}''$ ($\frac{6\frac{1}{2}}{16}''$ or half of $\frac{13}{16}''$).

Quarter Size. Fig. 2.37 (b). Use the 3″ scale in which $3'' = 1'$. The subdivided portion to the left of zero represents one foot compressed to actually 3″ in length, and is divided into inches, then half inches, quarter inches, and finally eighth inches. Thus the entire portion representing one foot would actually measure three inches; therefore, $3'' = 1'$. To set off anything less than 12″, start at zero and measure to the left.

To set off $10\frac{1}{8}''$, read off 9″ from zero to the left, then add $1\frac{1}{8}''$ and set off the total $10\frac{1}{8}''$, as shown. To set off more than 12″, for example, $1'\text{-}9\frac{3}{8}''$ (see your scale), find the 1′ mark to the right of zero and the $9\frac{3}{8}''$ mark to the left of zero; the required distance is the distance between these marks, and represents $1'\text{-}9\frac{3}{8}''$.

Eighth Size. Fig. 2.37 (b). Use the $1\frac{1}{2}''$ scale in which $1\frac{1}{2}'' = 1'$. The subdivided portion to the right of zero represents 1′, and is divided into inches, then half inches, and finally quarter inches. The entire portion, representing 1′, actually is $1\frac{1}{2}''$; therefore: $1\frac{1}{2}'' = 1'$. To set off anything less than 12″, start at zero and measure to the right.

To set off $7\frac{1}{4}''$ (see figure), read off 7″ from zero to the right, then add $\frac{1}{4}''$, and set off the total $7\frac{1}{4}''$, as shown.

To set off more than twelve inches, for example $3'\text{-}10\frac{3}{4}''$ (see your scale), find the 3′ mark to the left of zero, and the $10\frac{3}{4}''$ mark to the right of zero; the required distance is the distance between these marks, and represents $3'\text{-}10\frac{3}{4}''$, being one-eighth of $3'\text{-}10\frac{3}{4}''$.

Double Size. Use the full-size scale, and multiply every dimension mentally by two. To set off 1″, measure 2″; to set off $3\frac{1}{4}''$, measure $6\frac{1}{2}''$, and so on. The double-size scale is occasionally used to represent small objects. In such cases, a small actual-size outline view should be shown near the bottom of the sheet to help the shop man visualize the actual size of the object.

Other Sizes. Fig. 2.37. The other scales besides those described above are used chiefly by architects. Machine drawings are customarily made only double size, full size, $\frac{1}{2}$ size, $\frac{1}{4}$ size, or $\frac{1}{8}$ size.

Special Methods. The $\frac{3}{8}''$ scale can be used conveniently to set off thirty-seconds of an inch full size, since each small subdivision on this scale equals $\frac{1}{32}''$. Similarly, the $\frac{3}{16}''$ scale can be used to set off sixty-fourths of an inch full size, since each small subdivision equals $\frac{1}{64}''$. If it is desired, for example, to set off the

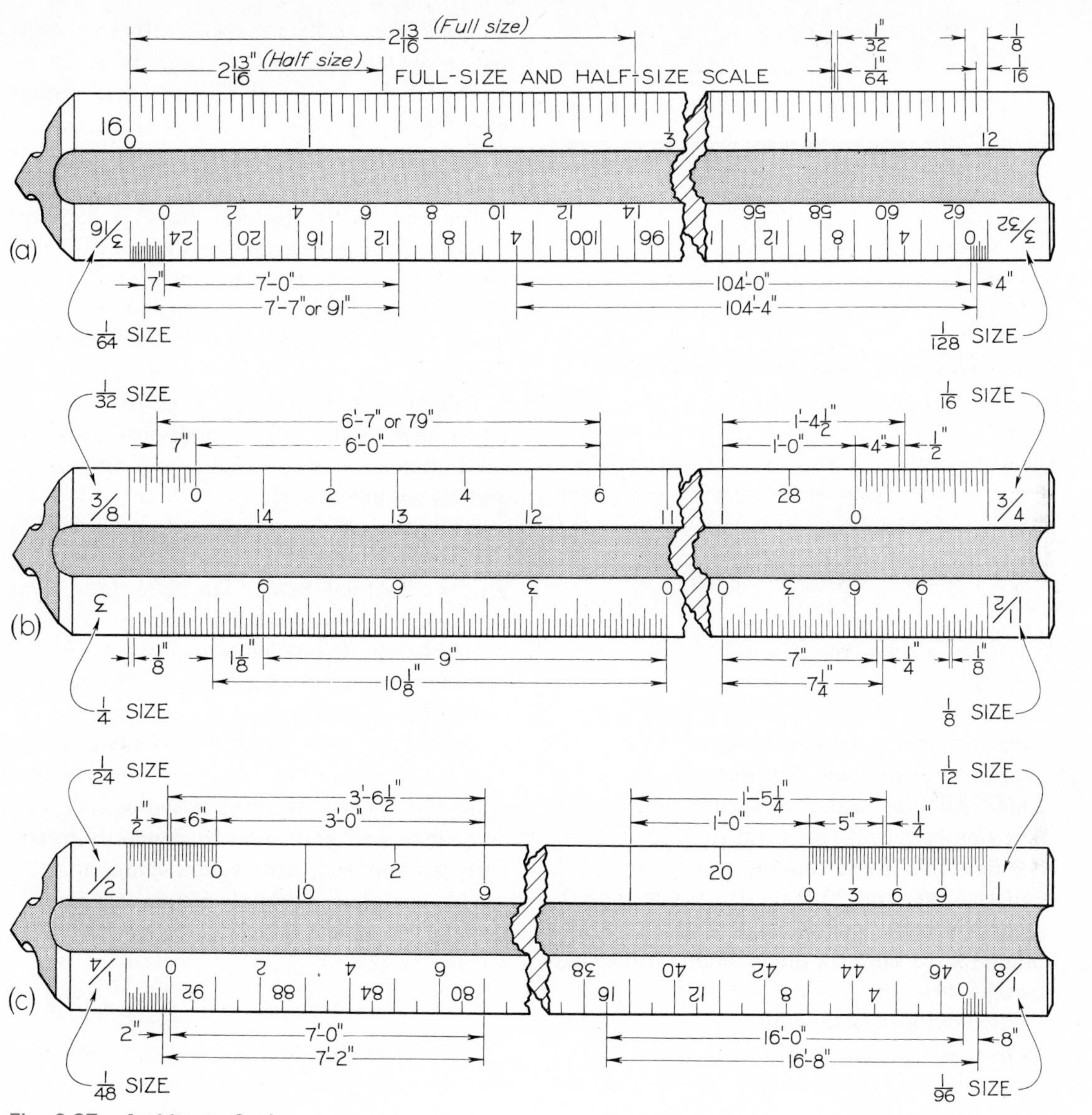


Fig. 2.37 **Architects Scales.**

radius of a $\frac{13}{64}''$ drill hole, set off $6\frac{1}{2}$ (half of 13) small divisions on the $\frac{3}{16}''$ scale.

Do not abuse the scale by using it as a straightedge, hammering thumbtacks with it, pricking holes in it with dividers to take off dimensions, or using it in ways other than its intended use.

## 2.27 Engineers Scale.

Fig. 2.35 (b). The *engineers scale* is graduated in the decimal system. It is also frequently called the *civil engineers scale* because it was originally used mainly in civil engineering. The name *chain scale* also persists because it was derived from the surveyors' chain composed of 100 links, used for land measurements. The name *engineers scale* is perhaps best, because the scale is used generally by engineers of all kinds.

The engineers scale is graduated in units of one inch divided into 10, 20, 30, 40, 50, and 60 parts. Thus, the engineers scale is conve-

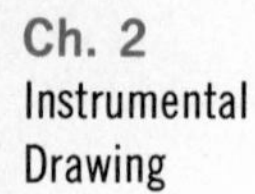


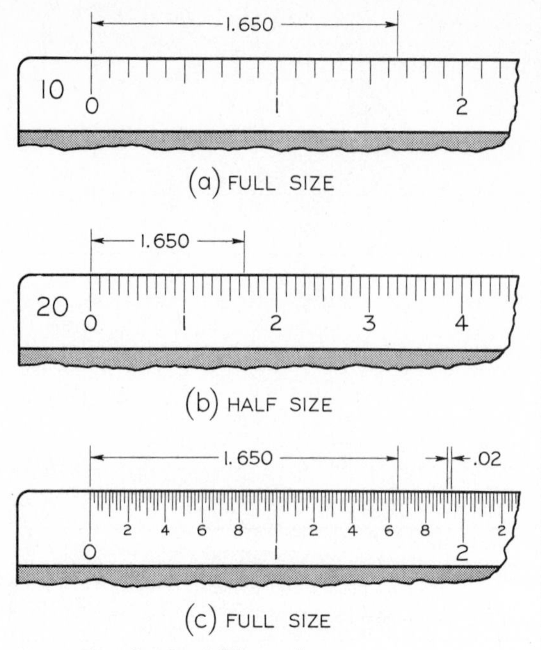

Fig. 2.38 **Decimal Dimensions.**

nient in machine drawing to set off dimensions expressed in decimals. For example, to set off 1.650″ full size, Fig. 2.38 (a), use the 10-scale and simply set off one main division plus $6\frac{1}{2}$ subdivisions. To set off the same dimension half size, use the 20-scale, (b), since the 20-scale is exactly half the size of the 10-scale. Similarly, to set off a dimension quarter size, use the 40-scale.

The engineers scale is used also in drawing maps to scales of 1″ = 50′, 1″ = 500′, 1″ = 5 miles, etc., and in drawing stress diagrams or other graphical constructions to such scales as 1″ = 20 lb and 1″ = 4000 lb.

## 2.28 Mechanical Engineers Scale.

Fig. 2.35 (c). The objects represented in machine drawing vary in size from small parts, an inch or smaller in size, to machines of large dimensions. By drawing these objects full size, half size, quarter size, or eighth size, the drawings will readily come within the limits of the standard-size sheets. For this reason the mechanical engineers scales are divided into units representing inches to full size, half size, quarter size, or eighth size. To make a drawing of an object to a scale of one-half size, for example, use the mechanical draftsmans scale marked half size, which is graduated so that every $\frac{1}{2}$″ represents 1″. Thus, the half-size scale is simply a full-size scale compressed to one-half size.

These scales are also very useful in dividing dimensions. For example, to draw a $3\frac{11}{16}$″ dia. circle full size, we need half of $3\frac{11}{16}$″ to use as radius. Instead of using arithmetic to find half of $3\frac{11}{16}$″, it is easier to set off $3\frac{11}{16}$″ on the half-size scale.

Triangular combination scales are available which include the full and half-size mechanical draftsmans scales, several architects scales, and an engineers scale.

## 2.29 Decimal Scale.

Fig. 2.35 (d). The increasing use of decimal dimensions has brought about the development of a scale specifically for that use, approved by the ANSI.* On the full-size scale, each inch is divided into fiftieths of an inch, or .02″, as shown in Fig. 2.38 (c); and on the half- and quarter-size scales the inches are compressed to half size or quarter size, and then are divided into ten parts, so that each subdivision stands for .1″.

The complete decimal system of dimensioning, in which this scale is used, is described in §11.10.

## 2.30 Metric Scale.

Fig. 2.35 (e). The *metric scale* is used when the meter is the linear measurement standard, as is the case in most foreign countries and in scientific work in the United States. The metric system has the advantages of the decimal system and is rapidly becoming the world standard of measure. Great Britain and Canada are in the process of changing to the metric system of weights and measure. The United States is the only industrial nation not committed to the metric system. A three-year study by the National Bureau of Standards is expected to recommend the metric system so that the United States may increase its participation in the

* ANSI Z75.1-1955.

world markets. Since the cost of conversion of machines, tools, and so on, to the metric system would be enormous, it appears that the United States is moving rapidly to the use of the decimilized inch as a prelude to acceptance of the metric system. For metric equivalents, see inside front cover and Appendix 26.

## 2.31 To Specify the Scale on a Drawing.

For machine drawings, the recommended practice is to letter FULL SIZE, 1.00 = 1.00, 1 = 1, or 1/1; HALF SIZE, .50 = 1.00, $\frac{1}{2}$ = 1, or 1/2; QUARTER SIZE, .25 = 1.00, $\frac{1}{4}$ = 1, or 1/4; EIGHTH SIZE, .125 = 1.00, $\frac{1}{8}$ = 1, or 1/8; TWICE SIZE, 2.00 = 1.00, 2 = 1, or 2/1; or TEN TIMES SIZE, 10.00 = 1.00, 10 = 1, or 10/1. For examples of how scales are shown on machine drawings, see Figs. 14.26 and 14.27.

For drawings of buildings, piping, and other structures in which the dimensions are in feet and inches, the architects scales should be given in terms of inches to feet, such as 3″ = 1′-0 (quarter size) or $\frac{1}{2}$″ = 1′-0 (twenty-fourth size).

Map scales are indicated in terms of fractions, as Scale $\frac{1}{62500}$, or graphically as 400 0 400 800 Ft. See also Fig. 3.38 and Fig. 23.2.

## 2.32 Accurate Measurements.

Accurate drafting depends considerably upon the correct use of the scale in setting off distances. Do not take measurements directly off the scale with the dividers or compass, as damage will result to the scale. Place the scale on the drawing with the edge parallel to the line on which the measurement is to be made and, with a sharp pencil having a conical point, make a short dash at right angles to the scale and opposite the correct graduation mark, as shown in Fig. 2.39 (a). If extreme accuracy is required, a tiny prick mark may be made at the required point with the needle point or stylus, as shown at (b), or with one leg of the dividers.

*Avoid cumulative errors* in the use of the scale. If a number of distances are to be set off end-to-end, all should be set off at one setting of the scale by adding each successive measurement to the preceding one, if possible. Avoid setting off the distances individually by moving the scale to a new position each time, since slight errors in the measurements may accumulate and give rise to a large error.

## 2.33 Drawing Instruments.

*Drawing instruments* are generally sold in sets, in velvet-lined cases, but they may be purchased separately. The principal parts of high-grade instruments are usually made of nickel silver, which has a silvery luster, is corrosion-resistant, and can be readily machined into desired shapes. Tool steel is used for the blades of ruling pens, for spring parts, for divider points, and for the various screws.

In technical drawing, accuracy, neatness, and speed are essential, §2.2. These objectives are not likely to be obtained with cheap or inferior drawing instruments. For the student or the professional draftsman it is advisable,

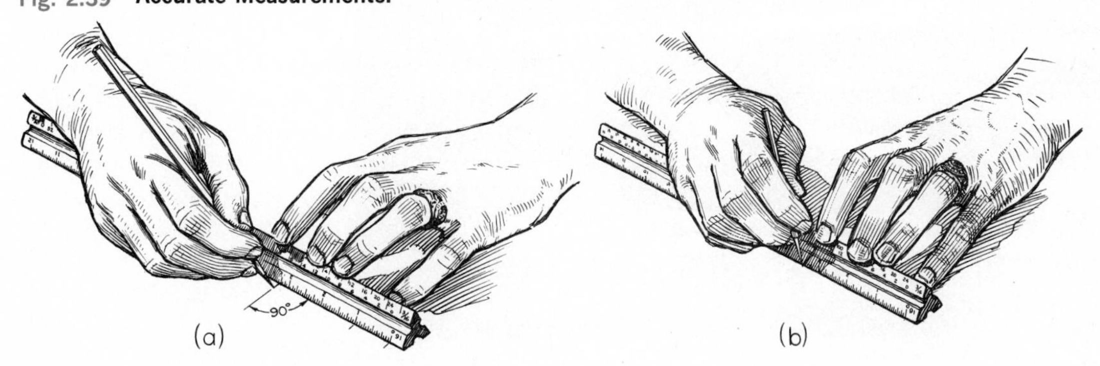


Fig. 2.39 **Accurate Measurements.**

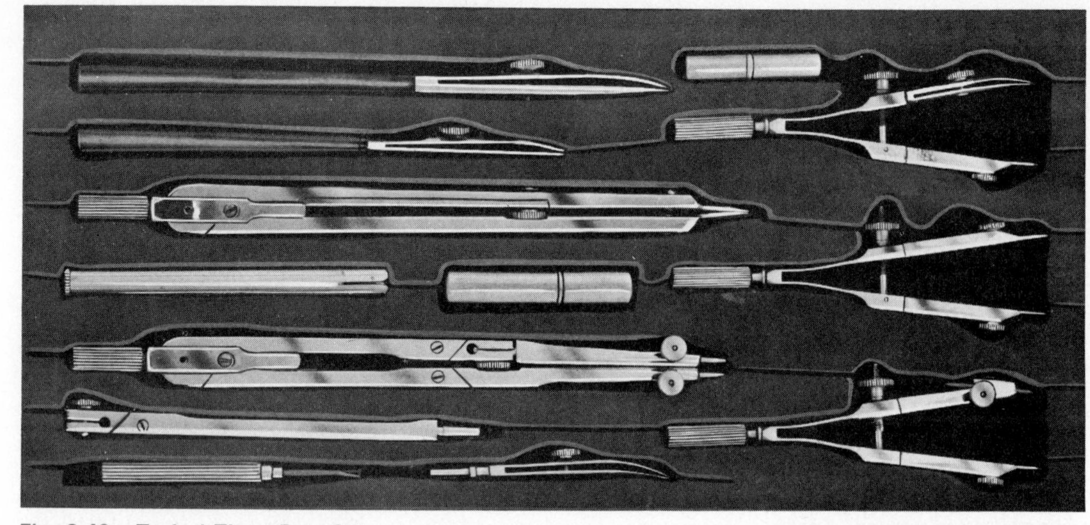

Fig. 2.40 **Typical Three-Bow Set.** *Courtesy Gramercy Guild Group, Inc.*

and in the end more economical, to purchase the best instruments that can be afforded. Good instruments will satisfy the most rigid requirements, and the satisfaction, saving in time, and improved quality of work that good instruments can produce will more than justify the higher price.

Unfortunately, the qualities of high-grade instruments are not likely to be recognized by the beginner, who is not familiar with the performance characteristics required and who is apt to be attracted by elaborate sets containing a large number of shiny low-quality instruments. Therefore, the student should obtain the advice of his drafting instructor, of an experienced draftsman, or of a reliable dealer.

A typical set of traditional style drawing instruments is shown in Fig. 2.40. This set contains a compass, dividers, bow pencil, bow pen, bow dividers, two ruling pens, and various auxiliary parts.

**2.34 Giant Bow Set.** Formerly it was general practice to make pencil drawings on detail paper and then to make an inked tracing from it on tracing cloth. As reproduction methods and transparent tracing papers were improved,

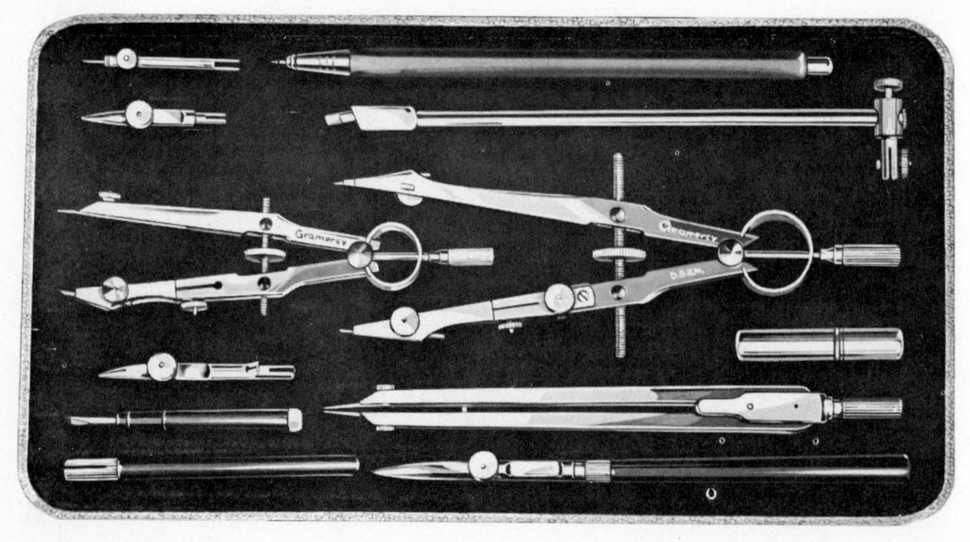


Fig. 2.41 **Giant Bow Set.**
*Courtesy Gramercy Guild Group, Inc.*

it was found that a great deal of time could be saved by making drawings directly in pencil with dense black lines on the tracing paper and making prints or photocopies therefrom, thus doing away with the preliminary pencil drawing on detail paper. Today, though inked tracings are still made when a fine appearance is necessary and where the greater cost is justified, the overwhelming proportion of drawings are made directly in pencil on tracing paper, vellum, polyester films, or pencil tracing cloth.

These developments have brought about the present giant bow sets that are offered now by all the major manufacturers, Fig. 2.41. The sets contain various combinations of instruments, but all feature a large bow compass in place of the traditional large compass. The large bow instrument is much sturdier and is capable of taking the heavy pressure necessary to produce dense black lines without losing the setting.

Most of the large bows are of the center-wheel type, Fig. 2.42 (a). Several manufacturers now offer different varieties of quick-acting bows. The large bow compass shown at (b) can be adjusted to the approximate setting by simply opening or closing the legs in the same manner as for the old style compass. Final adjustment of the setting is made with the center wheel.

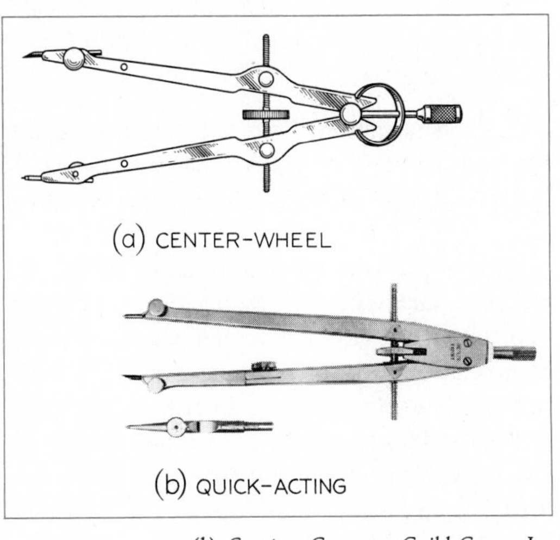


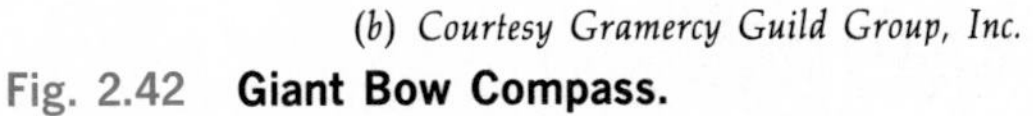


*(b) Courtesy Gramercy Guild Group, Inc.*

Fig. 2.42 **Giant Bow Compass.**

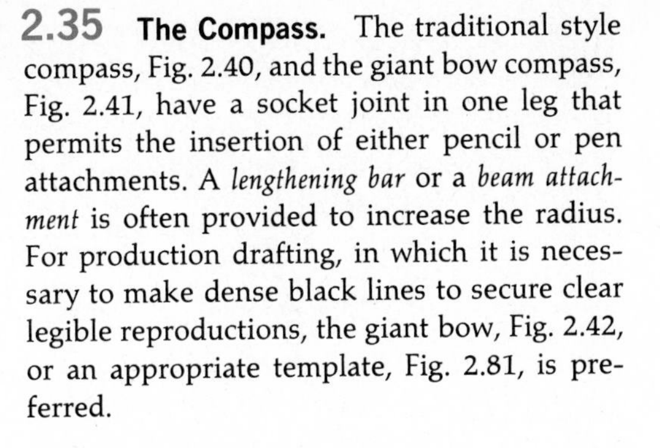

## 2.35 The Compass.

The traditional style compass, Fig. 2.40, and the giant bow compass, Fig. 2.41, have a socket joint in one leg that permits the insertion of either pencil or pen attachments. A *lengthening bar* or a *beam attachment* is often provided to increase the radius. For production drafting, in which it is necessary to make dense black lines to secure clear legible reproductions, the giant bow, Fig. 2.42, or an appropriate template, Fig. 2.81, is preferred.

## 2.36 Using the Compass.

These instructions apply generally both to the old style and the giant bow compasses. The compass, with pencil and inking attachments, is used for drawing circles of approximately 1″ radius or larger, Fig. 2.43. Most compass needle points have a plain end for use when the compass is converted into dividers, and a shoulder end for use as a compass. Adjust the needle point with the shoulder end out and so that the small point extends *slightly* farther than the pencil lead or pen nibs, Fig. 2.45 (d).

To draw a penciled circle, Fig. 2.43: (1) set off the required radius on one of the center lines, (2) place the needle point at the exact intersection of the center lines, (3) adjust the

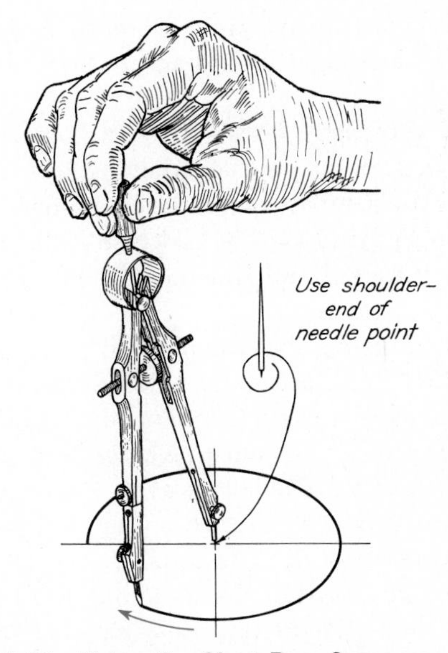


Fig. 2.43 **Using the Giant Bow Compass.**

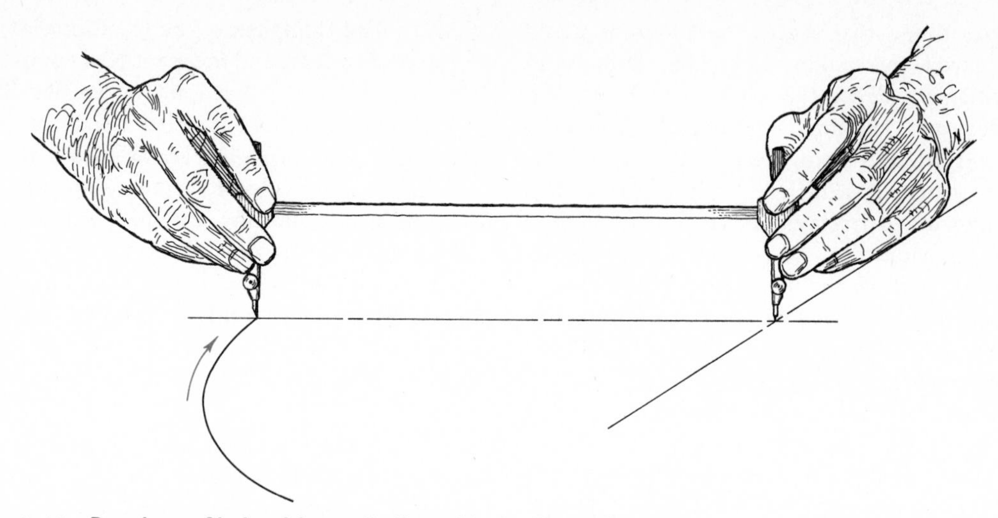

Fig. 2.44 **Drawing a Circle of Large Radius with the Beam Compass.**

compass to the required radius (1″ or more), (4) lean the compass forward and draw the circle clockwise while rotating the handle between the thumb and forefinger. To obtain sufficient weight of line, it may be necessary to repeat the movement several times.

Any error in radius will result in a doubled error in diameter; hence, it is best to draw a trial circle first on scrap paper or on the backing sheet and then check the diameter with the scale.

On drawings having circular arcs and tangent straight lines, draw the arcs first, whether in pencil or in ink, as it is much easier to connect a straight line to an arc than the reverse.

For very large circles a beam compass is preferred, or use the lengthening bar to increase the compass radius. Use both hands, as shown in Fig. 2.44, but be careful not to jar the instrument and thus change the adjustment.

When using the compass to draw construction lines, use a 4H to 6H lead so that the lines will be very dim. For required lines, the arcs and circles must be black, and softer leads must be used. However, since heavy pressure cannot be exerted on the compass as it can on a pencil, it is usually necessary to use a compass lead that is about one grade softer than the pencil used for the corresponding line work. For example, if an F pencil is used for visible lines drawn with the pencil, then an HB might be found suitable for the compass work. The hard leads supplied with the compass are usually unsatisfactory for most line work except construction lines. In summary, use leads in the compass that will produce arcs and circles that match the straight pencil lines.

It is necessary to exert pressure on the compass to produce heavy "reproducible" circles, and this tends to enlarge the compass center hole in the paper, especially if there are a number of concentric circles. In such cases, use a *horn center,* or center tack, in the hole, and place the needle point in the hole in the tack.

## 2.37 Sharpening the Compass Lead.

Various forms of compass lead points are illustrated in Fig. 2.45. At (a) a single elliptical face has been formed by rubbing on the sandpaper pad, as shown in Fig. 2.46. At (b) the point is narrowed by small side cuts. At (c) two long cuts and two small side cuts have been made so as to produce a point similar to that on a screwdriver. At (d) the cone point is prepared by chucking the lead in a mechanical pencil and shaping it in a pencil pointer. Avoid using leads that are too short to be exposed as shown.

In using the compass, *never use the plain end*

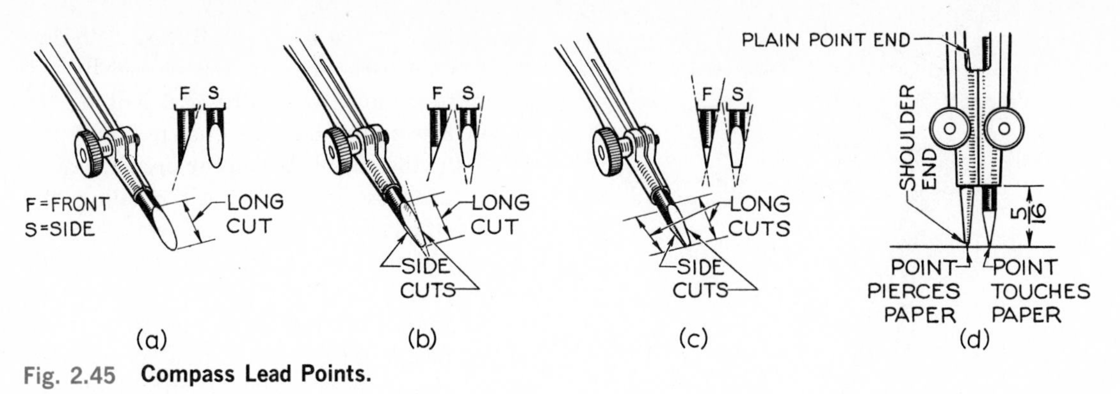


Fig. 2.45 **Compass Lead Points.**

Fig. 2.46 **Sharpening Compass Lead.**

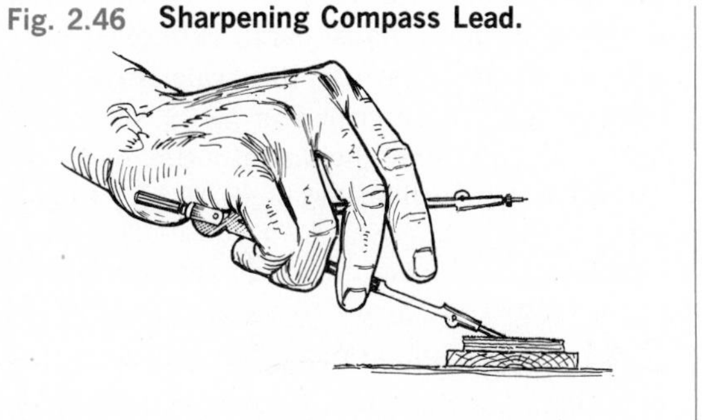

*of the needle point.* Instead, use the shoulder end, as shown in Fig. 2.45 (d), adjusted so that the tiny needle point extends about half-way into the paper when the compass lead just touches the paper.

## 2.38 Beam Compass.

The *beam compass,* or *trammel,* Fig. 2.47, is used for drawing arcs or circles larger than can be drawn with the regular compass and for transferring distances too great for the regular dividers. Besides steel points, pencil and pen attachments are provided. The beams may be made of nickel silver, steel, or wood, and are procurable in various lengths.

## 2.39 Dividers.

The dividers are similar to the compass in construction, and are made in square, flat, and round forms.

The friction adjustment should be loose enough to permit easy manipulation with one hand, as shown in Fig. 2.48. If the pivot joint is too tight, the legs of the compass tend to spring back instead of stopping at the desired point when the pressure of the fingers is released. To adjust the tension, use the small screwdriver.

Most dividers are provided with a *hair spring* so that minute adjustments can be made by turning the small thumbscrew.

Fig. 2.47 **Beam Compass.**

*Courtesy Gramercy Guild Group, Inc.*

Fig. 2.48 **Adjusting the Dividers.**

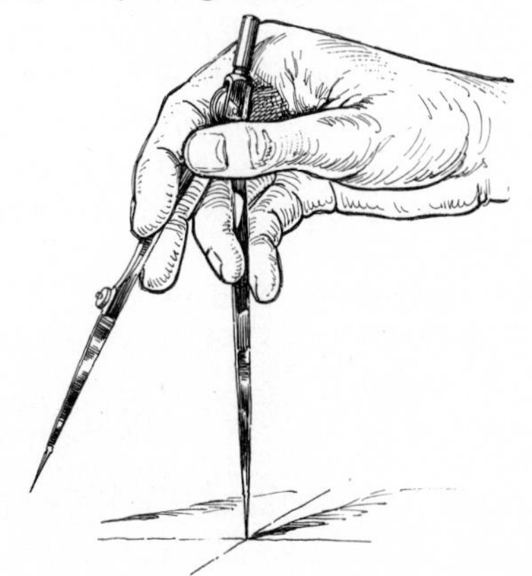

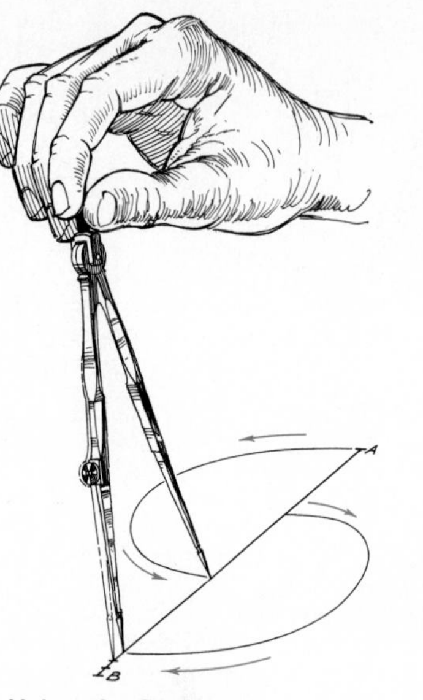


Fig. 2.49 **Using the Dividers.**

## 2.40 Using the Dividers.

The dividers, as the name implies, are used for *dividing* distances into a number of equal parts. They are used also for *transferring distances* or for *setting off* a series of equal distances. The dividers are used for spaces of approximately 1″ or more. For less than 1″ spaces, use the bow dividers, Fig. 2.52 (a). *Never use the large dividers for small spaces when the bow dividers can be used; the latter are more accurate.*

To divide a given distance into a number of equal parts, Fig. 2.49, the method is one of trial and error. Adjust the dividers with the fingers of the hand that holds them, to the approximate unit of division, estimated by eye. Rotate the dividers counterclockwise through 180°, and then clockwise through 180°, and so on, until the desired number of units has been stepped off. If the last prick of the dividers falls short of the end of the line to be divided, increase the distance between the divider points proportionately. For example, to divide the line AB, Fig. 2.49, into three equal parts, the dividers are set by eye to approximately one-third the length AB. When it is found that the trial radius is too small, the distance between the divider points is increased by one-third the remaining distance. If the last prick of the dividers is beyond the end of the line, a similar decreasing adjustment is made.

The student should avoid *cumulative errors,* which may result when the dividers are used to set off a series of distances end to end. To set off a large number of equal divisions, say 15, first set off 3 equal large divisions and then divide each into 5 equal parts. Wherever possible in such cases, use the scale instead of the dividers, as described in §2.32, or set off the total and then divide into the parts by means of the parallel-line method, §§4.15 and 4.16.

## 2.41 Proportional Dividers.

For enlarging or reducing a drawing, proportional dividers, Fig. 2.50, are convenient. They may be used also for dividing distances into a number of equal parts, or for obtaining a percentage reduction of a distance. For this purpose, points of division are marked on the instrument so as to secure the required subdivisions readily. Some instruments are calibrated to obtain special ratios, such as $1:\sqrt{2}$, the diameter of a circle to the side of an equal square, and feet to meters.

## 2.42 The Bow Instruments.

The bow instruments are classified as the *bow dividers, bow pen,* and *bow pencil.* A combination pen and pencil bow, usually with center-wheel adjustment, Fig. 2.51, and separate instruments with either side-wheel or center-wheel adjustment,

Fig. 2.50 **Proportional Dividers.**

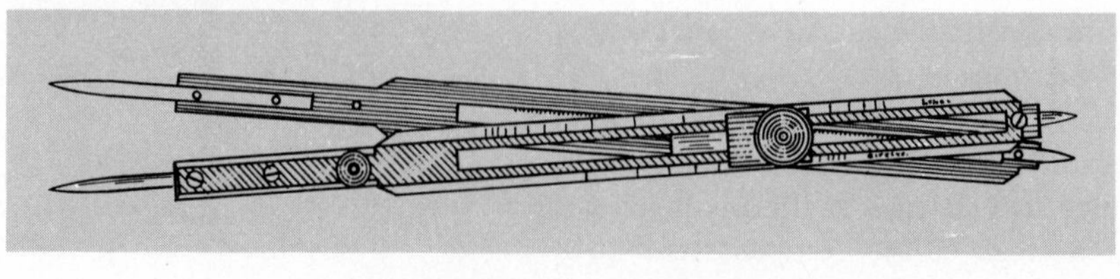

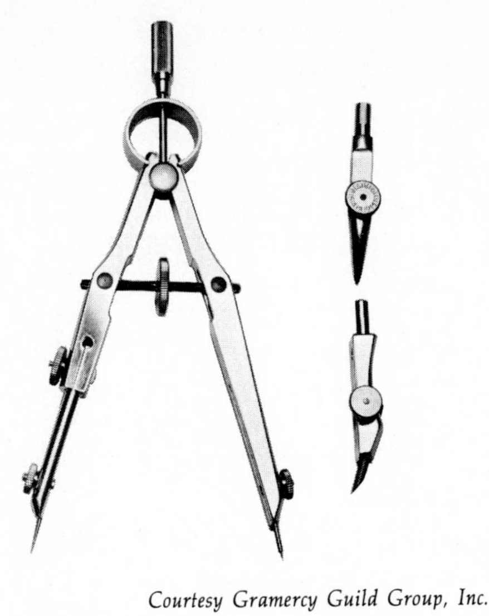

Courtesy Gramercy Guild Group, Inc.

Fig. 2.51 **Combination Pen and Pencil Bow.**

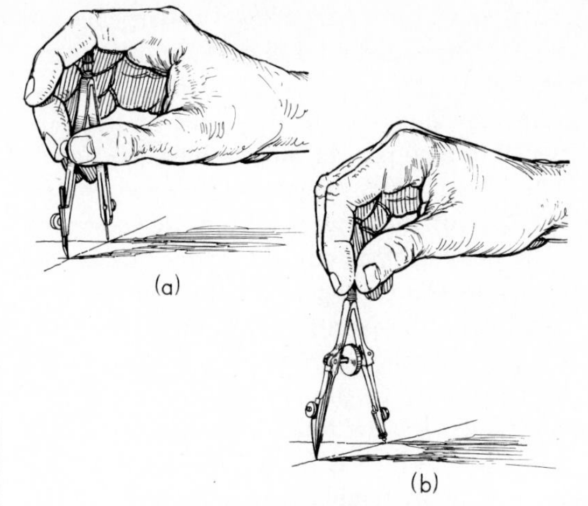


Fig. 2.53 **Using the Bow Instruments.**

Fig. 2.52, are available. The choice is a matter of personal preference.

**2.43 Using the Bow Instruments.** The bow pencil and bow pen are used for drawing circles of approximately 1″ radius or smaller. The bow dividers are used for the same purpose as the large dividers, but for smaller (approximately 1″ or less) spaces and more accurate work.

Whether the center-wheel or side-wheel instrument is used, the adjustment should be made with the thumb and finger of the hand that holds the instrument, Fig. 2.53 (a). The instrument is manipulated by twirling the head between the thumb and fingers, (b).

The lead is sharpened in the same manner as for the large compass, §2.37, except that for small radii, the inclined cut may be turned *inside* if preferred, Fig. 2.54 (a). For general use, the lead should be turned to the outside, as shown at (b). In either case, always keep the compass lead sharpened. *Avoid stubby compass*

Fig. 2.52 **Bow Instruments with Side Wheel.**

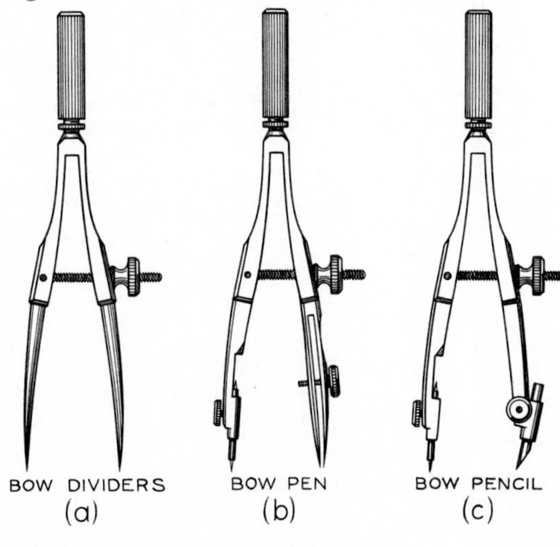


Fig. 2.54 **Compass-Lead Points.**

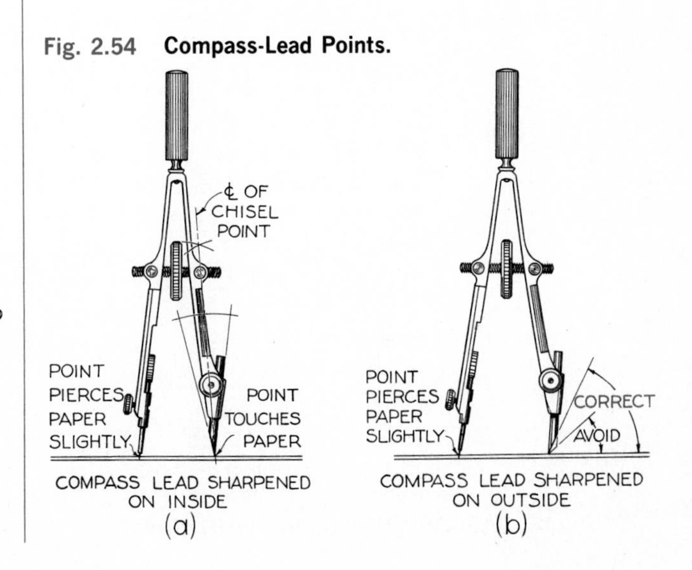

*leads,* which cannot be properly sharpened. At least $\frac{1}{4}''$ of lead should extend from the compass at all times.

In adjusting the needle point of the bow pencil or bow pen, be sure to have the needle extending slightly longer than the pen or lead, Fig. 2.45 (d), the same as for the large compass.

In drawing small circles, greater care is necessary in sharpening and adjusting the lead and the needle point, and especially in accurately setting the desired radius. If a $\frac{1}{4}''$-diameter circle is to be drawn, and if the radius is "off" only $\frac{1}{32}''$, the total error on diameter is 25 percent, which is far too much error.

Appropriate templates may be used also for drawing small circles. See Fig. 2.81.

## 2.44 Drop Spring Bow Pencil and Pen.

These compasses, Fig. 2.55, are designed for drawing multiple identical small circles, such as drill holes or rivet heads. A central pin is made to move easily up and down through a tube to which the pen or pencil unit is attached. To use the instrument, hold the knurled head of the tube between the thumb and second finger, placing the first finger on top of the knurled head of the pin. Place the point of the pin at the desired center, lower the pen or pencil until it touches the paper, and twirl the instrument clockwise with the thumb and second finger. Then lift the tube

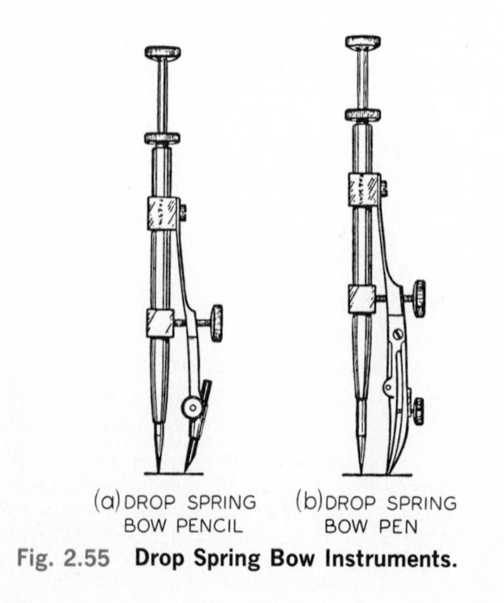

(a) DROP SPRING BOW PENCIL (b) DROP SPRING BOW PEN

Fig. 2.55 **Drop Spring Bow Instruments.**

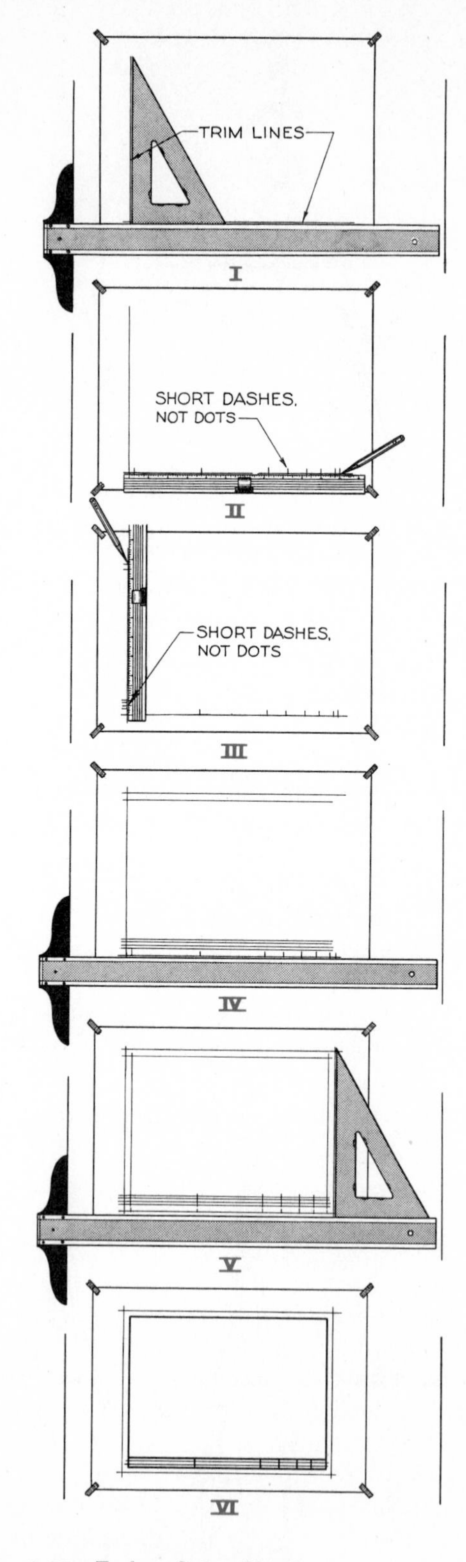


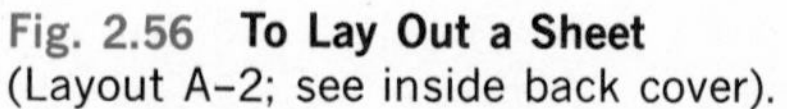

Fig. 2.56 **To Lay Out a Sheet** (Layout A-2; see inside back cover).

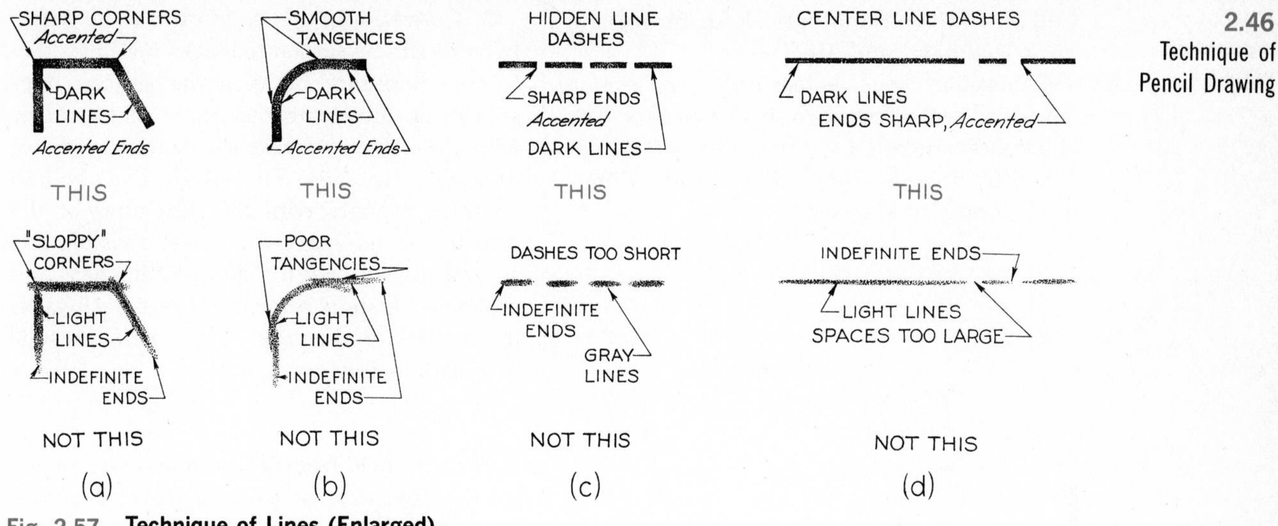


Fig. 2.57 **Technique of Lines (Enlarged).**

independently of the pin, and finally lift the entire instrument.

## 2.45 To Lay Out a Sheet.

Fig. 2.56. (See also Layout A-2, inside back cover.) After the sheet has been attached to the board, as explained in §2.7, proceed as follows:

I. Using the T-square, draw a horizontal *trim line* near the lower edge of the paper; and then using the triangle, draw a vertical trim line near the left edge of the paper. Both should be *light construction lines.*

II. Place the scale along the lower trim line with the full-size scale up. Draw short light dashes *perpendicular* to the scale at the required distances. See Fig. 2.39 (a).

III. Place the scale along the left trim line with the full-size scale to the left, and mark the required distances with short light dashes perpendicular to the scale.

IV. Draw horizontal construction lines with the aid of the T-square through the marks at the left of the sheet.

V. Draw vertical construction lines, *from the bottom upward,* along the edge of the triangle through the marks at the bottom of the sheet.

VI. Retrace the border and the title strip to make them heavier. Notice that the layout is made independently of the edges of the paper.*

## 2.46 Technique of Pencil Drawing.

By far the greater part of commercial drafting is executed in pencil. Most prints or photocopies are made from pencil tracings, and all ink tracings must be preceded by pencil drawings. It should therefore be evident that skill in drafting chiefly implies skill in pencil drawing.

*Technique* is a style or quality of drawing imparted by the individual draftsman to his work. It is characterized by crisp black linework and lettering. Technique in lettering is discussed in §3.12.

1. DARK ACCENTED LINES. The pencil lines of a finished pencil drawing or tracing should be very dark, Fig. 2.57. Dark crisp lines are necessary to give punch or snap to the drawing. Ends of lines should be accented by a little extra pressure on the pencil, (a). Curves should be as dark as other lines, (b). Hidden-line dashes and center-line dashes should be carefully estimated as to length and spacing,

*In industrial drafting rooms the sheets are available, cut to standard sizes, with border and title strips already printed. Drafting supply houses can supply such papers, printed to order, to schools for little or no extra cost.

and should be of uniform width throughout their length, (c) and (d).

Dimension lines, extension lines, section lines, and center lines also should be dark. The difference between these lines and visible lines is mostly in width—there is very little difference, if any, in blackness.

Fig. 2.58 **Testing Density of Lines.**

A simple way to determine whether your lines on tracing paper or cloth are dense black is to hold the tracing up to the light, Fig. 2.58. Lines that are not opaque black will not print clearly by blueprinting or otherwise.

Construction lines should be made with a sharp, hard pencil and *should be so light that they need not be erased* when the drawing is completed.

2. Contrast in Lines. Contrast in pencil lines should be similar to that of ink lines; that is, the difference between the various lines should be mostly in the *widths* of the lines, with little if any difference in the degree of darkness, Fig. 2.59. The visible lines should contrast strongly with the thin lines of the drawing. If necessary, draw over a visible line several times to get the desired thickness and darkness. A short retracing stroke backwards (to the left), producing a jabbing action, results in a darker line.

## 2.47 Pencil Tracing.

While some pencil tracings are made of a drawing placed underneath the tracing paper (usually when a great deal of erasing and changing is necessary on the original drawing), most drawings today are made directly in pencil on tracing paper, pencil tracing cloth, films, or vellum. These are not tracings but pencil drawings, and the methods and technique are the same as previously described for pencil drawing.

In making a drawing directly on a tracing medium, a smooth sheet of heavy white drawing paper should be placed underneath. Such a sheet is known as a *backing sheet*. The whiteness of the backing sheet improves the visibility of the lines, and the hardness of the surface makes it possible to exert pressure on the pencil and produce dense black lines without excessive grooving of the paper.

These tracings, or drawings, are intended to be reproduced by blueprinting or by other kindred processes, Chapter 15, and all lines must be dark and cleanly drawn.

Fig. 2.59 **Contrast of Lines (Enlarged).**

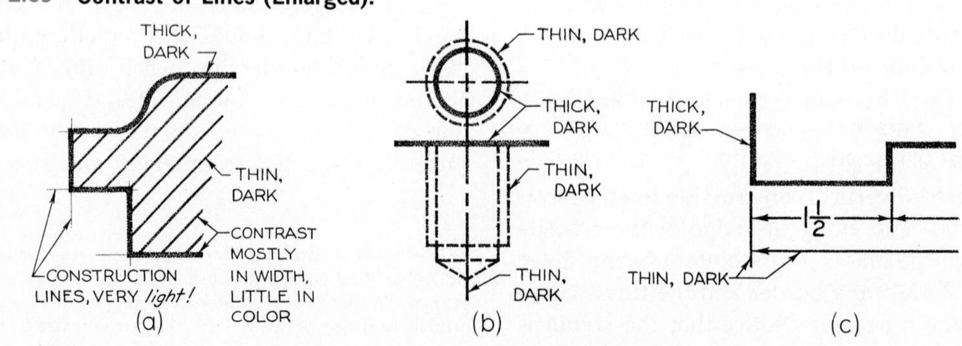

**2.48 Technical Fountain Pens.** The *technical fountain pen,* Fig. 2.60, with the tube and needle point, is available in several line widths. Many prefer this type of pen, for the line widths are fixed and it is suitable for freehand and mechanical lettering and line work. The pen requires an occasional filling and a minimum of skill to use. For uniform line work, the pen should be used perpendicular to the paper. For best results, follow the manufacturer's recommendations for operation and cleaning.

**2.49 Ruling Pens.** The *ruling pen,* Fig. 2.61, should be of the highest quality, with blades of high-grade tempered steel sharpened properly at the factory. The nibs should be sharp, but not sharp enough to cut the paper. See §2.55 for sharpening the ruling pen.

The *detail pen,* capable of holding a considerable quantity of ink, is extremely useful for drawing long heavy lines, (b), and is preferred by some for general use.

For methods of using the ruling pen, see §2.52.

**2.50 Special Pens.** The *contour pen,* Fig. 2.62 (a), is used for tracing freehand curves, such as contour lines on maps. The *railroad pen,* (b), is used for drawing two parallel lines straight or moderately curved, as for roads and railroads.

**2.51 Drawing Ink.** Drawing ink, Fig. 2.63 (a), is composed chiefly of carbon in colloidal suspension and gum. The fine particles of carbon give the deep, black luster to the ink, and the gum makes it waterproof and quick to dry. The ink bottle should not be left uncovered, as evaporation will cause the ink to thicken. Thickened ink may be thinned by adding a few drops of a solution of four parts of aqua ammonia to one part of distilled water. A convenient pen-filling inkstand, which requires the use of only one hand, is shown at (b).

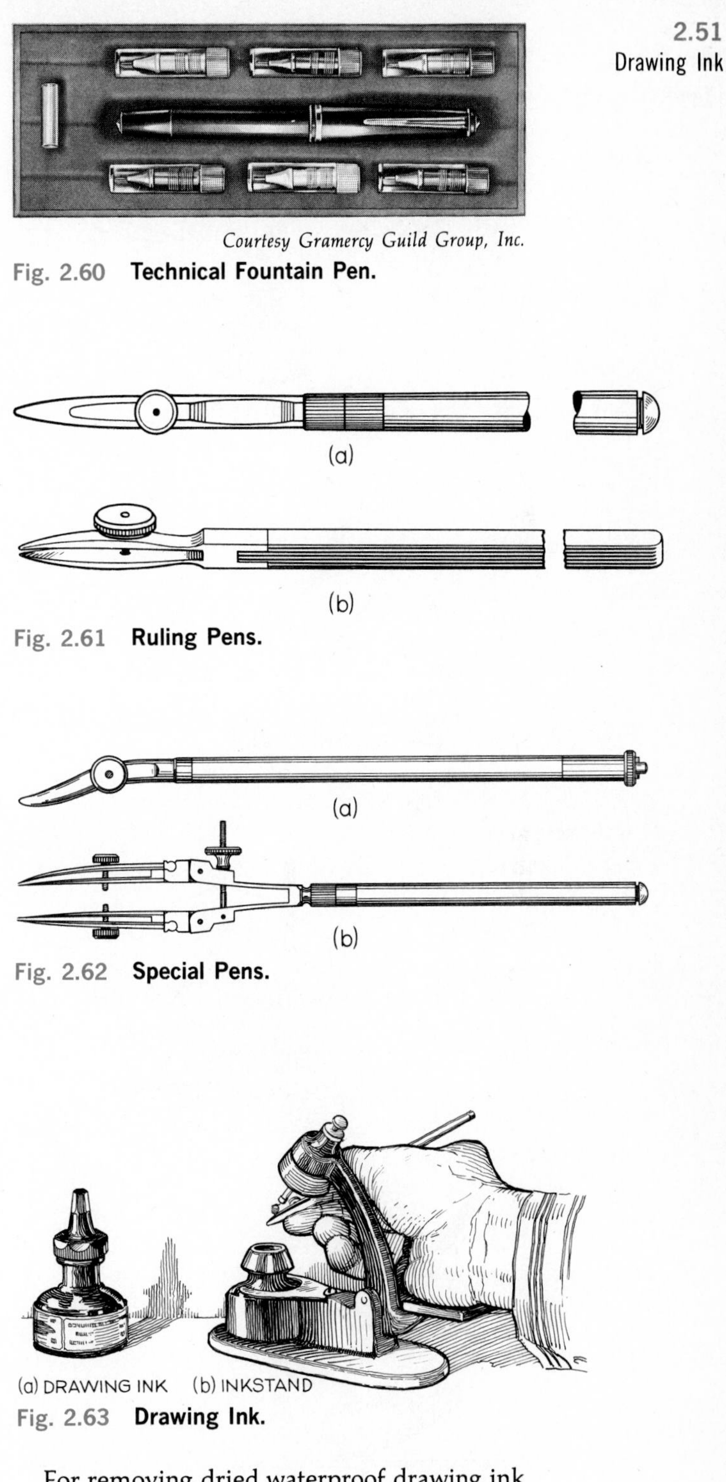


*Courtesy Gramercy Guild Group, Inc.*

Fig. 2.60 **Technical Fountain Pen.**

Fig. 2.61 **Ruling Pens.**

Fig. 2.62 **Special Pens.**

Fig. 2.63 **Drawing Ink.**

For removing dried waterproof drawing ink from pens or instruments, pen-cleaning fluids are available at dealers.*

*Higgins Pen Cleaner or Leroy Pen Cleaning Fluid.

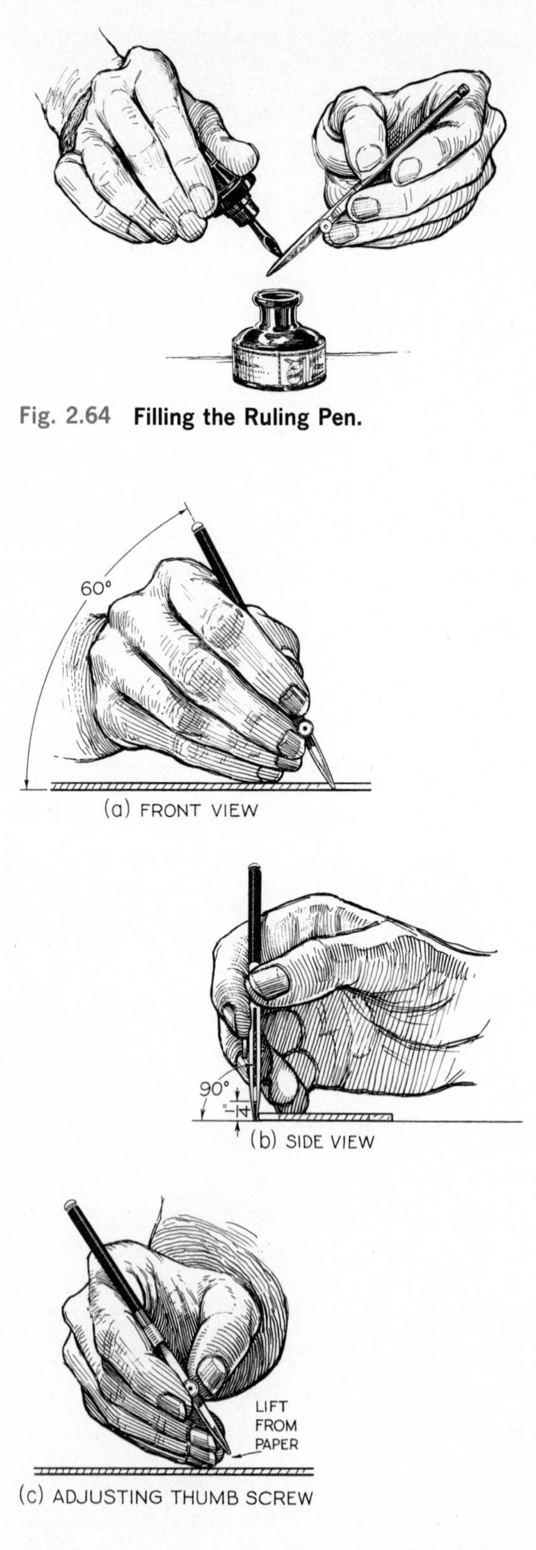


Fig. 2.64 **Filling the Ruling Pen.**

Fig. 2.65 **Position of Hand in Using the Ruling Pen.**

**2.52 Use of the Ruling Pen.** The ruling pen, Fig. 2.60, is used to ink lines drawn with instruments, never to ink freehand lines or freehand lettering. The proper method of filling the pen is shown in Fig. 2.64. The hands may be steadied by touching the little fingers together. Twisting instead of pulling the stopper from a new bottle of ink, or one that has not been used for some time, will often save the stopper from being broken. After the pen has been filled, the ink should stand about $\frac{1}{4}''$ deep in the pen.

Horizontal lines and vertical lines are drawn in the same manner as for the corresponding pencil lines, Figs. 2.21 (a) and 2.22 (a).

Practically all the difficulties encountered in the use of the ruling pen may be attributed to (1) incorrect position of the pen, (2) lack of allowance for the quick-drying properties of drawing ink, and (3) improper control of thickness of lines and incorrect junctures.

1. Position of the Pen. The pen should lean at an angle of about 60° with the paper in the direction in which the line is being drawn and in a vertical plane containing the line, Fig. 2.65 (a) and (b). In general, the more the pen is leaned toward the paper, the thicker the line will be; and the more nearly vertical the pen is held, the thinner the line will be. The thumbscrew is faced away from the straight edge, and is adjusted, (c), with the thumb and forefinger of the same hand that holds the instrument. The correct position of the pen and the resulting correct line are shown in Fig. 2.66 (a).

If the nibs are pulled tightly against the T-square or triangle, the effect is to close the nibs and thus reduce the thickness of the lines, (b). If the pen is held as shown at (c), the ink will come in contact with the T-square and paper at the same time and will run under the T-square and cause a blot on the drawing. The same result may occur if, in filling the pen, ink is deposited on the outside of the nib that touches the T-square. If the pen is held as shown at (d), the outside nib of the ruling pen may not touch the paper, and the line is apt to be ragged.

When the line has been correctly drawn, care must be exercised not to touch the wet

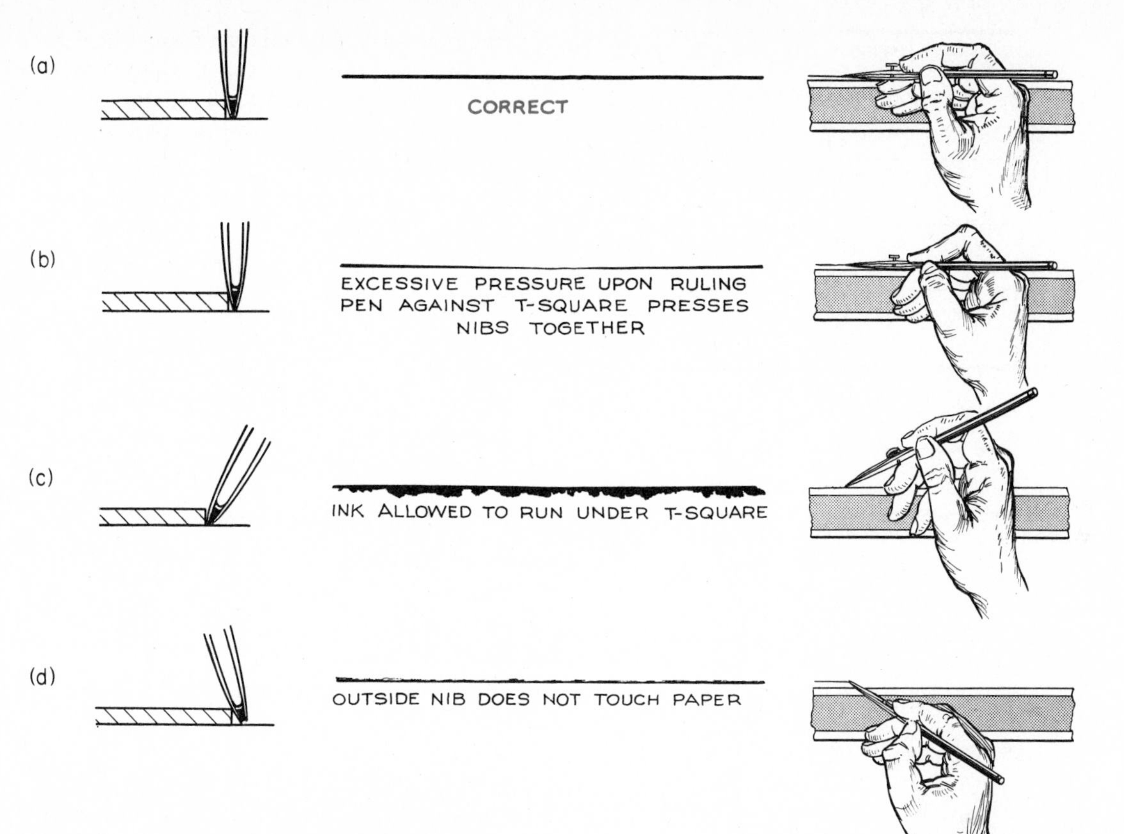


Fig. 2.66 **Errors in Using the Ruling Pen.**

ink when removing the T-square or triangle. The triangle or T-square should be carefully moved away from the line before being picked up. If more than $\frac{1}{4}''$ of ink is placed in the pen, the ink will flow too readily, thus increasing the danger of a blot.

Some draftsmen prefer to place another triangle under the one being used, as shown in Fig. 2.67, to raise the first triangle above the paper and thus prevent ink from running under the edge. This is especially useful if the lines are heavy and blots may easily occur.

2. Correct Use of Drawing Ink. One of the most common difficulties is that the pen will not "feed." A clogged pen often may be started by touching it to drafting tape or the back of a finger. If the pen will not make a fine line, the nibs have been screwed too close together, the ink has been allowed to partially dry and thicken in the pen, the ink in the bottle is too thick from age and exposure to air, or the pen is dull and needs sharpening.

The pen should never be filled until the draftsman is ready to use it, because the ink dries quickly when not flowing from the pen. *Ink should never be allowed to dry in any instrument. Never lay a ruling pen down with ink in it.* Some drawing inks have an acid content that will "pit" a ruling pen if left to dry in the pen repeatedly. The student should clean the pen frequently by slipping a stiff blotter or a folded cloth between the nibs. *Sandpaper should never be used to remove dry ink.* Dry ink should be removed by scraping very lightly with a pen knife. Ruling pens constructed so that the

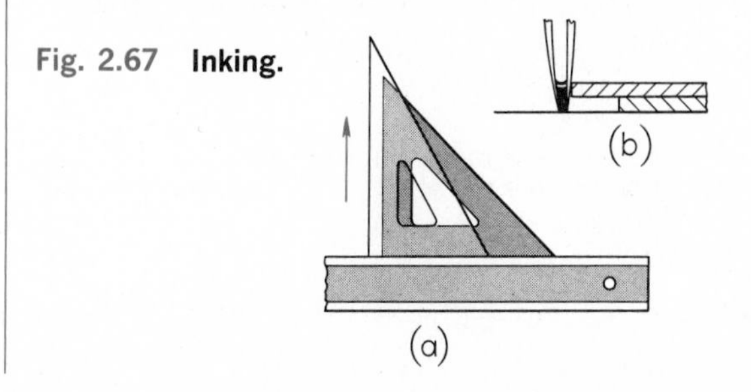


Fig. 2.67 **Inking.**

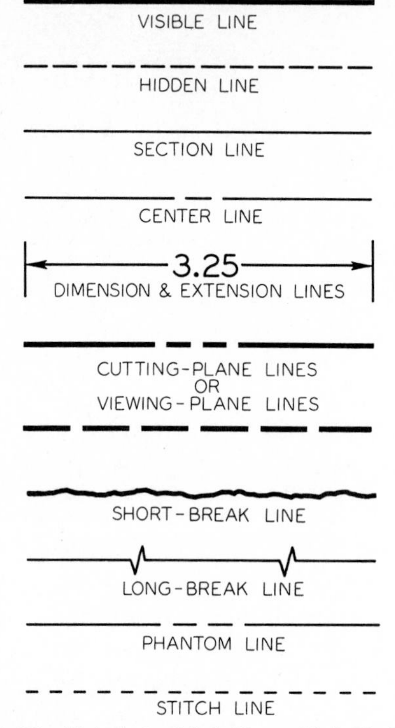


Fig. 2.68 **Alphabet of Ink Lines (Full Size).**

nibs will separate for cleaning are available in a number of good designs.

The stopper should always be kept in the bottle when it is not in use, since exposure of the ink to the air causes it to become thick and difficult to use.

3. How to Control Thickness of Lines. The various widths of lines used for inked drawings or tracings are shown in Fig. 2.68. The draftsman must first develop a trained eye to distinguish fine variations, and must also acquire skill in producing the desired widths. The student must remember that the thumbscrew alone does not control the width of the line. The following factors affect the width of a line with a given setting of the thumbscrew.

*Factors that tend to make line heavier:*

1. Excess ink in the pen.
2. Slow movement of the pen.
3. Dull nibs.
4. Caked particles of ink on the nibs.
5. Leaning the pen more toward the paper.
6. Soft working surface.

*Factors that tend to make line finer:*

1. Small amount of ink in the pen.
2. Rapid movement of the pen.
3. Sharp nibs.
4. Fresh ink and clean pen.
5. Pen more nearly vertical.
6. Hard working surface.

Before making a new ink line on a drawing, the thickness of line should be tested by drawing a test line on a separate piece of paper under the same conditions. *Never test the pen freehand, or on a different kind of paper.* Always use a straightedge, and use identical paper.

If excess ink is in the pen or if wet lines are allowed to intersect previously drawn lines that are still wet, teardrop ends and rounded corners will result, Fig. 2.69 (a) and (b).

The ruling pen is used in inking irregular curves, as well as straight lines, as shown in

Fig. 2.69 **Ink Lines.**

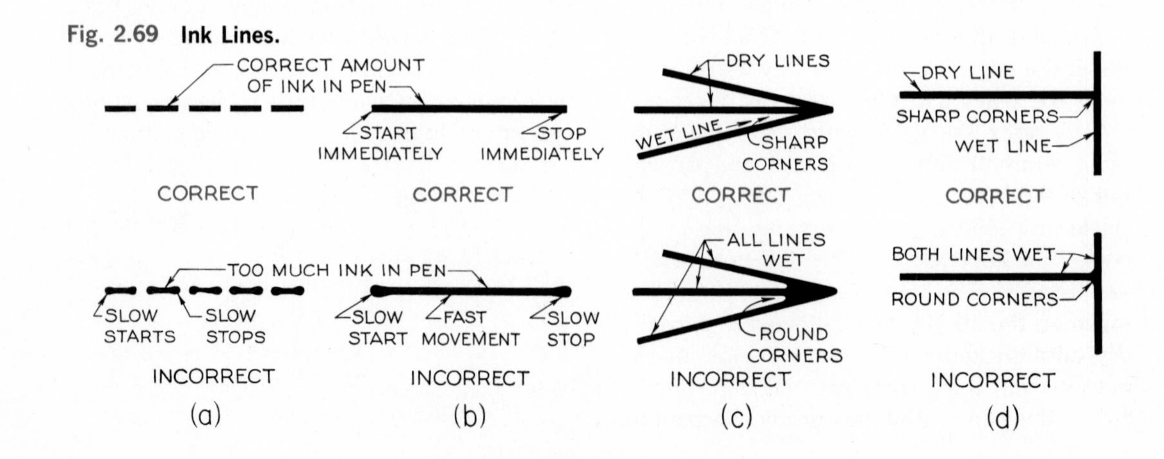

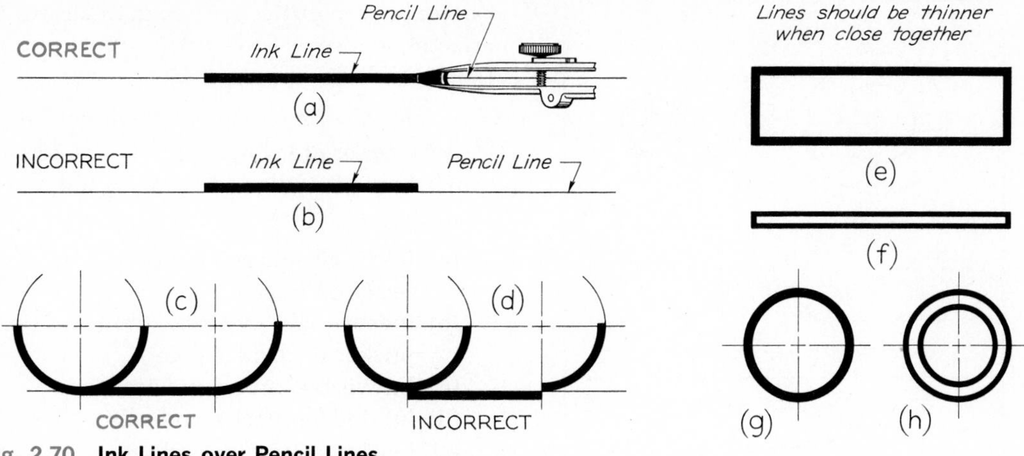


Fig. 2.70 **Ink Lines over Pencil Lines.**

Figs. 2.78 and 2.79. The pen should be held more nearly vertical when used with an irregular curve than when used with the T-square or a triangle. The ruling pen should lean only slightly in the direction in which the line is being drawn. It should be in a vertical plane containing a tangent to the curve at the position of the pen.

Some draftsmen insert a triangle under the irregular curve, back from the line, in order to raise the curve from the paper and prevent ink running under the edge. Another effective method is to glue several thin pieces of plastic to the faces of the curve or to the triangles.

## 2.53 Tracing in Ink.

To make a tracing in ink, tracing paper, §2.63, or tracing cloth, §2.64, is fastened over the drawing, and the copy is made by tracing the lines. When a drawing is important enough to warrant the use of ink, it is generally made on tracing cloth. Although the glazed side of the cloth formerly was intended as the working surface, most draftsmen prefer the dull side, because it takes ink better and can be marked with a pencil. It is common practice to make the pencil drawing directly upon the tracing cloth and then trace it with ink, thus eliminating the traditional pencil drawing on detail paper.

Before the ink is applied, the cloth should be dusted with a small quantity of *pounce,* which should be rubbed in lightly with a soft fabric and then thoroughly removed with a clean cloth. Instead of the special drafting pounce, any slightly abrasive powder, such as talcum or chalk dust (calcium carbonate), applied with an ordinary blackboard eraser, may be used.

A greater difference in the widths of lines is necessary on tracings than on pencil drawings, because the contrast between blue and white on blueprints is not so great as that between black and white on drawings. Visible lines should be very bold. Extension lines, dimension lines, section lines, and center lines should be fine, but strong enough to insure positive printing.

In inking or tracing a pencil line, the ink line should be centered over the pencil line, as shown in Fig. 2.70 (a), and not along one side as at (b). If this is done correctly in the case of tangent arcs, the line thicknesses will overlap at the points of tangency, as at (c), resulting in smooth tangencies. Incorrect practice is shown in exaggerated form at (d). Tangent points should be constructed in pencil, §§4.35 to 4.42, to assist in making smooth connections.

Make visible lines full thickness if the lines are spaced well apart, as shown at (e) and (g). When they are close together, the lines should be made thinner, (f) and (h).

Pencil guide lines for lettering should be drawn directly upon the tracing paper or cloth, since guide lines on the drawing underneath

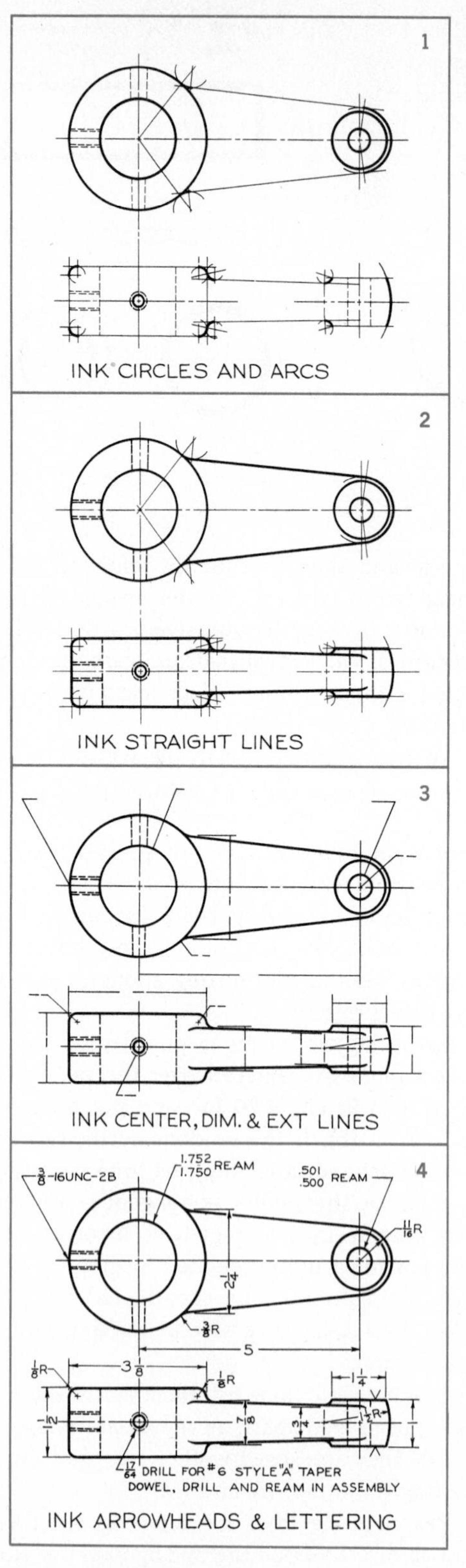


Fig. 2.71 **Order of Inking.**

cannot be seen distinctly enough to furnish an accurate guide for letter heights. For conventional ink lines, see Fig. 2.68.

## 2.54 Order of Inking.

Fig. 2.71. A definite order should be followed in inking a drawing or tracing, as follows:

1. (a) Mark all tangent points in pencil directly on tracing.
   (b) Indent all compass centers (with pricker or divider point).
   (c) Ink visible circles and arcs.
   (d) Ink hidden circles and arcs.
   (e) Ink irregular curves, if any.
2. (1st: horizontal; 2nd: vertical; 3rd: inclined)
   (a) Ink visible straight lines.
   (b) Ink hidden straight lines.
3. (1st: horizontal; 2nd: vertical; 3rd: inclined) Ink center lines, extension lines, dimension lines, leader lines, and section lines (if any).
4. (a) Ink arrowheads and dimension figures.
   (b) Ink notes, titles, etc. (pencil guide lines directly on tracing.)

Some draftsmen prefer to ink center lines before indenting the compass centers because of the possibility of ink going through the holes and causing blots on the back of the sheet.

## 2.55 To Sharpen the Ruling Pen.

If a ruling pen is subjected to frequent or extended use, its nibs will become so worn that good lines cannot be drawn with it. The correct point is shown in Fig. 2.72 (a), and a characteristic worn point is shown at (b). The nibs at (c) are too pointed, and as a result the ink tends to hang suspended in the pen and not touch the paper. The nibs at (d) are too rounded, or blunt, in which case the ink tends to run out of the pen onto the paper too readily. Another indication of a defective ruling pen is a scratchy contact with the paper, or the necessity of pressing the points firmly into the paper to get the ink to flow. The reason for this condition is illustrated in Fig. 2.73 (b), which shows one nib much longer than the other.

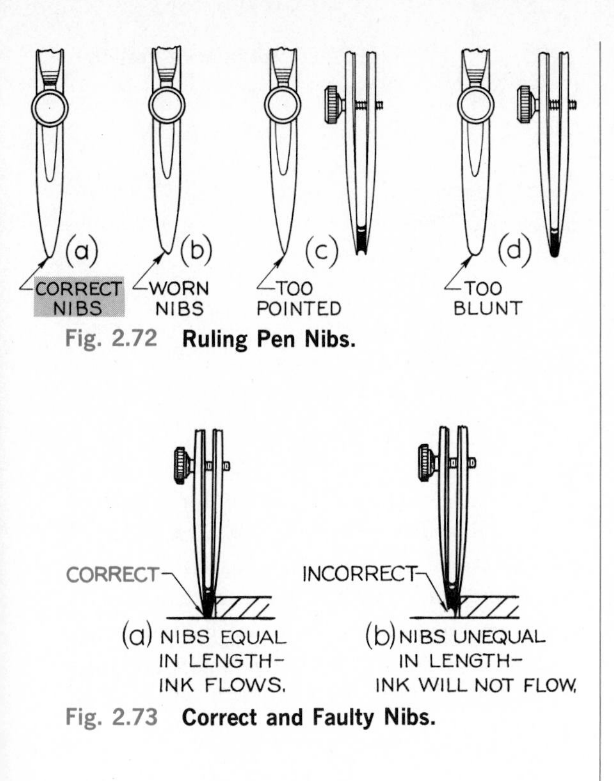

Fig. 2.72 **Ruling Pen Nibs.**

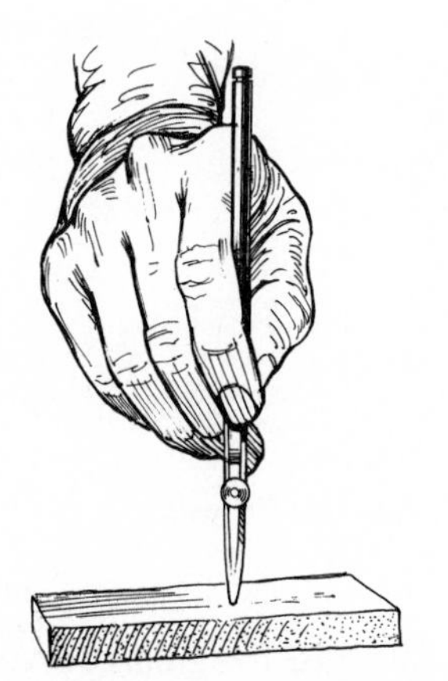

Fig. 2.73 **Correct and Faulty Nibs.**

Fig. 2.74 **Equalizing the Lengths of Nibs.**

Fig. 2.75 **Sharpening the Ruling Pen.**

A hard Arkansas oil stone is excellent for sharpening ruling pens. If the nibs of the ruling pen are unequal in length, they should first be equalized by moving the pen, with the nibs together, across the stone lightly with an oscillating movement from left to right, as shown in Fig. 2.74. To sharpen the nibs, they should be opened and each nib sharpened on the outside, as shown in Fig. 2.75, rolling the pen slightly from side to side to preserve the convex surface of the nib. Great care must be exercised to prevent oversharpening one nib and thus shortening it. The bright points, indicating dullness, should be carefully watched, and the nibs should be sharpened until the bright points disappear. Finally, to make sure that one nib has not been shortened, a few very light strokes should again be taken, as in Fig. 2.74. No attempt should ever be made to sharpen the inside of the nibs, for this always results in a slight convexity, which will ruin the pen.

**2.56 Ink Erasing.** Mistakes are certain to occur in inking, and correct methods of erasing should be considered a part of the technique. For general ink or pencil erasing, the Weldon Roberts India or the Eberhard Faber Ruby eraser is recommended. Ink erasers are usually gritty and too abrasive, and their use tends to destroy the surface of the paper. If this occurs, it may be impossible to ink over the erased area. Best results are obtained if a smooth hard surface, such as a triangle, is placed under the area being erased.

An application of pounce or chalk dust will improve the surface and prevent running of the ink. The erasing shield, Fig. 2.18, should be used to protect lines adjacent to the area to be erased.

When an ink blot is made, the excess ink should be taken up with a blotter, or smeared with the finger if a blotter is not available, and not allowed to soak into the paper. When the spot is thoroughly dry, the remaining ink can be erased easily.

For cleaning untidy drawings or for removing the original pencil lines from an inked drawing, a sponge rubber or Artgum is useful.

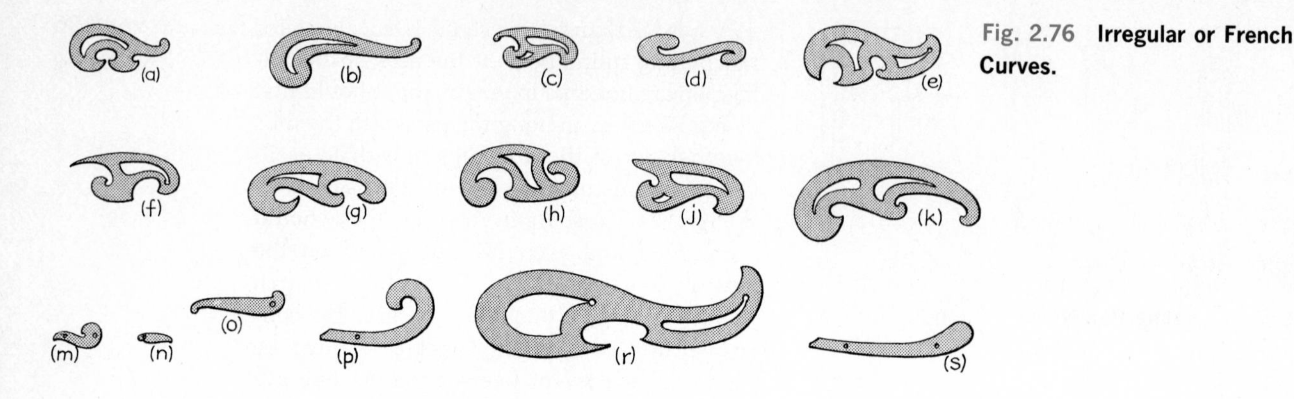


Fig. 2.76 **Irregular or French Curves.**

The Artgum is recommended for general use. Pencil lines or dirt can be removed from tracing cloth by rubbing lightly with a cloth moistened with carbon tetrachloride (Carbona) or benzine (Energine). Use either with care in a well-ventilated area.

When erasure on cloth damages the surface, it may be restored by rubbing the spot with soapstone and then applying pounce or chalk dust. If the damage is not too great, an application of the powder will be sufficient.

When a gap in a thick ink line is made by erasing, the gap should be filled in with a series of fine lines that are allowed to run together. A single heavy line is difficult to match and is more likely to run and cause a blot.

In commercial drafting rooms, the electric erasing machine, Fig. 2.19, is usually available to save the time of the draftsman.

## 2.57 Irregular Curves.

When it is required to draw mechanical curves other than circles or circular arcs, an *irregular* or *French curve* is generally employed. Many different forms and sizes of curves are manufactured, as suggested by the more common forms illustrated in Fig. 2.76.

The curves are composed largely of successive segments of the geometric curves, such as the ellipse, parabola, hyperbola, and involute. The best curves are made of highly transparent plastic. Among the many special types of curves that are available are hyperbolas, parabolas, ellipses, logarithmic spirals, ship curves, railroad curves.

*Adjustable curves,* Fig. 2.77, are also available. The curve shown at (a) consists of a core of lead, enclosed by a coil spring attached to a flexible strip. The one at (b) consists of a *spline,* to which *ducks* (weights) are attached. The spline can be bent to form any desired curve, limited only by the elasticity of the material. An ordinary piece of solder wire can be used very successfully by bending the wire to the desired curve.

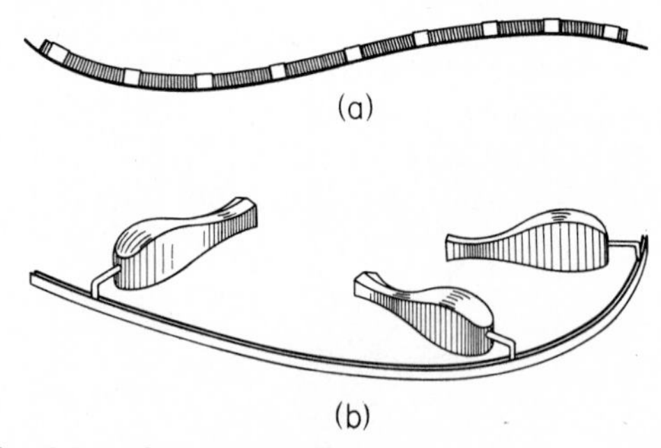

Fig. 2.77 **Adjustable Curves.**

## 2.58 Using the Irregular Curve.

The irregular curve is a device for the *mechanical drawing of curved lines and should not be applied directly to the points,* or used for purposes of producing an initial curve. The proper use of the irregular curve requires skill, especially when the lines are to be drawn in ink. After points have been plotted through which the curve is to pass, a light pencil line should be sketched freehand smoothly through the points.

To draw a mechanical line over the freehand line with the aid of the irregular curve, it is

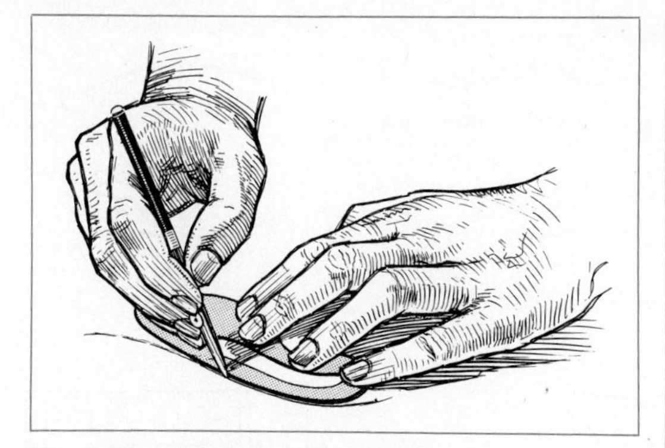
Fig. 2.78 **Using the Irregular Curve.**

only necessary to match the various segments of the irregular curve with successive portions of the freehand curve and to draw the line with pencil or ruling pen along the edge of the curve, Fig. 2.78. It is very important that the irregular curve match the curve to be drawn for some distance at each end beyond the segment to be drawn for any one setting of the curve, as shown in Fig. 2.79. When this rule is observed, the successive sections of the curve will be tangent to each other, without any abrupt change in the curvature of the line. In placing the irregular curve, the short-radius end of the curve should be turned toward the short-radius part of the curve to be drawn; that is, the portion of the irregular curve used should have the same curvilinear tendency as the portion of the curve to be drawn. This will prevent abrupt changes in direction.

The draftsman should change his position with respect to the drawing when necessary, so that he always works on the edge of the curve away from him; that is, he should avoid working on the near side of the curve.

When plotting points to establish the path of a curve, it is desirable to plot more points, and closer together, where sharp turns in the curve occur.

Free curves may also be drawn with the compass, as shown in Fig. 4.42.

For symmetrical curves, such as an ellipse, Fig. 2.80, use the same segment of the irregular curve in two or more opposite places. For example, at (a) the irregular curve is matched to the curve and the line drawn from 1 to 2. Light pencil dashes are then drawn directly on the irregular curve at these points (the curve will take pencil marks well if it is lightly "frosted" by rubbing with a hard pencil eraser). At (b) the irregular curve is turned over and matched so that the line may be drawn from 2 to 1. In similar manner, the same segment is used again at (c) and (d). The ellipse is completed by filling in the gaps at the ends by using the irregular curve or, if desired, the compass.

Fig. 2.79 **Settings of Irregular Curve.**

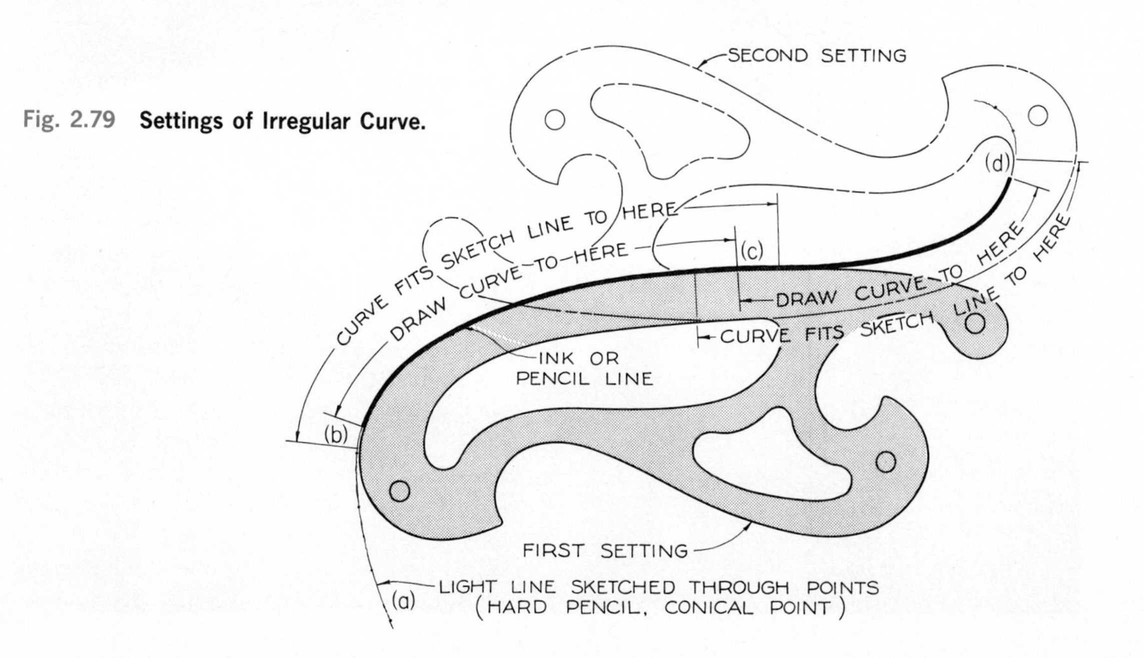

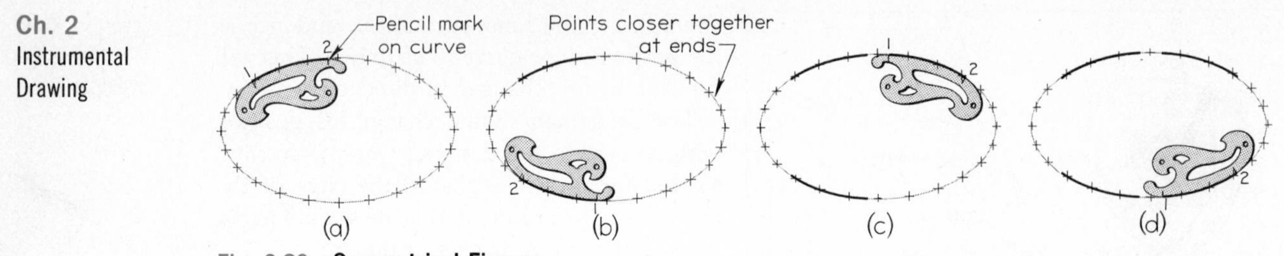


Fig. 2.80 **Symmetrical Figures.**

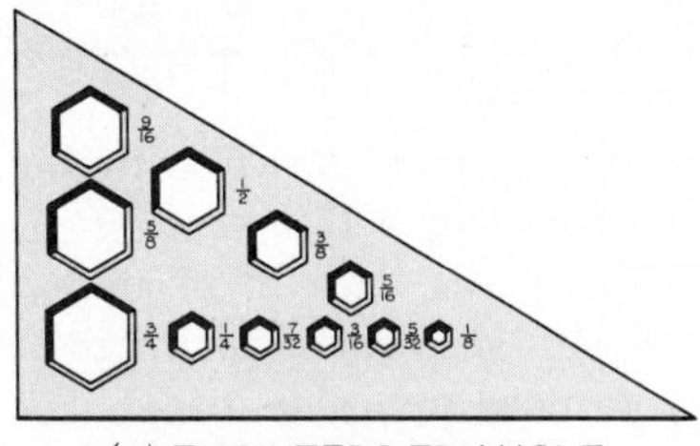

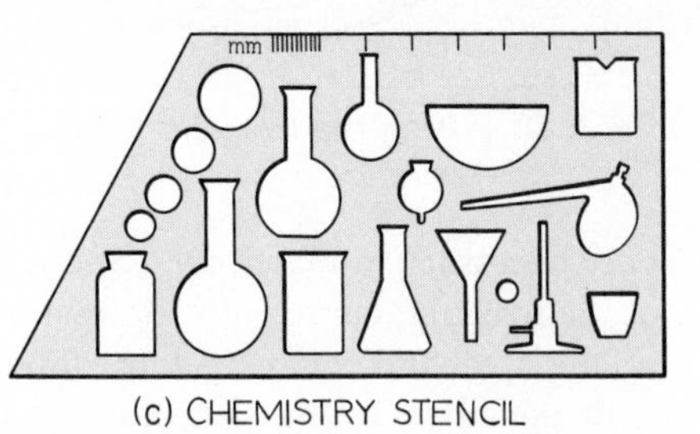


Fig. 2.81 **Drafting Devices.**

Fig. 2.82 **Drafting Machine.** *Courtesy Keuffel & Esser Co.*

**2.59 Templates.** A great variety of *templates* is available for specialized needs. A template may be found for drawing almost any ordinary drafting symbols or repetitive features. The *Engineers Triangle,* Fig. 2.81 (a), is useful for drawing hexagons or for bolt heads and nuts; the *Draftsquare,* (b), is convenient for drawing the curves on bolt heads and nuts, for drawing circles, thread forms, and so forth; and the *Chemistry Stencil,* (c), is useful for drawing chemical apparatus in schematic form.

Ellipse templates, §4.57, are perhaps more widely used than any other type. Circle templates are useful for drawing small circles quickly and for drawing fillets and rounds, and are used extensively in tool and die drawings.

**2.60 Drafting Machines.** The *drafting machine,* Fig. 2.82, is an ingenious device that replaces the T-square, triangles, scales, and protractor. The links, or bands, are arranged so that the controlling head is always in any desired fixed position regardless of where it is placed on the board; thus the horizontal straightedge will remain horizontal if so set. The controlling head is graduated in degrees (including a vernier on certain machines), which allows the straightedges, or scales, to be set and locked at any angle. There are automatic stops at the more frequently used angles, such as 15°, 30°, 45°, 60°, 75°, and 90°.

Drafting machines* have been greatly improved in recent years. The chief advantage of the drafting machine is that it speeds up drafting. Since its parts are made of metal, their accurate relationships are not subject to change, whereas T-squares, triangles, and working edges of drawing boards must be checked and corrected frequently. Drafting machines for left-handers are available from the manufacturers.

*Universal Drafting Machine Co., Charles Bruning Co., Keuffel & Esser Co., Eugene Dietzgen Co., Frederick Post Co., and V & E Manufacturing Co. are some of the manufacturers of drafting machines.

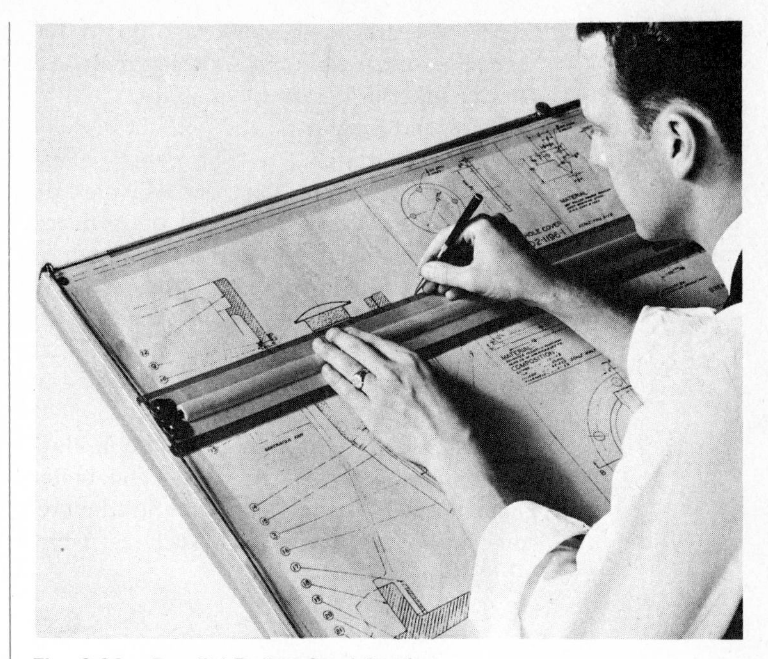

Fig. 2.83 **Parallel-Ruling Straightedge.** *Courtesy Eugene Dietzgen Co.*

**2.61 Parallel-Ruling Straightedge.** For large drawings, the long T-square becomes unwieldy, and considerable inaccuracy may result from the "give" or swing of the blade. In such case the *parallel-ruling straightedge,* Fig. 2.83, is recommended. The ends of the straightedge are controlled by a system of cords and pulleys which permit the straightedge to be moved up or down on the board while maintaining a horizontal position.

**2.62 Drawing Papers.** *Drawing paper,* or *detail paper,* is used whenever a drawing is to be made in pencil but not for reproduction. For working drawings and for general use, the preferred paper is light cream or buff in color, and is available in rolls of widths 24″ and 36″ and in cut sheets of standard sizes such as 9″ × 12″, 12″ × 18″, 18″ × 24″, or $8\frac{1}{2}$″ × 11″, 11″ × 17″, 17″ × 22″. Most industrial drafting rooms use standard sheets with printed borders and title strips, and since the cost for printing adds so little to the price per sheet, many schools have also adopted printed sheets.

The best drawing papers have up to 100 percent pure rag stock, have strong fibers that afford superior erasing qualities, folding strength, and toughness, and will not discolor or grow brittle with age. The paper should have a fine grain or tooth that will pick up the graphite and produce clean, dense black lines. However, if the paper is too rough, it will wear down the pencil excessively, and will produce ragged, grainy lines. The paper should have a hard surface so that it will not groove too easily when pressure is applied to the pencil.

For ink work, as for catalog and book illustrations, white papers are used. The better papers, such as Bristol Board and Strathmore, come in several thicknesses such as 2-ply, 3-ply, 4-ply.

## 2.63 **Tracing Papers.**

*Tracing paper* is a thin transparent paper upon which drawings are made for the purpose of reproducing by blueprinting or by other similar processes. Tracings are usually made in pencil, but may be made in ink. Most tracing papers will take pencil or ink, but some are especially suited to one or to the other.

Tracing papers are of two kinds: (1) those treated with oils, waxes, or similar substances to render them more transparent, called *vellums;* (2) those not so treated, but which may be quite transparent, owing to the high quality of the raw materials and the methods of manufacture. Some treated papers deteriorate rapidly with age, becoming brittle in many cases within a few months, but some excellent vellums are available. Untreated papers made entirely of good rag stock will last indefinitely and will remain tough. For a discussion of tracing methods, see §§2.52 and 2.53.

## 2.64 **Tracing Cloth.**

*Tracing cloth* is a thin transparent muslin fabric, (cotton, not "linen" as commonly supposed) sized with a starch compound or plastic to provide a good working surface for pencil or ink. It is much more expensive than tracing paper. Tracing cloth is available in rolls of standard widths, such as 30″, 36″, and 42″, and also in sheets of standard sizes, with or without printed borders and title forms.

For pencil tracings, special pencil tracing cloths are available. Many concerns make their drawings in pencil directly on this cloth, dispensing entirely with the preliminary pencil drawing on detail paper, thus saving a great deal of time. These cloths generally have a surface that will produce dense black lines when hard pencils are used. Hence, these drawings do not easily smudge and will stand up well with handling.

## 2.65 **Polyester Films and Coated Sheets.**

The polyester film is a superior drafting material available in rolls and sheets of standard size. It is made by bonding a mat surface to one or both sides of a clear *Mylar* polyester sheet. The transparency and printing qualities are very good, the mat drawing surface is excellent for pencil or ink, erasures leave no ghost marks, and the film has high dimensional stability. Its resistance to cracking, bending, or tearing make it virtually indestructible, if given reasonable care. The film is rapidly replacing cloth and is competing with vellum in some applications. Some companies have found it more economical to make their drawings directly in ink on the film.

Large coated sheets of aluminum, which provides good dimensional stability, are often used in the aircraft and auto industry for full-scale layouts that are scribed into the coating with a steel point rather than a pencil. The layouts are reproduced from the sheets photographically.

## 2.66 **Standard Sheets.**

Two systems of sheet sizes are approved by the ANSI* as follows:

| | | | |
|---|---|---|---|
| A | $8\frac{1}{2}$″ × 11″ | A | 9″ × 12″ |
| B | 11″ × 17″ | B | 12″ × 18″ |
| C | 17″ × 22″ | C | 18″ × 24″ |
| D | 22″ × 34″ | D | 24″ × 36″ |
| E | 34″ × 44″ | E | 36″ × 48″ |

* ANSI Y14.1-1957.

The use of the basic sheet size $8\frac{1}{2}'' \times 11''$, and multiples thereof, permits filing of small tracings and of folded prints in standard files with or without correspondence. These sizes can be cut without waste from the standard 36″ rolls of paper or cloth.

The alternate system based on size $9'' \times 12''$ is widely used in the automotive industry and has the advantage of slightly larger areas.

For title blocks, revision blocks, and list of materials blocks, see inside the back cover of this book. See also §14.5.

## 2.67 **Instrumental Drawing Problems.**

All of the following constructions, Figs. 2.84 to 2.94, are to be drawn in pencil on Layout A–2 (see inside the back cover of this book). The steps in drawing this layout are shown in Fig. 2.56. Draw all construction lines *lightly,* using a hard pencil (4H to 6H), and all required lines dense black with a softer pencil (F to H). If construction lines are drawn properly—that is, *lightly*—they need not be erased.

The drawings in Figs. 2.89 to 2.94 are to be drawn in pencil, preferably on tracing paper or vellum; then prints should be made to show the effectiveness of the students' technique. If ink tracings are required, the originals may be drawn on detail paper and then traced on vellum or tracing cloth. For any assigned problem, the instructor may require that all dimensions and notes be lettered in order to afford further lettering practice.

The problems of Chapter 4, "Geometrical Constructions," provide excellent additional practice to develop skill in the use of drawing instruments.

Problems in convenient form for solution may be found in *Technical Drawing Problems, Series 1,* by Giesecke, Mitchell, Spencer, and Hill; in *Technical Drawing Problems, Series 2,* by Spencer and Hill; and in *Technical Drawing Problems, Series 3,* by Spencer and Hill. These three workbooks are designed to accompany this text and all are published by Macmillan Publishing Co., Inc.

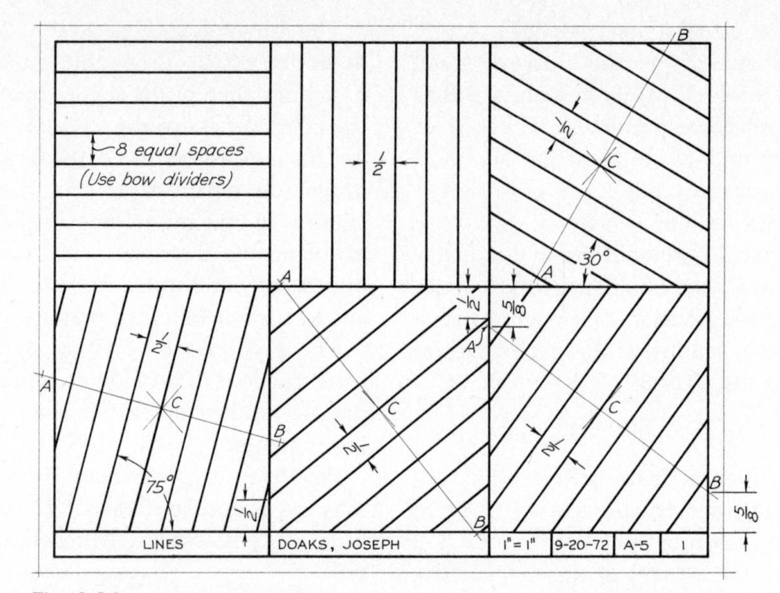


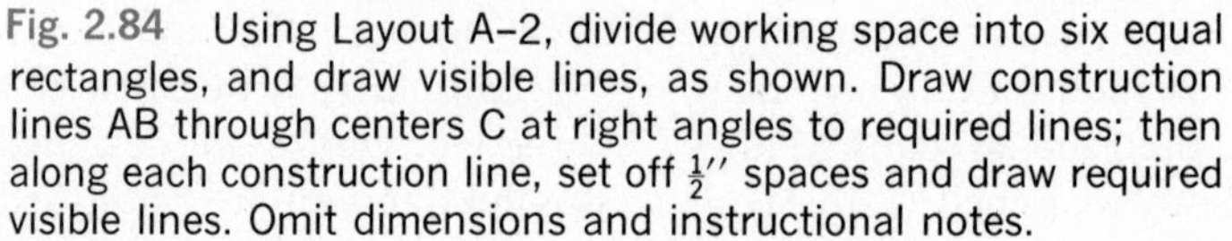
Fig. 2.84 Using Layout A–2, divide working space into six equal rectangles, and draw visible lines, as shown. Draw construction lines AB through centers C at right angles to required lines; then along each construction line, set off $\frac{1}{2}''$ spaces and draw required visible lines. Omit dimensions and instructional notes.

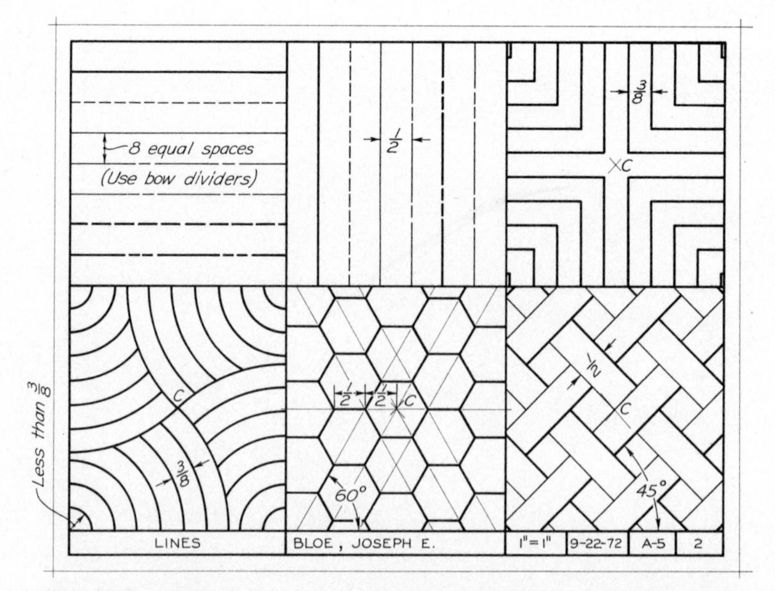


Fig. 2.85 Using Layout A–2, divide working space into six equal rectangles, and draw lines as shown. In first two spaces, draw conventional lines to match those in Fig. 2.15. In remaining spaces, locate centers C by diagonals, and then work constructions out from them. Omit dimensions and instructional notes.

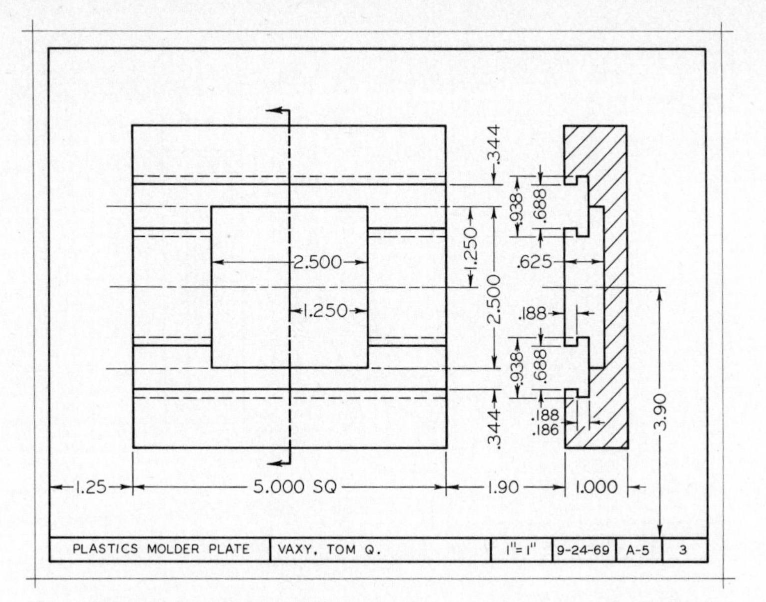


Fig. 2.86 Using Layout A–2, draw views in pencil, as shown. Omit all dimensions.

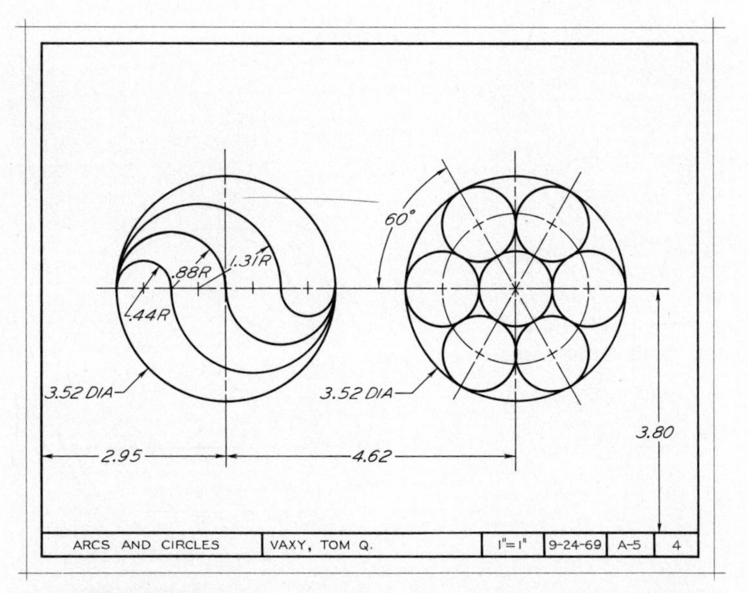


Fig. 2.87 Using Layout A–2, draw figures in pencil, as shown. Use bow pencil for all arcs and circles within its radius range. Omit all dimensions.

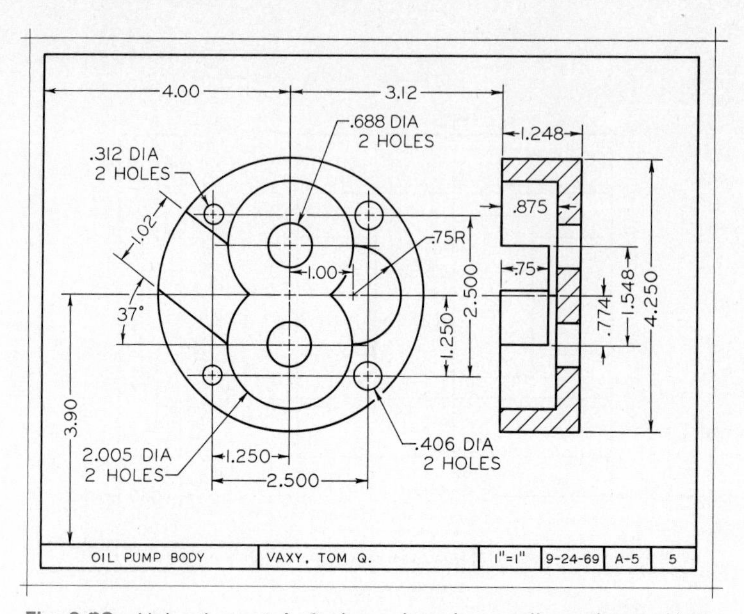


Fig. 2.88 Using Layout A–2, draw views in pencil, as shown. Use bow pencil for all arcs and circles within its radius range. Omit all dimensions.

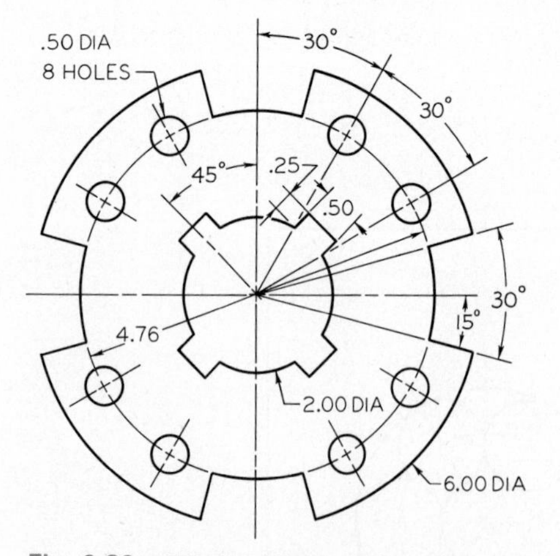


Fig. 2.89 **Friction Plate.** Using Layout A–2, draw in pencil. Omit dimensions and notes.

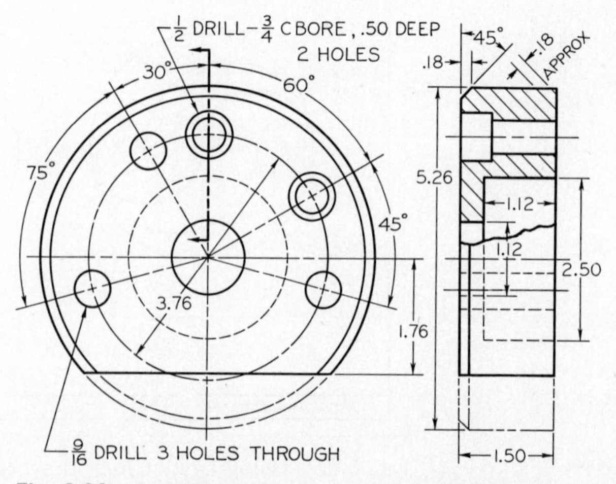


Fig. 2.90 **Seal Cover.** Using Layout A–2, draw views in pencil. Omit dimensions and notes. See §7.8.

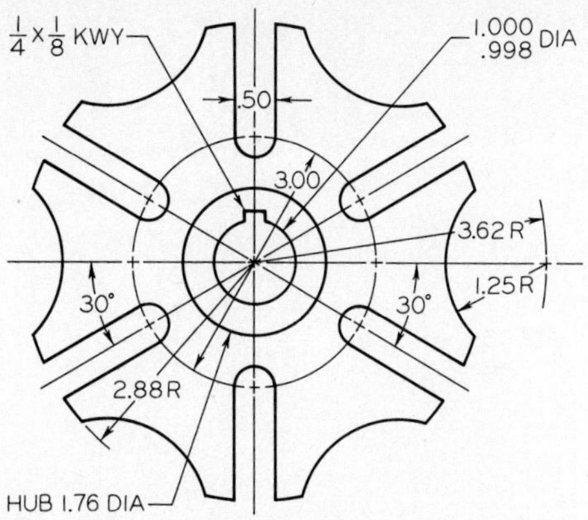

Fig. 2.91 **Geneva Cam.** Using Layout A-2, draw in pencil. Omit dimensions and notes.

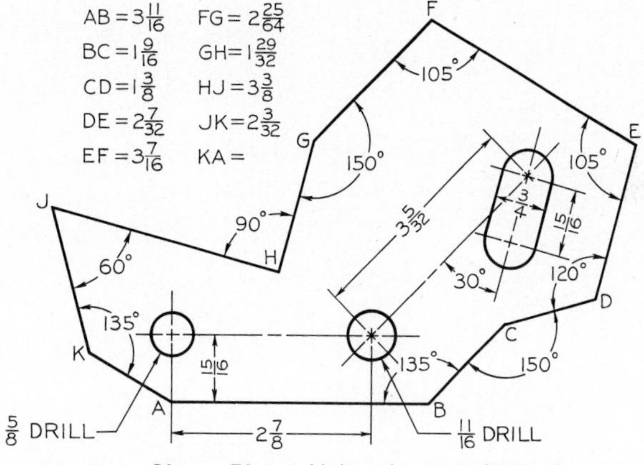


Fig. 2.92 **Shear Plate.** Using Layout A-2, draw accurately in pencil. Give length of KA. Omit other dimensions and notes.

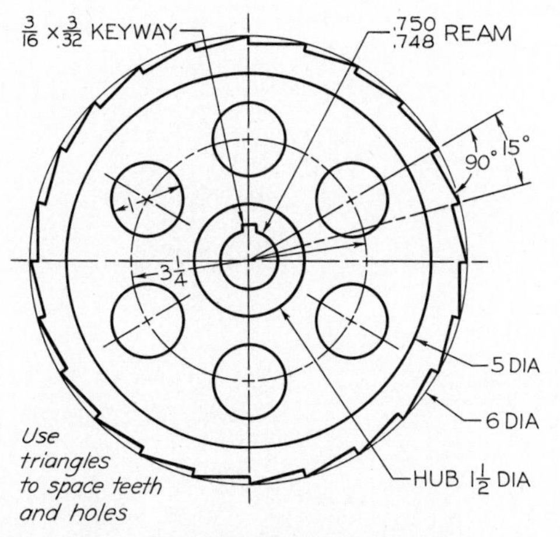


Fig. 2.93 **Ratchet Wheel.** Using Layout A-2, draw in pencil. Omit dimensions and notes.

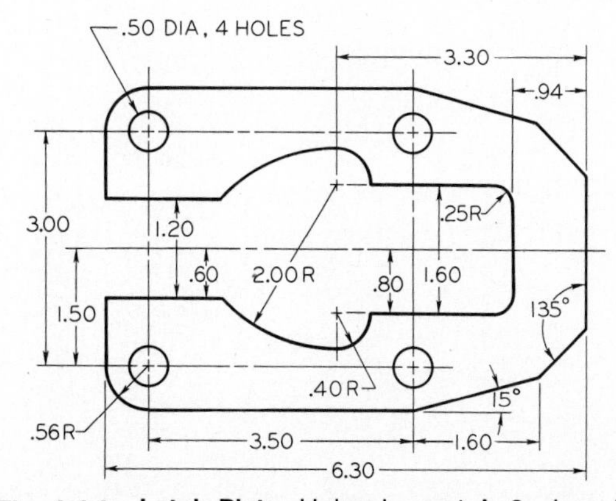


Fig. 2.94 **Latch Plate.** Using Layout A-2, draw in pencil. Omit dimensions and notes.

# 3

# lettering*

**3.1 Origin of Letter Forms.** The designs of modern alphabets had their origin in Egyptian hieroglyphics, Fig. 1.3, which were developed into a cursive hieroglyphic or hieratic writing. This was adopted by the Phoenicians and was developed by them into an alphabet of twenty-two letters. This Phoenician alphabet was later adopted by the Greeks, but it evolved into two distinct types in different sections of Greece: an Eastern Greek type, used also in Asia Minor, and a Western Greek type, used in the Greek colonies in and near Italy. In this manner the Western Greek alphabet became the Latin alphabet about 700 B.C. The Latin alphabet came into general use throughout the Old World.

Originally the Roman capital alphabet consisted of twenty-two characters, and these have remained practically unchanged to this day. They may still be seen on Trajan's Column and other Roman monuments. The letter V was used for both U and V until the tenth century. The last of the twenty-six characters, J, was adopted at the end of the fourteenth century as a modification of the letter I. The dot over the lowercase j still indicates its kinship to the i; in Old English the two letters are very similar. The numerous modern styles of letters were derived from the design of the original Roman capitals.

Before the invention of printing by Gutenberg in the fifteenth century, all letters were made by hand, and their designs were modified and decorated according to the taste of the individual writer. In England these letters

* *Lettering,* not "printing," is the correct term for making letters by hand. *Printing* means the production of printed material on a printing press.

ABCDEFGH abcdefgh GOTHIC All letters having the elementary strokes of even width are classified as Gothic
Made with Style A or B Speedball Pen

ABCDEFGH abcdefghij Roman All letters having elementary strokes "accented" or consisting of heavy and light lines, are classified as Roman.
Made with Style C or D Speedball Pen

ABCDEFGHI abcdefghijklm Italic- All slanting letters are classified as Italics- These may be further designated as Roman-Italics, Gothic Italics or Text Italic.
Made with Style C or D Speedball Pen

ABCDEFG abcdefghijkl Text—This term includes all styles of Old English, German text, Bradley text or others of various trade names — Text styles are too illegible for commercial purposes.
Made with Style C or D Speedball Pen

Fig. 3.1 **Classification of Letter Styles.**

became known as Old English. The early German printers adopted the Old English letters, and they are still in use in Germany. The early Italian printers used Roman letters, which were later introduced into England, where they gradually replaced the Old English letters. The Roman capitals have come down to us virtually in their original form.

**3.2 Letter Styles.** A general classification of letter styles is shown in Fig. 3.1. They were all made with Speedball pens, as indicated, and are therefore largely single-stroke letters.

If the letters are drawn in outline and filled in, they are referred to as *filled-in* letters, Figs. 3.39 and 3.42. The plainest and most legible style is the GOTHIC, from which our single-stroke engineering letters are derived. The term ROMAN refers to any letter having wide downward strokes and thin connecting strokes, as would result from the use of a wide pen, while the ends of the strokes are terminated with spurs called *serifs*. Roman letters include Old Roman and Modern Roman, and may be vertical or inclined. Inclined letters are also referred to as *Italic*, regardless of the letter style; those shown in Fig. 3.1 are inclined Modern Roman. *Text* letters are often loosely referred to as Old English, although these letters as well as the other similar letters, such as German Text, are actually Gothic. The Commercial Gothic shown at the top of Fig. 3.1 is a relatively modern development, which originates from the earlier Gothic forms. German Text is the only commercially used form of medieval Gothic in use today.

For more extensive and detailed information regarding the styles of letters, see §§3.27 to 3.29.

Fig. 3.2 **Condensed and Extended Letters.**

CONDENSED LETTERS
EXTENDED LETTERS
*Condensed Letters*
*Extended Letters*

**3.3 Extended and Condensed Letters.** To meet design or space requirements, letters may be narrower and spaced closer together, in which case they are called *compressed* or *condensed* letters. If the letters are wider than normal, they are referred to as *extended* letters, Fig. 3.2.

**3.4 Lightface and Boldface Letters.** Letters also vary as to the thickness of the stems or strokes. Letters having very thin stems are called LIGHTFACE, while those having heavy stems are called **BOLDFACE,** Fig. 3.3.

LIGHTFACE
**BOLDFACE**

Fig. 3-3 **Lightface and Boldface Letters.**

**3.5 Single-Stroke Gothic Letters.** During the latter part of the nineteenth century the development of industry and of technical drawing in the United States made evident a need for a simple legible letter that could be executed with single strokes of an ordinary pen. To meet this need C. W. Reinhardt, formerly Chief Draftsman for the *Engineering News,* developed alphabets of capital and lower-case inclined and "upright" letters,* based upon the old Gothic letters, Fig. 3.38. For each letter he worked out a systematic series of strokes. The single-stroke Gothic letters used on the technical drawings today are based upon Reinhardt's work.

**3.6 Standardization of Lettering.** The first step toward standardization of technical lettering was made by Reinhardt when he developed single-stroke letters with a systematic series of strokes, §3.5. However, since that time there has been an unnecessary and confusing diversity of lettering styles and forms, and the American National Standards Institute in 1935 suggested letter forms that are now generally considered as standard. The present Standard (ANSI Y14.2–1973) is practically the same as that given in 1935, except that vertical lower-case letters have since been added.

The letters in this chapter and throughout this text conform to the American National Standard. Vertical letters are perhaps slightly more legible than inclined letters, but are more difficult to execute. Both vertical and inclined letters are standard, and the engineer or draftsman may be called upon to use either. Students should, therefore, learn to execute both forms well, though they may give more attention to the style they like and in which they can do better.

Lettering on drawings must be legible and suitable for easy and rapid execution. The single-stroke Gothic letters shown in Figs. 3.18 and 3.19 meet these requirements. Either vertical or inclined lettering may be used but only one style should be used throughout a drawing. Drawings for microfilm reproduction require well-spaced lettering to prevent "fill-

*Published in the *Engineering News* about 1893, and in book form in 1895.

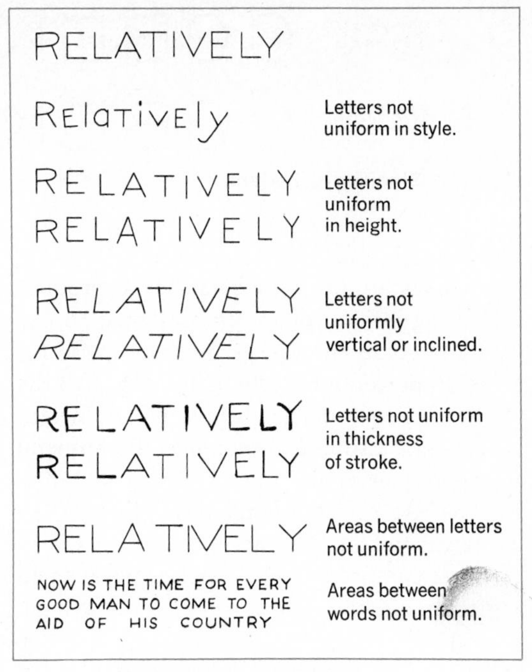


Fig. 3.4 **Uniformity in Lettering.**

ins." Background areas between letters in words should appear approximately equal, and words should be clearly separated by a space equal to the height of the lettering. Only when special emphasis is necessary should the lettering be underlined.

It is not desirable to vary the size of the lettering according to the size of the drawing except when a drawing is to be reduced in reproduction.

**3.7 Uniformity.** In any style of lettering, *uniformity* is essential. Uniformity in height, proportion, inclination, strength of lines, spacing of letters, and spacing of words insures a pleasing appearance, Fig. 3.4.

Uniformity in height and inclination is promoted by the use of light guide lines, §3.14. Uniformity in strength of lines can be obtained only by the skillful use of properly selected pencils and pens, §§3.10 and 3.11.

**3.8 Optical Illusions.** Good lettering involves artistic design, in which the white and

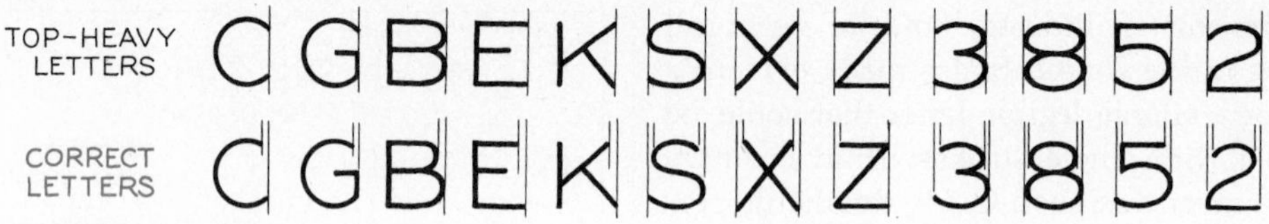

TOP-HEAVY LETTERS C G B E K S X Z 3 8 5 2

CORRECT LETTERS C G B E K S X Z 3 8 5 2

Fig. 3.5 **Stability of Letters.**

black areas are carefully balanced to produce a pleasing effect. Letters are designed to *look* well and some allowances must be made for errors in perception. Note that in Fig. 3.18 the width of the standard H is less than its height to eliminate a square appearance; the numeral 8 is narrower at the top to give it stability; and the width of the letter W is greater than its height for the very acute angles in the W give it a compressed appearance. Such acute angles should be avoided in good letter design.

**3.9 Stability.** If the upper portions of certain letters and numerals are equal in width to the lower portions, the characters appear top-heavy. To correct this, the upper portions are reduced in size where possible, thereby producing the effect of *stability* and a more pleasing appearance, Fig. 3.5.

If the central horizontal strokes of the letters B, E, F, and H are placed at midheight, they will appear to be below center. To overcome this optical illusion, these strokes should be drawn slightly above the center.

**3.10 Lettering Pencils.** Pencil letters can be best made with a medium-soft pencil with a conical point, Fig. 2.12 (c). First, sharpen the pencil to a needle point; then dull the point *very slightly* by marking on paper while holding the pencil vertically and rotating the pencil to round off the point. An F or H pencil is suitable for use on ordinary drawing paper of smooth surface. Between letters, rotate the pencil occasionally to new positions in order to keep the point symmetrical.

Today the majority of drawings are finished in pencil and reproduced as blueprints, ammonia prints, or other reproductions. To reproduce well by any process, the pencil lettering must be **dense black,** as should all other final lines on the drawing. The right pencil to use depends largely upon the amount of tooth in the paper, the rougher papers requiring the harder pencils. The lead should be soft enough to produce jet-black lettering, yet hard enough to prevent excessive wearing down of the point, crumbling of the point, and smearing of the graphite.

**3.11 Lettering Pens.** The choice of a pen for lettering is determined by the size and style of the letters, the thickness of stroke desired, and the personal preference of the draftsman. These conditions vary so much that it is impossible to specify any certain pen to use. The student who is zealous in his efforts to develop his ability to letter will learn by experience which pen is best suited to his purpose. Fig. 3.6 shows a variety of the best pen points in a range from the *tit-quill,* the

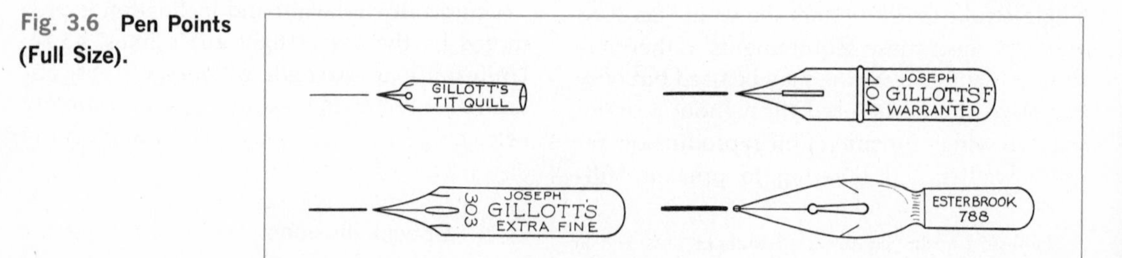


Fig. 3.6 **Pen Points (Full Size).**

finest, to the *ball-pointed,* the coarsest. The widths of the lines made by the several pens are shown full size. The medium widths, represented by the Gillott's 303 and 404 (or equivalent) are most widely used for lettering notes and dimensions on drawings, in which case the letters are usually $\frac{1}{8}''$ high. For lettering $\frac{3}{16}''$ to $\frac{1}{4}''$ high, as for titles, the ball-pointed pens are commonly used.

A very flexible pen should not be used for lettering, because the downward strokes are apt to be wider. A good lettering pen is one with which it is easy to make a stroke of uniform width. New pen points have a thin film of oil, which should be removed with a cloth—not burned off with a match flame. The best results are secured from a pen that has been used for some time, that is, "broken in" with use. Hence, when a pen point has proved satisfactory, it should be carefully wiped after using and taken care of as a valuable instrument.

Letters more than $\frac{1}{2}''$ in height generally require a special pen, Fig. 3.7. The *Speedball* pens are excellent for Gothic letters, Fig. 3.1, and are often used for titles and for the large drawing numbers in the corner of the title block, Figs. 14.15 to 14.17. Other styles of Speedball pens are suitable for Roman or text letters. These pens have the additional advantage of being low in cost. The *Barch-Payzant Lettering Pen* is available in eleven sizes ranging from 000 (very coarse) to 8 (very fine). The size 8 pen produces a stroke fine enough to be used for the usual lettering on technical drawings, being satisfactory for letters from $\frac{1}{8}''$ to $\frac{3}{16}''$ high.

Fig. 3.7 **Special Pens for Freehand Lettering.**

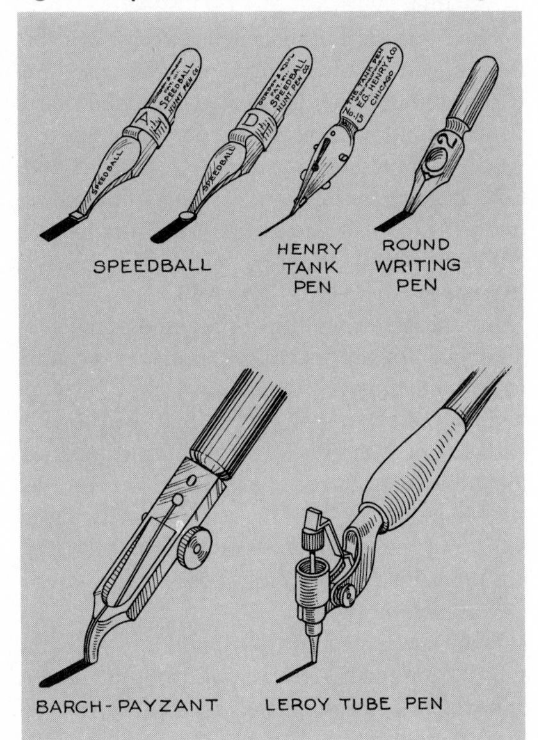


The *Henry Tank Pen* is available in both plain and ball points, and has a simple device under the pen to hold ink and prevent the nibs from spreading.

The *Leroy Pen* is also available in a wide range of sizes, and is highly recommended. It can be used in a regular pen staff for freehand lettering, or in the scriber for mechanical lettering, Fig. 3.31.

*Any lettering pen must be kept clean.* Drawing ink corrodes the point of the pen if allowed to dry and builds up the width of the point, so that it has to be cleaned anyway. To remove dried drawing ink from any instrument, scrape carefully with a knife or use a special pen-cleaning fluid.*

Esterbrook and Venus fountain pens, with removable points for cleaning, are available for use with drawing ink. The Koh-I-Noor "Rapidograph" is a different type in which the point is a small tube. See §2.48. A small automatic plunger rod keeps the ink flowing. All of these pens should be frequently cleaned with cleaning fluid to keep them in service.

## 3.12 Technique of Lettering.

*Any normal person can learn to letter if he is persistent and intelligent in his efforts.* While it is true that "practice makes perfect," it must be understood that practice alone is not enough; it must be accompanied by *continuous effort to improve.*

Lettering is freehand drawing and not writing. Therefore, the six fundamental strokes and their direction for freehand drawing are

*Higgins Ink Co. or Keuffel & Esser Co.

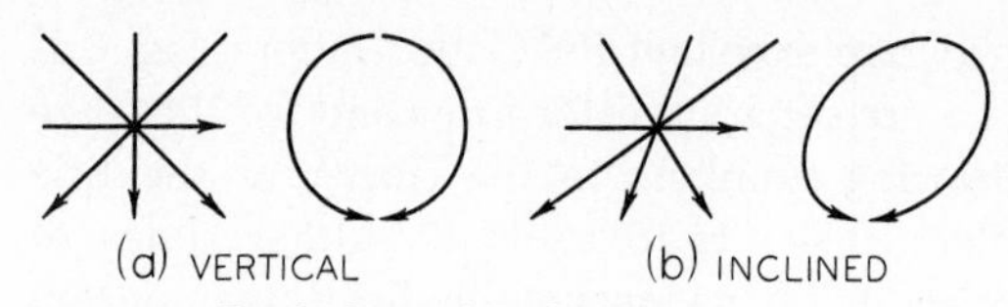


Fig. 3.8 **Basic Lettering Strokes.**

basic to lettering, Fig. 3.8. The horizontal strokes are drawn to the right, and all vertical, inclined, and curved strokes are drawn downward.

Good lettering is always accomplished by conscious effort and is never done well otherwise, though good muscular coordination is of great assistance. Ability to letter has little relationship to writing ability; excellent letterers are often poor writers.

There are three necessary steps in learning to letter:

1. Knowledge of the proportions and forms of the letters and the order of the strokes. No one can make a good letter who does not have a clear mental image of the correct form of the letter.
2. Knowledge of composition—the spacing of letters and words. Rules governing composition should be thoroughly mastered, §3.24.
3. Persistent practice, with *continuous effort to improve.*

Pencil lettering should be executed with a fairly soft pencil, such as an F or H for ordinary paper; and the strokes should be *dark* and *sharp,* not gray and blurred. Many draftsmen acquire "snap" in their lettering by accenting or "bearing down" at the beginning and the end of each stroke. Beginners should be careful not to overdo this trick or to try it without first acquiring the ability to form letters of correct shape. After a few letters are made, the pencil will tend to become dull. In order to wear the lead down uniformly and thereby to keep the lettering sharp, turn the pencil frequently to a new position.

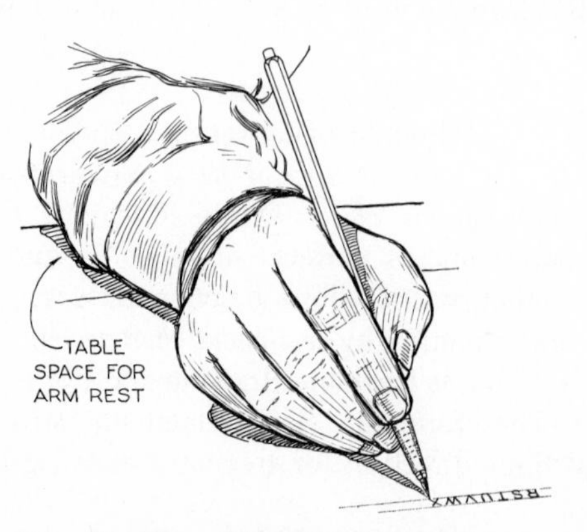


Fig. 3.9 **Position of Hand in Lettering.**

The correct position of the hand in lettering is shown in Fig. 3.9. In general, draw vertical strokes downward or toward you with a finger movement, and draw horizontal strokes from left to right with a wrist movement without turning the paper.

The forearm should be approximately at right angles to the line of lettering, and *resting on the board*—never suspended in mid-air. If the board is small, revolve it counterclockwise about 45°, until the line of lettering is approximately perpendicular to the forearm. If the board is larger and cannot be moved, shift your body around toward the left side of the board to approximate this position as nearly as possible.

Since practically all pencil lettering will be reproduced by blueprinting or otherwise, the letters should be **dense black.** Avoid hard pencils that, even with considerable pressure, produce gray lines. Use a fairly soft pencil and keep it sharp by frequent dressing of the point on the sandpaper pad or file. An example (full size) of pencil lettering exhibiting correct technique is shown in Fig. 3.10.

In ink lettering, most beginners have a tendency toward excessive pressure on the pen point, thus producing strokes of varying widths. Select a pen point that will make strokes of correct thickness without spreading the nibs. The correct position of the pen is shown in Fig. 3.9. Move the pen with light uniform pressure, and allow the ink to *flow off* the point instead of being forced off by pressure upon the point.

Some draftsmen transfer ink to the pen with a quill from the ink bottle, but this is generally unnecessary unless some ink-holding device (such as that on the Henry Tank Pen, Fig. 3.7)

THE IMPORTANCE OF GOOD LETTERING CANNOT BE OVER-EMPHASIZED. THE LETTERING CAN EITHER MAKE OR BREAK AN OTHERWISE GOOD DRAWING.

*PENCIL LETTERING SHOULD BE DONE WITH A FAIRLY SOFT SHARP PENCIL AND SHOULD BE CLEAN-CUT AND DARK. ACCENT THE ENDS OF THE STROKES.*

Fig. 3.10 **Pencil Lettering (Full Size).**

is used. Excess ink, however, should be removed from the point by lightly touching it against the opening of the bottle.

Before an inked tracing is lettered, all guide lines should be drawn in pencil directly upon the tracing paper or cloth if the guidelines underneath are too indistinct to serve their purpose well as a guide for inking.

## 3.13 Left-Handers.

All evidence indicates that the left-handed draftsman is just as skillful as the right-hander, and this includes skill in lettering. The most important step in learning to letter is to learn the correct shapes and proportions of letters, and these can be learned as well by the left-hander as by anyone else. The left-hander does have a problem of developing a system of strokes that seems most suitable for himself. The strokes shown in Figs. 3.18 and 3.19 are for right-handers. The left-hander should experiment with each letter to find out which strokes are best for him. The habits of left-handers vary so much that it is futile to suggest a standard system of strokes for all left-handers.

The left-hander, in developing his own system of strokes, should decide upon strokes he can make best with the pen, and he should then use the same strokes for pencil lettering. Pen strokes can be drawn in the direction the pen is leaned, or at right angles to this, or in curved paths between the two. The pen should never be jabbed in the direction contrary to the way the pen is leaned, as the point has a tendency to dig into the paper. The strokes should, therefore, be those which are in harmony with the natural and intended use of the pen point.

The regular left-hander assumes a natural position exactly opposite to that of the right-hander, but he will be able to use many of the same strokes as shown for right-handers, with perhaps some minor differences. As prescribed for right-handers, he will draw all vertical strokes downward, and may also draw all horizontal strokes from left to right. He may, however, prefer to draw horizontal strokes from right to left, and he should do this if it seems more natural to him. Also he may wish to change the order of drawing horizontal strokes, so that in the case of the E, for example, the top stroke would be drawn first and the bottom stroke drawn last. If this is done, the pen or pencil will not tend to hide strokes already drawn. Curved strokes will be essentially the same as for right-handers, with perhaps some adjustments of the starting and ending points of the curves.

The hooked-wrist left-hander has a more serious problem, and each such person will have to adopt a system that seems best for his own particular habits. Vertical strokes may be drawn downward as for right-handers, but many hooked-wrist left-handers will find it easier to draw vertical strokes upward. Horizontal strokes will most certainly be drawn from right to left with a finger movement, for the pencil or pen will dig into the paper if pushed in the other direction. Furthermore, the order of horizontal strokes will be to do those at the bottom first, and those at the top last, as described above for the letter E. Since a sheet is lettered from the top downward, the

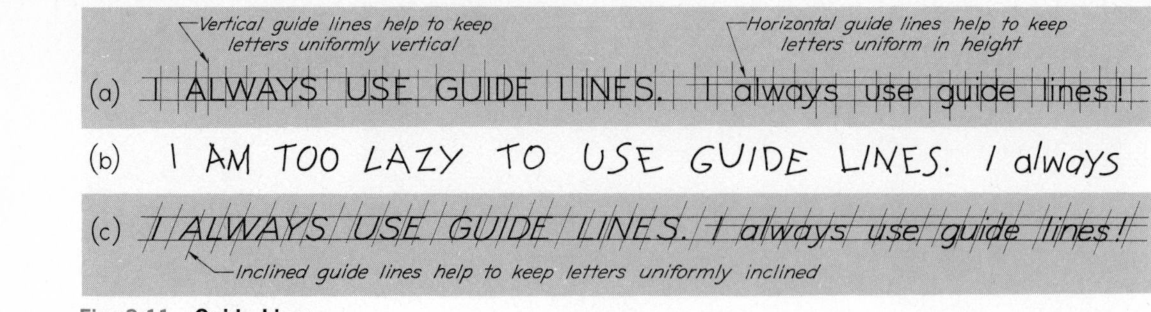


Fig. 3.11 **Guide Lines.**

hooked-wrist left-hander must move his hand over lines of lettering previously made. Therefore, a piece of paper should be placed over the lettered areas so that smearing of the graphite cannot occur.

If you are left-handed, advise your instructor at once. On examinations in which lettering is tested, use strokes that you have found most suitable for your own use, and letter a statement in the margin to the effect that you are left-handed.

## 3.14 Guide Lines.

Extremely light horizontal guide lines are necessary to regulate the height of letters. In addition, light vertical or inclined guide lines are needed to keep the letters uniformly vertical or inclined. Guide lines are absolutely essential for good lettering, and should be regarded as a welcome aid, not as an unnecessary requirement. Paradoxically, the better draftsman always uses guide lines, while the unskilled letterer who needs them most is inclined to slight this important step. See Fig. 3.11.

Make guide lines for finished pencil lettering *so lightly that they need not be erased,* as indeed they cannot be after the lettering has been completed. Guide lines should be barely visible at arm's length. Use a relatively hard pencil, such as a 4H to 6H, with a long, sharp, conical point, Fig. 2.12 (c). If the letters are inked, the guide lines may be removed with the Artgum after the ink is dry, Fig. 2.17 (b).

In preparation for ink lettering, complete guide lines should be drawn, and the letters first drawn lightly in pencil. Experienced letterers often draw the complete guide lines, and then letter directly in ink, without first penciling the letters.

## 3.15 Guide Lines for Capital Letters.

Guide lines for vertical capital letters are shown in Fig. 3.12. On working drawings, capital letters are commonly made $\frac{1}{8}''$ high, with the space between lines of lettering from three-fifths to the full height of the letters. The vertical guide lines are not used to space the letters (as this should always be done by eye while lettering), but only to keep the letters uniformly vertical, and they should accordingly be drawn at random. Where several lines of letters are to be made, these vertical guide lines should be continuous from top to bottom of the lettered area, as shown.

Guide lines for inclined capital letters are shown in Fig. 3.13. The spacing of horizontal

Fig. 3.12 **Guide Lines for Vertical Capital Letters.**

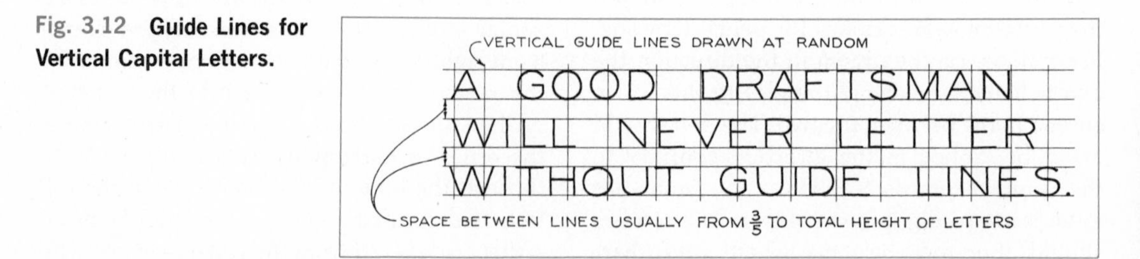

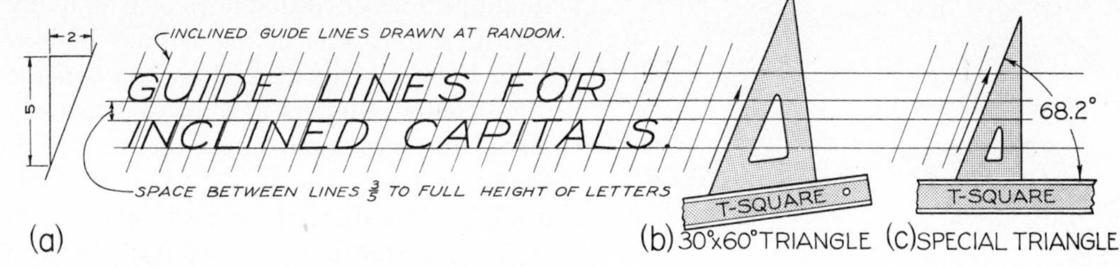


Fig. 3.13 **Guide Lines for Inclined Capital Letters.**

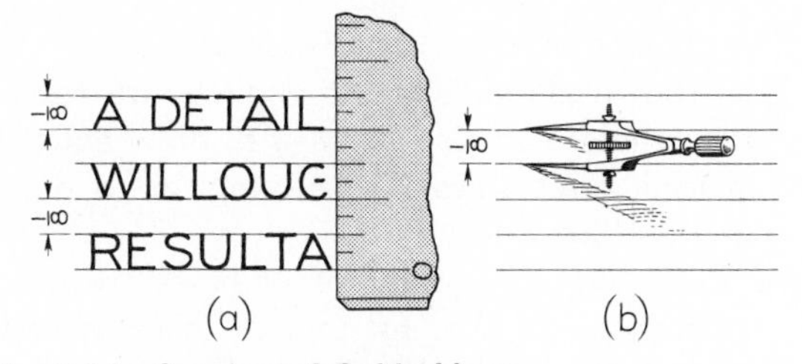


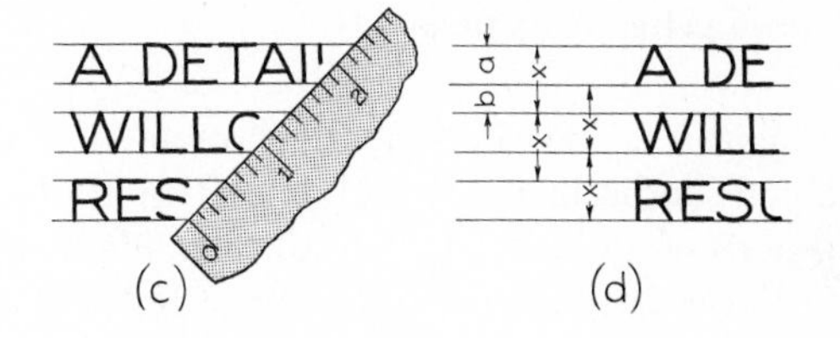


Fig. 3.14 **Spacing of Guide Lines.**

guide lines is the same as for vertical capital lettering. The American National Standard slope of 2 in 5 (or 68.2° with horizontal) may be established by drawing a "slope triangle," as shown at (a), and drawing the guide lines at random with the T-square and triangle, as shown at (b). Special triangles for the purpose may be used, as shown at (c), or the lines may be drawn with the Braddock-Rowe Lettering Triangle, Fig. 3.16, or the Ames Lettering Guide, Fig. 3.17.

A simple method of spacing horizontal guide lines is to use the scale, as shown in Fig. 3.14 (a), and merely set off a series of $\frac{1}{8}''$ spaces, making both the letters and the spaces between lines of letters $\frac{1}{8}''$ high. Another method of setting off equal spaces, $\frac{1}{8}''$ or otherwise, is to use the bow dividers, as shown at (b).

If it is desired to make the spaces between lines of letters less than the height of the letters, the methods shown at (c) and (d) will be convenient. At (c) the scale is placed diagonally, the letters in this case being four units high and the spaces between lines of lettering being three units. If the scale is rotated clockwise about the zero mark as a pivot, the height of the letters and the spaces between lines of letters diminish but remain proportional. If the scale is moved counterclockwise, the spaces are increased. The same unequal spacing may be accomplished with the bow dividers, as shown at (d). Let distance $x = a + b$, and set off x-distances, as shown.

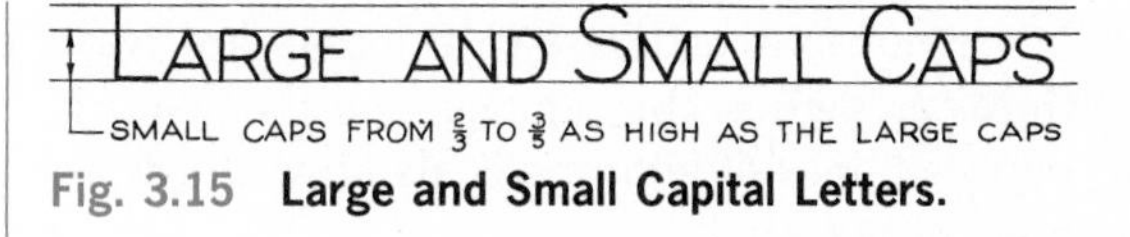


Fig. 3.15 **Large and Small Capital Letters.**

When large and small capitals are used in combination, the small capitals should be three-fifths to two-thirds as high as the large capitals, Fig. 3.15. This is in conformity with the guide-line devices described below, §§3.16 and 3.17.

## 3.16 Lettering Triangles.

*Lettering triangles,* which are available in a variety of shapes and sizes, are provided with sets of holes in which the pencil is inserted and the guide lines produced by moving the triangle with the pencil point along the T-square. The *Braddock-Rowe Lettering Triangle,* Fig. 3.16, is convenient for drawing guide lines for lettering and dimension figures, and also for drawing section lines. In addition, the triangle is used as a utility 45°

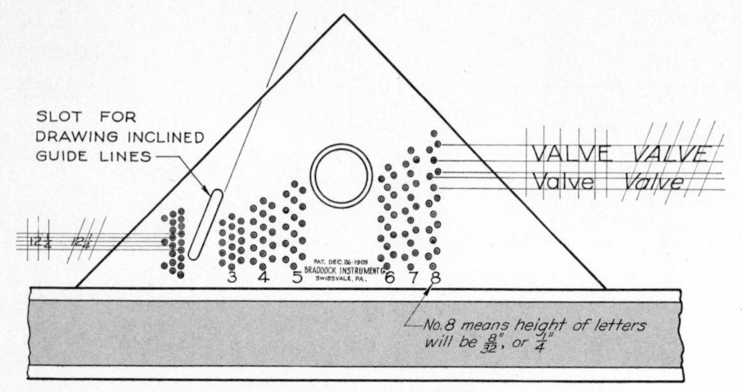


Fig. 3.16 **Braddock-Rowe Lettering Triangle.**

triangle. The numbers at the bottom of the triangle indicate heights of letters in thirty-seconds of an inch. Thus, to draw guide lines for $\frac{1}{8}''$ capitals, use the No. 4 set of holes. For lower-case letters, draw guide lines from every hole; for capitals, omit the second hole in each group. The spacing of holes is such that the lower portions of lower-case letters are two-thirds as high as the capitals, and the spacing between lines of lettering is also two-thirds as high as the capitals.

The column of holes at the extreme left is used to draw guide lines for dimension figures $\frac{1}{8}''$ high and fractions $\frac{1}{4}''$ high, and also for section lines $\frac{1}{16}''$ apart.

## 3.17 Ames Lettering Guide.

The *Ames Lettering Guide,* Fig. 3.17, is an ingenious transparent plastic device composed of a frame holding a disc containing three columns of holes. The vertical distances between the holes may be adjusted quickly to the desired spacing for guide lines or section lines by simply turning the disc to one of the settings indicated at the bottom of the disc. These numbers indicate heights of letters in thirty-seconds of an inch. Thus, for $\frac{1}{8}''$ high letters, the No. 4 setting would be used. The center column of holes is used primarily to draw guide lines for numerals and fractions, the height of the whole number being two units and the height of the fraction four units. The No. 4 setting of the disc will provide guide lines for $\frac{1}{8}''$ whole numbers, with fractions twice as high, or $\frac{1}{4}''$, as shown at (a). Since the spaces are equal, these holes can also be used to draw equally spaced guide lines for lettering, or to draw section lines.

The two outer columns of holes are used to draw guide lines for capitals or lower-case letters, the column marked three-fifths being used where it is desired to make the lower portions of lower-case letters three-fifths the total height of the letters, and the column marked two-thirds being used where the lower portion is to be two-thirds the total height of the letters. In each case, for capitals, the middle hole of each set is not used. The two-thirds and three-fifths also indicate the spaces between lines of letters.

The sides of the guide are used to draw vertical or inclined guide lines, as shown at (b) and (c).

## 3.18 Vertical Capital Letters and Numerals.

Fig. 3.18. For convenience in learning the proportions of the letters and numerals, each character is shown in a grid 6 units high.

Fig. 3.17 **Ames Lettering Guide.**

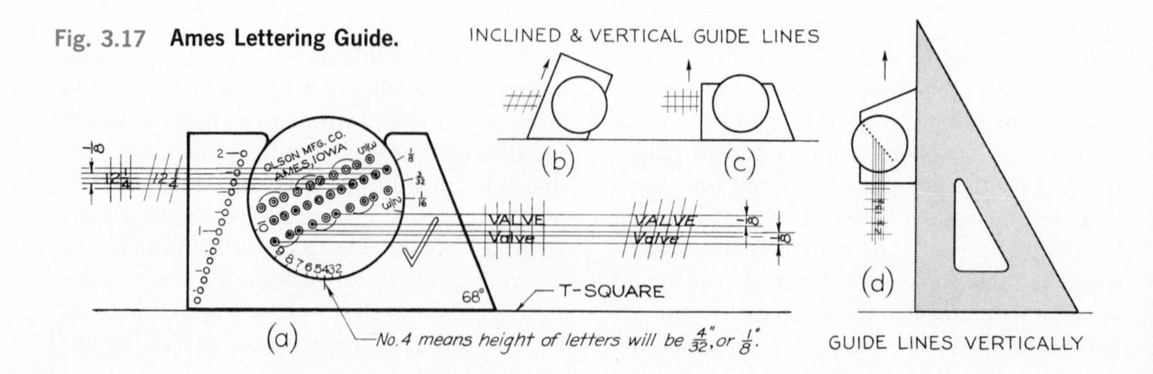

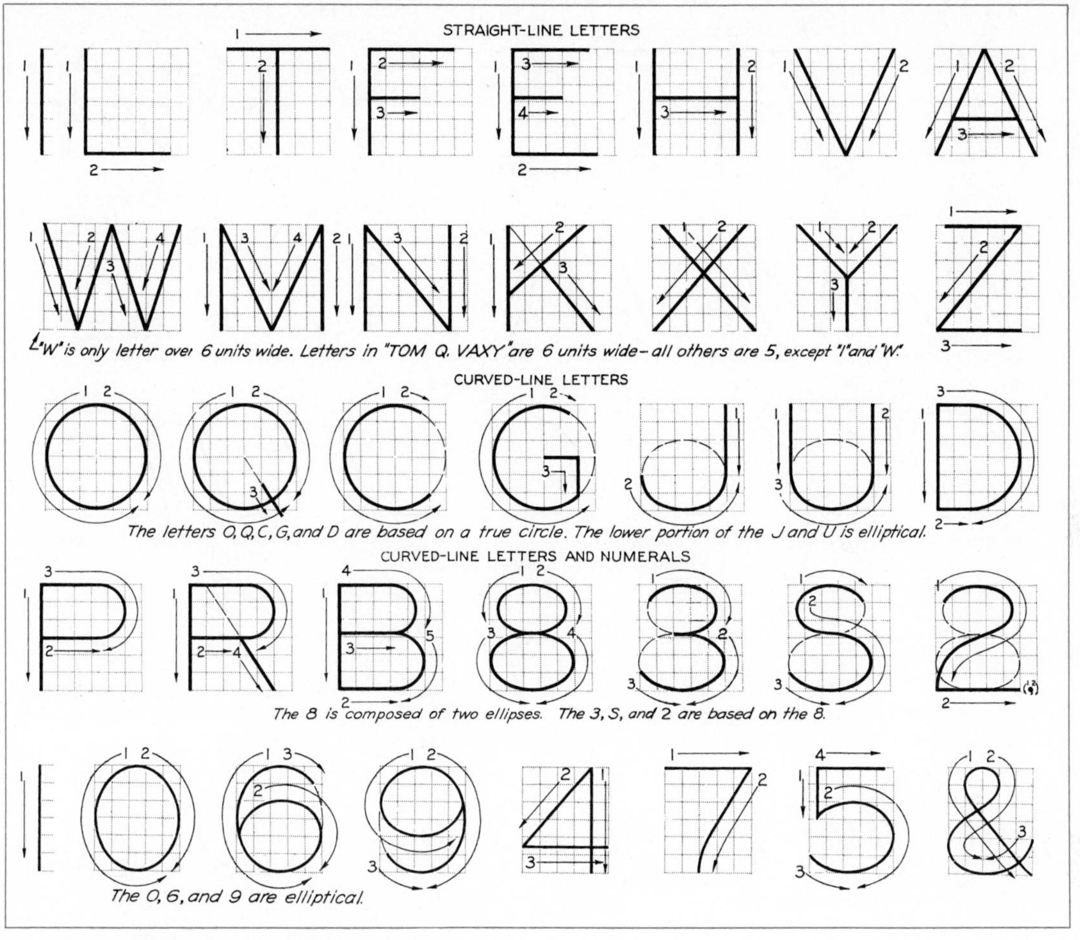

STRAIGHT-LINE LETTERS
"W" is only letter over 6 units wide. Letters in "TOM Q. VAXY" are 6 units wide–all others are 5, except "I" and "W."
CURVED-LINE LETTERS
The letters O, Q, C, G, and D are based on a true circle. The lower portion of the J and U is elliptical.
CURVED-LINE LETTERS AND NUMERALS
The 8 is composed of two ellipses. The 3, S, and 2 are based on the 8.
The 0, 6, and 9 are elliptical.


Fig. 3.18 **Vertical Capital Letters and Numerals.**

Numbered arrows indicate the order and direction of strokes. The widths of the letters can be easily remembered. The letter I, or the numeral 1, has no width. The W is 8 units wide ($1\frac{1}{3}$ times the height) and is the widest letter in the alphabet. All the other letters or numerals are either 5 or 6 units wide, and it is easy to remember the 6-unit letters because when assembled they spell TOM Q. VAXY. All numerals, except the 1, are 5 units wide.

All horizontal strokes are drawn to the right, and all vertical, inclined, and curved strokes are drawn downward, Fig. 3.8. Most of the strokes are natural and easy to remember. All the strokes and proportions should be thoroughly learned in the beginning, and it is recommended that this be done by practice sketching the vertical capital letters on cross-section paper, making the letters 6 squares high.

As shown in Fig. 3.18, the letters are classified as *straight-line letters* or *curved-line letters.* On the third row, the letters O, Q, C, and G are all based on the circle. The lower portions of the J and U are semiellipses, and the right sides of the D, P, R, and B are semicircular. The 8, 3, S, and 2 are all based on the figure 8, which is composed of a small ellipse over a larger ellipse. The 6 and 9 are based on the elliptical zero. The lower part of the 5 is also elliptical in shape.

## 3.19 Inclined Capital Letters and Numerals.

Fig. 3.19. The order and direction of the strokes and the proportions of the inclined

STRAIGHT-LINE LETTERS
"W" is only letter over 6 units wide. Letters in "TOM Q. VAXY" are 6 units wide - all others are 5, except "I" and "W".
CURVED-LINE LETTERS
The letters O, Q, C, G, and D are based on a true ellipse.
The lower portion of the J and U is elliptical.
CURVED-LINE LETTERS & NUMERALS
The 8 is composed of two ellipses. The 3, S, and 2 are based on the 8.
The O, 6, and 9 are elliptical.

Fig. 3.19 **Inclined Capital Letters and Numerals.**

capital letters and numerals are the same as those for the vertical characters. The methods of drawing guide lines for inclined capital letters are given in §3.15, and for numerals in §3.20.

Inclined capitals may be regarded as oblique projections, §§17.1 and 17.2, of vertical capitals. In the inclined letters, the circular parts become elliptical, the major axes of the ellipses based on the *O* making an angle of 45° with horizontal. The letters are classified as *straight-line letters* or *curved-line letters,* most of the curves being elliptical in shape. Therefore, skill in inclined lettering depends somewhat upon the ability to form smooth ellipses that appear to "lean" properly to the right.

The letters having sloping sides, such as the *V, A, W, X,* and *Y,* are also a source of difficulty for many beginners. The letters should be made symmetrically about an imaginary inclined center line. If this is done, the left side of the *V* and the right side of the *A* will be practically vertical, while the opposite sides will slope at less than 60° with horizontal.

## 3.20 Guide Lines for Whole Numbers and Fractions.

Complete guide lines should be drawn for whole numbers and fractions, especially by beginners. This means that both horizontal and vertical guide lines, or horizontal and inclined guide lines, should be drawn. Even the expert letterer will be able to do better lettering if he uses guide lines, and for this reason he is more likely to use them than the beginner who considers them

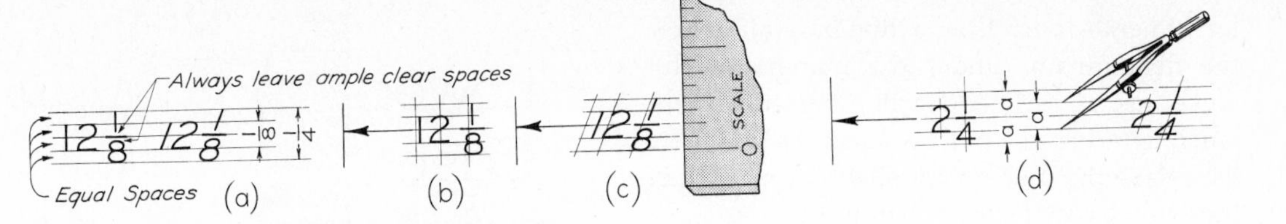


Fig. 3.20 **Guide Lines for Dimension Figures.**

"too much trouble." The guide lines, of course, should be drawn very lightly, with a hard pencil, 4H to 6H.

Draw five equally spaced guide lines for whole numbers and fractions, Fig. 3.20. Thus, fractions are twice the height of the corresponding whole numbers. Make the numerator and the denominator each about three-fourths as high as the whole number, to allow ample clear space between them and the fraction bar, as shown.

For dimensioning, the most commonly used height for whole numbers is $\frac{1}{8}$'', and for fractions $\frac{1}{4}$'', as shown at (a) and (c). These spaces may be easily set off directly with the scale, as shown. After the horizontal guide lines have been drawn, add vertical or inclined guide lines spaced at random, (b) and (c).

Another simple method is to set the bow dividers at $\frac{1}{8}$'' as shown at (d); then, with one point on the dimension line, set off $\frac{1}{8}$'' above the line and swing down and set off $\frac{1}{8}$'' below the line, to establish the top and bottom of the fraction; then center the $\frac{1}{8}$'' on the dimension line by eye to establish the height of the whole number. Draw the guide lines with the aid of the T-square.

If the Braddock-Rowe Triangle is used, the column of holes at the left produces five guide lines, each $\frac{1}{16}$'' apart, Fig. 3.21.

If the Ames Lettering Guide, Fig. 3.17, is used with the No. 4 setting of the disc, the same five guide lines, each $\frac{1}{16}$'' apart, may be drawn from the central column of holes.

Fig. 3.21 **Use of Braddock-Rowe Triangle.**

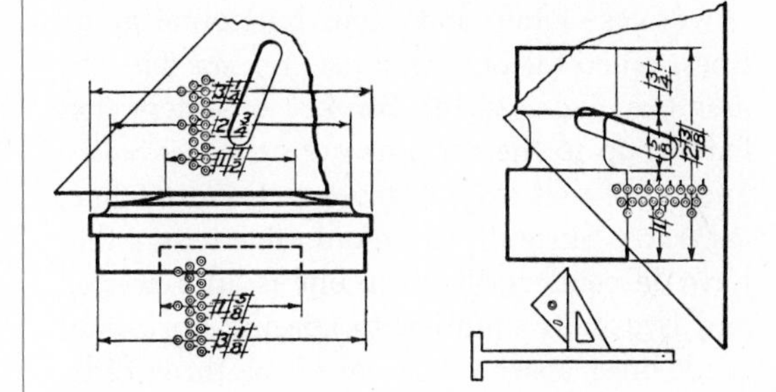

The experienced letterer may dispense with the drawing of guide lines for dimension figures, particularly where the most finished work is not required, by preparing a small card with marks indicating heights of numerals, Fig. 3.22, and then holding the card in place while lettering without actual guide lines.

Some of the most common errors in lettering fractions are illustrated in Fig. 3.23. Never

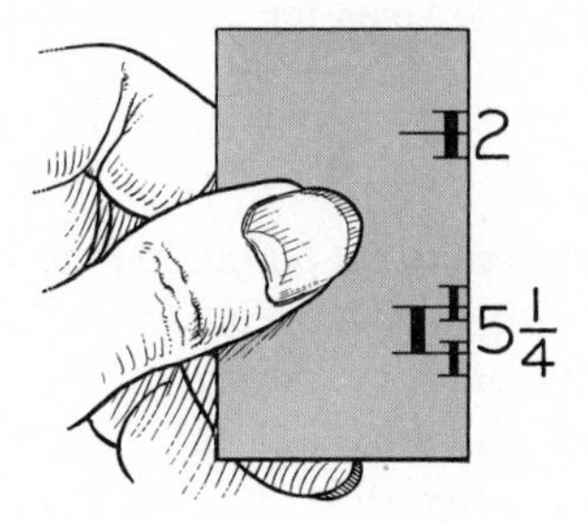

*Courtesy Prof. Albert Jorgensen*

Fig. 3.22 **Letter-Height Indicator.**

Fig. 3.23 **Common Errors.**

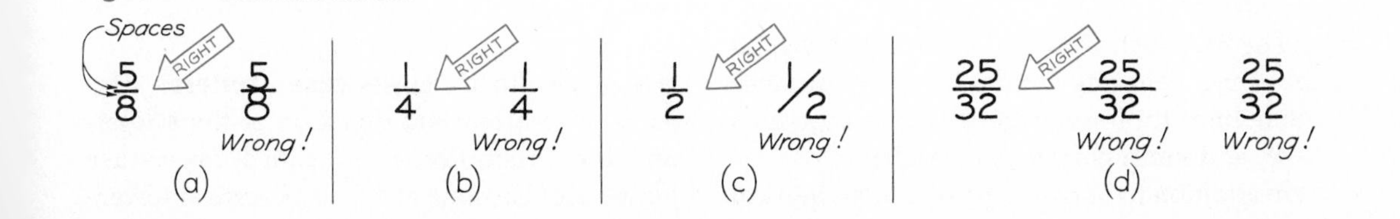

let numerals touch the fraction bar, (a). Center the denominator under the numerator, (b). Never use an inclined fraction bar, (c), except when lettering in a narrow space, as in a parts list. Make the fraction bar slightly longer than the widest part of the fraction, (d).

## 3.21 Guide Lines for Lower-case Letters.

Lower-case letters have four horizontal guide lines, called the *cap line, waist line, base line,* and *drop line,* Fig. 3.24 (a). Strokes of letters that extend up to the cap line are called *ascenders,* and those that extend down to the drop line, *descenders.* Since there are only five letters that have descenders, the drop line is little needed and is usually omitted. In spacing horizontal guide lines, space *a* may vary from three-fifths to two-thirds of space *b*. Spaces *c* are equal, as shown.

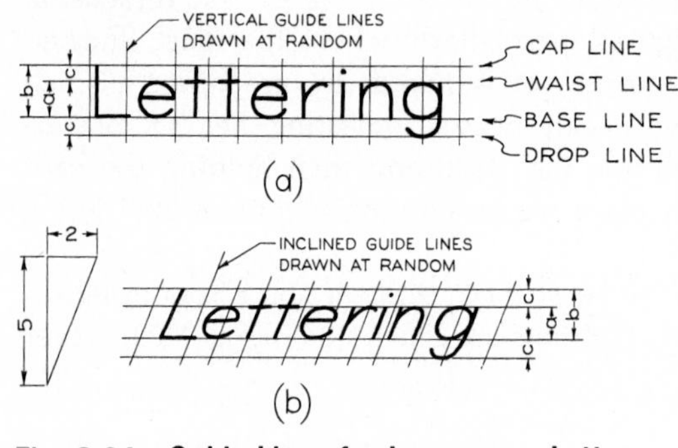


**Fig. 3.24 Guide Lines for Lower-case Letters.**

If it is desired to set off guide lines for letters $\frac{3}{16}''$ high with the scale (using the two-thirds ratio), it is only necessary to set off equal spaces each $\frac{1}{16}''$, Fig. 3.25 (a). The lower portion of the letter thus would be $\frac{1}{8}''$, and the space between lines of letters would also be $\frac{1}{8}''$. If the scale is placed at an angle, the spaces will diminish but remain equal, (b). Thus, this method may be easily used for various heights of lettering.

The Braddock-Rowe Triangle, Fig. 3.16, and the Ames Lettering Guide, Fig. 3.17, produce guide lines for lower-case letters as described here, and are highly recommended.

In addition to horizontal guide lines, vertical or inclined guide lines, drawn at random, should always be used to keep the letters uniformly vertical or inclined, Fig. 3.24.

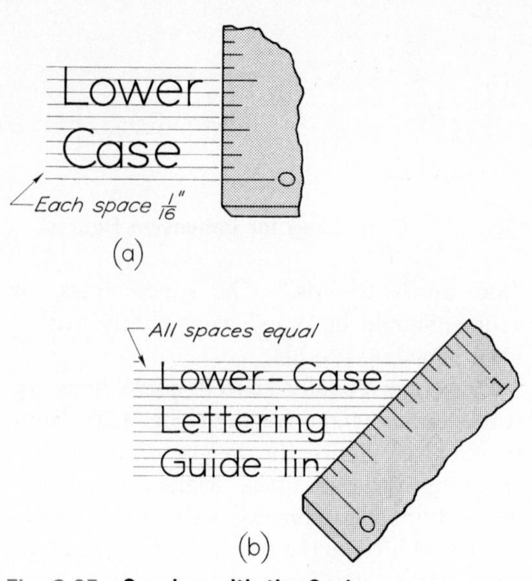


**Fig. 3.25 Spacing with the Scale.**

## 3.22 Vertical Lower-case Letters.

Fig. 3.26. Vertical lower-case letters are used largely on map drawings, and very seldom on machine drawings. The shapes are based upon a repetition of the circle or circular arc and the straight line, with some variations. The lower part of the letter is usually two-thirds the height of the capital letter.

Stroke 3 of the e is slightly above mid-height. The crosses on the f and t are on the waist line and are symmetrical with respect to strokes 1. The curved strokes of h, m, n, and r intersect strokes 1 approximately two-thirds of the distance from the base line to the waist line.

The descenders of the g, j, and y terminate in curves that are tangent to the drop line, while those of p and q terminate in the drop line without curves.

## 3.23 Inclined Lower-case Letters.

Fig. 3.27. The order and direction of the strokes and the proportions of inclined lower-case letters are the same as those of vertical lower-

Fig. 3.26 **Vertical Lower-case Letters.**

Fig. 3.27 **Inclined Lower-case Letters.**

case letters. The inclined lower-case letters may be regarded, like the inclined capital letters, as oblique projections of vertical letters, in which all circles in the vertical alphabet become ellipses in the inclined alphabet. As in the inclined capital letters, all ellipses have their major axes inclined at an angle of 45° with horizontal.

The forms of the letters c, o, s, v, w, x, and z are almost identical with those of the corresponding capitals.

The slope of the letters is the same as for inclined capitals, or 68.2° with horizontal. The slope may be determined by drawing a "slope triangle" of 2 in 5, as shown in Fig. 3.24 (b), or with the aid of the inclined slot in the Braddock-Rowe Triangle, Fig. 3.16, or with the Ames Lettering Guide, Fig. 3.17 (b).

## 3.24 Spacing of Letters and Words.

Uniformity in spacing of letters is a matter of equalizing spaces by eye. *The background areas between letters, not the distances between them, should be approximately equal.* In Fig. 3.28 (a) the actual distances are equal, but the letters do not ap-

Fig. 3.28 **Spacing Between Letters.**

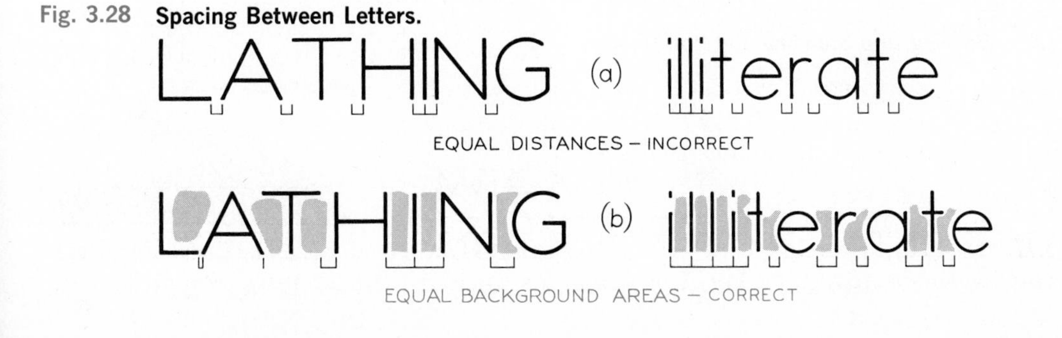

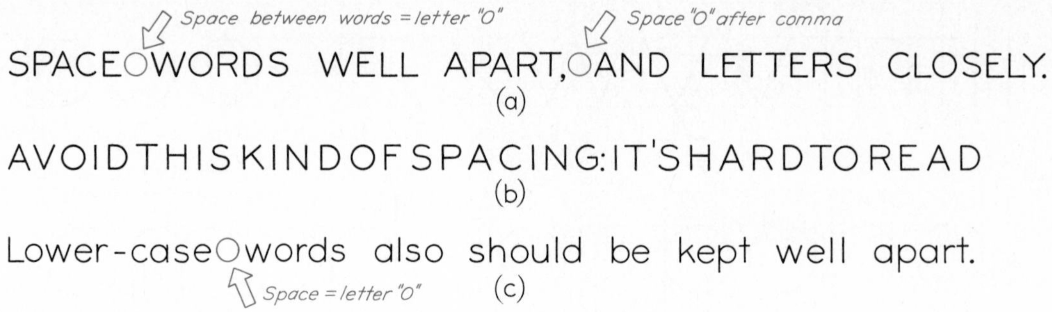

Space between words = letter "O"　　Space "O" after comma

SPACE○WORDS WELL APART,○AND LETTERS CLOSELY.

(a)

AVOIDTHISKINDOFSPACING:IT'SHARDTOREAD

(b)

Lower-case○words also should be kept well apart.

Space = letter "O"　(c)

Fig. 3.29 **Spacing Words.**

pear equally spaced. At (b) the distances are intentionally unequal, but the background areas between letters are approximately equal, and the result is an even and pleasing spacing.

Some combinations, such as LT and VA, may even have to be slightly overlapped to secure good spacing. In some cases the width of a letter may be decreased. For example, the lower stroke of the L may be shortened when followed by A.

*Space words well apart, but space letters closely within words.* Make each word a compact unit well-separated from adjacent words. For either upper-case or lower-case lettering, make the spaces between words approximately equal to a capital O, Fig. 3.29. Avoid spacing letters too far apart and words too close together, as shown at (b). Samples of good spacing are shown in Fig. 3.10.

When it is necessary to letter to a stop-line as in Fig. 3.30 (a), space each letter from *right to left,* as shown in step II, estimating the widths of the letters by eye. Then letter from *left to right,* as shown at III, and finally erase the spacing marks.

When it is necessary to space letters symmetrically about a center line, Fig. 3.30 (b), which is frequently the case in titles, Figs. 3.35 to 3.37, number the letters as shown, with the space between words considered as one letter. Then place the middle letter on center, making allowance for narrow letters (I's) or wide letters (W's) on either side. The X at (b) is placed slightly to the left of center to compensate for the letter I, which has no width. Check with the dividers to make sure that distances *a* are exactly equal.

Another method is to letter roughly a trial line of lettering along the bottom edge of a scrap of paper, place it in position immediately above, as shown at (c), and then letter the line in place. Be sure to use guide lines for the trial lettering.

If the lettering is being done on tracing paper or cloth, the trial letters can be placed underneath, arranged for lettering to a stop line or "on center," and then lettered directly over or with slight improvement as may be desired, Fig. 3.35.

## 3.25 Lettering Devices.

The *Leroy Lettering Instrument,* Fig. 3.31, is perhaps the most widely used lettering device. A guide pin follows grooved letters in a template, and the inking point moves on the paper. By adjusting

Fig. 3.30 **Spacing to a Stop-line and "on Center."**

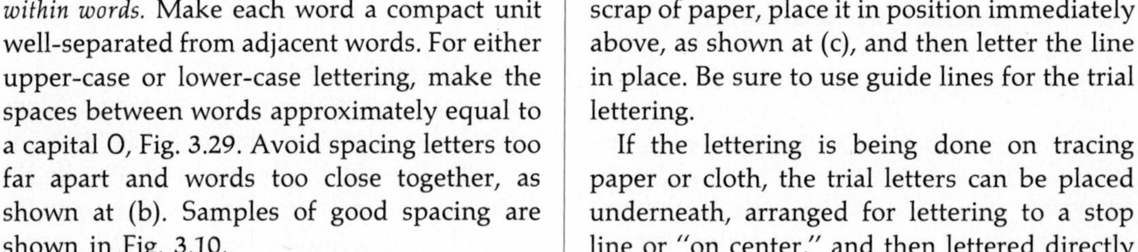

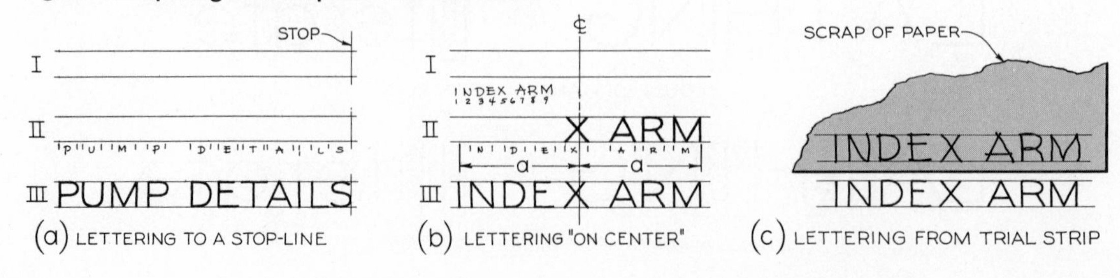


(a) LETTERING TO A STOP-LINE　(b) LETTERING "ON CENTER"　(c) LETTERING FROM TRIAL STRIP

*Courtesy Keuffel & Esser Co.*

Fig. 3.31 **Leroy Lettering Instrument.**

the arm on the instrument, the letters may be made vertical or inclined. A number of templates and sizes of pens is available, including templates for a wide variety of "built-up" letters similar to those made by the Varigraph and Letterguide, described below. Inside each pen is a cleaning pin used to keep the small tube open. These pins are easily broken, especially the small ones, when the pen is not promptly cleaned. To clean a pen, draw it across a blotter until all ink has been absorbed; then insert the pin and remove it and wipe it with a cloth. Repeat this until the pin remains clean. If the ink has dried, the pens may be cleaned with Leroy pen-cleaning fluid, available at dealers.

The *Wrico Lettering Guide,* Fig. 3.32, consists of a scriber and templates similar to the Leroy system. Wrico letters more closely resemble American National Standard letters than do the Leroy letters.

Fig. 3.32 **Wrico Lettering Guide.**

*Courtesy Wood-Regan Instrument Co., Inc.*

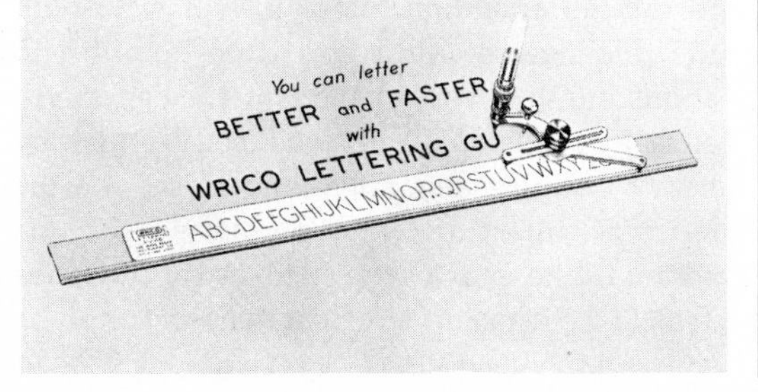


The *Varigraph* is a more elaborate device for making a wide variety of either single-stroke letters or "built-up" letters. As shown in Fig. 3.33, a guide pin (lower right) is moved along the grooves in a template, and the pen (upper left) forms the letters. The *Letterguide* scriber, Fig. 3.34, is a much simpler instrument, which also makes a large variety of styles and sizes of letters when used with the various templates available. It also operates with a guide pin moving in the grooved letters of the template, while the pen, which is mounted on an adjustable arm, makes the letters in outline.

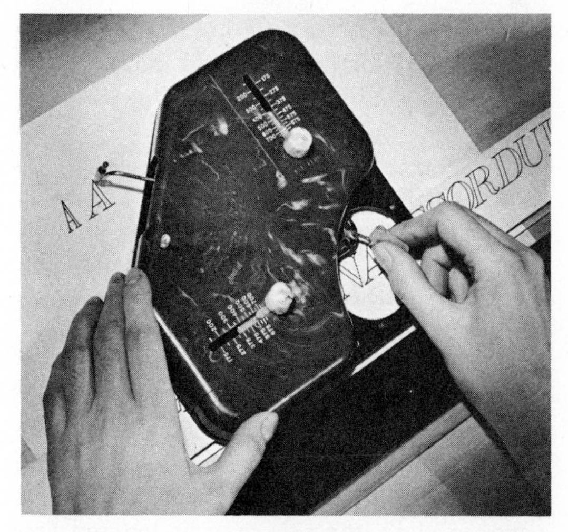

*Courtesy The Varigraph Co., Inc.*

Fig. 3.33 **The Varigraph.**

Fig. 3.34 **Letterguide.**

*Courtesy Letterguide*

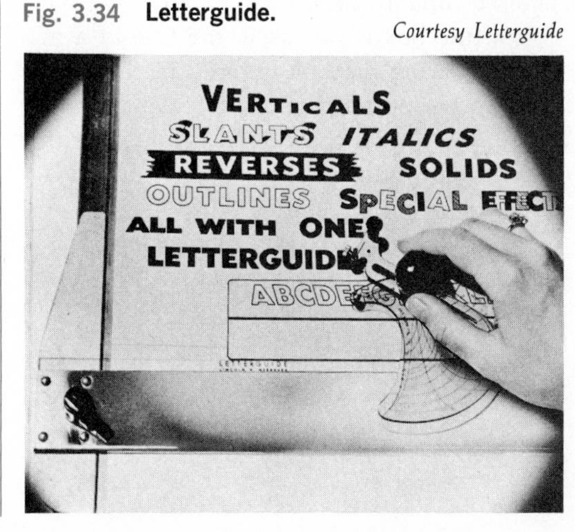

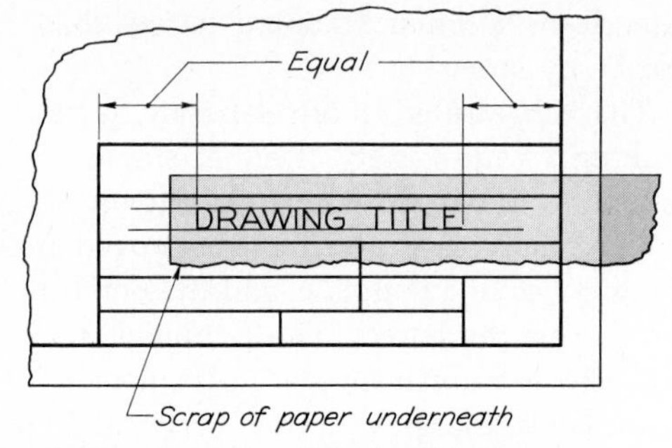


Fig. 3.35 **Centering Title in Title Box.**

Various forms of press-on lettering and special lettering devices (typewriters, etc.) are available. In whatever way the lettering is applied to the drawing and whatever style of lettering is used, the lettering must meet the requirements for legibility and microfilm reproduction.

## 3.26 Titles.

The composition of titles on machine drawings is relatively simple. In most cases, the title and related information are lettered in "title boxes" or "title strips," which are printed directly on the drawing paper, tracing paper, or cloth, Figs. 14.24 and 14.26 to 14.28. The main drawing title is usually centered in a rectangular space. This may be done by the method shown in Fig. 3.30 (b); or if the lettering is being done on tracing paper or cloth, the title may be lettered first on scrap paper and then placed underneath the tracing, as shown in Fig. 3.35, and then lettered directly over.

If a title box is not used, the title of a machine drawing may be lettered in the lower-right corner of the sheet as a "balanced title," Fig. 3.36. A balanced title is simply one that is arranged symmetrically about an imaginary center line. These titles take such forms as the rectangle, the oval, the inverted pyramid, or any other simple symmetrical form.

On display drawings, or on highly finished maps or architectural drawings, titles may be composed of filled-in letters, usually Gothic or Roman, Fig. 3.37.

In any kind of title, the most important words are given the most prominence by making the lettering larger, heavier, or both. Other data, such as scale and date, are not so important and should not be prominently displayed.

Fig. 3.36 **Balanced Machine-Drawing Title.**

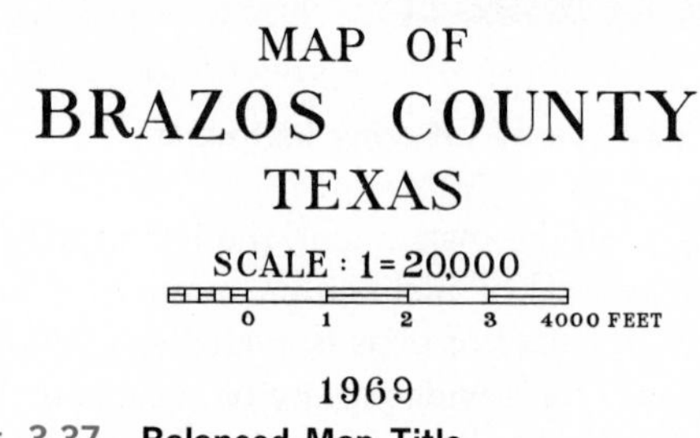

Fig. 3.37 **Balanced Map Title.**

## 3.27 Gothic Letters.

Fig. 3.38. Among the many forms of Gothic styles, including Old English and German Gothic, the so-called *sans-serif* Gothic letter is the only one of interest to engineers. It is from this style that the modern single-stroke engineering letters, discussed in the early part of this chapter, are derived. While they are admittedly not as beautiful as many other styles, they are very legible and comparatively easy to make.

Sans-serif Gothic letters should be used on drawings where legibility and not beauty is the determining factor. They should be drawn in outline and then filled in, Fig. 3.39, with the thickness of the stems from one-fifth to about one-tenth the height of the letter. An attractive letter may be produced by making heavy outlines and not filling in, as for the letter H shown at (a). A slight spur may be added to the ends of the stem, as for the letter T at (a). An example of condensed Gothic is shown at (b).

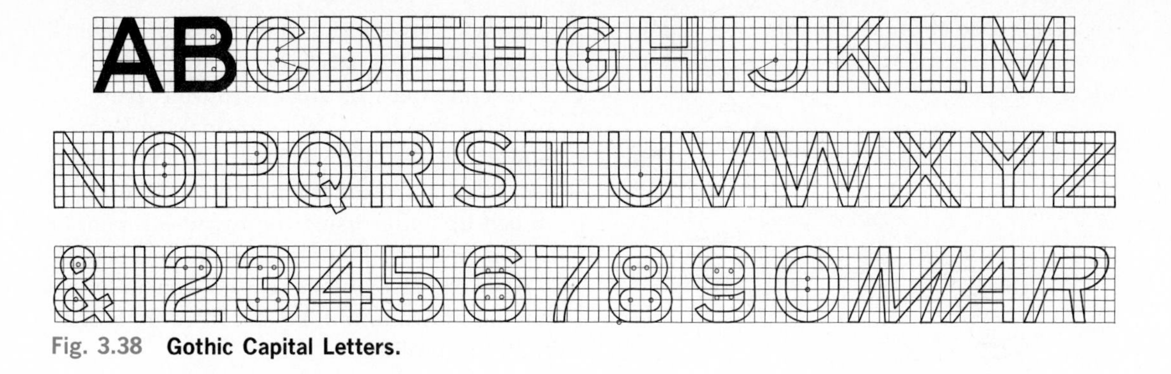

Fig. 3.38 **Gothic Capital Letters.**

Fig. 3.39 **Gothic Letter Construction.**

Fig. 3.40 **Old Roman Capitals, with Numerals and Lower-case of Similar Design.**

**3.28 Old Roman Letters.** Fig. 3.40. The Old Roman letter is the basis of all of our letters, and is still regarded as the most beautiful. The letters on the base of Trajan's Column in Rome are regarded by many as the finest example of Old Roman letters.

The Old Roman letter is used mostly by architects. Because of its great beauty, it is used almost exclusively on buildings and for inscriptions on bronze or stone. Full-size "details" of the letter are usually drawn for such inscriptions.

Originally, this letter was made on manuscripts with a broad-point reed pen, Fig. 1.8;

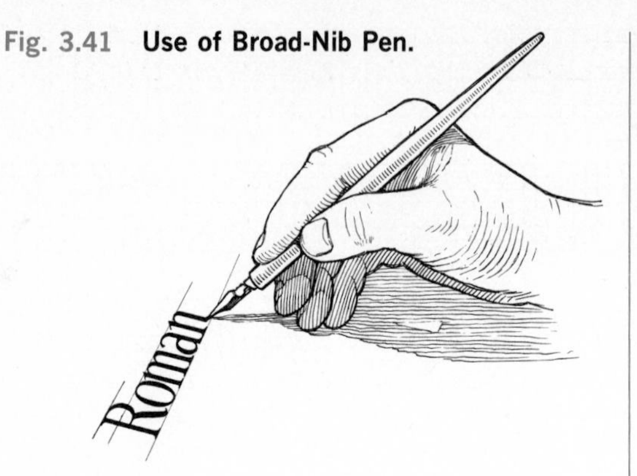

Fig. 3.41 **Use of Broad-Nib Pen.**

the wide stems were produced by downward strokes, and the narrow portions by horizontal strokes. A brief examination of any Roman letter will show why certain strokes are wide, while others are narrow.

Several styles of steel broad-nib pens are available and are suitable for making Roman, Gothic, or Text letters, Figs. 3.1 and 3.41. If necessary, an ordinary pen may be used to "touch up" after using the broad-nib pen, or to add fine-line flourishes, as in Text letters.

As in the case of Gothic, Old Roman letters may be drawn in outline and filled in, or may be left in outline.

## 3.29 Modern Roman Letters.

Figs. 3.42 and 3.43. The Modern Roman, or simply "Roman," letters were evolved during the

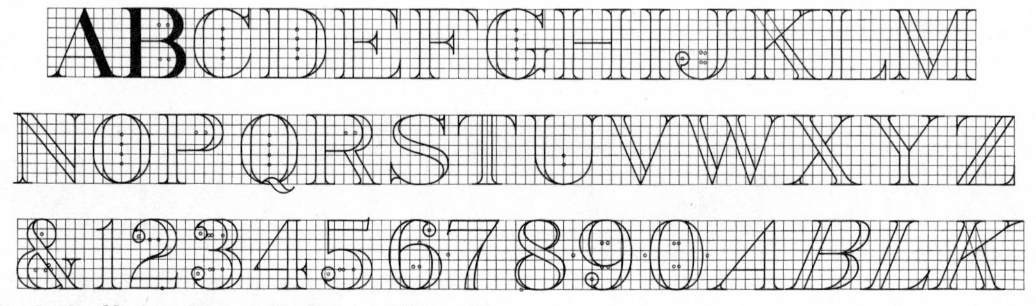

Fig. 3.42 **Modern Roman Capitals and Numerals.**

Fig. 3.43 **Lower-case Modern Roman Letters.**

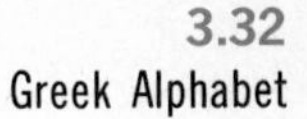


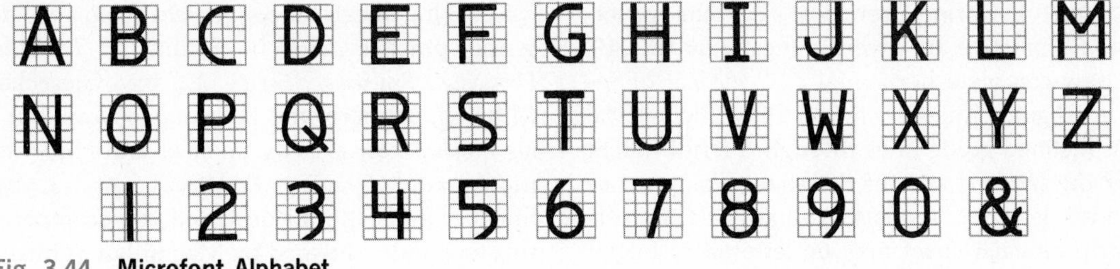


Fig. 3.44 **Microfont Alphabet.**

eighteenth century by the type founders; the letters used in most modern newspapers, magazines, and books are of this style. The text of this book is set in Modern Roman Letters. These letters are often used on maps, especially for titles. They may be drawn in outline and then filled in, as shown in Fig. 3.42, or they may be produced with one of the broad-nib pens shown in Fig. 3.7.

If drawn in outline and filled in, the straight lines may be drawn with the ruling pen, and the circular curves drawn with the bow pen. The fillets and other noncircular curves are drawn freehand. The thickness of the stem, or broad stroke, varies widely, the usual thickness being from one-sixth to one-eighth of the height of the letter.

A typical example of the use of Modern Roman in titles is shown in Fig. 3.37. Their use on maps is discussed below.

Lower-case Modern Roman letters are shown in Fig. 3.43. The lower-case italics are known as *stump letters.* They are easily made freehand, and are often used on maps and on Patent Office drawings, §14.26.

## 3.30 Lettering on Maps.

Modern Roman letters are generally used on maps, as follows:

1. *Vertical Capitals.* Names of states, countries, townships, capital cities, large cities, and titles of maps. See §3.26.
2. *Vertical Lower-case.* (First letter of each word a capital.) Names of small towns, villages, post offices, and so forth.
3. *Inclined Capitals.* Names of oceans, bays, gulfs, sounds, large lakes, and rivers.
4. *Inclined Lower-case, or "Stump Letters."* (First letter of each word a capital.) Names of rivers, creeks, small lakes, ponds, marshes, brooks, and springs.

Prominent land features, such as mountains, plateaus, and canyons, are lettered in vertical Gothic, Fig. 3.18, while the names of small land features, such as small valleys, islands, and ridges, are lettered in vertical lower-case Gothic, Fig. 3.26. Names of railroads, tunnels, highways, bridges, and other public structures are lettered in inclined Gothic capitals, Fig. 3.19.

## 3.31 Microfont Alphabet.

The microfont alphabet, Fig. 3.44, is a recent adaptation of the single-stroke Gothic characters developed by the National Microfilm Association. It is designed for general usage and increased legibility in reproduction. Only the vertical style is shown. The inclined style is under development.

Fig. 3.45 **Greek Alphabet.**

| | | | | | |
|---|---|---|---|---|---|
| Α α | alpha | Ι ι | iota | Ρ ρ | rho |
| Β β | beta | Κ κ | kappa | Σ ς | sigma |
| Γ γ | gamma | Λ λ | lambda | Τ τ | tau |
| Δ δ | delta | Μ μ | mu | ϒ υ | upsilon |
| Ε ϵ | epsilon | Ν ν | nu | Φ φ | phi |
| Ζ ζ | zeta | Ξ ξ | xi | Χ χ | chi |
| Η η | eta | Ο ο | omicron | Ψ ψ | psi |
| Θ θ | theta | Π π | pi | Ω ω | omega |

## 3.32 Greek Alphabet.

Greek letters are often used as symbols in both mathematics and technical drawing by the engineer. A Greek alphabet, showing both upper-case and lower-case letters, is given for reference purposes in Fig. 3.45.

**3.33 Lettering Exercises.** Layouts for lettering practice are given in Figs. 3.46 to 3.49. Draw complete horizontal and vertical or inclined guide lines *very lightly.* Draw the vertical or inclined guide lines through the full height of the lettered area of the sheet. For practice in ink lettering, the last two lines and the title strip on each sheet may be lettered in ink, if assigned by the instructor. Omit all dimensions.

Lettering sheets in convenient form for lettering practice may be found in *Technical Drawing Problems, Series 1,* by Giesecke, Mitchell, Spencer, and Hill; in *Technical Drawing Problems, Series 2,* by Spencer and Hill; and in *Technical Drawing Problems, Series 3,* by Spencer and Hill, all designed to accompany this text and published by Macmillan Publishing Co., Inc.

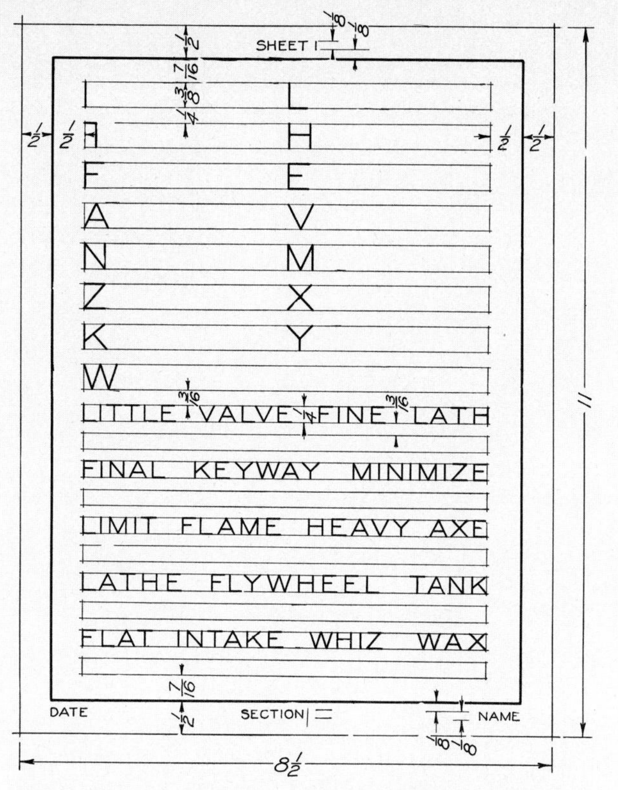


Fig. 3.46 Lay out sheet, add vertical or inclined guide lines, and fill in vertical or inclined capital letters as assigned.

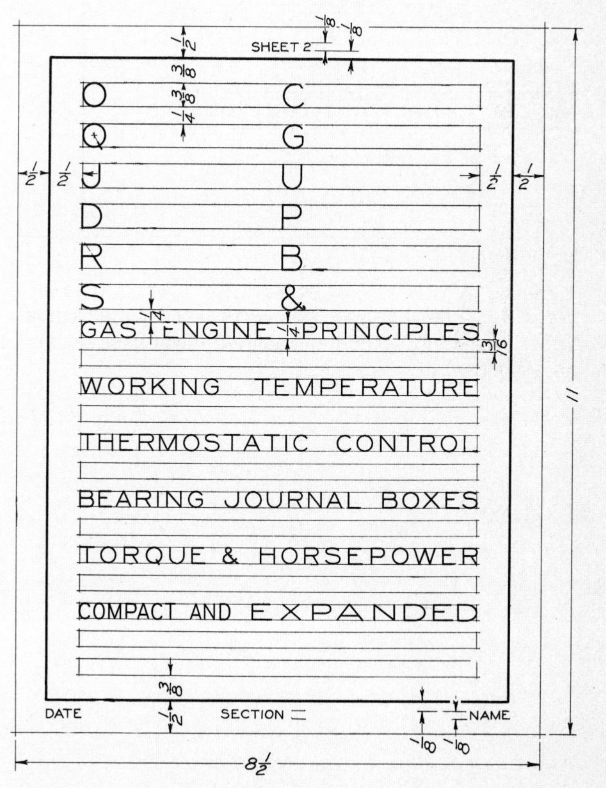


Fig. 3.47 Lay out sheet, add vertical or inclined guide lines, and fill in vertical or inclined capital letters as assigned.

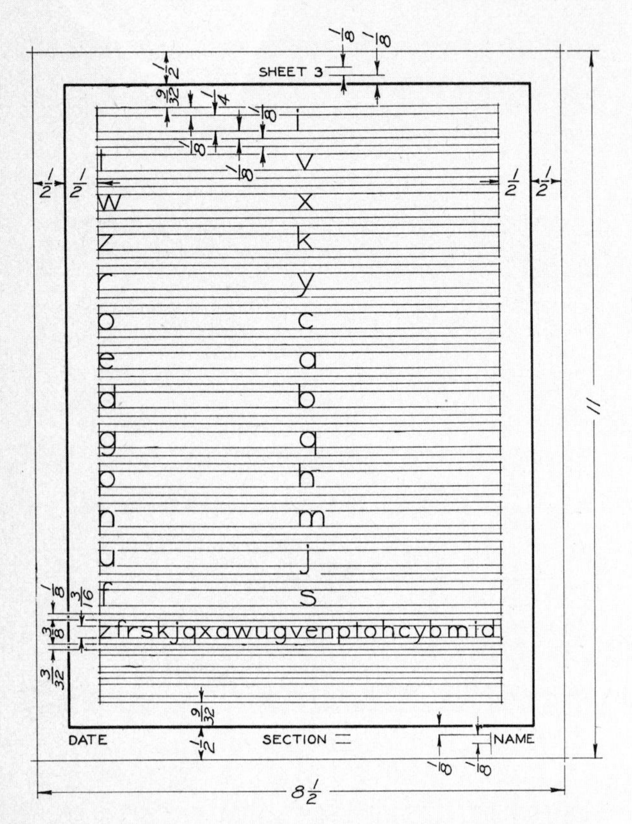


Fig. 3.48 Lay out sheet, add vertical or inclined guide lines, and fill in vertical or inclined lower-case letters as assigned.

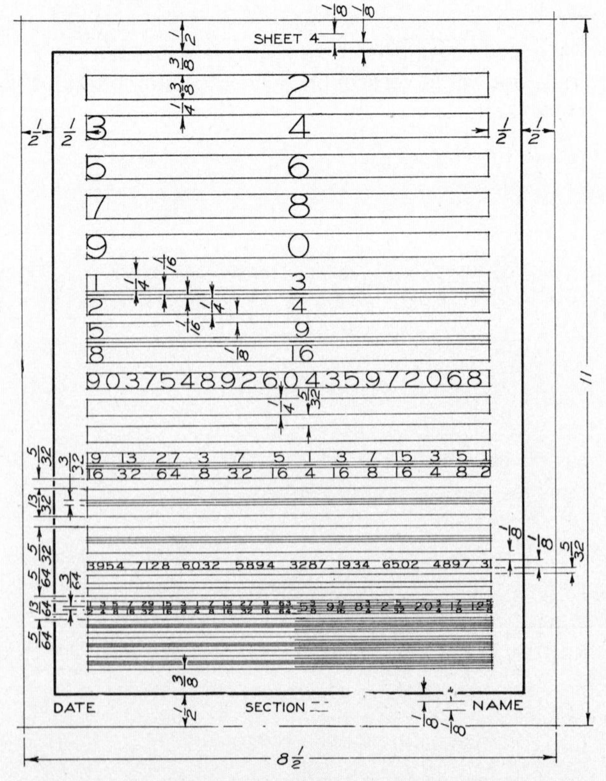


Fig. 3.49 Lay out sheet, add vertical or inclined guide lines, and fill in vertical or inclined numerals as assigned.

# 4

# geometric constructions

**4.1 Geometric Constructions.** Many of the constructions used in technical design drawing are based upon plane geometry, and every draftsman or engineer should be sufficiently familiar with them to be able to apply them to the solutions of problems. Pure geometry problems may be solved only with the compass and a straightedge, and in some cases these methods may be used to advantage in technical drawing. However, the draftsman or designer has at his disposal the T-square, triangles, dividers, and other equipment, such as drafting machines, which in many cases enable him to obtain accurate results more quickly by what we may term "draftsmen's methods." Therefore, many of the solutions in this chapter are draftsmen's adaptations of the principles of pure geometry.

This chapter is designed to present definitions of terms and geometric constructions of importance in technical drawing, suggest simplified methods of construction, point out practical applications, and afford opportunity for practice in accurate instrumental drawing. In the latter sense, the problems at the end of this chapter may be regarded as a continuation of those at the end of Chapter 2.

In drawing these constructions, accuracy is most important. Use a sharp medium-hard lead (H to 3H) in your pencil and compasses. Draw construction lines extremely light—so light that they can hardly be seen when your drawing is held at arm's length. Draw all given and required lines medium to thin but dark.

**4.2 Points and Lines.** Fig. 4.1. A *point* represents a location in space or on a drawing,

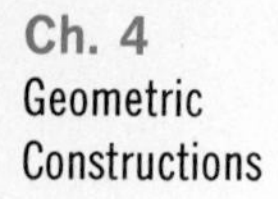

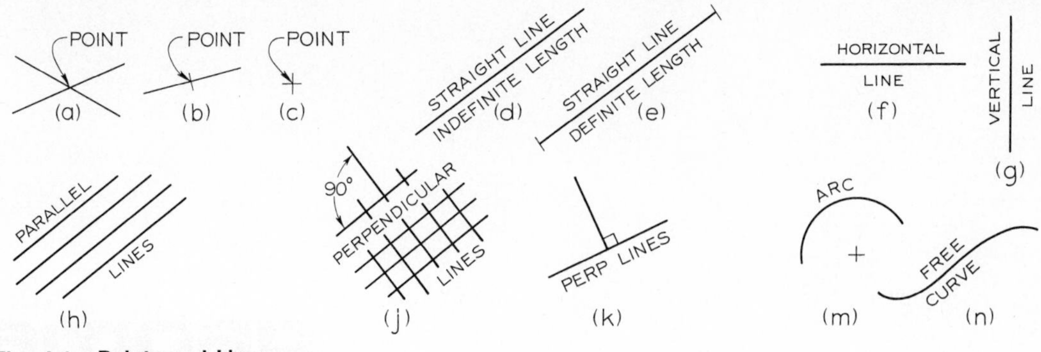


Fig. 4.1 **Points and Lines.**

and has no width, height, or depth. A point is represented by the intersection of two lines, (a), by a short crossbar on a line, (b), or by a small cross, (c). Never represent a point by a simple dot on the paper.

A line is defined by Euclid as "that which has length without breadth." A *straight line* is the shortest distance between two points and is commonly referred to simply as a "line." If the line is indefinite in extent, the length is a matter of convenience, and the end points are not fixed, (d). If the end points of the line are significant, they must be marked by means of small mechanically drawn crossbars, (e). Other common terms are illustrated from (f) to (h). Either straight lines or curved lines are parallel if the shortest distance between them remains constant. A common symbol for parallel lines is ||, and for perpendicular lines is ⊥ (singular) or ⊥s (plural). Two perpendicular lines may be marked with a "box" to indicate perpendicularity, as shown at (k). Such symbols may be used on sketches, but not on production drawings.

## 4.3 Angles.

Fig. 4.2. An *angle* is formed by two intersecting lines. A common symbol for angle is ∠ (singular) or ∠s (plural). There are 360 *degrees* (360°) in a full circle, as shown at (a). A degree is divided into 60 *minutes* (60′), and a minute is divided into 60 *seconds* (60″). Thus, 37° 26′ 10″ is read: 37 degrees, 26 minutes, and 10 seconds. When minutes alone are indicated, the number of minutes should be preceded by 0°, as 0° 20′.

The different kinds of angles are illustrated in (b) to (e). Two angles are *complementary,* (f), if they total 90°, and are *supplementary,* (g), if they total 180°. Most angles used in technical drawing can be drawn easily with the T-square and triangles, Fig. 2.26. To draw odd angles, use the protractor, Fig. 2.27. For considerable accuracy, use a *vernier protractor,* or the *tangent, sine,* or *chord* methods, §4.21.

## 4.4 Triangles.

Fig. 4.3. A *triangle* is a plane figure bounded by three straight sides, and the sum of the interior angles is always 180°. A right triangle, (d), has one 90° angle, and the square of the hypotenuse is equal to the sum of the squares of the two sides, (e). As shown at (f), any triangle inscribed in a semicircle is a *right triangle* if the hypotenuse coincides with the diameter.

Fig. 4.2 **Angles.**

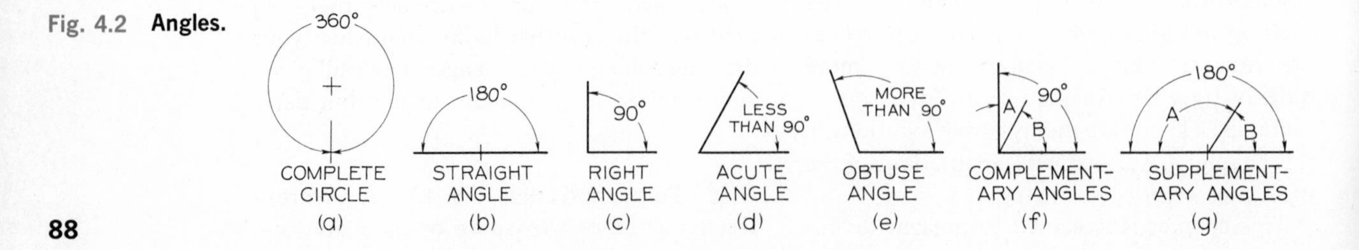

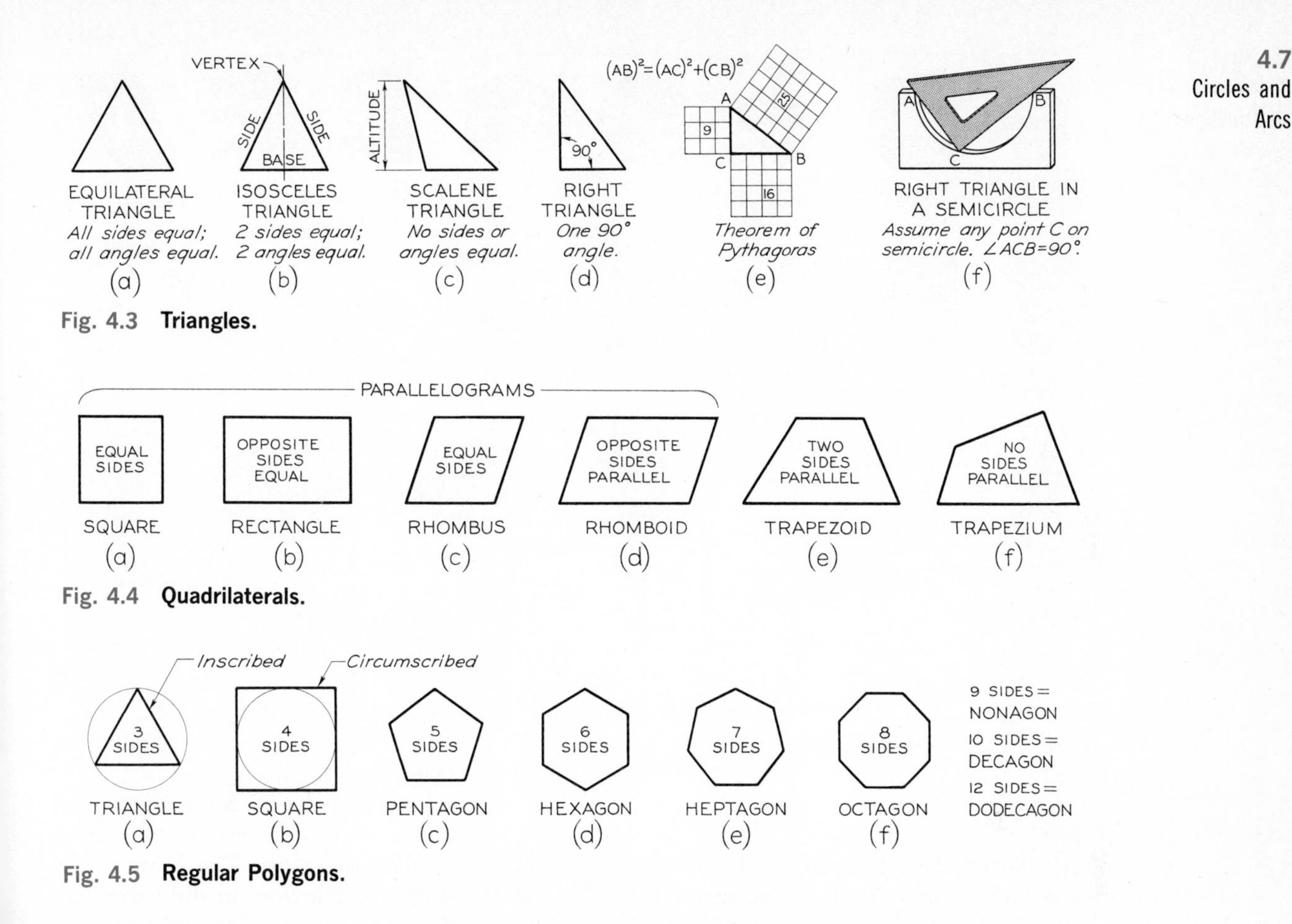


**Fig. 4.3 Triangles.**

**Fig. 4.4 Quadrilaterals.**

**Fig. 4.5 Regular Polygons.**

**4.5 Quadrilaterals.** Fig. 4.4. A *quadrilateral* is a plane figure bounded by four straight sides. If the opposite sides are parallel, the quadrilateral is also a *parallelogram.*

**4.6 Polygons.** Fig. 4.5. A *polygon* is any plane figure bounded by straight lines. If the polygon has equal angles and equal sides, it can be *inscribed* in or *circumscribed* around a circle, and is called a *regular polygon.*

**4.7 Circles and Arcs.** Fig. 4.6. A *circle,* (a), is a closed curve all points of which are the same distance from a point called the *center. Circumference* refers to the circle or to the distance around the circle. This distance equals

**Fig. 4.6 The Circle.**

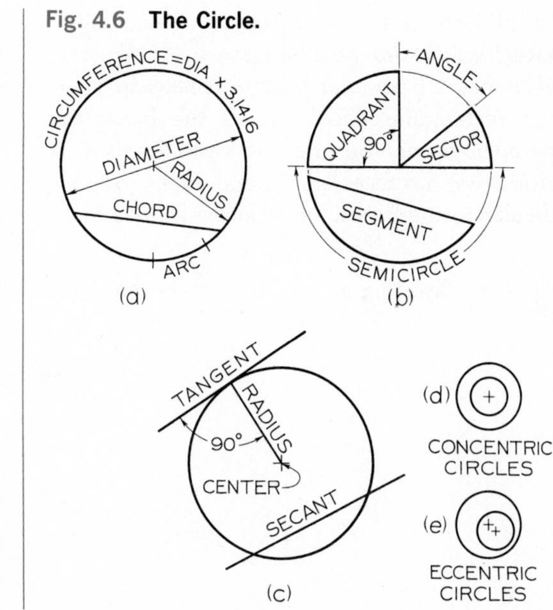

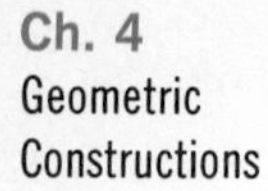


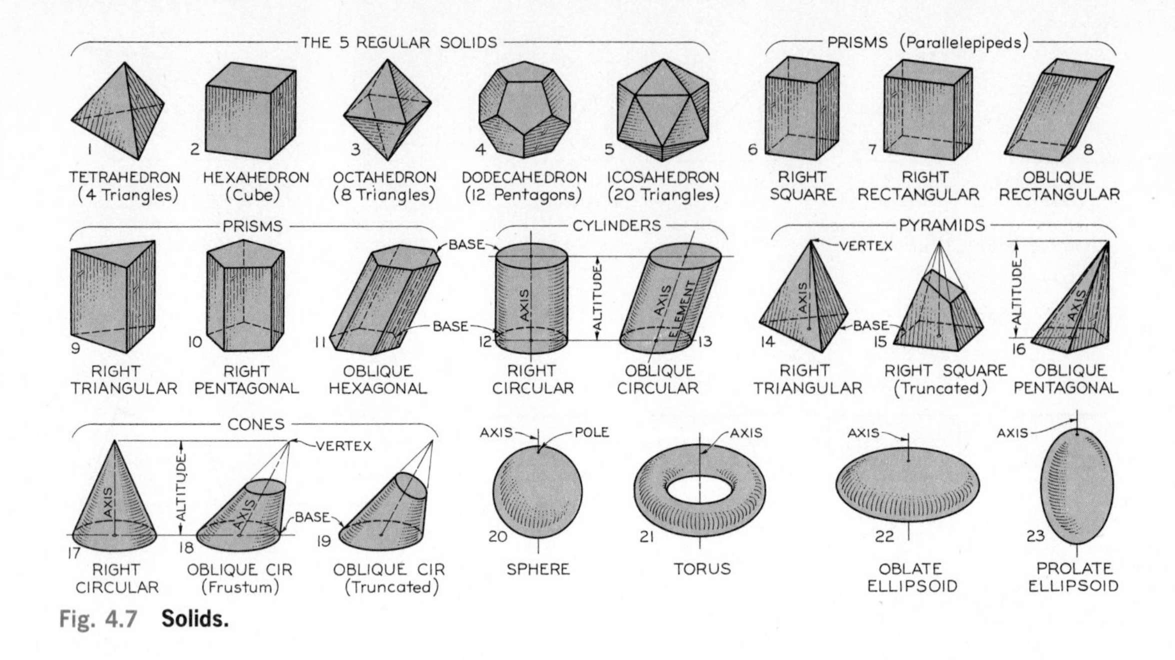


Fig. 4.7 **Solids.**

the diameter multiplied by $\pi$ (called *pi*) or 3.1416. Other definitions are illustrated in the figure.

## 4.8 Solids.

**4.8 Solids.** Fig. 4.7. Solids bounded by plane surfaces are *polyhedra.* The surfaces are called *faces,* and if these are equal regular polygons, the solids are *regular polyhedra.*

A *prism* has two *bases,* which are parallel equal polygons, and three or more lateral faces, which are parallelograms. A *triangular prism* has a triangular base; a *rectangular prism* has rectangular bases, etc. If the bases are parallelograms, the prism is a *parallelepiped.* A *right prism* has faces and lateral edges perpendicular to the bases; an *oblique prism* has faces and lateral edges oblique to the bases. If one end is cut off to form an end not parallel to the bases, the prism is said to be *truncated.*

A *pyramid* has a polygon for a base and triangular lateral faces intersecting at a common point called the *vertex.* The center line from the center of the base to the vertex is the *axis.* If the axis is perpendicular to the base, the pyramid is a *right pyramid;* otherwise it is an *oblique pyramid.* A *triangular pyramid* has a triangular base, a *square pyramid* has a square base, etc. If a portion near the vertex has been cut off, the pyramid is *truncated,* or referred to as a *frustum.*

A *cylinder* is generated by a straight line, called the *generatrix,* moving in contact with a curved line and always remaining parallel to

Fig. 4.8 **Bisecting a Line or a Circular Arc** (§4.9).

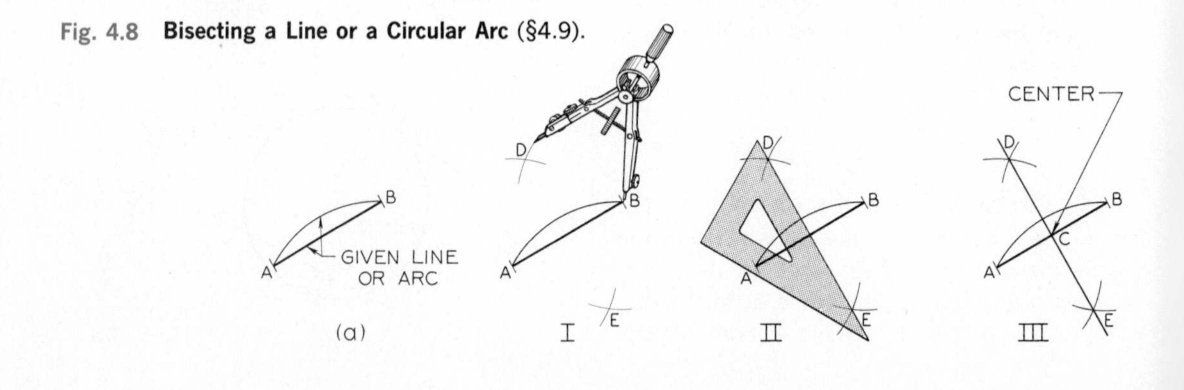

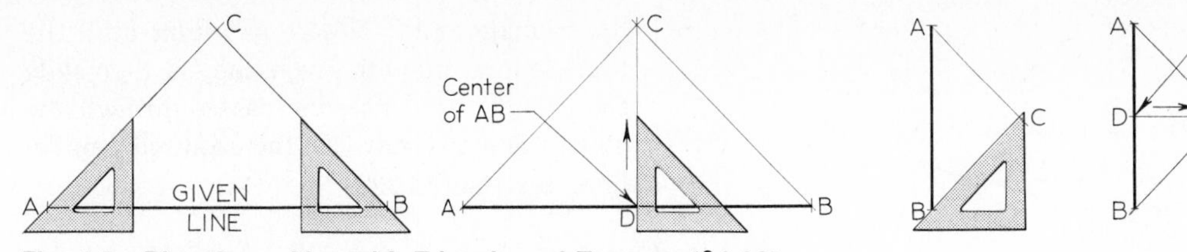


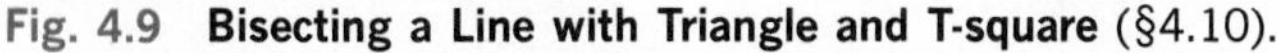

Fig. 4.9 **Bisecting a Line with Triangle and T-square** (§4.10).

its previous position or to the axis. Each position of the generatrix is called an *element* of the cylinder.

A *cone* is generated by a straight line moving in contact with a curved line, and passing through a fixed point, the vertex of the cone. Each position of the generatrix is an *element* of the cone.

A *sphere* is generated by a circle revolving about one of its diameters. This diameter becomes the *axis* of the sphere, and the ends of the axis are *poles* of the sphere.

A *torus* is generated by a circle (or other curve) revolving about an axis which is eccentric to the curve.

## 4.9 To Bisect a Line or a Circular Arc.

Fig. 4.8. Given line or arc AB, as shown at (a), to be bisected.

I. From A and B draw equal arcs with radius greater than half AB.

II. and III. Join intersections D and E with a straight line to locate center C.

## 4.10 To Bisect a Line with Triangle and T-square.

Fig. 4.9. From end points A and B, draw construction lines at 30°, 45°, or 60° with the given line; then through their intersection C, draw line perpendicular to the given line to locate the center D, as shown.

To divide a line with the dividers, see §2.40.

Fig. 4.10 **Bisecting an Angle** (§4.11).

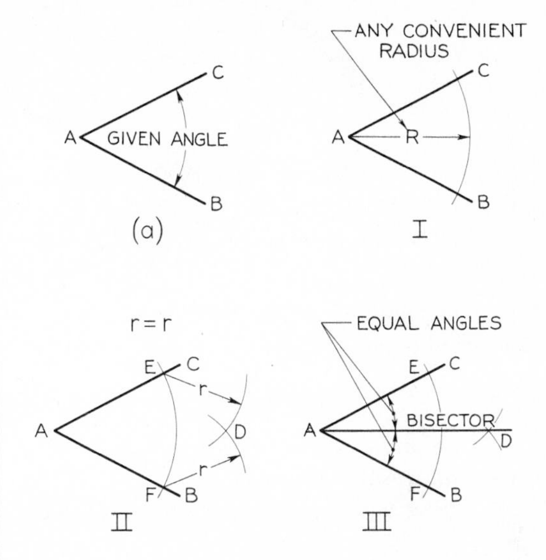


## 4.11 To Bisect an Angle.

Fig. 4.10. Given angle BAC, as shown at (a), to be bisected.

I. Strike large arc R.

II. Strike equal arcs r with radius slightly larger than half BC, to intersect at D.

III. Draw line AD, which bisects angle.

## 4.12 To Transfer an Angle.

Fig. 4.11. Given angle BAC, as shown at (a), to be transferred to the new position at A'B'.

Fig. 4.11 **Transferring an Angle** (§4.12).

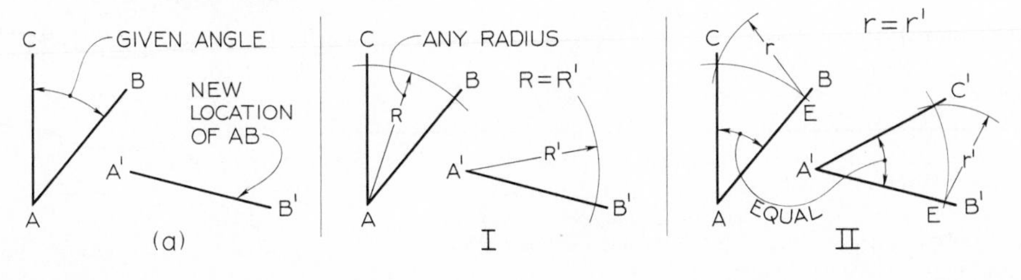

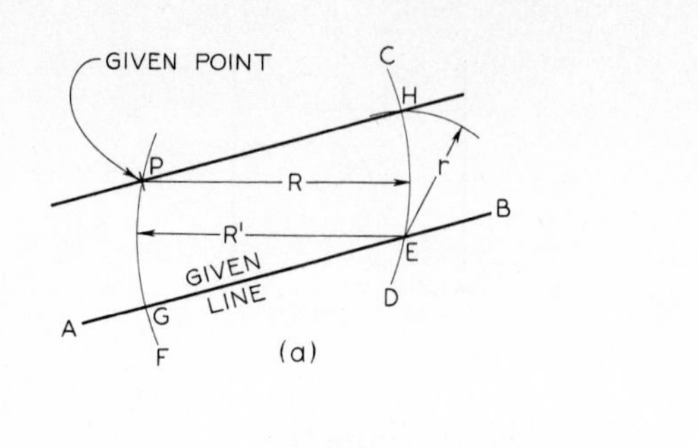


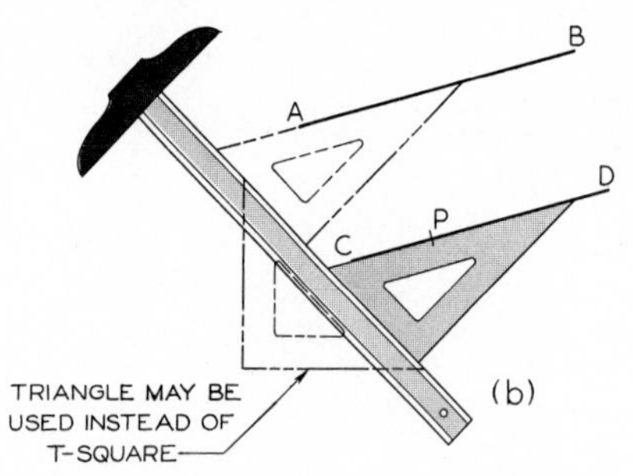


Fig. 4.12 **Drawing a Line Through a Point Parallel to a Line** (§4.13).

I. Use any convenient radius R, and strike arcs from centers A and A′.

II. Strike equal arcs r, and draw side A′C′.

## 4.13 To Draw a Line Through a Point and Parallel to a Line.

*Fig. 4.12 (a).* With given point P as center, and any convenient radius R, strike arc CD to intersect the given line AB at E. With E as center and the same radius, strike arc R′ to intersect the given line at G. With PG as radius, and E as center, strike arc r to locate point H. The line PH is the required line.

*Fig. 4.12 (b). Preferred by Draftsmen.* Move the triangle and T-square as a unit until the triangle lines up with given line AB; then slide the triangle until its edge passes through the given point P. Draw CD, the required parallel line. See also §2.22.

## 4.14 To Draw a Line Parallel to a Line and at a Given Distance.

Let AB be the line and CD the given distance.

*Fig. 4.13 (a).* With points E and F near A and B respectively as centers, and CD as radius, draw two arcs. The line GH, tangent to the arcs, is the required line.

*Fig. 4.13 (b). Preferred by Draftsmen.* With any point E of the line as center and CD as radius, strike an arc JK. Move the triangle and T-square as a unit until the triangle lines up with the given line AB; then slide the triangle until its edge is tangent to the arc JK, and draw the required line GH.

*Fig. 4.13 (c).* With centers selected at random on the curved line AB, and with CD as radius, draw a series of arcs; then draw the required line tangent to these arcs as explained in §2.57.

## 4.15 To Divide a Line into Equal Parts.

Fig. 4.14.

I. Draw light construction line at any convenient angle from one end of line.

II. With dividers or scale, set off from intersection of lines as many equal divisions as needed, in this case three.

III. Connect last division point to other end of line, using triangle and T-square, as shown.

IV. Slide triangle along T-square and draw

Fig. 4.13 **Drawing a Line Parallel to a Line at a Given Distance** (§4.14).

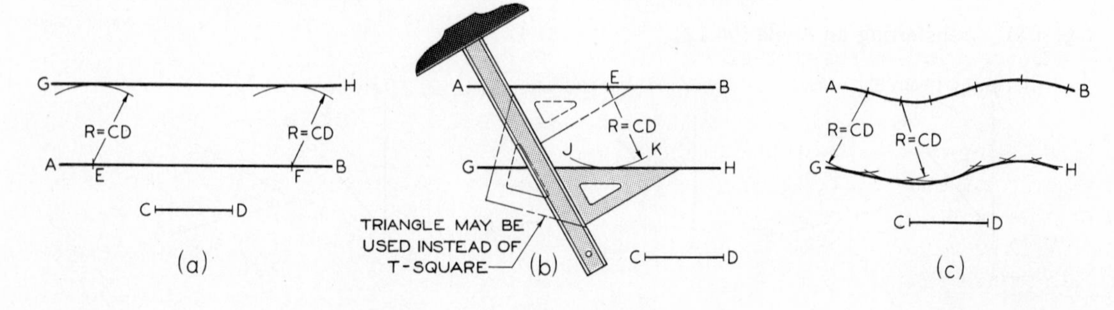

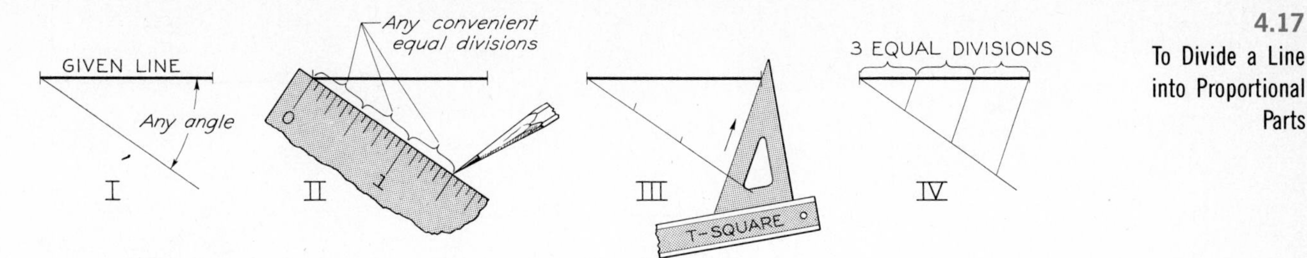


Fig. 4.14 **Dividing a Line into Equal Parts** (§4.15).

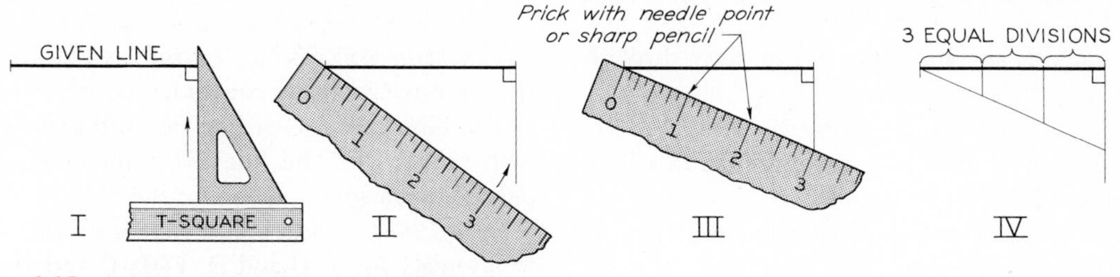


Fig. 4.15 **Dividing a Line into Equal Parts** (§4.16).

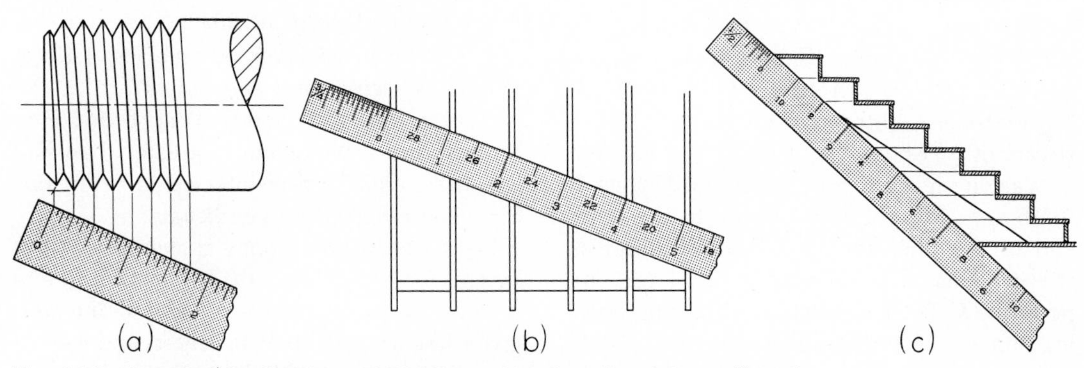


Fig. 4.16 **Practical Applications of Dividing a Line into Equal Parts** (§4.16).

parallel lines through other division points, as shown.

## 4.16 To Divide a Line into Equal Parts.

Fig. 4.15. *Preferred by Draftsmen.*

I. Draw vertical construction line at one end of given line.

II. Set zero of scale at other end of line.

III. Swing scale up until third unit falls on vertical line, and make tiny dots at each point, or prick points with dividers.

IV. Draw vertical construction lines through each point.

Some practical applications of this method are shown in Fig. 4.16.

## 4.17 To Divide a Line into Proportional Parts.

Fig. 4.17 (a) and (b). Let it be required to divide the line AB into three parts proportional to 2, 3, and 4.

*Fig. 4.17 (a). Preferred by Draftsmen.* Draw a vertical line from point B. Select a scale of convenient size for a total of 9 units and set the zero of the scale at A. Swing the scale up until the ninth unit falls on the vertical line. Along the scale, set off points for 2, 3, and

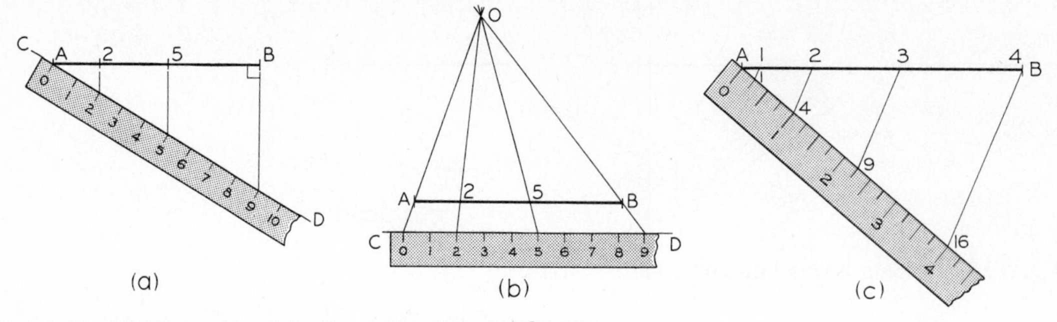


**Fig. 4.17 Dividing a Line into Proportional Parts** (§4.17).

4 units, as shown. Draw vertical lines through these points.

*Fig. 4.17 (b).* Draw a line CD parallel to AB and at any convenient distance. On this line, set off 2, 3, and 4 units, as shown. Draw lines through the ends of the two lines to intersect at the point O. Draw lines through O and the points 2 and 5 to divide AB into the required proportional parts.

Constructions of this type are useful in the preparation of graphs (Chapter 26).

*Fig. 4.17 (c).* Given AB, to divide into proportional parts, in this case proportional to the square of X, where X = 1, 2, 3, . . . . Set zero of scale at end of line and set off divisions 4, 9, 16, . . . . Join the last division to the other end of the line, and draw parallel lines as shown. This method may be used for any power of X. The construction is used in drawing nomographic charts (Chapter 27).

## 4.18 To Draw a Line Through a Point and Perpendicular to a Line.

Given the line AB and a point P.

When the Point Is Not on the Line. Fig. 4.18 (a). From P draw any convenient inclined line, as PD. Find center C of line PD, and draw arc with radius CP. The line EP is the required perpendicular.

*Fig. 4.18 (b).* With P as center, strike an arc to intersect AB at C and D. With C and D as centers, and radius slightly greater than half CD, strike arcs to intersect at E. The line PE is the required perpendicular.

When the Point Is on the Line. Fig. 4.18 (c). With P as center and any radius, strike arcs to intersect AB at D and G. With D and G as centers, and radius slightly greater than half DG, strike equal arcs to intersect at F. The line PF is the required perpendicular.

*Fig. 4.18 (d).* Select any convenient unit of length, for example $\frac{1}{4}''$. With P as center, and 3 units as radius, strike an arc to intersect given line at C. With P as center, and 4 units as radius, strike arc DE. With C as center, and 5 units as radius, strike an arc to intersect DE at F. The line PF is the required perpendicular.

This method is frequently used in laying off

**Fig. 4.18 Drawing a Perpendicular to a Line and Through a Point** (§4.18).

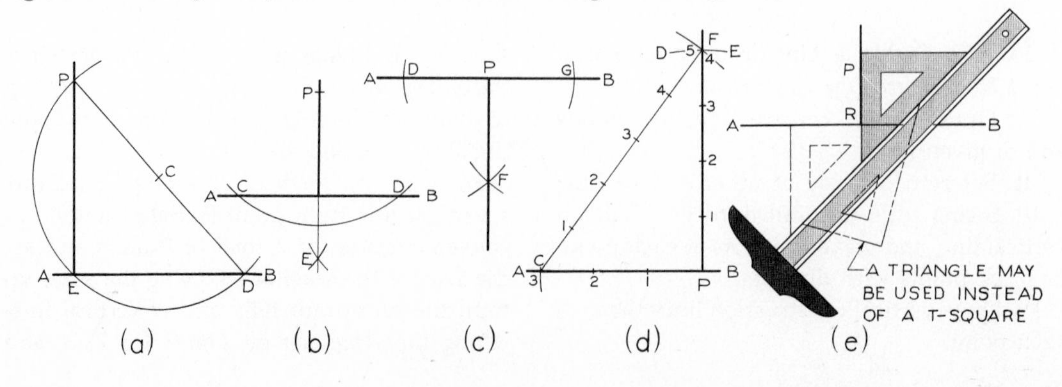

rectangular foundations of large machines, buildings, or other structures. For this purpose a steel tape may be used and distances of 30, 40, and 50 feet measured as the three sides of the right triangle.

*Fig. 4.18 (e). Preferred by Draftsmen.* Move the triangle and T-square as a unit until the triangle lines up with AB; then slide the triangle until its edge passes through the point P (whether P is on or off the line), and draw the required perpendicular.

## 4.19 To Draw a Triangle with Sides Given.

Fig. 4.19. Given the sides A, B, and C, as shown at (a):

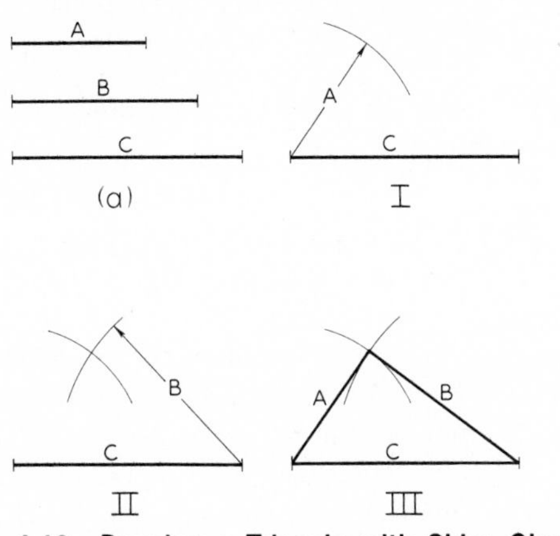


Fig. 4.19 **Drawing a Triangle with Sides Given** (§4.19).

I. Draw one side, as C, in desired position, and strike arc with radius equal to given side A.

II. Strike arc with radius equal to given side B.

III. Draw sides A and B from intersection of arcs, as shown.

## 4.20 To Draw a Right Triangle with Hypotenuse and One Side Given.

Fig. 4.20. Given sides S and R. With AB as a diameter equal to S, draw semicircle. With A as center, and R as radius, draw an arc intersecting the semicircle at C. Draw AC and CB to complete the right triangle.

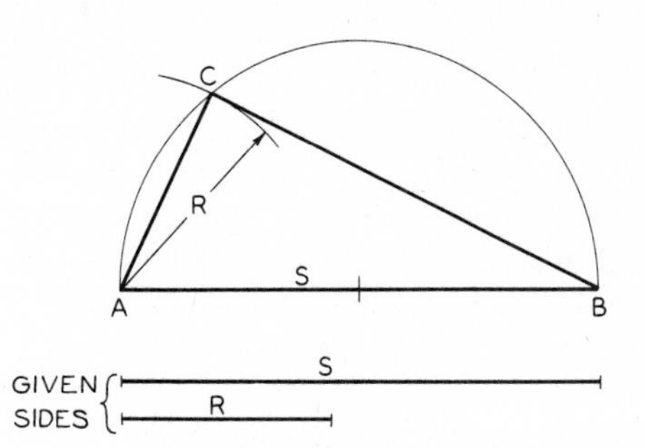


Fig. 4.20 **Drawing a Right Triangle** (§4.20).

## 4.21 To Lay Out an Angle.

Fig. 4.21. Many angles can be laid out directly with the triangle, Fig. 2.25; or they may be laid out with the protractor, Fig. 2.26. Other methods, where considerable accuracy is required, are as follows:

Tangent Method. Fig. 4.21 (a). The tangent of angle $\theta$ is $\frac{Y}{X}$, and $Y = X \tan \theta$. To construct the angle, assume a convenient value for X, preferably 10″, as shown. Find the tangent of angle $\theta$ in a table of natural tangents, multiply by 10, and set off $Y = 10 \tan \theta$.

Fig. 4.21 **Laying Out Angles** (§4.21).

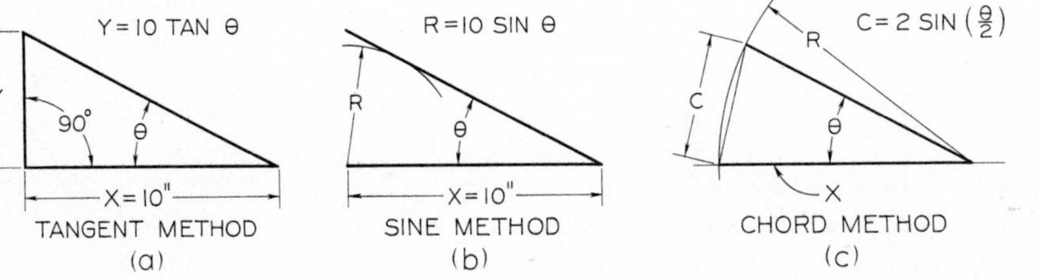

*Example:* To set off $31\frac{1}{2}°$, find the natural tangent of $31\frac{1}{2}°$, which is .6128. Then Y = 10″ (.6128) = 6.128″.

SINE METHOD. Fig. 4.21 (b). Draw line X to any convenient length, preferably 10″ as shown. Find the sine of angle $\theta$ in a table of natural sines, multiply by 10″, and strike arc R = 10″ sin $\theta$. Draw the other side of the angle tangent to the arc, as shown.

*Example:* To set off $25\frac{1}{2}°$, find the natural sine of $25\frac{1}{2}°$, which is .4305. Then R = 10″ (.4305) = 4.305″.

CHORD METHOD. Fig. 4.21 (c). Draw line X to any convenient length, draw arc with any convenient radius R, say 10″. Find the chordal length C in a table of chords (see a machinists' handbook), and multiply the value by 10″, since the table is made for a radius of 1″.

*Example:* To set off 43° 20′, the chordal length C for 1″ radius, as given in a table of chords = .7384, and if R = 10″, then C = 7.384″. If a table is not available, the chord C may be calculated by the formula $C = 2 \sin \frac{\theta}{2}$.

*Example:* Half of 43° 20′ = 21° 40′. The sine of 21° 40′ = .3692. C = 2 (.3692) = .7384 for a 1″ radius. For a 10″ radius, C = 7.384″.

## 4.22 To Draw an Equilateral Triangle.

Given side AB.

*Fig. 4.22 (a).* With A and B as centers and AB as radius, strike arcs to intersect at C. Draw lines AC and BC to complete the triangle.

*Fig. 4.22 (b). Preferred by Draftsmen.* Draw lines through points A and B making angles of 60° with the given line and intersecting at C, as shown.

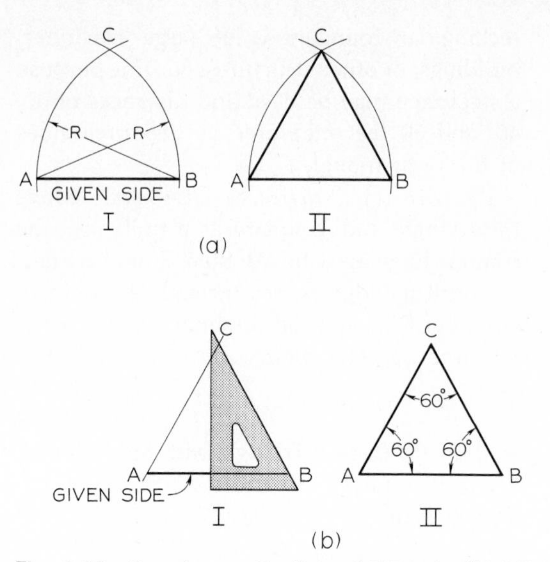


**Fig. 4.22 Drawing an Equilateral Triangle** (§4.22).

## 4.23 To Draw a Square. Fig. 4.23.

*Fig. 4.23 (a).* Given one side AB. Through point A, draw a perpendicular, Fig. 4.18 (c). With A as center, and AB as radius, draw the arc to intersect the perpendicular at C. With B and C as centers, and AB as radius, strike arcs to intersect at D. Draw lines CD and BD.

*Fig. 4.23 (b). Preferred by Draftsmen.* Given one side AB. Using the T-square and 45° triangle, draw lines AC and BD perpendicular to AB, and the lines AD and BC at 45° with AB. Draw line CD with the T-square.

*Fig. 4.23 (c). Preferred by Draftsmen.* Given the circumscribed circle (distance "across corners"). Draw two diameters at right angles to

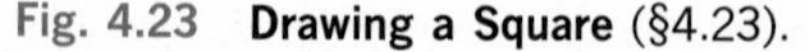

**Fig. 4.23 Drawing a Square** (§4.23).

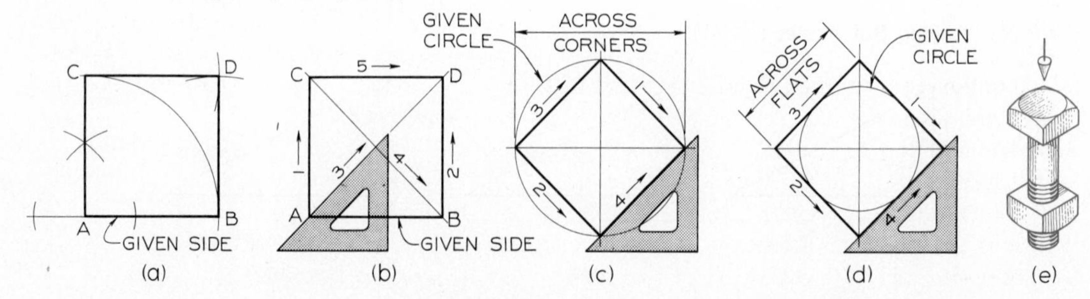

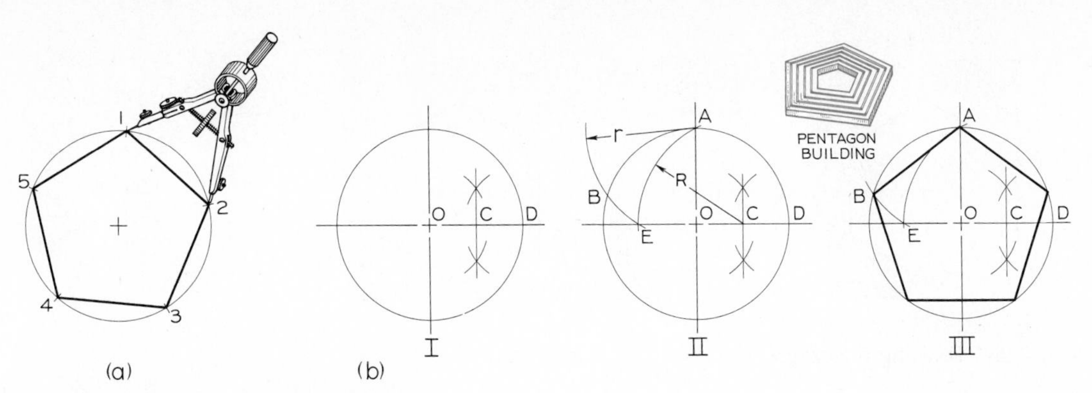

Fig. 4.24 **Drawing a Pentagon** (§4.24).

each other. The intersections of these diameters with the circle are vertexes of an inscribed square.

*Fig. 4.23 (d). Preferred by Draftsmen.* Given the inscribed circle (distance "across flats," as in drawing bolt heads). Using the T-square and 45° triangle, draw the four sides tangent to the circle. See Fig. 13.32.

## 4.24 To Draw a Regular Pentagon.

Given the circumscribed circle.

*Fig. 4.24 (a). Preferred by Draftsmen.* Divide the circumference of the circle into five equal parts with the dividers, and join the points with straight lines.

GEOMETRICAL METHOD. Fig. 2.24 (b).

I. Bisect radius OD at C.

II. With C as center, and CA as radius, strike arc AE. With A as center, and AE as radius, strike arc EB.

III. Draw line AB; then set off distances AB around the circumference of the circle, and draw the sides through these points.

## 4.25 To Draw a Hexagon.

Given the circumscribed circle.

*Fig. 4.25 (a).* Each side of a hexagon is equal to the radius of the circumscribed circle. Therefore, using the compass or dividers and the radius of the circle, set off the six sides of the hexagon around the circle, and connect the points with straight lines. As a check on the accuracy of the construction, make sure that opposite sides of the hexagon are parallel.

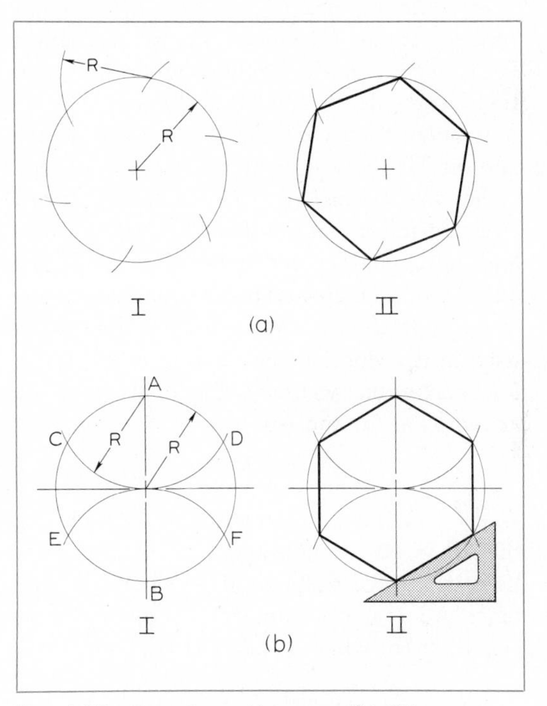

Fig. 4.25 **Drawing a Hexagon** (§4.25).

*Fig. 4.25 (b). Preferred by Draftsmen.* This construction is a variation of the one shown at (a). Draw vertical and horizontal center lines. With A and B as centers and radius equal to that of the circle, draw arcs to intersect the circle at C, D, E, and F, and complete the hexagon as shown.

## 4.26 To Draw a Hexagon.

Given the circumscribed or inscribed circle. *Both Preferred by Draftsmen.*

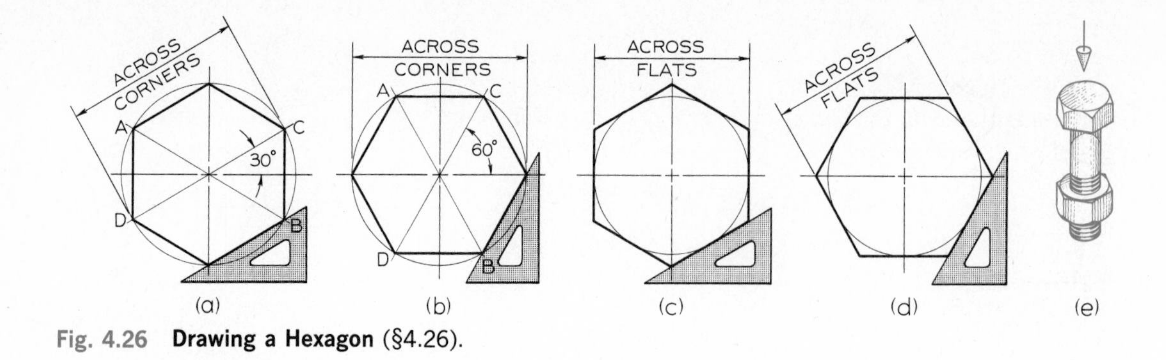


Fig. 4.26 **Drawing a Hexagon** (§4.26).

*Fig. 4.26 (a) and (b).* Given the circumscribed circle (distance "across corners"). Draw vertical and horizontal center lines, and then diagonals AB and CD at 30° or 60° with horizontal; then with the 30° × 60° triangle and the T-square, draw the six sides as shown.

*Fig. 4.26 (c) and (d).* Given the inscribed circle (distance "across flats"). Draw vertical and horizontal center lines; then with the 30° × 60° triangle and the T-square, draw the six sides tangent to the circle. This method is used in drawing bolt heads and nuts, §13.31. For maximum accuracy, diagonals may be added as at (a) and (b).

## 4.27 To Draw a Hexagon.

Fig. 4.27. Using the 30° × 60° triangle and the T-square, draw lines in the order shown at (a) where the distance AB ("across corners") is given, or as shown at (b) where a side CD is given.

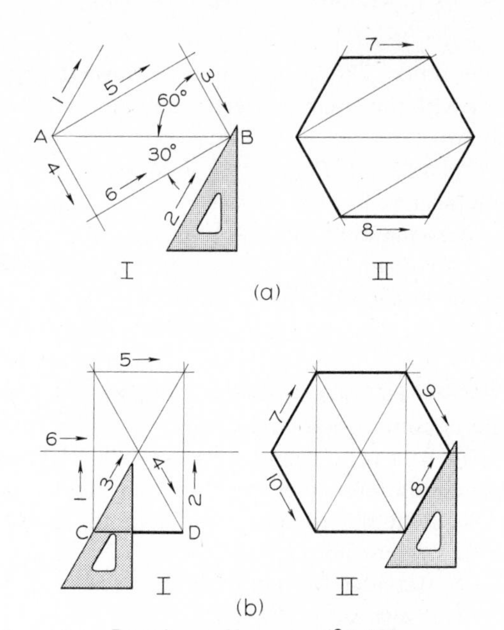


Fig. 4.27 **Drawing a Hexagon** (§4.27).

## 4.28 To Draw an Octagon.

*Fig. 4.28 (a). Preferred by Draftsmen.* Given inscribed circle, or distance "across flats." Using the T-square and 45° triangle, draw the eight sides tangent to the circle, as shown.

*Fig. 4.28 (b).* Given circumscribed square, or distance "across flats." Draw diagonals of

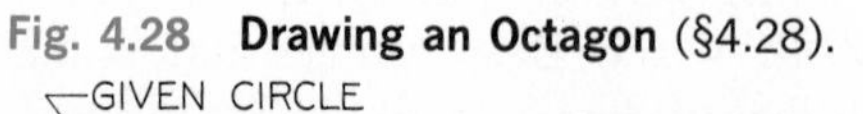

Fig. 4.28 **Drawing an Octagon** (§4.28).

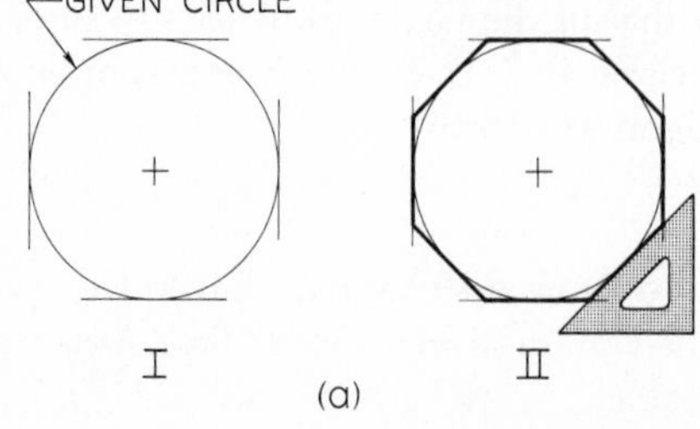


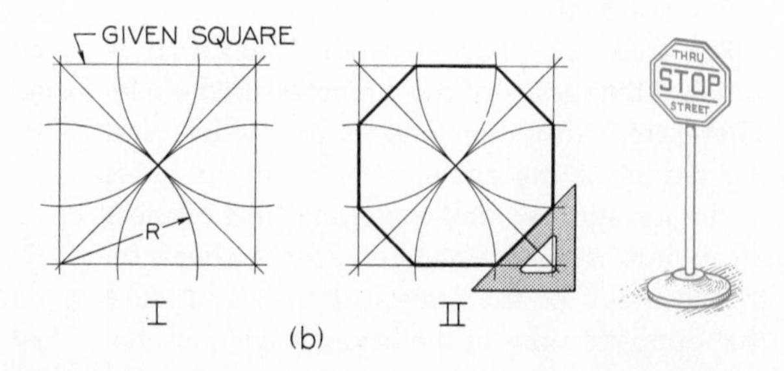

square; then with the corners of the given square as centers, and with half the diagonal as radius, draw arcs cutting the sides as shown at I. Using the T-square and 45° triangle, draw the eight sides as shown at II.

## 4.29 To Transfer Plane Figures by Geometric Methods.

To Transfer a Triangle to a New Location. Fig. 4.29 (a) and (b). Set off any side, as AB, in the new location, (b). With the ends of the line as centers and the lengths of the other sides of the given triangle, (a), as radii, strike two arcs to intersect at C. Join C to A and B to complete the triangle.

To Transfer a Polygon by the Triangle Method. Fig. 4.29 (c). Divide the polygon into triangles as shown, and transfer each triangle as explained above.

To Transfer a Polygon by the Rectangle Method. Fig. 4.29 (d). Circumscribe a rectangle about the given polygon. Draw a congruent rectangle in the new location and locate the vertexes of the polygon by transferring location measurements a, b, c, etc., along the sides of the rectangle to the new rectangle. Join the points thus found to complete the figure.

To Transfer Irregular Figures. Fig. 4.29 (e). Figures composed of rectangular and circular forms are readily transferred by enclosing the elementary features in rectangles and determining centers of arcs and circles. These may then be transferred to the new location.

To Transfer Figures by Offset Measurements. Fig. 4.29 (f). *Offset location measurements* are frequently useful in transferring figures composed of free curves. When the figure has been enclosed by a rectangle, the sides of the rectangle are used as reference lines for the location of points along the curve.

To Transfer Figures by a System of Squares. Fig. 4.29 (g). Figures involving free curves are easily copied, enlarged, or reduced by the use of a system of squares. For example, to enlarge a figure to double size, draw the containing rectangle and all small squares double their original size. Then draw the lines through the corresponding points in the new set of squares. See also Fig. 5.18.

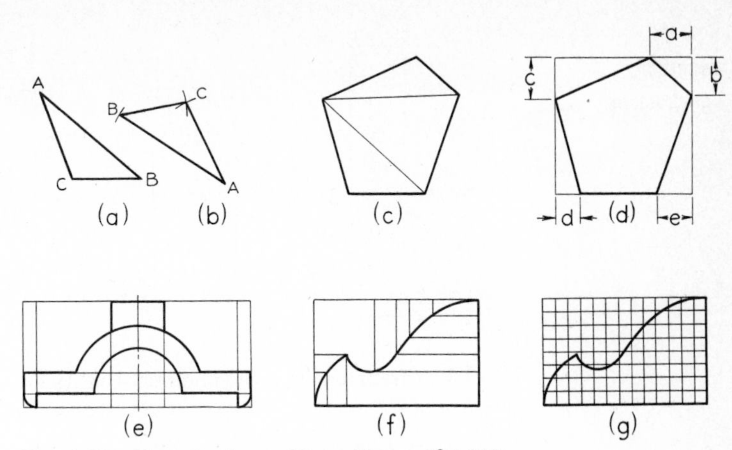


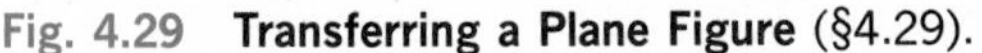


Fig. 4.29 **Transferring a Plane Figure** (§4.29).

## 4.30 To Transfer Drawings by Tracing-Paper Methods.

To transfer a drawing to an opaque sheet, the following procedures may be used:

Prick-Point Method. Lay tracing paper over the drawing to be transferred. With a sharp pencil, make a small dot directly over each important point on the drawing. Encircle each dot so as not to lose it. Remove the tracing paper, place it over the paper to receive the transferred drawing, and maneuver the tracing paper into the desired position. With a needle point (such as a point of the dividers), prick through each dot. Remove the tracing paper and connect the prick-points to produce the lines as on the original drawing.

To transfer arcs or circles, it is only necessary to transfer the center and one point on the circumference. To transfer a free curve, transfer as many prick-points on the curve as desired.

Tracing Method. Lay tracing paper over the drawing to be transferred, and make a pencil tracing of it. Turn the tracing paper over and mark over the lines with short strokes of a soft pencil so as to provide a coating of graphite over every line. Turn tracing face up and fasten in position where drawing is to be transferred. Trace over all lines of the tracing, using a hard pencil. The graphite on the back acts as a carbon paper and will produce dim but definite lines. Heavy in the dim lines to complete the transfer.

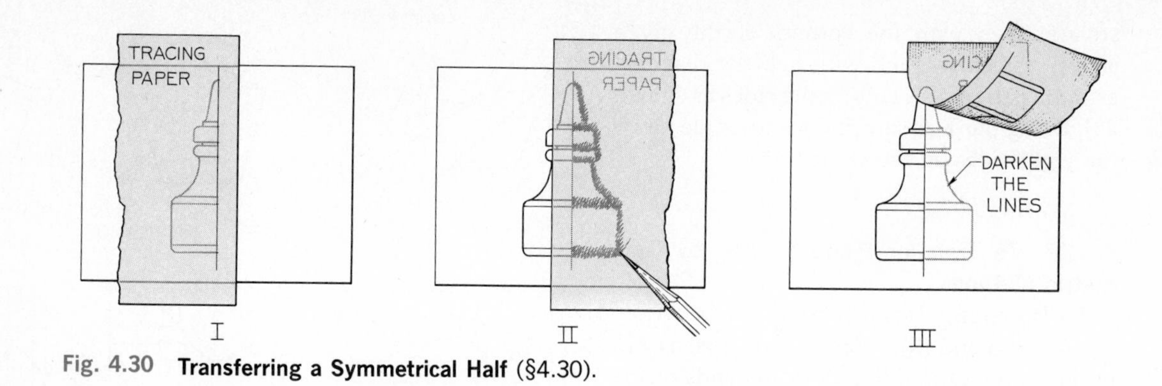


Fig. 4.30 **Transferring a Symmetrical Half** (§4.30).

*Fig. 4.30.* If one half of a symmetrical object has been drawn, as for the ink bottle at I, the other half may be easily drawn with the aid of tracing paper as follows:

I. Trace the half already drawn.

II. Turn tracing paper over and maneuver to the position for the right half. Then trace over the lines freehand or mark over the lines with short strokes as shown.

III. Remove the tracing paper, revealing the dim imprinted lines for the right half. Heavy in these lines to complete the drawing.

## 4.31 To Enlarge or Reduce a Drawing.

*Fig. 4.31 (a).* The construction shown is an adaptation of the parallel-line method, Figs. 4.14 and 4.15, and may be used whenever it is desired to enlarge or reduce any group of dimensions to the same ratio. Thus if full-size dimensions are laid off along the vertical line, the enlarged dimensions would appear along the horizontal line, as shown.

*Fig. 4.31 (b).* To enlarge or reduce a rectangle (say, a sheet of drawing paper), a simple method is to use the diagonal, as shown.

*Fig. 4.31 (c).* A simple method of enlarging or reducing a drawing is to make use of radial lines, as shown. The original drawing is placed underneath a sheet of tracing paper, and the enlarged or reduced drawing is made directly on the tracing paper.

## 4.32 To Draw a Circle Through Three Points. 

Fig. 4.32 (a).

I. Let A, B, and C be the three given points not in a straight line. Draw lines AB and BC, which will be chords of the circle. Draw perpendicular bisectors EO and DO, Fig. 4.8, intersecting at O.

II. Through center O, draw required circle through the points.

## 4.33 To Find the Center of a Circle.

Fig. 4.32 (b). Draw any chord AB, preferably horizontal as shown. Draw perpendiculars

Fig. 4.31 **Enlarging or Reducing** (§4.31).

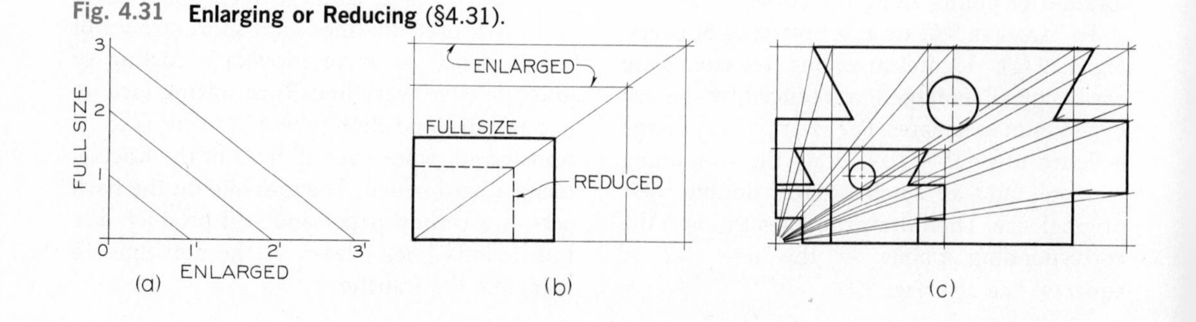

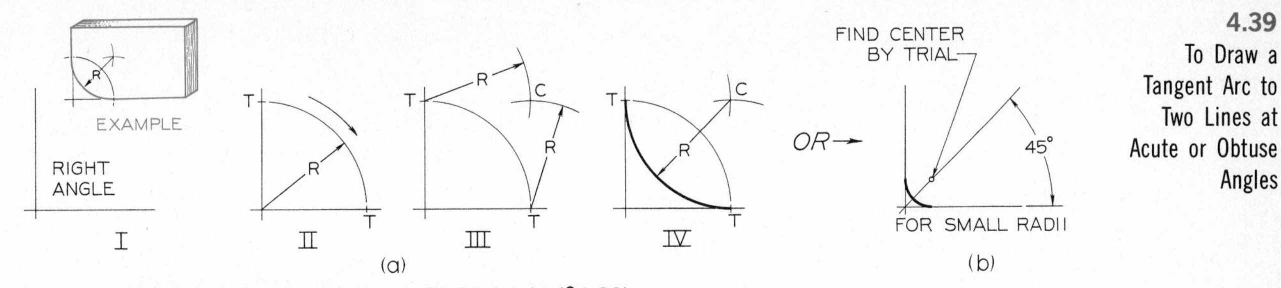


Fig. 4.37 **Drawing a Tangent Arc in a Right Angle** (§4.38).

erect a perpendicular to the line to intersect DE at C, the center of the required tangent arc.

*Fig. 4.36* (*c*). Given arc with center Q, point P, and radius R. From P strike arc with radius R. From Q strike arc with radius equal to that of the given arc plus R. The intersection C of the arcs is the center of the required tangent arc.

## 4.38 To Draw a Tangent Arc to Two Lines at Right Angles.

Fig. 4.37 (a).

I. Given two lines at right angles to each other.

II. With given radius R, strike arc intersecting given lines at tangent points T.

III. With given radius R again, and with points T as centers, strike arcs intersecting at C.

IV. With C as center and given radius R, draw required tangent arc.

For Small Radii. Fig. 4.37 (b). For small radii, such as $\frac{1}{8}$R for fillets and rounds, it is not practicable to draw complete tangency constructions. Instead, draw a 45° bisector of the angle and locate the center of the arc by trial along this line, as shown.

## 4.39 To Draw a Tangent Arc to Two Lines at Acute or Obtuse Angles.

Fig. 4.38 (a) or (b).

I. Given two lines not making 90° with each other.

II. Draw lines parallel to given lines at distance R from them, to intersect at C, the required center.

III. From C drop perpendiculars to the given lines respectively to locate points of tangency T.

IV. With C as center and with given radius R, draw required tangent arc between the points of tangency.

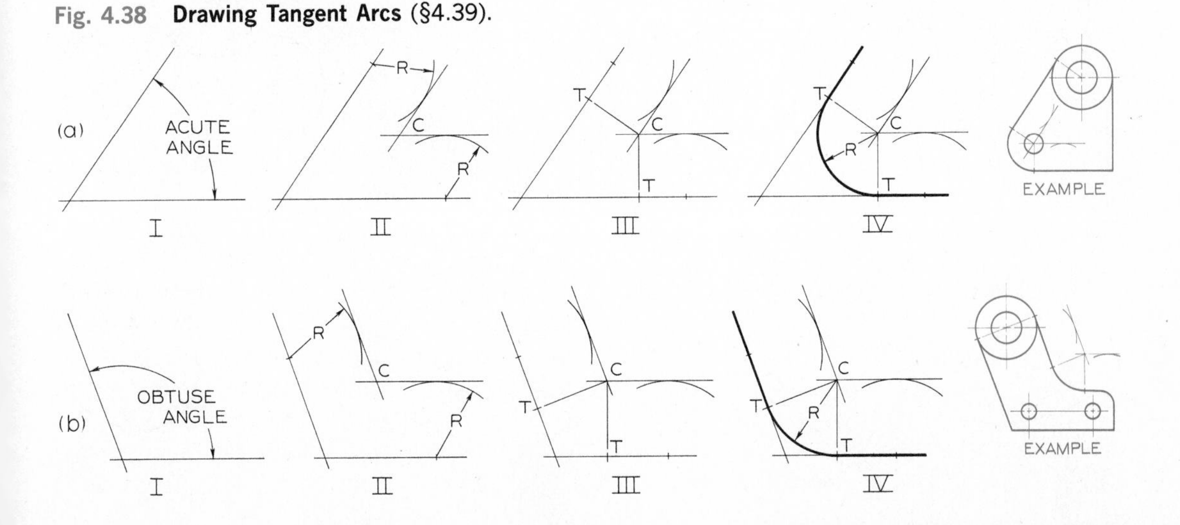


Fig. 4.38 **Drawing Tangent Arcs** (§4.39).

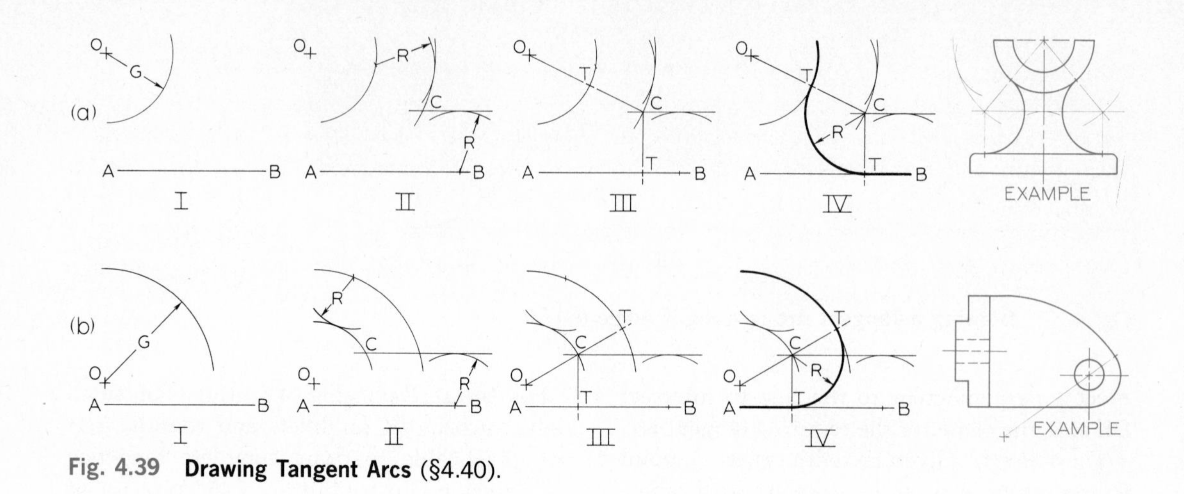


Fig. 4.39 **Drawing Tangent Arcs** (§4.40).

## 4.40 To Draw Tangent Arc to an Arc and a Straight Line. Fig. 4.39 (a) or (b).

I. Given arc with radius G and straight line AB.

II. Draw straight line and an arc parallel respectively to the given straight line and arc at the required radius distance R from them, to intersect at C, the required center.

III. From C drop a perpendicular to the given straight line to obtain one point of tangency T. Join the centers C and O with a straight line to locate the other point of tangency T.

IV. With center C and given radius R, draw required tangent arc between the points of tangency.

## 4.41 To Draw an Arc Tangent to Two Arcs. Fig. 4.40 (a) or (b).

Fig. 4.40 **Drawing an Arc Tangent to Two Arcs** (§4.41).

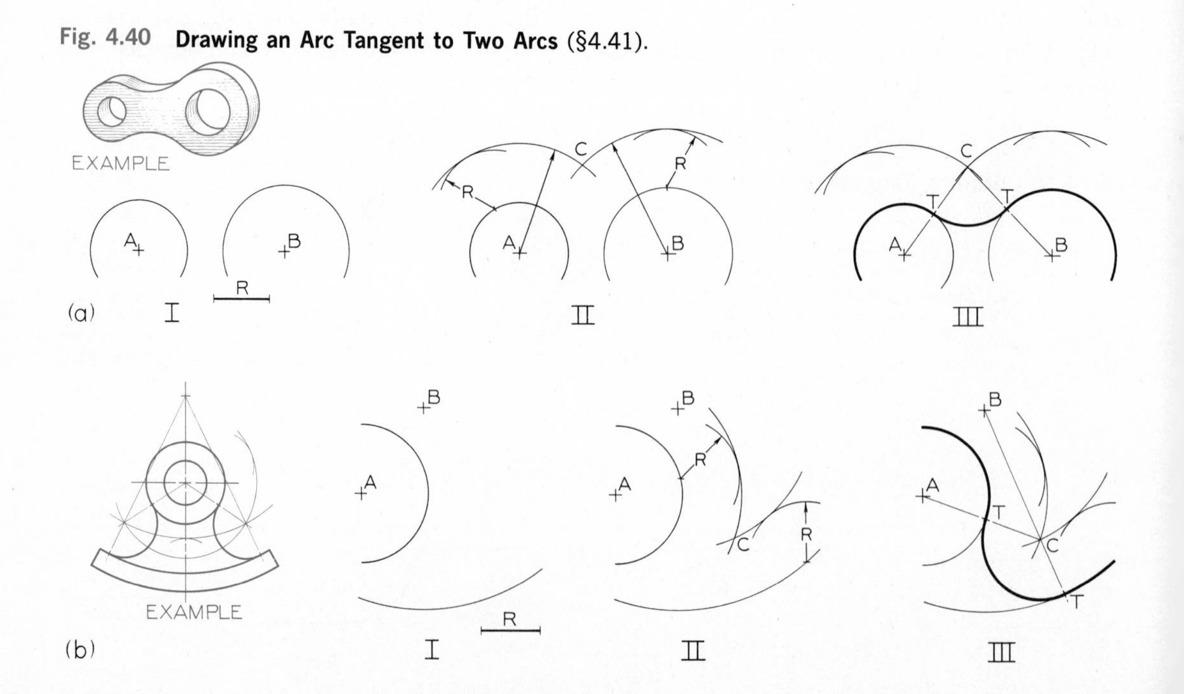

I. Given arcs with centers A and B, and required radius R.

II. With A and B as centers, draw arcs parallel to the given arcs and at a distance R from them; their intersection C is the center of the required tangent arc.

III. Draw lines of centers AC and BC to locate points of tangency T, and draw required tangent arc between the points of tangency, as shown.

## 4.42 To Draw an Arc Tangent to Two Arcs and Enclosing One or Both.

The Required Arc Encloses Both Given Arcs. Fig. 4.41 (a). With A and B as centers, strike arcs HK − r (given radius minus radius of small circle) and HK − R (given radius minus radius of large circle) intersecting at G, the center of the required tangent arc. Lines of centers GA and GB (extended) determine points of tangency T.

The Required Arc Encloses One Given Arc. Fig. 4.41 (b). With C and D as centers, strike arcs HK + r (given radius plus radius of small circle) and HK − R (given radius minus radius of large circle) intersecting at G, the center of the required tangent arc. Lines of centers GC and GD (extended) determine points of tangency T.

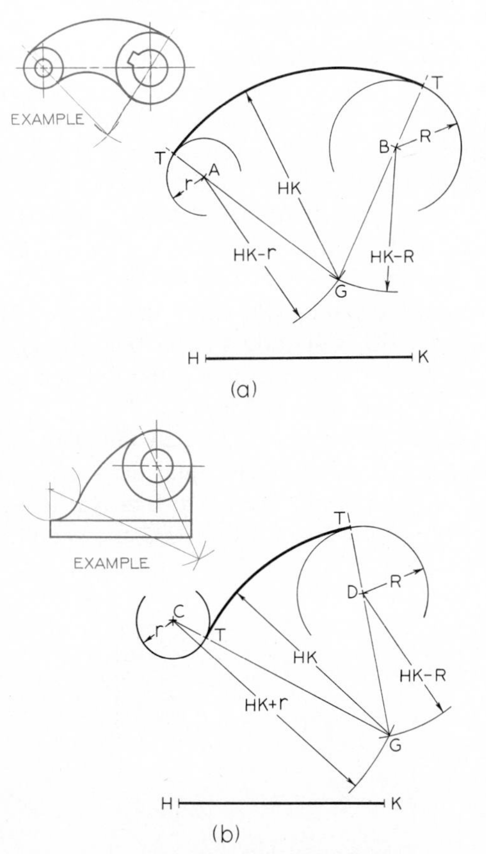


Fig. 4.41 **Drawing Tangent Arcs** (§4.42).

## 4.43 To Draw a Series of Tangent Arcs Conforming to a Curve.

Fig. 4.42. First sketch lightly a smooth curve as desired. By trial, find a radius R and a center C, producing an arc AB which closely follows that portion of the curve. The successive centers D, E, etc., will be on lines joining the centers with the points of tangency, as shown.

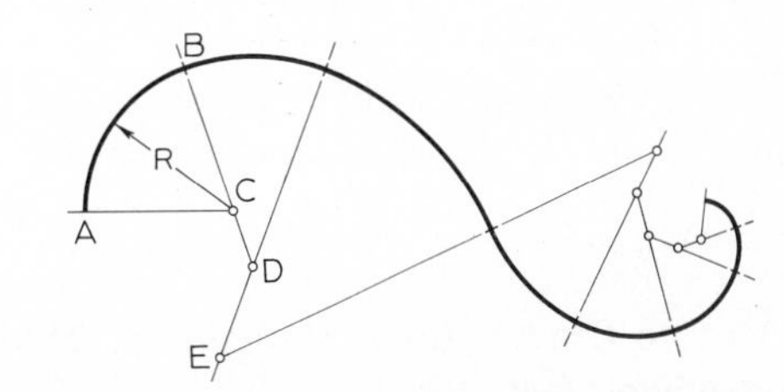


Fig. 4.42 **A Series of Tangent Arcs** (§4.43).

## 4.44 To Draw an Ogee Curve.

Connecting Two Parallel Lines. Fig. 4.43 (a). Let NA and BM be the two parallel lines. Draw AB, and assume inflection point T (at midpoint if two equal arcs are desired). At A and B erect perpendiculars AF and BC. Draw perpendicular bisectors of AT and BT. The intersections F and C of these bisectors and the perpendiculars, respectively, are the centers of the required tangent arcs.

*Fig. 4.43 (b).* Let AB and CD be the two parallel lines, with point B as one end of the curve, and R the given radii. At B erect perpendicular to AB, make BG = R, and draw arc as shown. Draw line SP parallel to CD at distance R from CD. With center G, draw arc of radius 2R, intersecting line SP at O. Draw

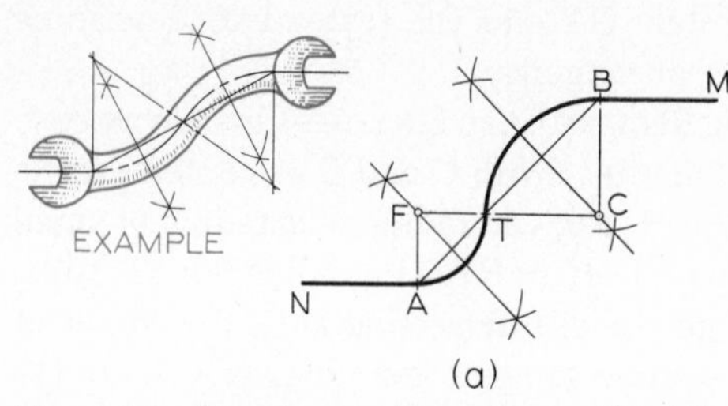


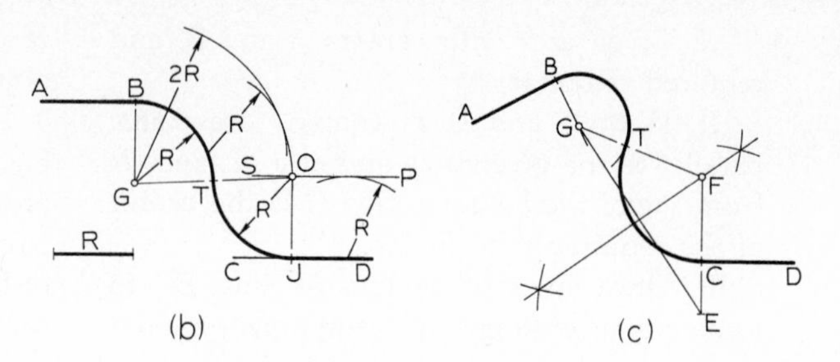


Fig. 4.43 **Drawing an Ogee Curve** (§4.44).

perpendicular OJ to locate tangent point J, and join centers G and O to locate point of tangency T. Using centers G and O and radius R, draw the two tangent arcs as shown.

Connecting Two Nonparallel Lines. Fig. 4.43 (c). Let AB and CD be the two nonparallel lines. Erect perpendicular to AB at B. Select point G on the perpendicular so that BG equals any desired radius, and draw arc as shown. Erect perpendicular to CD at C and make CE = BG. Join G to E and bisect it. The intersection F of the bisector and the perpendicular CE, extended, is the center of the second arc. Join centers of the two arcs to locate tangent point T, the inflection point of the curve.

## 4.45 To Draw a Curve Tangent to Three Intersecting Lines. Fig. 4.44 (a) and (b).

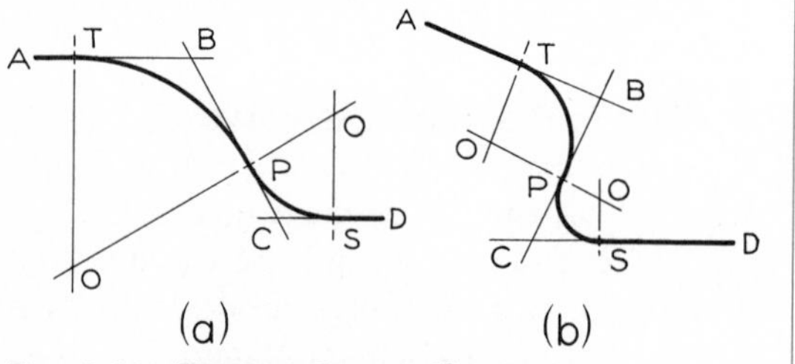


Fig. 4.44 **Tangent Curves** (§4.45).

Let AB, BC, and CD be the given lines. Select point of tangency P at any point on line BC. Make BT equal to BP, and CS equal to CP, and erect perpendiculars at the points P, T, and S. Their intersections O are the centers of the required tangent arcs.

## 4.46 To Rectify a Circular Arc.

To *rectify* an arc is to lay out its true length along a straight line. The constructions are approximate, but well within the range of accuracy of drawing instruments.

To Rectify a Quadrant of a Circle, AB. Fig. 4.45 (a). Draw AC tangent to the circle and BC at 60° to AC, as shown. The line AC is almost equal to the arc AB, the difference in length being about 1 in 240.

To Rectify Arc AB. Fig. 4.45 (b). Draw tangent at B. Draw chord AB and extend it to C, making BC equal to half AB. With C as center and radius CA, strike the arc AD. The tangent BD is slightly shorter than the given arc AB. For an angle of 45° the difference in length is about 1 in 2866.

*Fig. 4.45 (c).* Use the bow dividers, and beginning at A, set off equal distances until the division point nearest to B is reached. At this point, reverse the direction and set off an

Fig. 4.45 **Rectifying Circular Arcs** (§§4.46 and 4.47).

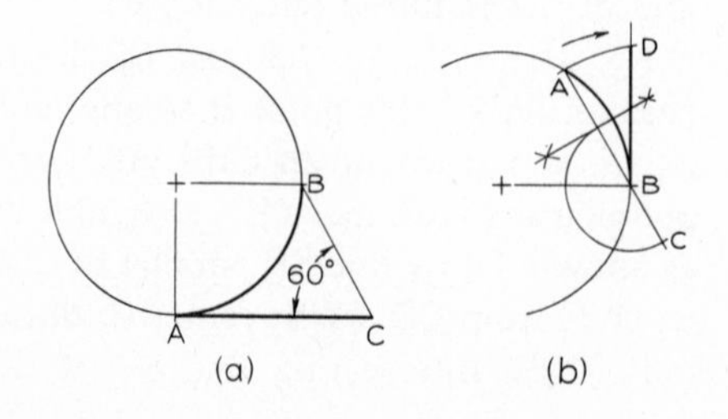


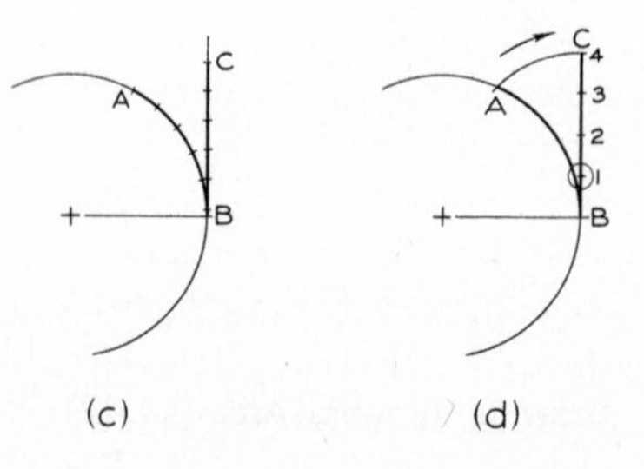

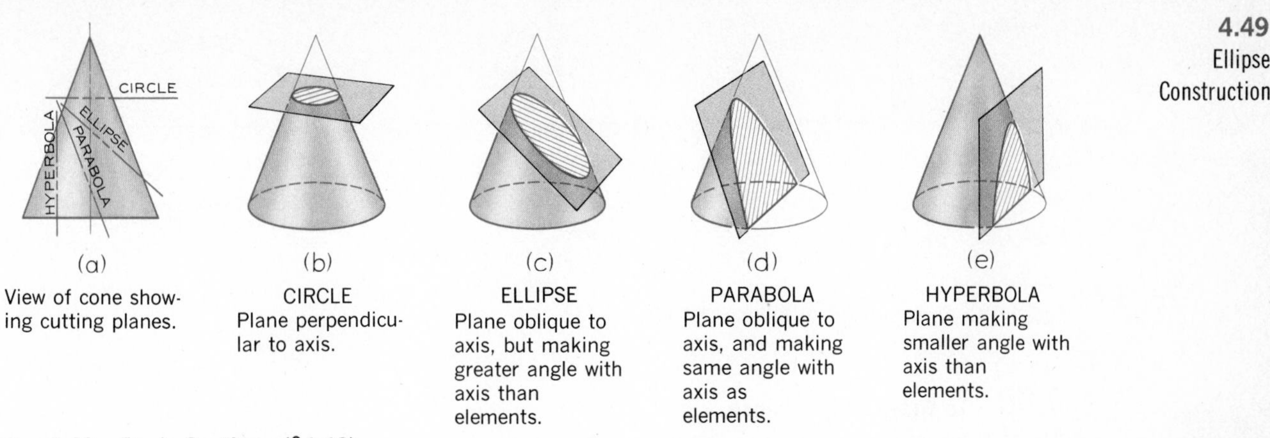


(a) View of cone showing cutting planes.

(b) CIRCLE Plane perpendicular to axis.

(c) ELLIPSE Plane oblique to axis, but making greater angle with axis than elements.

(d) PARABOLA Plane oblique to axis, and making same angle with axis as elements.

(e) HYPERBOLA Plane making smaller angle with axis than elements.

**Fig. 4.46 Conic Sections** (§4.48).

equal number of distances along the tangent to determine point C. The tangent BC is slightly shorter than the given arc AB. If the angle subtended by each division is 10°, the error is approximately 1 in 830.

*Note:* If the angle $\theta$ subtending an arc of radius R is known, the length of the arc is $2\pi R\frac{\theta}{360^\circ} = 0.01745R\theta$.

## 4.47 To Set Off a Given Length Along a Given Arc.

*Fig. 4.45 (c).* Reverse the method described above so as to transfer distances from the tangent line to the arc.

*Fig. 4.45 (d).* To set off the length BC along the arc BA, draw BC tangent to the arc at B. Divide BC into four equal parts. With center at 1, the first division point, and radius 1C, draw the arc CA. The arc BA is practically equal to BC for angles less than 30°. For 45° the difference is approximately 1 in 3232, and for 60° it is about 1 in 835.

## 4.48 The Conic Sections.

Fig. 4.46. The *conic sections* are curves produced by planes intersecting a right circular cone. Four types of curves are produced: the *circle, ellipse, parabola,* and *hyperbola,* according to the position of the planes, as shown. These curves were studied in detail by the ancient Greeks, and are of great interest in mathematics, as well as in technical drawing. For equations, see any text on analytic geometry.

**Fig. 4.47 Ellipse Constructions** (§4.49).

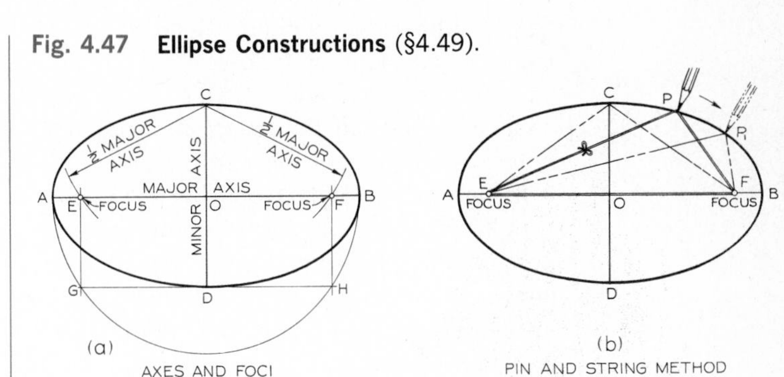


## 4.49 Ellipse Construction.

The long axis of an ellipse is the *major axis,* and the short axis is the *minor axis,* Fig. 4.47 (a). The *foci* E and F are found by striking arcs with radius equal to half the major axis and with center at the end of the minor axis. Another method is to draw a semicircle with the major axis as diameter, then to draw GH parallel to the major axis, and GE and HF parallel to the minor axis, as shown.

*An ellipse may be generated by a point moving so that the sum of its distances from two points (the foci) is constant and equal to the major axis.* For example, Fig. 4.47 (b), an ellipse may be constructed by placing a looped string around the foci E and F, and around C, one end of the minor axis, and moving the pencil point P along its maximum orbit while the string is kept taut.

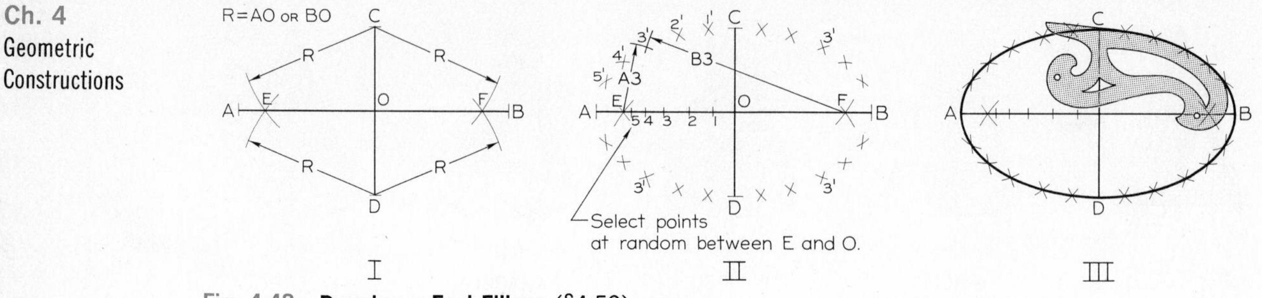


Fig. 4.48 **Drawing a Foci Ellipse** (§4.50).

## 4.50 To Draw a Foci Ellipse.

Fig. 4.48. Let AB be the major axis, and CD the minor axis. This method is the geometrical counterpart of the pin-and-string method. Keep the construction very light, as follows:

I. To find foci E and F, strike arcs R with radius equal to half the major axis and with centers at the ends of the minor axis.

II. Between E and O on the major axis, mark at random a number of points (spacing those on the left more closely), equal to the number of points desired in each quadrant of the ellipse. In this figure, five points were deemed sufficient. For large ellipses, more points should be used—enough to insure a smooth, accurate curve. Begin construction with any one of these points, such as 3. With E and F as centers and radii A3 and B3, respectively (from the ends of the major axis to point 3), strike arcs to intersect at four points 3′, as shown. Using the remaining points 1, 2, 4, and 5, for each find four additional points on the ellipse in the same manner.

III. Sketch the ellipse lightly through the points; then heavy in the final ellipse with the aid of the irregular curve, Fig. 2.79.

## 4.51 To Draw a Trammel Ellipse.

Fig. 4.49. A "long trammel" or a "short trammel" may be prepared from a small strip of stiff paper or thin cardboard, as shown. In both cases, set off on the edge of the trammel distances equal to the semimajor and semiminor axes. In one case these distances overlap; in the other they are end to end. To use either method, place the trammel so that two of the points are on the respective axes, as shown;

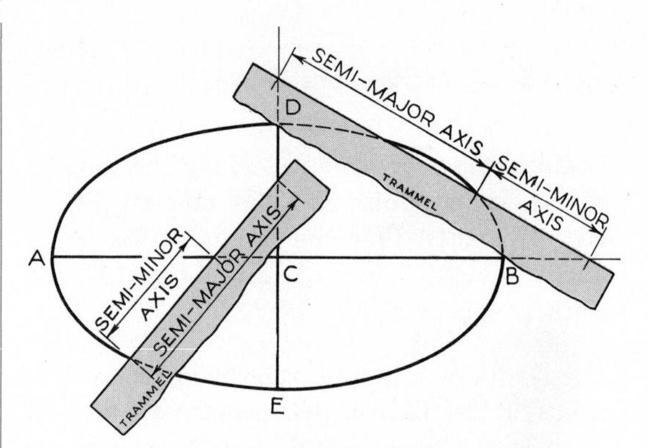

Fig. 4.49 **Drawing a Trammel Ellipse** (§4.51).

the third point will then be on the curve and can be marked with a small dot. Find additional points by moving the trammel to other positions, always keeping the two points exactly on the respective axes. Extend the axes to use the long trammel. Find enough points to insure a smooth and symmetrical ellipse. Sketch the ellipse lightly through the points; then heavy in the ellipse with the aid of the irregular curve, Fig. 2.79.

## 4.52 To Draw a Concentric-Circle Ellipse.

Fig. 4.50. If a circle is viewed so that the line of sight is perpendicular to the plane of the circle, as shown for the silver dollar at (a), the circle will appear as a circle, in true size and shape. If the circle is viewed at an angle, as shown at (b), it will appear as an ellipse. If the circle is viewed edgewise, it appears as a straight line, as shown at (c). The case shown at (b) is the basis for the construction of an ellipse by the *concentric-circle method,* as follows (keep the construction very light):

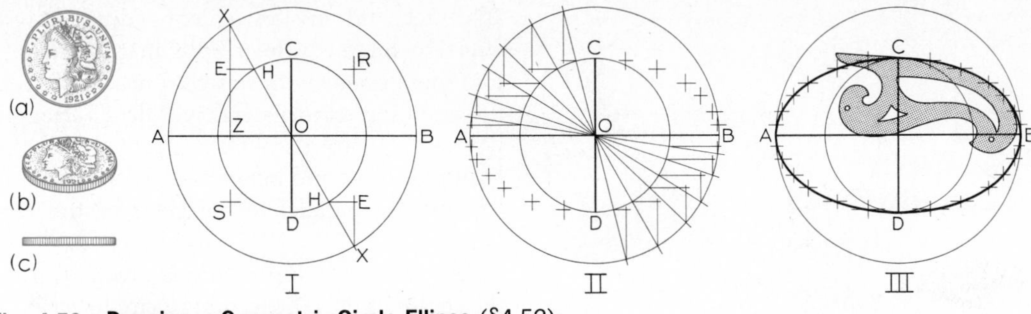


Fig. 4.50 **Drawing a Concentric-Circle Ellipse** (§4.52).

I. Draw circles on the two axes as diameters, and draw any diagonal XX through center O. From the points X, in which the diagonal intersects the large circle, draw lines XE parallel to the minor axis; and from points H, in which it intersects the small circle, draw lines HE parallel to the major axis. The intersections E are points on the ellipse. Two additional points, S and R, can be found by extending lines XE and HE, giving a total of four points from the one diagonal XX.

II. Draw as many additional diagonals as needed to provide a sufficient number of points for a smooth and symmetrical ellipse, each diagonal accounting for four points on the ellipse. Notice that where the curve is sharpest (near the ends of the ellipse), the points are constructed closer together to better determine the curve.

III. Sketch the ellipse lightly through the points; then heavy in the final ellipse with the aid of the irregular curve, Fig. 2.79.

*Note:* It is evident at I, Fig. 4.50, that the ordinate EZ of the ellipse is to the corresponding ordinate XZ of the circle as b is to a, where b represents the semiminor axis and a the semimajor axis. Thus, the area of the ellipse is equal to the area of the circumscribed circle multiplied by $\frac{b}{a}$; hence it is equal to $\pi ab$.

## 4.53 To Draw an Ellipse on Conjugate Diameters. Oblique Circle Method.

Fig. 4.51. Let AB and DE be the given conjugate diameters. *Two diameters are conjugate when each is parallel to the tangents at the extremities of the other.*

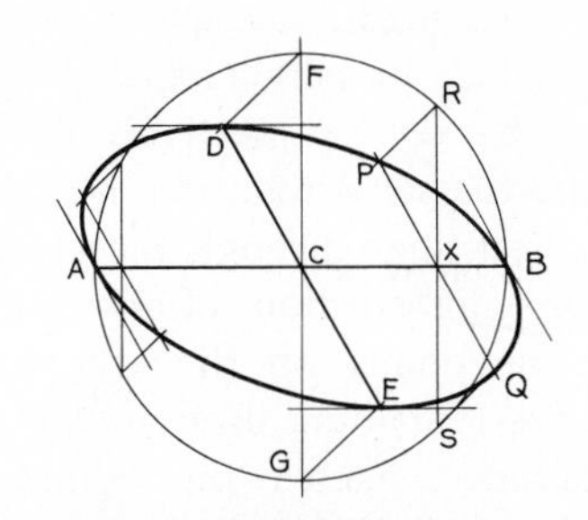

Fig. 4.51 **Oblique-Circle Ellipse** (§4.53).

With center at C and radius CA, draw a circle; draw the diameter GF perpendicular to AB, and draw lines joining the points D and F, and G and E.

Assume that the required ellipse is an oblique projection of the circle just drawn; the points D and E of the ellipse are the oblique projections of the points F and G of the circle, respectively; and similarly, the points P and Q are the oblique projections of the points R and S, respectively. The points P and Q are determined by assuming the point X at any point on AB and drawing the lines RS and PQ, and RP and SQ, parallel respectively to GF and DE, and FD and GE.

Determine at least five points in each quadrant (more for larger ellipses) by assuming additional points on the major axis and proceeding as explained for point X. Sketch the ellipse lightly through the points; then heavy in the final ellipse with the aid of the irregular curve, Fig. 2.79.

## 4.54 To Draw a Parallelogram Ellipse.

Fig. 4.52 (a) and (b). Given the major and minor

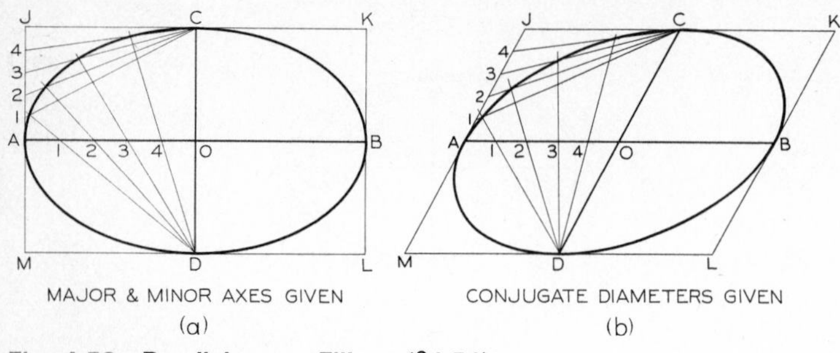


Fig. 4.52 **Parallelogram Ellipse** (§4.54).

axes, or the conjugate diameters AB and CD, draw a rectangle or parallelogram with sides parallel to the axes, respectively. Divide AO and AJ into the same number of equal parts, and draw *light* lines through these points, as shown. The intersection of like-numbered lines will be points on the ellipse. Locate points in the remaining three quadrants in a similar manner. Sketch the ellipse lightly through the points; then heavy in the final ellipse with the aid of the irregular curve, Fig. 2.79.

## 4.55 To Find the Axes of an Ellipse, with Conjugate Diameters Given.

*Fig. 4.53 (a).* Conjugate diameters AB and CD and the ellipse are given. With intersection O of the conjugate diameters (center of ellipse) as center, and any convenient radius, draw a circle to intersect the ellipse in four points. Join these points with straight lines, as shown; the resulting quadrilateral will be a rectangle whose sides are parallel, respectively, to the required major and minor axes. Draw the axes EF and GH parallel to the sides of the rectangle.

*Fig. 4.53 (b).* Ellipse only is given. To find the center of the ellipse, draw a circumscribing rectangle or parallelogram about the ellipse; then draw diagonals to intersect at center O, as shown. The axes are then found as shown at (a).

*Fig. 4.53 (c).* Conjugate diameters AB and CD only are given. With O as center and CD as diameter, draw a circle. Through center O and perpendicular to CD, draw line EF. From points E and F, where this perpendicular intersects the circle, draw lines FA and EA to form angle FAE. Draw the bisector AG of this angle. The major axis JK will be parallel to this bisector, and the minor axis LM will be perpendicular to it. The length AH will be one half the major axis, and HF one half the minor axis. The resulting major and minor axes are JK and LM, respectively.

Fig. 4.53 **Finding the Axes of an Ellipse** (§4.55).

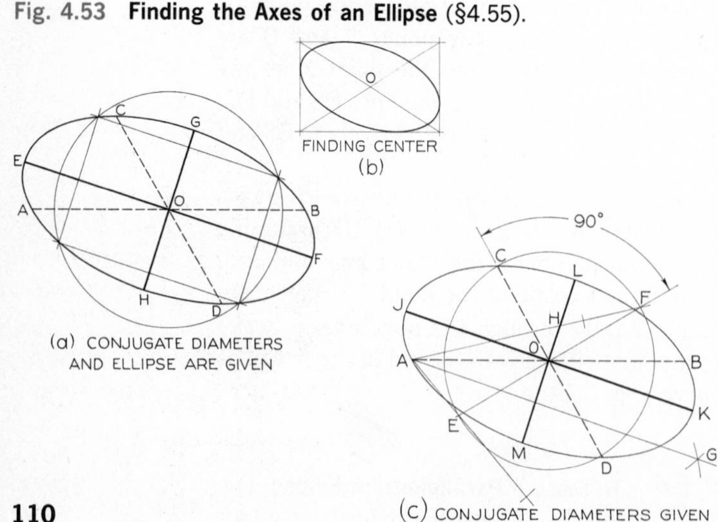


## 4.56 To Draw a Tangent to an Ellipse.

Concentric Circle Construction. Fig. 4.54 (a). To draw a tangent at any point on the ellipse, as E, draw the ordinate at E to intersect the circle at V. Draw a tangent to the circle at V, §4.35, and produce it to intersect the major axis produced at G. The line GE is the required tangent.

To draw a tangent from a point outside the ellipse, as P, draw the ordinate PY and extend it. Draw DP, intersecting the major axis at X. Draw FX and extend it to intersect the ordinate through P at Q. Then, from similar triangles, QY:PY = OF:OD. Draw tangent to the circle from Q, §4.35, find the point of tangency R, and draw the ordinate at R to intersect the ellipse at Z. The line ZP is the required tangent. As a check on the drawing, the tangents RQ and ZP should intersect at a point on the major axis extended. Two tangents to the ellipse can be drawn from point P.

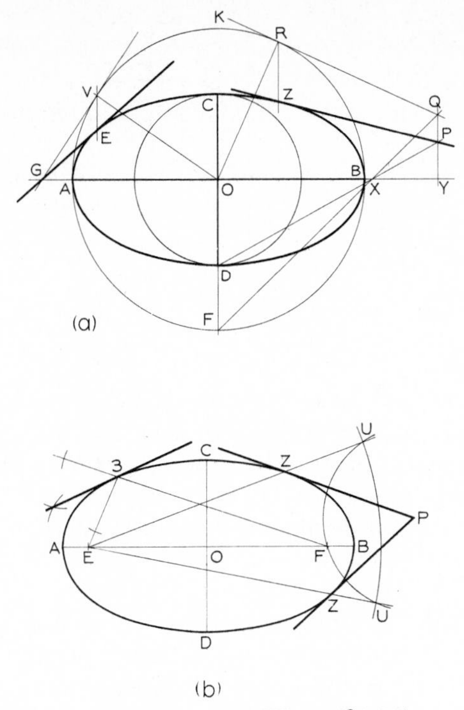


Fig. 4.54 **Tangents to an Ellipse** (§4.56).

Foci Construction. Fig. 4.54 (b). To draw a tangent at any point on the ellipse, such as point 3, draw the focal radii E3 and F3, extend one, and bisect the exterior angle, as shown. The bisector is the required tangent.

To draw a tangent from any point outside the ellipse, such as point P, with center at P and radius PF, strike an arc as shown. With center at E and radius AB, strike an arc to intersect the first arc at points U. Draw the lines EU to intersect the ellipse at the points Z. The lines PZ are the required tangents.

## 4.57 **Ellipse Templates.**

To save time in drawing ellipses, and to insure uniform results, *ellipse templates,* Fig. 4.55 (a), are often used. These are plastic sheets with elliptical openings in a wide variety of sizes, and usually come in sets of six or more sheets.

Ellipse guides are usually designated by the *ellipse angle,* the angle at which a circle is viewed to appear as an ellipse. In Fig. 4.55 (b) the angle between the line of sight and the edge view of the plane of the circle is found to be about 49°; hence the 50° ellipse template is indicated. Ellipse templates are generally available in ellipse angles at 5° intervals, as 15°, 20°, 25°, etc. On this 50° template a variety of sizes of 50° ellipses is provided, and it is only necessary to select the one that fits. If the ellipse angle is not easily determined, you can always look for the ellipse that is approximately as long and as "fat" as the ellipse to be drawn.

A simple construction for finding the ellipse angle when the views are not available is shown at (c). Using center O, strike arc BF; then draw CE parallel to the major axis. Draw diagonal OE, and measure angle EOB with the protractor, §2.19. Use the ellipse template nearest to this angle; in this case a 35° template is selected.

Since it is not practicable to have ellipse openings for every exact size that may be

Fig. 4.55 **Using the Ellipse Template** (§4.57).

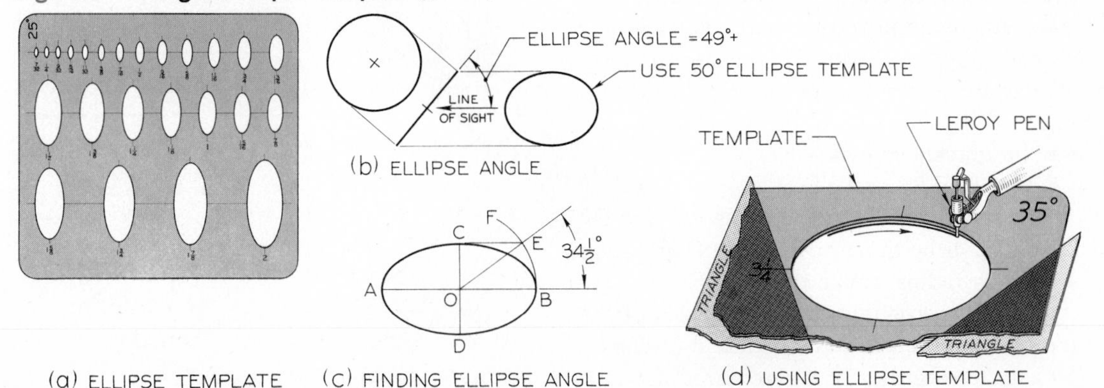

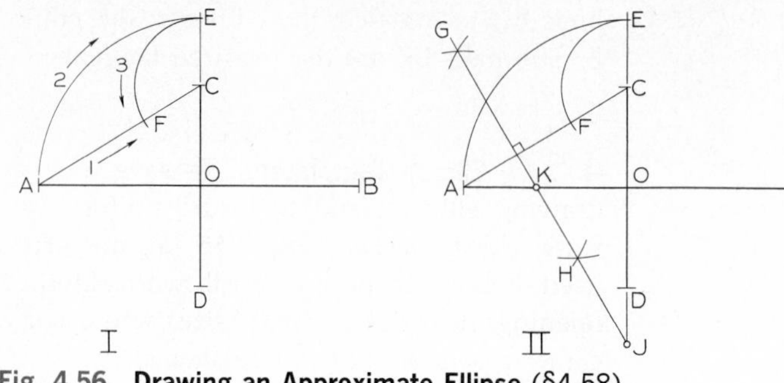


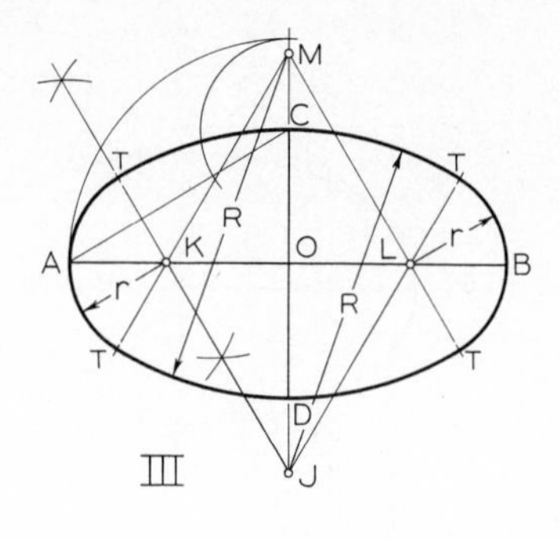


Fig. 4.56 **Drawing an Approximate Ellipse** (§4.58).

required, it is often necessary to use the template somewhat in the manner of an irregular curve. For example, if the opening is too long and too "fat" for the required ellipse, one end may be drawn and then the template shifted slightly to draw the other end. Similarly, one long side may be drawn and then the template shifted slightly to draw the opposite side. In such cases, leave gaps between the four segments, to be filled in freehand or with the aid of an irregular curve. When the differences between the ellipse openings and the required ellipse are small, it is only necessary to lean the pencil or pen slightly outward or inward from the guiding edge to offset the differences.

For inking the ellipses, the Leroy, Rapidograph, or Wrico pens are recommended. The Leroy pen is shown in Fig. 4.55 (d). Place triangles under the ellipse template, as shown, so as to lift the template from the paper and prevent ink from spreading under the template; or better still, place a larger opening of another ellipse guide underneath.

## 4.58 To Draw an Approximate Ellipse.

Fig. 4.56. For many purposes, particularly where a small ellipse is required, the approximate circular-arc method is perfectly satisfactory. Such an ellipse is sure to be symmetrical and may be quickly drawn.

Given axes AB and CD.

I. Draw line AC. With O as center and OA as radius, strike the arc AE. With C as center and CE as radius, strike the arc EF.

II. Draw perpendicular bisector GH of the line AF; the points K and J, where it intersects the axes, are centers of the required arcs.

III. Find centers M and L by setting off OL = OK and OM = OJ. Using centers K, L, M, and J, draw circular arcs as shown. The points of tangency T are at the junctures of the arcs on the lines joining the centers.

## 4.59 To Draw a Parabola.

The curve of intersection between a right circular cone and a plane parallel to one of its elements, Fig.

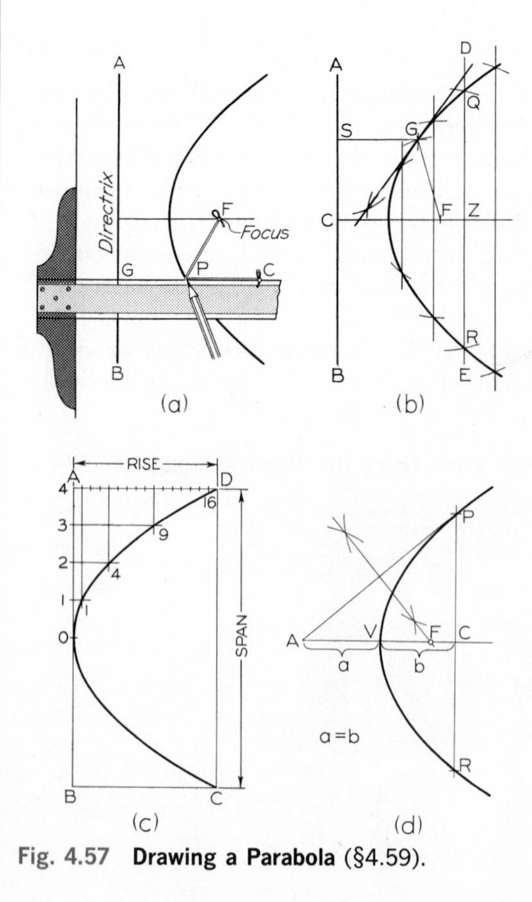


Fig. 4.57 **Drawing a Parabola** (§4.59).

4.46 (d), is a *parabola. A parabola may be generated by a point moving so that its distances from a fixed point, the focus, and from a fixed line, the directrix, remain equal.* For example:

*Fig. 4.57 (a).* Given focus F and directrix AB. A parabola may be generated by a pencil guided by a string, as shown. Fasten the string at F and C; its length is GC. The point C is selected at random, its distance from G depending on the desired extent of the curve. Keep the string taut and the pencil against the T-square, as shown.

*Fig. 4.57 (b).* Given focus F and directrix AB. Draw a line DE parallel to the directrix and at any distance CZ from it. With center at F and radius CZ, strike arcs to intersect the line DE in the points Q and R, which are points on the parabola. Determine as many additional points as are necessary to draw the parabola accurately, by drawing additional lines parallel to line AB and proceeding in the same manner.

A tangent to the parabola at any point G bisects the angle formed by the focal line FG and the line SG perpendicular to the directrix.

*Fig. 4.57 (c).* Given the rise and span of the parabola. Divide AO into any number of equal parts, and divide AD into a number of equal parts amounting to the square of that number. From line AB, each point on the parabola is offset by a number of units equal to the square of the number of units from point O. For example, point 3 projects 9 units (the square of 3). This method is generally used for drawing parabolic arches.

*Fig. 4.57 (d).* Given points P, R, and V of a parabola, to find the focus F. Draw tangent at P, making a = b. Draw perpendicular bisector of AP, which intersects the axis at F, the focus of the parabola.

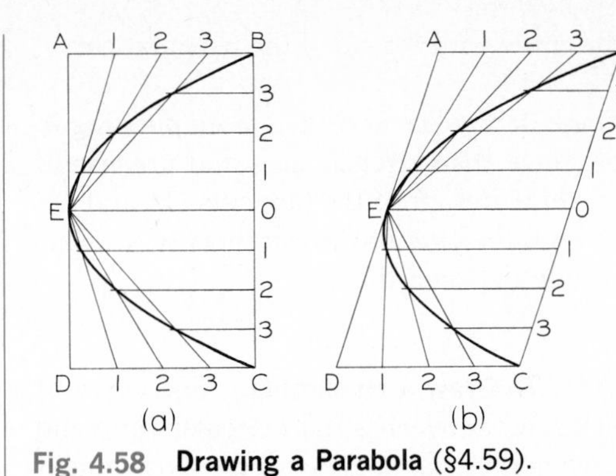


**Fig. 4.58 Drawing a Parabola** (§4.59).

*Fig. 4.58 (a)* or *(b).* Given rectangle or parallelogram ABCD. Divide BC into any even number of equal parts, and divide the sides AB and DC each into half as many parts, and draw lines as shown. The intersections of like-numbered lines are points on the parabola.

Practical Applications. The parabola is used for reflecting surfaces for light and sound, for vertical curves in highways, for forms of arches, and approximately for forms of the curves of cables for suspension bridges. It is also used to show the bending moment at any point on a uniformly loaded beam or girder.



## 4.60 To Join Two Points by a Parabolic Curve.

Fig. 4.59. Let X and Y be the given points. Assume any point O, and draw tangents XO and YO. Divide XO and YO into the same number of equal parts, number the division points as shown, and connect corresponding points. These lines are tangents of the required parabola and form its envelope. Sketch a light smooth curve, and then heavy

**Fig. 4.59 Parabolic Curves** (§4.60).

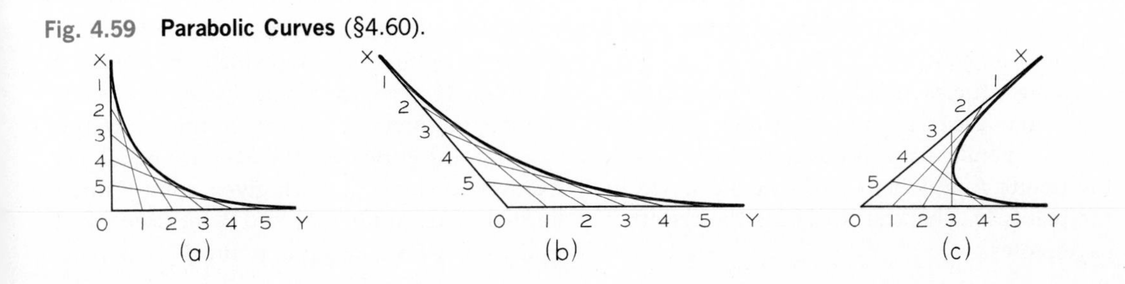

in the curve with the aid of the irregular curve, §2.57.

These parabolic curves are more pleasing in appearance than circular arcs and are useful in machine design. If the tangents OX and OY are equal, the axis of the parabola will bisect the angle between them.

## 4.61 To Draw a Hyperbola.

The curve of intersection between a right circular cone and a plane making an angle with the axis smaller than that made by the elements, Fig. 4.46 (e), is a *hyperbola. A hyperbola is generated by a point moving so that the difference of its distances from two fixed points, the foci, is constant and equal to the transverse axis of the hyperbola.*

*Fig. 4.60 (a).* Let F and F be the foci and AB the transverse axis. The curve may be generated by a pencil guided by a string, as shown. Fasten a string at F′ and C; its length is FC minus AB. The point C is chosen at pleasure; its distance from F depends on the desired extend of the curve.

Fasten the straightedge at F. If it is revolved about F, with the pencil point moving against it and with the string taut, the hyperbola may be drawn as shown.

*Fig. 4.60 (b).* To construct the curve geometrically, select any point X on the transverse axis produced. With centers at F and F′ and BX as radius, strike the arcs DE. With the same centers and AX as radius, strike arcs to intersect the arcs first drawn in the points Q, R, S, and T, which are points of the required hyperbola. Find as many additional points as necessary to draw the curves accurately by selecting other points similar to point X along the transverse axis, and proceeding as described for point X.

To draw the tangent to a hyperbola at a given point P, bisect the angle between the focal radii FP and F′P. The bisector is the required tangent.

To draw the asymptotes HCH of the hyperbola, draw a circle with the diameter FF′ and erect perpendiculars to the transverse axis at the points A and B to intersect the circle in the points H. The lines HCH are the required asymptotes.

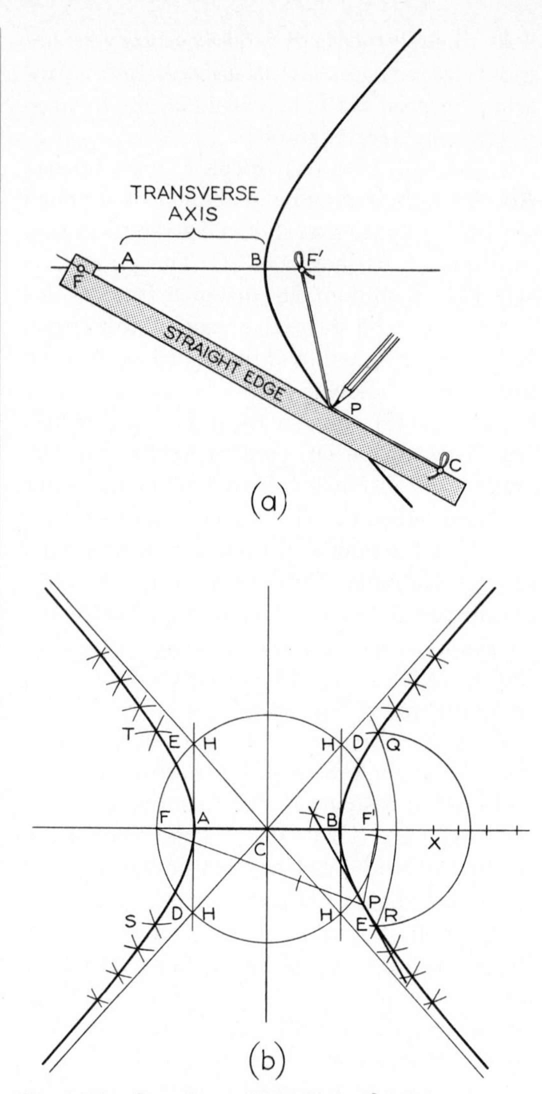


**Fig. 4.60 Drawing a Hyperbola** (§4.61).

## 4.62 To Draw an Equilateral Hyperbola.

Fig. 4.61. Let the asymptotes OB and OA, at right angles to each other, and the point P on the curve be given.

*Fig. 4.61 (a).* In an equilateral hyperbola the asymptotes, at right angles to each other, may be used as the axes to which the curve is referred. If a chord of the hyperbola is extended to intersect the axes, the intercepts between the curve and the axes are equal. For example, a chord through given point P intersects the axes at points 1 and 2, intercepts P–1 and 2–3 are equal, and point 3 is a point

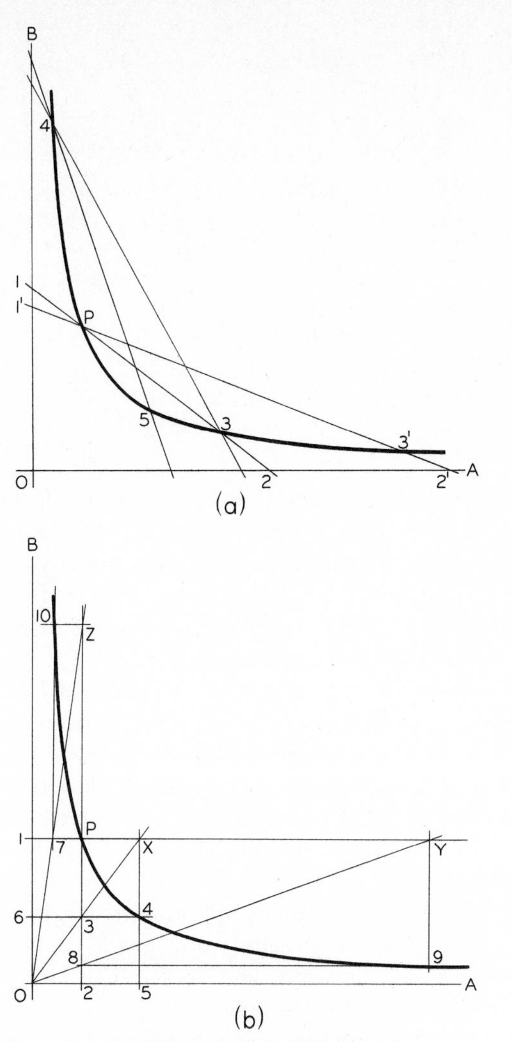


Fig. 4.61 **Equilateral Hyperbola** (§4.62).

on the hyperbola. Likewise, another chord through P provides equal intercepts P–1′ and 3′–2′, and point 3′ is a point on the curve. All chords need not be drawn through given point P, but as new points are established on the curve, chords may be drawn through them to obtain more points. After enough points are found to insure an accurate curve, the hyperbola is drawn with the aid of the irregular curve, §2.57.

*Fig. 4.61 (b).* In an equilateral hyperbola, the coordinates are related so that their products remain constant. Through given point P, draw lines 1–P–Y and 2–P–Z parallel, respectively, to the axes. From the origin of coordinates O, draw any diagonal intersecting these two lines at points 3 and X. At these points draw lines parallel to the axes, intersecting at point 4, a point on the curve. Likewise, another diagonal from O intersects the two lines through P at points 8 and Y, and lines through these points parallel to the axes intersect at point 9, another point on the curve. A third diagonal similarly produces point 10 on the curve, and so on. Find as many points as necessary for a smooth curve, and draw the parabola with the aid of the irregular curve, §2.57. It is evident from the similar triangles O–X–5 and O–3–2 that lines P–1 × P–2 = 4–5 × 4–6.

The equilateral hyperbola can be used to represent varying pressure of a gas as the volume varies, because the pressure varies inversely as the volume; that is, pressure × volume is constant.

## 4.63 To Draw a Spiral of Archimedes.

Fig. 4.62. To find points on the curve, draw lines through the pole C, making equal angles with each other, such as 30° angles, and beginning with any one line, set off any distance, such as $\frac{1}{16}$″; set off twice that distance on the next line, three times on the third, and so on. Through the points thus determined, draw a smooth curve, using the irregular curve, §2.57.

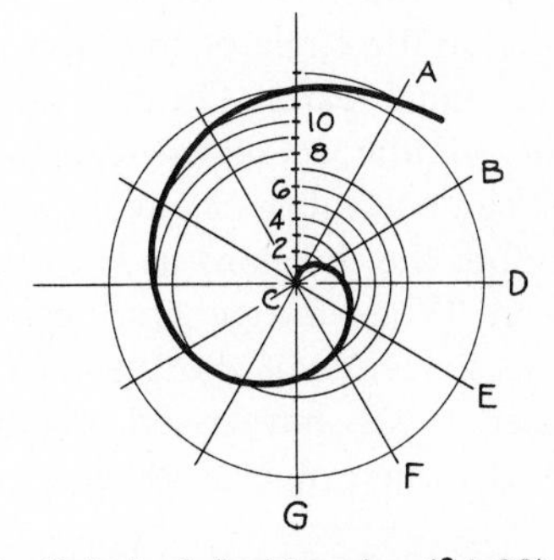


Fig. 4.62 **Spiral of Archimedes** (§4.63).

## 4.64 To Draw a Helix.

Fig. 4.63. *A helix is generated by a point moving around and along the surface of a cylinder or cone with a uniform angular velocity about the axis, and with a uniform*

Fig. 4.63 **Helix** (§4.64).

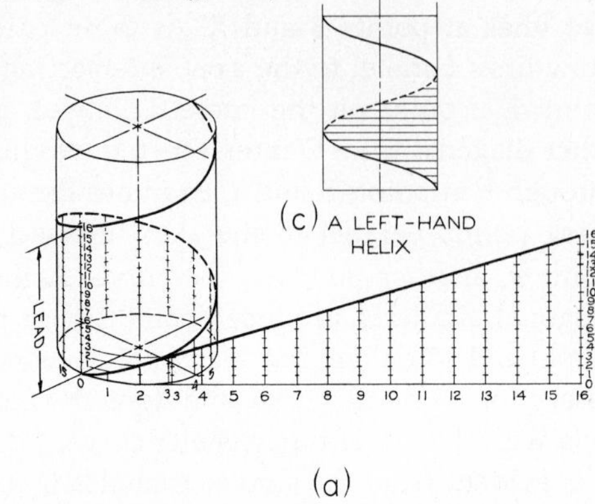


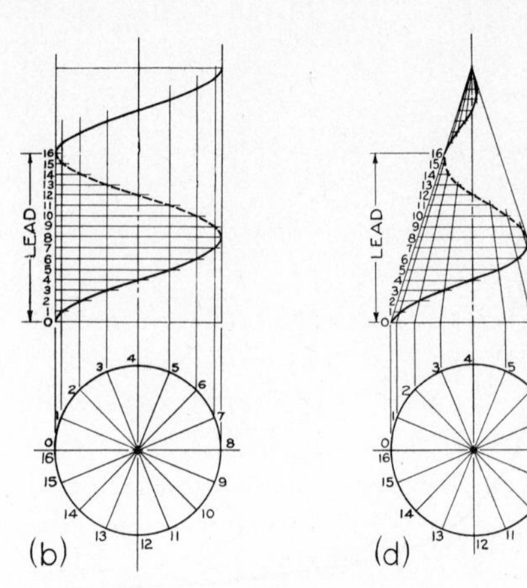


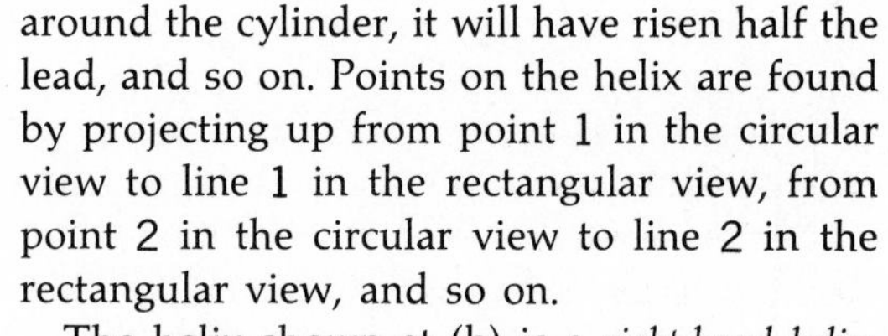


*linear velocity in the direction of the axis.* A cylindrical helix is generally known simply as a *helix.* The distance measured parallel to the axis traversed by the point in one revolution is called the *lead.*

If the cylindrical surface upon which a helix is generated is rolled out onto a plane, the helix becomes a straight line as shown in Fig. 4.63 (a), and the portion below the helix becomes a right triangle, the altitude of which is equal to the lead of the helix and the length of the base equal to the circumference of the cylinder. Such a helix can, therefore, be defined as the shortest line that can be drawn on the surface of a cylinder connecting two points not on the same element.

To draw the helix, draw two views of the cylinder upon which the helix is generated, (b), and divide the circle of the base into any number of equal parts. On the rectangular view of the cylinder, set off the lead and divide it into the same number of equal parts as the base. Number the divisions as shown, in this case sixteen. When the generating point has moved one-sixteenth of the distance around the cylinder, it will have risen one-sixteenth of the lead; when it has moved half-way around the cylinder, it will have risen half the lead, and so on. Points on the helix are found by projecting up from point 1 in the circular view to line 1 in the rectangular view, from point 2 in the circular view to line 2 in the rectangular view, and so on.

The helix shown at (b) is a *right-hand helix.* In a *left-hand helix,* (c), the visible portions of the curve are inclined in the opposite direction, that is, downward to the right. The helix shown at (b) can be converted into a left-hand helix by interchanging the visible and hidden lines.

The helix finds many applications in industry, as in screw threads, worm gears, conveyors, spiral stairways, and so on. The stripes of a barber pole are helical in form.

The construction for a right-hand conical helix is shown at (d).

## 4.65 To Draw an Involute.

Fig. 4.64. The path of a point on a string, as the string unwinds from a line, a polygon, or a circle, is an *involute.*

To Draw an Involute of a Line. Fig. 4.64 (a). Let AB be the given line. With AB as

Fig. 4.64 **Involutes** (§4.65).

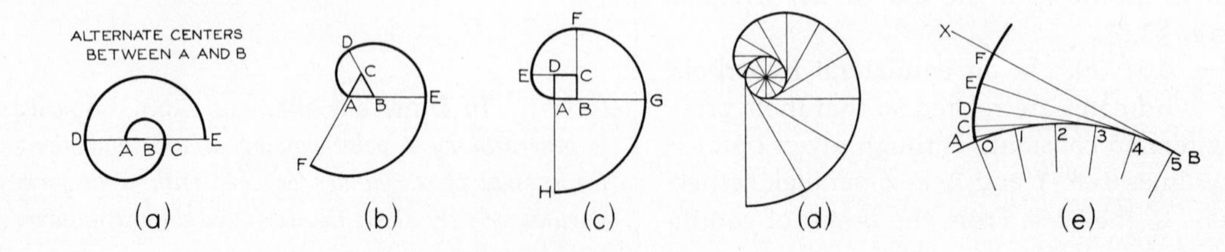

radius and B as center, draw the semicircle AC. With AC as radius and A as center, draw the semicircle CD. With BD as radius and B as center, draw the semicircle DE. Continue similarly, alternating centers between A and B, until a figure of the required size is completed.

To Draw an Involute of a Triangle. Fig. 4.64 (b). Let ABC be the given triangle. With CA as radius and C as center, strike the arc AD. With BD as radius and B as center, strike the arc DE. With AE as radius and A as center, strike the arc EF. Continue similarly until a figure of the required size is completed.

To Draw an Involute of a Square. Fig. 4.64 (c). Let ABCD be the given square. With DA as radius and D as center, draw the 90° arc AE. Proceed as for the involute of a triangle until a figure of the required size is completed.

To Draw an Involute of a Circle. Fig. 4.64 (d). A circle may be regarded as a polygon with an infinite number of sides. The involute is constructed by dividing the circumference into a number of equal parts, drawing a tangent at each division point, setting off along each tangent the length of the corresponding circular arc, Fig. 4.45 (c), and drawing the required curve through the points set off on the several tangents.

*Fig. 4.64 (e).* The involute may be generated by a point on a straight line which is rolled on a fixed circle. Points on the required curve may be determined by setting off equal distances 0–1, 1–2, 2–3, and so forth, along the circumference, drawing a tangent at each division point, and proceeding as explained for (d).

The involute of a circle is used in the construction of involute gear teeth. See §20.4. In this system, the involute forms the face and a part of the flank of the teeth of gear wheels; the outlines of the teeth of racks are straight lines.

## 4.66 To Draw a Cycloid.

Fig. 4.65. *A cycloid may be generated by a point P in the circumference of a circle that rolls along a straight line.*

Given the generating circle and the straight line AB tangent to it, make the distances CA and CB each equal to the semicircumference

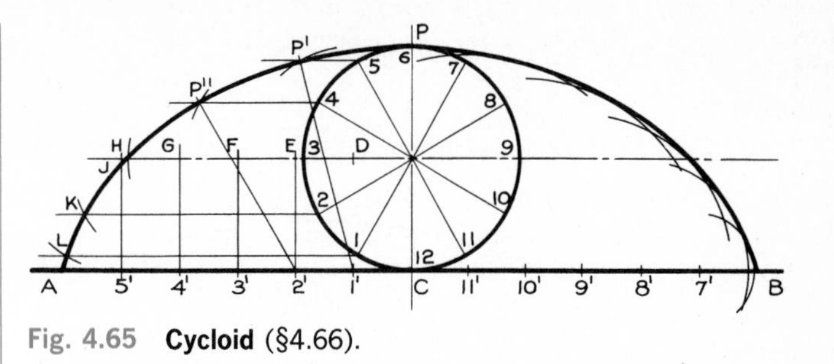


Fig. 4.65 **Cycloid** (§4.66).

of the circle, Fig. 4.45 (c). Divide these distances and the semicircumference into the same number of equal parts, six for instance, and number them consecutively as shown. Suppose the circle to roll to the left; when point 1 of the circle reaches point 1′ of the line, the center of the circle will be at D, point 7 will be the highest point of the circle, and the generating point 6 will be at the same distance from the line AB as point 5 is when the circle is in its central position. Hence, to find the point P′, draw a line through point 5 parallel to AB and intersect it with an arc drawn from the center D with a radius equal to that of the circle. To find point P″, draw a line through point 4 parallel to AB, and intersect it with an arc drawn from the center E, with a radius equal to that of the circle. Points J, K, and L are found in a similar manner.

Another method that may be employed is shown in the right half of the figure. With center at 11′ and the chord 11–6 as radius, strike an arc. With 10′ as center and the chord 10–6 as radius, strike an arc. Continue similarly with centers 9′, 8′, and 7′. Draw the required cycloid tangent to these arcs.

The student may use either method; the second is the shorter one and is preferred. It is evident, from the tangent arcs drawn in the manner just described, that the line joining the generating point and the point of contact for the generating circle is a normal of the cycloid; the lines 1′–P′ and 2′–P″, for instance, are normals; this property makes the cycloid suitable for the outlines of gear teeth.

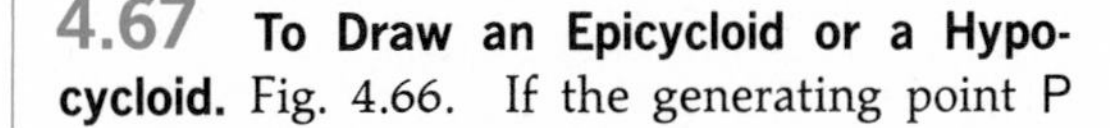

## 4.67 To Draw an Epicycloid or a Hypocycloid.

Fig. 4.66. If the generating point P

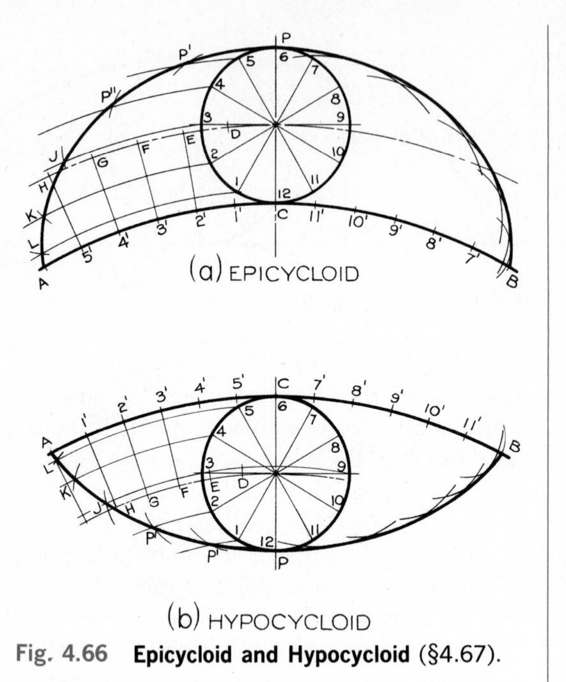


**Fig. 4.66 Epicycloid and Hypocycloid** (§4.67).

is on the circumference of a circle that rolls along the convex side of a larger circle, (a), the curve generated is an *epicycloid.* If the circle rolls along the concave side of a larger circle, (b), the curve generated is a *hypocycloid.* These curves are drawn in a manner similar to the cycloid, Fig. 4.65. These curves, like the cycloid, are used to form the outlines of certain gear teeth and are, therefore, of practical importance in machine design.

## 4.68 Geometric Construction Problems.

Fig. 4.67. The following geometric constructions should be made very accurately, with a hard pencil (2H to 4H) having a long sharp conical point. Draw given and required lines dark and medium in thickness, and draw construction lines *very light.* Do not erase construction lines. Indicate points and lines as described in §4.2.

Geometric construction problems in convenient form for solution may be found in *Technical Drawing Problems, Series 1,* by Giesecke, Mitchell, Spencer and Hill; in *Technical Drawing Problems, Series 2,* by Spencer and Hill; and in *Technical Drawing Problems, Series 3,* by Spencer and Hill, all designed to accompany this text and published by Macmillan Publishing Co., Inc.

Listed below are a large number of problems from which the instructor will make assignments. Use Layout A–2 (see inside of back cover) divided into four parts, as shown in Fig. 4.67. Additional sheets, with other problems selected from Figs. 4.68 to 4.79 and drawn on the same sheet layout, may be assigned by the instructor.

The student should exercise care in setting up each problem so as to make the best use of the space available, to present the problem to best advantage, and to produce a pleasing appearance. Letter the principal points of all constructions in a manner similar to the various illustrations in this chapter.

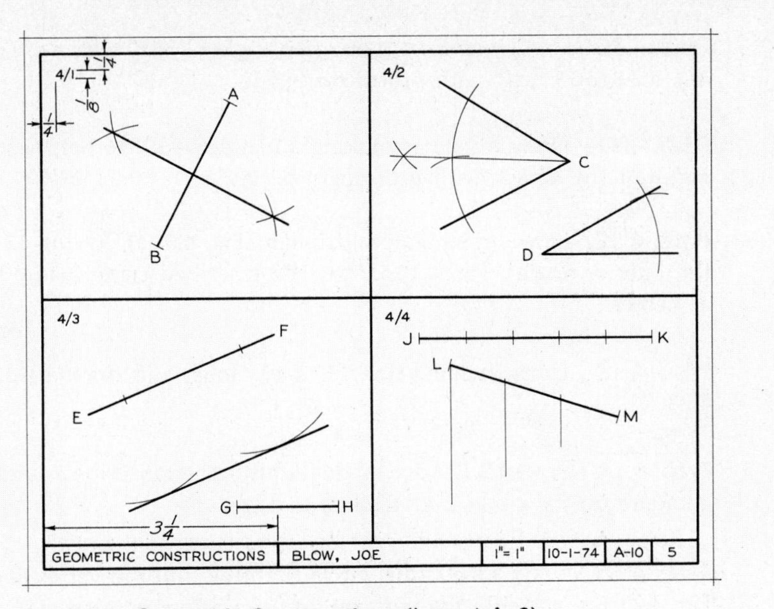


Fig. 4.67 **Geometric Constructions** (Layout A–2).

*The first four problems below are shown in Fig. 4.67.*

**Prob. 4.1** Draw an inclined line AB 2.50″ long and bisect it, Fig. 4.8.

**Prob. 4.2** Draw any angle with vertex at C. Bisect it, Fig. 4.10, and transfer one half in new position at D, Fig. 4.11.

**Prob. 4.3** Draw an inclined line EF and assume distance GH = $1\frac{5}{8}$″. Draw a line parallel to EF and at the distance GH from it, Fig. 4.13 (a).

**Prob. 4.4** Draw the line JK 3.75″ long and divide it into five equal parts with the dividers, §2.40. Draw a line LM 2.81″ long and divide it into three equal parts by the parallel-line method, Fig. 4.15.

**Prob. 4.5** Draw a line OP 3.62″ long and divide it into three proportional parts to 3, 5, and 9, Fig. 4.17 (a).

**Prob. 4.6** Draw a line $3\frac{7}{16}$″ long and divide it into parts proportional to the square of X where X = 1, 2, 3, and 4, Fig. 4.17 (c).

**Prob. 4.7** Draw a triangle having sides 3.00″, 3.25″, and 2.50″, Fig. 4.19. Bisect the three interior angles, Fig. 4.10. The bisectors should meet at a point. Draw the inscribed circle with the point as center.

**Prob. 4.8** Draw a right triangle having the hypotenuse $2\frac{1}{2}$″ and one leg $1\frac{9}{16}$″, Figs. 4.3 and 4.20, and draw a circle through the three vertexes, Fig. 4.32.

**Prob. 4.9** Draw an inclined line QR 84 mm long. Select a point P on the line 32 mm from Q, and erect a perpendicular, Fig. 4.18 (c). Assume a point S about 44.5 mm from the line, and erect a perpendicular from S to the line, Fig. 4.18 (b).

**Prob. 4.10** Draw two lines making an angle of $35\frac{1}{2}°$ with each other, using the tangent method, Fig. 4.21 (a). Check with protractor, §2.19.

**Prob. 4.11** Draw two lines making an angle of 33°16′ with each other, using the sine method, Fig. 4.21 (b). Check with protractor, §2.19.

**Prob. 4.12** Draw an equilateral triangle, Fig. 4.3 (a), having 63.5 mm sides, Fig. 4.22 (a). Bisect the interior angles, Fig. 4.10. Draw the inscribed circle, using the intersection of the bisectors as center.

**Prob. 4.13** Draw inclined line TU 2.19″ long, and draw a square on TU as a given side, Fig. 4.23 (a).

**Prob. 4.14** Draw a 2.12″ dia. circle (lightly); then inscribe a square in the circle and circumscribe a square on the circle, Fig. 4.23 (c) and (d).

**Prob. 4.15** Draw a 2.50″ dia. circle (lightly), find the vertexes of a regular inscribed pentagon, Fig. 4.24 (a), and join the vertexes to form a five-pointed star.

**Prob. 4.16** Draw a 2.50″ dia. circle (lightly), inscribe a hexagon, Fig. 4.25 (b), and circumscribe a hexagon, Fig. 4.26 (d).

**Prob. 4.17** Draw a square (lightly) with 63.5 mm sides, Fig. 4.23 (b), and inscribe an octagon, Fig. 4.28 (b).

**Prob. 4.18** Draw a triangle similar to that in Fig. 4.29 (a), having sides 2.00″, 1.50″, and 2.88″ long, and transfer the triangle to a new location and turned 180° similar to that in Fig. 4.29 (b). Check by prick-point method, §4.30.

**Prob. 4.19** In center of space, draw a rectangle 3.50″ wide and 2.40″ high. Show construction for reducing this rectangle until it is 2.30″ wide and again when 2.75″ wide, Fig. 4.31 (b).

**Prob. 4.20** Draw three points arranged approximately as in Fig. 4.32 (a), and draw a circle through the three points.

**Prob. 4.21** Draw a $2\frac{1}{4}$″ dia. circle. Assume a point S on the left side of the circle and draw a tangent at that point, Fig. 4.34 (a). Assume a point T to the right of the circle 2″ from its center, and draw two tangents to the circle through the point, Fig. 4.34 (b).

**Prob. 4.22** Through center of space, draw horizontal center line; then draw two circles 2″ dia. and $1\frac{1}{4}$″ dia., respectively, with centers $2\frac{1}{8}$″ apart. Locate the circles so that the construction will be centered in the space. Draw "open belt" tangents to the circles, Fig. 4.35 (a).

**Prob. 4.23** Same as Prob. 4.22, except draw "crossed belt" tangents to the circle, Fig. 4.35 (b).

**Prob. 4.24** Draw vertical line VW 1.30″ from left side of space. Assume point P 1.75″ farther to the right and 1.00″ down from top of space. Draw a 2.20″ dia. circle through P, tangent to VW, Fig. 4.36 (a).

**Prob. 4.25** Draw vertical line XY $1\frac{3}{8}''$ from left side of space. Assume point P $1\frac{3}{4}''$ farther to the right and 1″ down from top of space. Assume point Q on line XY and 2″ from P. Draw circle through P and tangent to XY at Q, Fig. 4.36 (b).

**Prob. 4.26** Draw $2\frac{1}{2}''$ dia. circle with center C $\frac{5}{8}''$ directly to left of center of space. Assume point P at the lower right and $2\frac{3}{8}''$ from C. Draw an arc with 1″ radius through P and tangent to the circle, Fig. 4.36 (c).

**Prob. 4.27** Draw a vertical line and a horizontal line, each 63.5 mm long, Fig. 4.37 (I). Draw arc with 38 mm radius, tangent to the lines.

**Prob. 4.28** Draw horizontal line $\frac{3}{4}''$ up from bottom of space. Select a point on the line 2″ from the left side of space, and through it draw a line upward to the right at 60° to horizontal. Draw arcs with $1\frac{3}{8}''$ radius within obtuse angle and acute angle, respectively, tangent to the two lines, Fig. 4.38.

**Prob. 4.29** Draw two intersecting lines making an angle of 60° with each other similar to Fig. 4.38 (a). Assume a point P on one line at a distance of $1\frac{3}{4}''$ from the intersection. Draw an arc tangent to both lines with one point of tangency at P, Fig. 4.33.

**Prob. 4.30** Draw vertical line AB $1\frac{1}{4}''$ from left side of space. Draw arc of $1\frac{5}{8}''$ radius with center 3″ to right of line and in lower right portion of space. Draw arc of 1″ radius tangent to AB and to the arc, Fig. 4.39.

**Prob. 4.31** With centers .75″ up from bottom of space, and 3.40″ apart, draw arcs of radius 1.75″ and .94″, respectively. Draw arc of 1.25″ radius tangent to the two arcs, Fig. 4.40.

**Prob. 4.32** Draw two circles as in Prob. 4.22. Draw arc of $2\frac{3}{4}''$ radius tangent to upper sides of circles and enclosing them, Fig. 4.41 (a). Draw arc of 2″ radius tangent to the circles but enclosing only the smaller circle, Fig. 4.41 (b).

**Prob. 4.33** Draw two parallel inclined lines 44.5 mm apart. Choose a point on each line and connect them with an ogee curve tangent to the two parallel lines, Fig. 4.43 (a).

**Prob. 4.34** Draw an arc of $2\frac{1}{8}''$ radius that subtends an angle of 90°. Find the length of the arc by two methods, Fig. 4.45 (a) and (c). Calculate the length of the arc and compare with the lengths determined graphically. See note at end of §4.46.

**Prob. 4.35** Draw major axis 4″ long (horizontally) and minor axis $2\frac{1}{2}''$ long, with their intersection at the center of the space. Draw ellipse by foci method with at least five points in each quadrant, Fig. 4.48.

**Prob. 4.36** Draw axes as in Prob. 4.35, but draw ellipse by trammel method, Fig. 4.49.

**Prob. 4.37** Draw axes as in Prob. 4.35, but draw ellipse by concentric-circle method, Fig. 4.50.

**Prob. 4.38** Draw axes as in Prob. 4.35, but draw ellipse by parallelogram method, Fig. 4.52 (a).

**Prob. 4.39** Draw conjugate diameters intersecting at center of space. Draw 3.50″ diameter horizontally, and 2.76″ diameter at 60° with horizontal. Draw oblique-circle ellipse, Fig. 4.51. Find at least 5 points in each quadrant.

**Prob. 4.40** Draw conjugate diameters as in Prob. 4.39, but draw ellipse by parallelogram method, Fig. 4.52 (b).

**Prob. 4.41** Draw axes as in Prob. 4.35, but draw approximate ellipse, Fig. 4.56.

**Prob. 4.42** Draw a parabola with a vertical axis, and the focus $\frac{1}{2}$″ from the directrix, Fig. 4.57 (b). Find at least 9 points on the curve.

**Prob. 4.43** Draw a hyperbola with a horizontal transverse axis 1.00″ long and the foci 1.50″ apart, Fig. 4.60 (b). Draw the asymptotes.

**Prob. 4.44** Draw horizontal line near bottom of space, and vertical line near left side of space. Assume point P $\frac{5}{8}$″ to right of vertical line and $1\frac{1}{4}$″ above horizontal line. Draw equilateral hyperbola through P and with reference to the two lines as asymptotes. Use either method of Fig. 4.61.

**Prob. 4.45** Using the center of the space as the pole, draw a spiral of Archimedes with the generating point moving in a counterclockwise direction and away from the pole at the rate of 1″ in each convolution, Fig. 4.62.

**Prob. 4.46** Through center of space, draw horizontal center line, and on it construct a right-hand helix 2″ dia., $2\frac{1}{2}$″ long, and with a lead of 1″, Fig. 4.63. Draw only a half-circular end view.

**Prob. 4.47** Draw the involute of an equilateral triangle with 12.7 mm sides, Fig. 4.64 (b).

**Prob. 4.48** Draw the involute of a $\frac{3}{4}$″ dia. circle, Fig. 4.64 (d).

**Prob. 4.49** Draw a cycloid generated by a 1.20″ dia. circle rolling along a horizontal straight line, Fig. 4.65.

**Prob. 4.50** Draw an epicycloid generated by a $1\frac{1}{4}$″ dia. circle rolling along a circular arc having a radius of $2\frac{1}{2}$″, Fig. 4.66 (a).

**Prob. 4.51** Draw a hypocycloid generated by a $1\frac{1}{4}$″ dia. circle rolling along a circular arc having a radius of $2\frac{1}{2}$″, Fig. 4.66 (b).

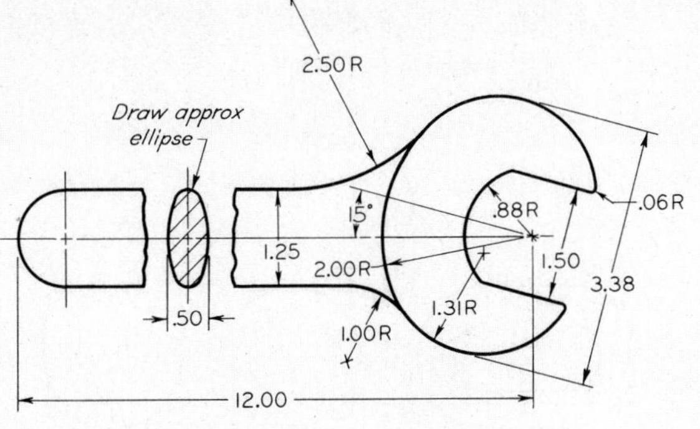


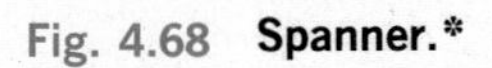
Fig. 4.68 **Spanner.***

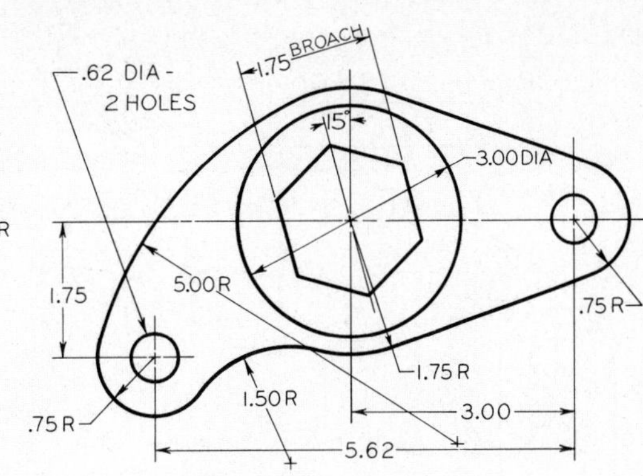


Fig. 4.69 **Rocker Arm.***

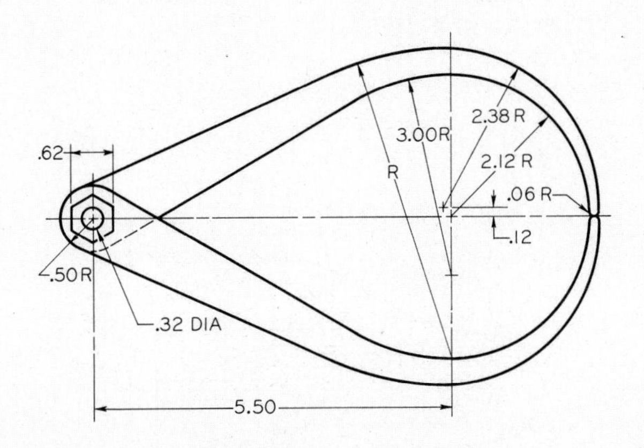


Fig. 4.70 **Outside Caliper.***

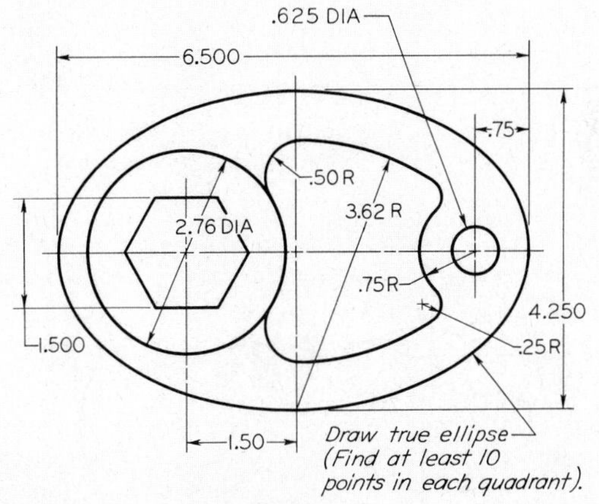


Fig. 4.71 **Special Cam.***

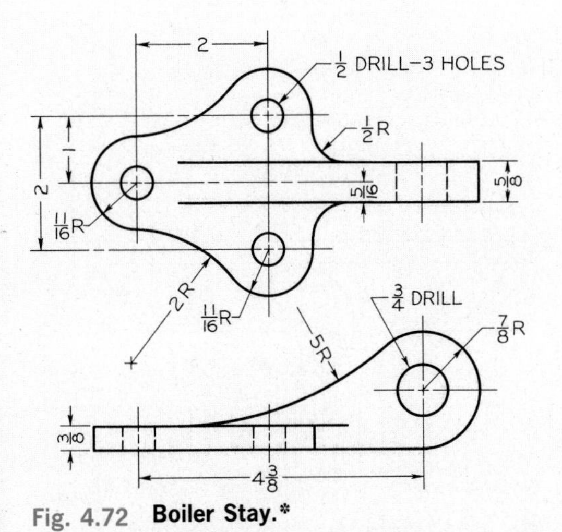


Fig. 4.72 **Boiler Stay.***

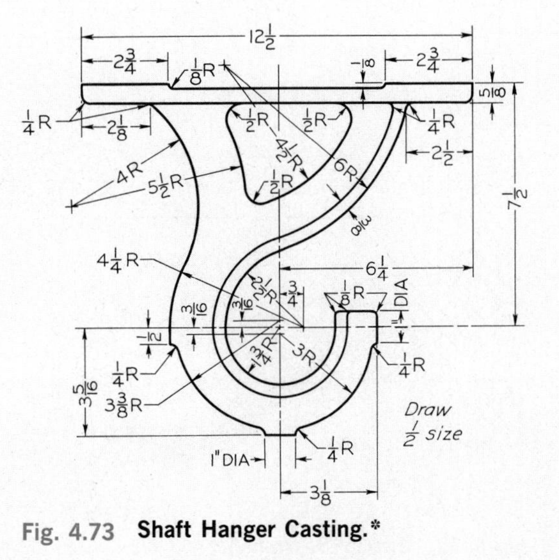


Fig. 4.73 **Shaft Hanger Casting.***

* Using Layout A-2, draw assigned problem with instruments. Omit dimensions and notes unless assigned by instructor.

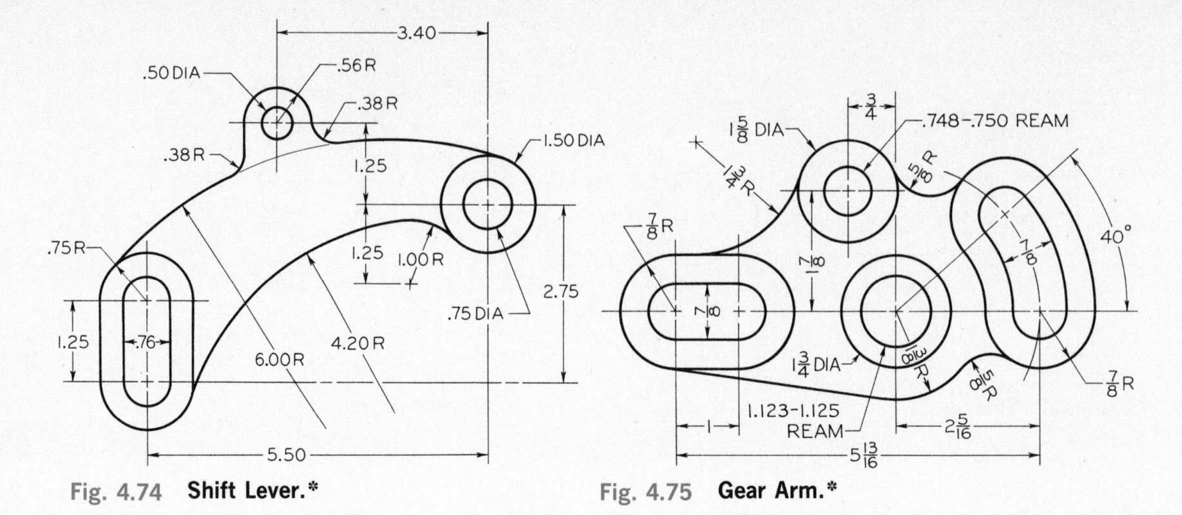


Fig. 4.74 **Shift Lever.***

Fig. 4.75 **Gear Arm.***

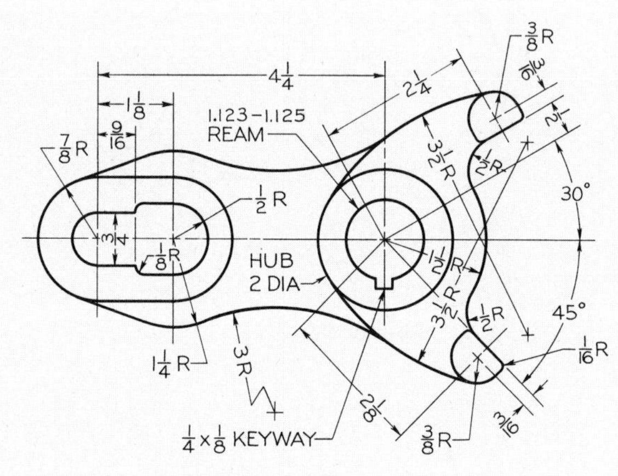


Fig. 4.76 **Form Roll Lever.***

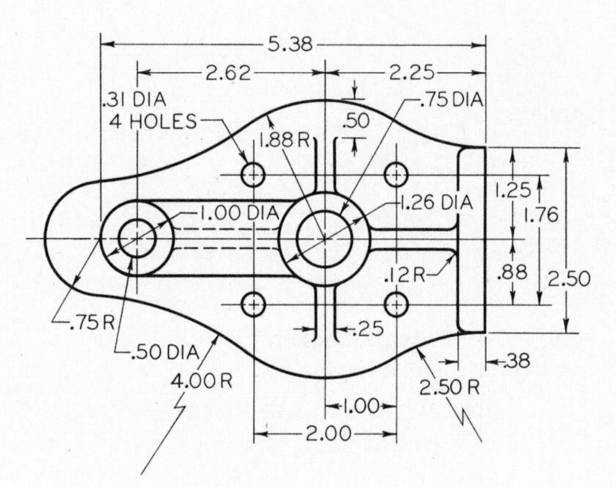


Fig. 4.77 **Press Base.***

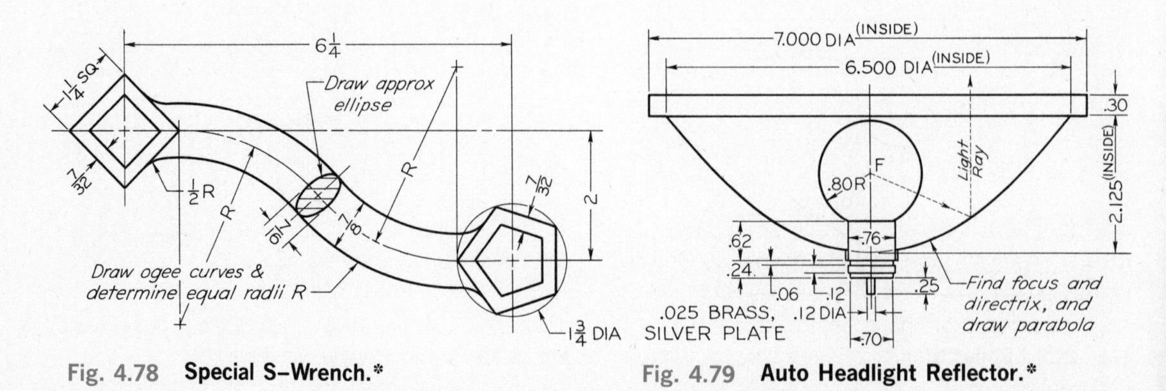


Fig. 4.78 **Special S-Wrench.***

Fig. 4.79 **Auto Headlight Reflector.***

*Using Layout A-2, draw assigned problem with instruments. Omit dimensions and notes unless assigned by instructor.

# 5

# sketching and shape description

**5.1 Importance of Technical Sketching.** The importance of freehand drawing or sketching in engineering and design cannot be overestimated. To the person who possesses a complete knowledge of drawing as a language, the ability to execute quick, accurate, and clear sketches of ideas and designs constitutes a valuable means of expression. The old Chinese saying that "one picture is worth a thousand words" is not without foundation.

Most original design ideas find their first expression through the medium of a freehand sketch, §14.2. Freehand sketching is a valuable means of amplifying and clarifying, as well as recording, verbal explanations. Executives sketch freehand daily to explain their ideas to subordinates. Engineers often prepare their designs and turn them over to their detailers or draftsmen in this convenient form, Fig. 5.1. Freehand sketches are of great assistance to the designer in organizing his thoughts and recording his ideas. They are an effective and economical means of formulating various solutions to a given problem so that a choice can be made between them at the outset. Often much time can be lost if the designer starts his scaled layout before adequate preliminary study with the aid of sketches. Information concerning changes in design or covering replacement of broken parts or lost drawings is usually conveyed through sketches. Many engineers consider the ability to render serviceable sketches of even greater value to them than skill in instrumental drawing. The draftsman or engineer will find daily use for this valuable means of formulating, expressing, and recording ideas in his work.

The degree of perfection required in a given sketch depends upon its use. Sketches hurriedly made to supplement oral description

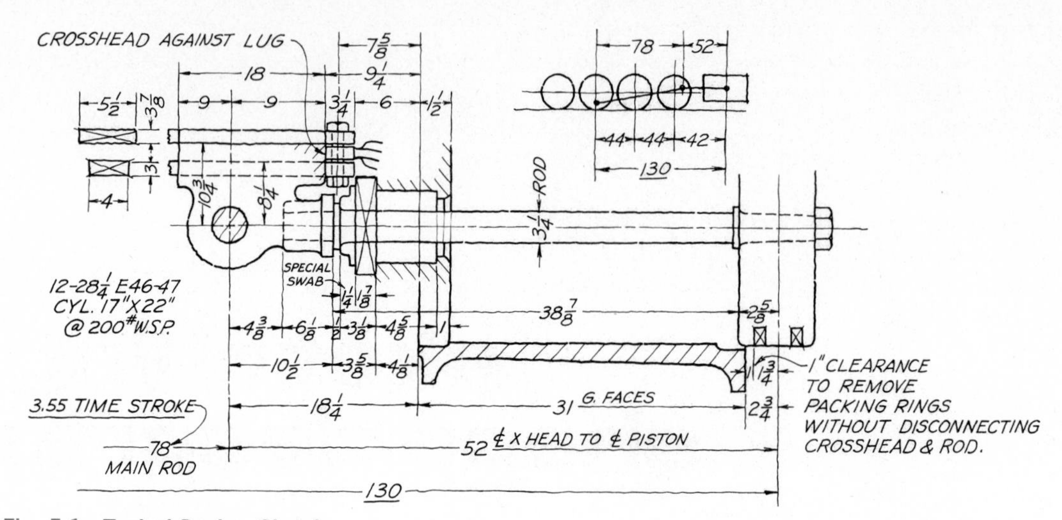


Fig. 5.1 **Typical Design Sketch.**

may be rough and incomplete. On the other hand, if a sketch is the medium of conveying important and precise information to engineers or to craftsmen, it should be executed as carefully as possible under the circumstances.

The term "freehand sketch" is too often understood to mean a crude or sloppy freehand drawing in which no particular effort has been made. On the contrary, a freehand sketch should be made with care and with attention to correct line widths.

## 5.2 Sketching Materials.

One of the advantages of freehand sketching is that it requires only pencil, paper, and eraser—items that anyone has for ready use.

When sketches are made in the field, where an accurate record is required, a small notebook or sketching pad is frequently used. Often clip boards are employed to hold the paper.

Cross-section paper is helpful to the sketcher, especially to the beginner or to the person who cannot sketch reasonably well without guide lines. It is available in rolls, sheets, and in pad form. Paper with 4, 5, 8, or 10 squares per inch is recommended. Such paper is convenient for sketching to scale since values can be assigned to the squares, and the squares counted to secure proportional distances, as shown in Fig. 5.2.

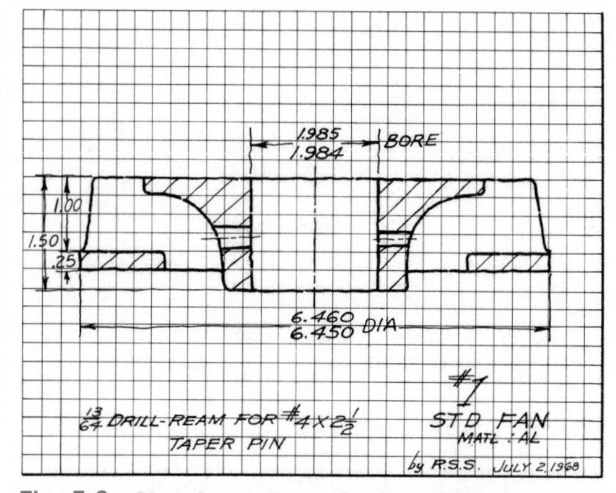


Fig. 5.2 **Sketch on Cross-Section Paper.**

Where multiple copies of a sketch are needed, a special tracing paper with cross-section lines that do not reproduce on a blueprint can be used. Also, sketching pads of plain tracing paper are available, accompanied by a master cross-section sheet. The draftsman places a blank sheet over the master grid and sketches while following the squares, which he can see through the transparent sheet.

An excellent procedure is to draw, with instruments, a master cross-section sheet, using India ink, making the squares $\frac{1}{4}''$ or $\frac{1}{8}''$, as desired. Ordinary bond typewriter paper is then placed over the master sheet and the sketch made thereon. Such a sketch is not only more uniform and "true," but shows up better because the cross-section lines are absent.

An excellent plain white paper for sketching is *ledger paper,* which is heavier than bond and may be purchased at low cost from a stationer or printing establishment.

For isometric sketching, a specially ruled *isometric paper* is available, Fig. 5.45.

Soft pencils, such as HB or F, should be used for freehand sketching. For carefully made sketches, two erasers are recommended, an Artgum and an orginary soft pencil eraser, Fig. 2.16.

**5.3 Types of Sketches.** Since technical sketches are made of three-dimensional objects, the form of the sketch conforms approximately to one of the four standard types of projection, as shown in Fig. 5.3. In *multiview* projection, (a), the object is described by its necessary views, as discussed in §§5.11 to 5.24. Or the object may be shown pictorially in a single view, by *axonometric, oblique,* or *perspective,* (b) to (d), as discussed in §§5.25 to 5.31.

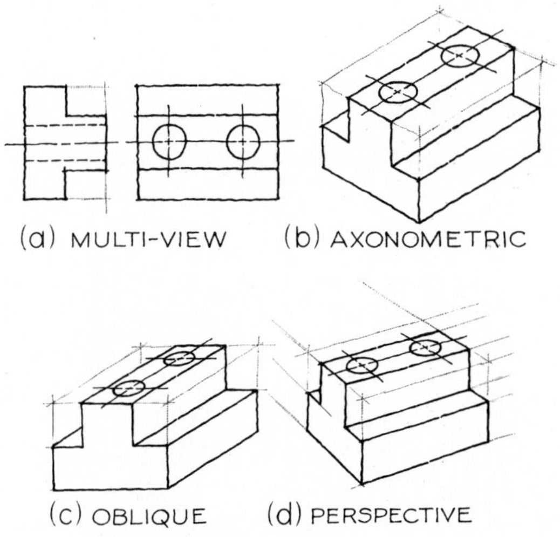

Fig. 5.3 **Types of Projection.**

**5.4 Scale.** *Sketches usually are not made to any scale.* Objects should be sketched in their correct proportions as accurately as possible, by eye. However, cross-section paper provides a ready scale (by counting squares) that may be used, if only to assist in sketching to correct proportions. The size of the sketch is purely optional, depending upon the complexity of the object and the size of paper available. Small objects are often sketched oversize so as to show the necessary details clearly.

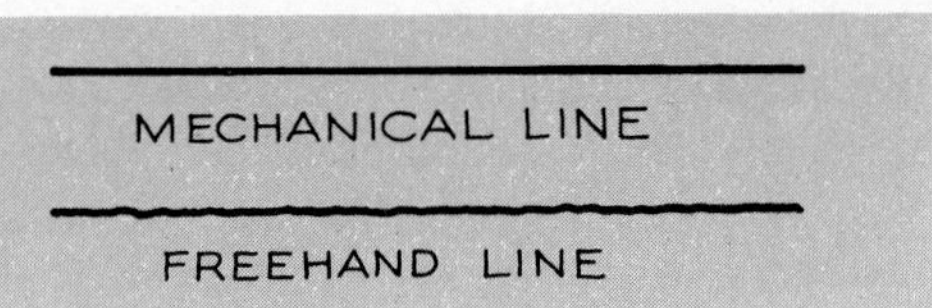

Fig. 5.4 **Comparison of Lines.**

**5.5 Technique of Lines.** The chief difference between an instrumental drawing and a freehand sketch lies in the character or *technique* of the lines. A good freehand line should not be rigidly straight or exactly uniform, as an instrumental line. While the effectiveness of an instrumental line lies in exacting uniformity, the quality of freehand line lies in its *freedom* and *variety,* Figs. 5.4 and 5.7.

Conventional lines, drawn instrumentally, are shown in Fig. 2.15, and the corresponding freehand renderings are shown in Fig. 5.5. The freehand construction line is a very light rough

Fig. 5.5 **Sketch Lines.**
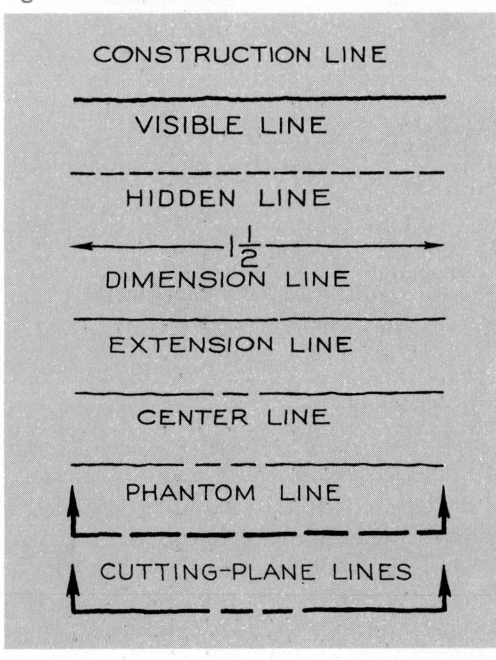

line in which some strokes may overlap. All other lines should be dark and clean-cut. Accent the ends of all dashes, and maintain a sharp contrast between the line thicknesses. Especially, make visible lines **heavy** so the outline will stand out clearly, and make hidden lines, center lines, dimension lines, and extension lines *thin.*

## 5.6 Sharpening Sketching Pencils.

Use a soft pencil, such as HB or F, or a mechanical pencil with an HB or F lead, and sharpen it to a conical point, as shown in Fig. 2.11 (c). Use this sharp point for center lines, dimension lines, and extension lines. For visible lines, hidden lines, and cutting-plane lines, round off the point slightly to produce the desired thickness of line, Fig. 5.6. Make all lines dark, with the exception of construction lines, which should be very light.

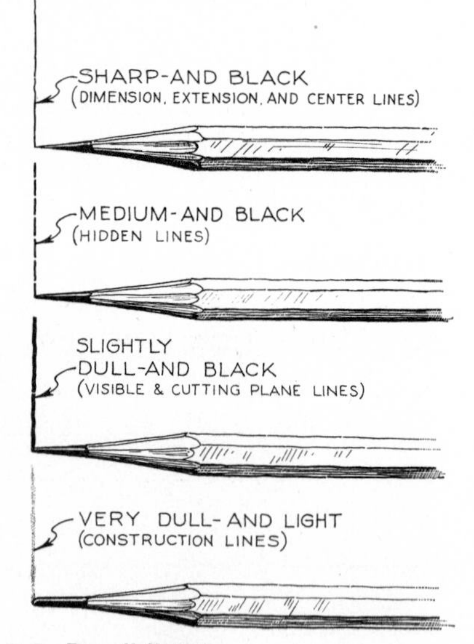


**Fig. 5.6 Pencil Points.**

## 5.7 Straight Lines.

Since the majority of lines on the average sketch are straight lines, it is necessary to learn to make them well. Hold the pencil naturally about $1\frac{1}{2}''$ back from the point, and approximately at right angles to the line to be drawn. Draw horizontal lines from left to right with a free and easy wrist-and-arm movement, Fig. 5.7. Draw vertical

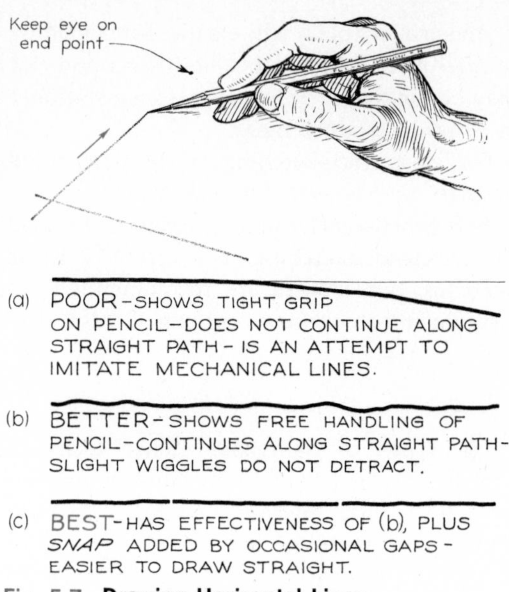

Keep eye on
end point
(a) POOR-SHOWS TIGHT GRIP ON PENCIL-DOES NOT CONTINUE ALONG STRAIGHT PATH-IS AN ATTEMPT TO IMITATE MECHANICAL LINES.
(b) BETTER-SHOWS FREE HANDLING OF PENCIL-CONTINUES ALONG STRAIGHT PATH-SLIGHT WIGGLES DO NOT DETRACT.
(c) BEST-HAS EFFECTIVENESS OF (b), PLUS SNAP ADDED BY OCCASIONAL GAPS-EASIER TO DRAW STRAIGHT.


**Fig. 5.7 Drawing Horizontal Lines.**

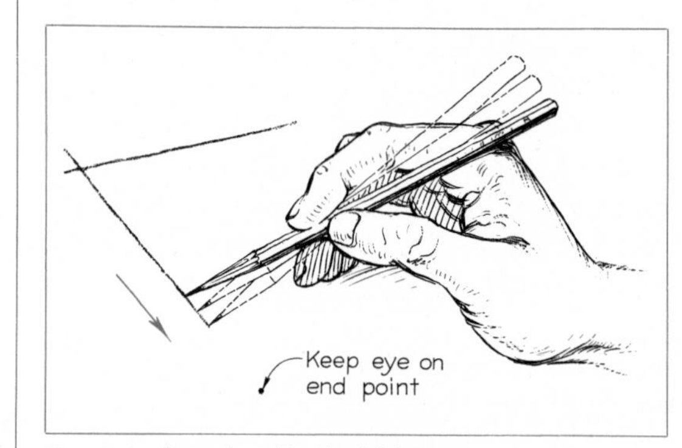


**Fig. 5.8 Drawing Vertical Lines.**

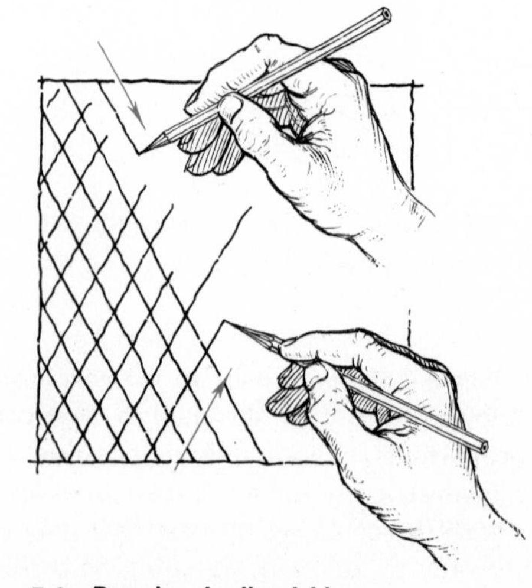

**Fig. 5.9 Drawing Inclined Lines.**

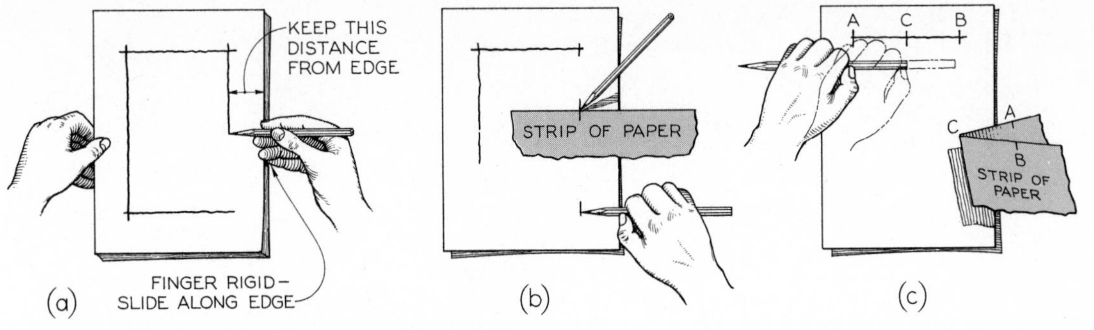


Fig. 5.10 **Blocking In Horizontal and Vertical Lines.**

lines downward with finger-and-wrist movements, Fig. 5.8.

Inclined lines may be made to conform in direction to horizontal or vertical lines by shifting position with respect to the paper or by turning the paper slightly; hence, they may be drawn with the same general movements, Fig. 5.9.

In sketching long lines, mark the ends of the line with light dots, then move the pencil back and forth between the dots in long sweeps, keeping the eye always on the dot toward which the pencil is moving, the point of the pencil touching the paper lightly, and each successive stroke correcting the defects of the preceding strokes. When the path of the line has been established sufficiently, apply a little more pressure, replacing the trial series with a distinct line. Then, dim the line with the Artgum and draw the final line clean-cut and dark, keeping the eye now on the point of the pencil.

An easy method of blocking in horizontal or vertical lines is to hold the hand and pencil rigidly and glide the finger tips along the edge of the pad or board, as shown in Fig. 5.10.

Another method, (b), is to mark the distance on the edge of a card or a strip of paper and transfer this distance at intervals, as shown; then draw the final line through these points. Or the pencil may be held as shown at the lower part of (b), and distance marks made on the paper at intervals by tilting the lead down to the paper. It will be seen that both methods of transferring distances are substitutes for the dividers and will have many uses in sketching.

To find the midpoint of a line AB at (c), hold the pencil in the left hand with the thumb gaging the estimated half-distance. Try this distance on the left and then on the right until the center is located by trial, and mark the center C, as shown. Another method is to mark the total distance AB on the edge of a strip of paper and then to fold the paper to bring points A and B together, thus locating center C at the crease. To find quarter points, the folded strip can be folded once more.

## 5.8 Circles and Arcs.

Small circles and arcs can be easily sketched in one or two strokes, as for the circular portions of letters, without any preliminary blocking in.

One method of sketching a larger circle, Fig. 5.11, is first to sketch lightly the enclosing

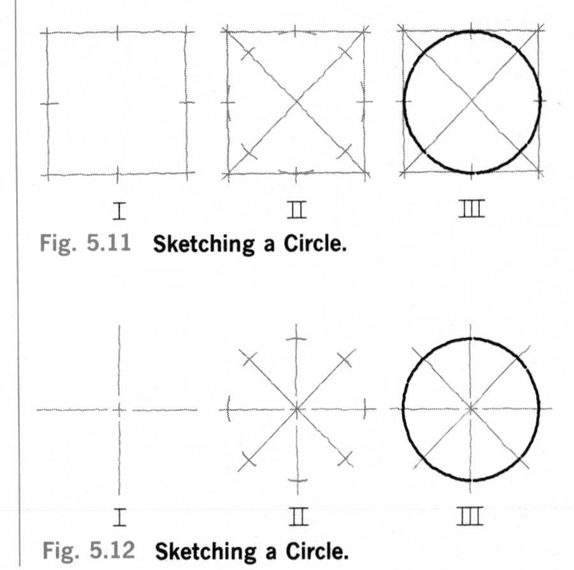


Fig. 5.11 **Sketching a Circle.**

Fig. 5.12 **Sketching a Circle.**

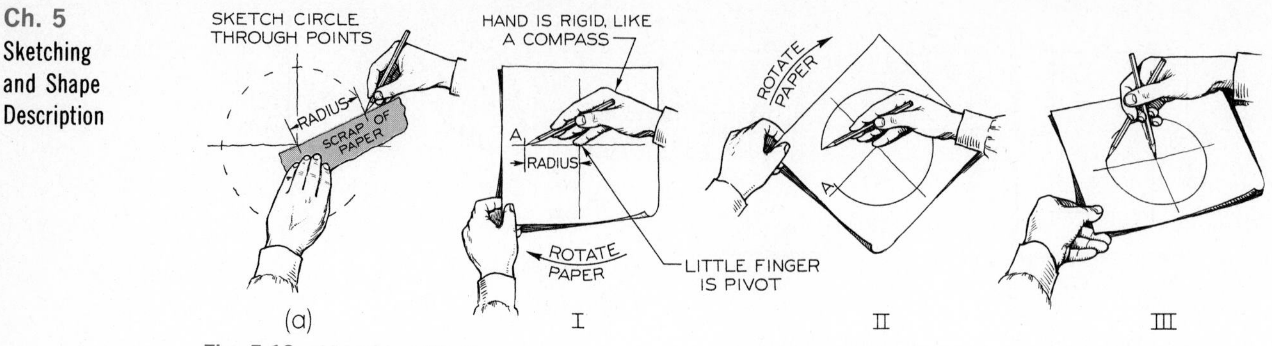


Fig. 5.13 **Sketching Circles.**

square, mark the midpoints of the sides, draw light arcs tangent to the sides of the square, and then heavy in the final circle.

Another method, Fig. 5.12, is to sketch the two center lines, add light 45° radial lines, sketch light arcs across the lines at the estimated radius distance from the center, and finally to sketch the required circle heavily.

In both methods, dim all construction lines with the Artgum before heavying in the final circle.

An excellent method, particularly for large circles, Fig. 5.13 (a), is to mark the estimated radius on the edge of a card or scrap of paper, to set off from the center as many points as desired, and to sketch the final heavy circle through these points.

The clever draftsman will prefer the method at I and II, in which the hand is used as a compass. Place the tip of the little finger, or the knuckle joint of the little finger, at the center; "feed" the pencil out to the desired radius, hold this position rigidly, and carefully revolve the paper with the other hand, as shown. If you are using a sketching pad, place the pad on your knee and revolve the entire pad on the knee as a pivot.

At III, two pencils are held rigidly like a compass and the paper is slowly revolved.

Methods of sketching arcs, Fig. 5.14, are adaptations of those used for sketching circles. In general, it is easier to sketch arcs with the hand and pencil on the concave side of the curve. In sketching tangent arcs, always keep in mind the actual geometric constructions, carefully approximating all points of tangency.

**5.9 Ellipses.** If a circle is viewed obliquely, Fig. 4.50 (b), it appears as an ellipse. With a

Fig. 5.14 **Sketching Arcs.**

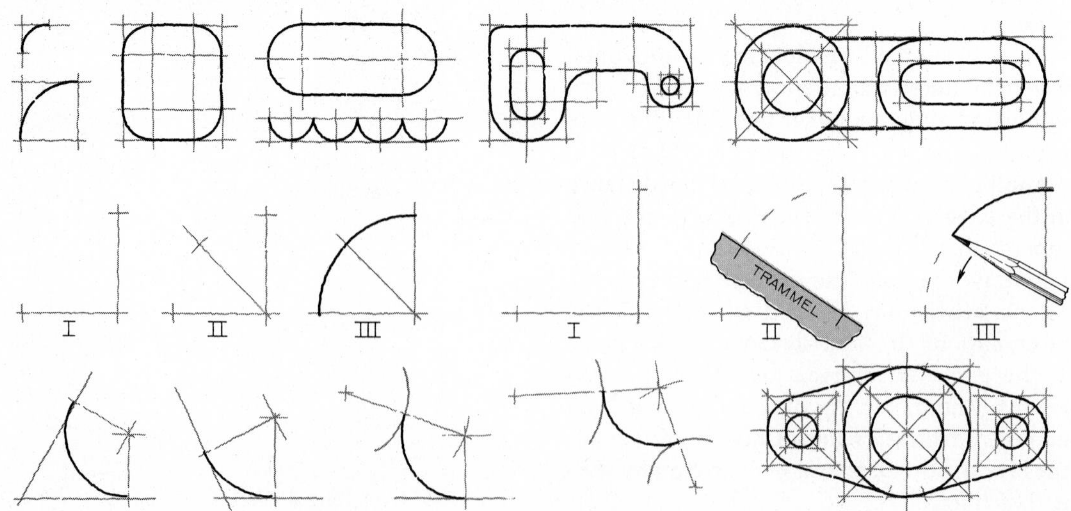

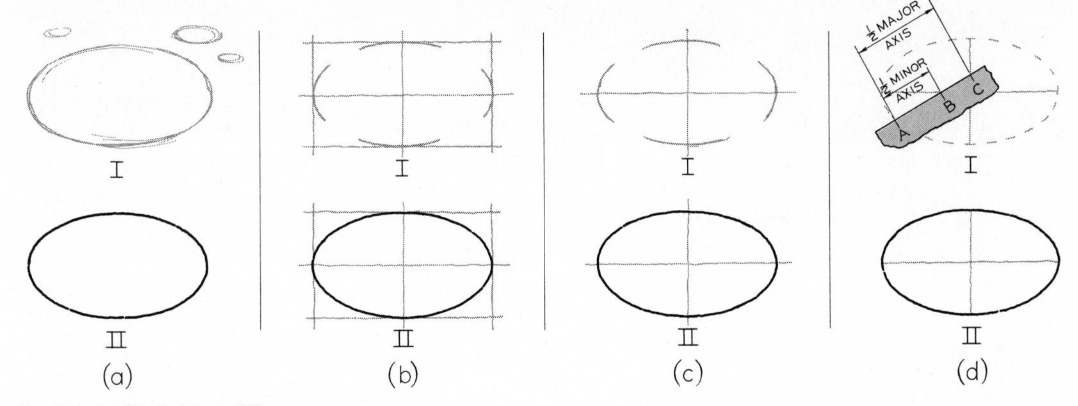


Fig. 5.15 **Sketching Ellipses.**

little practice, you can learn to sketch small ellipses with a free arm movement, Fig. 5.15 (a). Hold the pencil naturally, rest the weight on the upper part of the forearm, and move the pencil rapidly above the paper in the elliptical path desired; then lower the pencil so as to describe several light overlapping ellipses, as shown at I. Dim all lines with the Artgum and heavy in the final ellipse, II.

Another method, (b), is to sketch lightly the enclosing rectangle, I, mark the midpoints of the sides, and sketch light tangent arcs, as shown. Then, II, complete the ellipse lightly, dim all lines with the Artgum, and heavy in the final ellipse.

The same general procedure shown at (b) may be used in sketching the ellipse upon the given axes, as shown at (c).

The *trammel method,* (d), is excellent for sketching large ellipses. Prepare a "trammel" on the edge of a card or strip of paper, move it to different positions, and mark points on the ellipse at A. The trammel method is explained in §4.51. Sketch the final ellipse through the points, as shown.

For sketching isometric ellipses, see §5.13.

## 5.10 Proportions.

*The most important rule in freehand sketching is to keep the sketch in proportion.* No matter how brilliant the technique or how well the small details are drawn, if the proportions—especially the large over-all proportions—are bad, the sketch will be bad. First, the relative proportions of the height to the width must be carefully established; then as you proceed to the medium-size areas and the small details, constantly compare each new estimated distance with established distances already set down.

If you are working from a given picture, such as the night table in Fig. 5.16 (a), it is

Fig. 5.16 **Sketching a Night Table.**

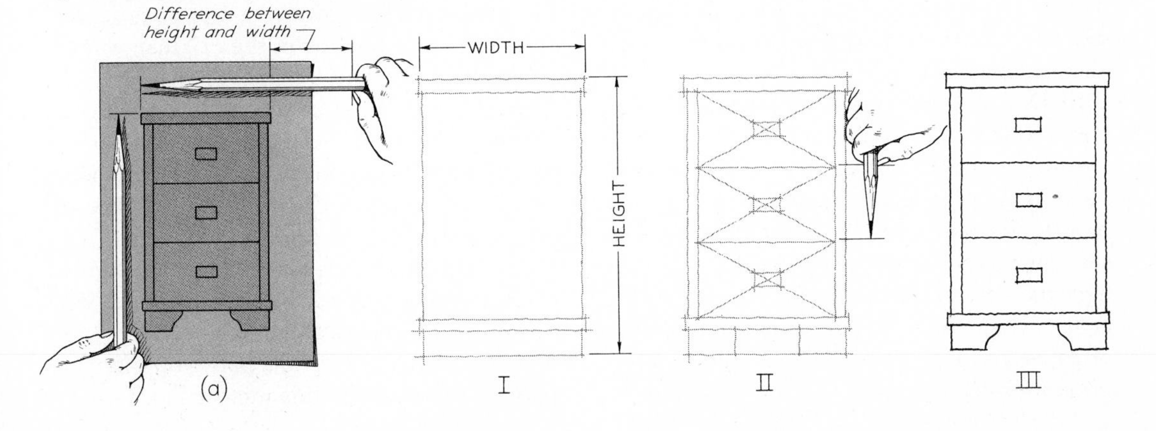

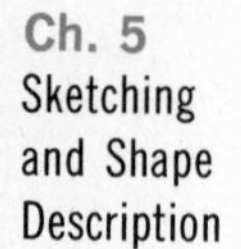


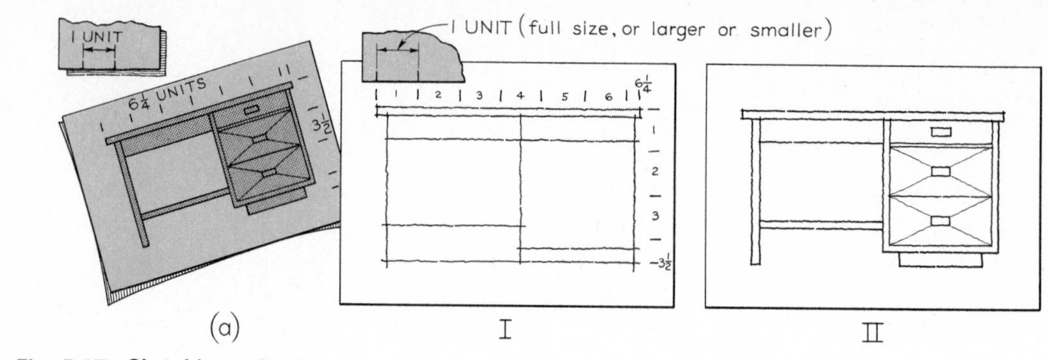


Fig. 5.17 **Sketching a Desk.**

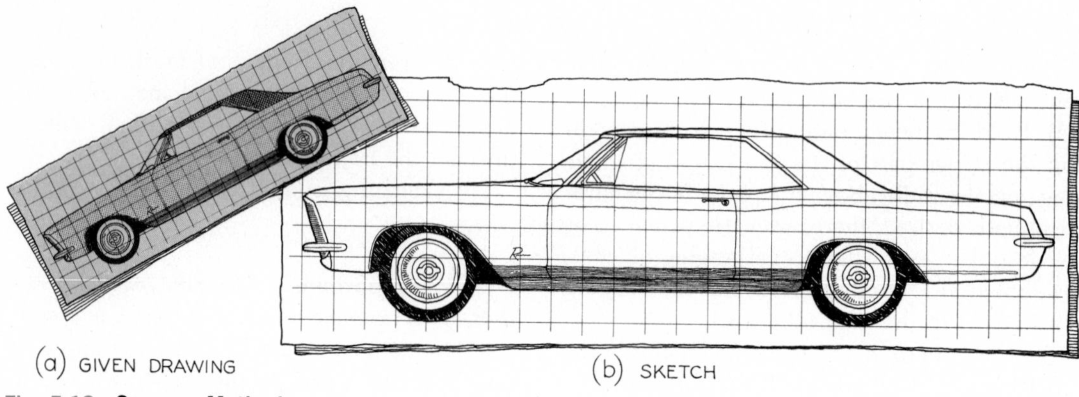

Fig. 5.18 **Squares Method.**

first necessary to establish the relative width compared to the height. One way is to use the pencil as a measuring stick, as shown. In this case, the height is about $1\frac{3}{4}$ times the width. Then:

I. Sketch the enclosing rectangle in the correct proportion. In this case, the sketch is to be slightly larger than the given picture.

II. Divide the available drawer space into three parts with the pencil by trial, as shown. Sketch light diagonals to locate centers of drawers, and block in drawer handles. Sketch all remaining details.

III. Dim all construction with Artgum or soft eraser, and heavy in all final lines.

Another method of estimating distances is illustrated in Fig. 5.17. On the edge of a card or strip of paper, mark an arbitrary unit. Then see how many units wide and how many units high the desk is. If you are working from the actual object, you could use a foot rule, a piece of paper, or the pencil itself as a unit to determine the proportions.

To sketch an object composed of many curves to the same scale or to a larger or smaller scale, the method of "squares" is recommended, Fig. 5.18. On the given picture, rule accurate grid lines to form squares of any convenient size. It is best to use a scale and some convenient spacing, such as $\frac{1}{2}''$. On the new sheet, rule a similar grid, making the spacing of the lines proportional to the original, but reduced or enlarged as desired. Make the final sketch by drawing the lines in and across the grid lines as in the original, as near as you can estimate by eye.

In sketching from an actual object, you can easily compare various distances on the object by using the pencil to compare measurements as shown in Fig. 5.19. While doing this, do not change your position, and always hold your pencil at arm's length. The length sighted can then be compared in similar manner with any other dimension of the object. If the object is small, such as a machine part, you can compare distances in the manner of Fig. 5.16, by

Fig. 5.19 **Estimating Dimensions.**

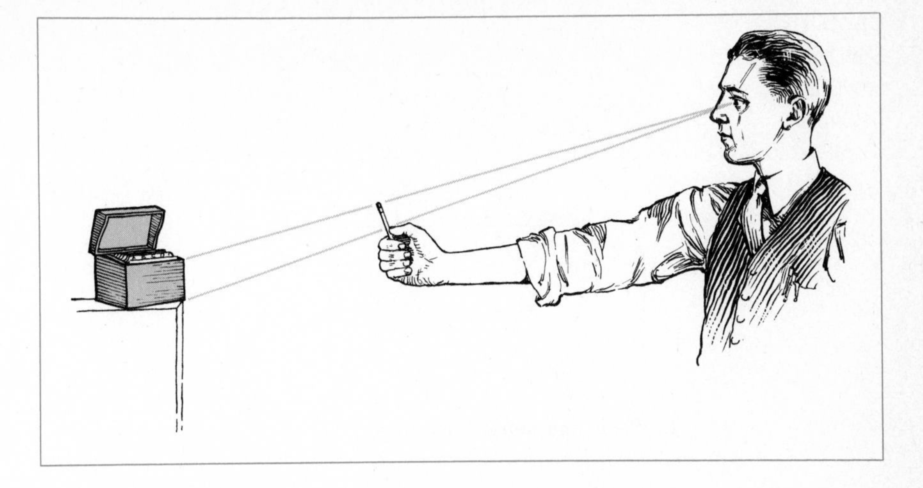

actually placing the pencil against the object itself.

In establishing proportions, the blocking-in method is recommended, especially for irregular shapes. The steps for blocking in and completing the sketch of a Shaft Hanger are shown in Fig. 5.20. As always, first give attention to the main proportions, next to the general sizes and direction of flow of curved shapes, and finally to the snappy lines of the completed sketch.

Fig. 5.20 **Blocking In an Irregular Object (Shaft Hanger).**

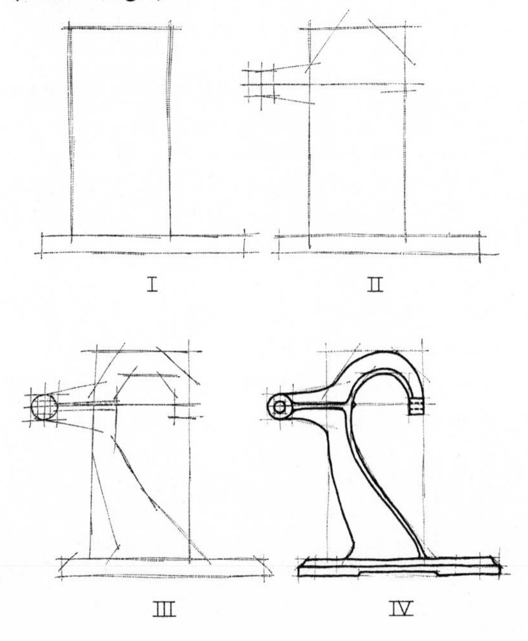


In making sketches from actual machine parts, it is necessary to use the measuring tools used in the shop, especially those needed to determine dimensions that must be relatively accurate. For a discussion of these methods, see §10.21.

## 5.11 Pictorial Sketching.

We shall now examine several simple methods of preparing *pictorial* sketches that will be of great assistance in learning the principles of multiview projection. A detailed and more scientific treatment of pictorial drawing is given in Chapters 16, 17, and 18.

## 5.12 Isometric Sketching.

To make an *isometric* sketch from an actual object, hold the object in your hand and tilt it toward you, as shown in Fig. 5.21 (a). In this position, the front corner will appear vertical, and the two receding bottom edges and those parallel to them, respectively, will appear at about 30° with horizontal, as shown. The steps in sketching are:

I. Sketch the enclosing box lightly, making AB vertical, and AC and AD approximately 30° with horizontal. These three lines are the *isometric axes.* Make AB, AC, and AD approximately equal in length to the actual corresponding edges on the object. Sketch the remaining lines parallel, respectively, to these three lines.

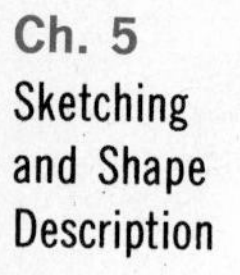

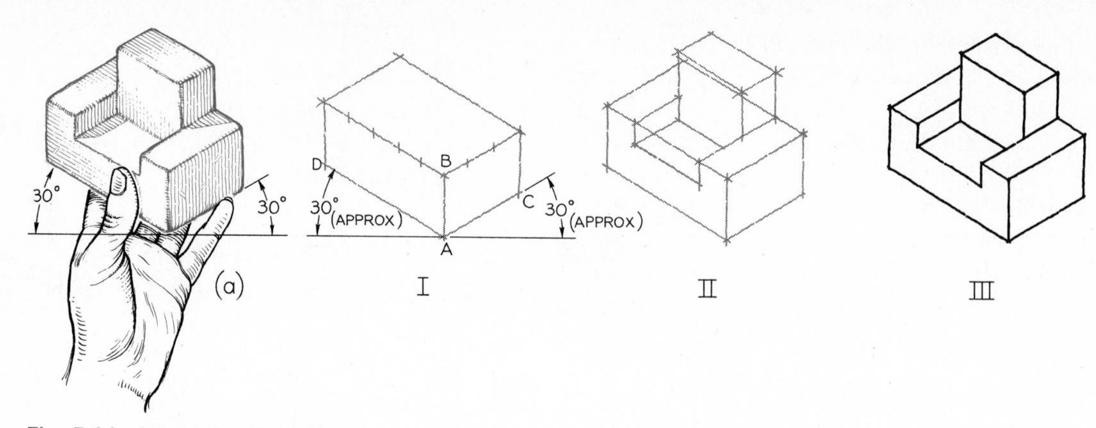


Fig. 5.21 **Isometric Sketching.**

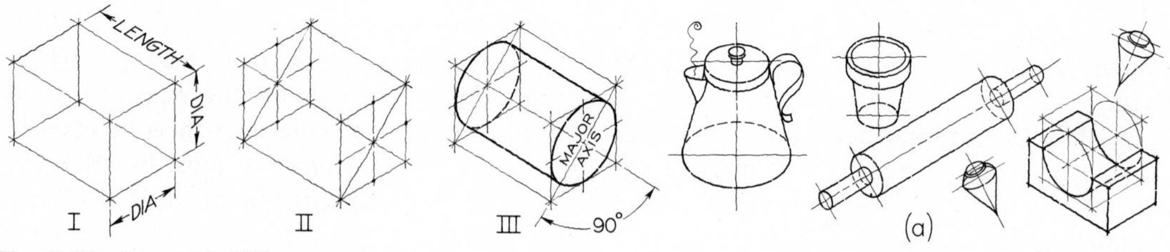

Fig. 5.22 **Isometric Ellipses.**

II. Block in the recess and the projecting block.

III. Dim all construction lines with the Artgum, and heavy in all final lines.

*Note:* The angle of the receding lines may be less than 30°, say 20° or 15°. Although the result will not be an isometric sketch, the sketch may be more pleasing and effective in many cases.

## 5.13 Isometric Ellipses.

As shown in Fig. 4.50 (b), a circle viewed at an angle appears as an ellipse. When objects having cylindrical or conical shapes are placed in the isometric or other oblique positions, the circles will be viewed at an angle and will appear as ellipses, Fig. 5.22.

The most important rule in sketching isometric ellipses is: *The major axis of the ellipse is always at right angles to the center line of the cylinder, and the minor axis is at right angles to the major axis and coincides with the center line.*

Two views of a block with a large cylindrical hole are shown in Fig. 5.23 (a). The steps in sketching the object are:

I. Sketch the block and the enclosing parallelogram for the ellipse, making the sides

Fig. 5.23 **Isometric Ellipses.**

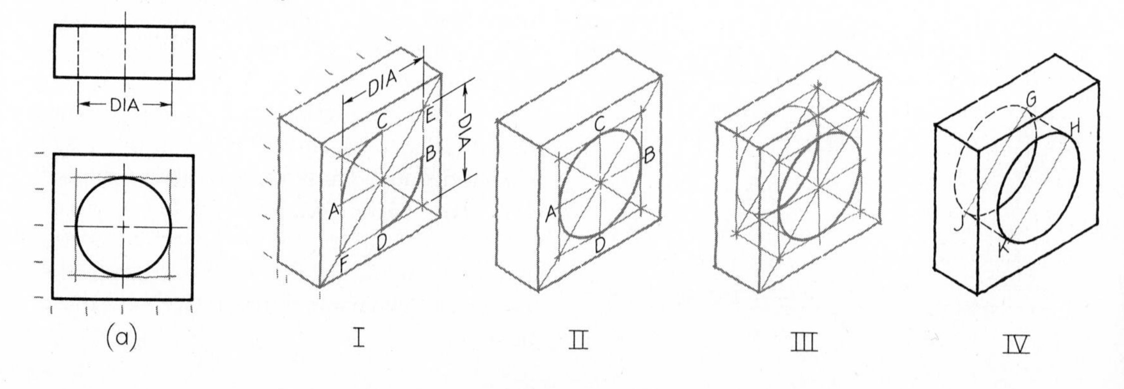

of the parallelogram parallel to the edges of the block and equal in length to the diameter of the hole. Draw diagonals to locate the center of the hole, and then draw center lines AB and CD. Points A, B, C, and D will be midpoints of the sides of the parallelogram, and the ellipse will be tangent to the sides at those points. The major axis will be on the diagonal EF, which is at right angles to the center line of the hole, and the minor axis will fall along the short diagonal. Sketch long flat elliptical sides CA and BD, as shown.

II. Sketch short small-radius arcs CB and AD to complete the ellipse. Avoid making the ends of the ellipse "squared off," or pointed like a football.

III. Sketch lightly the parallelogram for the back ellipse, and sketch the ellipse in the same manner as the front ellipse.

IV. Draw lines GH and JK tangent to the two ellipses. Dim all construction with the Artgum, and heavy in all final lines.

Another method for determining the back ellipse is shown in Fig. 5.24.

I. Select points at random on the front ellipse and sketch "depth lines" equal in length to the depth of the block.

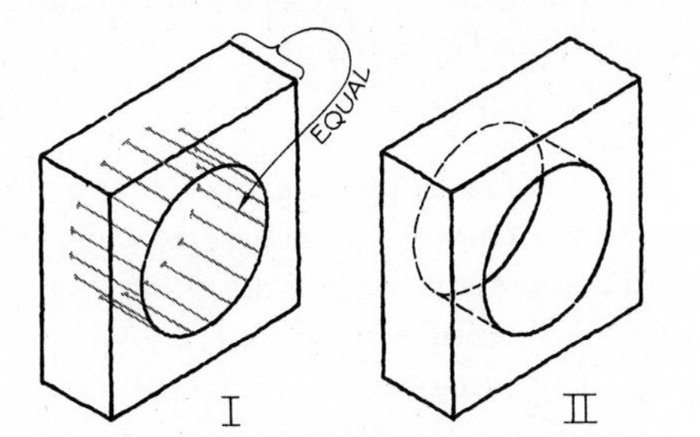

Fig. 5.24 **Isometric Ellipses.**

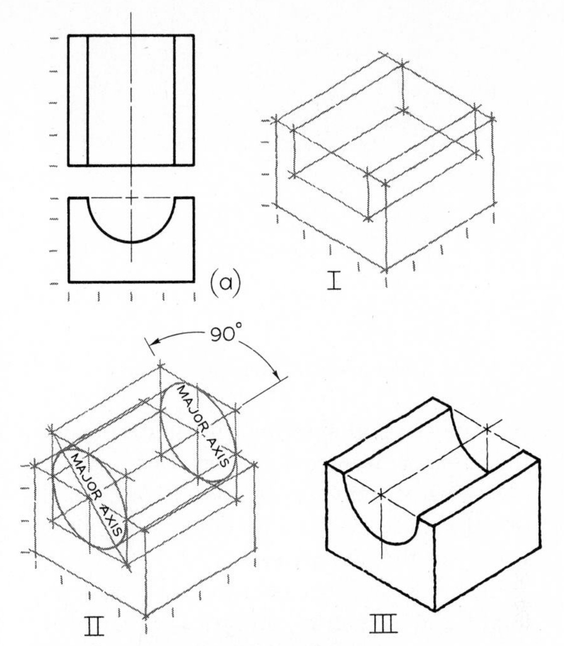

Fig. 5.25 **Sketching Semiellipses.**

II. Sketch the ellipse through the ends of the lines, as shown.

Two views of a Bearing with a semicylindrical opening are shown in Fig. 5.25 (a). The steps in sketching are:

I. Block in the object, including the rectangular space for the semicylinder.

II. Block in the box enclosing the complete cylinder. Sketch the entire cylinder lightly.

III. Dim all construction lines, and heavy in all final lines, showing only the lower half of the cylinder.

## 5.14 Sketching on Isometric Paper.

Two views of a Guide Block are shown in Fig. 5.26

Fig. 5.26 **Sketching on Isometric Paper.**

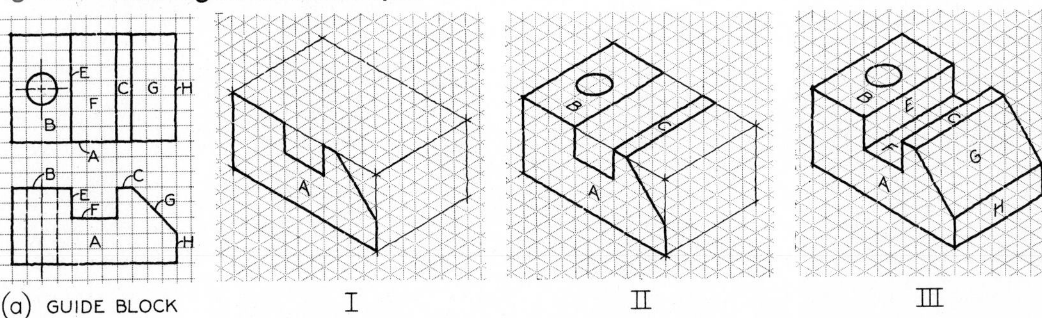

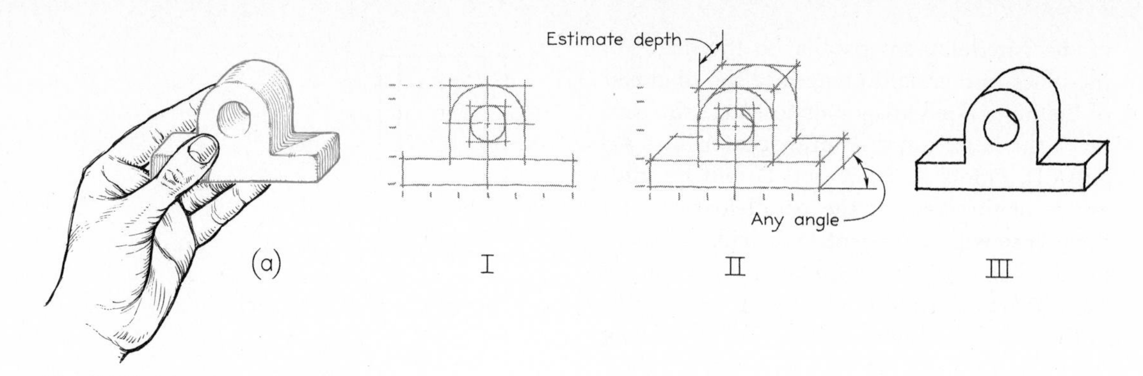


Fig. 5.27 **Sketching in Oblique.**

(a). The steps in sketching illustrate not only the use of isometric paper, but also the sketching of individual planes or faces of the object in order to build up pictorially a visualization of the given views.

I. Sketch isometric of enclosing box, counting off the isometric grid spaces to equal the corresponding squares on the given views. Sketch surface A, as shown.

II and III. Sketch additional surfaces B, C, E, and so forth, and the small ellipse, to complete the sketch.

## 5.15 Oblique Sketching.

Another simple method for sketching pictorially is *oblique* sketching, Fig. 5.27. Hold the object in your hand, as shown at (a).

I. Block in the front face of the object, as if you were sketching a front view.

II. Sketch receding lines parallel to each other and at any convenient angle, say 30° or 45° with horizontal, approximately. Cut off receding lines so that the depth appears correct. These lines may be full length, but a more natural appearance results if they are cut to three-quarters or one-half size, approximately. If they are full length, the sketch is a *cavalier* sketch. If half size, the sketch is a *cabinet* sketch. See §17.4.

III. Dim all construction and heavy in the final lines.

*Note:* Oblique sketching is a less suitable method for any object having circular shapes in or parallel to more than one plane of the object, because ellipses result when circular shapes are viewed obliquely. Therefore, place the object with most or all of the circular shapes toward you, so that they will appear as true circles and arcs in oblique sketching, as in Fig. 5.27.

## 5.16 Oblique Sketching on Cross-Section Paper.

Ordinary cross-section paper is suitable and convenient for oblique sketching. Two views of a Bearing Bracket are shown in Fig. 5.28 (a). The dimensions are determined simply by counting the squares.

Fig. 5.28 **Oblique Sketching on Cross-Section Paper.**

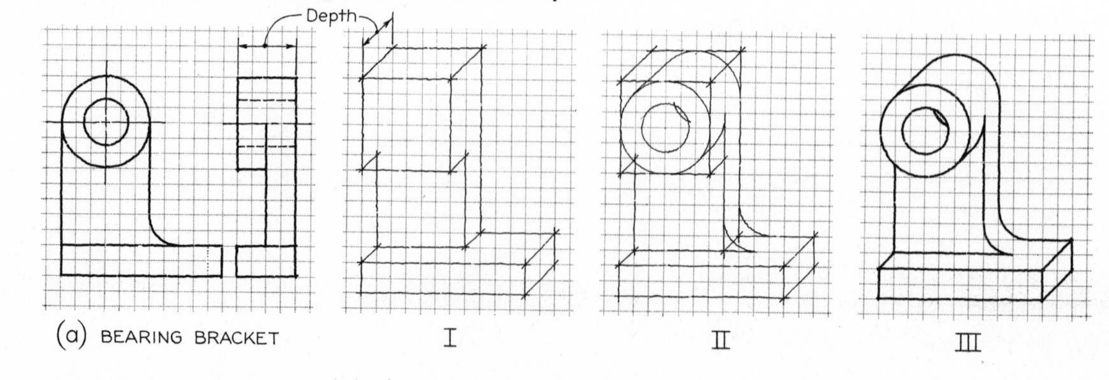

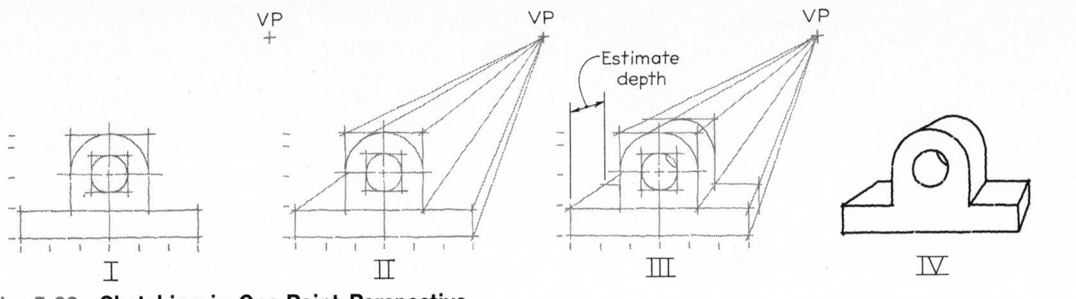


Fig. 5.29 **Sketching in One-Point Perspective.**

I. Sketch lightly the enclosing box construction. Sketch the receding lines at 45° diagonally through the squares. To establish the depth at a reduced scale, sketch the receding lines diagonally through half as many squares as the given number shown at (a).

II. Sketch all arcs and circles.

III. Heavy in all final lines.

## 5.17 Perspective Sketching.

The Bearing sketched in oblique in Fig. 5.27 can easily be sketched in *one-point perspective* (one vanishing point), as shown in Fig. 5.29.

I. Sketch the true front face of the object, just as in oblique sketching. Select the vanishing point (VP) for the receding lines. In most cases, it is desirable to place VP above and to the right of the picture, as shown, although it can be placed anywhere in the vicinity of the picture. But if it is placed too close to the center, the lines will converge too sharply, and the picture will be distorted.

II. Sketch the receding lines toward VP.

III. Estimate the depth to look well, and sketch in the back portion of the object. Note that the back circle and arc will be slightly smaller than the front circle and arc.

IV. Dim all construction with the Artgum or a soft eraser, and heavy in all final lines. Note the similarity between the perspective sketch and the oblique sketch in Fig. 5.27.

*Two-point perspective* (two vanishing points) is the most true to life of all pictorial methods, but requires some natural sketching ability or considerable practice for best results. A simple method is shown in Fig. 5.30 that can be used successfully by the nonartistic student:

Fig. 5.30 **Two-Point Perspective.**

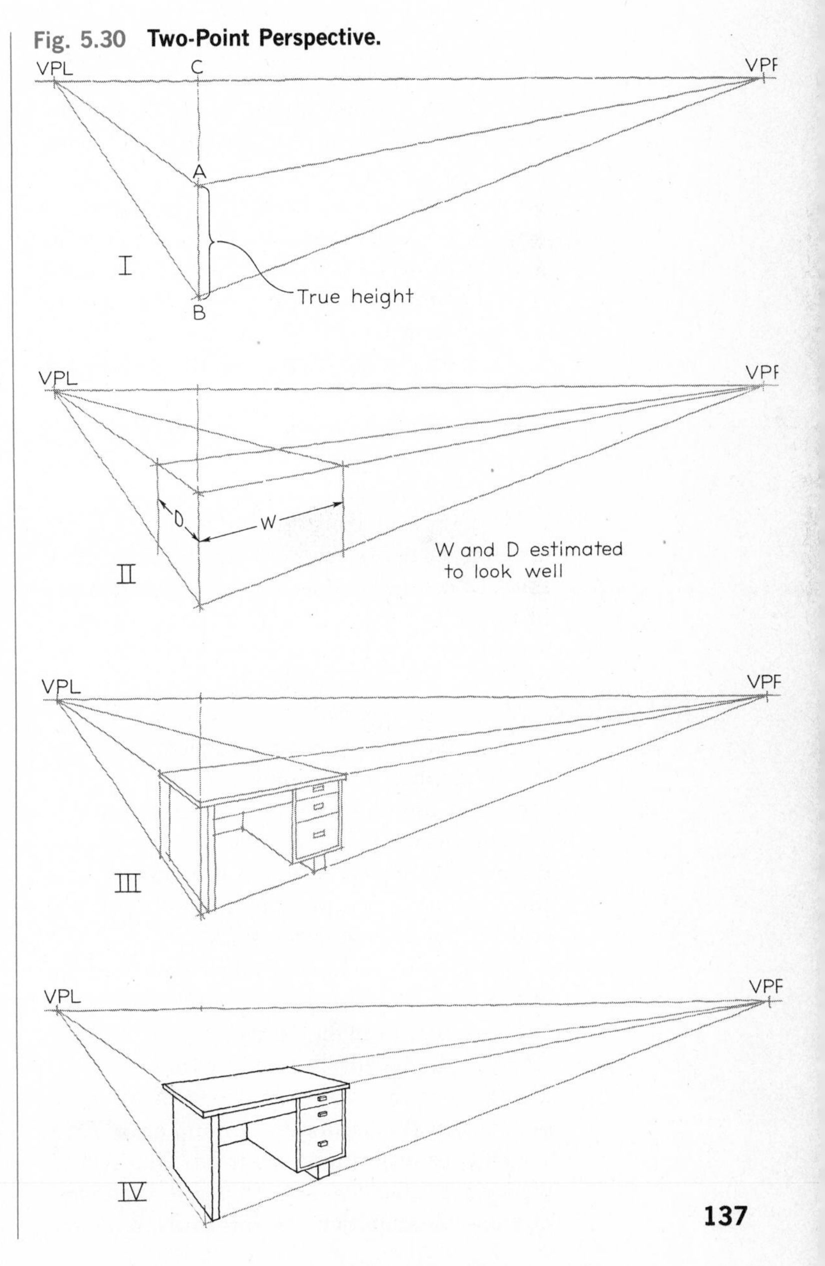

Fig. 5.31 **Front View of an Object.**

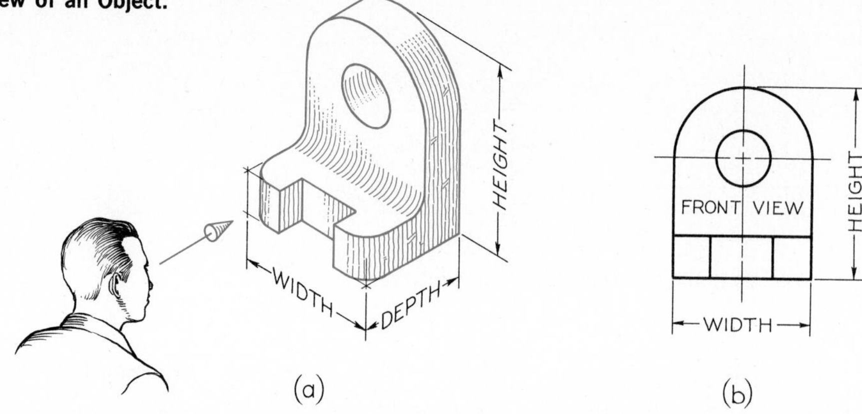


I. Sketch front corner of desk in true height, and locate two *vanishing points* (VPL and VPR) on a *horizon* line (eye level). The distance CA may vary—the greater it is, the higher the eye level will be and the more we will be looking down on top of the object. A good rule of thumb is to make C–VPL one third to one fourth of C–VPR.

II. Estimate depth and width, and sketch enclosing box.

III. Block in all details. Note that all parallel lines converge toward the same vanishing point.

IV. Dim the construction lines with the Artgum as necessary, and heavy in all final lines. Make the outlines thicker and the inside lines thinner, especially where they are close together.

## 5.18 Views of Objects.

A pictorial drawing or a photograph shows an object as it *appears* to the observer, but not as it *is*. Such a picture cannot describe the object fully, no matter from which direction it is viewed, because it does not show the exact shapes and sizes of the several parts.

In industry, a complete and clear description of the shape and size of an object to be made is necessary, to make certain that the object will be manufactured exactly as intended by the designer. In order to provide this information clearly and accurately, a number of *views*, systematically arranged, are used. This system of views is called *multiview projection*. Each view provides certain definite information if the view is taken in a direction perpendicular to a principal face or side of the object. For example, as shown in Fig. 5.31 (a), if the observer looks perpendicularly toward one face of the object, he obtains a true view of the shape and size of that side. This view as seen by the observer is shown at (b). (The observer is theoretically at an infinite distance from the object.)

An object has three principal dimensions: *width, height,* and *depth,* as shown at (a). In technical drawing, these fixed terms are used for dimensions taken in these directions, regardless of the shape of the object. The terms "length" and "thickness" are not used because they cannot be applied in all cases. Note at (b) that the front view shows only the height and width of the object and not the depth. In fact, *any one view of a three-dimensional object can show only two dimensions; the third dimension will be found in an adjacent view.*

## 5.19 Revolving the Object.

To obtain additional views, revolve the object as shown in Fig. 5.32. First, hold the object in the front-view position, as shown at (a).

To get the *top view,* (b), revolve the object so as to bring the *top of the object up and toward you.*

To get the *right-side view,* (c), revolve the object so as to bring the *right side to the right and toward you.*

To obtain views of any of the other sides, merely turn the object so as to bring those sides toward you.

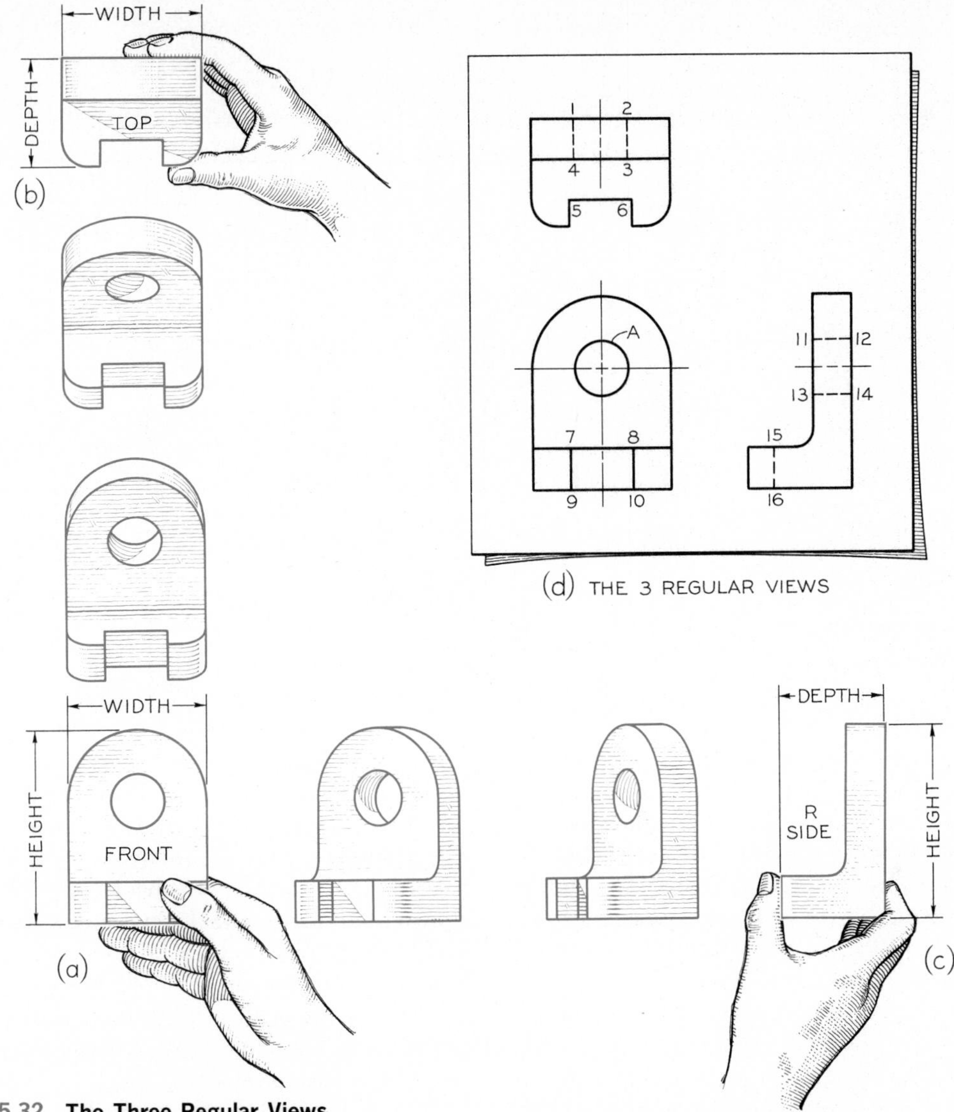


Fig. 5.32 **The Three Regular Views.**

The top, front, and right-side views, arranged closer together, are shown at (d). These are called the *three regular views,* because they are the views most frequently used.

At this stage we can consider spacing between views as purely a matter of appearance. The views should be spaced well apart and yet close enough to appear related to each other. The space between the front and top views may or may not be equal to the space between the front and side views. If dimensions (Chapter 11) are to be added to the sketch, sufficient space for them between views will have to be allowed.

An important advantage that a view has over a photograph of an object is that hidden features can be clearly shown by means of *hidden lines,* Fig. 2.15. In Fig. 5.32 (d), surface 7–8–9–10 in the front view appears as a visible line 5–6 in the top view, and as a hidden line 15–16 in the side view. Also, hole A, which appears as a circle in the front view, shows as hidden lines 1–4 and 2–3 in the top view, and 11–12 and 13–14 in the side view. For a complete discussion of hidden lines, see §5.25.

Also, at (d) note the use of center lines for the hole. See §5.26.

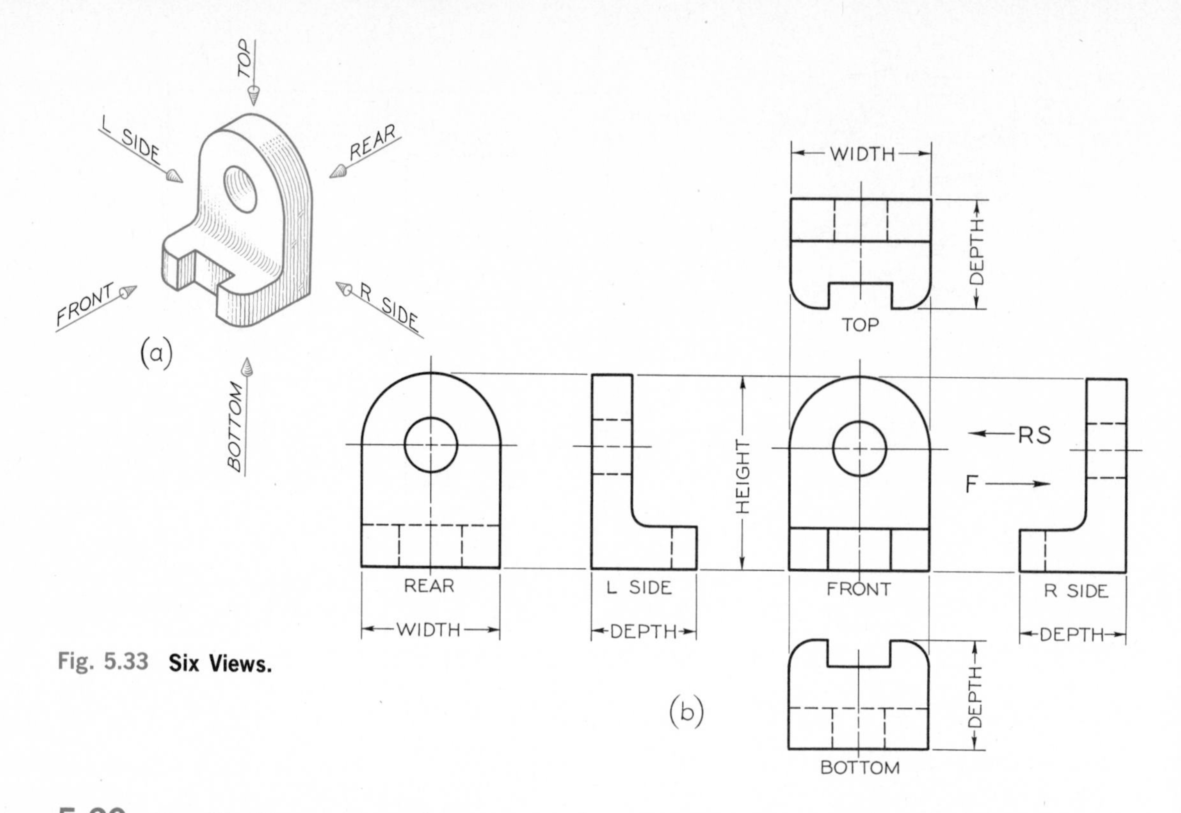


Fig. 5.33 **Six Views.**

## 5.20 The Six Views.

Any object can be viewed from six mutually perpendicular directions, as shown in Fig. 5.33 (a). Thus, six views may be drawn if necessary, as shown at (b). These six views are always arranged as shown,* which is the American National Standard arrangement of views. The *top, front,* and *bottom views* line up vertically, while the *rear, left-side, front,* and *right-side views* line up horizontally. To draw a view out of place is a very serious error, generally regarded as one of the worst mistakes one can make in this subject. See Fig. 5.47.

Note that the height is shown in the rear, left-side, front, and right-side views; the width is shown in the top, front, and bottom views; and the depth is shown in the four views that surround the front view, namely the left-side, top, right-side, and bottom views. In each view, two of the principal dimensions are shown, and the third is not shown. Observe also that in the four views that surround the front view, the front of the object is faced toward the front view.

*Except as explained in §6.8.

*Adjacent views are reciprocal.* If the front view, Fig. 5.33, is imagined to be the object itself, the right-side view is obtained by looking toward the right side of the front view, as shown by the arrow RS. Likewise, if the right-side view is imagined to be the object, the front view is obtained by looking toward the left side of the right-side view, as shown by the arrow F. The same relation exists between any two adjacent views.

Obviously, the six views may be obtained either by shifting the object with respect to the observer, as we have seen, Fig. 5.32, or by shifting the observer with respect to the object, Fig. 5.33. Another illustration of the second method is given in Fig. 5.34; showing six views of a house. The observer can walk around the house and view its front, sides, and rear, and he can imagine the top view as seen from an airplane and the bottom or "worm's-eye view" as seen from underneath.* Notice

*Architects usually draw the views of a building on separate sheets because of the large sizes of the drawings. When two or more views are drawn together, they are usually drawn in first-angle projection, §6.38.

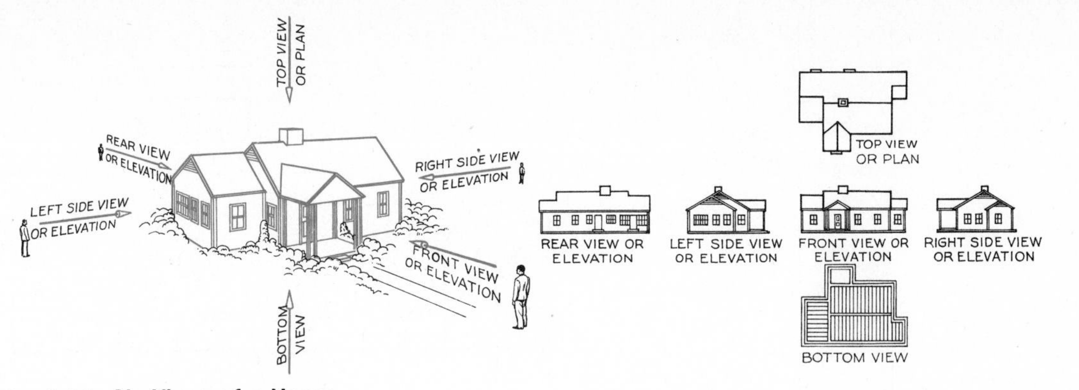


Fig. 5.34 **Six Views of a House.**

the use of the terms *plan,* for the top view, and *elevation,* for all views showing the height of the building. These terms are regularly used in architectural drawing and, occasionally, with reference to drawings in other fields.

## 5.21 Orientation of Front View.

Six views of an automobile are shown in Fig. 5.35. The view chosen for the front view in this case is the side, not the front of the automobile. In general, the front view should show the object in its operating position, particularly of familiar objects such as the house and automobile shown above. A machine part is often drawn in the position it occupies in the assembly. However, in most cases this is not important, and the draftsman may assume the object to be in any convenient position. For example, an automobile connecting rod is usually drawn horizontally on the sheet, Fig. 14.37. Also, it is customary to draw screws, bolts, shafts, tubes, and other elongated parts in a horizontal position, not only because they are usually manufactured in this position but also because they can be presented more satisfactorily on paper in this position.

## 5.22 Choice of Views.

*A drawing for use in production should contain only those views needed for a clear and complete shape description of the object.* These minimum required views are referred to as the *necessary views.* In selecting views, the draftsman should choose those that show best the essential contours or shapes and

Fig. 5.35 **Six Views of an Automobile.**

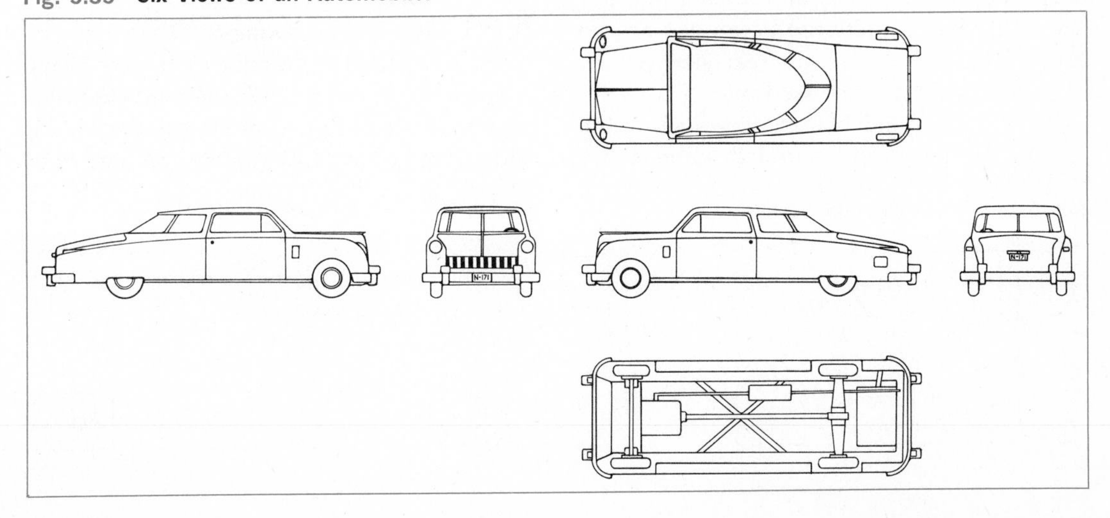

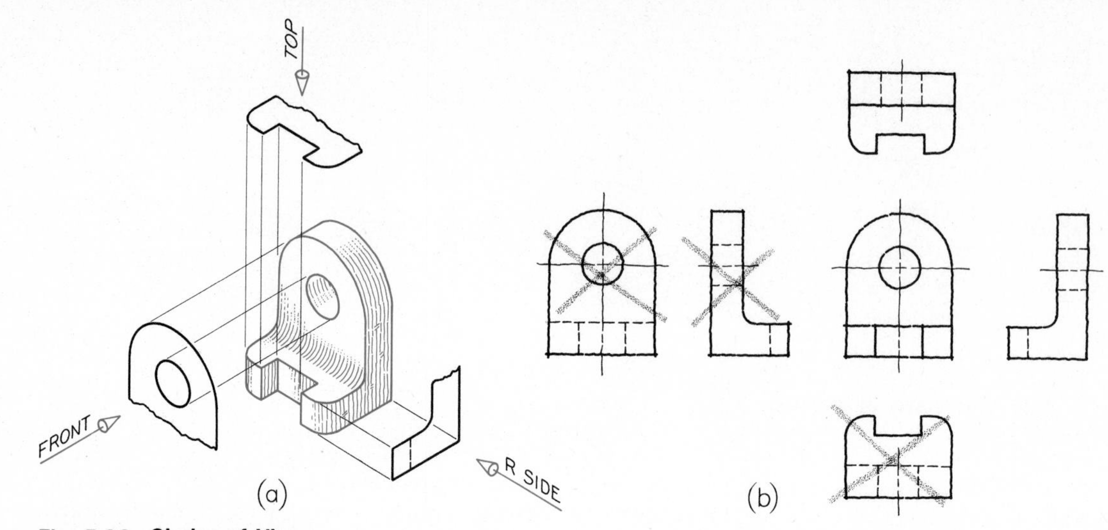


Fig. 5.36 **Choice of Views.**

should give preference to those with the least number of hidden lines.

As shown in Fig. 5.36 (a), there are three distinctive features of this object that need to be shown on the drawing:

1. Rounded top and hole, seen from the front.
2. Rectangular notch and rounded corners, seen from the top.
3. Right angle with filleted corner, seen from the side.

Another way to choose necessary views is to eliminate unnecessary views. At (b) a "thumbnail sketch" of the six views is shown. Both the front and rear views show the true shapes of the hole and the rounded top, but the front view is preferred because it has no hidden lines. Therefore, the rear view (which is seldom needed) is crossed out.

Both the top and bottom views show the rectangular notch and rounded corners, but the top view is preferred because it has fewer hidden lines.

Both the right-side and left-side views show the right angle with the filleted corner. In fact, in this case the side views are identical, except reversed. In such instances, it is customary to choose the right-side view.

The necessary views, then, are the three remaining views: the top, front, and right-side views. These are the three regular views referred to in connection with Fig. 5.32.

More complicated objects may require more than three views, or in many cases special views such as partial views, §6.9; sectional views, Chapter 7; auxiliary views, Chapter 8.

## 5.23 Two-View Drawings.

Often only two views are needed to describe clearly the shape of an object. In Fig. 5.37 (a), the right-side view shows no significant contours of the object, and is crossed out. At (b) the top and front

Fig. 5.37 **Two Necessary Views.**

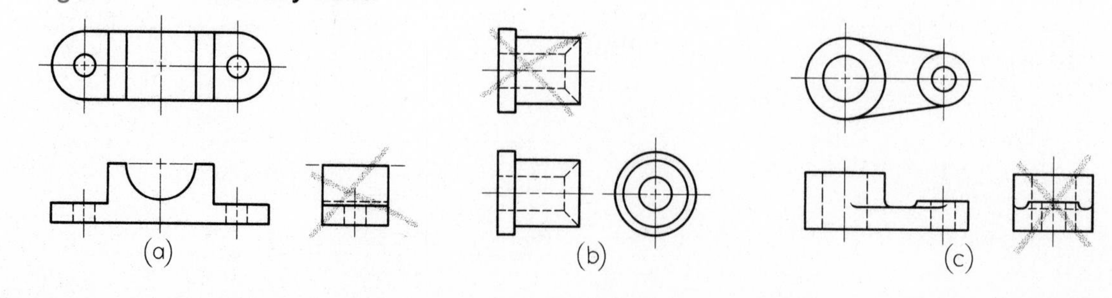

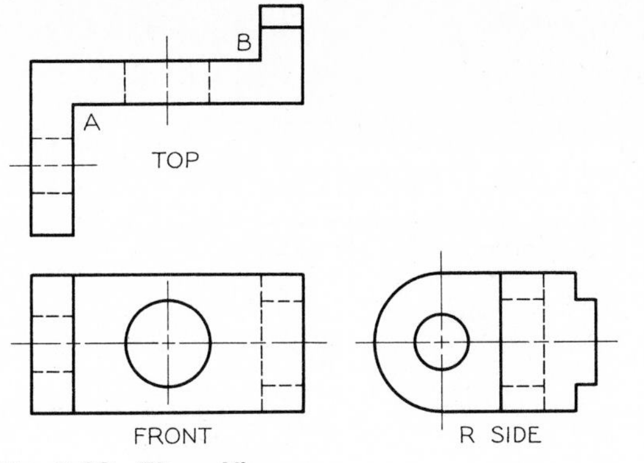


Fig. 5.38 **Three Views.**

views are identical, so the top view is eliminated. At (c), no additional information not already given in the front and top views is shown in the side view, so the side view is unnecessary.

The question often arises: What are the absolute minimum views required? For example, in Fig. 5.38, the top view might be omitted, leaving only the front and right-side views. However, it is more difficult to "read" the two views or visualize the object, because the characteristic "Z" shape of the top view is omitted. In addition, one must assume that

Fig. 5.39 **Choice of Right-Side View.**

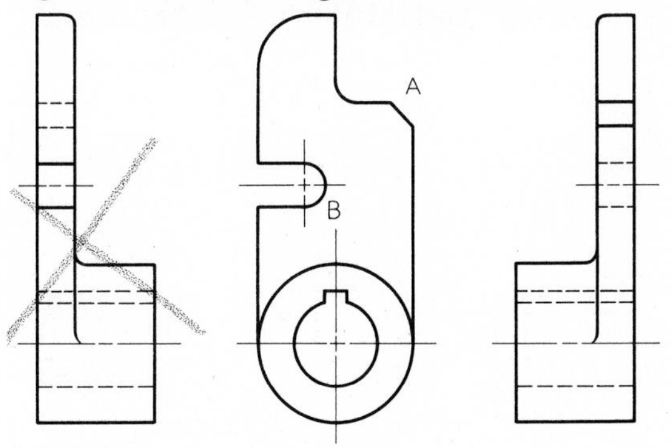


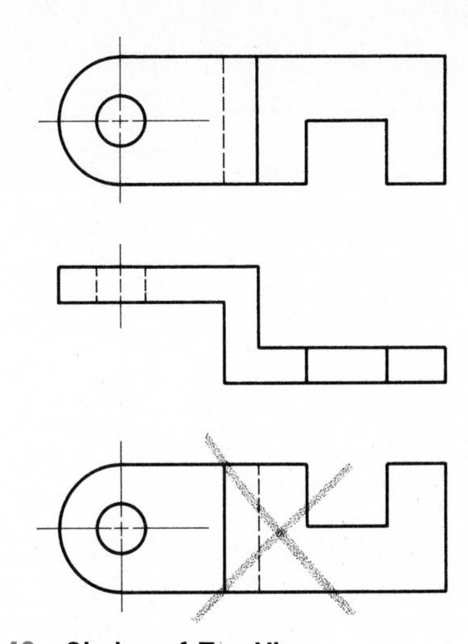

Fig. 5.40 **Choice of Top View.**

corners A and B (top view) are square and not filleted. In this example, all three views are necessary.

If the object requires only two views, and the left-side and right-side views are equally descriptive, the right-side view is customarily chosen, Fig. 5.39. If contour A were omitted, then the presence of slot B would make it necessary to choose the left-side view in preference to the right-side view.

If the object requires only two views, and the top and bottom views are equally descriptive, the top view is customarily chosen, Fig. 5.40.

If only two views are necessary, and the top view and right-side view are equally descriptive, the combination chosen is that which spaces best on the paper, Fig. 5.41.

## 5.24 One-View Drawings.

Frequently a single view supplemented by a note or lettered symbols is sufficient to describe clearly the

Fig. 5.41 **Choice of Views to Fit Paper.**

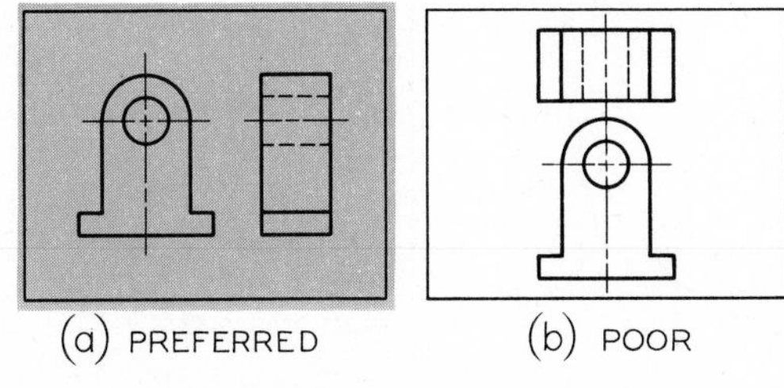

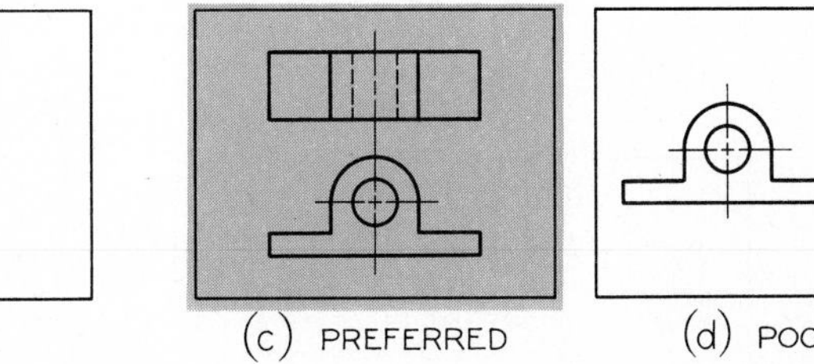

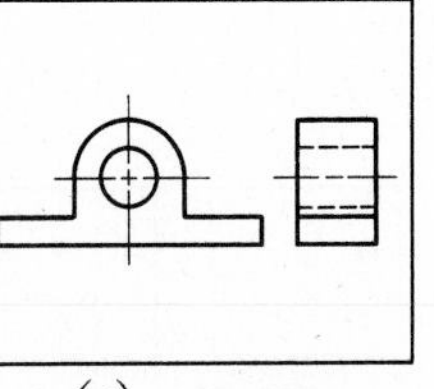

(a) PREFERRED (b) POOR (c) PREFERRED (d) POOR

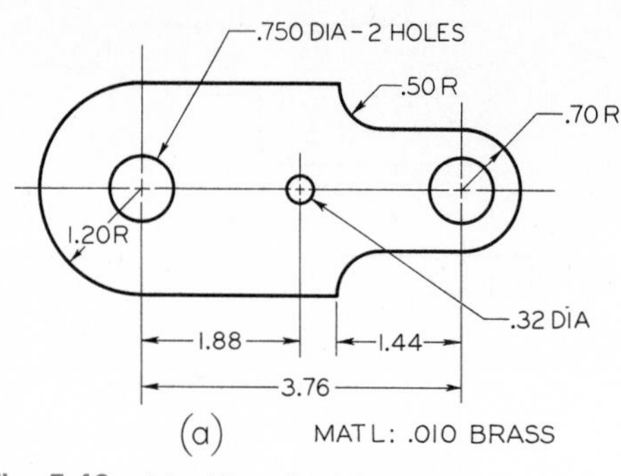


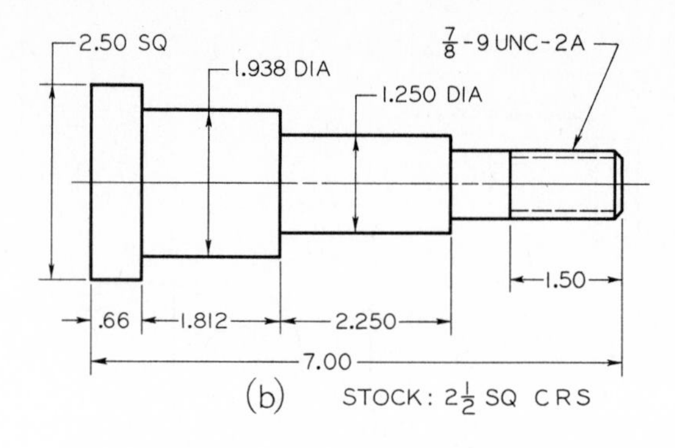


Fig. 5.42 **One-View Drawings.**

shape of a relatively simple object. In Fig. 5.42 (a), one view of the Shim, plus a note indicating the thickness as .010″, is sufficient. At (b), the left end is 2.50″ square, the next portion is 1.938″ diameter, the next is 1.250″ diameter, and the portion with the thread is $\frac{7}{8}''$ diameter, as indicated in the note. Nearly all shafts, bolts, screws, and similar parts can and should be represented by single views in this manner.

## 5.25 Hidden Lines.

Correct and incorrect practices in drawing hidden lines are illustrated in Fig. 5.43. In general, a hidden line should join a visible line except when it causes the visible line to extend too far, as shown at (a). In other words, *leave a gap whenever a hidden-line dash forms a continuation of a visible line.* Hidden lines should intersect to form L and T corners, as shown at (b). A hidden line preferably should "jump" a visible line when possible, (c). Parallel hidden lines should be drawn so that the dashes are staggered, in a manner similar to bricklaying, as at (d). When two or three hidden lines meet at a point, the dashes should join, as shown for the bottom of the drilled hole at (e), and for the top of a countersunk hole, (f). The example at (g) is similar to (a) in that hidden lines should not join visible lines when it makes the visible line extend too far. Correct and incorrect methods of drawing hidden arcs are shown at (h).

Poorly drawn hidden lines can easily spoil a drawing. Each dash should be carefully

Fig. 5.43 **Hidden-Line Practices.**

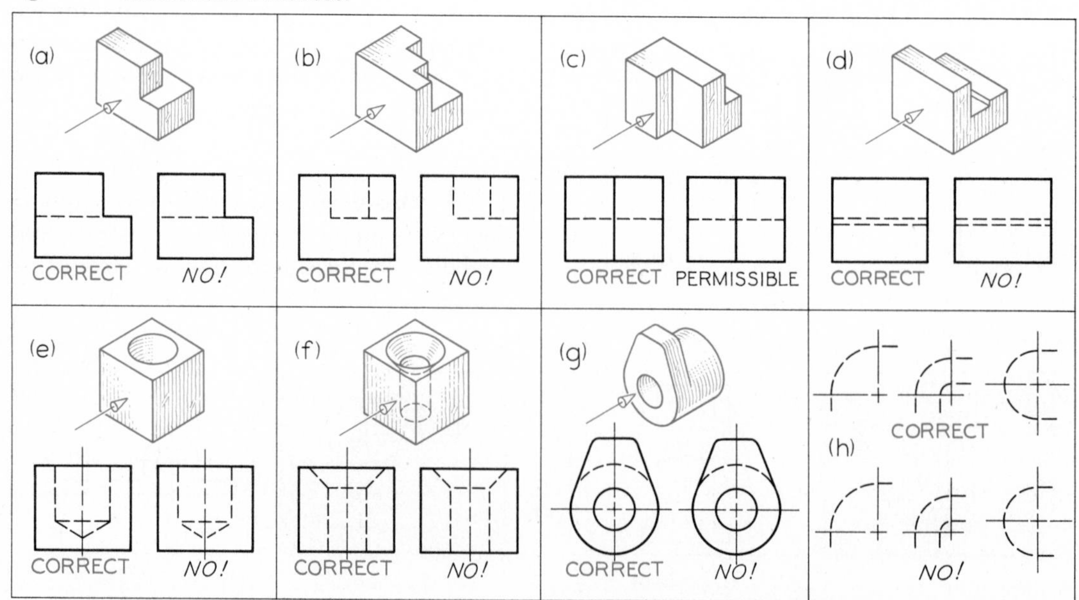

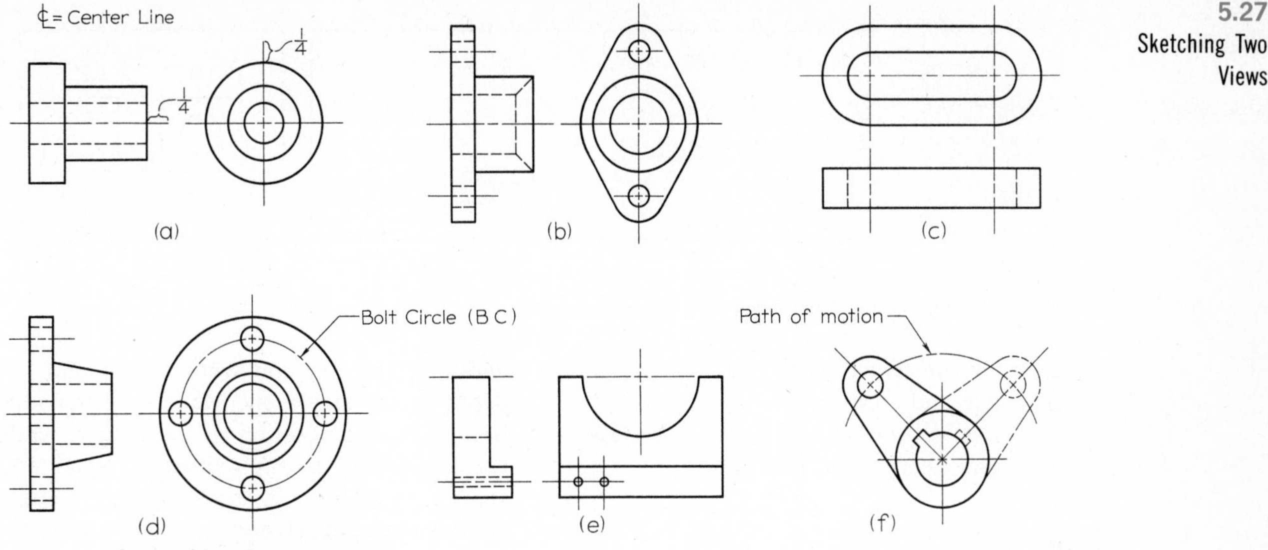


Fig. 5.44 **Center-Line Applications.**

drawn about $\frac{1}{8}''$ long and spaced only about $\frac{1}{32}''$ apart, by eye. Accent the beginning and end of each dash by pressing down on the pencil, whether drawn freehand or mechanically.

In general, views should be chosen that show features with visible lines, so far as possible. After this has been done, hidden lines should be used wherever necessary to make the drawing clear. Where they are not needed for clearness, hidden lines should be omitted, so as not to clutter the drawing any more than necessary and in order to save time. The beginner, however, would do well to be cautious about leaving out hidden lines until experience shows him when they can be safely omitted.

## 5.26 Center Lines.

*Center lines* (symbol: ℄ ) are used to indicate axes of symmetrical objects or features, bolt circles, and paths of motion. Typical applications are shown in Fig. 5.44. As shown at (a), a single center line is drawn in the longitudinal view and crossed center lines in the circular view. The small dashes should cross at the intersections of center lines. Center lines should extend uniformly about $\frac{1}{4}''$ outside the feature for which they are drawn.

The long dashes of center lines may vary from $\frac{3}{4}''$ to $1\frac{1}{2}''$ or more in length, depending upon the size of the drawing. The short dashes should be about $\frac{1}{8}''$ long, with spaces about $\frac{1}{16}''$. Center lines should always start and end with long dashes. Short center lines, especially for small holes, as at (e), may be made solid as shown. Always leave a gap as at (e) when a center line forms a continuation of a visible or hidden line. Center lines should be thin enough to contrast well with the visible and hidden lines, but dark enough to reproduce well.

Center lines are useful mainly in dimensioning and should be omitted from unimportant rounded or filleted corners and other shapes that are self-locating.

## 5.27 Sketching Two Views.

The Support Block in Fig. 5.45 (a) requires only two views. The steps in sketching are:

I. Block in lightly the enclosing rectangles for the two views. Sketch horizontal lines 1 and 2 to establish the height of the object, while making spaces A approximately equal. Sketch vertical lines 3, 4, 5, and 6 to establish the width and depth in correct proportion to the already established height, while making spaces B approximately equal, and space C equal to or slightly less than space B.

II. Block in smaller details, using diagonals

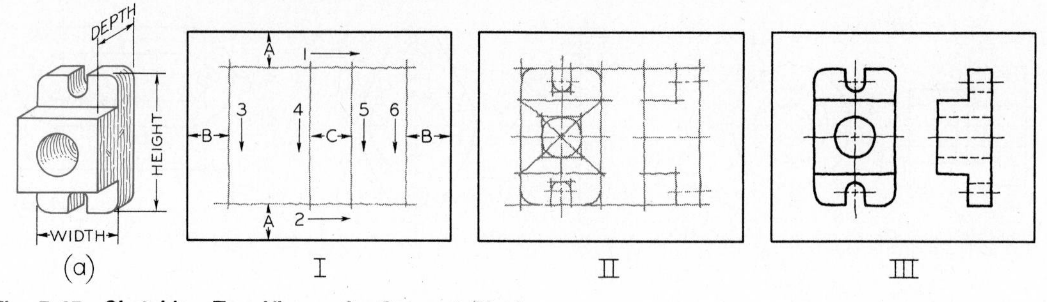


Fig. 5.45 **Sketching Two Views of a Support Block.**

to locate the center, as shown. Sketch lightly the circle and arcs.

III. Dim all construction lines with the Artgum, and heavy in all final lines.

## 5.28 Sketching Three Views.

A Lever Bracket requiring three views is shown in Fig. 5.46 (a). The steps in sketching the three views are as follows:

I. Block in the enclosing rectangles for the three views. Sketch horizontal lines 1, 2, 3, and 4 to establish the height of the front view and the depth of the top view, while making spaces A approximately equal, and space C equal to or slightly less than space A. Sketch vertical lines 5, 6, 7, and 8 to establish the width of the top and front views, and the depth of the side view. Make sure that this is in correct proportion to the height, while making spaces B approximately equal and space D equal to or slightly less than one space B. Note that spaces C and D are not necessarily equal, but are independent of each other. Similarly, spaces A and B are not necessarily equal. To transfer the depth dimension from the top view to the side view, use the edge of a card or strip of paper, as shown; or transfer the distance by using the pencil as a measuring stick, as shown in Fig. 5.10 (b) and (c). *Note that the depth in the top and side views must always be equal.*

II. Block in all details lightly.

III. Sketch all arcs and circles lightly.

Fig. 5.46 **Sketching Three Views of a Lever Bracket.**

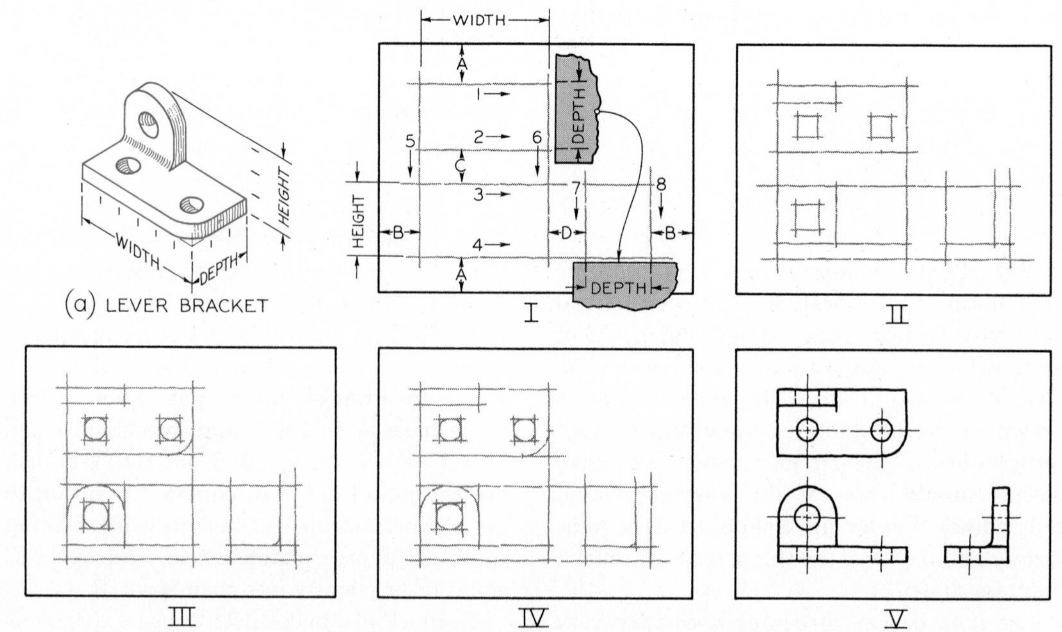

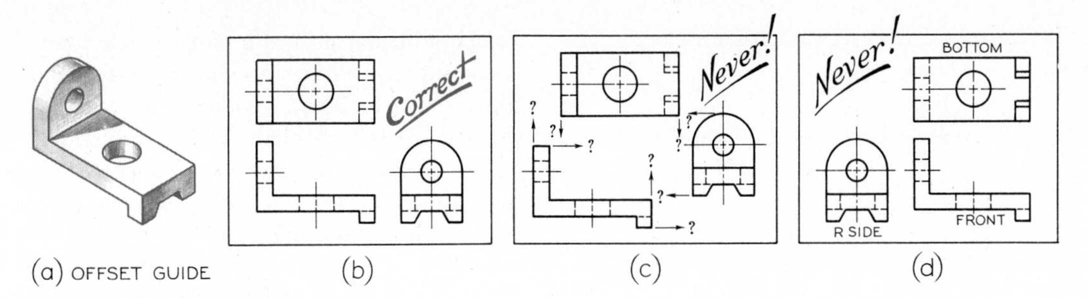

Fig. 5.47 **Position of Views.**

IV. Dim all construction lines with Artgum.

V. Heavy in all final lines so that the views will stand out clearly.

## 5.29 **Alignment of Views.**

Errors in arranging the views are so commonly made by students that it is necessary to repeat: the views must be drawn in accordance with the American National Standard arrangement, Fig. 5.33. In Fig. 5.47 (a) an Offset Guide is shown that requires three views. These three views, correctly arranged, are shown at (b). The top view must be directly above the front view, and the right-side view directly to the right of the front view—not out of alignment, as at (c). Also, never draw the views in reversed positions, with the bottom over the front view, or the right-side to the left of the front view, as shown at (d), even though the views do line up with the front view.

## 5.30 **Meaning of Lines.**

A visible line or a hidden line has three possible meanings, Fig. 5.48: (1) intersection of two surfaces, (2) edge view of a surface, and (3) contour view of a curved surface. Since *no shading is used on a working drawing,* it is necessary to examine all the views to determine the meaning of the lines. For example, the line AB at the top of the front view might be regarded as the edge view of a flat surface if we look at only the front and top views and do not observe the curved surface on top of the object as shown in the right-side view. Similarly, the vertical line CD in the front view might be regarded as the edge view of a plane surface if we look at only the front and side views. However, the top view shows that the line represents the intersection of an inclined surface.

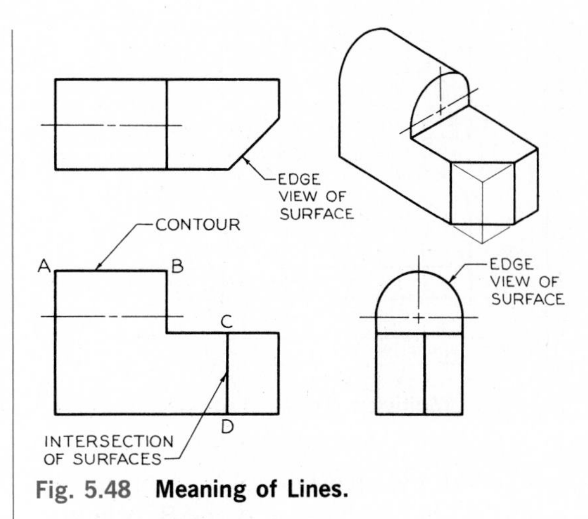

Fig. 5.48 **Meaning of Lines.**

## 5.31 **Precedence of Lines.**

Visible lines, hidden lines, and center lines often coincide on a drawing, and it is necessary for the draftsman to know which line to show. As shown in Fig. 5.49, a visible line always takes

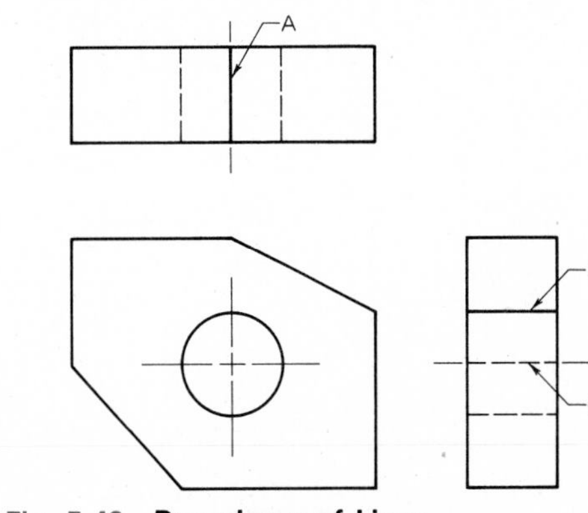

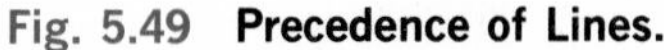
Fig. 5.49 **Precedence of Lines.**

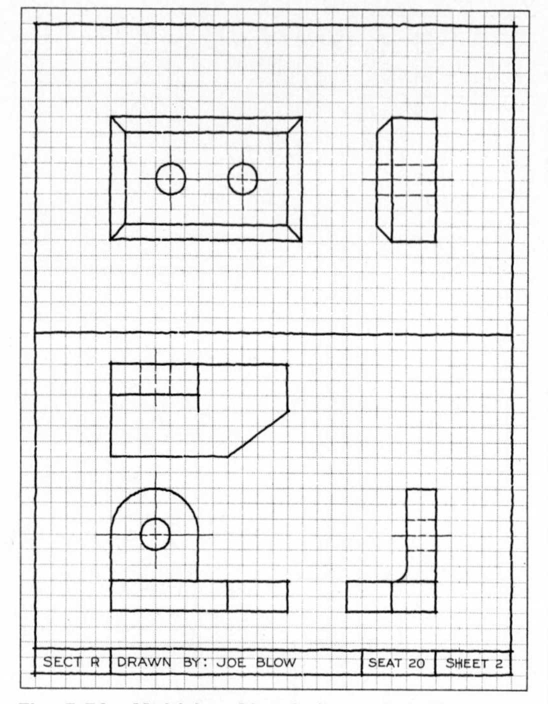


Fig. 5.50 **Multiview Sketch** (Layout A–1).

precedence over (covers up) a center line or a hidden line, as shown at A and B. A hidden line always takes precedence over a center line, as at C. Note that at A and C the ends of the center line are shown, but are separated from the view by short gaps.

## 5.32 Sketching Problems.

Figures 5.51 and 5.52 present a variety of objects from which the student is to sketch the necessary views. Using $8\frac{1}{2}'' \times 11''$ cross-section paper, sketch a border and title strip and divide the sheet into two parts as shown in Fig. 5.50. Sketch two assigned problems per sheet, as shown. On the problems in Fig. 5.51, "ticks" are given that indicate $\frac{1}{2}''$ or $\frac{1}{4}''$ spaces. Thus, measurements may be easily spaced off on cross-section paper having $\frac{1}{8}''$ or $\frac{1}{4}''$ grid spacings.

On the problems in Fig. 5.52, no indications of size are given. The student is to sketch the necessary views of assigned problems to fit the spaces comfortably, about as shown in Fig. 5.50. It is suggested that the student prepare a small paper scale, making the divisions equal to those on the paper scale in Prob. 1. This scale can be used to determine the approximate sizes. Let each division equal $\frac{1}{2}''$ on your sketch.

Missing-line and missing-view problems are given in Figs. 5.53 and 5.54, respectively. These are to be sketched, two problems per sheet, in the arrangement shown in Fig. 5.50. If the instructor so assigns, the missing lines or views may be sketched with a colored pencil. The problems given in Figs. 5.53 and 5.54 may be sketched in isometric on isometric paper or in oblique on cross-section paper.

Sketching problems in convenient form for solution are available in *Technical Drawing Problems, Series 1,* by Giesecke, Mitchell, Spencer, and Hill; *Technical Drawing Problems, Series 2,* by Spencer and Hill; and *Technical Drawing Problems, Series 3,* by Spencer and Hill, all designed to accompany this text and published by Macmillan Publishing Co., Inc.

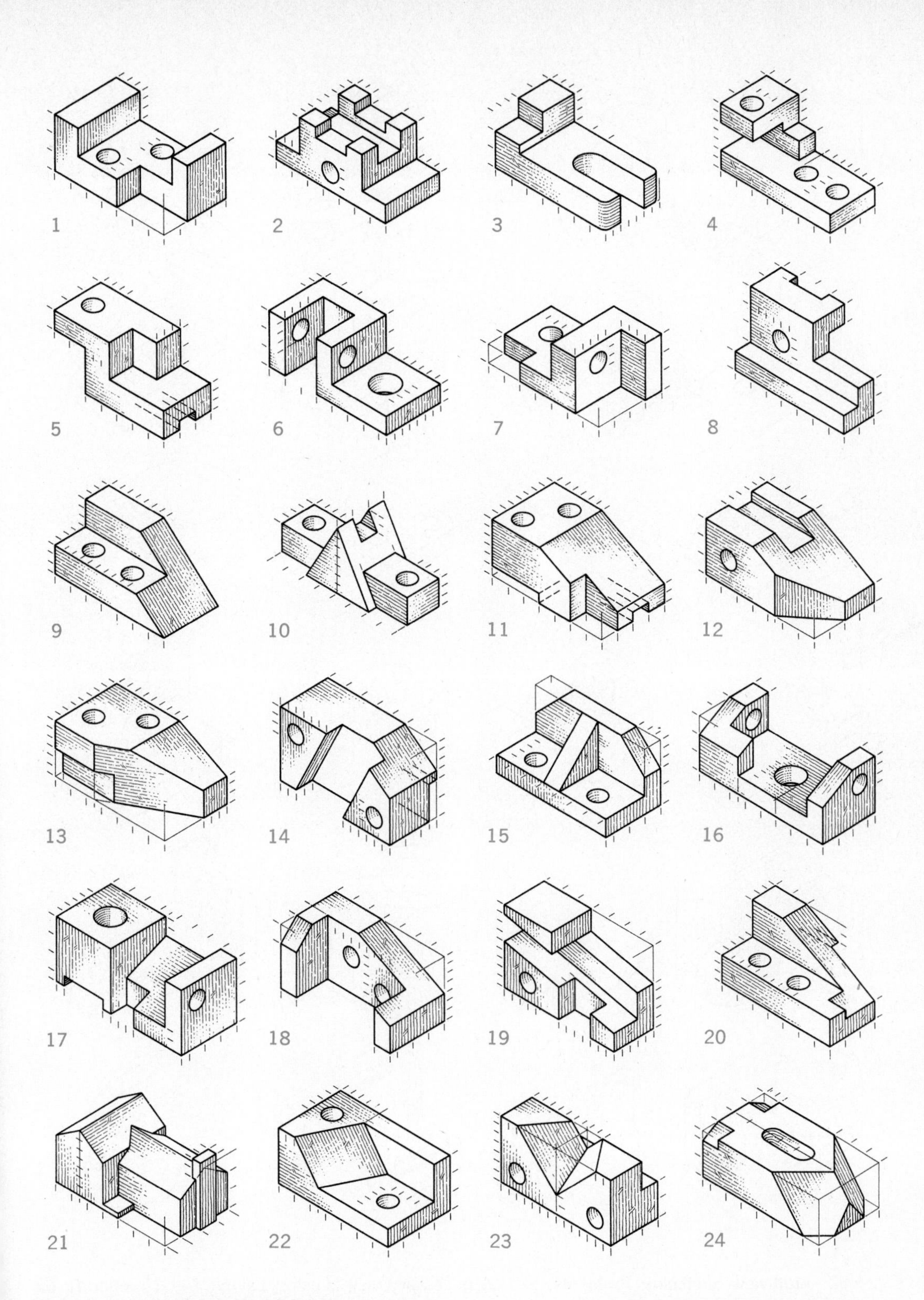


Fig. 5.51 **Multiview Sketching Problems.** Sketch necessary views, using Layout A–1 (freehand), on cross-section paper or plain paper, two problems per sheet as in Fig. 5.50. The units shown are $\frac{1}{2}''$ or $\frac{1}{4}''$. See §5.32. All holes are through holes.

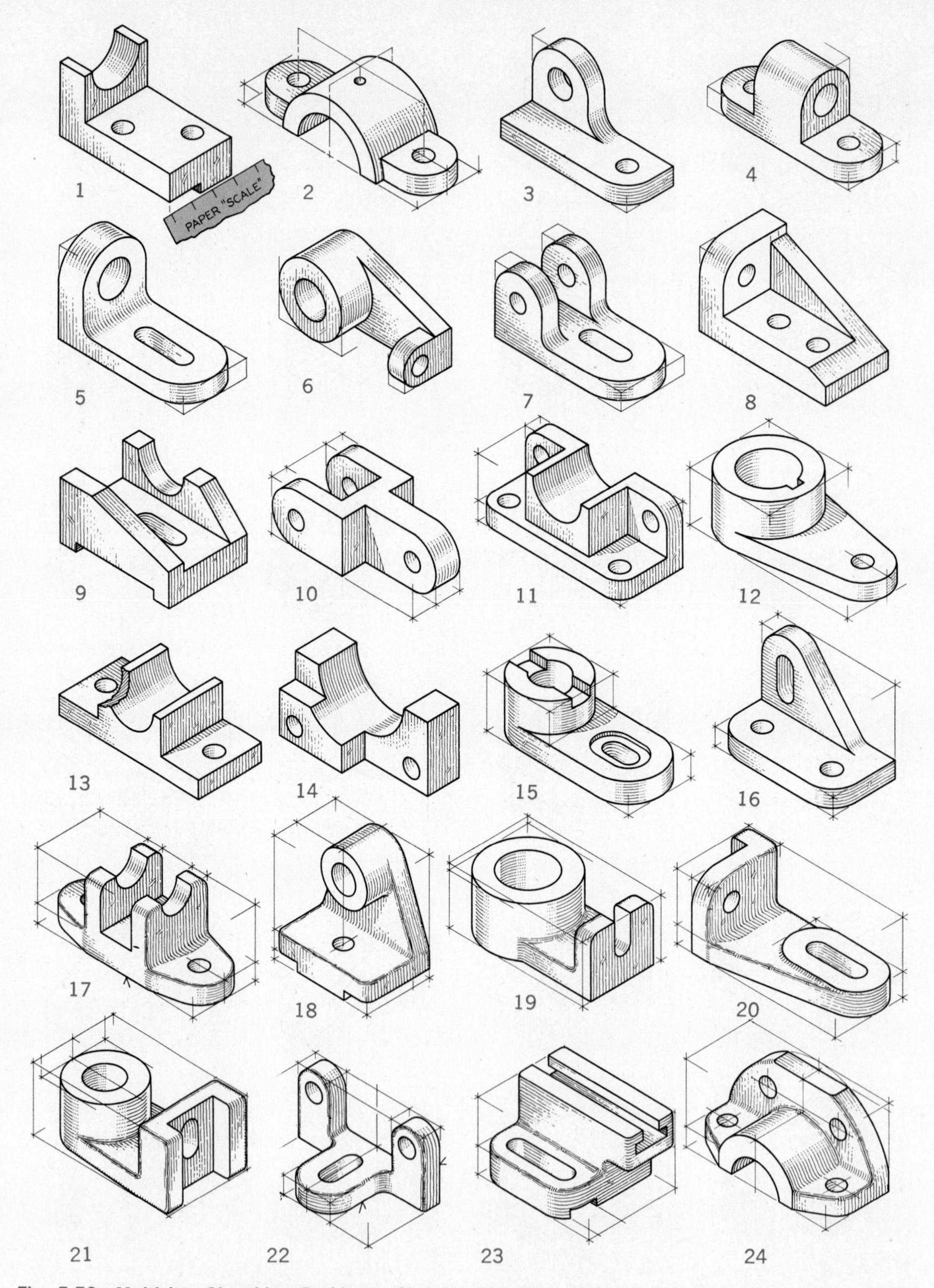


Fig. 5.52 **Multiview Sketching Problems.** Sketch necessary views, using Layout A–1 (freehand), on cross-section paper or plain paper, two problems per sheet as in Fig. 5.50. Prepare paper scale with divisions equal to those in Prob. 1, and apply to problems to obtain approximate sizes. Let each division = $\frac{1}{2}''$ on your sketch. For Probs. 17 to 24, study §§6.34 to 6.36.

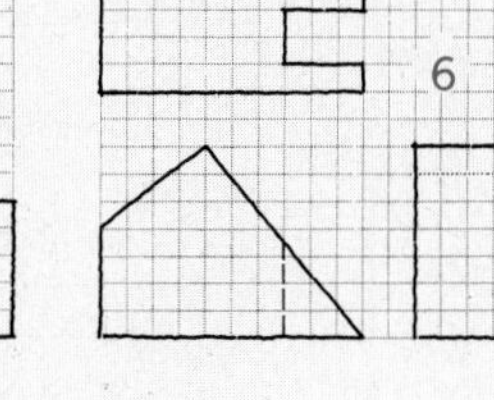

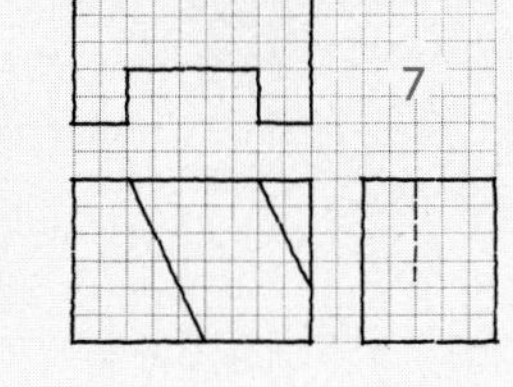

*Probs. 1–5: No inclined or oblique surfaces.*

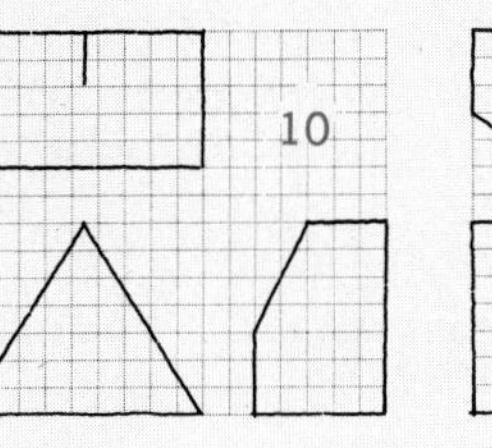

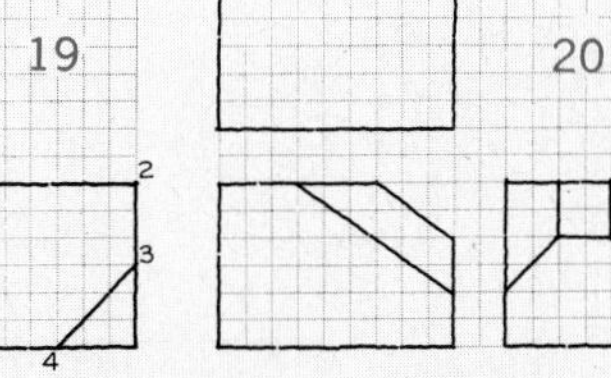

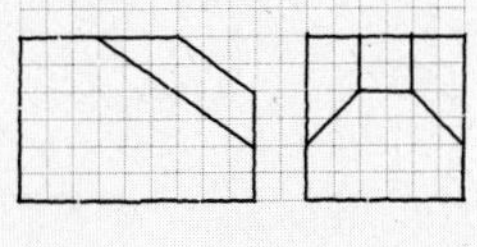

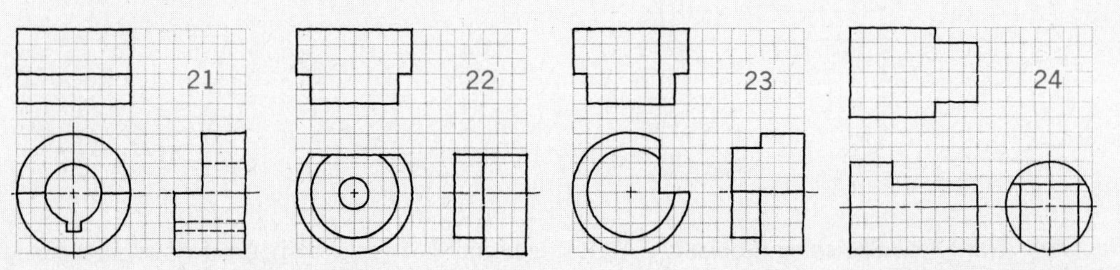

Fig. 5.53 **Missing-Line Sketching Problems.** (1) Sketch given views, using Layout A–1 (freehand), on cross-section paper or plain paper, two problems per sheet as in Fig. 5.50. Add missing lines. The squares are each $\frac{1}{4}''$. (2) Sketch in isometric on isometric paper or in oblique on cross-section paper.

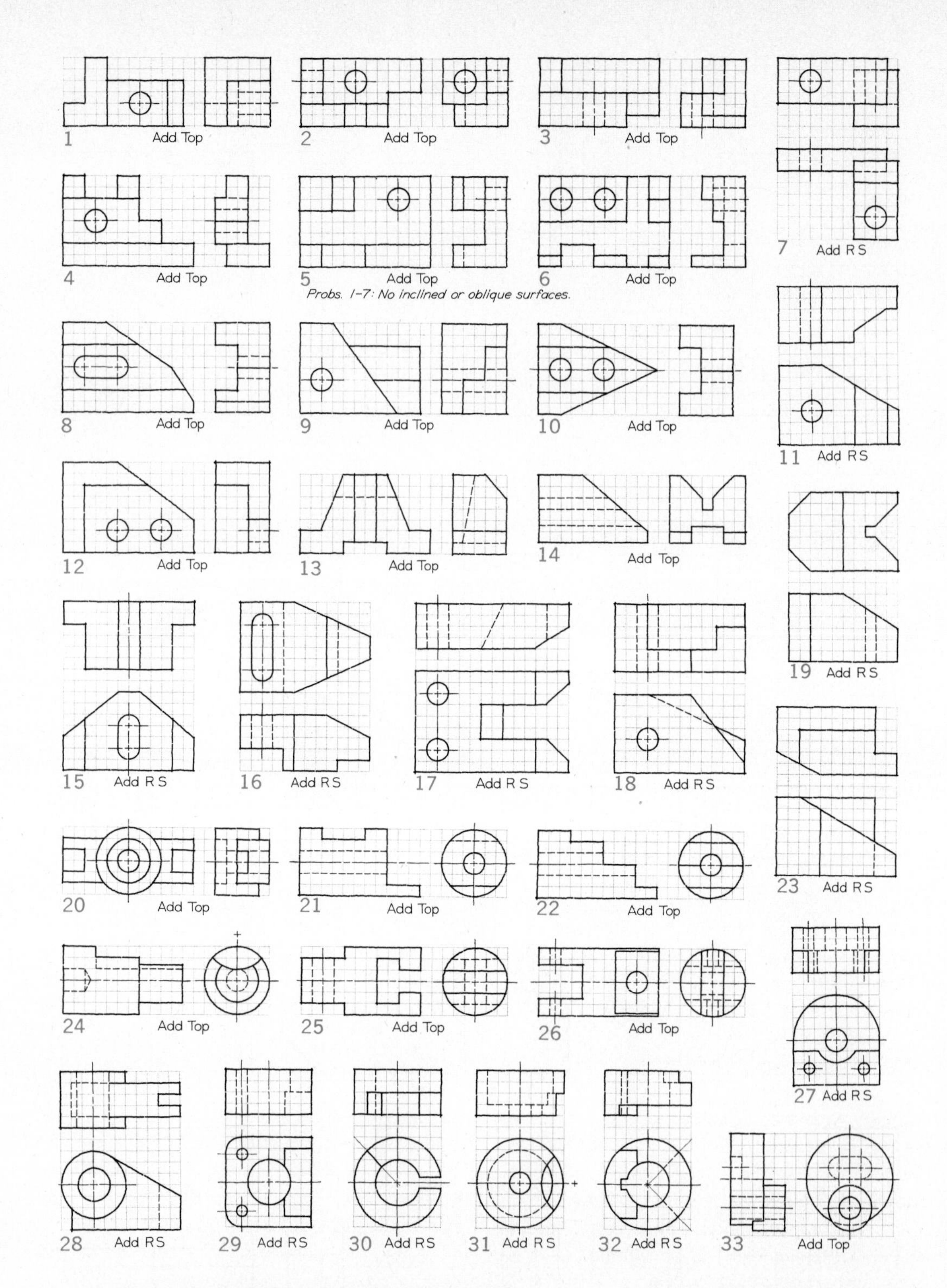


Fig. 5.54 **Third-View Sketching Problems.** (1) Using Layout A–1 (freehand), on cross-section paper or plain paper, two problems per sheet as in Fig. 5.50, sketch the two given views and add the missing views, as indicated. The given views are either front and right-side views or front and top views. Hidden holes with center lines are drilled holes. (2) Sketch in isometric on isometric paper or in oblique on cross-section paper.

# 6 multiview projection

**6.1 Multiview Projection.*** A view of a part for a design is known technically as a *projection.* A projection is a view conceived to be drawn or projected onto a plane known as the *plane of projection.*

The method of viewing the part to obtain a *multiview projection,* in this case a front view, is shown in Fig. 6.1 (a). Between the observer and the part, a transparent plane or pane of glass representing a plane of projection is placed parallel to the front surfaces of the part. Shown on the pane of glass in outline is how the design appears to the observer. Theoretically, the observer is at an infinite distance from the part or object, so that the *lines of sight* are parallel.

*See ANSI Y14.3-1957.

In more precise terms, this view is obtained by drawing perpendiculars, called *projectors,* from all points on the edges or contours of the part or object to the plane of projection, (b). The collective piercing points of these projectors, being infinite in number, form lines on the pane of glass, as shown at (c).

Thus, as shown at (c), a projector from point 1 on the object pierces the plane of projection at point 7, which is a view or projection of the point. The same procedure applies to point 2, whose projection is point 9. Since 1 and 2 are end points of a straight line on the object, the projections 7 and 9 are joined to give the projection of the line 7-9. Similarly, if the projections of the four corners 1, 2, 3, and 4 are found, the projections 7, 9, 10, and 8 may be joined by straight lines to form the projection of the rectangular surface.

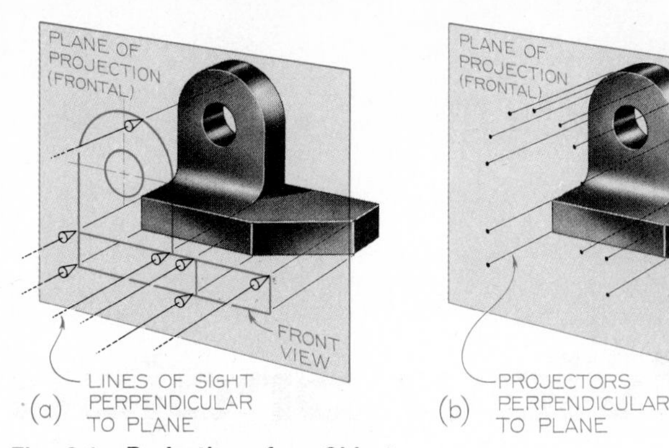


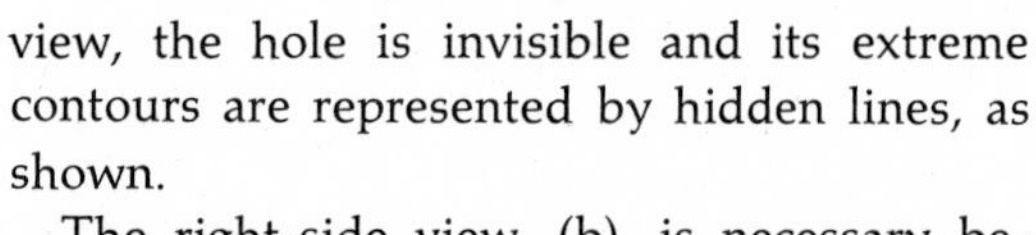


**Fig. 6.1 Projection of an Object.**

The same procedure can be applied to curved lines—for example, the top curved contour of the object. A point, 5, on the curve is projected to the plane at 6. The projection of an infinite number of such points, a few of which are shown at (b), on the plane of projection results in the projection of the curve. If this procedure of projecting points is applied to all edges and contours of the object, a complete view or projection of the object results. This view is necessary in the shape description because it shows the true curvature of the top and the true shape of the hole.

A similar procedure may be used to obtain the top view, Fig. 6.2 (a). This view is necessary in the shape description because it shows the true angle of the inclined surface. In this view, the hole is invisible and its extreme contours are represented by hidden lines, as shown.

The right-side view, (b), is necessary because it shows the right-angled characteristic shape of the object and shows the true shape of the curved fillet. Note how the cylindrical contour on top of the object appears when viewed from the side. The extreme or contour element 1–2 on the object is projected to give the line 3–4 on the view. The hidden hole is also represented by projecting the extreme elements.

The plane of projection upon which the front view is projected is called the *frontal plane,* that upon which the top view is projected, the *horizontal plane,* and that upon which the side view is projected, the *profile plane.*

**Fig. 6.2 Top and Right-Side Views.**



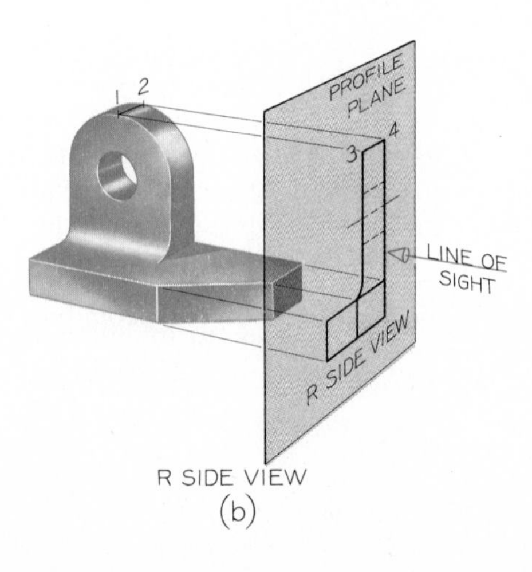

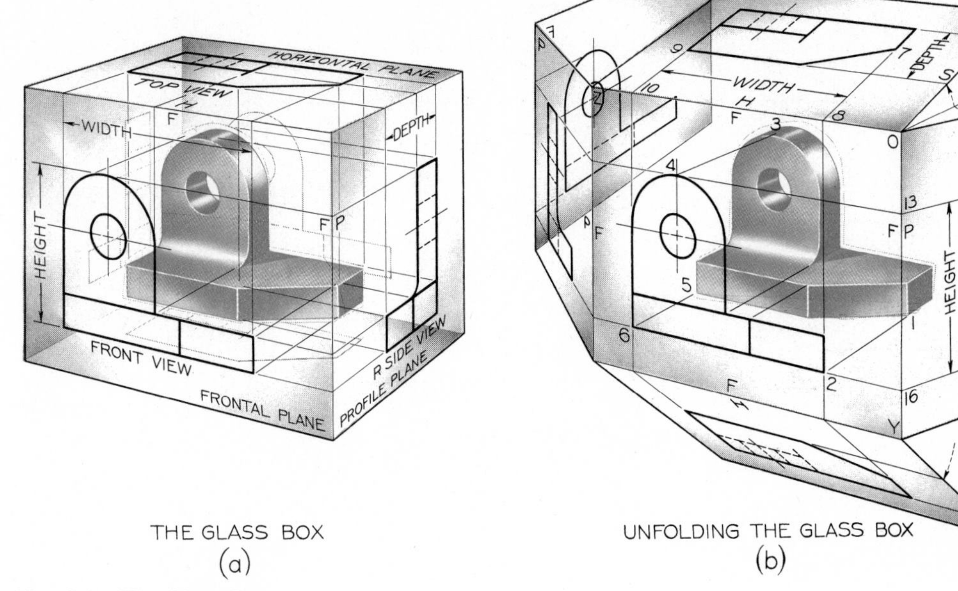


Fig. 6.3 **The Glass Box.**

**6.2 The Glass Box.** If planes of projection are placed parallel to the principal faces of the object, they form a "glass box," as shown in Fig. 6.3 (a). Notice that the observer is always *on the outside looking in,* so that he sees the object through the planes of projection. Since the glass box has six sides, six views of the object are obtained.

Note that the object has three principal dimensions: *width, height,* and *depth.* These are fixed terms used for dimensions in these directions, regardless of the shape of the object. See §5.18.

Since it is required to show the views of a solid or three-dimensional object on a flat sheet of paper, it is necessary to unfold the planes so that they will all lie in the same plane, Fig. 6.3 (b). All planes except the rear plane are hinged upon the frontal plane, the rear plane being hinged to the left-side plane.* Each plane revolves outwardly from the original box position until it lies in the frontal plane, which remains stationary. The hinge lines of the glass box are known as *folding lines.*

*Except as explained in §6.8.

The positions of these six planes, after they have been revolved, are shown in Fig. 6.4. Carefully identify each of these planes and corresponding views with its original position in the glass box, and repeat this mental procedure, if necessary, until the revolutions are thoroughly understood.

In Fig. 6.3 (b), observe that lines extend around the glass box from one view to another upon the planes of projection. These are the *projections of the projectors* from points on the object to the views. For example, the projector 1–2 is projected on the horizontal plane at 7–8 and on the profile plane at 16–17. When the top plane is folded up, lines 9–10 and 7–8 will become vertical and line up with 10–6 and 8–2, respectively. Thus, 9–10 and 10–6 form a single straight line 9–6, and 7–8 and 8–2 form a single straight line 7–2, as shown in Fig. 6.4. This explains why the top view is the same width as the front view and why it is placed directly above the front view. The same relation exists between the front and bottom views. Therefore, *the front, top, and bottom views all line up vertically and are the same width.*

In Fig. 6.3 (b), when the profile plane is folded out, lines 4–13 and 13–15 become a

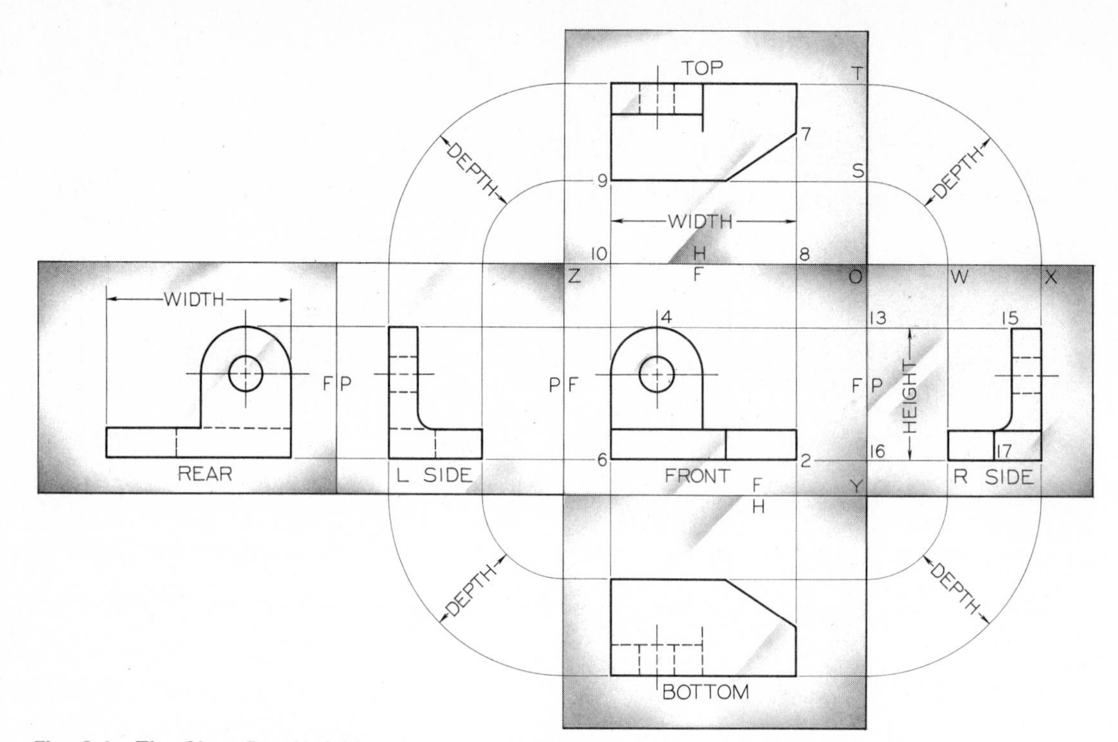


Fig. 6.4 **The Glass Box Unfolded.**

single straight line 4–15, and lines 2–16 and 16–17 become a single straight line 2–17 as shown in Fig. 6.4. The same relation exists between the front, left-side, and rear views. Therefore, *the rear, left-side, front, and right-side views all line up horizontally, and are the same height.*

In Fig. 6.3 (b), note that lines OS and OW and lines ST and WX are respectively equal. These equal lines are shown in the unfolded position in Fig. 6.4. Thus, it is seen that the top view must be the same distance from the folding line OZ as the right-side view is from the folding line OY. Similarly, the bottom view and the left-side view are the same distance from their respective folding lines as are the right-side view and the top view. Therefore, *the top, right-side, bottom, and left-side views are all equidistant from the respective folding lines, and are the same depth.* Note that in these four views that surround the front view, the front surfaces of the object are faced inward, or toward the front view. Observe also that the left-side and right-side views and the top and bottom views are the reverse of each other in outline shape. Similarly, the rear and front views are the reverse of each other.

## 6.3 **Folding Lines.**

Three views of the object discussed above are shown in Fig. 6.5 (a), with folding lines between the views. These folding lines correspond to the hinge lines of the glass box, as we have seen. The H/F folding line, between the top and front views, is the intersection of the horizontal and frontal planes. The F/P folding line, between the front and side views, is the intersection of the frontal and profile planes. See Figs. 6.3 and 6.4.

The distances X and Y, from the front view to the respective folding lines, are not necessarily equal, since they depend upon the relative distances of the object from the horizontal and profile planes. However, as explained in §6.2, distances $D_1$, from the top and side views to the respective folding lines, must always be equal. Therefore, the views may be any desired distance apart, and the folding lines may be drawn anywhere between them, so long as

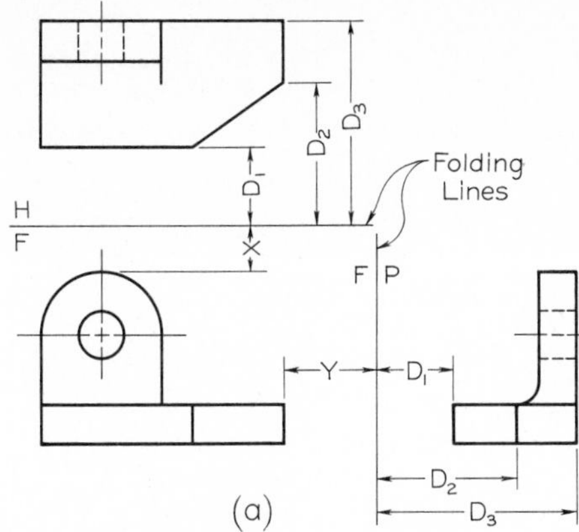


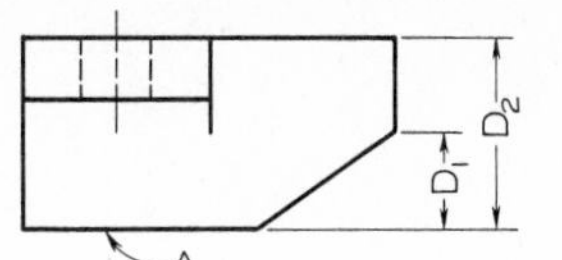


Folding lines omitted

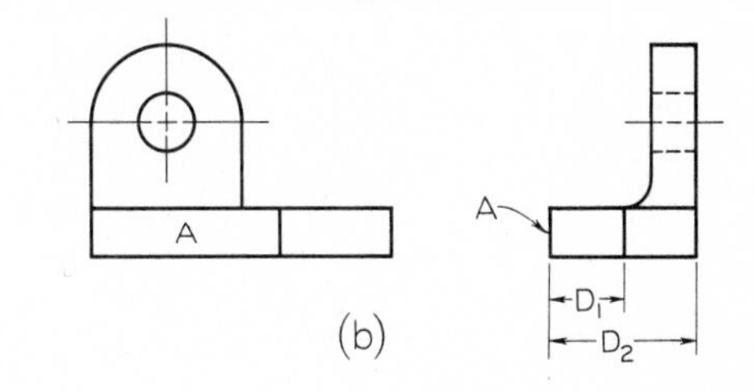

Fig. 6.5 **Folding Lines.**

distances $D_1$ are kept equal and the folding lines are at right angles to the projection lines between the views.

It will be seen that distances $D_2$ and $D_3$, respectively, are also equal, and the folding lines H/F and F/P are in reality reference lines for making equal *depth* measurements in the top and side views. Thus, any point in the top view is the same distance from H/F as the corresponding point in the side view is from F/P.

While it is necessary to understand the folding lines, particularly because they are useful in solving graphical problems in descriptive geometry, they are as a rule omitted in industrial drafting. The three views, with the folding lines omitted, are shown in Fig. 6.5 (b). Again, the distances between the top and front views and between the side and front views are not necessarily equal. Instead of using the folding lines as reference lines for setting off depth measurements in the top and side views, we use the front surface A of the object as a reference line. In this way, $D_1$, $D_2$, and all other depth measurements are made to correspond in the two views in the same manner as if folding lines were used.

## 6.4 Two-View Mechanical Drawing.

Let it be required to draw, full size with instruments on Layout A–2, the necessary views of the Operating Arm shown in Fig. 6.6 (a). In this case, as shown by the arrows, only the front and top views are needed.

I. Determine the spacing of the views. The width of the front and top views is 6″, and the width of the working space is $10\frac{1}{2}$″. As shown at (b), subtract 6″ from $10\frac{1}{2}$″ and divide the result by 2 to get the value of space A. To set off the spaces, place the scale horizontally along the bottom of the sheet and make short vertical marks.

The depth of the top view is $2\frac{1}{2}$″ and the height of the front view is $1\frac{3}{4}$″, while the height of the working space is $7\frac{5}{8}$″. Assume a space C, say 1″, between views that will look well and that will provide sufficient space for dimensions, if any.

As shown at (b), add $2\frac{1}{2}$″, 1″, and $1\frac{3}{4}$″, subtract the total from $7\frac{5}{8}$″, and divide the result by 2 to get the value of space B. To set off the spaces, place the scale vertically along the left side of the sheet with the full-size scale on the left, and make short marks perpendicular to the scale. See Fig. 2.56 (III).

II. Locate center lines from spacing marks. Construct arcs and circles lightly.

III. Draw horizontal and then vertical construction lines in the order shown. Allow construction lines to cross at corners.

IV. Add hidden lines and heavy in all final lines, clean-cut and dark. The visible lines should be heavy enough to make the views stand out. The hidden lines and center lines should be sharp in contrast to the visible lines, but dark enough to reproduce well. See §2.46 for technique of pencil drawing. Construction lines need not be erased if drawn lightly. If you are working on tracing paper, hold the

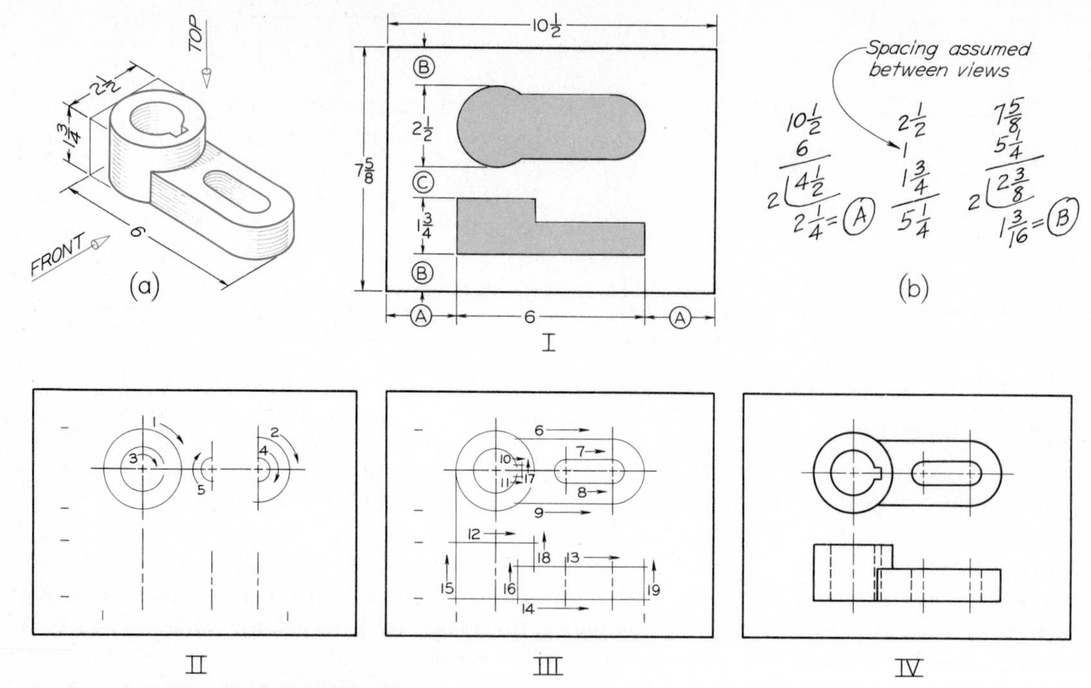


Fig. 6.6 **Two-View Mechanical Drawing.**

sheet up to the light to see if the density of your lines is sufficient to reproduce well, Fig. 2.58.

## 6.5 Transferring Depth Dimensions.

Since all depth dimensions in the top and side views must correspond point for point, accurate methods of transferring these distances, such as $D_1$ and $D_2$, Fig. 6.5 (b), must be used.

The 45° *mitre line* method, Fig. 6.7 (a), is a convenient method, especially when transferring a large number of points, as when plotting a curve, Fig. 6.35. Note that the right-side view may be moved to the right or left, or the top view may be moved upward or downward, by shifting the 45° line accordingly. It is not necessary to draw continuous lines between the top and side views via the mitre line. Instead, make short dashes across the mitre line and project from these.

In practice it is generally recommended, for the sake of accuracy, that the depth dimensions be transferred with the aid of the dividers, (b), or scale, (c). These methods are best when only a small number of very accurate

Fig. 6.7 **Transferring Depth Dimensions.**

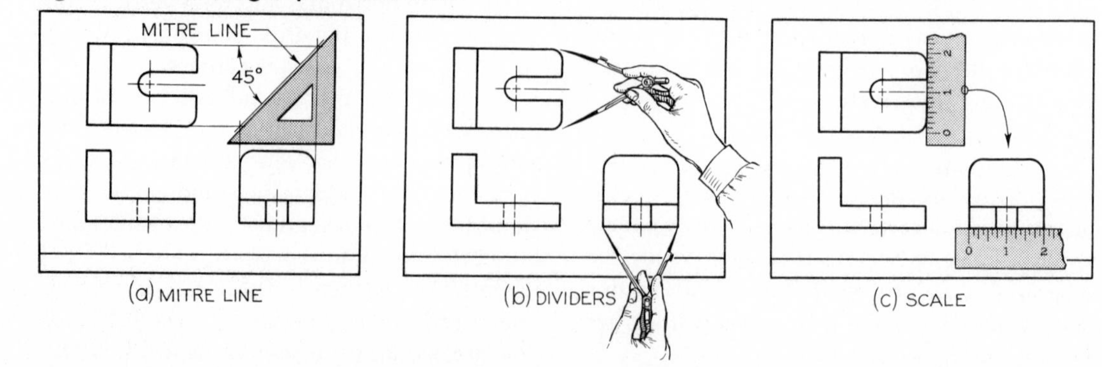

measurements are to be transferred, as is usually the case. The scale method is especially convenient when the drafting machine, Fig. 2.82, is used, because both vertical and horizontal scales are readily available.



## 6.6 Projecting a Third View.

In Fig. 6.8 (top) is a pictorial drawing of a given object, three views of which are required. Each corner of the object is given a number, as shown. At I, the top and front views are shown, with each corner properly numbered in both views. Each number appears twice, once in the top view and once again in the front view.

If a point is *visible* in a given view, the number is placed *outside* the corner, but if the point is *invisible*, the numeral is placed *inside* the corner. For example, at I point 1 is visible in both views, and is therefore placed outside the corners in both views. However, point 2 is visible in the top view and the number is placed outside, while in the front view it is invisible and is placed inside.

This numbering system, in which points are identified by the same numbers in all views, is useful in projecting known points in two views to unknown positions in a third view. Note that in this numbering system a given point has the same number in all views, and should not be confused with the numbering system used in Fig. 6.23 and others, in which a point has different numbers in each view.

In Fig. 6.8, before starting to project the right-side view, try to visualize the view as seen in the direction of the arrow (see pictorial drawing). Then construct the right-side view point by point, using a hard pencil and very light lines.

As shown at I, locate point 1 in the side view by projecting from point 1 in the top view and point 1 in the front view. In space II, project points 2, 3, and 4 in a similar manner to complete the vertical end surface of the object. In space III, project points 5 and 6 to complete the side view of the inclined surface 5-6-2-1. This completes the right-side view, since invisible points 9, 10, 8, and 7 are directly behind visible corners 5, 6, 4, and 3, respectively. Note that in the side view also,

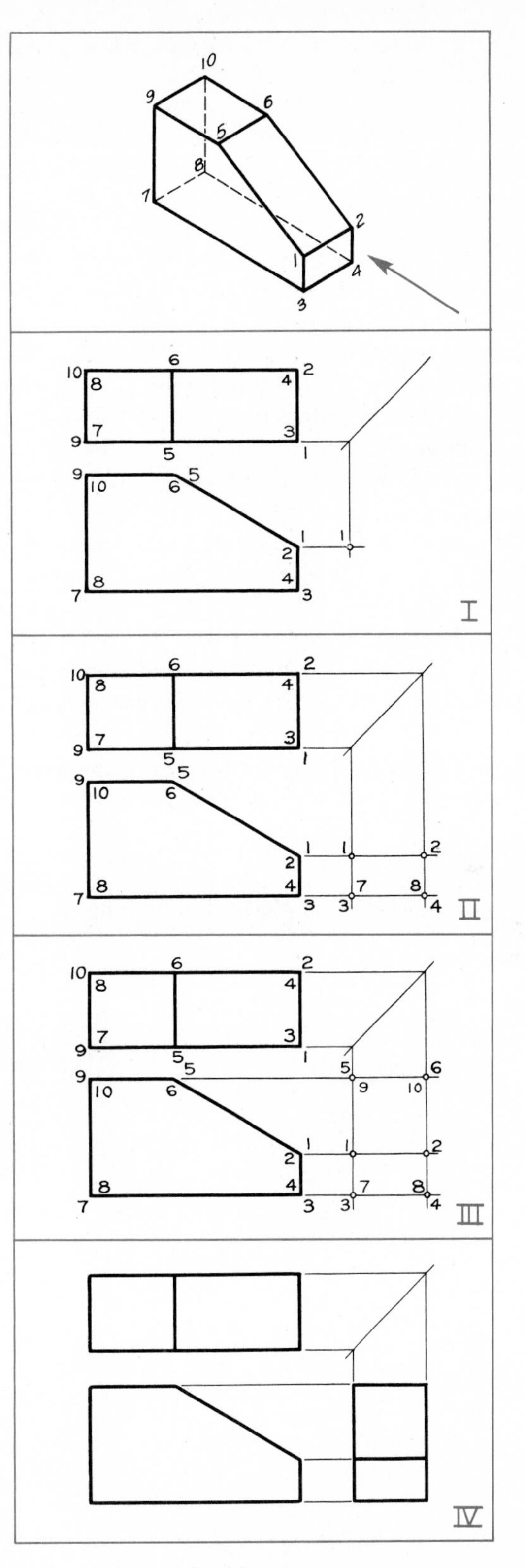


Fig. 6.8 **Use of Numbers.**

the invisible points are lettered *inside*, and the visible points *outside.*

As shown in space IV, the drawing is completed by heavying in the lines in the right-side view.

## 6.7 Three-View Mechanical Drawing.

Let it be required to draw, full size with instruments on Layout A–2, the necessary views of the V-Block in Fig. 6.9 (a). In this case, as shown by the arrows, three views are needed.

I. Determine the spacing of the views. The width of the front view is $4\frac{1}{4}''$, and the depth of the side view is $2\frac{1}{4}''$, while the width of the working space is $10\frac{1}{2}''$. Assume a space C between views, say $1\frac{1}{4}''$, that will look well, and that will allow sufficient space for dimensions, if any.

As shown at (b), add $4\frac{1}{4}''$, $1\frac{1}{4}''$, and $2\frac{1}{4}''$, subtract the total from $10\frac{1}{2}''$, and divide the result by 2 to get the value of space A. To set off these horizontal spacing measurements, place the scale along the bottom of the sheet and make short vertical marks.

The depth of the top view is $2\frac{1}{4}''$, and the height of the front view is $1\frac{3}{4}''$, while the height of the working space is $7\frac{5}{8}''$. Assume a space D between views, say $1''$. As shown in §6.3, space D need not be the same as space C. As shown at (b), add $2\frac{1}{4}''$, $1''$, and $1\frac{3}{4}''$, subtract the total from $7\frac{5}{8}''$, and divide the result by 2 to get the value of space B. To set off these vertical spacing measurements, place the scale along the left side of the sheet with the scale used on the left, and make short marks perpendicular to the scale. Allow for dimensions, if any.

II. Locate the center lines from the spacing marks. Construct lightly the arcs and circles.

III. Draw horizontal, then vertical, then inclined construction lines, in the order shown. Allow construction lines to cross at the corners. Do not complete one view at a time; construct the views simultaneously.

IV. Add hidden lines and heavy in all final lines, clean-cut and dark. A convenient method of transferring a hole diameter from the top view to the side view is to use the compass with the same setting used for draw-

Fig. 6.9 **Three-View Mechanical Drawing.**

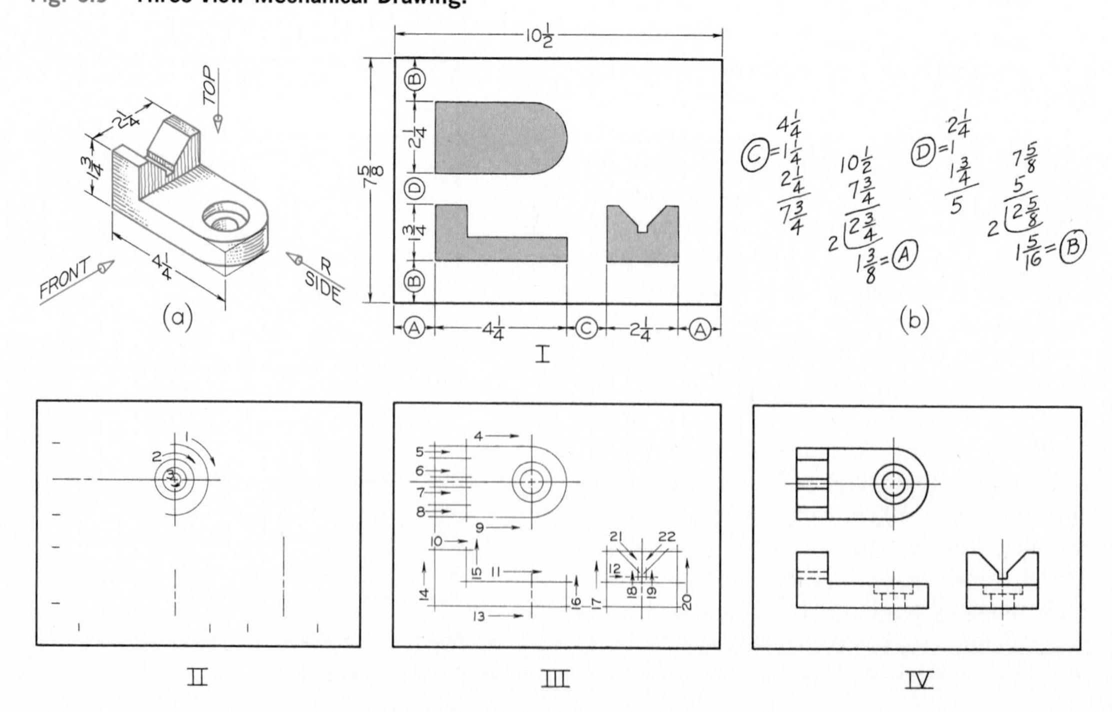

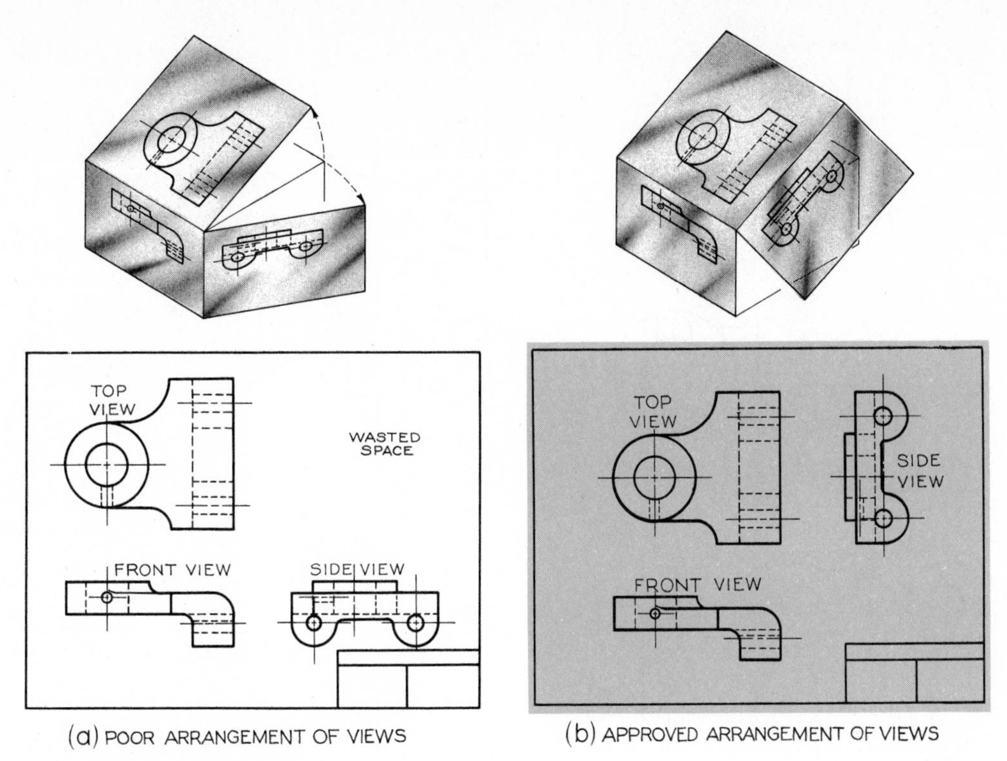


(a) POOR ARRANGEMENT OF VIEWS

(b) APPROVED ARRANGEMENT OF VIEWS

Fig. 6.10 **Position of Side View.**

ing the hole. The visible lines should be heavy enough to make the views stand out. The hidden lines and center lines should be sharp in contrast to the visible lines, but dark enough to reproduce well. Construction lines need not be erased if they are drawn lightly. If you are working on tracing paper, hold the sheet up to the light to see if the density of your lines is sufficient to reproduce well. See §2.46.

## 6.8 Alternate Positions of Views.

If three views of a wide flat object are drawn, using the conventional arrangement of views, Fig. 6.10 (a), a large wasted space is left on the paper, as shown. In such cases, the profile plane may be considered hinged to the horizontal plane instead of the frontal plane, as shown at (b). This places the side view beside the top view, which results in better spacing and in some cases makes the use of a reduced scale unnecessary.

It is also permissible in extreme cases to place the side view across horizontally from the bottom view, in which case the profile plane is considered hinged to the bottom plane of projection. Similarly, the rear view may be placed directly above the top view or under the bottom view, if necessary, in which case the rear plane is considered hinged to the horizontal or bottom plane, as the case may be, and then rotated into coincidence with the frontal plane.

## 6.9 Partial Views.

A view may not need to be complete but may show only what is necessary in the clear description of the object. Such a view is a *partial view*, Fig. 6.11. A break line, (a), may be used to limit the partial view; the contour of the part shown may limit the view, (b); or if symmetrical, a half-view may be drawn on one side of the center line (c), or a partial view, "broken out," may be drawn as at (d). The half shown at (c) and (d) should be the near side, as shown. For half-views in connection with sections, see Fig. 7.32.

Do not place a break line so as to coincide with a visible or hidden line.

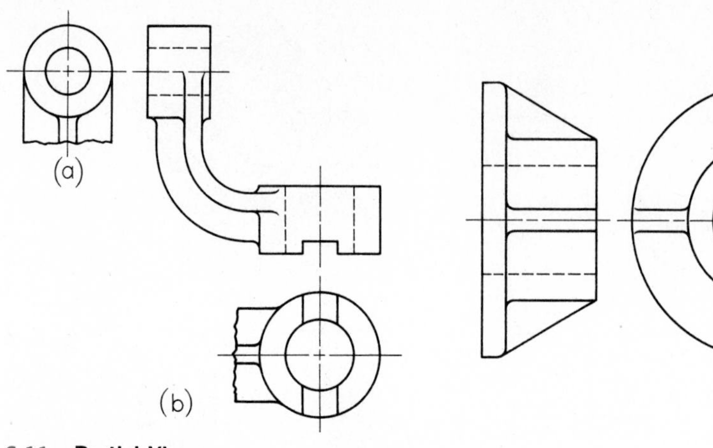


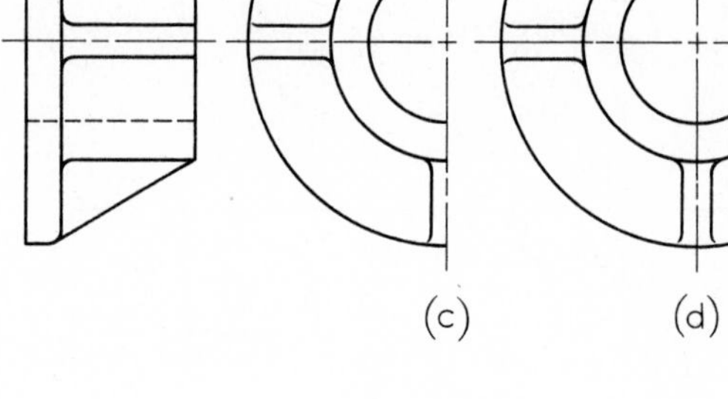


Fig. 6.11 **Partial Views.**

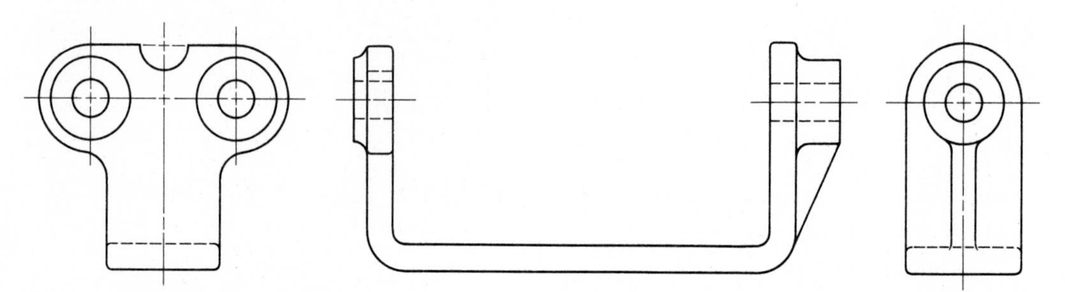

Fig. 6.12 **Incomplete Side Views.**

Occasionally the distinctive features of an object are on opposite sides, so that in either complete side view there would be a considerable overlapping of shapes, resulting in an unintelligible view. In such cases two side views are often the best solution, Fig. 6.12. Observe that the views are partial views, in both of which certain visible and invisible lines have been omitted for clearness.

**6.10 Revolution Conventions.** In some cases regular multiview projections are either awkward, confusing, or actually misleading. For example, Fig. 6.13 (a) shows an object that has three triangular ribs, three holes equally spaced in the base, and a keyway. The right-side view at (b) is a regular projection and is not recommended. The lower ribs appear in a foreshortened position, the holes do not appear in their true relation to the rim of the base, and the keyway is projected as a confusion of hidden lines.

The conventional method shown at (c) is preferred, not only because it is simpler to read, but requires less drafting time. Each of the features mentioned has been revolved in the front view to lie along the vertical center line from where it is projected to the correct side view at (c).

At (d) and (e) are shown regular views of a flange with many small holes. The hidden holes at (e) are confusing and take unnecessary time to draw. The preferred representation at (f) shows the holes revolved, and the drawing is clear.

Another example is shown in Fig. 6.14. As shown at (a), a regular projection results in a confusing foreshortening of an inclined arm. In order to preserve the appearance of symmetry about the common center, the lower arm is revolved to line up vertically in the front view so that it projects true length in the side view at (b).

Revolutions of the type discussed here are frequently used in connection with sectioning.

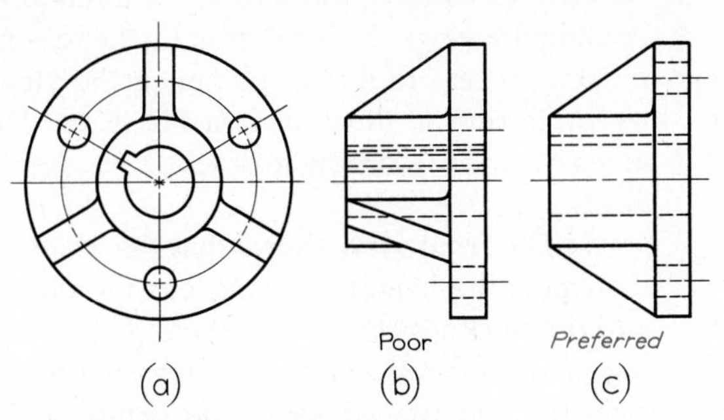
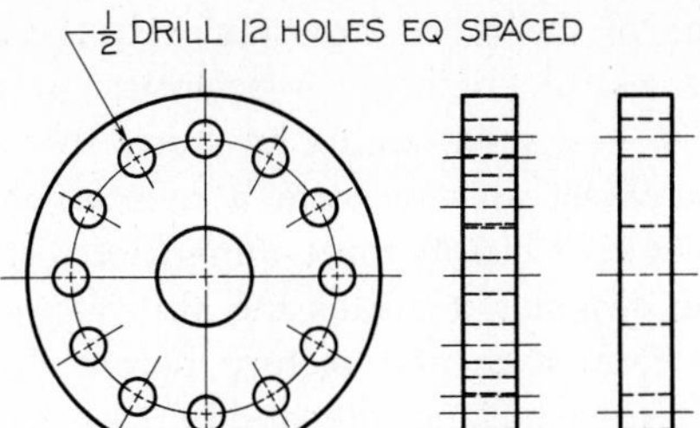

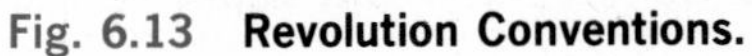

Fig. 6.13 **Revolution Conventions.**

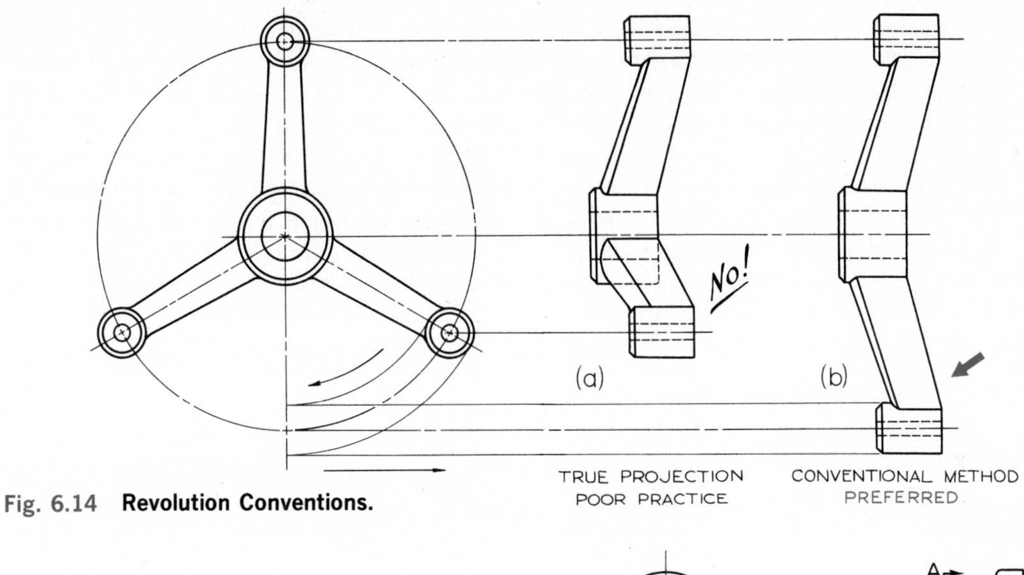


Fig. 6.14 **Revolution Conventions.**

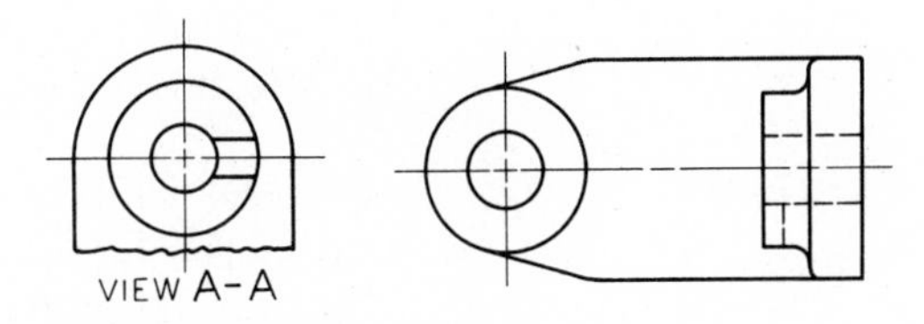

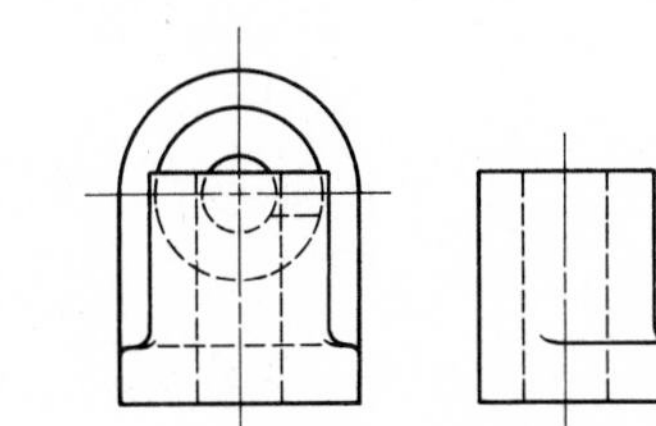
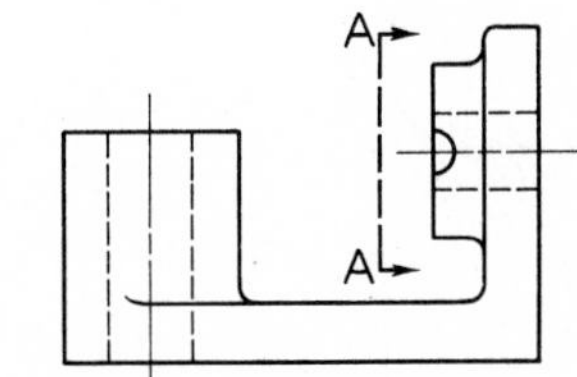


Fig. 6.15 **Removed View.**

Such sectional views are called *aligned sections*, §7.13.

## 6.11 Removed Views.

A *removed view*, Fig. 6.15, is a complete or a partial view removed to another place on the sheet so that it no longer is in direct projection with any other view. Such a view may be used to show some feature of the object more clearly, possibly to a larger scale, or to save drawing a complete regular view. A viewing-plane line is used to indicate the part being viewed, the arrows at the corners showing the direction of sight. See §7.5. The removed view should be labeled VIEW A–A or VIEW B–B, etc., the letters referring to those placed at the corners of the viewing-plane line.

## 6.12 Visualization

As stated in §1.10, the ability to *visualize* or *think in three dimensions* is one of the most important requisites of the successful engineer or scientist. In practice, this means the ability to study the views of an object and to form a mental picture of it—to *visualize* its three-dimensional shape. To the designer it means the ability to *synthesize* or form a mental picture before the object even exists, and the ability to express this image in terms of views. The engineer is the master planner in the construction of new machines, structures, or processes. The ability to visualize, and to use the language of drawing as a means of communication or recording of mental images, is indispensable.

Even the experienced engineer cannot look at a multiview drawing and instantly visualize the object represented (except for the simplest shapes), any more than he can grasp the ideas on a book page merely at a glance. It is necessary to *study* the drawing, to read the lines in a logical way, to piece together the little things until a clear idea of the whole emerges. How this is done is described in the following paragraphs.

## 6.13 Visualizing the Views.

A method of reading drawings that is essentially the reverse mental process to that of obtaining the views by projection is illustrated in Fig. 6.16. The given views of an Angle Bracket are shown at (a).

I. The front view shows that the object is L-shaped, the height and width of the object, and the thickness of the members. The meaning of the hidden and center lines is not yet clear, nor do we yet know the depth of the object.

II. The top view tells us that the horizontal member is rounded on the end and has a round hole. Some kind of slot is indicated at the left end. The depth and width of the object are shown.

III. The right-side view tells us that the left end of the object has rounded corners at the top and has an open-end slot in a vertical position. The height and depth of the object are shown.

Thus, each view provides certain definite information regarding the shape of the object.

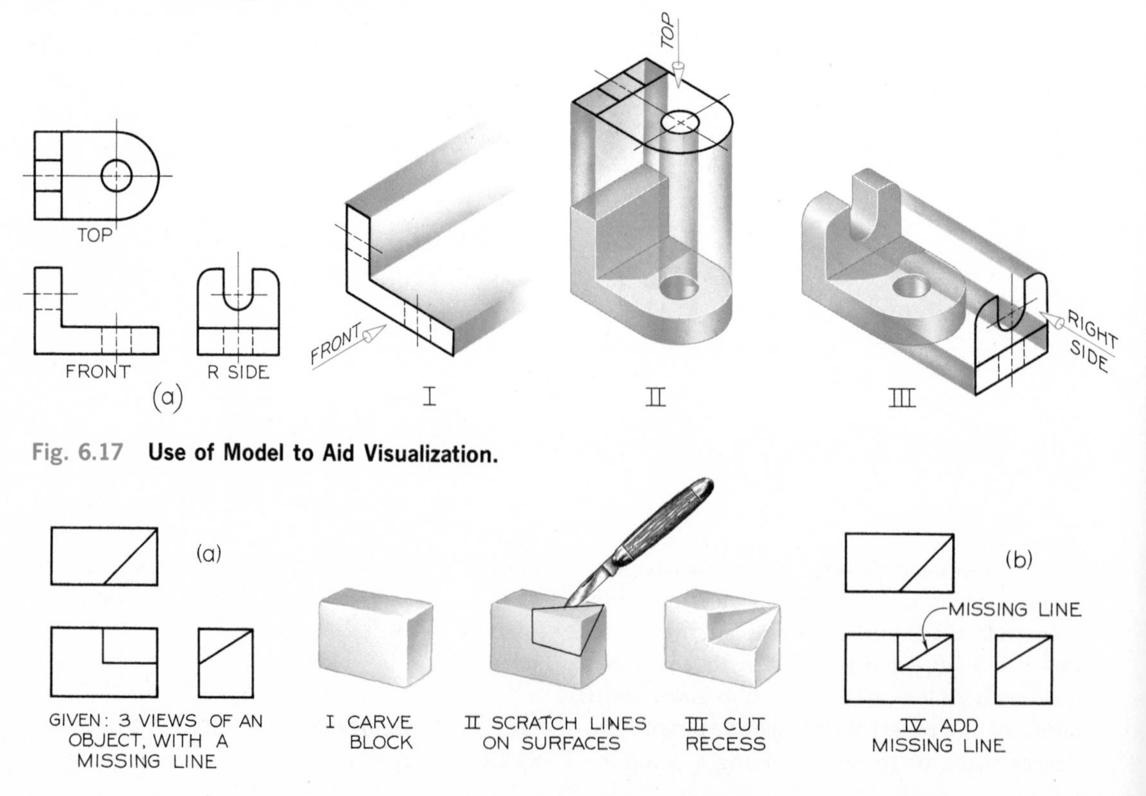


Fig. 6.16 **Visualizing from Given Views.**

Fig. 6.17 **Use of Model to Aid Visualization.**

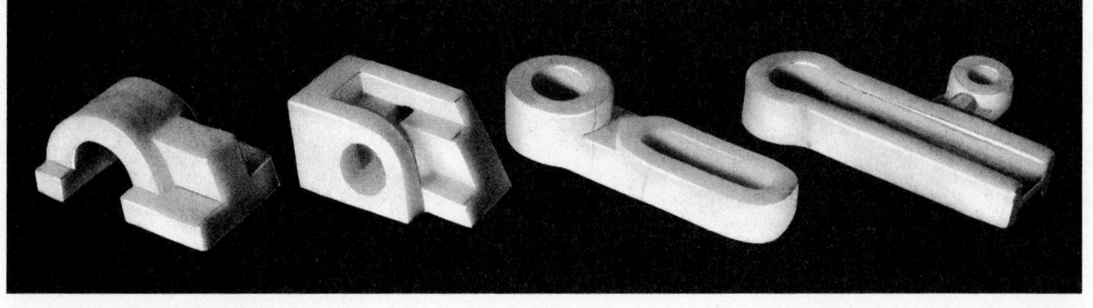

Fig. 6.18 **Soap Models.**

All views must be considered in order to visualize the object completely.

## 6.14 Models.

One of the best aids to visualization is an actual model of the object. Such a model need not be made accurately to scale, and may be made of any convenient material, such as modeling clay, laundry soap, wood, styrofoam, or any material that can be easily carved or cut.

A typical example of the use of soap or clay models is shown in Fig. 6.17, in which three views of an object are given, (a), and the student is to supply a missing line. The model is carved as shown in I, II, and III, and the "missing" line, discovered in the process, is added to the drawing as shown at (b).

Some typical examples of soap models are shown in Fig. 6.18.

## 6.15 Surfaces, Edges, and Corners.

In order to analyze and synthesize multiview projections, it is necessary to consider the component elements that make up most solids. A *surface* (plane) may be bounded by straight lines or curves, or a combination of them. A surface may be *frontal, horizontal,* or *profile,* according to the plane of projection to which it is parallel. See §6.1.

If a plane surface is perpendicular to a plane of projection, it appears as a line, *edge view* (EV), Fig. 6.19 (a). If it is parallel, it appears as a surface, *true size* (TS), (b). If it is situated at an angle, it appears as a surface, *foreshortened* (FS), (c). Thus, *a plane surface always projects as a line or a surface.*

The intersection of two plane surfaces produces an *edge,* or a straight line. Such a line is common to both surfaces and forms a boundary line for each. If an edge is perpendicular to a plane of projection, it appears as a point, Fig. 6.20 (a); otherwise it appears as a line, (b) and (c). If it is parallel to the plane of projection, it shows true length, (b); if not parallel, it shows foreshortened, (c). Thus, a *straight line always projects as a straight line or as*

Fig. 6.19 **Projections of Surfaces.**

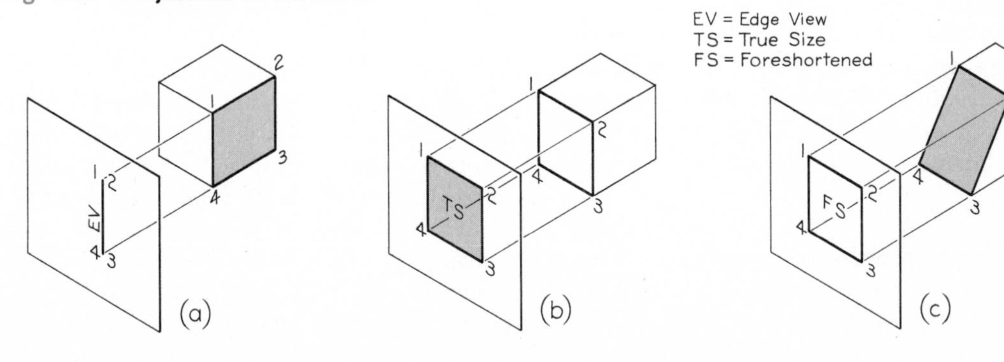

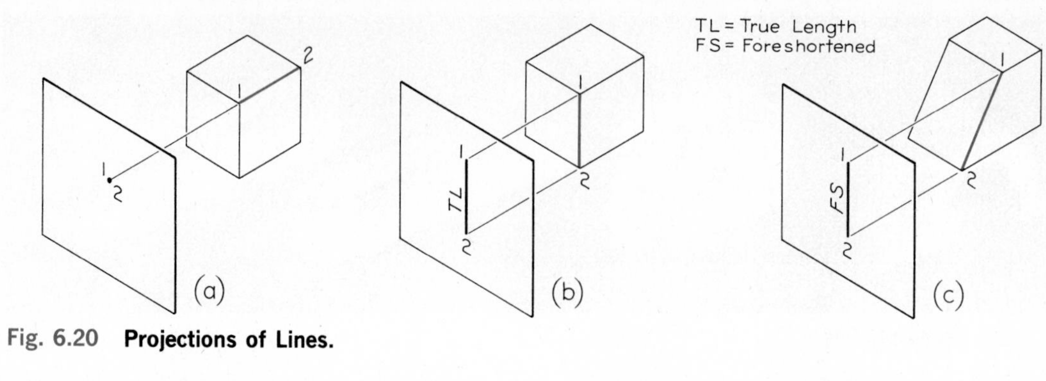


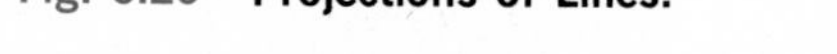
Fig. 6.20 **Projections of Lines.**

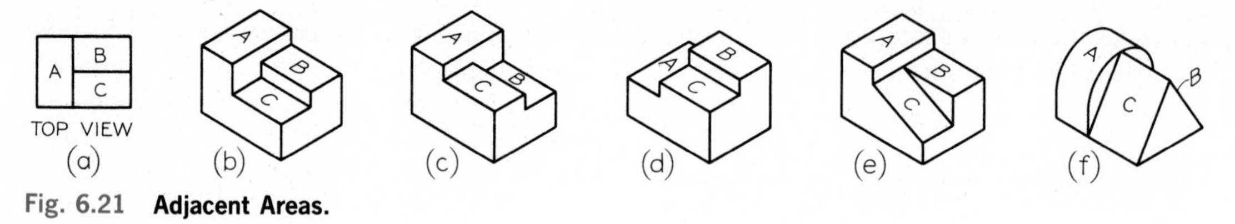


Fig. 6.21 **Adjacent Areas.**

*a point.* A line may be *frontal, horizontal,* or *profile,* according to the plane of projection to which it is parallel.

A *corner,* or point, is the common intersection of three or more surfaces or edges. A point always appears as a point in every view.

## 6.16 Adjacent Areas.

Consider a given top view, as shown at Fig. 6.21 (a). Lines divide the view into three areas. Each of these must represent a surface *at a different level.* Surface A may be high and B and C lower, as shown at (b). Or B may be lower than C, as shown at (c). Or B may be highest, with C and A each lower, (d). Or one or more surfaces may be inclined, as at (e). Or one or more surfaces may be cylindrical, as at (f), etc. Hence the rule: *No two adjacent areas can lie in the same plane.*

The same reasoning can apply, of course, to the adjacent areas in any given view. Since an area (surface) on a view can be interpreted in several different ways, it is necessary to observe other views also in order to determine which interpretation is correct.

## 6.17 Similar Shapes of Surfaces.

If a surface is viewed from several different positions, it will in each case be seen to have a certain number of sides and to have a certain characteristic shape. An L-shaped surface, Fig. 6.22 (a), will appear as an L-shaped figure in every view in which it does not appear as a line. A T-shaped surface, (b), a U-shaped surface, (c), or a hexagonal surface, (d), will in each case have the same number of sides and the same characteristic shape in every view in which it appears as a surface.

This repetition of shapes is one of our best means for analyzing the views.

## 6.18 Reading a Drawing.

Let it be required to read or visualize the object shown by three views in Fig. 6.23. Since no lines are curved, the object is made up of plane surfaces.

Surface 2–3–10–9–6–5 in the top view is an L-shaped surface of six sides. It appears in the side view at 16–17–21–20–18–19, and is L-shaped and six-sided. No such shape appears in the front view, but we note that points 2 and 5 line up with 11 in the front view, points 6 and 9 line up with 13, and points 3 and 10 line up with 15. Evidently line 11–15 in the front view is the edge view of the L-shaped surface.

Surface 11–13–12 in the front view is triangular in shape, but no corresponding triangles

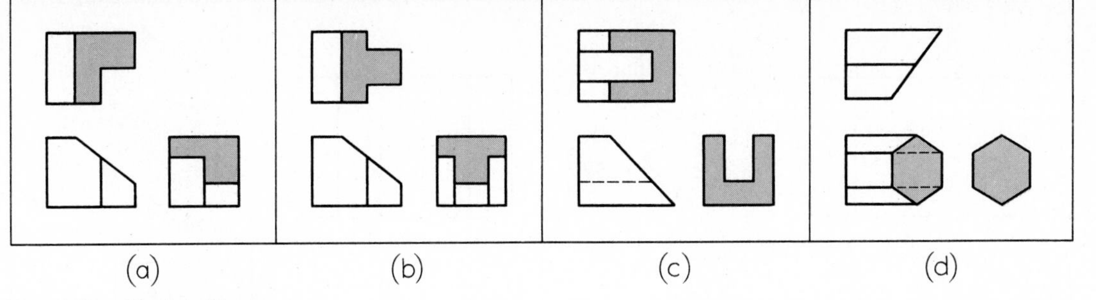


Fig. 6.22 **Similar Shapes.**

appear in either the top or the side view. We note that point 12 lines up with 8 and 4 and that point 13 lines up with 6 and 9. However, surface 11–13–12 of the front view cannot be the same as surface 4–6–9–8 in the top view because the former has three sides and the latter has four. Obviously, the triangular surface appears as a line 4–6 in the top view and as a line 16–19 in the side view.

Surface 12–13–15–14 in the front view is trapezoidal in shape. But there are no trapezoids in the top and side views, so the surface evidently appears in the top view as line 7–10, and in the side view as line 18–20.

The remaining surfaces can be identified in the same manner, whence it will be seen that the object is bounded by seven plane surfaces, two of which are rectangular, two triangular, two L-shaped, and one trapezoidal.

Note that the numbering system used in Fig. 6.23 is different from that in Fig. 6.8 in that different numbers are used for all points and there is no significance in a point being inside or outside a corner.

## 6.19 **Normal Surfaces.** 

*A normal surface is a plane surface that is parallel to a plane of projection.* It appears in true size and shape on the plane to which it is parallel, and as a vertical or a horizontal line on adjacent planes of projection.

In Fig. 6.24 four stages in machining a block of cold-rolled steel to produce the final Tool Block in space IV are shown. All surfaces are normal surfaces. In space I, normal surface A is parallel to the horizontal plane and appears true size in the top view at 2–3–7–6, as line

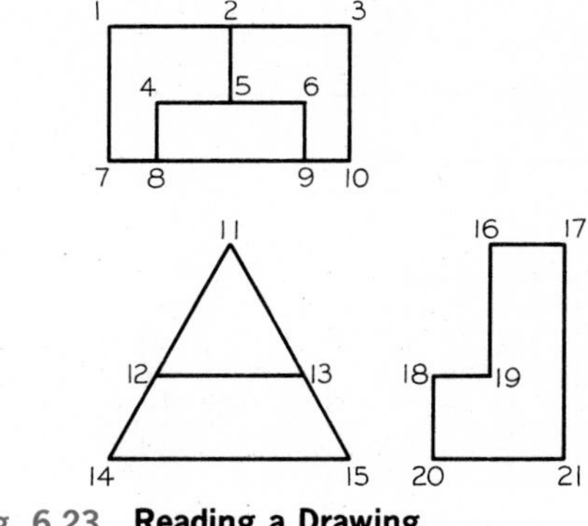

Fig. 6.23 **Reading a Drawing.**

9–10 in the front view, and as line 17–18 in the side view. Normal surface B is parallel to the profile plane and appears true size in the side view at 17–18–20–19, as line 3–7 in the top view, and as line 10–13 in the front view. Normal surface C, an inverted T-shaped surface, is parallel to the frontal plane and appears true size in the front view at 9–10–13–14–16–15–11–12, as line 5–8 in the top view, and as line 17–21 in the side view.

All other surfaces of the object may be visualized in a similar manner. In the four stages of Fig. 6.24, observe carefully the changes in the views produced by the machining operations, including the introduction of new surfaces, new visible edges and hidden edges, and the dropping out of certain lines as the result of a new cut.

The top view in space I is cut by lines 2–6 and 3–7, which means that there are three surfaces, 1–2–6–5, 2–3–7–6, and 3–4–8–7. In the front view, surface 9–10 is seen to be the highest, and surfaces 11–12 and 13–14 are at the same lower level. In the side view both

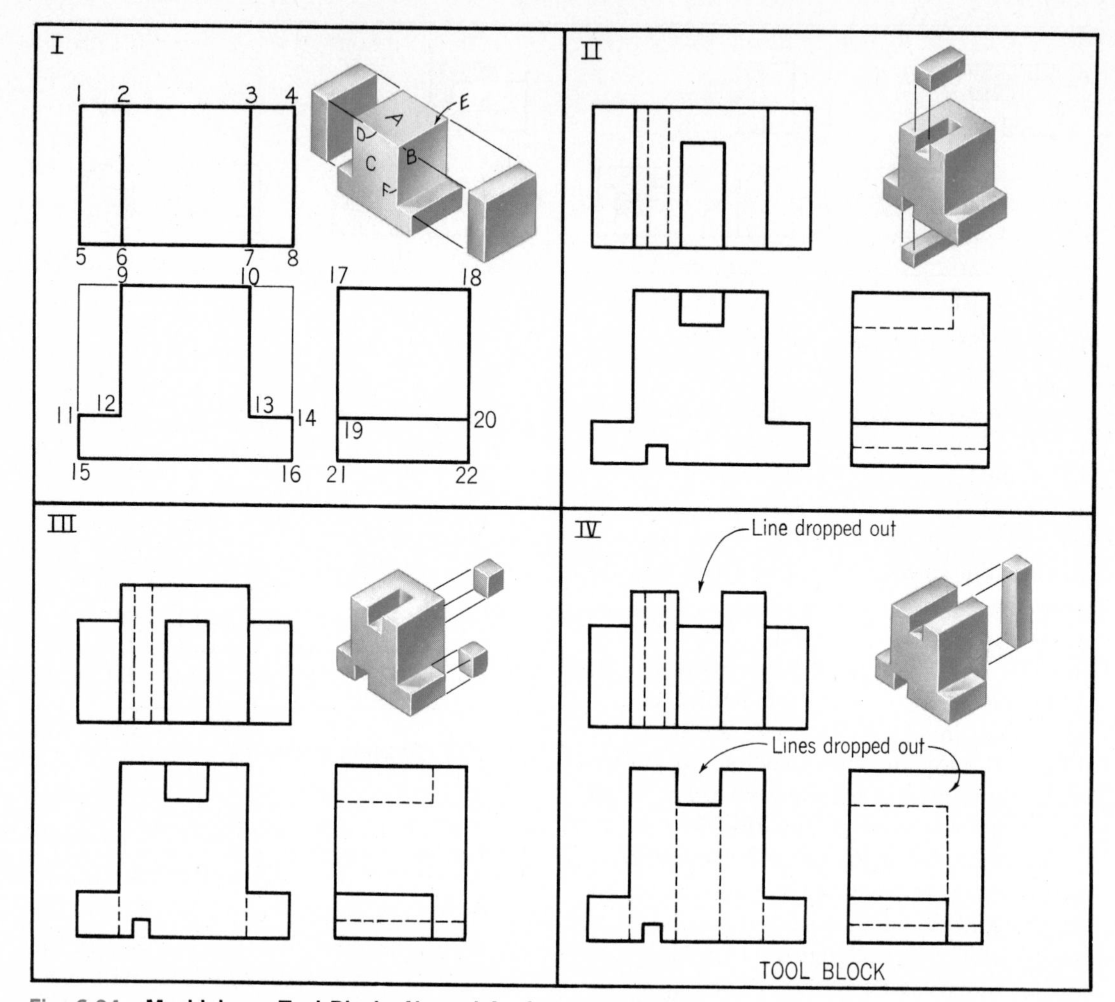


Fig. 6.24 **Machining a Tool Block—Normal Surfaces and Edges.**

of these latter surfaces appear as one line 19–20. Surface 11–12 might appear as a hidden line in the side view, but surface 13–14 appears as a visible line 19–20, which covers up the hidden line and takes precedence over it. See §5.31.

In space IV, how many normal surfaces are there altogether?

## 6.20 Normal Edges.

*A normal edge is a line that is perpendicular to a plane of projection.* It will appear as a point on the plane of projection to which it is perpendicular and as a line in true length on adjacent planes of projection. In space I of Fig. 6.24, edge D is perpendicular to the profile plane of projection and appears as point 17 in the side view. It is parallel to the frontal and horizontal planes of projection, and is shown true length at 9–10 in the front view and 6–7 in the top view. Edges E and F are perpendicular respectively to the frontal and horizontal planes of projection, and their views may be similarly analyzed.

In space II, how many normal edges are there?

## 6.21 Inclined Surfaces.

*An inclined surface is a plane surface that is perpendicular to one plane of projection but inclined to adjacent planes.* An inclined surface will project as a straight line on the plane to which it is perpendicular, and it will appear foreshortened (FS) on planes to which it is inclined, the degree of foreshortening being proportional to the angle of inclination.

In Fig. 6.25 four stages in machining a Locating Finger are shown, producing several inclined surfaces. In space I, inclined surface A is perpendicular to the horizontal plane of projection and appears as line 5–3 in the top view. It is shown as a foreshortened surface in the front view at 7–8–11–10, and in the side view at 12–13–16–15. Note that the surface is more foreshortened in the side view than in the front view because the plane makes a greater angle with the profile plane of projection than with the frontal plane of projection.

In space III, edge 23–24 in the front view is the edge view of an inclined surface that appears in the top view as 21–2–3–22 and in the side view as 25–14–27–26. Note that 25–14 is equal in length to 21–22, and that the surface has the same number of sides (four) in both views in which it appears as a surface.

In space IV, edge 29–23 in the front view is the edge view of an inclined surface that appears in the top view as visible surface 1–21–22–5–18 and in the side view as invisible surface 25–14–32–31–30. While the surface does not appear true size in any view, it does have the same characteristic shape and the same number of sides (five) in the views in which it appears as a surface.

In space IV, how many normal surfaces are there? Inclined surfaces?



Fig. 6.25 **Machining a Locating Finger—Inclined Surfaces.**

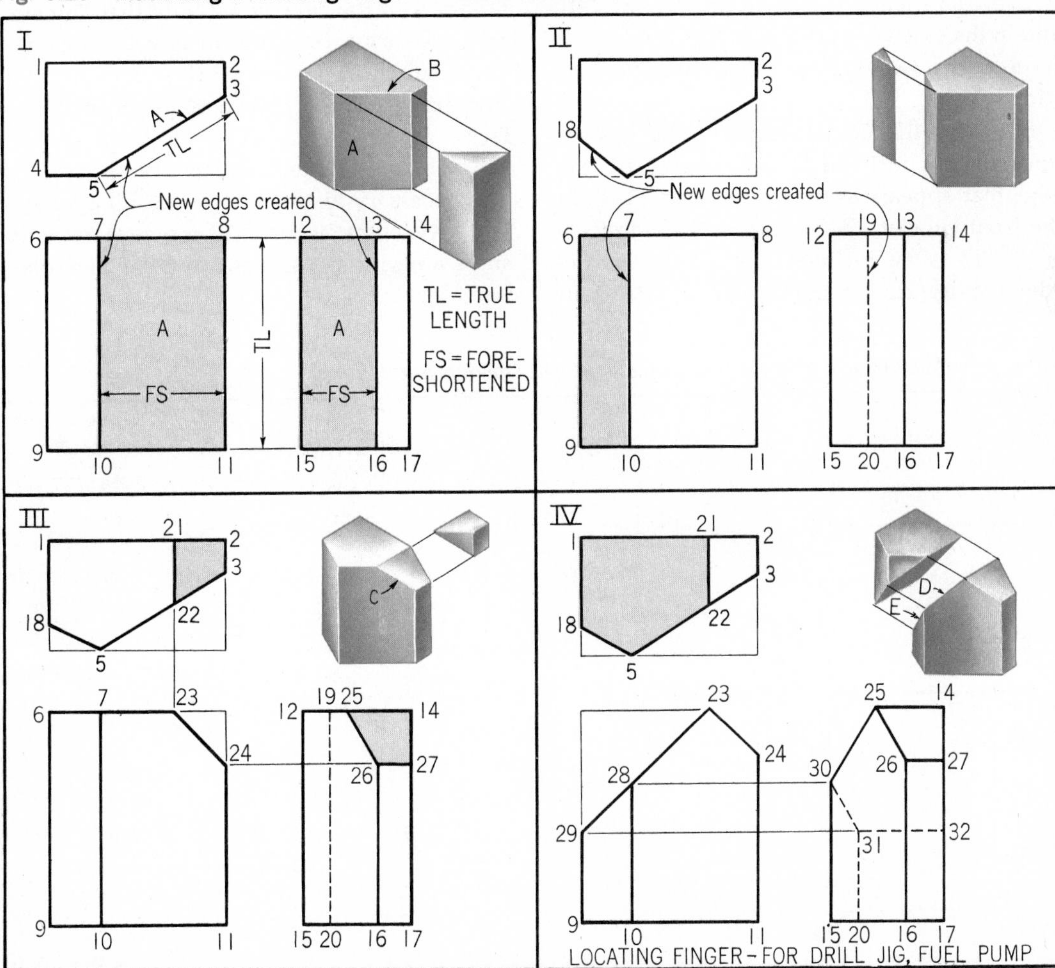

In order to obtain the true size of an inclined surface, it is necessary to construct an auxiliary view, §8.1, or to revolve the surface until it is parallel to a plane of projection, §9.1.

## 6.22 Inclined Edges.

*An inclined edge is a line that is parallel to a plane of projection but inclined to adjacent planes.* It will appear true length on the plane to which it is parallel, and foreshortened on adjacent planes, the degree of foreshortening being proportional to the angle of inclination. The true-length view of an inclined line is always inclined, while the foreshortened views are either vertical or horizontal lines.

In space I of Fig. 6.25, inclined edge B is parallel to the horizontal plane of projection, and appears true length in the top view at 5–3. It is foreshortened in the front view at 7–8, and in the side view at 12–13. Note that plane A produces two normal edges and two inclined edges.

In spaces III and IV, some of the sloping lines are not inclined lines. In space III, the edge that appears in the top view at 21–2, in the front view at 23–24, and in the side view at 14–27 is an inclined line. However, the edge that appears in the top view at 22–3, in the front view at 23–24, and in the side view at 25–26 is not an inclined line by the definition given here. Actually, it is an *oblique line,* §6.24.

In space IV, how many normal edges are there? Inclined edges?

## 6.23 Oblique Surfaces.

*An oblique surface is a plane that is oblique to all planes of projection.* Since it is not perpendicular to any plane, it cannot appear as a line in any view. Since it is not parallel to any plane, it cannot appear true size in any view. Thus, an oblique surface always appears as a foreshortened surface in all three views.

In space II of Fig. 6.26, oblique surface C appears in the top view at 25–3–6–26, and in the front view at 29–8–31–30. What are its numbers in the side view? Note that any surface appearing as a line in any view cannot be an oblique surface. How many inclined surfaces are there? How many normal surfaces?

To obtain the true size of an oblique surface, it is necessary to construct a secondary auxiliary view, §§8.21 and 8.22, or to revolve the surface until it is parallel to a plane of projection, §9.11.

Fig. 6.26 **Machining a Control Lever—Inclined and Oblique Surfaces.**

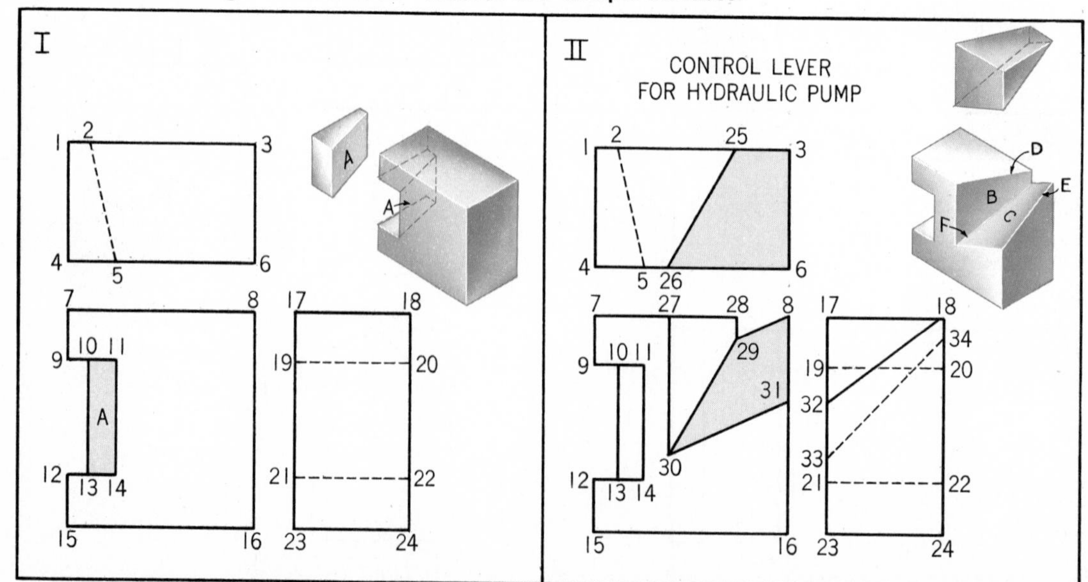

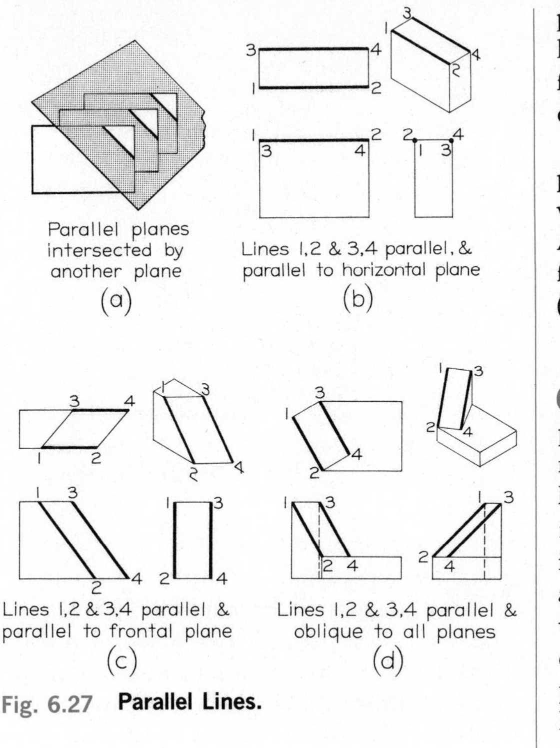


Fig. 6.27 **Parallel Lines.**

## 6.24 Oblique Edges.

*An oblique edge is a line that is oblique to all planes of projection.* Since it is not perpendicular to any plane, it cannot appear as a point in any view. Since it is not parallel to any plane, it cannot appear true length in any view. An oblique edge appears foreshortened and in an inclined position in every view.

In space II of Fig. 6.26, oblique edge F appears in the top view at 26–25, in the front view at 30–29, and in the side view at 33–34. Are there any other oblique lines in this figure? What are the oblique lines in Fig. 6.25 (IV)?

## 6.25 Parallel Edges.

If a series of parallel planes is intersected by another plane, the resulting lines of intersection will be parallel, Fig. 6.27 (a). At (b) the top plane of the object intersects the front and rear planes, producing the parallel edges 1–2 and 3–4. If two lines are parallel in space, their projections in any view are parallel. The example at (b) is a special case in which the two lines appear as points in one view and coincide as a single line in another and should not be regarded as an exception to the rule. Note that even in the pictorial drawings the lines are shown parallel.

Parallel inclined lines are shown at (c), and parallel oblique lines at (d).

In Fig. 6.28 it is required to draw three views of the object after a plane has been passed

Fig. 6.28 **Oblique Surface.**

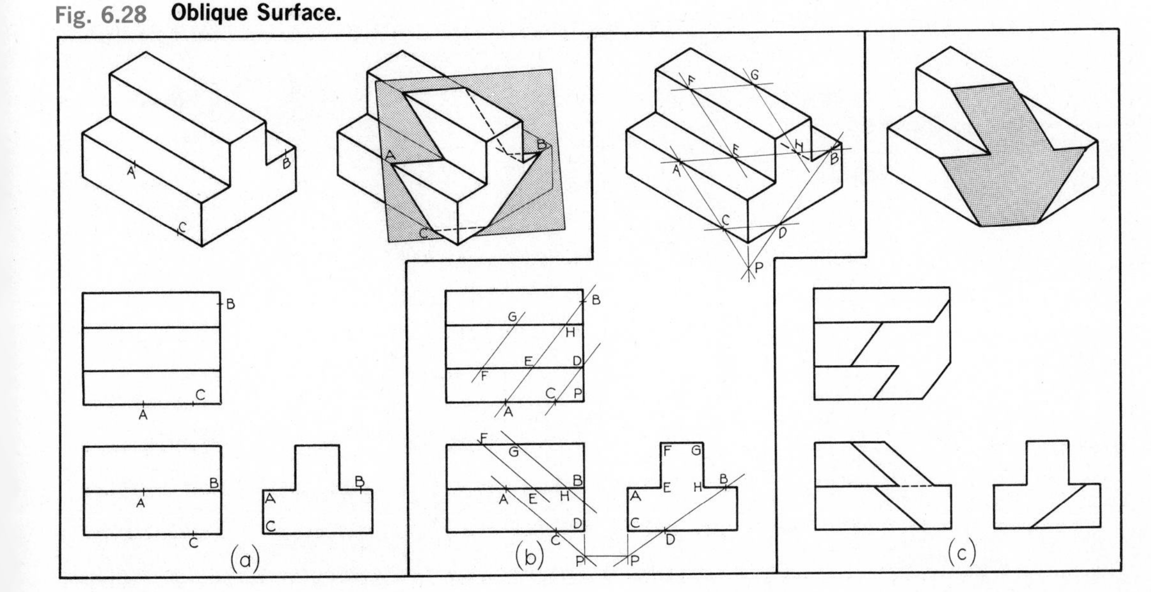

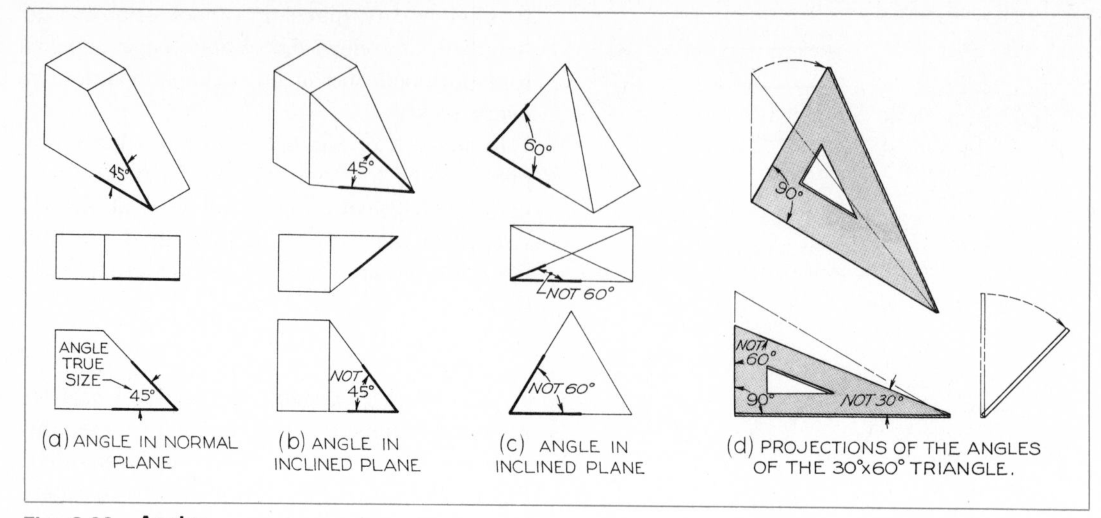


Fig. 6.29 **Angles.**

through the points A, B, and C. As shown at (b), only points that lie in the same plane are joined. In the front view, join points A and C, which are in the same plane, extending the line to P on the vertical front edge of the block extended. In the side view, join P to B, and in the top view, join B to A. Complete the drawing by applying the rule: *parallel lines in space will be projected as parallel lines in any view.* The remaining lines are thus drawn parallel to lines AP, PB, and BA.

**6.26 Angles.** If an angle is in a normal plane—that is, parallel to a plane of projection—the angle will be shown true size on the plane of projection to which it is parallel, Fig. 6.29 (a).

If the angle is in an inclined plane, (b) and (c), the angle may be projected either larger or smaller than the true angle, depending upon its position. At (b) the 45° angle is shown *oversize* in the front view, and at (c) the 60° angle is shown *undersize* in both views.

Fig. 6.30 **Curved Surfaces.**

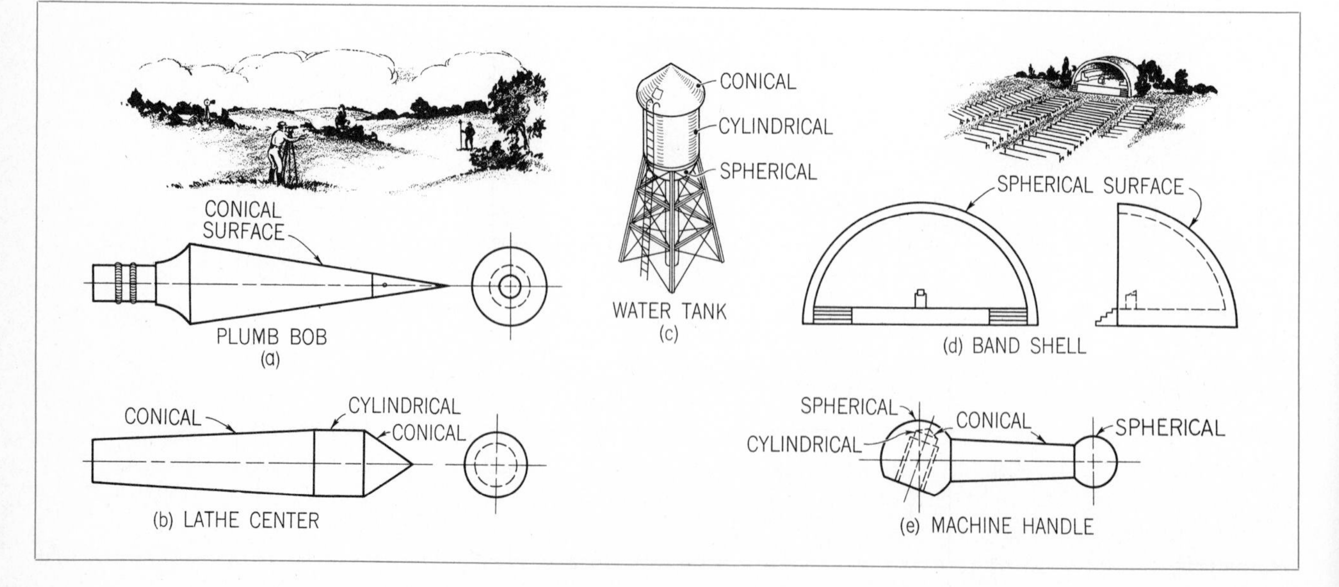

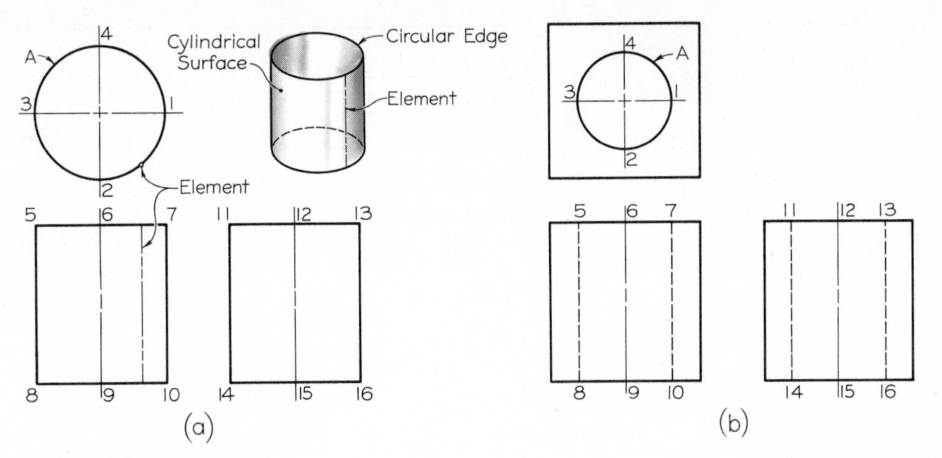


Fig. 6.31 **Cylindrical Surfaces.**

A 90° angle will be projected true size, even though it is in an inclined plane, provided one leg of the angle is a normal line, as shown at (d). In this figure, the 60° angle is projected *oversize* and the 30° angle *undersize.* Study these relations, using your own 30° × 60° triangle as a model.

## 6.27 Curved Surfaces.

Rounded surfaces are common in engineering practice because they are easily formed on the lathe, the drill press, and other machines using the principle of rotation either of the "work" or of the cutting tool. The most common are the cylinder, cone, and sphere, a few of whose applications are shown in Fig. 6.30. For other geometric solids, see Fig. 4.7.

## 6.28 Cyclindrical Surfaces.

Three views of a *right-circular cylinder,* the most common type, are shown in Fig. 6.31 (a). The single cylindrical surface is intersected by two plane (normal) surfaces, forming two curved lines of intersection or *circular edges* (the bases of the cylinder). These circular edges are the only actual edges on the cylinder.

The cylinder is represented on a drawing by its circular edges and the contour elements. An *element* is a straight line on the cylindrical surface, parallel to the axis, as shown in the pictorial view of the cylinder at (a). In this figure, at both (a) and (b), the circular edges appear in the top views as circles A, and in the front views as horizontal lines 5–7 and 8–10, and in the side views as horizontal lines 11–13 and 14–16.

The contour elements 5–8 and 7–10 in the front views appear as points 3 and 1 in the top views. The contour elements 11–14 and 13–16 in the side views appear as points 2 and 4 in the top views.

In Fig. 6.32 four possible stages in machining a Cap are shown, producing several cylindrical surfaces. In space I, the removal of the two upper corners forms cylindrical surface A which appears in the top view as surface 1–2–4–3, in the front view as arc 5, and in the side view as surface 8–9–Y–X.

In space II, a large reamed hole shows in the front view as circle 16, in the top view as cylindrical surface 12–13–15–14, and in the side view as cylindrical surface 17–18–20–19.

In space III, two drilled and counterbored holes are added, producing four more cylindrical surfaces and two normal surfaces. The two normal surfaces are those at the bottoms of the counterbores.

In space IV, a cylindrical cut is added, producing two cylindrical surfaces that appear edgewise in the front view as arcs 30 and 33, in the top view as surfaces 21–22–26–25 and 23–24–28–27, and in the side view as surfaces 36–37–40–38 and 41–42–44–43.

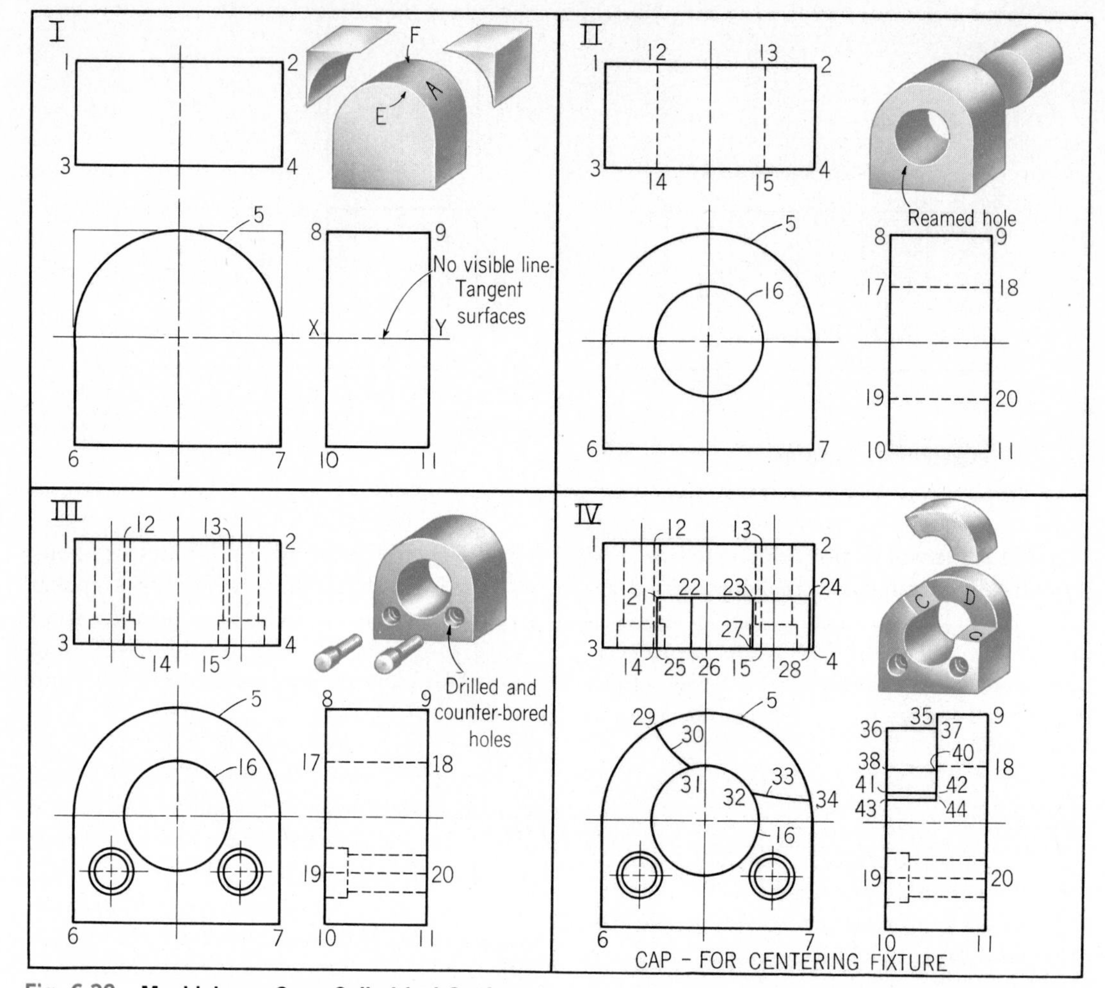


Fig. 6.32 **Machining a Cap—Cylindrical Surfaces.**

## 6.29 **Deformities of Cylinders.**

In shop practice, cylinders are usually machined or formed so as to introduce other surfaces, usually plane surfaces. In Fig. 6.33 (a) a cut is shown which introduces two normal surfaces. One surface appears as a line 3–4 in the top view, as surface 6–7–10–9 in the front view, and as line 13-16 in the side view. The other appears as line 15–16 in the side view, line 9–10 in the front view, and surface 3–4, arc 2 in the top view.

All elements touching arc 2, between 3 and 4 in the top view, become shorter as a result of the cut. For example, element A, which shows as a point in the top view, now becomes CD in the front view, and 15–17 in the side view. As a result of the cut, the front half of the cylindrical surface has changed from 5–8–12–11 to 5–6–9–10–7–8–12–11 (front view). The back half remains unchanged.

At (b) two cuts introduce four normal surfaces. Note that surface 7–8 (top view) is through the center of the cylinder, producing in the side view line 21–24, and in the front view surface 11–14–16–15 equal in width to the diameter of the cylinder. Surface 15–16 (front view) is read in the top view as 7–8–arc 4. Surface 11–14 (front view) is read in the top view as 5–6–arc 3–8–7–arc 2.

At (c) two cylinders on the same axis are shown, intersected by a normal surface parallel to the axis. Surface 17–20 (front view) is 23–25 in the side view, and 2–3–11–9–15–14–8–6 in the top view. A common error is to draw a visible line in the top view between 8 and 9. However, this would produce two surfaces 2–3–11–6 and 8–9–15–14 not in the same plane. In the front view, the larger surface appears as line 17–20 and the smaller as line 18–19. These lines coincide; hence they are all one surface, and there can be no visible line joining 8 and 9 in the top view.

The surface that appears in the front view at 17-18-arc 22-19-20-arc 21 appears in the top view at 5-12, which explains the hidden line 8-9 in the top view.

## 6.30 **Cylinders and Ellipses.**

If a cylinder is cut by an inclined plane, as in Fig. 6.34 (a), the inclined surface is bounded by an ellipse. The ellipse appears as circle 1 in the top view, as straight line 2–3 in the front view, and as ellipse ADBC in the side view. Note that circle 1 in the top view would remain a circle regardless of the angle of the cut. If the cut is 45° with horizontal, the ellipse will appear as a circle in the side view (see phantom lines) since the major and minor axes in that view would be equal. To find the true size and shape of the ellipse, an auxiliary view will be required, with the line of sight perpendicular to surface 2–3 in the front view, §8.12.

Since the major and minor axes AB and CD are known, the ellipse can be drawn by any of the methods in Figs. 4.48 to 4.50 and 4.52 (a) (true ellipses) or by the aid of an ellipse guide, Fig. 4.55.

If the cylinder is tilted forward, (b), the bases or circular edges 1–2 and 3–4 (side view) become ellipses in the front and top views. Points on the ellipses can be plotted

Fig. 6.33 **Deformities of Cylinders.**

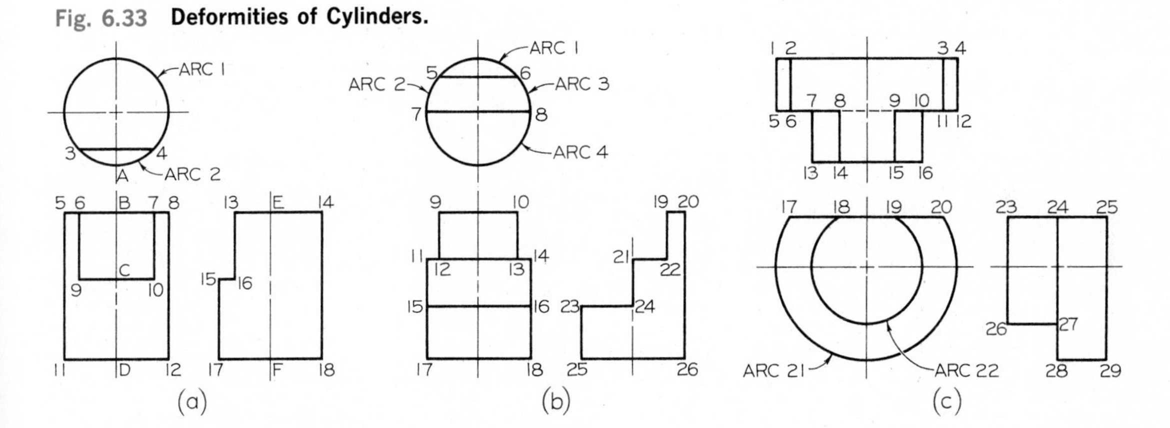


Fig. 6.34 **Cylinders and Ellipses.**

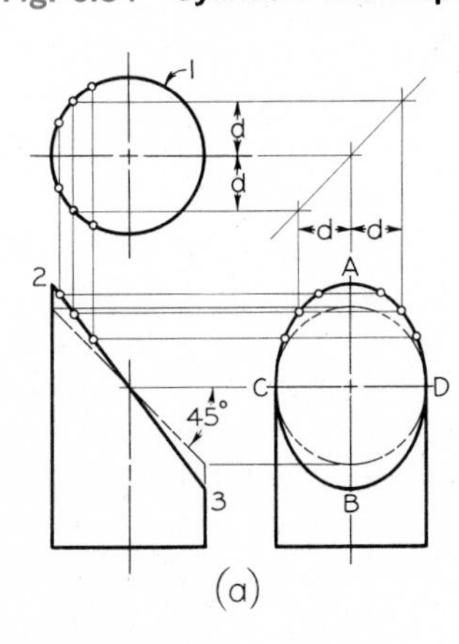


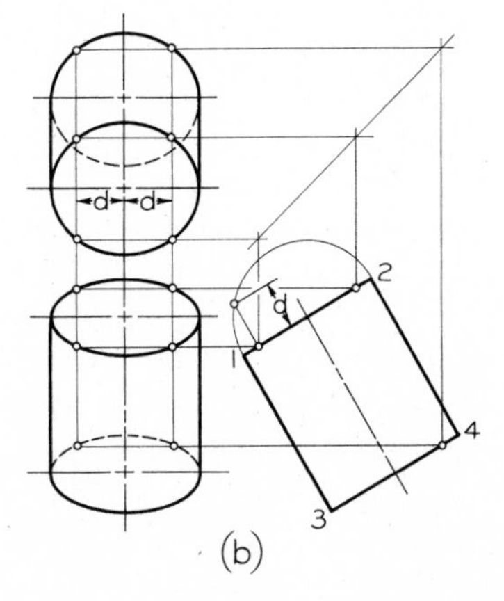


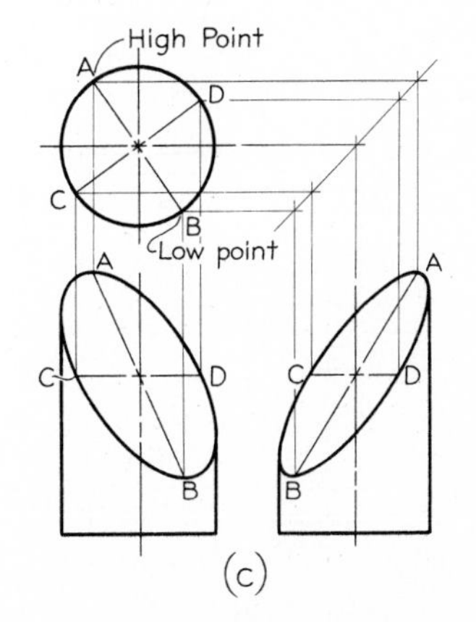

from the semicircular end view of the cylinder, as shown, distances d being equal. Since the major and minor axes for each ellipse are known, the ellipses can be drawn with the aid of an ellipse guide, or by any of the true ellipse methods, or by the approximate method.

If the cylinder is cut by an oblique plane, (c), the elliptical surface appears as an ellipse in two views. In the top view, points A and B are selected, diametrically opposite, as the high and low points in the ellipse, and CD is drawn perpendicular to AB. These are the projections of the major and minor axes, respectively, of the actual ellipse in space. In the front and side views, points A and B are assumed at the desired altitudes. Since CD appears true length in the top view, it will appear horizontal in the front and side views, as shown. These axes in the front and side views are the *conjugate axes* of the ellipses. The ellipses may be drawn upon these axes by the method of Fig. 4.51 or 4.52 (b), or by trial with the aid of an ellipse template, Fig. 4.55.

In Fig. 6.35, the intersection of a plane and a quarterround molding is shown at (a), and with a cove molding at (b). In both figures, assume points 1, 2, 3, . . . , at random in the side views in which the cylindrical surfaces appear as curved lines, and project the points to the front and top views, as shown. A sufficient number of points should be used to insure smooth curves. Draw the final curves through the points with the aid of the irregular curve, §2.58.

Fig. 6.35 **Plotting Elliptical Curves.**

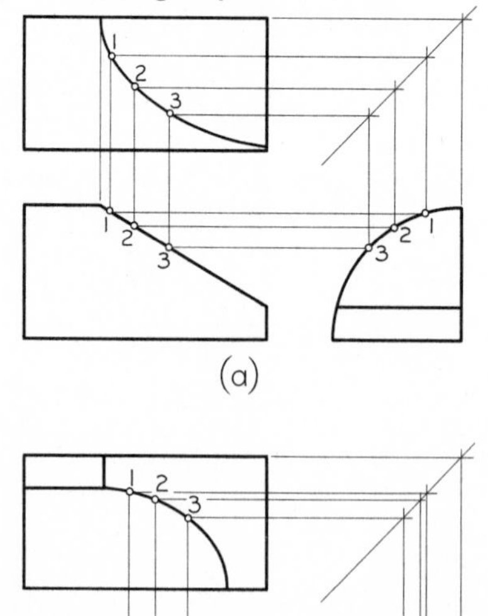


## 6.31 Space Curves.

The views of a space curve are established by the projections of points along the curve, Fig. 6.36. In this figure any points 1, 2, 3, . . . , are selected along the curve in the top view and then projected to the side view (or the reverse), and points are located in the front view by projecting downward from the top view and across from the side view. The resulting curve in the front view is drawn with the aid of the irregular curve, §2.58.

## 6.32 Intersections and Tangencies.

No line should be drawn where a curved surface is tangent to a plane surface, Fig. 6.37 (a), but when a curved surface *intersects* a plane surface, a definite edge is formed, (b). If curved surfaces are arranged as at (c), no lines appear in the top view, as shown. If the surfaces are arranged as at (d), a vertical surface in the front view produces a line in the top view.

Fig. 6.36 **Space Curve.**

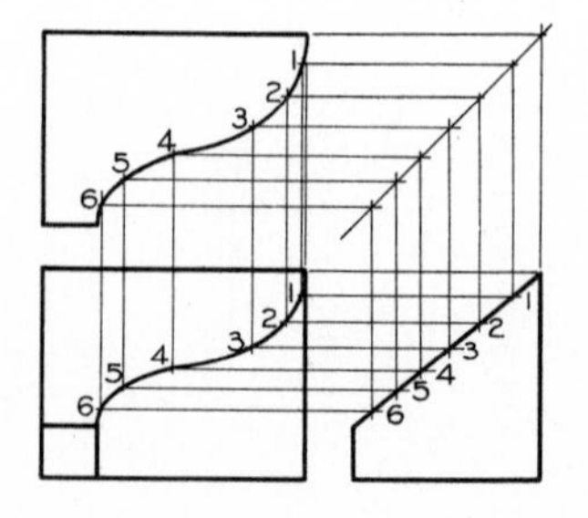

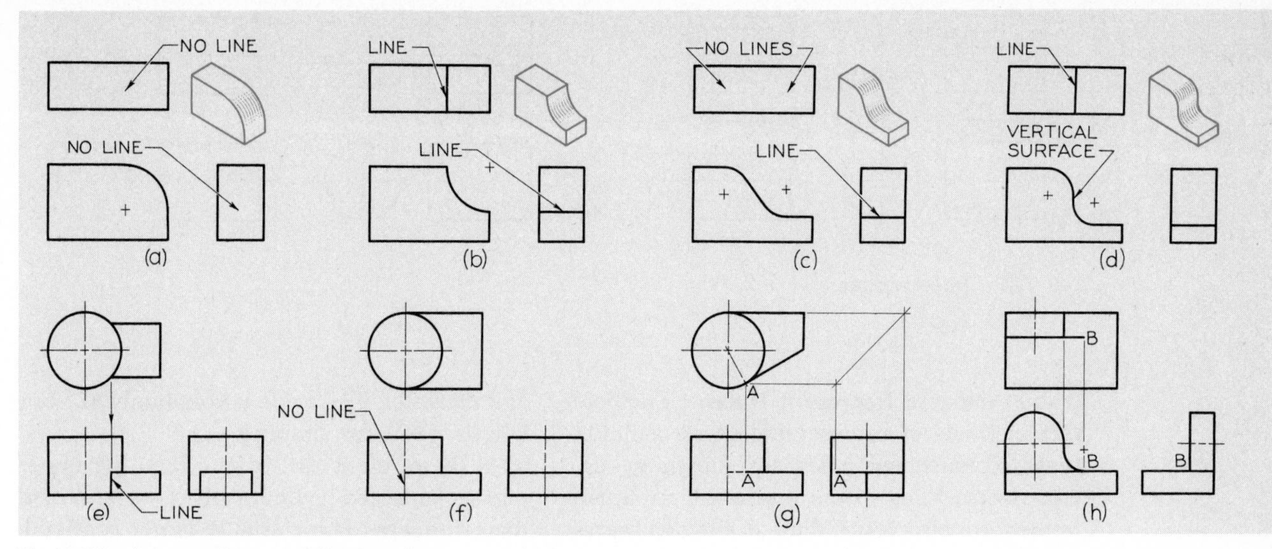


Fig. 6.37 **Intersections and Tangencies.**

Other typical intersections and tangencies of surfaces are shown from (e) to (h). To locate the point of tangency A in (g), refer to Fig. 4.34 (b).

The intersection of a small cylinder with a large cylinder is shown in Fig. 6.38 (a). The intersection is so small that it is not plotted, a straight line being used instead. At (b) the intersection is larger, but still not large enough to justify plotting the curve. The curve is approximated by drawing an arc whose radius r is the same as radius R of the large cylinder.

The intersection at (c) is significant enough to justify constructing the true curve. Points are selected at random in the circle in the side or top view, and these are then projected to the other two views to locate points on the curve in the front view, as shown. A sufficient number of points should be used, depending upon the size of the intersection, to insure a smooth and accurate curve. Draw the final curve with the aid of the irregular curve, §2.58.

At (d), the cylinders are the same diameter. The figure of intersection consists of two semi-ellipses that appear as straight lines in the front view.

If the intersecting cylinders are holes, the intersections would be similar to those for the external cylinders in Fig. 6.38. See also Fig. 7.34 (d).

In Fig. 6.39 (a), a narrow prism intersects a cylinder, but the intersection is insignificant and is ignored. At (b) the prism is larger and the intersection is noticeable enough to warrant construction, as shown. At (c) and (d) a keyseat and a small drilled hole, respectively, are shown; in both cases the intersection is not important enough to construct.

Fig. 6.38 **Intersections of Cylinders.**

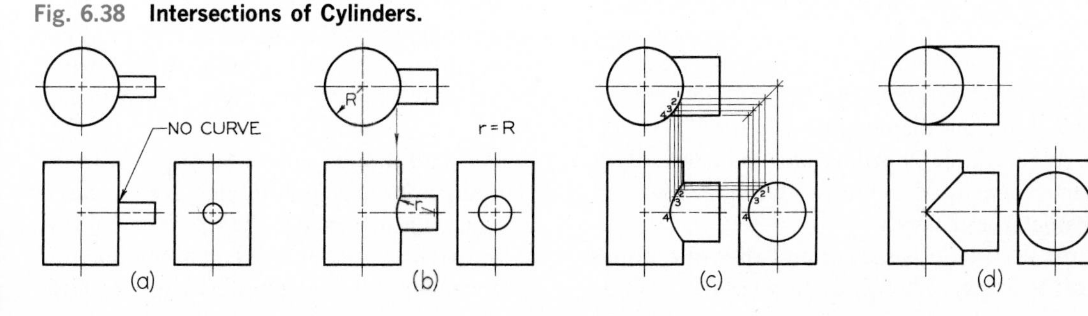

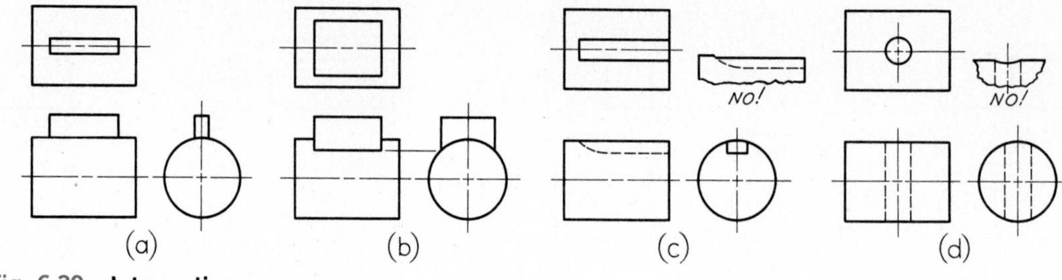


Fig. 6.39 **Intersections.**

**6.33 How to Represent Holes.** The correct methods of representing most common types of machined holes are shown in Fig. 6.40. Instructions to the machinist are given in the form of notes, and the draftsman represents the holes in conformity with these specifications. In general, the notes tell the machine operator what to do and in which order it is to be done. Hole sizes are always specified by diameter—never by radius. For each operation specified, the diameter is given first, followed by the method such as drill, ream, etc., as shown in (a) and (b).

The size of the hole may be specified as a diameter without the specific method such as drill, ream, etc., since the selection of the method will depend upon available production facilities. See (h) to (j).

A drilled hole is a *through* hole if it goes through a member. If the hole has a specified depth, as shown at (a), the hole is called a *blind* hole. The depth includes the cylindrical portion of the hole only. The point of the drill leaves a conical bottom in the hole, drawn approximately with the 30° × 60° triangle, as shown. For drill sizes, see Appendix 11 (Twist Drill Sizes). For abbreviations, see Appendix 4.

A through drilled or reamed hole is drawn as shown at (b). The note tells how the hole is to be produced—in this case by reaming. Note that tolerances are ignored in actually laying out the diameter of a hole.

At (c) a hole is drilled and then the upper part is enlarged cylindrically to a specified diameter and depth.

At (d) a hole is drilled and then the upper part is enlarged conically to a specified angle and diameter. The angle is commonly 82° but is drawn 90° for simplicity.

At (e) a hole is drilled and then the upper part is enlarged cylindrically to a specified diameter. The depth usually is not specified, but is left to the shop to determine. For average cases, the depth is drawn $\frac{1}{16}$″.

For complete information about how holes are made in the shop, see §10.20. For further information on notes, see §11.24.

**6.34 Fillets and Rounds.** A rounded interior corner is called a *fillet,* and a rounded exterior corner a *round,* Fig. 6.41 (a). Sharp corners should be avoided in designing parts to be cast or forged not only because they are difficult to produce but also because, in the case of interior corners, they are a source of weakness and failure. See §10.6 for shop processes involved.

Two intersecting rough surfaces produce a rounded corner, (b). If one of these surfaces is machined, (c), or if both surfaces are machined, (d), the corner becomes sharp. Therefore, on a drawing a rounded corner means that both intersecting surfaces are rough, and a sharp corner means that one or both surfaces are machined. On working drawings, fillets and rounds are never shaded. The presence of the curved surfaces is indicated only where they appear as arcs, except as shown in Fig. 6.45.

Fillets and rounds should be drawn with the bow pencil or bow pen if they are $\frac{1}{8}$″ radius or larger. Those smaller than $\frac{1}{8}$″ R should be made carefully freehand. As an aid in drawing these smaller arcs, some prefer to use the ends

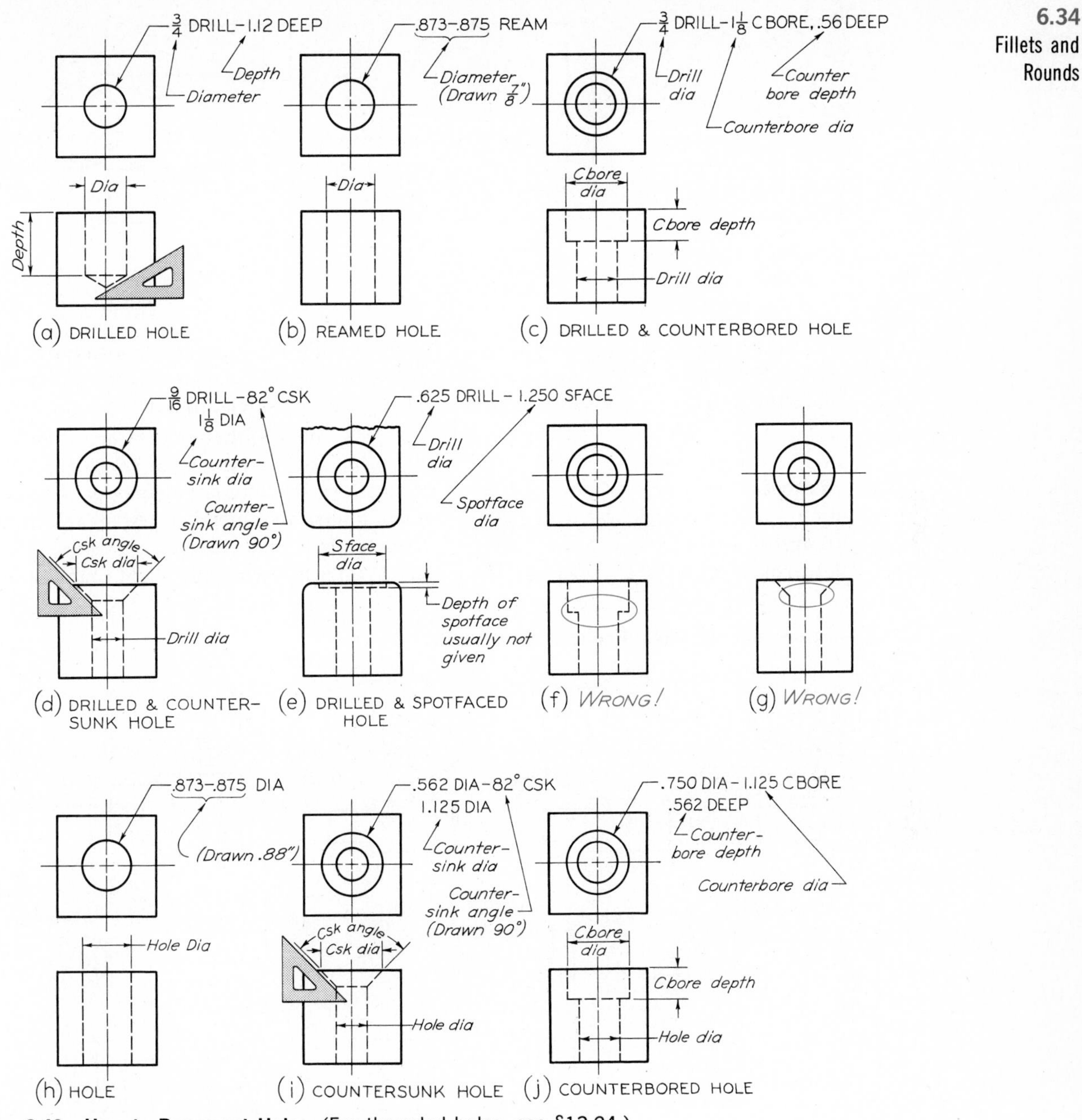


**Fig. 6.40 How to Represent Holes.** (For threaded holes, see §13.24.)

**Fig. 6.41 Rough and Finished Surfaces.**

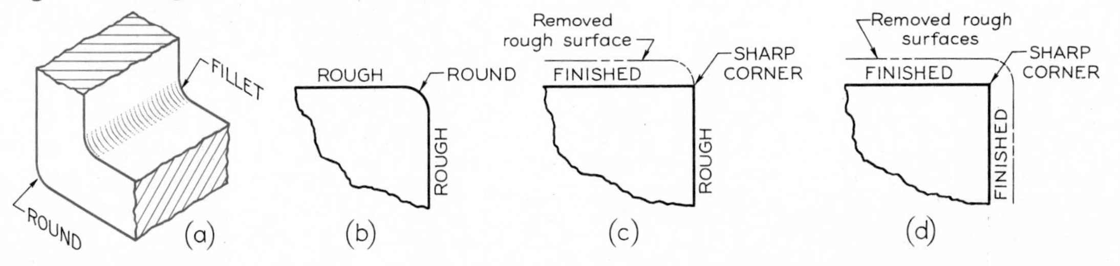

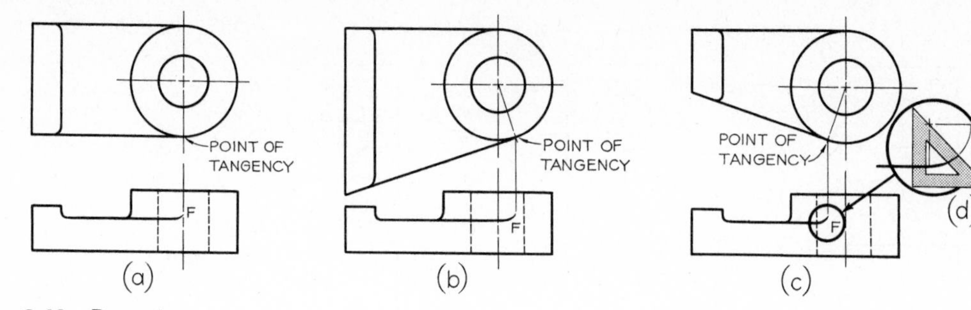


Fig. 6.42 **Runouts.**

of the slots in the erasing shield, the filleted corners of the triangle, a special fillets-and-rounds template, or a circle template.

**6.35 Runouts.** The correct method of representing fillets in connection with plane surfaces tangent to cylinders is shown in Fig. 6.42. These small curves are called *runouts.* Note that the runouts F have a radius equal to that of the fillet and a curvature of about one-eighth of a circle, (d).

Typical filleted intersections are shown in Fig. 6.43. The runouts from (a) to (d) differ because of the different shapes of the horizontal intersecting members. At (e) and (f) the runouts differ because the top surface of the web at (e) is flat, with only slight rounds along

Fig. 6.43 **Conventional Fillets, Rounds, and Runouts.**

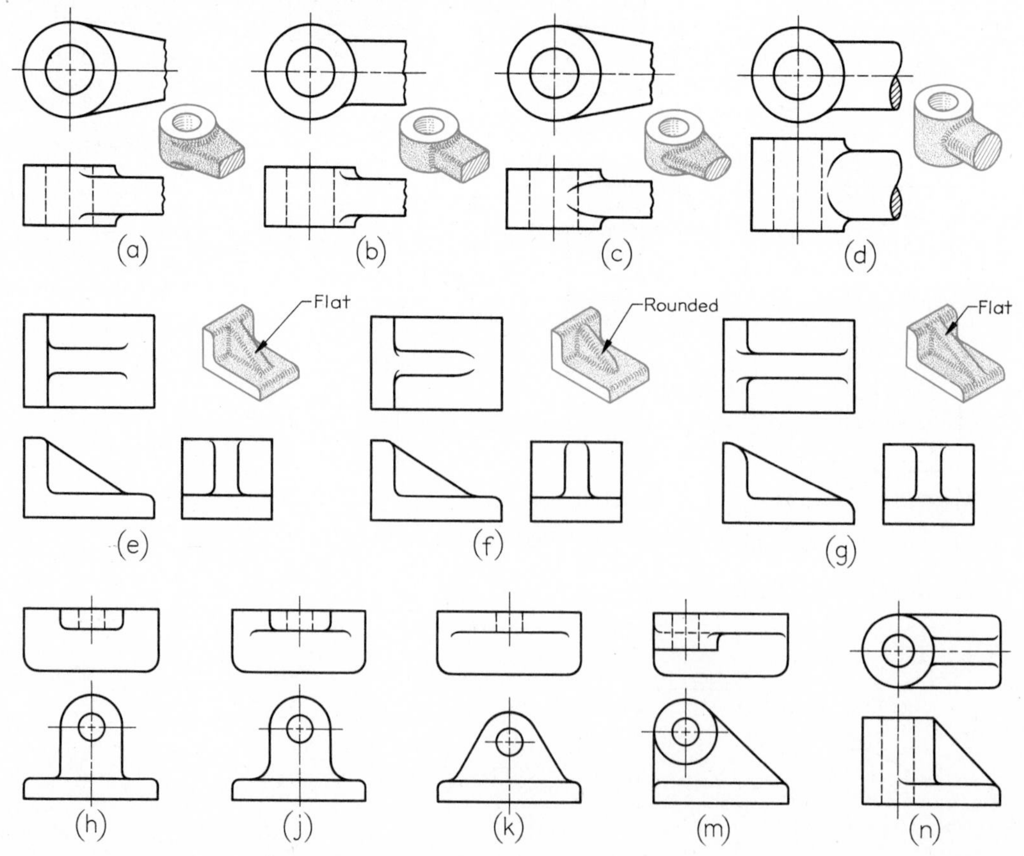

the edge, while the top surface of the web at (f) is considerably rounded. When two different sizes of fillets intersect, as at (g) and (j), the direction of the runout is dictated by the larger fillet, as shown.

## 6.36 Conventional Edges.

Rounded and filleted intersections eliminate sharp edges and sometimes make it difficult to present a clear shape description. In fact, true projection in some cases may be actually misleading, as in Fig. 6.44 (a), in which the side view of the railroad rail is quite blank. A more clear representation results if lines are added for rounded and filleted edges, as shown at (b) and (c). The added lines are projected from the actual intersections of the surfaces as if the fillets and rounds were not present.

In Fig. 6.45, two top views are shown for each given front view. The upper top views are nearly devoid of lines that contribute to the shape descriptions, while the lower top views, in which lines are used to represent the rounded and filleted edges, are quite clear. Note, in the lower top views at (a) and (c), the use of small freehand Y's where rounded or filleted edges meet a rough surface. If such an edge intersects a finished surface, no Y is shown.

## 6.37 Right-Hand and Left-Hand Parts.

In industry many individual parts are located symmetrically so as to function in pairs. These opposite parts are often exactly alike, as for example, the hub caps used on the left and right sides of the automobile. In fact, whenever possible, for economy's sake the designer will design identical parts for use on both the right and left. But opposite parts often cannot be exactly alike, such as a pair of gloves or a pair of shoes. Similarly, the right-front fender of an automobile cannot be the same shape as the left-front fender. Therefore, a left-hand part is not simply a right-hand part turned around; the two parts will be opposite and not interchangeable.

**Fig. 6.44 Conventional Representation of a Rail.**

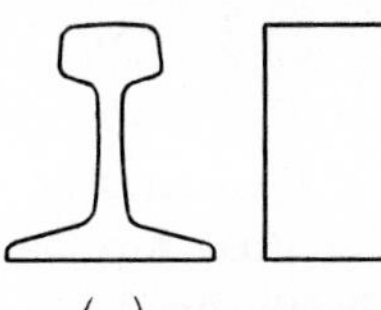

(a) TRUE PROJECTION POOR PRACTICE

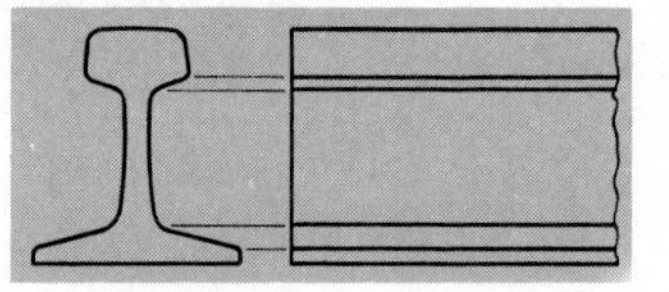

(b) CONVENTIONAL DRAWING PREFERRED IN LARGE SIZES

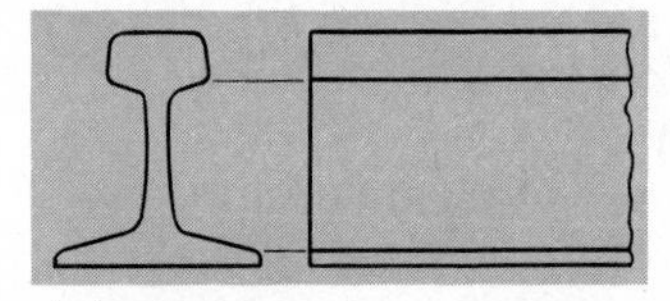

(c) CONVENTIONAL DRAWING PREFERRED IN SMALL SIZES

**Fig. 6.45 Conventional Edges.**

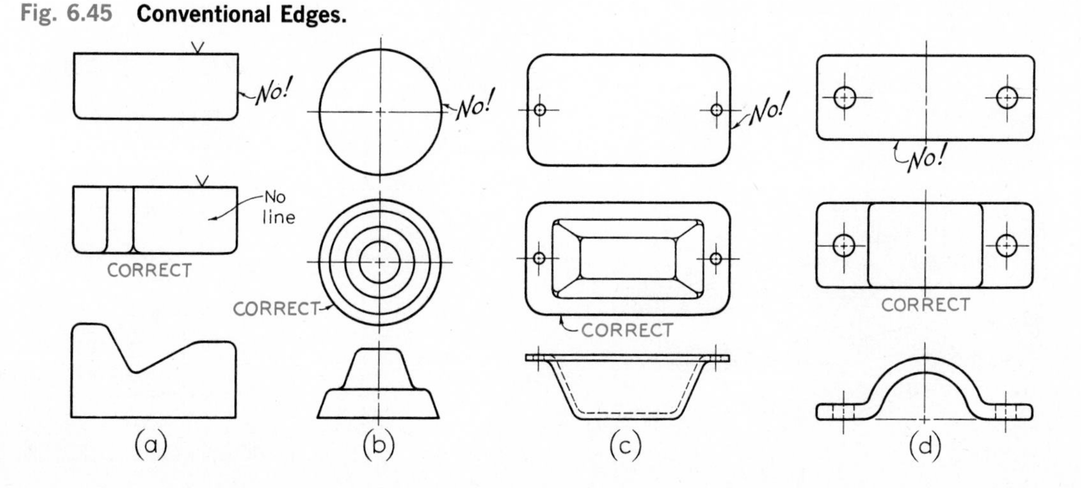

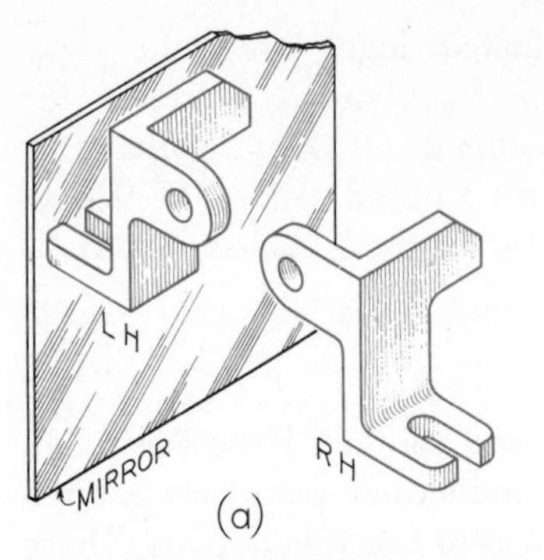


Fig. 6.46 **Right-Hand and Left-Hand Parts.**

A left-hand part is referred to as an LH part, and a right-hand part as an RH part. In Fig. 6.46 (a), the image in the mirror is the "other hand" of the part shown. If the part in front of the mirror is an RH part, the image shows the LH part. No matter how the object is turned, the image will show the LH part. At (b) and (c) are shown LH and RH drawings of the same object, and it will be seen that the drawings are also symmetrical with respect to a reference-plane line between them.

If you hold a drawing faced against a window pane or a light table so that the lines can be seen through the paper, you can trace the reverse image of the part on the back or on tracing paper, which will be a drawing of the opposite part. Or if you run a tracing upside down through the blueprint machine, the print will be reversed, and although a mirror will be needed to read the lettering, the print will be that of the opposite part.

It is customary in most cases to draw only one of two opposite parts, and to label the one that is drawn with a note, such as LH PART SHOWN, RH OPPOSITE. If the opposite-hand shape is not clear, a separate drawing must be made for it and properly identified.

## 6.38 First-Angle Projection.

If the vertical and horizontal planes of projection are considered indefinite in extent and intersecting at 90° with each other, the four dihedral angles produced are the *first, second, third,* and *fourth* angles, Fig. 6.47 (a). The profile plane intersects these two planes and may extend into all angles. If the object is placed below the horizontal plane and behind the vertical plane as in the glass box, Fig. 6.3, the object is said to be in the third angle. In this case, as we have seen, the observer is always "outside, looking in," so that for all views the lines of

Fig. 6.47 **First-Angle Projection.**

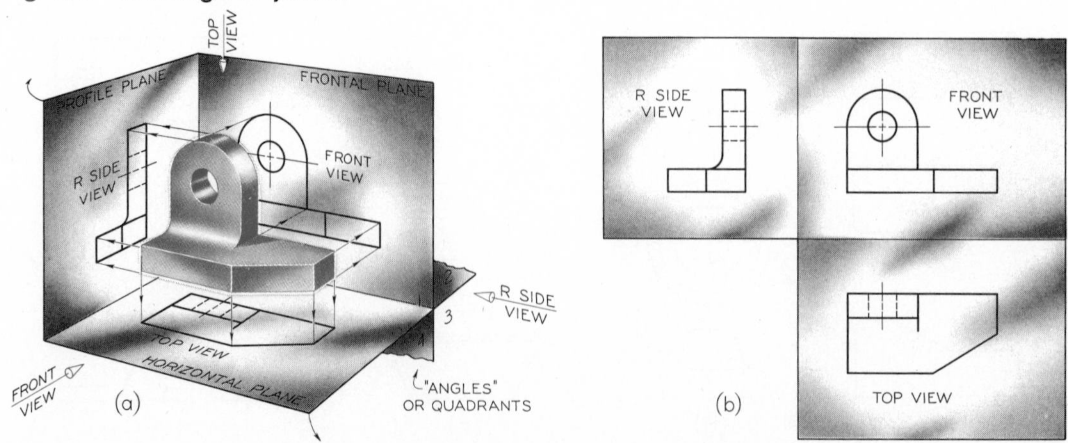

sight proceed from the eye *through the planes of projection and to the object.*

If the object is placed above the horizontal plane and in front of the vertical plane, the object is in the first angle. In this case, the observer always looks *through the object and to the planes of projection.* Thus, the right-side view is still obtained by looking toward the right side of the object, the front by looking toward the front, and the top by looking down toward the top; but the views are projected from the object onto a plane in each case. When the planes are unfolded, as at (b), the right-side view falls at the left of the front view, and the top view falls below the front view, as shown. Thus, the only ultimate difference between third-angle and first-angle projection is in the arrangement of the views. The views themselves are the same in both systems. Compare the views in Figs. 6.47 (b) and 6.5 (b).

In the United States and Canada and to some extent in England, third-angle projection is standard, while in most of the rest of the world, first-angle projection is used. First-angle projection was originally used all over the world, including the United States, but in this country it was abandoned about eighty-five years ago.

## 6.39 Multiview Projection Problems.

The following problems are intended primarily to afford practice in instrumental drawing, but any of them may be sketched freehand on cross-section paper or plain paper. Sheet layouts, Figs. 6.48 and 6.49, or inside back cover, are suggested, but the instructor may prefer a different sheet size or arrangement.

Dimensions may or may not be required by the instructor. If they are assigned, the student should study §§11.1 to 11.25. In the given problems, whether in multiview or in pictorial form, it is often not possible to give dimensions in the preferred places or, occasionally, in the standard manner. The student is expected to move dimensions to the preferred locations and otherwise to conform to the dimensioning practices recommended in Chapter 11.

For the problems in Figs. 6.53 to 6.112, it is suggested that the student make a thumbnail sketch of the necessary views, in each case, and obtain his instructor's approval before starting the mechanical drawing.

For additional problems, see Fig. 8.29. Draw top views instead of auxiliary views.

Problems in convenient form for solution are available in *Technical Drawing Problems, Series 1,* by Giesecke, Mitchell, Spencer, and Hill; in *Technical Drawing Problems, Series 2,* by Spencer and Hill; and *Technical Drawing Problems, Series 3,* by Spencer and Hill, all designed to accompany this text and published by Macmillan Publishing Co., Inc.

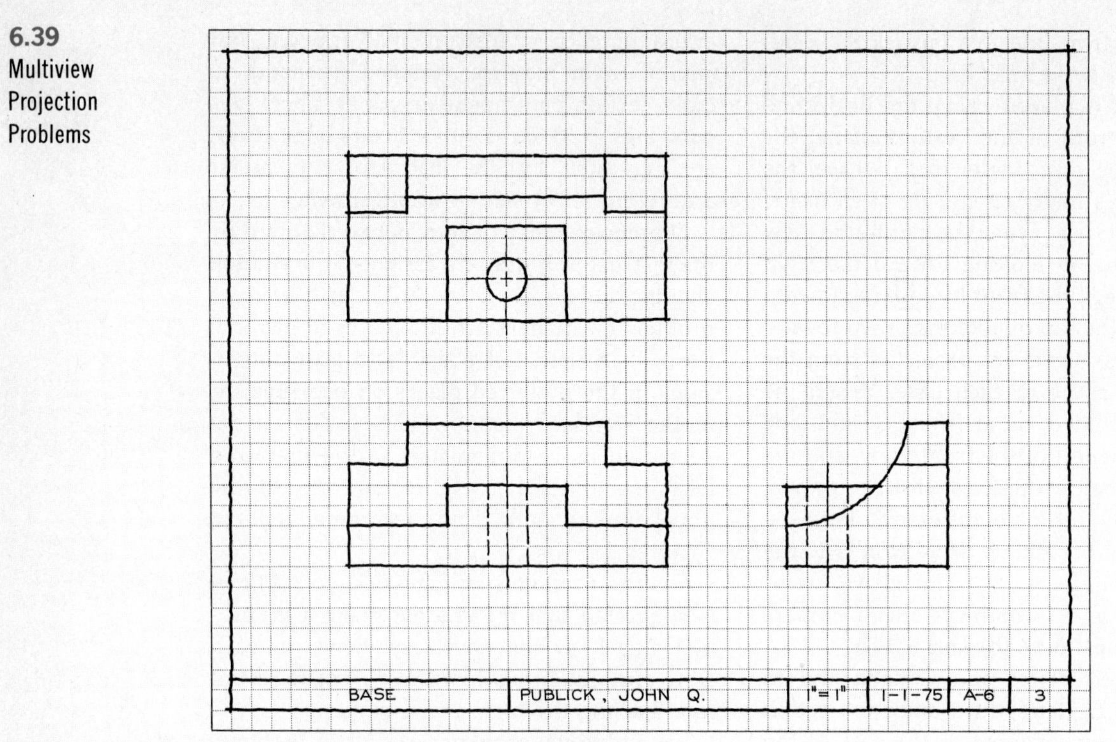

Fig. 6.48 **Freehand Sketch** (Layout A-2).

Fig. 6.49 **Mechanical Drawing** (Layout A-3).

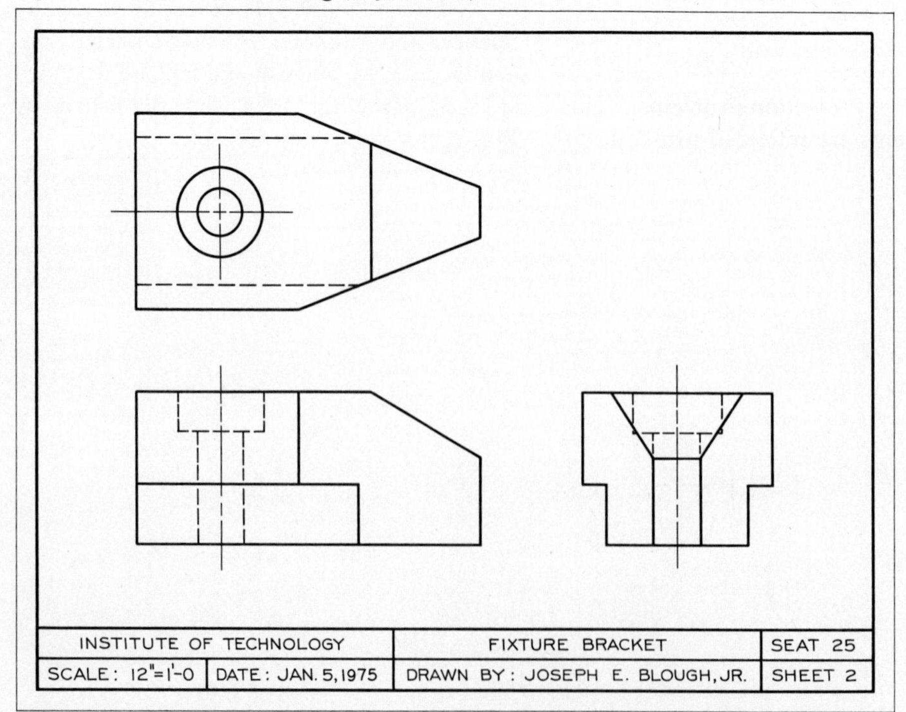

INSTITUTE OF TECHNOLOGY
FIXTURE BRACKET
SEAT 25
SCALE: 12"=1'-0
DATE: JAN. 5, 1975
DRAWN BY: JOSEPH E. BLOUGH, JR.
SHEET 2

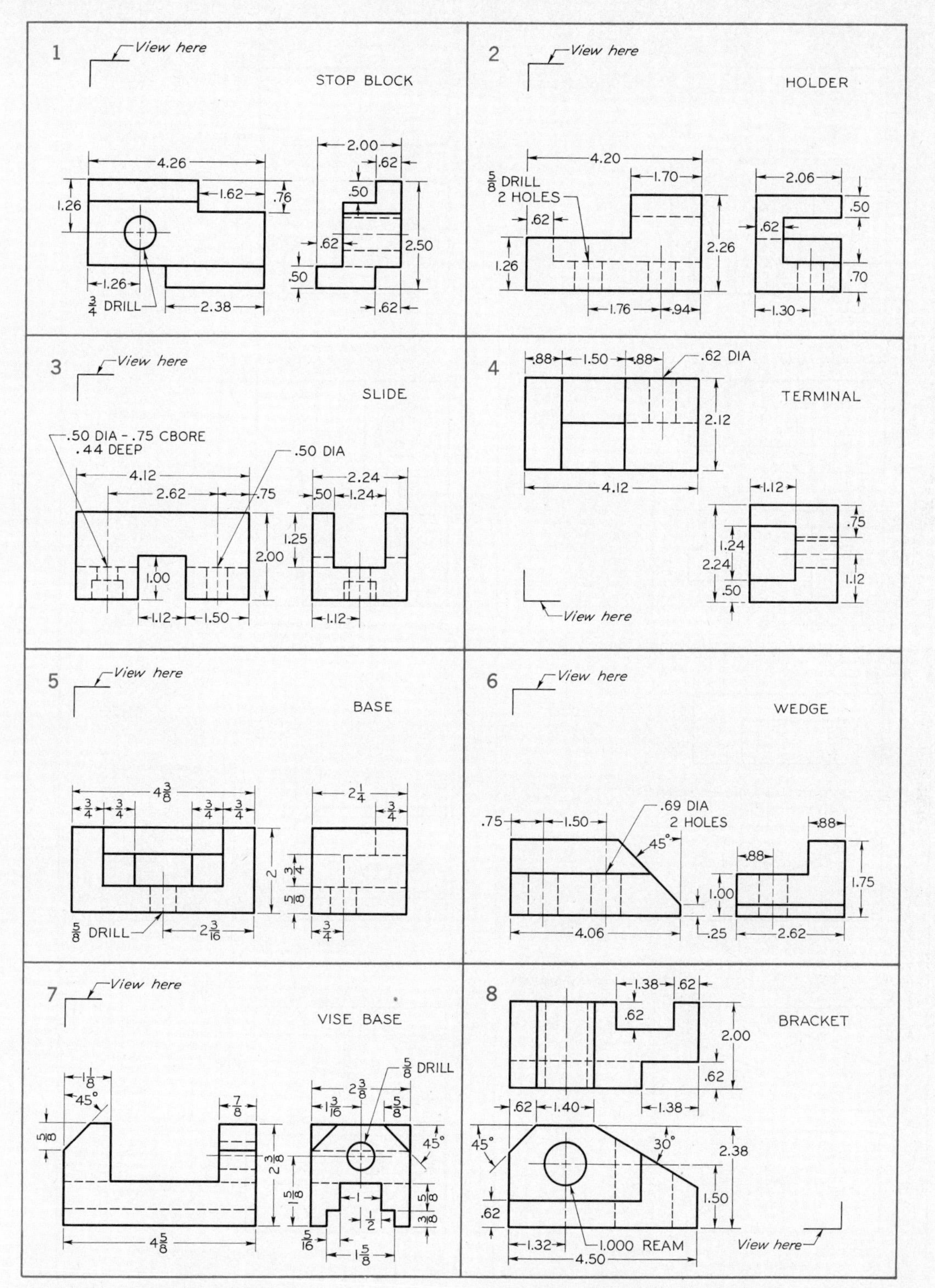


Fig. 6.50 **Missing-View Problems.** Using Layout A–2 or A–3, sketch or draw mechanically the given views, and add the missing view, as shown in Figs. 6.48 and 6.49. If dimensions are required, study §§11.1 to 11.25. Move dimensions to better locations where possible. In Probs. 1 to 5, all surfaces are normal surfaces.

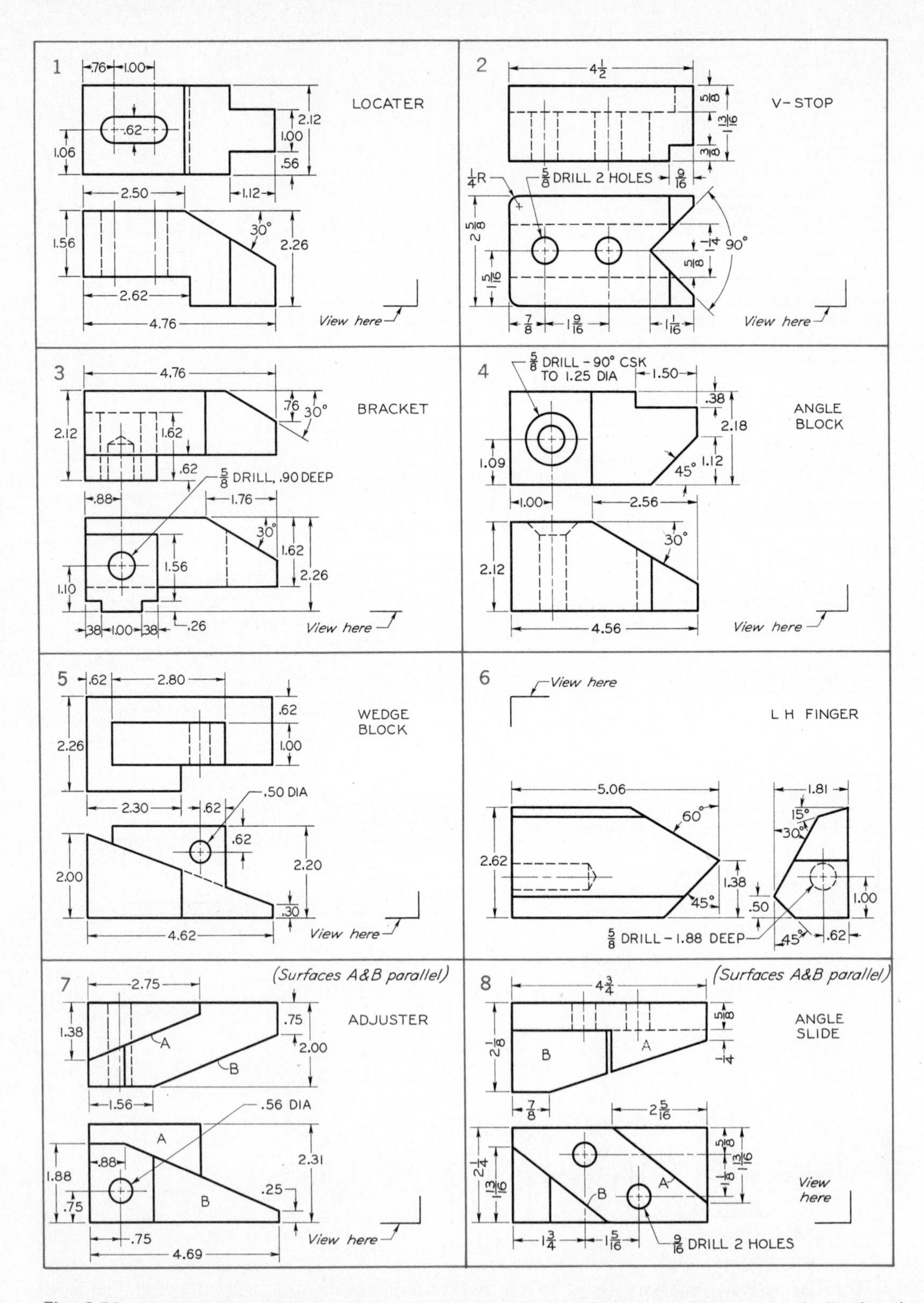


Fig. 6.51 **Missing-View Problems.** Using Layout A–2 or A–3, sketch or draw mechanically the given views, and add the missing view, as shown in Figs. 6.48 and 6.49. If dimensions are required, study §§11.1 to 11.25. Move dimensions to better locations where possible.

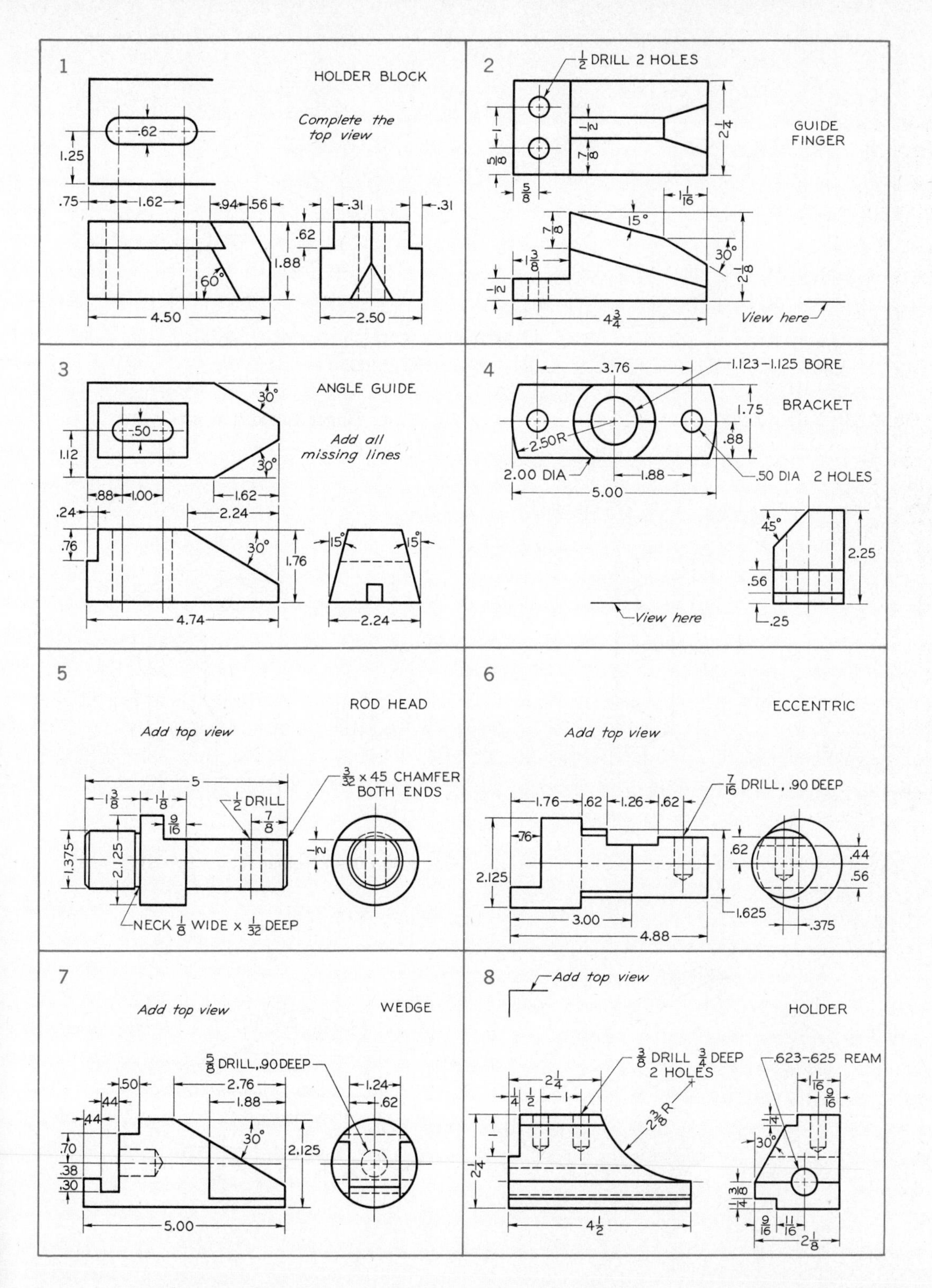


Fig. 6.52 **Missing-View Problems.** Using Layout A–2 or A–3, sketch or draw mechanically the given views, and add the missing view, as shown in Figs. 6.48 and 6.49. If dimensions are required, study §§11.1 to 11.25. Move dimensions to better locations where possible.

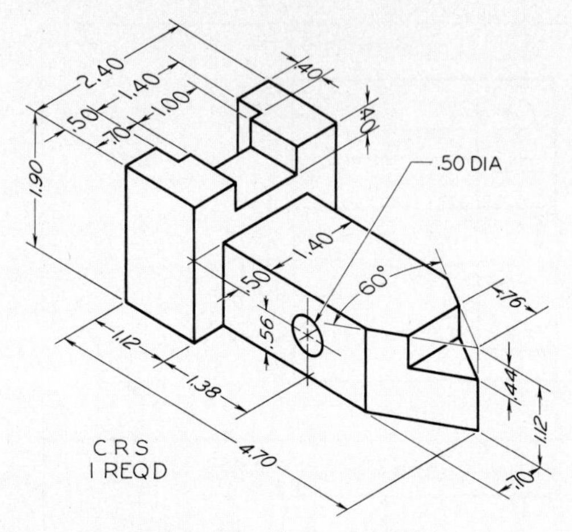


Fig. 6.53 **Safety Key** (Layout A-3).*

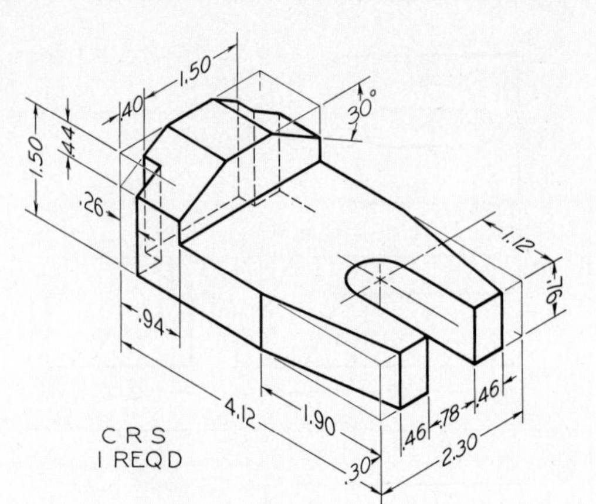


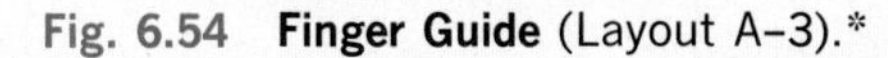

Fig. 6.54 **Finger Guide** (Layout A-3).*

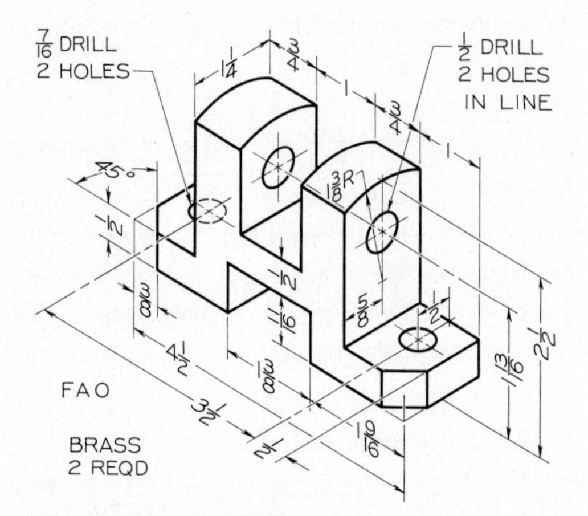


Fig. 6.55 **Rod Support** (Layout A-3).*

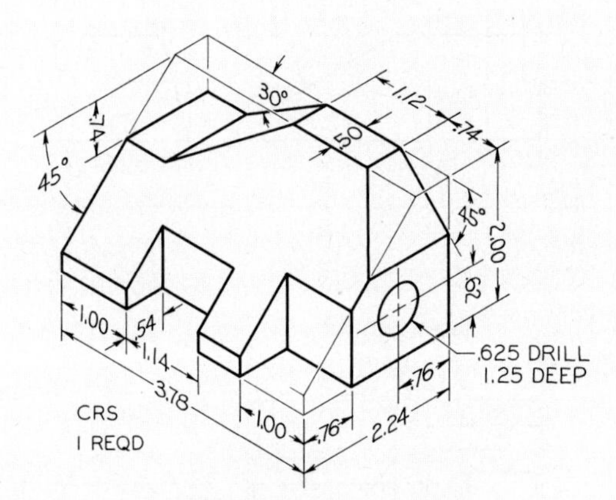


Fig. 6.56 **Tool Holder** (Layout A-3).*

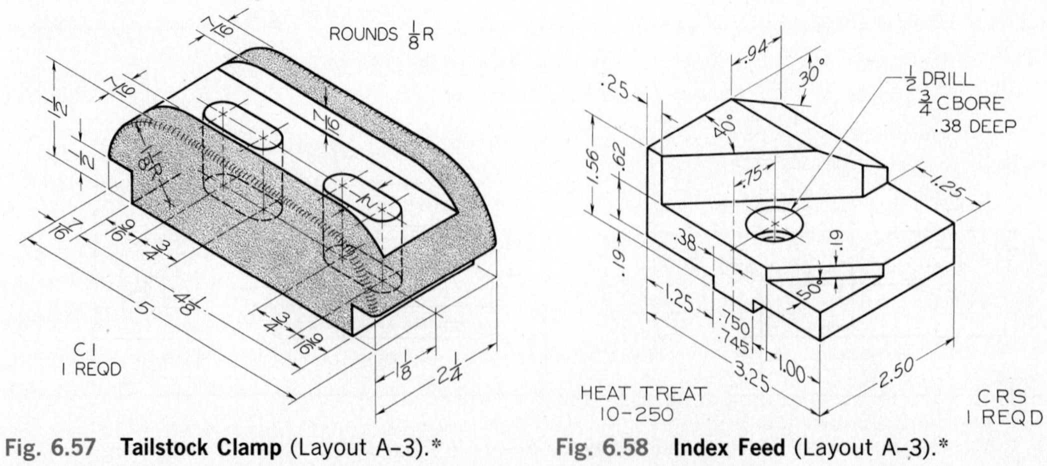


Fig. 6.57 **Tailstock Clamp** (Layout A-3).*

Fig. 6.58 **Index Feed** (Layout A-3).*

* Draw or sketch necessary views. If dimensions are required, study §§11.1 to 11.25.

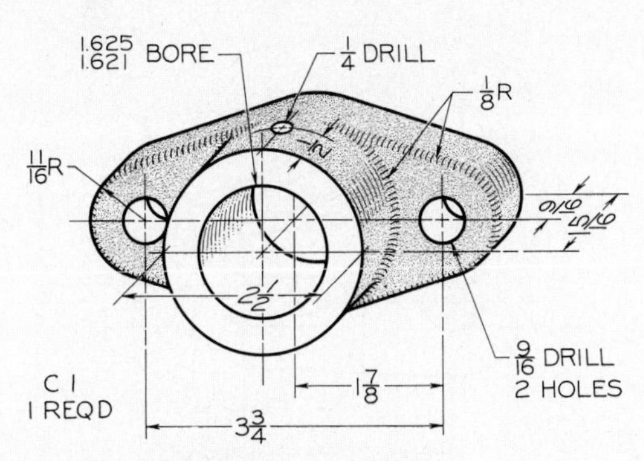

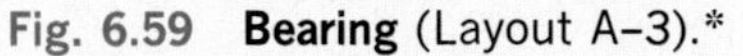


Fig. 6.59 **Bearing** (Layout A-3).*

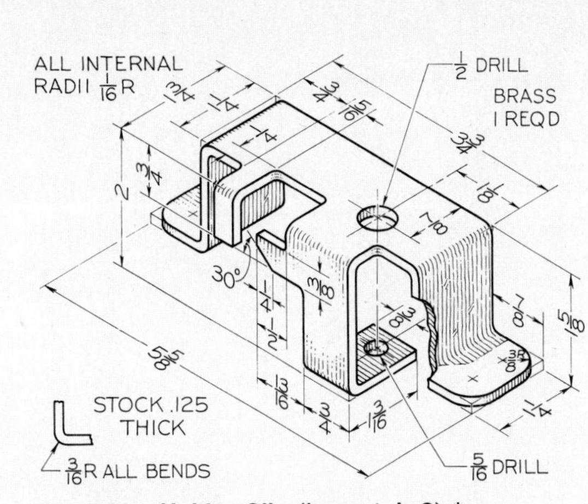


Fig. 6.60 **Holder Clip** (Layout A-3).*

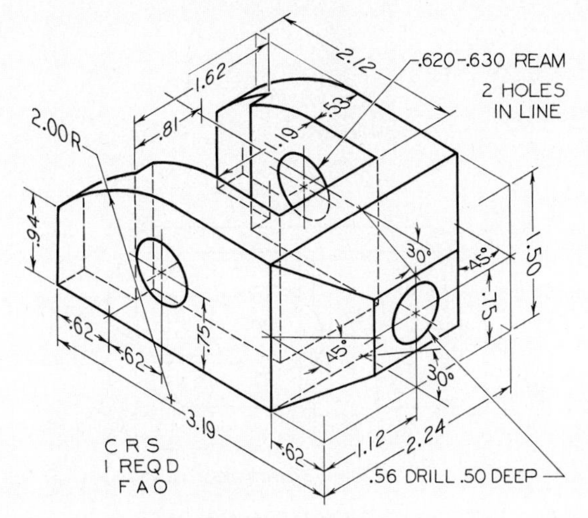


Fig. 6.61 **Cam** (Layout A-3).*

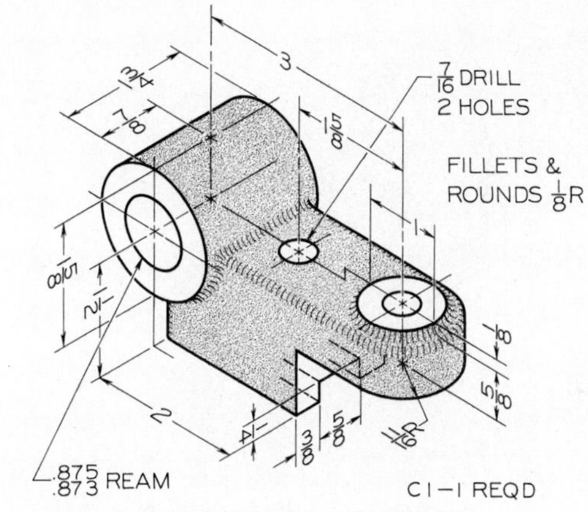

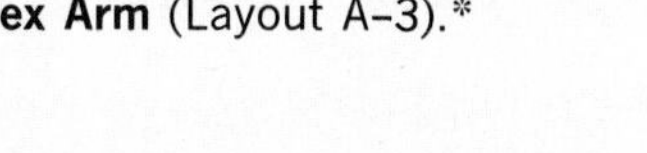


Fig. 6.62 **Index Arm** (Layout A-3).*

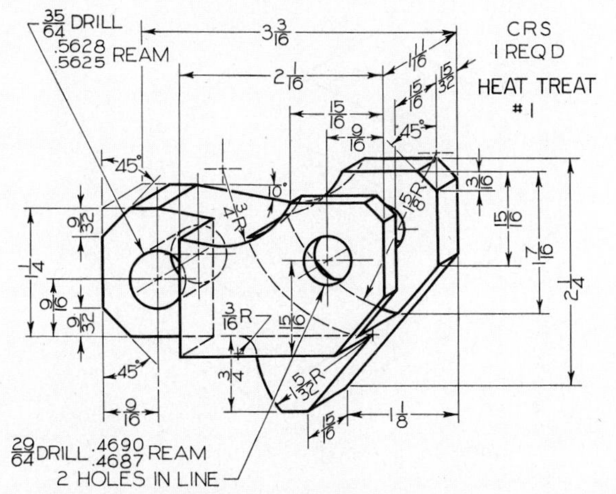


Fig. 6.63 **Roller Lever** (Layout A-3).*

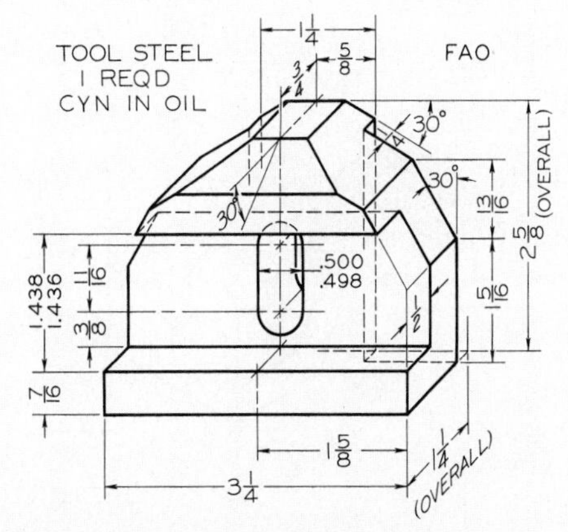


Fig. 6.64 **Support** (Layout A-3).*

* Draw or sketch necessary views. If dimensions are required, study §§11.1 to 11.25.

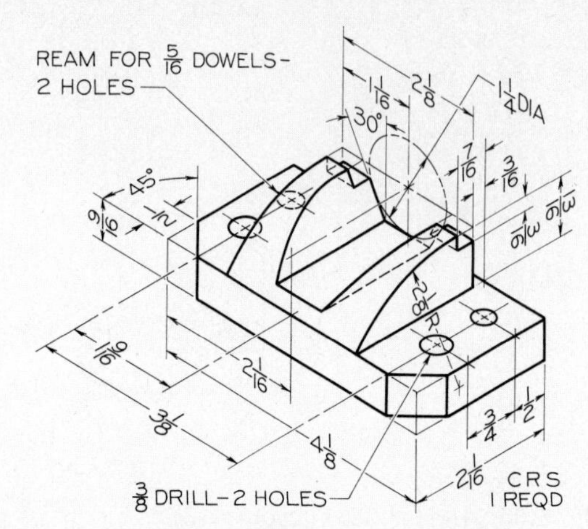


Fig. 6.65 **Locating Finger** (Layout A–3).*

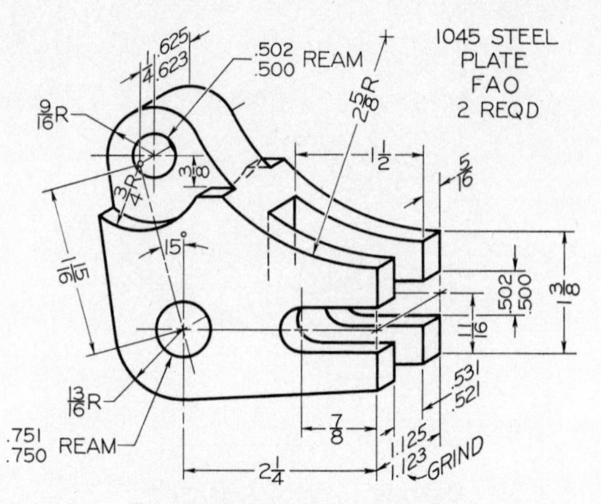


Fig. 6.66 **Toggle Lever** (Layout A–3).*

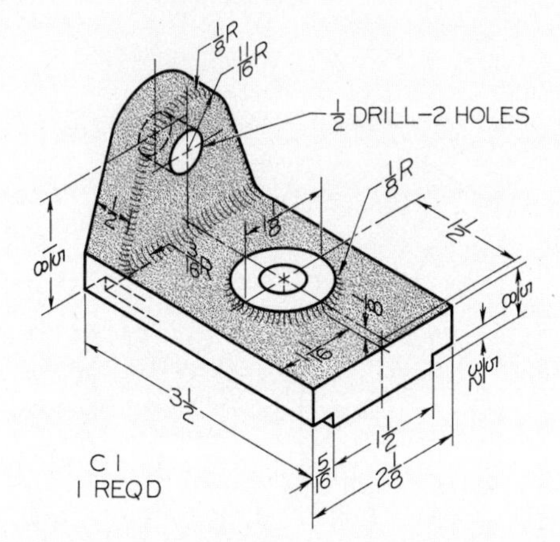


Fig. 6.67 **Cut-off Holder** (Layout A–3).*



Fig. 6.68 **Index Slide** (Layout A–3).*

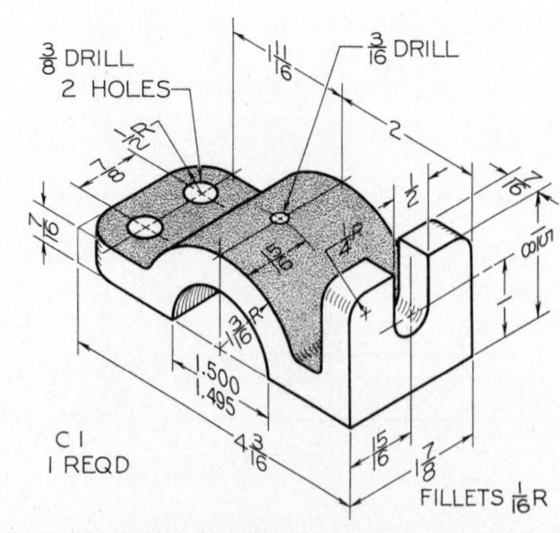

Fig. 6.69 **Frame Guide** (Layout A–3).*

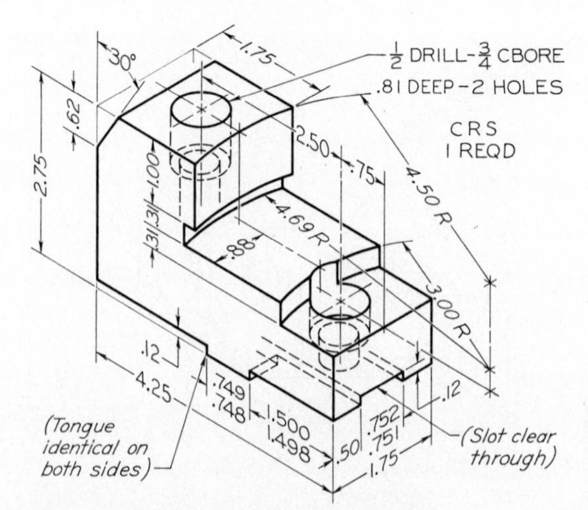


Fig. 6.70 **Chuck Jaw** (Layout A–3).*



* Draw or sketch necessary views. If dimensions are required, study §§11.1 to 11.25.

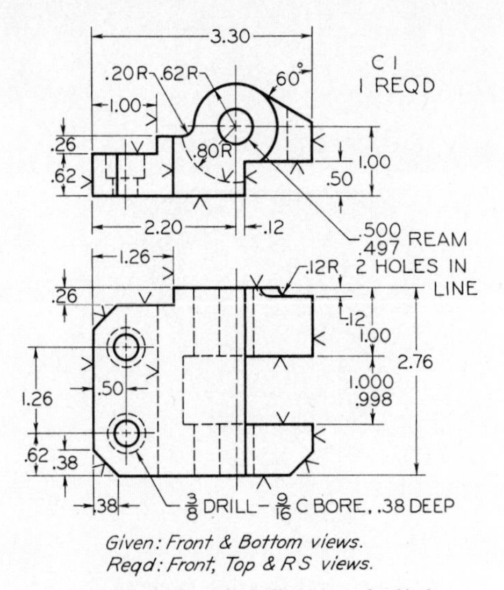


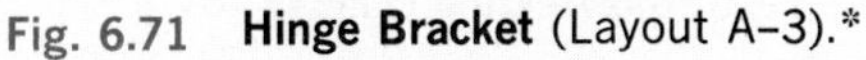
Fig. 6.71 **Hinge Bracket** (Layout A–3).*

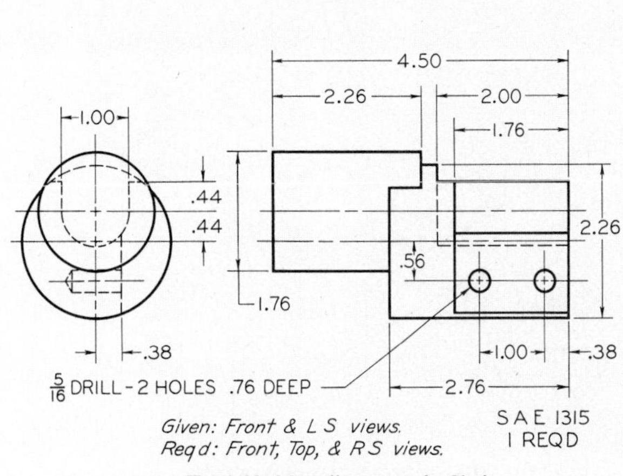


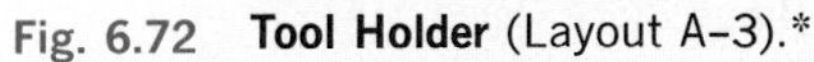
Fig. 6.72 **Tool Holder** (Layout A–3).*

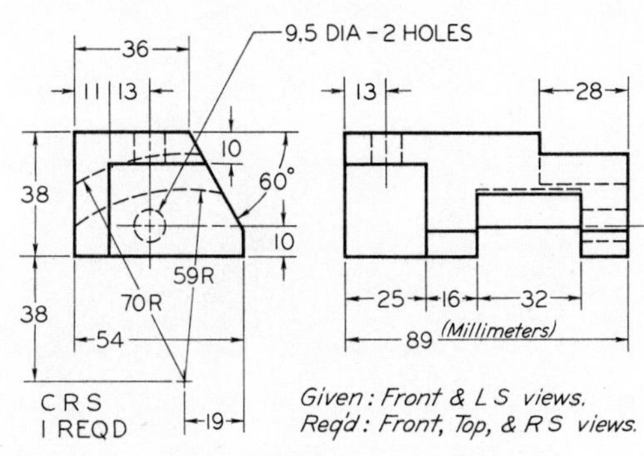

Fig. 6.73 **Shifter Block** (Layout A–3).*

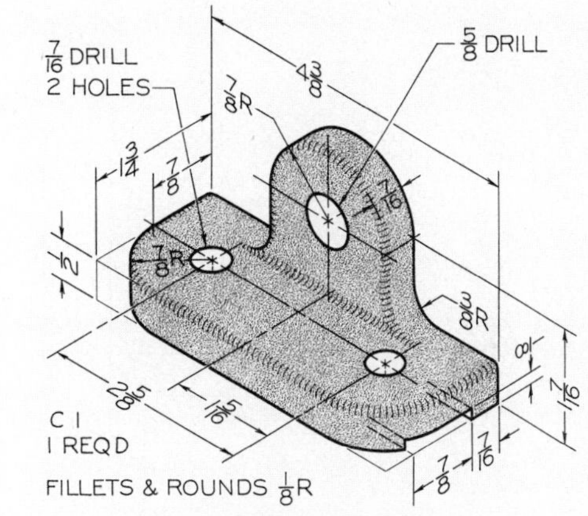


Fig. 6.74 **Cross-feed Stop** (Layout A–3).*

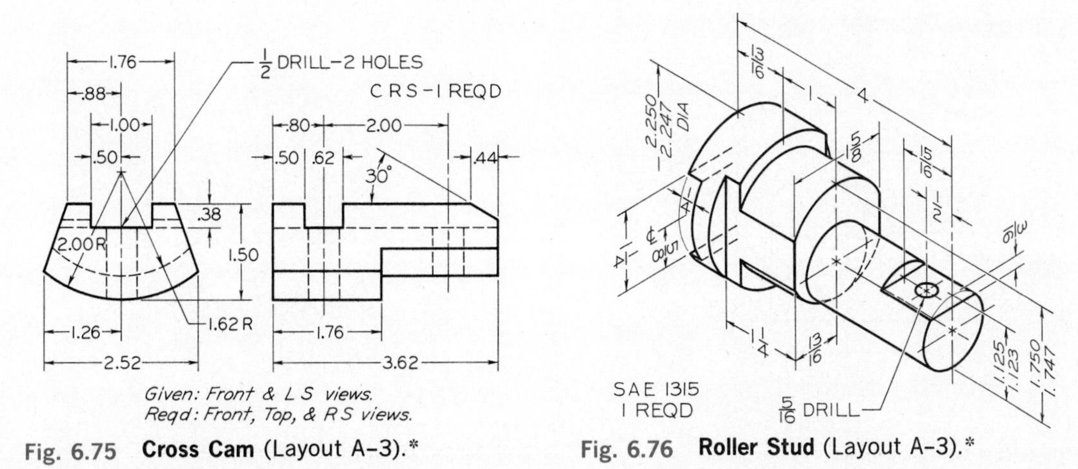


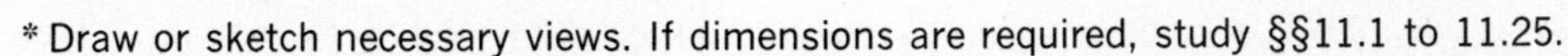
Fig. 6.75 **Cross Cam** (Layout A–3).*

Fig. 6.76 **Roller Stud** (Layout A–3).*

* Draw or sketch necessary views. If dimensions are required, study §§11.1 to 11.25.

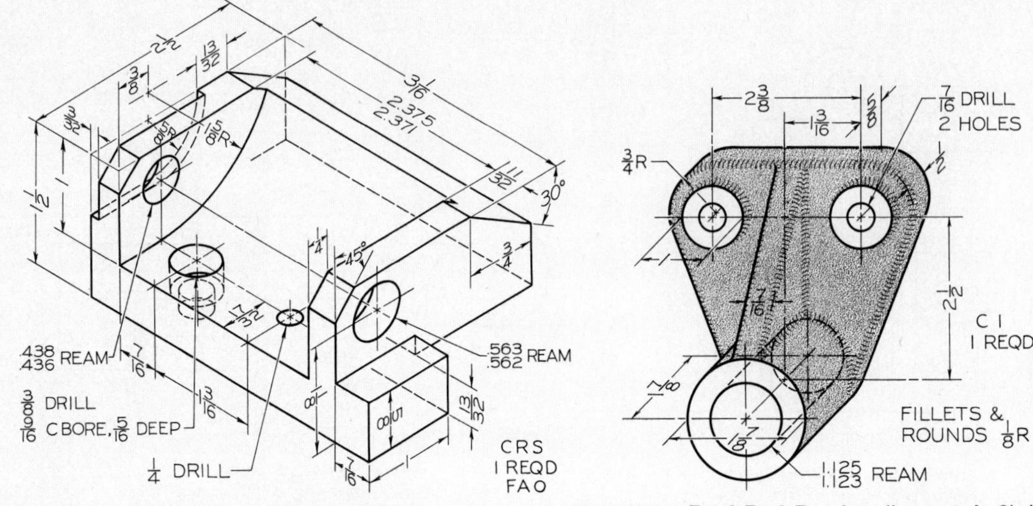


Fig. 6.77 **Hinge Block** (Layout A–3).*

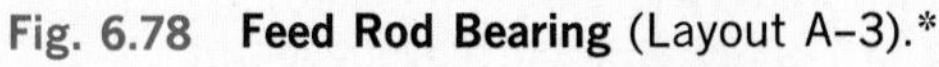

Fig. 6.78 **Feed Rod Bearing** (Layout A–3).*

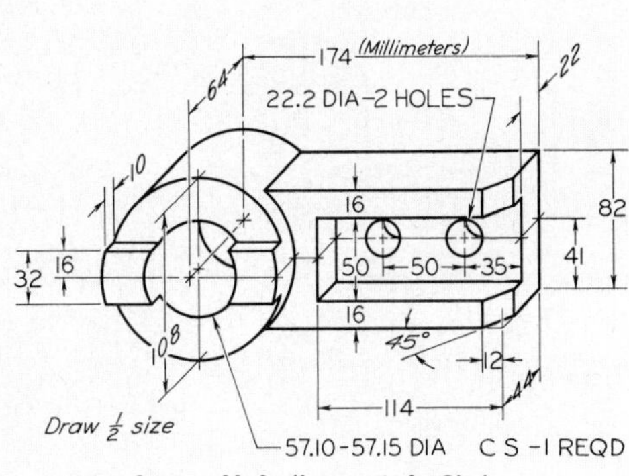


Fig. 6.79 **Lever Hub** (Layout A–3).*



Fig. 6.80 **Vibrator Arm** (Layout A–3).*

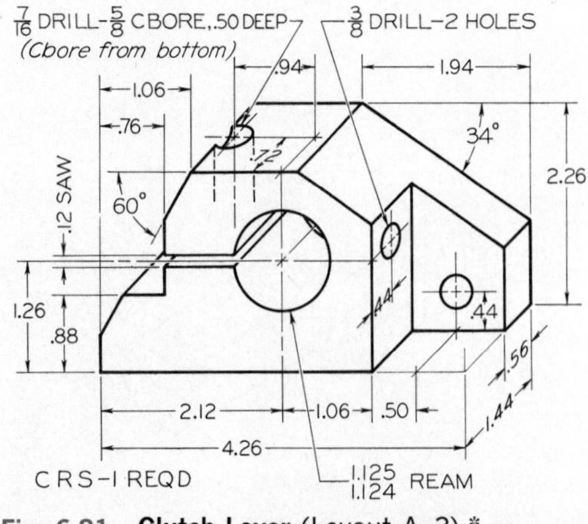


Fig. 6.81 **Clutch Lever** (Layout A–3).*

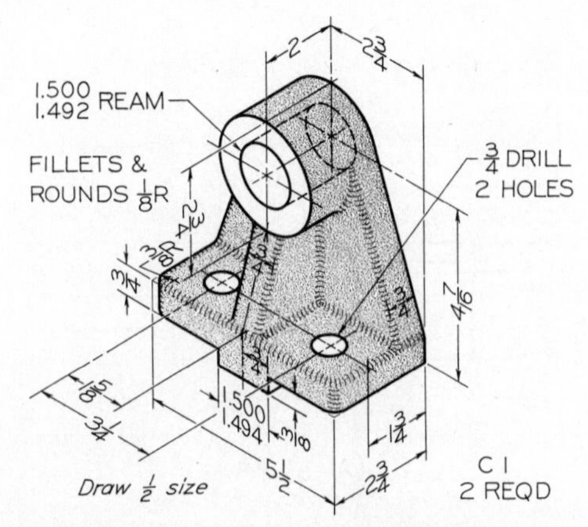


Fig. 6.82 **Counter Bearing Bracket** (Layout A–3).*

* Draw or sketch necessary views. If dimensions are required, study §§11.1 to 11.25.

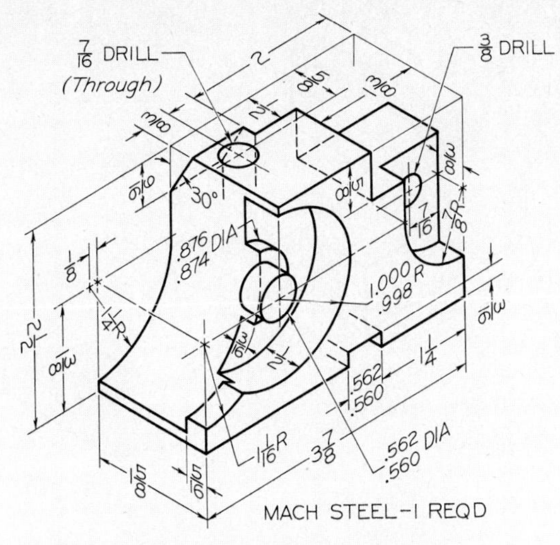


Fig. 6.83 **Tool Holder** (Layout A-3).*

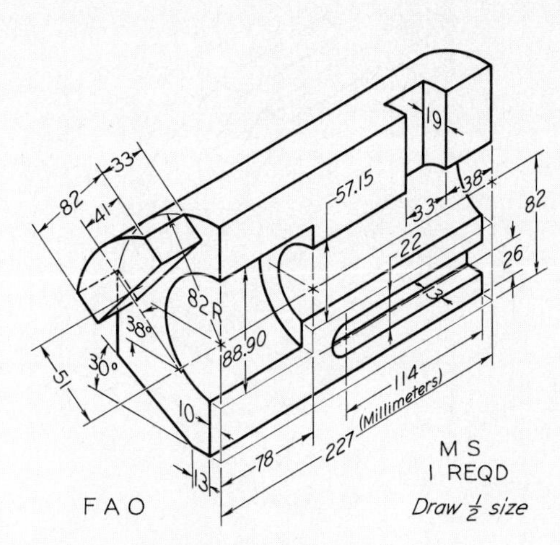


Fig. 6.84 **Control Block** (Layout A-3).*

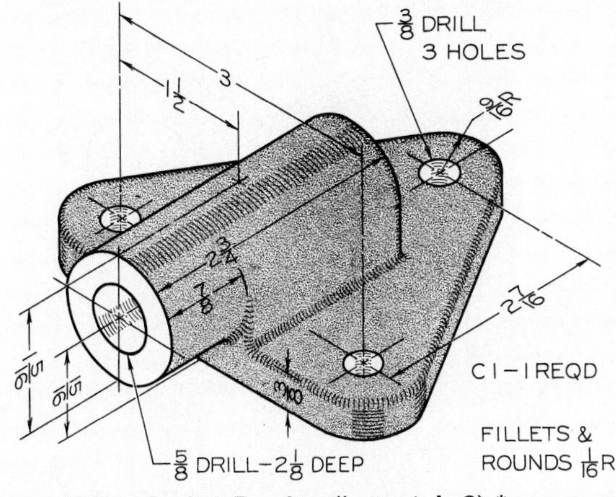

Fig. 6.85 **Socket Bearing** (Layout A-3).*

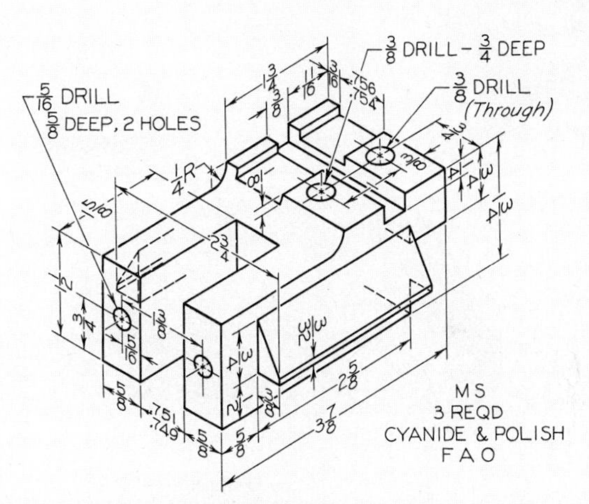


Fig. 6.86 **Tool Holder** (Layout A-3).*

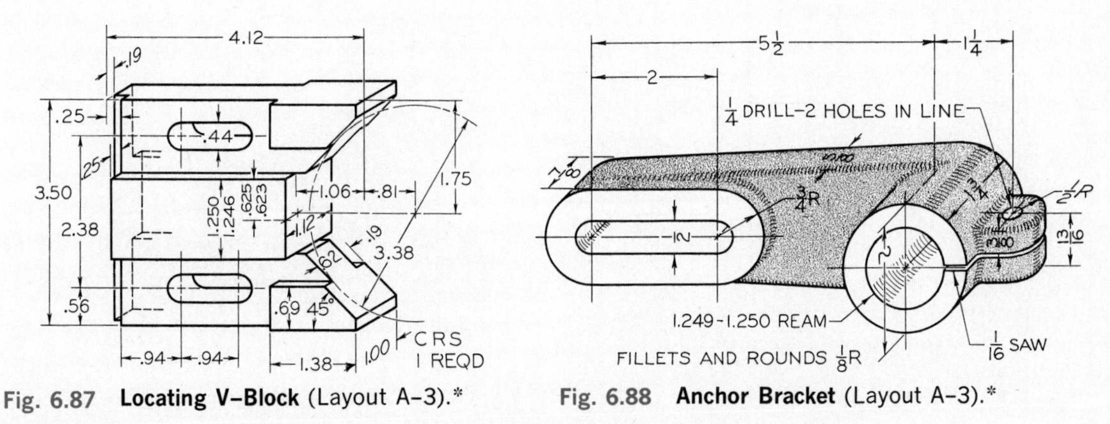


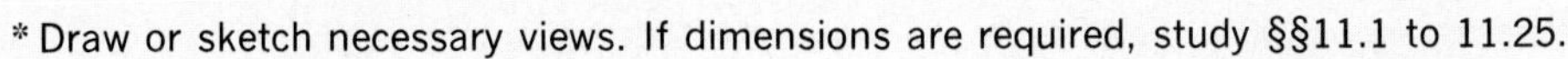

Fig. 6.87 **Locating V-Block** (Layout A-3).*

Fig. 6.88 **Anchor Bracket** (Layout A-3).*

* Draw or sketch necessary views. If dimensions are required, study §§11.1 to 11.25.

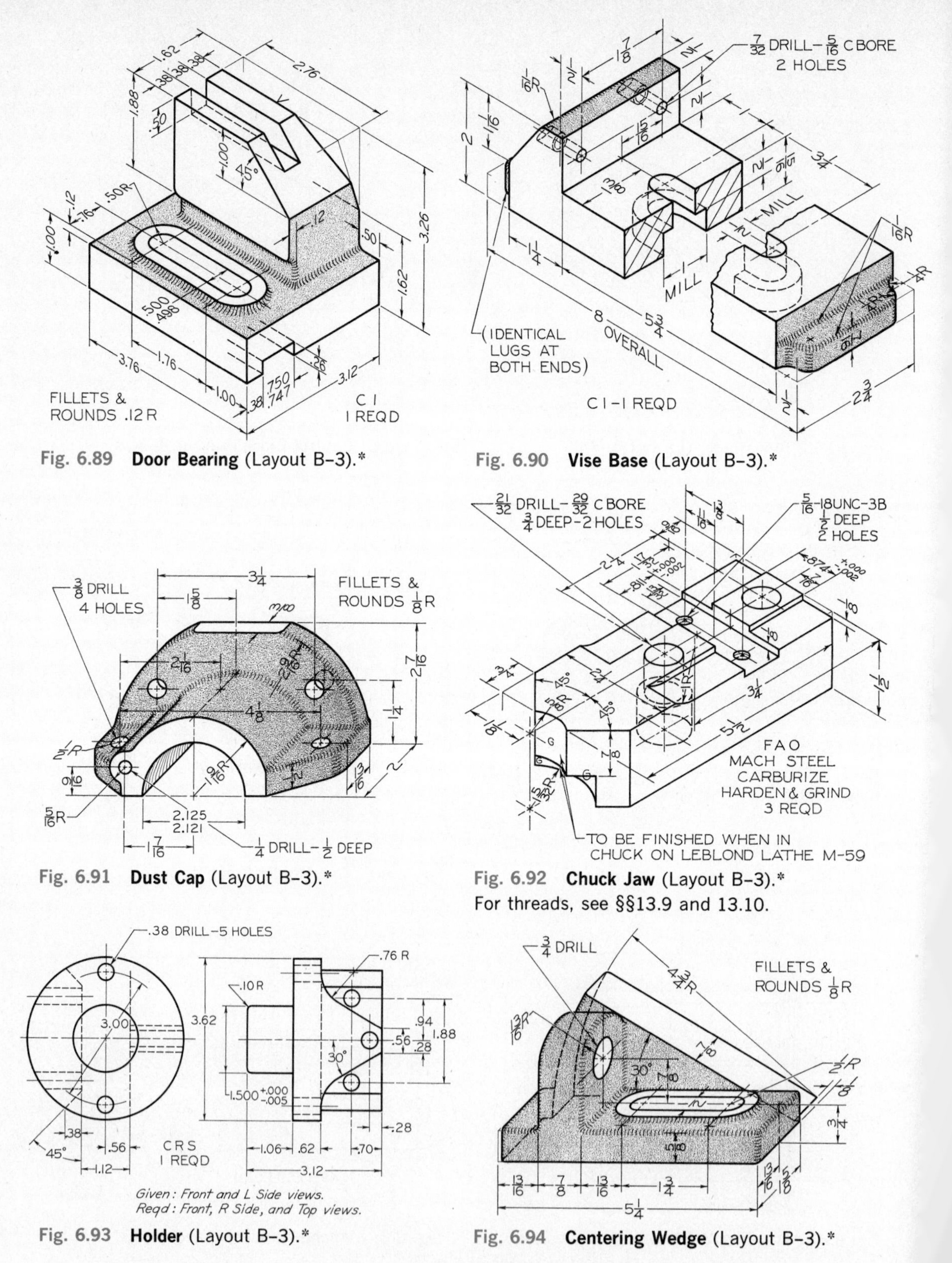


**Fig. 6.89** **Door Bearing** (Layout B-3).*

**Fig. 6.90** **Vise Base** (Layout B-3).*

**Fig. 6.91** **Dust Cap** (Layout B-3).*

**Fig. 6.92** **Chuck Jaw** (Layout B-3).*
For threads, see §§13.9 and 13.10.

**Fig. 6.93** **Holder** (Layout B-3).*

**Fig. 6.94** **Centering Wedge** (Layout B-3).*



* Draw or sketch necessary views. If dimensions are required, study §§11.1 to 11.25.

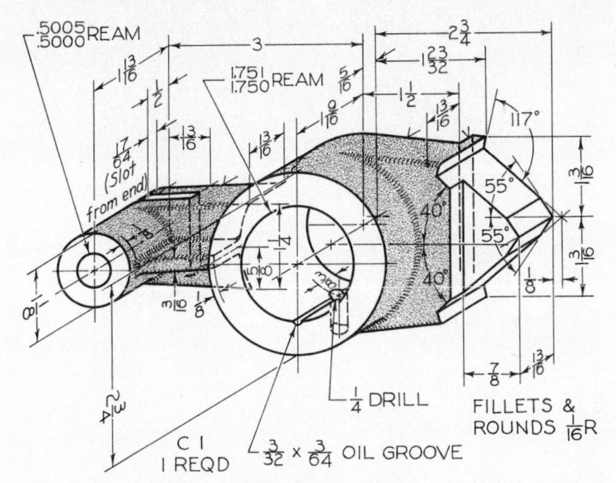


Fig. 6.95 **Motor Switch Lever.** Draw or sketch necessary views (Layout B–3).

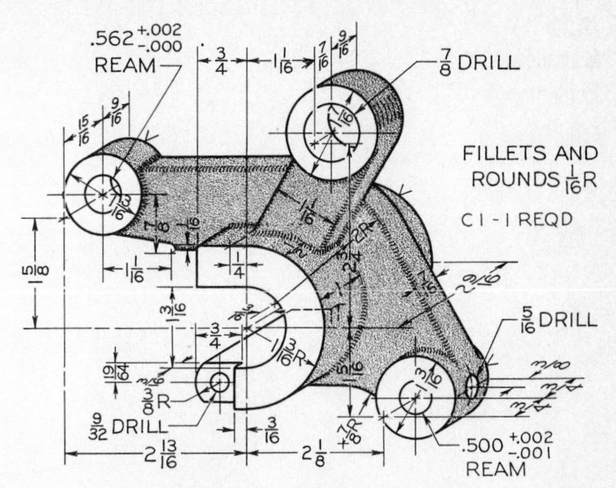


Fig. 6.96 **Socket Form Roller—LH.** Draw or sketch necessary views (Layout B–4).

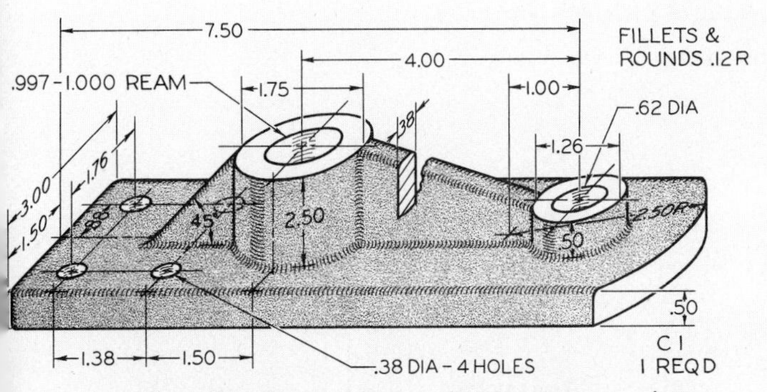


Fig. 6.97 **Stop Base.** Draw or sketch necessary views (Layout B–3).

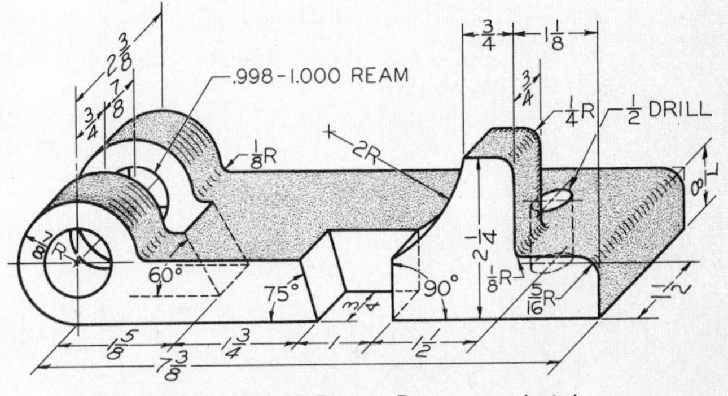


Fig. 6.98 **Hinge Base.** Draw or sketch necessary views (Layout B–3).

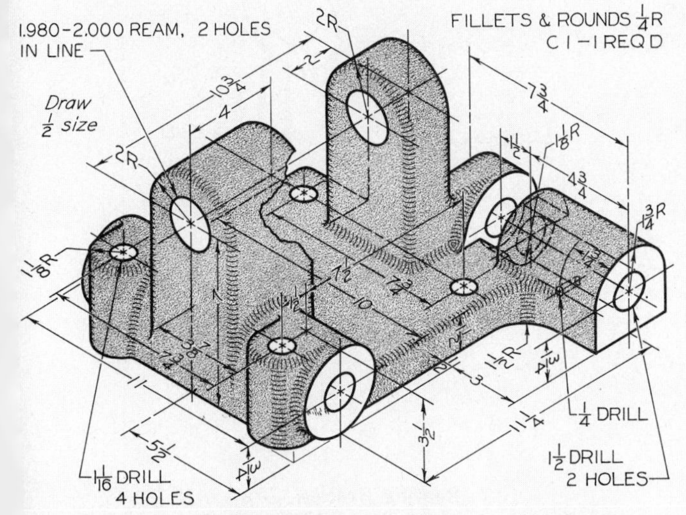


Fig. 6.99 **Automatic Stop Base.** Draw or sketch necessary views (Layout C–3).

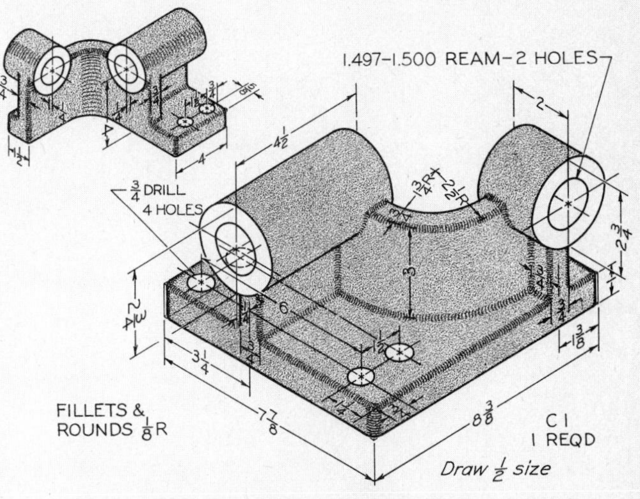


Fig. 6.100 **Lead Screw Bracket.** Draw or sketch necessary views (Layout C–3).

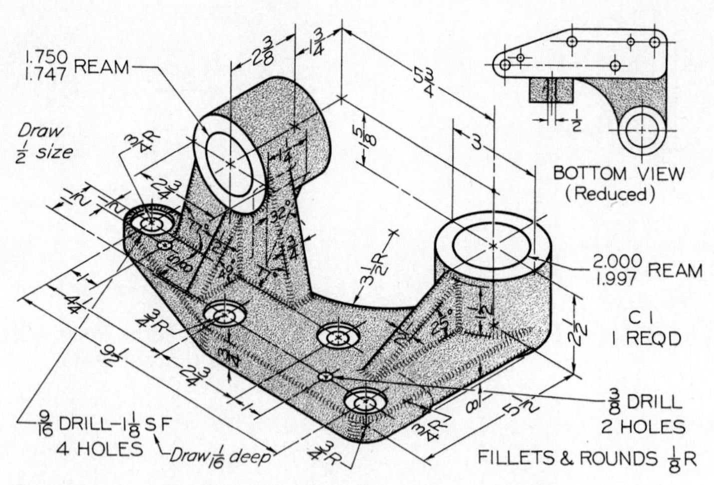


Fig. 6.101 **Lever Bracket.** Draw or sketch necessary views (Layout C–3).

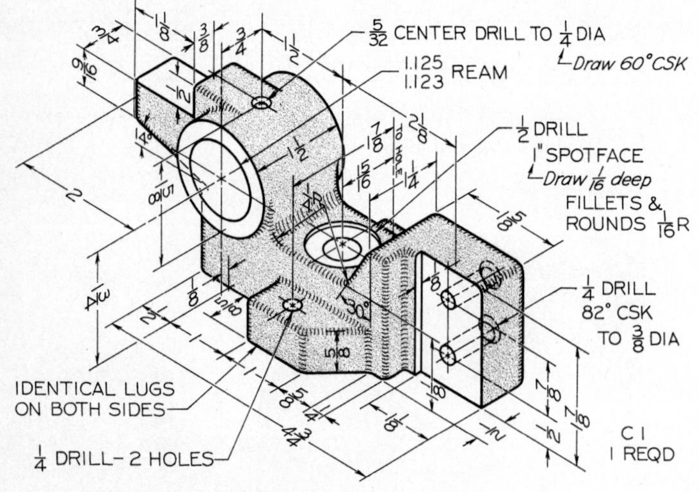


Fig. 6.102 **Gripper Rod Center.** Draw or sketch necessary views (Layout B–3).

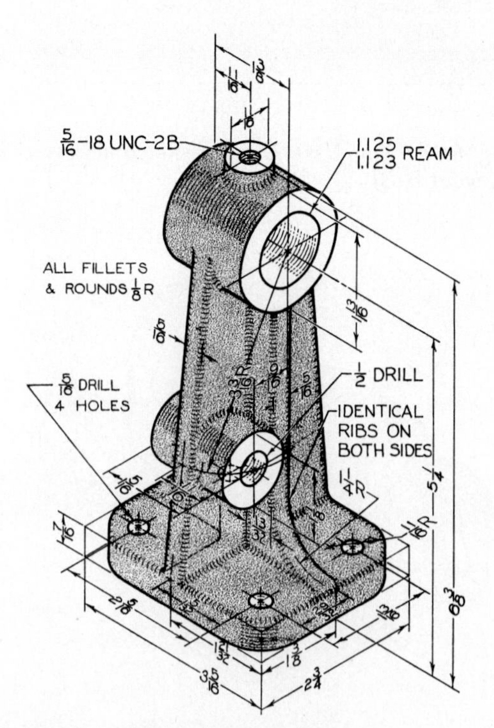


Fig. 6.103 **Bearing Bracket.** Draw or sketch necessary views (Layout B–3). For threads, see §§13.9 and 13.10.

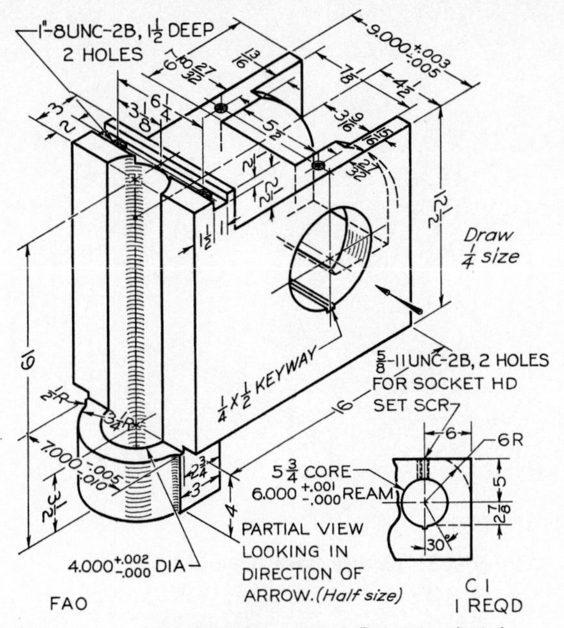


Fig. 6.104 **Link Arm Connector.** Draw or sketch necessary views (Layout B-3). For threads, see §§13.9 and 13.10.

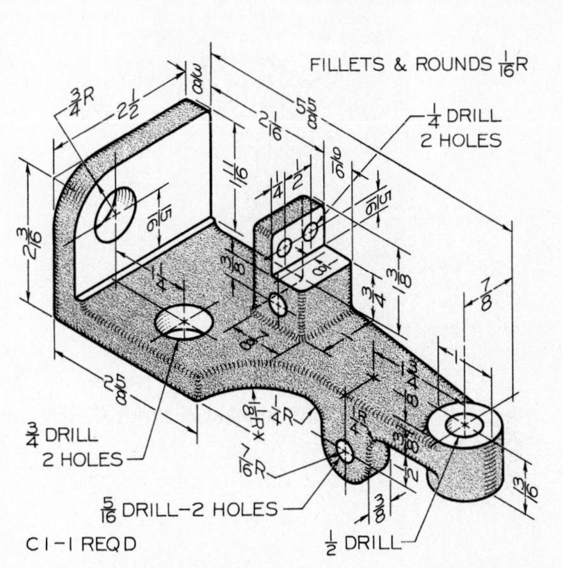


Fig. 6.105 **Mounting Bracket.** Draw or sketch necessary views (Layout B-3).

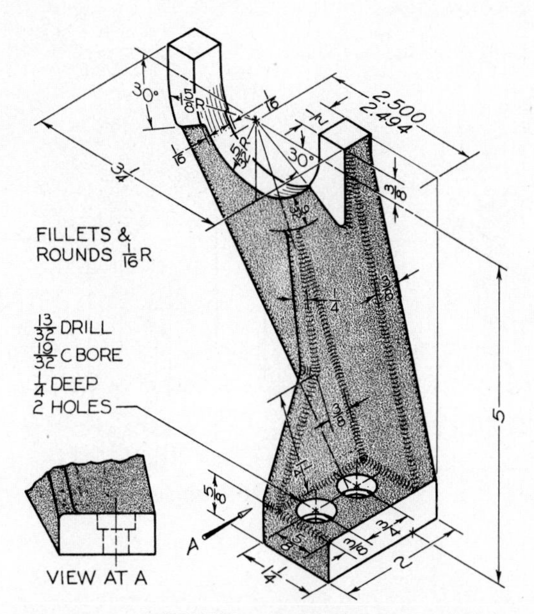


Fig. 6.106 **LH Shifter Fork.** Draw or sketch necessary views (Layout B-3).

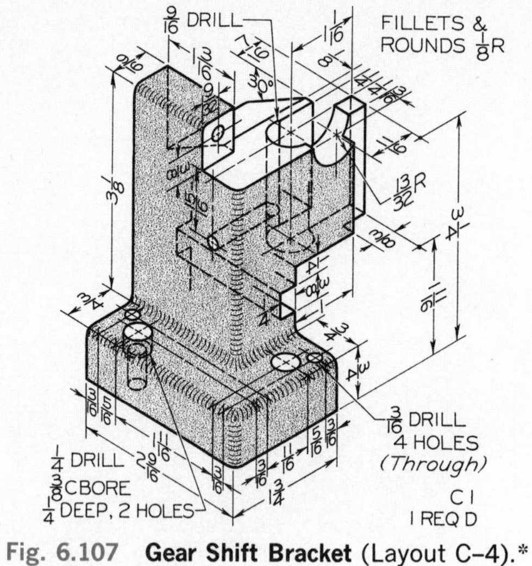


Fig. 6.107 **Gear Shift Bracket** (Layout C-4).*

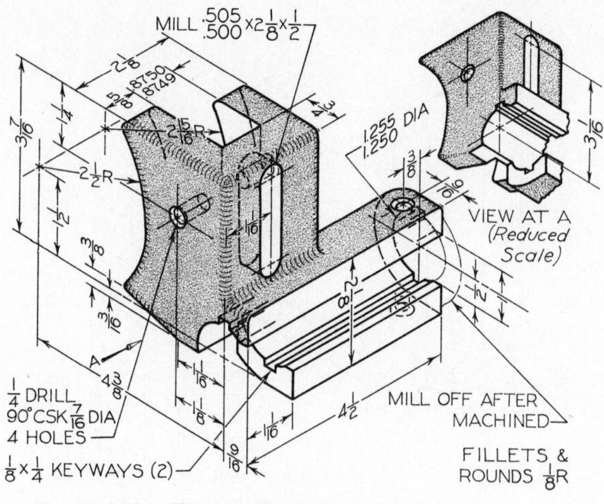


Fig. 6.108 **Fixture Base** (Layout C-4).*

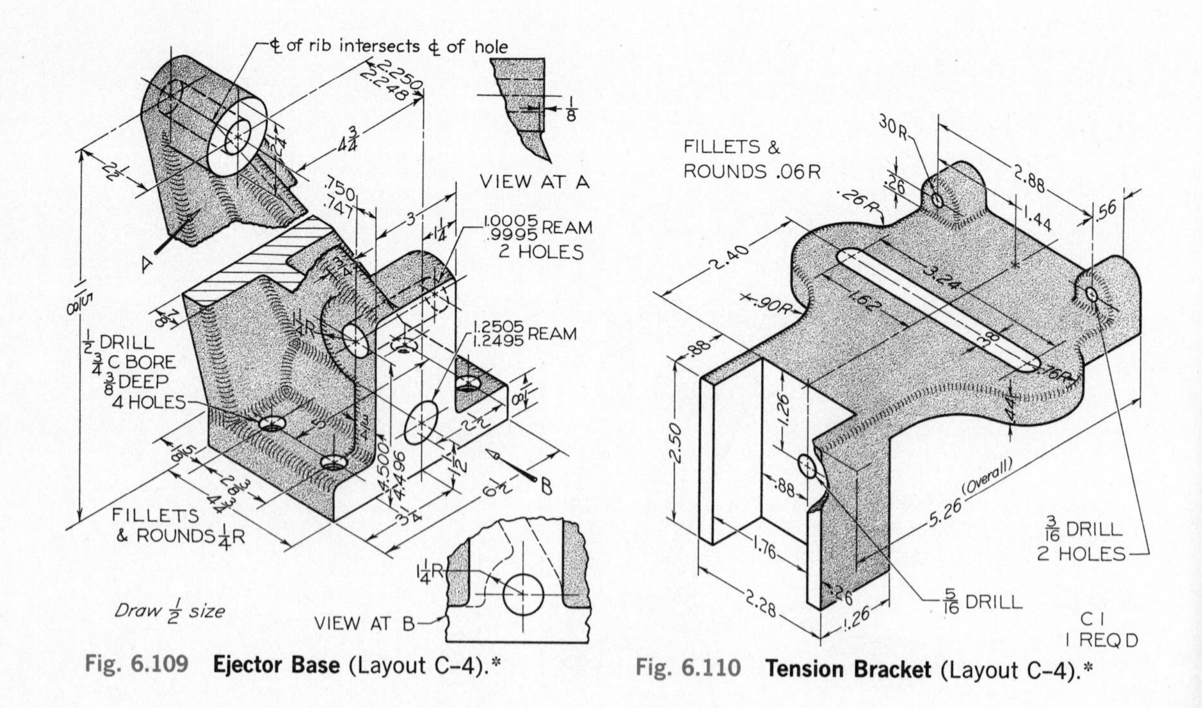


Fig. 6.109 **Ejector Base** (Layout C-4).*

Fig. 6.110 **Tension Bracket** (Layout C-4).*

* Draw or sketch necessary views. If dimensions are required, study §§11.1 to 11.25.

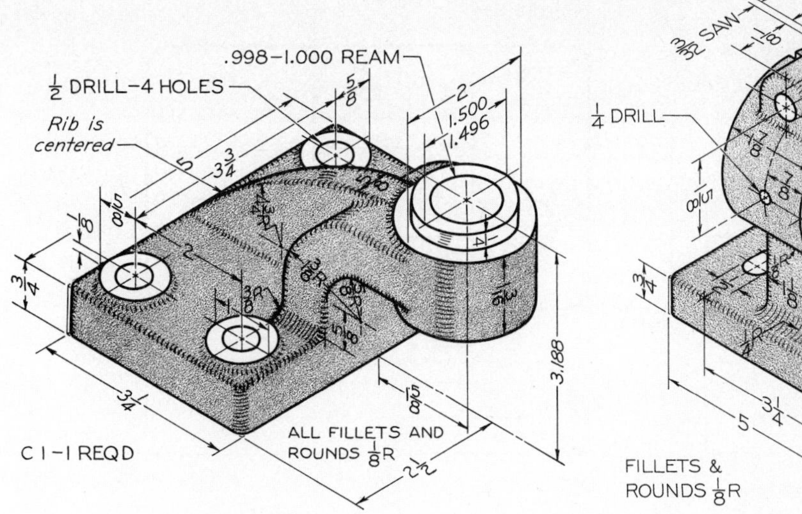


Fig. 6.111 **Offset Bearing** (Layout C–4).*

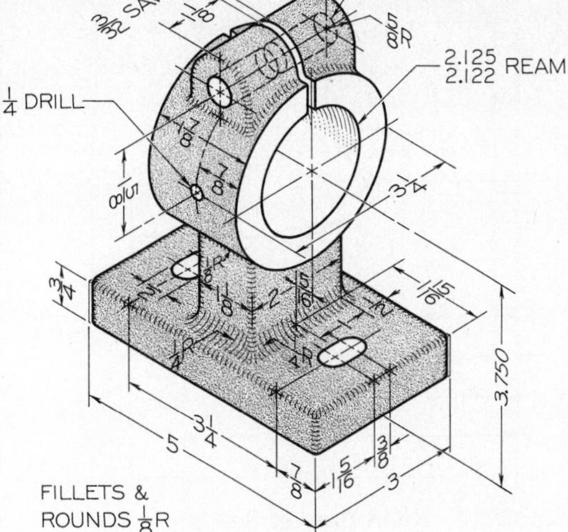


Fig. 6.112 **Feed Guide** (Layout C–4).*

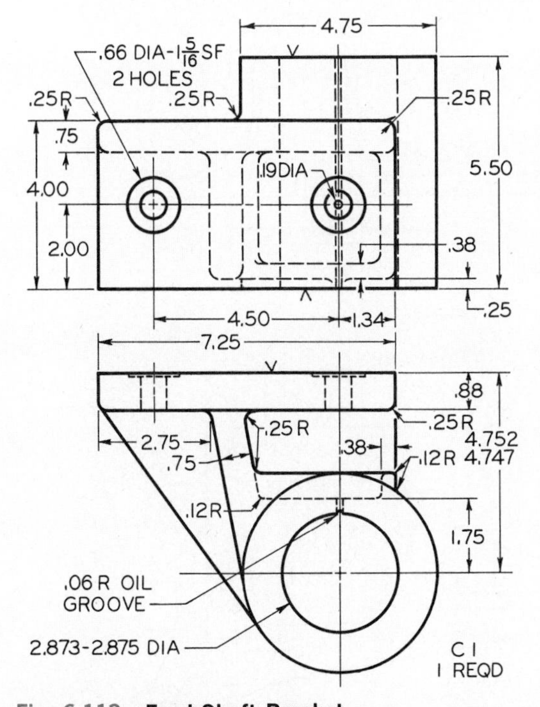


Fig. 6.113 **Feed Shaft Bracket.**
Given: Front and top views.
Required: Front, top, and right-side views, half size (Layout B–3).

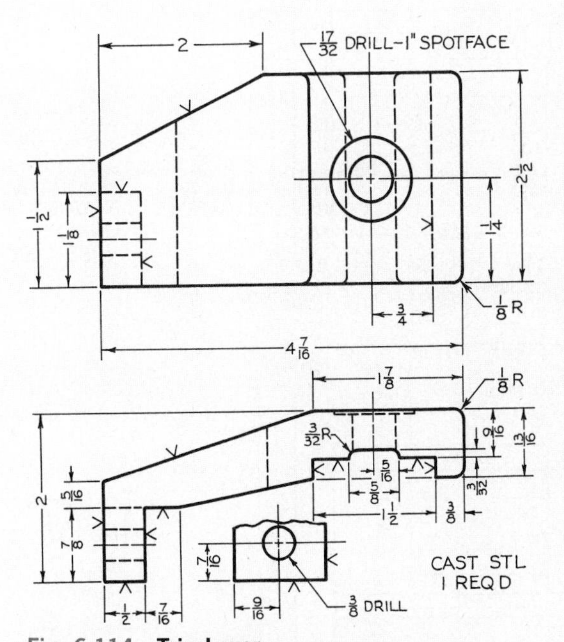


Fig. 6.114 **Trip Lever.**
Given: Front, top, and partial side views.
Required: Front, bottom, and left-side views, drawn completely (Layout B–3).

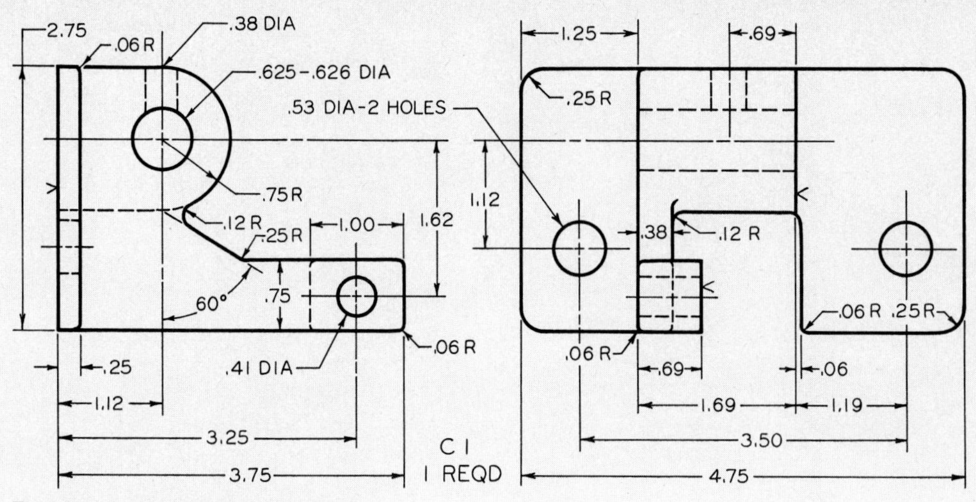


**Fig. 6.115 Knurl Bracket Bearing.**
Given: Front and left-side views.
Required: Take front as top view on new drawing, and add front and right-side views (Layout B–3).

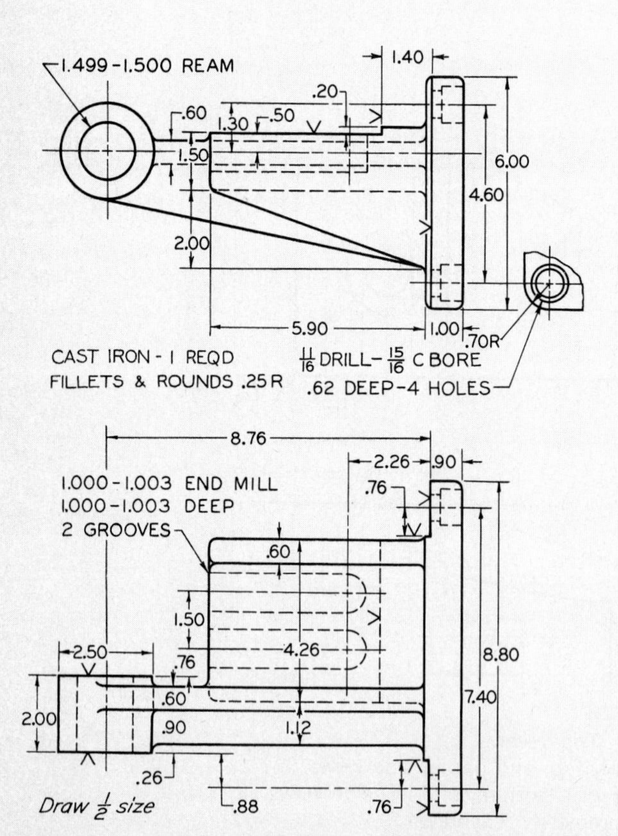


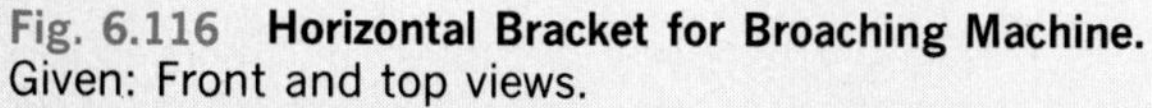

**Fig. 6.116 Horizontal Bracket for Broaching Machine.**
Given: Front and top views.
Required: Take top as front view in new drawing; then add top and left-side views (Layout C–4).

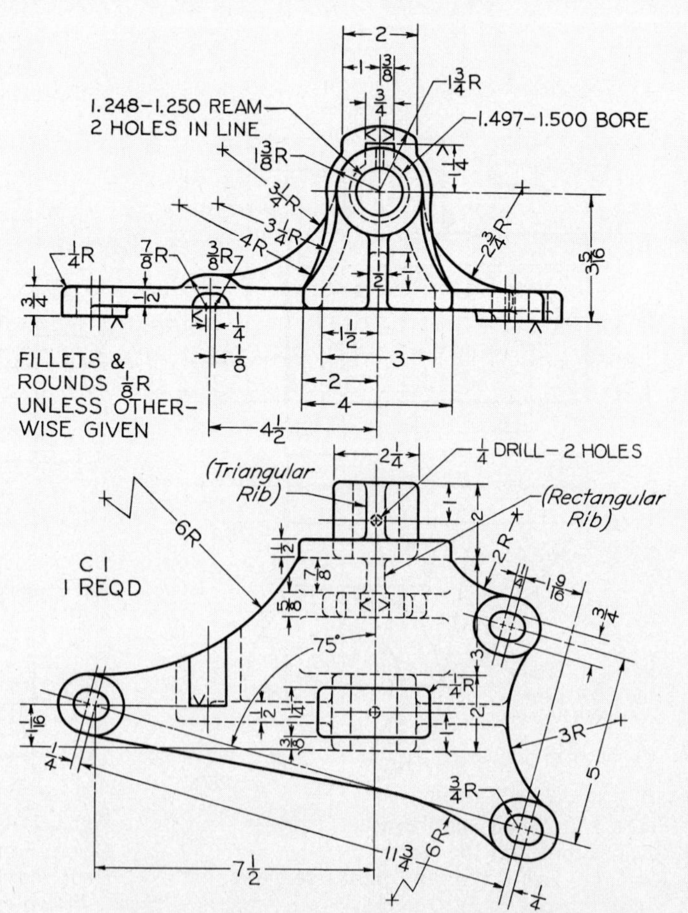


**Fig. 6.117 Boom Swing Bearing for a Power Crane.**
Given: Front and bottom views.
Required: Front, top, and left-side views (Layout C–4).

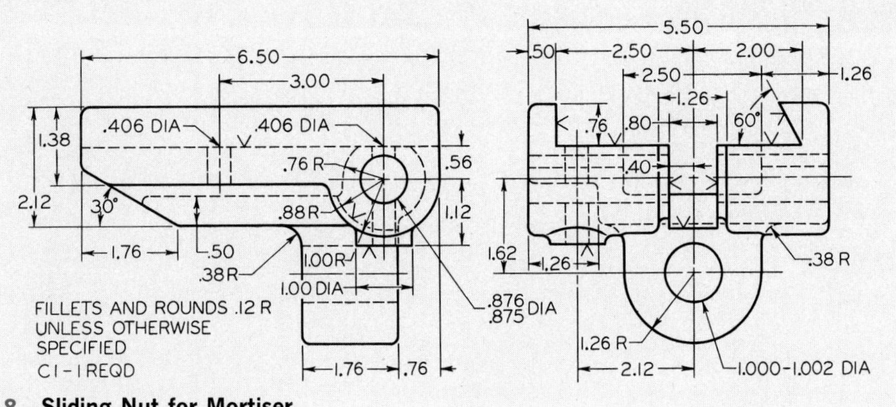


Fig. 6.118 **Sliding Nut for Mortiser.**
Given: Top and right-side views.
Required: Front, top, and left-side views, full size (Layout C–4).

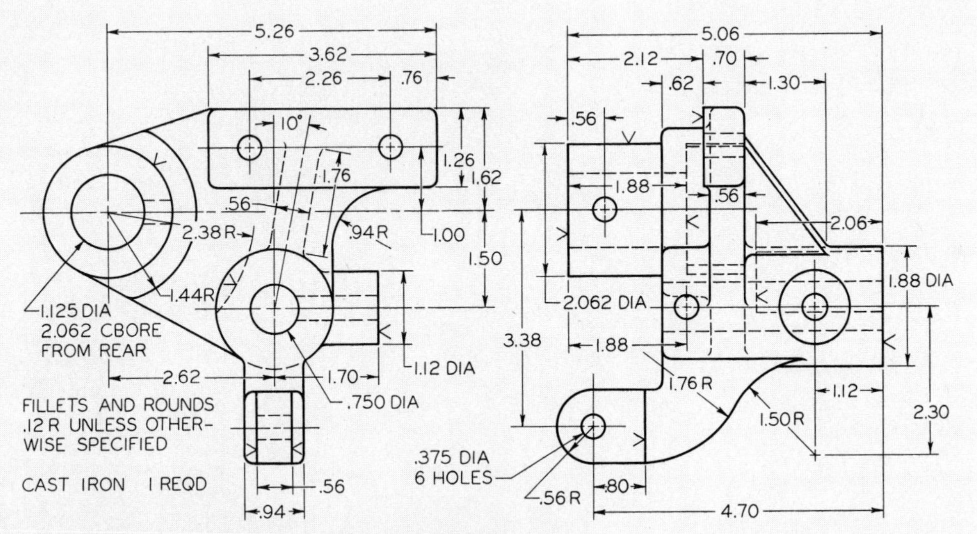

Fig. 6.119 **Power Feed Bracket for Universal Grinder.**
Given: Front and right-side views.
Required: Front, top, and left-side views, full size (Layout C–4).

# 7

# sectional views

**7.1 Sectional Views.*** The basic method of representing parts for designs by views, or projections, has been explained in previous chapters. By means of a limited number of carefully selected views, the external features of the most complicated designs can be thus fully described.

However, we are frequently confronted with the necessity of showing more or less complicated interiors of parts that cannot be shown clearly by means of hidden lines. We accomplish this by slicing through the part much as one would cut through an apple or a melon. A cutaway view of the part is then drawn; it is called a *sectional view*, a *cross section*, or simply a *section*.

More exactly, a *cutting plane*, §7.5, is assumed to be passed through the part for the design, as shown in Fig. 7.1 (a). Then, at (b) the cutting plane is removed and the two halves drawn apart, exposing the interior construction. In this case, the direction of sight is toward the left half, as shown, and for purposes of the section the right half is mentally discarded. The sectional view will be in the position of a right-side view.

*See ANSI Y14.2-1973, ANSI Y14.3-1957.

**7.2 Full Sections.** The sectional view obtained by passing the cutting plane fully through the object is called a *full section*, Fig. 7.2 (c). A comparison of this sectional view with the left-side view, (a), emphasizes the advantage in clearness of the former. The left-side view would naturally be omitted. In

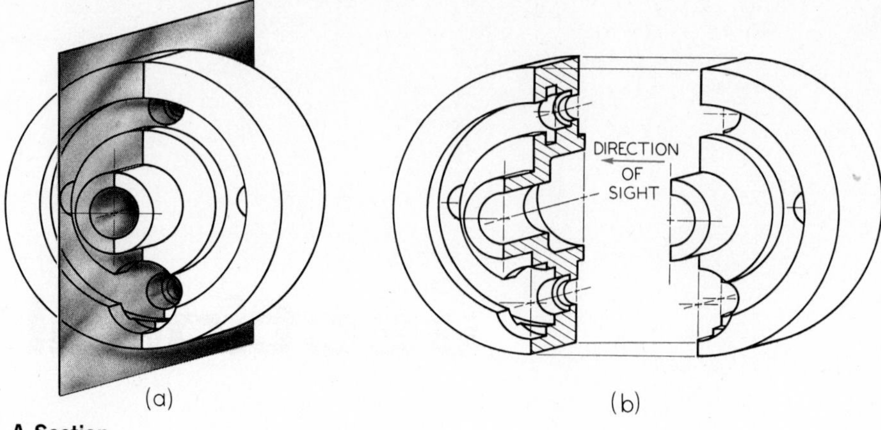


Fig. 7.1 **A Section.**

the front view, the cutting plane appears as a line, called a *cutting-plane line,* §7.5. The arrows at the ends of the cutting-plane line indicate the direction of sight for the sectional view.

Note that in order to obtain the sectional view, the right half is only *imagined* to be removed and not actually shown removed anywhere except in the sectional view itself. In the sectional view, the section-lined areas are those portions that have been in actual contact with the cutting plane. Those areas are *crosshatched* with thin parallel section lines spaced carefully by eye. In addition, the visible parts behind the cutting plane are shown but not crosshatched.

As a rule, the location of the cutting plane is obvious from the section itself, and, therefore, the cutting-plane line is omitted. It is shown in Fig. 7.2 for illustration only. Cutting-plane lines should, of course, be used wherever necessary for clearness, as in Figs. 7.21, 7.22, 7.24, and 7.25.

## 7.3 Lines in Sectioning.

A correct front view and sectional view are shown in Fig. 7.3 (a) and (b). In general, *all visible edges and con-*

Fig. 7.2 **Full Section.**

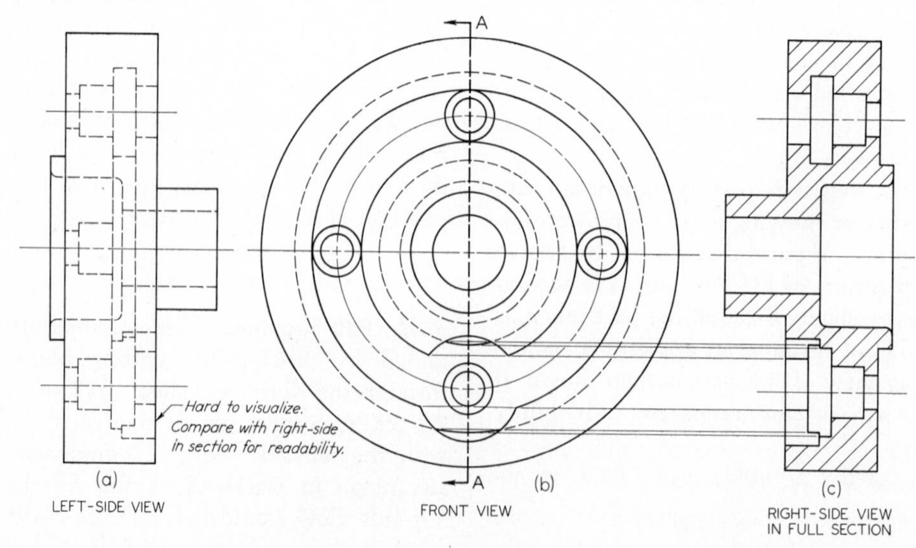

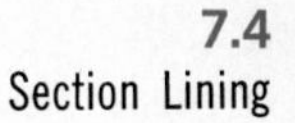

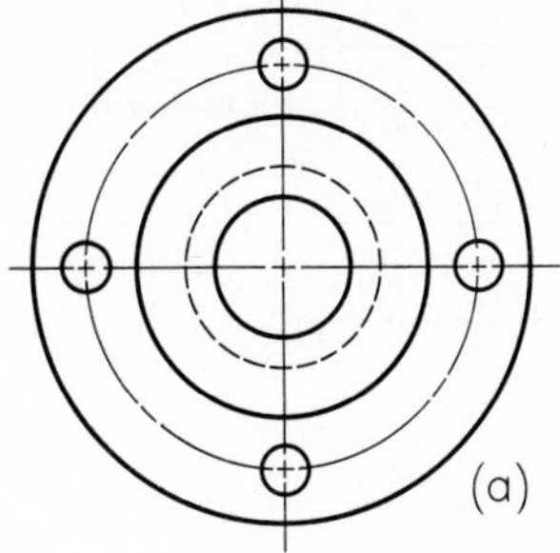

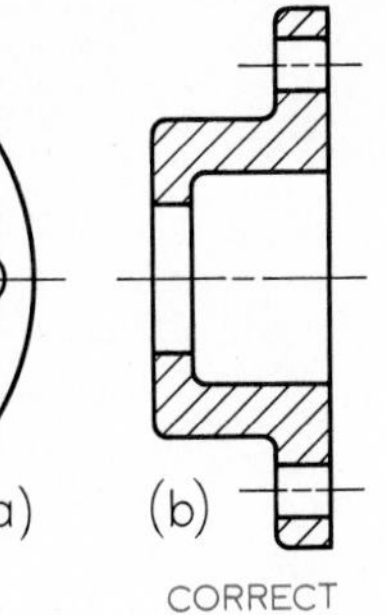


**Fig. 7.3 Lines in Sectioning.**

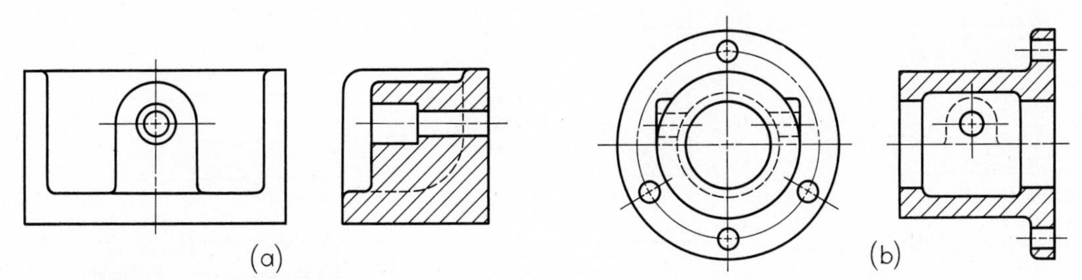


**Fig. 7.4 Hidden Lines in Sections.**

*tours behind the cutting plane should be shown;* outherwise a section will appear to be made up of disconnected and unrelated parts, as shown at (c). Occasionally, however, visible lines behind the cutting plane are not necessary for clearness and should be omitted.

Sections are used primarily to replace hidden-line representation; hence, as a rule, *hidden lines should be omitted in sectional views.* As shown in Fig. 7.3 (d), the hidden lines do not clarify the drawing; they tend to confuse, and they take unnecessary time to draw. Sometimes hidden lines are necessary for clearness and should be used in such cases, especially if their use will make it possible to omit a view, Fig. 7.4.

A section-lined area is always completely bounded by a visible outline—never by a hidden line as in Fig. 7.3 (e), since in every case the cut surfaces and their boundary lines will be visible. Also, a visible line can never cut across a section-lined area.

In a sectional view of a part, alone or in assembly, the section lines in all sectioned areas must be parallel, not as shown in Fig. 7.3 (f). The use of section lining in opposite directions is an indication of different parts, as when two or more parts are adjacent in an assembly drawing, Fig. 14.41.

## 7.4 Section Lining.

In the past, different section-lining symbols, Fig. 7.5, have been used to indicate the material to be used in producing the object. These symbols represented the general types only, such as cast-iron, brass, and steel. Now, however, there are so many different materials, and each general type has so many subtypes, that a general name or symbol is not enough. For example, there are hundreds of different kinds of steel alone. Since detailed specifications of material must be lettered in the form of a note or in the title strip, the ANSI recommends that on detail drawings (single parts) the general-purpose (cast-iron) section lining be used for all materials.

Symbolic section lining may be used in assembly drawings in cases where it is desirable to distinguish the different materials; otherwise, the general-purpose symbol is used for all parts. For assembly sections, see §14.22.

The correct method of drawing section lines is shown in Fig. 7.6 (a). Draw the section lines

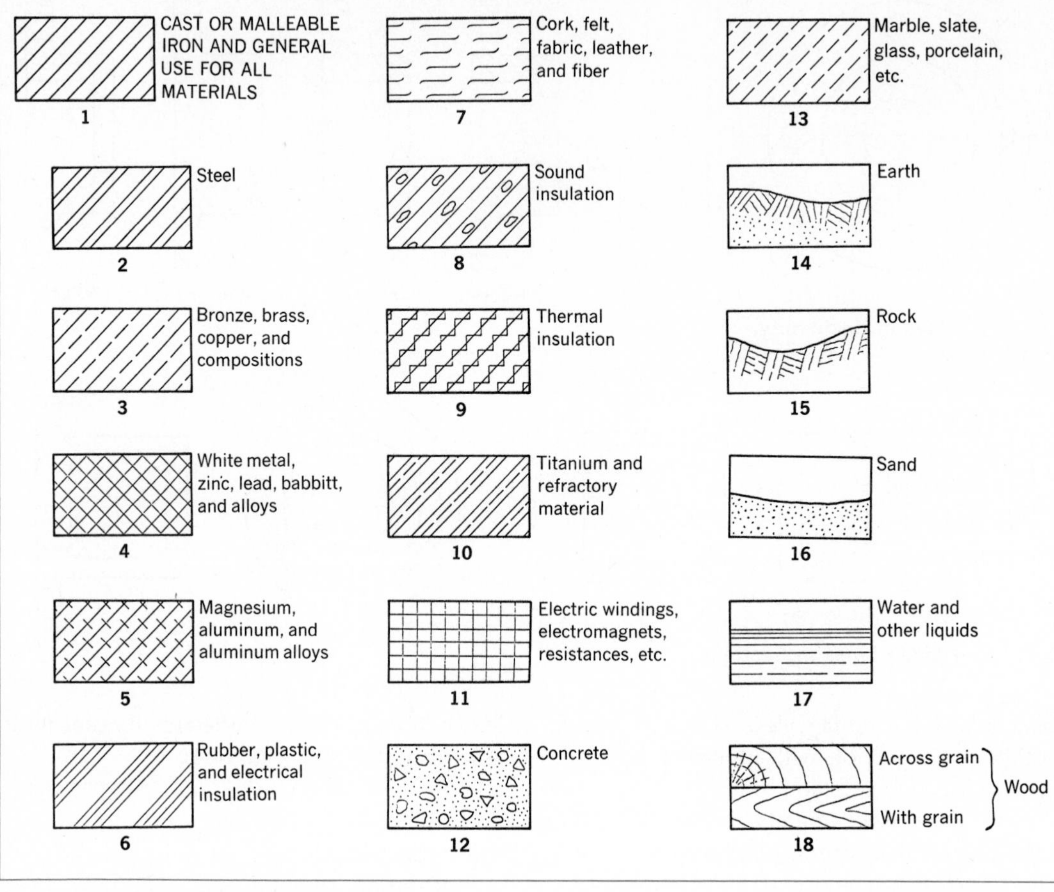


Fig. 7.5 **Symbols for Section Lining.**

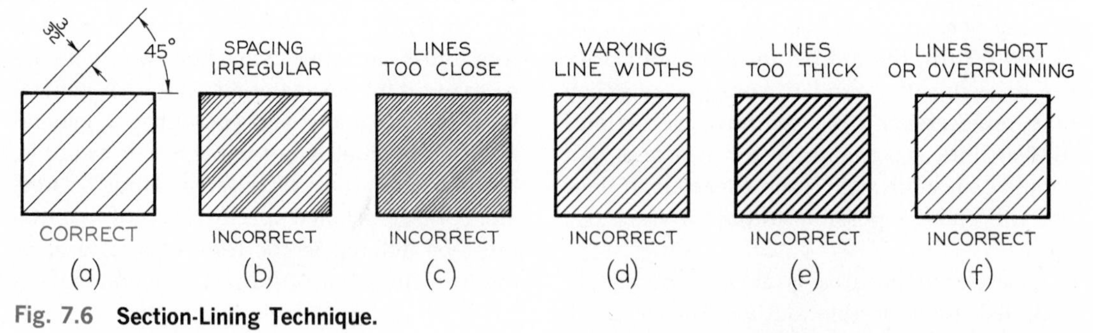


Fig. 7.6 **Section-Lining Technique.**

Fig. 7.7 **Direction of Section Lines.**

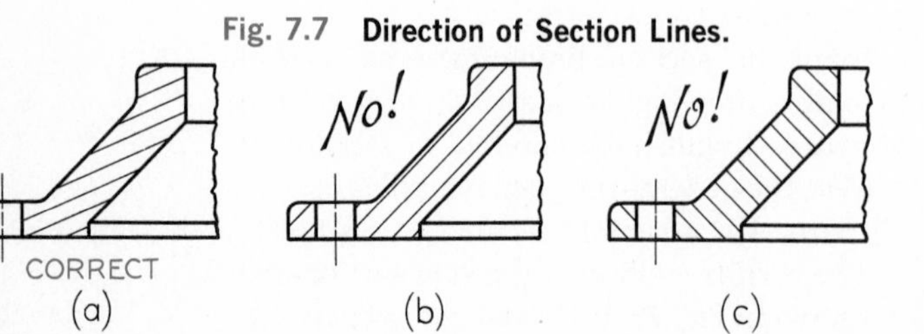


with a sharp medium-grade pencil (H or 2H) with a conical point as shown in Fig. 2.12 (c). Always draw the lines at 45° with horizontal as shown, unless there is some advantage in using a different angle. Space the section lines as uniformly as possible by eye from about $\frac{1}{16}''$ to $\frac{1}{8}''$ or more apart, depending on the size of the drawing or of the sectioned area. *For average drawings, space the lines about $\frac{3}{32}''$ or more*

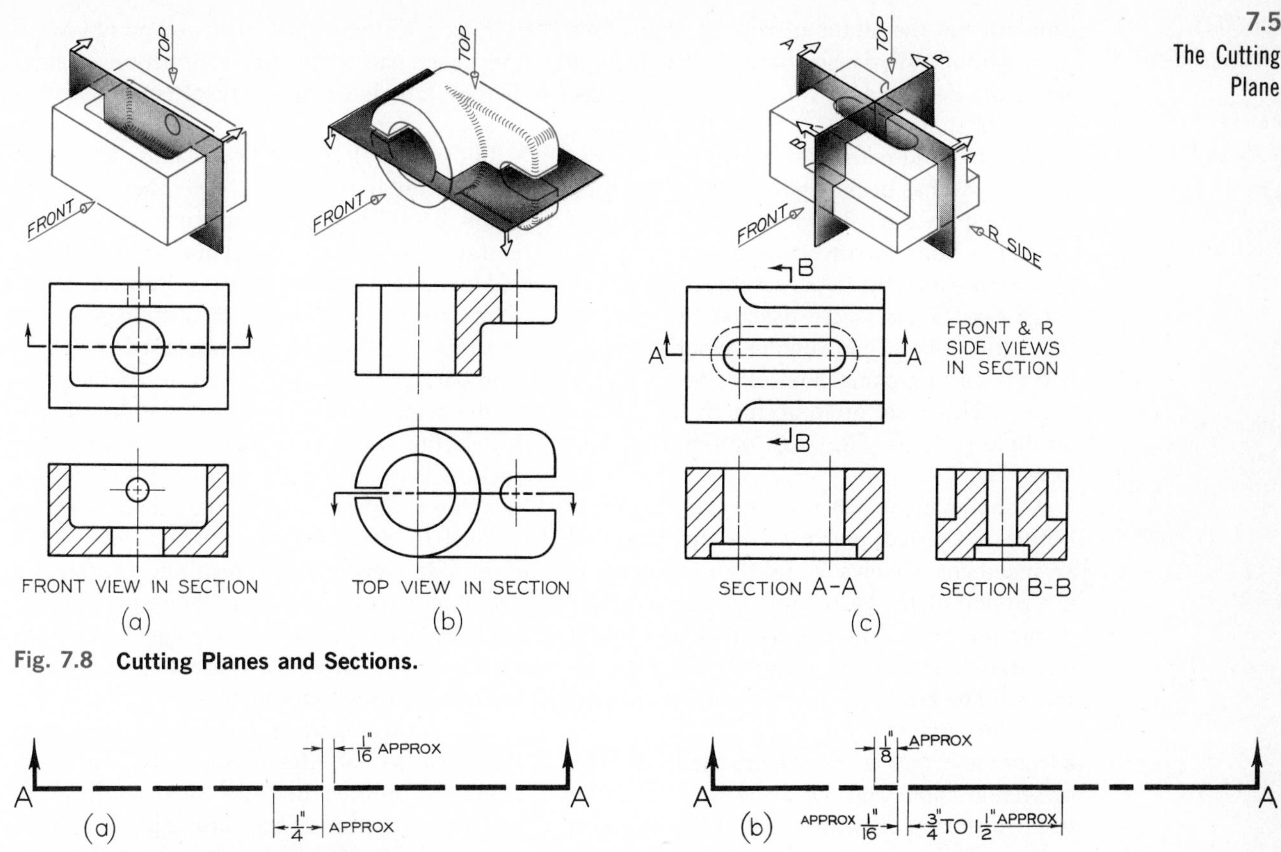


**Fig. 7.8 Cutting Planes and Sections.**

**Fig. 7.9 Cutting-Plane Lines (Full Size).**

*apart.* As a rule, space the lines as generously as possible and yet close enough to distinuish clearly the sectioned areas.

After the first few lines have been drawn, look back repeatedly at the original spacing to avoid gradually increasing or decreasing the intervals, Fig. 7.6 (b). Beginners almost invariably draw section lines too close together, (c). This is very tedious because with small spacing the least inaccuracy in spacing is conspicuous.

Section lines should be uniformly thin, never varying in thickness, as at (d). There should be a marked contrast in thickness of the visible outlines and the section lines. Section lines should not be too thick, as at (e). Also avoid running section lines beyond the visible outlines or stopping the lines too short, as at (f).

If section lines drawn at 45° with horizontal would be parallel or perpendicular (or nearly so) to a prominent visible outline, the angle should be changed to 39°, 60°, or some odd angle, Fig. 7.7.

Dimensions should be kept off sectioned areas, but when this is unavoidable the section lines should be omitted where the dimension figure is placed. See Fig. 11.13.

Section lines may be drawn adjacent to the boundaries of the sectioned areas (outline sectioning) providing clarity is not sacrificed. See Fig. 14.99.

## 7.5 The Cutting Plane.

The *cutting plane* is indicated in a view adjacent to the sectional view, Fig. 7.8. In this view, the cutting plane appears edgewise or as a line, called the *cutting-plane line.* Alternate styles of cutting-plane lines are shown in Fig. 7.9. The first form, Fig. 7.9 (a), composed of equal dashes each about $\frac{1}{4}''$ or more long plus the arrowheads, is the standard in the automotive industry. This form without the dashes between the ends is especially desirable on complicated drawings. The form shown in (b), composed of alternate long dashes and pairs of short dashes plus the arrowheads, has been in general use for a long time. Both lines are drawn the same thickness as visible lines. Arrowheads indicate the direction in which the cutaway object is viewed. Large capital letters are used at the ends of the cutting-plane line when necessary to identify the cutting-plane line with the indicated section. This most often occurs in the case of multiple sections, Fig. 7.25, or removed sections, Fig. 7.21.

As shown in Fig. 7.8, sectional views occupy normal projected positions in the standard arrangement of views. At (a) the cutting plane is a frontal plane, §6.15, and appears as a line in the top view. The front half of the object (lower half in the top view) is imagined removed. The arrows at the ends of the cutting-plane line point in the direction of sight for a front view; that is, away from the front view or section. Note that the arrows do not point in the direction of withdrawal of the removed portion. The resulting full section may be referred to as the "front view in section," since it occupies the front view position.

In Fig. 7.8 (b), the cutting plane is a horizontal plane, §6.15, and appears as a line in the front view. The upper half of the object is imagined removed. The arrows point toward the lower half in the same direction of sight as for a top view, and the resulting full section is a "top view in section."

In Fig. 7.8 (c), two cutting planes are shown, one a frontal plane and the other a profile plane, §6.15, both of which appear edgewise in the top view. Each section is completely independent of the other and drawn as if the other were not present. For section A–A, the front half of the object is imagined removed. The back half then is viewed in the direction of the arrows for a front view, and the resulting section is a "front view in section." For section B–B, the right half of the object is imagined removed. The left half then is viewed in the direction of the arrows for a right-side view, and the resulting section is a "right-side view in section." The cutting-plane lines are preferably drawn through an exterior view, in this case the top view, as shown, instead of a sectional view.

The cutting-plane lines in Fig. 7.8 are shown for purposes of illustration only. They are generally omitted in cases such as these, in which the location of the cutting plane is obvious. When a cutting-plane line coincides with a center line, the cutting-plane line takes precedence.

Correct and incorrect relations between cutting-plane lines and corresponding sectional views are shown in Fig. 7.10.

Fig. 7.10 **Cutting Planes and Sections.**

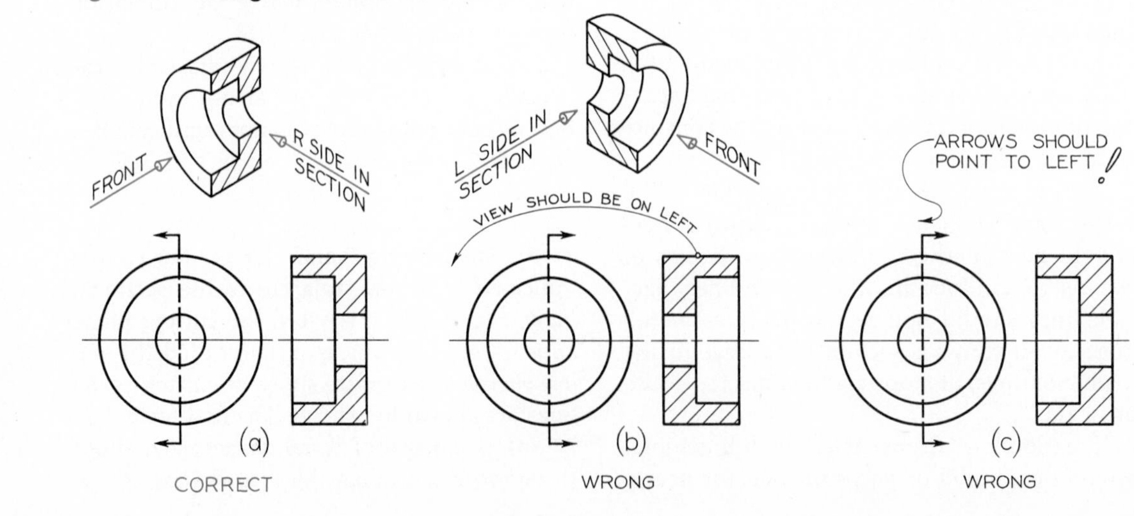

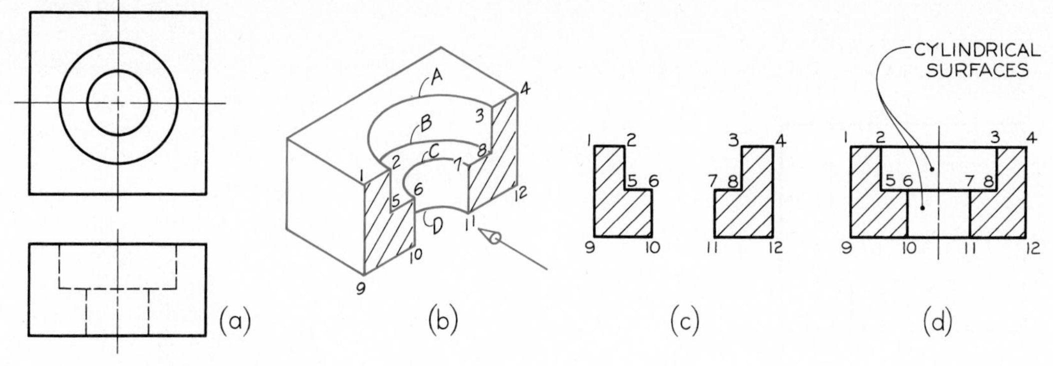


Fig. 7.11 **Visualizing a Section.**

**7.6 Visualizing a Section.** Two views of an object to be sectioned, having a drilled and counterbored hole, are shown in Fig. 7.11 (a). The cutting plane is assumed along the horizontal center line in the top view, and the front half of the object (lower half of the top view) is imagined removed. A pictorial drawing of the remaining back half is shown at (b). The two cut surfaces produced by the cutting plane are 1-2-5-6-10-9 and 3-4-12-11-7-8. However, the corresponding section at (c) is incomplete because certain visible lines are missing.

If the section is viewed in the direction of sight, as shown at (b), arcs A, B, C, and D will be visible. As shown at (d), these arcs will appear as straight lines 2-3, 6-7, 5-8, and 10-11. These lines may also be accounted for in other ways. The top and bottom surfaces of the object appear in the section as lines 1-4 and 9-12. The bottom surface of the counterbore appears in the section as line 5-8. Also, the semicylindrical surfaces for the back half of the counterbore and of the drilled hole will appear as rectangles in the section at 2-3-8-5 and 6-7-11-10.

The front and top views of a Collar are shown in Fig. 7.12 (a), and a right-side view in full section is required. The cutting plane is understood to pass along the center lines AD and EL. If the cutting plane were drawn, the arrows would point to the left in conformity with the direction of sight (see arrow) for the right-side view. The right side of the object is imagined removed and the left half will be viewed in the direction of the arrow, as shown pictorially at (d). The cut surfaces will appear edgewise in the top and front views along AD and EL; and since the direction of sight for the section is at right angles to them, they will appear in true size and shape in the sectional view. Each sectioned area will be completely enclosed by a boundary of visible lines. The sectional view will show, in addition to the cut surfaces, all visible parts behind the cutting plane. No hidden lines will be shown.

Whenever a surface of the object (plane or cylindrical) appears as a line and is intersected by a cutting plane that also appears as a line, a new edge (line of intersection) is created that will appear as a *point* in that view. Thus, in the front view, the cutting plane creates new edges appearing as points at E, F, G, H, J, K, and L. In the sectional view, (b), these are horizontal lines 31-32, 33-34, 35-36, 37-38, 39-40, 41-42, and 43-44.

Whenever a surface of the object appears as a surface (that is, not as a line) and is cut by a cutting plane that appears as a line, a new edge is created that will appear as a line in the view, coinciding with the cutting-plane line, and as a line in the section.

In the top view, D is the *point view* of a vertical line KL in the front view and 41-43 in the section at (b). Point C is the point view of a vertical line HJ in the front view and 37-39 in the section. Point B is the point view of two vertical lines EF and GH in the front view, and 31-33 and 35-38 in the section. Point A is the point view of three vertical lines

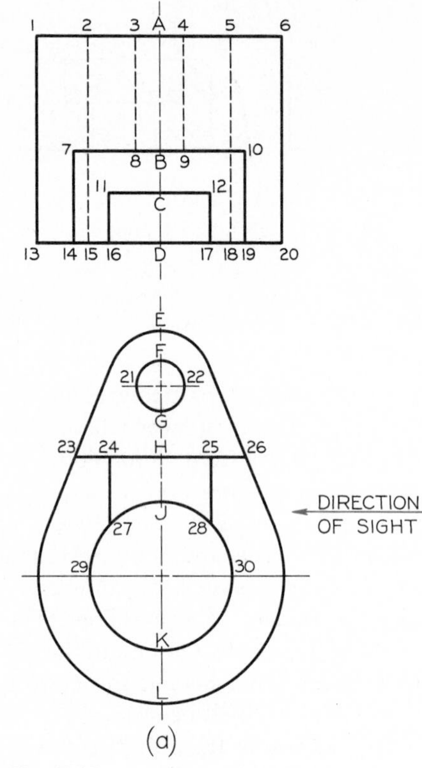


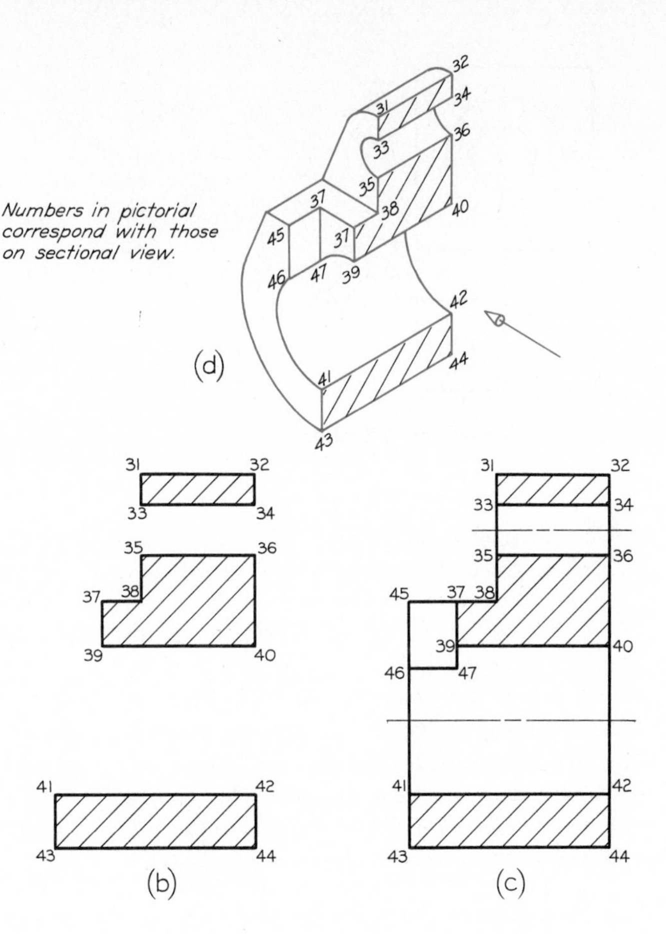


Fig. 7.12 **Drawing a Full Section.**

EF, GJ, and KL in the front view, and 32–34, 36–40, and 42–44 in the section. This completes the boundaries of three sectioned areas 31-32-34-33, 35-36-40-39-37-38, and 41-42-44-43. It is only necessary now to add the visible lines beyond the cutting plane.

The semicylindrical left half F-21-G of the small hole (front view) will be visible as a rectangle in the sections at 33-34-36-35, as shown at (c). The two semicircular arcs will appear as straight lines in the section at 33–35 and 34–36.

Surface 24-27, appearing as a line in the front view, appears as a line 11-16 in the top view, and as surface 45-37-47-46, true size, in the section at (c).

Cylindrical surface J-29-K, appearing as an arc in the front view, appears in the top view as 2-A-C-11-16-15, and in the section as 46-47-39-40-42-41. Thus, arc 27-29-K (front view) appears in the section, (c), as straight lines 46-41; and arc J-29-K appears as straight line 40-42.

All cut surfaces here are part of the same object; hence the section lines must all run in the same direction, as shown.

## 7.7 Half Sections.

If the cutting plane passes halfway through the object, the result is a half section, Fig. 7.13. A half section has the advantage of exposing the interior of one half of the object and retaining the exterior of the other half. Its usefulness is, therefore, largely limited to symmetrical objects. It is not widely used in detail drawings (single parts) because of this limitation of symmetry and also because of difficulties in dimensioning internal shapes that are shown in part in the sectioned half, Fig. 7.13 (b).

In general, hidden lines should be omitted from both halves of a half section. However,

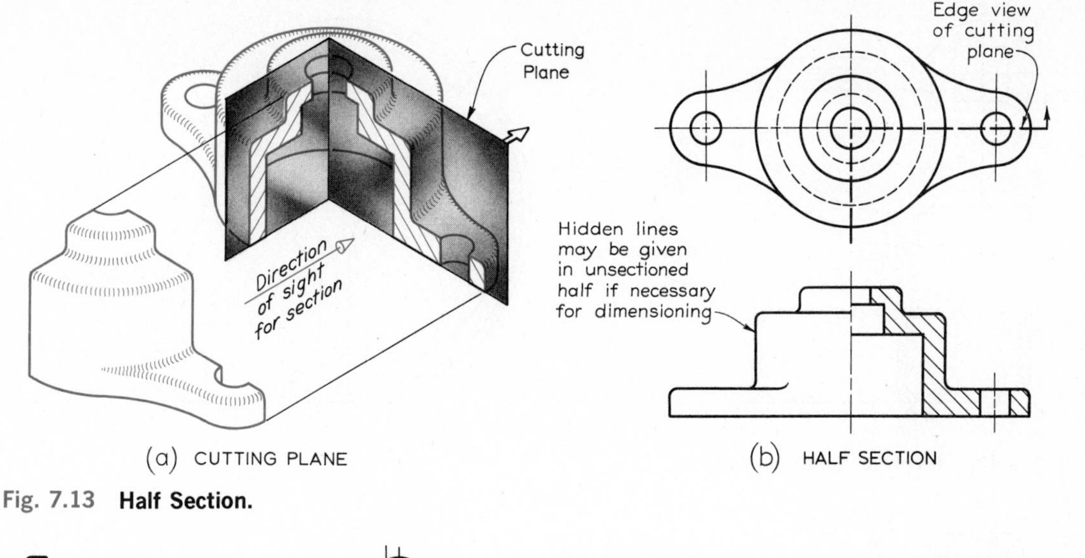


Fig. 7.13 **Half Section.**

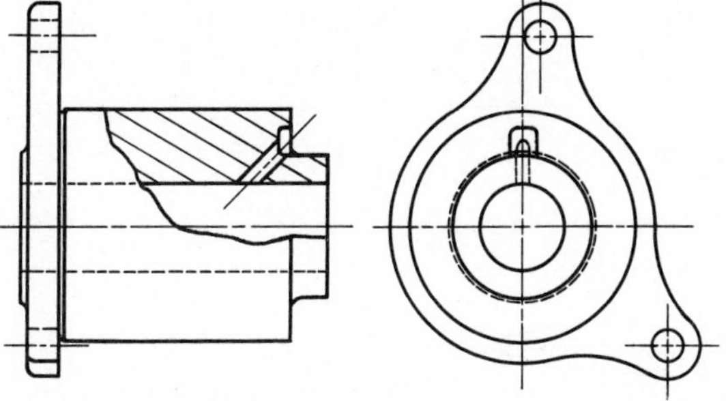

Fig. 7.14 **Broken-Out Section.**

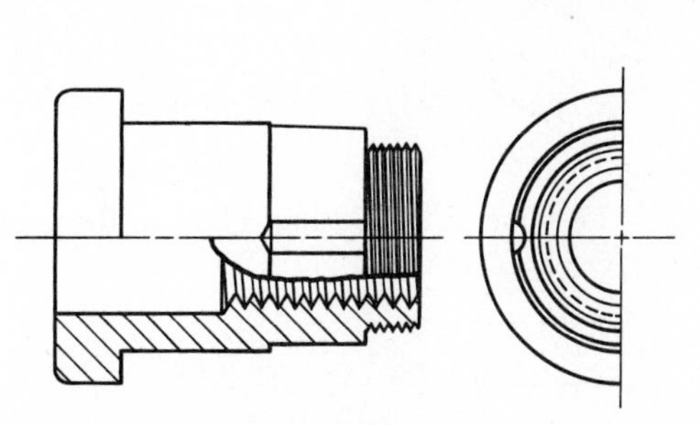

Fig. 7.15 **Break Around Keyway.**

they may be used in the unsectioned half if necessary for dimensioning.

The greatest usefulness of the half section is in assembly drawing, Fig. 14.41, in which it is often necessary to show both internal and external construction on the same view, but without the necessity of dimensioning.

As shown in Fig. 7.13 (b), a center line is used to separate the halves of the half section. The American National Standard recommends a center line for the division line between the sectioned half and the unsectioned half of a half-sectional view, although in some cases the same overlap of the exterior portion, as in a broken-out section, is preferred. See Fig. 7.33 (b). Either form is acceptable.

**7.8 Broken-Out Sections.** It often happens that only a partial section of a view is needed to expose interior shapes. Such a section, limited by a *break line,* Fig. 2.15, is called a *broken-out section.* In Fig. 7.14, a full or half section is not necessary, a small broken-out section being sufficient to explain the construction. In Fig. 7.15, a half section would have caused the removal of half the keyway. The keyway is preserved by breaking out around it. Note that in this case the section is limited partly by a break line and partly by a center line.

**7.9 Revolved Sections.** The shape of the cross section of a bar, arm, spoke, or other elongated object may be shown in the longitudinal view by means of a *revolved section,* Fig. 7.16. Such sections are made by assuming a plane perpendicular to the center line or axis of the bar or other object, as shown in Fig. 7.17 (a), then revolving the plane through 90°

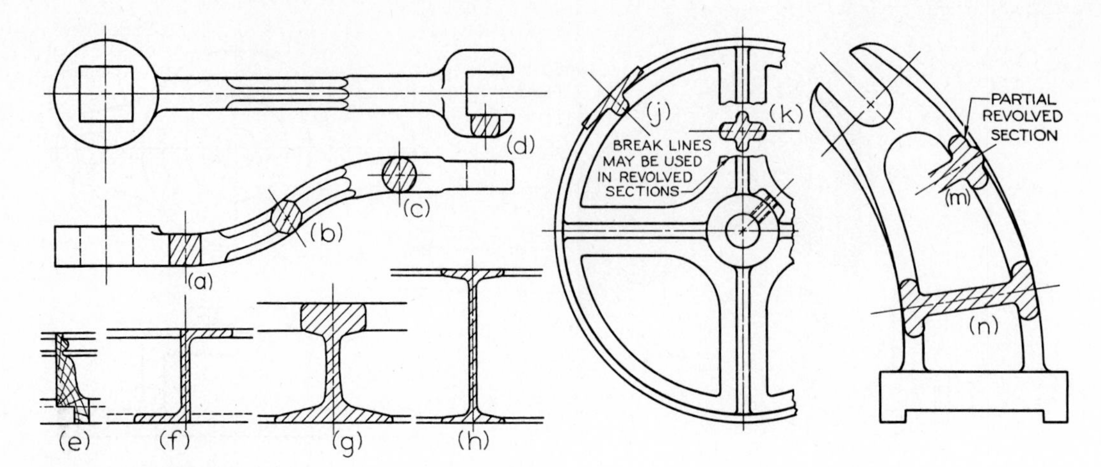


Fig. 7.16 **Revolved Sections.**

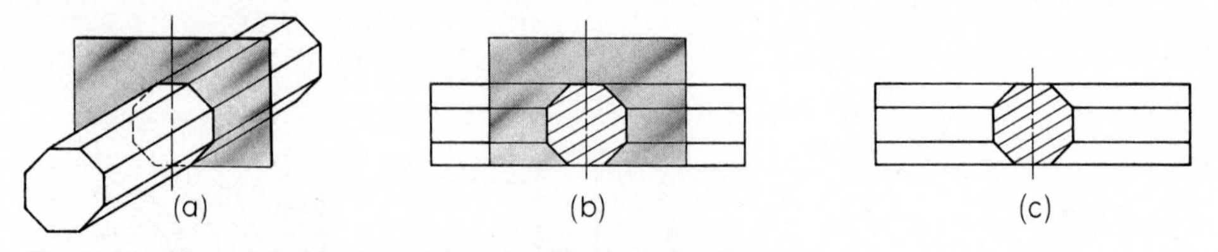


Fig. 7.17 **Use of the Cutting Plane in Revolved Sections.**

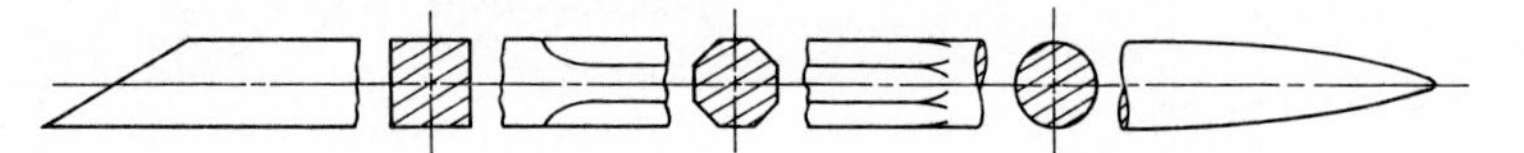

Fig. 7.18 **Conventional Breaks Used with Revolved Sections.**

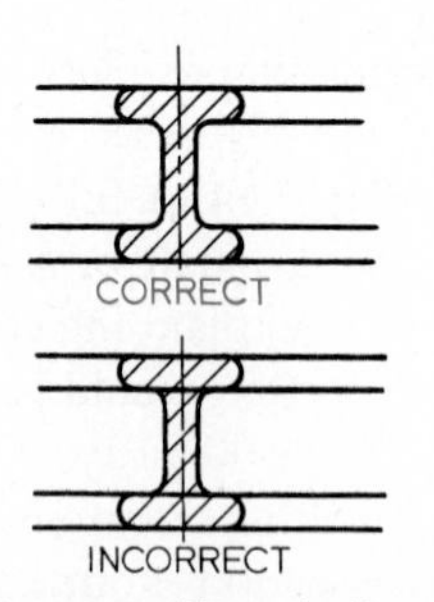


Fig. 7.19 **A Common Error in Drawing Revolved Sections.**

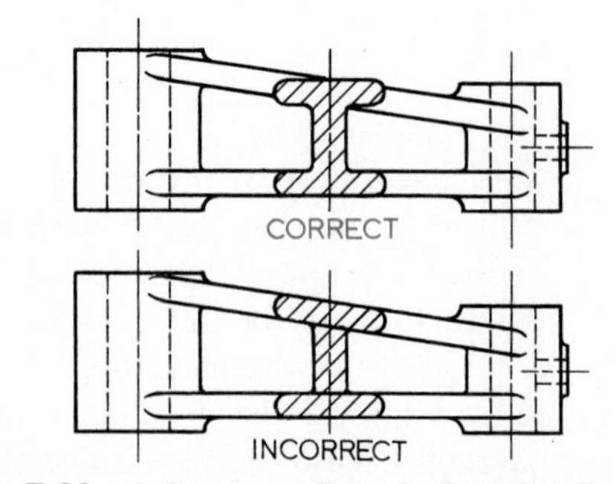


Fig. 7.20 **A Common Error in Drawing Revolved Sections.**

about a center line at right angles to the axis, as at (b) and (c).

The visible lines adjacent to a revolved section may be broken out if desired, as shown in Figs. 7.16 (k) and 7.18.

The superimposition of the revolved section requires the removal of all original lines covered by it, Fig. 7.19. The true shape of a revolved section should be retained after the revolution of the cutting plane, regardless of the direction of the lines in the view, Fig. 7.20.

## 7.10 Removed Sections.

A removed section is one not in direct projection from the

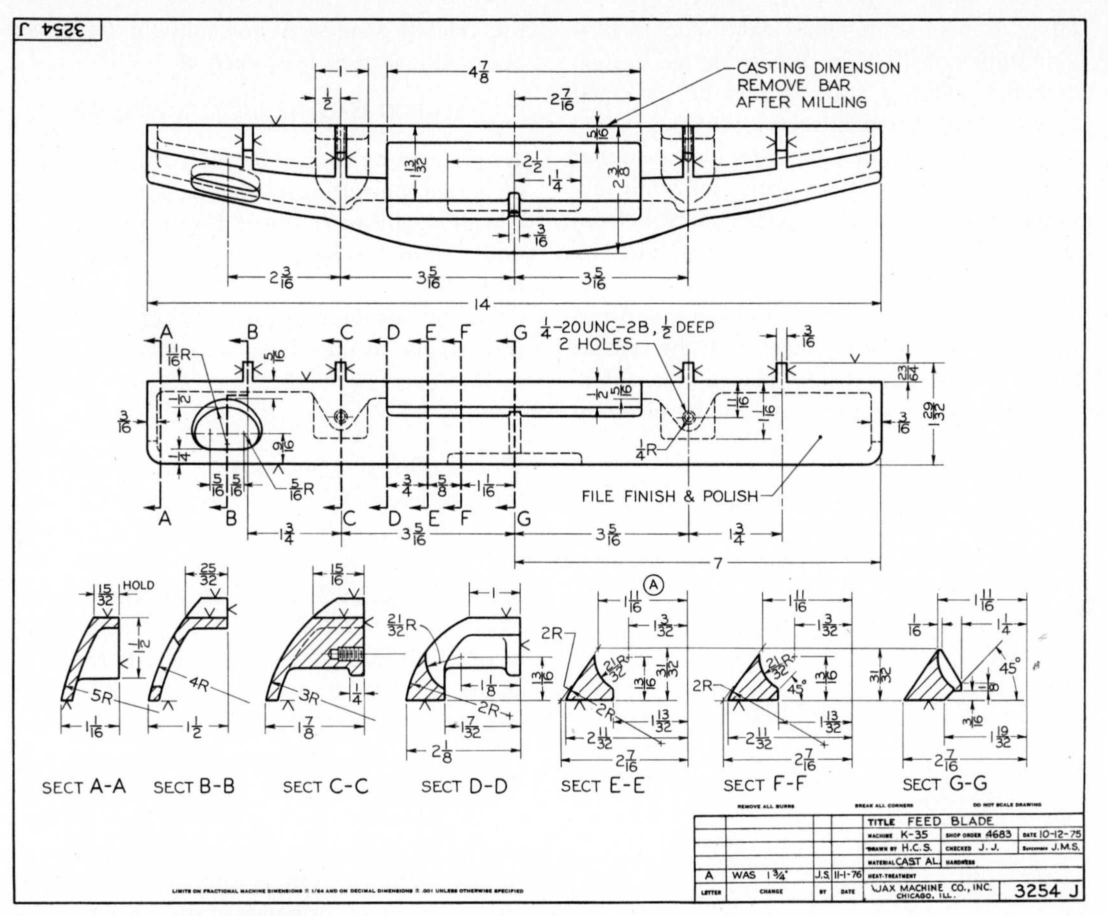


Fig. 7.21 **Removed Sections.**

view containing the cutting plane—that is, it is not positioned in agreement with the standard arrangement of views. This displacement from the normal projection position should be made without turning the section from its normal orientation.

Removed sections, Fig. 7.21, should be labeled, such as SECTION A–A and SECTION B–B, corresponding to the letters at the ends of the cutting-plane line. They should be arranged in alphabetical order from left to right on the sheet. Section letters should be used in alphabetical order, but letters I, O, and Q should not be used because they are easily confused with the numeral 1 or the zero.

A removed section is often a partial section. Such a removed section, Fig. 7.22, is frequently drawn to an enlarged scale, as shown. This is often desirable in order to show clear

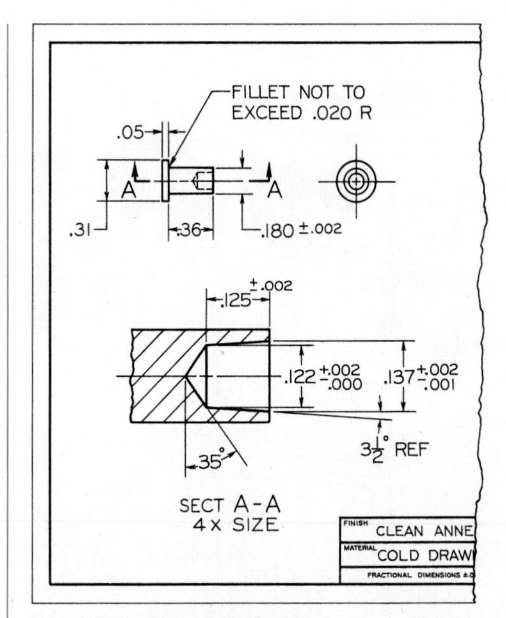


Fig. 7.22 **Removed Section.**

delineation of some small detail and to provide sufficient space for dimensioning. In such case the enlarged scale should be indicated beneath the section title.

A removed section should be placed so that it no longer lines up in projection with any other view. It should be separated clearly from the standard arrangement of views. See Fig. 12.9.

Whenever possible removed sections should be on the same sheet with the regular views. If a section must be placed on a different sheet, cross references should be given on the related sheets. A note should be given below the section title, such as

SECTION B–B ON SHEET 4, ZONE A3

A similar note should be placed on the sheet on which the cutting-plane line is shown, with a leader pointing to the cutting-plane line and referring to the sheet on which the section will be found.

Sometimes it is convenient to place removed sections on center lines extended from the section cuts, Fig. 7.23.

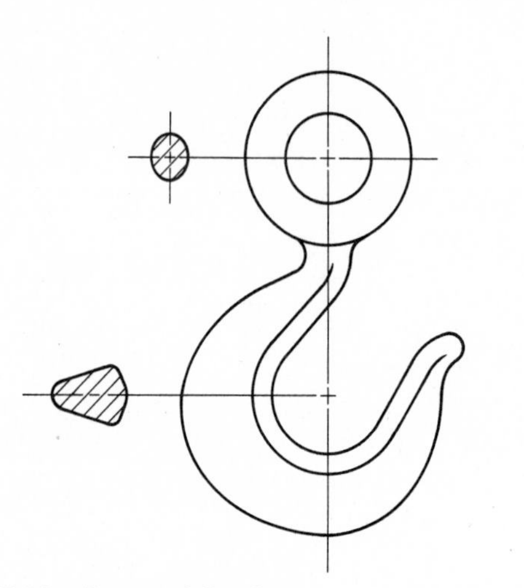

Fig. 7.23 **Removed Sections.**

## 7.11 **Offset Sections.**

In sectioning through irregular objects, it is often desirable to show several features that do not lie in a straight line, by "offsetting" or bending the cutting plane. Such a section is called an *offset section.* In Fig. 7.24 (a) the cutting plane is offset in several places in order to include the hole at the left end, one of the parallel slots, the rectangular recess, and one of the holes at the right end. The front portion of the object is then imagined to be removed, (b). The path of the cutting plane is shown by the cutting-plane line in the top view at (c), and the resulting offset section is shown in the front view. The offsets or bends in the cutting plane are all 90° and are *never shown in the sectional view.*

Fig. 7.24 **Offset Section.**

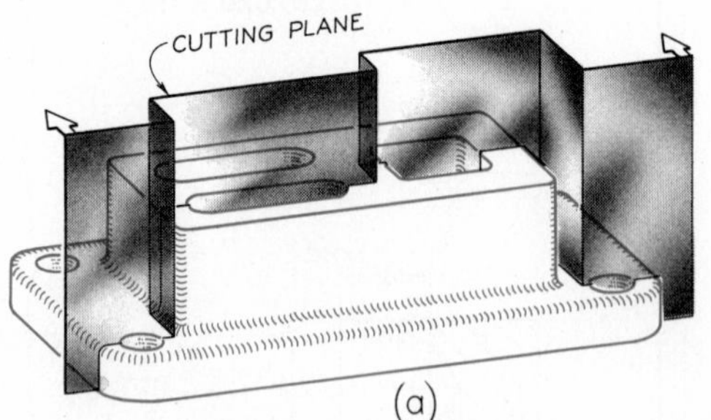


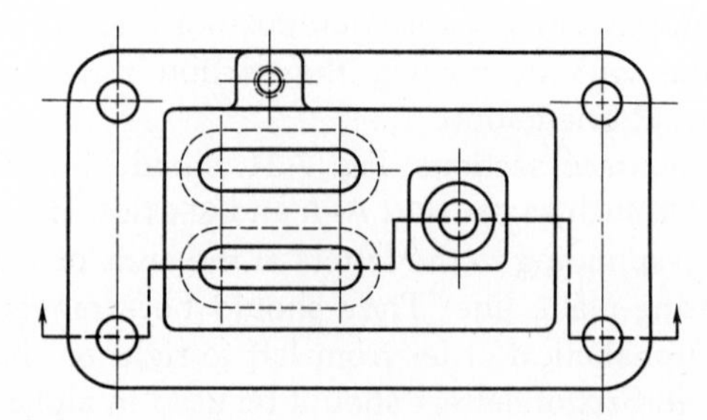

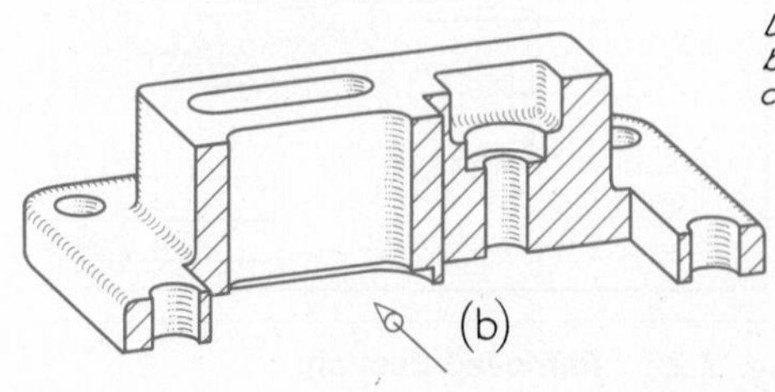


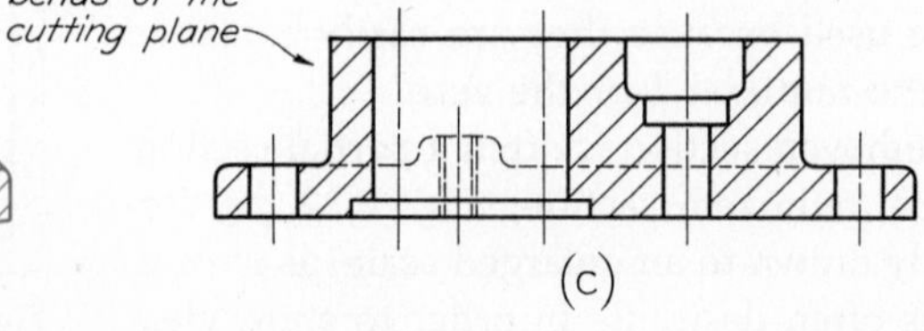

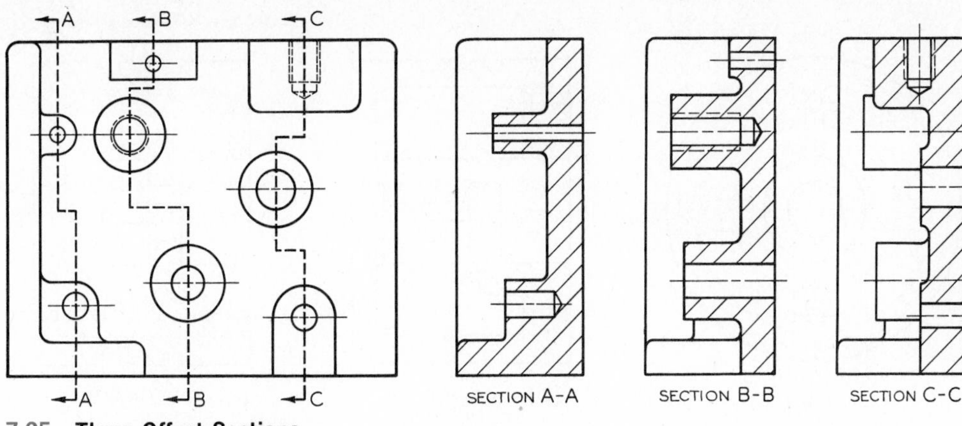


Fig. 7.25 **Three Offset Sections.**

Figure 7.24 also illustrates an example in which hidden lines in a section eliminate the need for an additional view. In this case, an extra view would be needed to show the small boss on the back if hidden lines were not shown.

An example of multiple offset sections is shown in Fig. 7.25.

## 7.12 Ribs in Section.

To avoid a false impression of thickness and solidarity, ribs, webs, gear teeth, and other similar flat features are not sectioned even though the cutting plane passes along the center plane of the feature. For example, in Fig. 7.26, the cutting plane A–A passes flatwise through the vertical web, or rib, and the web is not section-lined, (a). *Such thin features should not be section-lined, even though the cutting plane passes through them.* The incorrect section is shown at (b). Note the false impression of thickness or solidity resulting from section lining the rib.

If the cutting plane passes *crosswise* through a rib or any thin member, as does the plane B–B in Fig. 7.26, the member should be section-lined in the usual manner, as shown in the top view at (c).

In some cases, if a rib is not section-lined when the cutting plane passes through it flatwise, it is difficult to tell whether the rib is actually present, as for example, ribs A in Fig. 7.27 (a) and (b). It is difficult to distinguish spaces B as open spaces and spaces A as ribs. In such cases, double-spaced *section lining* of

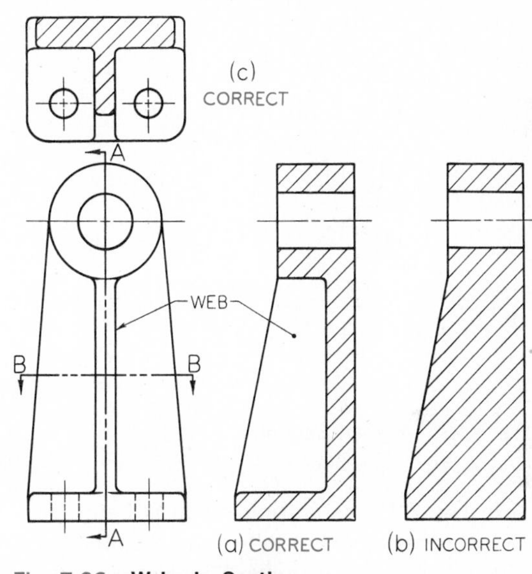


Fig. 7.26 **Webs in Section.**

the ribs should be used, (c). This consists simply in continuing alternate section lines through the ribbed areas, as shown.

## 7.13 Aligned Sections.

In order to include in a section certain angled elements, the cutting plane may be bent so as to pass through those features. The plane and feature are then imagined to be revolved into the original plane. For example, in Fig. 7.28, the cutting plane was bent to pass through the angled arm, and then revolved to a vertical position (aligned), from where it was projected across to the sectional view.

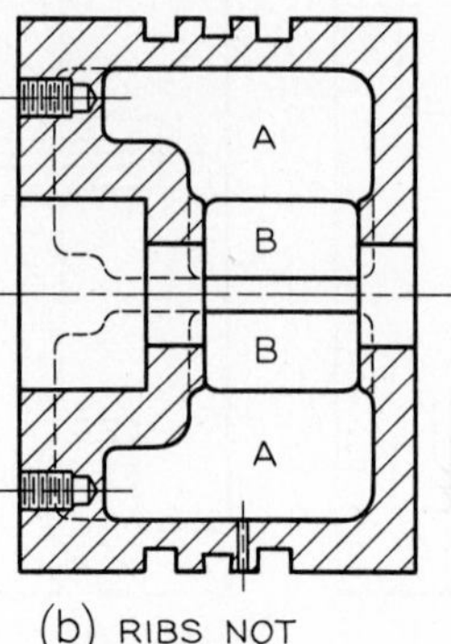


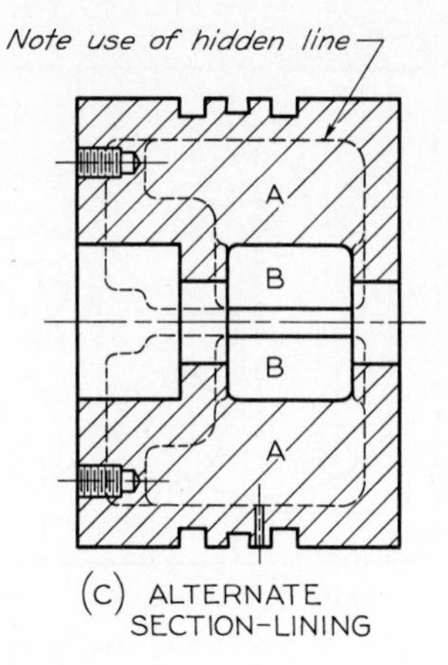


Fig. 7.27 **Alternate Section Lining.**

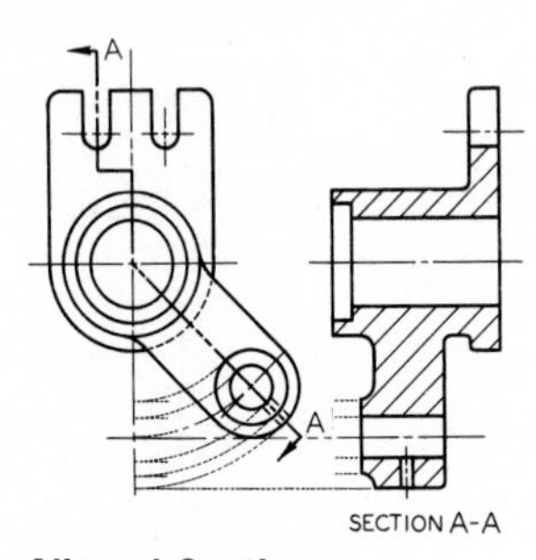


Fig. 7.28 **Aligned Section.**

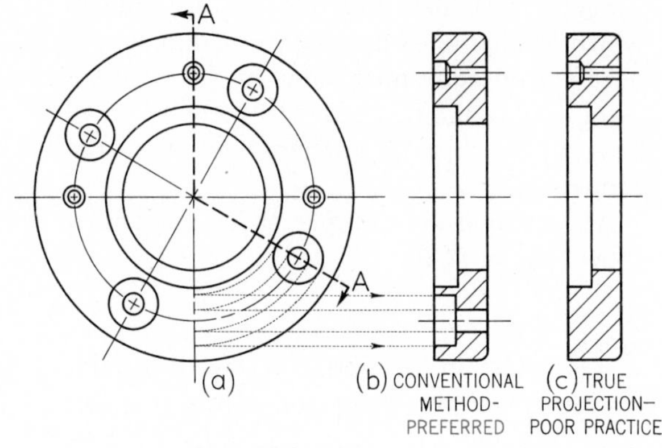


Fig. 7.29 **Aligned Section.**

In Fig. 7.29 the cutting plane is bent so as to include one of the drilled and counterbored holes in the sectional view. The correct section view at (b) gives a clearer and more complete description than does the section at (c), which was taken along the vertical center line of the front view—that is, without any bend in the cutting plane.

In such cases, the angle of revolution should always be less than 90°.

The student is cautioned *not to revolve* features when clearness is not gained. In some cases the revolving features will result in a loss of clarity. Examples in which revolution should not be used are Fig. 7.39, Probs. 17 and 18.

In Fig. 7.30 (a) is an example in which the projecting lugs were not sectioned on the same basis that ribs are not sectioned. At (b) the projecting lugs are located so that the cutting plane would pass through them crosswise; hence they are sectioned.

Another example involving rib sectioning and also aligned sectioning is shown in Fig. 7.31. In the circular view, the cutting plane is offset in circular-arc bends to include the upper hole, and upper rib, the keyway and center hole, the lower rib, and one of the lower holes. These features are then imagined to be revolved until they line up vertically, and then projected from that position to obtain the section at (b). Note that the ribs are not sectioned. If a regular full section of the object were drawn, without the use of conventions discussed here, the resulting section, (c), would be both incomplete and confusing and, in addition, would take more time to draw.

In sectioning a pulley or any spoked wheel, it is standard practice to revolve the spokes if necessary (if there is an odd number) and not

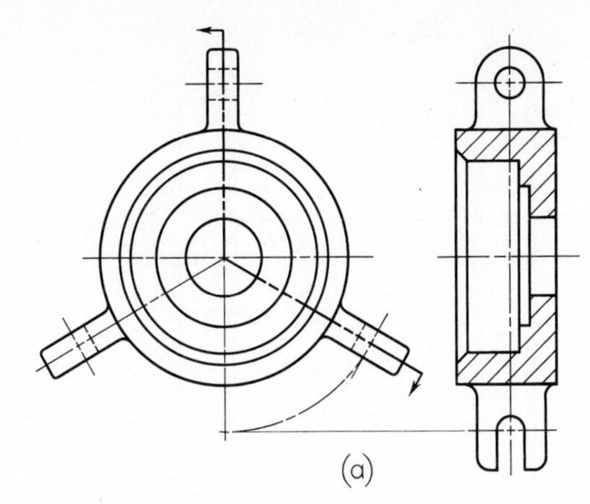


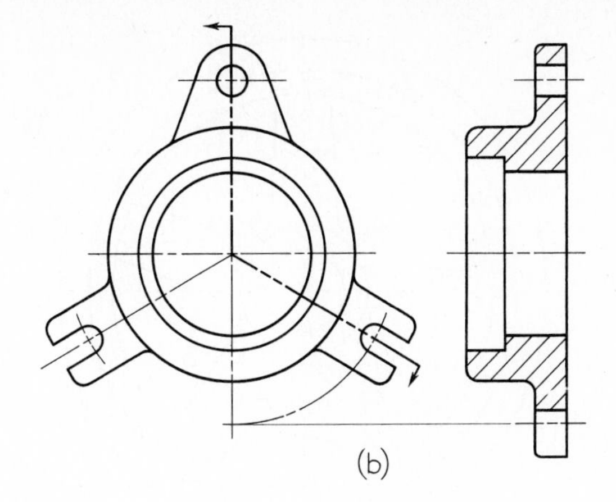


Fig. 7.30 **Aligned Sections.**

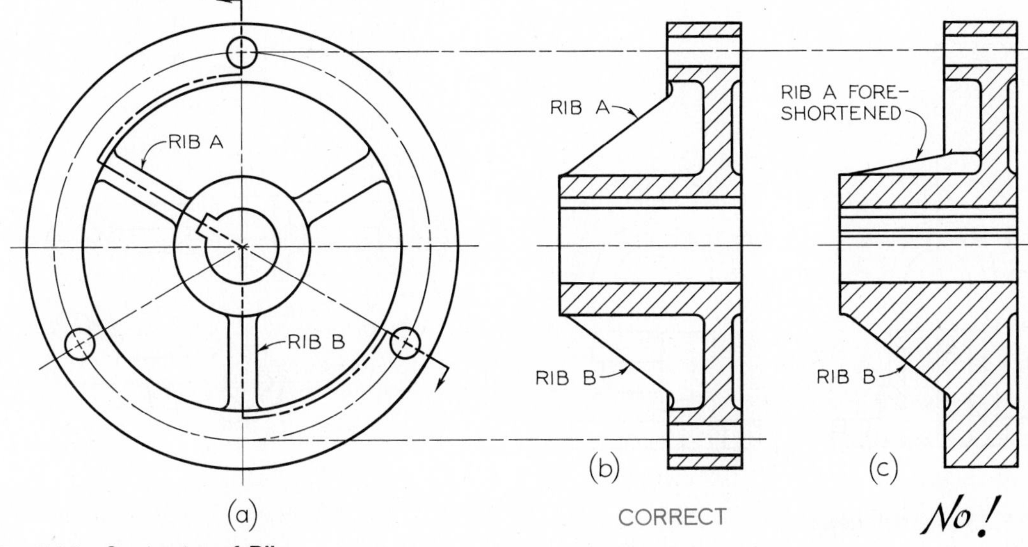


Fig. 7.31 **Symmetry of Ribs.**

to section-line the spokes, Fig. 7.32 (b). If the spoke is sectioned, as shown at (c), the section gives a false impression of continuous metal. If the lower spoke is not revolved, it will be foreshortened in the sectional view in which it presents an "amputated" and wholly misleading appearance.

Figure 7.32 also illustrates correct practice in omitting visible lines in a sectional view. Notice that spoke B is omitted at (b). If it were included, (c), the spoke would be foreshortened, difficult and time consuming to draw, and confusing to the reader of the drawing.

**7.14 Partial Views.** If space is limited on the paper or if it is necessary to save drafting time, *partial views* may be used in connection with sectioning, Fig. 7.33. *Half views* are shown at (a) and (b) in connection with a full section and a half section, respectively. Note that in each case the back half of the object in the circular view is shown, in conformity with the idea of removing the front portion of the object in order to expose the back portion for viewing in section. See also §6.9.

Another method of drawing a partial view is to break out much of the circular view,

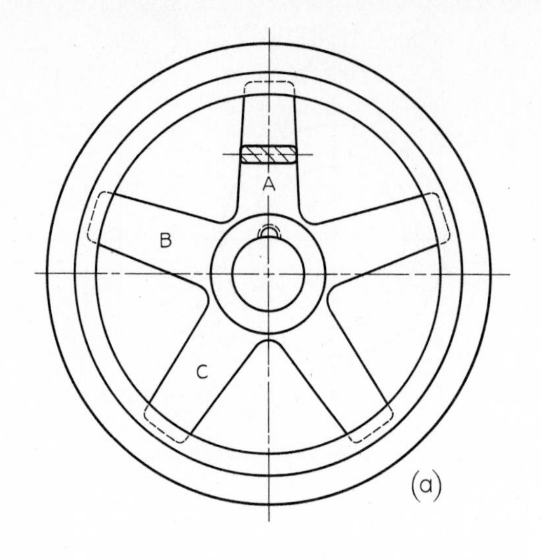


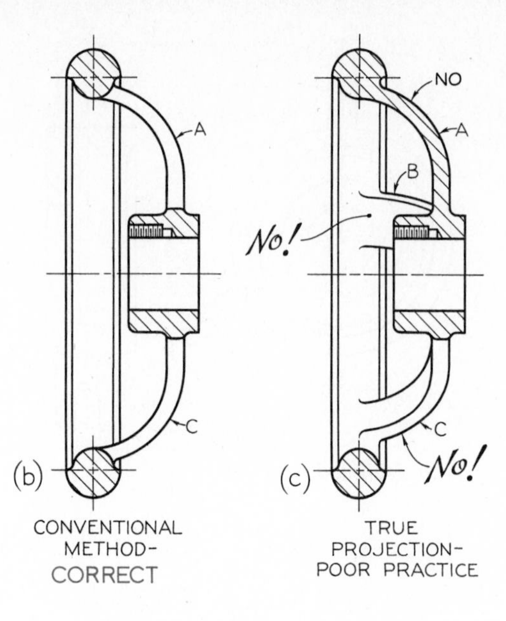


Fig. 7.32 **Spokes in Section.**

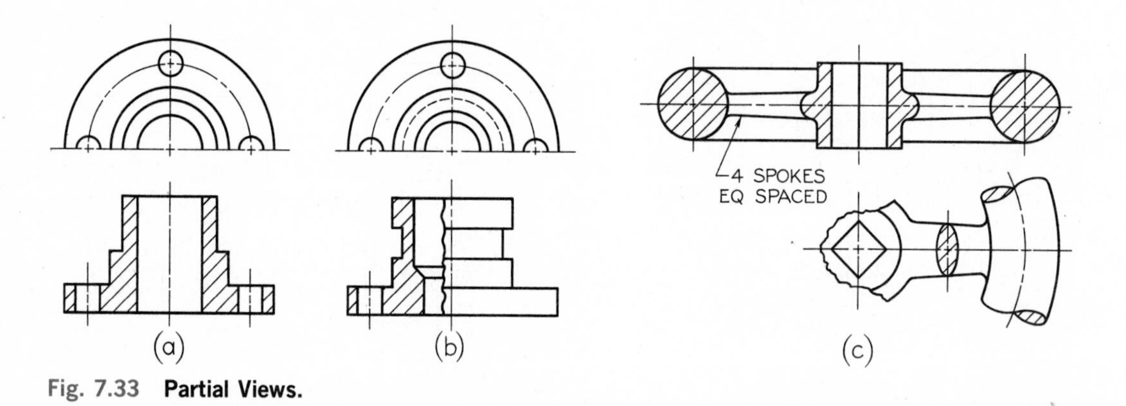


Fig. 7.33 **Partial Views.**

retaining only those features that are needed for minimum representation, Fig. 7.33 (c).

## 7.15 Intersections in Sectioning.

Where an intersection is small or unimportant in a section, it is standard practice to disregard the true projection of the figure of intersection as shown in Fig. 7.34 (a) and (c). Larger figures of intersection may be projected, as shown at (b), or approximated by circular arcs, as shown for the smaller hole at (d). Note that the larger hole K is the same diameter as the vertical hole. In such cases the curves of intersection (ellipses) appear as straight lines, as shown. See also Figs. 6.38 and 6.39.

## 7.16 Conventional Breaks.

In order to shorten a view of an elongated object, conventional breaks are recommended, as shown in Fig. 7.35. For example, the two views of a garden rake are shown in Fig. 7.36 (a), drawn to a small scale in order to get it on the paper. At (b) the handle was "broken," a long central portion removed, and the rake then drawn to a larger scale, producing a much more clear delineation.

Parts thus broken must have the same section throughout, or if tapered they must have a uniform taper. Note at (b) the full-length dimension is given, just as if the entire rake were shown.

The breaks used on cylindrical shafts or tubes are often referred to as "S-breaks" and in the industrial drafting room are usually drawn entirely freehand or partly freehand and partly with the irregular curve or the compass. By these methods, the result is often very crude, especially when attempted by beginners. Simple methods of construction for use by the student or the industrial draftsmen

Fig. 7.34 **Intersections.**

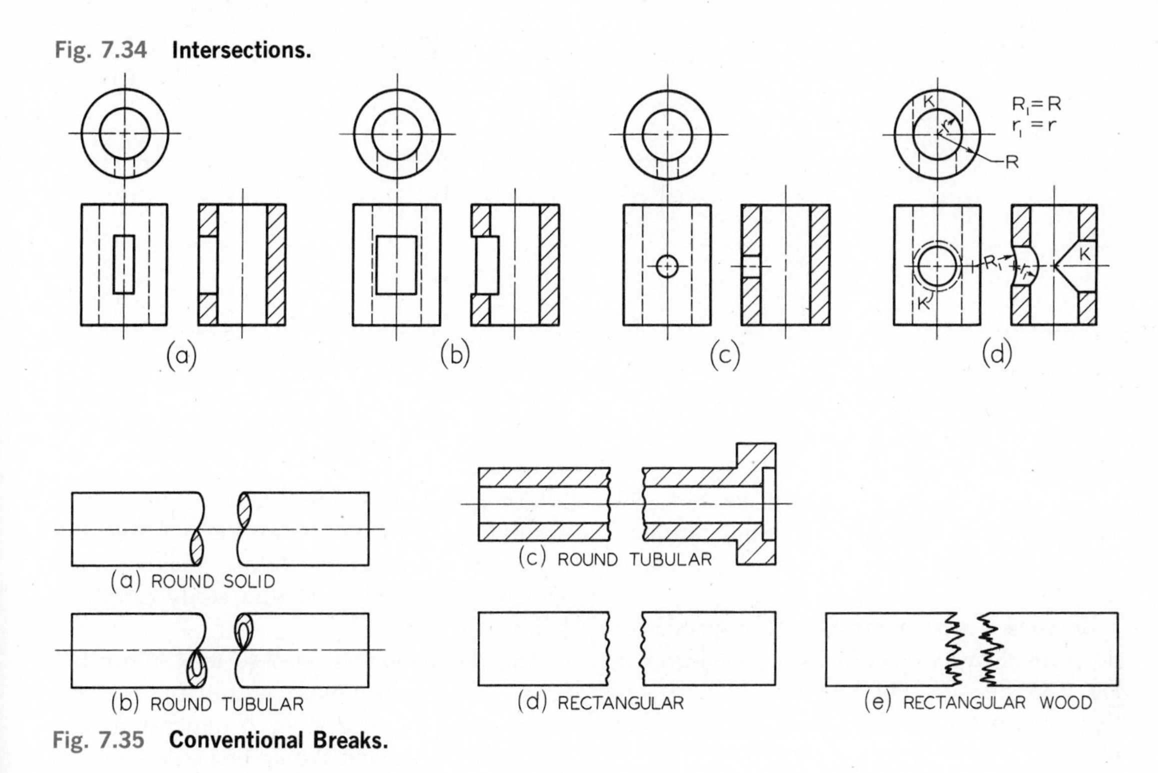


Fig. 7.35 **Conventional Breaks.**

Fig. 7.36 **Use of Conventional Breaks.**

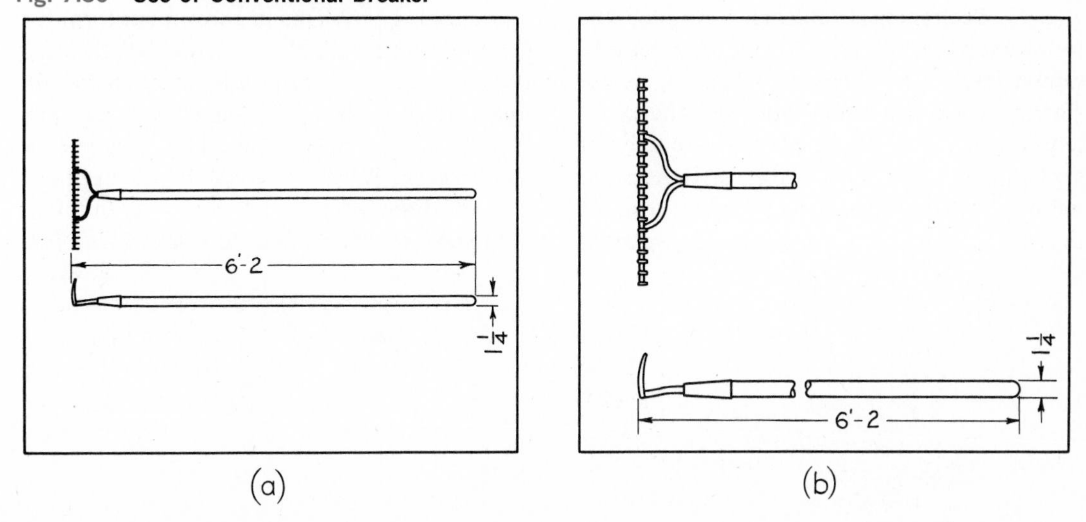

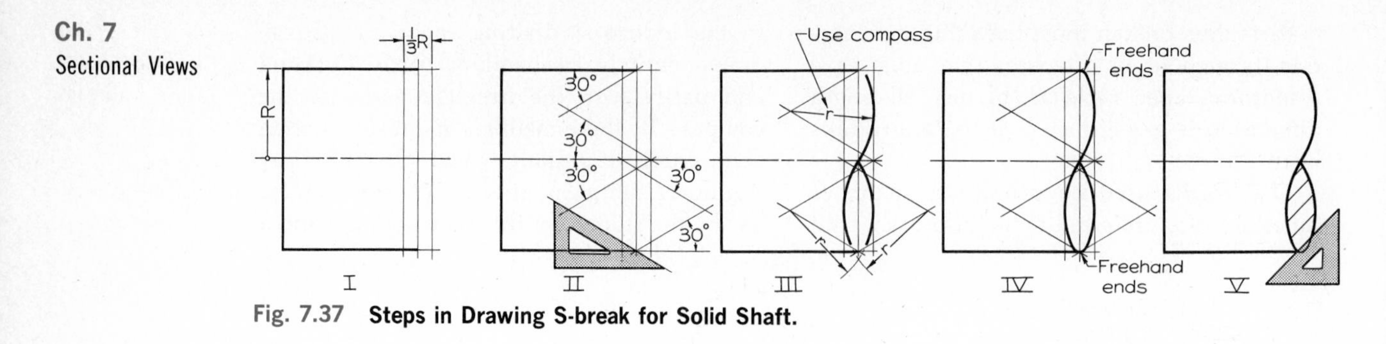


Fig. 7.37 **Steps in Drawing S-break for Solid Shaft.**

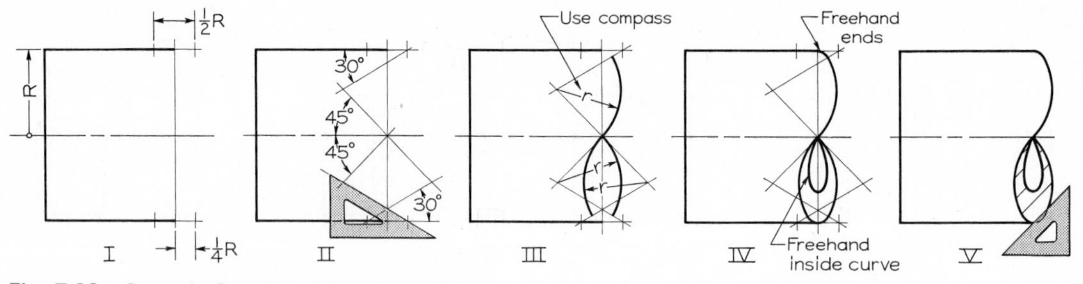


Fig. 7.38 **Steps in Drawing S-break for Tubing.**

are shown in Figs. 7.37 and 7.38 and will always produce a professional result. Excellent S-breaks are also obtained with an S-break template.

Breaks for rectangular metal and wood sections are always drawn freehand, as shown in Fig. 7.35. See also Fig. 7.18, which illustrates the use of breaks in connection with revolved sections.

## 7.17 Sectioning Problems.

Any of the following problems, Figs. 7.39 to 7.69, may be drawn freehand or with instruments, as assigned by the instructor. However, the problems in Fig. 7.39 are especially suitable for sketching on $8\frac{1}{2}'' \times 11''$ cross-section paper with $\frac{1}{4}''$ grid squares. Two problems can be drawn on one sheet, using Layout A–1 similar to Fig. 5.50, with borders drawn freehand. If desired, the problems may be sketched on plain drawing paper. Before making any sketches, the student should study carefully §§5.1 to 5.10.

The problems in Figs. 7.40 to 7.59 are intended to be drawn mechanically, but may be drawn freehand, if desired. If dimensions are required, the student should first study §§11.1 to 11.25. If an ink tracing is required, the student is referred to §§2.51 to 2.54 and 2.56.

Sectioning problems in convenient form for solution are available in *Technical Drawing Problems, Series 1,* by Giesecke, Mitchell, Spencer, and Hill; in *Technical Drawing Problems, Series 2,* by Spencer and Hill; and in *Technical Drawing Problems, Series 3,* by Spencer and Hill, all designed to accompany this text and published by Macmillan Publishing Co., Inc.

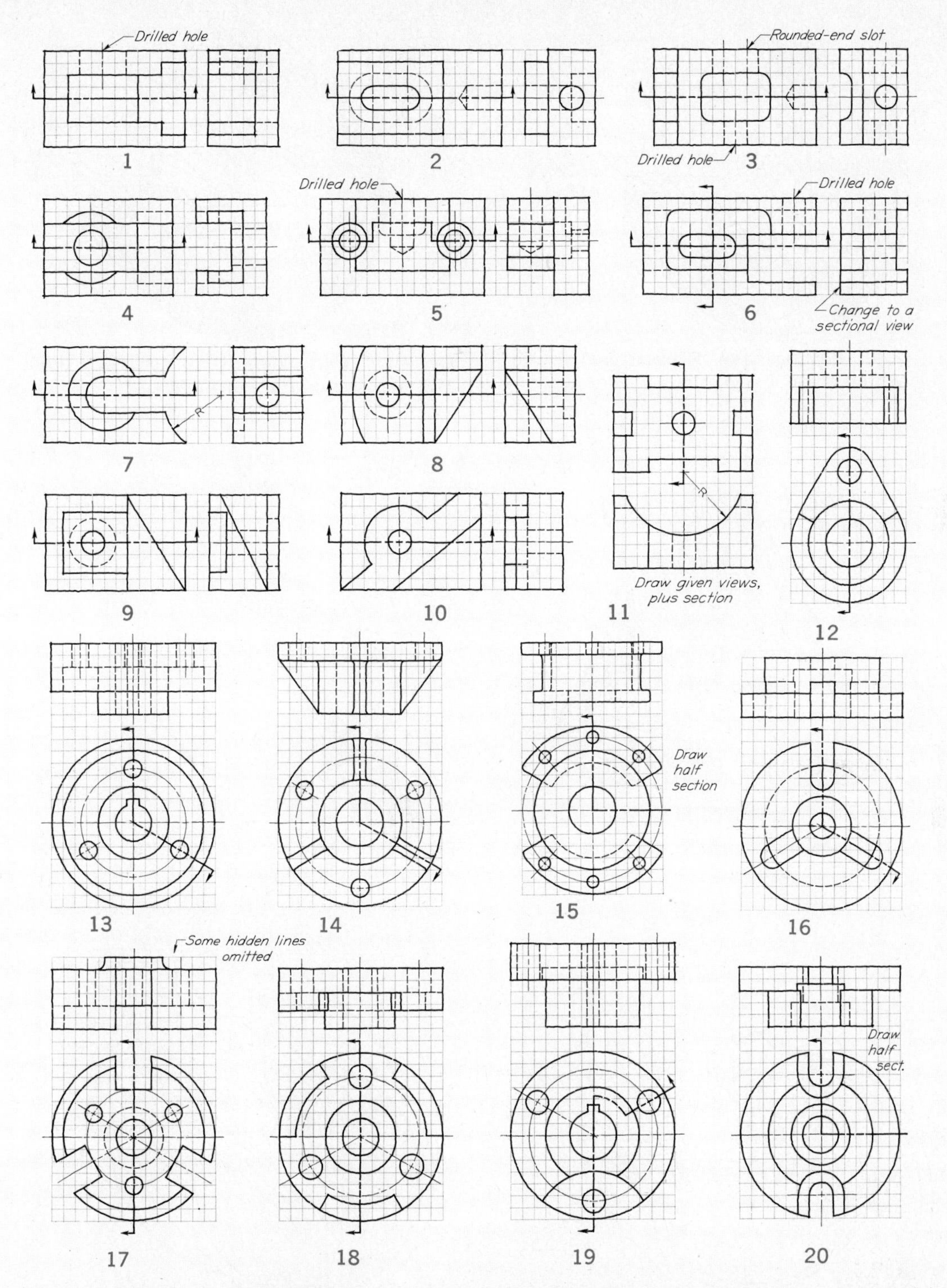


Fig. 7.39 **Freehand Sketching Problems.** Using Layout A–1 (freehand) on cross-section paper or plain paper, two problems per sheet as in Fig. 5.50, sketch views with sections as indicated. Each grid square = $\frac{1}{4}''$. In Probs. 1 to 10, top and right-side views are given. Draw front sectional views and then move right-side views to line up horizontally with front sectional views. In Probs. 12 to 20, draw given front views plus sectional views, omitting given top views. Omit cutting planes except in Probs. 5 and 6.

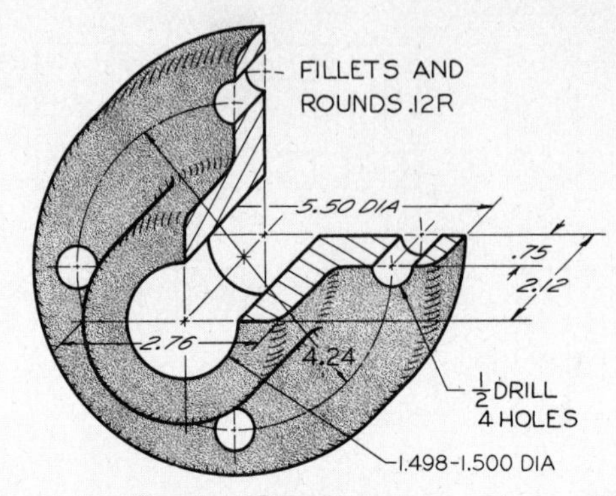


Fig. 7.40 **Bearing.** Draw necessary views, with full section (Layout A–3).

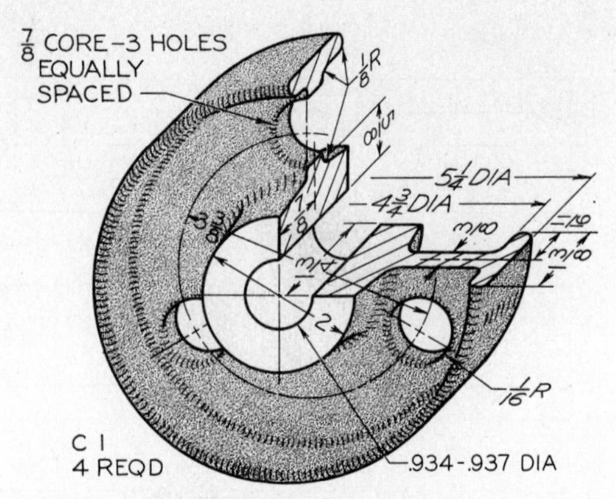


Fig. 7.41 **Truck Wheel.** Draw necessary views, with half section (Layout A–3).

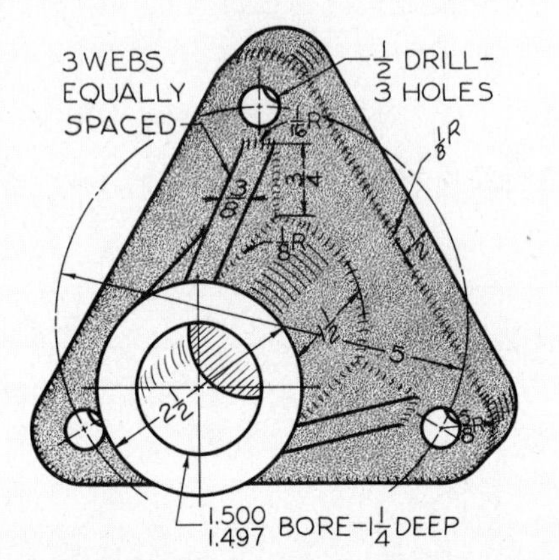


Fig. 7.42 **Column Support.** Draw necessary views, with full section (Layout A–3).

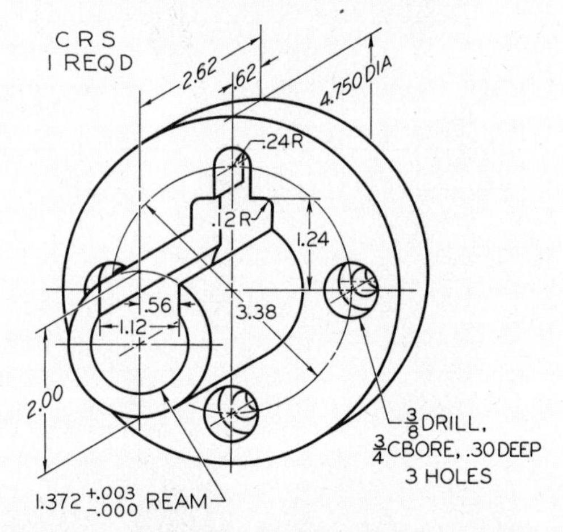


Fig. 7.43 **Centering Bushing.** Draw necessary views, with full section (Layout A–3).

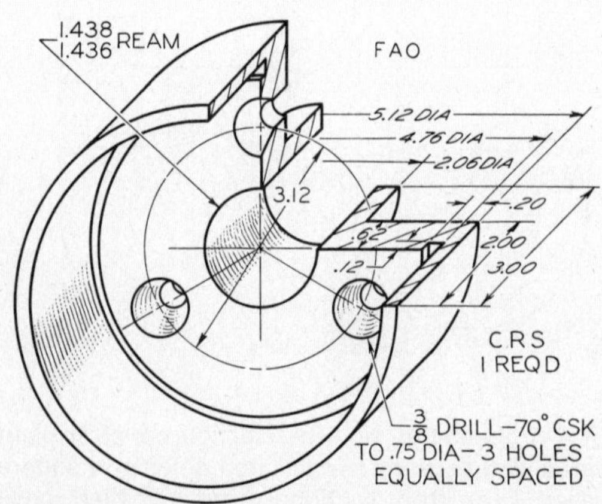


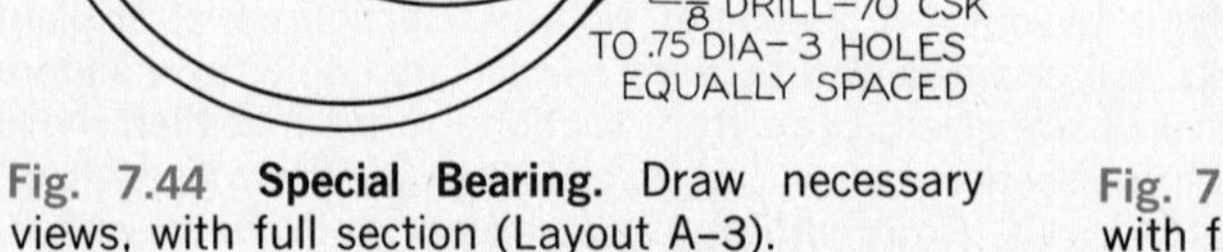


Fig. 7.44 **Special Bearing.** Draw necessary views, with full section (Layout A–3).

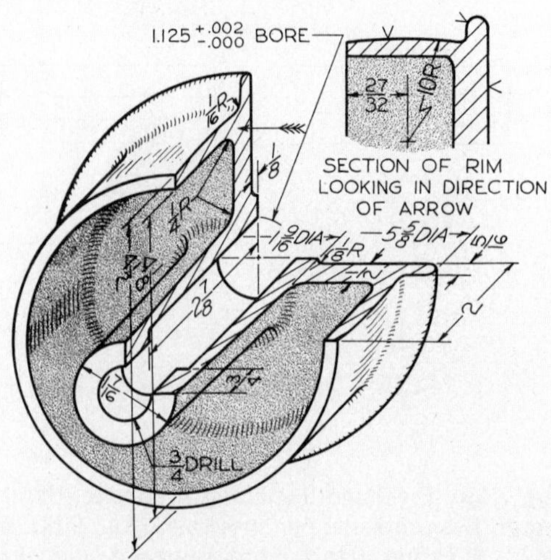


Fig. 7.45 **Idler Pulley.** Draw necessary views, with full section (Layout A–3).

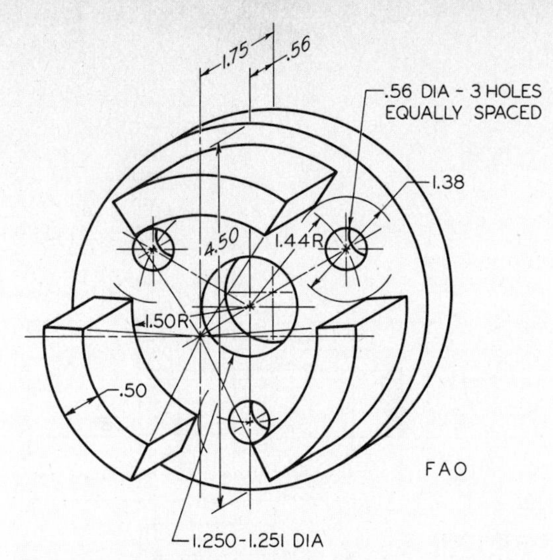


Fig. 7.46 **Cup Washer.** Draw necessary views, with full section (Layout A–3).

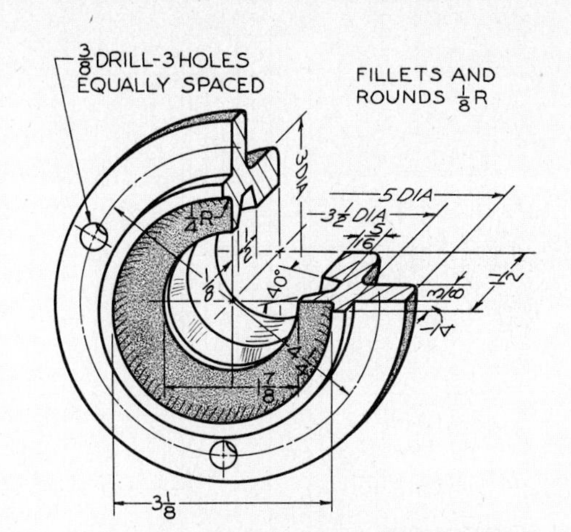


Fig. 7.47 **Fixed Bearing Cup.** Draw necessary views, with full section (Layout A–3).

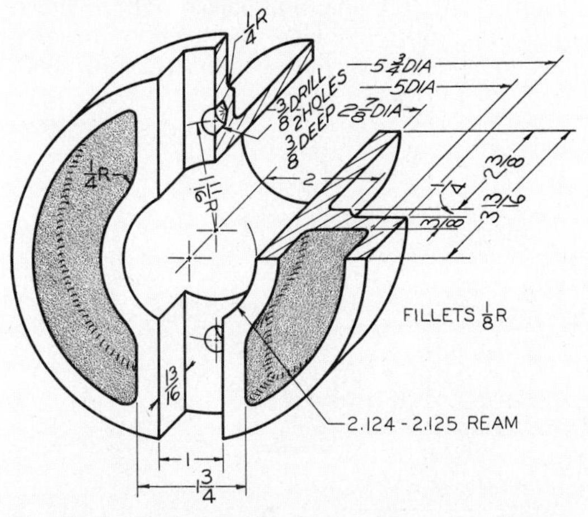


Fig. 7.48 **Stock Guide.** Draw necessary views, with half section (Layout B–4).

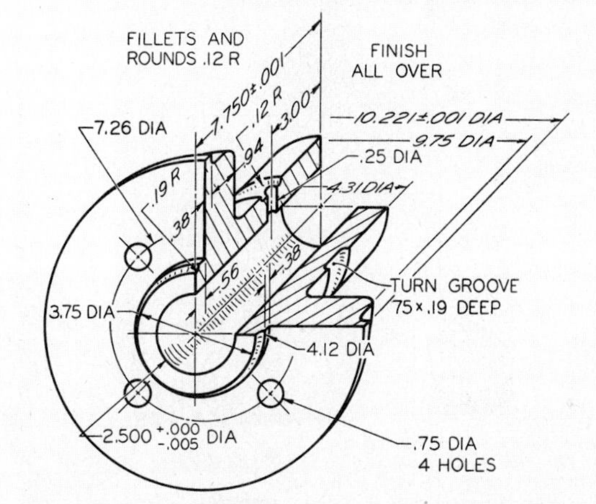


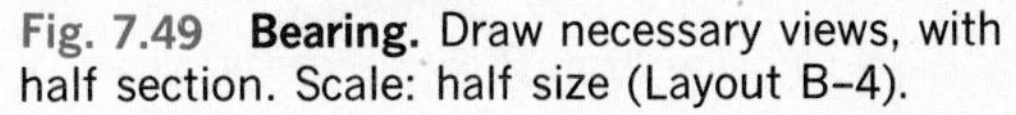

Fig. 7.49 **Bearing.** Draw necessary views, with half section. Scale: half size (Layout B–4).

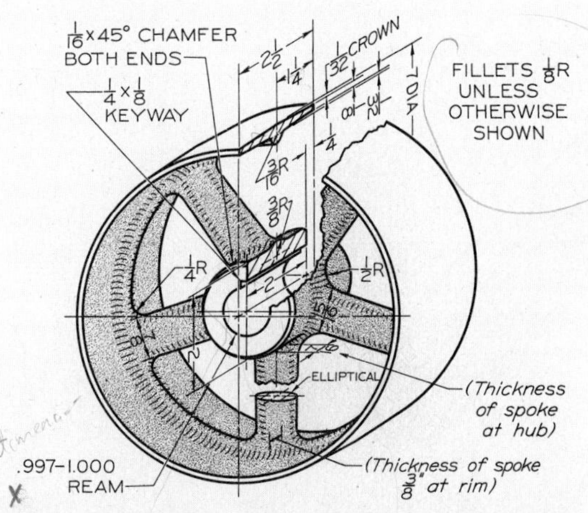


Fig. 7.50 **Pulley.** Draw necessary views, with full section, and revolved section of spoke (Layout B–4).

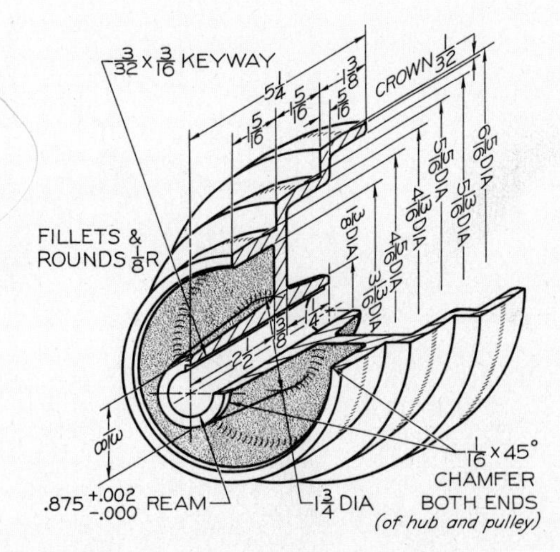


Fig. 7.51 **Step-Cone Pulley.** Draw necessary views, with full section (Layout B–4).

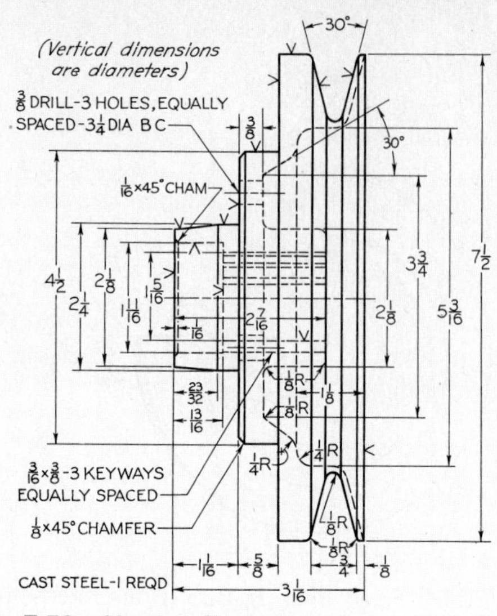

Fig. 7.52 **Sheave.** Draw two views, including half section (Layout B–4).

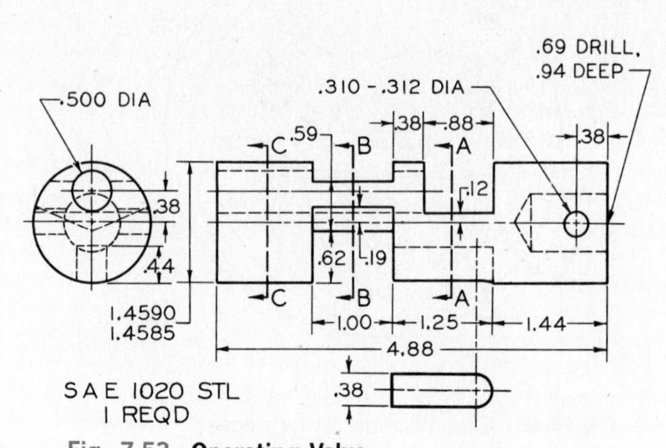

Fig. 7.53 **Operating Valve.** Given: Front, left-side, and partial bottom views. Required: Front, right-side, and full bottom views, plus indicated removed sections (Layout B–4).

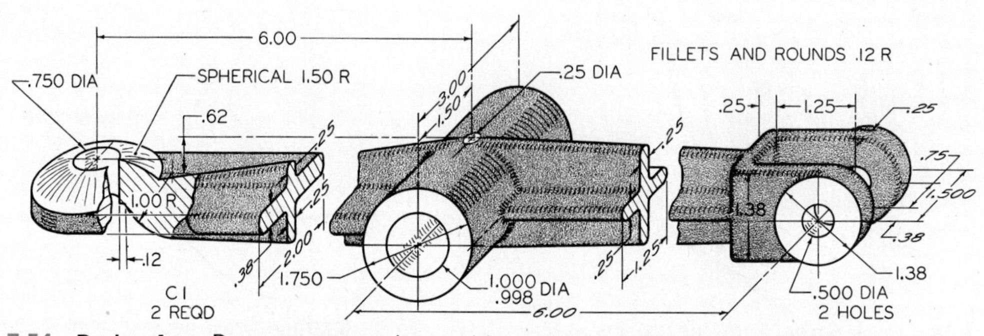

Fig. 7.54 **Rocker Arm.** Draw necessary views, with revolved sections (Layout B–4).

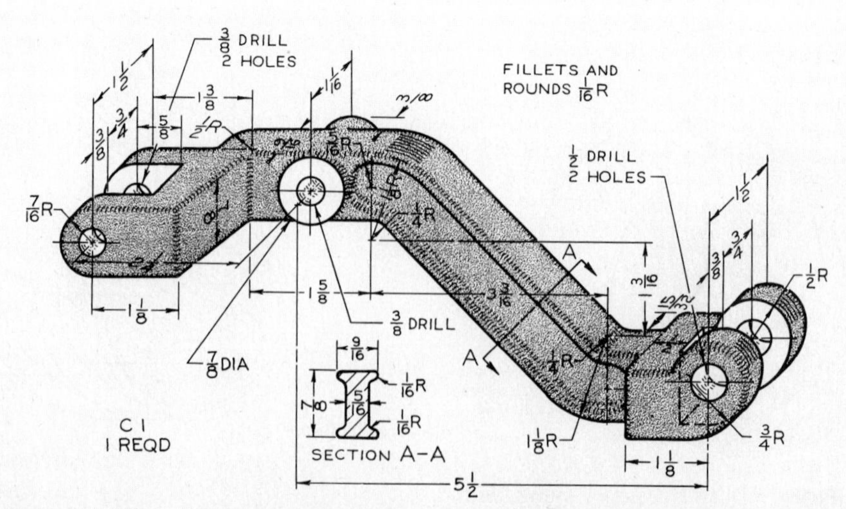

Fig. 7.55 **Dash Pot Lifter.** Draw necessary views, using revolved section instead of removed section (Layout B–4).

Fig. 7.56 **Adjuster Base.**
Given: Front and top views.
Required: Front and top views and sections A–A, B–B, and C–C. Show all visible lines (Layout B-4).

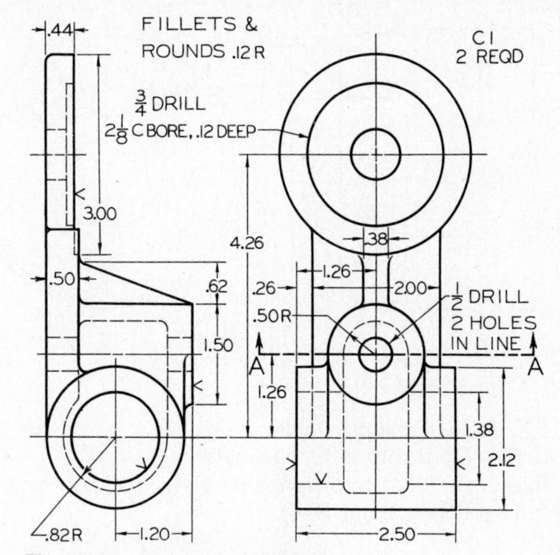


Fig. 7.57 **Mobile Housing.**
Given: Front and left-side views.
Required: Front view, right-side view in full section, and removed section A–A (Layout B-4).

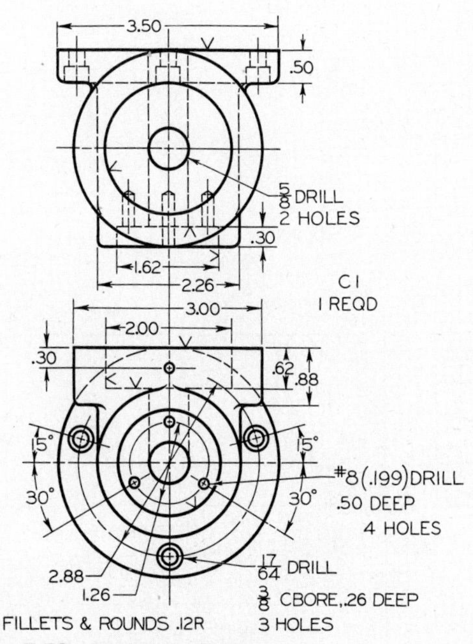


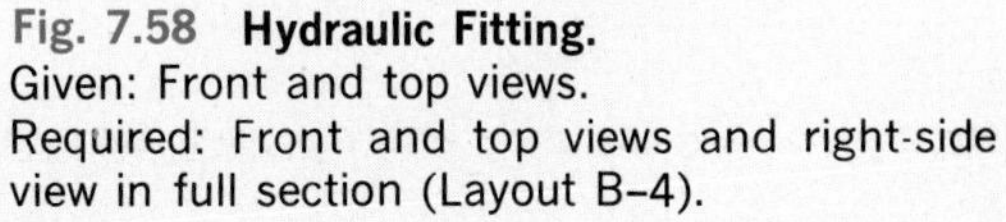


Fig. 7.58 **Hydraulic Fitting.**
Given: Front and top views.
Required: Front and top views and right-side view in full section (Layout B-4).

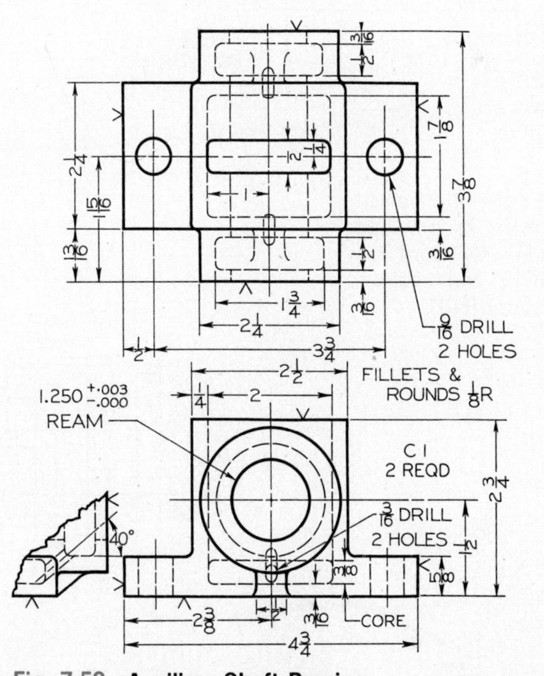


Fig. 7.59 **Auxiliary Shaft Bearing.**
Given: Front and top views.
Required: Front and top views and right-side view in full section (Layout B-4).

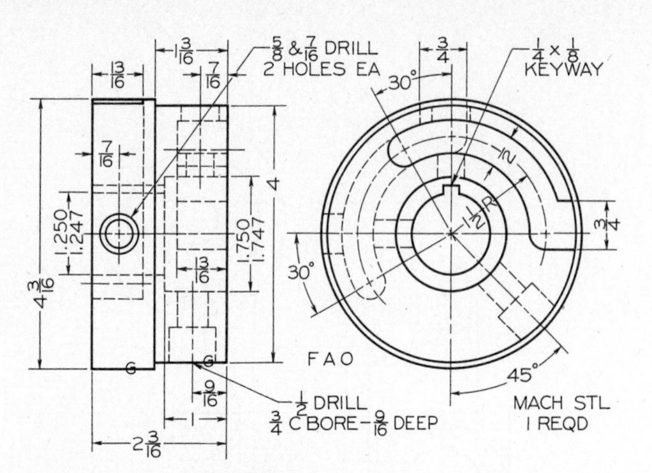


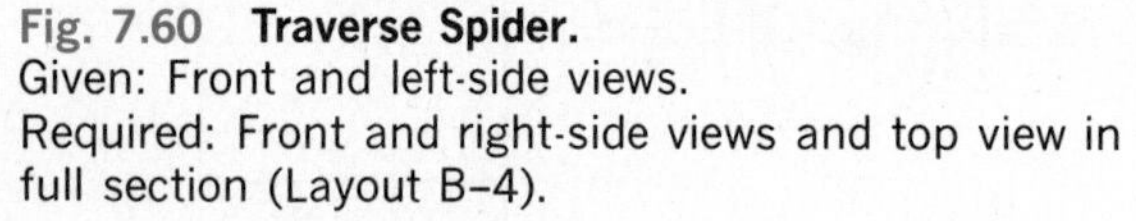
Fig. 7.60 **Traverse Spider.**
Given: Front and left-side views.
Required: Front and right-side views and top view in full section (Layout B–4).

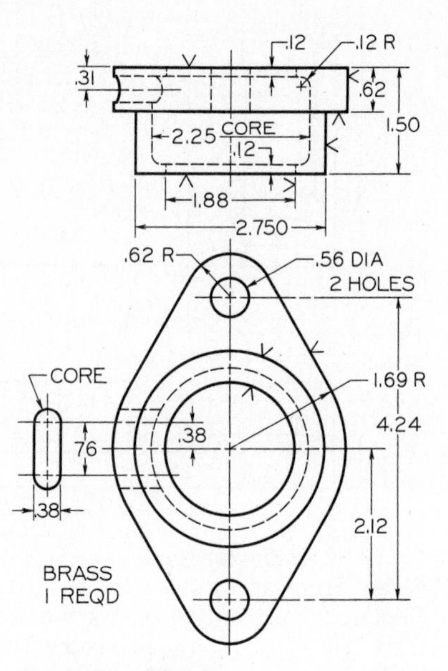


Fig. 7.61 **Gland.**
Given: Front, top, and partial left-side views.
Required: Front view and right-side view in full section (Layout A–3).

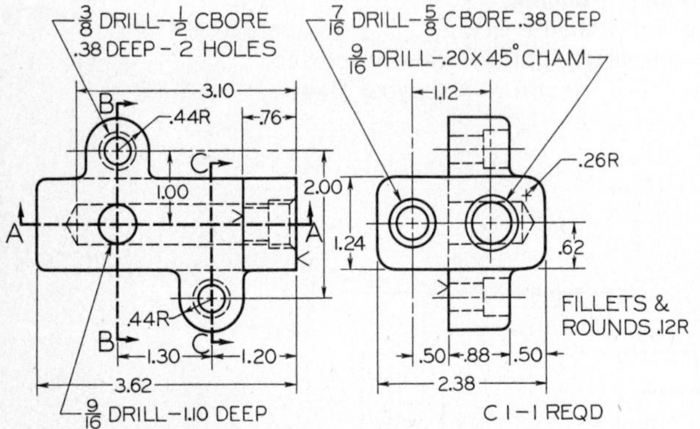


Fig. 7.62 **Bracket.**
Given: Front and right-side views.
Required: Take front as new top; then add right-side view, front view in full section A–A, and sections B–B and C–C (Layout B–4).

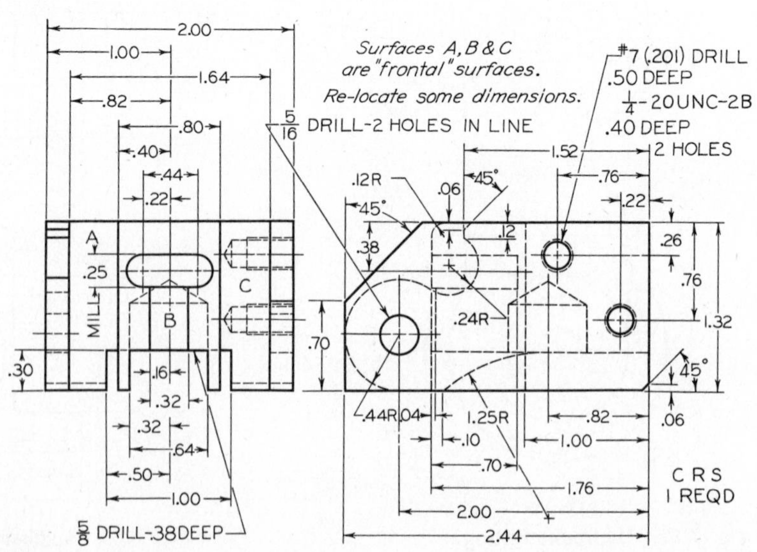


Fig. 7.63 **Cocking Block.**
Given: Front and right-side views.
Required: Take front as new top view; then add new front view, left-side view, and right-side view in full section. Draw double size on Layout C–4.

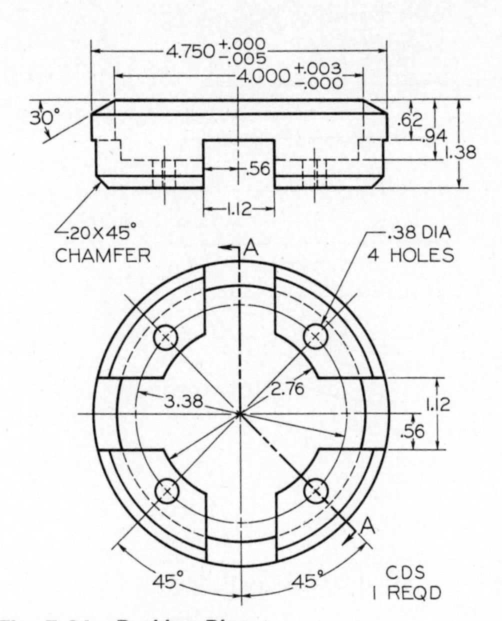


Fig. 7.64 **Packing Ring.**
Given: Front and top views.
Required: Front view and section A–A (Layout A–3).

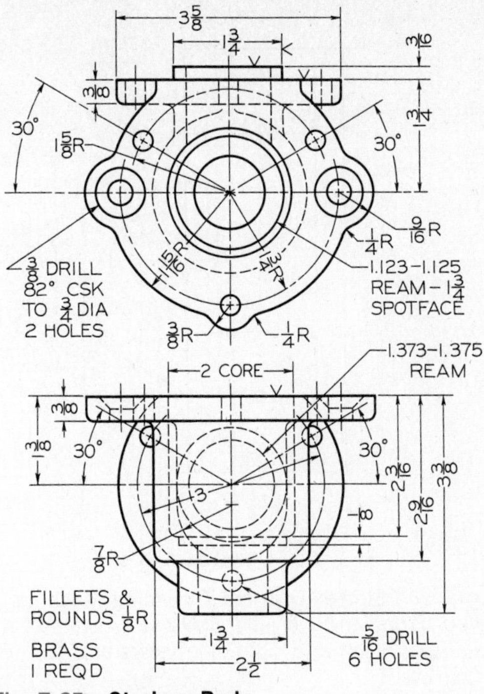


Fig. 7.65 **Strainer Body.**
Given: Front and bottom views.
Required: Front and top views and right-side view in full section (Layout C–4).

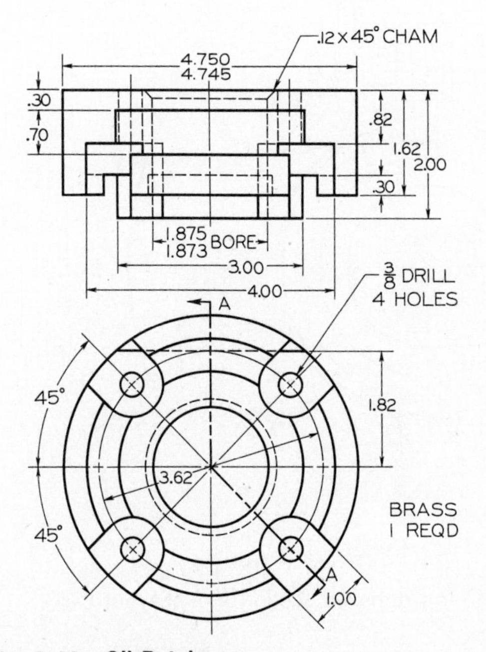


Fig. 7.66 **Oil Retainer.**
Given: Front and top views.
Required: Front view and section A–A (Layout B–4).

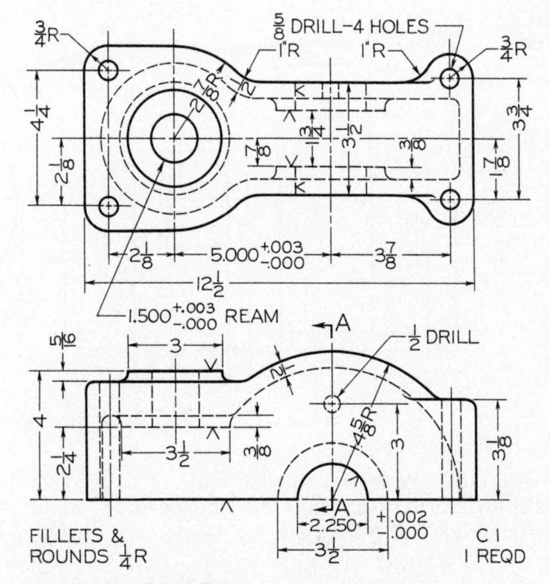


Fig. 7.67 **Gear Box.**
Given: Front and top views.
Required: Front in full section, bottom view, and right-side section A–A. Half size (Layout B–4).

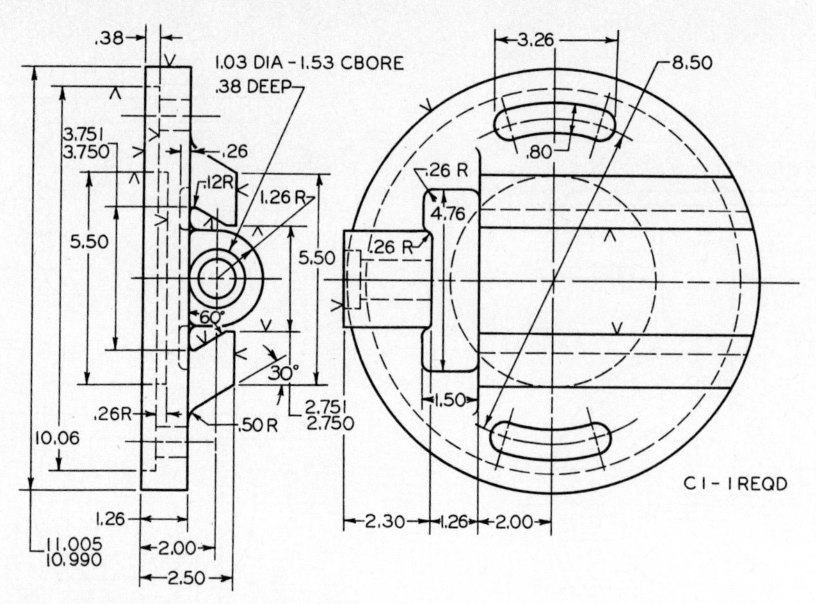


Fig. 7.68 **Slotted Disk for Threading Machine.**
Given: Front and left-side views.
Required: Front and right-side views and top full-section view. Half size (Layout B-4).

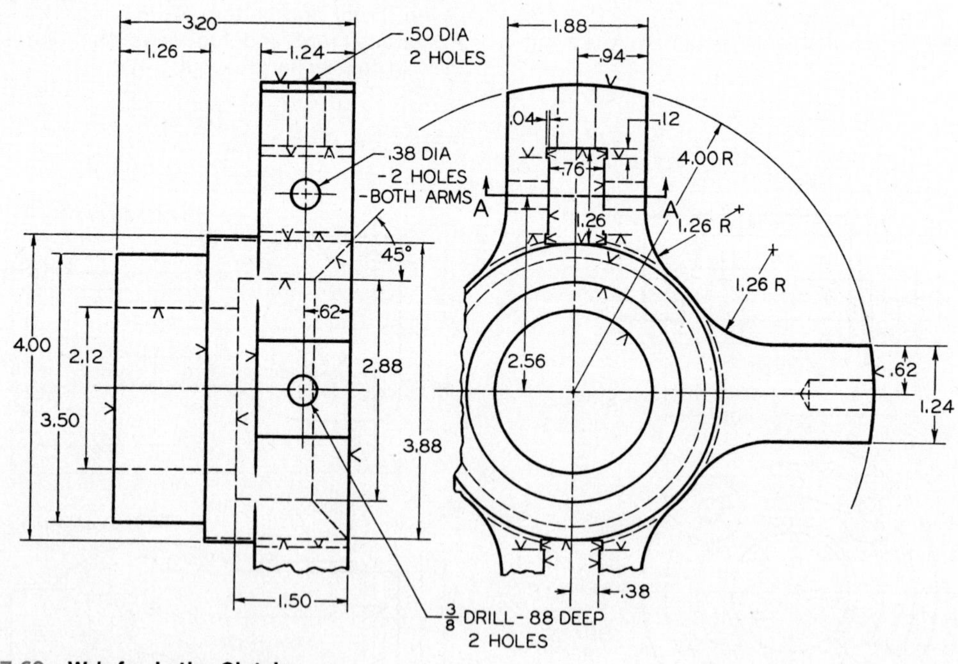


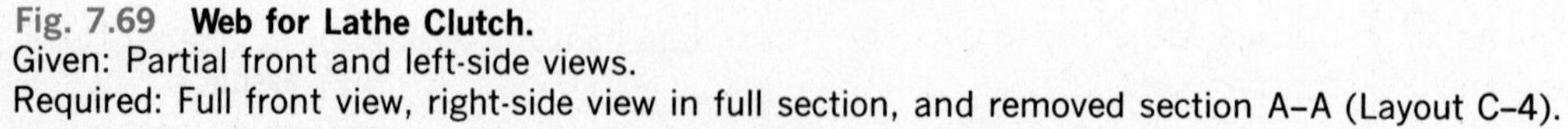
Fig. 7.69 **Web for Lathe Clutch.**
Given: Partial front and left-side views.
Required: Full front view, right-side view in full section, and removed section A–A (Layout C-4).

# 8
# auxiliary views

**8.1 Auxiliary Views.** Many objects are of such shape that their principal faces cannot always be assumed parallel to the regular planes of projection. For example, in Fig. 8.1 (a), the base of the design for the bearing is shown in its true size and shape, but the rounded upper portion is situated at an angle with the planes of projection and does not appear in its true size and shape in any of the three regular views.

In order to show the true circular shapes, it is necessary to assume a direction of sight perpendicular to the planes of those curves, as shown at (b). The resulting view is called an *auxiliary view.* This view, together with the top view, completely describes the object. The front and right-side views are not necessary.

A view obtained by projection on any plane other than the horizontal, frontal, and profile projection planes is known as an auxiliary view. A *primary* auxiliary view is projected on a plane that is perpendicular to one of the principal planes of projection and is inclined to the other two. A *secondary* auxiliary view is projected from a primary view and on a plane inclined to all three principal projection planes. See §8.19.

**8.2 The Auxiliary Plane.** In Fig. 8.2 (a), the object has an inclined surface that does not appear in its true size and shape in any regular view. The auxiliary plane is assumed parallel to the inclined surface P, that is, perpendicular to the line of sight, which is at right angles to that surface. The auxiliary plane is then perpendicular to the frontal plane of projection and hinged to it.

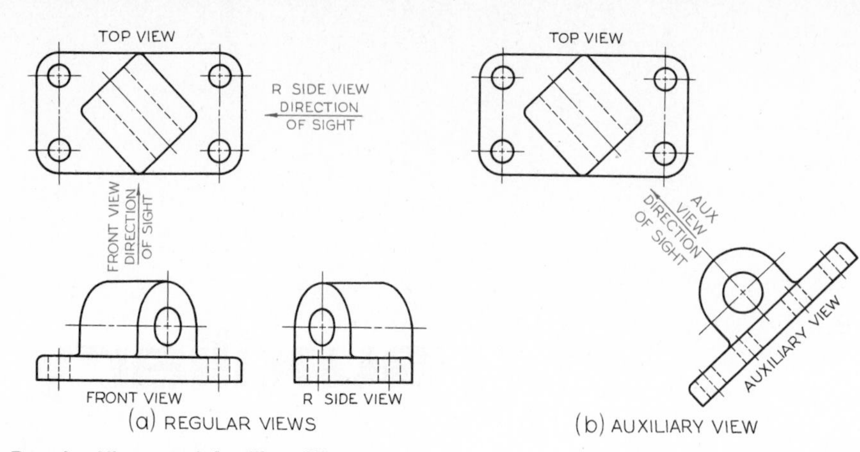


Fig. 8.1 **Regular Views and Auxiliary Views.**

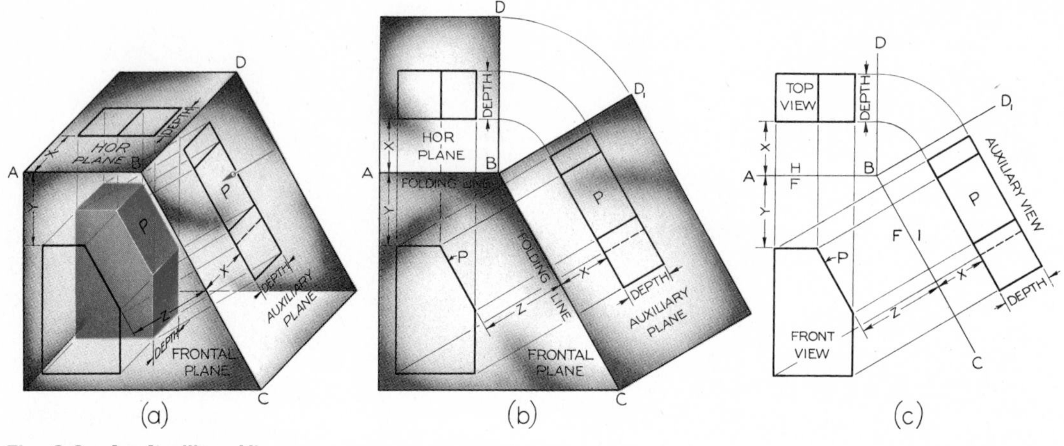


Fig. 8.2 **An Auxiliary View.**

When the horizontal and auxiliary planes are unfolded to lie in the plane of the front view, as shown at (b), the *folding lines* represent the hinge lines joining the planes. The drawing is simplified as shown at (c) by retaining the folding lines, H/F and F/1, and omitting the planes. As will be shown later, the folding lines may themselves be omitted in the actual drawing. The inclined surface P is shown in its true size and shape in the auxiliary view, the long dimension of the surface being projected directly from the front view and the *depth* from the top view.

It should be observed that the positions of the folding lines depend upon the relative positions of the planes of the glass box at (a). If the horizontal plane is moved upward, the distance Y is increased. If the frontal plane is brought forward, the distances X are increased but remain *equal.* If the auxiliary plane is moved to the right, the distance Z is increased. Note that both top and auxiliary views show the *depth* of the object.

## 8.3 To Draw an Auxiliary View, Using Folding Lines.

**8.3 To Draw an Auxiliary View, Using Folding Lines.** As shown in Fig. 8.2 (c), the folding lines are the hinge lines of the glass box. Distances X must be equal, since they both represent the distance of the front surface of the object from the frontal plane of projection.

Although distances X must remain equal, distances Y and Z, from the front view to the respective folding lines, may or may not be equal.

The steps in drawing an auxiliary view with the aid of the folding lines, shown in Fig. 8.3, are described as follows:

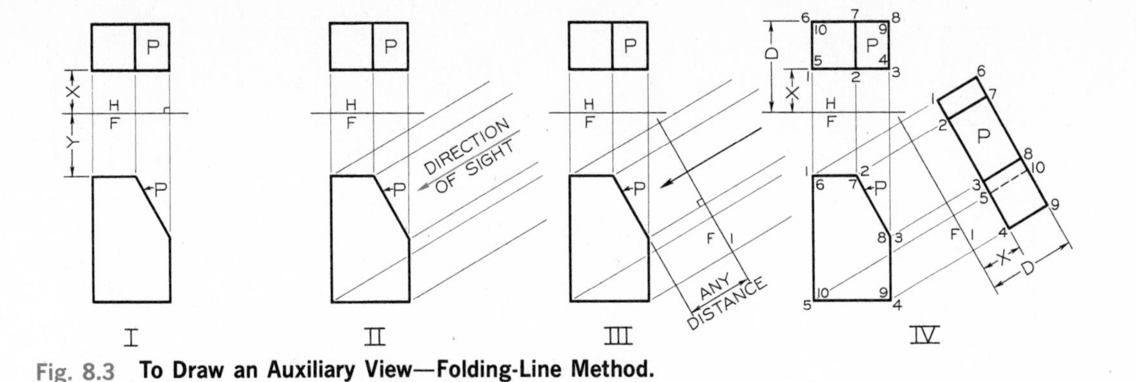

Fig. 8.3 **To Draw an Auxiliary View—Folding-Line Method.**

I. The front and top views are given. It is required to draw an auxiliary view showing the true size and shape of inclined surface P. Draw the folding line H/F between the views at right angles to the projection lines. Distances X and Y may or may not be equal, as desired.

*Note:* In the following steps, manipulate the triangle (either triangle) as shown in Fig. 8.4 to draw lines parallel or perpendicular to the inclined face.

II. Assume arrow, indicating direction of sight, perpendicular to surface P. Draw light projection lines from the front view parallel to the arrow, or perpendicular to surface P.

III. Draw folding line F/1 for the auxiliary view at right angles to the projection lines and at any convenient distance from the front view.

IV. Draw the auxiliary view, using the numbering system explained in §6.6. Locate all points the same distances from folding line F/1 as they are from folding line H/F in the top view. For example, points 1 to 5 are distance X from the folding lines in both the top and auxiliary views, and points 6 to 10 are distance D from the corresponding folding lines. Since the object is viewed in the direction of the arrow, it will be seen that edge 5–10 will be hidden in the auxiliary view.

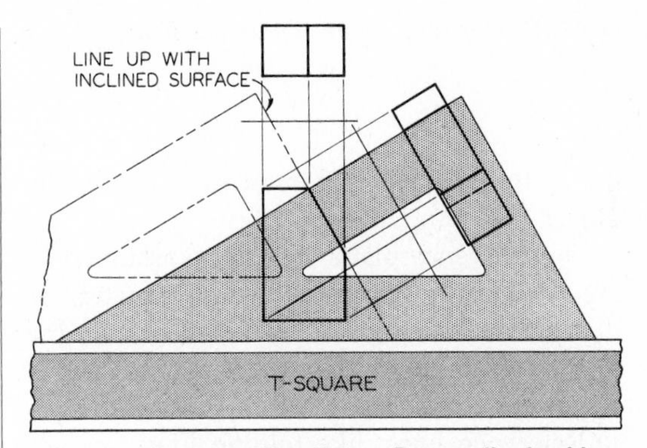

Fig. 8.4 **Drawing Parallel or Perpendicular Lines.**

## 8.4 Reference Planes.

In Figs. 8.2 (c) and 8.3 the folding lines are edge views of the front plane of projection. In effect, the frontal plane is used as a *reference plane,* or *datum plane,* for transferring distances (*depth measurements*) from the top view to the auxiliary view.

Instead of using one of the planes of projection as a reference plane, it is often more convenient to assume a reference plane inside the glass box parallel to the plane of projection and touching or cutting through the object. For example, Fig. 8.5 (a), a reference plane is assumed to coincide with the front surface of the object. This plane appears edgewise in the top and auxiliary views, and the two reference lines are then used in the same manner as folding lines. Dimensions D, to the reference lines, are equal. The advantage of the reference-plane method is that fewer measurements are required, since some points of the object lie in the reference plane.

The reference plane may coincide with the front surface of the object as at (a), or it may cut through the object as at (b) if the object is symmetrical, or the reference plane may coincide with the back surface of the object as at

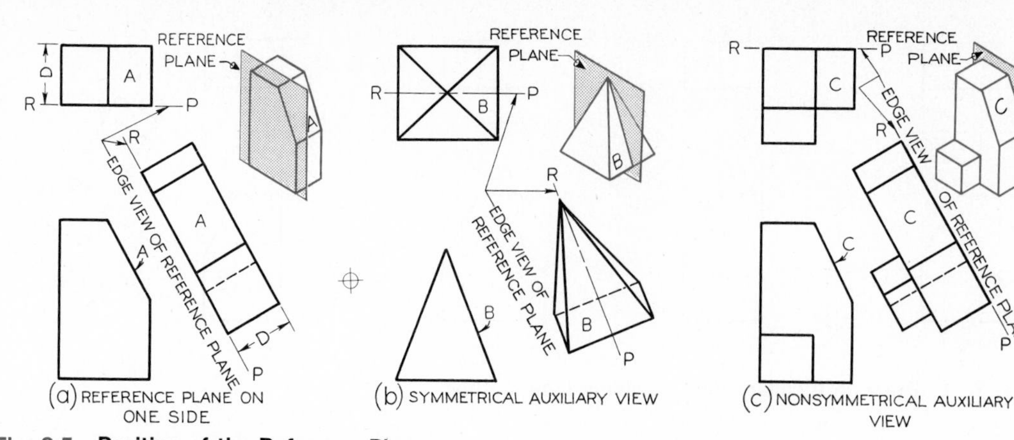


**Fig. 8.5 Position of the Reference Plane.**

(c), or through any intermediate point of the object.

The reference plane should be assumed in the position most convenient for transferring distances with respect to it. Remember the following:

1. Reference lines, like folding lines, are always at right angles to the projection lines between the views.
2. A reference plane appears as a line in two alternate views, never in adjacent views.
3. Measurements are always made at right angles to the reference lines, or parallel to the projection lines.
4. In the auxiliary view, all points are at the same distances from the reference line as the corresponding points are from the reference line in the *second previous view,* or alternate view.

## 8.5 To Draw an Auxiliary View, Using Reference Plane.

To object shown in Fig. 8.6 (a) is numbered as explained in §6.6. To draw the auxiliary view, proceed as follows:

I. Draw two views of the object, and assume an arrow indicating the direction of sight for the auxiliary view of surface A.

II. Draw projection lines parallel to the arrow.

III. Assume reference plane coinciding with back surface of object as shown at (a). Draw reference lines in the top and auxiliary views at right angles to the projection lines; *these are the edge views of the reference plane.*

IV. Draw auxiliary view of surface A. It will be true size and shape because the direction of sight was taken perpendicular to that surface. Transfer depth measurements from the top view to the auxiliary view with dividers or scale. Each point in the auxiliary view will be on its projection line from the front view and the same distance from the reference line as it is in the top view to the corresponding reference line.

V. Complete the auxiliary view by adding other visible edges and surfaces of the object. Each numbered point in the auxiliary view lies on its projection line from the front view and is the same distance from the reference line as it is in the top view.

Note that two surfaces of the object appear as lines in the auxiliary view. Which surfaces are these? (Give numbers.) Does the bottom surface of the object appear true size and shape in the auxiliary view? Why? Before you draw the bottom surface in the auxiliary view, how do you know its general configuration and exact number of sides? Which edges of surface 2-5-6-3 are foreshortened and which are true length in the auxiliary view? Does surface 5-8-9-6 appear true size in the auxiliary view? Which edges of surface 5-8-9-6 appear true size in the front view? In the auxiliary view?

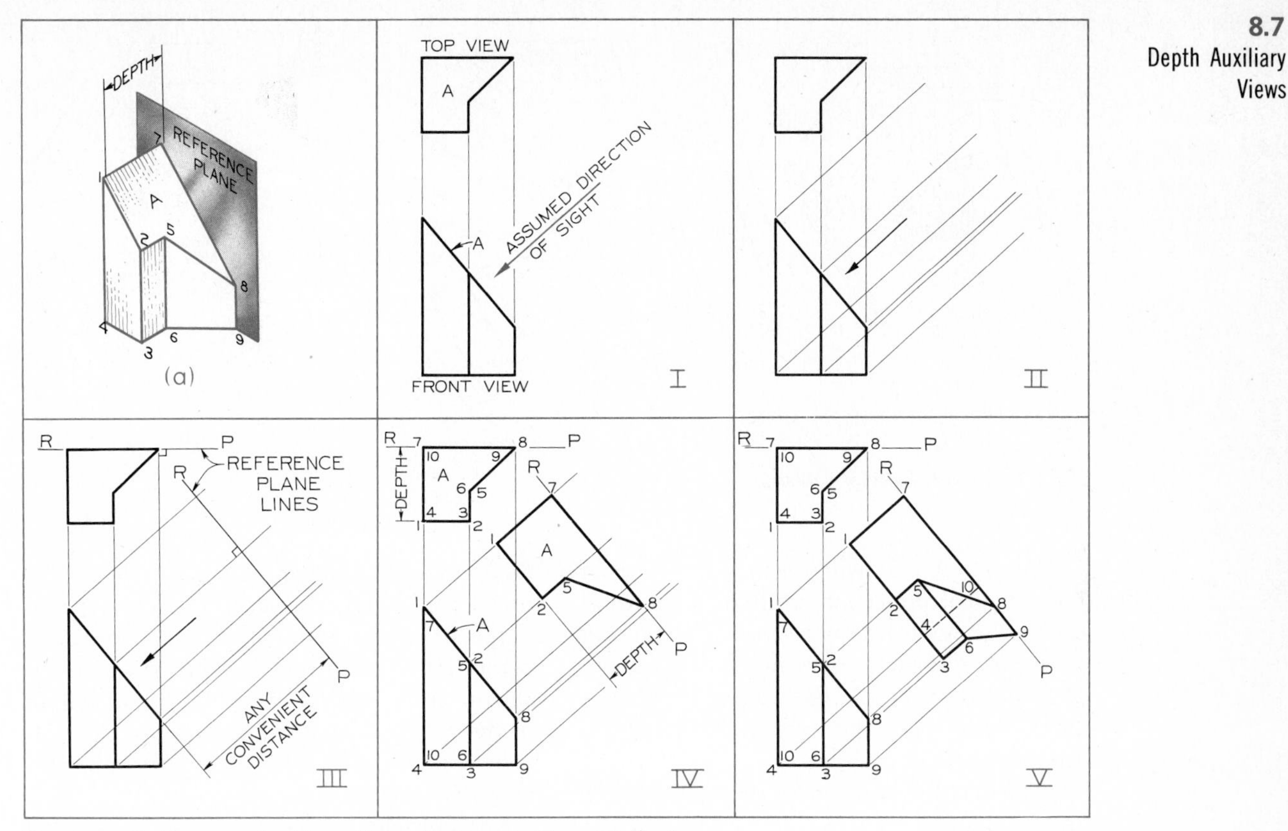


Fig. 8.6 **To Draw an Auxiliary View—Reference-Plane Method.**

## 8.6 Classification of Auxiliary Views.

Auxiliary views are classified and named according to the principal dimensions of the object shown in the auxiliary view. For example, the auxiliary view in Fig. 8.6 is a *depth auxiliary view* because it shows the principal dimension of the object, *depth.* Any auxiliary view projected from the front view will show the depth of the object and is a depth auxiliary view.

Similarly, any auxiliary view projected from the top view is a *height auxiliary view,* and any auxiliary view projected from the side view (either side) is a *width auxiliary view.* What kind of auxiliary view is Fig. 8.1 (b)? Fig. 8.13 (b)? Fig. 8.27? Fig. 8.33?

## 8.7 Depth Auxiliary Views.

An infinite number of auxiliary planes can be assumed perpendicular to, and hinged to, the frontal plane (F) of projection. Five such planes are shown in Fig. 8.7 (a), the horizontal plane being included to show that it is similar to the others. In all of these views the principal dimension, *depth,* is shown; hence all of the auxiliary views are *depth auxiliary views.*

The unfolded auxiliary planes are shown at (b), which also shows how the depth dimension may be projected from the top view to all auxiliary views. The arrows indicate the directions of sight for the several views, and the projection lines are respectively parallel to these arrows. The arrows may be assumed but need not be actually drawn, since the projection lines determine the direction of sight. The folding lines are perpendicular to the arrows and to the corresponding projection lines. Since the auxiliary planes can be assumed at any distance from the object, it follows that the folding lines may be any distance from the front view.

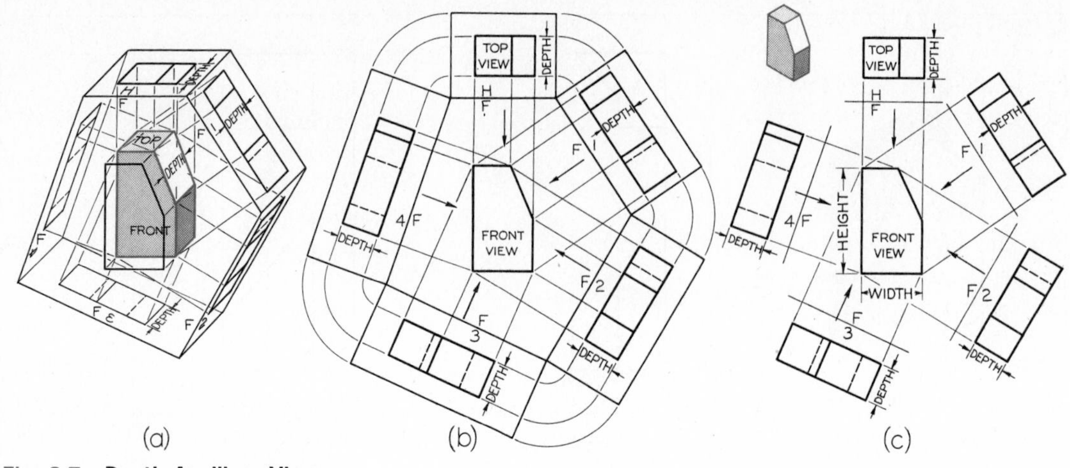


Fig. 8.7 **Depth Auxiliary Views.**

The complete drawing, with the outlines of the planes of projection omitted, is shown at (c). This shows the drawing as it would appear on paper, in which use is made of reference planes as described in §8.4, all depth dimensions being measured perpendicular to the reference line in each view.

Note that the front view shows the *height* and the *width* of the object, *but not the depth.* The depth is shown in all views that are projected from the front view; hence, this rule; *The principal dimension shown in an auxiliary view is that one which is not shown in the adjacent view from which the auxiliary view was projected.*

## 8.8 Height Auxiliary Views.

An infinite number of auxiliary planes can be assumed perpendicular to, and hinged to, the horizontal plane (H) of projection, several of which are shown in Fig. 8.8 (a). The front view and all of the auxiliary views show the principal dimension, *height.* Hence, all of the auxiliary views are *height auxiliary views.*

The unfolded planes are shown at (b), and the complete drawing, with the outlines of the planes of projection omitted, is shown at (c). All reference lines are perpendicular to the corresponding projection lines, and all height dimensions are measured parallel to the pro-

Fig. 8.8 **Height Auxiliary Views.**

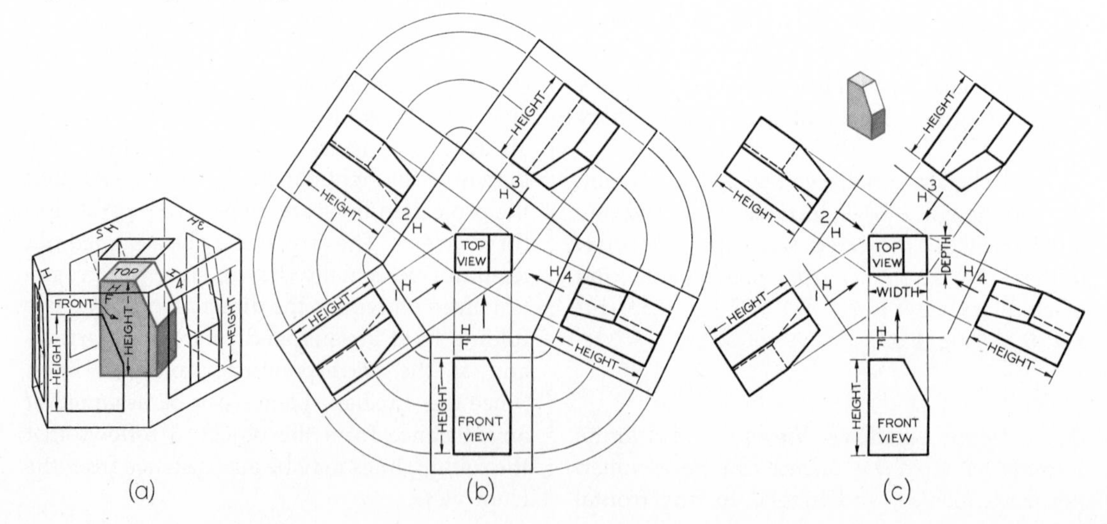

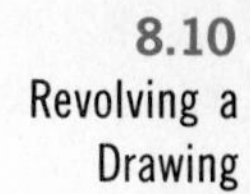


jection lines, or perpendicular to the reference lines, in each view. Note that in the view projected from, which is the top view, the only dimension *not shown* is height.

## 8.9 Width Auxiliary Views.

An infinite number of auxiliary planes can be assumed perpendicular to, and hinged to, the profile plane (P) of projection, several of which are shown in Fig. 8.9 (a). The front view and all of the auxiliary views show the principal dimension, *width*. Hence, all of the auxiliary views are *width auxiliary views.*

The unfolded planes are shown at (b), and the complete drawing, with the outlines of the planes of projection omitted, is shown at (c). All reference lines are perpendicular to the corresponding projection lines, and all width dimensions are measured parallel to the projection lines, or perpendicular to the reference lines, in each view. Note that in the right-side view, from which the auxiliary views are projected, the only dimension *not shown* is width.

## 8.10 Revolving a Drawing.

In Fig. 8.10 (a) is a drawing showing top, front, and auxiliary views. At (b) the drawing is shown revolved, as indicated by the arrows, until the auxiliary view and the front view line up horizontally.

**Fig. 8.9 Width Auxiliary Views.**

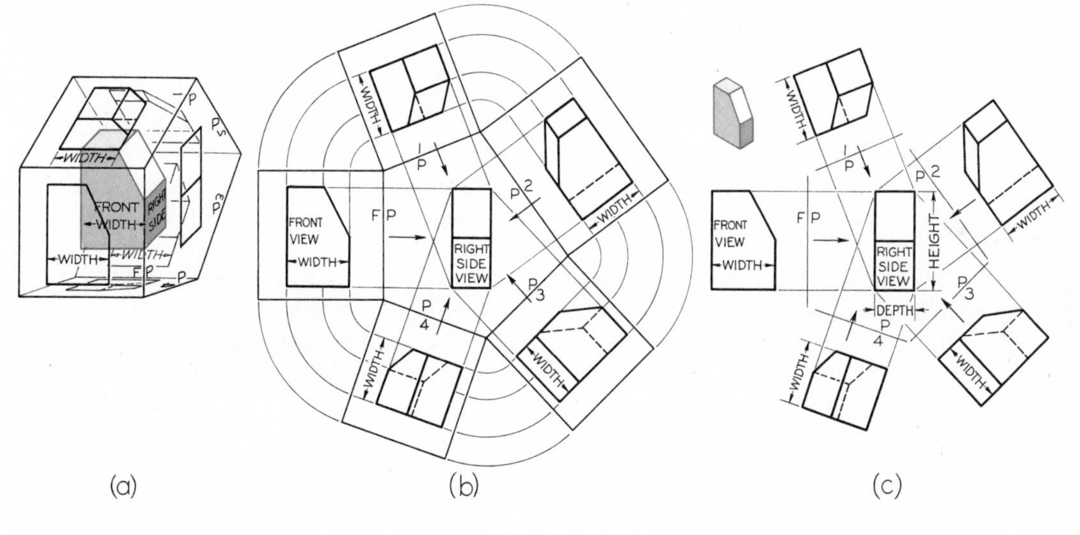


**Fig. 8.10 Revolving a Drawing.**

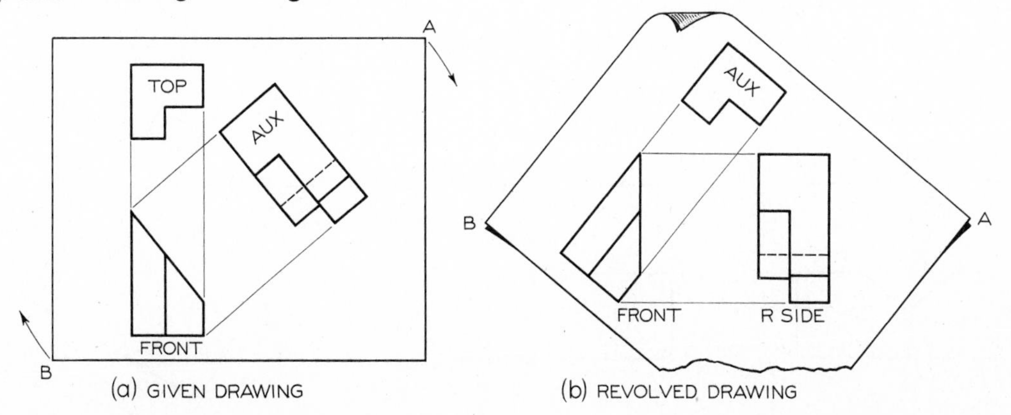

Although the views remain exactly the same, the names of the views must be changed; the auxiliary view now becomes a right-side view, and the top view becomes an auxiliary view. Some students find it easier to visualize and draw an auxiliary view when revolved to the position of a regular view in this manner. In any case, it must be understood that an auxiliary view basically is like any other view.

## 8.11 Dihedral Angles.

The angle between two planes is a *dihedral angle.* One of the principal uses of auxiliary views is to show dihedral angles in true size, mainly for dimensioning purposes. In Fig. 8.11 (a) is shown a block with a V-groove situated so that the true dihedral angle between inclined surfaces A and B is shown in the front view. Why does this view show the true angle?

Assume a line in a plane. For example, draw a straight line on a sheet of paper; then hold the paper so as to view the line as a point. You will observe that when the line appears as a point, the plane containing the line appears as a line. Hence, this rule: *To get the edge view of a plane, get the point view of any line in that plane.*

In Fig. 8.11 (a), line 1–2 is the line of intersection of planes A and B. Now, line 1–2 lies in both planes at the same time; therefore, a point view of this line will show both planes as lines, and the angle between them is the dihedral angle between the planes. Hence, this rule: *To get the true angle between two planes, get the point view of the line of intersection of the planes.*

At (b), the line of intersection 1–2 does not appear as a point in the front view; hence, planes A and B do not appear as lines, and the true dihedral angle is not shown. Assuming that the actual angle is the same as at (a), does the angle show larger or smaller than at (a)? The drawing at (b) is unsatisfactory. The true angle does not appear because the direction of sight (see arrow) is not parallel to the line of intersection 1–2.

At (c) the direction of sight arrow is taken parallel to line 1–2, producing an auxiliary view in which line 1–2 appears as a point, planes A and B appear as lines, and the true dihedral angle is shown. *To draw a view showing a true dihedral angle, assume the direction of sight parallel to the line of intersection between the planes of the angle.*

## 8.12 Plotted Curves.

As shown in §6.30, if a cylinder is cut by an inclined plane, the inclined surface is elliptical in shape. In Fig. 6.34 (a), such a surface is produced, but the ellipse does not show true size and shape because the plane of the ellipse is not seen at right angles in any view.

In Fig. 8.12 (a), the line of sight is taken perpendicular to the edge view of the inclined surface, and the resulting ellipse is shown in true size and shape in the auxiliary view. The major axis is found by direct projection from the front view, and the minor axis is equal to the diameter of the cylinder. The left end of the cylinder (a circle) will appear as an ellipse

Fig. 8.11 Dihedral Angles.

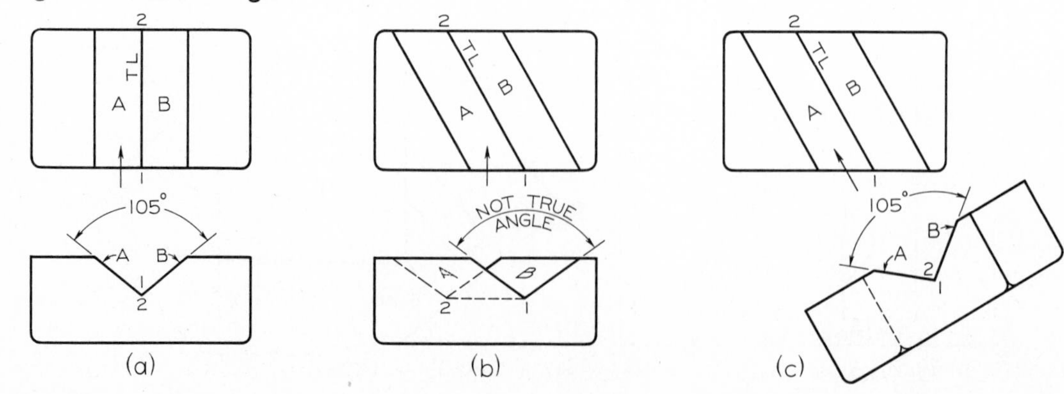

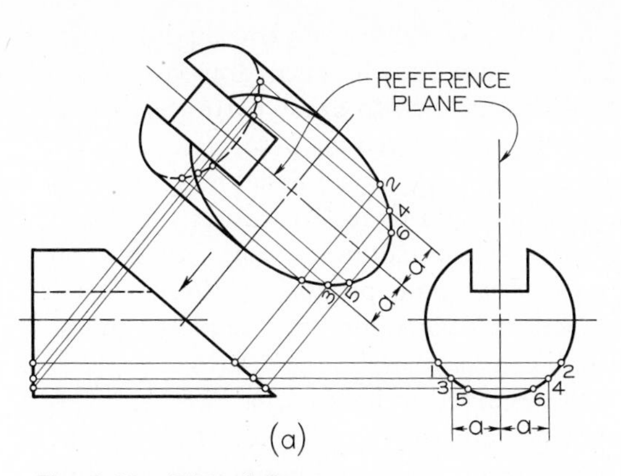


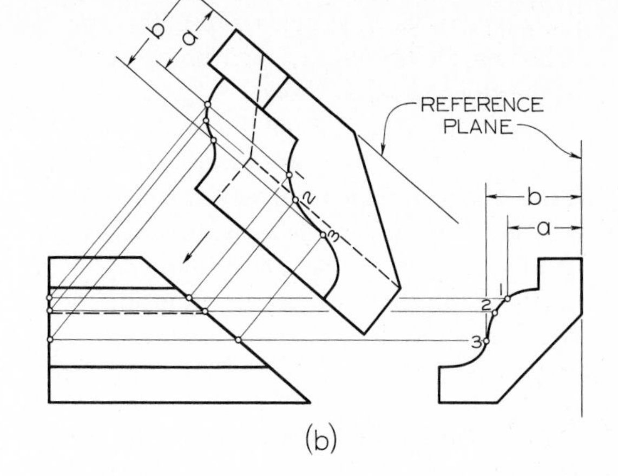


Fig. 8.12 **Plotted Curves.**

in the auxiliary view, the major axis of which is equal to the diameter of the cylinder.

Since this is a symmetrical object, the reference plane is assumed through the center, as shown. To plot points on the ellipses, select points on the circle of the side view, and project them across to the inclined surface or to the left-end surface, and then upward to the auxiliary view. In this manner, two points can be projected each time, as shown for points 1–2, 3–4, and 5–6. Distances a are equal and are transferred from the side view to the auxiliary view with the aid of dividers. A sufficient number of points must be projected to establish the curves accurately. Use the irregular curve as described in §2.58.

Since the major and minor axes are known, any of the true ellipse methods of Figs. 4.48 to 4.50 and 4.52 (a) may be used. Or if an approximate ellipse is adequate for the job in hand, the method of Fig. 4.56 can be used. But the quickest and easiest method is to use an ellipse template, as explained in §4.57.

In Fig. 8.12 (b), the auxiliary view shows the true size and shape of the inclined cut through a piece of molding. The method of plotting points is similar to that above.

## 8.13 Reverse Construction.

In order to complete the regular views, it is often necessary to construct an auxiliary view first. For example, in Fig. 8.13 (a) the upper portion of

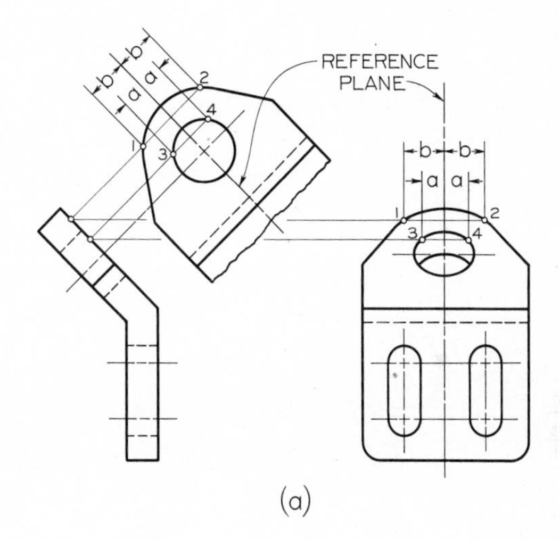


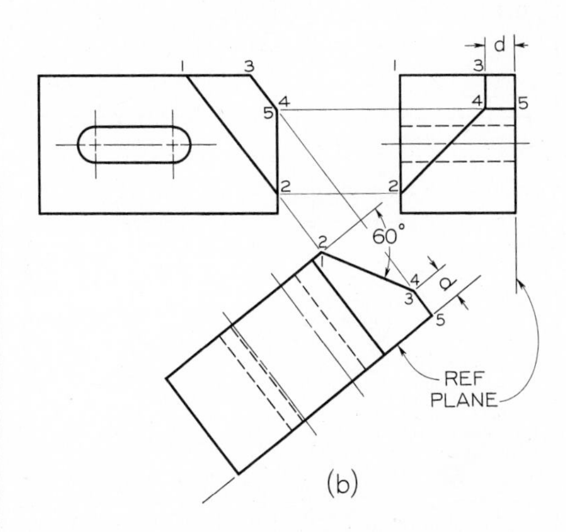


Fig. 8.13 **Reverse Construction.**

the right-side view cannot be constructed until the auxiliary view is drawn and points established on the curves and then projected back to the front view, as shown.

At (b), the 60° angle and the location of line 1–2 in the front view are given. In order to locate line 3–4 in the front view, the lines 2–4, 3–4, and 4–5 in the side view, it is necessary first to construct the 60° angle in the auxiliary view and project back to the front and side views, as shown.

## 8.14 Partial Auxiliary Views.

The use of an auxiliary view often makes it possible to eliminate one or more regular views and thus to simplify the shape description, as shown in Fig. 8.1 (b).

In Fig. 8.14 three complete auxiliary-view drawings are shown. Such drawings take a great deal of time to draw, particularly when ellipses are involved as is so often the case, and the completeness of detail may add nothing in clearness or may even detract from it because of the clutter of lines. However, in these cases some portion of every view is needed—no view can be completely eliminated, as was done in Fig. 8.1 (b).

As described in §6.9, *partial views* are often sufficient, and the resulting drawings are considerably simplified and easier to read. Similarly, as shown in Fig. 8.15, partial regular

Fig. 8.14 **Primary Auxiliary Views.**

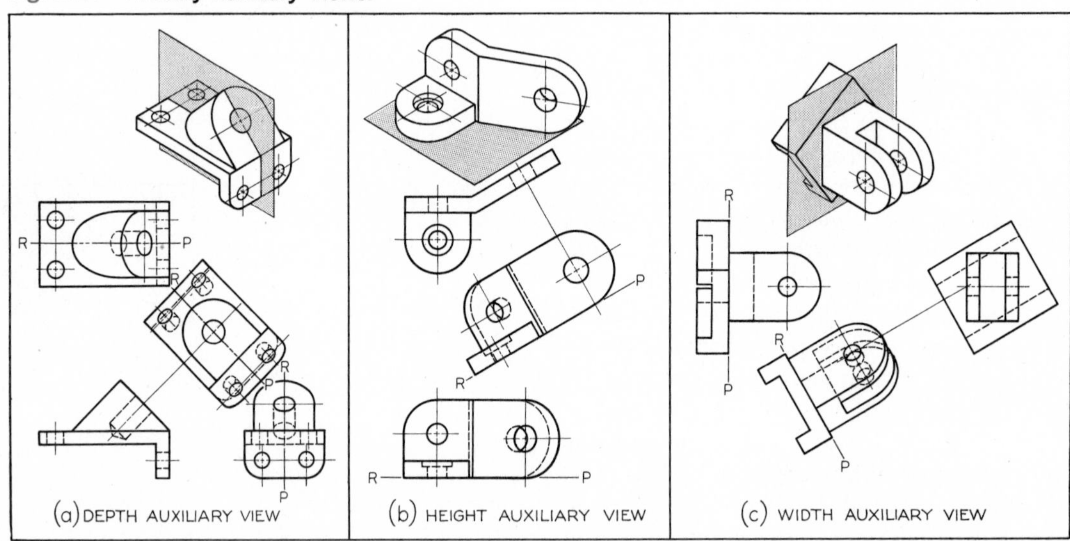


Fig. 8.15 **Partial Views.**

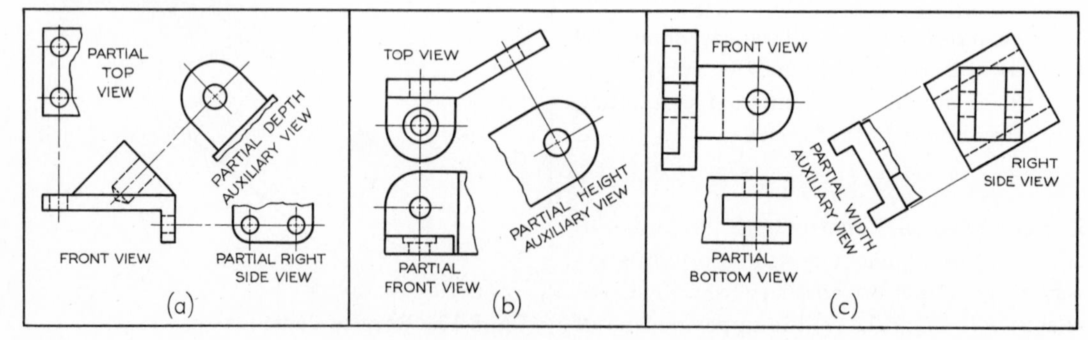

views and partial auxiliary views are used with the same result. Usually a break line is used to indicate the imaginary break in the views. *Do not draw a break line coinciding with a visible line or a hidden line.*

In order to clarify the relation of views, the auxiliary views should be connected to the views from which they are projected, either with a center line, or with one or two projection lines. This is particularly important with regard to partial views that often are small and easily appear to be "lost" and not related to any view.

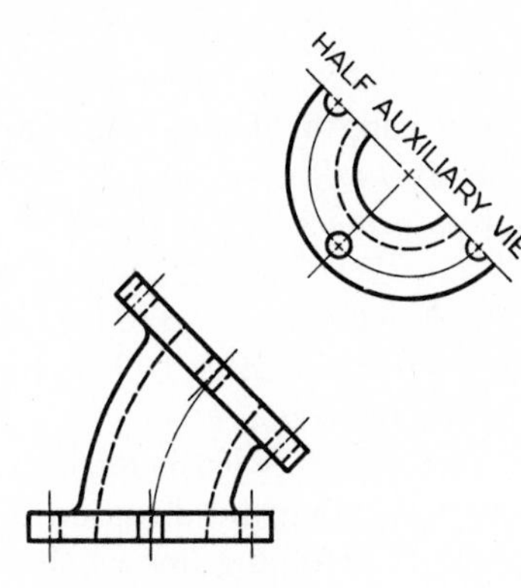


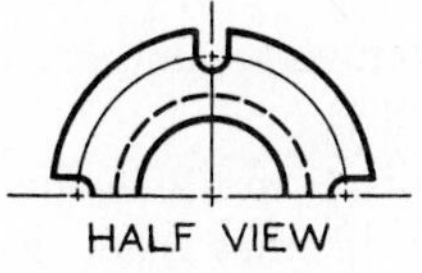


Fig. 8.16 **Half Views.**

## 8.15 Half Auxiliary Views.

If an auxiliary view is symmetrical, and if it is necessary to save space on the drawing or to save time in drafting, only a half of the auxiliary view may be drawn, as shown in Fig. 8.16. In this case, a half of a regular view is also shown, since the bottom flange is also symmetrical. See §§6.9 and 7.14. Note that in each case the *near half* is shown.

## 8.16 Hidden Lines in Auxiliary Views.

In practice, hidden lines should be omitted in auxiliary views, as in ordinary views, §5.25, unless they are needed for clearness. The beginner, however, should show all hidden lines, especially if the auxiliary view of the entire object is shown. Later, in advanced work, it will become clearer as to when hidden lines can be omitted.

## 8.17 Auxiliary Sections.

An *auxiliary section* is simply an auxiliary view in section. In Fig. 8.17 (a), note the cutting-plane line and the terminating arrows that indicate the direction of sight for the auxiliary section. Observe that the section lines are drawn at approximately 45° with visible outlines. In drawing an auxiliary section, the entire portion of the object behind the cutting plane may be shown, as at (a), or the cut surface alone, as at (b) and (c).

Fig. 8.17 **Auxiliary Sections.**

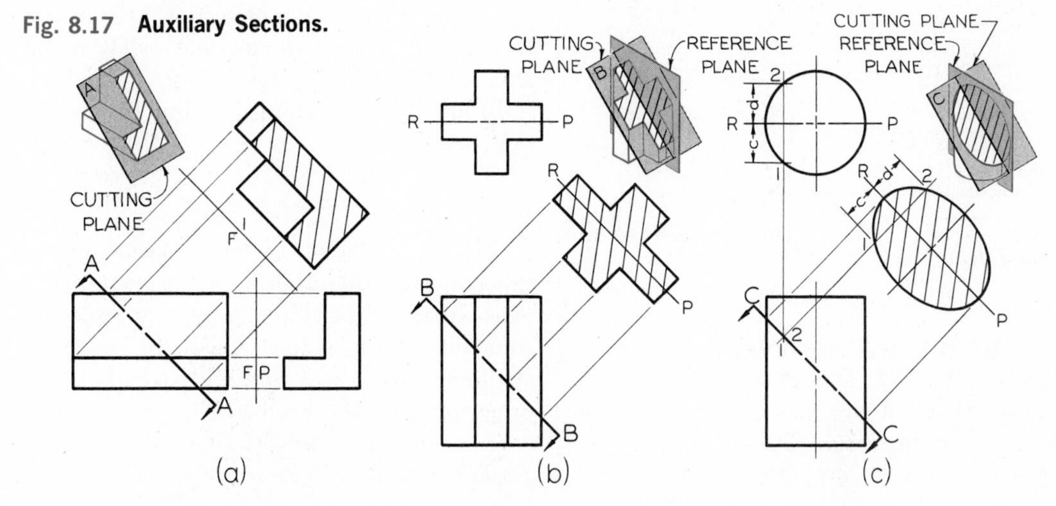

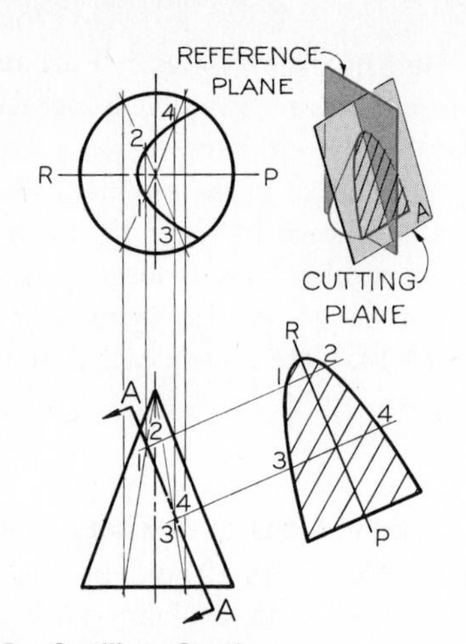


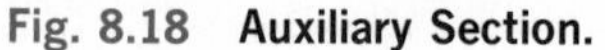

Fig. 8.18 **Auxiliary Section.**

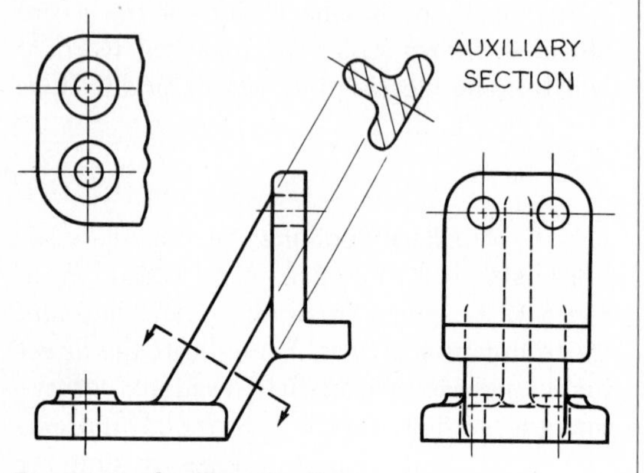


Fig. 8.19 **Auxiliary Section.**

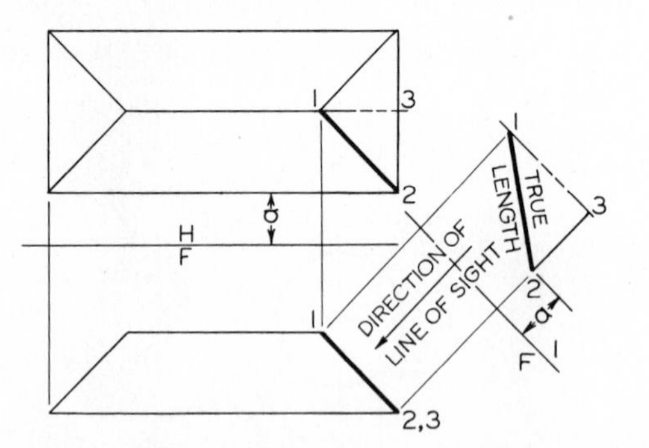


Fig. 8.20 **True Length of a Line by Means of an Auxiliary View.**

An auxiliary section through a cone is shown in Fig. 8.18. This is one of the conic sections, §4.48, in this case a *parabola.* The parabola may be drawn by other methods, Figs. 4.57 and 4.58, but the method shown here is by projection. In Fig. 8.18, elements of the cone are drawn in the front and top views. These intersect the cutting plane at points 1, 2, 3, and so on. These points are established in the top view by projecting upward to the top views of the corresponding elements. In the auxiliary section, all points on the parabola are the same distance from the reference plane RP as they are in the top view.

A typical example of an auxiliary section in machine drawing is shown in Fig. 8.19. Here, there is not sufficient space for a *revolved section,* §7.9, although a *removed section,* §7.10, could have been used instead of an auxiliary section.

## 8.18 True Length of Line—Auxiliary-View Method.

A line will show in *true length* when projected to a projection plane parallel to the line.

In Fig. 8.20, let it be required to find the true length of the hip rafter 1–2 by means of a depth auxiliary view.

I. Assume an arrow perpendicular to 1–2 (front view) indicating the direction of sight, and place the H/F folding line as shown.

II. Draw the F/1 folding line perpendicular to the arrow and at any convenient distance from 1–2 (front view), and project the points 1 and 3 toward it.

III. Set off the points 1 and 2 in the auxiliary view at the same distance from the folding line as they are in the top view. The triangle 1–2–3 in the auxiliary view shows the true size and shape of the roof section 1–2–3, and the distance 1–2 in the auxiliary view is the true length of the hip rafter 1–2.

To find the true length of a line by revolution, see §9.10.

## 8.19 Successive Auxiliary Views.

Up to this point we have dealt with *primary auxiliary views*—that is, single auxiliary views projected from one of the regular views. In Fig. 8.21, auxiliary view 1 is a primary auxiliary view, projected from the top view.

From primary auxiliary view 1, a *secondary auxiliary view* 2 can be drawn; then from it a

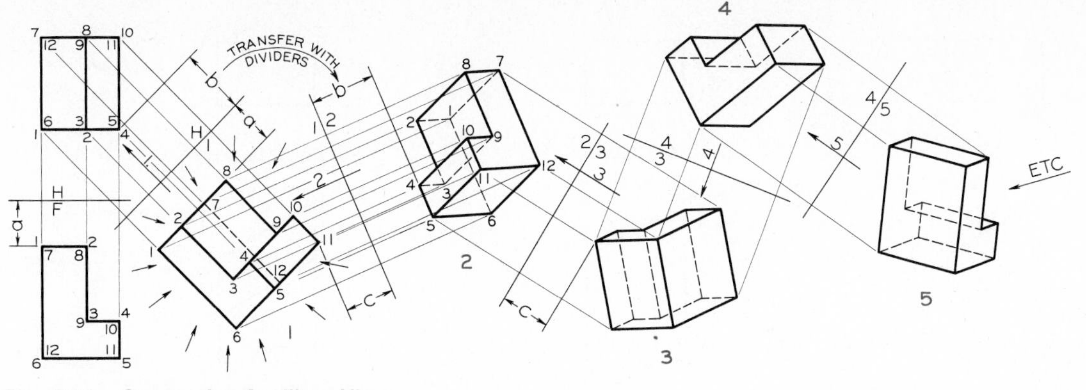


Fig. 8.21 **Successive Auxiliary Views.**

third auxiliary view 3, and so on. An infinite number of such successive auxiliary views may be drawn, a process which may be likened to the "chain reaction" in nuclear explosions.

However, secondary auxiliary view 2 is not the only one that can be projected from primary auxiliary view 1 and thus start an independent "chain reaction." As shown by the arrows around view 1, an infinite number of secondary auxiliary views, with different lines of sight, may be projected. *Any auxiliary view projected from a primary auxiliary view is a secondary auxiliary view.* Furthermore, any succeeding auxiliary view may be used to project an infinite number of "chains" of views from it.

In this example, folding lines are more convenient than reference-plane lines. In auxiliary view 1, all numbered points of the object are the same distance from folding line H/1 as they are in the front view from folding line H/F. These distances, such as distance a, are transferred from the front view to the auxiliary view with the aid of dividers.

To draw the secondary auxiliary view 2, drop the front view from consideration, and center attention on the sequence of three views: the top view, view 1, and view 2. Draw arrow 2 toward view 1 in the direction desired for view 2, and draw light projection lines parallel to the arrow. Draw folding line 1/2 perpendicular to the projection lines and at any convenient distance from view 1. Locate all numbered points in view 2 from folding line 1/2 at the same distances they are in the top view from folding line H/1, using the dividers to transfer distances. For example, transfer distance b to locate points 4 and 5. Connect points with straight lines, and determine visibility. The corner nearest the observer (11) for view 2 will be visible, and the one farthest away (1) will be hidden, as shown.

To draw views 3, 4, and so on, repeat the above procedure, remembering that each time we will be concerned only with a sequence of three views. In drawing any auxiliary view, the paper may be revolved so as to make the last two views line up as regular views.

## 8.20 Uses of Auxiliary Views.

Generally, auxiliary views are used to show the true shape or true angle of features that appear distorted in the regular views. Auxiliary views have the following four uses:

1. True length of line (TL), §8.18.
2. Point view of line, §8.11.
3. Edge view of plane (EV), §8.21.
4. True size of plane (TS), §8.21.

## 8.21 True Size of an Oblique Surface—Folding Line Method.

A typical requirement of a secondary auxiliary view is to show the true size and shape of an oblique surface, such as surface 1–2–3–4 in Fig. 8.22. In this case folding lines are used, but the same results can be obtained with reference lines. Proceed as follows:

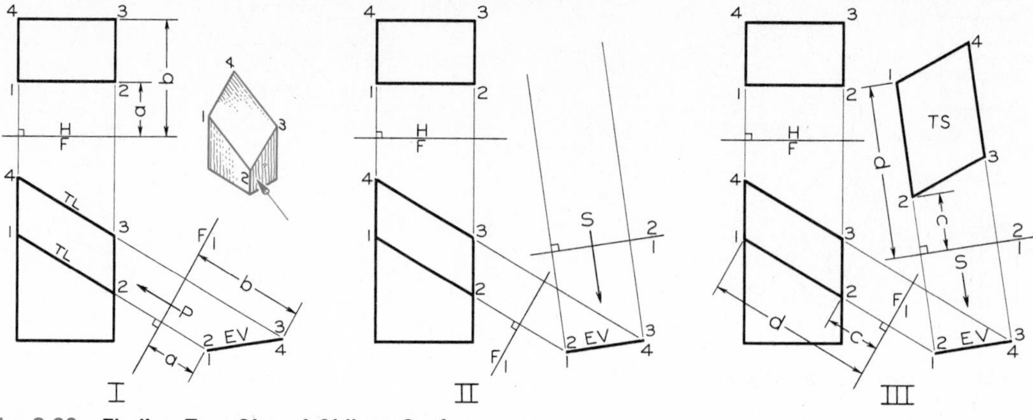


Fig. 8.22 **Finding True Size of Oblique Surface.**

I. Draw primary auxiliary view showing surface 1–2–3–4 as a line. As explained in §8.11, the edge view (EV) of a plane is found by getting the point view of a line in that plane. To get the point view of a line, the line of sight must be assumed parallel to the line. Therefore, draw arrow P parallel to lines 1–2 and 3–4, which are true length (TL) in the front view, and draw projection lines parallel to the arrow. Draw folding line H/F between the top and front views and F/1 between the front and auxiliary views, perpendicular to the respective projection lines. All points in the auxiliary view will be the same distance from the folding line F/1 as they are in the top view from folding line H/F. Lines 1–2 and 3–4 will appear as points in the auxiliary view, and plane 1–2–3–4 will therefore appear *edgewise,* that is, as a line.

II. Draw arrow S perpendicular to the edge view of plane 1–2–3–4 in the primary auxiliary view, and draw projection lines parallel to the arrow. Draw folding line 1/2 perpendicular to these projection lines and at a convenient distance from the primary auxiliary view.

III. Draw secondary auxiliary view. Locate each point (transfer with dividers) the same distance from the folding line 1/2 as it is in the front view to the folding line F/1, as for example, dimensions c and d. The true size (TS) of the surface 1–2–3–4 will be shown in the secondary auxiliary view, since the direction of sight, arrow S, was taken perpendicular to it.

## 8.22 True Size of an Oblique Surface Reference-Plane Method.

In Fig. 8.23 (a), it is required to draw an auxiliary view in which triangular surface 1–2–3 will appear in true size and shape. In order for the true size of the surface to appear in the secondary auxiliary view, arrow S must be assumed perpendicular to the edge view of that surface; so it is necessary to have the edge view of surface 1–2–3 in the primary auxiliary view first. In order to do this, the direction of sight, arrow P, must be parallel to a line in surface 1–2–3 that appears true length (TL) in the front view. Hence, arrow P is drawn parallel to line 1–2 of the front view, line 1–2 will appear as a point in the primary auxiliary view, and surface 1–2–3 must therefore appear edgewise in that view.

In this case it is convenient to use reference lines and to assume the reference plane X (for drawing the primary auxiliary view) coinciding with the back surface of the object, as shown. For the primary auxiliary view, all depth measurements, as a in the figure, are transferred with dividers from the top view with respect to the reference line X–X.

For the secondary auxiliary view, reference plane Y is assumed cutting through the object for convenience in transferring measurements. All measurements perpendicular to Y–Y in the secondary auxiliary view are the same as between the reference plane and the corresponding points in the front view. Note that corresponding measurements must be *inside* (toward the central view in the sequence of

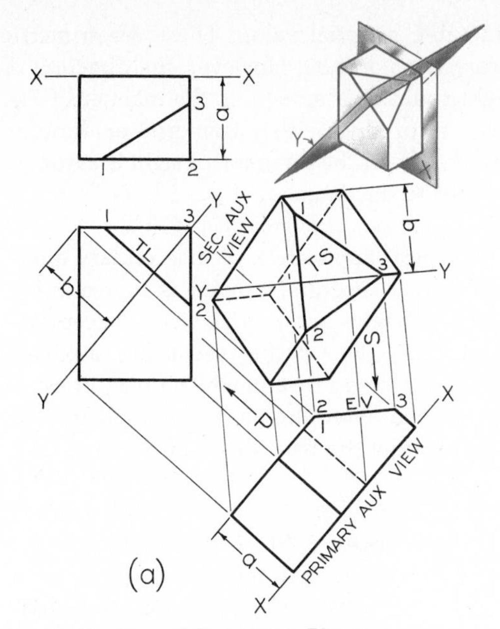


Fig. 8.23 **Use of Reference Planes.**

three views) or *outside* (away from the central view). For example, dimension b is on the side of Y–Y *away* from the primary auxiliary view in both places.

In Fig. 8.23 (b) it is required to find the true size and shape of surface 1-2-3-4-5-6-7, and not to draw the complete secondary auxiliary view. The method is similar to that described above.

## 8.23 Oblique Direction of Sight Given.

In Fig. 8.24 two views of a block are given, with two views of an arrow indicating the direction in which it is desired to look at the object to obtain a view. Proceed as follows:

I. *Draw primary auxiliary view of both the object and the assumed arrow,* which will show the true length *of the arrow.* In order to do this, assume a horizontal reference plane X–X in the front and auxiliary views, as shown. Then assume a direction of sight perpendicular to the given arrow. In the front view, the butt end of the arrow is a distance *a* higher than the arrow point, and this distance is transferred to the primary auxiliary view as shown. All *height* measurements in the auxiliary view correspond to those in the front view.

II. *Draw secondary auxiliary view,* which will show the arrow as a point. This can be done because the arrow shows in true length in the primary auxiliary view, and projection lines for the secondary auxiliary view are drawn parallel to it. Draw reference line Y–Y, for the secondary auxiliary view, perpendicular to these projection lines. In the top view, draw

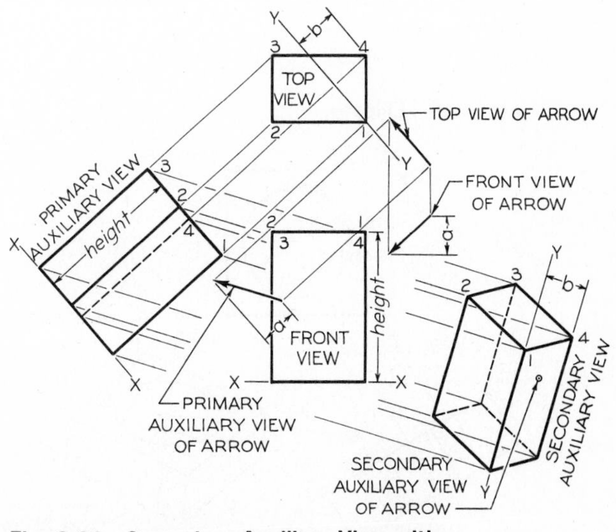


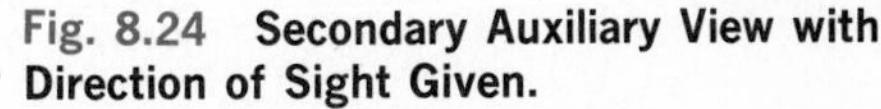

Fig. 8.24 **Secondary Auxiliary View with Direction of Sight Given.**

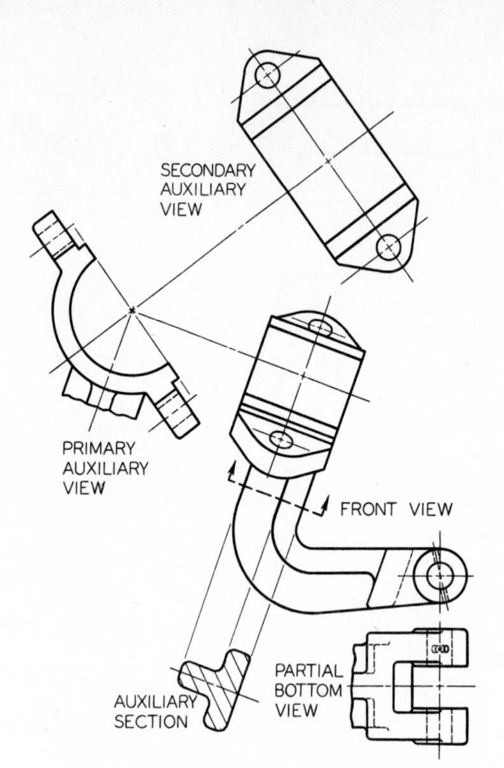


Fig. 8.25 **Secondary Auxiliary View.**

Y–Y perpendicular to the projection lines to the primary auxiliary view. All measurements, such as b, with respect to Y–Y correspond in the secondary auxiliary view and the top view.

It will be observed that the secondary auxiliary views of Figs. 8.23 (a) and 8.24 have considerable pictorial value. These are trimetric projections, §16.32. However, the direction of sight could be assumed, in the manner of Fig. 8.24, to produce either isometric or dimetric projections. If the direction of sight is assumed parallel to the diagonal of a cube, the resulting view is an *isometric projection,* §16.4.

A typical application of a secondary auxiliary view in machine drawing is shown in Fig. 8.25. All views are partial views, except the front view. The partial secondary auxiliary view illustrates a case in which break lines are not needed. Note the use of an auxiliary section to show the true shape of the arm.

## 8.24 Ellipses.

As shown in §6.30, if a circle is viewed obliquely, the result is an ellipse. This often occurs in successive auxiliary views, because of the variety of directions of sight. In Fig. 8.26 (a) the hole appears as a true circle in the top view. The circles appear as straight lines in the primary auxiliary view, and as ellipses in the secondary auxiliary view. In the latter, the major axis AB of the ellipse is parallel to the projection lines and equal in length to the true diameter of the circle as shown in the top view. The minor axis CD is perpendicular to the major axis, and its foreshortened length is projected from the primary auxiliary view.

The ellipse can be completed by projecting

Fig. 8.26 **Ellipses.**

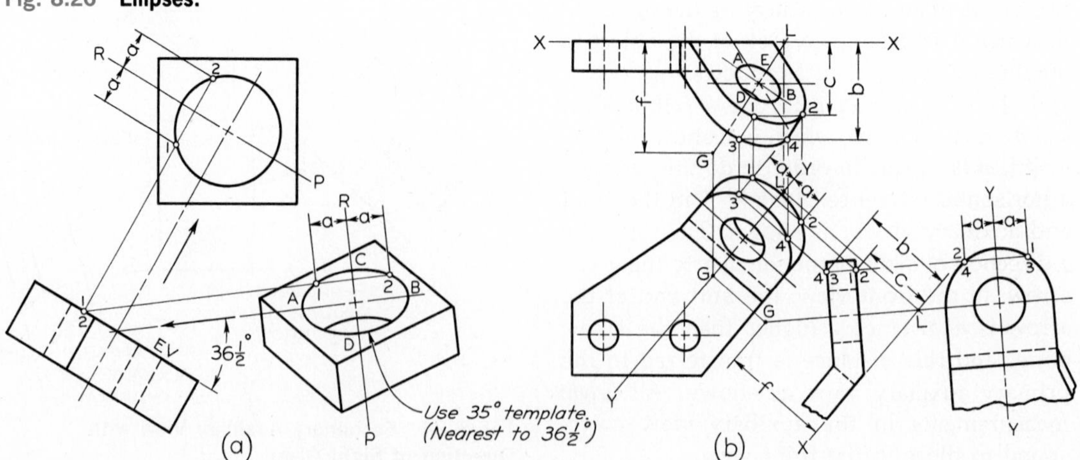

points, such as 1 and 2, symmetrically located about the reference plane RP coinciding with CD with distances a equal in the top and secondary auxiliary views as shown, and finally, after a sufficient number of points have been plotted, by applying the irregular curve, §2.58.

Since the major and minor axes are easily found, any of the true-ellipse methods of Figs. 4.48 to 4.50 and 4.52 (a) may be used, or an approximate ellipse, Fig. 4.56, may be found sufficiently accurate for a particular drawing. Or the ellipses may be easily and rapidly drawn with the aid of an ellipse template, §4.57. The "angle" of ellipse to use is the one that most closely matches the angle between the direction of sight arrow and the plane (EV) containing the circle, as seen in this case in the primary auxiliary view. Here the angle is $36\frac{1}{2}°$, so a 35° ellipse is selected.

At (b) successive auxiliary views are shown in which the true circular shapes appear in the secondary auxiliary view, and the elliptical projections in the front and top views. It is necessary to construct the circular shapes in the secondary auxiliary view, then to project plotted points back to the primary auxiliary view, the front view, and finally to the top view, as shown in the figure for points 1, 2, 3, and 4. The final curves are then drawn with the aid of the irregular curve.

If the major and minor axes are found, any of the true-ellipse methods may be used; or better still, an ellipse template, §4.57, may be employed. The major and minor axes are easily established in the front view, but in the top view they are more difficult to find. The major axis AB is at right angles to the center line GL of the hole, and equal in length to the true diameter of the hole. The minor axis ED is at right angles to the major axis. Its length is found by plotting several points in the vicinity of one end of the minor axis, or by using descriptive geometry to find the angle between the line of sight and the inclined surface, and by this angle selecting the ellipse guide required.

## 8.25 Auxiliary-View Problems.

The problems in Figs. 8.27 to 8.58 are to be drawn mechanically or freehand. If partial auxiliary views are not assigned, the auxiliary views are to be complete views of the entire object, including all necessary hidden lines.

It is often difficult to space properly the views of an auxiliary-view drawing. In some cases it may be necessary to make a trial blocking out on a preliminary sheet before starting the actual drawing. Allowances for dimensions must be made if dimensions are to be included. In such case, the student should study §§11.1 to 11.25.

Auxiliary-view problems in convenient form for solution are available in *Technical Drawing Problems, Series 1,* by Giesecke, Mitchell, Spencer, and Hill; *Technical Drawing Problems, Series 2,* by Spencer and Hill; and in *Technical Drawing Problems, Series 3,* by Spencer and Hill, all designed to accompany this text and published by Macmillan Publishing Co., Inc.

## 8.25 Auxiliary-View Problems

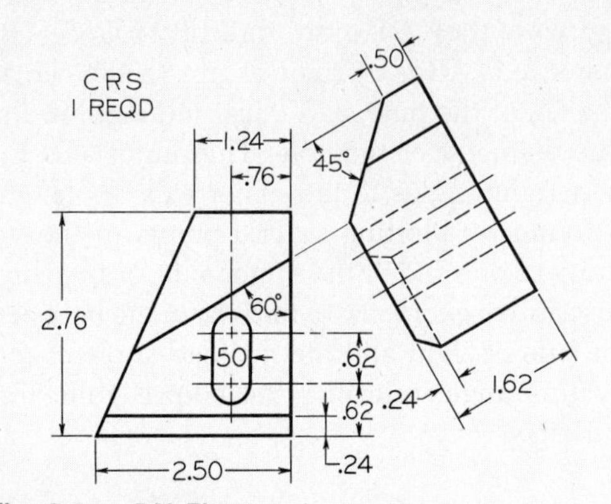


Fig. 8.27 **RH Finger.**
Given: Front and auxiliary views.
Required: Complete front, auxiliary, left-side, and top views (Layout A-3).

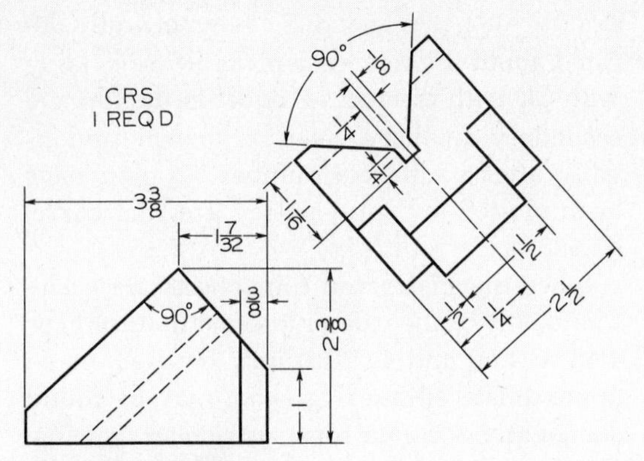


Fig. 8.28 **V-Block.**
Given: Front and auxiliary views.
Required: Complete front, top, and auxiliary views (Layout A-3).

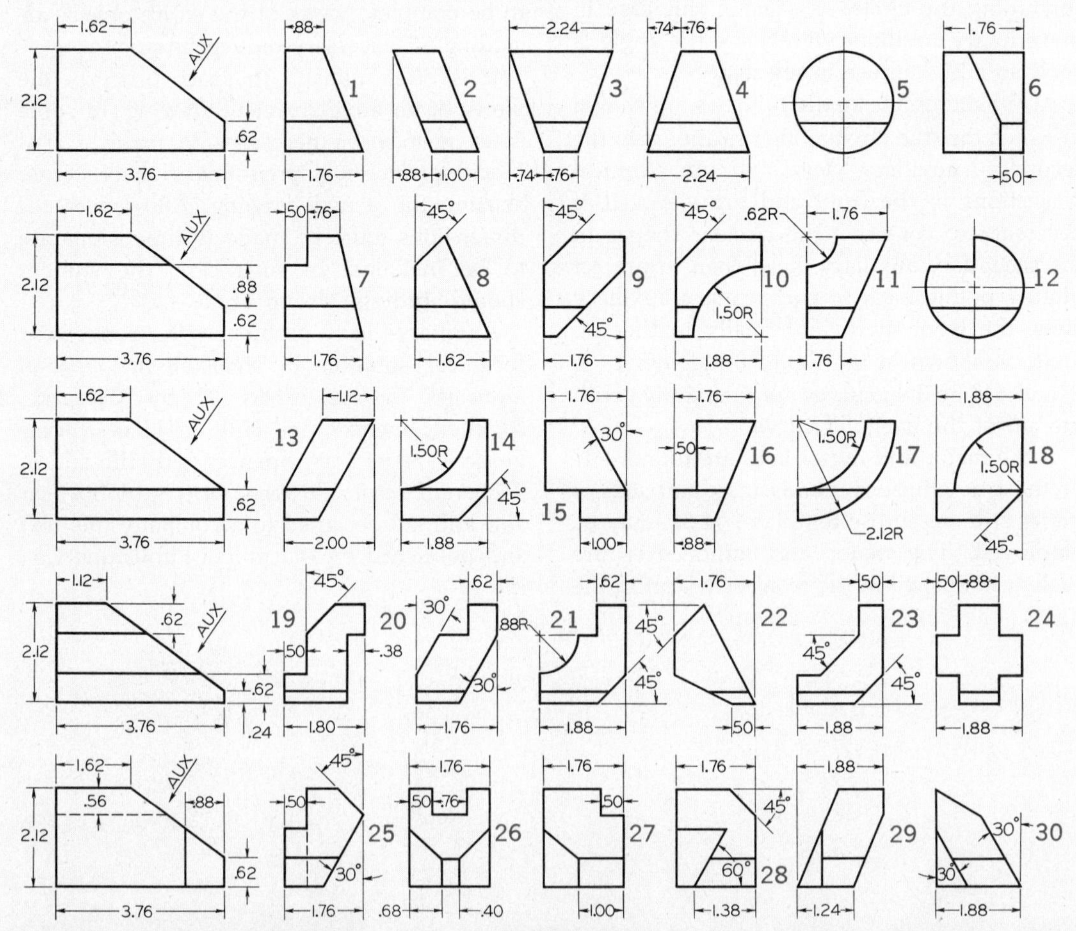


Fig. 8.29 **Auxiliary-View Problems.** Make free hand sketch or mechanical drawing of selected problem as assigned by instructor. Draw given front and right-side views, and add complete auxiliary view, including all hidden lines (Layout A-3). If assigned, design your own right-side view consistent with given front view, and then add complete auxiliary view.

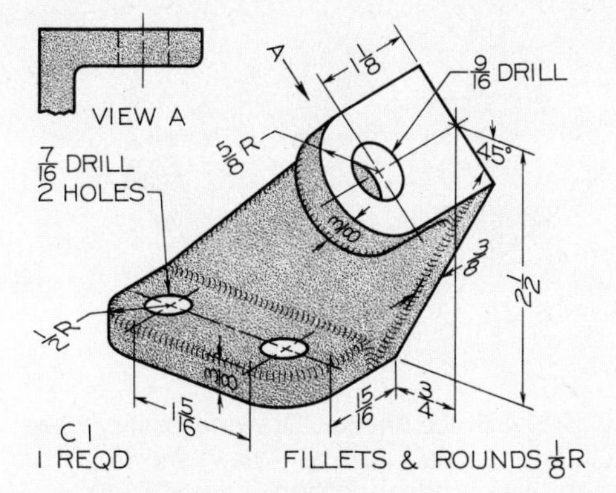


Fig. 8.30 **Anchor Bracket.** Draw necessary views or partial views (Layout A–3).

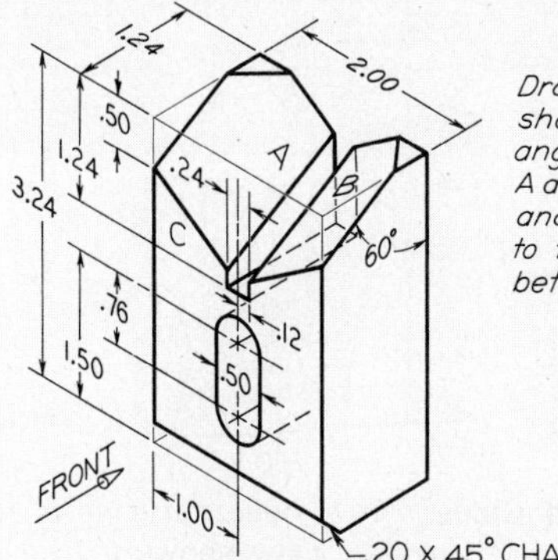


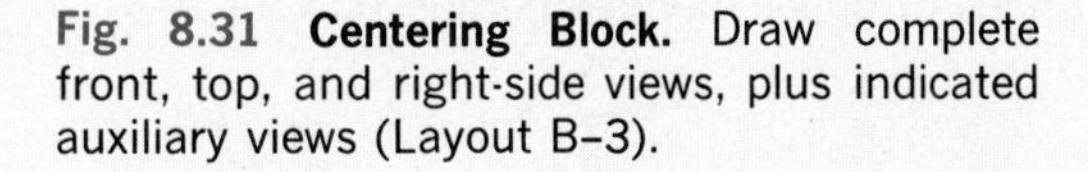

Fig. 8.31 **Centering Block.** Draw complete front, top, and right-side views, plus indicated auxiliary views (Layout B–3).

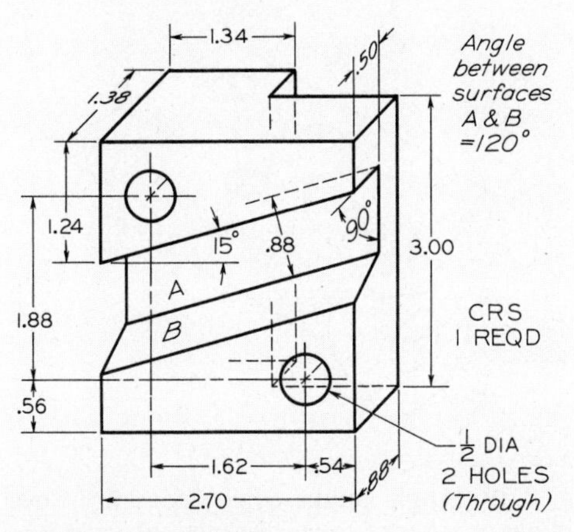


Fig. 8.32 **Clamp Slide.** Draw necessary views completely (Layout B–3).

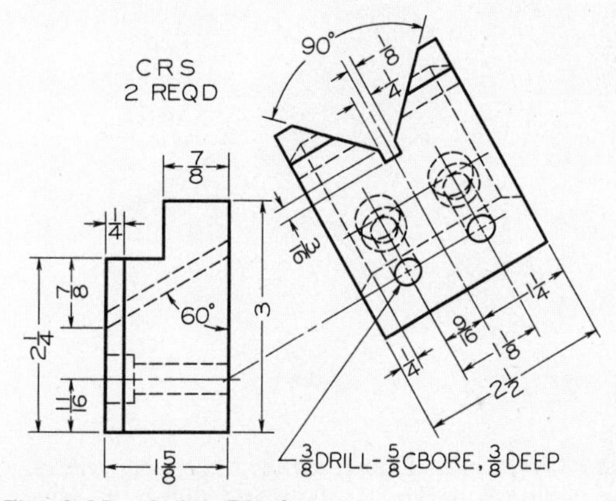


Fig. 8.33 **Guide Block.**
Given: Right-side and auxiliary views.
Required: Right-side, auxiliary, plus front and top views—all complete (Layout B–3).

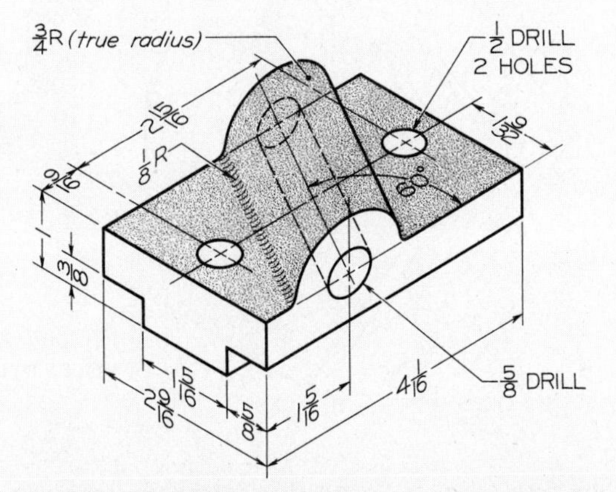


Fig. 8.34 **Angle Bearing.** Draw necessary views, incliuding a complete auxiliary view (Layout A–3).

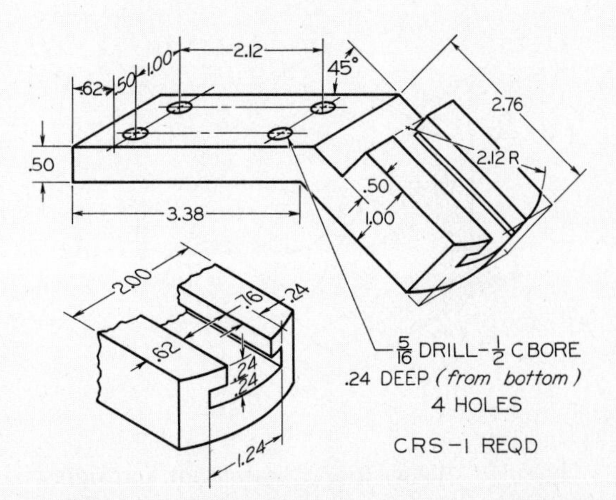


Fig. 8.35 **Guide Bracket.** Draw necessary views or partial views (Layout B–3).

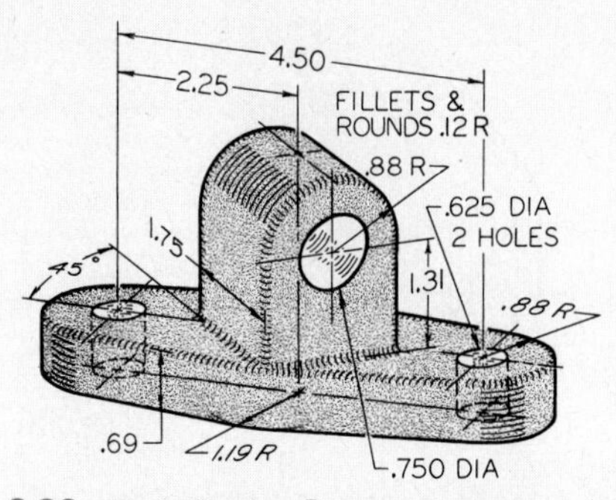


Fig. 8.36 **Rod Guide.** Draw necessary views, including complete auxiliary view showing true shape of upper rounded portion (Layout B–4).

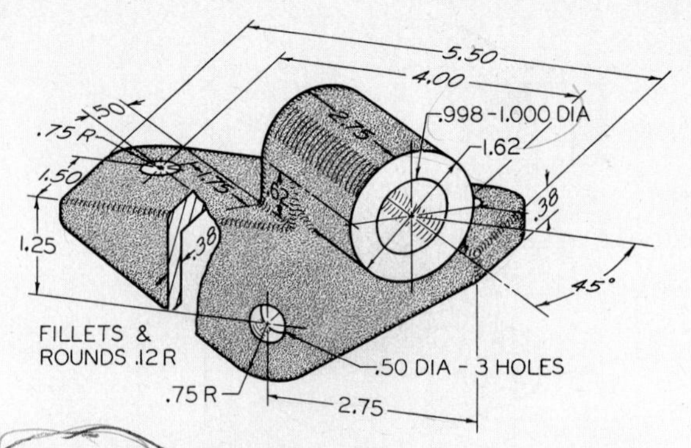


Fig. 8.37 **Brace Anchor.** Draw necessary views, including partial auxiliary view showing true shape of cylindrical portion (Layout B–4).

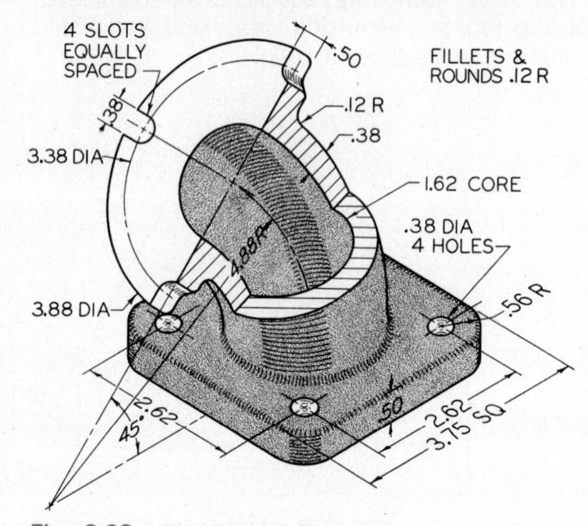


Fig. 8.38 **45° Elbow.** Draw necessary views, including a broken section and two half views of flanges (Layout B–4).

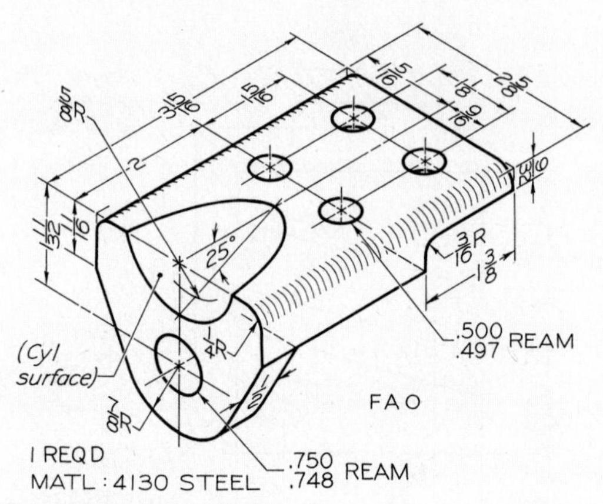


Fig. 8.39 **Angle Guide.** Draw necessary views, including a partial auxiliary view of cylindrical recess (Layout B–4).

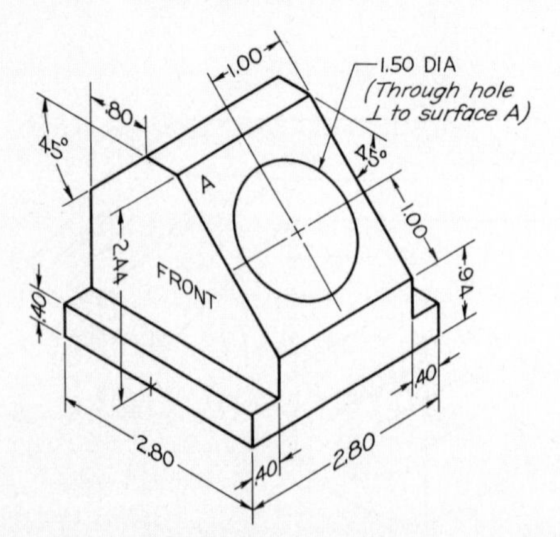


Fig. 8.40 **Holder Block.** Draw front and right-side views (2.80″ apart) and complete auxiliary view of entire object showing true shape of surface A and all hidden lines (Layout A–3).

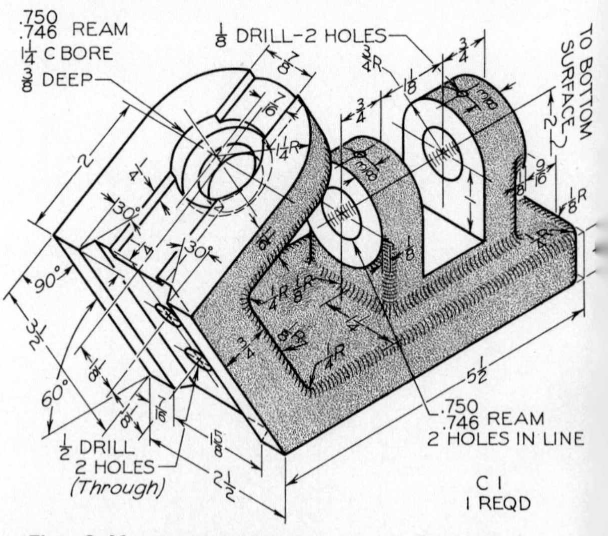


Fig. 8.41 **Control Bracket.** Draw necessary views, including partial auxiliary views and regular views (Layout C–4).

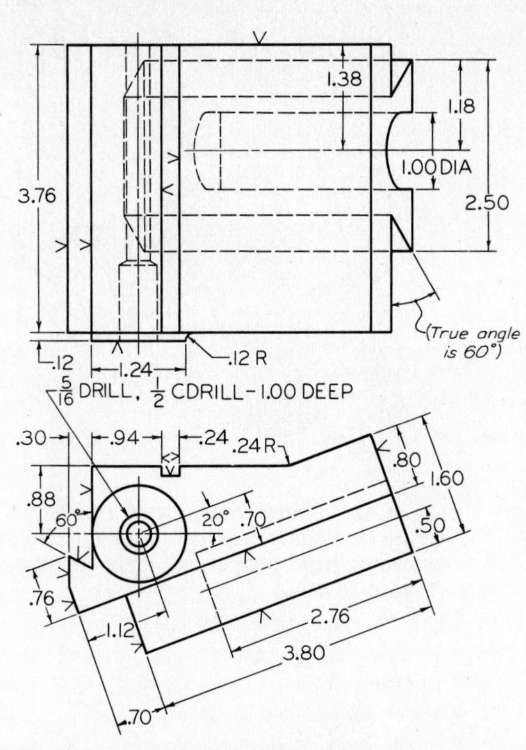


Fig. 8.42 **Tool Holder Slide.** Draw given views, and add complete auxiliary view showing true curvature of slot on bottom (Layout B–4).

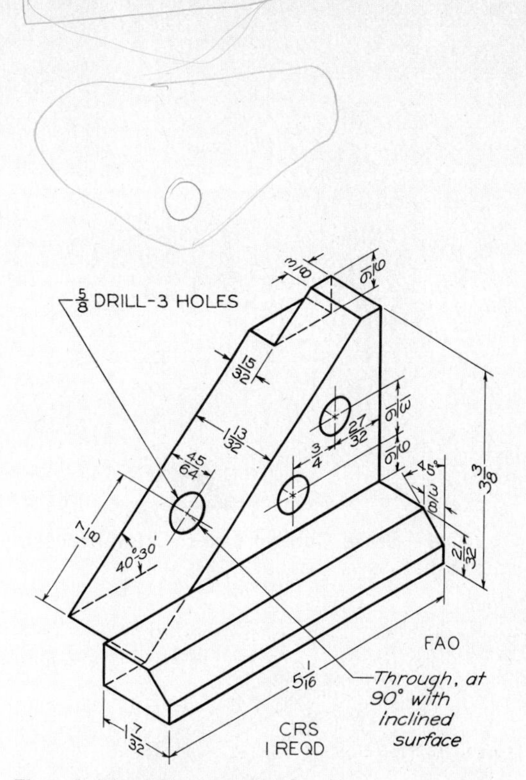


Fig. 8.43 **Adjuster Block.** Draw necessary views, including complete auxiliary view showing true shape of inclined surface (Layout B–4).

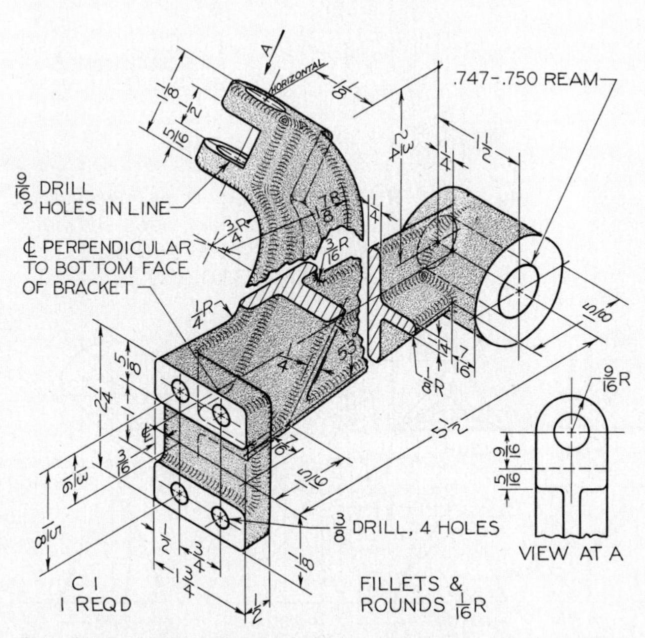


Fig. 8.44 **Guide Bearing.** Draw necessary views and partial views, including two partial auxiliary views (Layout C–4).

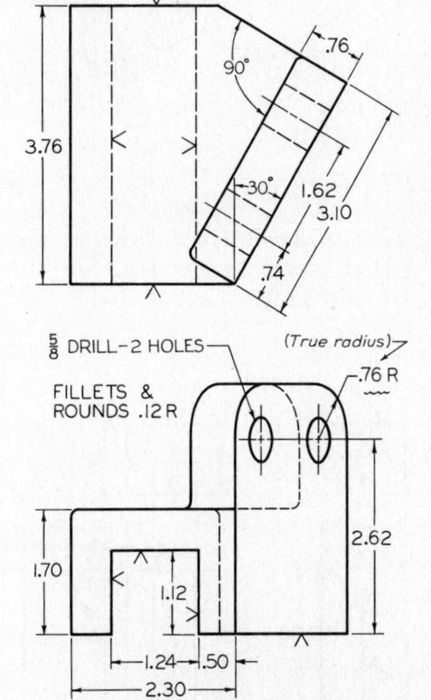


Fig. 8.45 **Drill Press Bracket.** Draw given views and add complete auxiliary view showing true shape of inclined face (Layout B–4).

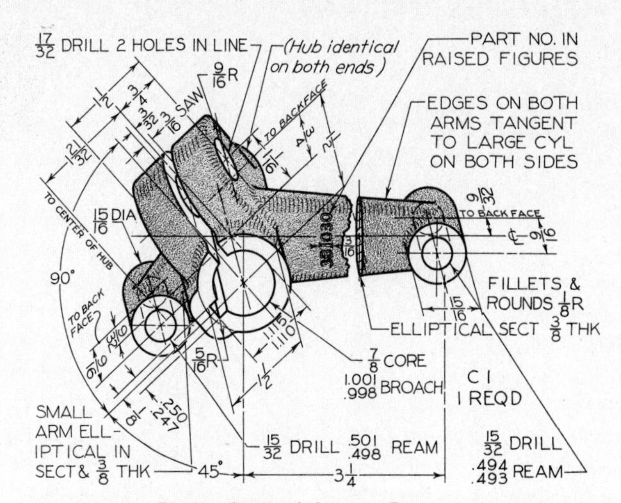


Fig. 8.46 **Brake Control Lever.** Draw necessary views and partial views (Layout B-4).

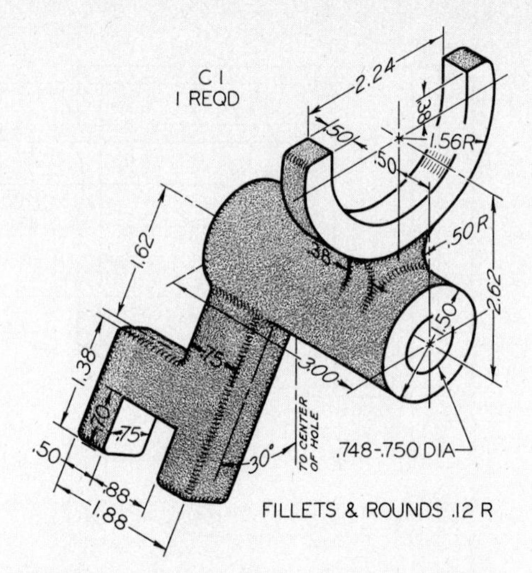


Fig. 8.47 **Shifter Fork.** Draw necessary views, including partial auxiliary view showing true shape of inclined arm (Layout B-4).

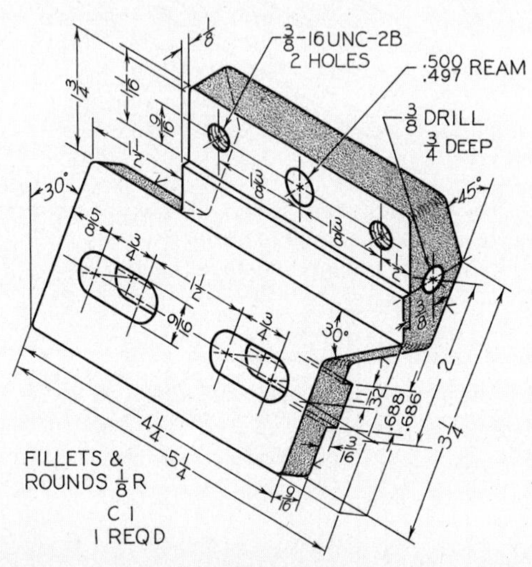


Fig. 8.48 **Cam Bracket.** Draw necessary views or partial views as needed. For threads, see §§13.9 and 13.10 (Layout B-4).

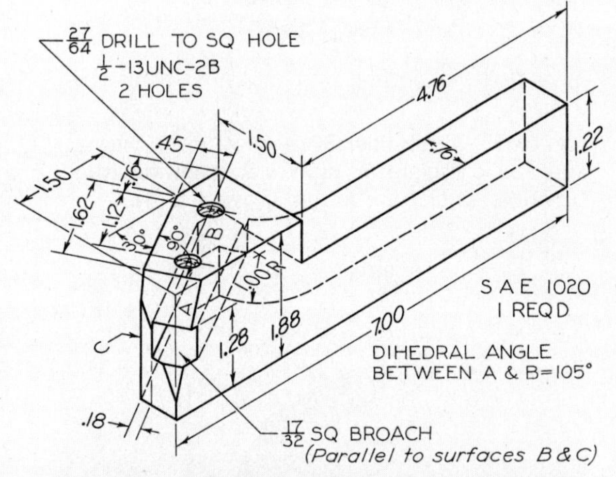


Fig. 8.49 **RH Tool Holder.** Draw necessary views, including partial auxiliary views showing 105° angle and square hole true size. For threads, see §§13.9 and 13.10 (Layout B-4).

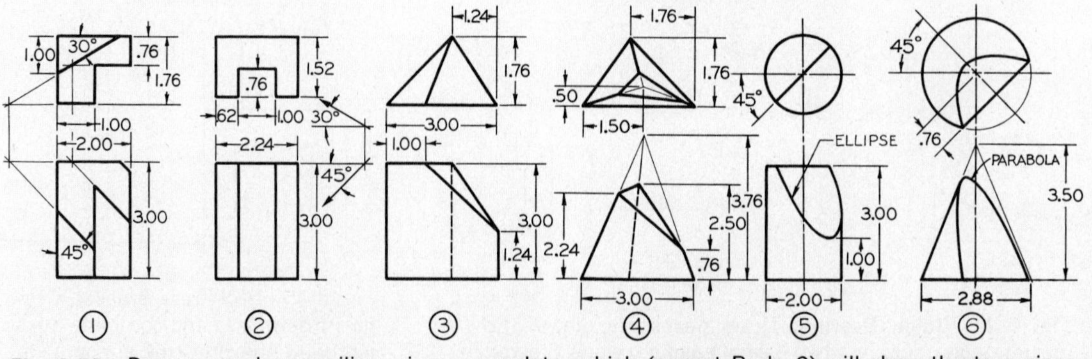


Fig. 8.50 Draw secondary auxiliary views, complete, which (except Prob. 2) will show the true sizes of the inclined surfaces. In Prob. 2, draw secondary auxiliary view as seen in direction of arrow (Layout B-3).

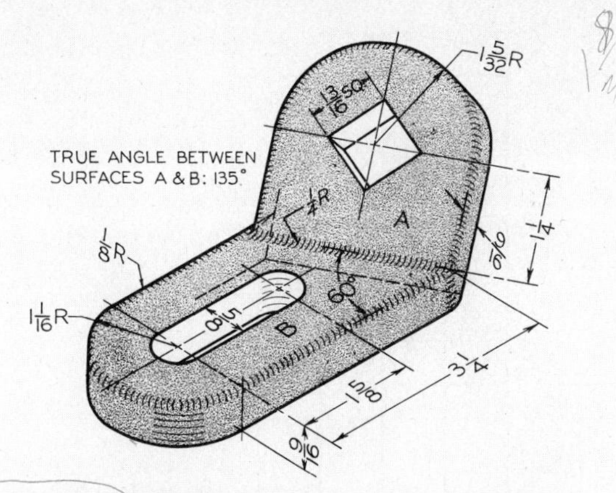


**Fig. 8.51 Control Bracket.** Draw necessary views including primary and secondary auxiliary views so that the latter shows true shape of oblique surface A (Layout B–4).

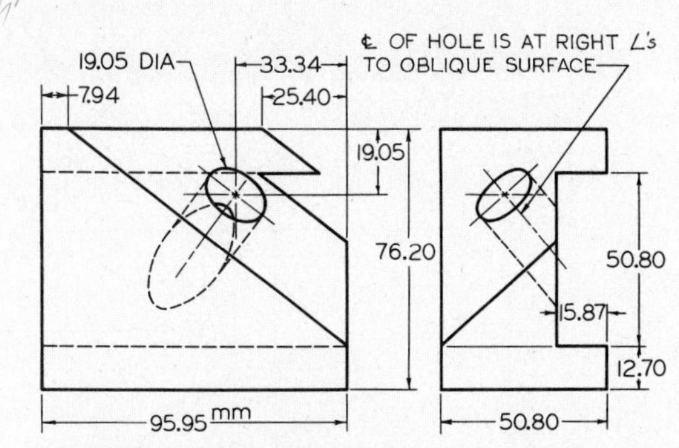


**Fig. 8.52 Holder Block.** Draw given views and primary and secondary auxiliary views so that the latter shows true shape of oblique surface (Layout B–4).

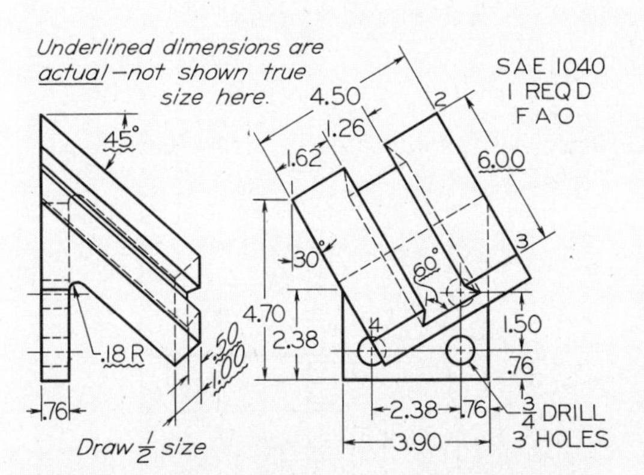


**Fig. 8.53 Dovetail Slide.** Draw complete given views and auxiliary views, including view showing true size of surface 1–2–3–4 (Layout B–4).

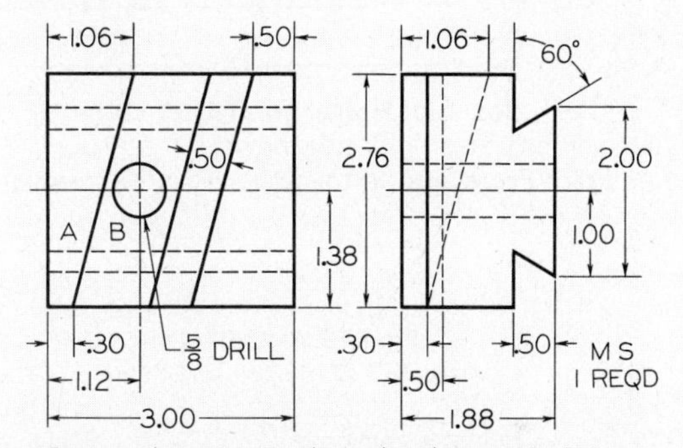

Draw primary aux. view showing angle between
planes A and B; then secondary auxiliary view
showing true size of surface A.


**Fig. 8.54 Dovetail Guide.** Draw given views plus complete auxiliary views as indicated (Layout B–4).

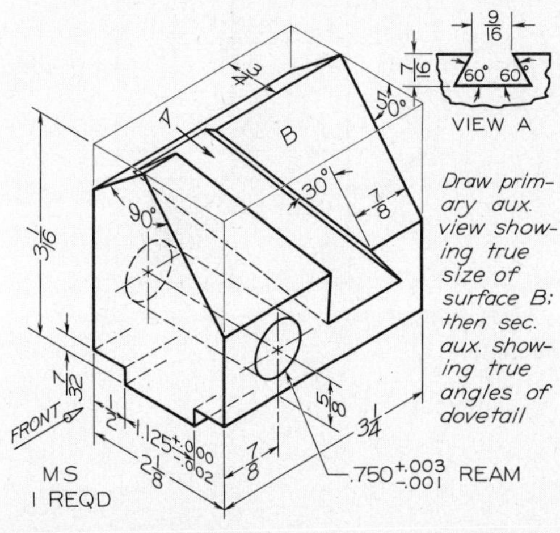


**Fig. 8.55 Adjustable Stop.** Draw complete front and auxiliary views plus partial right-side view. Show all hidden lines (Layout C–4).

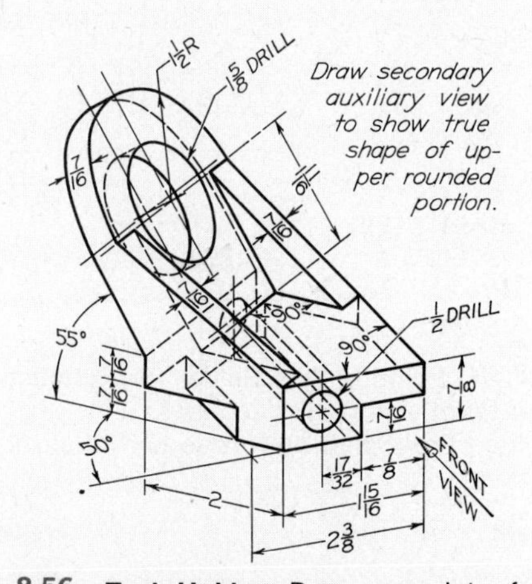


**Fig. 8.56 Tool Holder.** Draw complete front view, and primary and secondary auxiliary views as indicated (Layout B–4).

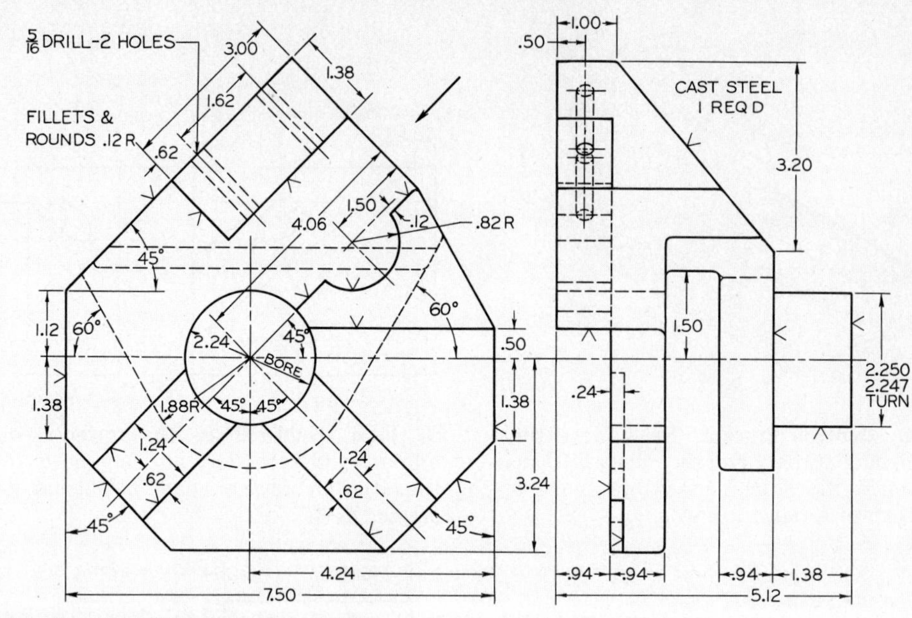


Fig. 8.57 **Box Tool Holder for Turret Lathe.**
Given: Front and right-side views.
Required: Front and left-side views, and complete auxiliary view as indicated by arrow (Layout C–4).

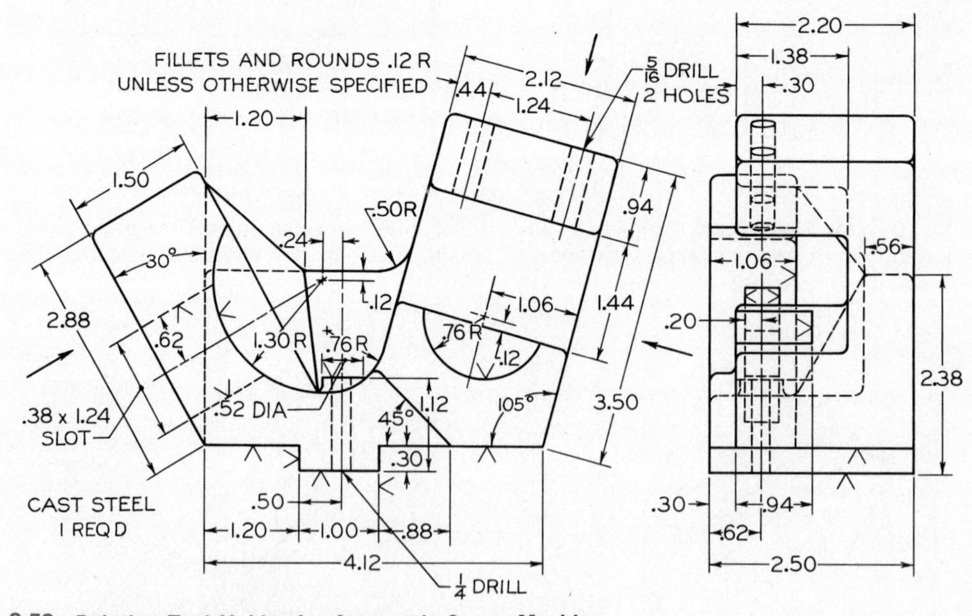


Fig. 8.58 **Pointing Tool Holder for Automatic Screw Machine.**
Given: Front and right-side views.
Required: Front view and three partial auxiliary views (Layout C–4).

# 9

# revolutions

**9.1 Revolutions Compared with Auxiliary Views.** To obtain an auxiliary view, the observer shifts his position with respect to the object, as shown by the arrow in Fig. 9.1 (a). The auxiliary view shows the true size and shape of surface A. Exactly the same view of the object also can be obtained by shifting the object with respect to the observer, as shown at (b). Here the object is revolved until surface A appears in its true size and shape in the right-side view. The *axis of revolution* is assumed perpendicular to the frontal plane of projection, as shown.

Note that the view in which the axis of revolution appears as a point (in this case the front view) *revolves but does not change shape,* and that in the views in which the axis is shown as a line in true length the *dimensions of the object parallel to the axis do not change.*

To make a revolution drawing, the view on the plane of projection that is perpendicular to the axis of revolution is drawn first, since it is the only view that remains unchanged in size and shape. This view is drawn revolved either *clockwise* or *counterclockwise* about a point that is the end view, or point view, of the axis of revolution. This point may be assumed at any convenient point on or outside the view. The other views are then projected from this view.

The axis of revolution is usually considered perpendicular to one of the three principal planes of projection. Thus, an object may be revolved about an axis perpendicular to the horizontal, frontal, or profile planes of projection, and the views drawn in the new positions. Such a process is called a *primary revolution.* If this drawing is then used as a basis for

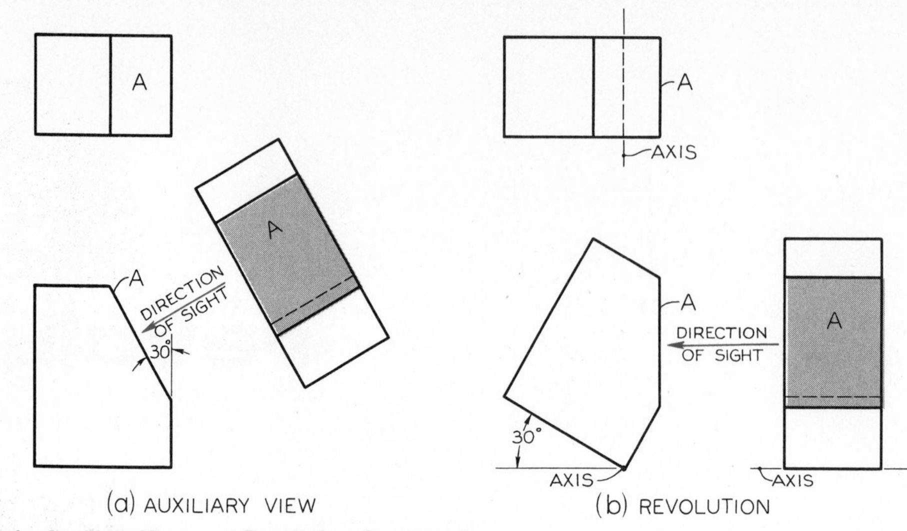


Fig. 9.1 **Auxiliary View and Revolution Compared.**

another revolution, the operation is called *successive revolutions.* Obviously, this process may be continued indefinitely, which reminds us of the "chain reaction" in successive auxiliary views, Fig. 8.21.

**9.2 Revolution About Axis Perpendicular to Frontal Plane.** A primary revolution is illustrated in Fig. 9.2. An imaginary axis XY is assumed, about which the object is to revolve to the desired position. In this case the axis is selected perpendicular to the frontal plane of projection, and during the revolution all points of the object describe circular arcs parallel to that plane. The axis may pierce the object at any point or may be exterior to it. In space II, the front view is drawn *revolved* (but not changed in shape) through the angle desired (30° in this case), and the top and side views are obtained by projecting from the front view. The *depth* of the top view and the side view is found by projecting from the top view of the first unrevolved position (space I), *because the depth, since it is parallel to the axis, remains unchanged.* If the front view of the revolved position is drawn directly without first drawing the normal unrevolved position, the depth of the object, as shown in the revolved top and side views, may be drawn to known dimensions. No difficulty should be encountered by the student who understands how to obtain projections of points and lines, §6.6.

Note the similarity between the top and side views in space II of Fig. 9.2 and some of the auxiliary views of Fig. 8.7 (c).

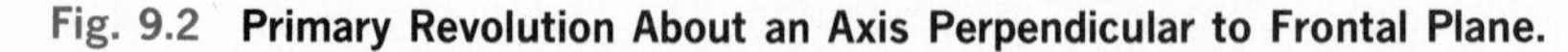

Fig. 9.2 **Primary Revolution About an Axis Perpendicular to Frontal Plane.**

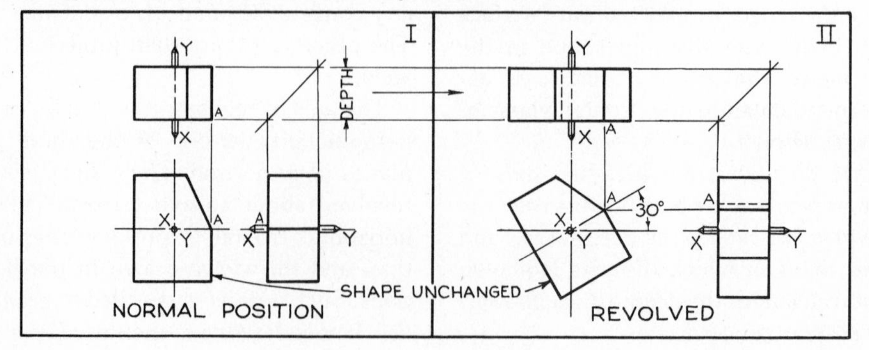

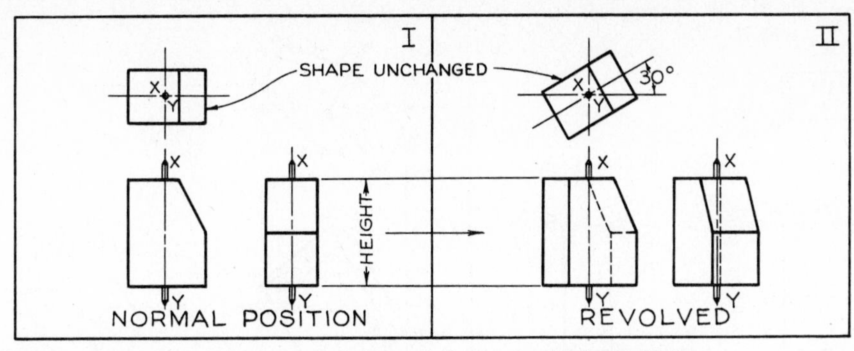


Fig. 9.3 **Primary Revolution About an Axis Perpendicular to Horizontal Plane.**

## 9.3 Revolution About Axis Perpendicular to Horizontal Plane.

A revolution about an axis perpendicular to the horizontal plane of projection is shown in Fig. 9.3. An imaginary axis XY is assumed perpendicular to the top plane of projection, the top view is drawn revolved (but not changed in shape) to the desired position (30° in this case), and the other views are obtained by projecting from this view. During the revolution, all points of the object describe circular arcs parallel to the horizontal plane. The *heights* of all points in the front and side views in the revolved position remain unchanged, since they are measured parallel to the axis, and may be drawn by projecting from the initial front and side views of space I.

Note the similarity between the front and side views in space II of Fig. 9.3 and some of the auxiliary views of Fig. 8.8 (c).

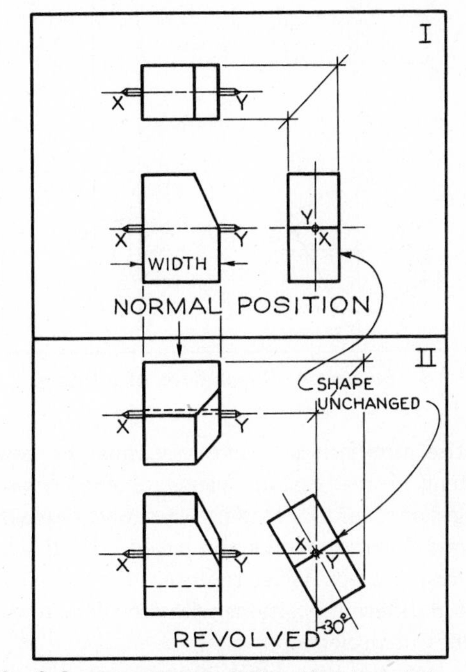


Fig. 9.4 **Primary Revolution About an Axis Perpendicular to Profile Plane.**

## 9.4 Revolution About Axis Perpendicular to Profile Plane.

A revolution about an axis XY perpendicular to the profile plane of projection is illustrated in Fig. 9.4. During the revolution, all points of the object describe circular arcs parallel to the profile plane of projection. The widths of the top and front views in the revolved position remain unchanged, since they are measured parallel to the axis, and may be obtained by projection from the top and front views of space I, or may be set off by direct measurement.

Note the similarity between the top and front views in space II of Fig. 9.4 and some of the auxiliary views of Fig. 8.9 (c).

## 9.5 Successive Revolutions.

It is possible to draw an object in an infinite number of revolved positions by making successive revolutions. Such a procedure, Fig. 9.5, limited to three or four stages, offers excellent practice in multiview projection. While it is possible to make several revolutions of a simple object without the aid of a system of numbers, it is absolutely necessary in successive revolutions to assign a number or a letter to every corner of the object. See §6.6.

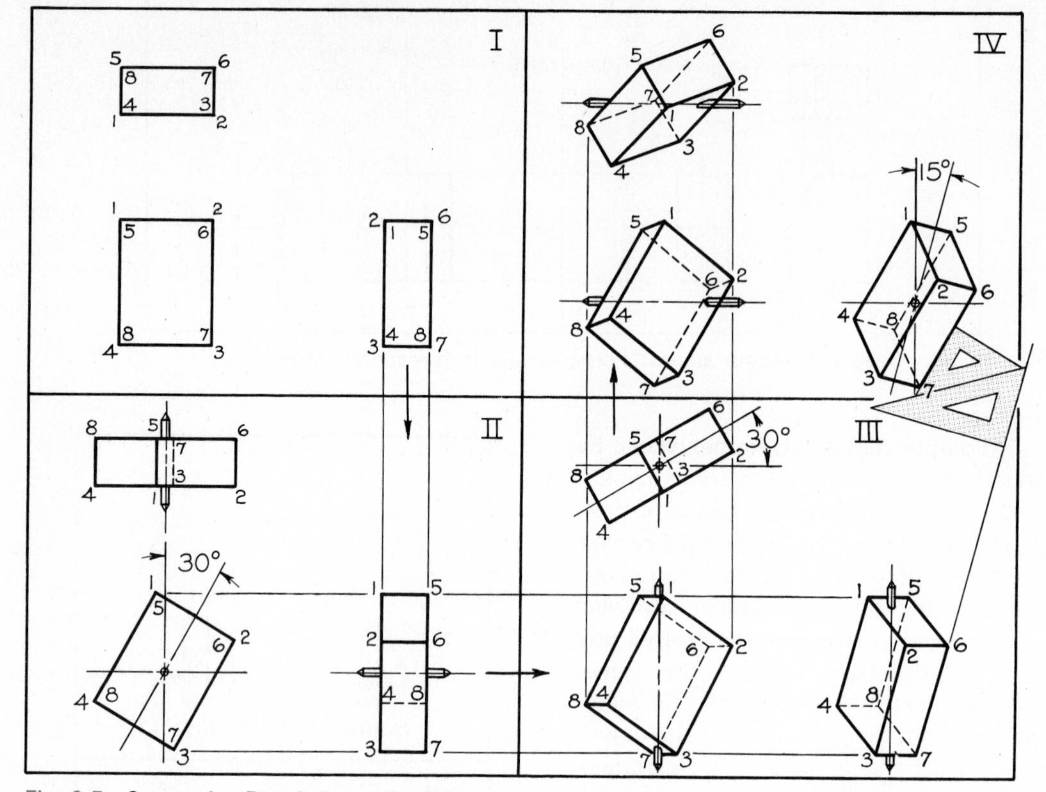


Fig. 9.5 **Successive Revolutions of a Prism.**

The numbering or lettering must be consistent in the various views of the several stages of revolution. Figure 9.5 shows four sets of multiview drawings numbered I, II, III, and IV, respectively. These represent the same object in different positions with reference to the planes of projection.

In space I, the object is represented in its normal position, with its faces parallel to the planes of projection. In space II, the object is represented after it has been revolved clockwise through an angle of 30° about an axis perpendicular to the frontal plane. The drawing in space II is placed under space I so that the side view, whose width remains unchanged, can be projected from space I to space II as shown.

During the revolution, all points of the object describe circular arcs parallel to the frontal plane of projection and remain at the same distance from that plane. The side view, therefore, may be projected from the side view of space I and the front view of space II. The top view may be projected in the usual manner from the front and side views of space II.

In space III, the object is taken as represented in space II and is revolved counterclockwise through an angle of 30° about an axis perpendicular to the horizontal plane of projection. During the revolution, all points describe *horizontal* circular arcs and remain at the same distance from the horizontal plane of projection. The top view is copied from space II but is revolved through 30°. The front and side views are obtained by projecting from the front and side views of space II and from the top view of space III.

In space IV, the object is taken as represented in space III and revolved clockwise through 15° about an axis perpendicular to the profile plane of projection. During the revolution, all points of the object describe circular arcs parallel to the profile plane of projection and remain at the same distance from that plane. The side view is copied, §4.29, from the side view of space III but revolved through 15°. The front and top views are projected from the side view of space IV and from the top and front views of space III.

Another convenient method of copying a view in a new revolved position is to use tracing paper as described in §4.30. Either a tracing can be made and transferred by rubbing, or the prick points may be made and transferred, as shown.

In spaces III and IV of Fig. 9.5, each view is an axonometric projection, §16.3. An isometric projection can be obtained by revolution, as shown in Fig. 16.3, and a dimetric projection, §16.29, can be constructed in a similar manner. If neither an isometric nor a dimetric projection is specifically sought, the successive revolution will produce a trimetric projection, §16.32, as shown in Fig. 9.5.

## 9.6 Revolution of a Point—Normal Axis.

Examples of the revolution of a point about a straight-line axis are often found in design problems that involve pulleys, gears, cranks, linkages, etc. For example, in Fig. 9.6 (a), as the disk is revolved, point 3 moves in a circular path lying in a plane perpendicular to the axis 1–2. This relationship is represented in the two views at (b). Note in this instance that the axis is normal or perpendicular to the frontal plane of projection, resulting in a front view that shows a point view of the axis and a true-size view of the circular path of revolution for point 3. The top view shows the path of revolution in edge view and perpendicular to the true-length view of the axis. Similar two-view relationships would occur if the axis were perpendicular or normal to either the horizontal or profile planes of projection.

The clockwise revolution through 150° for point 3 is illustrated at (c).

## 9.7 Revolution of a Point—Inclined Axis.

In Fig. 9.7 (a) the axis of revolution for point 3 is positioned parallel to the frontal plane and inclined to the horizontal and profile projection planes. Since the axis 1–2 is true length in the front view, the edge view of the path of revolution can be located as at (b). In order

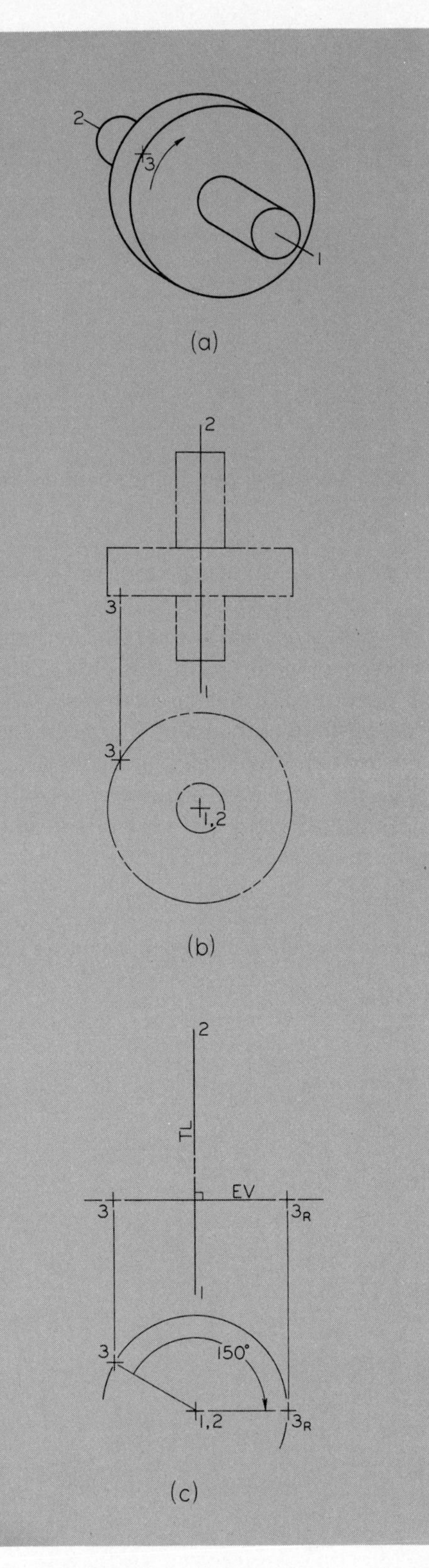


Fig. 9.6 **Revolution of a Point About a Normal Axis.**

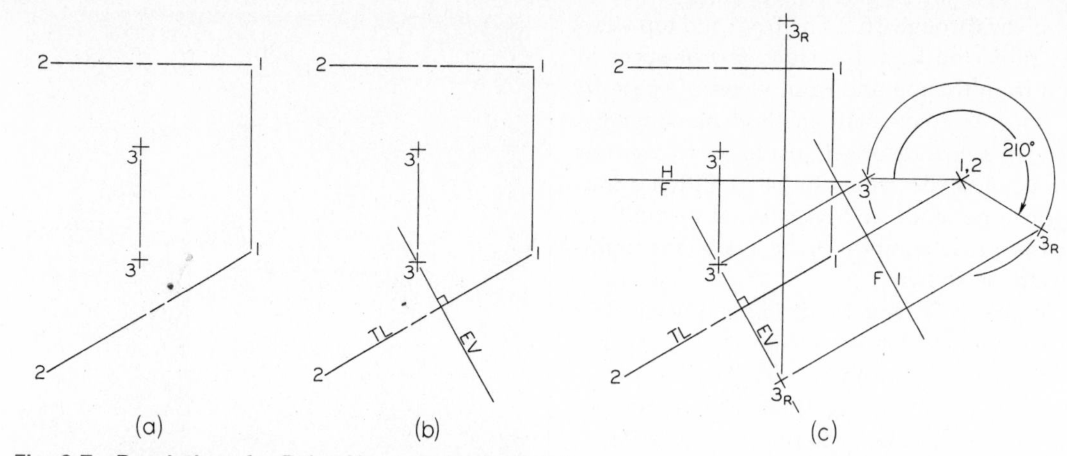


Fig. 9.7 **Revolution of a Point About an Inclined Axis.**

to establish the circular path of revolution for point 3, an auxiliary view showing the axis in point view is required as at (c). The required revolution of point 3 (in this case, 210°) is now performed in this circular view. The revolved position of the point is projected to the given front and top views, as shown.

Note the similarity of the relationships of the front view and auxiliary view and the constructions shown in Fig. 9.6 (c).

**9.8 Revolution of a Point—Oblique Axis.** In Fig. 9.8 (a), the axis of revolution for point 3 is oblique to all principal planes of projection and, therefore, is shown neither in true length nor as a point view in the top, front, or profile views. To establish the necessary true length and point view of the axis 1–2 in adjacent views, two successive auxiliary views are required, as shown at (b). The required revolved position of point 3 can now be lo-

Fig. 9.8 **Revolution of a Point About an Oblique Axis.**

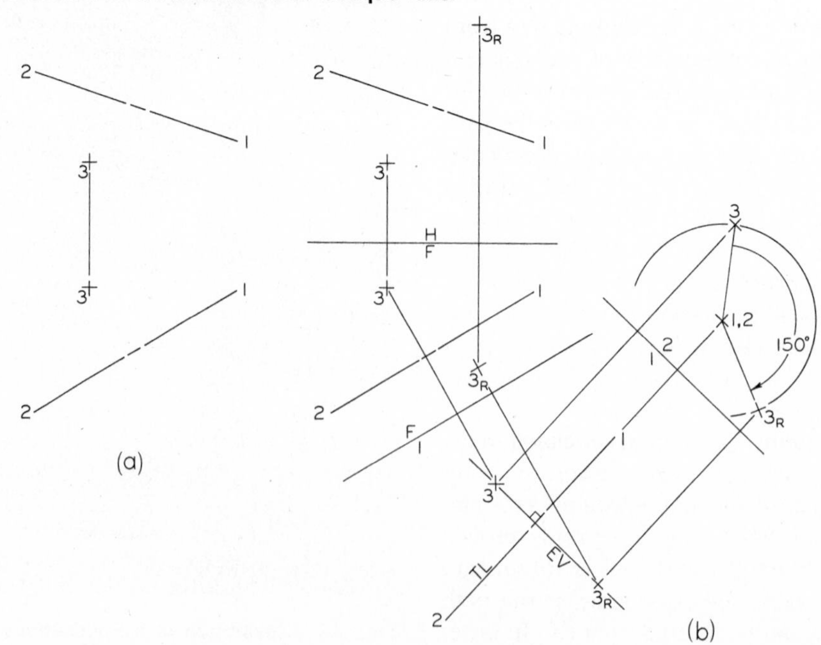

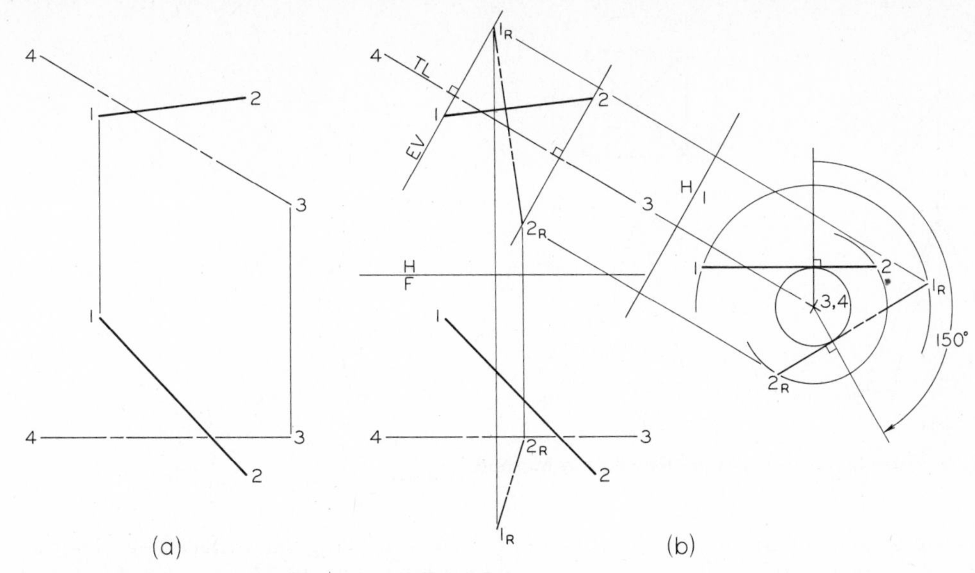


**Fig. 9.9 Revolution of a Line About an Inclined Axis.**

cated and then projected back to complete the given front and top views.

## 9.9 Revolution of a Line.

The procedure for the revolution of a line about an axis is very similar to that required for the revolution of a point, §9.6. All points on a line must revolve through the same angle, or the revolved line becomes altered.

In Fig. 9.9 (a), the line 1–2 is to be revolved thru 150° about the inclined axis 3–4.

Since the axis 3–4 is given in true length in the top view, an auxiliary view is required to provide a point view of the axis, as shown at (b). The necessary revolution can then be made about point view 3–4. In order to insure that all points on the line rotate through the same number of degrees, note that a construction circle tangent to line 1–2 is drawn, and a perpendicular through the tangency point becomes the reference for measurement of the angle of rotation. The circular arc paths for points 1 and 2 locate the points $1_R$ and $2_R$, as the revolved position of the line is drawn perpendicular to the radial line subtending the 150° arc of revolution, and tangent to the smaller circle. The alternate-position line is used to distinguish the revolved-position line from the original given line.

## 9.10 True Length of a Line—Revolution Method.

If a line is parallel to one of the planes of projection, its projection on that plane is equal in length to the line, Fig. 6.20. In Fig. 9.10 (a), the element AB of the cone is oblique to the planes of projection; hence its projections are foreshortened. If AB is revolved about the axis of the cone until it coincides with either of the contour elements, for example $AB_R$, it will be shown in its true length in the front view for it will then be parallel to the frontal plane of projection.

Likewise, at (b), the edge of the pyramid CD is shown in its true length $CD_R$ when it has been revolved about the axis of the pyramid until it is parallel to the frontal plane of projection. At (c), the line EF is shown in its true length at $EF_R$ when it has been revolved about a vertical axis until it is parallel to the frontal plane of projection.

The true length of a line may also be found by constructing a right triangle or a true-length diagram, as shown at (d), whose base is equal to the top view of the line, and whose altitude is the difference in elevation of the ends. The hypotenuse of the triangle is equal to the true length of the line.

In these cases the lines are revolved until parallel to a plane of projection. The true length of a line may also be found by leaving

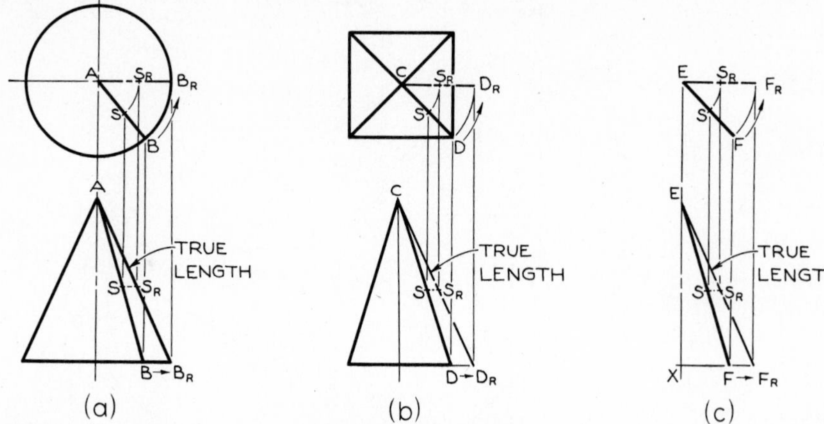

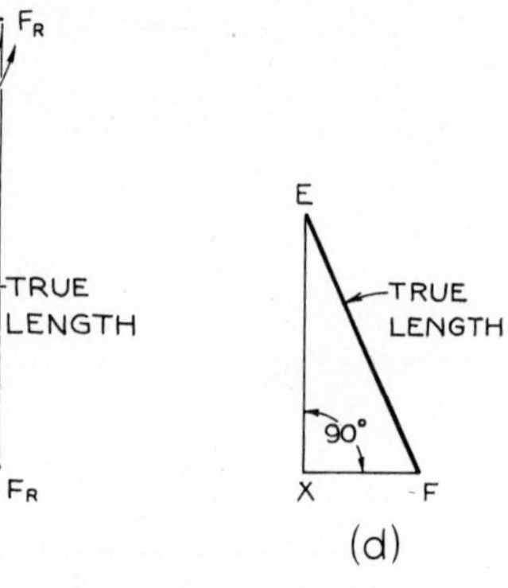


Fig. 9.10 **True Length of a Line—Revolution Method.**

the line stationary but shifting the position of the observer—that is, the method of auxiliary views, §8.18.

## 9.11 True Size of a Plane Surface—Revolution Method.

If a surface is parallel to one of the planes of projection, its projection on that plane is true size, Fig. 6.19. In Fig. 9.11 (a), the inclined surface 1-2-3-4 is foreshortened in the top and side views and appears as a line in the front view. Line 2-3 is taken as the axis of revolution, and the surface is revolved clockwise in the front view to the position $4_R$-3 and projected to the side view at $4_R$-$1_R$-2-3, which is the true size of the surface. In this case the surface was revolved until parallel to the profile plane of projection.

At (b), triangular surface 1-2-3 is revolved until parallel to the horizontal plane of projection so that the surface appears true size in the top view, as shown.

At (c), the true size of the oblique surface cannot be found by a simple primary revolution. The true size can be found by two successive revolutions or by a combination of an auxiliary view and a primary revolution. The latter is shown at (c). First, draw an auxiliary view that will show the edge view (EV) of the plane. See Fig. 8.22. Second, revolve the edge view of the surface until it is parallel to the folding line F/1, as shown. All points in the front view, except those in the axis of revolution, will describe circular arcs parallel to the reference plane F/1. These arcs will appear in the front view as lines parallel to the folding

Fig. 9.11 **True Size of Plane Surface—Revolution Method.**

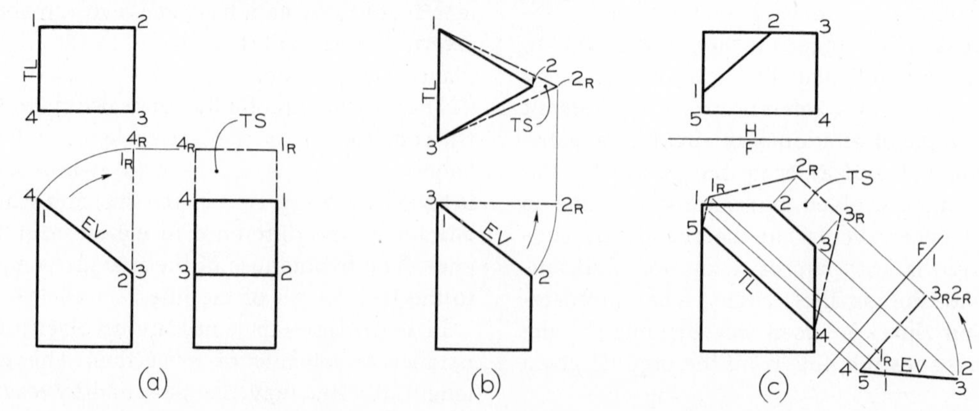

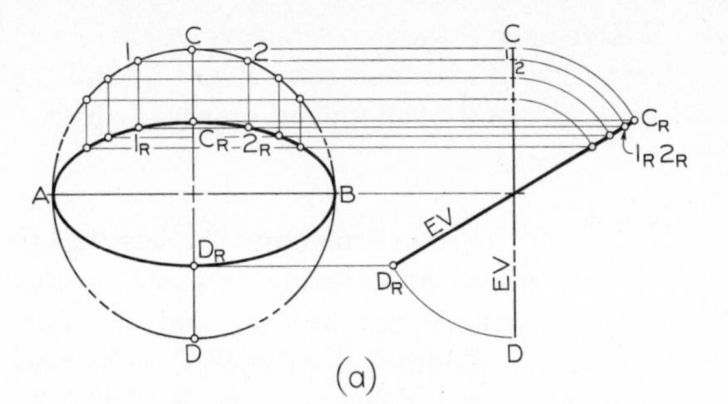


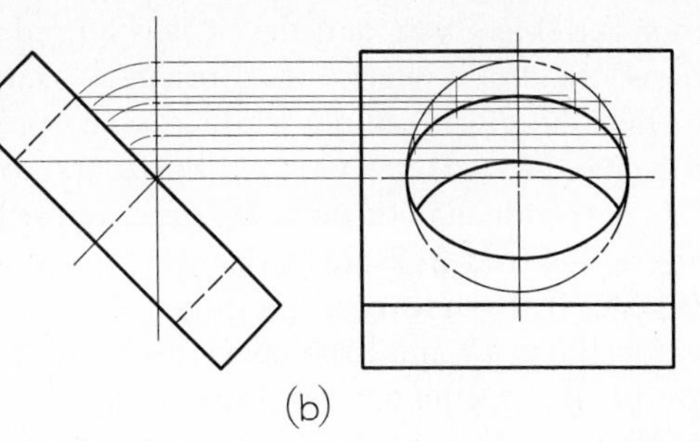


Fig. 9.12 **Revolution of a Circle.**

line, such as $2\text{–}2_R$ and $3\text{–}3_R$. The true size of the surface is found by connecting the points with straight lines.

## 9.12 Revolution of Circles.

As shown in Fig. 4.50 (a) to (c), a circle, when viewed obliquely, appears as an ellipse. In that case the coin is revolved by the fingers. The geometric construction of this revolution is shown in Fig. 9.12 (a). In the front view the circle appears as ACBD, and in the side view as line CD, which is really the edge view of the plane containing the circle. In the side view, CD (the side view of the circle) is revolved through any desired acute angle to $C_RD_R$.

To find points on the ellipse, draw a series of horizontal lines across the circle in the front view. Each line will cut the circle at two points, as 1 and 2. Project these points across to the vertical line representing the unrevolved circle; then, revolve each point and project horizontally to the front view to establish points on the ellipse. Plot as many points as necessary to secure a smooth curve.

An application of this construction to the representation of a revolved object with a large hole is shown at (b).

## 9.13 Counterrevolution.

The reverse procedure to revolution is *counterrevolution.* For example, if the three views of space II in Fig. 9.2 are given, the object can be drawn in the unrevolved position of space I by counterrevolution. The front view is simply counterrevolved back to its normal upright position, and the top and side views are drawn as shown. Similarly, in Fig. 9.5, the object may

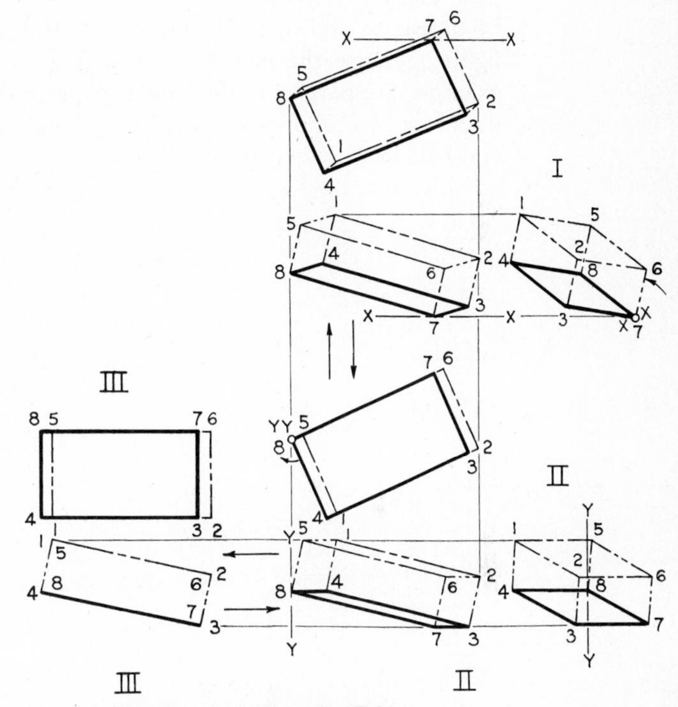


Fig. 9.13 **Counterrevolution of a Prism.**

be counterrevolved from its position of space IV to its unrevolved position of space I by simply reversing the process.

In practice, it sometimes becomes necessary to draw the views of an object located on or parallel to a given oblique surface. In such an oblique position, it is very difficult to draw the views of the object because of the foreshortening of lines. The work is greatly simplified by counterrevolving the oblique surface to a simple position, completing the drawing, and then revolving to the original given position.

An example is shown in Fig. 9.13. Assume that the oblique surface 8–4–3–7 (three views in space I) is given, and that it is required to draw the three views of a prism $\frac{1}{2}''$ high, having the given oblique surface as its base. Revolve the surface about any horizontal axis XX, perpendicular to the side view, until the edges 8–4 and 3–7 are horizontal, as shown in space II. Then revolve the surface about any vertical axis YY until the edges 8–7 and 4–3 are parallel to the frontal plane, as shown in space III. In this position the given surface is perpendicular to the frontal plane, and the front and top views of the required prism can be drawn, as shown by phantom lines in the figure, because the edges 4–1 and 3–2, for example, are parallel to the frontal plane and, therefore, are shown in their true lengths, one-half inch. When the two views in space III have been drawn, counterrevolve the object from III to II and then from II to I to find the required views of the given object in space I.

**9.14 Revolution Problems.** In Figs. 9.14 to 9.20 are problems covering primary revolutions, successive revolutions, and counterrevolutions. Additional problems, in convenient form for solution, are available in *Technical Drawing Problems, Series 1,* by Giesecke, Mitchell, Spencer, and Hill; *Technical Drawing Problems, Series 2,* by Spencer and Hill; and in *Technical Drawing Problems, Series 3,* by Spencer and Hill, all designed to accompany this text and published by Macmillan Publishing Co., Inc.

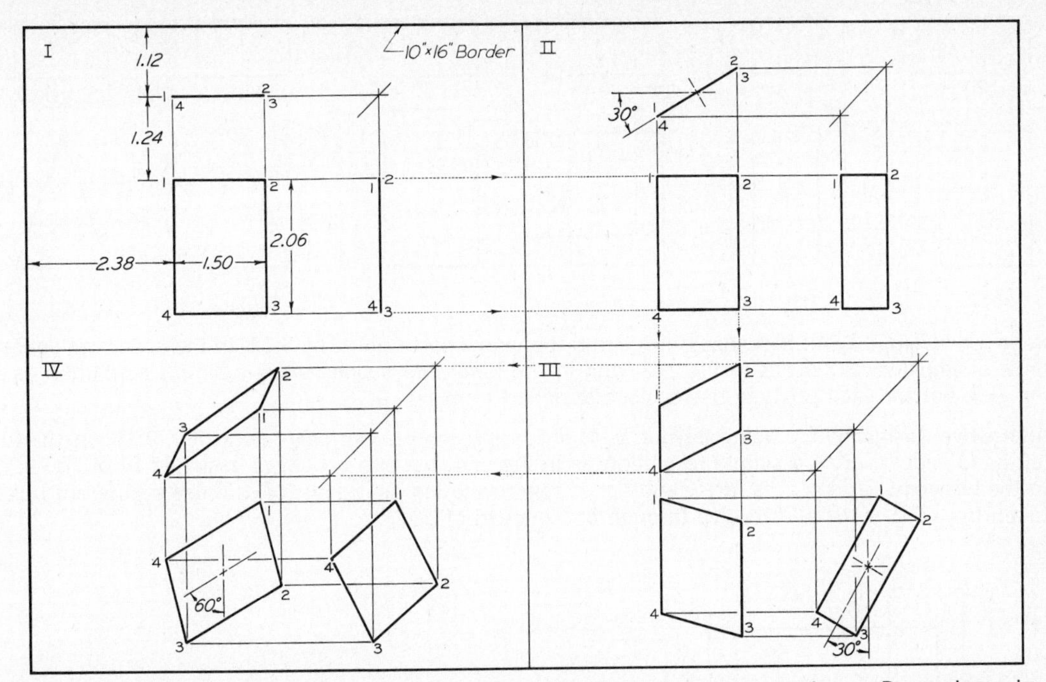


Fig. 9.14 Using size B sheet, divide working area into four equal parts, as shown. Draw given views of rectangle, and then the primary revolution in space II, followed by successive revolutions in spaces III and IV. Number points as shown. Omit dimensions. Letter date below border at the left and your name correspondingly at the right.

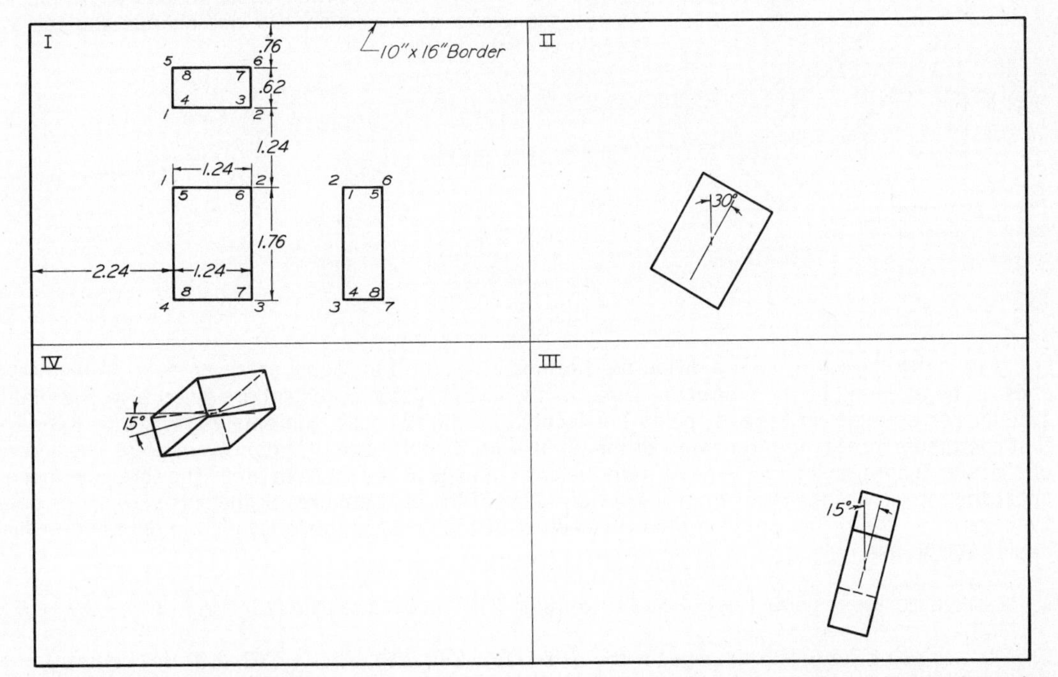


Fig. 9.15 Using size B sheet, divide working area into four equal parts, as shown. Draw given views of prism as shown in space I; then draw three views of the revolved prism in each succeeding space, as indicated. Number all corners. Omit dimensions. Letter date below border at the left and your name correspondingly at the right.

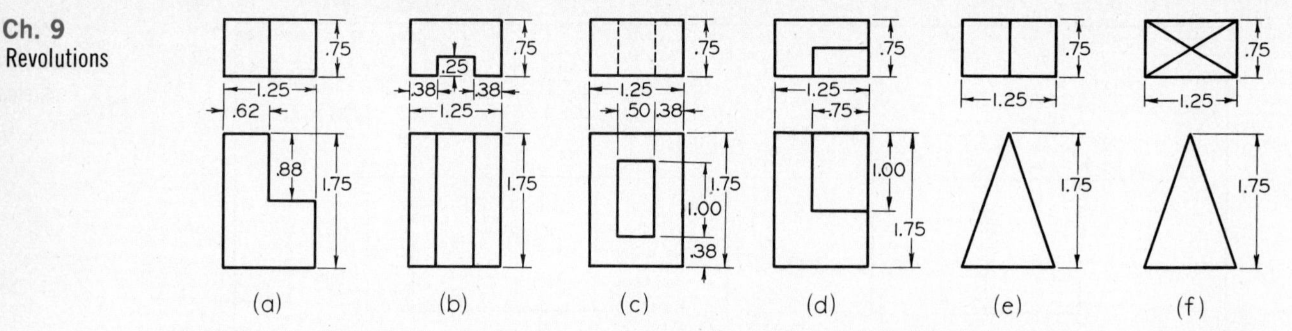


Fig. 9.16 Using Layout B–4 sheet, divide into four equal parts as in Fig. 9.14. In the upper two spaces, draw a simple revolution as in Fig. 9.2, and in the lower two spaces, draw a simple revolution as in Fig. 9.3; but for each problem use a block assigned from Fig. 9.16.

**Alternative Assignment:** Using Layout B–4, divide into four equal parts as in Fig. 9.14. In the two left-hand spaces, draw a simple revolution as in Fig. 9.4, but use an object assigned from Fig. 9.16. In the two right-hand spaces, draw another simple revolution as in Fig. 9.4, but use a different object taken from Fig. 9.16, and revolve through 45° instead of 30°.

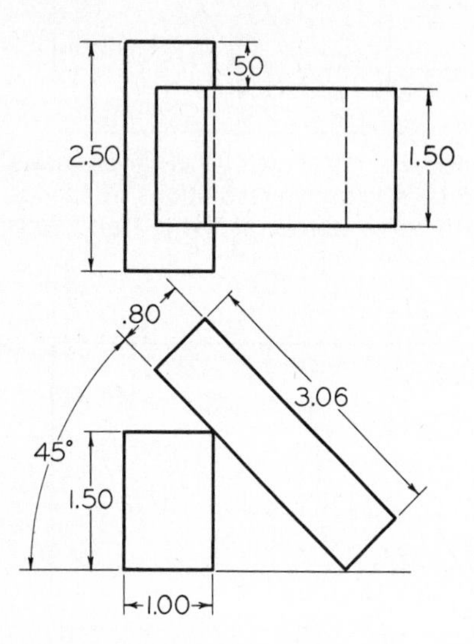


Fig. 9.17 Using Layout A–2 or A–3, draw three views of the blocks, but revolved 30° clockwise about an axis perpendicular to the top plane of projection. Do not change the relative positions of the blocks.

Fig. 9.18 Use Layout A–1, and divide the working area into four equal areas for four problems per sheet to be assigned by the instructor. Data for the layout of each problem are given by a coordinate system. For example, in Prob. 1, point 1 is located by the full-scale coordinates (1.1, 1.5, 3.0). The first coordinate locates the front view of the point from the left edge of the problem area. The second one locates the front view of the point from the bottom edge of the problem area. The third one locates either the top view of the point from the bottom edge of the problem area or the side view of the point from the left edge of the problem area. Inspection of the given problem layout will determine which application to use.

1. Revolve clockwise point 1(1.1, 1.5, 3.0) through 210° about the axis 2(2.0, 2.3, 3.7) – 3(2.0, 0.3, 3.7).
2. Revolve point 3(1.6, 1.5, 2.1) about the axis 1(1.1, 2.5, 2.9) – 2(1.1, 0.3, 2.9) until point 3 is at the farthest distance behind the axis.
3. Revolve point 3(0.8, 0.3, 3.3) about the axis 1(0.4, 0.7, 4.8) – 2(2.2, 0.7, 3.0) through 210° and to the rear of line 1–2.
4. Revolve point 3(0.2, 2.1, 2.1) about the axis 1(0.4, 0.5, 2.8) – 2(0.9, 2.6, 2.8) to its extreme position to the left in the front view.

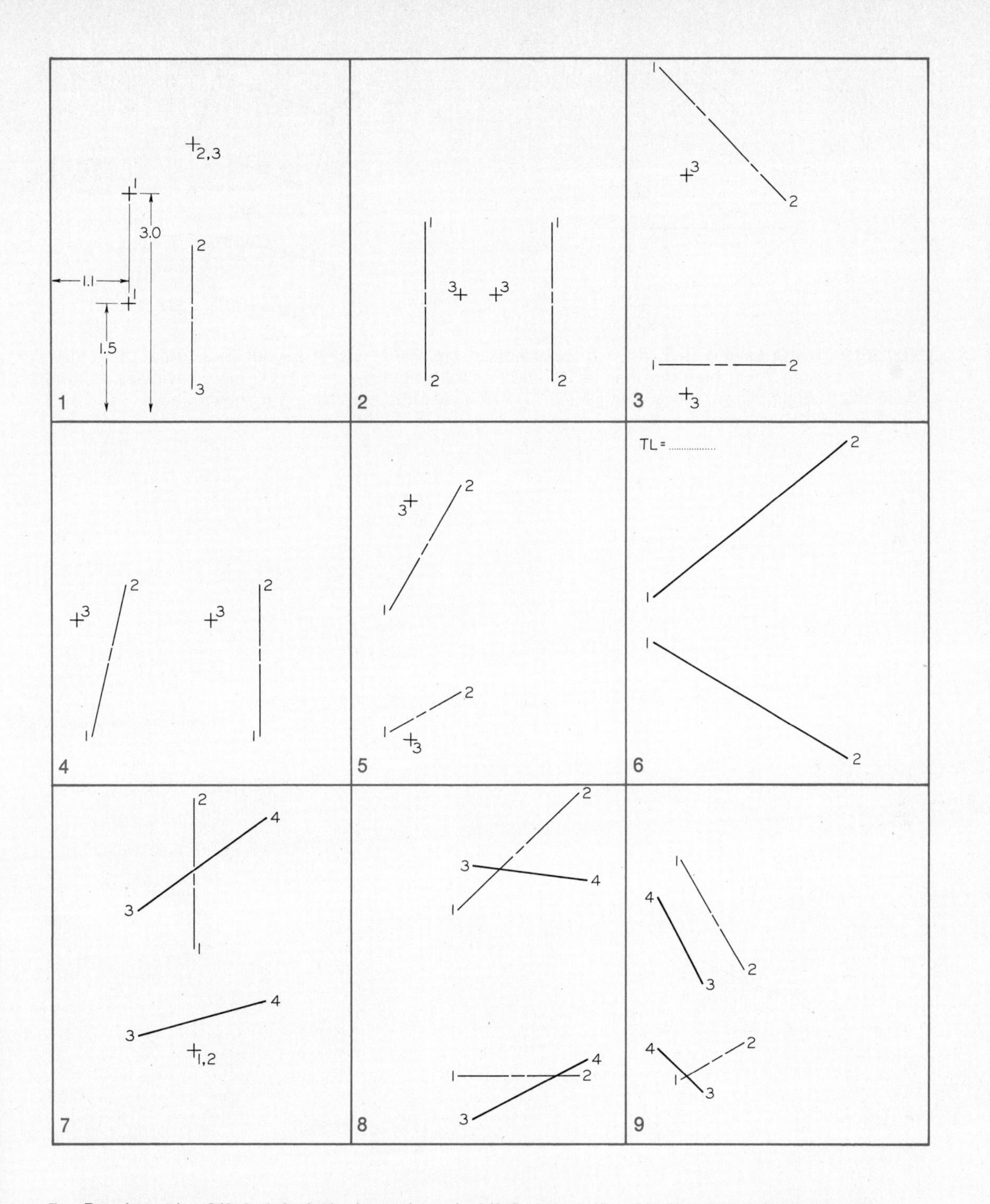


**5.** Revolve point 3(0.6, 0.3, 3.9) about the axis 1(0.3, 0.4, 2.4) – 2(1.3, 1.0, 4.1) through 180°.
**6.** By revolution find the true length of line 1(0.3, 1.9, 2.5) – 2(3.1, 0.3, 4.7). Scale: 1″ = 10′-0.
**7.** Revolve line 3(1.2, 1.5, 3.2) – 4(3.0, 2.0, 4.5) about axis 1(2.0, 1.3, 2.7) – 2(2.0, 1.3, 4.8) until line 3–4 is shown true length and below the axis 1–2. Scale: 1″ = 30′-0.
**8.** Revolve line 3(2.1, 0.3, 3.8) – 4(3.7, 1.1, 3.6) about the axis 1(1.9, 0.9, 3.2) – 2(3.6, 0.9, 4.8) until line 3–4 is in true length and above the axis.
**9.** Revolve line 3(1.1, 0.6, 2.1) – 4(0.5, 1.2, 3.3) about the axis 1(0.8, 0.8, 3.8) – 2(1.7, 1.3, 2.3) until line 3–4 is level and above the axis.

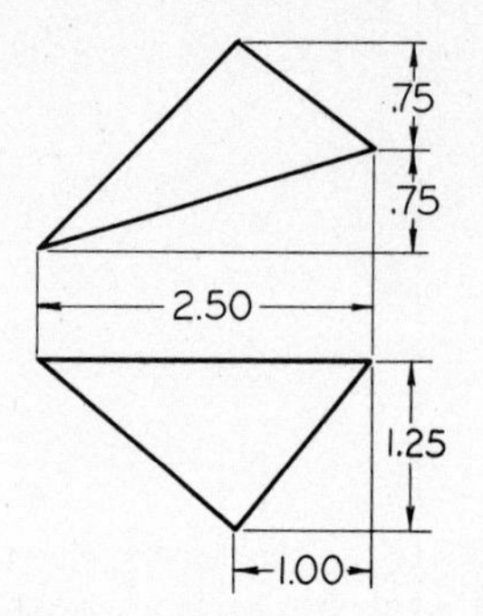


**Fig. 9.19** Using Layout B–3, draw three views of a right prism 1.50″ high that has as its lower base the triangle shown above. See §9.13.

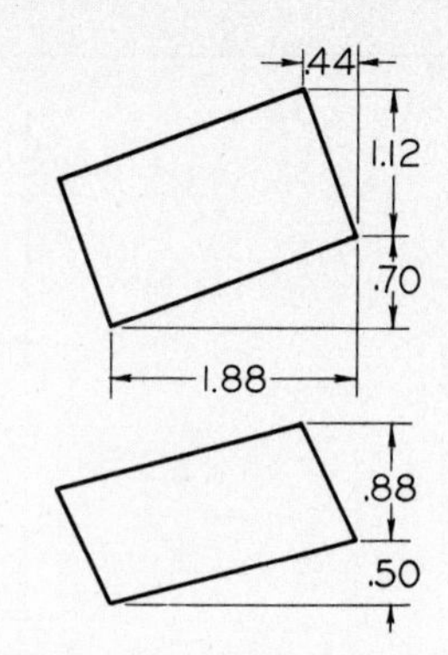


**Fig. 9.20** Using Layout B–3, draw three views of a right pyramid 2.00″ high, having as its lower base the parallelogram shown above. See §9.13.

# 10

# shop processes

by J. George H. Thompson* and John Gilbert McGuire†

**10.1 Introduction.** The test of the usefulness of any working drawing is whether the object described can be satisfactorily produced without further information than that furnished on the drawing. The drawing must furnish information as to shape, size, material, and finish and, if necessary, suggest the shop processes required to produce the desired object. It is the purpose of this chapter to provide the young engineer with some information about certain fundamental shop terms and processes and to assist him in using this information on his drawings. Even the more sophisticated techniques of automatic and numerically controlled shop processes are in essence basic machine tools and processes with programmed operation sequence and control, §10.23.

The student engineer should seize every opportunity to visit shops in order to learn what can be done on various machines and thus be able to make practical specifications on his drawings. The draftsman in any organization must have a thorough knowledge of what can be done in his shop before he can properly indicate on his drawing the machining operations, heat-treatment, finish, and the accuracy desired on each part.

**10.2 Shop Processes.** The shop starts with what might be called *raw stock* and modifies this until it agrees with the detail drawing. The shape of the raw stock may have to be altered. For example, an automobile body shell may be pressed out of sheet steel

*Professor of Mechanical Engineering, Texas A&M University.

†Assistant Dean of Engineering, Texas A&M University.

on a massive press before it conforms with the drawing. The size may also have to be changed; for example, a 2″ diameter bar may have to be turned down on a lathe until it becomes 1.774″ diameter.

Changing the shape and size of the material of which a part is being made requires one or more of the following processes: (1) removing part of the original material, (2) adding more material, and (3) redistributing original material. Cutting, as turning on a lathe, or punching holes by means of a power press, removes material. Welding, brazing, soldering, metal spraying, and electrochemical plating add material. Forging, pressing, drawing, extruding, and spinning redistribute material.

The shop is also concerned with the characteristics of the material from which the part is being produced. Frequently, the characteristics of the material may have to be altered to agree with the properties specified by the designer on the drawing. For example, the part may be required to have a Brinell hardness* of 400. Since steel is not ordinarily supplied in this degree of hardness, it would become necessary to harden such a part by heat-treatment, §10.30.

## 10.3 Manufacturing Methods and the Drawing.

Before the draftsman prepares a drawing for the production of a part, he should consider what manufacturing processes are to be used, inasmuch as these determine the representation of the detailed features of the part, the choice of dimensions, and the machining accuracy. Principal types are (1)

* A Brinell number indicates hardness. The number is obtained by measuring the diameter of indentation produced by a standard steel ball loaded according to specified conditions. A steel of "150 Brinell" is moderately soft. A "250 to 400 Brinell" is as hard as can ordinarily be cut with single-pointed steel cutting tools. Harder materials can be machined by grinding or by the use of nonferrous cutting tools, such as tungsten carbides.

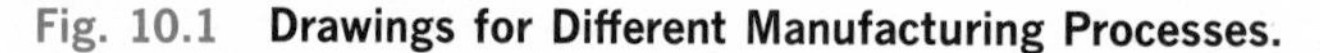

Fig. 10.1 **Drawings for Different Manufacturing Processes.**

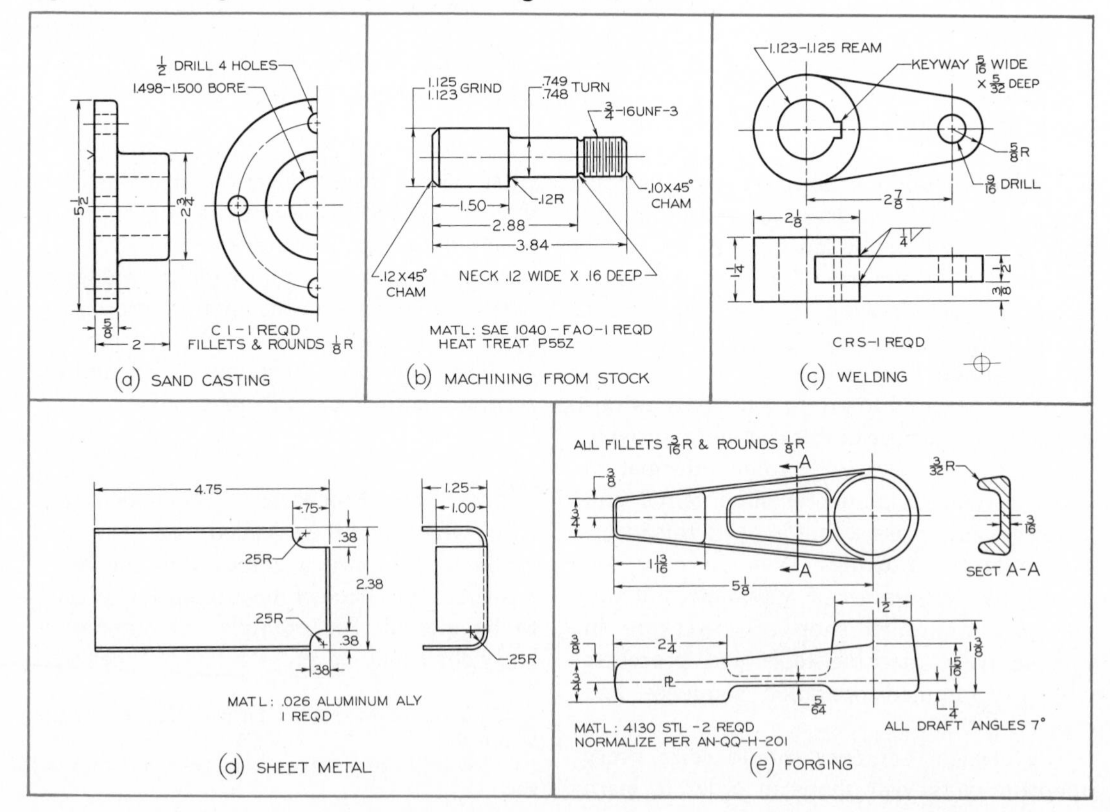

*casting,* (2) *machining from standard stock,* (3) *welding,* (4) *forming from sheet stock,* and (5) *forging.* A knowledge of these processes, along with a thorough understanding of the intended use of the part, will help determine the basic manufacturing processes. Drawings that reflect these manufacturing methods are shown in Fig. 10.1.

In sand casting, Fig. 10.1 (a), all unfinished surfaces remain rough, and all rough corners are filleted or rounded. Sharp corners appear where two surfaces intersect if at least one is a finished surface, §6.34. Finish marks are shown on the edge views of finished surfaces. See §§11.16 and 11.17. Pattern draft is usually not shown on a drawing. Dimensions are given for the patternmaker and the machinist on the same drawing, §11.27.

In drawings of parts machined from standard stock, (b), most surfaces are machined. In some cases, as on shafting, the surface existing on the raw stock is often accurate enough without further finishing. Corners are usually sharp, but fillets and rounds are machined when necessary. For example, an interior corner may be machined with a radius to provide greater strength.

On welding drawings, (c), the several pieces are cut to size, brought together and then welded. Welding symbols are used to indicate the welds required. Generally there are no fillets and rounds. Certain surfaces may be machined after welding, or in some cases before welding. Notice that lines are shown where the separate pieces are joined.

On sheet-metal drawings, (d), the thickness of material is uniform and is usually given in the material specification note. Bend radii and bend reliefs at corners are specified according to standard practice. For dimensions, whole numbers and common fractions may be used. In the aircraft and automotive industries, and in many others, the complete decimal system, §11.10, may be used as shown in the figure. Allowances of extra material for joints may be required.

For forged parts, separate drawings are usually made for the diemaker and for the machinist. Thus, a forging drawing, (e), provides only the information to produce the forging, and the dimensions given are those needed by the diemaker. See §10.28. All corners are rounded and filleted, and are so shown on the drawing. The draft is drawn to scale, and is usually specified by degrees in a note, as shown.

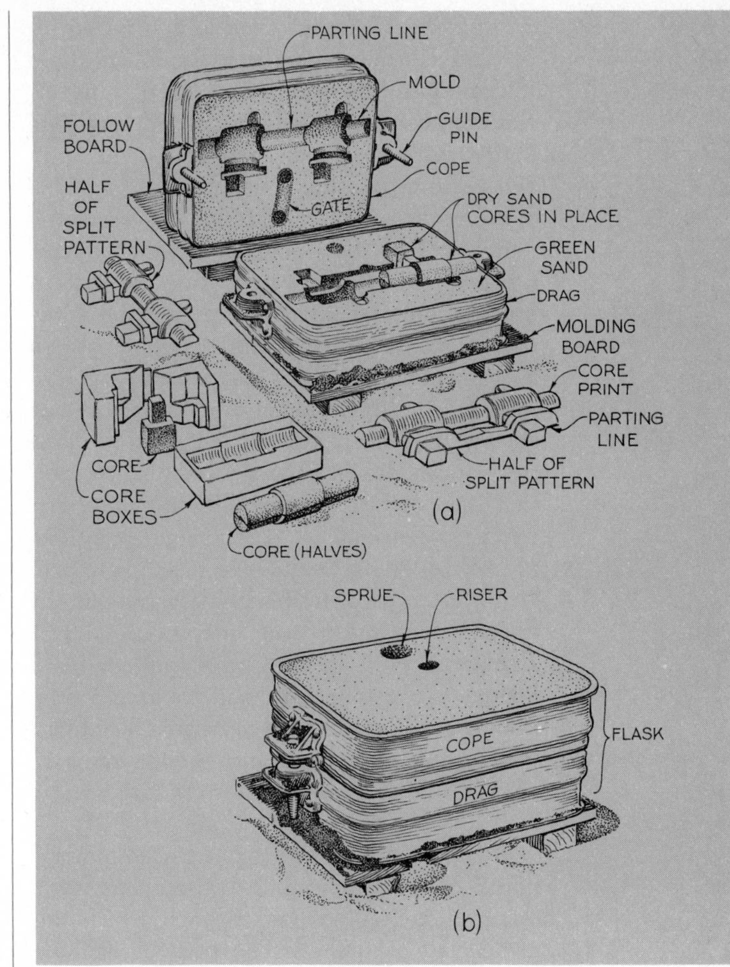


Fig. 10.2 **Sand Molding of Table Bracket** (Fig. 14.30).

## 10.4 Sand Casting.

While there are a number of processes used to produce castings, sand molds are used far more extensively than any other types of molds. Sand molds are made by ramming sand around a model, called a pattern, and then carefully withdrawing it, leaving a cavity to receive the molten metal, as shown in Fig. 10.2 (a). The sand is contained in a two-part box called a *flask,* (b), the upper part of which is called the *cope* and the lower the *drag*. For more complex work, one or more intermediate boxes, called *cheeks*, may be introduced between the cope and the drag. The pattern must be of such shape that it will "pull away" from both the cope and the drag. The plane of separation of the two halves of the pattern marks the *parting line* on the pattern. On each side of the parting line the pattern must be tapered slightly to permit the withdrawal of the pattern from the sand, unless a segmented pattern is used. This taper is known as *draft*. Although draft is usually not shown, and dimensions are not given for it on the working drawing, the design must be such that draft can be properly built into the pattern by the patternmaker. When large quantities of castings are to be made, a matchplate pattern is often used. In this process, the pattern is split along the parting line, mounted and aligned on both sides of an aluminum plate that fits between the cope and the drag. Withdrawal of the matchplate pattern is facilitated by the guide pins on the cope and the integrity of the mold is protected despite rapid withdrawal of the pattern.

A *sprue stick*, or round peg, is placed in position during the ramming process and then removed to leave a hole through which the metal may be poured. The part of the hole adjacent to the casting is called the *gate*, and the vertical part the *sprue*, Fig. 10.2. On some molds another hole, known as the *riser*, is provided to allow gases to escape and to provide a reservoir of metal that feeds back to the casting as it cools and shrinks.

Since shrinkage occurs when metal cools, patterns are made slightly oversize. The patternmaker accomplishes this by using a *shrink rule* whose units are oversize according to the shrinkage characteristics of the metal used. For example, cast iron shrinks about $\frac{1}{8}''$ per linear foot as it cools, while steel shrinks about $\frac{3}{16}''$ per linear foot. Allowance for shrinkage is not shown on the working drawing but is taken care of entirely by the pattern shop. The patternmaker must refer to the drawing for the kind of metal to be used in the casting.

A *core print* is a projection added to a pattern for the purpose of forming a cavity in a mold into which a corresponding portion of a *core* will rest, as shown in Fig. 10.2, thus forming an anchor to hold the core in place. Cores are made of sand and are used to provide certain hollow portions of the casting. The most common use of a core is to extend through a casting to form a *cored hole*. When it is necessary to form the sand into shapes that would ordinarily not permit the necessary adhesion and strength, or in instances where the shape of the casting would interfere with the removal of the pattern, a *dry sand core* is used. Dry sand cores, Fig. 10.2, are made by ramming a prepared mixture of sand and a binding substance into a *core box*; the core is then removed and baked in a *core oven* to make it sufficiently rigid. A *green sand core* (not baked) is used when it is practical to make the core along with the mold, as for example the central hole shown in Fig. 10.3.

It is from considerations of shrinkage and of draft that one arrives at a general rule that small holes (even if so placed as to draw out of the sand) are better drilled from a solid casting, and that large holes are better cored and then bored.

## 10.5 The Pattern Shop and the Drawing.

The pattern shop receives the working drawing showing the object in its completed state, including all dimensions and finish marks. Usually the same drawing is used by the pattern shop and the machine shop; hence, it should contain all dimensions and notes needed by both shops, as shown in Fig. 10.3. Some companies follow the practice of giving all dimensions for the pattern shop or forge shop in pencil and those for the machine shop in ink. On the blueprint, the difference is easily distinguishable. If the part is large and complicated, two separate drawings are sometimes made, one showing the pattern dimensions, and the other the machine dimensions.

Pattern dimensions, §11.27, which are used to make a pattern, should not be closer than $\frac{1}{16}''$, while dimensions for the machine shop must often reflect a tolerance of a few thousandths of an inch or less. Therefore, to dimension correctly, a careful study of a part

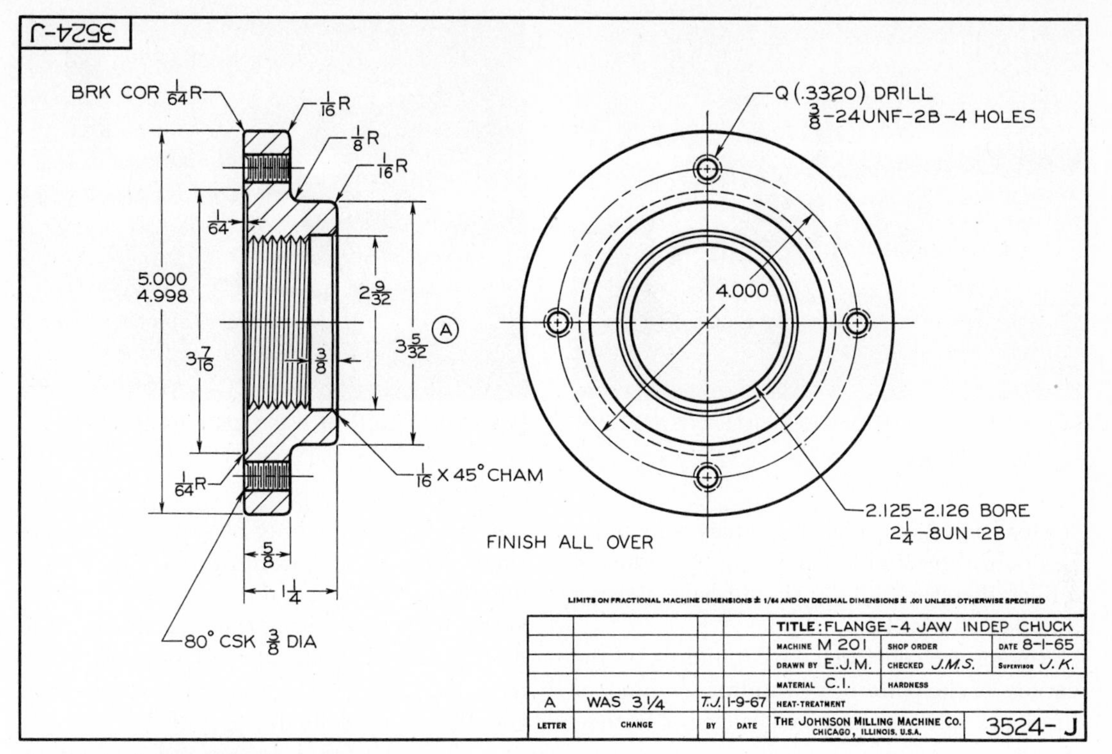


**Fig. 10.3 A Detail Working Drawing.**

must be made to determine which dimensions are to be used by the patternmaker and which by other craftsmen, §10.7.

Finish marks, §11.17, are as important to the patternmaker as to the machinist because additional material on each surface to be machined must be provided. For small and medium-sized castings, $\frac{1}{16}''$ to $\frac{1}{8}''$ is usually sufficient; larger allowances are made if there is probability of distortion or warping. On the Flange, Fig. 10.3, it is necessary for the patternmaker to provide material for finish on all surfaces.

Sometimes the patternmaker works directly from the drawing; in other cases, he may find it desirable to make his own pattern layout. Except for simple objects, it is common practice for the pattern shop to prepare full-size (to shrink rule) pattern layout drawings on white pine boards. Wood is used instead of paper because it is more durable and because paper stretches and shrinks excessively. The pattern is then checked against this pattern layout drawing on which draft, shrinkage, core print dimensions, and other such information are clearly shown.

## 10.6 Fillets and Rounds.

*Fillets* (inside rounded corners) and *rounds* (outside rounded corners) must be taken care of on the pattern, in order to provide for strength and appearance in the casting, Fig. 10.4. Crystals of cooling metal tend to arrange themselves perpendicular to the exterior surfaces as indicated at (b) and (c). If the corners of a casting are rounded as shown at (c), a much stronger casting results. It has been demonstrated that

**Fig. 10.4 Fillets and Rounds.**

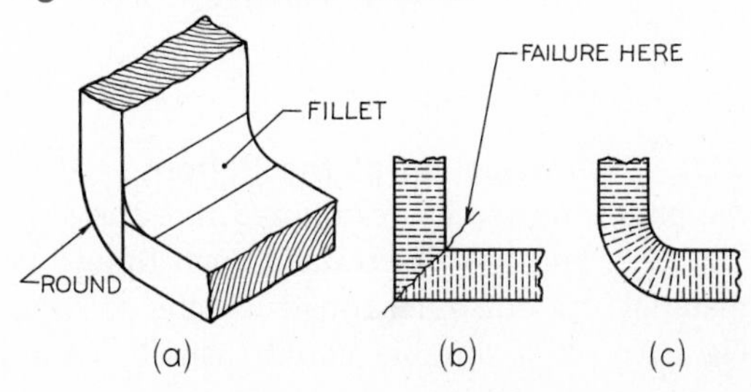

(a) APPLICATION OF LEATHER FILLET TO PATTERN

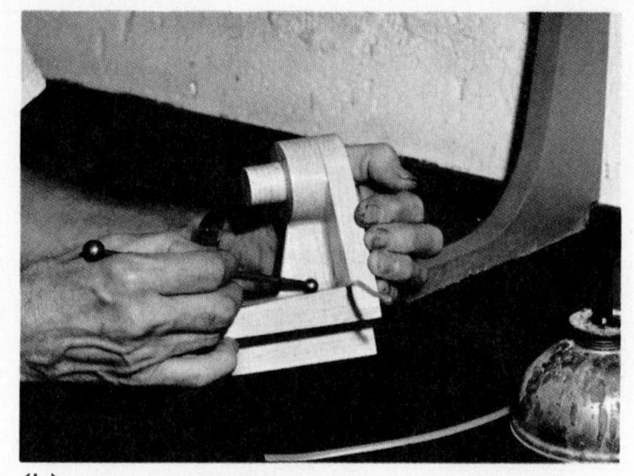

(b) APPLICATION OF WAX FILLET TO PATTERN

Fig. 10.5 **Fillets.**

the load a part can safely carry may sometimes be doubled by simply increasing the radius of a fillet; therefore, a fillet should have as large a radius as possible.

Rounds are constructed by rounding off the corners of the pattern (by sanding, planing, or turning) while fillets are constructed of leather, wax, or wood. Preformed leather, with a range of radii, is available for this purpose. A leather fillet of proper radius is fastened with glue and firmly pressed in place with the use of an unheated spherical tool, as shown in Fig. 10.5 (a). Wax fillets are applied in much the same way as leather but with the use of a hot spherical tool and without the use of glue, as shown at (b). For some cylindrical patterns it is possible to form (on the lathe) the proper radius for a fillet as an integral part of the wood pattern.

All fillets and rounds should be shown on the drawing, either freehand or drawn to scale with the use of a compass or with a circle guide, as in Fig. 10.3. For a general discussion of fillets and rounds from the standpoint of representation on the drawing, see §§6.34 to 6.36. For dimensioning of fillets and rounds, see §11.16.

## 10.7 Construction of the Pattern.

After the patternmaker has examined the drawing, Fig. 10.3, and has determined such things as material and finish, he constructs the pattern, Fig. 10.6, of a durable wood (usually white pine or mahogany), or of plastic, using dimensions given on the working drawing or pattern layout drawing and making proper allowances for metal shrinkage, machining, and draft. At (a) a cylindrical block of wood is first mounted on the face plate of a wood lathe and then turned to the proper diameter (5″ plus allowance for shrinkage and machining). At (b) the hub diameter $3\frac{5}{32}''$ on working drawing) is checked. It should be noted that the $\frac{1}{8}''$ fillet between the hub and the Flange has been formed on the pattern. Also, it may be noted at (c) that the flange thickness ($\frac{5}{8}''$ on the working drawing) is checked with a shrink rule, and at (d) the center hole is being sanded on a spindle sander to provide a 1° draft taper to the hole previously turned on the lathe. If a quantity of molds were to be made from the pattern, the pattern stock would probably be glued up from several pieces of wood with the grain arranged to minimize warpage. For a large number of molds, a *master pattern* would be constructed of wood and a duplicate working pattern cast in aluminum for actual use in the foundry. In this way the master pattern would be protected from excessive wear. In this case the patternmaker would employ "double" shrinkage, one for aluminum and one for the material to be cast.

## 10.8 The Foundry.

After the pattern has been completed, as described above, it is sent to the foundry for casting. The steps in pre-

(a) TURNING TO REQUIRED DIAMETER

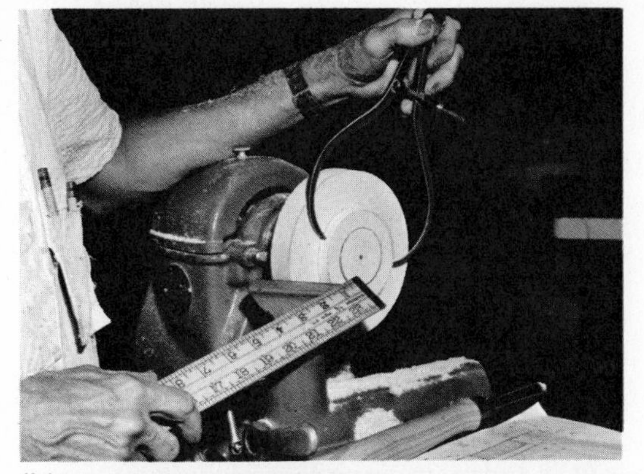
(b) CHECKING DIAMETER OF HUB

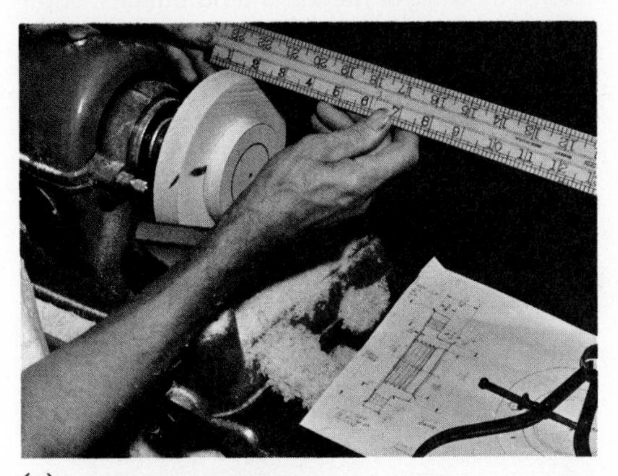
(c) CHECKING THICKNESS OF FLANGE

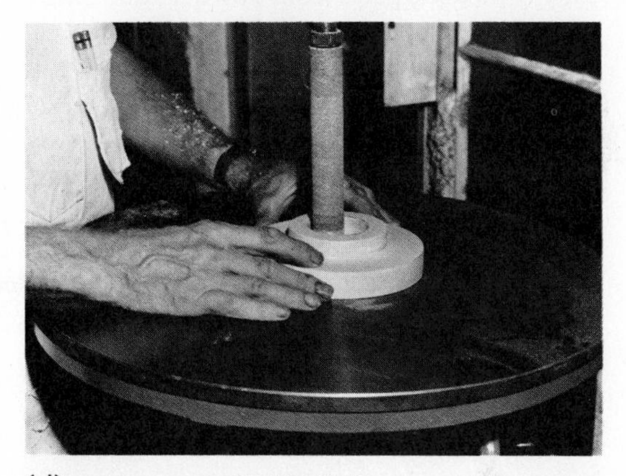
(d) SANDING DRAFT IN HOLE

Fig. 10.6 **Making a Pattern.**

paring a sand mold for the Flange, Fig. 10.3, are illustrated in Fig. 10.7. At (a) sand has been rammed around the pattern, and then at (b) the pattern has been carefully removed, leaving a cavity for molten metal. After the cope has been returned to position on the drag at (c), molten metal is poured through the sprue (hole leading to cavity) into the mold. After sufficient cooling time, the casting is removed from the broken mold, as shown at (d).

Although sand molds are the least expensive to make, sometimes plaster of Paris molds are used when it is desired to produce castings with smoother surfaces. Stainless steel, for example, can be successfully cast in plaster of Paris molds.

The *investment casting process* (also called the "lost wax process") produces castings of great dimensional accuracy and detail, and cast sections .015" in thickness may be produced. In this process, a wax pattern is melted after the mold has been formed. Thus, none of the details of the mold are injured by the removal of the pattern, and for that reason shapes may be molded that would be impossible if a pattern had to be drawn out of the mold.

A casting process of importance is *centrifugal casting.* Molten metal is poured into a mold that is already rotating and that may continue to rotate until the metal has solidified. Centrifugal force produces a less porous casting than that produced in a sand mold. This process is extensively used in the manufacture of cast iron pipe, and for steel gears and discs.

*Die casting* is the process of forcing molten metal into metallic dies by a force greater than atmospheric pressure. These castings are accurate in size and shape and possess surfaces superior in appearance and accuracy to those produced by other casting processes. Die castings, as a rule, require little or no machining. Although die castings generally cannot be heat-treated or put into service at high temperatures, this process is the fastest of the casting processes, and consequently the least expensive for mass-produced items, such as aluminum automobile engine blocks, carburetors, door handles, and home appliances.

## 10.9 Powder Metallurgy.

Powder metallurgy is an old process (dating back to 300 B.C.) but is today an important technique of producing metallic parts from powders of a single metal, of two or more metals, or of a combination of metals and nonmetals. It consists of successive steps of mixing powders, compressing them at high pressures (30 or 40 tons per square inch) into a preliminary shape, and heating them at high temperatures, but below the melting point of the principal metal. This process bonds the individual particles and produces a part with smooth surfaces that are quite accurate in dimension (±.001″ per inch in the direction perpendicular to the direction of the applied pressure and approximately ±.005″ per inch in the direction of pressure).

Some of the advantages of this process include the elimination of scrap, elimination of machining, suitability to mass production, better control of composition and structure of a part, and the ability to produce parts made from a mixture of metals and nonmetals. Some of the disadvantages include high cost of dies, lower physical properties, higher cost of raw materials, and the limitation on design of a part. Powders when pressed do not flow around corners, nor do they transmit pressures

Fig. 10.7 **Preparation of Mold and Pouring Molten Metal.**

(a) SAND RAMMED AROUND PATTERN

(b) PATTERN REMOVED FROM MOLD

(c) POURING OF MOLTEN METAL

(d) REMOVAL OF CASTING FROM MOLD

Fig. 10.8 **Engine Lathe.**

as a liquid; therefore, a part designed for casting will normally require redesigning in order to be made by powder metallurgy.

## 10.10 The Machine Shop and the Drawing.

The casting is sent from the foundry to the machine shop for machining. The drawing is used in the machine shop to obtain information for the machining operations. In some cases, a special drawing is made for the machine shop, and in rare instances, a drawing is produced for each machining operation. Also, in some production shops, a special set of instructions indicate routing among machines and individual operations for each particular machine. As a rule, however, working drawings used in the preceding shops are also used in the machine shop to bring the product to completion.

It is important to note that machined surfaces are usually held to much greater accuracy than are cast surfaces. Dimensions should reflect the desired accuracy for machining.

## 10.11 Machine Tools

Some of the more common machine tools are the engine lathe, drill press, milling machine, shaper, planer, grinding machine, and boring mill. Brief descriptions of these machines follow.

## 10.12 Engine Lathe.

The *engine lathe,* Fig. 10.8, is one of the most versatile machines used in the machine shop, and on it are performed such operations as turning, boring, reaming, facing, threading, and knurling. Figure 10.21 shows how a workpiece may be held in the lathe by means of a *chuck* (essentially a rotating vise). The *cutting tool* is fastened in the *tool holder* of the lathe and fed mechanically into the work as required. Several lathe operations are illustrated in Figs. 10.21, 10.22, 10.24 and 10.25.

## 10.13 Drill Press.

The *drill press,* Fig. 10.9, is one of the most used machine tools in the shop. Some of the operations that may be performed on this machine are drilling, reaming, boring, spot facing, counterboring, and countersinking. The *sensitive drill press* is used for light work and is fed by hand. The *heavy-duty drill press* is used for heavy work and is fed mechanically at the required speed. The *radial drill press,* shown in Fig. 10.9, with its adjustable head and spindle, is very versatile and is especially suitable for large work. A *multiple-spindle drill press* supports a number of spindles driven from the same shaft and is used in mass production. A table of twist drills is given in Appendix 11.

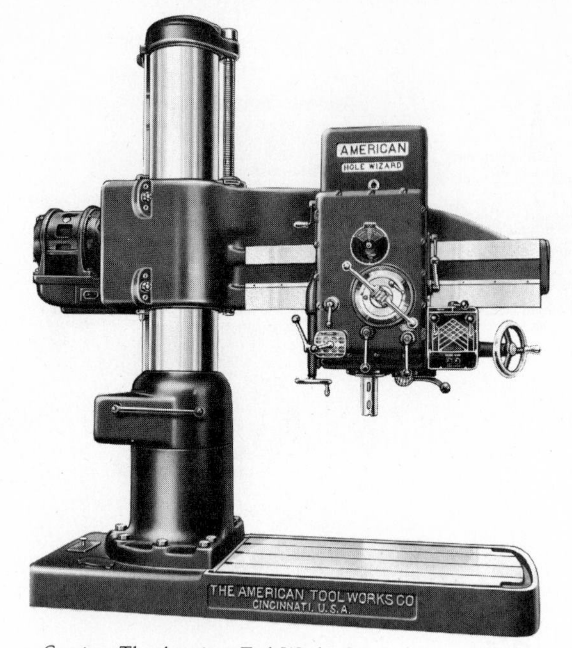


*Courtesy The American Tool Works Co.*

**Fig. 10.9 Radial Drill Press.**

## 10.14 Milling Machine.

On the *milling machine,* Fig. 10.10, cutting is accomplished by feeding the work into a rotating cutter. Milling machines are economical metal removers because of their high efficiencies. Their versatility allows their use for machining slabs, cutting contours such as keyways, or even gear manufacture, as shown in Fig. 10.11. Large-scale production of gears, however, is more often done by special machine tools, such as gear hobbers. Large plane milling machines, the basic components of which are similar to the planer shown in Fig. 10.15, are also important production machines. In such plane mills, the single point cutting tools shown are replaced with rotating milling cutters, Fig. 10.12. In addition, it is practical to drill, ream, and bore on the milling machine.

## 10.15 Shaper.

On the *shaper,* Figs. 10.13 and 10.14, work is held in a vise while a single-pointed cutting tool mounted in a reciprocating head is forced to move in a straight line past the stationary work. Between succeeding strokes of the tool, the vise that holds the work is fed mechanically into the path of the tool for the next cut alongside the one just completed.

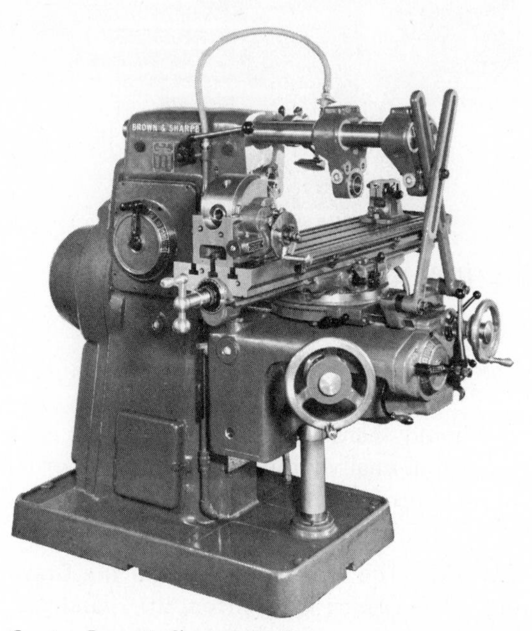


*Courtesy Brown & Sharpe Mfg. Co.*

**Fig. 10.10 Milling Machine.**

**Fig. 10.11 Cutting Teeth on Gear in Milling Machine.**

*Courtesy Brown & Sharpe Mfg. Co.*

Fig. 10.12 **Typical Milling Cutters.**

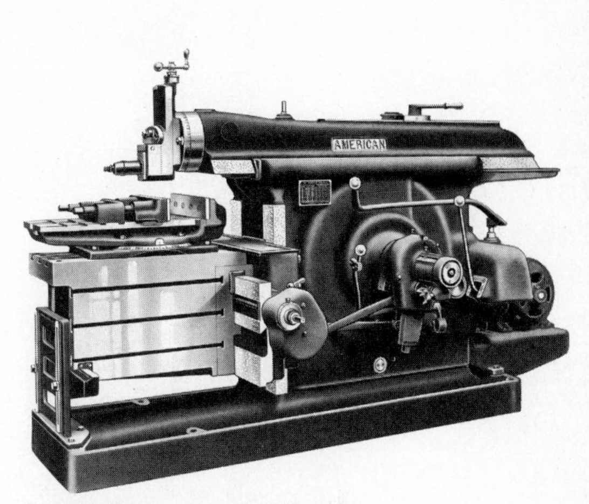


*Courtesy The American Tool Works Co.*

Fig. 10.13 **Shaper.**

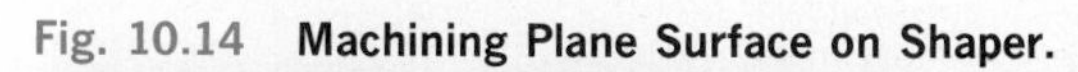

Fig. 10.14 **Machining Plane Surface on Shaper.**

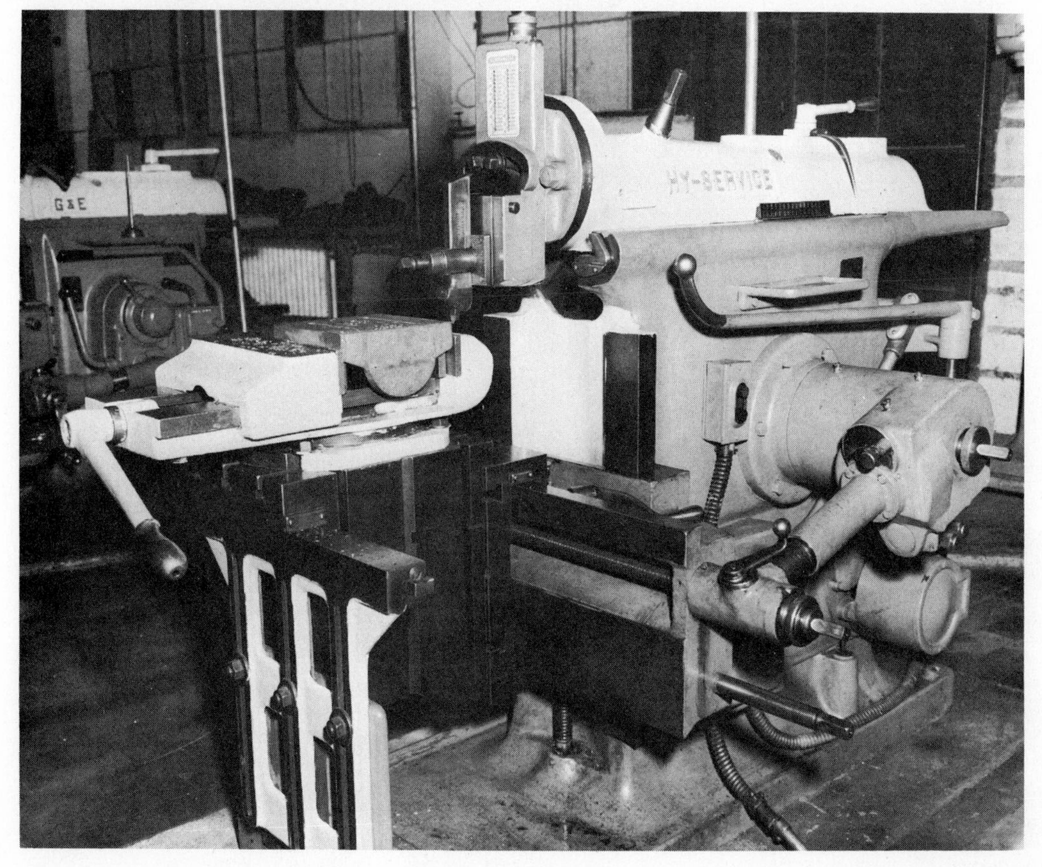

Fig. 10.15 **Planer.**

*Courtesy Cincinnati Planer Co.*

Fig. 10.16 **Grinding Machine.** *Courtesy Brown & Sharpe Mfg. Co.*

Fig. 10.17 **Vertical Boring Mill.**

*Courtesy Cincinnati Planer Co.*

**10.16 Planer.** On the *planer,* Fig. 10.15, work is fixed to a table that is moved mechanically so that the cutting takes place between stationary tools and moving work. This machine, with its reciprocating bed, and a tool head that is adjustable both horizontally and vertically, is used principally for machining large plane surfaces, or surfaces on a large number of pieces, as shown in the figure. In the extremely large planers, the table is stationary and the tool actually is carried along a track past the work in order to cut.

**10.17 Grinding Machine.** A *grinding machine,* Fig. 10.16, is used for removing a relatively small amount of material to bring the work to a very fine and accurate finish. In grinding, the work is fed mechanically against the grinding wheel, and the depth of cut may be varied from .001″ to .00025″. Two principal types of grinding machines exist. Surface grinders reciprocate the work on a table that is simultaneously fed transverse to the grinding wheel. Cylindrical grinders may require mounting of the workpiece on centers, as shown in Fig. 10.16, or they may be of the centerless type. Also, specially formed grinding wheels are often used to cut gear teeth, threads, and other shapes.

**10.18 Boring Mill.** The *vertical boring mill,* Fig. 10.17, is used for facing, turning, and boring heavy work. Castings weighing up to twenty tons are commonly handled on this machine. The vertical boring mill has a large rotating table and a nonrotating cutting toll that moves mechanically into the work.

The *horizontal boring machine* and the *jig borer* are similar to the milling machine in that the cutting action is between a rotating tool and nonrotating work. The horizontal boring machine is suitable for accurate boring, reaming, facing, counterboring, and milling of pieces larger than could be handled on the milling machine. The jig borer is a precision machine that somewhat resembles a drill press in its basic features of a rotating vertical spindle supporting a cutting tool and a stationary horizontal table for holding the work. The *precision jig borer,* however, is equipped with a table that can be locked in position while a hole is being cut but may be moved between cutting operations so as to locate one hole with respect to another. The work is not moved with respect to the table; instead, the table may be accurately positioned in two perpendicular directions by means of two accurate lead screws or by micrometer measuring bars.

Since conditions for boring vary, it is difficult to give a figure for accuracy of a bored hole. On S.I.P. (Swiss) jig-boring machines, holes are said to be bored within .00008″ of true size and are said to be located to within .0001″ accuracy. Diamond-boring machines* are said to produce holes to the following limits: .00001″ for out-of-round, .0001″ for straightness, and .00001″ for size. Thus, the boring operation at its best is capable of very accurate results. However, the tables in Appendixes 5 to 9 are more representative of current practice.

**10.19 Broaching.** *Broaching* is illustrated in Fig. 10.18. A broach is shown in an arbor press cutting a keyway in a small gear. Broaching is similar to a single-stroke filing operation. As the broach is forced through the work, each succeeding tooth bites deeper and deeper into the metal, thus enlarging and forming the keyway as the broach passes through the hole. Another typical broach, together with the corresponding drawing calling for the use of this broach, is shown in Fig. 10.19.

Figures 10.18 and 10.19 illustrate *internal broaching.* Square, cylindrical, hexagonal, and other shaped holes are produced by this process. An initial hole, produced by drilling, punching, or otherwise, is necessary in order to permit the broach to enter the work. Thus, internal broaching only enlarges and improves the shape of a hole that has already been produced by some other operation.

In addition to internal broaching, *surface broaching* is used to produce flat surfaces on such parts as engine blocks.

* Diamond is the material used for the cutting tool.

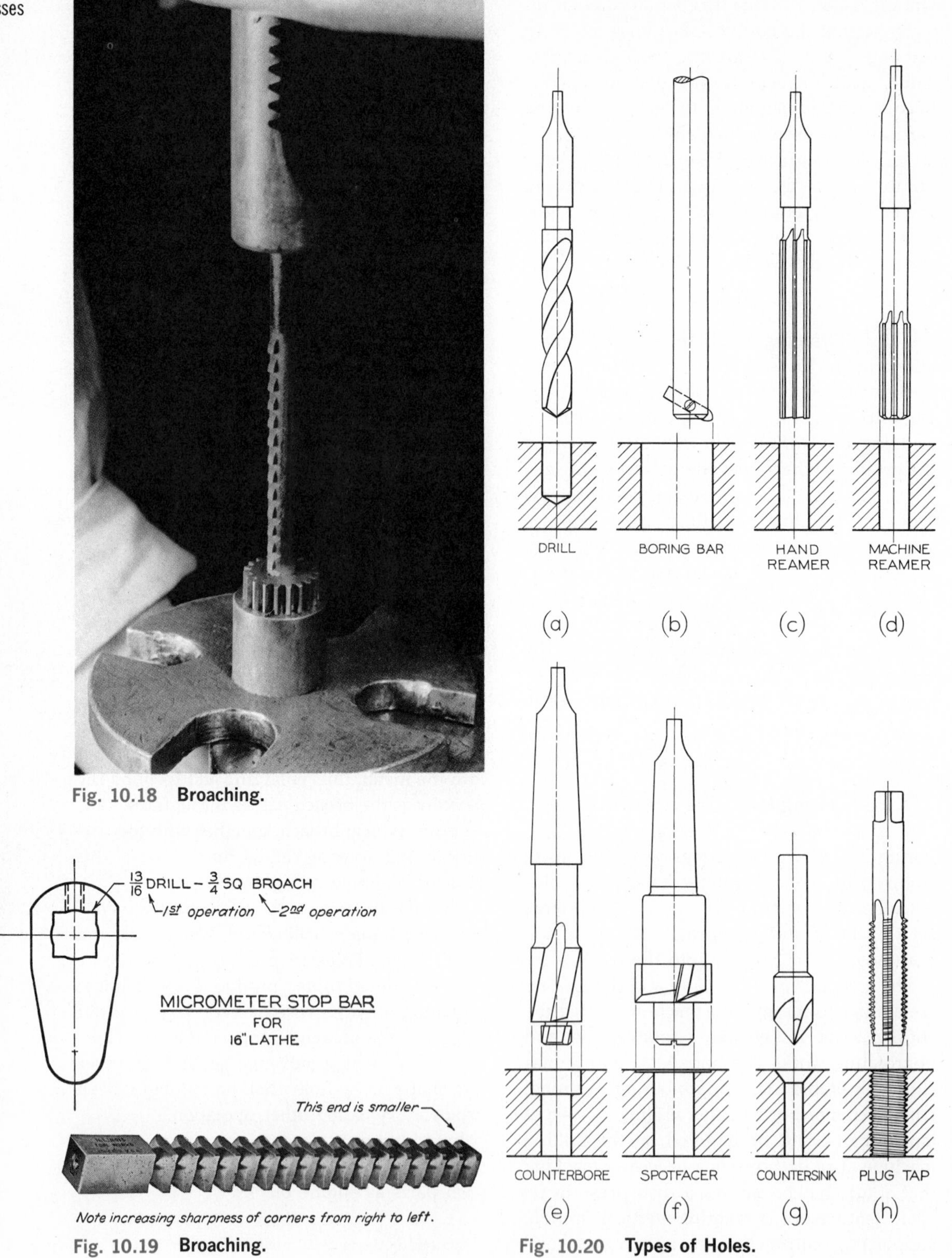


Fig. 10.18 **Broaching.**

*Note increasing sharpness of corners from right to left.*

Fig. 10.19 **Broaching.**

Fig. 10.20 **Types of Holes.**

Broaching is distinctly a mass-production process in which a large expenditure of money is usually involved both for the broaches themselves and for the special machines developed for their use in mass production. Work produced by broaching is of a high order of precision. It is common to produce holes that are within .0005″ of the desired diameter with this process.

## 10.20 Holes.

Holes are produced in metal by coring, piercing (punching), flame cutting, or drilling. The first three of these methods are very rough. Drilling, Figs. 10.20 (a) and 10.21, while producing a hole superior in finish to coring, piercing, or flame cutting, does not produce a hole of extreme accuracy in roundness, straightness, or size. Drills frequently cut holes slightly larger than their norminal size. A twist drill is somewhat flexible, which makes it tend to follow a path of least resistance. For work that demands greater accuracy, drilling is followed by boring, Fig. 10.20 (b), or by reaming, (c) or (d), and Fig. 10.22. When a drilled hole is to be finished by boring, it is drilled undersize and the boring tool, which is supported by a relatively rigid bar, generates a hole that is round and straight. Boring is also used in finishing a cored hole. Also, reaming is used for enlarging ($\frac{1}{64}$″ approx.) and improving the surface quality of a drilled or bored hole. Good practice is to drill, bore, and then ream to produce an accurate and finely finished hole. Standard reamers are available in $\frac{1}{64}$″ increments of diameter.

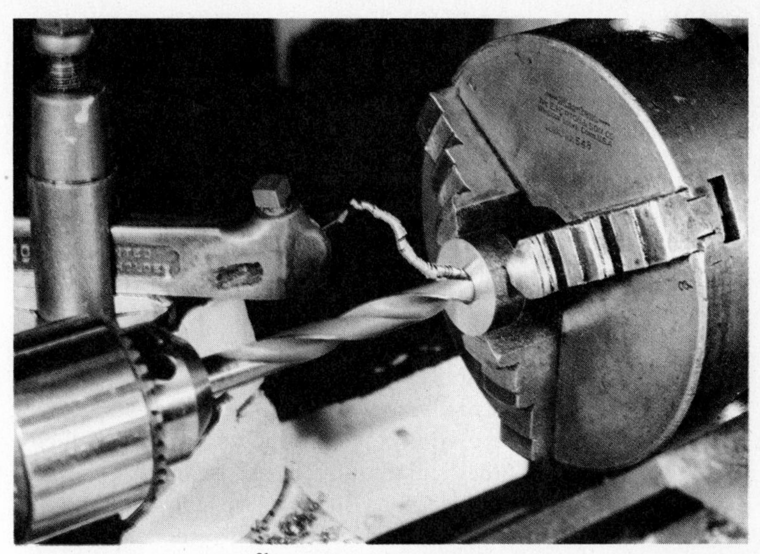

Fig. 10.21 **Drilling $\frac{31}{64}$″ Hole in Gear Blank.**

Fig. 10.22 **Reaming to Finish $\frac{1}{2}$″ Hole.**

*Counterboring,* Fig. 10.20 (e), is the cutting of an enlarged cylindrical portion of a previously produced hole, usually to receive the head of a fillester-head or socket-head screw, Figs. 13.34 and 13.35.

*Spotfacing* is similar to counterboring, but quite shallow, usually about $\frac{1}{16}$″ deep or deep enough to clean a rough surface, Fig. 10.20 (f), or to finish the top of a boss to form a bearing surface. Although the depth of a spotface is commonly drawn $\frac{1}{16}$″ deep, the actual depth is usually left to the shop. It is also good practice to include a note, "spotface to clean." A spotface provides an accurate bearing surface for the underside of a bolt or screw head.

*Countersinking,* Fig. 10.20 (g), is the process of cutting a conical taper at one end of a hole, usually to receive the head of a flat-head screw, Figs. 13.33 and 13.34, or to provide a seat for a lathe center, as shown on part No. 9 of Fig. 14.90. Actually, this operation is an example of drilling and countersinking, which is accomplished in one operation with a tool called a *combined drill and countersink,* made especially for the purpose.

*Tapping* is the threading of small holes by the use of one or more taps, Fig. 10.20 (h). Before tapping may be accomplished, a hole must be drilled, as shown in Fig. 13.26.

The location of holes is a subject that is as important as the production of the holes themselves. Details are beyond the scope of this chapter, but it should be mentioned that the layout of the part to be machined is usually a necessary operation before cutting begins.

For the production of holes, the layout is customarily accomplished by first treating the surface with copper sulphate. Prussian blue, or other suitable substances so that scratches will be clearly visible. Sharp-pointed instruments (scribers) are used to scratch center lines on the work. The surface plate or layout table is often used for such layouts. Frequently the work is clamped to a toolmaker's angle and the center lines scratched on with a vernier height gage.

Of the several methods commonly used to locate work relative to the cutting tool, the use of toolmaker's buttons is probably the most common. Buttons are small hardened-steel rings that may be temporarily attached to the work so that the centers of the rings are precisely where the centers of the holes are to be. The work is then set up with the center of one of the buttons in the center of rotation of, say, the faceplate of an engine lathe. The button is then unscrewed and a hole drilled, bored, and reamed in proper location. By repeating this process, the required holes may be produced in their proper locations.

**Fig. 10.23 Measuring Devices Used in the Shop.**

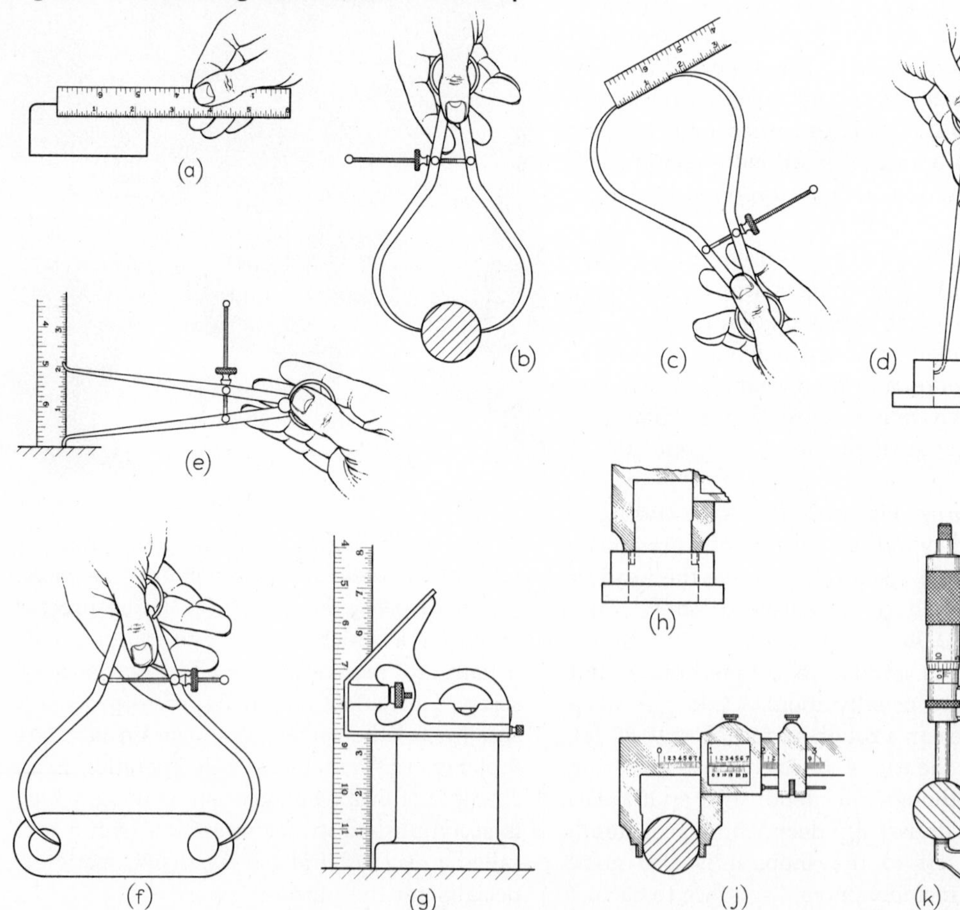

## 10.21 Measuring Devices in the Shop.

Inasmuch as the machinist uses various measuring devices depending upon the kind of dimensions shown on the drawing, it is evident that to dimension correctly, the engineering draftsman must have at least a working knowledge of the common measuring tools. The *machinist's steel rule,* or *scale,* is the most commonly used measuring tool in the shop, Fig. 10.23 (a). The smallest division on one scale of this rule is $\frac{1}{64}''$, and such a scale is used for common fractional dimensions. Also, many machinist's rules have a decimal scale with the smallest division of .01″, which is used for dimensions given on the drawing by the decimal system, §11.10. For checking the nominal size of outside diameters, the *outside spring caliper* and steel scale are used as shown at (b) and (c). Likewise, the *inside spring caliper* is used for checking nominal dimensions, as shown at (d) and (e). Another use for the outside caliper, (f), is to check the nominal distance between holes (center-to-center). The *combination square* may be used for checking height, as shown at (g), and for a variety of other measurements.

For dimensions that require more precise measurements, the *vernier caliper,* (h) and (j), or the *micrometer caliper,* (k), may be used. It is common practice to check measurements to .001″ with these instruments and in some instances they are used to measure directly to .0001″.

Most measuring devices in the shop are adjustable so they can each be employed to measure any size within their range of designed usage. There is also a need for measuring devices designed to be used for only one particular dimension. These are called *fixed gages* because their setting is fixed and cannot be changed in the shop.

A common type of fixed gage consists of two carefully finished rounds. One might think of each of these rounds as being 1″ in diameter and $1\frac{1}{2}''$ long. Let one of these diameters be slightly larger than the other. One can see that, for a certain range of hole sizes, the smaller round will enter the hole but the larger will not. If the larger round diameter is made slightly greater than the largest acceptable hole diameter and if the diameter of the smaller round is made slightly less than the smallest acceptable hole diameter, then the large round *will never go* into any acceptable hole but the small round *will go* into any acceptable hole. A fixed gage consisting of two such rounds is called a "go–no go" gage. There are, of course, many other kinds of "go–no go" gages.

The subject of gages and gaging is a specialized field and involves so many technical considerations that large companies employ highly trained men to attend to nothing but this one feature of their operations.

## 10.22 Machine Operations.

To demonstrate how the machinist uses drawings, the operations called for on the working drawing in Fig. 10.3 will be shown in the following steps. This drawing was also used by the pattern shop and the foundry, Figs. 10.6 and 10.7.

*Step I.* After the casting has been properly cleaned, it is then chucked in the lathe, as shown in Fig. 10.24 (a). The drawing specifies that the casting is to be finished on all surfaces; therefore, the machinist must produce a smooth finish on the flat surface at the end of the cylinder. This is called *facing.* After this surface has been machined, it serves as a control surface in locating parallel surfaces.

*Step II.* The next machine operation on this casting is that of boring the central hole, as shown at (b). The machinist reads 2.125–2.126 from the note 2.125–2.126 BORE $2\frac{1}{4}$–8 UN–2B on the drawing to determine the required size of the bored hole. This note indicates that he is allowed a tolerance of .001″, and the finished hole diameter must be between 2.125″ and 2.126″. Although a preliminary check of this inside diameter may be made, as shown at (c), using an inside spring caliper and an outside micrometer caliper, the desired tolerance would suggest that the hole diameter be measured directly with an inside micrometer caliper or checked with a "go–no go" gage.

*Step III.* After boring is completed, the internal thread is cut with a 60° thread cutting

(a) FACING

(b) BORING

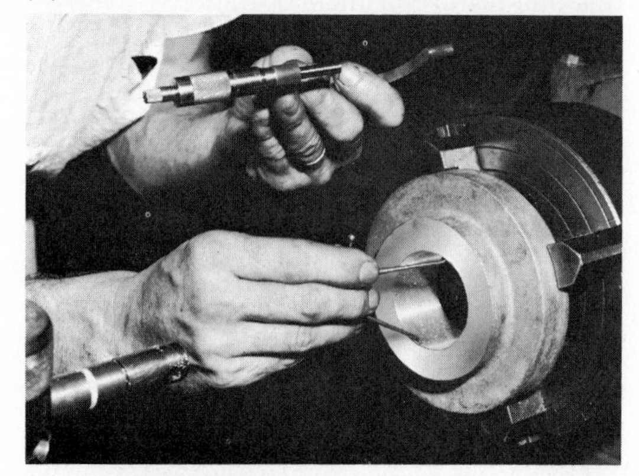

(c) CHECKING BORE

(d) THREAD CUTTING

(e) CHECKING WITH THREAD GAGE

(f) CHECKING RELIEF DEPTH

Fig. 10.24 **Machining and Checking.**

tool, as shown at (d). The $2\frac{1}{4}$ in the thread note $2\frac{1}{4}$–8 UN–2B, Fig. 10.3, indicates that the major diameter of the thread must be $2\frac{1}{4}''$; 8 UN indicates 8 Unified threads per inch; and 2B indicates a Class 2 internal thread. The finished thread is checked with a thread plug gage, as shown at (e).

*Step IV.* The threads are now relieved by boring $2\frac{9}{32}''$ diameter by $\frac{3}{8}''$ deep, and checked as shown at (f) and Fig. 10.25 (a). These dimensions are checked with a steel scale, since the required tolerance is only $\pm\frac{1}{64}''$ on all fractional dimensions of this drawing, as shown in Fig. 10.3.

(a) CHECKING RELIEF DIAMETER

(b) CHECKING CHAMFER DEPTH

(c) CHECKING HUB LENGTH

(d) CHECKING HUB DIAMETER

(e) CHECKING ROUND

(f) CHECKING FLANGE DIAMETER

Fig. 10.25 **Checking.**

*Step V.* The next step is to cut the $\frac{1}{16} \times 45°$ chamfer. The cutting tool is set at a 45° angle, and the depth of cut is checked with a depth gage set at $\frac{1}{16}''$, as shown in Fig. 10.25 (b).

*Step VI.* The next associated group of machine operations includes facing the Flange, turning the hub diameter, and turning the $\frac{1}{8}''$ radius between the hub and flange. The $\frac{5}{8}''$ length of the hub is checked with the depth gage, as shown at (c). The $3\frac{5}{32}''$ hub diameter is checked at (d), with a spring caliper and steel scale, because this dimension is not critical. The $\frac{1}{16}''$ radius on the hub is checked with a radius gage, as shown at (e).

After this group of operations is completed, the Flange is removed from the chuck of the lathe and screwed onto the lathe spindle.

*Step VII.* As shown at (f), the outside diameter of the Flange is turned to size and checked with a micrometer caliper. The outside diameter dimension $^{5.000}_{4.998}$ is critical, as the Flange must fit in a mating part of the 4-Jaw Independent Chuck.

*Step VIII.* The back surface is now faced, Fig. 10.26 (a), to provide a smooth flat surface and to bring the over-all length of the Flange to $1\frac{1}{4}''$. Again, since this dimension is not critical, it is checked with the spring caliper and

**Fig. 10.26 Machining and Checking.**

(a) FACING BACK SURFACE

(b) CHECKING RELIEF DIAMETER

(c) LAYING OUT SMALL HOLES

(d) DRILLING SMALL HOLES

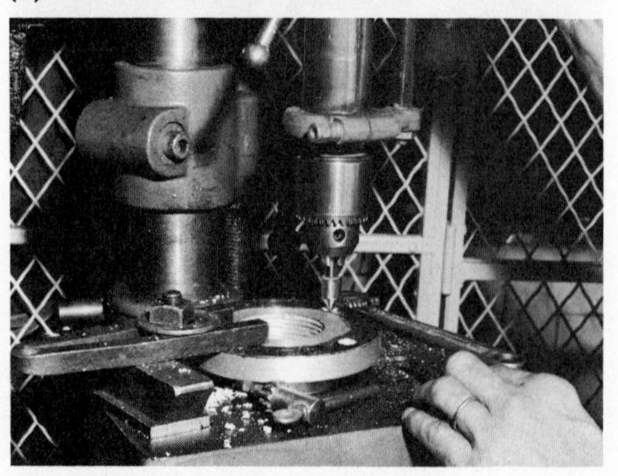

(e) COUNTERSINKING SMALL HOLES

(f) TAPPING SMALL HOLES

**Fig. 10.27 Numerically Controlled Milling Machine.** *Courtesy Metalworking News, Fairchild Publications, Inc.*

steel scale. The facing tool is used to break the back surface corner $\frac{1}{64}$ R and to round the forward corner $\frac{1}{16}$ R as dimensioned in Fig. 10.3.

*Step IX.* After the back surface is faced, it is then relieved, Fig. 10.26 (b), to a diameter of $3\frac{7}{16}''$ and to a depth of $\frac{1}{64}''$, as dimensioned in Fig. 10.3. By relieving the back surface of the Flange in this manner, a better seat is insured between the Flange and its mating part in the 4-Jaw Independent Chuck.

*Step X.* The next machining operation is to locate and drill the four small holes, as shown in Fig. 10.26 (c) and (d). After locating the centers 90° apart on a 4" diameter circle, the machinist selects a "Q" diameter (0.3320) drill, and drills the four holes.

*Step XI.* The next step is to countersink, Fig. 10.26 (e), the four small holes with an 80° countersink to a diameter equal to the major diameter of the threads. The countersinking facilitates the tapping of the holes.

*Step XII.* The last step in machining the Flange is to tap the four small holes as shown at (f). The $\frac{3}{8}$ in the drawing note $\frac{3}{8}$–24 UNF–2B indicates that the major diameter of the tapped holes must be $\frac{3}{8}''$; 24 UNF indicates 24 Unified Fine threads per inch of thread length; and 2B indicates a Class 2 fit internal thread.

## 10.23 Automation.

*Automation* is the term applied to systems of automatic machines and processes. These machines and processes are essentially the basic machines and processes with mechanisms to control the sequence of operation, movement of tool, flow, and so forth, with little attention from an operator once the equipment has been "set up." Among such automatic devices is a numerically controlled machine tool with a programmed cutting sequence, which provides the engineer with a method of producing accurately machined parts with complex geometrical surfaces, Fig. 10.27. Actually, all machines previously described in this chapter are available as numerically controlled machines, also. Other automatic or semiautomatic devices include various machine tools that follow directions punched on a tape, computers that make thousands of intricate mathematical calculations in a matter of seconds or a fraction of a second, and gages measuring electrically in millionths of an inch.

Automation does not mean that there will be less thinking for the engineer, but more. Therefore, the engineering student is going to need even more intensive grounding in basic fundamentals than previously. In addition, the enterprising student will seize opportunities for keeping informed on current industrial developments.

## 10.24 Chipless Machining.

Several manufacturing processes do not employ the cutting action described in most of the previous sections.

Chemical milling removes material through chemical reactions at the surface of the work piece. An etchant, acidic or basic, is agitated over an immersed workpiece that is masked where no metal removal is desired. For a material with high homogeneity, the tolerances obtainable in such an operation are $\pm.003''$.

Electrodischarge machining utilizes high energy electric sparks that build up high charge densities on the surface of the work material. Thermal stresses, exceeding the strength of the work material, cause minute particles to break away from the work. Though this process is slow, it allows intricate shapes to be cut in hard materials, such as tungsten carbides.

## 10.25 Stock Forms.

Many standardized structural shapes are available in stock sizes for the fabrication of parts or structures. Among these are bars of various shapes, flat stock, rolled structural shapes, and extrusions, Fig. 10.28. The manufacturing processes of rolling, drawing, and extruding often add a great deal of toughness and strength to a metal, and these stock forms so processed are very useful in the manufacture of small machined parts, such as screws, and for the fabrication of welded and riveted structures.

## 10.26 Welding.

*Welding* is a process of joining metals by fusion into a single homogeneous mass. Arc welding, gas welding, resistance welding, and atomic hydrogen welding are commonly used, as well as the old method of forge welding, which is still used to some extent.

Welded structures are built up in most cases from stock forms, particularly plate, tubing, and angles. Often both heat-treating and machine-shop operations must be performed on welded machine parts. Since welding frequently distorts a part or structure enough to alter permanently the dimensions to which it has been cut in the shop, it is usual practice not to weld accurate work after finish machining has been done unless the volume of the welded material is very small in comparison to the volume of the part.

## 10.27 Jigs and Fixtures.

A general-purpose machine tool may have its effectiveness on a specific job increased by means of jigs or fixtures. A *jig* is a device that holds the work and guides the tool; it is usually not rigidly fixed to a machine. A *drilling jig*, Fig. 10.29, is a common device by means of which holes on many duplicate parts may be drilled

Fig. 10.28 **Common Stock Forms.**

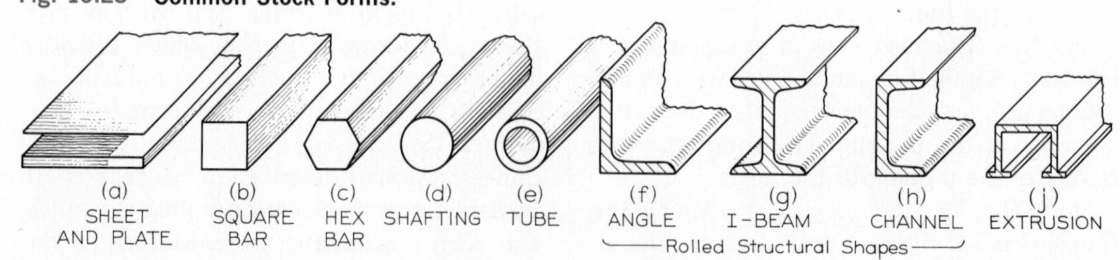

Fig. 10.29 **Use of Drilling Jig.**

exactly alike. A *fixture* is rigidly attached to the machine, becoming in reality an extension of it, and holds the work in position for the cutting tools without acting as a guide for them. Drawings of one of the important fixtures used in the production of a connecting rod, Figs. 14.35 to 14.37, are shown in Figs. 14.120 to 14.122. This fixture was built at considerable expense for the single purpose of holding the connecting rod in the exact position required for the efficient and speedy execution of a single operation.

Jigs and fixtures are usually designed in a tooling department by tool designers. Usually they are built by machinists of much better than average skill, using especially accurate equipment. Such tooling devices are commonly held to tolerances one-tenth of those applied to the parts to be produced on these jigs and fixtures.

Jigs and fixtures may be grouped into two general classes: manufacturing tooling and assembly tooling. *Manufacturing tooling* consists of devices used in producing individual parts. An example of this is a fixture for holding a connecting rod in a milling machine when the ends are being faced by straddle milling.

*Assembly tooling* consists of devices to hold work and guide tools as parts are being assembled. For example, the center section of the wing of a large airplane was assembled in a special jig in which parts were held in place, drilled when together, and riveted. The assembly jig was built so that the component parts could fit together only when correctly located, and so that no measurement was ever made and no blueprint ever referred to by workers assembling the wing.

Large quantities of precision products may be produced and assembled by less skilled labor than would otherwise be required if a relatively small group of highly skilled workers first produce the required manufacturing tooling and assembly tooling.

## 10.28 Hot and Cold Working of Metals.

Let us now review some of the ideas we have already discussed separately in several of the preceding sections. Recall that in each of the casting operations, §§10.4 and 10.8, the casting assumes its shape by filling a mold as a liquid and does not appreciably change its shape as it solidifies. Notice that in many manufacturing processes, however, the shape of the part is changed after the metal has solidified. If these processes involve slowly pressing the part (squeezing) or rapidly and repeatedly striking the part (hammering), then the term *mechanical working* is used to describe the operation.

The practical effects of mechanical working depend a great deal on the temperature at which the metal is squeezed or hammered. The terms *hot working* and *cold working* are used widely in engineering today, and the student will find in his metallurgy courses and machine design courses that he will devote a great deal of effort to the study of the hot working and cold working of metals.

## 10.29 Forging.

*Forging* is the process of shaping metal to a desired form by means of pressing or hammering. Generally, forging is hot forging, in which the metal is heated to a required temperature before forging. Some softer metals can be forged without heating, and this is cold forging.

When metal is worked under pressure, as in forging, the material is compressed and

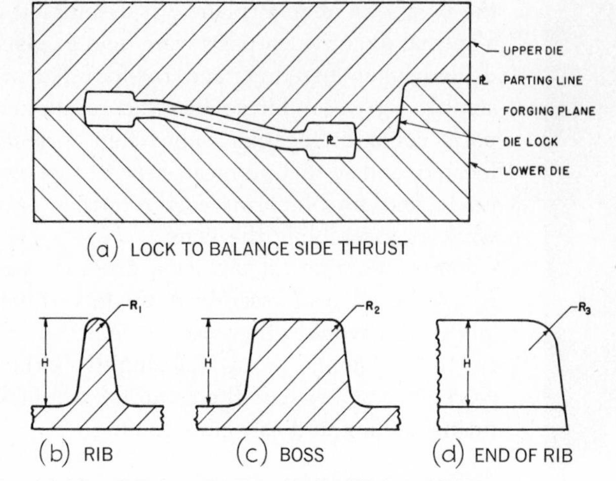


(a) LOCK TO BALANCE SIDE THRUST

(b) RIB (c) BOSS (d) END OF RIB

| H | $R_1$ | $R_2$ | $R_3$ |
|---|---|---|---|
| 1/4 | 1/16 | 1/16 | 3/16 |
| 1/2 | 1/16 | 1/16 | 3/16 |
| 1 | 1/8 | 1/8 | 3/8 |
| 2 | 3/16 | 1/4 | 1/2 |
| 3 | 1/4 | 5/16 | 3/4 |
| 4 | 5/16 | 7/16 | 1 |
| 5 | 3/8 | 1/2 | 1 1/8 |
| 6 | 7/16 | 5/8 | 1 1/4 |
| 7 | 1/2 | 11/16 | 1 1/2 |

(e) CORNER RADII

**Fig. 10.30 Forging Design** (ANSI Y14.9–1958).

greatly strengthened. Complicated shapes are extremely expensive because of costly dies. Forgings are made of steel, copper, brass, bronze, aluminum, magnesium, and more recently, the new light-weight titanium.

Closed-die forgings, or drop forgings, are formed between dies that are machined to the desired shapes. Open-die forgings, or hand forgings, are formed between flat dies and manipulated by hand during forging to obtain the required shape.

Usually the first step in the design of a forging is to locate the *parting plane,* which separates the upper and lower dies. As a rule, this surface follows the center line of the piece, as shown in Fig. 10.30 (a), and should pass through the largest portion of the object.

*Draft,* or taper, must be provided on all forgings. Although draft is not usually shown on drawings for sand casting, it is always shown on forging drawings. A typical forging drawing is shown in Fig. 14.35. For outside surfaces, the standard draft angle is 7° for all materials. For inside surfaces in aluminum and magnesium, the draft is also 7°, but in steel the angle is usually 10°.

Small fillets and rounds in forgings reduce die life and decrease the quality of the forging. Minimum rounds or corner radii are shown in Fig. 10.30 (b) to (e), and minimum fillet radii in Fig. 10.31. Whenever possible, larger radii than those shown should be used.

The minimum thickness of a web depends on the area at the parting line. Minimum web-thickness values are shown in the following table.

| Area, in.$^2$ | Thickness |
|---|---|
| 4 | 3/32 |
| 20 | 1/8 |
| 80 | 3/16 |
| 200 | 1/4 |
| 400 | 5/16 |
| 600 | 3/8 |
| 800 | 7/16 |
| 1000 | 1/2 |

Ribs will increase in thickness as the height increases. Minimum values conform to radius $R_1$ in Fig. 10.30 (b).

The working drawing, as it comes to the shop, may show the completed or machined part, in which case the necessary forging and machining allowances are made by the diemaker. However, since dies are so expensive and the forging itself is so important, it is common practice to make separate forging and machining drawings, as shown in Figs. 14.35 and 14.36. It is standard practice to show visible lines for rounded corners as if the corners were sharp. See §6.36. Note that the parting plane is represented by a center line.

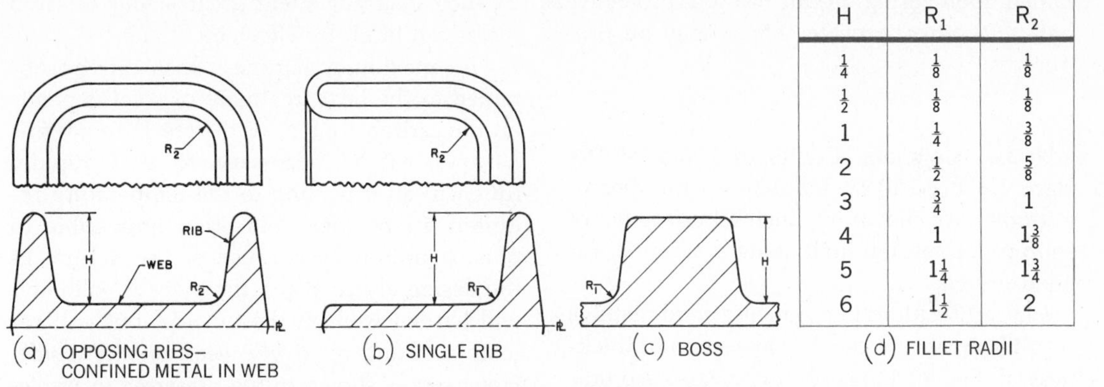


| H | $R_1$ | $R_2$ |
| --- | --- | --- |
| $\frac{1}{4}$ | $\frac{1}{8}$ | $\frac{1}{8}$ |
| $\frac{1}{2}$ | $\frac{1}{8}$ | $\frac{1}{8}$ |
| 1 | $\frac{1}{4}$ | $\frac{3}{8}$ |
| 2 | $\frac{1}{2}$ | $\frac{5}{8}$ |
| 3 | $\frac{3}{4}$ | 1 |
| 4 | 1 | $1\frac{3}{8}$ |
| 5 | $1\frac{1}{4}$ | $1\frac{3}{4}$ |
| 6 | $1\frac{1}{2}$ | 2 |

(a) OPPOSING RIBS–CONFINED METAL IN WEB (b) SINGLE RIB (c) BOSS (d) FILLET RADII

Fig. 10.31 **Minimum Radii for Forgings** (ANSI Y14.9–1958).

## 10.30 Heat-Treating.

We have seen above that heat, along with heavy forces, often plays an important role in the making of a part. Sometimes the heat itself, without any mechanical working, is enough to improve a piece significantly. When heat alone is so employed and when the object is to change the properties of the metal rather than merely to change its shape, we use the term *heat-treating.*

Let us mention briefly some of the heat-treating terms the draftsman will use on his drawings. *Annealing* and *normalizing* soften the metal and release internal stresses developed in the manufacturing processes. They involve heating to the critical temperature range and then slowly cooling. *Hardening* requires heating to above the critical temperature followed by rapid cooling—quenching in oil, water, or brine, or in some instances in air (air-hardened steels). *Tempering* reduces internal stresses caused by hardening. Tempering also improves the toughness and ductility. *Surface hardening* is a way of hardening the surface of a steel part while leaving the inside of the piece soft. Surface hardening is accomplished by *carburizing,* followed by heat-treatment; by *cyaniding,* followed by heat-treatment; by *nitriding;* by *induction hardening;* and by *flame hardening.*

## 10.31 Plastics Processing.

Specialized manufacturing processes readily and inexpensively convert plastics from the raw material into the finished product, often in a single operation. Therefore, in recent decades plastics have enjoyed an increasing utilization in home appliances, aircraft, automobiles, and recreational vehicles.

Plastics are conveniently divided into two major types, thermosetting and thermoplastic. The former type retains its hardness regardless of temperature, while the latter will have a hardness inversely proportional to temperature.

Thermoplastic material may be molded into parts through injection molding. The material is preheated to its melting temperature range and subsequently injected into a cool die cavity where solidification takes place. Ejection follows rapidly and high production rates are thus insured. Because of high die costs, the process is only economically suited for mass production. Since thermoplastics soften with increased temperature, the extrusion process offers a capability for producing long, constant-cross-section parts, such as beams and cable.

Hot-compression molding may be used for most plastics. A powder loaded into a hot die cavity is followed by a plunger under high pressure. The pressure is maintained during the curing cycle, after which the plunger is removed and the part ejected.

Laminated plastics, consisting of paper or rough-weave cloth impregnated with resin, may be cured in a mold; a vacuum existing between the mold and the laminated plate will insure that the shape of the mold will be retained upon curing. In this fashion, protective hats and other intricate shapes may be produced.

## 10.32 Do's and Don'ts of Practical Design.

In Figs. 10.32 and 10.33 a number of examples are shown in which a knowledge of shop processes and limitations is essential for good design.

Many difficulties in producing good castings result from abrupt changes in section or thickness. In Fig. 10.32 (a) rib thicknesses are uniform so that the metal will flow easily to all parts. Fillet radii are equal to the rib thickness—a good general rule to follow. When it is necessary to join a thin member to a thicker member, the thin member should be thickened as it approaches the intersection, as shown at (b).

At (c), (g), and (h) coring is used to produce walls of more uniform sections. At (d) an abrupt change in sections is avoided by making thinner walls and leaving a collar, as shown.

At (e) and (f) are shown examples in which the preferred design will tend to allow the castings to cool without introducing internal stresses. The less desirable design is more likely to crack as it cools, since there is no "give" in the design. Curved spokes are preferable to straight spokes, and an odd number of spokes is better than an even number because direct stresses along opposite spokes are avoided.

The design of a part may cause unnecessary trouble and expense for the pattern shop and foundry without any gain in the usefulness of the design. For example, in the "poor" designs at (j) and (k), one-piece patterns would not withdraw from the sand, and two-piece patterns would be necessary. In the "preferred" examples, the design is just as useful and is conducive to economical work in the pattern shop and foundry.

As shown in Fig. 10.33 (a), a narrower piece of stock sheet metal can be used for certain designs that can be linked or overlapped. In this case, the stampings may be overlapped if dimension W is increased slightly, as shown. By such an arrangement, great savings in scrap metal can often by effected.

The maximum hardness that can be obtained in the heat treatment of steel depends on the carbon content of the steel. To get this hardness, it is necessary to cool rapidly (quench) after heating to the temperature required. In practice it is often impossible to quench uniformly because of the design. In the design above at (b), the piece is solid and will harden well on the outside, but will remain soft and relatively weak on the inside. However, as shown in the preferred example, a hollow piece can be quenched from both the outside and the inside. Thus it is possible for a hardened hollow shaft to be stronger than a hardened solid shaft of the same diameter.

As shown at (c), the addition of a rounded groove, called a *neck,* around a shaft next to a shoulder will eliminate a practical difficulty in precision grinding. It is not only more expensive to grind a sharp internal corner, but such sharp corners often lead to cracking and failure.

The design at the right at (d) eliminates a costly "reinforced" weld, which would be required by the design on the left. The strong virgin metal with a generous radius is present at the point at which the stress is likely to be most severe. It is possible to make the design on the left as strong as the design on the right, but it is more expensive and requires expert skill and special equipment.

It is difficult to drill into a slanting surface, as shown at the left at (e). The drilling will be greatly facilitated if a boss is provided, as shown at the right.

At (f), the design at the left requires accurately boring or reaming a blind hole all the way to a flat bottom, which is difficult and expensive. It is better to drill deeper than the hole is to be finished, as shown at the right, in order to provide room for tool clearance and for chips.

In the upper example at (g), the drill and counterbore cannot be used for the hole in the center piece because of the raised portion at the right end. In the approved example, the end is redesigned to provide access for the drill and counterbore.

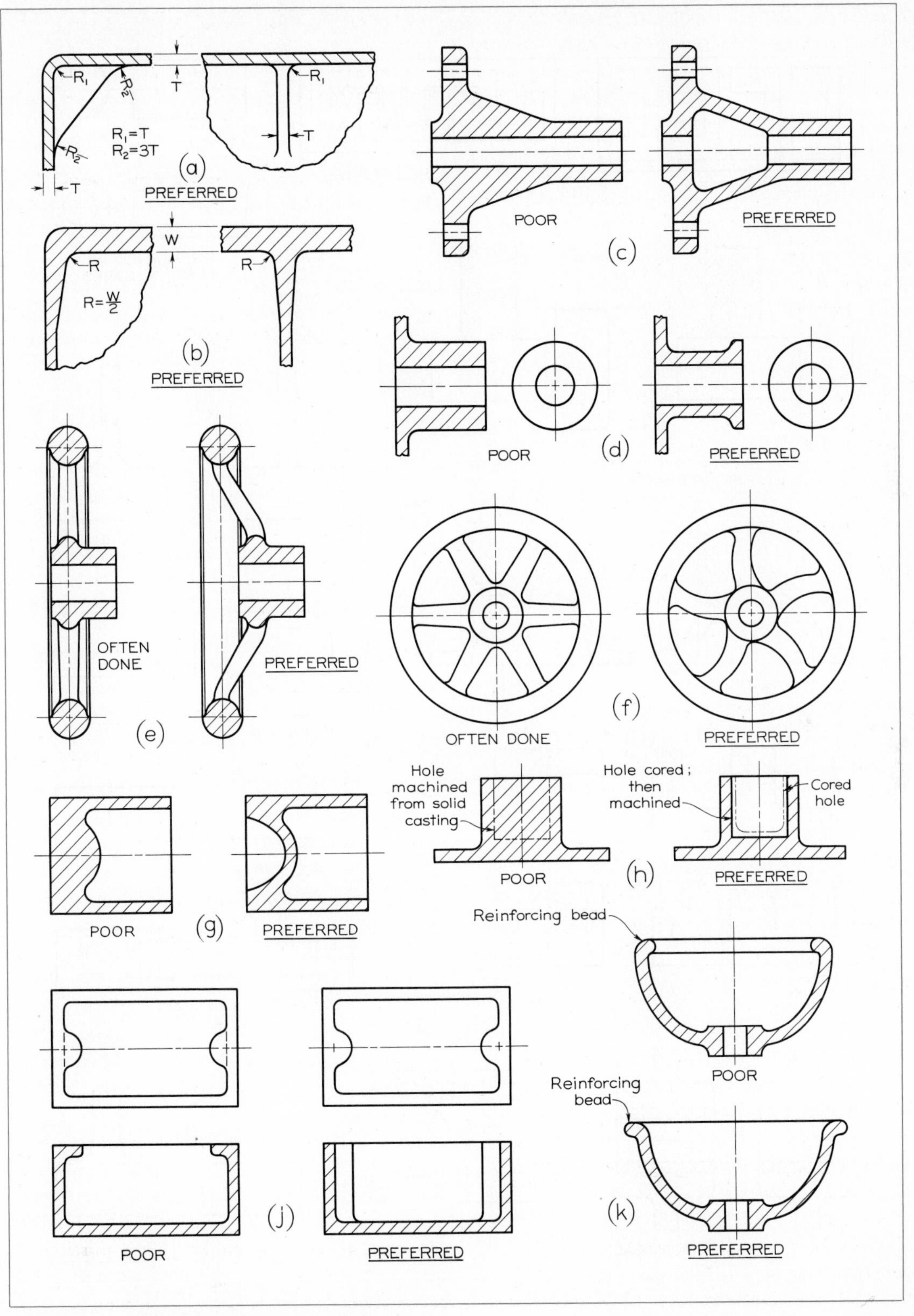


Fig. 10.32 **Casting Design Do's and Don'ts.**

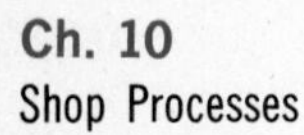


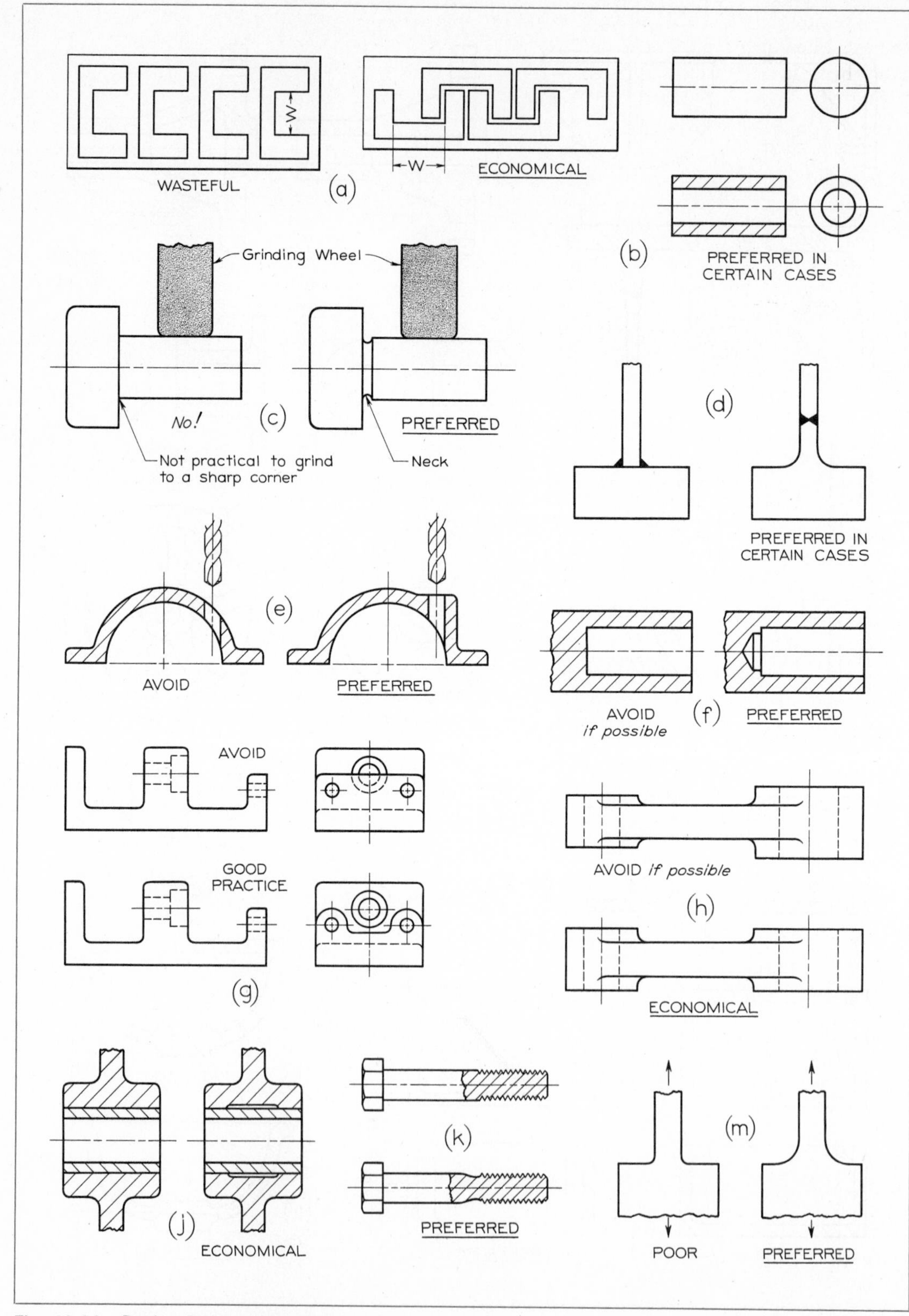


Fig. 10.33 **Design Do's and Don'ts.**

In the design above at (h), the ends are not the same height. As a result, each flat surface must be machined separately. In the design below, the ends are the same height, the surfaces are in line horizontally, and only two machining operations are necessary. It is always good design to simplify and limit the machining as much as possible.

The design at the left at (j) requires that the housing be bored for the entire length in order to receive a pressed bushing. If the cored recess is made as shown, machining time can be decreased. This assumes that average loads would be applied in use.

At (k), the lower bolt is encircled by a rounded groove no deeper than the root of the thread. This makes a gentle transition from the small diameter at the root of the threads and the large diameter of the body of the bolt, producing less stress concentration and a stronger bolt. In general, sharp internal corners should be avoided as points of stress concentration and possible failure.

At (m) a $\frac{1}{4}''$ steel plate is being pulled, as shown by the arrows. Increasing the radius of the inside corners increases the strength of the plate by distributing the load over a greater area.

# 11 dimensioning

**11.1 Historical Measurements.** We have all heard of "rule of thumb." Actually, at one time an inch was defined as the width of a thumb, and a foot was simply the length of a man's foot. In old England, an inch used to be "three barley corns, round and dry." In the time of Noah and the Ark, a *cubit* was the length of a man's forearm, or about 18″.

In 1793, when Napoleon was rising to power, France adopted the *meter* and from this evolved the decimalized metric system (1 meter = 39.37″; 1″ = 25.4 mm). In the meantime England was setting up a more accurate determination of the *yard*, which was legally defined in 1824 by act of Parliament. A foot was one third of a yard, and an inch was one thirty-sixth of a yard. From these specifications, graduated rulers, scales, and many types of measuring devices have been developed to achieve even more accuracy of measurement and inspection.

Until this century, common fractions were considered adequate for dimensions; but as designs became more complicated, and as it became necessary to have interchangeable parts in order to support mass production, more accurate specifications were required and it became necessary to turn to the decimal system. Today, decimals are widely used in industry. See §§11.9 and 11.10.

The rapid growth of worldwide science and commerce fostered an international system of units (SI), based on the meter and suitable for measurements in physical science and engineering. The six basic units of measure are the meter (length), kilogram (mass), second (time), ampere (electric current), degree Kelvin (temperature), and candela (luminous intensity).

Several industries have found it desirable to use a dual dimensioning system that includes both decimal inches and millimeters, so that their drawings may be utilized domestically as well as abroad. The metric system is now under serious consideration in this country, and eventually the United States will adopt the system since practically all other countries either have adopted it or are committed to do so. See inside the front cover of this book for a table for converting fractional inches to decimal inches or millimeters. For the International System of Units and their United States equivalents, see Appendix 26.

## 11.2 Size Description.*

In addition to a complete *shape description* of an object, as discussed in previous chapters, a drawing of the design must also give a complete *size description;* that is, it must be *dimensioned.*

In the early years of machine manufacturing, the designing and production functions were closely allied under one roof. In many cases these processes were even carried out by the same individual. Design drawings, mostly of the assembly type, were scaled by the workmen to obtain the dimensions. It was up to the shop man to make the parts correctly and to see to it that they would fit together and operate properly. If any question arose, he could always consult the designer who would be nearby. Under these conditions it was not necessary for drawings to carry detailed dimensions and notes.

The need for *interchangeability* of parts is the basis for the development of modern methods of size description. Drawings today must be dimensioned so that workmen in widely separated places can make mating parts that will fit properly when brought together for final assembly in the factory or when used as repair or replacement parts by the customer, §11.26.

The increasing need for precision manufacturing and the necessity to control sizes for interchangeability has shifted responsibility for size control from the machinist to the designing engineer and the draftsman. The workman no longer exercises judgment in engineering matters, but only in the proper execution of instructions given on the drawings. Therefore, it is necessary for engineers and draftsmen to be familiar with materials and methods of construction and with requirements of the shops. The engineering student or the draftsman should seize every opportunity to familiarize himself with the fundamental shop processes, especially *patternmaking, foundry, forging,* and *machine-shop practice,* discussed in the previous chapter.

The drawing should show the object in its completed condition, and should contain all necessary information to bring it to that final state. Therefore, in dimensioning a drawing, the designer and the draftsman should keep in mind the finished piece, the shop processes required, and above all the function of the part in assembly. Whenever possible—that is, when there is no conflict with functional dimensioning—dimensions should be given that are convenient for the workman or the production engineer. These dimensions should be given so that it will not be necessary to calculate, scale, or assume any dimensions. Do not give dimensions to points or surfaces that are not accessible to the man in the shop.

Dimensions should not be duplicated or superfluous, §11.30. Only those dimensions should be given that are needed to produce and inspect the part exactly as intended by the designer. The student often makes the mistake of giving the dimensions he used *to make the drawing.* These are not necessarily the dimensions required. There is much more to the theory of dimensioning, as we shall see.

## 11.3 Scale of Drawing.

Drawings should be made to scale, and the scale should be indicated in the title block even though the workman is never expected to scale the drawing or print for a needed dimension. See §2.31 for indications of scales.

A wavy line should be drawn under any dimension that is not to scale, Fig. 11.15, or the abbreviation NTS (Not to Scale) should be indicated. This procedure may be necessary when a change is made in the drawing that

*See ANSI Y14.5-1966.

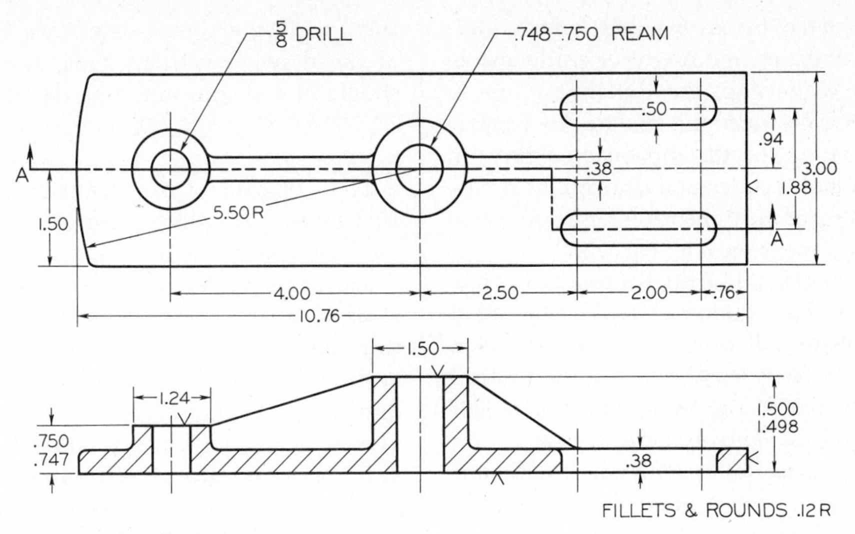


Fig. 11.1 **Dimensioning Technique.**

is not important enough to justify making an entirely new drawing.

## 11.4 Learning to Dimension.

Dimensions are given in the form of linear distances, angles, or notes. Ability to dimension properly requires the following:

1. The student must learn the *technique of dimensioning:* the character of the lines, the spacing of dimensions, the making of arrowheads, and so forth. A typical dimensioned drawing is shown in Fig. 11.1. Note the strong contrast between the visible lines of the object and the thin lines used for the dimensions.

2. The student must learn the rules of *placement of dimensions* on the drawing. These practices assure a logical and practical arrangement with maximum legibility.

3. The student should learn the *choice of dimensions.* Formerly, manufacturing processes were considered the governing factor in dimensioning. Now function is considered first and shop processes second. The proper procedure is to dimension tentatively for function, then review the dimensioning to see if any improvements from the standpoint of production can be made without adversely affecting the functional dimensioning. A "geometric breakdown," §11.20, will assist the beginner in selecting dimensions. In most cases dimensions thus determined will be functional, but this method should be accompanied by a logical analysis of the functional requirements.

## 11.5 Lines Used in Dimensioning.

A *dimension line,* Fig. 11.2 (a), is a thin, dark, solid

Fig. 11.2 **Dimensioning Technique.**

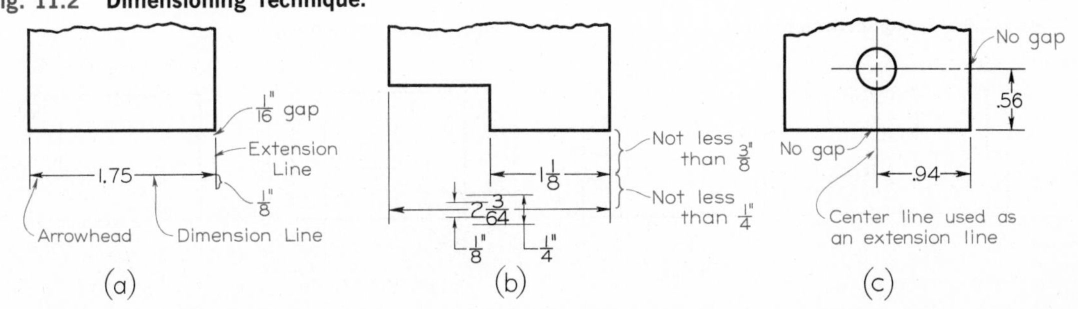

line terminated by arrowheads, which indicates the direction and extent of a dimension. In machine drawing, the dimension line is broken, usually near the middle, to provide an open space for the dimension figure. In structural and architectural drawing, it is customary to place the dimension figure above an unbroken dimension line, Fig. 22.9.

As shown in Fig. 11.2 (b), the dimension line nearest the object outline should be spaced at least $\frac{3}{8}''$ away. All other parallel dimension lines should be at least $\frac{1}{4}''$ apart, and more if space is available. *The spacing of dimension lines should be uniform throughout the drawing.*

An *extension line,* (a), is a thin, dark, solid line that "extends" from a point on the drawing to which a dimension refers. The dimension line meets the extension lines at right angles except in special cases, as in Fig. 11.6 (a). A gap of about $\frac{1}{16}''$ should be left where the extension line would join the object outline. The extension line should extend about $\frac{1}{8}''$ beyond the outermost arrowhead, (a) and (b).

A *center line* is a thin, dark line composed of alternate long and short dashes, and is used to represent axes of symmetrical parts and to denote centers. As shown in Fig. 11.2 (c), center lines are commonly used as extension lines in locating holes and other features. When so used, the center line crosses over other lines of the drawing without gaps. A center line should always end in a long dash.

## 11.6 Placement of Dimension and Extension Lines.

A correct example of the placement of dimension lines and extension lines is shown in Fig. 11.3 (a). The shorter dimensions are nearest to the object outline. Dimension lines should not cross extension lines as at (b), which results from placing the shorter dimensions outside. Note that it is perfectly satisfactory to cross extension lines, as shown at (a). They should never be shortened, as at (c). A dimension line should never coincide with, or form a continuation of, any line of the drawing, as shown at (d). Avoid crossing dimension lines wherever possible.

Dimensions should be lined up and grouped together as much as possible, as shown in Fig. 11.4 (a), and not as at (b).

In many cases, extension lines and center lines must cross visible lines of the object, Fig. 11.5 (a). When this occurs, gaps should not be left in the lines, as at (b).

Dimension lines are normally drawn at right angles to extension lines, but an exception may be made in the interest of clearness, as

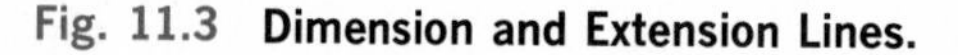
**Fig. 11.3 Dimension and Extension Lines.**

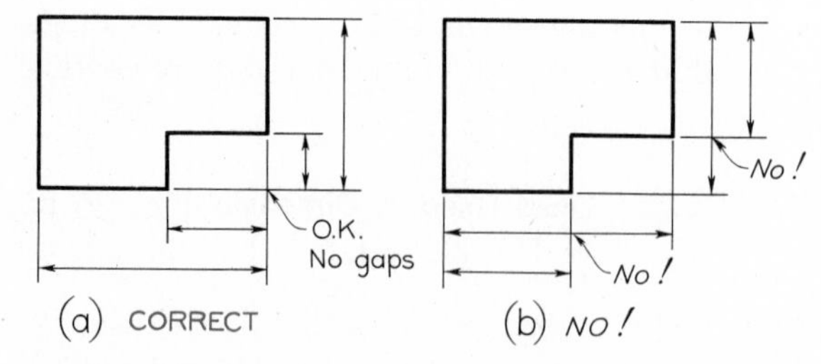


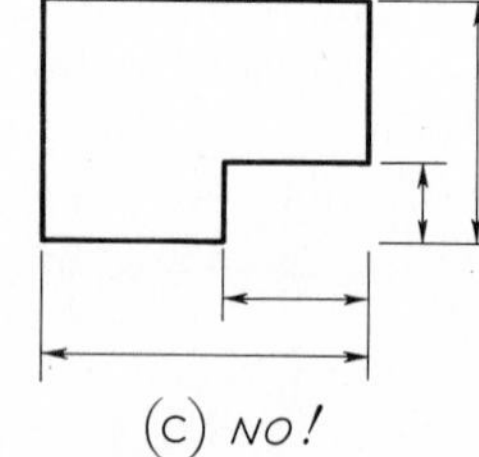


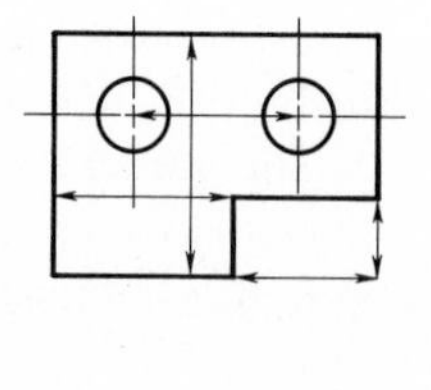


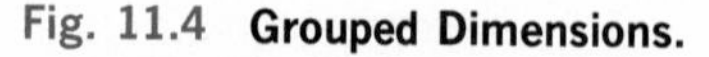
**Fig. 11.4 Grouped Dimensions.**

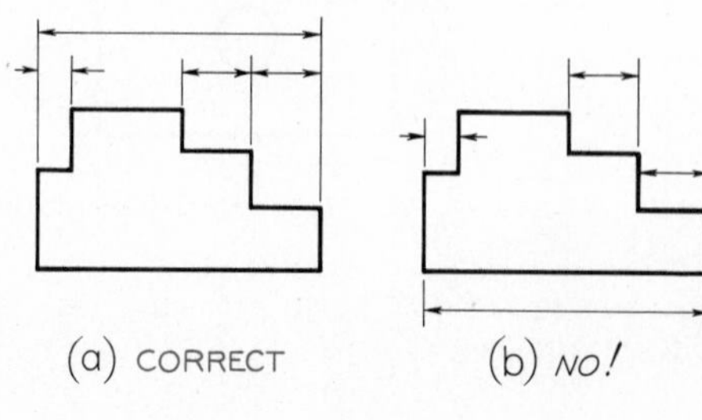

**Fig. 11.5 Crossing Lines.**

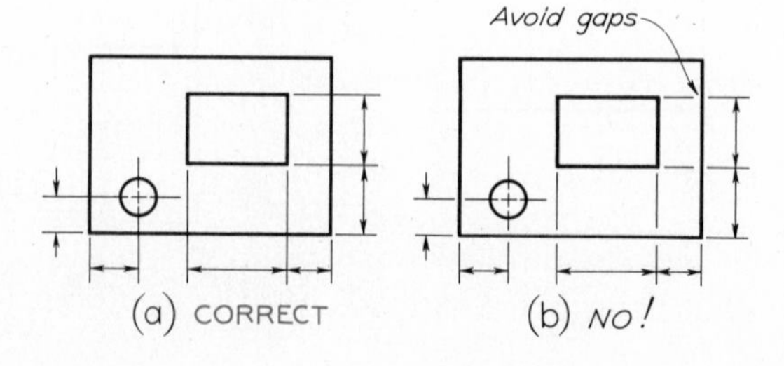

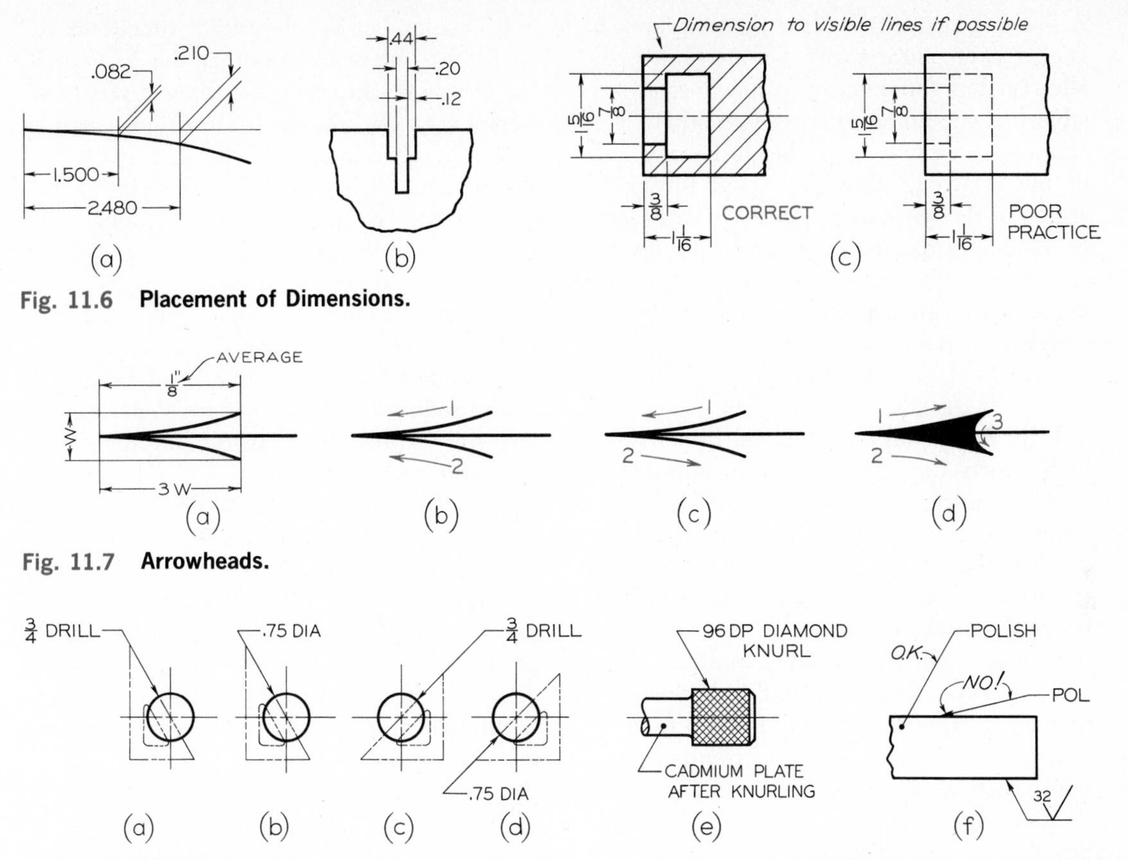


**Fig. 11.6 Placement of Dimensions.**

**Fig. 11.7 Arrowheads.**

**Fig. 11.8 Leaders.**

shown in Fig. 11.6 (a). In crowded conditions, gaps in extension lines near arrowheads may be left, in order to clarify the dimensions, as shown at (b). In general, avoid dimensioning to hidden lines, as shown at (c).

**11.7 Arrowheads.** *Arrowheads,* Fig. 11.7, indicate the extent of dimensions. They should be uniform in size and style throughout the drawing, and not varied according to the size of the drawing or the length of dimensions. Arrowheads should be drawn freehand and the length and width should be in a ratio of 3:1. The length of the arrowhead should be equal to the height of the dimension whole numbers. For average use, make arrowheads about $\frac{1}{8}''$ long and very narrow, (a). Use strokes toward the point or away from the point as desired, (b) to (d). The method at (b) is easier when the strokes are drawn toward the draftsman. For best appearance, fill in the arrowhead as at (d). A suitable pen for inking arrowheads is the Gillott's 303.

**11.8 Leaders.** A *leader,* Fig. 11.8, is a thin solid line leading from a note or dimension, and terminated by an arrowhead or a dot touching the part to which attention is directed. Arrowheads should always terminate on a line such as the edge of a hole; dots should be within the outline of the object. A leader should generally be an inclined straight line, if possible, except for the short horizontal shoulder ($\frac{1}{4}''$, approx.) extending from mid-height of the lettering *at the beginning or end of a note.* The shoulder is optional, but if used it should not be drawn so as to underline the lettering.

A leader to a circle should be radial, so that if extended it would pass through the center.

A drawing presents a more pleasing appearance if leaders near each other are drawn parallel. Leaders should cross as few lines as possible, and should never cross each other. They should not be drawn parallel to nearby lines of the drawing, allowed to pass through a corner of the view, drawn unnecessarily long, or drawn horizontally or vertically on the sheet. A leader should be drawn at a large angle and terminate with the appropriate arrowhead or dot as shown at (f).

## 11.9 Fractional and Decimal Dimensions.

In the early days of machine manufacturing in this country, the workman would scale the undimensioned design drawing to obtain any needed dimensions, and it was his responsibility to see to it that the parts fitted together properly. When blueprinting came into use, workman in widely separated localities could use the same drawings, and it became the practice to dimension the drawings, more or less completely, in inches and common fractions, such as $\frac{1}{4}''$, $\frac{1}{8}''$, $\frac{1}{16}''$, $\frac{1}{32}''$, and $\frac{1}{64}''$. The smallest dimension the workman was supposed to measure directly was $\frac{1}{64}''$, which was the smallest division on the machinist's scale.* (However, a good machinist could "split" sixty-fourths with ease.) When close fits were required, the drawing would carry a note, such as "running fit" or "drive fit," and the workman would make considerably finer adjustment of size than $\frac{1}{64}''$. Workmen were skilled, and it should not be thought that very accurate and excellent fits were not obtained. Hand-built machines were often beautiful examples of precision workmanship.

This system of units and common fractions is still used in architectural and structural work where close accuracy is relatively unimportant, and where the steel tape or framing square is used to set off measurements. Architectural and structural drawings are therefore dimensioned in this manner.

As industry has progressed, there has been greater and greater demand for more accurate specifications of the important functional dimensions—more accurate than the $\frac{1}{64}''$ permitted by the machinists scale. Since it was cumbersome to use still smaller fractions, such as $\frac{1}{128}$ or $\frac{1}{256}$, it became the practice to give decimal dimensions, such as 4.2340 and 3.815, for the dimensions requiring accuracy. Along with this, many of the dimensions, such as pattern dimensions, forging dimensions, relatively unimportant machine dimensions, standard nominal sizes of materials, punched holes, drilled holes, threads, keyways, and other features produced by tools that are so designated are still expressed in whole numbers and common fractions.

Thus, a given drawing today may be dimensioned entirely with whole numbers and common fractions, or entirely with decimals, or with a combination of the two. However, practice is rapidly moving toward the adoption of the decimal system as recommended by ANSI, for decimals can be added, subtracted, multiplied, and divided more easily than fractions. In addition the decimal system is compatible with the metric system, the calibrations of machine-tool controls, the requirements for numerically controlled machine tools, and for computer-programmed digital plotting. For an example of a computer-made drawing, see Fig. 14.15.

For inch-millimeter equivalents of decimal and common fractions, see inside the front cover of this book. For additional metric equivalents, see Appendix 26. For rounding off decimals, see §11.10.

## 11.10 Decimal Systems.

As shown in §11.9, the ever-increasing requirement for accuracy has brought greater use of decimals, while common fractions were continued for dimensions that did not require great accuracy. The use of both common fractions and decimals on the same drawings has caused a great deal of confusion, and there is today a definite trend toward the use of a decimal system for all dimensions.

A decimal system, based upon the inch as a unit of measure, has most of the advantages of the nearly universal metric system and is

*Decimal scales, graduated in fiftieths of an inch, are coming into increasing use, Fig. 11.9.

compatible with most measuring devices and machine tools in use today. American industries have found that the decimalized inch system rather than the metric system could be adopted without the necessity of scrapping all measuring devices and undergoing the upheaval and the enormous cost of making the change from the inch system.

In 1932, the Ford Motor Company adopted the decimal system. The shop scale adopted, Fig. 11.9, is divided on one edge into inches

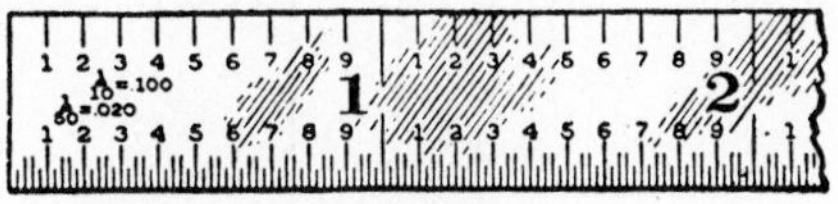

**Fig. 11.9 Ford Special Rule.**

and tenths and on the other edge into inches, tenths, and fiftieths. Thus, the smallest division is one fiftieth or .02″, two divisions are .04″, and so on, so that when necessary to halve any measurement, the result will still be a two-place decimal. This scale has now been widely adopted, especially in the automotive and aircraft industries.

Although dimensions may be expressed as decimals or common fractions, decimal dimensioning is preferred by the American National Standard Institute. "*Complete decimal dimensioning* employs decimals for all dimensions and designations. *Combination dimensioning* employs decimals for all dimensions except the designations of nominal sizes of parts or features such as bolts, screw threads, keyseats, or other standardized fractional designations."* In these systems, a two-place decimal is used when a common fraction is regarded as sufficiently accurate. When common fractions are replaced by decimals, computations are simplified, as decimals can be added, subtracted, multiplied, or divided much more easily.

Two-place decimals are used when tolerance limits of ±.01, or more, can be permitted. Three or more decimal places are used for tolerance limits less than ±.01. In a two-place decimal, the second place preferably should be an even digit (for example: .02, .04, and .06 are preferred to .01, .03, or .05) so that when the dimension is divided by two, as is necessary in determining the radius from a diameter, the result will be a decimal of two places. However, odd two-place decimals are used when required for design purposes, such as in dimensioning points on a smooth curve, or when strength or clearance is a factor.

*In the combination dimensioning system, common fractions are continued* to indicate nominal sizes of materials, drilled holes, punched holes, threads, keyways, and other standard features.

*Examples:*

$\frac{1}{4}$-20 UNC-2A; $\frac{5}{16}$ DRILL; STOCK $1\frac{1}{4} \times 1\frac{1}{2}$.

In the complete decimal system, decimals are used for everything, including standard nominal sizes, as .250-20 UNC-2B, or .750 HEX.

A typical example of the use of the complete decimal system is shown in Fig. 11.10.

When a decimal value is to be rounded off to a lesser number of places than that available, the method is as follows:*

The last figure to be retained should not be changed when the figure beyond the last figure to be retained is less than 5.

*Example:* 3.46325, if cut off to three places, should be 3.463.

The last figure to be retained should be increased by 1 when the figure beyond the last figure to be retained is greater than 5.

*Example:* 8.37652, if cut off to three places, should be 8.377.

The last figure to be retained should be unchanged if it is even, or increased by 1 if odd, when followed by exactly 5.

*Example:* 4.365 becomes 4.36 when cut off to two places. Also 4.355 becomes 4.36 when cut off to two places.

Shop scales and drafting scales for use in the decimal system are standardized by the ANSI.† The drafting scale is known as the

* ANSI Y14.5-1966.

* ANSI Z25.1-1940 (R1961).
† ANSI Z75.1-1955.

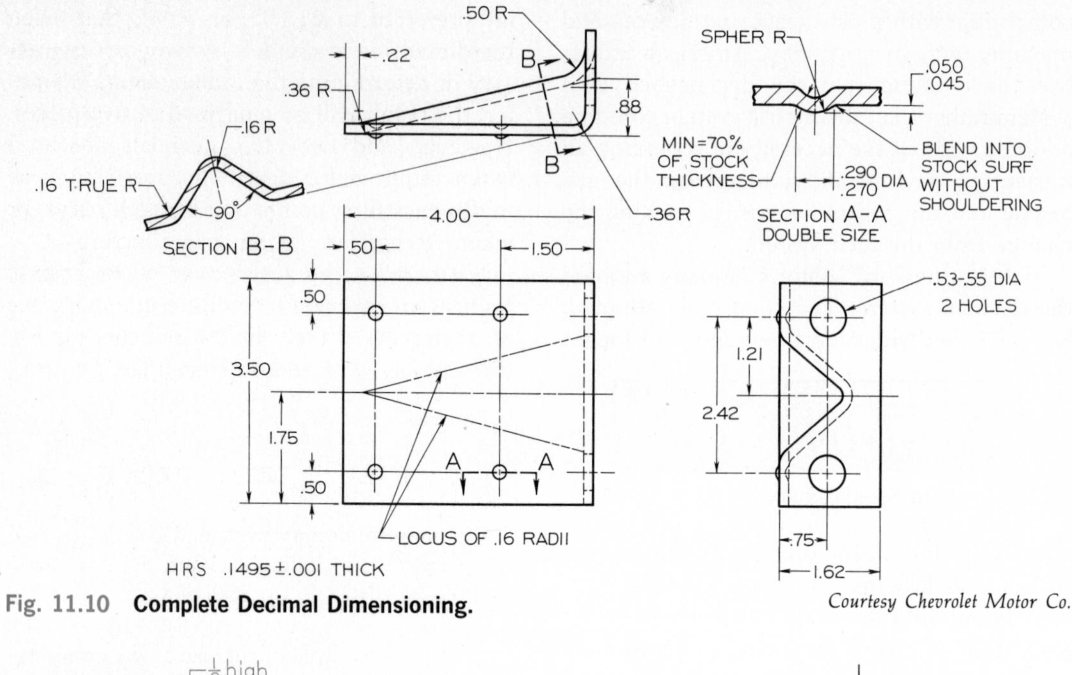


Fig. 11.10 **Complete Decimal Dimensioning.** *Courtesy Chevrolet Motor Co.*

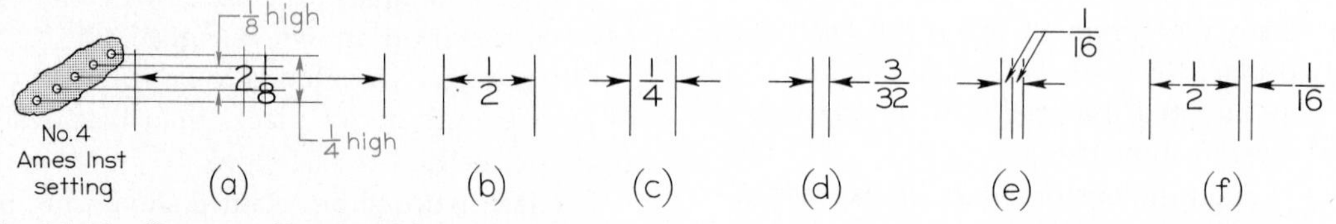


Fig. 11.11 **Common-Fraction Dimension Figures.**

*decimal scale,* and is discussed in §2.29. See inside the front cover for 2-, 3-, and 4-place decimal equivalent table.

The use of the decimal system means not only an expensive change-over of measuring equipment, but also a change-over in thinking on the part of draftsmen and designers. They must discontinue thinking in terms of units and common fractions, and think in terms of tenths, fiftieths, and hundredths of an inch. However, once the new system is installed, it is obvious that the advantages in computation, in checking, and in simplified dimensioning technique will be considerable.

## 11.11 Dimension Figures.

The importance of good lettering of dimension figures cannot be overstated. The shop produces according to the directions on the drawing, and to save time and prevent costly mistakes, all lettering should be perfectly legible. A complete discussion of numerals is given in §§3.18 to 3.20.

As shown in Fig. 11.11 (a), the standard height for whole numbers is $\frac{1}{8}''$, and for fractions double that, or $\frac{1}{4}''$. Beginners should use guide lines, as shown in Figs. 3.20 to 3.22. The numerator and denominator of a fraction should be clearly separated from the fraction bar, and the fraction bar should always be horizontal; as in Fig. 3.23 (c). An exception to this may be made in crowded places, such as parts lists, but never in dimensioning.

Legibility should never be sacrificed by crowding dimension figures into limited spaces. For every such case there is a practical and effective method, as shown in Fig. 11.11. At (a) and (b) there is enough space for the dimension line, the numeral, and the arrowheads. At (c) there is only enough room for the figure, and the arrowheads are placed out-

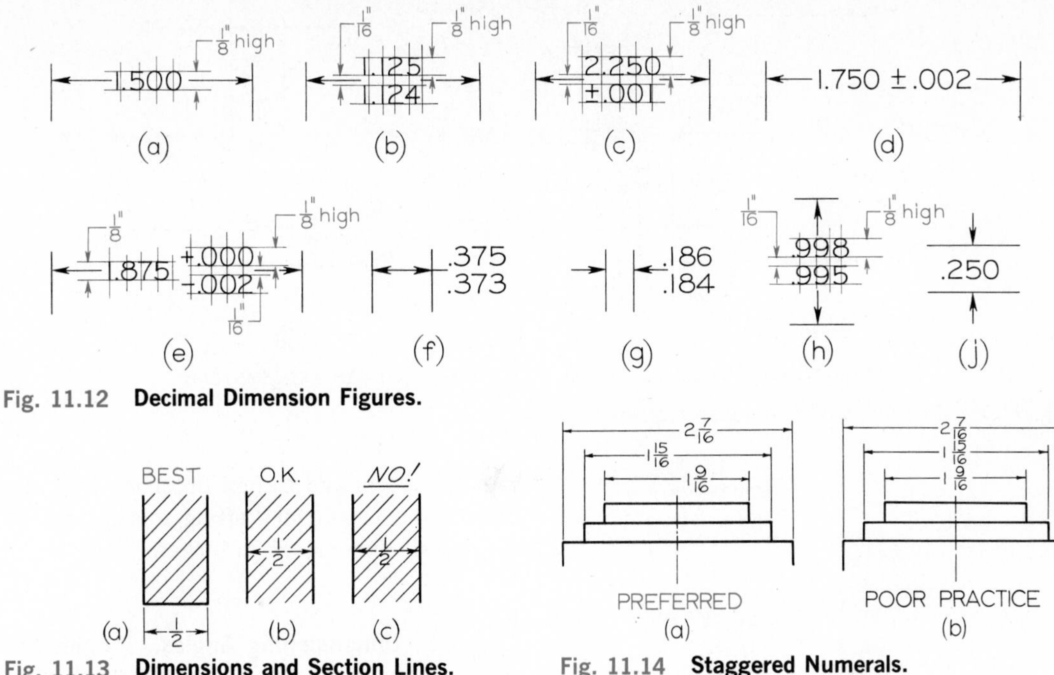


Fig. 11.12 **Decimal Dimension Figures.**

Fig. 11.13 **Dimensions and Section Lines.**

Fig. 11.14 **Staggered Numerals.**

side. At (d) both the arrowheads and the figure are placed outside. Other methods are shown at (e) and (f).

If necessary, a removed partial view may be drawn to an enlarged scale to provide the space needed for clear dimensioning, Fig. 7.22.

Methods of lettering decimal dimension figures are shown in Fig. 11.12. All numerals are $\frac{1}{8}''$ high whether on one line or on two lines. As shown at (b), the space between lines of numerals is $\frac{1}{16}''$, or $\frac{1}{32}''$ on each side of the dimension line. To draw guide lines with the Braddock-Rowe Triangle, Fig. 3.16, use the "fraction" holes at the left side of the triangle. For the Ames Lettering Guide, Fig. 3.17, use the No. 4 setting and the center column of holes.

Make all decimal points bold, allowing ample space. Never letter a dimension figure over any line of the drawing, but break the line if necessary. Place dimension figures outside a sectioned area if possible, Fig. 11.13 (a). When a dimension must be placed on a sectioned area, leave an opening in the section lining for the dimension figure, (b).

In a group of parallel dimension lines, the numerals should be staggered, as in Fig. 11.14 (a), and not stacked up one above the other, as at (b).

## 11.12 Direction of Dimension Figures.

Two systems of reading direction for dimension figures are approved by ANSI. In the preferred *unidirectional system,* Fig. 11.15 (a), all dimension figures and notes are lettered horizontally on the sheet and are read from the bottom of the drawing. The unidirectional system has been extensively adopted in the aircraft, automotive, and other industries because it is easier to use and read, especially on large drawings. In the *aligned system,* Fig. 11.15 (b), all dimension figures are aligned with the dimension lines so that they may be read from the bottom or from the right side of the sheet. Dimension lines in this system should not run in the directions included in the shaded area of Fig. 11.16, if avoidable. Notes should always be lettered horizontally on the sheet.

## 11.13 Feet and Inches.

*Inches* are indicated by the symbol " placed slightly above and to the right of the numeral, thus: $2\frac{1}{2}''$. *Feet*

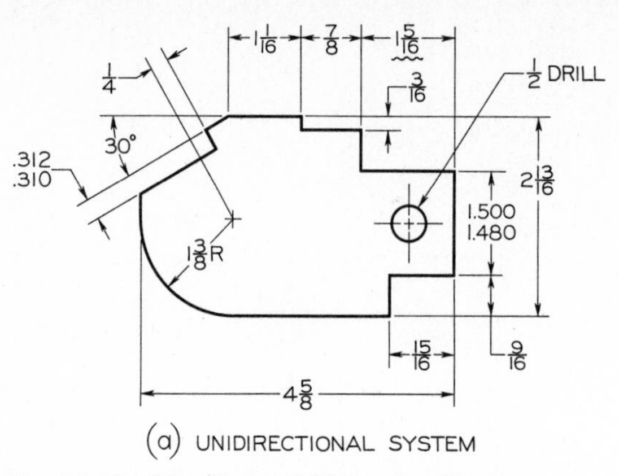


(a) UNIDIRECTIONAL SYSTEM

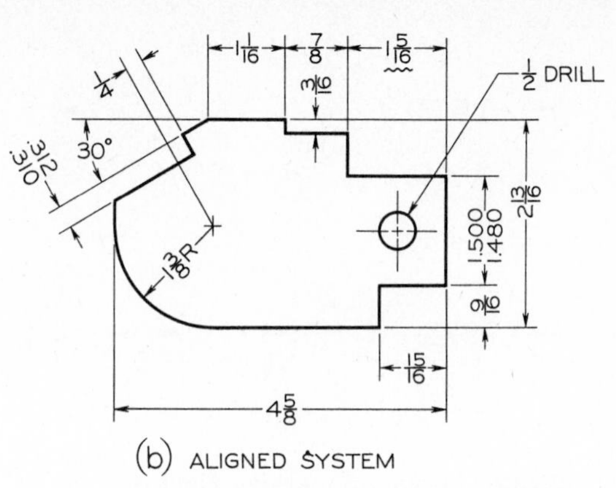


(b) ALIGNED SYSTEM

**Fig. 11.15 Directions of Dimension Figures.**

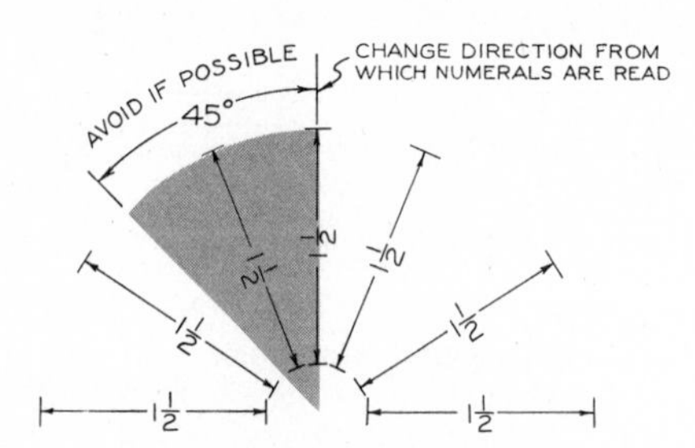


**Fig. 11.16 Directions of Dimensions.**

are indicated by the symbol ′ similarly placed, thus: 3′–0, 5′–6, 10′–0¼. It is customary in such expressions to omit the inch marks.

It is standard practice to omit inch marks when all dimensions on a drawing are in inches, except when there is a possibility of misunderstanding. For example, 1 VALVE should be 1″ VALVE, and 1 DRILL should be 1″ DRILL.

In some industries all dimensions, regardless of size, are given in inches; in others dimensions up to 72″ inclusive are given in inches, and those greater are given in feet and inches. In structural and architectural drafting, all dimensions of 1′ or over are usually expressed in feet and inches. In locomotive, aircraft, and sheet-metal drafting, it is customary to give all dimensions in inches.

## 11.14 Dimensioning Angles.

Angles are dimensioned by means of coordinate dimensions of the two legs of a right triangle. Fig. 11.17 (a), or by means of a linear dimension and an angle in degrees, (b). The coordinate method is suitable for work requiring a high degree of accuracy. Variations of angle (in degrees) are hard to control because the amount of variation increases with the distance from the vertex of the angle. Methods of indicating various angles are shown from (c) to (f). Tolerances of angles are discussed in §12.16.

When degrees alone are indicated, the symbol ° or the abbreviation DEG is used. When minutes alone are given, the number should be preceded by 0°. *Example:* 0° 23′. If desired, an angle may be given in degrees and decimal fractions of a degree, as 49.5°. In all cases, whether in the aligned system or in the unidirectional system, the dimension figures are

**Fig. 11.17 Angles.**

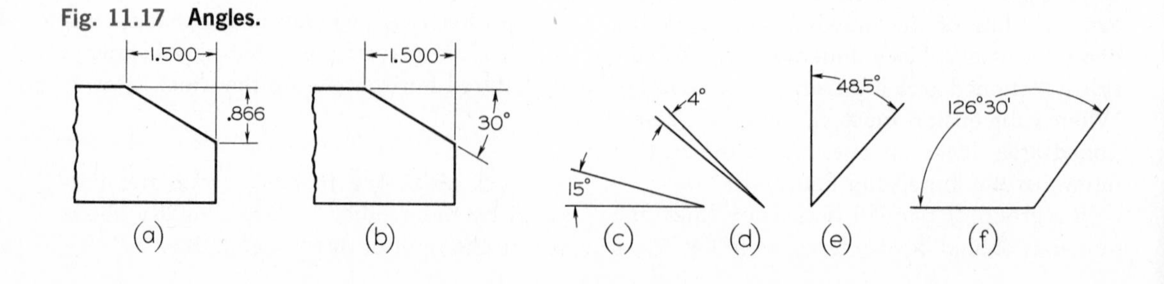

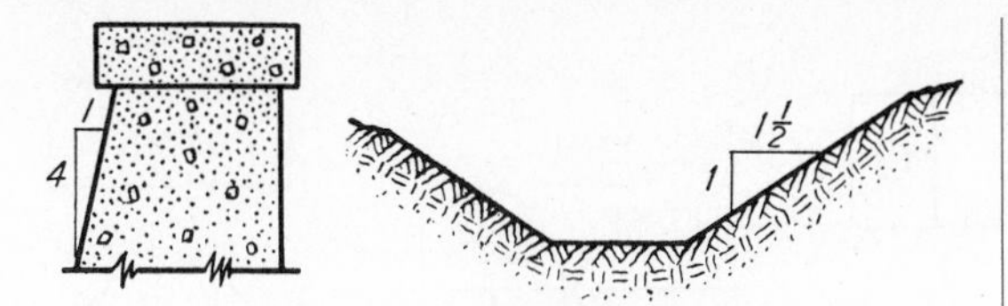


Fig. 11.18 **Angles in Civil Engineering Projects.**

lettered on horizontal guide lines. For a general discussion of angles, see §4.3.

In civil engineering drawings, *slope* represents the angle with the horizontal, while *batter* is the angle referred to the vertical. Both are expressed by making one member of the ratio equal to 1, as shown in Fig. 11.18. *Grade,* as of a highway, is similar to slope but is expressed in percentage of rise per 100′ of run. Thus a 20′ rise in a 100′ run is a grade of 20 percent.

In structural drawings, angular measurements are made by giving the ratio of "run" to "rise," with the larger size being 12″. These right triangles are referred to as *bevels.*

## 11.15 Dimensioning Arcs.

A circular arc is dimensioned in the view in which its true shape is shown by giving the numeral denoting its radius, followed by the abbreviation R, as shown in Fig. 11.19. The centers may be indicated by small crosses to clarify the drawing but not for small or unimportant radii. Crosses should not be drawn for undimensioned arcs. As shown at (a) and (b), when there is room enough, both the numeral and the arrowhead are placed inside the arc. At (c) the arrowhead is left inside, but the numeral had to be moved outside. At (d) both the arrowhead and the numeral had to be moved outside. At (e) is shown an alternate method to (c) or (d) to be used when section lines or other lines are in the way. Note that in the unidirectional system, all of these numerals would be lettered horizontally on the sheet.

For a long radius, as shown at (f), when the center falls outside the available space, the dimension line is drawn toward the actual center; but a false center may be indicated and the dimension line "jogged" to it, as shown.

## 11.16 Fillets and Rounds.

Individual fillets and rounds are dimensioned as any arc, as shown in Fig. 11.19 (b) to (e). If there are only a few and they are obviously the same size, as in Fig. 11.41 (5), one typical radius is sufficient. However, fillets and rounds are often quite numerous on a drawing and most of them are likely to be some standard size, as $\frac{1}{8}$″R or $\frac{1}{4}$″R. In such cases it is customary to give a note in the lower portion of the drawing to cover all such uniform fillets and rounds, thus: FILLETS .24 R AND ROUNDS .12 R UNLESS OTHERWISE SPECIFIED, or ALL CASTING RADII .24 R UNLESS NOTED, or simply ALL FILLETS AND ROUNDS .12 R.

For a discussion of fillets and rounds in the shop, see §10.6.

## 11.17 Finish Marks.

A *finish mark* is used to indicate that a surface is to be machined, or finished, as on a rough casting or forging. To the patternmaker or diemaker, a finish mark means that allowance of extra metal in the rough workpiece must be provided for the machining. See §10.5. On drawings of parts to be machined from rolled stock, finish marks are generally unnecessary, for it is obvious that the surfaces are finished. Similarly, it is

Fig. 11.19 **Dimensioning Arcs.**

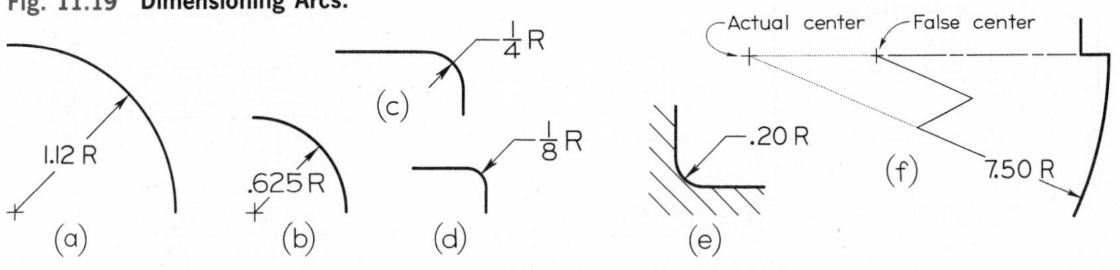

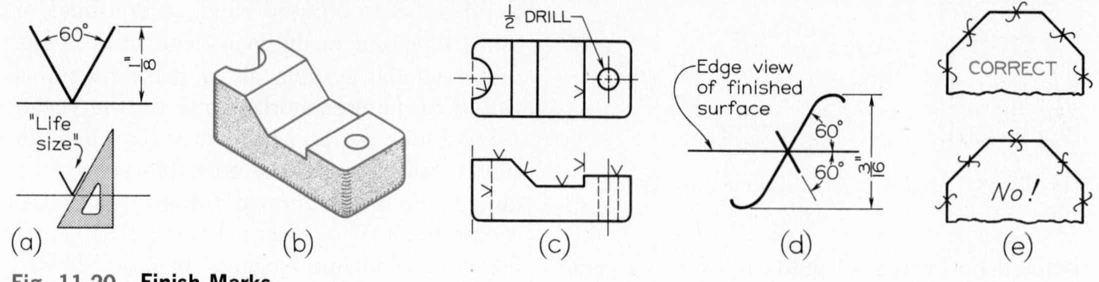


Fig. 11.20 **Finish Marks.**

not necessary to show finish marks when a shop operation is specified in a note that indicates machining, such as drilling, reaming, boring, countersinking, counterboring, and broaching, or when the dimension implies a finished surface, such as .245–.250 DIA.

Two styles of finish marks are used, the preferred newer V symbol and the older *f* symbol. These symbols indicate an ordinary machine finish. The V symbol, Fig. 11-20 (a), is like a capital V, made about $\frac{1}{8}''$ high in conformity with the dimension figures. For best results it should be drawn with the aid of a template or the 30° × 60° triangle.

At (b) is shown a simple casting having several finished surfaces, and at (c) are shown two views of the same casting, showing how the finish marks are indicated on a drawing. The *finish mark is shown only on the edge view of a finished surface, and is repeated in any other view in which the surface appears as a line, even if the line is a hidden line.* The point of the V should point inward toward the body of metal in a manner similar to that of a tool bit. When it is necessary to control the surface texture of finished surfaces beyond that of an ordinary machine finish, the V symbol is used as a base for the more elaborate surface quality symbols as discussed in §11.42.

The several kinds of finishes are detailed in machine shop practice manuals. The following terms are commonly used: *finish all over, rough finish, file finish, sand blast, pickle, scrape, lap, hone, grind, polish, burnish, buff, chip, spotface, countersink, counterbore, core, drill, ream, bore, tap, broach, knurl,* etc.

The old symbol *f* is still used, though it is executed in a variety of forms. The preferred form, Fig. 11.20 (d) and (e), is shown on the edge views of finished surfaces as described above for the V-type finish marks.

If a part is to be finished all over, finish marks should be omitted, and a general note, such as FINISH ALL OVER or FAO, should be lettered on the lower portion of the sheet.

## 11.18 Dimensions On or Off Views.

*Dimensions should not be placed upon a view unless the clearness of the drawing is promoted thereby.* The

Fig. 11.21 **Dimensions On or Off the Views.**

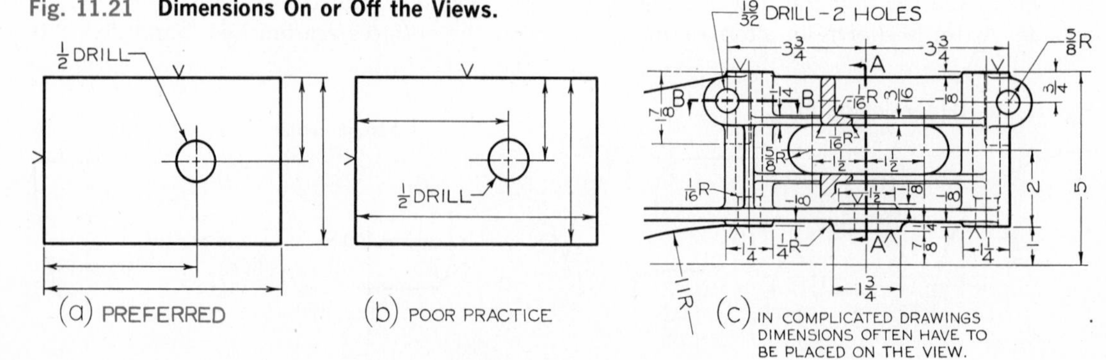

ideal form is shown in Fig. 11.21 (a), in which all dimensions are placed outside the view. Compare this with the evidently poor practice shown at (b). This is not to say that a dimension should never be placed on a view, for in many cases, particularly in complicated drawings, this is necessary, as shown at (c). Certain radii and other dimensions are given on the views, but in each case investigation will reveal a good reason for placing the dimension on the view. *Place dimensions outside of views, except where directness of application and clarity are gained by placing them on the views where they will be closer to the features dimensioned.* When a dimension must be placed in a sectioned area or on the view, leave an opening in the sectioned area or a break in the lines for the dimension figures, Figs. 11.13 (b) and 11.21 (c).

## 11.19 Contour Dimensioning.

Views are drawn to describe the shapes of the various features of the object, and dimensions are given to define exact sizes and locations of those shapes. It follows that *dimensions should be given where the shapes are shown,* that is, in the views where the contours are delineated, as shown in Fig. 11.22 (a). Incorrect placement of the dimensions is shown at (b).

If individual dimensions are attached directly to the contours that show the shapes being dimensioned, this will automatically prevent the attachment of dimensions to hidden lines, as shown for the depth .38 of the slot at (b). It will also prevent the attachment of dimensions to a visible line, the meaning of which is not clear in a particular view, such as dimension .76 for the height of the base at (b).

Although the placement of notes for holes follows the contour rule wherever possible, as shown at (a), the diameter of an external cylindrical shape is preferably given in the rectangular view where it can be readily found near the dimension for the length of the cylinder, as shown in Figs. 11.23 (b), 11.26, and 11.27.

## 11.20 Geometric Breakdown.

Engineering structures are composed largely of simple geometric shapes, such as the prism, cylinder, pyramid, cone, and sphere, as shown in Fig. 11.23 (a). They may be exterior (positive) or interior (negative) forms. For example, a steel shaft is a positive cylinder, and a round hole is a negative cylinder.

These shapes result directly from the design

Fig. 11.22 **Contour Dimensioning.**

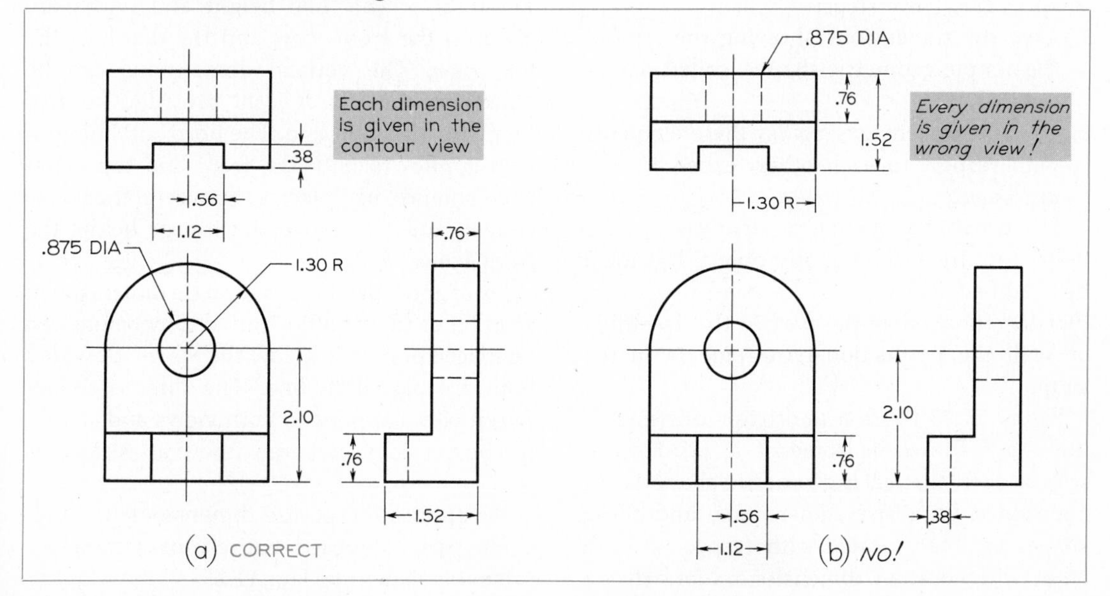

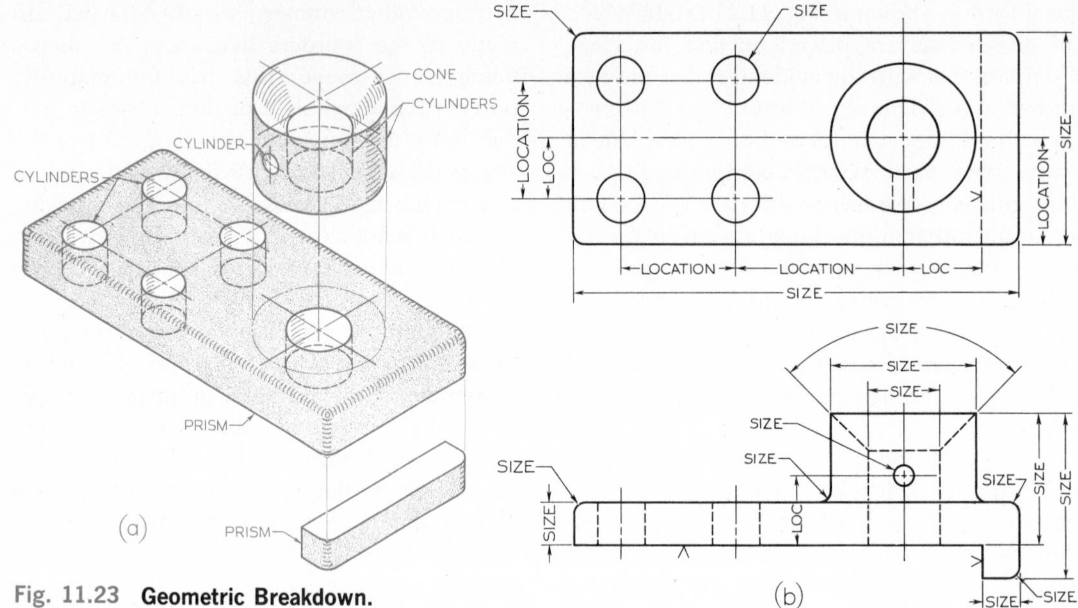


Fig. 11.23 **Geometric Breakdown.**

necessity to keep forms as simple as possible and from the requirements of the fundamental shop operations. Forms having plane surfaces are produced by planing, shaping, milling, and so forth, while forms having cylindrical, conical, or spherical surfaces are produced by turning, drilling, reaming, boring, countersinking, and other rotary operations. See Chapter 10, "Shop Processes."

The dimensioning of engineering structures involves two basic steps:

1. Give the dimensions showing the *sizes* of the simple geometric shapes, called *size dimensions;*
2. Give the dimensions *locating* these elements with respect to each other, called *location dimensions.*

The process of geometric analysis is very helpful in dimensioning any object, but must be modified when there is a conflict with either the function of the part in the assembly or with the production requirements in the shop.

Figure 11.23 (b) is a multiview drawing of the object shown in isometric at (a). Here it will be seen that each geometric shape is dimensioned with size dimensions, and these shapes are then located with respect to each other with location dimensions. Note that a *location dimension locates a three-dimensional geometric element* and not just a surface; otherwise, all dimensions would have to be classified as location dimensions.

## 11.21 Size Dimensions—Prisms.

The right rectangular prism, Fig. 4.7, is probably the most common geometric shape. Front and top views are dimensioned as shown in Fig. 11.24 (a) or (b). The height and width are given in the front view and the depth in the top view. The vertical demensions can be placed on the left or right provided both of them are placed in line. The horizontal dimension applies to both the front and top views and should be placed between them, as shown, and not above the top or below the front view.

Front and side views should be dimensioned as at (c) or (d). The horizontal dimensions can be placed above or below the views, provided both are placed in line. The dimension between views applies to both views and should not be placed elsewhere without a special reason.

An application of size dimensions to a machine part composed entirely of rectangular prisms is shown in Fig. 11.25.

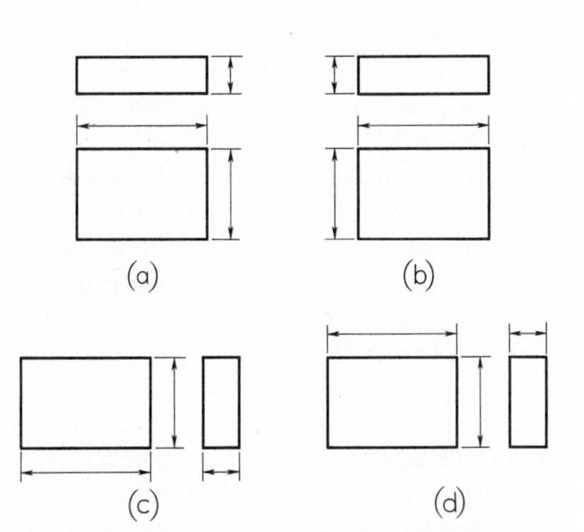

Fig. 11.24 **Dimensioning Rectangular Prisms.**

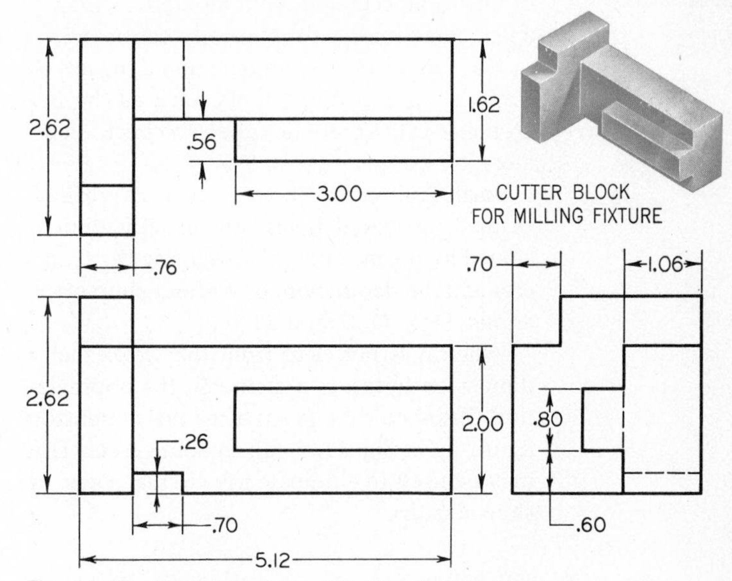

Fig. 11.25 **Dimensioning a Machine Part Composed of Prismatic Shapes.**

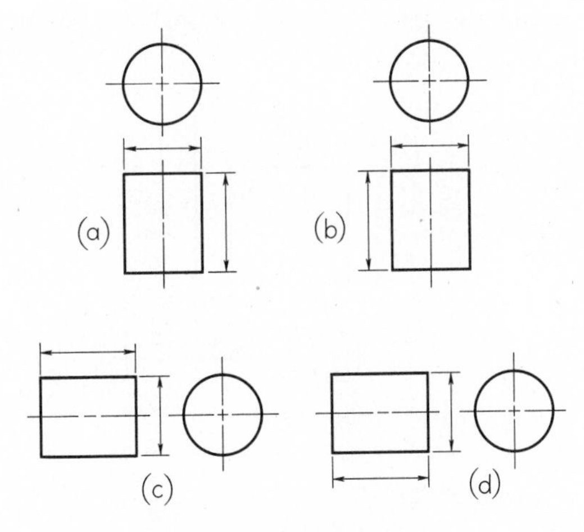

Fig. 11.26 **Dimensioning Cylinders.**

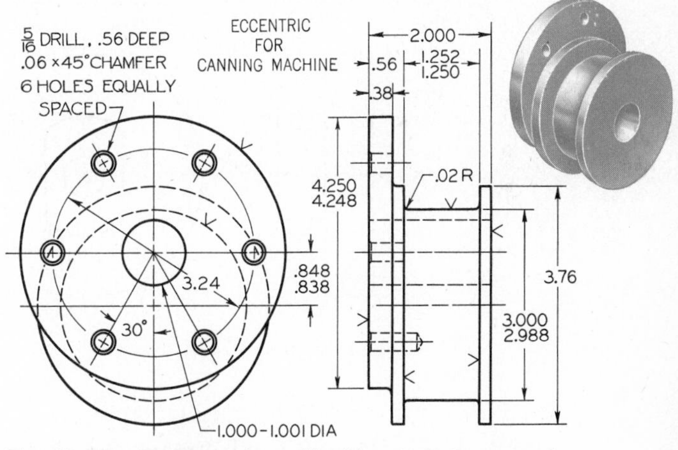

Fig. 11.27 **Dimensioning a Machine Part Composed of Cylindrical Shapes.**

## 11.22 Size Dimensions—Cylinders.

The right circular cylinder is the next most common geometric shape and is commonly seen as a shaft or a hole. The general method of dimensioning a cylinder is to give both its diameter and its length in the rectangular view, Fig. 11.26. If the cylinder is drawn in a vertical position, the length or altitude of the cylinder may be given at the right as at (a), or on the left as at (b). If the cylinder is drawn in a horizontal position, the length may be given above the rectangular view as at (c) or below as at (d). An application showing the dimensioning of cylindrical shapes is shown in Fig. 11.27.

The ANSI approves the use of a diagonal diameter in the circular view, in addition to the method shown in Fig. 11.26, but the authors do not recommend this method except in special cases when clearness is gained thereby. The use of several diagonal diameters on the same center is definitely to be discouraged, since the result is usually confusing.

The radius of a cylinder should never be given, since measuring tools, such as the micrometer caliper, are designed to check diameters.

Small cylindrical holes, such as drilled, reamed, or bored holes, are usually dimensioned by means of notes specifying the diameter and the depth, with or without shop operations, Figs. 11.27 and 11.30.

When it is not clear from the views that a dimension indicates a diameter, the abbreviation DIA should be given after the dimension figure, as in Fig. 11.28 (a). In some cases, DIA may be used to eliminate the circular view, as shown at (b).

## 11.23 Size Dimensions—Miscellaneous Shapes.

**Shapes.** A triangular prism is dimensioned, Fig. 11.29 (a) by giving the height, width, and

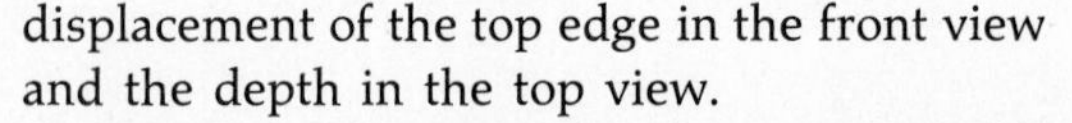

displacement of the top edge in the front view and the depth in the top view.

A rectangular pyramid is dimensioned, (b), by giving the heights in the front view, and the dimensions of the base and the centering of the vertex in the top view. If the base is square, (c), it is necessary to give the dimensions for only one side of the base, provided it is labeled SQ as shown.

A cone is dimensioned, (d), by giving its altitude and the diameter of the base in the triangular view. A frustum of a cone may be dimensioned, (e), by giving the vertical angle and the diameter of one of the bases. Another method is to give the length and the diameters of both ends in the front view. Still another is to give the diameter at one end and the amount of taper per foot in a note, §11.33.

At (f) is shown a two-view drawing of a plastic knob. The main body is spherical and is dimensioned by giving its diameter, followed by the abbreviation SPHER. A bead around the knob is in the shape of a torus,

6.24
3.12 DIA
"DIA" not necessary
(a)
USE OF "DIA" TO INDICATE CIRCULAR SHAPE

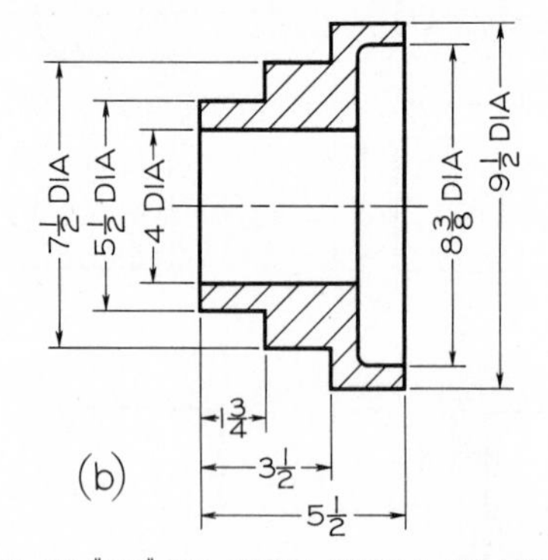

USE OF "DIA" TO OMIT CIRCULAR VIEW

**Fig. 11.28 Use of DIA in Dimensioning Cylinders.**

**Fig. 11.29 Dimensioning Various Shapes.**

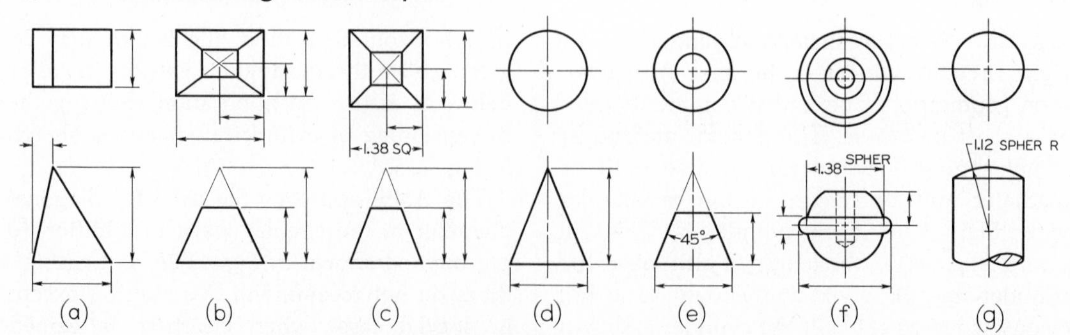

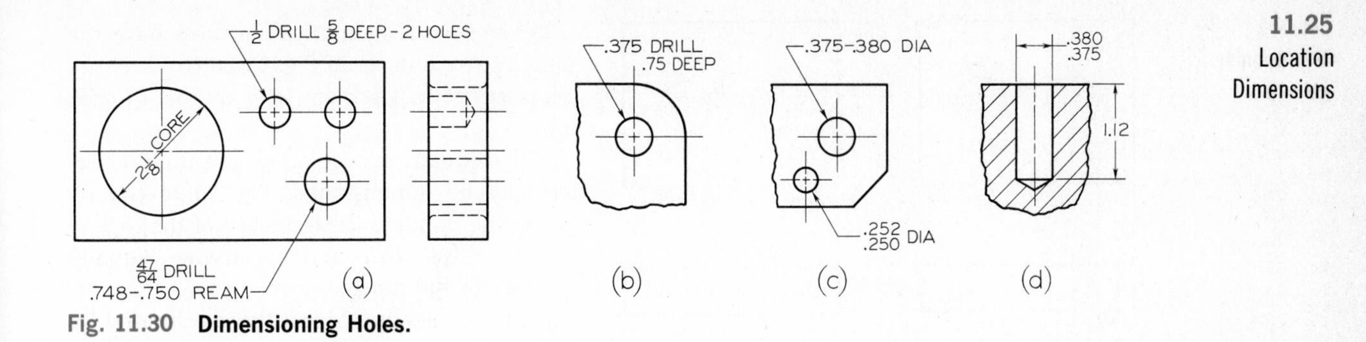


Fig. 11.30 **Dimensioning Holes.**

Fig. 4.7, and it is dimensioned by giving the thickness of the ring and the outside diameter, as shown. At (g) a spherical end is dimensioned by a radius, followed by SPHER R.

Internal shapes corresponding to the external shapes in Fig. 11.29 would be dimensioned in a similar manner.

## 11.24 Size Dimensioning of Holes.

Holes that are to be drilled, bored, reamed, punched, cored, etc., are usually specified by standard notes, as shown in Figs. 6.40, 11.30 (a), and 11.42. The order of items in a note corresponds to the order of procedure in the shop in producing the hole. Two or more holes are dimensioned by a single note, the leader pointing to one of the holes, as shown at the top of Fig. 11.30 (a).

As illustrated in Figs. 6.40 and 11.30, the leader of a note should, as a rule, point to the circular view of the hole. It should point to the rectangular view only when clearness is promoted thereby. When the circular view of the hole has two or more concentric circles, as for counterbored, countersunk, or tapped holes, the arrowhead should touch the outer circle, Fig. 11.42 (c) to (j).

*Notes should always be lettered horizontally on the paper, and guide lines should always be used.*

The use of decimal fractions instead of common fractions to designate drill sizes has gained wide acceptance,* Fig. 11.30 (b). For

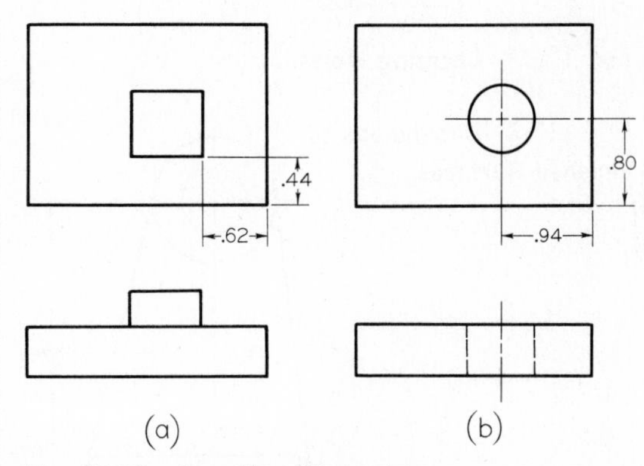


Fig. 11.31 **Location Dimensions.**

numbered or letter-size drills, Appendix 11, it is recommended that the decimal size be given in this manner, or given in parentheses, thus: #28 (.1405) DRILL, or "P" (.3230) DRILL.

On drawings of parts to be produced in large quantity for interchangeable assembly, dimensions and notes may be given without specification of the shop process to be used. Only the dimensions of the hole are given, without reference to whether the holes are to be drilled, reamed, or punched, as in Fig. 11.30 (c) and (d). It should be realized that even though shop operations are omitted from a note, the tolerances indicated would tend to dictate the shop processes required.

## 11.25 Location Dimensions.

After the geometric shapes composing a structure have been dimensioned for *size,* as discussed above, *location dimensions* must be given to show the relative positions of these geometric shapes, as shown in Fig. 11.23. As shown in Fig. 11.31

*Although drills are still usually listed fractionally in most manufacturers' catalogs, many companies have supplemented drill and wire sizes with a decimal value. In many cases the number, letter, or common fraction has been replaced by the decimal size.

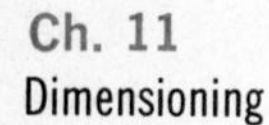


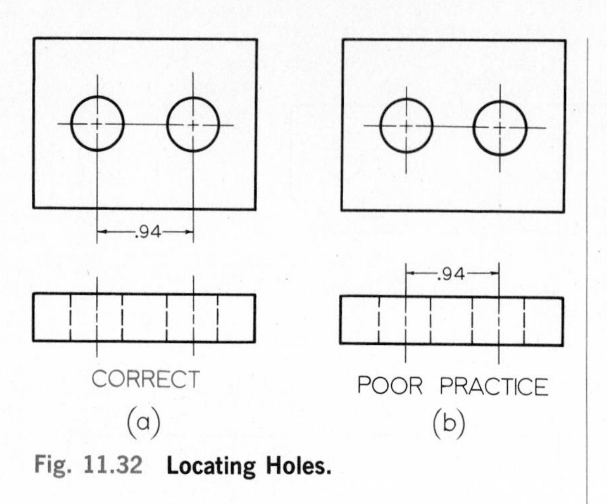


**Fig. 11.32 Locating Holes.**

**Fig. 11.33 Dimensions to Finished Surfaces.**

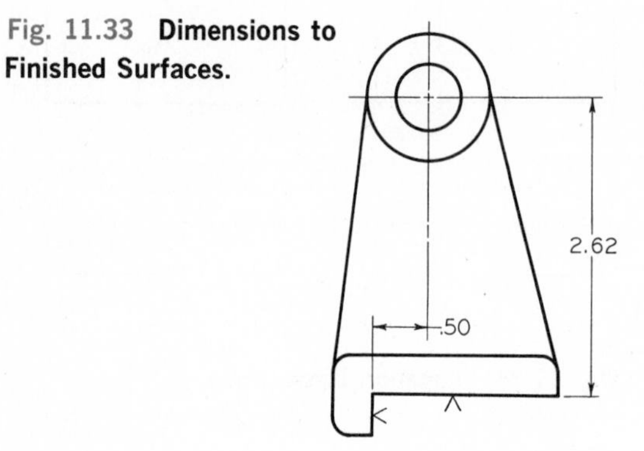


(a), rectangular shapes, whether in the form of solids or of recesses, are located with reference to their faces. As shown at (b), cylindrical or conical holes or bosses, or other symmetrical shapes, are located with reference to their center lines.

As shown in Fig. 11.32, location dimensions for holes are preferably given in the circular view of the holes.

Location dimensions should lead to finished surfaces wherever possible, Fig. 11.33, because rough castings and forgings vary in size, and unfinished surfaces cannot be relied upon for accurate measurements. Of course, the *starting dimension,* used in locating the first machined surface on a rough casting or forging, must necessarily lead from a rough surface, or from a center or a center line of the rough piece. See Figs. 14.115 and 14.116, parts 219–12 and 219–6.

In general, location dimensions should be built from a finished surface as a datum plane, or from an important center or center line.

When several cylindrical surfaces have the same center line, as in Fig. 11.28 (b), it is not necessary to locate them with respect to each other.

Holes equally spaced about a common center may be dimensioned, Fig. 11.34 (a), by giving the diameter (diagonally) of the *circle of centers,* or *bolt circle,* and specifying "equally spaced" in the note.

Holes unequally spaced, (b), are located by means of the bolt circle diameter plus angular measurements with reference to *only one* of the center lines, as shown.

Where greater accuracy is required, coordinate dimensions should be given, as at (c). In this case, the diameter of the bolt circle is marked REF to indicate that it is to be used only as a *reference dimension.* Reference dimensions are given for information only. They are not intended to be measured and do not govern the shop operations. They represent calculated dimensions and are often useful in showing the intended design sizes. See Fig. 11.35 (c).

When several nonprecision holes are located on a common arc, they are dimensioned, Fig. 11.35 (a), by giving the radius and the angular measurements from a *base line,* as shown. In this case, the base line is the horizontal center line.

At (b) the three holes are on a common center line. One dimension locates one small hole from the center; the other gives the distance between the small holes. Note the omission of a dimension at X. This method is used when (as is usually the case) the distance between the small holes is the important consideration. If the relation between the center hole and each of the small holes is more important, then include the distance at X, and mark the over-all dimension REF.

At (c) is another example of coordinate dimensioning. The three small holes are on a bolt circle whose diameter is marked REF, for reference purposes only. From the main center, the small holes are located in two mutually perpendicular directions.

Another example of locating holes by means of linear measurements is shown at (d). In this case, one such measurement is made at an

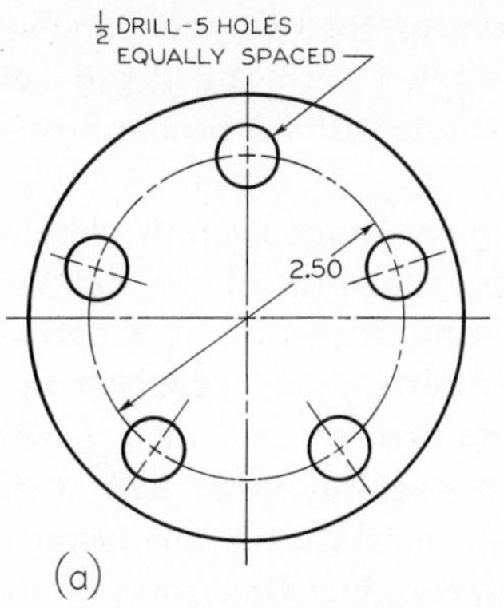


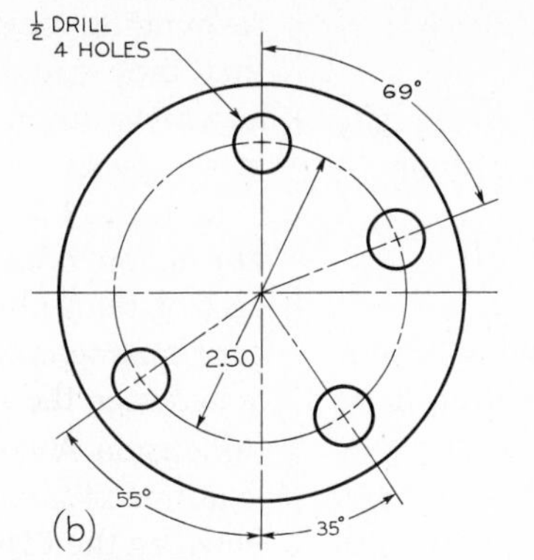


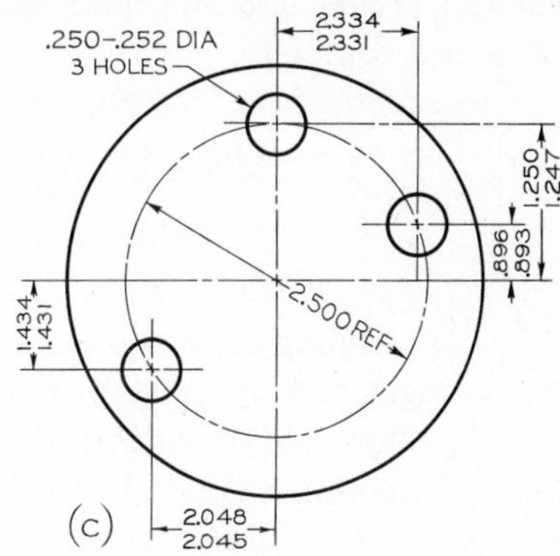


Fig. 11.34 **Locating Holes About a Center.**

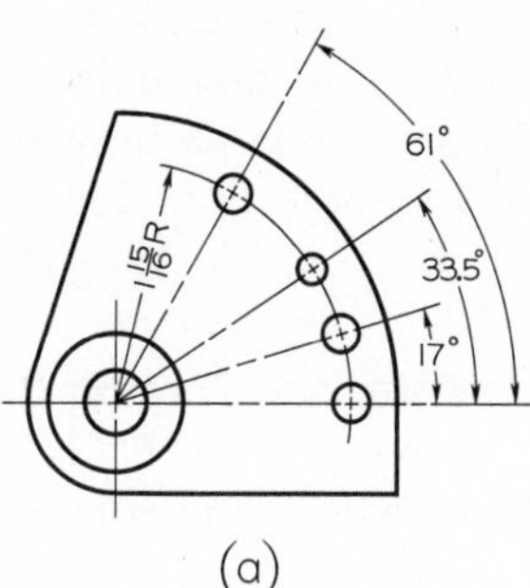


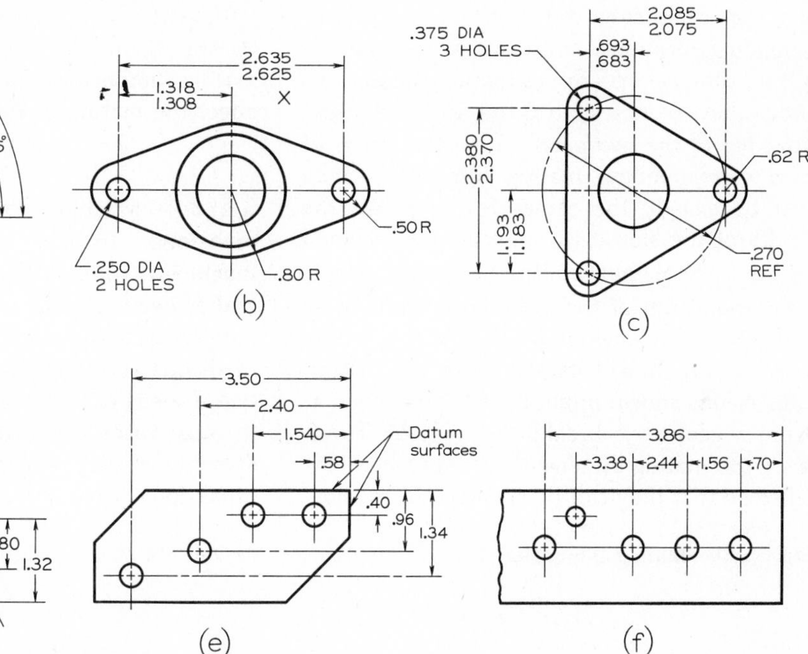


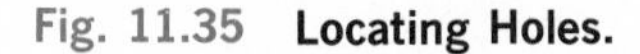


Fig. 11.35 **Locating Holes.**

angle to the coordinate dimensions because of the direct functional relationship of the two holes.

At (e) the holes are located from two *base lines* or *datums.* When all holes are located from a common datum, the sequence of measuring and machining operations is controlled, overall tolerance accumulations are avoided, and proper functioning of the finished part is assured as intended by the designer. The datum surfaces selected must be more accurate than any measurement made from them, must be accessible during manufacture, and must be arranged so as to facilitate tool and fixture design. Thus it may be necessary to specify accuracy of the datum surfaces in terms of straightness, roundness, flatness, and so forth. See §12.14.

At (f) is shown a method of giving, in a single line, all of the dimensions from a common datum. Each dimension except the first has a single arrowhead and is accumulative in value. The final and longest dimension is separate and complete.

These methods of locating holes are equally applicable to locating pins or other symmetrical features.

## 11.26 Mating Dimensions.

In dimensioning a single part, its relation to mating parts must be taken into consideration. For example, in Fig. 11.36 (a), a Guide Block fits into a slot in a Base. Those dimensions common to both parts are *mating dimensions,* as indicated.

These mating dimensions should be given on the multiview drawings in the corresponding locations, as shown at (b) and (c). Other dimensions are not mating dimensions since they do not control the accurate fitting together of the two parts. The actual *values* of two corresponding mating dimensions may not be exactly the same. For example, the width of the slot at (b) may be dimensioned $\frac{1}{32}''$ or several thousandths of an inch larger than the width of the Block at (c), but these are mating dimensions figured from a single basic width. It will be seen that the mating dimensions shown might have been arrived at from a geometric breakdown, §11.20. However, the mating dimensions need to be identified so that they can be specified in the corresponding locations on the two parts, and so that they can be given with the degree of accuracy commensurate with the proper fitting of the parts.

In Fig. 11.37 (a) the dimension A should appear on both the drawings of the Bracket and of the Frame and, therefore, is a necessary mating dimension. At (b), which shows a redesign of the Bracket into two parts, dimension A is not used on either part, as it is not necessary to control closely the distance between the cap screws. But dimensions F are now essential mating dimensions and should appear correspondingly on the drawings of both parts. The remaining dimensions E, D, B, and C, at (a), are not considered to be mating dimensions, since they do not directly affect the mating of the parts.

## 11.27 Machine, Pattern, and Forging Dimensions.

In Fig. 11.36 (a), the Base is machined from a rough casting; the patternmaker needs certain dimensions to make the pattern, and the machinist needs certain dimensions for the machining. In some cases one dimension will be used by both. Again, in most cases, these dimensions will be the same as those resulting from a geometric breakdown, §11.20, but it is important to

Fig. 11.36 **Mating Dimensions.**

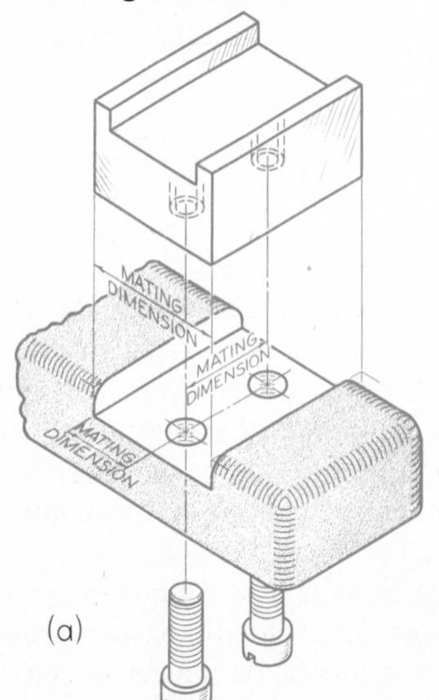


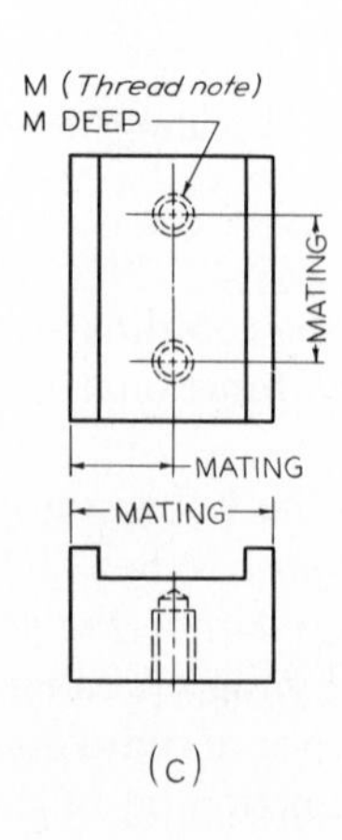

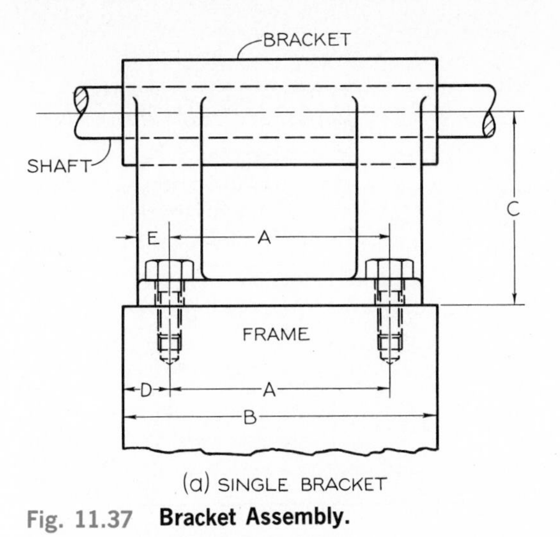

(a) SINGLE BRACKET

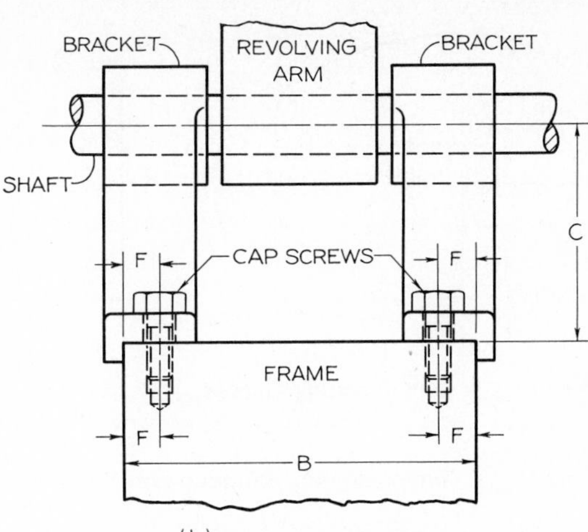

(b) DOUBLE BRACKET

Fig. 11.37 **Bracket Assembly.**

identify them in order to assign values to them intelligently.

The same part is shown in Fig. 11.38, with the machine dimensions and pattern dimensions identified by the letters M and P. The patternmaker is interested only in the dimensions he needs to make the pattern, and the machinist, in general, is concerned only with the dimensions he needs to machine the part. Frequently, a dimension that is convenient for the machinist is not convenient for the patternmaker, or vice versa. Since the patternmaker uses the drawing only once, while making the pattern, and the machinist refers to it continuously, the dimensions should be given primarily for the convenience of the machinist.

If the part is large and complicated, two separate drawings are sometimes made, one showing the pattern dimensions and the other the machine dimensions. The usual practice, however, is to prepare one drawing for both the patternmaker and the machinist. See §10.5.

For forgings, it is common practice to make separate forging drawings and machining drawings. A forging drawing of a connecting rod, showing only the dimensions needed in the forge shop, is shown in Fig. 14.35. A machining drawing of the same part, but containing only the dimensions needed in the machine shop, is shown in Fig. 14.36. See also Figs. 14.115 and 14.116.

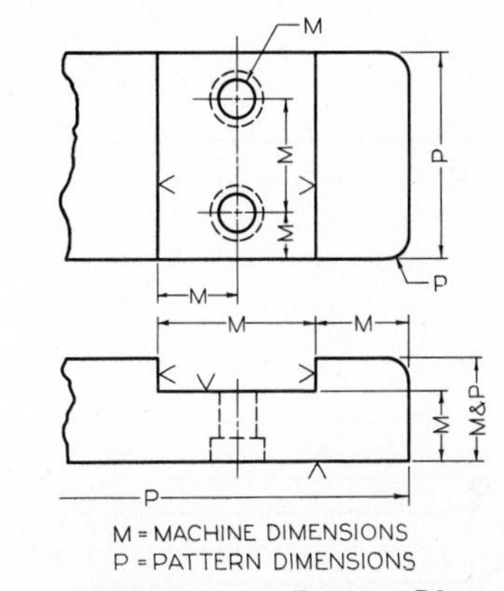

Fig. 11.38 **Machine and Pattern Dimensions.**

Unless a decimal system is used, §11.10, the pattern dimensions are always nominal, usually to the nearest $\frac{1}{16}''$, and given in whole numbers and common fractions. If a machine dimension is given in whole numbers and common fractions, the machinist is usually allowed a tolerance (permissible variation in size) of $\frac{1}{64}''$, corresponding to his steel scale, which has $\frac{1}{64}''$ divisions. Some companies specify a tolerance of .010″ on all common fractions. If greater accuracy is required, the dimensions are given in decimal form to three or more places, §§11.10 and 12.1.

## 11.28 Dimensioning of Curves.

Curved shapes may be dimensioned by giving a group of radii, as shown in Fig. 11.39 (a). Note that

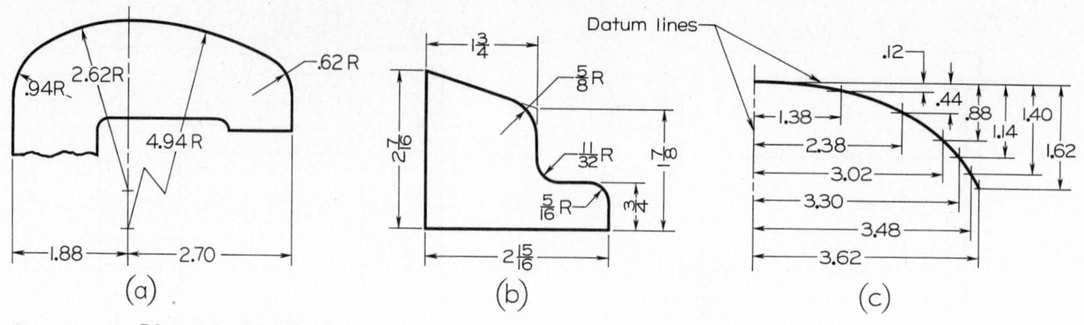


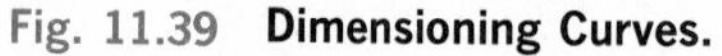
Fig. 11.39 **Dimensioning Curves.**

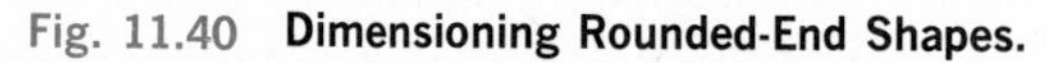
Fig. 11.40 **Dimensioning Rounded-End Shapes.**

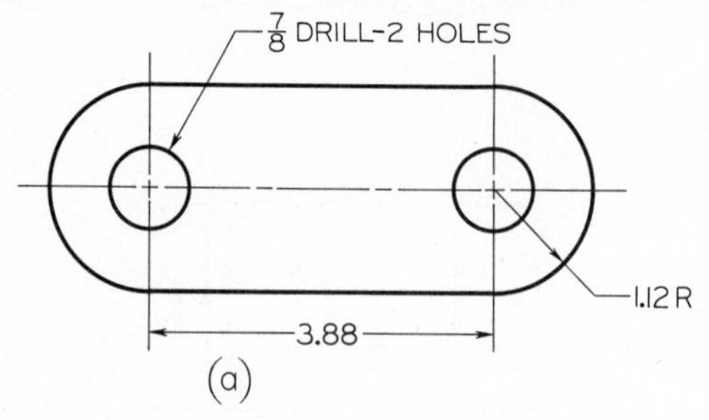


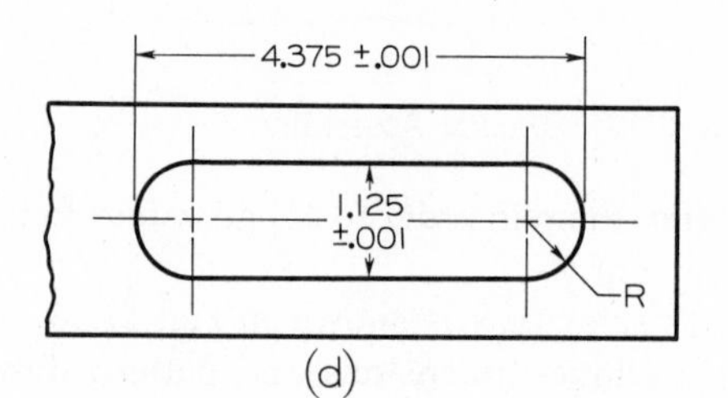


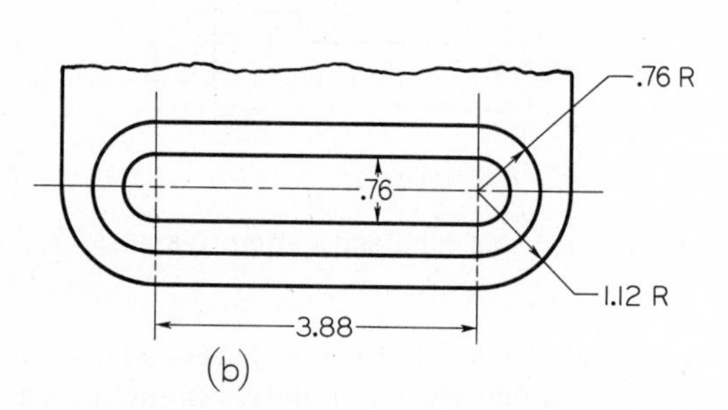


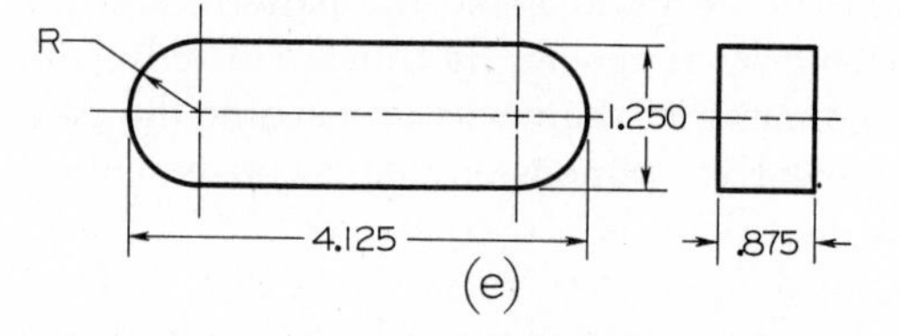


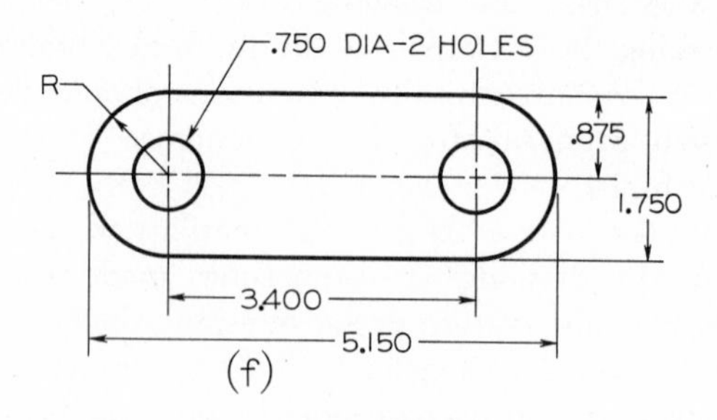


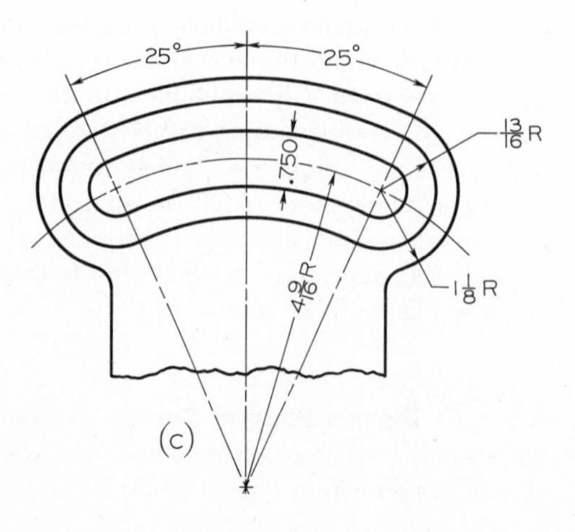


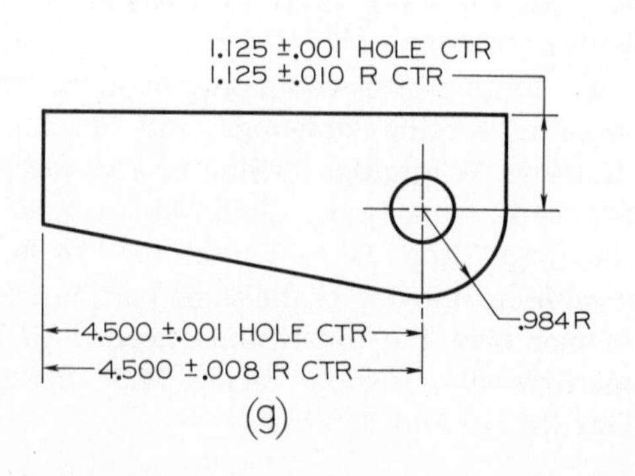

in dimensioning the 4.94R arc whose center is inaccessible, the center may be moved inward along a center line, and a jog made in the dimension line. See also Fig. 11.19 (f). Another method is to dimension the outline envelope of a curved shape so that the various radii are self-locating from "floating centers," as at (b). Either a circular or a noncircular curve may be dimensioned by means of coordinate dimensions referred to datums, as shown at (c). See also Fig. 11.6 (a).

### 11.29 Dimensioning of Rounded-End Shapes.

The method used for dimensioning rounded-end shapes depends upon the degree of accuracy required, Fig. 11.40. When precision is not necessary, the methods used are those that are convenient for the shop, as at (a), (b), and (c).

At (a) the link, to be cast or to be cut from sheet metal or plate, is dimensioned as it would be laid out in the shop, by giving the center-to-center distance and the radii of the ends. Note that only one such radius dimension is necessary.

At (b) the pad on a casting, with a milled slot, is dimensioned from center to center for the convenience of both the patternmaker and the machinist in layout. An additional reason for the center-to-center distance is that it gives the total travel of the milling cutter, which can be easily controlled by the machinist. The width dimension indicates the diameter of the milling cutter; hence it is incorrect to give the radius of a machined slot. On the other hand, a cored slot (see §10.4) should be dimensioned by radius in conformity with the patternmaker's layout procedure.

At (c) the semicircular pad is laid out in a similar manner to the pad at (b), except that angular dimensions are used. Angular tolerances, §12.16, can be used if necessary.

When accuracy is required, the methods shown at (d) to (g) are recommended. Over-all lengths of rounded-end shapes are given in each case, and radii are indicated, but without specific values. In the example at (f), the center-to-center distance is required because of necessity for accurate location of the holes.

At (g) the hole location is more critical than the location of the radius; hence the two are located independently, as shown.

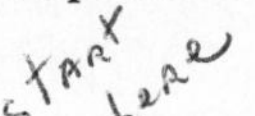


### 11.30 Superfluous Dimensions.

Although it is necessary to give all dimensions, the draftsman should avoid giving unnecessary or superfluous dimensions, Fig. 11.41. Dimensions should not be repeated on the same view or on different views, nor should the same information be given in two different ways.

Figure 11.41 (2) illustrates a type of superfluous dimensioning that should generally be avoided, especially in machine drawing where accuracy is important. The workman should not be allowed a choice between two dimensions. *Avoid "chain" dimensioning,* in which a complete series of detail dimensions is given, together with an over-all dimension. In such cases, one dimension of the chain should be omitted, as shown, so that the machinist is obliged to work from one surface only. This is particularly important in tolerance dimensioning, §12.1, where an accumulation of tolerances can cause serious difficulties. See also §12.10.

Some inexperienced draftsmen have the habit of omitting both dimensions, such as those at the right at (2), on the theory that the holes are symmetrically located and will be understood to be centered. One of the two location dimensions should be given.

As shown at (5), when one dimension clearly applies to several identical features, it need not be repeated. This is generally true of fillets and rounds and of other noncritical features. For example, the radii of the rounded ends in Fig. 11.40 (a) to (c) need not be repeated; and in Fig. 11.1 both ribs are obviously the same thickness so it is unnecessary to repeat the .38″ dimension.

### 11.31 Notes.

It is usually necessary to supplement the direct dimensions with notes, Fig. 11.42. They should be brief and should be carefully worded so as to be capable of only one interpretation. *Notes should always be lettered*

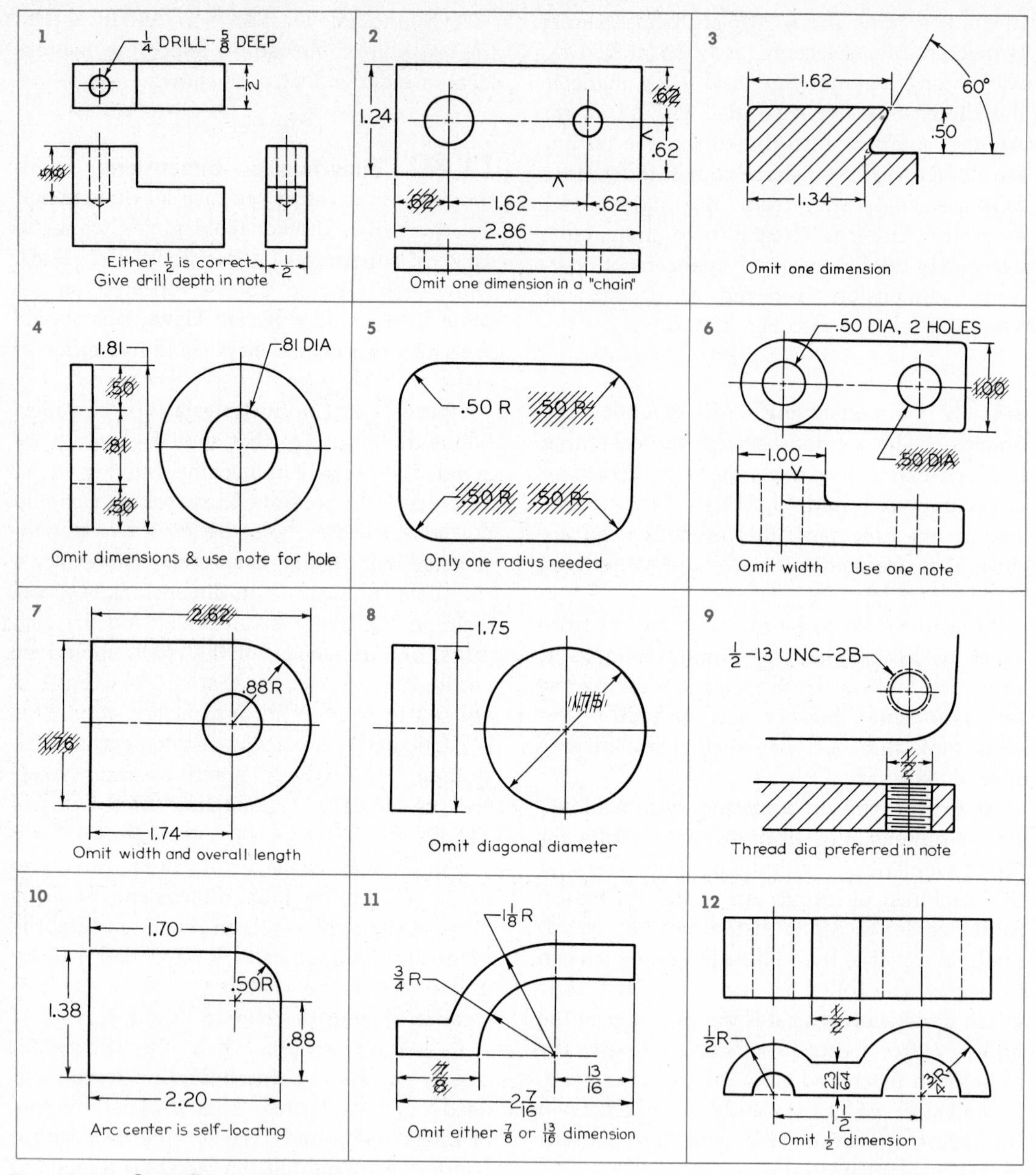


Fig. 11.41 **Superfluous Dimensions.**

*horizontally on the sheet, with guide lines, and arranged in a systematic manner.* They should not be lettered in crowded places. Avoid placing notes between views, if possible. They should not be lettered closely enough to each other to confuse the reader, or close enough to another view or detail to suggest application to the wrong view. Leaders should be as short as possible and cross as few lines as possible. They should never run through a corner of a view or through any specific points or intersections.

Notes are classified as *general notes* when they apply to an entire drawing and as *local notes* when they apply to specific items.

General Notes. General notes should be lettered in the lower right-hand corner of the drawing, above or to the left of the title block,

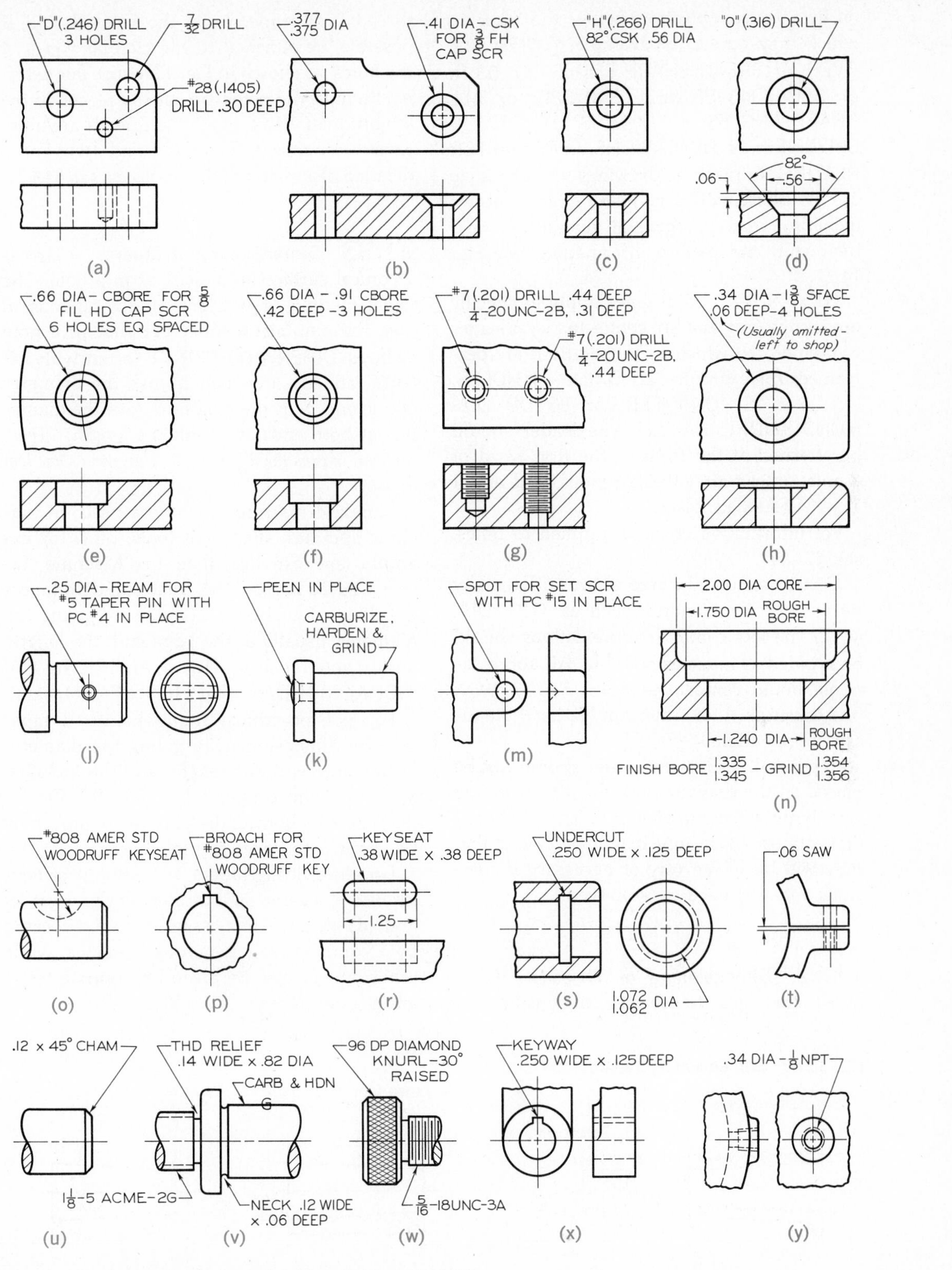


Fig. 11.42 **Local Notes.** See also Figs. 6.40 and 11.30.

or in a central position below the view to which they apply; for example, FINISH ALL OVER; or BREAK SHARP EDGES TO .03 R; or SAE 3345–BRINELL 340–380; or ALL DRAFT ANGLES 3° UNLESS OTHERWISE SPECIFIED; or DIMENSIONS APPLY AFTER PLATING. In machine drawings, the title strip or title block will carry many general notes, including material, general tolerances, heat treatment, and pattern information. See Fig. 14.27.

LOCAL NOTES. Local notes apply to specific operations only and are connected by a leader to the point at which such operations are performed; for example, .25 DRILL—4 HOLES; or .06 × 45° CHAMFER; or 96 DP DIAMOND KNURL, RAISED. The leader should be attached at the front of the first word of a note, or just after the last word, and not at any intermediate place.

For information on notes applied to holes, see §11.24.

Certain commonly used abbreviations may be used freely in notes, such as THD, DIA, MAX. The less common abbreviations should be avoided as much as possible. All abbreviations should conform to ANSI Z32.13–1950. See Appendix 4 for American National Standard abbreviations.

In general, leaders and notes should not be placed on the drawing until the dimensioning is substantially completed. If notes are lettered first, almost invariably it will be found that they will be in the way of necessary dimensions and will have to be moved.

## 11.32 Dimensioning of Threads.

Local notes are used to specify dimensions of threads. For tapped holes the notes should, if possible, be attached to the circular views of the holes, as shown in Fig. 11.42 (g). For external threads, the notes are usually placed in the longitudinal views where the threads are more easily recognized, as at (v) and (w). For a detailed discussion of thread notes, see §13.21.

## 11.33 Dimensioning of Tapers.

A *taper* is a conical surface on a shaft or in a hole. The usual method of dimensioning a taper is to give the amount of taper per foot in a note such as TAPER 2.00 PER FT (often with TO GAGE added), and then to give the diameter at one end, plus the length, or give the diameters at both ends and omit the length. *"Taper per foot" means the difference in diameter in one foot of length.*

Standard machine tapers are used on machine spindles, shanks of tools, pins, for example, and are described in "Machine Tapers," ANSI B5.10–1963. Such standard tapers are dimensioned on a drawing by giving the diameter, usually at the large end, the length, and a note, such as NO. 4 AMERICAN NATIONAL STANDARD TAPER. See Fig. 11.43 (a).

For not-too-critical requirements, a taper may be dimensioned by giving the diameter at the large end, the length, and the included angle, all with proper tolerances, (b). Or the diameters of both ends, plus the length, may be given with necessary tolerances.

For close-fitting tapers, the amount of *taper per inch on diameter* is indicated as shown at (c) and (d). A gage line is selected and located by a comparatively generous tolerance, while other dimensions are given appropriate tolerances as required.

Fig. 11.43 **Dimensioning Tapers.**

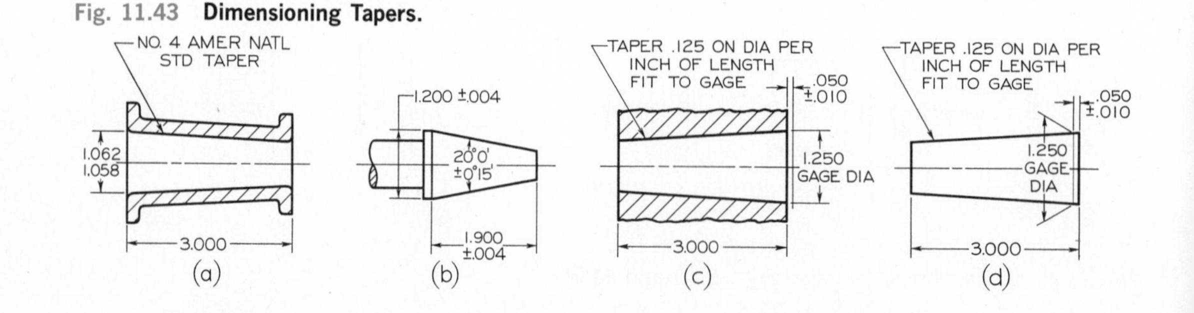

## 11.34 Dimensioning of Chamfers.

A *chamfer* is a beveled or sloping edge, and it is dimensioned by giving the length of the offset and the angle, as in Fig. 11.44 (a). A 45° chamfer also may be dimensioned in a manner similar to that shown at (a), but usually it is dimensioned by note without or with the word CHAMFER as at (b).

## 11.35 Shaft Centers.

Shaft centers are required on shafts, spindles, and other conical or cylindrical parts for turning, grinding, and other operations. Such a center may be dimensioned as shown in Fig. 11.45. Normally the centers are produced by a combined drill and countersink. See Appendix 38 for a table of shaft center sizes.

## 11.36 Dimensioning Keyways.

Methods of dimensioning keyways for Woodruff keys and stock keys are shown in Fig. 11.46. Note, in both cases, the use of a dimension to center the keyway in the shaft or collar. The preferred method of dimensioning the depth of a keyway is to give the dimension from the bottom of the keyway to the opposite side of the shaft or hole, as shown. The method of computing such a dimension is shown at (d). Values for A may be found in machinists' handbooks.

For general information about keys and keyways, see §13.34.

## 11.37 Dimensioning of Knurls.

A *knurl* is a roughened surface to provide a better handgrip or to be used for a press fit between two parts. For handgripping purposes, it is necessary only to give the pitch of the knurl, the type of knurling, and the length of the knurled area, Fig. 11.47(a) and (b). To dimension a knurl for a press fit, the toleranced

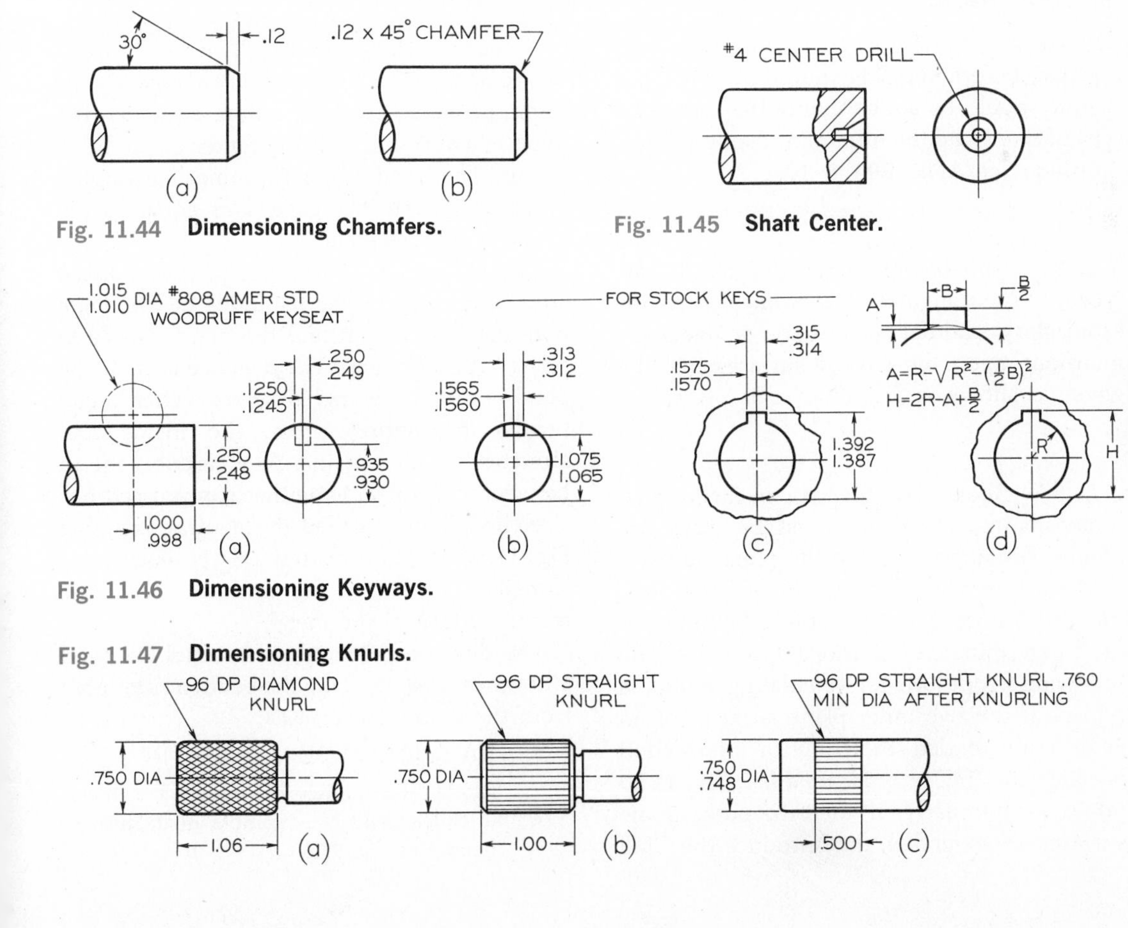


Fig. 11.44 **Dimensioning Chamfers.**

Fig. 11.45 **Shaft Center.**

Fig. 11.46 **Dimensioning Keyways.**

Fig. 11.47 **Dimensioning Knurls.**

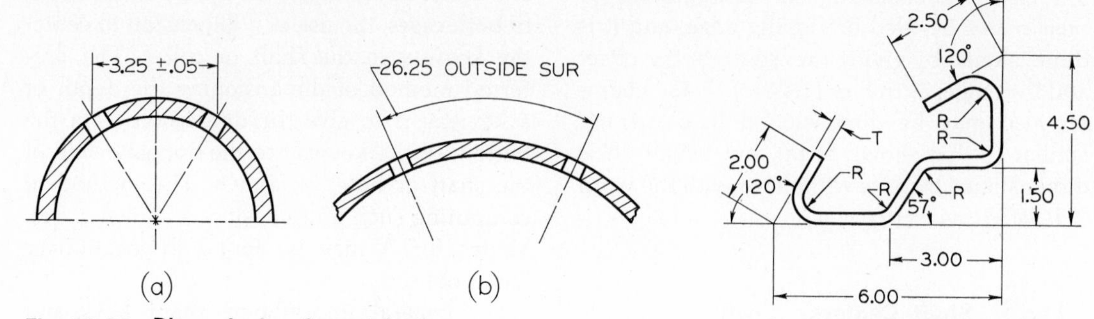


Fig. 11.48 **Dimensioning Curved Surfaces.**

Fig. 11.49 **Profile Dimensioning.**

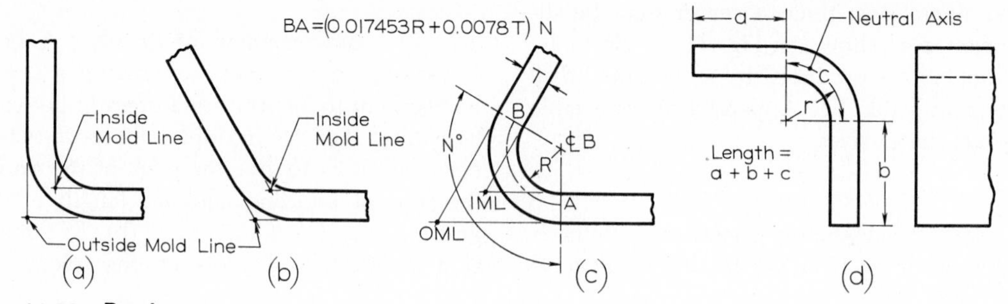


Fig. 11.50 **Bends.**

diameter before knurling should be given, (c). A note should be added giving the pitch and type of knurl and the minimum diameter after knurling. See ANSI B94. 6–1966.

**11.38 Dimensioning Along Curved Surfaces.** When angular measurements are unsatisfactory, chordal dimensions or linear dimensions upon the curved surfaces may be given, as shown in Fig. 11.48.

**11.39 Sheet-Metal Bends.** In sheet-metal dimensioning, allowance must be made for bends. The intersection of the plane surfaces adjacent to a bend is called the *mold line,* and this line, rather than the center of the arc, is used to terminate dimensions, Fig. 11.49. The following procedure for calculating bends is typical. If the two inner plane surfaces of an angle are extended, their line of intersection is called the IML or *inside mold line,* Fig. 11.50 (a) to (c). Similarly, if the two outer plane surfaces are extended, they produce the OML or *outside mold line.* The *center line of bend* (℄ B) refers primarily to the machine on which the bend is made, and is at the center of the bend radius. The bend radius for annealed dural is taken as double the gage, and equal to the gage for other metals.

The length, or *stretchout,* of the pattern equals the sum of the flat sides of the angle plus the distance around the bend measured along the *neutral axis.* The distance around the bend is called the *bend allowance.* When metal bends, it compresses on the inside and stretches on the outside. At a certain zone in between, the metal is neither compressed nor stretched. This is called the neutral axis. See Fig. 11.50 (d). The neutral axis is usually assumed to be .44 of the thickness from the inside surface of the metal.

The developed length of material, or bend allowance (BA), to make the bend is computed from the empirical formula:

$$BA = (0.017453R + 0.0078T)N$$

where R = radius of bend, T = metal thickness, and N = number of degrees of bend. See Fig. 11.50 (c)

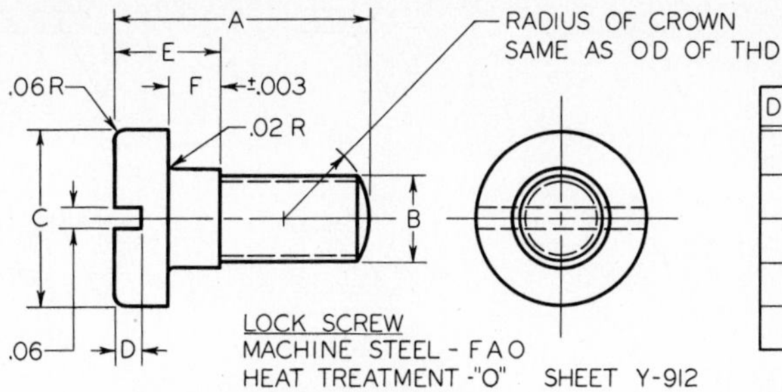


| DETAIL | A | B | C | D | E | F | UNC THD | STOCK | LBS |
|---|---|---|---|---|---|---|---|---|---|
| 1 | .62 | .38 | .62 | .06 | .25 | .135 | $\frac{5}{16}$ - 18 | $\frac{3}{4}$ DIA | .09 |
| 2 | .88 | .38 | .62 | .09 | .38 | .197 | $\frac{5}{16}$ - 18 | $\frac{3}{4}$ DIA | .12 |
| 3 | 1.00 | .44 | .75 | .12 | .38 | .197 | $\frac{3}{8}$ - 16 | $\frac{7}{8}$ DIA | .19 |
| 4 | 1.25 | .50 | .88 | .12 | .50 | .260 | $\frac{7}{16}$ - 14 | 1" DIA | .30 |
| 5 | 1.50 | .56 | 1.00 | .16 | .62 | .323 | $\frac{1}{2}$ - 13 | $1\frac{1}{8}$ DIA | .46 |

Fig. 11.51 **Tabular Dimensioning.**

## 11.40 Tabular Dimensions.

A series of objects having like features but varying in dimensions may be represented by one drawing, Fig. 11.51. Letters are substituted for dimension figures on the drawing, and the varying dimensions are given in tabular form. The dimensions of many standard parts are given in this manner in the various catalogs and handbooks.

## 11.41 Standards.

Dimensions should be given, wherever possible, to make use of readily available materials, tools, parts, and gages. The dimensions for many commonly used machine elements, such as bolts, screws, nails, keys, tapers, wire, pipes, sheet metal, chains, belts, ropes, pins, and rolled metal shapes, have been standardized, and the draftsman must obtain these sizes from company standards manuals, from published handbooks, from American National Standards, or from manufacturers' catalogs. Tables of some of the more common items are given in the Appendix of this text.

Such standard parts are not delineated on detail drawings unless they are to be altered for use, but are drawn conventionally on assembly drawings and are listed in parts lists, §14.14. Common fractions are generally used to indicate the nominal sizes of standard parts or tools. If the complete decimal system is used, all such sizes ordinarily are expressed by decimals; for example, .250 DRILL instead of $\frac{1}{4}$ DRILL.

## 11.42 Surface Roughness, Waviness, and Lay.

The modern demands of the automobile, the airplane, and other machines for parts that can stand heavier loads and higher speeds with less friction and wear have increased the need for accurate control of surface quality by the designer. Simple finish marks are no longer enough to specify surface finish on such parts.

The ANSI* recommends a system of symbols for use on drawings that is now broadly used by American industry. These symbols define *roughness, waviness,* and *lay. Surface roughness* refers to the small peaks and valleys that will be found on any surface, however smooth in appearance. *Waviness* refers to the larger undulations of a surface upon which the roughness features are superimposed. *Lay* refers to the direction of the toolmarks, scratches, or grains of a surface.

Surface finish is intimately related to the functioning of a surface, and proper specification of finish of such surfaces as bearings and seals is very important. If the finish of a surface is not important, it may not be necessary to specify the surface quality. It is to be used where needed only, since the cost of producing a surface becomes greater as the quality of surface called for is finer. Generally speaking, the ideal finish is the roughest one that will do the job satisfactorily.

When surface quality is to be specified, the V finish mark, §11.17, is used as a base, and the right side is extended upward as on a check mark, with a horizontal bar across the

*ANSI B46.1-1962.

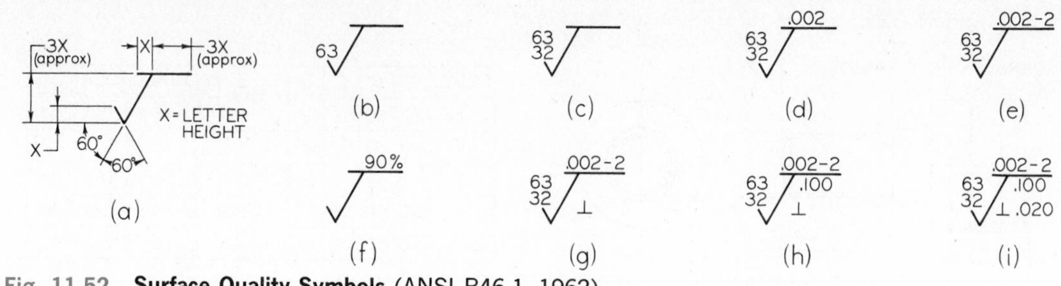


Fig. 11.52 **Surface Quality Symbols** (ANSI B46.1–1962).

top. The recommended proportions for the construction of the symbol are shown in Fig. 11.52 (a). When it is necessary to specify only maximum roughness height, and the width between ridges or direction of tool marks is unimportant, the symbol at (b) is used, but the horizontal crossbar may be omitted if desired. The roughness height is indicated by the numeral in microinches (1 microinch = 1 one-millionth inch) representing the *arithmetic-average deviation* of the surface from the mean line in a profile. The arithmetic average is an expression of the average amount of deviation of peaks and valleys from a mean line. The higher the number of microinches, the rougher the surface. Surface roughness in terms of arithmetic average can be easily measured with a surfagage or other tracer-type electrical instrument. This method is superior to the method of comparison, visually and by touch, with sample surfaces having accurately measured surface irregularities, although the latter method has a definite usefulness. When it is desired to specify maximum and minimum average roughness height, the upper and lower numbers are placed as at (c). The preferred roughness height values in microinches are as follows:

| | | | |
|---|---|---|---|
| 1 | 10 | 50 | 250 |
| 2 | 13 | 63 | 320 |
| 3 | 16 | 80 | 400 |
| 4 | 20 | 100 | 500 |
| 5 | 25 | 125 | 600 |
| 6 | 32 | 160 | 800 |
| 8 | 40 | 200 | 1000 |

When maximum waviness height is to be specified, the numerical value, in inches, is placed above the horizontal bar as at (d). Standard values to be used are (in inches) as follows:

| | | |
|---|---|---|
| 0.00002 | 0.0003 | 0.005 |
| 0.00003 | 0.0005 | 0.008 |
| 0.00005 | 0.0008 | 0.010 |
| 0.00008 | 0.001 | 0.015 |
| 0.0001 | 0.002 | 0.020 |
| 0.0002 | 0.003 | 0.030 |

The maximum waviness width, in inches, is given to the right of the waviness height number as at (e). In this example the waviness width is 2″.

Minimum requirements for contact or bearing area with a mating part or reference surface may be specified by a percentage value placed above the horizontal crossbar as shown at (f).

If it is further desired to specify the lay, it will be indicated by the addition of a symbol at the right of the V as in (g) to (i). The lay symbols are shown in Fig. 11.53.

When it is necessary to indicate the "roughness-width cut-off rating," this is given in inches underneath the horizontal crossbar, as shown at Fig. 11.52 (h). This number expresses the "maximum width in inches of surface irregularities to be included in the measurement of roughness height."* Standard values to be used are (in inches) 0.003, 0.010, 0.030, 0.100, 0.300, and 1.000. When no value is specified, the 0.030 value is assumed.

"Roughness width" is the maximum allowed spacing, in inches, between repetitive units of the surface pattern. The numeral is placed under the right portion of the crossbar, as shown at Fig. 11.52 (i).

*ANSI B46.1–1962.

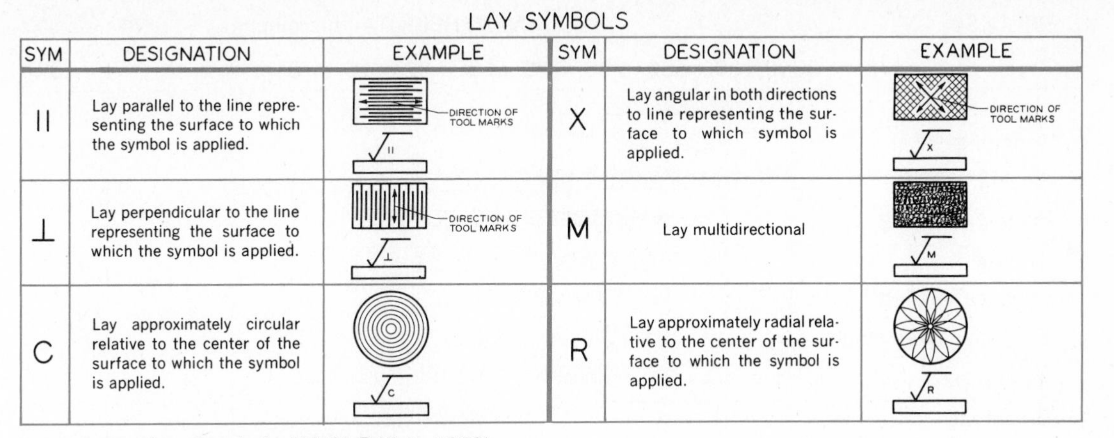

LAY SYMBOLS

| SYM | DESIGNATION | EXAMPLE | SYM | DESIGNATION | EXAMPLE |
|---|---|---|---|---|---|
| \|\| | Lay parallel to the line representing the surface to which the symbol is applied. | DIRECTION OF TOOL MARKS | X | Lay angular in both directions to line representing the surface to which symbol is applied. | DIRECTION OF TOOL MARKS |
| ⊥ | Lay perpendicular to the line representing the surface to which the symbol is applied. | DIRECTION OF TOOL MARKS | M | Lay multidirectional | |
| C | Lay approximately circular relative to the center of the surface to which the symbol is applied. | | R | Lay approximately radial relative to the center of the surface to which the symbol is applied. | |

Fig. 11.53 **Lay Symbols** (ANSI B46.1-1962).

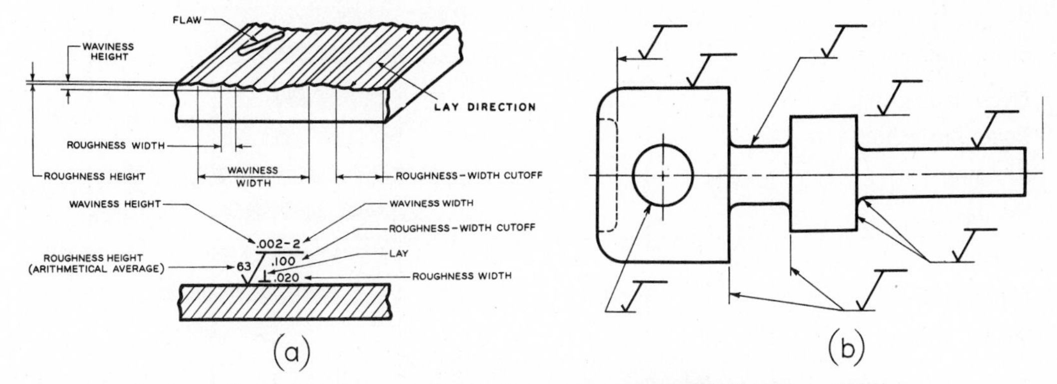


Fig. 11.54 **Surface Characteristics and Applications of Symbols** (ANSI B46.1-1962).

The relation of symbols to surface characteristics is illustrated in Fig. 11.54 (a). Only those numbers required to specify the surface adequately for the function should be included in the symbol.

The symbol is always made in the standard upright position—never at an angle or upside down, etc. Applications are shown at (b).

The typical range of surface roughness values that may be obtained from the various production methods is shown in Fig. 11.55. Preferred roughness-height values are shown at the top of the chart.

## 11.43 Dimensioning for Numerical Control.

In general, the basic dimensioning practices are compatible with the data requirements for tape-controlled automatic production machines. However, to make the best use of this type of production, the designer and draftsman should first consult the manufacturing machine manuals before making the drawings for production. Certain considerations should be noted:

1. A set of three mutually perpendicular datum or reference planes is usually required for coordinate dimensioning. These planes either must be obvious or clearly identified. See Fig. 11.56.
2. The designer selects as origins for dimensions those surfaces or other features most important to the functioning of the part. Enough of these features are selected to position the part in relation to the set of mutually perpendicular planes. All related dimensions on the part are then made from these planes.
3. All dimensions should be in decimals.
4. Angles should be given by coordinate dimensions rather than in degrees.

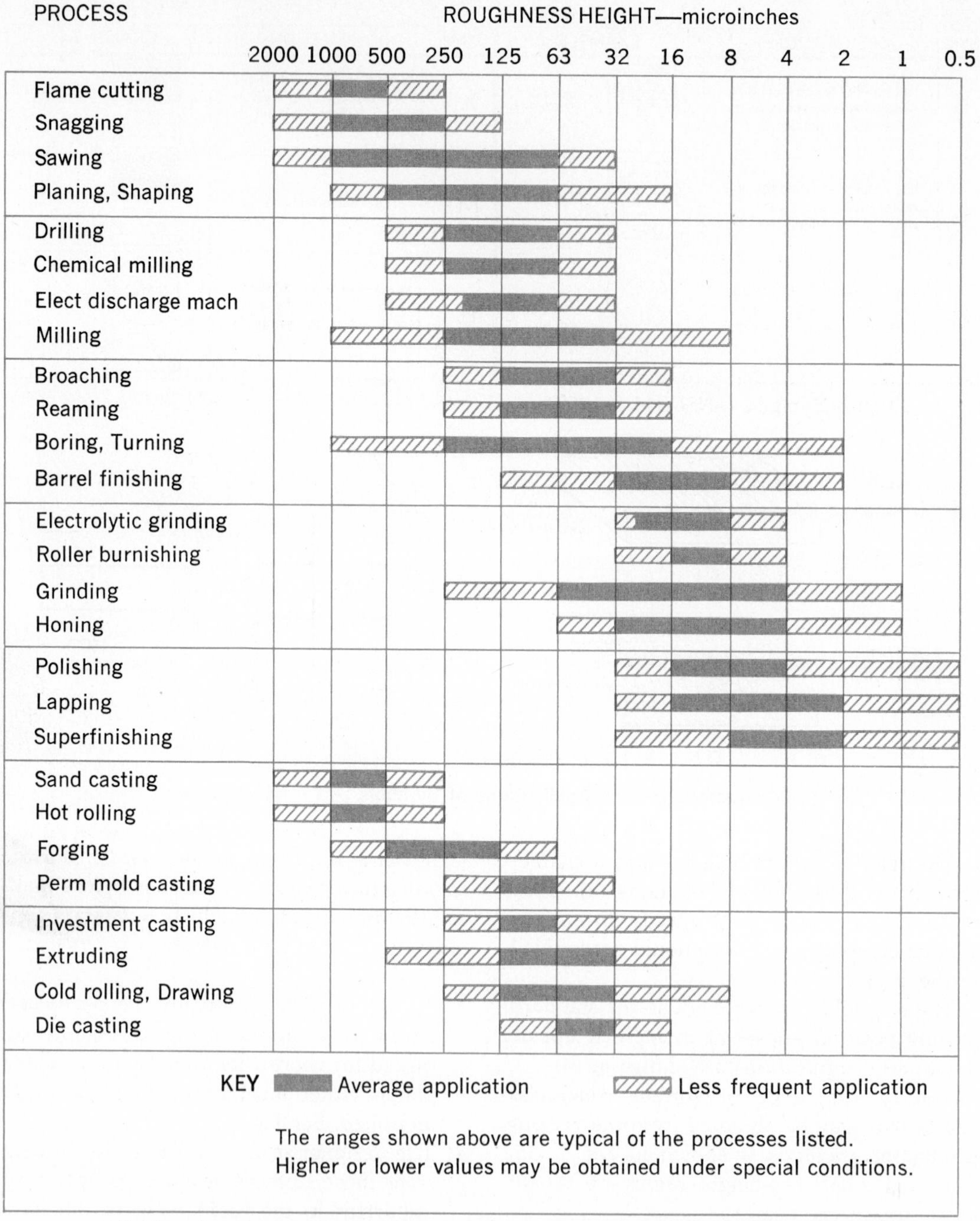

PROCESS

ROUGHNESS HEIGHT—microinches

2000 1000 500 250 125 63 32 16 8 4 2 1 0.5

Flame cutting
Snagging
Sawing
Planing, Shaping
Drilling
Chemical milling
Elect discharge mach
Milling
Broaching
Reaming
Boring, Turning
Barrel finishing
Electrolytic grinding
Roller burnishing
Grinding
Honing
Polishing
Lapping
Superfinishing
Sand casting
Hot rolling
Forging
Perm mold casting
Investment casting
Extruding
Cold rolling, Drawing
Die casting

KEY Average application Less frequent application

The ranges shown above are typical of the processes listed.
Higher or lower values may be obtained under special conditions.

Fig. 11.55 **Surface Roughness Produced by Common Production Methods** (ANSI B46.1-1962).

5. Standard tools, such as drills, reamers, and taps, should be specified wherever possible.
6. All tolerances should be determined by the design requirements of the part, not by the capability of the manufacturing machine.

**11.44 Do's and Don'ts of Dimensioning.** The following check list summarizes briefly most of the situations in which a beginning draftsman is likely to make a mistake in dimensioning. The student should check his

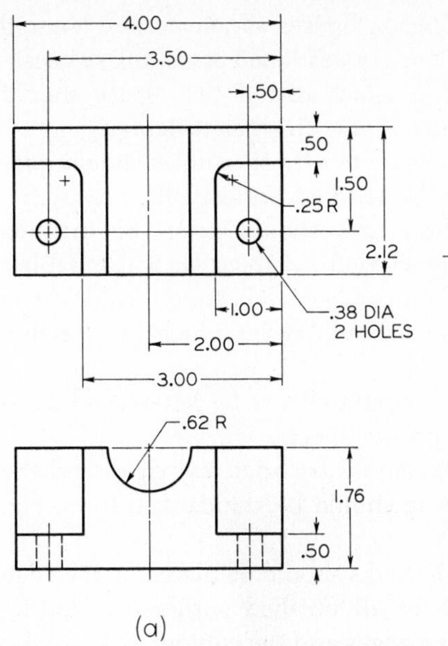


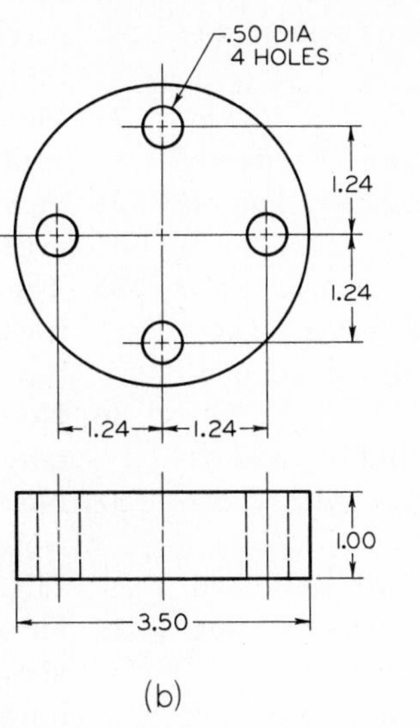


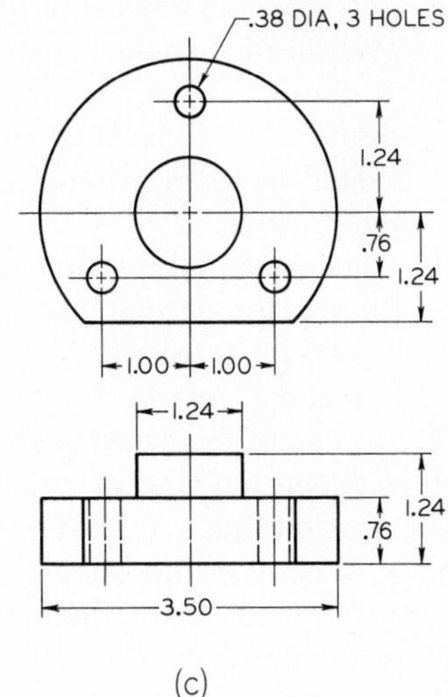

Fig. 11.56 **Coordinate Dimensioning.**

drawing by this list before submitting it to his instructor.

1. Each dimension should be given clearly, so that it can be interpreted in only one way.
2. Dimensions should not be duplicated or the same information be given in two different ways, and no dimensions should be given except those needed to produce or inspect the part.
3. Dimensions should be given between points or surfaces that have a functional relation to each other or that control the location of mating parts.
4. Dimensions should be given to finished surfaces or important center lines in preference to rough surfaces wherever possible.
5. Dimensions should be so given that it will not be necessary for the machinist to calculate, scale, or assume any dimension.
6. Dimensions should be attached to the view where the shape is best shown (contour rule).
7. Dimensions should be placed in the views where the features dimensioned are shown true shape.
8. Avoid dimensioning to hidden lines wherever possible.
9. Dimensions should not be placed upon a view unless clearness is promoted and long extension lines are avoided.
10. Dimensions applying to two adjacent views should be placed between views, unless clearness is promoted by placing them outside.
11. The longer dimensions should be placed outside all intermediate dimensions, so that dimension lines will not cross extension lines.
12. In machine drawing, omit all inch marks, except when necessary for clearness; for example, 1″ VALVE.
13. Do not expect the workman to assume a feature is centered (as a hole on a plate), but give a location dimension from one side. However, if a hole is to be centered on a symmetrical rough casting, mark the center line ℄ and omit the locating dimension from the center line.
14. A dimension should be attached to only one view (extension lines not connecting two views).
15. Detail dimensions should "line up" in chain fashion.

16. Avoid a complete chain of detail dimensions; better omit one, otherwise add REF (reference) to one detail dimension or the over-all dimension.
17. A dimension line should never be drawn through a dimension figure. A figure should never be lettered over any line of the drawing.
18. Dimension lines should be spaced uniformly throughout the drawing. They should be at least $\frac{3}{8}''$ from the object outline and $\frac{1}{4}''$ apart.
19. No line of the drawing should be used as a dimension line or coincide with a dimension line.
20. A dimension line should never be joined end-to-end (chain fashion) with any line of the drawing.
21. Dimension lines should not cross, if avoidable.
22. Dimension lines and extension lines should not cross, if avoidable (extension lines may cross each other).
23. When extension lines cross extension lines or visible lines, no break in either line should be made.
24. A center line may be extended and used as an extension line, in which case it is still drawn like a center line.
25. Center lines should generally not extend from view to view.
26. Leaders for notes should be straight, not curved, and pointing to the circular views of holes wherever possible.
27. Leaders should slope at 45°, or 30°, or 60° with horizontal but may be made at any odd angle except vertical or horizontal.
28. Leaders should extend from the beginning or end of a note, the horizontal "shoulder" extending from the mid-height of the lettering.
29. Dimension figures should be approximately centered between the arrowheads, except that in a "stack" of dimensions, the figures should be "staggered."
30. Dimension figures should be about $\frac{1}{8}''$ high for whole numbers and $\frac{1}{4}''$ high for fractions.
31. Dimension figures should never be crowded or in any way made difficult to read.
32. Dimension figures should not be lettered over lines or sectioned areas unless necessary, in which case a clear space should be left for the dimension figures.
33. Dimension figures for angles should generally be lettered horizontally.
34. Fraction bars should never be inclined except in confined areas, such as in tables.
35. The numerator and denominator of a fraction should never touch the fraction bar.
36. Notes should always be lettered horizontally on the sheet.
37. Notes should be brief and clear, and the wording should be standard in form, Fig. 11.42.
38. Finish marks should be placed on the edge views of all finished surfaces, including hidden edges and the contour and circular views of cylindrical surfaces.
39. Finish marks should be omitted on holes or other features where a note specifies a machining operation.
40. Finish marks should be omitted on parts made from rolled stock.
41. If a part is finished all over, omit all finish marks, and use the general note: FINISH ALL OVER, or FAO, not "*f*"AO or "*f*"ALL OVER.
42. A cylinder is dimensioned by giving both its diameter and length in the rectangular view, except when notes are used for holes. A diagonal diameter in the circular view may be used in cases where clearness is gained thereby.
43. Holes to be bored, drilled, reamed, etc., are size-dimensioned by notes in which the leaders preferably point toward the circular views of the holes. Indications of shop processes may be omitted from notes.
44. Drill sizes are preferably expressed in decimals. Particularly for drills designated by number or letter, the decimal size must also be given.
45. In general, a circle is dimensioned by its diameter, an arc by its radius.
46. Avoid diagonal diameters, except for very large holes and for circles of centers. They may be used on positive cylinders when clearness is gained thereby.

47. A diameter dimension figure should be followed by DIA except when it is obviously a diameter.
48. The letter R should always follow a radius dimension figure. The radial dimension line should have only one arrowhead, and it should touch the arc.
49. Cylinders should be located by their center lines.
50. Cylinders should be located in the circular views, if possible.
51. Cylinders should be located by coordinate dimensions in preference to angular dimensions where accuracy is important.
52. When there are several rough noncritical features obviously the same size (fillets, rounds, ribs, etc.), it is necessary to give only typical dimensions, or to use a note.
53. When a dimension is not to scale, it should be underscored with a wavy line or marked NTS or NOT TO SCALE.
54. Mating dimensions should be given correspondingly on drawings of mating parts.
55. Pattern dimensions should be given in two-place decimals or in common whole numbers and fractions to the nearest $\frac{1}{16}''$.
56. Decimal dimensions should be used when accuracy greater than $\frac{1}{64}''$ is required on a machine dimension.
57. Avoid cumulative tolerances, especially in limit dimensioning, described in §12.10.

## 11.45 Dimensioning Problems.

It is expected that most of the student's practice in dimensioning will be in connection with working drawings assigned from other chapters. However, a limited number of special dimensioning problems are available here in Figs. 11.57 and 11.58. The problems are designed for Layout A-3 ($8\frac{1}{2}'' \times 11''$), and are to be drawn with instruments and dimensioned to a full-size scale.

Dimensioning problems in convenient form for solution may be found in *Technical Drawing Problems, Series 1,* by Giesecke, Mitchell, Spencer, and Hill; in *Technical Drawing Problems, Series 2,* by Spencer and Hill; and in *Technical Drawing Problems, Series 3,* by Spencer and Hill, all designed to accompany this text and published by Macmillan Publishing Co., Inc.

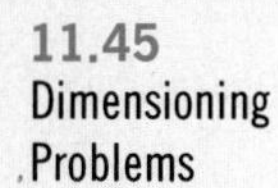


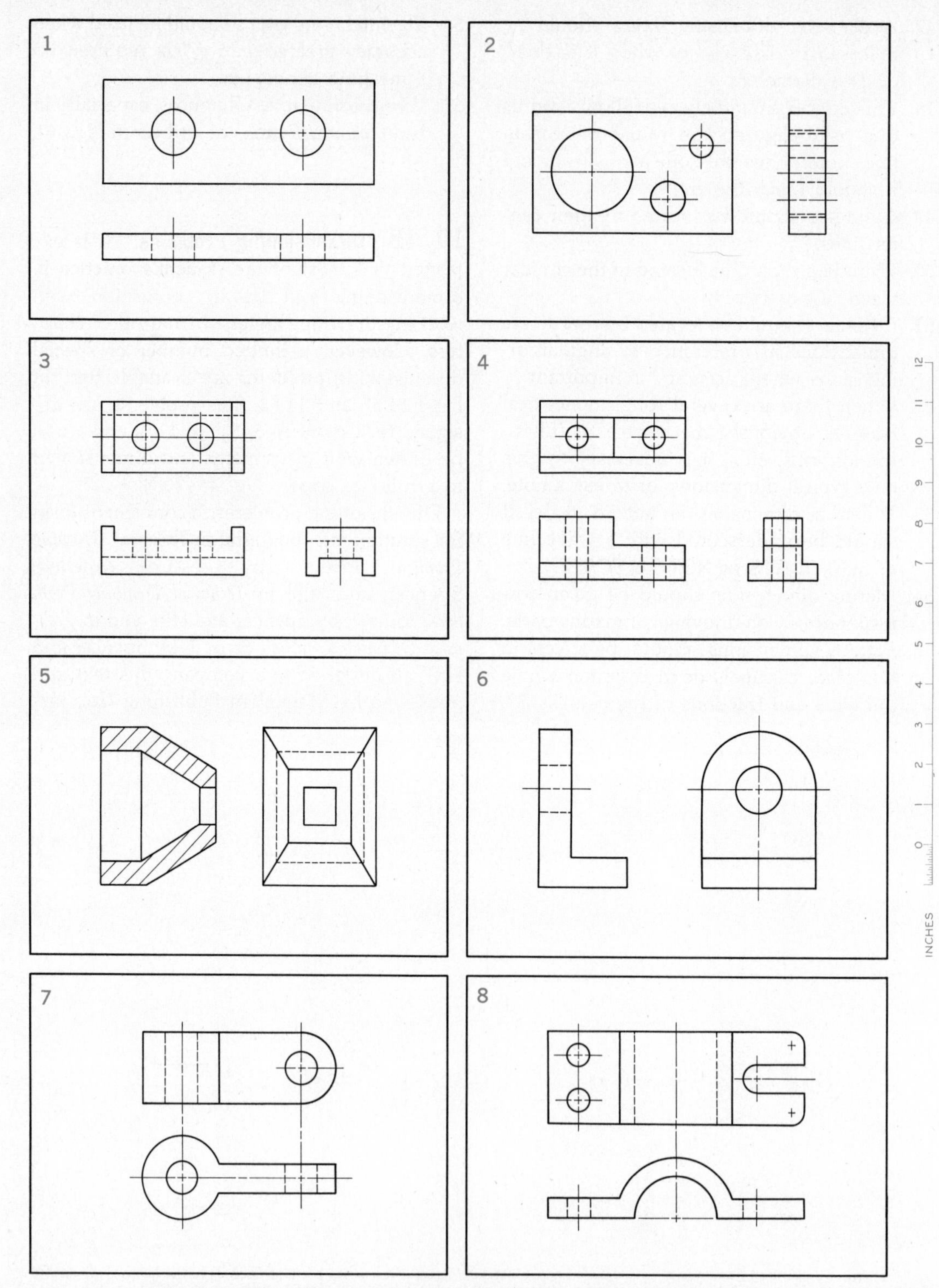


Fig. 11.57 Using Layout A-3, draw assigned problem with instruments. To obtain sizes, place bow dividers on the views on this page and transfer to scale at the side to obtain values in inches (nearest $\frac{1}{16}''$). Dimension drawing completely in fractions or two-place decimals—inches or millimeters (see inside front cover) as assigned, full size.

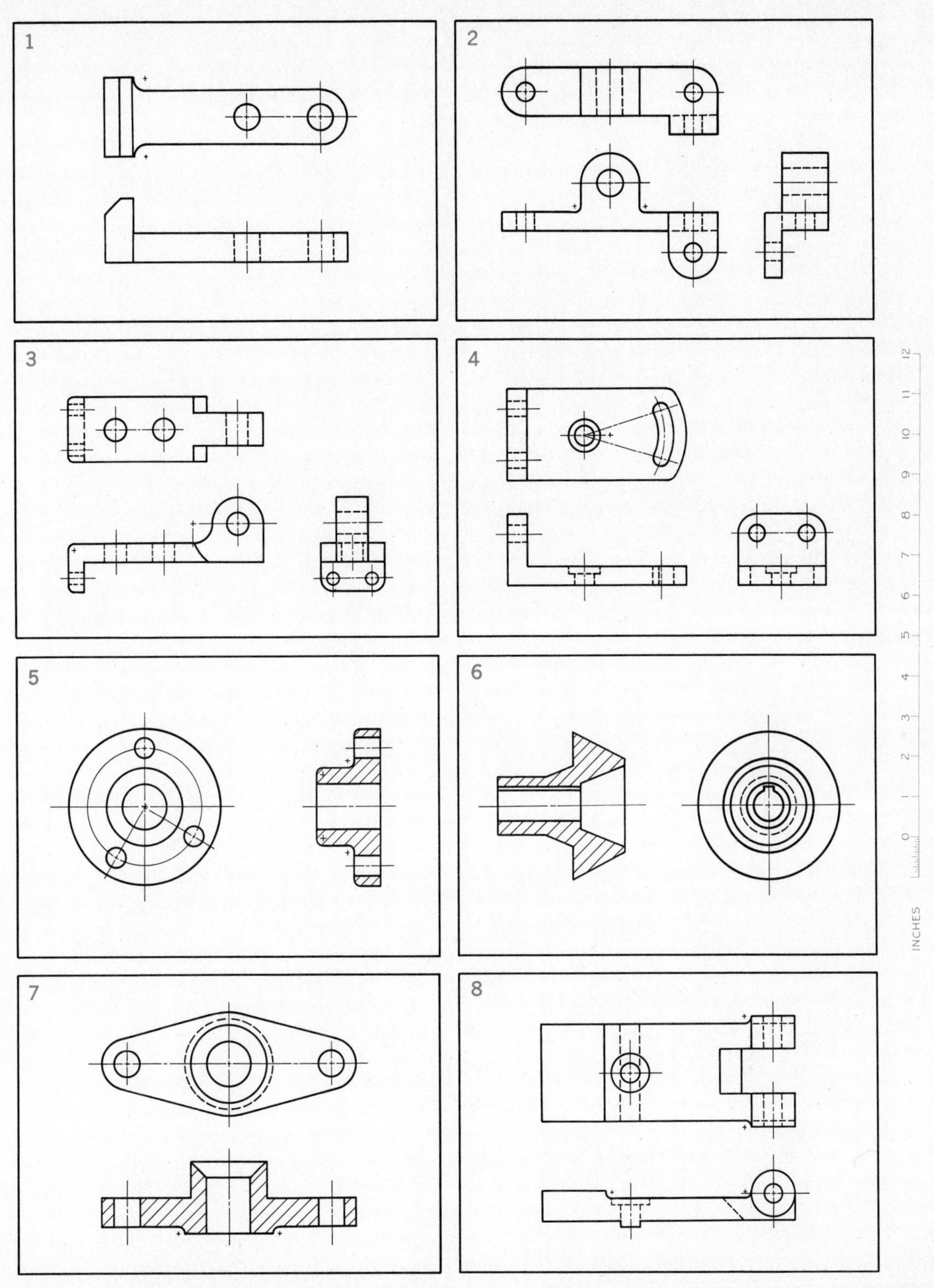


Fig. 11.58 Using Layout A-3, draw assigned problem with instruments. To obtain sizes, place bow dividers on the views on this page and transfer to scale at the side to obtain values in inches (nearest $\frac{1}{16}''$). Dimension drawing completely in fractions or two-place decimals—inches or millimeters (see inside front cover) as assigned, full size.

# 12
# tolerancing

**12.1 Dimensioning.** *Interchangeable manufacturing,* by means of which parts can be made in widely separated localities and then be brought together for assembly, where the parts will all fit together properly, is an essential element of mass production. Without interchangeable manufacturing, modern industry could not exist, and without effective size-control by the engineer, interchangeable manufacturing could not be achieved.

For example, an automobile manufacturer not only subcontracts the manufacture of many parts of a design to other companies, but he must be concerned with parts for replacement. All parts in each category must be near enough alike so that any one of them will fit properly in any assembly. It might be thought that if the dimensions are given on the blueprint, such as $4\frac{5}{8}$ and 2.632, all the parts will be exactly alike and will naturally fit properly in the machine. But, unfortunately, *it is impossible to make anything to exact size.* Parts can be made to very close dimensions, even to a few millionths of an inch (e.g., gage blocks), but such accuracy is extremely expensive.

However, exact sizes are not needed, only varying degrees of accuracy according to functional requirements. A manufacturer of children's tricycles would soon go out of business if he insisted on making the parts with jet-engine accuracy, as no one would be willing to pay the price. So what is needed is a means of specifying dimensions with whatever degree of accuracy may be required. The answer to the problem is the specification of a *tolerance* on each dimension.

**12.2 Tolerancing.*** *Tolerance is the amount of variation permitted in the size of a part or in the location of points or surfaces.* For example, a dimension given as 1.625 ± .002 means that it may be (on the manufactured part) 1.627″ or 1.623″, or anywhere between these dimensions. The tolerance, or total amount of variation "tolerated," is .004″. Thus, it becomes the function of the draftsman or designer to specify the allowable error that may be tolerated for a given dimension and still permit the satisfactory functioning of the part. Since greater accuracy costs more money, he will not specify the closest tolerance, but instead will specify as generous a tolerance as possible.

The old method of indicating dimensions of two mating parts was to give the nominal dimensions of the two parts in common whole numbers and fractions and to indicate by a note the kind of fit that was desired, as shown in Fig. 12.1 (a), and then to depend upon the workman to produce the parts so that they would fit together and function properly. Other types of fit included "drive fit," "sliding fit," "tunking fit," and "force fit."

In the example shown at (a), the machinist would make the hole close to $1\frac{1}{4}$″ diameter and would then make the shaft, say, .003″ less in diameter. It would not matter if the hole were several thousands more or less than 1.250″; he could make the shaft about .003″ less and obtain the desired fit. But this method would not work in quantity production, since the sizes would vary considerably and would not be interchangeable; that is, any given shaft would not fit properly in any hole.

In order to control the dimensions of quantities of the two parts so that any two mating parts will be interchangeable, it is necessary to assign tolerances to the dimensions of the parts, as shown at (b). The diameter of the hole may be machined not less than 1.250″ and not more than 1.251″, these two figures representing the *limits* and the difference between them, .001″, being the *tolerance.* Likewise, the shaft must be produced between the limits of 1.248″ and 1.247″, the tolerance on the shaft being the difference between these, or .001″.

*See ANSI Y14.5-1966.

A pictorial illustration of the dimensions in Fig. 12.1 (b) is shown in Fig. 12.2 (a). The maximum shaft is shown solid, and the minimum shaft is shown in phantom. The difference in diameters, .001″, is the tolerance on the shaft. Similarly, the tolerance on the hole is the difference between the two limits shown, or .001″. The loosest fit, or maximum clearance, occurs when the smallest shaft is in the largest hole, as shown at (b). The tightest fit, or minimum clearance, occurs when the largest shaft is in the smallest hole, as shown at (c). The difference between these, .002″, is the *allowance.* The average clearance is .003″, which is the same difference as allowed in the example of Fig. 12.1 (a); thus any shaft will fit any hole interchangeably.

When parts are required to fit properly in assembly but not to be interchangeable, the size of one part need not be toleranced, but indicated to be made to fit at assembly, Fig. 12.3.

**12.3 Definitions of Terms.*** At this point it is well to fix in mind the definitions of certain terms:

*ANSI Y14.5-1966.

Fig. 12.1 **Fits Between Mating Parts.**

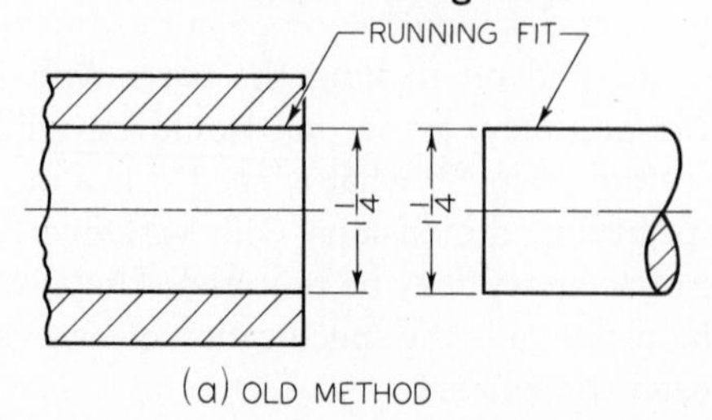

(a) OLD METHOD

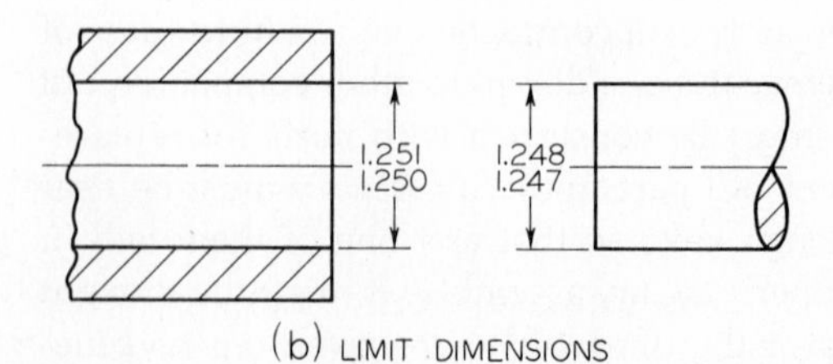

(b) LIMIT DIMENSIONS

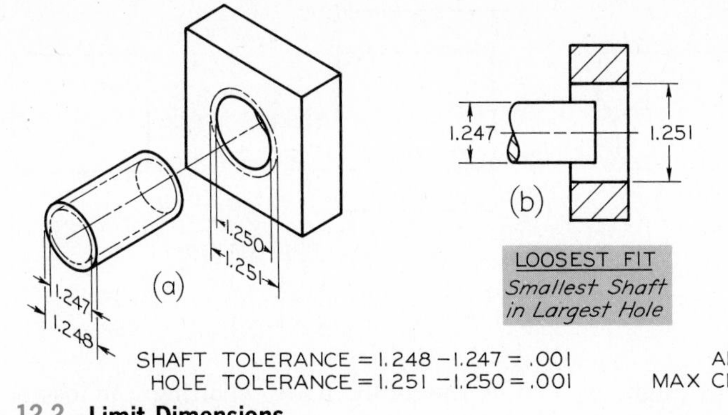
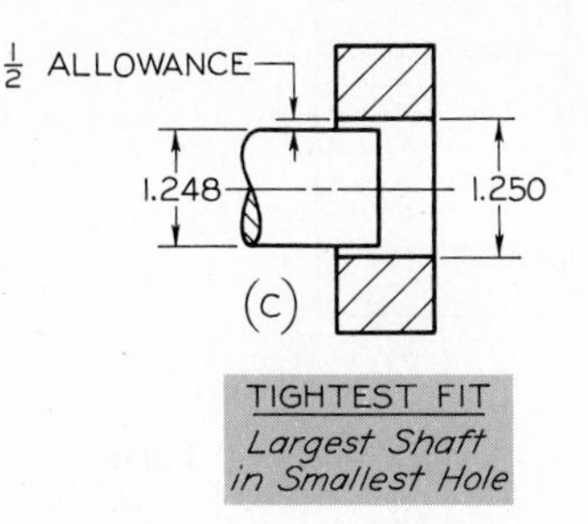


Fig. 12.2 **Limit Dimensions.**

*Nominal size*—The designation which is used for the purpose of general identification. In Fig. 12.1 (b), the nominal size of both hole and shaft is $1\frac{1}{4}''$, or 1.25″ in a decimal system of dimensioning.

*Basic size*—The size from which limits of size are derived by the application of allowances and tolerances. It is the exact theoretical size from which limits are figured. In Fig. 12.1 (b) the basic size is the decimal equivalent of the nominal size $1\frac{1}{4}''$, or 1.250″.

*Tolerance*—The total amount by which a given dimension may vary, or the difference between the limits. In Fig. 12.2 (a) the tolerance on either the shaft or hole is the difference between the limits, or .001″.

*Limits*—The maximum and minimum sizes indicated by a toleranced dimension. In Fig. 12.2 (a) the limits for the hole are 1.250″ and 1.251″, and for the shaft are 1.248″ and 1.247″.

*Allowance*—The minimum clearance space (or maximum interference) intended between the maximum material limits of mating parts. In Fig. 12.2 (c) the allowance is the difference between the smallest hole, 1.250″, and the largest shaft, 1.248″, or .002″. Allowance, then, represents the tightest permissible fit, and is simply the smallest hole minus the largest shaft. For clearance fits, this difference will be positive, while for interference fits it will be negative.

## 12.4 Fits Between Mating Parts.

"Fit is the general term used to signify the range of

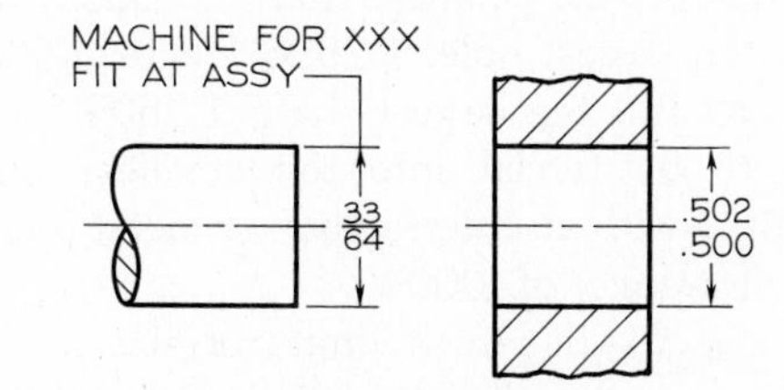


Fig. 12.3 **Noninterchangeable Fit.**

tightness or looseness which may result from the application of a specific combination of allowances and tolerances in the design of mating parts."* There are four general types of fits between parts:

1. *Clearance fit*—In which an internal member fits in an external member (as a shaft in a hole), and always leaves an air space or clearance between the parts. In Fig. 12.1 (b) the largest shaft is 1.248″ and the smallest hole is 1.250″, which permits a minimum air space of .002″ between the parts. This space is the allowance, and in a clearance fit it is always positive.

2. *Interference fit*—In which the internal member is larger than the external member such that there is always an actual interference of metal. In Fig. 12.4 (a) the smallest shaft is 1.2513″, and the largest hole is 1.2506″, so that there is an actual interference of metal amounting to at least .0007″. Under maximum material conditions the interference would be .0019″. This interference is the allowance, and in an interference fit it is always negative.

* ANSI Y14.5-1966.

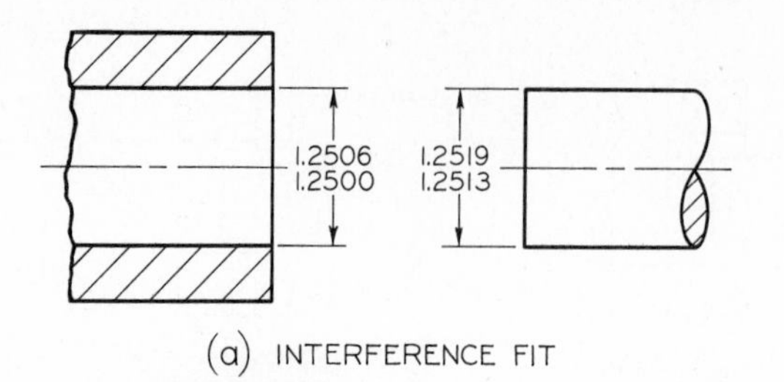


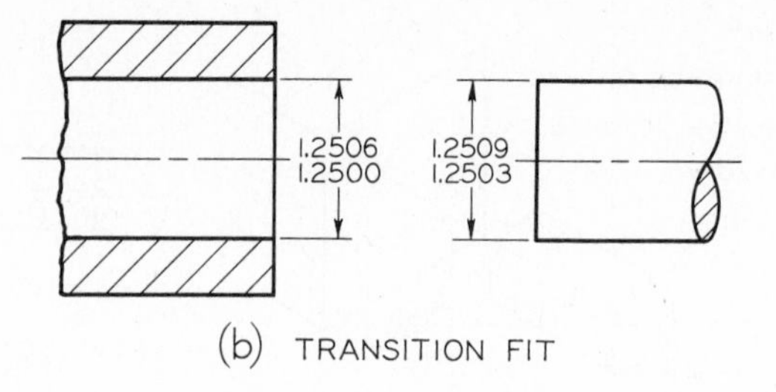


Fig. 12.4 **Fits Between Parts.**

3. *Transition fit*—In which the fit might be either a clearance fit or an interference fit. In Fig. 12.4 (b) the smallest shaft, 1.2503″, will fit in the largest hole, 1.2506″, with .0003″ to spare. But the largest shaft, 1.2509″, will have to be forced into the smallest hole, 1.2500″, with an interference of metal (negative allowance) of .0009″.

4. *Line fit*—In which limits of size are so specified that a clearance or surface contact may result when mating parts are assembled.

## 12.5 Selective Assembly.

If allowances and tolerances are properly given, mating parts can be completely interchangeable. But for close fits, it is necessary to specify very small allowances and tolerances, and the cost may be very high. In order to avoid this expense, either manual or computer controlled *selective assembly* is often used. In selective assembly, all parts are inspected and classified into several grades according to actual sizes, so that "small" shafts can be matched with "small" holes, "medium" shafts with "medium" holes, and so on. In this way, very satisfactory fits often may be obtained at much less expense than by machining all mating parts to very accurate dimensions. Since a transition fit may or may not represent an interference of metal, interchangeable assembly generally is not as satisfactory as selective assembly.

## 12.6 Basic Hole System.

Standard reamers, broaches, and other standard tools are often used to produce holes, and standard plug gages are used to check the actual sizes. On the other hand, shafting can easily be machined to any size desired. Therefore, toleranced dimensions are commonly figured on the so-called *basic hole system.* In this system, the *minimum hole is taken as the basic size,* an allowance is assigned, and tolerances are applied on both sides of, and away from, this allowance.

In Fig. 12.5 (a) the minimum size of the hole, .500″, is taken as the basic size. An allowance of .002″ is decided upon and subtracted from the basic hole size, giving the maximum shaft, .498″. Tolerances of .002″ and .003″, respectively, are applied to the hole and shaft to obtain the maximum hole of .502″ and the minimum shaft of .495″. Thus the minimum clearance between the parts becomes .500″ − .498″ = .002″ (smallest hole minus largest shaft), and the maximum clearance is .502″ − .495″ = .007″ (largest hole minus smallest shaft).

In the case of an interference fit, the maximum shaft size would be found by *adding the desired allowance* (maximum interference) to the basic hole size. In Fig. 12.4 (a), the basic size is 1.2500″. The maximum interference decided upon was .0019″, which added to the basic size gives 1.2519″, the largest shaft size.

The basic hole size can be changed to the basic shaft size by subtracting the allowance for a clearance fit, or adding it for an interference fit. The result is the largest shaft size, which is the new basic size.

## 12.7 Basic Shaft System.

In some branches of industry, such as textile machinery manufacturing, in which use is made of a great deal of cold-finished shafting, the *basic shaft system* is often used. This system should be used only when there is a reason for it. For example, it is advantageous when several parts having different fits, but one nominal size, are required on a single shaft. In this system *the maximum shaft is taken as the basic size,* an allowance for each mating part is assigned, and tolerances are applied on both sides of, and away from, this allowance.

In Fig. 12.5 (b) the maximum size of the shaft, .500″, is taken as the basic size. An allowance of .002″ is decided upon and added to the basic shaft size, giving the minimum hole, .502″. Tolerances of .003″ and .001″, respectively, are applied to the hole and shaft to obtain the maximum hole .505″ and the minimum shaft .499″. Thus the minimum clearance between the parts is .502″ − .500″ = .002″ (smallest hole minus largest shaft), and the maximum clearance is .505″ − .499″ = .006″ (largest hole minus smallest shaft).

In the case of an interference fit, the minimum hole size would be found by *subtracting the desired allowance from the basic shaft size.*

The basic shaft size may be changed to the basic hole size by adding the allowance for a clearance fit, or subtracting it for an interference fit. The result is the smallest hole size, which is the new basic size.

## 12.8 Specification of Tolerances.

A tolerance of a decimal dimension must be given in decimal form to the same number of places. A tolerance of a dimension given in common-fraction form must be given as a common fraction. See Fig. 12.8.

*General tolerances* on decimal dimensions in which tolerances are not given may also be covered in a printed note, such as DECIMAL DIMENSIONS TO BE HELD TO ±.001. Thus if a dimension 3.250 is given, the worker machines between the limits 3.249 and 3.251. See Fig. 14.14. Every dimension on a drawing should have a tolerance, either direct or by general tolerance note, except that commercial material is often assumed to have the tolerances set by commercial standards.

It is customary to indicate an over-all general tolerance for all common-fraction dimensions by means of a printed note in or just above the title block, for example, ALL FRACTIONAL DIMENSIONS $\pm\frac{1}{64}$ UNLESS OTHERWISE SPECIFIED; or HOLD FRACTIONAL DIMENSIONS TO $\pm\frac{1}{64}$ UNLESS OTHERWISE NOTED. See Fig. 14.14. General angular tolerances also may be given, as: ANGULAR TOLERANCE ±1°.

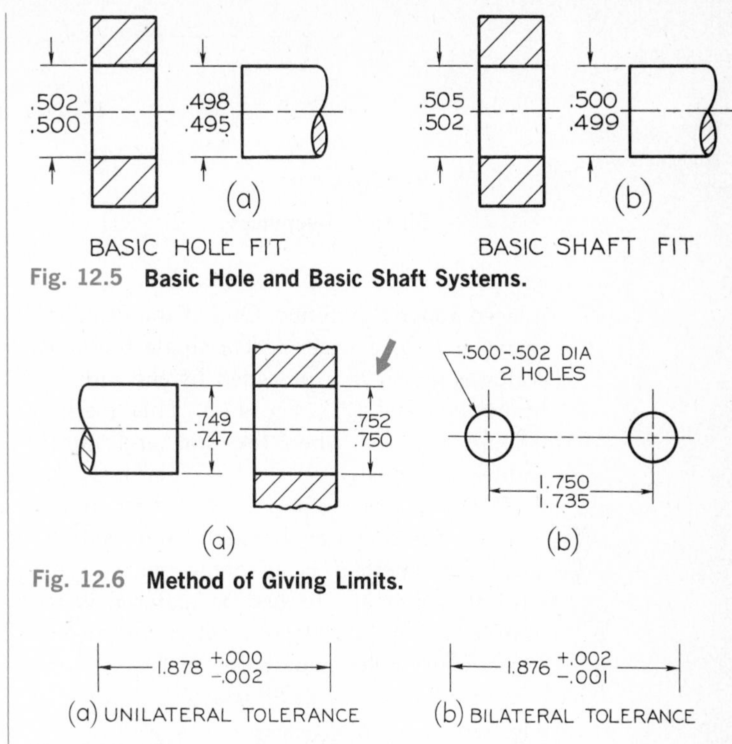


Fig. 12.5 **Basic Hole and Basic Shaft Systems.**



Fig. 12.6 **Method of Giving Limits.**

Fig. 12.7 **Tolerance Expression.**

Several methods of expressing tolerances in dimensions are approved by ANSI* as follows:

1. *Limit Dimensioning.* In this preferred method, the maximum and minimum limits of size and location are specified, as shown in Fig. 12.6. The high limit is placed above the low limit, thus: $^{.749}_{.747}$. In note form, the low limit is given first, thus: .500−.502 DIA.

2. *Plus and Minus Dimensioning.* In this method the basic size is followed by a plus and minus expression of tolerance resulting in either a unilateral or bilateral tolerance as in Fig. 12.7. If two unequal tolerance numbers are given, one plus and one minus, the plus is

* ANSI Y14.5-1966.

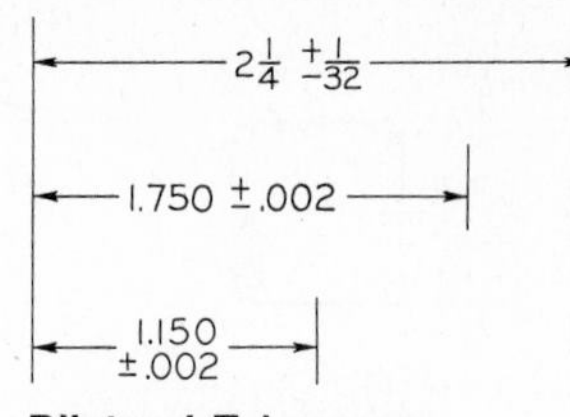


Fig. 12.8 **Bilateral Tolerances.**

placed above the minus. One of the numbers may be zero, if desired. If a single tolerance value is given, it is preceded by the plus-or-minus symbol (±), Fig. 12.8. This method should be used when the plus and minus values are equal.

The *unilateral system* of tolerances allows variations in only one direction from the basic size. This method is advantageous when a critical size is approached as material is removed during manufacture, as in the case of close-fitting holes and shafts. In Fig. 12.7 (a) the basic size is 1.878″. The tolerance .002″ is all in one direction—toward a smaller size. If this is a shaft diameter, the basic size 1.878″ is the size nearest the critical size because it is nearest to the tolerance zone; hence, the tolerance is taken *away* from the critical size. A unilateral tolerance is always all plus or all minus; that is, either the plus or the minus value must be zero. However, the zeros should not be omitted.

The *bilateral system* of tolerances allows variations in both directions from the basic size. Bilateral tolerances are usually given with location dimensions or with any dimensions that can be allowed to vary in either direction. In Fig. 12.7 (b) the basic size is 1.876″, and the actual size may be larger by .002″ or smaller by .001″. If it is desired to specify an equal variation in both directions, the combined plus-or-minus symbol (±) is used with a single value, as shown in Fig. 12.8.

Fig. 12.9 **Limit Dimensions.**

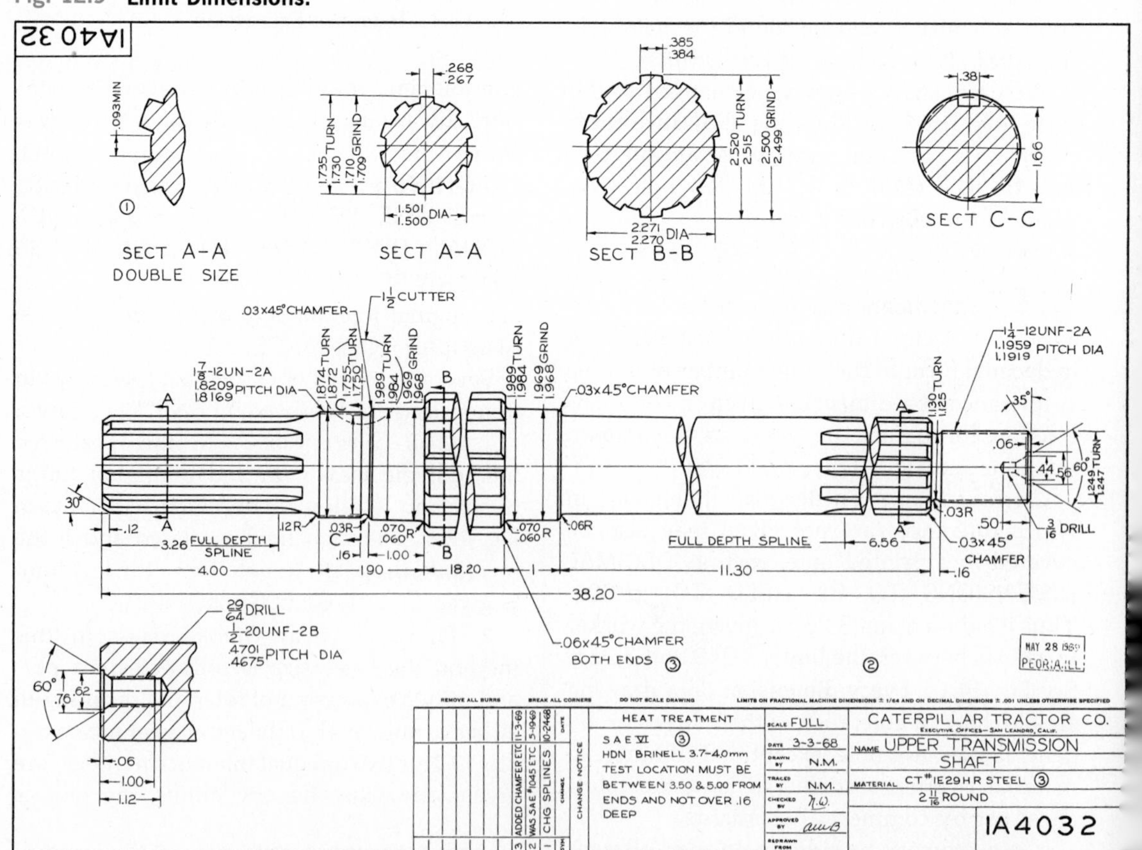

A typical example of limit dimensioning is given in Fig. 12.9.

3. *Single Limit Dimensioning.* It is not always necessary to specify both limits. MIN or MAX is often placed after a number to indicate minimum or maximum dimensions desired where other elements of design determine the other unspecified limit. For example, a thread length may be dimensioned thus: |←— 1.500 —→| MIN FULL THD or a radius dimensioned: .05 R MAX—↘ Other applications include depths of holes, chamfers, etc.

4. *Angular tolerances* are usually bilateral and in terms of degrees, minutes, and seconds.

*Examples:* 25° ± 1°, 25° 0′ ± 0° 15′, or 25° ± .25°. See also §12.16.

## 12.9 American National Standard Limits and Fits.

The American National Standards Institute (ANSI) has issued the ANSI B4.1-1967, "Preferred Limits and Fits for Cylindrical Parts," defining terms and recommending preferred standard sizes, allowances, tolerances, and fits.

This standard gives

> a series of standard types and classes of fits on a unilateral hole basis such that the fit produced by mating parts in any one class will produce approximately similar performance throughout the range of sizes. These tables prescribe the fit for any given size, or type of fit; they also prescribe the standard limits for the mating parts which will produce the fit.*

The tables are designed for the basic hole system, §12.6. See Appendixes 5 to 9.

Letter symbols to identify the five types of fits are:

- RC Running or Sliding Clearance Fits
- LC Locational Clearance Fits
- LT Transition Clearance or Interference Fits
- LN Locational Interference Fits
- FN Force or Shrink Fits

* ANSI B4.1-1967.

These letter symbols, plus a number indicating the class of fit within each type, are used to indicate a complete fit. Thus, FN 4 means a Class 4 Force Fit. The fits are described as follows:*

### Running and Sliding Fits

Running and sliding fits, for which description of classes of fits and limits of clearance are given [Appendix 5], are intended to provide a similar running performance, with suitable lubrication allowance, throughout the range of sizes. The clearances for the first two classes, used chiefly as slide fits, increase more slowly with diameter than the other classes, so that accurate location is maintained even at the expense of free relative motion.

### Locational Fits

Locational fits [Appendixes 6 to 8] are fits intended to determine only the location of the mating parts; they may provide rigid or accurate location, as with interference fits, or provide some freedom of location, as with clearance fits. Accordingly they are divided into three groups: clearance fits, transition fits, and interference fits.

### Force Fits

Force or shrink fits [Appendix 9] constitute a special type of interference fit, normally characterized by maintenance of constant bore pressures throughout the range of sizes. The interference therefore varies almost directly with diameter, and the difference between its minimum and maximum value is small, to maintain the resulting pressures within reasonable limits.

In the tables for each class of fit, the range of nominal sizes of shafts or holes is given in inches. To simplify the tables and reduce the space required to present them, the other values are given in thousandths of an inch. Minimum and maximum limits of clearance are given, the top number being the least clearance, or the allowance, and the lower number the maximum clearance, or greatest looseness of fit. Then, under the heading "Standard Limits" are given the limits for the hole and for the shaft that are to be applied algebraically to the basic size to obtain the limits of size for the parts, using the basic hole system.

* ANSI B4.1-1967.

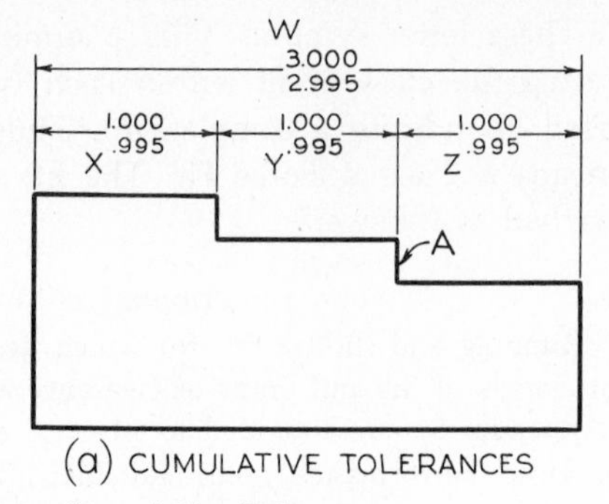

(a) CUMULATIVE TOLERANCES

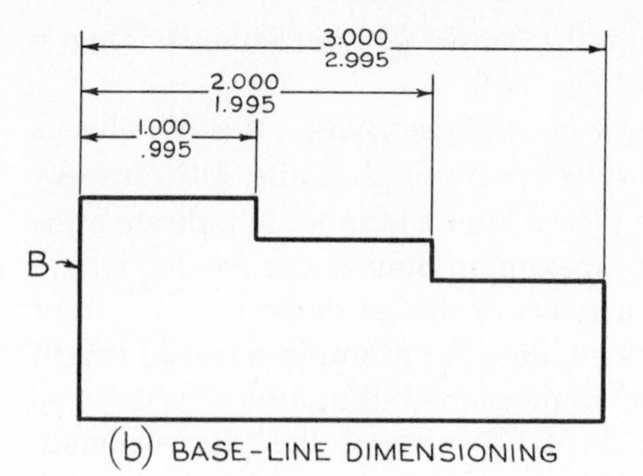

(b) BASE-LINE DIMENSIONING

Fig. 12.10 **Cumulative Tolerances.**

For example, take a 2.0000″ basic diameter with a Class RC 1 fit. This fit is given in Appendix 5. In the column headed "Nominal Size Range, Inches," find 1.97–3.15, which embraces the 2.0000″ basic size. Reading to the right we find under "Limits of Clearance" the values 0.4 and 1.2, representing the maximum and minimum clearance between the parts *in thousandths of an inch.* To get these values in inches, simply multiply by one thousandth, thus: $\frac{4}{10} \times \frac{1}{1000} = .0004$. Or, to convert 0.4 thousandths to inches, simply move the decimal point three places to the left, thus: .0004″. Therefore, for this 2.0000″ diameter, with a Class RC 1 fit, the minimum clearance, or allowance, is .0004″, and the maximum clearance, representing the greatest looseness, is .0012″.

Reading farther to the right we find under "Standard Limits" the value +0.5, which converted to inches is .0005″. Add this to the basic size thus: 2.0000″ + .0005″ = 2.0005″, the upper limit of the hole. Since the other value given for the hole is zero, the lower limit of the hole is the basic size of the hole, or 2.0000″. The hole would then be dimensioned as follows:

$$\begin{matrix}2.0005\\2.0000\end{matrix} \quad \text{or} \quad 2.0000\begin{matrix}+.0005\\-.0000\end{matrix}$$

The limits for the shaft are read as −.0004″ and −.0007″. To get the limits of the shaft, subtract these values from the basic size, thus:

2.0000″ − .0004″ = 1.9996″ (upper limit)
2.0000″ − .0007″ = 1.9993″ (lower limit)

The shaft would then be dimensioned as follows:

$$\begin{matrix}1.9996\\1.9993\end{matrix} \quad \text{or} \quad 1.9996\begin{matrix}+.0000\\-.0003\end{matrix}$$

**12.10 Accumulation of Tolerances.** In tolerance dimensioning, it is very important to consider the effect of one tolerance on another. When the location of a surface in a given direction is affected by more than one tolerance figure, the tolerances are *cumulative.* For example, in Fig. 12.10 (a), if dimension Z is omitted, surface A would be controlled by both dimensions X and Y, and there could be a total variation of .010 instead of the variation of .005 permitted by dimension Y, which is the dimension directly applied to surface A. Further, if the part is made to all the minimum tolerances of X, Y, and Z, the total variation in the length of the part will be .015, and the part can be as short as 2.985. However, the tolerance on the over-all dimension W is only .005, permitting the part to be only as short as 2.995. The part is superfluously dimensioned.

In some cases, for functional reasons, it may be desired to hold all three small dimensions X, Y, and Z closely without regard to the over-all length. In such a case the over-all dimension is just a *reference dimension* and should be marked REF. In other cases it may be desired to hold two small dimensions X and Y and the over-all closely without regard to dimension Z. In that case, dimension Z should be omitted, or marked REF.

As a rule, it is best to dimension each surface so that it is affected by only one dimension. This can be done by referring all dimensions to a single datum surface, such as B, as shown at (b). See also Fig. 11.35 (d) to (f).

## 12.11 **Tolerances and Shop Processes.**

As has been repeatedly stated in this chapter, tolerances should be as coarse as possible and still permit satisfactory use of the part. If this is done, great savings can be effected as a result of the use of less expensive tools, lower labor and inspection costs, and reduced scrapping of material.

Figure 12.11 shows a chart of tolerances in relation to shop processes that may be used as a guide by the draftsman in selecting appropriate tolerances. See also Chapter 10 for detailed information on shop processes and measuring devices.



## 12.12 **Positional Tolerances.**

In §11.25 are shown a number of examples of the traditional methods of locating holes—that is, by means of rectangular coordinates or angular dimensions. Each dimension has a tolerance, either given directly or indicated by a general note.

For example, in Fig. 12.12 (a) is shown a hole located from two surfaces at right angles to each other. As shown at (b), the center may lie anywhere within a square tolerance zone the sides of which are equal to the tolerances. Thus, the total variations along either diagonal of the square will be greater than the indicated tolerance.

If four holes are dimensioned with rectangular coordinates as in Fig. 12.13 (a), acceptable patterns for the square tolerance zones for the holes are shown at (b) and (c). The locational tolerances are actually greater than indicated by the dimensions.

Fig. 12.11 **Tolerances Related to Shop Processes.** *Mil-Std-8C*

| Range of Sizes | | Tolerances | | | | | | | | |
|---|---|---|---|---|---|---|---|---|---|---|
| From | To & Incl. | | | | | | | | | |
| .000 | .599 | .00015 | .0002 | .0003 | .0005 | .0008 | .0012 | .002 | .003 | .005 |
| .600 | .999 | .00015 | .00025 | .0004 | .0006 | .001 | .0015 | .0025 | .004 | .006 |
| 1.000 | 1.499 | .0002 | .0003 | .0005 | .0008 | .0012 | .002 | .003 | .005 | .008 |
| 1.500 | 2.799 | .00025 | .0004 | .0006 | .001 | .0015 | .0025 | .004 | .006 | .010 |
| 2.800 | 4.499 | .0003 | .0005 | .0008 | .0012 | .002 | .003 | .005 | .008 | .012 |
| 4.500 | 7.799 | .0004 | .0006 | .001 | .0015 | .0025 | .004 | .006 | .010 | .015 |
| 7.800 | 13.599 | .0005 | .0008 | .0012 | .002 | .003 | .005 | .008 | .012 | .020 |
| 13.600 | 20.999 | .0006 | .001 | .0015 | .0025 | .004 | .006 | .010 | .015 | .025 |

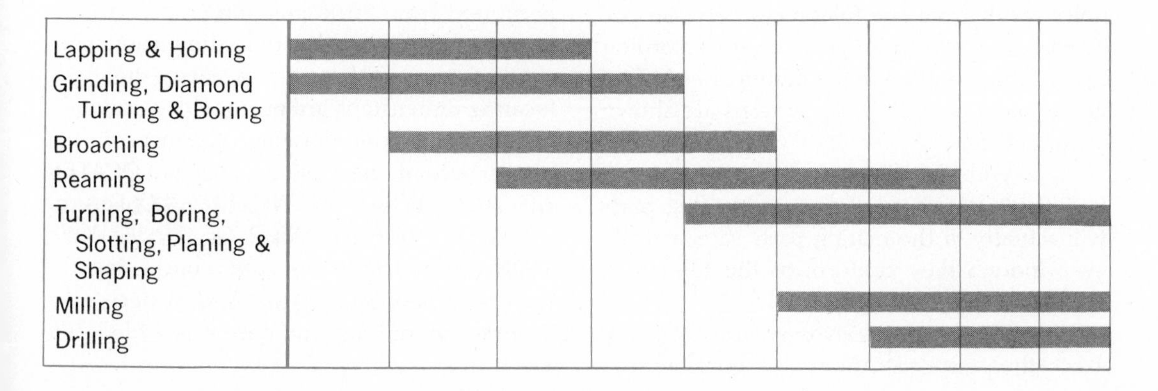

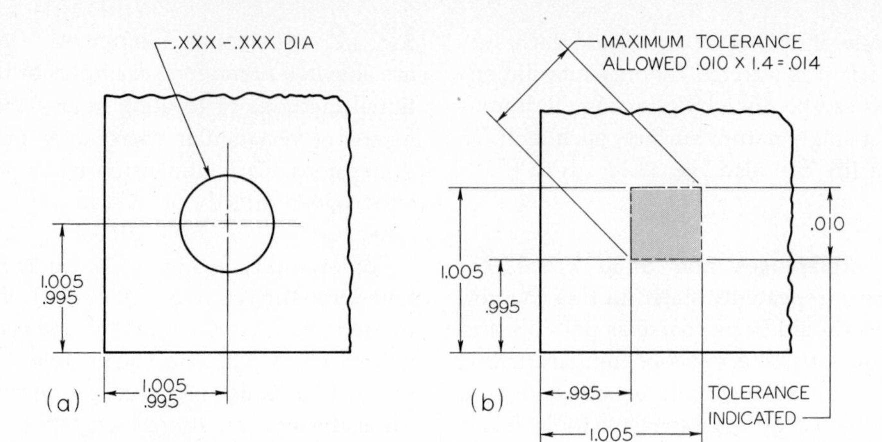


Fig. 12.12 **Tolerance Zones** (ANSI Y14.5–1966).

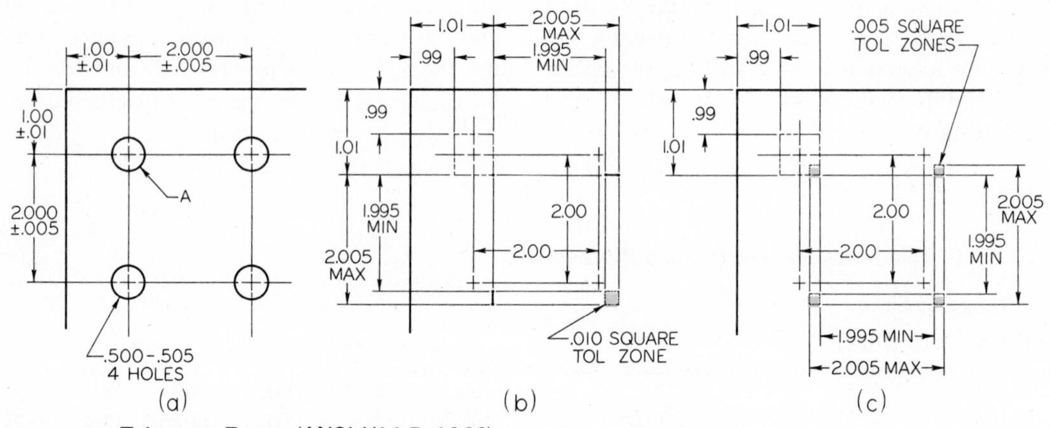


Fig. 12.13 **Tolerance Zones** (ANSI Y14.5–1966).

In Fig. 12.13 (a), hole A is selected as a datum, and the other three are located from it. The square tolerance zone for hole A results from the tolerances on the two rectangular coordinate dimensions locating hole A. The sizes of the tolerance zones for the other three holes result from the tolerances between the holes, while their locations will vary according to the actual location of the datum hole A. Two of the many possible zone patterns are shown at (b) and (c).

Thus, with the dimensions shown at (a), it is difficult to say whether the resulting parts will actually fit the mating parts satisfactorily even though they conform to the tolerances shown on the drawing.

These disadvantages are overcome by giving exact theoretical locations by untoleranced dimensions and then specifying by a note how far actual positions may be displaced from these locations. This is called *true-position dimensioning.* It will be seen that the tolerance zone for each hole will be a circle, the size of the circle depending upon the amount of variation permitted from "true position."

A complete true-position dimension consists of two or three parts. First, the basic locating dimensions are given with some indication that general drawing tolerances do not apply. Second, expressions such as LOCATED AT TRUE POSITION WITHIN .XXX DIA or LOCATED WITHIN .XXX R OF TRUE POSITION are added to the note concerning the feature to be located. And third, if necessary, a phrase identifying the datum is added. See Fig. 12.14.

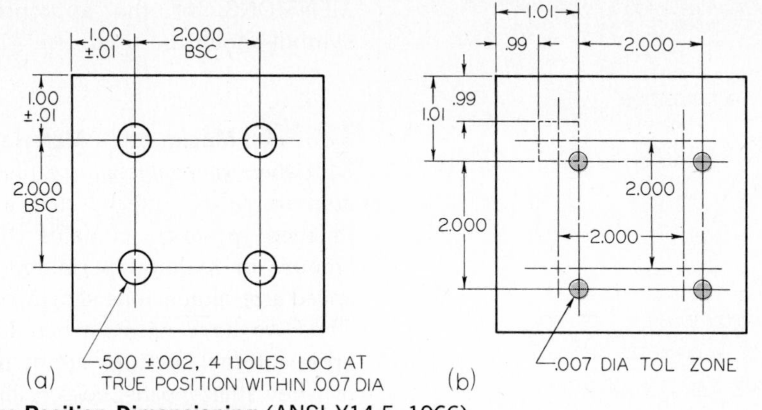


Fig. 12.14 **True-Position Dimensioning** (ANSI Y14.5–1966).

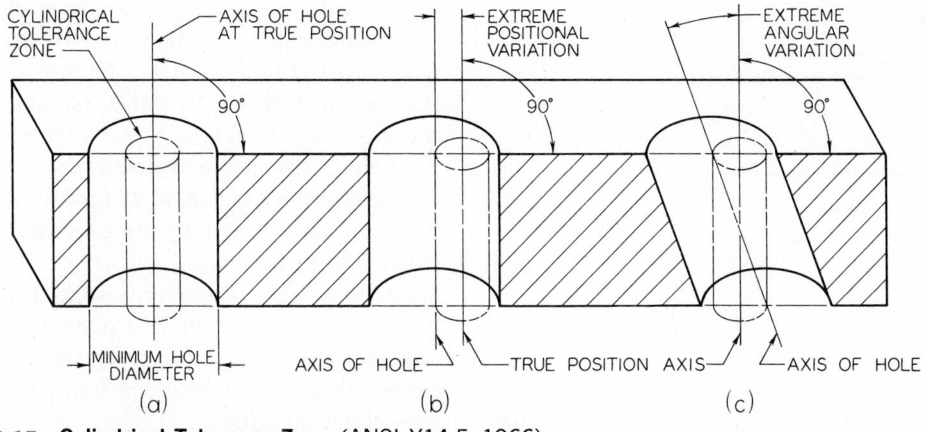


Fig. 12.15 **Cylindrical Tolerance Zone** (ANSI Y14.5–1966).

To prevent misunderstanding, true position should always be established with respect to a datum. In simple arrangements, the choice of datum is obvious and it does not require identification, Fig. 12.14 (a). When necessary for clearness, a phrase such as IN RELATION TO DATUM A should be added to the true-position note or the appropriate tolerance symbol should be used. See §12.17.

Actually, the "circular tolerance zone" is a cylindrical tolerance zone, and the axis of the hole must be within the cylinder, Fig. 12.15. The center line of the hole may coincide with the center line of the cylindrical tolerance zone, (a), or it may be parallel to it but displaced so as to remain within the tolerance cylinder, (b), or it may be inclined while remaining within the tolerance cylinder, (c). In this last case we see that the true-position tolerance also defines the limits of squareness variation.

In terms of the cylindrical surface of the hole, the true-position specification indicates that all elements on the hole surface must be on or outside a cylinder whose diameter is equal to the minimum diameter (MMC; §12.13) or the maximum diameter of the hole minus the true position tolerance (diameter, or twice the radius), with the center line of the cylinder located at true position, Fig. 12.16.

The use of basic untoleranced dimensions to locate features at true position avoids one of the chief difficulties in tolerancing—the accumulation of tolerances, §12.10, even in a chain of dimensions, Fig. 12.17.

While features, such as holes and bosses,

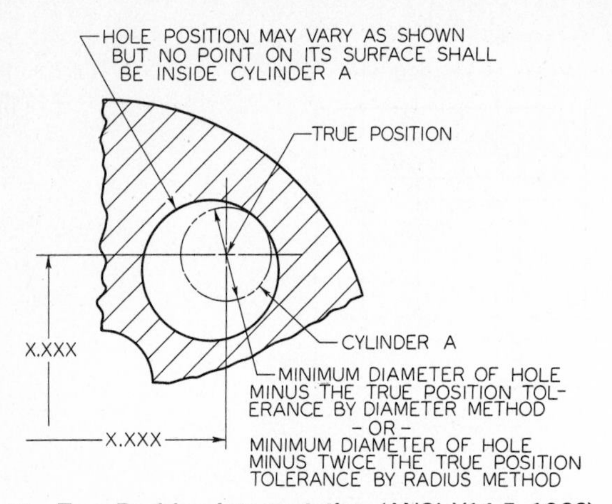


Fig. 12.16 **True Position Interpretation** (ANSI Y14.5–1966).

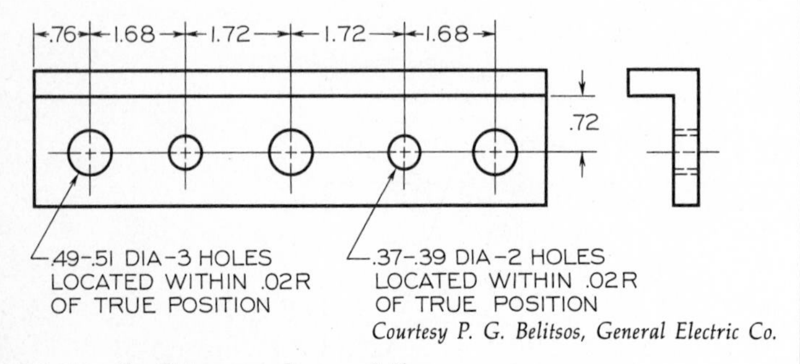


*Courtesy P. G. Belitsos, General Electric Co.*

Fig. 12.17 **No Tolerance Accumulation.**

may vary in any direction from the true-position axis, other features, such as slots, may vary on either side of a true-position plane. The note may be worded in either of two ways:

1. 6 SLOTS LOCATED AT TRUE POSITION WITHIN .010 WIDE ZONE.
2. 6 SLOTS LOCATED WITHIN .005 EITHER SIDE OF TRUE POSITION.

It has been quite common in some industries to use the note: LOCATED WITHIN .005 OF TRUE POSITION, which means the same as 2 above.

Since the exact locations of the true positions of the tolerances are given by untoleranced dimensions, it is important to prevent the application of general tolerances to these. A note should be added to the drawing such as: GENERAL TOLERANCES DO NOT APPLY TO BASIC TRUE-POSITION DIMENSIONS, or the appropriate tolerance symbol should be used. See §12.17.

## 12.13 Maximum Material Condition.

*Maximum material condition,* usually abbreviated to MMC, means that a feature of a finished product contains the maximum amount of material permitted by the toleranced size dimensions shown for that feature. Thus we have MMC when holes, slots, or other internal features are at minimum size, or when shafts, pads, bosses, and other external features are at their maximum size. We have MMC for both mating parts when the largest shaft is in the smallest hole and there is the least clearance between the parts.

In assigning positional tolerance to a hole, it is necessary to consider the size limits of the hole. If the hole is at MMC (smallest size), the positional tolerance is not affected; but if the hole is larger, the available positional tolerance is greater. In Fig. 12.18 (a) two half-inch holes are shown. If they are exactly .500 dia. (MMC, or smallest size) and are exactly 2.000 apart, they should receive a gage, (b), made of two round pins fixed in a plate 2.000 apart and .500 in diameter. However, the center-to-center distance between the holes may vary from 1.995 to 2.005.

If the .500 dia. holes are at their extreme positions, (c), the pins in the gage would have to be .005 smaller, or .495 diameter, to enter the holes. Thus, if the .500 dia. holes are located at the maximum distance apart, the .495 dia. gage pins would contact the inner sides of the holes; and if the holes are located at the minimum distance apart, the .495 dia. pins would contact the outer surfaces of the holes, as shown. If gagemakers' tolerances are not considered, the gage pins would have to be .495 dia. and exactly 2.000 apart if the holes are .500 dia., or MMC.

If the holes are .505 dia.—that is, at maximum size, as at (d)—they will be accepted by the same .495 dia. gage pins at 2.000 apart if the inner sides of the holes contact the inner sides of the gage pins, and the outer sides of the holes contact the outer sides of the gage pins, as shown. Thus the holes may be 2.010

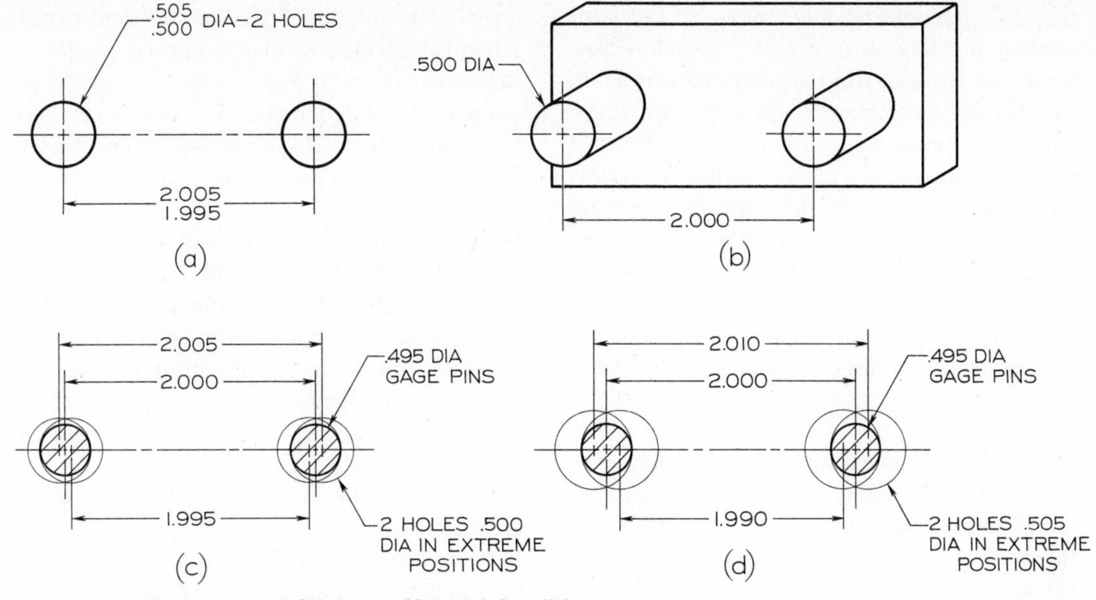


Fig. 12.18 **Maximum and Minimum Material Conditions.**

apart, which is beyond the tolerance permitted for the center-to-center distance between the holes. Similarly, the holes may be as close together as 1.990 from center to center, which again is outside the specified positional tolerance.

Thus, when the holes are not at MMC—that is, when they are at maximum size—a greater positional tolerance becomes available. Since all features may vary in size, it is necessary to be made clear on the drawing at what condition of size the true-position tolerance applies. In all but a few exceptional cases, the additional positional tolerance available, when holes are larger than minimum size, is acceptable and desirable. Parts thus accepted can be freely assembled whether the holes or other features are within the specified positional tolerance or not. This practice has been recognized and used in manufacturing for years, as is evident from the use of fixed-pin gages, which have been commonly used to inspect parts and control the least favorable condition of assembly. Thus it has become common practice for both manufacturing and inspection to assume that positional tolerance applies to MMC, and that greater positional tolerance becomes permissible when the part is not at MMC.

However, this practice should be followed with caution. The designer should be conscious of the additional tolerance and should permit it when it is practical and desirable. The ANSI* recommends that to avoid possible misinterpretation as to whether maximum material condition (MMC) or regardless of feature size (RFS) applies, it should be clearly stated on the drawing:

1. By the addition of MMC or RFS or modifiers to each applicable tolerance; or
2. By the use of a general note; or
3. By suitable coverage in a document referenced on the drawing.

When MMC or RFS is not specified on the drawing with respect to an individual tolerance, datum reference or both, the following rules shall apply:

1. True-position tolerances and related datum references apply at MMC.
2. Angularity, parallelism, perpendicularity, concentricity, and symmetry tolerances, including related datum references, apply RFS. No element of the actual feature shall extend beyond the envelope of the perfect form at MMC.

* ANSI Y14.5–1966.

## 12.14 Geometric Tolerances.*

*Geometric tolerance* or "tolerance of form" specifies how far actual surfaces are permitted to vary from the perfect geometry implied by drawings. The term "geometric" refers to the various geometric forms, as a plane, a cylinder, a cone, a square, or a hexagon. Theoretically, these are perfect forms, but since it is impossible to produce perfect forms, it may be necessary to specify the amount of variation permitted. Geometric tolerances define conditions of straightness, flatness, parallelism, perpendicularity, angularity, symmetry, concentricity, and roundness.

Methods of indicating geometric tolerances by means of notes, as recommended by ANSI,* are shown in Figs. 12.19 and 12.20. At the right of each example, the meaning of the tolerance note or dimension is illustrated. In these examples, tolerances of form are considered individually, not taking into account the effects of combinations with other tolerances of position, size, or form.

An example of combined expressions is shown in Fig. 12.20 (j), in which a hole is to be held more closely parallel to a flat surface than its positional tolerance alone would indicate. As shown at the right, the requirement that the center line be anywhere within the .010 dia. positional tolerance cylinder is further restricted by the specification that the center line must lie between two planes parallel to datum surface A, and .003 apart. The diagonal line between the two planes is one possible position in this case.

The tolerances on size and on form and specifications of surface roughness may overlap. But for a given part, variation in form must not exceed the size tolerance; therefore, tolerance of form need not be stated unless it is smaller than the size tolerance. As an example, if a size tolerance of ±.005 is permitted, any tolerance of form under .010 should be stated. Or if the flatness tolerance is .001, a surface roughness smoother than $^{1000}\surd$ should be indicated.

Since it is very expensive to maintain accurate geometric tolerances in the shop, tolerances of form should not be specified except when established shop procedures cannot be depended upon to produce the necessary accuracy. When geometric tolerances are not indicated on the drawing, the actual part is understood to be acceptable if it is within the dimensional limits shown, regardless of variations in form. Some extreme variations are shown in Fig. 12.20 (k) to (p). However, in the case of fabricated bars, sheets, tubing, etc., established industry standards prescribe acceptable conditions of straightness, flatness, and so on, and these standards are understood to hold if geometric tolerances are not shown on the drawing.

Frequently the amount of geometric tolerance can vary according to the actual sizes of the finished parts. In Fig. 12.20 (r) the note indicates that when the shaft is at maximum diameter (MMC), it may have an error of .005 in straightness. This allowable variation at MMC is illustrated at (s). If .005 variation is allowed at MMC, then .015 is available when the part is at minimum diameter, as shown at (t). This additional tolerance should be permitted only when it does not interfere with functional requirements.

In Fig. 12.20 (u) the pin at maximum diameter (MMC) may have a .002 perpendicularity error. This allowable variation of .002 is shown at (v), in which the gage hole is .627 (MMC shaft plus perpendicularity tolerance). But if .002 perpendicularity error is permitted at MMC, then .003 error is available if the part is at minimum diameter as shown at (w).

## 12.15 Zone Tolerances for Contours.

A "zone" tolerance may be given when a uniform amount of variation can be permitted along a contour. The drawing is constructed to show the desired contour fully defined by dimensions without tolerances. At a conspicuous place along the contour, one or two phantom lines are drawn with dimension lines and arrowheads to indicate the location of the tolerance zone. The value of the tolerance is given by a note. See Fig. 12.21 (a). The distance between the contour and phantom lines is usually exaggerated for clarity. In cases

* ANSI Y14.5-1966.

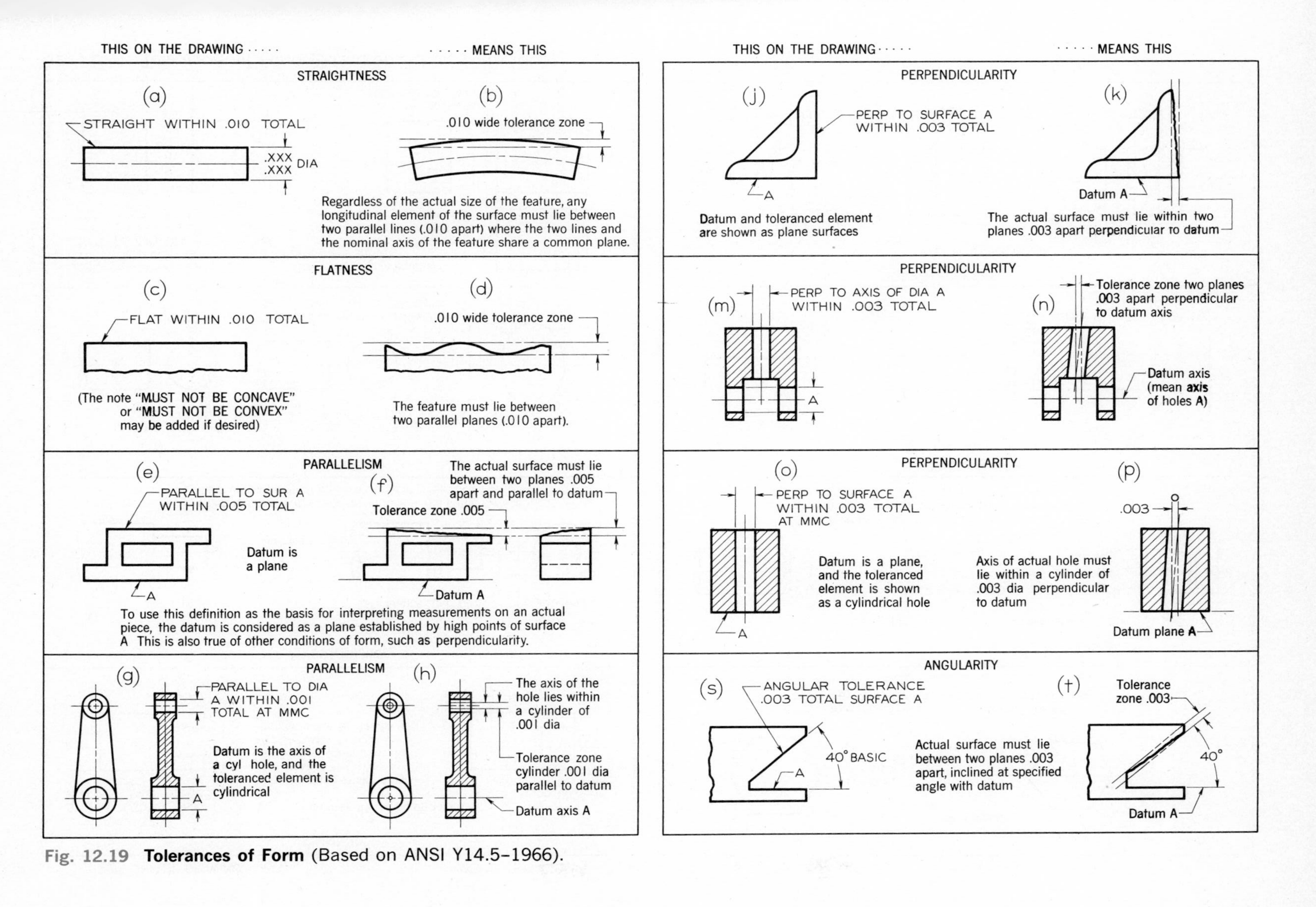
THIS ON THE DRAWING . . . . .
. . . . . MEANS THIS
STRAIGHTNESS
(a)
STRAIGHT WITHIN .010 TOTAL
.XXX
.XXX DIA
(b)
.010 wide tolerance zone
Regardless of the actual size of the feature, any longitudinal element of the surface must lie between two parallel lines (.010 apart) where the two lines and the nominal axis of the feature share a common plane.
FLATNESS
(c)
FLAT WITHIN .010 TOTAL
(The note "MUST NOT BE CONCAVE" or "MUST NOT BE CONVEX" may be added if desired)
(d)
.010 wide tolerance zone
The feature must lie between two parallel planes (.010 apart).
PARALLELISM
(e)
PARALLEL TO SUR A WITHIN .005 TOTAL
Datum is a plane
A
(f)
The actual surface must lie between two planes .005 apart and parallel to datum
Tolerance zone .005
Datum A
To use this definition as the basis for interpreting measurements on an actual piece, the datum is considered as a plane established by high points of surface A This is also true of other conditions of form, such as perpendicularity.
PARALLELISM
(g)
PARALLEL TO DIA A WITHIN .001 TOTAL AT MMC
Datum is the axis of a cyl hole, and the toleranced element is cylindrical
A
(h)
The axis of the hole lies within a cylinder of .001 dia
Tolerance zone cylinder .001 dia parallel to datum
Datum axis A
THIS ON THE DRAWING . . . . .
. . . . . MEANS THIS
PERPENDICULARITY
(j)
PERP TO SURFACE A WITHIN .003 TOTAL
A
Datum and toleranced element are shown as plane surfaces
(k)
Datum A
The actual surface must lie within two planes .003 apart perpendicular to datum
PERPENDICULARITY
(m)
PERP TO AXIS OF DIA A WITHIN .003 TOTAL
A
(n)
Tolerance zone two planes .003 apart perpendicular to datum axis
Datum axis (mean axis of holes A)
PERPENDICULARITY
(o)
PERP TO SURFACE A WITHIN .003 TOTAL AT MMC
Datum is a plane, and the toleranced element is shown as a cylindrical hole
A
(p)
.003
Axis of actual hole must lie within a cylinder of .003 dia perpendicular to datum
Datum plane A
ANGULARITY
(s)
ANGULAR TOLERANCE .003 TOTAL SURFACE A
40° BASIC
A
(t)
Tolerance zone .003
Actual surface must lie between two planes .003 apart, inclined at specified angle with datum
40°
Datum A

**Fig. 12.19 Tolerances of Form** (Based on ANSI Y14.5–1966).

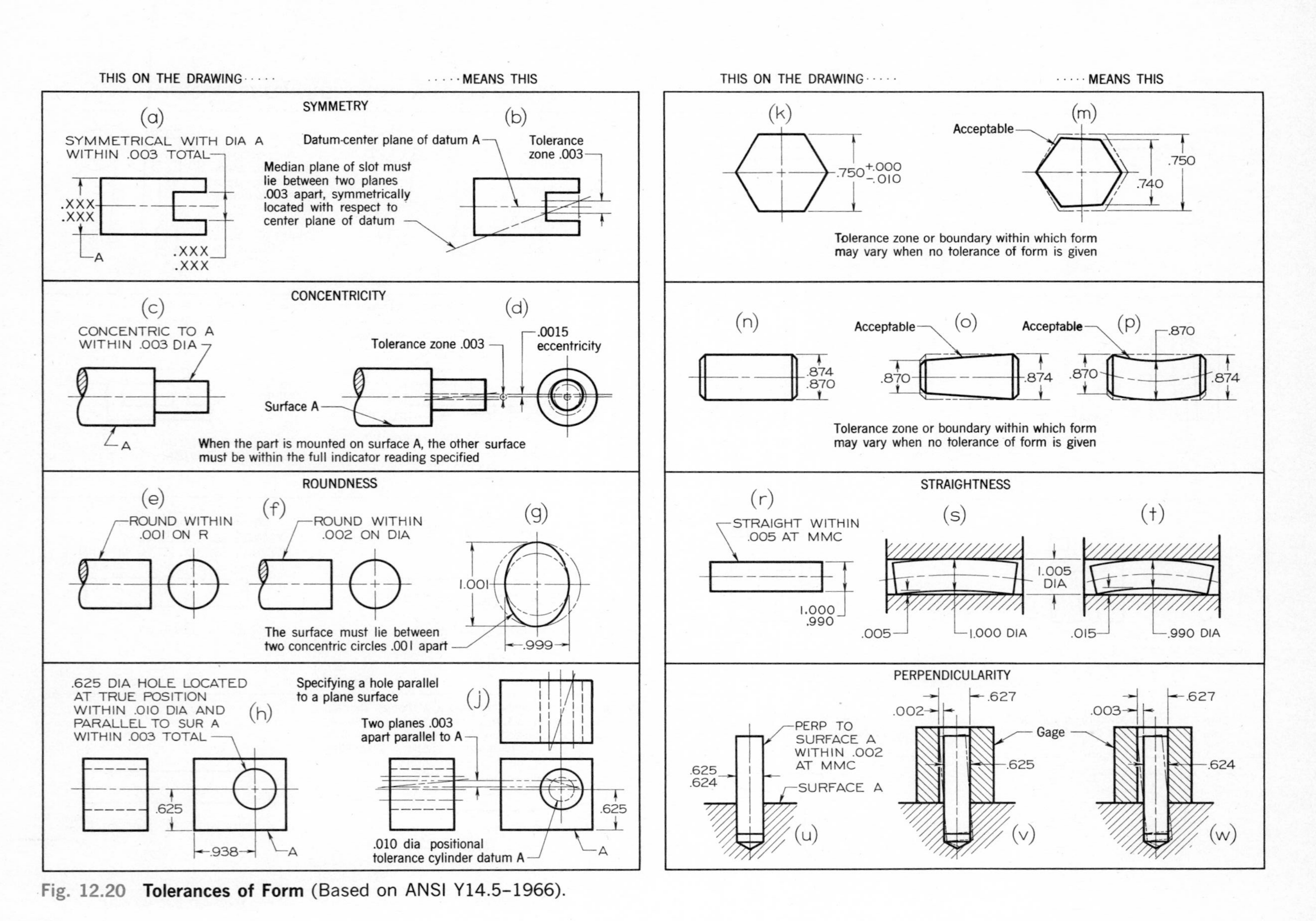


Fig. 12.20 **Tolerances of Form** (Based on ANSI Y14.5–1966).

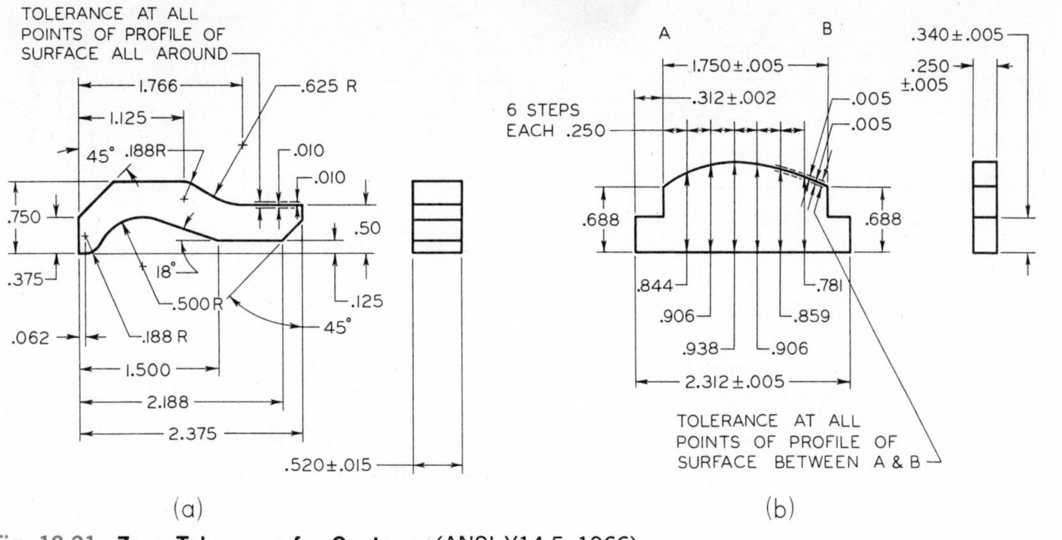


Fig. 12.21 **Zone Tolerances for Contours** (ANSI Y14.5-1966).

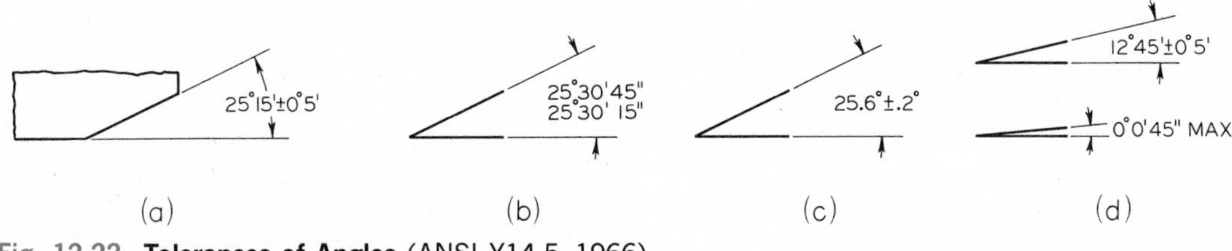


Fig. 12.22 **Tolerances of Angles** (ANSI Y14.5-1966).

where some limits on a drawing are given by zone tolerance while others are given by a general tolerance, it is necessary to indicate the extent of the zone tolerance, as at (b). The tolerance zone may be symmetrical about the contour lines (bilateral), or it may be all on either side (unilateral).

## 12.16 Tolerances of Angles.*

Bilateral tolerances are usually given on angular dimensions where accuracy must be maintained, as illustrated in Fig. 12.22. In figuring tolerances on angles, it should be kept in mind that the width of the tolerance zone increases as the distance from the vertex of the angle increases. The tolerance should be figured after considering the total allowable displacement at the point farthest from the vertex of the angle, and a tolerance specified that will not exceed this. The use of angular tolerances may be avoided by using gages. Taper turning is often handled by machining to fit a gage or by fitting to the mating part.

If an angular surface is located by a linear and an angular dimension, Fig. 12.23 (a), the surface must lie within a tolerance zone as shown at (b). The angular zone will be wider as the distance from the vertex increases. In order to avoid the accumulation of tolerances, i.e., to decrease the tolerance zone, the *basic angle* tolerancing method of (c) is recommended by ANSI. The angle is marked BASIC and no angular tolerance is specified. The tolerance zone is now defined by two parallel planes, resulting in improved angular control, (d). See also Fig. 12.19 (s) and (t) for tolerancing angularity as tolerance of form.

## 12.17 Symbols for Tolerances of Position and Form.

Feature-control symbols may be used in lieu of notes to specify positional and form tolerances to the extent that they are

*ANSI Y14.5-1966.

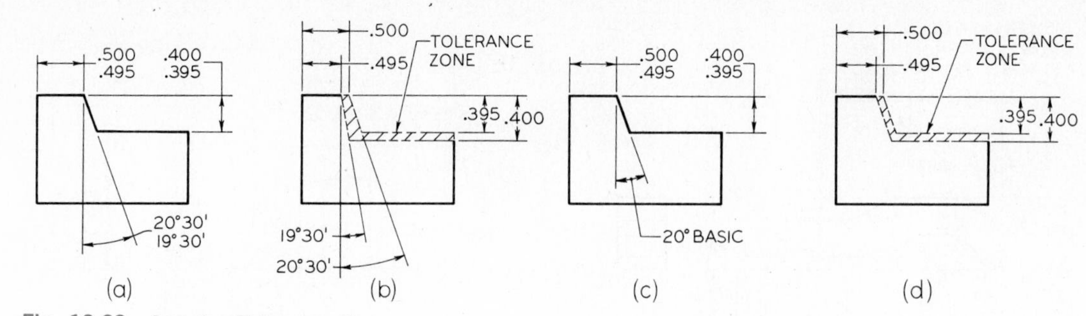


Fig. 12.23 **Angular Tolerance Zones.**

| GEOMETRIC CHARACTERISTIC SYMBOLS | | | |
|---|---|---|---|
| Characteristic | Symbol | Characteristic | Symbol |
| **Form Tolerances** | | For Related Features | |
| For Single Feature | | Parallelism[b] | ∥ |
| Flatness | ⏥ | Perpendicularity (Squareness) | ⊥ |
| Straightness | — | Angularity | ∠ |
| Roundness (Circularity) | ○ | Runout[c] | ↗ |
| Cylindricity | ⌭ | **Positional Tolerances** | |
| Profile of any line[a] | ⌒ | True position | ⊕ |
| Profile of any surface[a] | ⌓ | Concentricity[d] | ◎ |
| | | Symmetry[e] | ⌯ |

[a] Although included under "Form Tolerances," profile tolerances control size as well as form.
[b] Parallel lines may be shown oblique.
[c] Although included under "Form Tolerances," a runout tolerance controls position as well as form.
[d] Where concentricity RFS applies, it is preferred that the runout symbol be used. Where concentricity at MMC applies, it is preferred that the true-position symbol be used. Optionally, the inner circle of the symbol may be filled solid.
[e] Where symmetry applies, it is preferred that the true-position symbol be used.

Fig. 12.24 **Symbols for Tolerances of Position and Form** (ANSI Y14.5–1966).

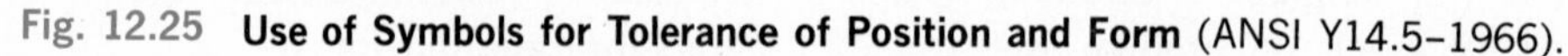

Fig. 12.25 **Use of Symbols for Tolerance of Position and Form** (ANSI Y14.5–1966).

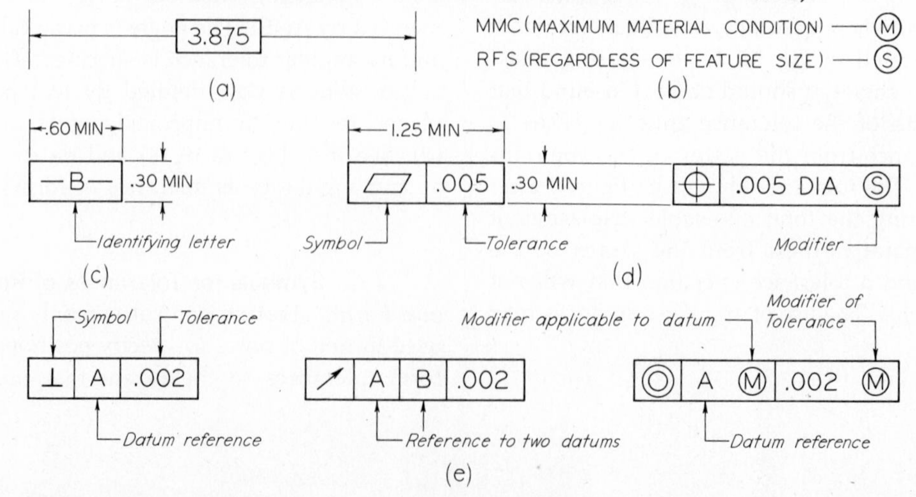

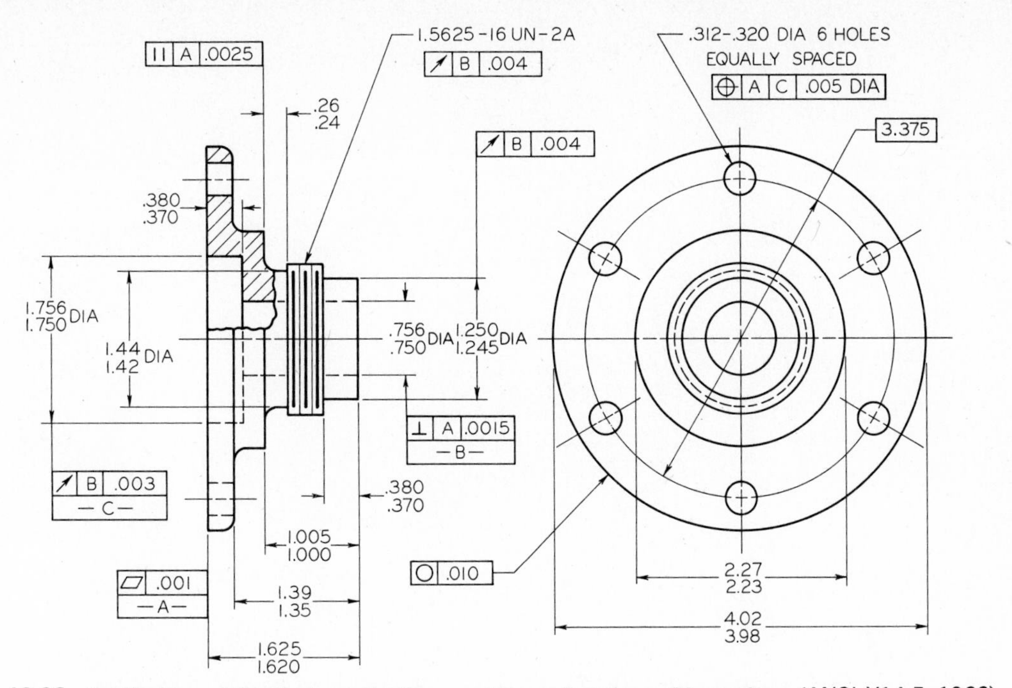


Fig. 12.26 **Application of Symbols to Position and Form Tolerance Dimensions** (ANSI Y14.5–1966).

applicable, and in other cases a note should be used to describe the precise requirement. The symbols for the geometric characteristic to which the tolerance applies are given in Fig. 12.24.

Basic (BSC) and true-position dimensions are indicated by enclosure in a frame as in Fig. 12.25 (a). Symbols for MMC and RFS are given at (b). Each datum that must be identified on the drawing shall be given an identifying letter of the alphabet except I, O, or Q. The datum letter shall be preceded and followed by a dash and enclosed in a rectangular frame, as shown at (c). Feature-control symbols are indicated at (d). Feature-control symbols incorporating datum references are given at (e).

Figure 12.26 illustrates the use of feature-control symbols in lieu of notes with position and form tolerance dimensions.

# 13

# threads, fasteners, and springs

**13.1 Screw Threads.*** The concept of the screw thread seems to have occurred first to Archimedes, the third-century B.C. mathematician, who wrote briefly on spirals and invented or designed several simple devices applying the screw principle. By the first century B.C. the screw was a familiar element, but was crudely cut from wood or filed by hand on a metal shaft. Nothing more was heard of the screw thread in Europe until the fifteenth century, though the Greeks and the Arabs had preserved their knowledge of it. Leonardo da Vinci understood the screw principle, and he has left sketches showing how to cut screw threads by machine. In the sixteenth century, screws appeared in German watches, and screws were used to fasten suits of armour. In 1569 the screw-cutting lathe was invented by the Frenchman Besson, but the method did not take hold, and for another century and a half, nuts and bolts continued to be made largely by hand. In the eighteenth century, screw manufacturing got started in England during the Industrial Revolution.

In these early times, there was no such thing as standardization. The nuts made by one manufacturer would not fit the bolts of another. In 1841 Sir Joseph Whitworth started crusading for a standard screw thread, and soon the Whitworth thread was accepted throughout England. In 1864 in the United States, a committee named by the Franklin Institute adopted a thread proposed by William Sellers of Philadelphia; but the Sellers' nuts would not screw on a Whitworth bolt, or vice versa, the thread angles being different.

*See ANSI Y14.6-1957.

In 1935 the American Standard thread, with the same 60° V form of the old Sellers' thread, was adopted in this country. Still there was no standardization between countries. In peacetime it was a nuisance; in World War I it was a serious inconvenience; and in World War II the obstacle was so great that the Allies decided to do something about it, and talks began between the Americans, British, and Canadians (the so-called ABC countries). In 1948 an agreement was reached on the unification of American and British screw threads. The new thread was called the *Unified screw thread,* and it represents a compromise between the American Standard and Whitworth systems, allowing complete interchangeability of threads in the three countries.

Today screw threads are vital to our industrial life. They are designed for hundreds of different purposes, the three basic applications being: (1) to *hold parts* together, (2) to *adjust parts* with reference to each other, and (3) to *transmit power.*

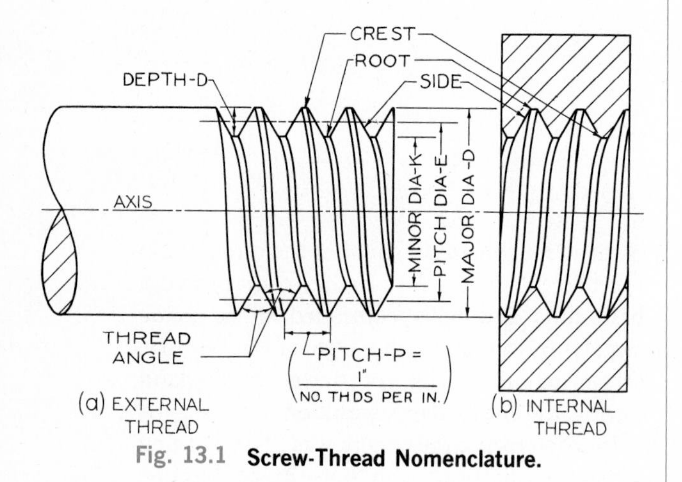


Fig. 13.1 **Screw-Thread Nomenclature.**

## 13.2 Definitions of Terms.

The following definitions apply to screw threads generally, Fig. 13.1.

*Screw thread*—A ridge of uniform section in the form of a helix, §4.64, on the external or internal surface of a cylinder.

*External thread*—A thread on the outside of a member, as on a shaft.

*Internal thread*—A thread on the inside of a member, as in a hole.

*Major diameter*—The largest diameter of a screw thread (applies to both internal and external threads).

*Minor diameter*—The smallest diameter of a screw thread (applies to both internal and external threads).

*Pitch*—The distance from a point on a screw thread to a corresponding point on the next thread measured parallel to the axis. The pitch P is equal to 1 divided by the number of threads per inch.

*Pitch diameter*—The diameter of an imaginary cylinder passing through the threads so as to make equal the widths of the threads and the widths of the spaces cut by the cylinder.

*Lead*—The distance a screw thread advances axially in one turn.

*Angle of thread*—The angle included between the sides of the thread measured in a plane through the axis of the screw.

*Crest*—The top surface joining the two sides of a thread.

*Root*—The bottom surface joining the sides of two adjacent threads.

*Side*—The surface of the thread that connects the crest with the root.

*Axis of screw*—The longitudinal center line through the screw.

*Depth of thread*—The distance between the crest and the root of the thread measured normal to the axis.

*Form of thread*—The cross section of thread cut by a plane containing the axis.

*Series of thread*—Standard number of threads per inch for various diameters.

## 13.3 Screw-Thread Forms.

Various forms of threads, Fig. 13.2, are used to hold parts together, to adjust parts with reference to each other, or to transmit power. The 60° *Sharp-V thread* was originally called the United States Standard thread, or the Sellers thread. For purposes of certain adjustments, the Sharp-V thread is useful with the increased friction resulting from the full thread-face. It is also used on brass pipe work.

The *American National thread** with flattened

* ANSI B1.1–1960.

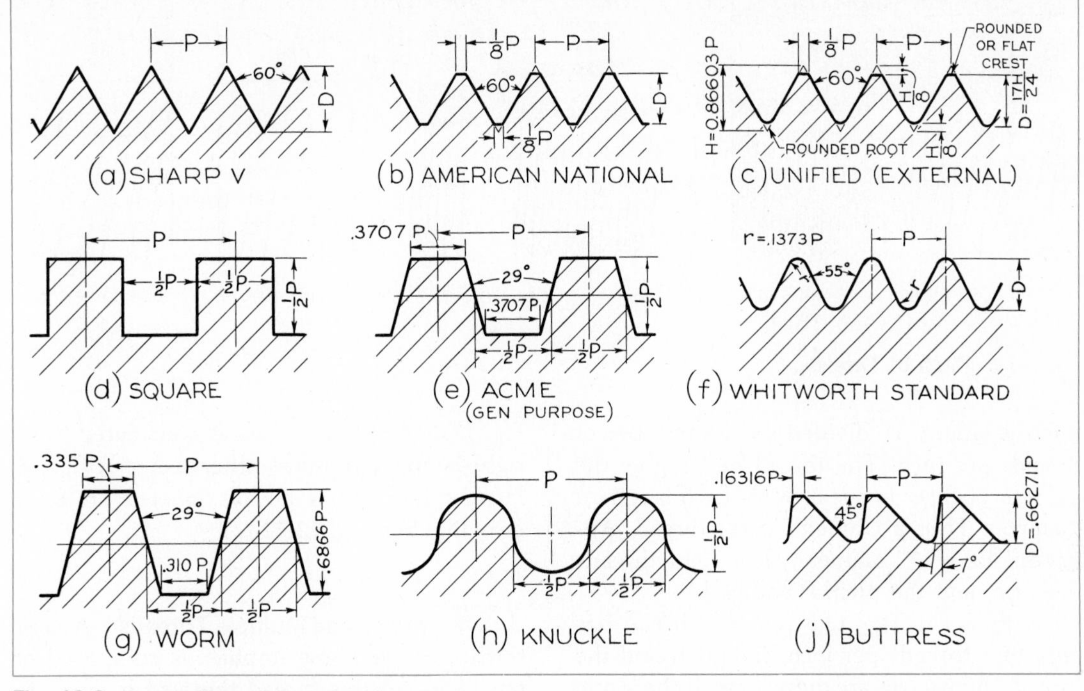


Fig. 13.2 **Screw-Thread Forms.**

roots and crests is a stronger thread. This form replaced the Sharp-V thread for general use and is still used for existing design.

The *Unified thread** is the new standard thread agreed upon by the United States, Canada, and Great Britain in 1948. The crest of the external thread may be flat or rounded, and the root is rounded; otherwise the thread form is essentially the same as the American National. The transition of industry from the American National thread to the Unified thread for new design has been nearly completed.

The *Square thread* is theoretically the ideal thread for power transmission, since its face is nearly at right angles to the axis; but owing to the difficulty of cutting it with dies and because of other inherent disadvantages, such as the fact that split nuts will not readily disengage, the Square thread has been displaced to a large extent by the Acme thread. The Square thread is not standardized.

The *Acme thread* is a modification of the Square thread, and has largely replaced it. It is stronger than the Square thread, is easier to cut, and has the advantage of easy disengagement from a split nut, as on the lead screw of a lathe.

The *Whitworth thread* has been the British standard and is being replaced by the Unified thread. Its uses correspond to those of the American National thread.

The *Standard Worm thread* is similar to the Acme thread, but is deeper. It is used on shafts to carry power to worm wheels.

The *Knuckle thread* is usually rolled from sheet metal but is sometimes cast, and is used in modified forms in electric bulbs and sockets, bottle tops, etc.

The *Buttress thread* is designed to transmit power in one direction only, and is used in the breech locks of large guns, in jacks, in airplane propeller hubs, and in other mechanisms of similar requirements.

## 13.4 Thread Pitch.

The *pitch* of a thread, of whatever form, is the distance parallel to the axis between corresponding points on adjacent threads, Figs. 13.1 (a) and 13.2. The

*ANSI B1.1-1960.

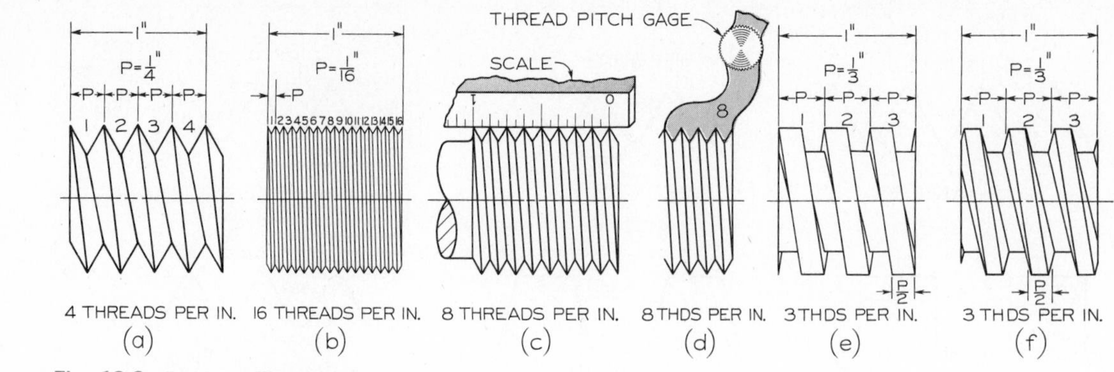


Fig. 13.3 **Pitch of Threads.**

pitch is simply 1″ divided by the number of threads per inch. The thread tables give the number of threads per inch for each standard diameter. Thus a Unified Coarse thread, Appendix 10, of 1″ diameter, has eight threads per inch, and the pitch P equals $\frac{1}{8}$″.

As shown in Fig. 13.3 (a), if a thread has only four threads per inch, the pitch and the threads themselves are quite large. If there are, say, sixteen threads per inch, the pitch is only $\frac{1}{16}$″, and the threads are relatively small, as shown at (b).

The pitch or number of threads per inch can easily be measured with an ordinary scale, (c), or with a *thread pitch gage,* (d).

It will be seen at (e) and (f) that the pitch for Square and Acme threads includes a thread ridge and a space.

## 13.5 Right-Hand and Left-Hand Threads.

A right-hand thread, when viewed toward an end, winds in a clockwise and receding direction. Thus a right-hand thread is one that advances into the nut when turned clockwise, and a left-hand thread is one that advances into the nut when turned counterclockwise, Fig. 13.4. A thread is always considered to be right-hand (RH) unless otherwise specified. A left-hand thread is always labeled LH on a drawing. See Fig. 13.20 (a).

## 13.6 Single and Multiple Threads.

A *single* thread, as the name implies, is composed of one ridge; in this thread the lead is equal to the pitch. *Multiple* threads are composed of two or more ridges running side by side. As shown in Fig. 13.5 (a) to (c), the *slope line* is the hypotenuse of a right triangle whose short side equals $\frac{1}{2}$P for single threads, P for double threads, $1\frac{1}{2}$P for triple threads, etc. This applies to all forms of threads. In *double* threads, the lead is twice the pitch; in *triple* threads the lead is three times the pitch, etc. On a drawing of a single or triple thread, a root is opposite a crest; in the case of a double or quadruple thread, a root is drawn opposite a root. Therefore, in one turn, a double thread advances twice as far as a single thread; a triple thread advances three times as far, etc.

RH Double Square and RH Triple Acme threads are shown at (d) and (e), respectively.

Multiple threads are used wherever quick

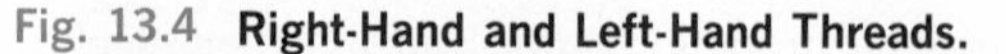
Fig. 13.4 **Right-Hand and Left-Hand Threads.**

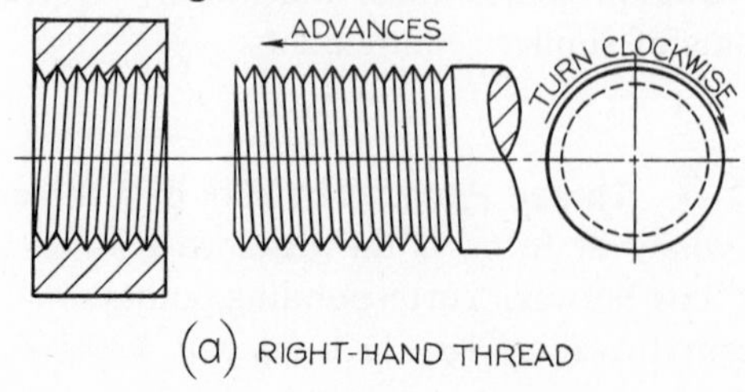


(a) RIGHT-HAND THREAD

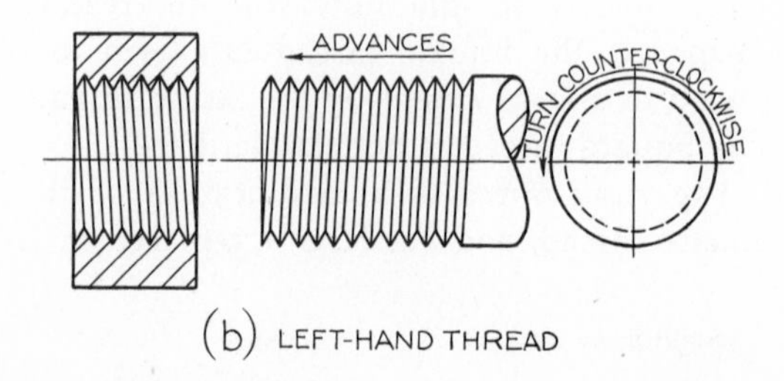


(b) LEFT-HAND THREAD

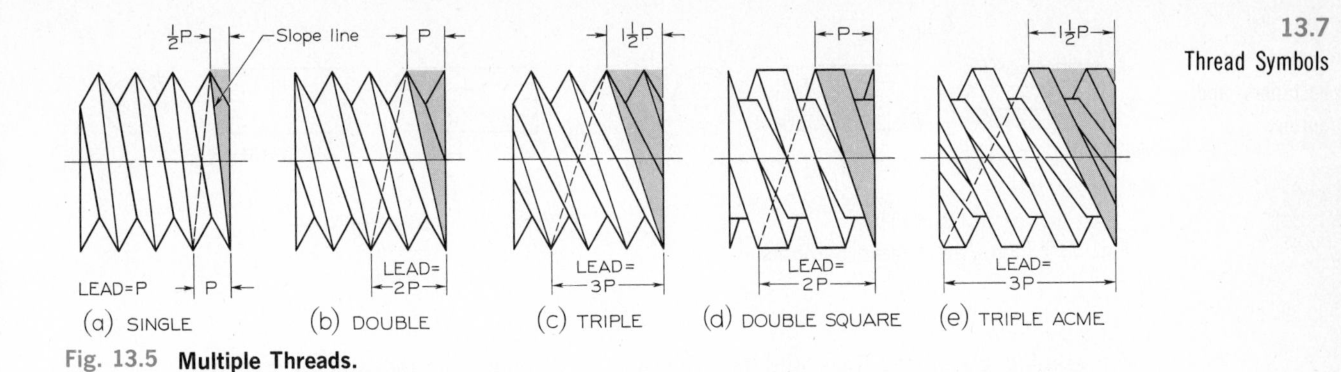


Fig. 13.5 **Multiple Threads.**

motion, but not great power, is desired, as on fountain pens, toothpaste caps, valve stems, etc. The threads on a valve stem are frequently multiple threads, to impart quick action in opening and closing the valve. Multiple threads on a shaft can be recognized and counted by observing the number of thread endings on the end of the screw.

## 13.7 **Thread Symbols.**

There are three conventions in general use for showing screw threads on drawings—the *schematic,* the *simplified,* and the *detailed.** For clarity of representation and where good judgment dictates, the detailed, schematic, and simplified thread symbols may be combined on a single drawing.

Two sets of thread symbols, the schematic and the simplified, are used to represent threads of small diameter, under approximately 1″ diameter on the drawing. The symbols are the same for all forms of threads, such as Unified, Square, and Acme. Although both of these symbols are widely used, the simplified form is more common in industry but should be avoided where there is any possibility of confusion with other drawing detail.

The detailed representation is a close approximation of the appearance of a screw thread. The true projection of the helical curves of a screw thread, Fig. 13.1, presents a pleasing appearance, but this does not compensate for the laborious task of plotting the helices, §4.64. Consequently, the true projection is used rarely in practice.

When the diameter of the thread on the drawing is over approximately 1″, a pleasing drawing may be made by the *detailed representation* method, in which the true profiles of the threads (any form of thread) are drawn; but the helical curves are replaced by straight lines, as shown in Fig. 13.6.* Whether the crests or roots are flat or rounded, they are

* ANSI Y14.6-1957.

* A thread 1½″ dia., if drawn half size, would be less than 1″ dia. on the drawing and hence would be too small for this method of representation.

Fig. 13.6 **Detailed American National and Unified Threads.**

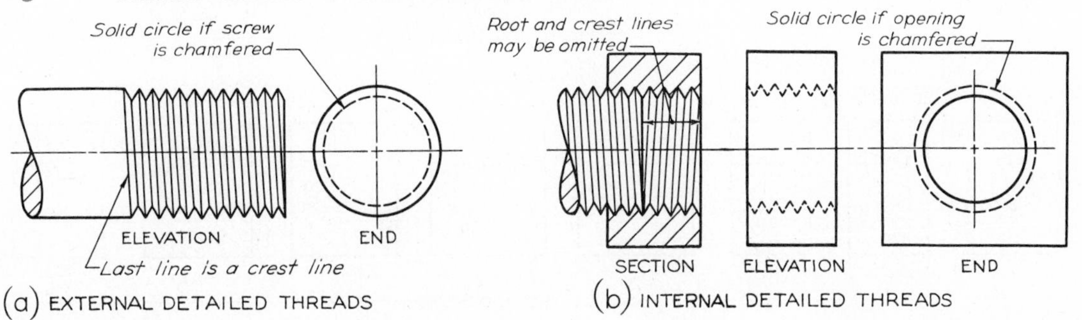

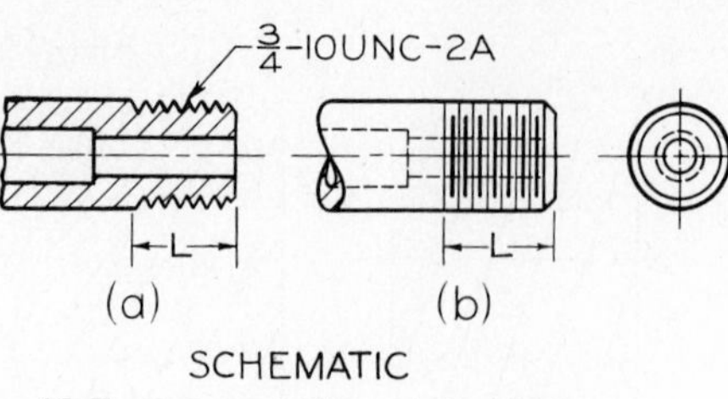


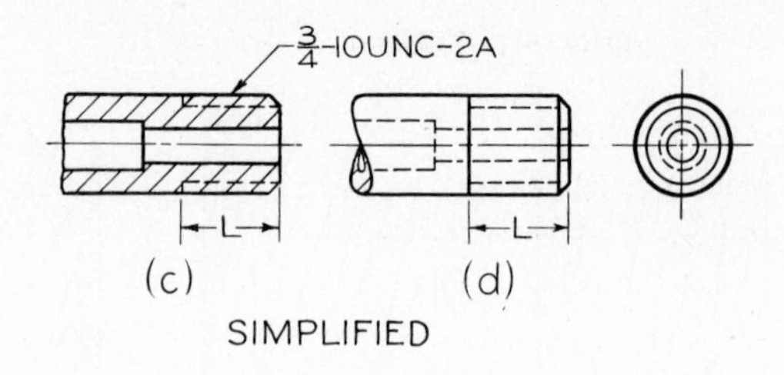


Fig. 13.7 **External Thread Symbols.**

represented by single lines and not double lines as in Fig. 13.1; consequently, American National and Unified threads are drawn in exactly the same way.

**13.8 External Thread Symbols.** External thread symbols are shown in Fig. 13.7. When the schematic form is shown in section, (a), it is necessary to show the V's; otherwise no threads would be evident. However, it is not necessary to show the V's to scale or according to the actual slope of the crest lines. To draw the V's, use the schematic thread depth, Fig. 13.9 (a), and let the pitch be determined by the 60° V's.

Schematic threads in elevation, Fig. 13.7 (b), are indicated by alternate long and short lines at right angles to the center line, the root lines being preferably thicker than the crest lines. Although theoretically the crest lines would be spaced according to actual pitch, the lines would often be very crowded and tedious to draw, thus defeating the purpose of the symbol, to save drafting time. Therefore, in practice the experienced draftsman spaces the crest lines carefully by eye, as he does section lines, and then adds the heavy root lines spaced by eye half-way between the crest lines. In general, the spacing should be proportionate for all diameters. For convenience in drawing, proportions for the schematic symbol are given in Fig. 13.9.

Simplified external thread symbols are shown in Fig. 13.7 (c) and (d). The threaded portions are indicated by hidden lines parallel to the axis at the approximate depth of the thread, whether in section or in elevation. Use the schematic depth of thread as given in the table in Fig. 13.9 (a), to draw these lines, (d).

Fig. 13.8 **Internal Thread Symbols.**

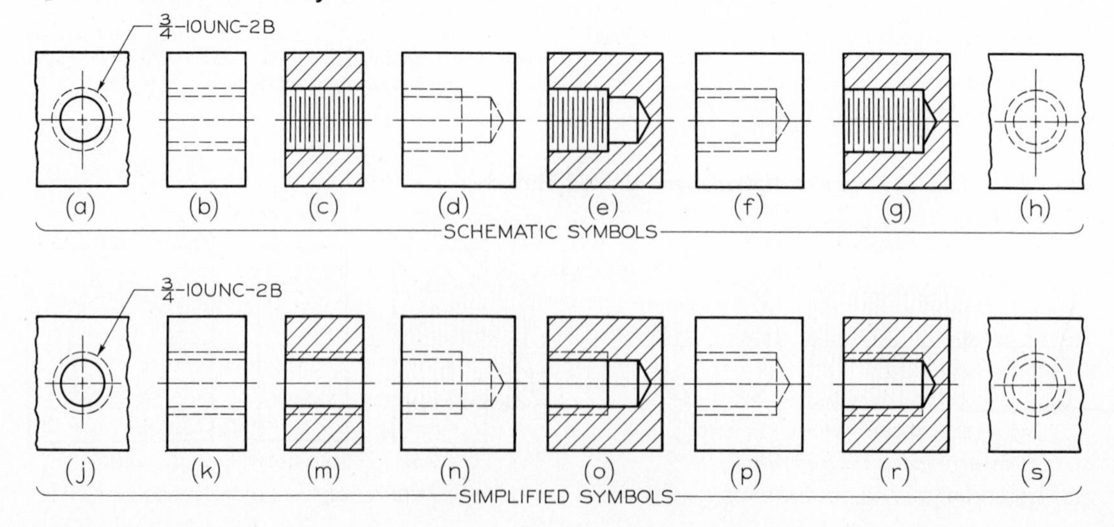

## 13.9 Internal Thread Symbols.

Internal thread symbols are shown in Fig. 13.8. The schematic thread in section is exactly the same as the external symbol in Fig. 13.7 (b). Hidden threads, by either method, are represented by pairs of hidden lines. The hidden dashes should be staggered, as shown. Note that the only differences between the schematic and simplified internal thread symbols occur in the sectional views.

In the case of blind tapped holes, the drill depth normally is drawn at least three schematic pitches beyond the thread length, as shown in Fig. 13.8 (d), (e), (n), and (o). The representations at (f) and (p) are used to represent the use of a bottoming tap, when the length of thread is the same as the depth of drill. See also §13.24.

## 13.10 To Draw Schematic Thread Symbols.

In Fig. 13.9 (a) is shown a table of values of depth and pitch to use in drawing the thread symbols. These values are selected to produce a well-proportioned symbol and to be convenient to set off with the scale. The experienced draftsman will carefully space the lines by eye, but the student should use the scale. Note that the values of D and P are for the diameter *on the drawing.* Thus a $1\frac{1}{2}''$ diameter thread at half scale would be $\frac{3}{4}''$ diameter on the drawing, and values of D and P for a $\frac{3}{4}''$ major diameter would be used.

SCHEMATIC SYMBOLS. The steps for drawing the schematic symbols for an external thread in elevation and in section are shown in Fig. 13.9 (b). Note that when the pitches P are set off in II, the final crest line for a full pitch may fall beyond the actual thread length as shown. The completed schematic symbol for an external thread in elevation is shown at III. The completed schematic symbol for an external thread in section is shown at IV. The schematic thread depth is used for drawing the V's, and the pitch is established by the 60° V's.

The steps for drawing the schematic symbols for an internal thread in elevation and in section are shown at (c). Here again the symbol thread length may be slightly longer than the actual given thread length. If the tap drill depth is known or given, the drill is drawn to that depth, as shown. If the thread note omits this information, as is often done in practice, the draftsman merely draws the hole three thread pitches (schematic) beyond the thread length. The tap drill diameter is represented approximately, as shown, and not to actual size. The completed schematic symbol for an internal thread in elevation is shown at II. Pairs of hidden lines represent the threads, and the hidden-line dashes are staggered. The completed schematic symbol for an internal thread in section is shown at III. The schematic internal thread in section is represented in the same manner as for the schematic external thread.

SIMPLIFIED SYMBOLS. The steps for drawing the simplified symbols for an external thread in elevation and in section are shown in Fig. 13.9 (d). The thread depth from the table at (a) is used for establishing the pairs of hidden lines that represent the threads in elevation and in section. No pitch measurement is needed. The completed symbols are shown at III and IV.

The steps for drawing the simplified symbol for an internal thread in section are shown in Fig. 13.9 (e). The simplified representation for the internal thread in elevation is identical to that used for schematic representation, as the threads are indicated by pairs of hidden lines as shown at II. The simplified symbol for an internal thread in section is shown at III. The major diameter of the thread is represented by hidden lines across the sectioned area.

## 13.11 Detailed Representation—American National and Unified Threads.

The detailed representation for American National and Unified threads is the same, since the flats, if any, are disregarded. The steps in drawing are shown in Fig. 13.10:

I. Draw center line and lay out length and major diameter.

II. Find the number of threads per inch in Appendix 10. This number depends upon the major diameter of the thread, whether the

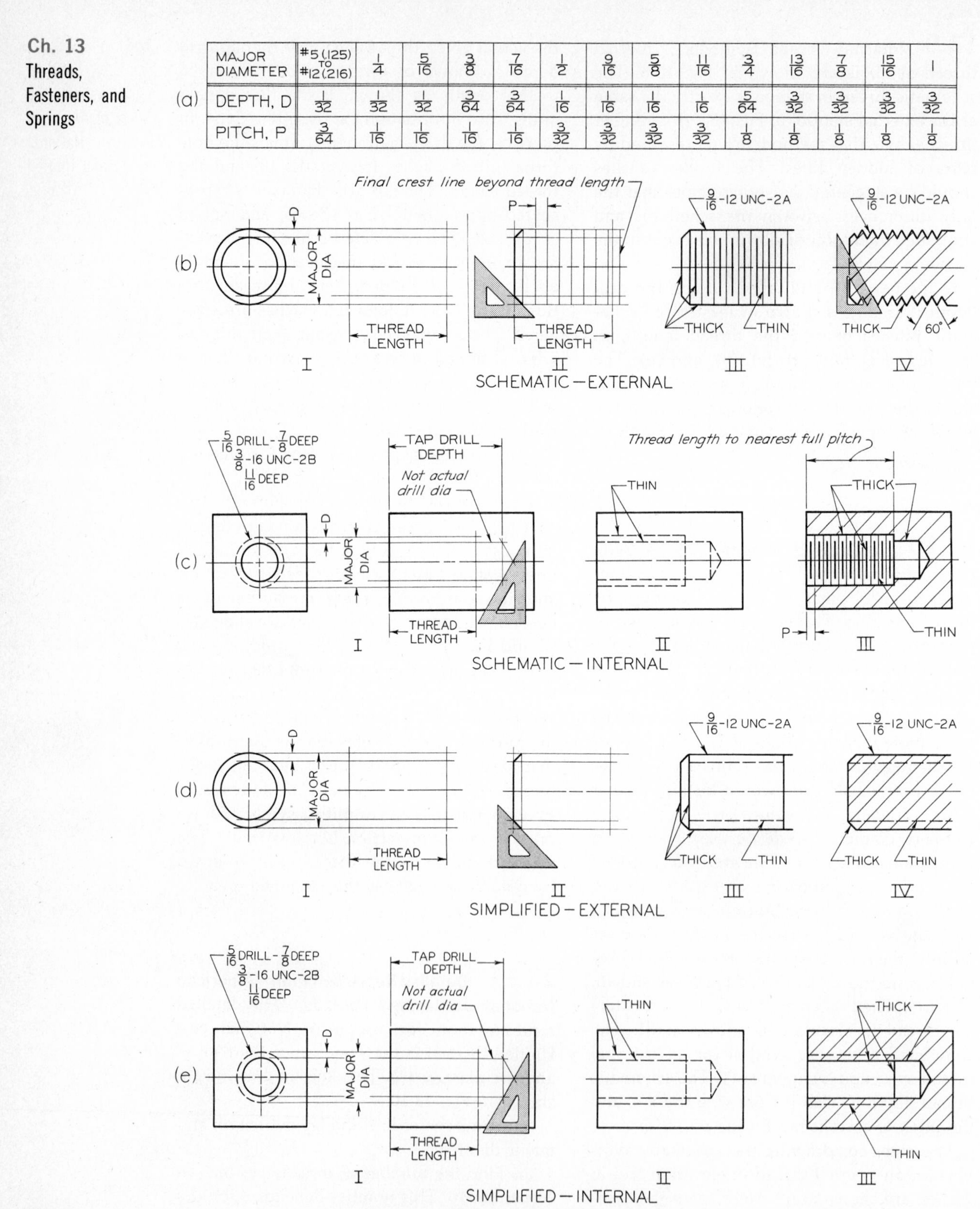

(a)

| MAJOR DIAMETER | #5 (.125) TO #12 (.216) | $\frac{1}{4}$ | $\frac{5}{16}$ | $\frac{3}{8}$ | $\frac{7}{16}$ | $\frac{1}{2}$ | $\frac{9}{16}$ | $\frac{5}{8}$ | $\frac{11}{16}$ | $\frac{3}{4}$ | $\frac{13}{16}$ | $\frac{7}{8}$ | $\frac{15}{16}$ | 1 |
|---|---|---|---|---|---|---|---|---|---|---|---|---|---|---|
| DEPTH, D | $\frac{1}{32}$ | $\frac{1}{32}$ | $\frac{1}{32}$ | $\frac{3}{64}$ | $\frac{3}{64}$ | $\frac{1}{16}$ | $\frac{1}{16}$ | $\frac{1}{16}$ | $\frac{1}{16}$ | $\frac{5}{64}$ | $\frac{3}{32}$ | $\frac{3}{32}$ | $\frac{3}{32}$ | $\frac{3}{32}$ |
| PITCH, P | $\frac{3}{64}$ | $\frac{1}{16}$ | $\frac{1}{16}$ | $\frac{1}{16}$ | $\frac{1}{16}$ | $\frac{3}{32}$ | $\frac{3}{32}$ | $\frac{3}{32}$ | $\frac{3}{32}$ | $\frac{1}{8}$ | $\frac{1}{8}$ | $\frac{1}{8}$ | $\frac{1}{8}$ | $\frac{1}{8}$ |



Fig. 13.9 **To Draw Thread Symbols—Schematic and Simplified.**

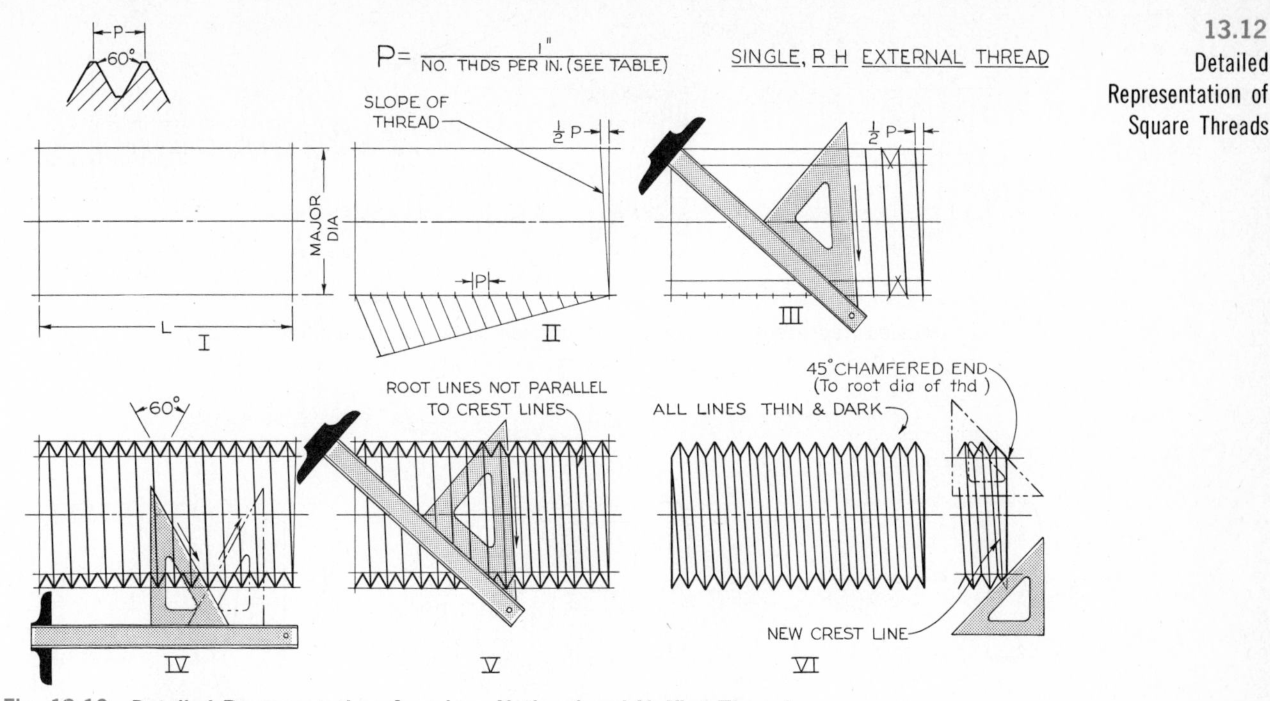


Fig. 13.10 **Detailed Representation–American National and Unified Threads.**

thread is internal or external. Find P (pitch) by dividing 1 by the number of threads per inch, §13.4. Establish the slope of the thread by offsetting the slope line $\frac{1}{2}$P for single threads, P for double threads, $1\frac{1}{2}$P for triple threads, and so on* For right-hand threads, the slope line slopes upward to the left; for left-hand threads, the slope line slopes upward to the right. If the number of threads per inch conforms to the scale, the pitch can be set off directly. For example, eight threads per inch can easily be set off with the architects scale, and ten threads per inch with the engineers scale. Otherwise, use the bow dividers or use the parallel-line method shown in Fig. 13.10(II).

III. From the pitch-points, draw crest lines parallel to slope line. These should be dark thin lines. Slide triangle on T-square (or another triangle) to make lines parallel. Draw two V's to establish depth of thread, and draw guide lines for root of thread, as shown.

IV. Draw 60° V's finished weight. These V's should stand vertically; that is, they should not "lean" with the thread.

V. Draw root lines dark at once. Root lines will *not* be parallel to crest lines. Slide triangle on straightedge to make root lines parallel.

VI. When the end is chamfered (usually 45° with end of shaft, sometimes 30°), the chamfer extends to the thread depth. The chamfer creates a new crest line, which is then drawn between the two new crest points. It is not parallel to the other crest lines. In the final drawing, all thread lines should be approximately the same weight—thin, but dark.

The corresponding internal detailed threads, in section, are drawn as shown in Fig. 13.11. Notice that for LH threads the lines slope upward to the left, as shown at (a), while for RH threads the lines slope upward to the right, as at (b). Make all final thread lines medium-thin but dark.

*These offsets are the same in terms of P for any form of thread.

## 13.12 Detailed Representation of Square Threads.

The steps in drawing the detailed representation of an external Square thread

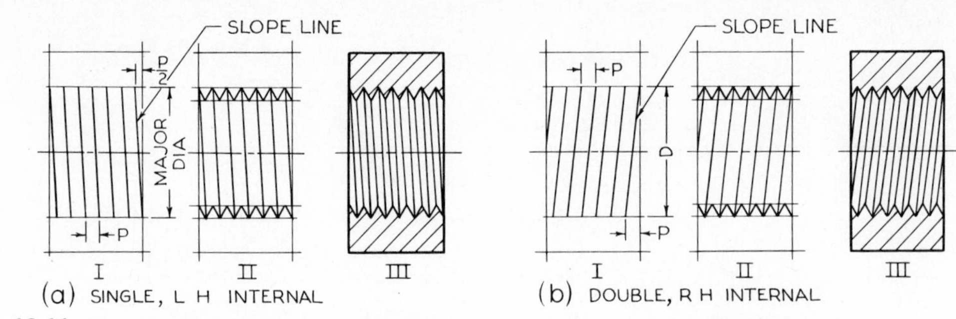


Fig. 13.11 **Detailed Representation—Internal American National and Unified Threads.**

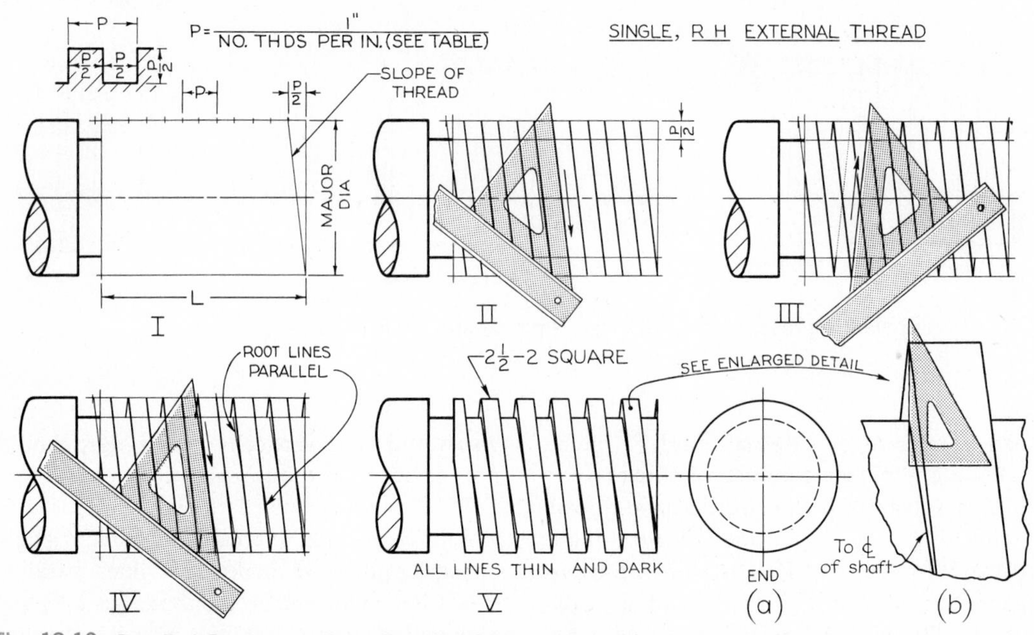


Fig. 13.12 **Detailed Representation—External Square Threads.**

when the major diameter is over 1″ (approx.) on the drawing are shown in Fig. 13.12.

I. Draw center line, and lay out length and major diameter of thread. Determine P by dividing 1 by the number of threads per inch. See Appendix 17. For a single RH thread, the lines slope upward to the left, and the slope line is offset $\frac{P}{2}$ *as for all single threads of any form.* On the upper line, set off spaces equal to $\frac{P}{2}$, as shown, using a scale if possible; otherwise use the bow dividers or the parallel-line method to space the points.

II. From the $\frac{P}{2}$ points on the upper line, draw crest lines parallel to slope line, dark and fairly thin. Draw guide lines for root of thread, making the depth $\frac{P}{2}$ as shown.

III. Draw parallel visible back edges of threads.

IV. Draw parallel visible root lines. Note enlarged detail at (b).

V. Accent the lines. All lines should be thin and dark.

Note the end view of the shaft at (a). The root circle is hidden; no attempt is made to show the true projection. If the end is chamfered, a solid circle would be drawn instead of the hidden circle.

An assembly drawing, showing an external Square thread partly screwed into the nut, is shown in Fig. 13.13. The detail of the Square thread at A is the same as shown in Fig. 13.12. But when the external and internal threads are

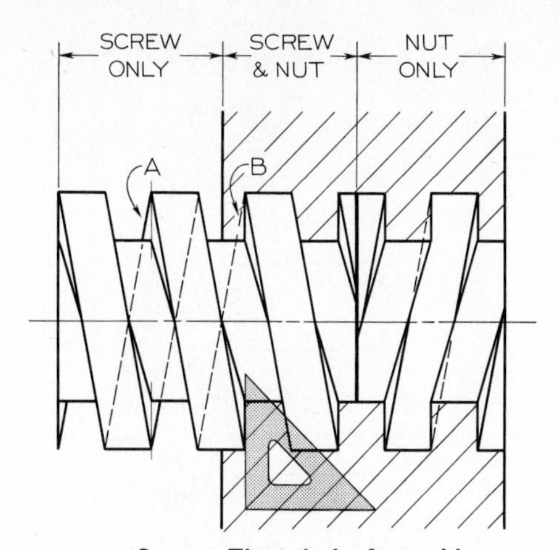


Fig. 13.13 **Square Threads in Assembly.**

assembled, the thread in the nut overlaps and covers up half of the V, as shown at B.

The internal thread construction is the same as in Fig. 13.14. Note that the thread lines representing the back half of the internal threads (since the thread is in section), slope in the opposite direction from those on the front side of the screw.

Steps in drawing a single internal Square thread in section are shown in Fig. 13.14. Note in step II that a crest is drawn opposite a root. This is the case for both single and triple threads. For double or quadruple threads, a crest is opposite a crest. Thus the construction in step I is the same for any multiple of thread. The differences are developed in step II where the threads and spaces are distinguished and outlined.

The same internal thread is shown in elevation (external view) in Fig. 13.14 (a). The profiles of the threads are drawn in their normal position, but with hidden lines; and the sloping lines are omitted for simplicity. The end view of the same internal thread is shown at (b). Note that the hidden and solid circles are opposite those for the end view of the shaft. See Fig. 13.12 (a).

## 13.13 Detailed Representation of Acme Thread.

The steps in drawing the detailed representation of Acme threads when the major diameter is over 1″ (approx.) on the drawing are shown in Fig. 13.15.

I. Draw center line, and lay out length and major diameter of thread. Determine P by dividing 1 by the number of threads per inch. See Appendix 17. Draw construction lines for the root diameter, making the thread depth $\frac{P}{2}$. Draw construction lines halfway between crest and root guide lines.

II. On the intermediate construction lines, lay off $\frac{P}{2}$ spaces, as shown. Set off spaces directly with scale is possible (for example, if $\frac{P}{2} = \frac{1}{10}$″, use the engineers scale); otherwise, use bow dividers or parallel-line method.

III. Through alternate points, draw construction lines for sides of threads (draw 15° instead of $14\frac{1}{2}$°).

Fig. 13.14 **Detailed Representation—Internal Square Threads.**

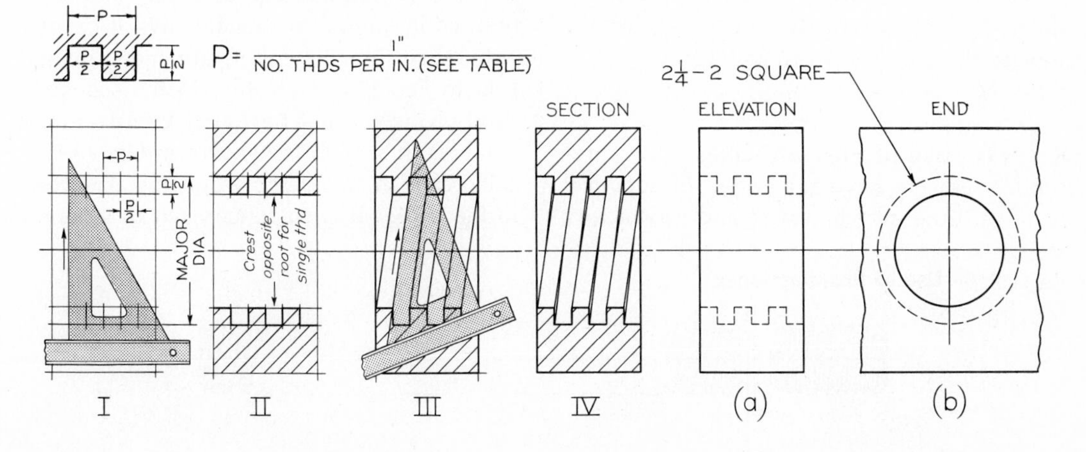

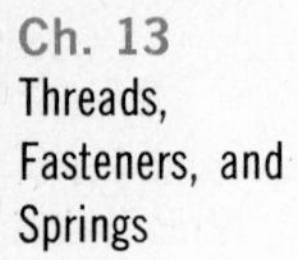

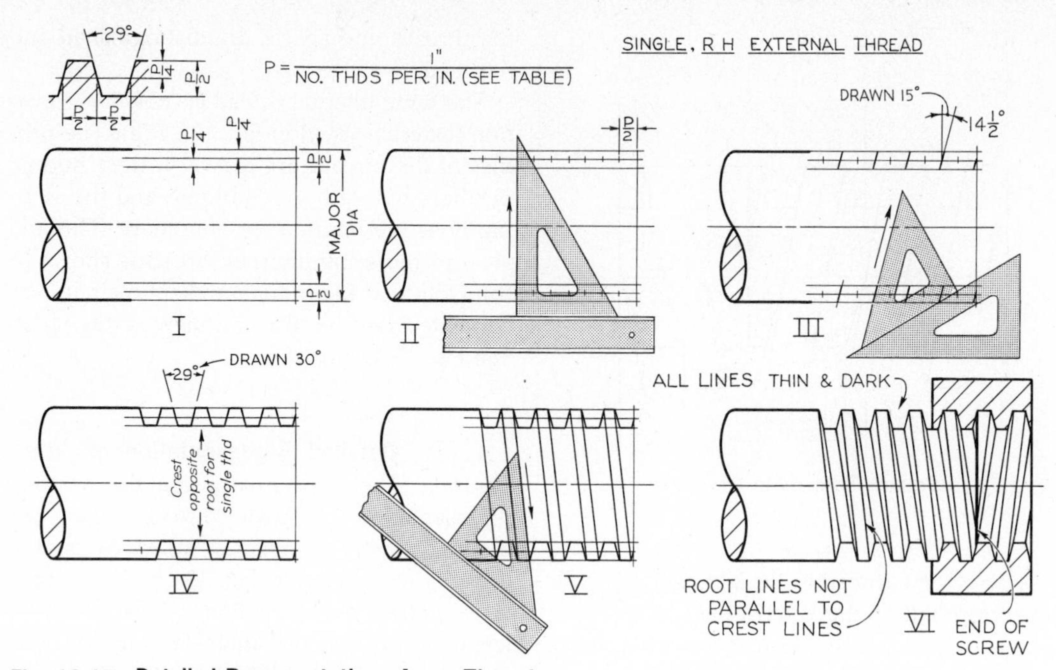


Fig. 13.15 **Detailed Representation—Acme Threads.**

IV. Draw construction lines for other sides of threads. Note that for single and triple threads, a crest is opposite a root, while for double and quadruple threads, a crest is opposite a crest. Heavy in tops and bottoms of threads.

V. Draw parallel crest lines, final weight at once.

VI. Draw parallel root lines, final weight at once, and heavy in the thread profiles. All lines should be thin and dark. Note that the internal threads in the back of the nut slope in the opposite direction to the external threads on the front side of the screw.

End views of Acme threaded shafts and holes are drawn exactly like those for the Square thread, Figs. 13.12 and 13.14.

## 13.14 Use of Phantom Lines.

In representing objects having a series of identical features, time may be saved and satisfactory representation effected by the use of phantom lines, as in Fig. 13.16. Threaded shafts thus represented may be shortened without the use of conventional breaks, but must be correctly dimensioned. The same methods may be applied to springs, Fig. 13.45 (d). This use of phantom lines is limited almost entirely to detail drawings.

## 13.15 Threads in Section.

Detailed representations of large threads in section are shown in Figs. 13.6, 13.11, 13.13, 13.14 and 13.15. As indicated by the note in Fig. 13.6 (b), the root lines and crest lines may be omitted in internal sectional views, if desired.

External thread symbols are shown in section in Fig. 13.7. Note that in the schematic symbol, the V's must be drawn. Internal thread symbols in section are shown in Fig. 13.8.

Threads in an assembly drawing are shown in Fig. 13.17. It is customary not to section a

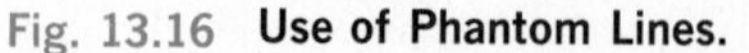

Fig. 13.16 **Use of Phantom Lines.**

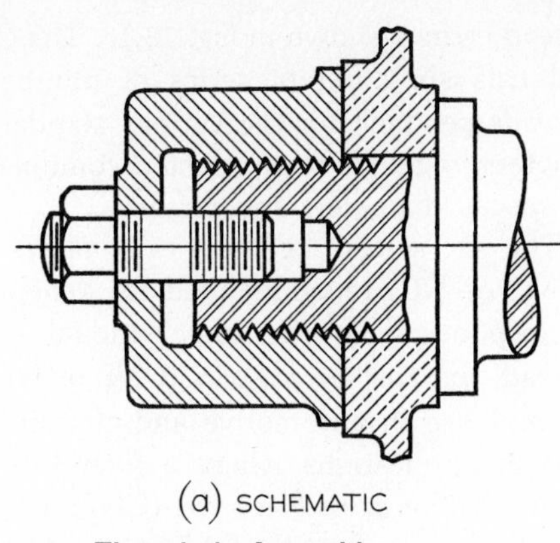

(a) SCHEMATIC

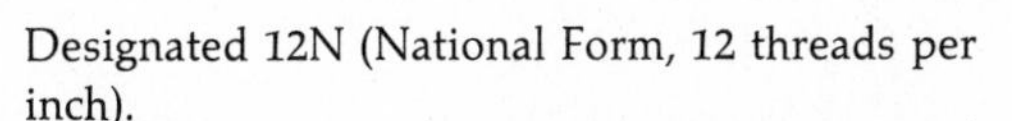

(b) SIMPLIFIED

Fig. 13.17 **Threads in Assembly.**

stud or a nut, or any solid part, unless necessary to show some internal shapes. See §14.22. Note that when external and internal threads are sectioned in assembly, the V's are required to show the threaded connection.

## 13.16 American National Thread.

The old American National thread was adopted in 1935. The *form,* or profile, Fig. 13.2 (b), is the same as the old Sellers' profile, or U.S. Standard, and is known as the *National Form.* The methods of representation are the same as for the Unified thread. American National threads are being replaced by the Unified threads, but the old threads will be frequently encountered on drawings for a long time.

Five *series* of threads were embraced in the old standard,* as follows:

1. *Coarse thread*—A general-purpose thread for holding purposes. Designated NC (National Coarse).
2. *Fine thread*—A greater number of threads per inch; used extensively in automotive and aircraft construction. Designated NF (National Fine).
3. *8-Pitch thread*—All diameters have 8 threads per inch. Used on bolts for high-pressure pipe flanges, cylinder-head studs, and similar fasteners. Designated 8N (National Form, 8 threads per inch).
4. *12-Pitch thread*—All diameters have 12 threads per inch; used in boiler work and for thin nuts on shafts and sleeves in machine construction. Designated 12N (National Form, 12 threads per inch).
5. *16-Pitch thread*—All diameters have sixteen threads per inch; used where necessary to have a fine thread regardless of diameter, as on adjusting collars and bearing retaining nuts. Designated 16N (National Form, sixteen threads per inch).

*ASA B1.1-1935.

## 13.17 American National Thread Fits.

The old standard* also established for general use four classes of screw thread *fits* between mating threads (as between bolt and nut). These fits are produced by the application of tolerances listed in the standards, and are described as follows:

*Class 1 fit*—Recommended only for screw thread work where clearance between mating parts is essential for rapid assembly and where shake or play is not objectionable.

*Class 2 fit*—Represents a high quality of commercial thread product and is recommended for the great bulk of interchangeable screw thread work.

*Class 3 fit*—Represents an exceptionally high quality of commercially threaded product and is recommended only in cases where the high cost of precision tools and continual checking are warranted.

*Class 4 fit*—Intended to meet very unusual requirements more exacting than those for which Class 3 is intended. It is a selective fit if initial assembly by hand is required. It is not, as yet, adaptable to quantity production.

*ASA B1.1-1935.

The class of fit desired on a thread is indicated by the number of the fit in the thread note, as shown in §13.21.

## 13.18 Unified Extra Fine Threads.*

*The Unified Extra Fine thread series* has many more threads per inch for given diameters than any series of the American National or Unified. The form of thread is the same as the American National. These small threads are used in thin metal where the length of thread engagement is small, in cases where close adjustment is required, and where vibration is great. It is designated UNEF (Extra Fine).

## 13.19 Unified Threads.*

The new Unified thread constitutes the present American National Standard. Some of the earlier American National threads are still included in the new standard. See Appendix 10. The Unified thread form is shown in Fig. 13.18. The standard lists six different series of numbers of threads per inch for the various standard diameters, together with selected combinations of special diameters and pitches.

The six series are the *Coarse thread series* (UNC or NC) recommended for general use corresponding to the old National Coarse thread; the *Fine thread series* (UNF or NF), for general use in automotive and aircraft work and in applications where a finer thread is required; the *Extra Fine series* (UNEF or NEF), which is the same as the SAE Extra Fine series, used particularly in aircraft and aeronautical equipment and generally for threads in thin walls; and the *8-, 12-, and 16-Pitch thread series* (8UN or 8N, 12UN or 12N, and 16UN or 16N), recommended for the uses corresponding to the old 8-, 12-, and 16-Pitch American National threads. In addition, there are three special thread series: UNS, NS, and UN, which involve special combinations of diameter, pitch, and length of engagement.

* ANSI B1.1-1960.

**Fig. 13.18 Unified Screw Thread Form** (ANSI B1.1-1960).

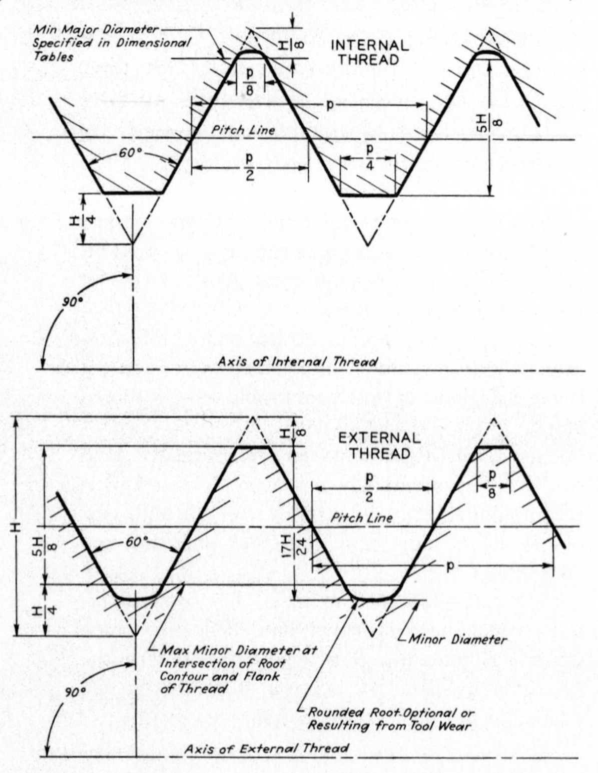


## 13.20 Unified Thread Fits.

The tables of the standard* specify tolerances and allowances defining the several classes of fit (degree of looseness or tightness) between mating threads. In the symbols for fit, the letter A refers to external threads, and B to internal threads. In the new standard, Classes 1A and 1B take the place of old Class 1, and have generous tolerances facilitating rapid assembly and disassembly; Classes 2A and 2B are used in the normal production of screws, bolts, and nuts, as well as a variety of general applications; and Classes 3A and 3B are newly toleranced classes for highly accurate and close-fitting requirements. Class 4 of the old standard has been dropped because of its infrequent and specialized use.

## 13.21 Thread Notes.

Thread notes for American National threads are shown in Fig. 13.19. These same symbols are used in correspondence, on shop and storeroom cards, and

* ANSI B1.1-1960.

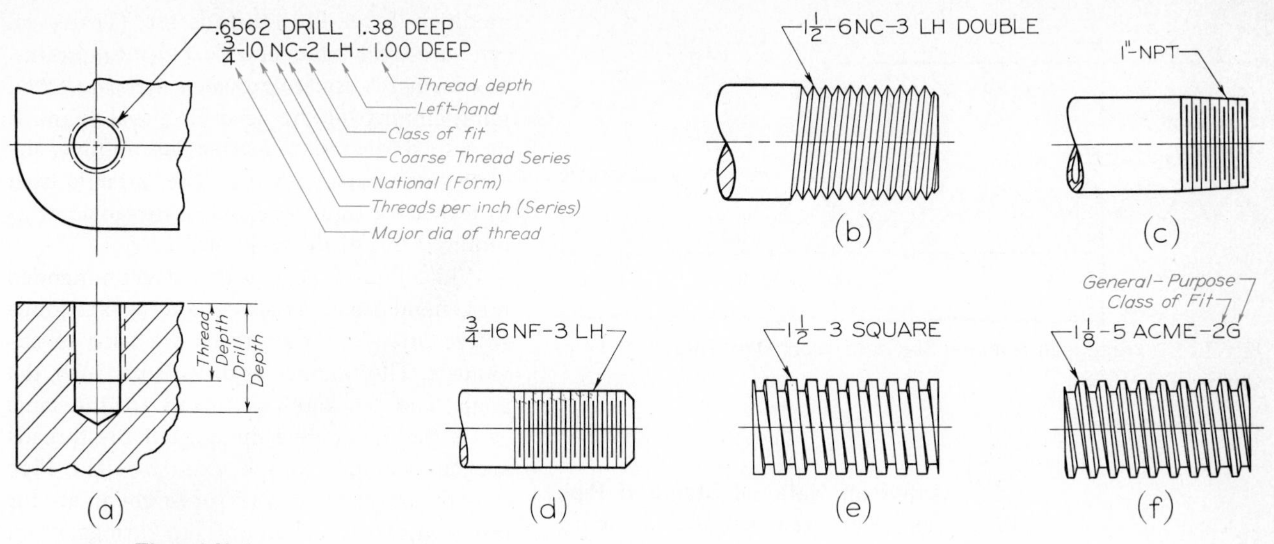


Fig. 13.19 **Thread Notes.**

in specifications for parts, taps, dies, tools, and gages. A thread note for a blind tapped hole is shown at (a). In a complete note, the tap drill and depth should be given, though in practice they are often omitted and left to the shop. For tap drill sizes, see Appendix 10. If the LH symbol is omitted, the thread is understood to be RH. If the thread is a multiple thread, the word DOUBLE, TRIPLE, or QUADRUPLE should precede the thread depth; otherwise the thread is understood to be single. Thread notes for holes are preferably attached to the circular views of the holes, as shown.

Thread notes for external threads are preferably given in the longitudinal view of the threaded shaft, as shown from (b) to (f). Examples of 8-, 12-, and 16-Pitch threads, not shown in the figure, are 2–8N–2, 2–12N–2, and 2–16N–2. A sample special thread designation is $1\frac{1}{2}$–7N–LH.

General-purpose Acme threads are indicated by the letter G, and centralizing Acme threads by the letter C. Typical thread notes are: $1\frac{3}{4}$–4 ACME–2G or $1\frac{3}{4}$–6 ACME–4C.

Thread notes for Unified threads are shown in Fig. 13.20. Unified threads are distinguished from American National threads by the insertion of the letter U before the series letters, and by the letters A or B (for external or internal, respectively) after the numeral designating the class of fit. If the letters LH are omitted, the thread is understood to be RH. Some typical thread notes are as follows:

$\frac{1}{4}$–20 UNC–2A TRIPLE
$\frac{9}{16}$–18 UNF–2B
$1\frac{3}{4}$–16 UN–2A
$1\frac{5}{8}$–10 NS–2 PD 1.5600–1.5532

Fig. 13.20 **Unified Thread Notes.**

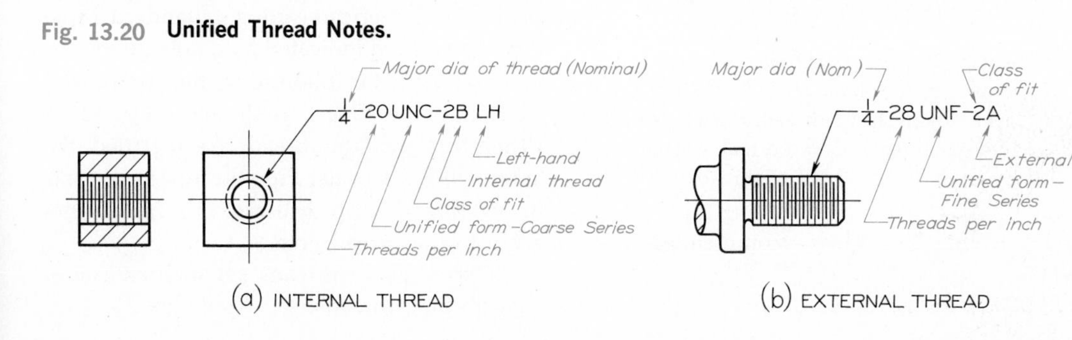


(a) INTERNAL THREAD

(b) EXTERNAL THREAD

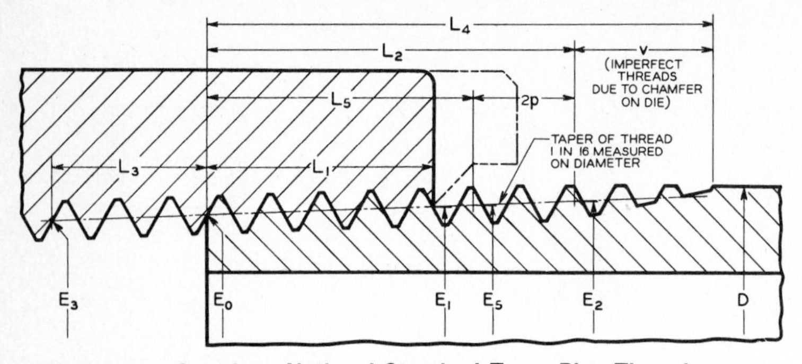


Fig. 13.21 **American National Standard Taper Pipe Thread** (ANSI B2.1-1968).

**13.22 American National Standard Pipe Threads.** The American National Standard for Pipe Threads, originally known as the Briggs Standard, was formulated by Robert Briggs in 1882. Two general types of pipe threads have been approved as American National Standard: *Taper Pipe threads* and *Straight Pipe threads.** The profile of the Taper Pipe thread is illustrated in Fig. 13.21. See Appendix 32. The taper of the thread is 1 in 16 or 0.75″ per foot measured on the diameter and along the axis. The angle between the sides of the thread is 60°. The depth of the sharp V is 0.8660p, and the basic maximum depth of the truncated thread is 0.800p, where p = pitch. The basic pitch diameters, $E_0$ and $E_1$, and the basic length of the effective external taper thread, $L_2$, are determined by the formulas:

$$E_0 = D - (0.050D + 1.1)\frac{1}{n}$$

$$E_1 = E_0 + 0.0625L_1$$

$$L_2 = (0.80D + 6.8)\frac{1}{n}$$

where D = basic O.D. of pipe; $E_0$ = pitch diameter of thread at end of pipe; $E_1$ = pitch diameter of thread at large end of internal thread; $L_1$ = normal engagement by hand; n = number of threads per inch.

The ANSI* has also recommended two modified Taper Pipe threads for: (1) dryseal pressure-tight joints and (2) rail fitting joints. The former is used to provide a metal-to-metal joint, eliminating the need for a sealer, and is used in refrigeration, marine, automotive, aircraft, and ordnance work. The latter is used to provide a rigid mechanical thread joint as required in rail fitting joints.

While Taper Pipe threads are recommended for general use, there are certain types of joints where Straight Pipe threads are used to advantage. The number of threads per inch, the angle, and the depth of thread are the same as on the Taper Pipe thread, but the threads are cut parallel to the axis. Straight Pipe threads are used for pressure-tight joints for pipe couplings, fuel and oil line fittings, drain plugs, free-fitting mechanical joints for fixtures, loose-fitting mechanical joints for locknuts, and loose-fitting mechanical joints for hose couplings.

Pipe threads are represented by detailed or symbolic methods in a manner similar to the representation of Unified and American National threads. The symbolic representation (schematic or simplified) is recommended for general use regardless of diameter, Fig. 13.22, the detailed method being approved only when the threads are large and when it is desired to show the profile of the thread as, for example, in a sectional view of an assembly.

As shown in Fig. 13.22, it is not necessary to draw the taper on the threads unless there is some reason to emphasize it, since the thread note indicates whether the thread is straight or tapered. If it is desired to show the taper, it should be exaggerated, as shown in Fig. 13.23, where the taper is drawn $\frac{1}{16}$″ per 1″ *on radius*, instead of the actual taper of $\frac{1}{16}$″ *on diameter.* American National Standard Taper Pipe threads are indicated by a note giving the nominal diameter followed by the letters NPT (National Pipe Taper), as shown in Fig. 13.22. When Straight Pipe threads are specified, the letters NPS (National Pipe Straight) are used. In practice, the tap drill size is normally not given in the thread note.

For numbers of threads per inch and other data on pipe threads, see Appendix 32.

*ANSI B2.1-1968 and ANSI B2.2-1968.

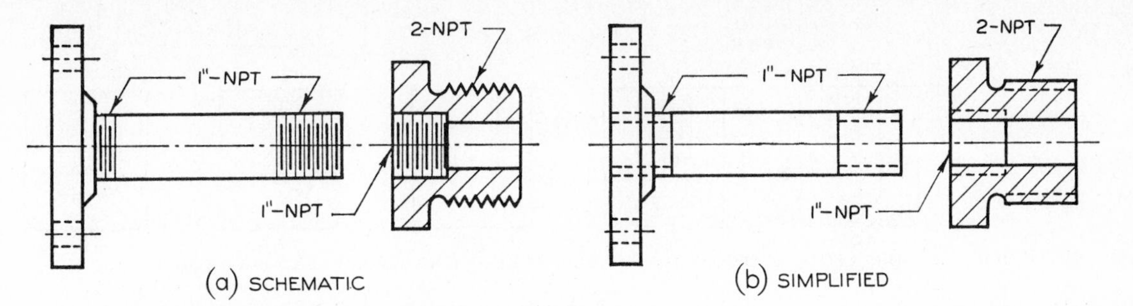


Fig. 13.22 **Conventional Representation of Pipe Threads.**

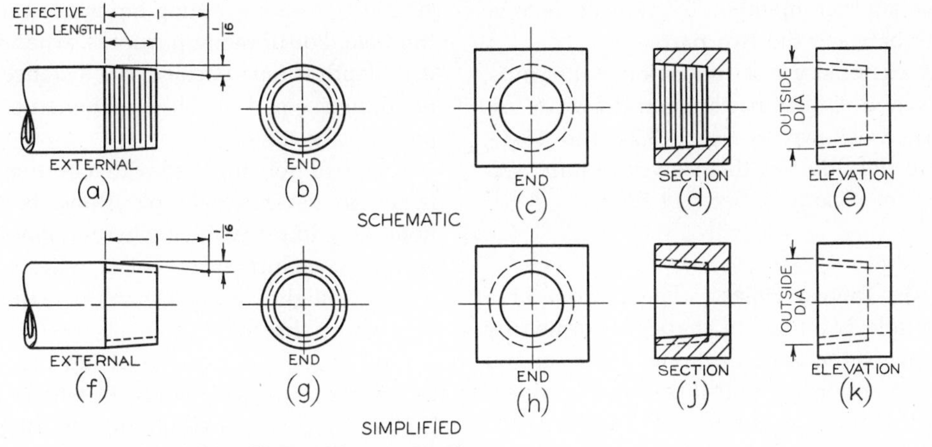


Fig. 13.23 **Conventional Pipe Thread Representation.**

**13.23 Bolts, Studs, and Screws.** The term *bolt* is generally used to denote a "through bolt" that has a head on one end and is passed through clearance holes in two or more aligned parts and is threaded on the other end to receive a nut to tighten and hold the parts together. See Fig. 13.24 (a) and §§13.25 and 13.26.

A hexagon head *cap screw,* (b), is similar to a bolt, except that it generally has a greater length of thread, for when it is used without a nut, one of the members being held together is threaded to act as a nut. It is screwed on with a wrench. Cap screws are not screwed into thin materials if strength is desired. See §13.29.

A *stud,* (c), is a steel rod threaded on both ends. It is screwed into place with a pipe wrench or, preferably, with a stud driver. As a rule, a stud is passed through a clearance hole in one member and screwed into another member, and a nut is used on the free end, as shown.

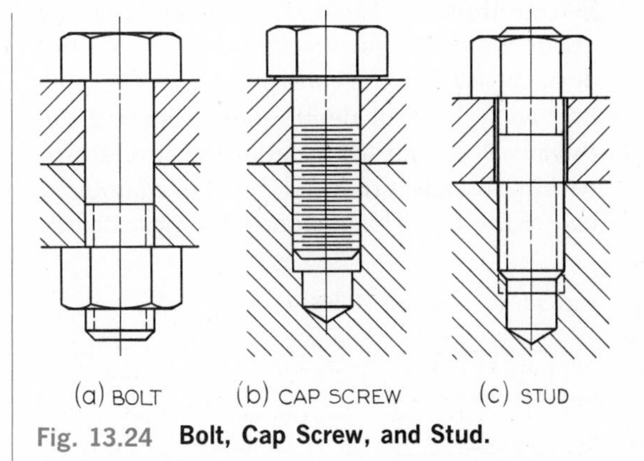


Fig. 13.24 **Bolt, Cap Screw, and Stud.**

A *machine screw,* Fig. 13.34, is similar to the slotted-head cap screws but, in general, is smaller. It may be used with or without a nut.

A *set screw,* Fig. 13.35, is a screw with or without a head that is screwed through one

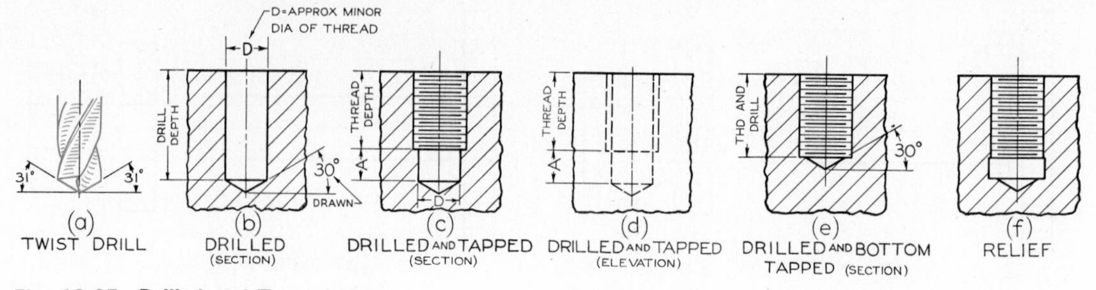


Fig. 13.25 **Drilled and Tapped Holes.**

member and whose special point is forced against another member to prevent relative motion between the two parts.

It is customary not to section bolts, nuts, screws, and similar parts when drawn in assembly, as shown in Figs. 13.24 and 13.33, because they do not themselves require sectioning for clearness. See §14.22.

## 13.24 Tapped Holes.

The bottom of a drilled hole is conical in shape, as formed by the point of the twist drill, Fig. 13.25 (a) and (b). When an ordinary drill is used in connection with tapping, it is referred to as a *tap drill.* On drawings, an angle of 30° is used to approximate the actual 31°.

The thread length is the length of full or perfect threads. The tap drill depth is the depth of the cylindrical portion of the hole and does not include the cone point, (b). The portion A of the drill depth shown beyond the threads at (c) and (d) includes the several imperfect threads produced by the chamfered end of the tap. This distance A varies according to drill size and whether a plug tap, Fig. 10.20 (h), or a bottoming tap is used to finish the hole. For drawing purposes, when the tap drill depth is not specified, the distance A may be drawn equal to three schematic thread pitches, Fig. 13.9.

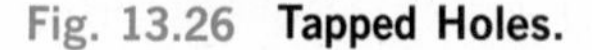
Fig. 13.26 **Tapped Holes.**

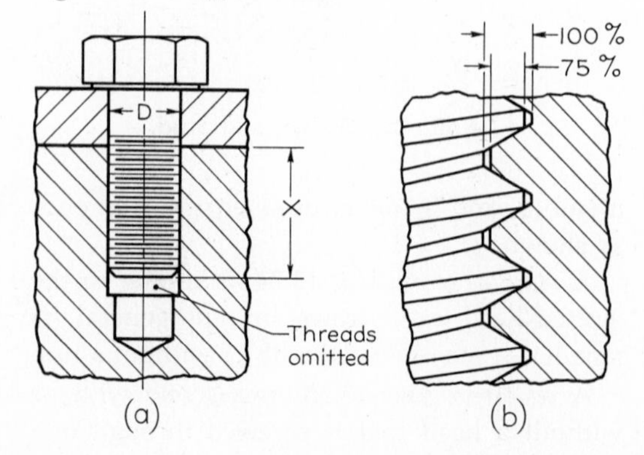


A tapped hole finished with a bottoming tap is drawn as shown at (e). Blind bottoming holes should be avoided wherever possible. A better procedure is to cut a relief with its diameter slightly greater than the major diameter of the thread, (f).

One of the chief causes of tap breakage is insufficient tap drill depth, in which the tap is forced against a bed of chips in the bottom of the hole. Therefore, the draftsman should never draw a blind hole when a through hole of not much greater length can be used; and when a blind hole is necessary, he should provide generous tap drill depth. Tap drill sizes for Unified and American National threads are given in Appendix 10.

The thread length in a tapped hole depends upon the major diameter and the material being tapped. In Fig. 13.26 (a), the minimum engagement length X, when both parts are steel, is equal to the diameter D of the thread. When a steel screw is screwed into cast iron, brass, or bronze, $X = 1\frac{1}{2}D$; and when screwed into aluminum, zinc, or plastic, $X = 2D$.

Since the tapped thread length contains only full threads, it is necessary to make this length only one or two pitches beyond the end of the engaging screw. In schematic representation, the threads are omitted in the bottoms of tapped holes so as to show the ends of the screws clearly. See Fig. 13.26.

In the early days of machine construction, it was the practice to make threads engage virtually 100 percent of the thread depth, but it was learned after a while that such a thread was only 5 percent stronger than a 75 percent thread and required about three times as much power to tap and resulted in many more broken taps. To produce a 75 percent tapped thread, Fig. 13.26 (b), the diameter of the drill is slightly greater than the root diameter of the internal thread. The tap drill sizes for Unified and American National threads given in Appendix 10 are for 75 percent threads. It is good practice to give the tap drill size in the thread note, §13.21.

When a bolt or a screw is passed through a clearance hole in one member, the hole may be drilled $\frac{1}{32}$″ larger than the screw up to $\frac{3}{8}$″ diameter and $\frac{1}{16}$″ larger for larger diameters. For more precise work, the clearance hole may be only $\frac{1}{64}$″ larger than the screw up to $\frac{3}{8}$″ diameter and $\frac{1}{32}$″ larger for larger diameters. Closer fits may be specified for special conditions. The clearance spaces on each side of a screw or bolt need not be shown on a drawing unless it is necessary to show that there is no thread engagement, in which case the clearance spaces are drawn about $\frac{3}{64}$″ wide for clarity.

## 13.25 American National Standard Bolts and Nuts.

American National Standard bolts and nuts* are produced in two forms: square and hexagon, Fig. 13.27. Square heads and nuts are chamfered at 30°, and hexagon heads and nuts are chamfered at 25°. Both are drawn at 30° for simplicity.

Bolt Types. Bolts are grouped according to use: *Regular* bolts are for general service and *Heavy* bolts are for heavier service or easier wrenching. Square bolts come only in the Regular type, while hexagon bolts, screws, and nuts and square nuts are standard in both types.

Finish. Square bolts and nuts, hexagon bolts, and hexagon flat nuts are *unfinished.* Unfinished bolts and nuts are not machined on any surface except for the threads. Traditionally, hexagon bolts and hexagon nuts have been available as unfinished, semifinished, or finished. According to the latest standards, hexagon cap screws and finished hexagon bolts have been consolidated into a single product—*Hex cap screws*—thus eliminating the regular semifinished hexagon bolt classification. Heavy semifinished hexagon bolts and heavy finished hexagon bolts also have been combined into a single product called *Heavy Hex screws.* Hexagon cap screws, Heavy hexagon screws and all hexagon nuts, except hexagon flat nuts, are considered *finished* to some degree and are characterized by a "washer face" machined or otherwise formed on the bearing surface. The washer face is $\frac{1}{64}$″ thick (drawn $\frac{1}{32}$″), and its diameter is equal to $1\frac{1}{2}$ times the body diameter D.

For nuts the bearing surface also may be a circular surface produced by chamfering. Hexagon screws and hexagon nuts have closer tolerances and may have a more finished appearance, but are not completely machined. There is no difference in the drawing for the degree of finish on finished screws and nuts.

Proportions. Sizes based on diameter D of the bolt body, which are either exact formula proportions or close approximations for drawing purposes, are as follows:

*Regular hexagon and square bolts and nuts:*

$$W = 1\tfrac{1}{2}D \qquad H = \tfrac{2}{3}D \qquad T = \tfrac{7}{8}D$$

Fig. 13.27 **Standard Bolts and Nuts.**

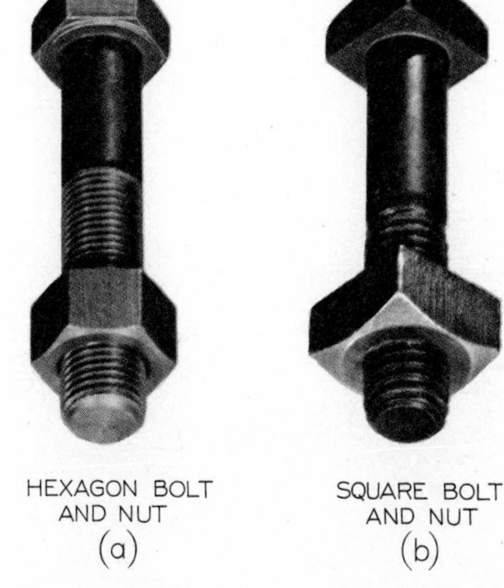


*ANSI B18.2.1-1965 and ANSI B18.2.2-1965. The standards cover several bolts and nuts. For complete details, see the standards.

*Heavy hexagon bolts and nuts and square nuts.**

$$W = 1\tfrac{1}{2}D + \tfrac{1}{8}'' \qquad H = \tfrac{2}{3}D \qquad T = D$$

where W = width across flats; H = head height; and T = nut height.

The washer face is always included in the head or nut height for finished hexagon screw heads and nuts.

THREADS. Unfinished bolts have Coarse threads, Class 2A, while hex cap screws (finished bolts) have Coarse, Fine, or 8-Pitch threads, Class 2A. Unfinished nuts have Coarse threads, Class 2B. Finished nuts have Coarse or Fine threads, Class 2B, while certain of these also may have 8-Pitch threads.

THREAD LENGTHS. *For bolts or screws up to 6″ in length:*

$$\text{Thread length} = 2D + \tfrac{1}{4}''.$$

*For bolts or screws over 6″ in length:*

$$\text{Thread length} = 2D + \tfrac{1}{2}''.$$

*There are no Heavy square bolts.

Fasteners too short for these formulas are threaded as close to the head as practicable. For drawing purposes, this may be taken as three thread pitches, approximately. The threaded end may be rounded or chamfered, but is usually drawn with a 45° chamfer from the thread depth, Fig. 13.28.

BOLT LENGTHS. Lengths have not been standardized because of the endless variety required by industry. The following increments (differences in successive lengths) are compiled from manufacturers' catalogs. These increments apply to stock sizes of cut-thread bolts. Long bolts of small diameter or short bolts of large diameter would have to be ordered "special."

*Square head bolts:*

Lengths $\frac{1}{2}''$ to $\frac{3}{4}''$, increment = $\frac{1}{8}''$
Lengths $\frac{3}{4}''$ to 5″, increment = $\frac{1}{4}''$
Lengths 5″ to 12″, increment = $\frac{1}{2}''$
Lengths 12″ to 30″, increment = 1″

Fig. 13.28 **Bolt Proportions** (Regular).

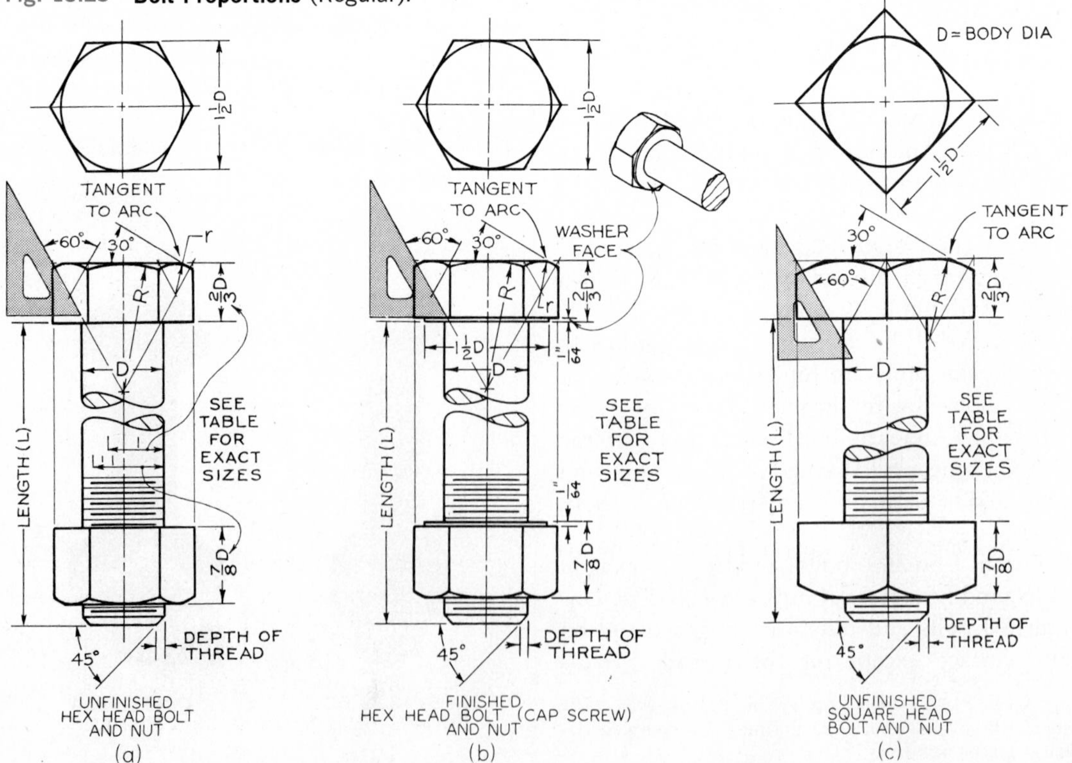

*Hexagon head bolts and screws:*

Lengths $\frac{3}{4}''$ to 8″, increment = $\frac{1}{4}''$
Lengths 8″ to 20″, increment = $\frac{1}{2}''$
Lengths 20″ to 30″, increment = 1″

For dimensions of standard bolts and nuts, see Appendix 13.

## 13.26 To Draw American National Standard Bolts.

In practice, standard bolts and nuts are not shown on detail drawings unless they are to be altered, but they appear so frequently on assembly drawings that a suitable but rapid method of drawing them must be used. They may be drawn from exact dimensions taken from tables* if accuracy is important, as in figuring clearances; but in the great majority of cases the conventional representation, in which proportions based upon the body diameter are used, will be sufficient, and a considerable amount of time may be saved. Three typical bolts illustrating the use of these proportions for the Regular bolts are shown in Fig. 13.28.

Although the curves produced by the chamfer on the bolt heads and nuts are hyperbolas, in practice these curves are always represented approximately by means of circular arcs, as shown in Fig. 13.28.

Generally, bolt heads and nuts should be drawn "across corners" in all views, regardless of projection, as shown in Figs. 14.82 to 14.84. This conventional violation of projection is used to prevent confusion between the square and hexagon heads and nuts and to show actual clearances. Only when there is a special reason should bolt heads and nuts be drawn across flats. In such cases, the conventional proportions shown in Fig. 13.29 are used.

Steps in drawing hexagon bolts (cap screws) and nuts are illustrated in Fig. 13.30, and those for square bolts and nuts in Fig. 13.31. Before you start drawing, the diameter of the bolt, the length from the under side of the bearing surface to the tip, the style of head (square or hexagon), and the type (Regular or Heavy), as well as the finish, must be known.

*See Appendix 13.

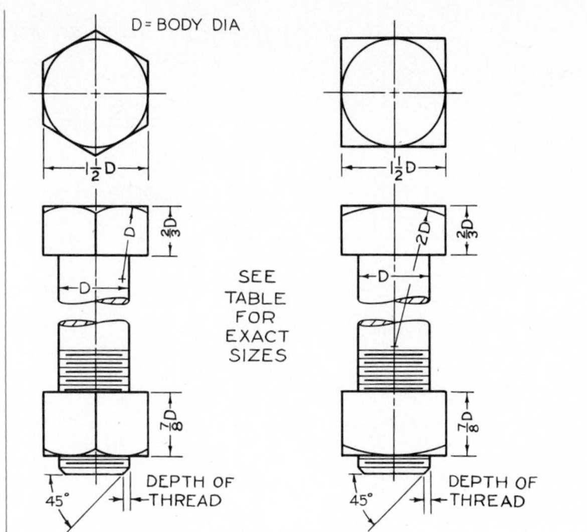


Fig. 13.29 **Bolts "Across Flats."**

If only the longitudinal view of a bolt is needed, it is necessary to draw only the lower half of the top views in Figs. 13.30 and 13.31 *with light construction lines* in order to project the corners of the hexagon or square to the front view. These construction lines may then be erased if desired.

The head and nut heights can be spaced off with the dividers on the shaft diameter and then transferred as shown in both figures, or the scale may be used as in Fig. 4.15. The heights should not be determined by arithmetic.

The $\frac{1}{64}''$ washer face has a diameter equal to the distance across flats of the bolt head or nut. It appears only on the finished hexagon screws or nuts, the $\frac{1}{64}''$ thickness being drawn at $\frac{1}{32}''$ for clearness. The $\frac{1}{32}''$ is included in the head or nut height.

Threads should be drawn in schematic or simplified form for body diameters of 1″ or less on the drawing, Fig. 13.9 (b) or (d), and by detailed representation for larger diameters, §§13.7 and 13.8. The threaded end of the screw should be chamfered at 45° from the schematic thread depth, Fig. 13.9 (a).

On drawings of small bolts or nuts (under $\frac{1}{2}''$ diameter approximately) where the chamfer is hardly noticeable, the chamfer on the head or nut may be omitted in the longitudinal view.

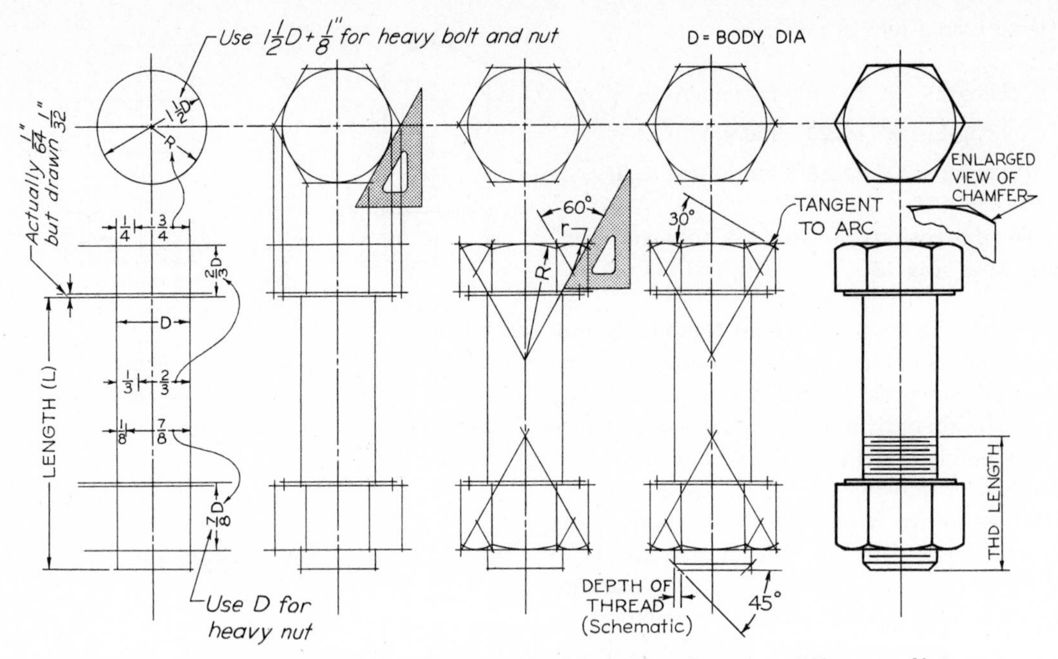


Fig. 13.30 **Steps in Drawing a Finished Hexagon-Head Bolt (Cap Screw) and Hexagon Nut.**

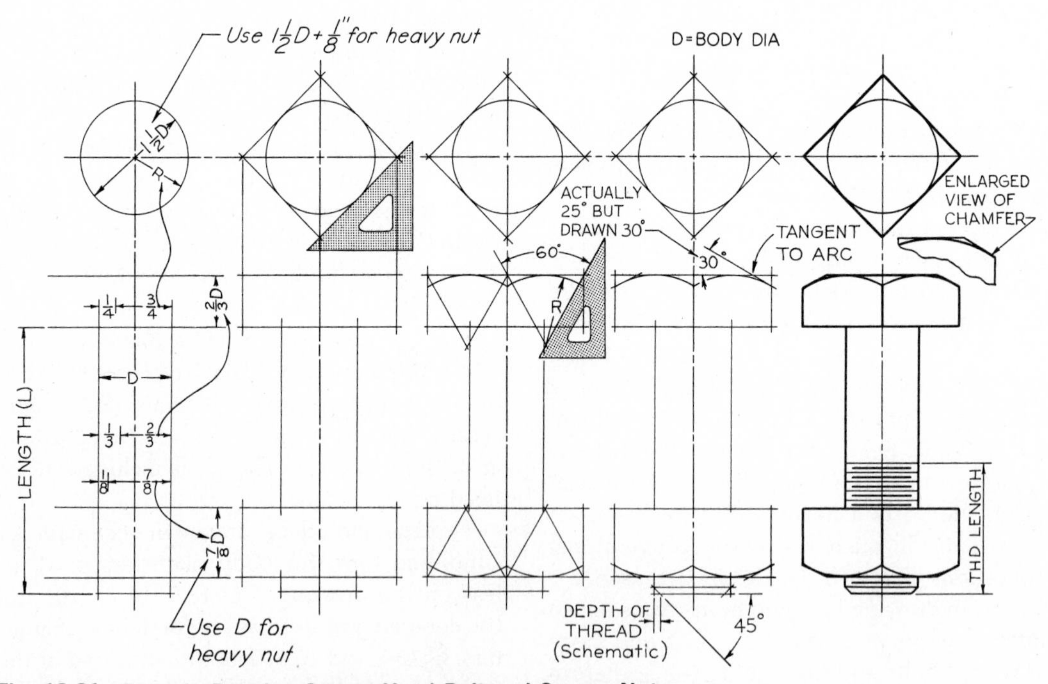


Fig. 13.31 **Steps in Drawing Square-Head Bolt and Square Nut.**

Many styles of templates are available for saving time in drawing bolt heads and nuts. One of these, the *Draftsquare*, is illustrated in Fig. 2.80 (b).

## 13.27 Specifications for Bolts and Nuts.

In specifying bolts in parts lists, in correspondence, or elsewhere, the following information must be covered in order:

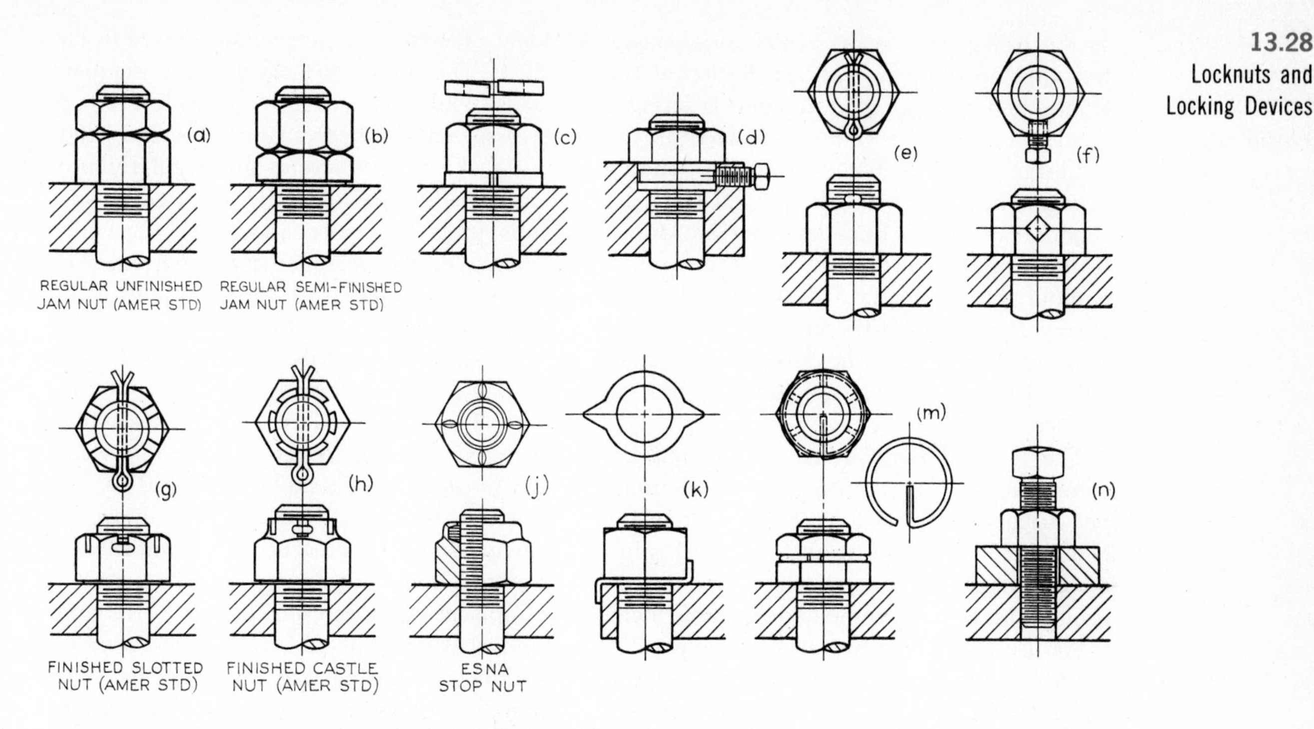


Hexagon Head Screws: Coarse, Fine, or 8-Thread Series, 2A. Thread length = 2D + $\frac{1}{4}''$ up to 6″ long and 2D + $\frac{1}{2}''$ if over 6″ long. For screws too short for formula, threads extend to within $2\frac{1}{2}$ threads of the head for diameters up to 1″. Screw lengths not standardized.

Slotted Head Screws: Coarse, Fine, or 8-Thread Series, 2A. Thread length = 2D + $\frac{1}{4}''$. Screw lengths not standardized. For screws too short for formula, threads extend to within $2\frac{1}{2}$ threads of the head.

Hexagon Socket Screws: Coarse or Fine threads, 3A. Coarse thread length = 2D + $\frac{1}{2}''$ where this would be over $\frac{1}{2}$L; otherwise thread length = $\frac{1}{2}$L. Fine thread length = $1\frac{1}{2}$D + $\frac{1}{2}''$ where this would be over $\frac{3}{8}$L; otherwise thread length = $\frac{3}{8}$L. Increments in screw lengths = $\frac{1}{8}''$ for screws $\frac{1}{4}''$ to 1″ long, $\frac{1}{4}''$ for screws 1″ to 3″ long, and $\frac{1}{2}''$ for screws $3\frac{1}{2}''$ to 6″ long.

Fig. 13.32 **Locknuts and Locking Devices.**

1. Nominal size of bolt body
2. Thread specification (see §13.21)
3. Length of bolt
4. Finish of bolt
5. Style of head
6. Name

*Example* (complete): $\frac{3}{4}$-10 UNC-2A × $2\frac{1}{2}$ HEXAGON CAP SCREW

*Example* (abbreviated): $\frac{3}{4}$ × $2\frac{1}{2}$ HEX CAP SCR

Nuts may be specified as follows:

*Example* (complete): $\frac{5}{8}$-11 UNC-2B SQUARE NUT

*Example* (abbreviated): $\frac{5}{8}$ SQ NUT

For either bolts or nuts, the word REGULAR is assumed if left out of the specification. If the Heavy series is intended, the word HEAVY should appear as the first word in the name of the fastener. Similarly, finish need not be mentioned if the fastener or nut is correctly named. See §13.25.

## 13.28 Locknuts and Locking Devices.

Many types of special nuts and devices to prevent nuts from unscrewing are available, some of the most common of which are illustrated in Fig. 13.32. The American National Standard *jam nuts,** (a) and (b), are the same

* ANSI B18.2.2-1965.

as the hexagon or hexagon flat nuts, except that they are thinner. The application at (b), where the larger nut is on top and is screwed on more tightly, is recommended. They are the same distance across flats as the corresponding hexagon nuts ($1\frac{1}{2}$D or $1\frac{1}{2}$D + $\frac{1}{8}''$). They are slightly over $\frac{1}{2}$D in thickness, but are drawn $\frac{1}{2}$D for simplicity. They are available with or without the washer face in the Regular and Heavy types. The tops of all are flat and chamfered at 30°, and the finished forms have either a washer face or a chamfered bearing surface.

The lock washer, shown at (c), and the cotter pin, (e), (g), and (h), are very common. See Appendixes 22 and 25. The set screw, (f), is often made to press against a plug of softer material, such as brass, which in turn presses against the threads without deforming them.

For use with cotter pins (see Appendix 25), the ANSI* recommends a Hex slotted nut, (g), and a Hex Castle nut, (h), as well as a Hex thick slotted nut and a Heavy Hex thick slotted nut.

## 13.29 American National Standard Cap Screws.†

The five types of American National Standard cap screws are shown in Fig. 13.33. The first four have these standard heads, while the socket head cap screws, (e), have several different shapes of round heads and sockets. Cap screws are regularly produced in finished form and are used on machine tools and other machines, for which accuracy and appearance are important. The ranges of sizes and exact dimensions are given in Appendixes 13 and 14.

Cap screws ordinarily pass through a clearance hole in one member and screw into another. The hole is drilled slightly larger than the screw in order to permit minimum clearance for assembly, as explained in §13.24. The clearance hole need not be shown on the drawing when the presence of the unthreaded clearance hole is obvious.

Cap screws are inferior to studs if frequent removal is necessary; hence they are used on machines requiring few adjustments. The slotted or socket-type heads are best under crowded conditions.

The actual standard dimensions may be used in drawing the cap screws whenever exact sizes are necessary, but this is seldom the case. In Fig. 13.33 the dimensions are given in terms of body diameter D, and they closely conform to the actual dimensions. The resulting drawings are almost exact reproductions

* ANSI B5.20–1958.
† ANSI B18.6.2–1956 and ANSI B18.3–1969.

Fig. 13.33 **American National Standard Cap Screws.** (See Appendixes 13 and 14.)

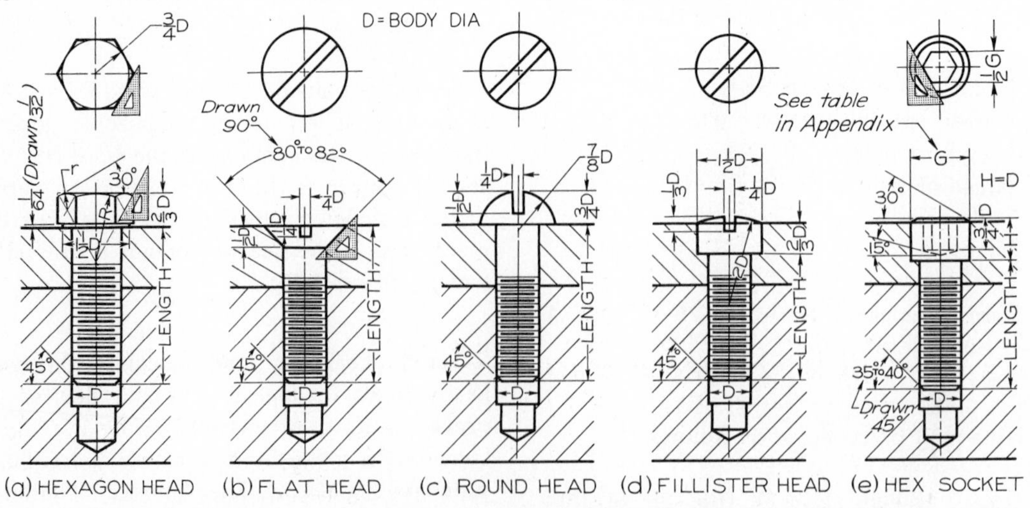


Threads: National Coarse or Fine, Class 2 fit. On screws 2″ long or less, threads extend to within 2 threads of head; on longer screws thread length = $1\frac{3}{4}''$. Screw lengths not standardized.

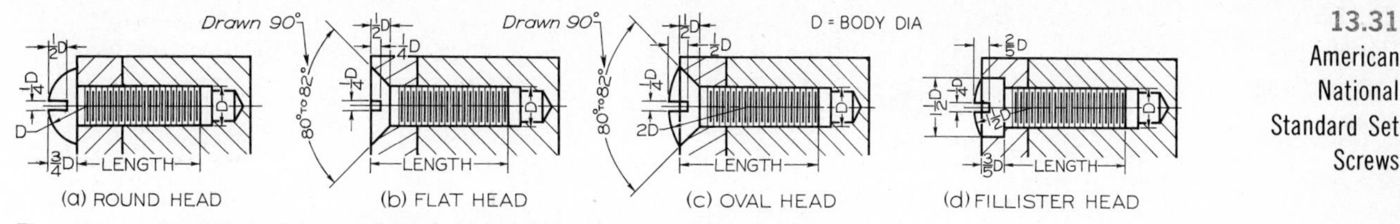


**Fig. 13.34 American National Standard Machine Screws.** See Appendix 15.

and are easy to draw. The hexagonal head cap screw is drawn in the manner shown in Fig. 13.30. The points are drawn chamfered at 45° from the schematic thread depth.

For correct representation of tapped holes, see §13.25. For information on drilled, countersunk, or counterbored holes, see §§6.33 and 10.20.

In an assembly section, it is customary not to section screws, bolts, shafts, or other solid parts whose center lines lie in the cutting plane. Such parts in themselves do not require sectioning and are, therefore, shown "in the round," Fig. 13.33 and §14.22.

Note that screwdriver slots are drawn at 45° in the circular views of the heads, without regard to true projection, and that threads in the bottom of the tapped holes are omitted so that the ends of the screws may be clearly seen. A typical cap screw note is as follows:

*Example* (complete): $\frac{3}{8}$–16 UNC–2A × $2\frac{1}{2}$ HEXAGON HEAD CAP SCREW

*Example* (abbreviated): $\frac{3}{8}$ × $2\frac{1}{2}$ HEX HD CAP SCR

## 13.30 American National Standard Machine Screws.

Machine screws are similar to cap screws but are in general smaller (0.060" to 0.750" dia.). The ANSI* has approved eight forms of heads, as shown in Appendix 15. The hexagon head may be slotted if desired. All others are available in either slotted or recessed-head forms. American National Standard machine screws are regularly produced with a naturally bright finish, not heat-treated, and are regularly supplied with plain-sheared ends, not chamfered.

Machine screws are particularly adapted to screwing into thin materials, and all the smaller-numbered screws are threaded nearly to the head. They are used extensively in firearms, jigs, fixtures, and dies. Machine screw nuts are used mainly on the round head and flat head types, and are hexagonal in form.

Exact dimensions of machine screws are given in Appendix 15, but they are seldom needed for drawing purposes. The four most common types of machine screws are shown in Fig. 13.34, where proportions based on diameter D conform closely to the actual dimensions and produce almost exact drawings. Clearance holes and counterbores should be made slightly larger than the screws, as explained in §13.24.

Note that the threads in the bottom of the tapped holes are omitted so that the ends of the screws will be clearly seen. Observe also that it is conventional practice to draw the screwdriver slots at 45° in the circular view without regard to true projection.

A typical machine screw note is as follows:

*Example* (complete): No. 10 (.1900)–32 NF-3 × $\frac{5}{8}$ FILLISTER HEAD MACHINE SCREW

*Example* (abbreviated): No. 10 (.1900) × $\frac{5}{8}$ FILL HD MACH SCR

## 13.31 American National Standard Set Screws.

The function of set screws, Fig. 13.35 (a), is to prevent relative motion, usually rotary, between two parts, such as the movement of the hub of a pulley on a shaft. A set screw is screwed into one part so that its point bears firmly against another part. If the point of the set screw is cupped, (e), or if a flat is milled on the shaft, (a), the screw will hold much more firmly. Obviously, set screws are not efficient when the load is heavy or is suddenly applied. Usually they are manufactured of steel and case hardened.

* ANSI B18.6.3-1962.

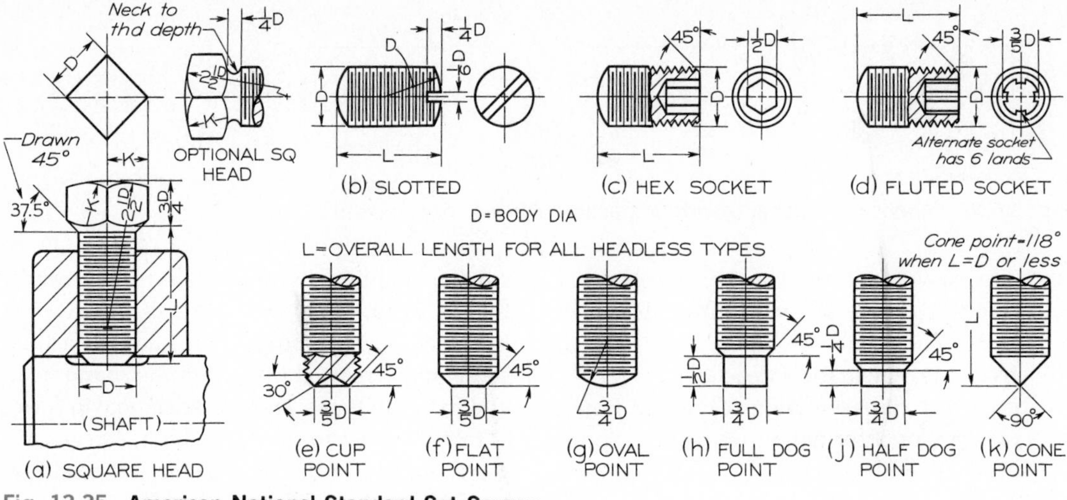


Fig. 13.35 **American National Standard Set Screws.**

The American National Standard square head set screw* is shown in Fig. 13.35 (a), and the American National Standard slotted headless set screw* at (b). Two American National Standard socket set screws† are illustrated at (c) and (d). American National Standard set screw points are shown from (e) to (k). The headless set screws have come into greater use because the projecting head of headed set screws has caused many industrial casualties; this has resulted in legislation prohibiting their use in many states.

Most of the dimensions in Fig. 13.35 are American National Standard formula dimensions, and the resulting drawings are almost exact representations.

Square head set screws have Coarse, Fine, or 8-Pitch threads, Class 2A, but are usually furnished with Coarse threads, since the square head set screw is generally used on the rougher grades of work. Slotted headless and socket set screws have Coarse or Fine threads, Class 3A.

Nominal diameters of socket set screws are No's. 0 to 10, 12, $\frac{1}{4}$, $\frac{5}{16}$, $\frac{3}{8}$, $\frac{7}{16}$, $\frac{1}{2}$, $\frac{9}{16}$, $\frac{5}{8}$, $\frac{3}{4}$, $\frac{7}{8}$, 1, $1\frac{1}{8}$, $1\frac{1}{4}$, $1\frac{3}{8}$, $1\frac{1}{2}$, $1\frac{3}{4}$, and 2. Square head set screws are No. 10 to $1\frac{1}{2}$ only, while slotted headless set screws are No. 5 to $\frac{3}{4}$ only, of this series of diameters.

* ANSI B18.6.2–1956.
† ANSI B18.3–1969.

Socket set screw lengths are standardized* as follows:

Lengths $\frac{1}{16}''$ through $\frac{3}{16}''$,[a] increment = $\frac{1}{32}''$
Lengths $\frac{1}{8}''$ through $\frac{1}{2}''$, increment = $\frac{1}{16}''$
Lengths $\frac{1}{2}''$ through 1″, increment = $\frac{1}{8}''$
Lengths 1″ through 2″, increment = $\frac{1}{4}''$
Lengths 2″ through 6″, increment = $\frac{1}{2}''$
Lengths over 6″, increment = 1″

[a] Applicable only to sizes 0 (.060″) to 3 (.099″) inclusive.

Square head set screw lengths are not standardized, but manufacturers list increments as follows:

Lengths $\frac{3}{8}''$ to 1″, increment = $\frac{1}{8}''$
Lengths 1″ to 4″, increment = $\frac{1}{4}''$
Lengths 4″ to 5″, increment = $\frac{1}{2}''$
Lengths 5″ & 6″, increment = 1″

Slotted headless set screw lengths are not standardized, but manufacturers list increments as follows:

Lengths $\frac{1}{8}''$ to $\frac{3}{8}''$, increment = $\frac{1}{16}''$
Lengths $\frac{3}{8}''$ to 1″, increment = $\frac{1}{8}''$
Lengths 1″ to 2″, increment = $\frac{1}{4}''$

* ANSI B18.3–1969.

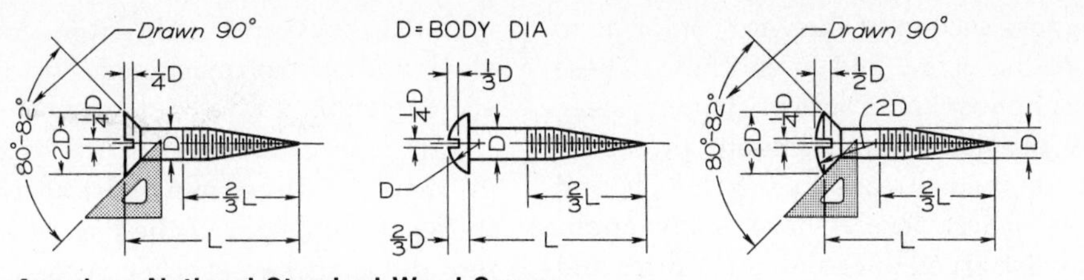


Fig. 13.36 **American National Standard Wood Screws.**

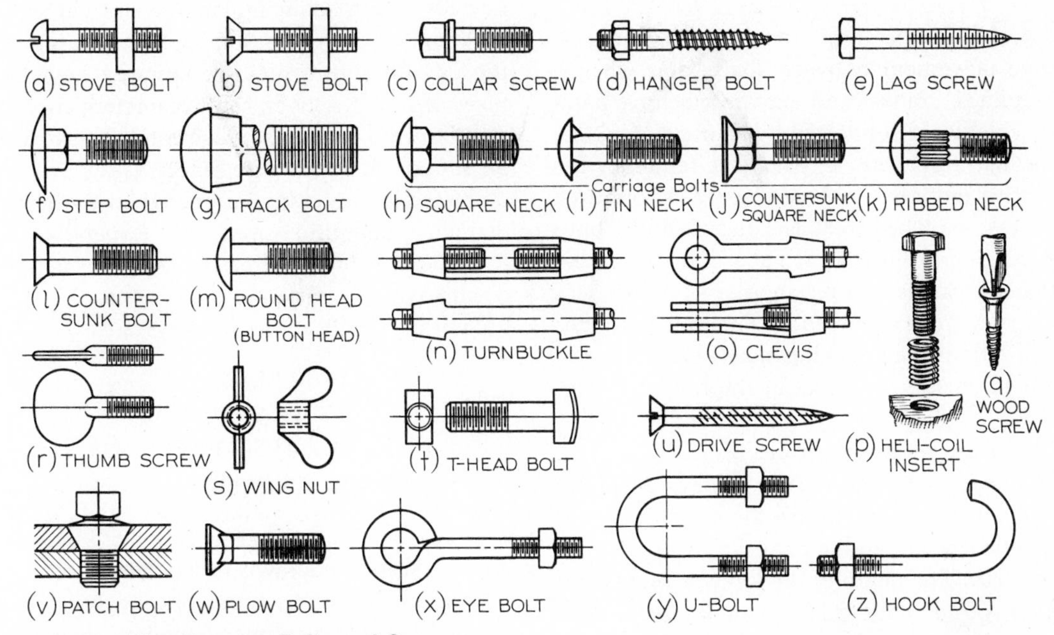


Fig. 13.37 **Miscellaneous Bolts and Screws.**

Set screws are specified as follows:

*Example* (complete): $\frac{3}{8}$-16 UNC-2a × $\frac{3}{4}$ SQUARE HEAD FLAT POINT SET SCREW

*Examples* (abbreviated):

$\frac{3}{8}$ × $1\frac{1}{4}$ SQ HD FL PT SS

$\frac{7}{16}$ × $\frac{3}{4}$ HEX SOC CUP PT SS

$\frac{1}{4}$-20 UNC-2A × $\frac{5}{8}$ SLOT. HDLS CONE PT SS

## 13.32 American National Standard Wood Screws.

Wood screws with three types of heads have been standardized,* Fig. 13.36. The dimensions shown closely approximate the actual dimensions and are more than sufficiently accurate for use on drawings.

Instead of the screwdriver slot, the Phillips recessed head is becoming more popular. Two styles of cross recesses have been standardized by the ANSI.* Many examples may be seen on the automobile. A special screwdriver is used, as shown in Fig. 13.37 (q), and results in rapid assembly without damage to the head.

*ANSI B18.6.1-1961.

## 13.33 Miscellaneous Fasteners.

Many other types of fasteners have been devised for specialized uses. Some of the more common types are shown in Fig. 13.37. A number of these are American National Standard round head bolts,† including carriage, button head, step, and countersunk bolts.

Aero-thread inserts, or Heli-coil inserts, as shown at (p), are shaped like a spring except

*ANSI B18.6.1-1961.
†ANSI B18.5-1952 (R1959).

that the cross section of the wire conforms to threads in the screw and in the hole. These are made of phosphor bronze or stainless steel, and they provide a hard, smooth protective lining for tapped threads in soft metals and in plastics. These inserts have many applications in aircraft engines and accessories and are coming into wider use.

## 13.34 Keys.

*Keys* are used to prevent relative movement between shafts and wheels, couplings, cranks, and similar machine parts attached to or supported by shafts, Fig. 13.38. For light duty, that is, when the tendency for relative motion is not very great, a round or *pin key* may be used. For heavy duty, only rectangular keys (flat or square) are suitable, and sometimes two rectangular keys are necessary for one connection. For even stronger connections, interlocking *splines* may be machined on the shaft and in the hole. See Fig. 12.9.

A *square key* is shown in Fig. 13.38 (a), and a *flat key* at (b). The widths of keys generally used are about one-fourth the shaft diameter. In either case, one-half the key is sunk into the shaft. The depth of the keyway or keyseat is measured on the side—not in the center, (a). Square and flat keys may have the top surface tapered $\frac{1}{8}''$ per foot, in which case they become square taper or flat taper keys.

A rectangular key that prevents rotary motion but permits relative longitudinal motion is a *feather key,* and is usually provided with *gib heads,* or otherwise fastened so it cannot slip out of the keyway. A *gib head key* is shown at (c). It is exactly the same as the square taper or flat taper key, except that a gib head, which provides for easy removal, is added. Square and flat keys are made from cold-finished stock and are not machined. For dimensions, see Appendix 16.

The *Pratt & Whitney key,* (d), is rectangular in shape, with semicylindrical ends. Two-thirds of the height of the P & W key is sunk into the shaft keyseat. See Appendix 20.

The American National Standard* *Woodruff key* is semicircular in shape, Fig. 13.39. The key fits into a semicircular key slot cut with a Woodruff cutter, as shown, and the top of the key fits into a plain rectangular keyway. Sizes of keys for given shaft diameters are not standardized, but for average conditions it will be found satisfactory to select a key whose diameter is approximately equal to the shaft diameter. For dimensions, see Appendix 18.

A *keyseat* is in a shaft; a *keyway* is in the hub or surrounding part.

Typical specifications for keys are:

$\frac{1}{4} \times 1\frac{1}{2}$ SQ KEY
No. 204 WOODRUFF KEY
$\frac{1}{4} \times \frac{1}{16} \times 1\frac{1}{2}$ FLAT KEY
No. 10 P & W KEY

Notes for nominal specifications of keyways and keyseats are shown in Fig. 11.42 (o), (p), (r) and (x). For production work, keyways and keyseats should be dimensioned as shown in Fig. 11.46.

## 13.35 Machine Pins.

American National Standard machine pins† include taper pins, straight pins, dowel pins, clevis pins, and cotter pins. For light work, the taper pin is effective for fastening hubs or collars to shafts, as shown

* ANSI B17.2-1967.
† ANSI B5.20-1958.

Fig. 13.38 **Square and Flat Keys.**

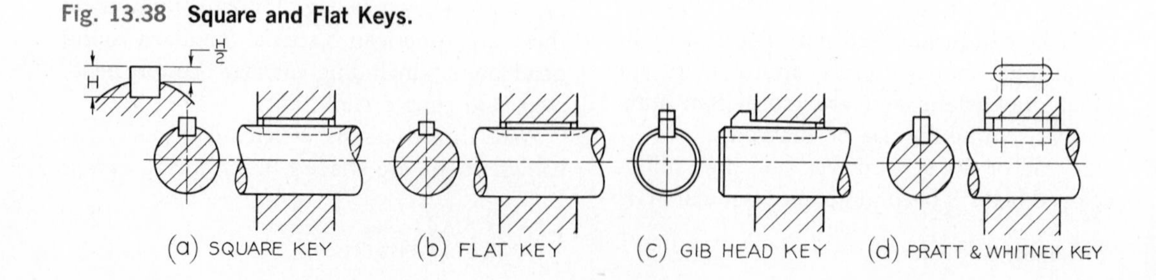

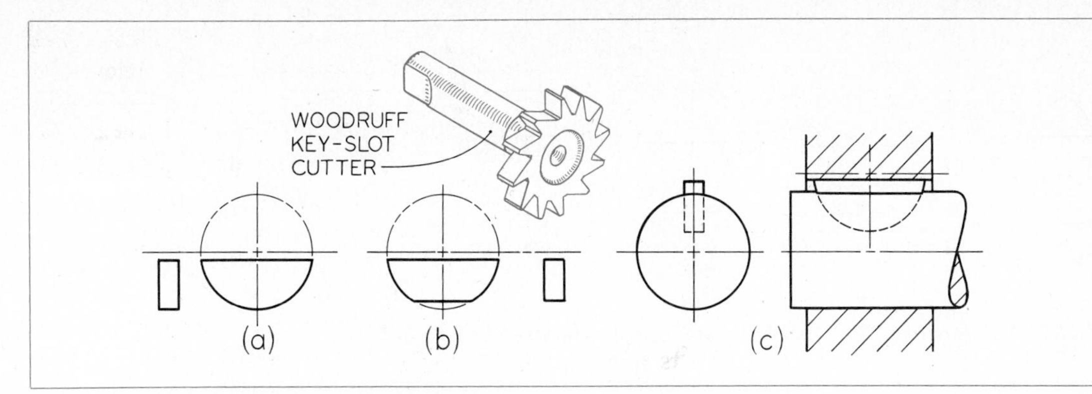


Fig. 13.39 **Woodruff Keys and Key-Slot Cutter.**

in Fig. 13.40, in which the hole through the collar and shaft is drilled and reamed when the parts are assembled. For slightly heavier duty, the taper pin may be used parallel to the shaft as for square keys. See Appendix 24.

Dowel pins are cylindrical or conical in shape and are used for a variety of purposes, chief of which is to keep two parts in a fixed position or to preserve alignment. The taper dowel pin is most commonly used and is recommended where accurate alignment is essential. Dowel pins are usually made of steel and are hardened and ground in a centerless grinder.

The clevis pin is used in a clevis and is held in place by a cotter pin. For the latter, see Appendix 25.

## 13.36 Rivets.

*Rivets* are regarded as permanent fastenings as distinguished from removable fastenings, such as bolts and screws. They are generally used to hold sheet metal or rolled steel shapes together and are made of wrought iron, soft steel, copper, or occasionally other metals.

To fasten two pieces of metal together, holes are punched, drilled, or punched and then reamed, slightly larger in diameter than the shank of the rivet. Rivet diameters in practice are made from $d = 1.2\sqrt{t}$ to $d = 1.4\sqrt{t}$, where d is the rivet diameter and t is the metal thickness. The larger size is used for steel and single-riveted joints, and the smaller may be used for multiple-riveted joints. In structural

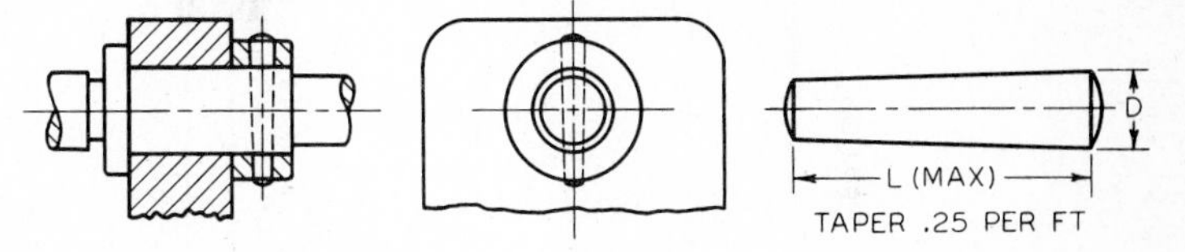


Fig. 13.40 **Taper Pin.**

work it is common practice to make the hole $\frac{1}{16}''$ larger than the rivet.

When the red-hot rivet is inserted, a "dolly bar," having a depression the shape of the driven head, is held against the head. A riveting machine is then used to drive the rivet and to form the head on the plain end. This action causes the rivet to swell and fill the hole tightly.

American National Standard large rivets are used in structural work of bridges, buildings, and in ship and boiler construction, and they are shown in their exact formula proportions in Fig. 13.41.* The button head and countersunk head types, (a) and (e), are the rivets most commonly used in structural work. The button head and cone head are commonly used in tank and boiler construction.

Typical riveted joints are illustrated in Fig. 13.42. Notice that the longitudinal view of each rivet shows the shank of the rivet with both heads made with circular arcs, and the circular view of each rivet is represented by only the outside circle of the head. In structural drafting, where there may be many such

* ANSI B18.4-1960.

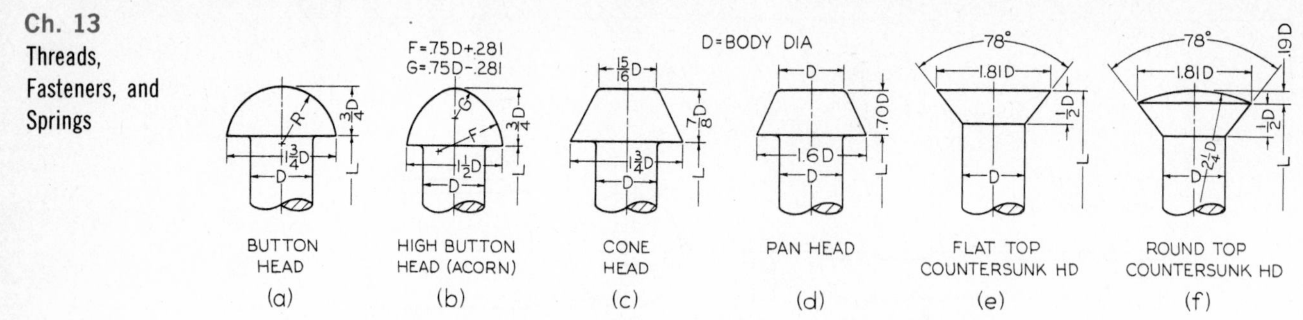


**Fig. 13.41 American National Standard Large Rivets** (ANSI B18.4–1960).

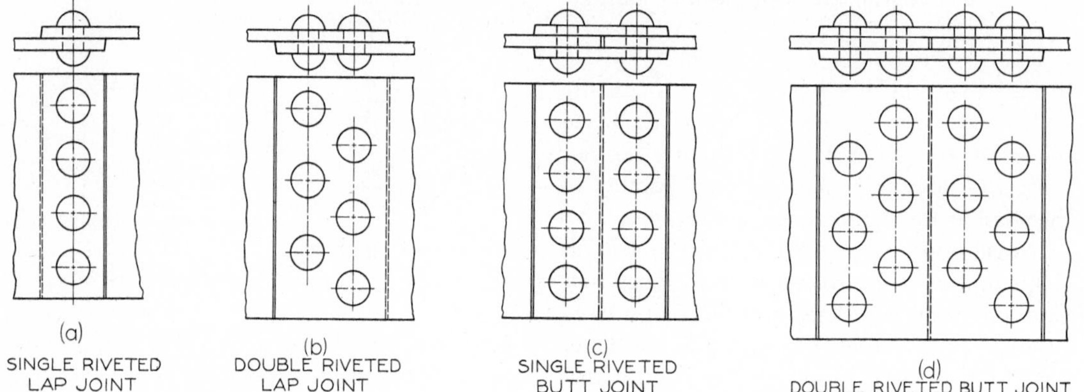


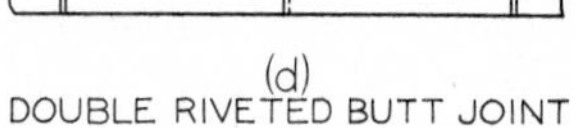


**Fig. 13.42 Common Riveted Joints.**

circles to draw, the drop spring bow, Fig. 2.54, is a convenient instrument.

Since many engineering structures are too large to be built in the shop, they are built in the largest units possible and then transported to the desired location. Trusses are common examples of this. The rivets driven in the shop are called *shop rivets,* and those driven on the job are called *field rivets.* However, heavy steel bolts are commonly used on the job for structural work. Solid black circles are used to represent field rivets, and other standard symbols are used to show other features, as shown in Fig. 13.43.

For light work, small rivets are used. American National Standard small solid rivets are illustrated with dimensions showing their standard proportions in Fig. 13.44* Included in the same standard are tinners', coppers', and belt rivets.

* ANSI B18.1-1965.

## 13.37 Springs.

A *spring* is a mechanical device designed to store energy when deflected and to return the equivalent amount of energy when released. Springs are classified as *helical springs,* Fig. 13.45, or *flat springs,* Fig. 13.50. Helical springs may be cylindrical or conical, but are usually the former.

There are three types of helical springs: *compression springs,* which offer resistance to a compressive force, Fig. 13.45 (a) to (e), *extension springs,* which offer resistance to a pulling force, Fig. 13.48, and *torsion springs,* which offer resistance to a torque load or twisting force, Fig. 13.49.

On working drawings, true projections of helical springs are never drawn because of the labor involved. As in the drawing of screw threads, the detailed and schematic methods, employing straight lines in place of helical curves, are used as shown in Fig. 13.45.

The elevation view of the square-wire spring is similar to the square thread with the

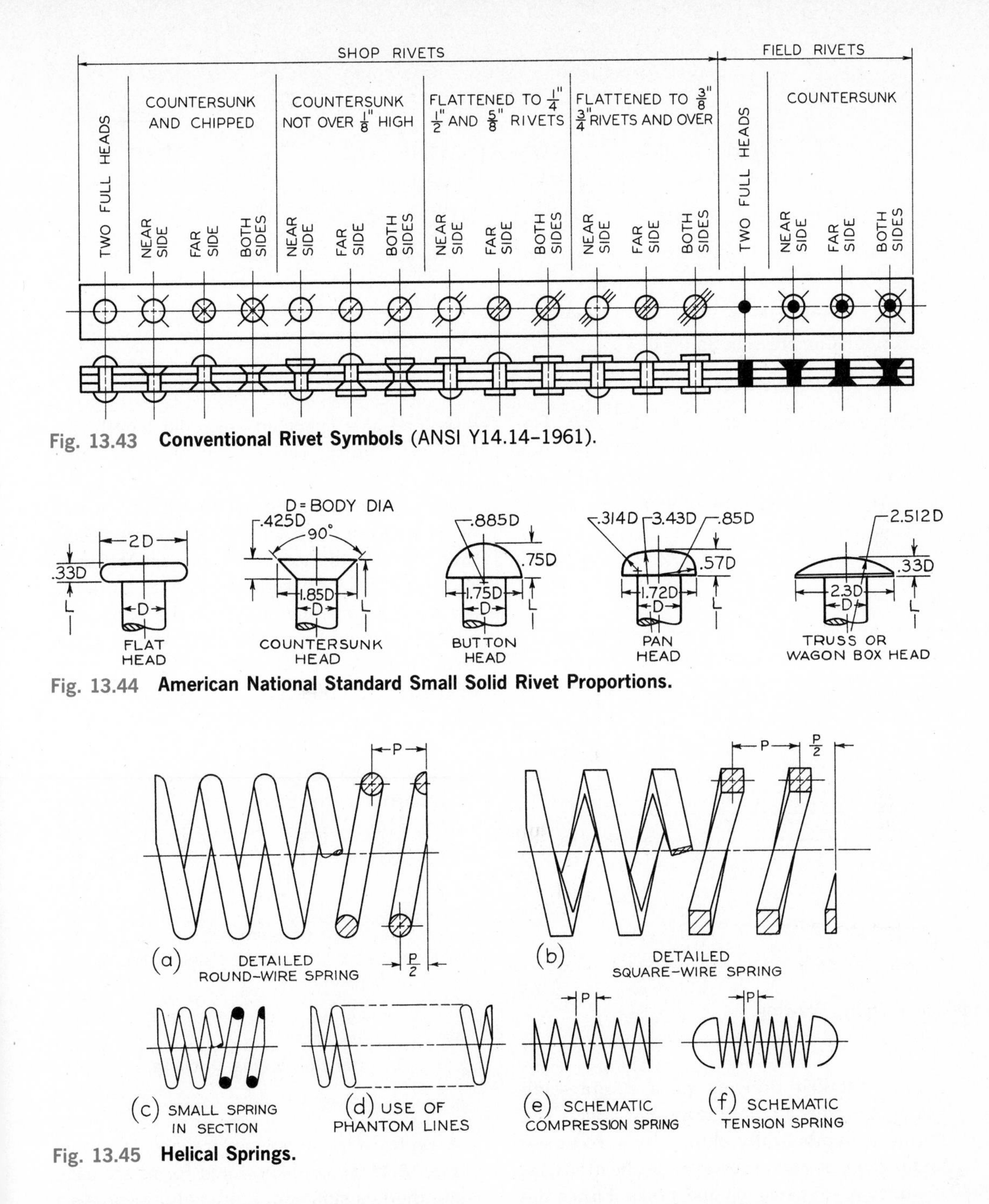


**Fig. 13.43** **Conventional Rivet Symbols** (ANSI Y14.14–1961).

**Fig. 13.44** **American National Standard Small Solid Rivet Proportions.**

**Fig. 13.45** **Helical Springs.**

core of the shaft removed, Fig. 13.12. Standard section lining is used if the areas in section are large, as shown in Fig. 13.45 (a) and (b). If these areas are small, the sectioned areas may be made solid black, (c). In cases where a complete picture of the spring is not necessary, phantom lines may be used to save time in drawing the coils, (d). If the drawing of the spring is too small to be represented by the outlines of the wire, it may be drawn by the schematic method, in which single lines are used, (e) and (f).

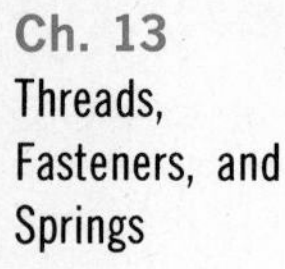


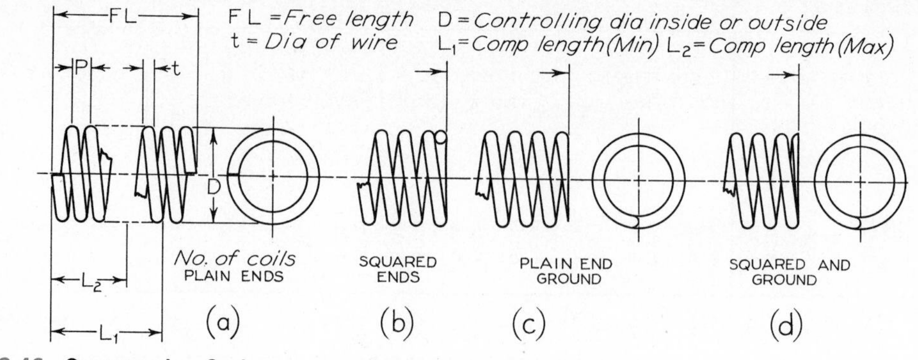


Fig. 13.46 **Compression Springs.**

Compression springs have *plain ends,* Fig. 13.46 (a), or *squared (closed) ends,* (b). The ends may be *ground* as at (c), or both *squared and ground* as at (d). Required dimensions are indicated in the figure.

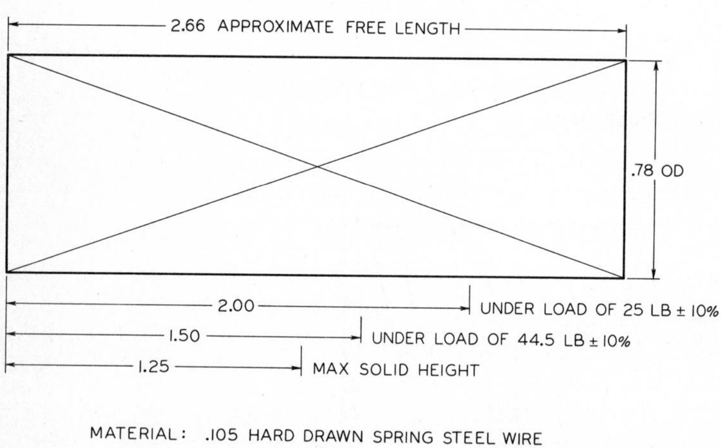


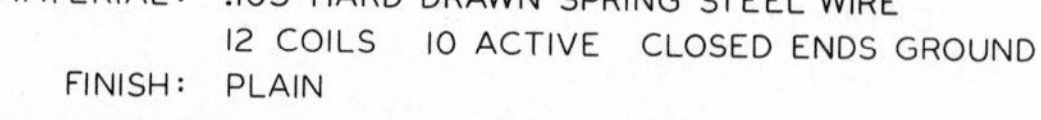


Fig. 13.47 **Compression Spring Drawing.**

In a detailed drawing of a compression spring, the coils are not drawn, Fig. 13.47. The spring is symbolically shown by a rectangle and diagonals, and necessary specifications are included as dimensions and notes. Either the ID or the OD is given, depending upon whether the spring works on a rod or in a hole.

Many companies use standard printed spring drawings with a printed form to be filled in by the draftsman, providing the necessary information as indicated in Fig. 13.47, plus load at a specified deflected length, the load rate, finish, type of service, and other data.

An extension spring may have any one of many types of ends, and it is therefore necessary to draw the spring or at least the ends and a few adjacent coils, Fig. 13.48. Note the use of phantom lines to show the continuity of coils. Printed forms are used when a given form of spring is produced with differences in verbal specification only.

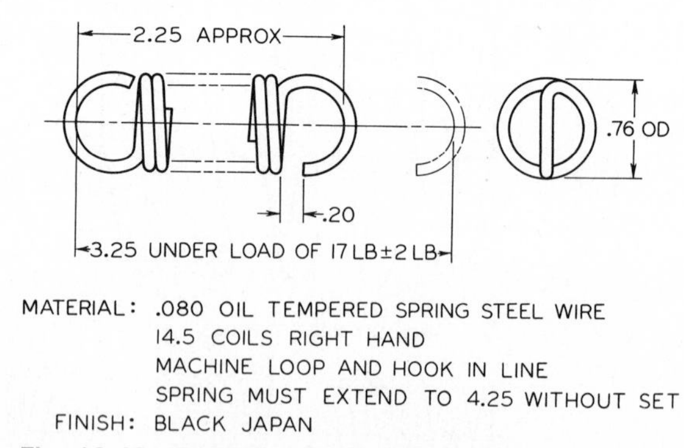

Fig. 13.48 **Extension Spring Drawing.**

A typical torsion spring drawing is shown in Fig. 13.49. Here also printed forms are used when there is sufficient uniformity in product to permit a common representation.

A typical flat spring drawing is shown in Fig. 13.50. Other types of flat springs are *power springs* (or flat coil springs), *Belleville springs* (like spring washers), and *leaf springs* (commonly used in automobiles).

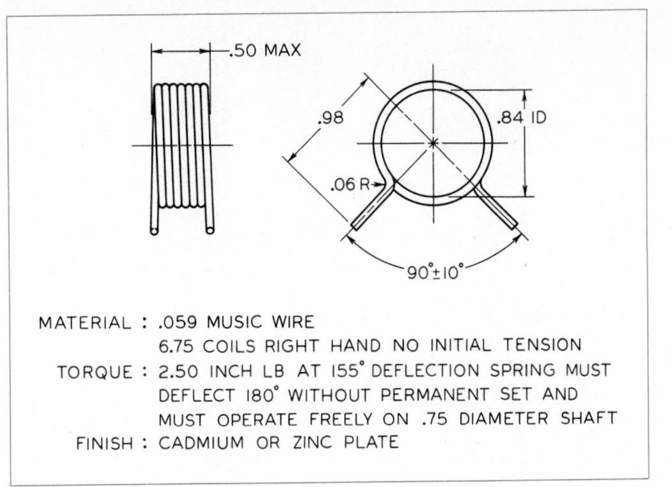


Fig. 13.49 **Torsion Spring Drawing.**

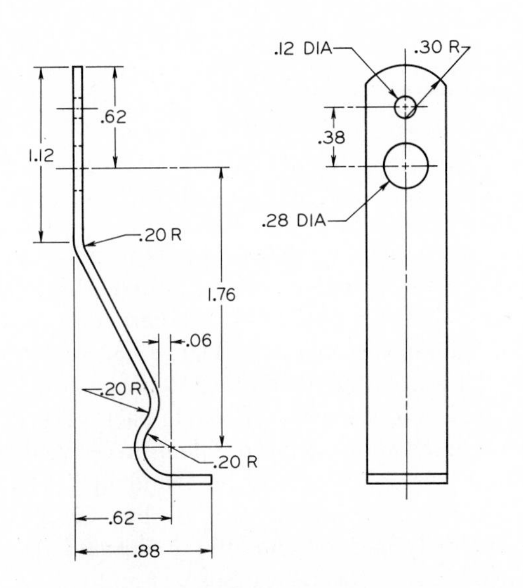


Fig. 13.50 **Flat Spring.**

**13.38 To Draw Helical Springs.** The construction for a schematic elevation view of a compression spring having six total coils is shown in Fig. 13.51 (a). Since the ends are closed, or squared, two of the 6 coils are "dead" coils, leaving only four full pitches to be set off along the top of the spring, as shown.

If there are $6\frac{1}{2}$ total coils, as at (b), the $\frac{P}{2}$ spacings will be on opposite side of the spring. The construction of an extension spring with 6 active coils and loop ends is shown at (c).

In Fig. 13.52 are shown the steps in drawing a sectional view and an elevation view of a compression spring by detailed representation. The given spring is shown pictorially at (a). At (b) a cutting plane has passed through the center line of the spring, and the front portion of the spring has been removed.

Fig. 13.51 **Schematic Spring Representation.**

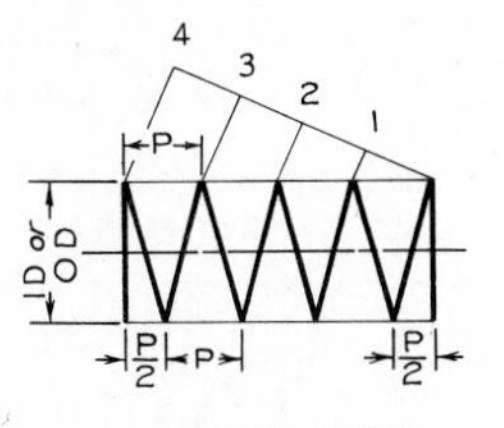

(a)

$6\frac{1}{2}$ TOTAL COILS
COMPRESSION SPRING

(b)

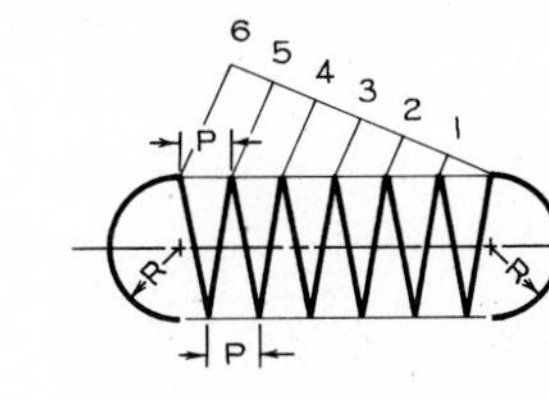


6 TOTAL COILS
EXTENSION SPRING

(c)

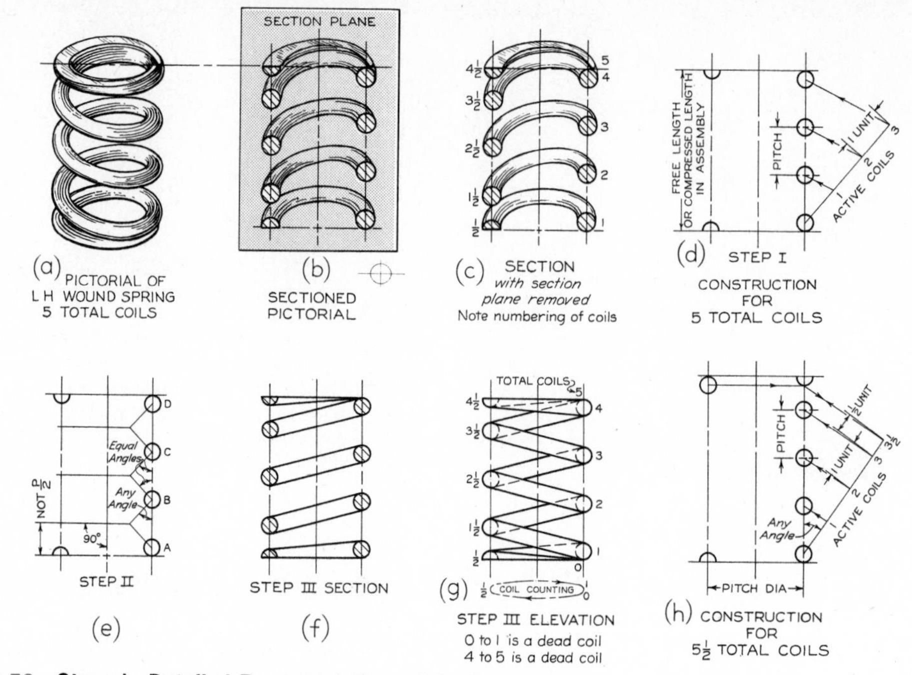


Fig. 13.52 **Steps in Detailed Representation of Spring.**

At (c) the cutting plane has been removed. Steps in constructing the spring through several stages to obtain the sectional view are shown at (d) to (f). The corresponding elevation view is shown at (g).

If there is a fractional number of coils, such as $5\frac{1}{2}$ coils at (h), note that the half-rounds of sectional wire are placed on opposite sides of the spring.

## 13.39 Thread and Fastener Problems.

It is expected that the student will make use of the information in this chapter in connection with working drawings at the end of the next chapter, where many different kinds of threads and fasteners are required. However, several problems are included here for specific assignment in this area, Figs. 13.53 to 13.56. All are to be drawn on tracing paper or detail paper, size B sheet (see inside back cover).

Thread and fastener problems in convenient form for solution may be found in *Technical Drawing Problems, Series 1,* by Giesecke, Mitchell, Spencer, and Hill; in *Technical Drawing Problems, Series 2,* by Spencer and Hill; and in *Technical Drawing Problems, Series 3,* by Spencer and Hill, all designed to accompany this text and published by Macmillan Publishing Co., Inc.

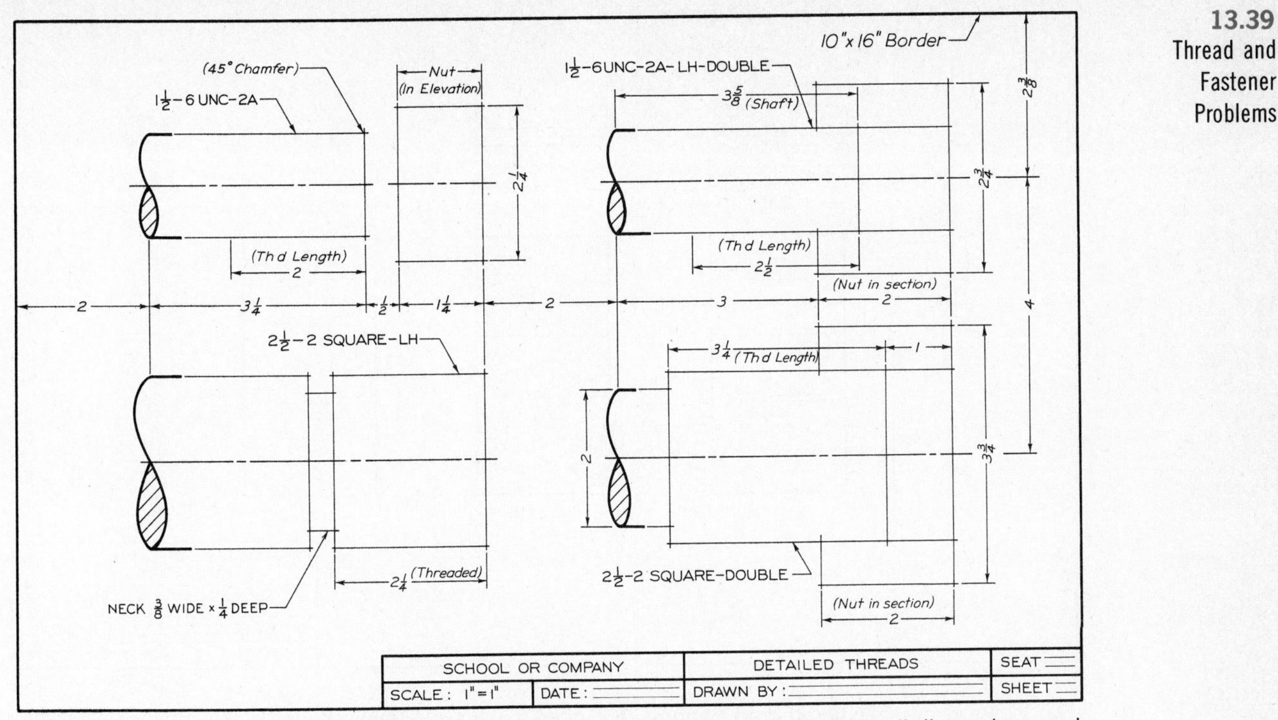


Fig. 13.53 Draw specified detailed threads arranged as shown, Layout B-3. Omit all dimensions and notes given in inclined letters. Letter only the thread notes and the title strip.

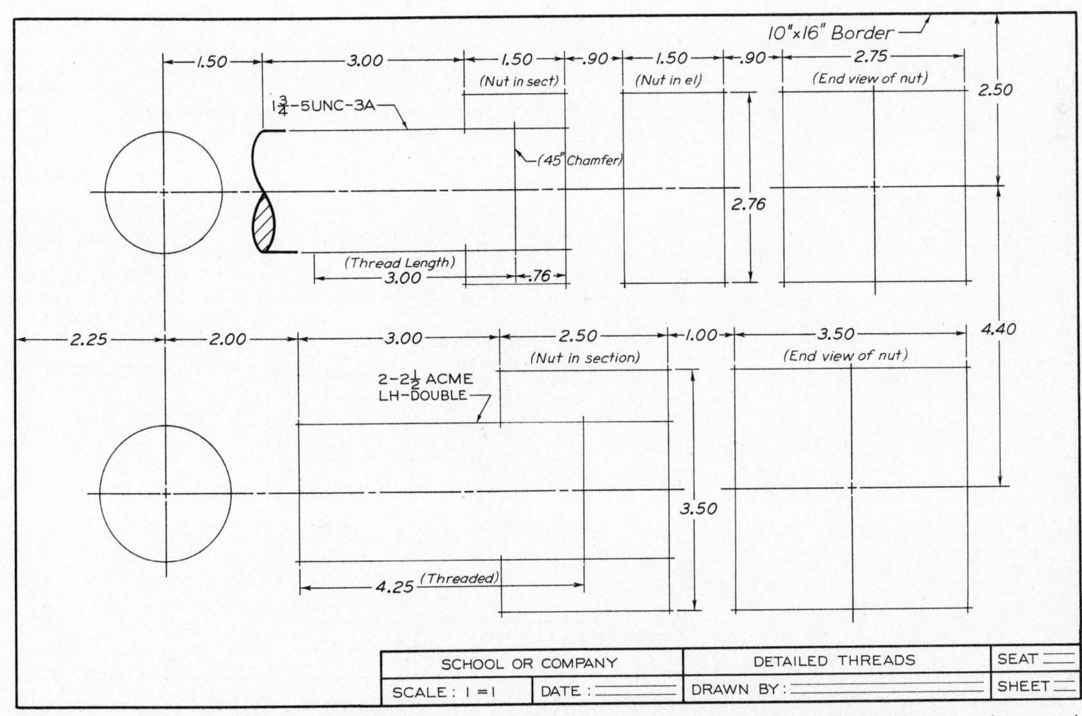


Fig. 13.54 Draw specified detailed threads, arranged as shown, Layout B-3. Omit all dimensions and notes given in inclined letters. Letter only the thread notes and the title strip.

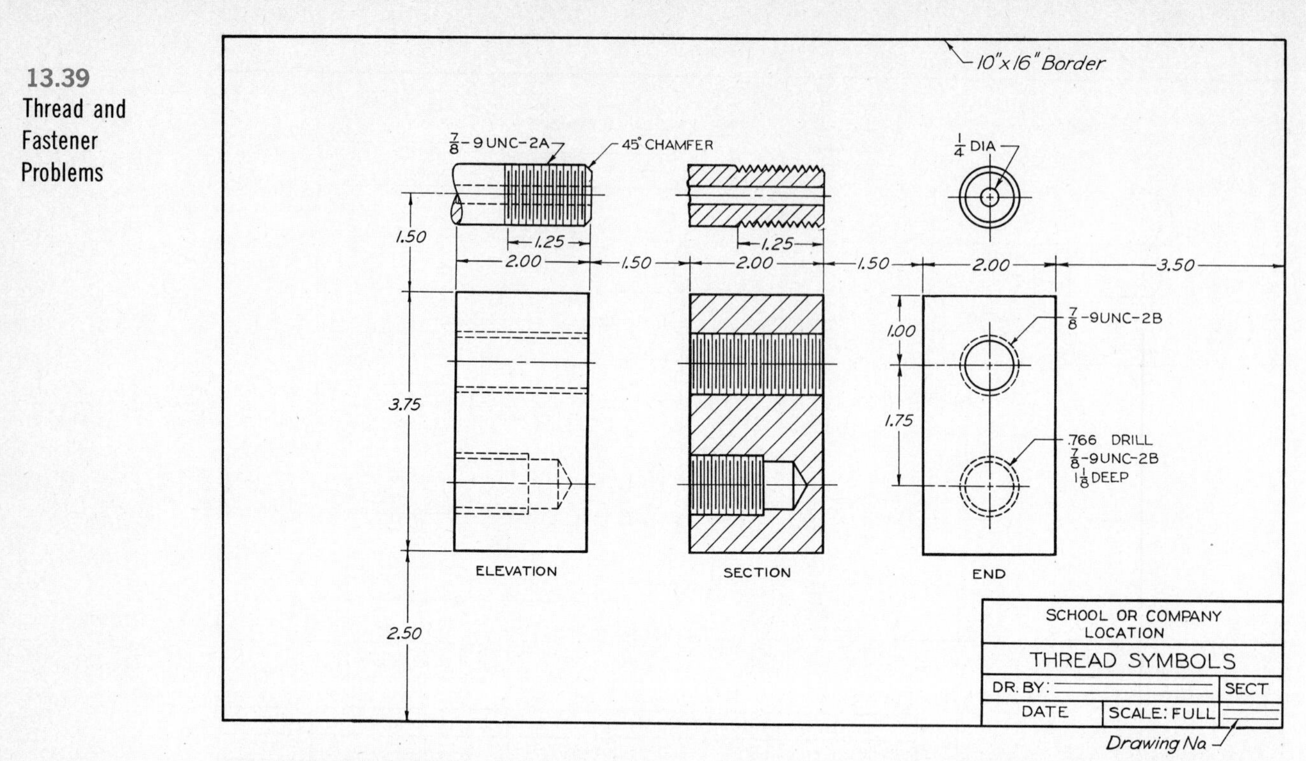


Fig. 13.55 Draw specified thread symbols, arranged as shown. Draw schematic or simplified symbols, as assigned by instructor, Layout B–5. Omit all dimensions and notes given in inclined letters. Letter only the drill and thread notes, the titles of the views, and the title strip.

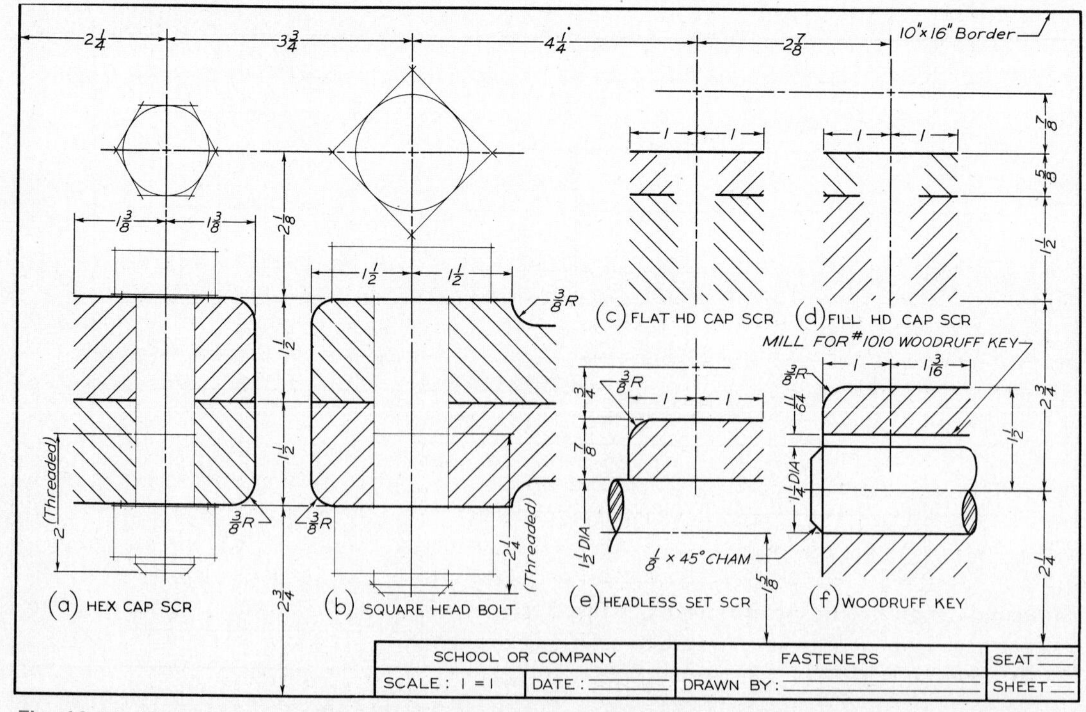


Fig. 13.56 Draw fasteners, arranged as shown, Layout B–3. At (a) draw $\frac{7}{8}$–9 UNC–2A × 4 Hex Cap Screw. At (b) draw $1\frac{1}{8}$–7 UNC–2A × $4\frac{1}{4}$ Sq Hd Bolt. At (c) draw $\frac{3}{8}$–16 UNC–2A × $1\frac{1}{2}$ Flat Hd Cap Screw. At (d) draw $\frac{7}{16}$–14 UNC–2A × 1 Fill Hd Cap Screw. At (e) draw $\frac{1}{2}$ × 1 Headless Slotted Set Screw. At (f) draw front view of No. 1010 Woodruff Key. Draw schematic or simplified thread symbols as assigned. Letter titles under each figure as shown.

# 14 design and working drawings

**14.1 Design and Engineering.** The many products, systems, and services that enrich our standard of living are largely the result of the design activities of those in engineering. It is principally this design activity that distinguishes engineering from science and research.

Although the engineer functions in a wide range of activities in industry, he most often participates in a team effort such as product application (sales), less familiar areas of research and development (health needs), or in his own particular area of specialization such as heat power (internal combustion engines). Even though engineering covers a broad range of activities, fundamentally the engineer is a *designer,* a creator, or a "builder."

There are two general types of design: *empirical design,* sometimes referred to as conceptual design, and *scientific design*. In scientific design, use is made of the principles of physics, mathematics, chemistry, mechanics, and other sciences in the new or revised design of devices, structures, or systems intended to function under specific conditions. In empirical design, much use is made of the information in handbooks, which in turn has been learned by experience. Nearly all technical design is a combination of scientific and empirical design. Therefore, a competent designer should have adequate engineering and scientific knowledge, and he should have access to the many handbooks related to his field. See Appendix 1. He should also have the ability to communicate verbally, symbolically, and graphically.

It is the design procedure that is an exciting and challenging effort in which the engineer

(a) POSTAL SCALE

(b) EDUCATIONAL TOYS

Fig. 14.1 **Working Models of Student Designs.**

relies heavily upon graphics as a means for *creating, recording, analyzing, and communicating to others* his design concepts or ideas. See §14.27 for suggested format of a student design report.

Normally, you as a student may not yet have acquired the necessary educational background and experience to undertake a sophisticated design. And you probably will not have access to the variety of experts on materials, production methods, business economics, legal considerations, or other vital areas that likely would be available in a large company. Nevertheless, you can attain a great deal of accomplishment through conscientious application of your design ability and available sources of information to the creation or improvement of some device, system, or service that *works* and is uniquely your own. The working models of the Postal Scale and the Educational Toys in Fig. 14.1 are examples of student design efforts.

**14.2 Design Concepts.** New ideas or design concepts must exist initially in the mind of the designer. In order to capture, preserve, and develop these ideas the designer makes liberal use of freehand sketches of views and pictorials, Chapter 5. These sketches are revised or redrawn as the concept is developed. All sketches should be preserved for reference and as a record of the development of the design.

At some point in the development of the idea, you will probably find it to your advantage to pool your ideas with those of others and begin working in a team effort; such a team may include others familiar with problems of materials, production, marketing, etc. In industry, the project becomes a team effort long before the product is produced and marketed. Obviously, the design process is not a haphazard operation of an inventor working in his basement, although it might well begin in that manner. Industry could not long survive if its products were determined in a haphazard manner. Hence nearly all successful companies support a well-organized design effort, and the vitality of the company depends to a large extent on the *planned* output of its designers.

Since it is important for you to be able to work effectively with others in a group or team, you must be able to express yourself clearly and concisely. Do not underestimate the importance of your communication skills, your ability to express your ideas verbally (written and spoken), symbolically (equations, formulas, etc.), and *graphically.* The graphical skills include the ability to present information and ideas clearly and effectively in the form of sketches, drawings, graphs, etc. This textbook is dedicated to helping you develop your communication skills in graphics.

## 14.3 The Design Process.

Design is the ability to combine scientific principles, resources, and often existing products into a solution of a problem. This ability to solve problems in design is the result of an organized and orderly approach to the problem known as the *design process.*

The design process leading to manufacturing, assembly, marketing, service, and the many activities necessary for a successful product is composed of several easily recognized phases. Although many industrial groups may identify them in their own particular way, a convenient procedure for the design of a new or improved product is in five stages as follows:

1. Identification of problem.
2. Concepts and ideas.
3. Compromise solutions.
4. Models or prototypes.
5. Production or working drawings.

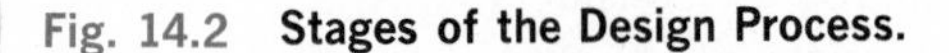

Ideally, the design moves through the stages as shown in Fig. 14.2, but if a particular stage proves unsatisfactory, it may be necessary to return to a previous stage and repeat the procedure as indicated by the dashed-line paths. This repetitive procedure is often referred to as *looping.*

Fig. 14.2 **Stages of the Design Process.**

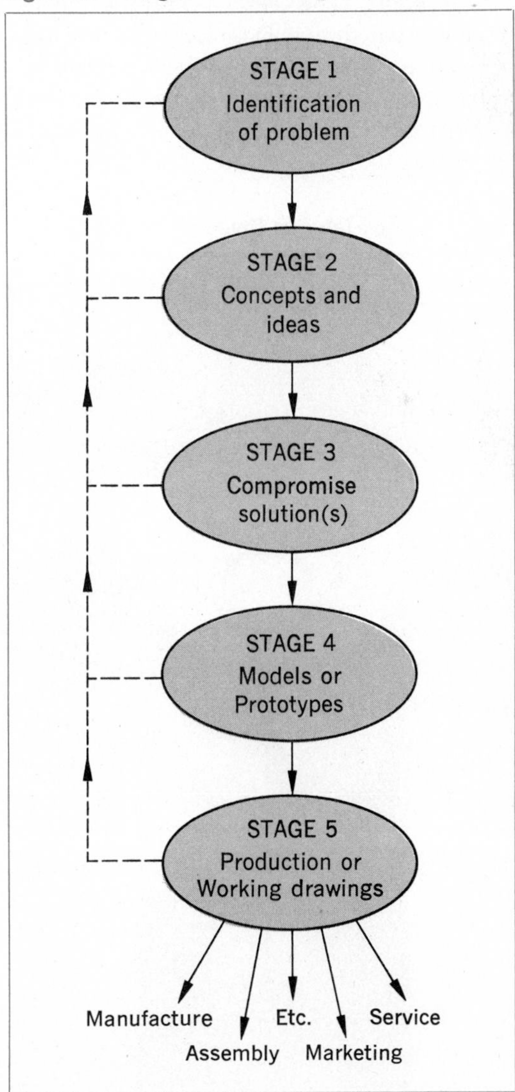

## 14.4 Stage 1—Identification of Problem.

The design activity begins with the recognition and determination of a need or want for a product, service, or system and the economic feasibility of fulfilling this need. Engineering design problems may range from the need for a simple and inexpensive container opener such as the pull tab commonly used on beverage cans, Fig. 14.3, to the more complex problems associated with the needs of air and ground travel, space exploration, environmental control, and so forth. Although the product may be very simple, such as the pull tab on a beverage can, the production tools and dies require considerable engineering and design effort. The new airport automated transit system design, Fig. 14.4, meets the need of moving people efficiently between the terminal areas. The system is capable of moving 3300 people every ten minutes.

Fig. 14.3 **Pull-Tab Can Opener.**

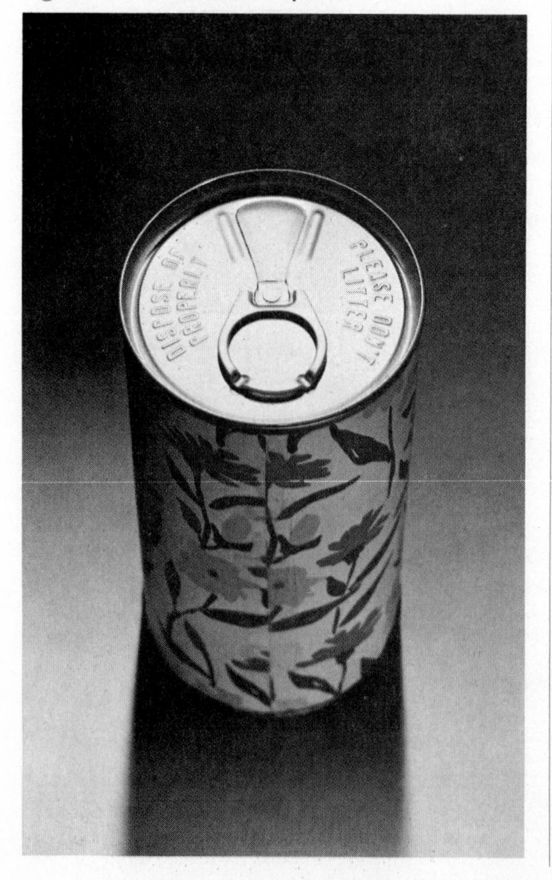


The Lunar Roving Vehicle, Fig. 14.5, is a solution to a need in the space program to explore larger areas of the lunar surface. This vehicle is the end result of a great deal of design work associated with the support systems and the related hardware.

At the problem identification stage, either the designer recognizes that there does exist a need requiring a design solution or, perhaps more often, he receives a directive to that effect from management. No attempt is made at this time to set goals or criteria for the solution.

Information concerning the identified problem becomes the basis for a problem proposal, which may be a paragraph or a multipage report presented for formal consideration. A proposal is a plan for action that will be followed to solve the problem. The proposal, if approved, becomes an agreement to follow the plan. In the classroom, the agreement is made between you and your instructor on the identification of the problem and your proposed plan of action.

Following approval of the proposal, further aspects of the problem are explored. Available information related to the problem is collected and parameters or guide lines for time, cost, function, etc., are defined within which you will work. For example: What is the design expected to do? What is the estimated cost limit? What is the market potential? What can it be sold for? When will the prototype be ready for testing? When must production drawings be ready? When will production begin? When will the product be available on the market?

The parameters of a design problem, including the time schedule, are established at this stage. Nearly all designs represent a compromise, and the amount of time budgeted to a project is no exception.

## 14.5 Stage 2—Concepts and Ideas.

At this stage, many ideas are collected, "wild" or otherwise, for possible solutions to the problem. The ideas are broad and unrestricted to permit the possibility of new and unique solutions. This compilation of ideas may be from

Fig. 14.4 **Airport Transit System.** *Courtesy Westinghouse Electric Corp.*

Fig. 14.5 **Lunar Roving Vehicle.** *Courtesy The Boeing Co.*

individuals or they may come from group or team "brainstorming" sessions wherein a suggested idea often generates many more ideas from the group. As the ideas are elicited, they are written down and/or recorded in graphic form (multiview or pictorial sketches) for future consideration and refinement.

The larger the collection of ideas, the greater are the chances of finding one or more ideas suitable for further refinement. All sources of ideas such as technical literature, reports, design and trade journals, patents and existing products, etc., are explored. The Greenfield Village Museum in Dearborn, Michigan, the Museum of Science and Industry in Chicago, trade exhibitions, large hardware and supply stores, mail order catalogs, etc., are all excellent sources for ideas. Another source is the user of an existing product. He often has suggestions for improvement. The potential user may be helpful with his reactions to the proposed solution.

No attempt is made to evaluate ideas at this stage. All notes and sketches are signed, dated, and retained for possible patent proof.

## 14.6 Stage 3—Compromise Solutions.

Various features of the many conceptual ideas generated in the preceding stages are selected after careful consideration and combined into one or more promising compromise solutions. At this point the best of the solutions is evaluated in detail, and attempts are made to simplify it and thereby make manufacture and performance more efficient.

The sketches of the design are often followed by a study of suitable materials and of motion problems that may be involved. What source of power is to be used—manual, electric motor, or what? What type of motion is needed? Is it necessary to translate rotary motion to linear motion or vice versa? Many of these problems are solved graphically by means of a schematic drawing in which various parts are shown in skeleton form. A pulley is represented by a circle, meshing gears by tangent pitch circles, an arm by a single line, paths of motion by center lines, and so on. At this time, too, certain basic calculations, such as those related to velocity and acceleration, may be made, if required.

These preliminary studies are followed by a *design layout,* made with instruments, or a *layout sketch* from which a draftsman makes an accurate to-scale instrumental drawing so that actual sizes and proportions can be clearly visualized, Fig. 14.6. At this time all parts are carefully designed for strength and function. Costs are constantly kept in mind, for no matter how well the device performs, it must be built to sell for a profit or the time and development costs will have been a loss.

During the layout process, great reliance is placed upon what has gone before. Experience provides a sense of proportion and size that permits the noncritical or more standard features to be designed by eye or with the aid of empirical data. Stress analysis and detailed computation may be necessary in connection with high speeds, heavy loads, or with special requirements or conditions.

As shown in Fig. 14.6, the layout is an assembly showing how parts fit together and the basic proportions of the various parts. Auxiliary views or sections are used, if necessary. Section lining may be used sparingly to save time. All lines should be sharp and the drawing made as accurately as possible, since most dimensions are omitted except for a few key ones that will be used in the detail or working drawings for production. Any notes or other information related to the detail drawing should be given on the layout.

Special attention is given to clearances of moving parts, to ease of assembly, and to serviceability. Standard parts are used wherever possible for it is less costly to use stock items. Most companies maintain some form of an *engineering standards manual,* which contains much of the empirical data and detailed information that is regarded as "company standard." Materials and costs are carefully considered. Although functional considerations must come first, manufacturing problems must be kept constantly in mind.

A great many design problems are concerned with the improvement of an existing product or with the redesign of a device from a different approach in which many of the

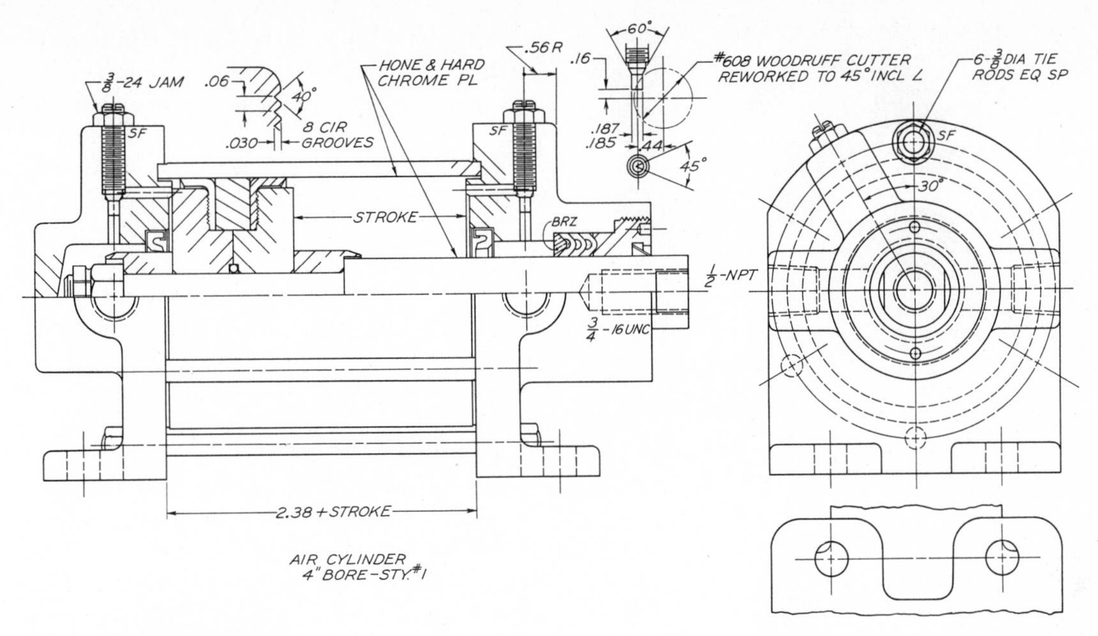


Fig. 14.6 **Design Layout.**

details will be similar to others previously used. For example, in Fig. 14.7, the Indicating Head is attached to a portable beam scale to add damping, sensitivity, and improved visibility to the weight readout. The original design of the housing was made of three sheet-metal parts with a plastic window, (b). The new two-piece design of a die cast housing and larger plastic window, (c), provides more resistance to abuse and a drop in the unit cost of the housing after the first 2400 units, to less than one third of the cost of the original sheet-metal design. Very often a change in material or a slight change in the shape of some part

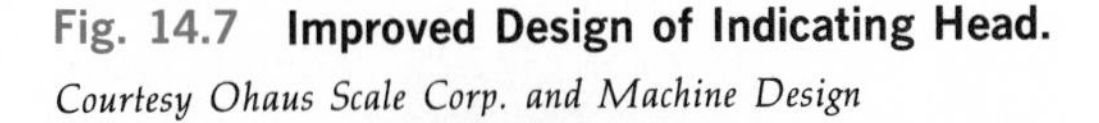

Fig. 14.7 **Improved Design of Indicating Head.**
*Courtesy Ohaus Scale Corp. and Machine Design*

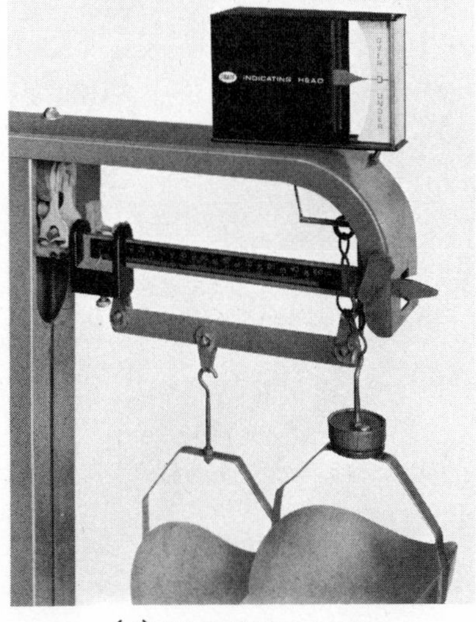

(a) INDICATING HEAD

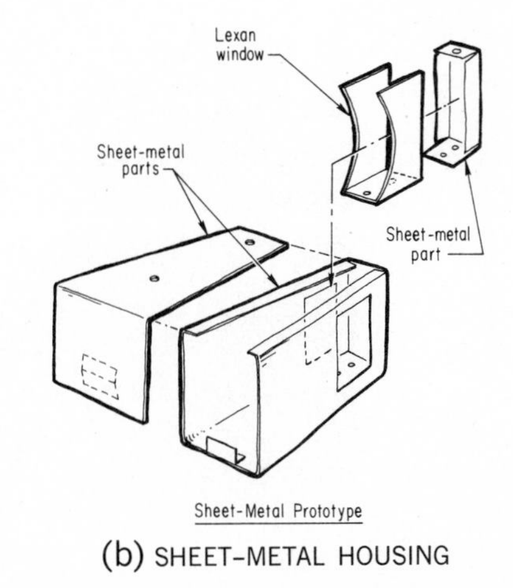


(b) SHEET-METAL HOUSING

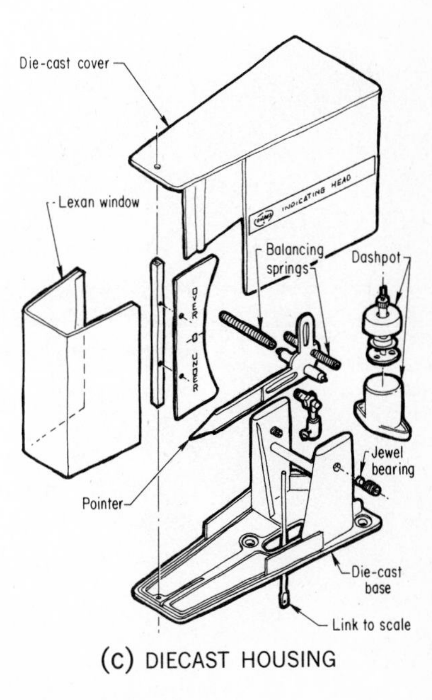


(C) DIECAST HOUSING

may be made without any loss of effectiveness and yet may save hundreds or thousands of dollars. The *ideal design* is the one that will do the job required at the lowest possible cost.

The Totalizer Wheel, Fig. 14.8, represents a cost-reducing redesign of an assembly in an adding machine. By using a new material (acetal resin) and the injection molding process, §10.31, the redesigned wheel replaces an assembly of the twenty-three parts shown and continues to act as an indexing gear, integral bearing, integral spring, position stop, and print wheel.

An example of design from a different approach is the Electric Wheel used in heavy duty four-wheel drive earth-moving equipment, Fig. 14.9. This self-contained wheel design eliminates the usual restrictive drive train components, such as the drive shaft, universal joints, differential gears, and transmission, and makes possible nonspin traction at each wheel. The motor in the wheel is powered by a heavy-duty diesel-electric generating system aboard the equipment.

*Courtesy Addmaster Corp. and E. I. du Pont de Nemours & Co., Inc.*

Fig. 14.8 **Redesigned Totalizer Wheel for Adding Machine.**

Fig. 14.9 **Electric Wheel.**

*Courtesy Marathon LeTourneau Co.*

Model L-700, LeTro-Loader, manufactured by Marathon LeTourneau Co., Equipment Division, Longview, Texas.

Fig. 14.10 **Electronic Organ.** *Courtesy Allen Organ Co.*

In a revised electronic organ design, the room-filling pipes and bellows have been replaced by a digital musical computer composed of aerospace microelectronics that requires about one cubic foot, Fig. 14.10. The computer contains some 48,000 transistors and enables the organ to be played in a virtually unlimited number of voices.

The designs for a large vacation-area complex include unique and unusual approaches or systems to the problem of housing and transportation. The fourteen-story A-frame hotel, as shown by the model, Fig. 14.11 (a), is serviced by water craft, surface vehicles, and special monorail trains that pass through the structure as indicated at (b). A pictorial of the train is shown at (c).

An improved design of a freight handling system for the transporting of subcompact automobiles includes the unique design of a special railroad car called Vert-A-Pack, Fig. 14.12. The sides of the five compartments on each side, which hold three cars each, are hinged at the bottom for use as ramps for efficient drive-on loading and drive-away unloading. The autos lock into place as the ramps are raised, and when the compartments are closed the autos are protected from vandalism and the weather.

**14.7 Stage 4—Models and Prototypes.** A model to scale is often constructed to study, analyze, and refine a design, Figs. 14.1, 14.11,

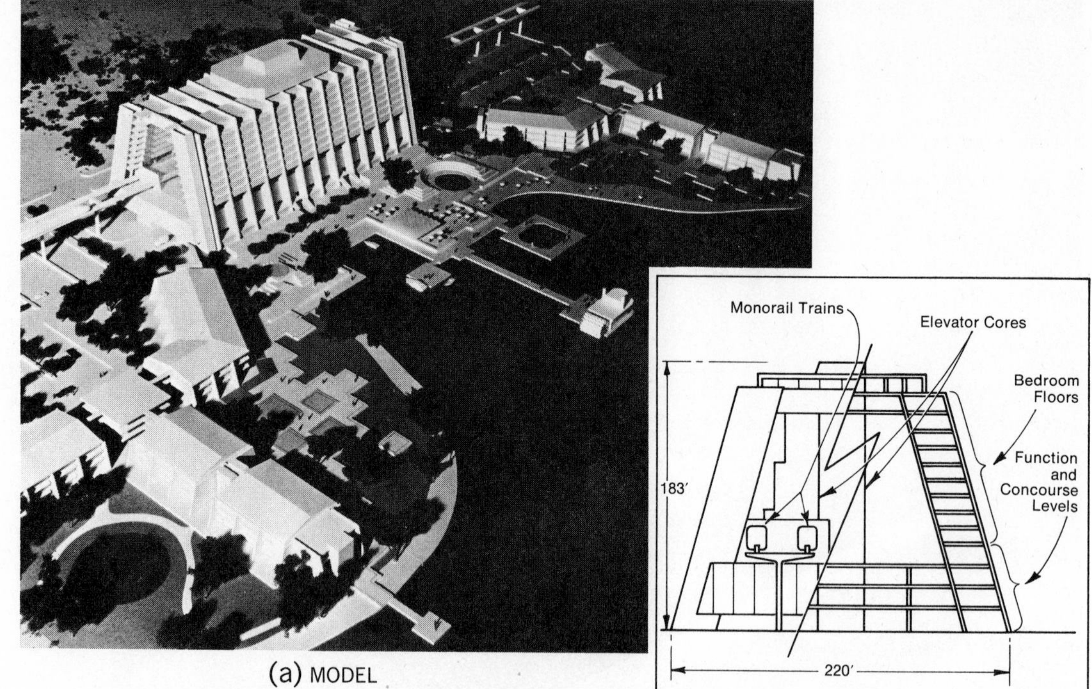


(a) MODEL

(b) ELEVATION

© *Walt Disney Productions*

(c) MONORAIL TRAIN

Fig. 14.11 **Recreation Park.**

*Courtesy Engineering News-Record*

and 14.17. The model of the Carveyor, Fig. 14.13, shows how it works and how people may be moved in such areas as airports, shopping centers, and college campuses. It is a loop system and is designed to carry up to 22,000 people per hour.

To instruct the model-shop craftsman in the construction of the prototype or model, dimensioned sketches and/or rudimentary working drawings are required. A full-size working model made to final specifications, except possibly for materials, is known as a *prototype.* The prototype is tested, modified where necessary, and the results noted in the revision of the sketches and working drawings.

If the prototype proves to be unsatisfactory, it may be necessary to return to a previous stage in the design process and repeat the procedures. It must be remembered that time and expense ceilings always limit the duration of this "looping." Eventually, a decision must be reached for the production model.

## 14.8 Stage 5—Working Drawings.

To produce or manufacture a product, a final set of production or working drawings are made, checked, and approved.

In industry the approved production design layouts are turned over to the engineering department for the production drawings. The draftsmen, or detailers, "pick off" the details from the layouts with the aid of the scale or dividers. The necessary views, §5.15, are drawn for each part to be made and complete dimensions and notes, Chapter 11, are added so that the drawings will describe these parts completely. These working drawings of the individual parts are also known as *detail drawings,* §14.10.

Unaltered standard parts, §11.42, do not require a detail drawing but are shown conventionally on the assembly drawing and listed with specifications in the parts list, §14.14.

A detail drawing of one of the parts from the design layout of Fig. 14.6 is shown in Fig. 14.14. For details concerning working drawings, see §§14.10 to 14.19.

After the parts have been detailed, an *assembly drawing* is made, showing how all the parts go together in the complete product. The assembly may be made by tracing the various details in place directly from the detail drawings, or the assembly may be traced from the original design layout; but if either is done,

*Courtesy American Iron and Steel Institute*

**Fig. 14.12 Railway Auto Transport System.**

**Fig. 14.13 Carveyor.**

*Courtesy Goodyear Tire and Rubber Co.*

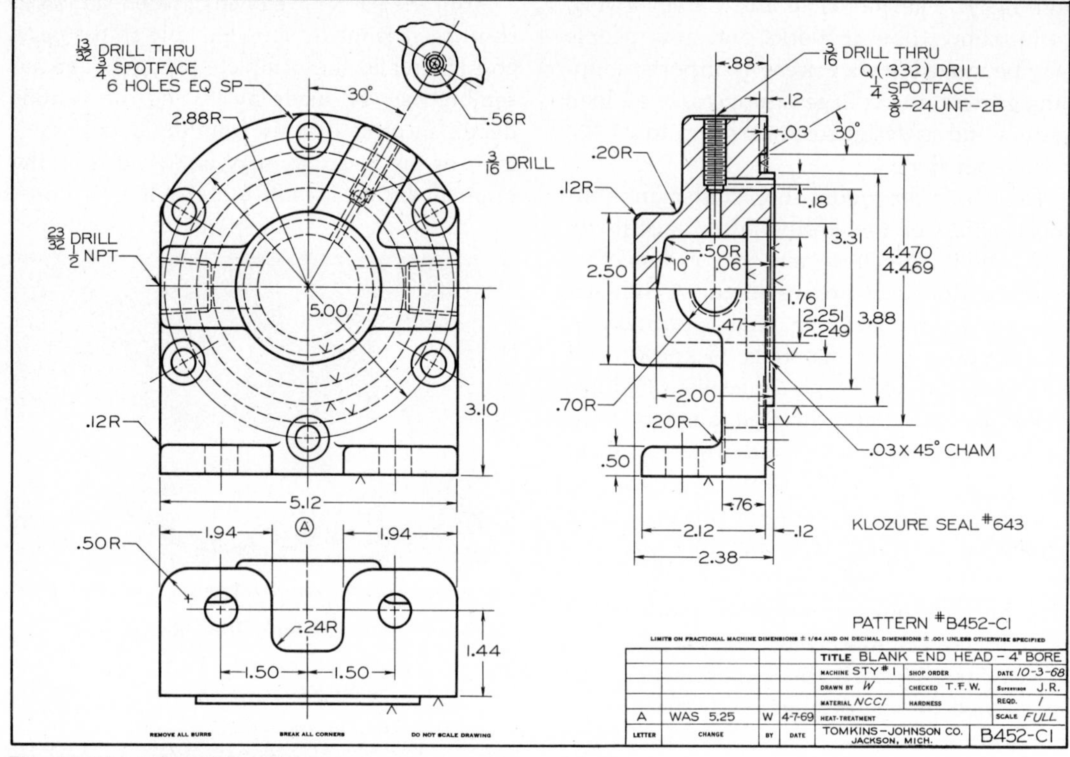


**Fig. 14.14 A Detail Drawing.**

**Fig. 14.15 Cordless Eraser with Recharging Console.** *Courtesy Pierce Corp.*

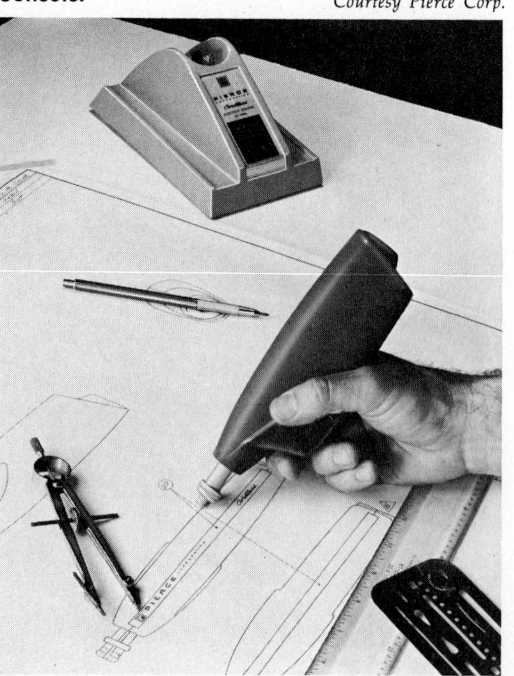

the value of the assembly for checking purposes, §14.25, will be largely lost. The various types of assemblies are discussed in §§14.20 to 14.25.

Finally, in order to protect the manufacturer, a *patent drawing,* which is often a form of assembly, is prepared and filed with the U.S. Patent Office. Patent drawings are line shaded, often lettered in script, and otherwise follow rules of the Patent Office, §14.26.

## 14.9 Design of a New Product.

An example of the design and development of a new product is that of the Cordless Electric Eraser shown in Figs. 2.19 and 14.15.

Stage 1—Identification of the Problem. In order to determine the feasibility of the *first* cordless eraser, opinions and ideas were solicited from many sources, including engineers, draftsmen, drafting teachers, drafting supply house managers, drafting supply store owners, and others. Price ranges and estimated sales were also carefully explored. This extensive survey indicated that there was a need and a potential market for a cordless eraser providing it was convenient to use, durable, lightweight, versatile, maintenance free, and competitively priced.

STAGE 2—CONCEPTS AND IDEAS. Various cord erasers on the market were examined, tested, and analyzed for possible improvements. Several cordless devices on the market were also studied. Power and speed requirements for efficient erasing of pencil and/or ink lines were determined. Several methods of holding the eraser were reviewed. Various eraser refills for electric erasers were tested. With this collection of information as a background, several questions now needed to be considered. Should a direct drive and a large motor or a reduction drive and a small motor be used? What power source should be used—replaceable or rechargeable batteries? What recharging arrangements are necessary for rechargeable battery power? What about safety precautions with 110 volt power for the charger? What materials are suitable? What bearings are available? What is a suitable way to chuck the eraser? Should long or short eraser plugs or only short ones be used? What standard parts are available for such items as the motor, bearings, batteries, recharging unit, and power cord?

How many ways can these components be arranged for the solution?

STAGE 3—COMPROMISE SOLUTION. The cordless version of the eraser with a charging unit in a separate stand or console was selected as the preferred goal. The power train of a small battery-driven motor with a pinion gear meshed with a larger spur gear on the main shaft provided adequate power and speed for the eraser chuck. The batteries provided power long enough for normal usage. Recharging would occur while the eraser was at rest on the charging console. Careful selection of components and materials could lead to a durable and lightweight unit.

The few simple components of the system could be arranged in several ways. Thus some flexibility in the final design was possible. Pictorials of several concepts, Fig. 14.16, were made and evaluated for balance, handling qualities, and appearance.

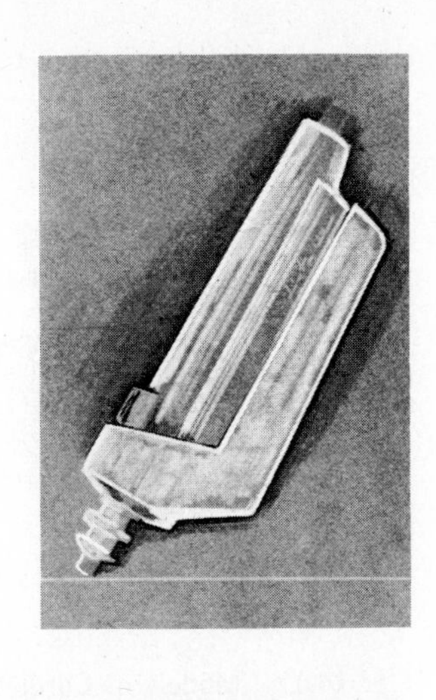

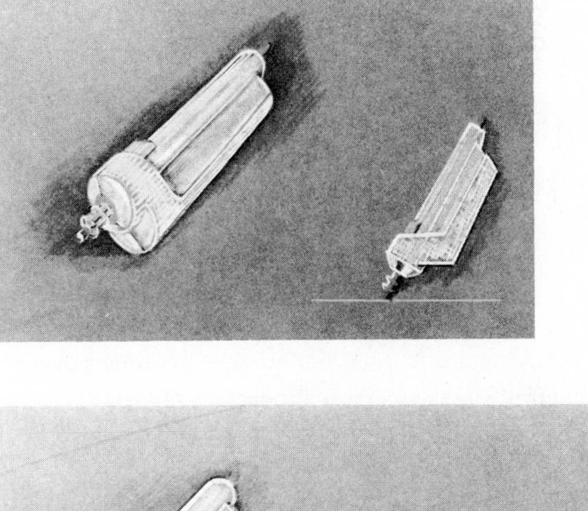

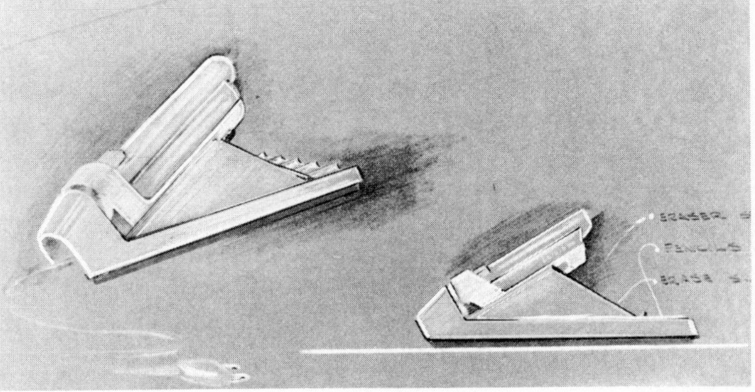

*Courtesy Pierce Corp.*

Fig. 14.16 **Preliminary Design Pictorials of Cordless Eraser.**

STAGE 4—PROTOTYPE. A prototype or working model, Fig. 14.17 (a), was built, tested, refined, and restyled into the final production

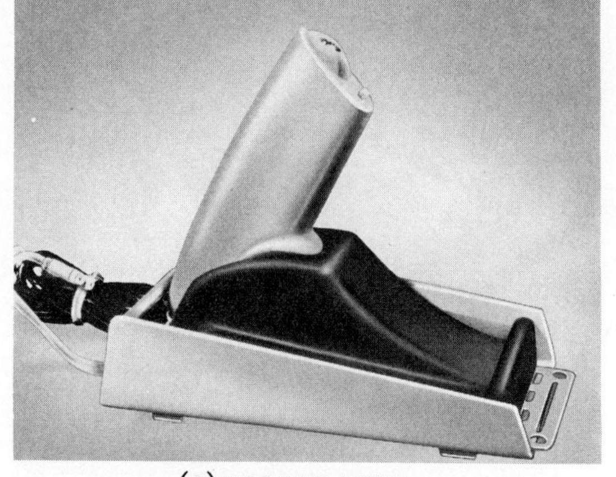

(a) WORKING MODEL

(b) PRODUCTION MODEL

*Courtesy Pierce Corp.*

**Fig. 14.17 Models of Cordless Eraser.**

*Courtesy Pierce Corp.*

**Fig. 14.18 Section of Cordless Eraser Production Model.**

design, (b). The section of the production model, Fig. 14.18, shows the selected arrangement of components; note that all the components are held in place in one half of the molded shell without additional parts or fasteners. The Main Shaft assembly is shown in Fig. 14.19. To achieve the goals of lightweight durability and a competitive selling price, the engineers selected a molding of reinforced nylon resin rather than a metal tubing. A manufacturing cost saving of over 50 percent for the main shaft alone was made possible by the elimination of several secondary operations. For example, in Fig. 14.19 note the following features:

1. No thread cutting required—threads are molded.
2. No groove cutting—molded hubs locate the bearings and also eliminate the need for retaining rings.
3. No slot cutting for chuck removal slot—slot is molded.
4. No external grinding of shaft is necessary—precision molding gives a diameter suitable for installation of the high-speed bearings.
5. No keyseat required—molded-in key eliminates need for a key.
6. No keyway required on gear—molded-in keyway. No key or retaining ring installations facilitate more simplified assembly.

The above considerations are typical of the attention given all components in the design.

STAGE 5—PRODUCTION DRAWINGS. Complete sets of detail and assembly drawings were made for the eraser and the charging console. The assembly drawing for the eraser is shown in Fig. 14.20 and the assembly drawing of the

**Fig. 14.19 Main Shaft Assembly.**

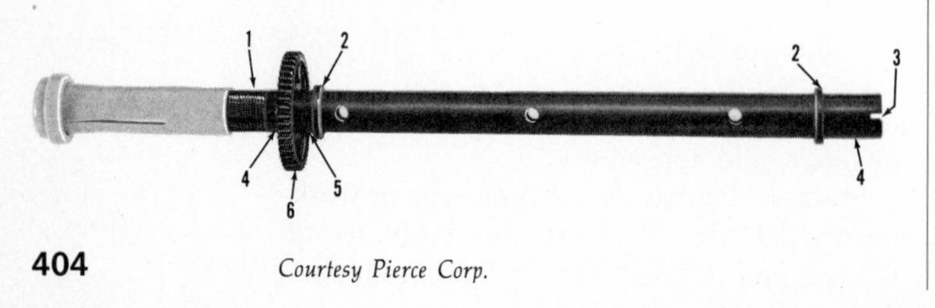


*Courtesy Pierce Corp.*

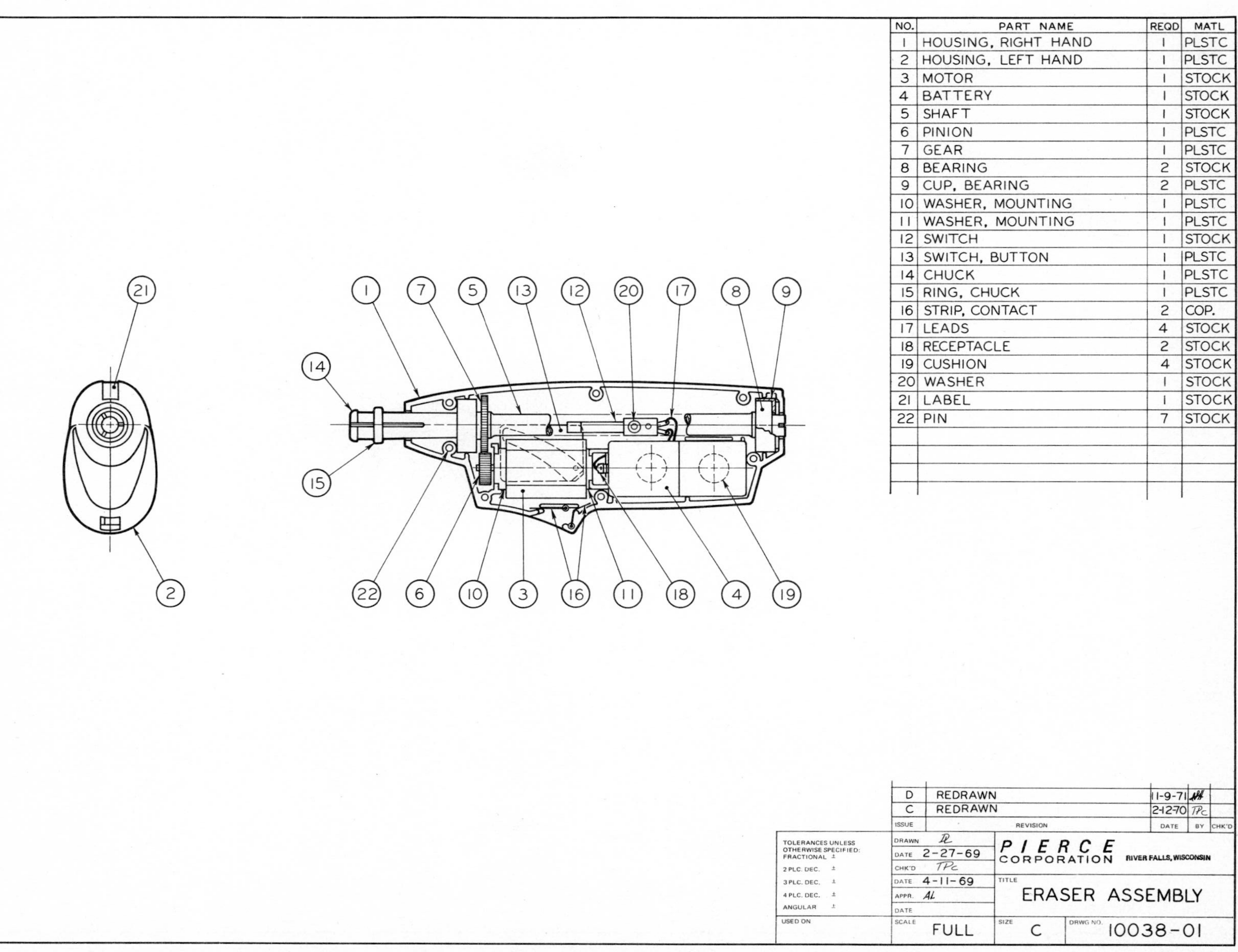

21
1
7
5
13
12
20
17
8
9
14
15
2
22
6
10
3
16
11
18
4
19
NO. | PART NAME | REQD | MATL
1 HOUSING, RIGHT HAND 1 PLSTC
2 HOUSING, LEFT HAND 1 PLSTC
3 MOTOR 1 STOCK
4 BATTERY 1 STOCK
5 SHAFT 1 STOCK
6 PINION 1 PLSTC
7 GEAR 1 PLSTC
8 BEARING 2 STOCK
9 CUP, BEARING 2 PLSTC
10 WASHER, MOUNTING 1 PLSTC
11 WASHER, MOUNTING 1 PLSTC
12 SWITCH 1 STOCK
13 SWITCH, BUTTON 1 PLSTC
14 CHUCK 1 PLSTC
15 RING, CHUCK 1 PLSTC
16 STRIP, CONTACT 2 COP.
17 LEADS 4 STOCK
18 RECEPTACLE 2 STOCK
19 CUSHION 4 STOCK
20 WASHER 1 STOCK
21 LABEL 1 STOCK
22 PIN 7 STOCK
D REDRAWN 11-9-71
C REDRAWN 2-12-70 TPC
ISSUE REVISION DATE BY CHK'D
TOLERANCES UNLESS OTHERWISE SPECIFIED:
FRACTIONAL ±
2 PLC. DEC. ±
3 PLC. DEC. ±
4 PLC. DEC. ±
ANGULAR ±
USED ON
DRAWN
DATE 2-27-69
CHK'D TPC
DATE 4-11-69
APPR. AL
DATE
PIERCE CORPORATION RIVER FALLS, WISCONSIN
TITLE
ERASER ASSEMBLY
SCALE FULL
SIZE C
DRWG NO. 10038-01


Courtesy Pierce Corp.

Fig. 14.20 **Assembly Drawing of Cordless Eraser.**

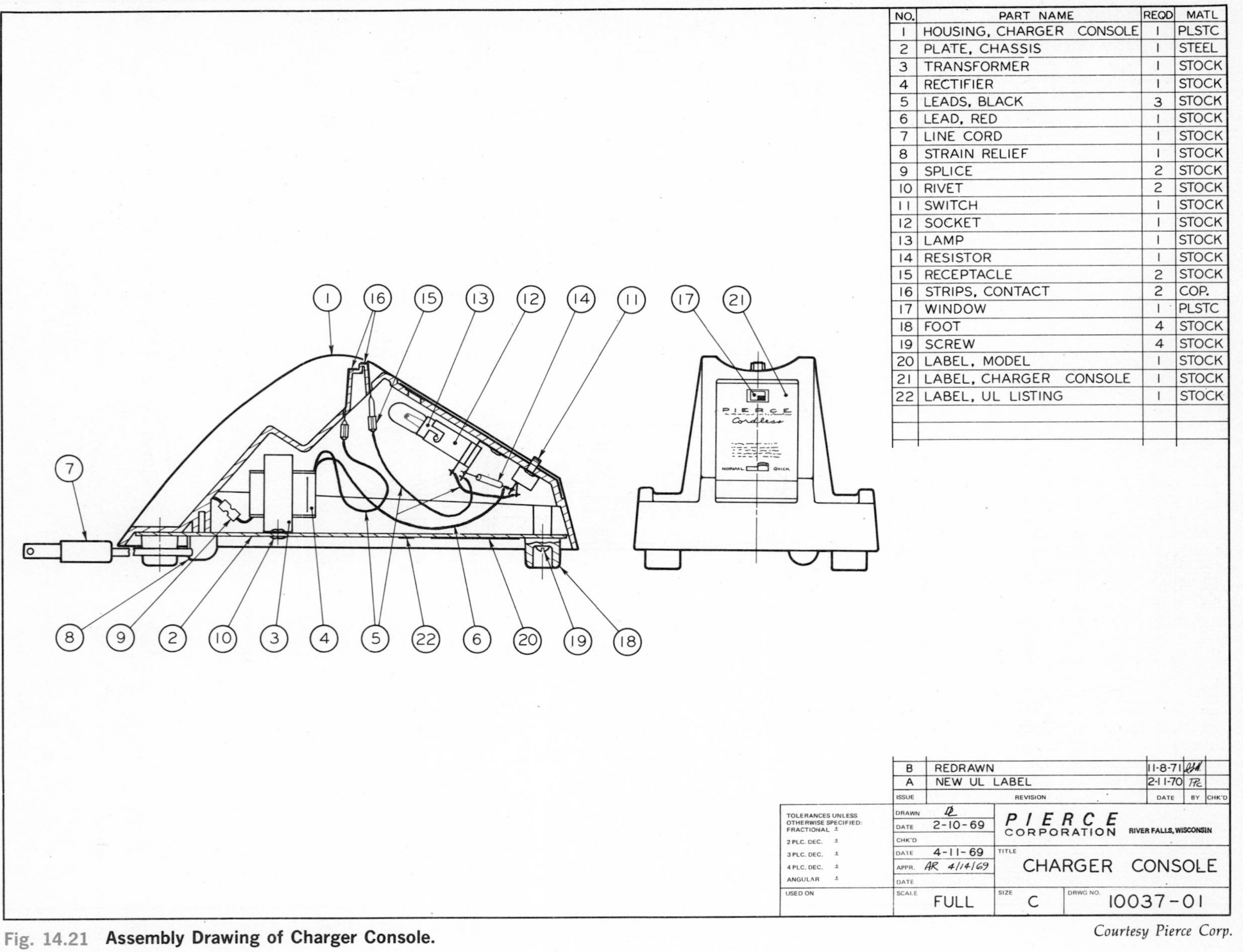


| NO. | PART NAME | REQD | MATL |
|---|---|---|---|
| 1 | HOUSING, CHARGER CONSOLE | 1 | PLSTC |
| 2 | PLATE, CHASSIS | 1 | STEEL |
| 3 | TRANSFORMER | 1 | STOCK |
| 4 | RECTIFIER | 1 | STOCK |
| 5 | LEADS, BLACK | 3 | STOCK |
| 6 | LEAD, RED | 1 | STOCK |
| 7 | LINE CORD | 1 | STOCK |
| 8 | STRAIN RELIEF | 1 | STOCK |
| 9 | SPLICE | 2 | STOCK |
| 10 | RIVET | 2 | STOCK |
| 11 | SWITCH | 1 | STOCK |
| 12 | SOCKET | 1 | STOCK |
| 13 | LAMP | 1 | STOCK |
| 14 | RESISTOR | 1 | STOCK |
| 15 | RECEPTACLE | 2 | STOCK |
| 16 | STRIPS, CONTACT | 2 | COP. |
| 17 | WINDOW | 1 | PLSTC |
| 18 | FOOT | 4 | STOCK |
| 19 | SCREW | 4 | STOCK |
| 20 | LABEL, MODEL | 1 | STOCK |
| 21 | LABEL, CHARGER CONSOLE | 1 | STOCK |
| 22 | LABEL, UL LISTING | 1 | STOCK |

Fig. 14.21 **Assembly Drawing of Charger Console.**

*Courtesy Pierce Corp.*

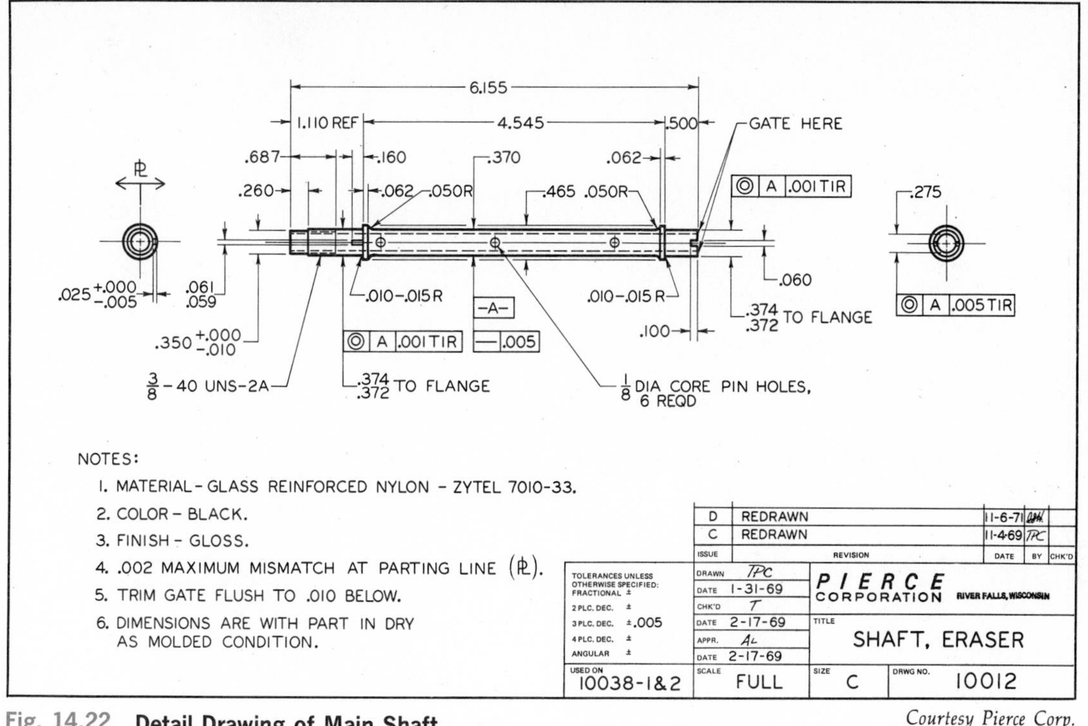


**Fig. 14.22 Detail Drawing of Main Shaft.** *Courtesy Pierce Corp.*

charger base is shown in Fig. 14.21. A detail of the shaft is given in Fig. 14.22. Standard parts were specified on separate sheets or in a parts list. (Space limitations do not permit including more of the drawings necessary for the product.)

## 14.10 Working Drawings.

Working drawings, which normally include assembly and details, are the specifications for the manufacture of a design. Therefore, they must be neatly made and carefully checked. The working drawings of the individual parts are also referred to as *detail drawings.* See §§14.11 to 14.18.

## 14.11 Number of Details per Sheet.

Two general methods are followed in industry regarding the grouping of details on sheets. If the machine or structure is small or composed of few parts, all the details may be shown on one large sheet, Fig. 14.23.

When larger or more complicated mechanisms are represented, the details may be drawn on several large sheets, several details to the sheet, and the assembly drawn on a separate sheet. Most companies have now adopted the practice of drawing only one detail per sheet, however simple or small, Fig. 14.22. The basic $8\frac{1}{2}'' \times 11''$ sheet is most commonly used for details, multiples of this size being used for larger details or the assembly. For American National Standard sheet sizes, see §2.66.

When several details are drawn on one sheet, careful consideration must be given to spacing. The draftsman should determine the necessary views for each detail, and *block in all views lightly before beginning to draw any view,* as shown in Fig. 14.23. Ample space should be allowed for dimensions and notes. A simple method to space the views is to cut out rectangular scraps of paper roughly equal to the sizes of the views, and to move these around on the sheet until a suitable spacing is determined. The corner locations are then marked

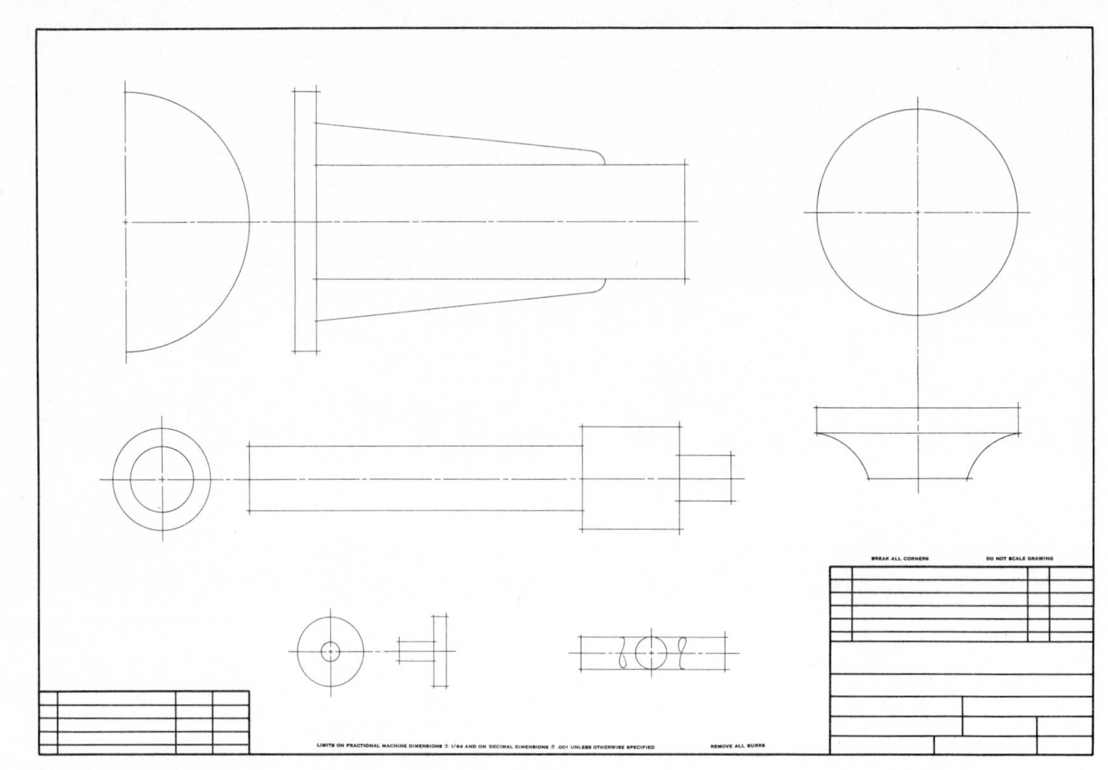

Fig. 14.23 **Blocking In the Views.** (See Fig. 14.77.)

on the sheet, and the scraps of paper are discarded.

The same scale should be used for all details on a single sheet, if possible. When this is not possible, the scales for the dissimilar details should be clearly noted under each.

## 14.12 Title and Record Strips.

The function of the title and record strip is to show, in an organized manner, all necessary information not given directly on the drawing with its dimensions and notes. Obviously, the type of title used depends upon the filing system in use, the processes of manufacture, and the requirements of the product. The following information should generally be given in the title form:

1. Name of the object represented.
2. Name and address of the manufacturer.
3. Name and address of the purchasing company, if any.
4. Signature of the draftsman who made the drawing, and the date of completion.
5. Signature of the checker, and the date of completion.
6. Signature of the chief draftsman, chief engineer, or other official, and the date of approval.
7. Scale of the drawing.
8. Number of the drawing.

Other information may be given, such as material, quantity, heat treatment, finish, hardness, pattern number, estimated weight, superseding and superseded drawing numbers, symbol of machine, and many other items, depending upon the plant organization and the peculiarities of the product. Some typical commercial titles are shown in Figs. 14.24 and 14.26 to 14.28.

The title form is usually placed in the lower right-hand corner of the sheet, Fig. 14.28, or along the bottom of the sheet, Fig. 14.27, because drawings are often filed in flat, horizontal drawers, and the title must be easily

SUPERSEDES DWG No. 17913-18233
SUPERSEDED BY DWG No.
PROPERTY OF HUGHES TOOL CO. HOUSTON, TEXAS
DRAWN BY McK — TRACED BY McK — CHECKED BY R.P.S. — DATE BEGUN 10-11-68 — DATE FINISHED 10-11-68
REVISIONS: A 10-15-68 McK; B 10-17-68 McK; C 10-19-68 BRA; D 10-21-68 R.P.S.; E 10-24-68 McK; F 1-14-69 BRA; G 1-31-69 BRA; H 2-2-69 RPS
FILE No. F/D — SCALE — — 18305

Fig. 14.24 **Title Strip.**

| Use | Letter Heights, in. (minimum) | | Drawing Size |
|---|---|---|---|
| | Freehand | Mechanical | |
| Drawing numbers in title block | $\frac{5}{16}$ (.312) | .350 | All |
| Drawing title | $\frac{1}{4}$ (.250) | .240 | |
| Section and tabulation letters | $\frac{1}{4}$ (.250) | .240 | |
| Zone letters and numerals in border | $\frac{3}{16}$ (.187) | .175 | |
| Dimension, tolerances, limits, notes, subtitles for special views, tables, revisions, and zone letters for the body of the drawing | $\frac{1}{8}$ (.125) | .120 | Up to and including 18″ × 24″ |
| | $\frac{5}{32}$ (.156) | .150 | Larger than 18″ × 24″ |

Fig. 14.25 **Lettering Heights.**

REPORT ALL ERRORS TO FOREMAN
NO. REQUIRED 1 — MATERIAL SAE 3115 — HEAT TREATMENT SEE NOTE — PART NAME FEED WORM SHAFT — DRAWN BY H.F. — UNIT 3134
DRAWN FOR SIMPLEX & DUPLEX (1200) — TRACED BY E.E.Z. — ALSO USED ON ABOVE MACHINES
REPLACED BY — REPLACES — OLD PART NO. 563-310 — ENGINEERING DEPARTMENT — CHECKED BY C.STB. — FIRST USED ON LOT — LAST USED ON LOT
KEARNEY & TRECKER CORPORATION MILWAUKEE, WISCONSIN, U. S. A. — APPROVED BY
ALTERATIONS — DATE OF CHG — JUN 25 1973 — SCALE FULL SIZE — DATE 7-10-71 — 17840 B

Fig. 14.26 **Title Strip.**

DO NOT SCALE THIS DRAWING FOR DIMENSIONS. MACHINE FRACTIONAL DIMENSIONS ± 1/64 ALL DIMENSIONS IN INCHES UNLESS OTHERWISE SPECIFIED.
2 CHGD MATL ETC 10-22-73
1 WAS #2345-ETC 5-21-73
SYM — CHANGE — DATE — CHANGE NOTICE
HEAT TREATMENT
SAE VIII
HDN ROCKWELL C-50-56
NOTE 3 TEST LOCATIONS
SCALE FULL — DATE 6-26-72 — DRAWN BY S.G. — TRACED BY L.R. — CHECKED BY n.w. — APPROVED BY amB. — REDRAWN FROM
CATERPILLAR TRACTOR CO.
EXECUTIVE OFFICES — SAN LEANDRO, CALIF.
NAME FIRST, FOURTH & THIRD SLIDING PINION
MATERIAL C.T. #1E36 STEEL ② ①
UPSET FORGING 3 $\frac{7}{8}$ ROUND MAX
1A4045

Fig. 14.27 **Title Strip.**

found. However, many filing systems are in use, and the location of the title form is completely governed by the system employed.

Lettering should be single-stroke vertical or inclined Gothic capitals, Figs. 3.18 and 3.19. The items in the title form should be lettered in accordance with their relative importance. The drawing number should receive greatest emphasis, closely followed by the name of the object and the name of the company. The date, scale, and draftsmen's and checkers' names are important, but they do not deserve prominence. Greater importance of items is indicated by heavier lettering, larger lettering, wider spacing of letters, or by a combination of these methods. See Fig. 14.25 for recommended letter heights.

Most companies have adopted standard title

forms and have them printed on standard-size sheets, so that the draftsmen need merely fill in the blank spaces, as shown in Figs. 14.24, 14.26, and 14.27.

Drawings constitute important and valuable information regarding the products of a manufacturer. Hence, carefully designed, well-kept, systematic files are generally maintained for the filing of drawings, §15.2.

## 14.13 Drawing Numbers.

Every drawing should be numbered. Some companies use serial numbers, such as 60412, or a number with a prefix or suffix letter to indicate the sheet size, as A60412 or 60412–A. A size A sheet would probably be the standard $8\frac{1}{2}'' \times 11''$ or $9'' \times 12''$, and the B size a multiple thereof. Many different numbering schemes are in use in which various parts of the drawing number indicate different things, such as model number of the machine and the general nature or use of the part. In general it is best to use a simple numbering system and not to load the number with too many indications.

The drawing number should be lettered $\frac{5}{16}''$ high in the lower-right and upper-left corners of the sheet, Fig. 14.39.

## 14.14 Parts Lists.

A bill of material, or *parts list*, consists of an itemized list of the several parts of a structure shown on a detail drawing or an assembly drawing. This list is often given on a separate sheet, Fig. 14.114, but is frequently lettered directly on the drawing, Fig. 14.38. The title strip alone is sufficient on detail drawings of only one part, Fig. 14.22, but a parts list is necessary on detail drawings of several parts, Fig. 14.28.

Parts lists on machine drawings contain the part numbers or symbols, a descriptive title of each part, the number required, the material specified, and frequently other information, such as pattern numbers, stock sizes of materials, and weights of parts.

Parts are listed in general order of size or importance. The main castings or forgings are listed first, parts cut from cold-rolled stock second, and standard parts such as fasteners, bushings, and roller bearings third. If the parts list rests on top of the title box or strip, the order of items should be from the bottom upward, Figs. 14.28 and 14.38, so that new items can be added later, if necessary. If the parts list is placed in the upper-right corner, the items should read downward.

Each detail on the drawing may be identified with the parts list by the use of a small circle containing the part number, placed adjacent to the detail, as in Fig. 14.28. One of the sizes in Fig. 14.29 will be found suitable, depending on the size of the drawing.

Standard parts, purchased or company produced, are not drawn, but are listed in the parts list, §11.41. Bolts, screws, bearings, pins, keys, etc., are identified by the part number from the assembly drawing and are specified by name and size or number.

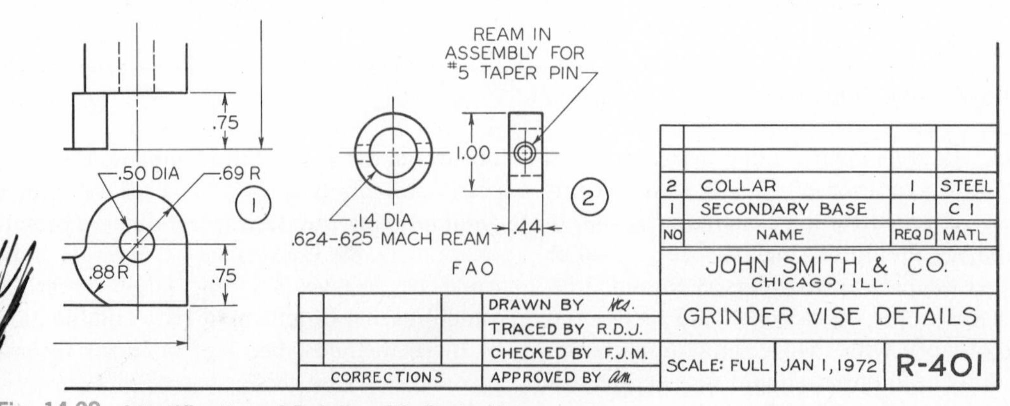


Fig. 14.28 **Identification of Details with Parts List.**

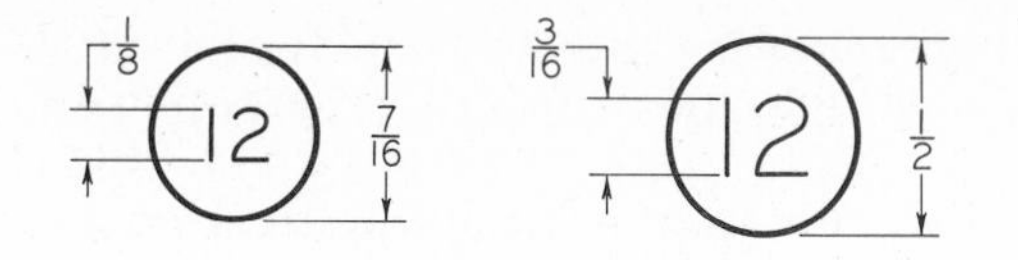

Fig. 14.29 **Identification Numbers.**

## 14.15 Zoning.

To facilitate locating an item on a large or complex drawing, regular ruled intervals are labeled along the margins, usually the right and lower margins only. The intervals on the horizontal margin are labeled from right to left with numerals and the intervals on the vertical margin are labeled from bottom to top with letters. See Fig. 14.44.

## 14.16 Checking.

The importance of accuracy in technical drawing cannot be overestimated. In commercial offices, errors sometimes cause tremendous unnecessary expenditures. *The draftsman's signature on a drawing identifies him, and he is held responsible for the accuracy of his work.*

In small offices, checking is usually done by the designer or by one of the draftsmen. In large offices, experienced engineers are employed who devote their entire time to checking.

The pencil drawing, upon completion, is carefully checked and signed by the draftsman who made it. The drawing is then checked by the designer for function, economy, practicability, and so on. Corrections, if any, are then made by the original draftsman.

The final checker should be able to discover all remaining errors. If his work is to be effective, he must proceed in a systematic way, studying the drawing with particular attention to the following points:

1. Soundness of design, with reference to function, strength, materials, economy, manufacturability, serviceability, ease of assembly and repair, lubrication, etc.
2. Choice of views, partial views, auxiliary views, sections, line work, lettering, etc.
3. Dimensions, with special reference to repetition, ambiguity, legibility, omissions, errors, and finish marks. Special attention should be given to tolerances.
4. Standard parts. In the interest of economy, as many parts as possible should be standard.
5. Notes, with special reference to clear wording and legibility.
6. Clearances. Moving parts should be checked in all possible positions to assure freedom of movement.
7. Title form information.

## 14.17 Drawing Revisions.

Changes on drawings are necessitated by changes in design, changes in tools, desires of customers, or by errors in design or in production. In order that the sources of all changes of information on drawings may be understood, verified, and accessible, an accurate record of all changes should be made on the drawings. The record should show the character of the change, by whom, when, and why made.

The changes are made by erasures directly on the original drawing or by means of erasure fluid on a reproduction print. See §15.7. Additions are simply drawn in on the original. The removal of information by crossing out is not recommended. If a dimension is not noticeably affected by a change, it may be underlined with a wavy line as shown in Fig. 11.15 to indicate that it is not to scale. In any case, prints of each issue or microfilms are kept on file to show how the drawing appeared before the revision. New prints are issued to supersede old ones each time a change is made.

If considerable change on a drawing is necessary, a new drawing may be made and the old one then stamped OBSOLETE and placed in the "obsolete" file. In the title block of the old drawing, under "SUPERSEDED BY . . . ," or "REPLACED BY . . . ," (see Figs. 14.24 and 14.26), the number of the new drawing is entered. On the new drawing, under "SUPERSEDES . . . ," or "REPLACES . . . ," the number of the old drawing is entered.

Various methods are used to reference the area on a drawing where the change is made, with the entry in the revision block. The most common is to place numbers or letters in small

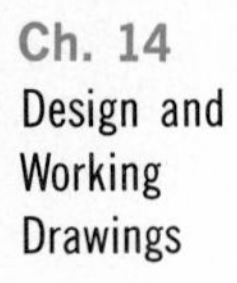

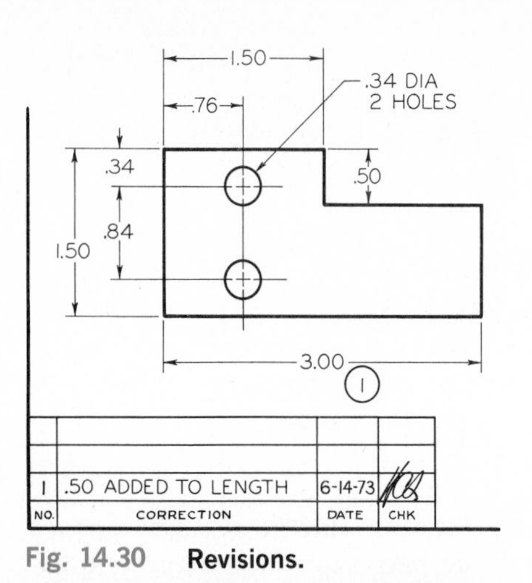


Fig. 14.30 **Revisions.**

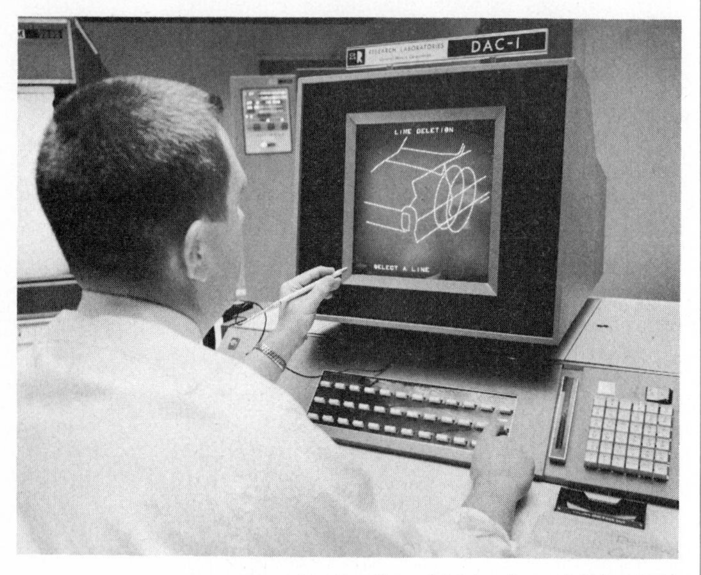


*Courtesy General Motors Research Laboratories*

Fig. 14.31 **Design Drawing with a Computer.**

circles near the places where the changes were made and to use the same numbers or letters in the revision block, Fig. 14.30. On zoned drawings, §14.15, the zone of the correction would be shown in the revision block. In addition, a brief description of the change should be made, and the date and the initials of the person making the change should be given.

## 14.18 Simplified Drafting.

Drafting time is a considerable element of the total cost of a product. Consequently, industry attempts to reduce drawing costs by simplifying its drafting practices, but without loss of clarity to the user.

The American National Standard Drafting Manual, composed of some 27 or more sections when completed and published by the American National Standards Institute, incorporates the best and most representative practices in this country, and the authors are in full accord with them. These standards advocate simplification in many ways; for example, partial views, half views, thread symbols, piping symbols, and single-line spring drawings. Any line or lettering on a drawing that is not needed for clarity should be omitted.

A summary of practices to simplify drafting is as follows:

1. Use word description in place of drawing wherever practicable.
2. Never draw an unnecessary view. Often a view can be eliminated by using abbreviations or symbols, as HEX, SQ, DIA, ℄, etc.
3. Draw partial views instead of full views wherever possible. Draw half views of symmetrical parts.
4. Avoid elaborate, pictorial, or repetitive detail as much as possible. Use phantom lines to avoid drawing repeated features, §13.14.
5. List rather than draw, when possible, standard parts such as bolts, nuts, keys, pins, etc.
6. Omit unnecessary hidden lines. See §5.25.
7. Use outline section lining in large sectioned areas wherever it can be done without loss of clarity.
8. Omit unnecessary duplication of notes and lettering.
9. Use symbolic representation wherever possible, as piping symbols, thread symbols, etc.
10. Draw freehand, or mechanically plus freehand, wherever practicable.
11. Avoid hand lettering as much as possible. For example, parts lists should be typed on a separate sheet.
12. Use laborsaving devices wherever feasible, such as templates, plastic overlays, etc.

Some industries have attempted to simplify their drafting practices even more. Until these practices are accepted generally by industry and in time find their way into the ANSI standards, the student should follow the ANSI standards as exemplified throughout this book. Fundamentals should come first—shortcuts perhaps later.

The old rule of thumb that one dollar on the drafting board saves ten dollars in the shop should not be taken lightly.

## 14.19 Computer Drawing.

Several systems designed to harness the versatility of the computer into a useful engineering and design tool have been developed. For example, in one system the freehand sketches made by the designer or draftsman with a special light pen on the face of a cathode-ray tube provide the input data for the memory units of the computer, Fig. 14.31. The designer can recall his sketches at will, change their shapes, erase them, and "think out" his design on the tube much the same way he would on a drawing board. The final drawing displayed on the tube in this method may be converted to a permanent record photographically.

Other computerized design and drawing systems require the input data to be numerical rather than graphical. Lines, circles, arcs, and curves, for example, are expressed mathematically, programmed in a computer language, and fed into the computer on punched cards or tapes. The computer processes the data and converts it to a punched or magnetic tape that in turn controls an electronic drafting machine, or X–Y plotter, Fig. 14.32. The X–Y

Fig. 14.32 **An Electronic Drafting Machine.** *Courtesy Fort Worth Division of General Dynamics Corporation*

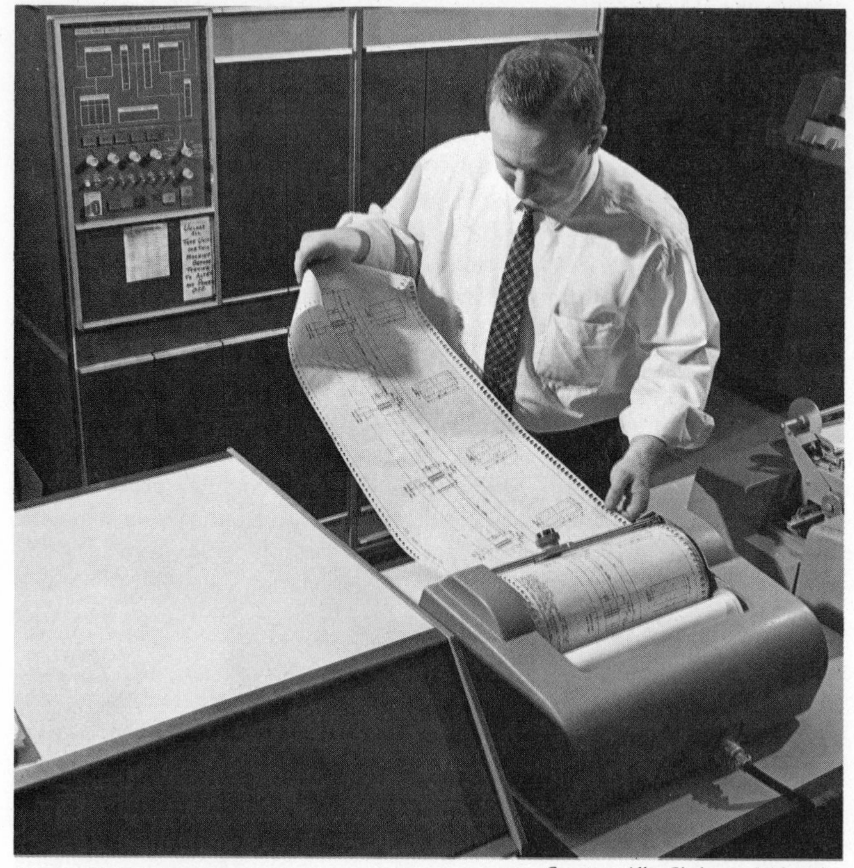

*Courtesy Allis-Chalmers, Milwaukee*

Fig. 14.33 **Drawing Made on a Computer.**

plotter is an electronically controlled pen or stylus that moves with respect to horizontal (X-axis) and vertical (Y-axis) references according to commands from the control tape. Thus by means of programmed numerical data supplied to the computer, the designer or draftsman is able to secure from an X–Y plotter a drawing complete with all required lines, lettering, arrowheads, and so forth, Fig. 14.33.

The procedure for the design and preparation of engineering and parts drawings by computer for the development of a new aircraft is shown in the work-flow diagram, Fig. 14.34. The information and design drawings supplied by the designers and engineers and programmed by the numerographics engineer to produce extremely accurate drawings is preserved on punch cards and tapes for use later to guide the numerically controlled programmed machine tools for the manufacture of the parts.

The engineers and draftsmen must have a thorough understanding of the graphic language in order to prepare the correct input data for the computer and to evaluate the output of the X–Y plotter. The computer will do only what it is programmed to do. Regardless of how complex or automated the method of making a drawing becomes, the time-proven mode of graphic expression is indispensable for purposes of communication, specifications, records, etc.

**14.20 Assembly Drawings.** An assembly drawing shows the assembled machine or structure, with all detail parts in their func-

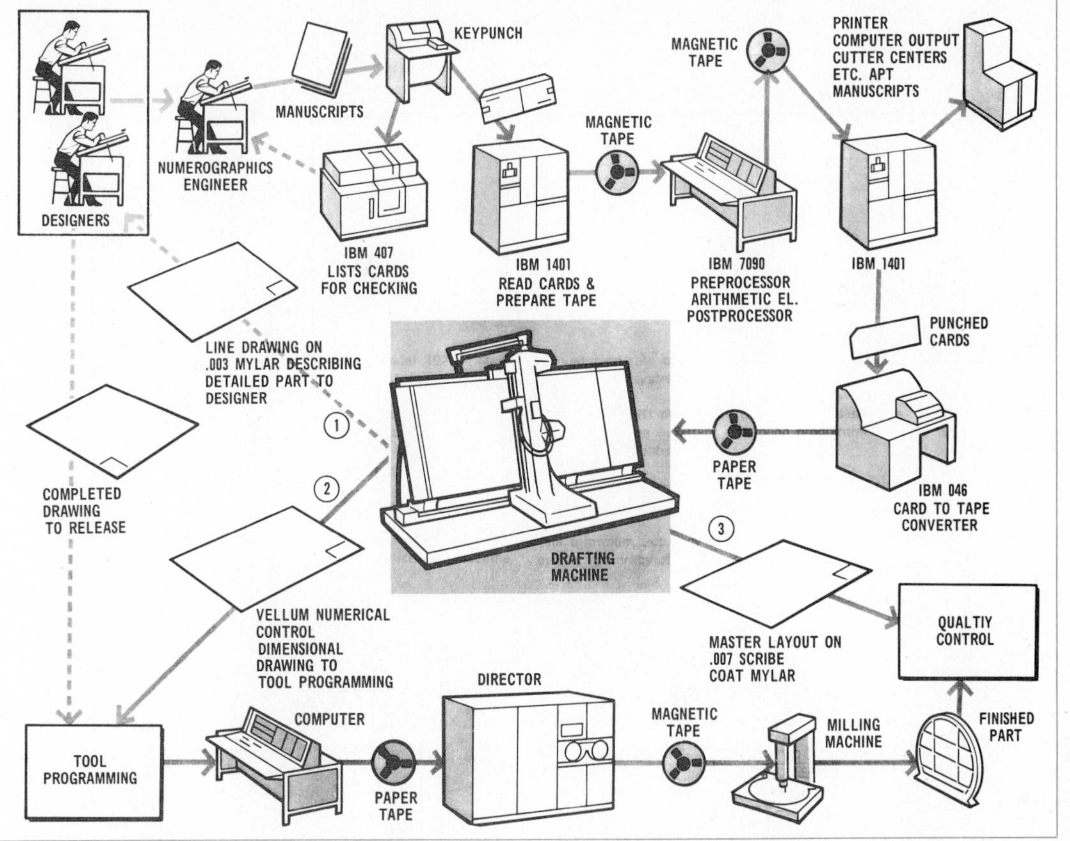


*Courtesy Fort Worth Division of General Dynamics Corporation*

Fig. 14.34 **Work Flow Diagram for Producing Detail and Assembly Drawings, and Machine Parts by Numerical Control.**

tional positions. Assembly drawings vary in character according to use, as follows: (1) Design Assemblies, or Layouts, discussed in §14.6, (2) General Assemblies, (3) Working-Drawing Assemblies, (4) Outline or Installation Assemblies, and (5) Check Assemblies.

## 14.21 General Assemblies.

A set of working drawings includes the *detail drawings* of the individual parts and the *assembly drawing* of the assembled unit. The detail drawings of an automobile connecting rod are shown in Figs. 14.35 and 14.36, and the corresponding assembly drawing is shown in Fig. 14.37. Such an assembly, showing only one unit of a larger machine, is often referred to as a *subassembly.*

An example of a complete general assembly appears in Fig. 14.38, which shows the assembly of a hand grinder. Another example of a subassembly is shown in Fig. 14.39.

1. Views. In selecting the views for an assembly drawing, the purpose of the drawing must be kept in mind: to show how the parts fit together in the assembly and to suggest the function of the entire unit, not to describe the shapes of the individual parts. The assembly worker receives the actual finished parts. If he should need some information about a part that he cannot get from the part itself, he can consult the detail drawings. Thus the assembly drawing purports to show *relationships* of parts, *not shapes.* The view or views selected should be the minimum views or partial views that will show how the parts fit together. In Fig. 14.37, only two views are necessary, while in Fig. 14.38 only one view is needed.

2. Sections. Since assemblies generally

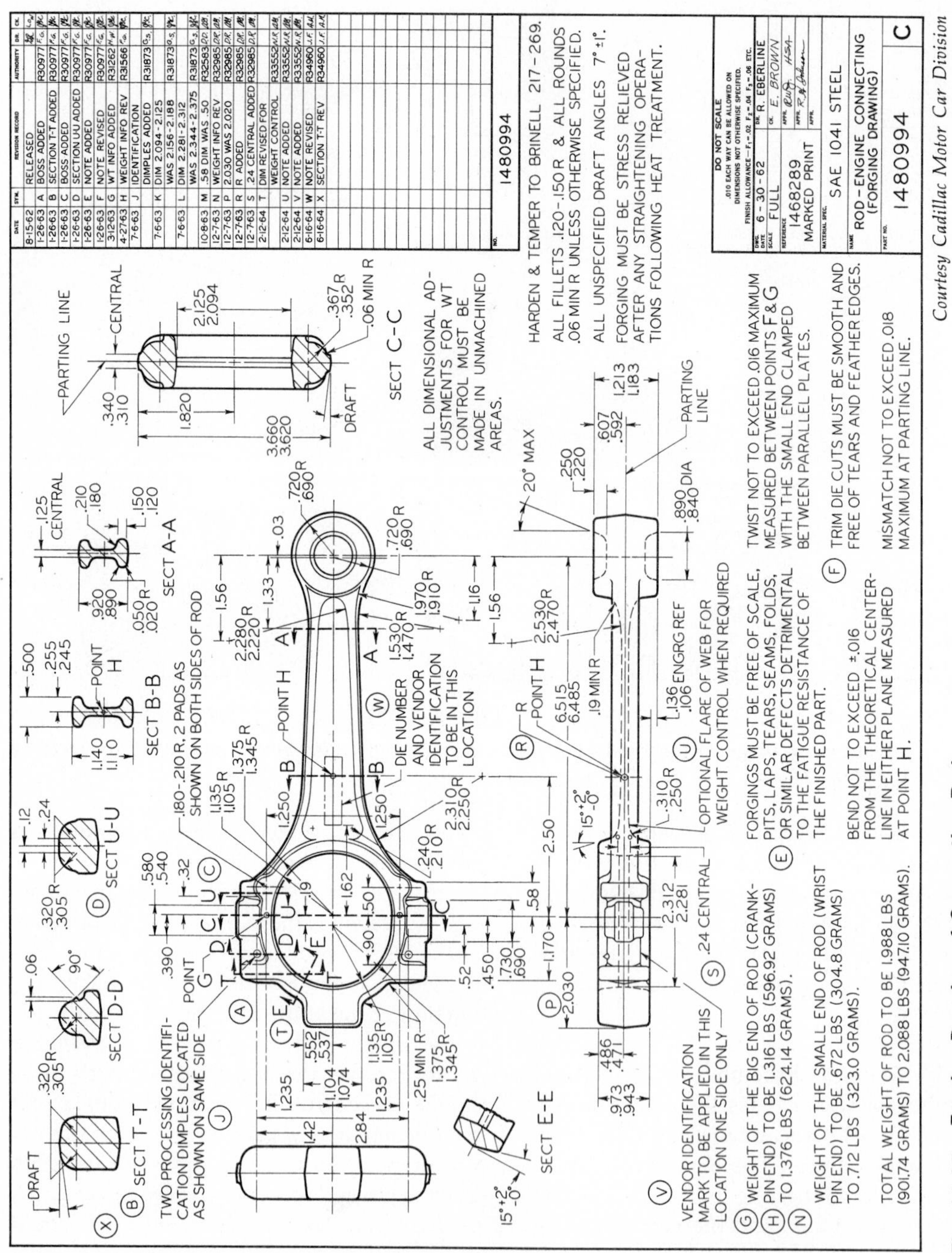


Courtesy Cadillac Motor Car Division

**Fig. 14.35 Forging Drawing of Connecting Rod.**

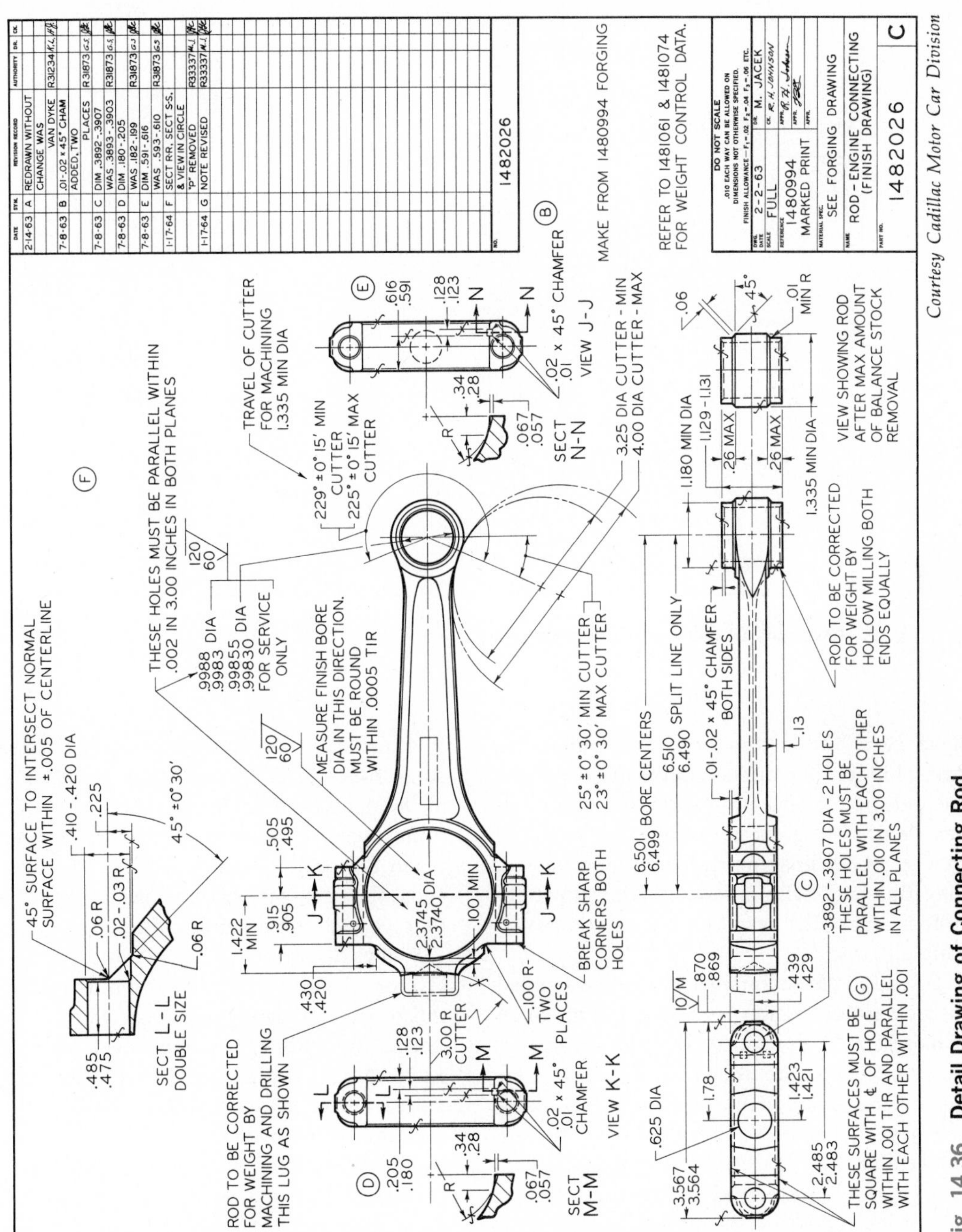


*Courtesy Cadillac Motor Car Division*

**Fig. 14.36 Detail Drawing of Connecting Rod.**

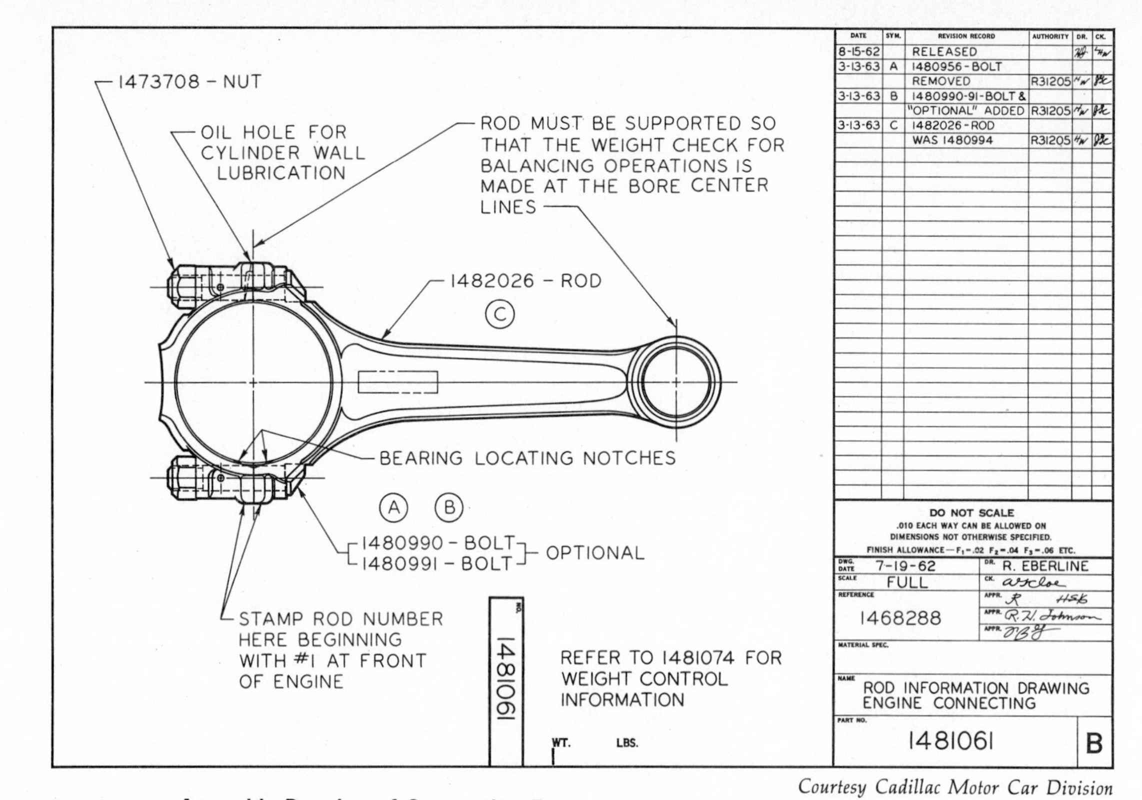


*Courtesy Cadillac Motor Car Division*

Fig. 14.37 **Assembly Drawing of Connecting Rod.**

have parts fitting into or overlapping other parts, hidden-line delineation is usually out of the question. Hence, in assemblies, sectioning can be used to great advantage. For example, in Fig. 14.38, imagine the right-side view in elevation and with interior parts shown by hidden lines. The result would be completely unintelligible.

Any kind of section may be used as needed. A broken-out section is shown in Fig. 14.38, a half section in Fig. 14.39, and several removed sections are shown in Fig. 14.35. For general information on assembly sectioning, see §14.22. For methods of drawing threads in sections, see §13.15.

3. Hidden Lines. As a result of the extensive use of sectioning in assemblies, hidden lines are often not needed. However, they should be used wherever necessary for clearness.

4. Dimensions. As a rule, dimensions are not given on assembly drawings, since they are given completely on the detail drawings. If dimensions are given, they are limited to some function of the object as a whole, such as the maximum height of a jack, or the maximum opening between the jaws of a vise. Or when machining is required in the assembly shop, the necessary dimensions and notes may be given on the assembly drawing.

5. Identification. The methods of identification of parts in an assembly are similar to those used in detail drawings where several details are shown on one sheet, as in Fig. 14.28. Circles containing the part numbers are placed adjacent to the parts, with leaders terminated by arrowheads touching the parts as in Fig. 14.38. The circles shown in Fig. 14.29 for detail drawings are, with the addition of radial leaders, satisfactory for assembly drawings. Note, in Fig. 14.38, that these circles are placed in orderly horizontal or vertical rows and not scattered over the sheet. Leaders are never allowed to cross, and adjacent leaders are parallel or nearly so.

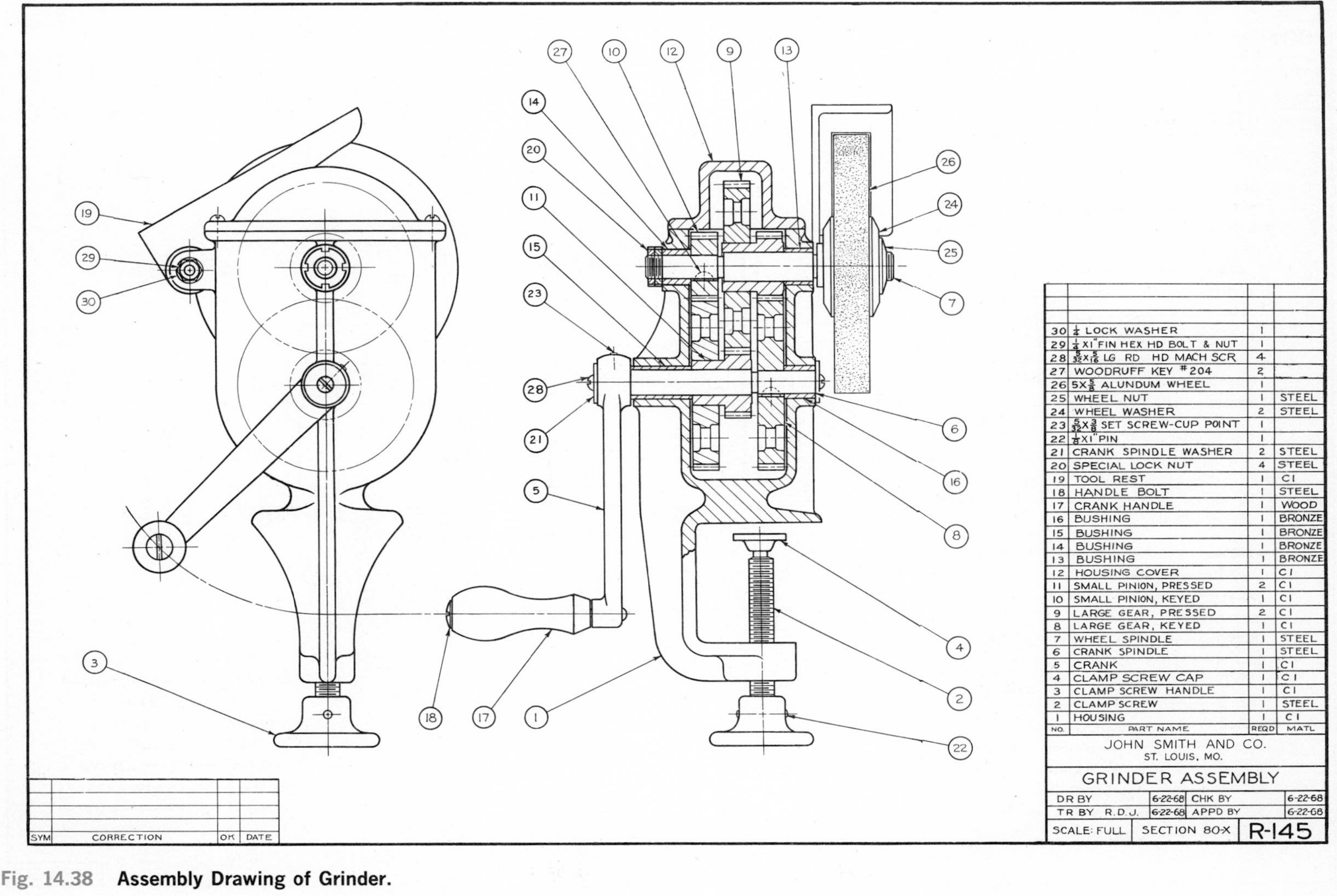

30 1/4 LOCK WASHER 1
29 1/4 x 1" FIN HEX HD BOLT & NUT 1
28 5/32 x 5/16 LG RD HD MACH SCR 4
27 WOODRUFF KEY #204 2
26 5 x 5/8 ALUNDUM WHEEL 1
25 WHEEL NUT 1 STEEL
24 WHEEL WASHER 2 STEEL
23 5/32 x 3/8 SET SCREW-CUP POINT 1
22 1/8 x 1" PIN 1
21 CRANK SPINDLE WASHER 2 STEEL
20 SPECIAL LOCK NUT 4 STEEL
19 TOOL REST 1 CI
18 HANDLE BOLT 1 STEEL
17 CRANK HANDLE 1 WOOD
16 BUSHING 1 BRONZE
15 BUSHING 1 BRONZE
14 BUSHING 1 BRONZE
13 BUSHING 1 BRONZE
12 HOUSING COVER 1 CI
11 SMALL PINION, PRESSED 2 CI
10 SMALL PINION, KEYED 1 CI
9 LARGE GEAR, PRESSED 2 CI
8 LARGE GEAR, KEYED 1 CI
7 WHEEL SPINDLE 1 STEEL
6 CRANK SPINDLE 1 STEEL
5 CRANK 1 CI
4 CLAMP SCREW CAP 1 CI
3 CLAMP SCREW HANDLE 1 CI
2 CLAMP SCREW 1 STEEL
1 HOUSING 1 CI
NO. PART NAME REQD MATL
JOHN SMITH AND CO.
ST. LOUIS, MO.
GRINDER ASSEMBLY
DR BY 6-22-68 CHK BY 6-22-68
TR BY R.D.J. 6-22-68 APPD BY 6-22-68
SCALE: FULL SECTION 80-X R-145
SYM CORRECTION OK DATE


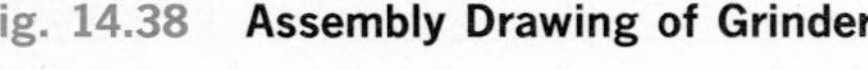
Fig. 14.38 **Assembly Drawing of Grinder.**

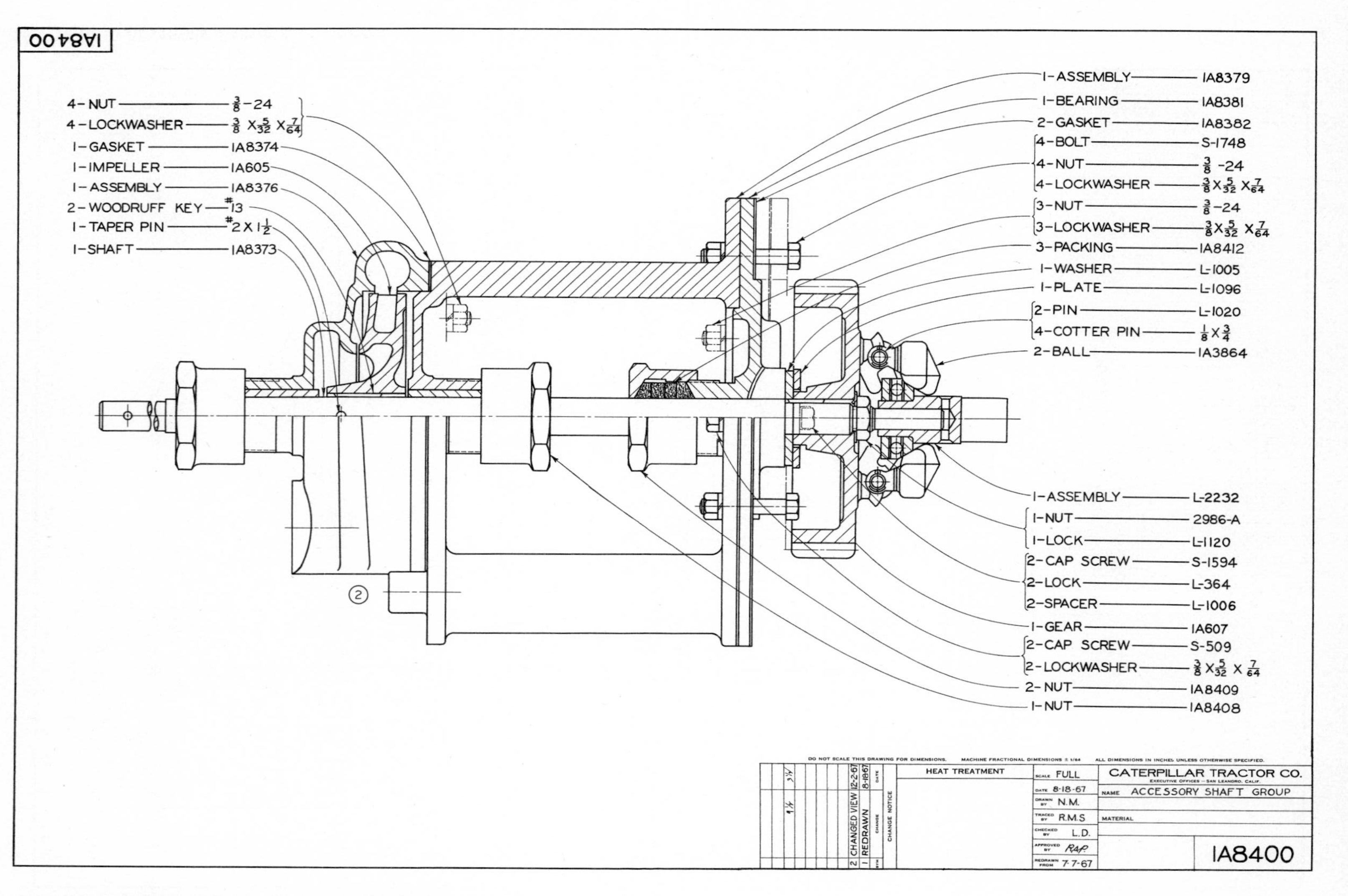


Fig. 14.39 Subassembly of Accessory Shaft Group.

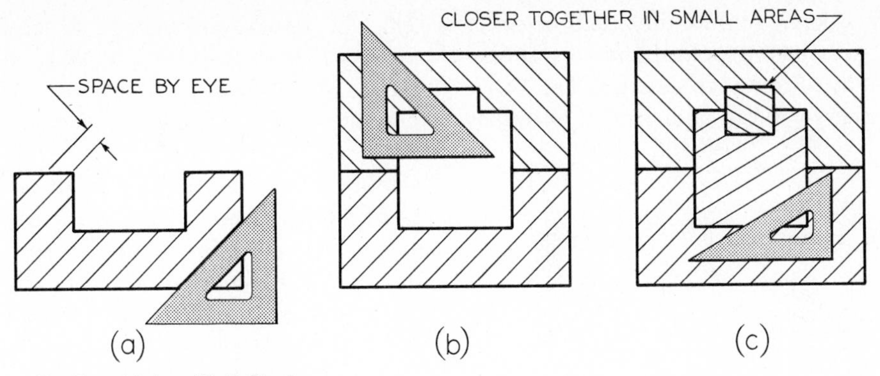


Fig. 14.40 **Section Lining (Full Size).**

The parts list includes the part numbers or symbols, a descriptive title of each part, the number required per machine or unit, the material specified, and frequently other information, such as pattern numbers, stock sizes, weights, etc. Frequently the parts list is lettered or typed on a separate sheet, as shown in Fig. 14.114.

Another method of identification is to letter the part names, numbers required, and part numbers, at the end of leaders as shown in Fig. 14.39. More commonly, however, only the part numbers are given, together with ANSI-approved straight-line leaders.

6. Drawing Revisions. Methods of recording changes are the same as those for detail drawings. See §14.17 and Figs. 14.24 to 14.26.

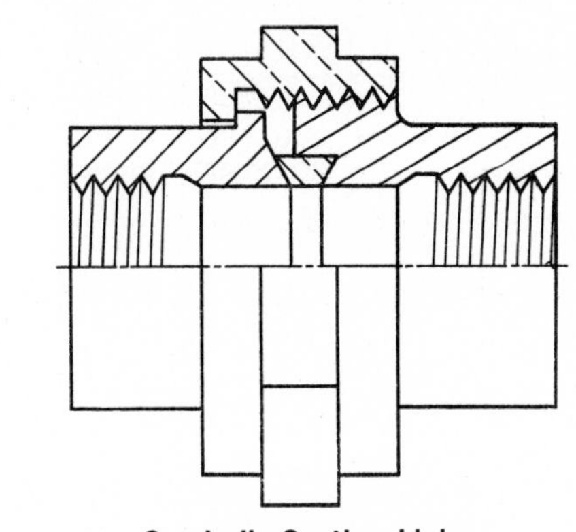

Fig. 14.41 **Symbolic Section Lining.**

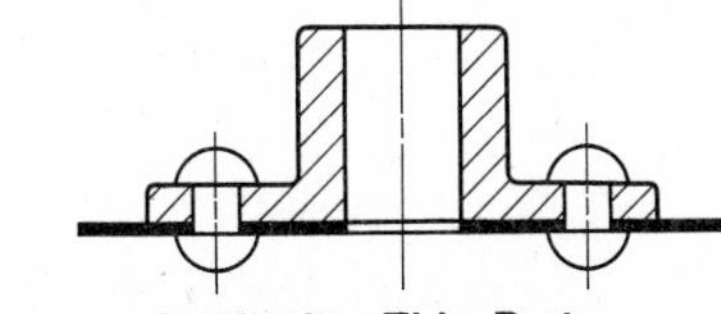

Fig. 14.42 **Sectioning Thin Parts.**

## 14.22 Assembly Sectioning.

In assembly sections it is necessary not only to show the cut surfaces but to distinguish between adjacent parts. This is done by drawing the section lines in opposing directions, as shown in Fig. 14.40. The first large area, (a), is section-lined at 45°. The next large area, (b), is section-lined at 45° in the opposite direction. Additional areas are then section-lined at other angles, as 30° or 60° with horizontal, as shown at (c). If necessary, "odd" angles may be used. Note at (c) that in small areas it is necessary to space the section lines closer together. The section lines in adjacent areas should not meet at the visible lines separating the areas.

For general use, the cast iron general-purpose section lining is recommended for assemblies. Wherever it is desired to give a general indication of the materials used, symbolic section lining may be used, as in Fig. 14.41. The American National Standard symbols for section lining are shown in Fig. 7.5.

In sectioning relatively thin parts in assembly, such as gaskets and sheet metal parts, section lining is ineffective, and such parts should be shown in solid black, Fig. 14.42.

Often solid objects, or parts which themselves do not require sectioning, lie in the path of the cutting plane. It is customary and standard practice to show such parts unsectioned, or "in the round." These include bolts, nuts, shafts, keys, screws, pins, ball or roller bearings, gear teeth, spokes, and ribs among others. Many of these are shown in Fig. 14.43. See how many you can find. Similar examples are shown in Figs. 14.38 and 14.39.

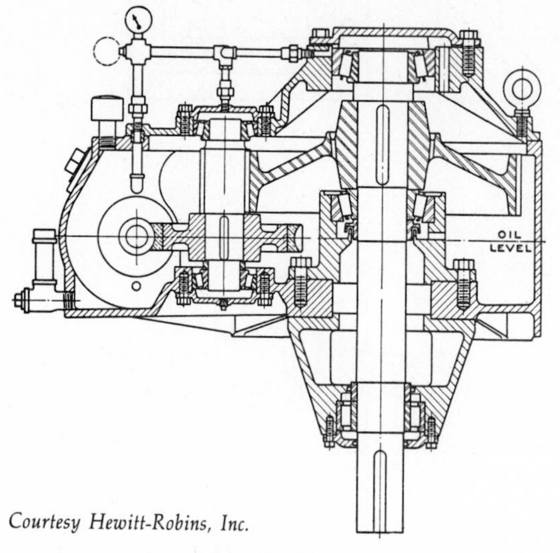


*Courtesy Hewitt-Robins, Inc.*

Fig. 14.43 **Assembly Section.**

## 14.23 Working-Drawing Assembly.

A *working-drawing assembly,* Fig. 14.44, is a combined detail and assembly drawing. Such drawings are often used in place of separate detail and assembly drawings when the assembly is simple enough for all of its parts to be shown clearly in the single drawing. In some cases, all but one or two parts can be drawn and dimensioned clearly in the assembly drawing, in which event these parts are detailed separately on the same sheet. This type of drawing is common in valve drawings, locomotive subassemblies, aircraft subassemblies, and in drawings of jigs and fixtures.

## 14.24 Installation Assemblies.

An assembly made specifically to show how to install or erect a machine or structure is an *installation assembly.* This type of drawing is also often called an *outline assembly,* because it shows only the outlines and the relationships of exterior surfaces. A typical installation assembly is

Fig. 14.44 **Working-Drawing Assembly of Drill Jig.**

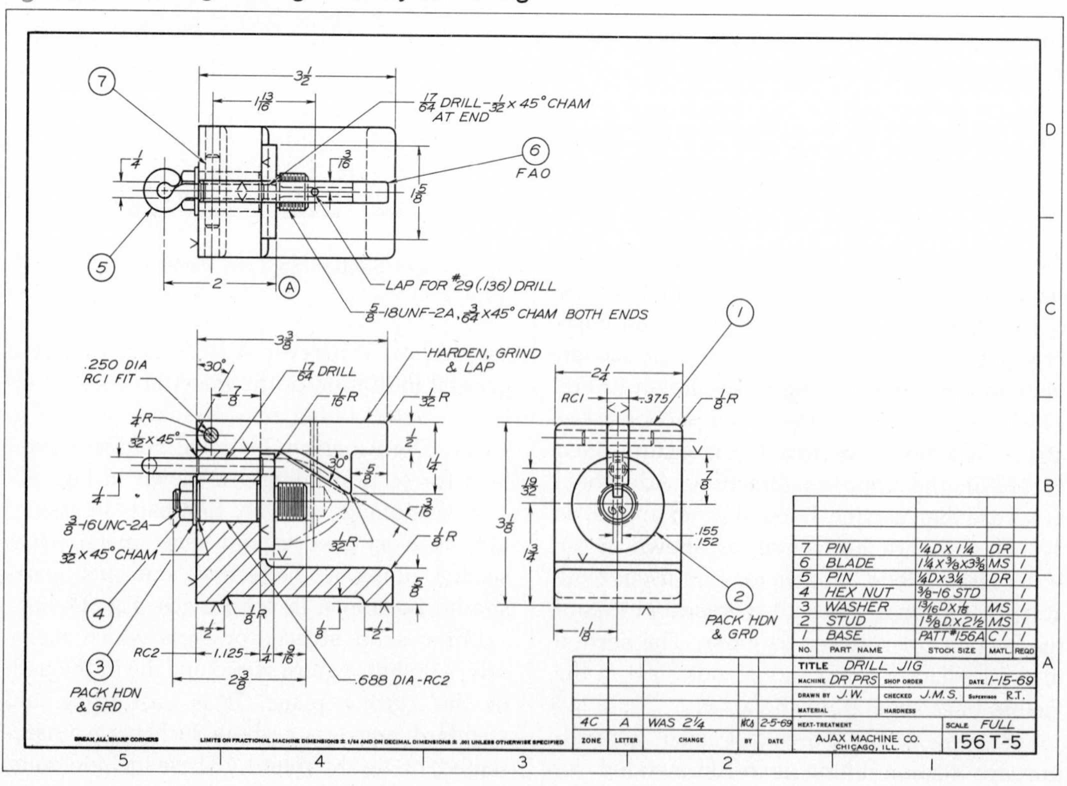

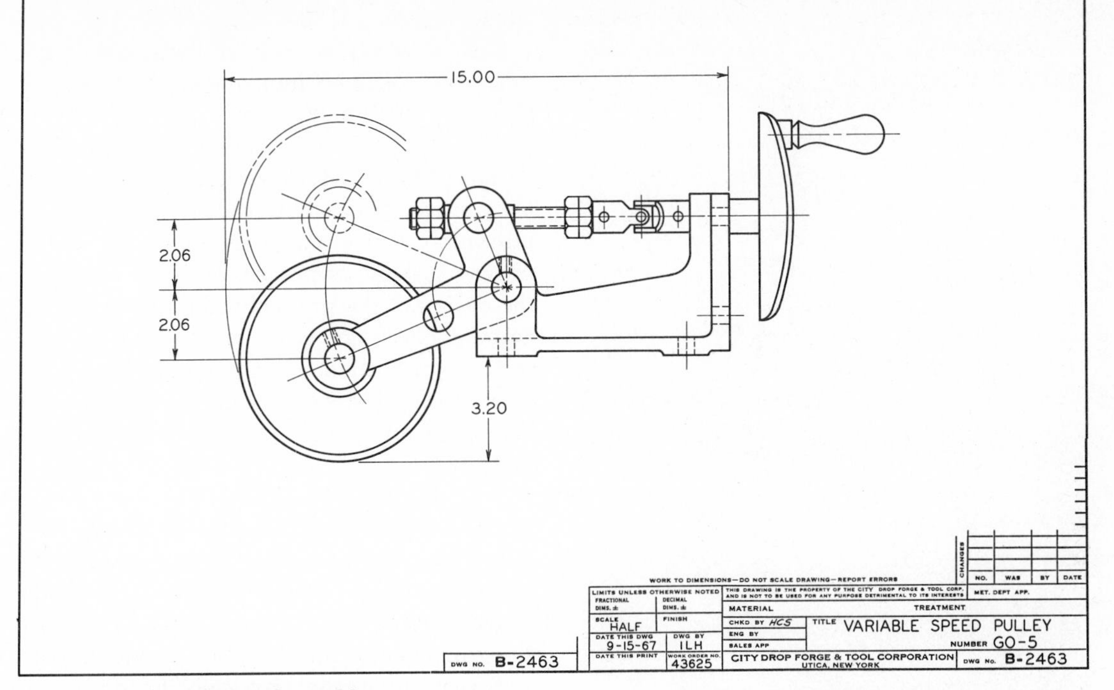


Fig. 14.45 **Installation Assembly.**

shown in Fig. 14.45. In aircraft drafting, an installation drawing (assembly) gives complete information for placing details or subassemblies in their final positions in the airplane.

## 14.25 Check Assemblies.

After all detail drawings of a unit have been made, it may be necessary to make a *check assembly,* especially if a number of changes were made in the details. Such an assembly is drawn accurately to scale in order to check graphically the correctness of the details and their relationship in assembly. After the check assembly has served its purpose, it may be converted into a general assembly drawing.

## 14.26 Patent Drawings.

The patent application for a machine or device must include drawings to illustrate and explain the invention. It is essential that all patent drawings be mechanically correct and constitute complete illustrations of every feature of the invention claimed. The strict requirements of the U.S. Patent Office in this respect serve to facilitate the examination of applications and the interpretation of patents issued thereon. A typical patent drawing is shown in Fig. 14.46.

The drawings for patent applications are pictorial and explanatory in nature; hence they are not detailed as are working drawings for production purposes. Center lines, dimensions, notes, and so forth are omitted. Views, features, and parts, for example, are identified by numbers which refer to the descriptions and explanations given in the specification section of the patent application.

Patent drawings are made with India ink on heavy, smooth, white paper, exactly 10″ × 15″ with 1″ borders on all sides. A space of not less than $1\frac{1}{4}$″ from the shorter border, which is the top of the drawing, is left blank for the heading of title, name, number, and other data to be added by the Patent Office.

All lines must be solid black and suitable for reproduction at a smaller size. Line shading is used whenever it improves readability.

The drawings must contain as many figures as necessary to show the invention clearly.

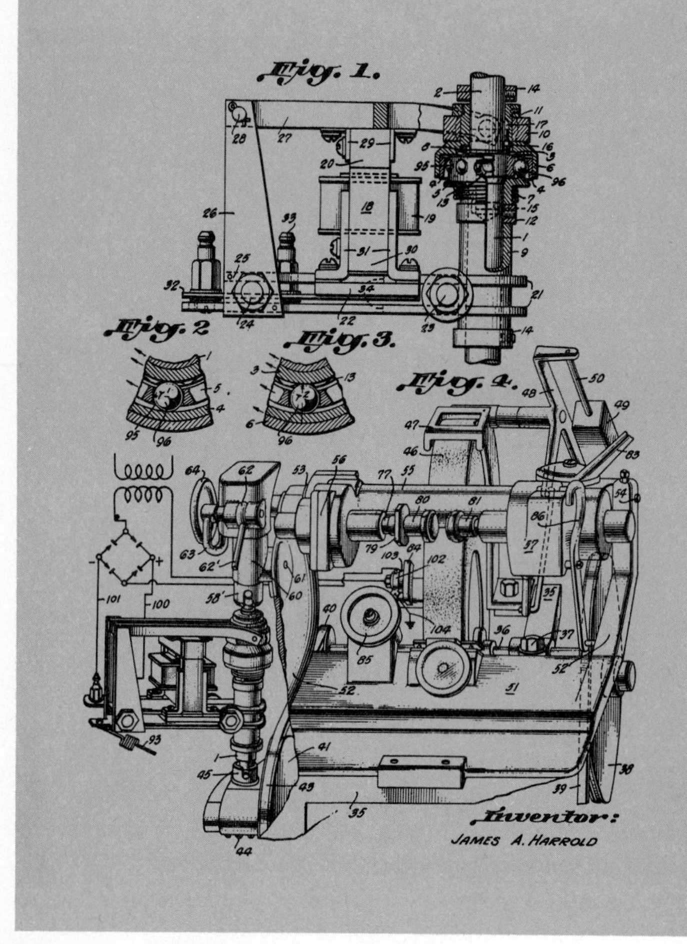


Fig. 14.46 **A Well-Executed Patent Drawing.**

There is no restriction on the number of sheets. The figures may be plan, elevation, section, pictorial, and detail views of portions or elements, and they may be drawn to an enlarged scale if necessary. The required signatures must be placed in the lower right-hand corner of the drawing, either inside or outside the border line.

Because of the strict requirements of the Patent Office, applicants are advised to employ competent draftsmen to make their drawings. To aid draftsmen in the preparation of drawings for submission in patent applications, the *Guide for Patent Draftsmen* was prepared and can be obtained from the Superintendent of Documents, U.S. Government Printing Office, Washington, D.C. 20402, at a cost of 15 cents per copy.

## 14.27 Design and Working Drawing Problems.

Design Problems. The following suggestions for project assignments are of a general and very broad nature and it is expected that they will help generate many ideas for specific design projects. Much design work attempts to improve an existing product or system by utilization of new materials, new techniques, or new systems or procedures. In addition to the design of the product itself, another large amount of design work is essential for the tooling, production, and handling of the product. Each student is encouraged to discuss with his instructor any ideas he may have for a project.

1. Design new or improved playground, recreational, or sporting equipment. For example, a new child's toy could be both recreational and educational.
2. Design new or improved health equipment. For example, the physically handicapped have need for special equipment.
3. Design security or safety devices. Fire, theft, or poisonous gases are a threat to life and property.
4. Design devices and/or systems for waste handling. Home and factory waste disposal needs serious consideration.
5. Design new or improved educational equipment. Both teacher and student would welcome more efficient educational aids.
6. Design improvements in our land, sea, and air transportation systems. Vehicles, controls, highways, and airports need further refinement.
7. Design new or improved devices for material handling. A dispensing device for a powdered product is an example.
8. Improve the design of an existing device or system.

The individual student design solution or a solution formulated by a group to a design problem should be in the form of a *report,* which should be typed or carefully lettered, assembled, and bound. Suitable folders or binders are usually available at the local school supply store. It is suggested that the report contain the following (or variations of them, as specified by your instructor):

1. A title sheet. The title of the design project should be placed in approximately the center of the sheet and in the lower right-hand corner place your name or the names of those in the group. The symbol PL should follow the name of the project leader.
2. Table of contents with page numbers.
3. Statement of the purpose of project with appropriate comments.
4. Preliminary design sketches, with comments on advantages and disadvantages of each, leading to the final selection of the *best* solution. All work should be signed and dated.
5. An accurately made pictorial and/or assembly drawing(s), if more than one part is involved in the design.
6. Detail working drawings, freehand or mechanical as assigned. The $8\frac{1}{2}'' \times 11''$ sheet size is preferred for convenient insertion in the report. Larger sizes may be bound in the report with appropriate folding.
7. A bibliography or credit for important sources of information, if applicable.

WORKING DRAWING PROBLEMS. The problems in Figs. 14.47 to 14.122 are presented to provide practice in making regular working drawings of the type used in industry. Many problems, especially those of the assemblies, offer an excellent opportunity for the student to exercise his ability to redesign or improve upon the existing design. Owing to the variations in sizes and in scales that may be used, the student is required to select his own sheet sizes and scales when these are not specified, subject to the approval of the instructor. Standard sheet layouts are shown inside the back cover of this book.

The statements for each problem are intentionally brief, so that the instructor may amplify or vary the requirements when making assignments. Many problems lend themselves to the complete decimal system or the metric system, while others are more suitable for a combination of fractional and decimal dimensions. Either unidirectional or aligned dimensioning may be assigned.

Some tracings in ink may be assigned, but this is left to the discretion of the instructor.

The student should clearly understand that in problems presented in pictorial form, the placement of dimensions and finish marks cannot always be followed in the drawing. *The dimensions given are in most cases those needed to make the parts, but owing to the limitations of pictorial drawing they are not in all cases the dimensions that should be shown on the working drawing.* In the pictorial problems the rough and finished surfaces are shown, but finish marks are usually omitted. The student should add all necessary finish marks. The student should place all dimensions in the preferred places in the final drawings.

Each problem should be preceded by a thumbnail sketch or a complete technical sketch, fully dimensioned. Any of the title blocks shown inside the back cover of this book may be used, or the student may design the title block if so assigned by the instructor.

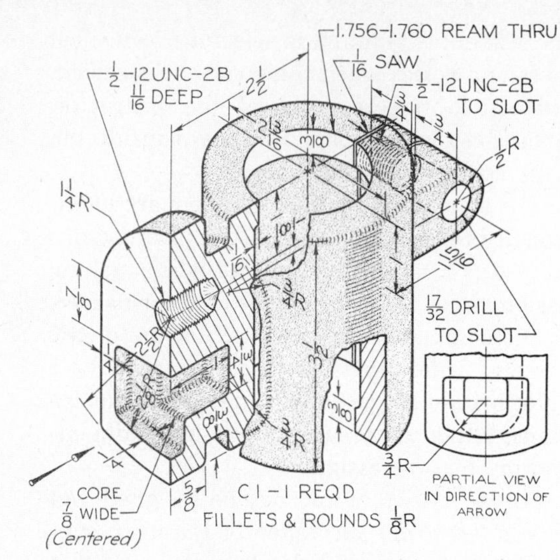


Fig. 14.47 **Table Bracket.** Make detail drawing. Use Size B sheet. If assigned, convert dimensions to decimal or metric system.

17/32 DRILL 5/8-11UNC-2B
45° CHAMFER × DEPTH OF THD
2 HOLES

5/8 DRILL
15/16 C BORE .88 DEEP
2 HOLES

C I – I REQD

ROUNDS .06R

Fig. 14.48 **RH Tool Post.** Make detail drawing. Use Size B sheet.

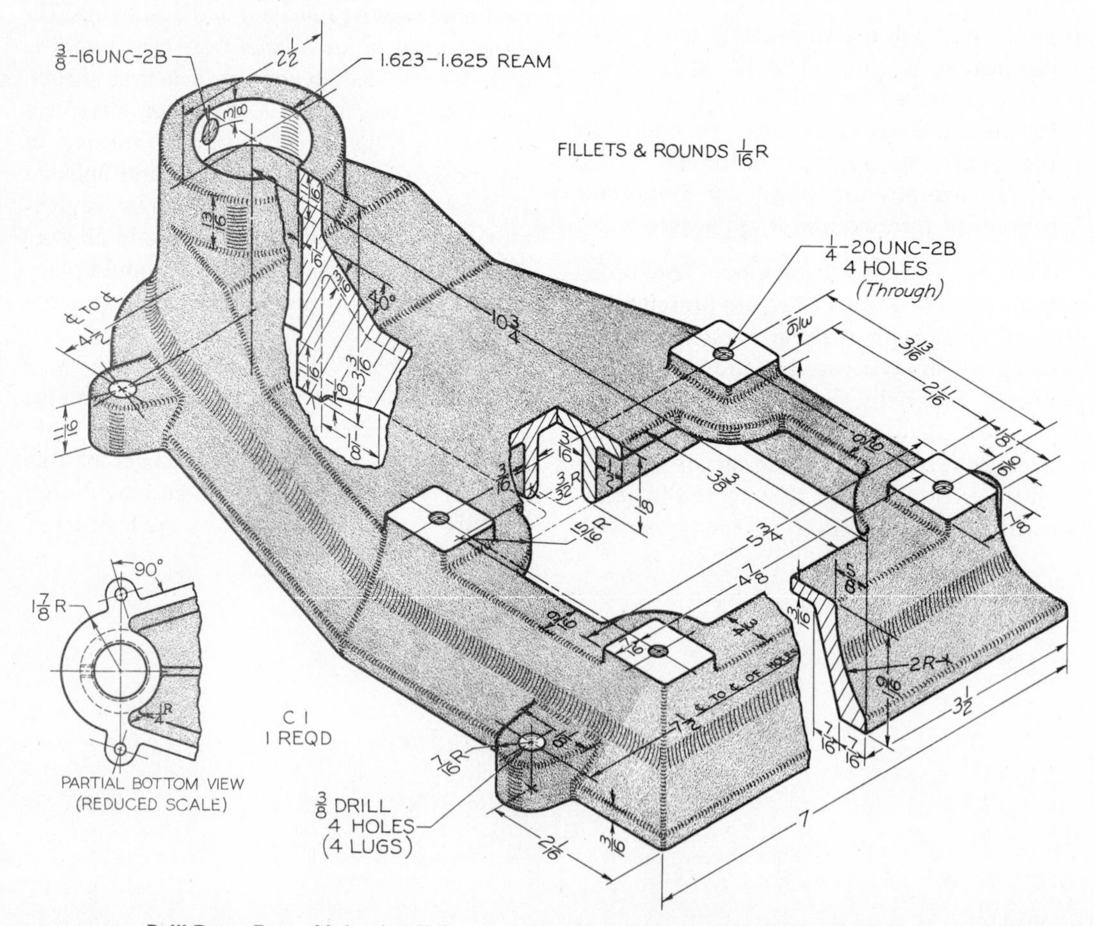


Fig. 14.49 **Drill Press Base.** Make detail drawing. Use Size C sheet. Use unidirectional two-place decimal dimensions.

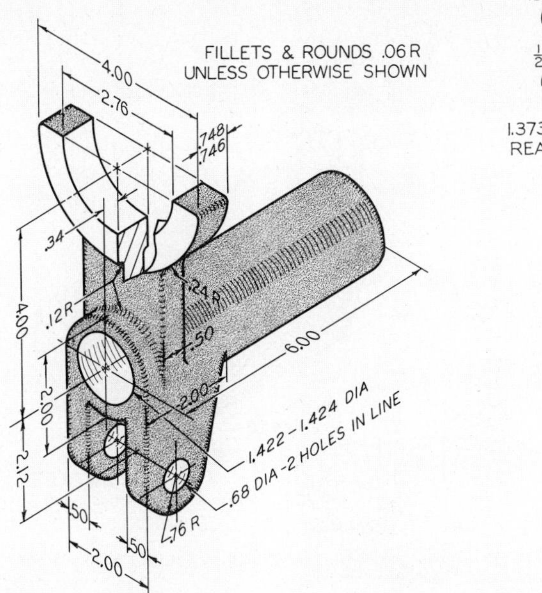


Fig. 14.50 **Shifter Fork.** Make detail drawing. Use Size B sheet.

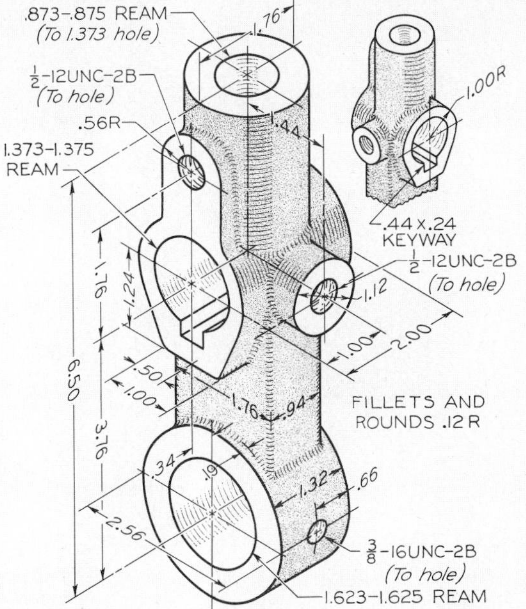


Fig. 14.51 **Idler Arm.** Make detail drawing. Use Size B sheet.

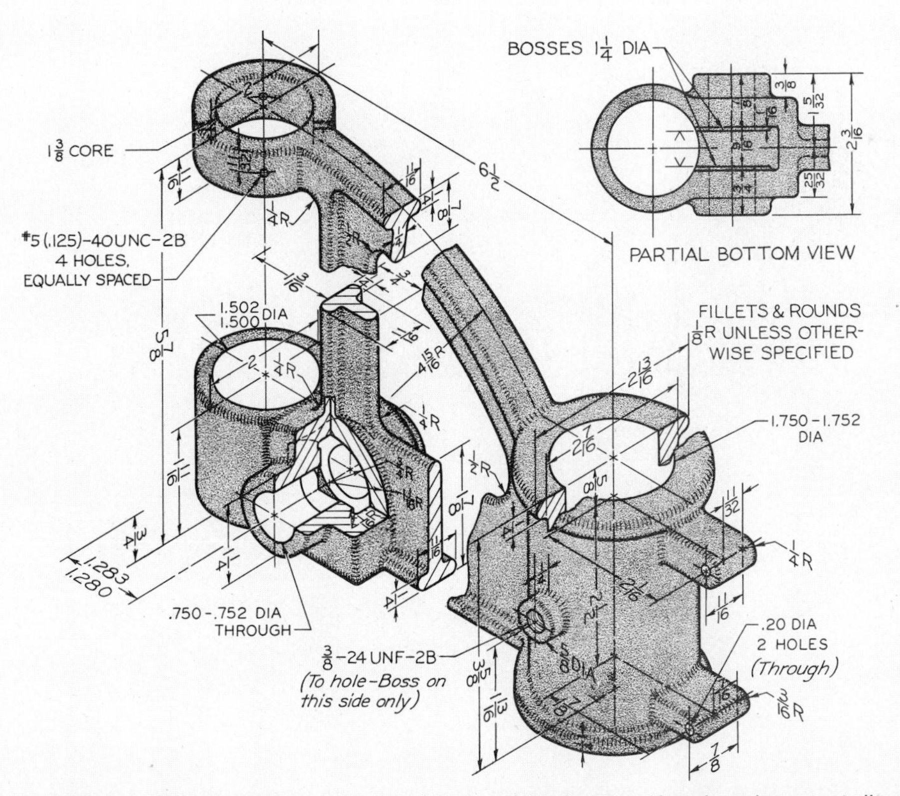


Fig. 14.52 **Drill Press Bracket.** Make detail drawing. Use Size C sheet. If assigned, convert dimensions to decimal or metric system.

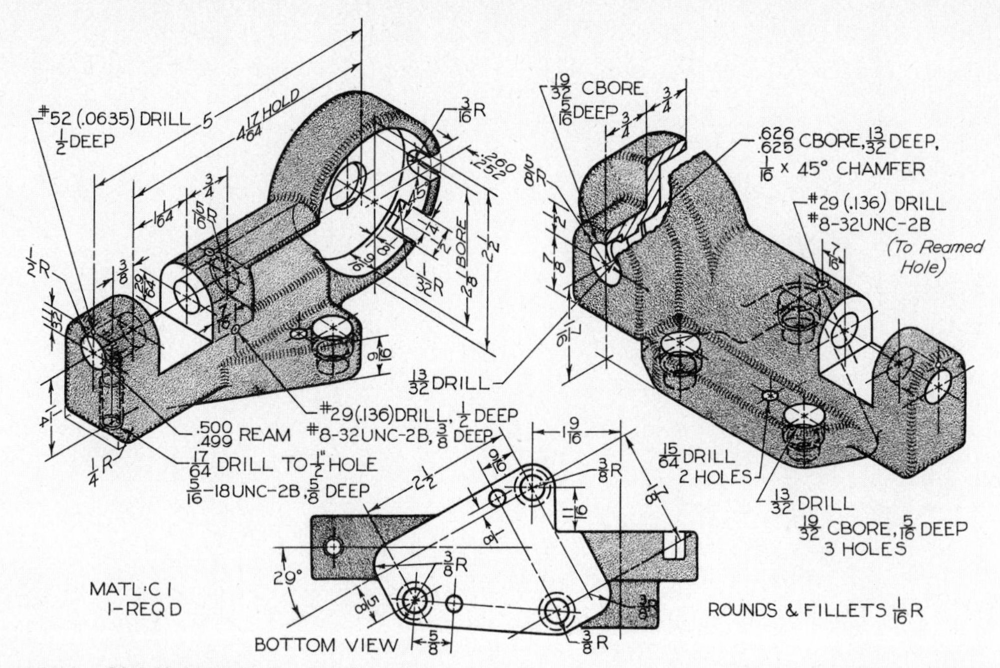


Fig. 14.53 **Dial Holder.** Make detail drawing. Use Size C sheet. If assigned, convert dimensions to decimal or metric system.

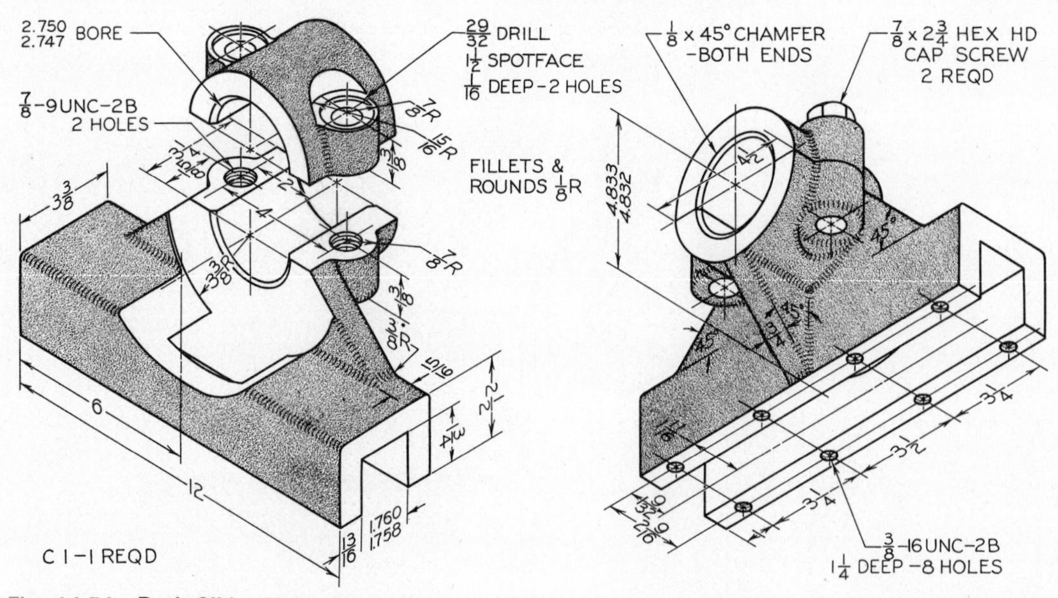


Fig. 14.54 **Rack Slide.** Make detail drawings. Draw half size on Size B sheet. If assigned, convert dimensions to decimal or metric system.

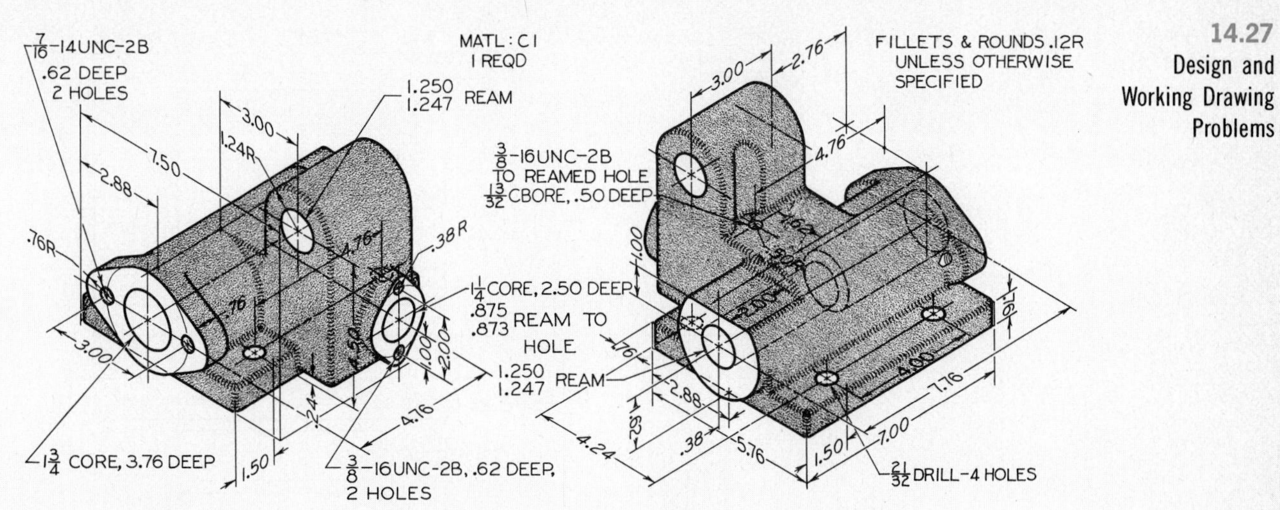


Fig. 14.55 **Automatic Stop Box.** Make detail drawing. Draw half size on Size B sheet.

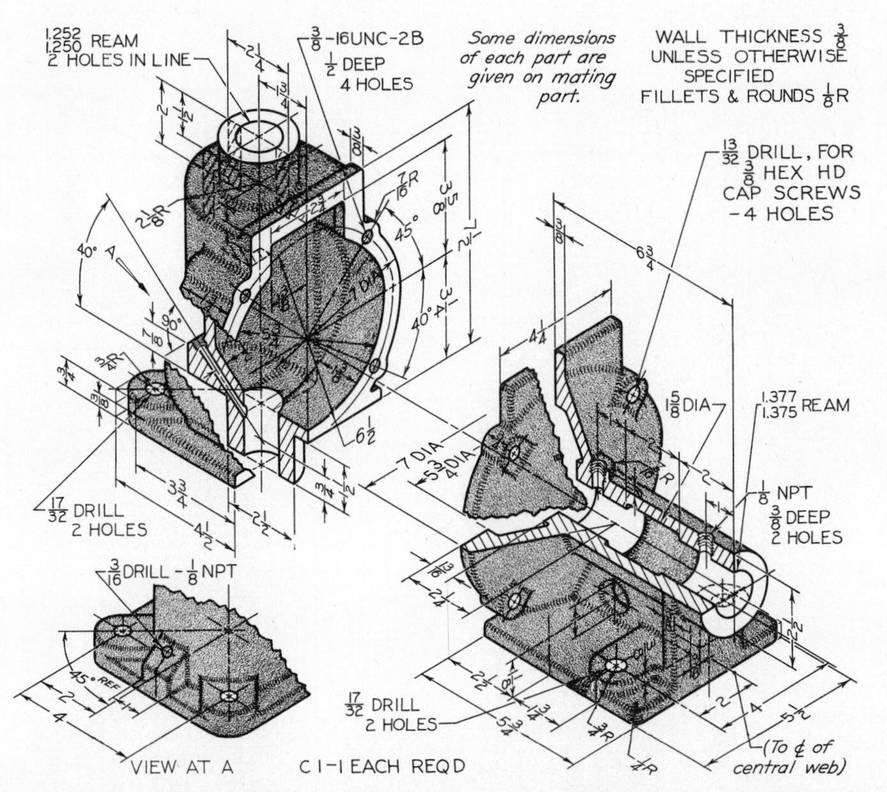


Fig. 14.56 **Conveyor Housing.** Make detail drawings. Draw half size on Size C sheets. If assigned, convert dimensions to decimal or metric system.

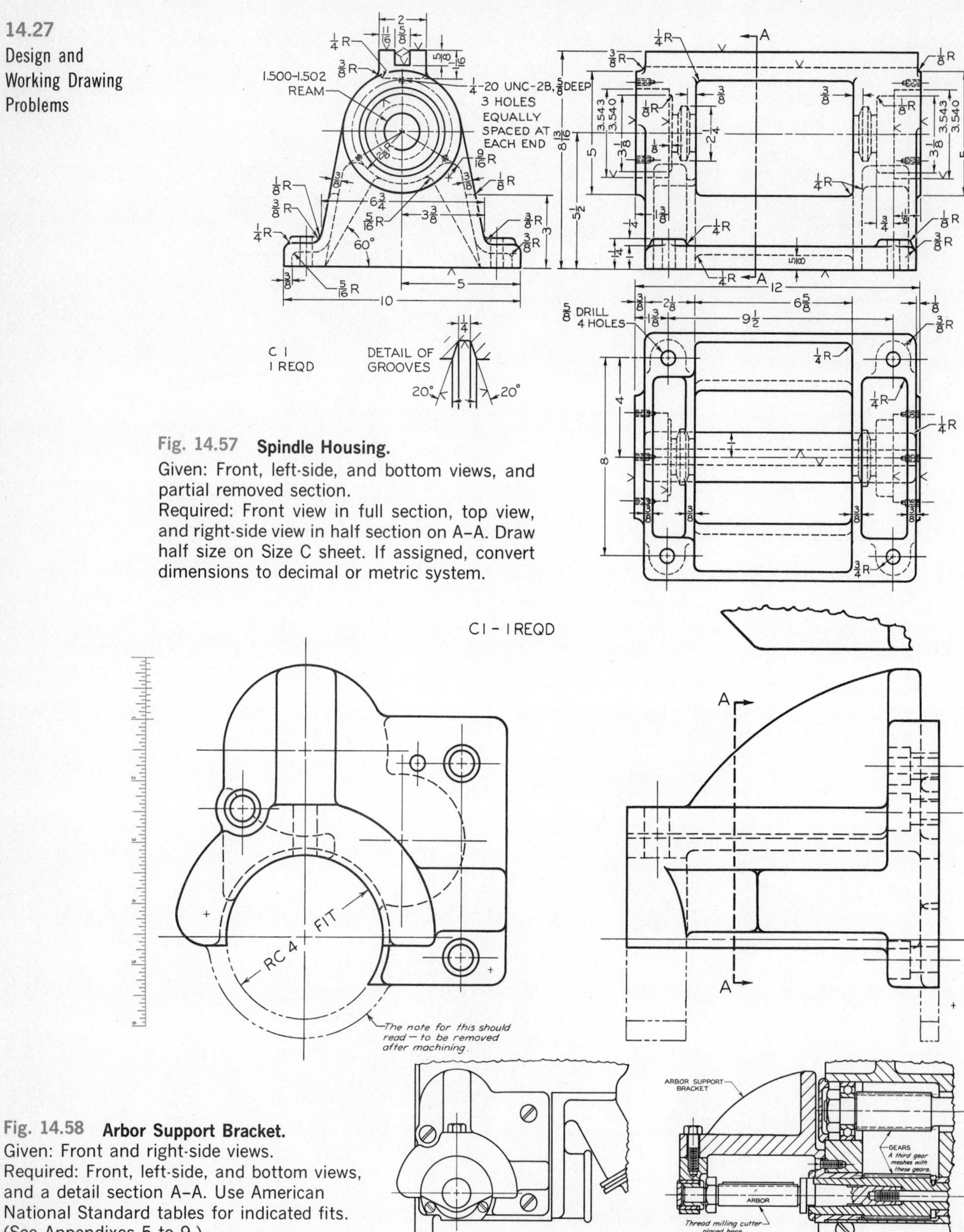


Fig. 14.57 **Spindle Housing.**
Given: Front, left-side, and bottom views, and partial removed section.
Required: Front view in full section, top view, and right-side view in half section on A–A. Draw half size on Size C sheet. If assigned, convert dimensions to decimal or metric system.

Fig. 14.58 **Arbor Support Bracket.**
Given: Front and right-side views.
Required: Front, left-side, and bottom views, and a detail section A–A. Use American National Standard tables for indicated fits. (See Appendixes 5 to 9.)

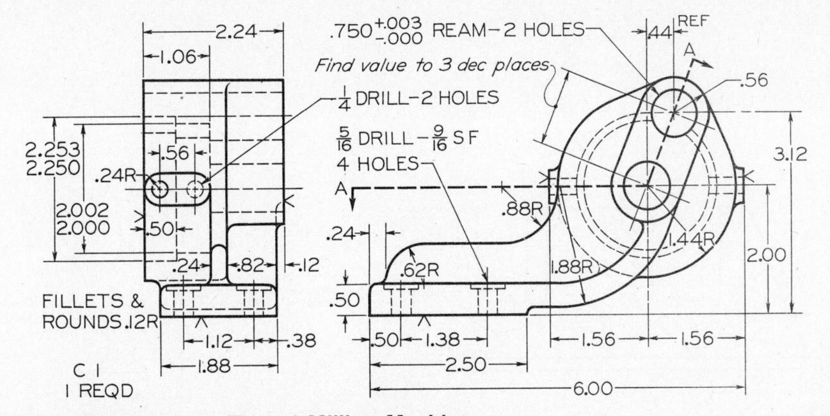


Fig. 14.59 **Pump Bracket for a Thread Milling Machine.**
Given: Front and left-side views.
Required: Front and right-side views, and top view in section on A–A. Draw full size on Size B sheet.

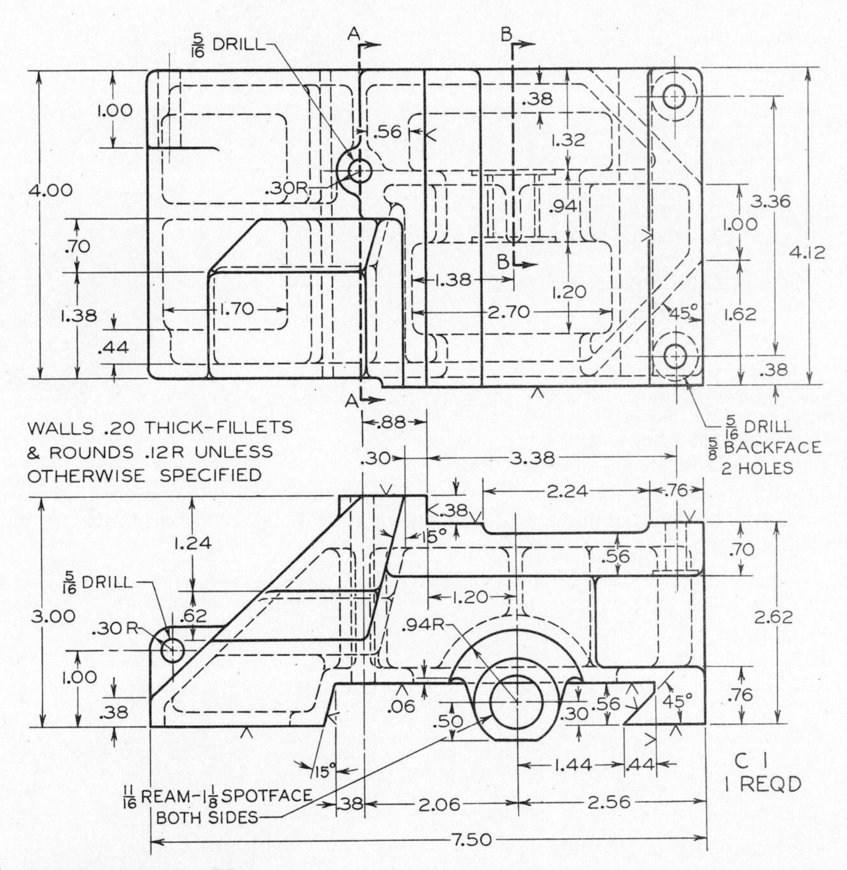


Fig. 14.60 **Support Base for Planer.**
Given: Front and top views.
Required: Front and top views, left-side view in full section A–A, and removed section B–B. Draw full size on Size C sheet.

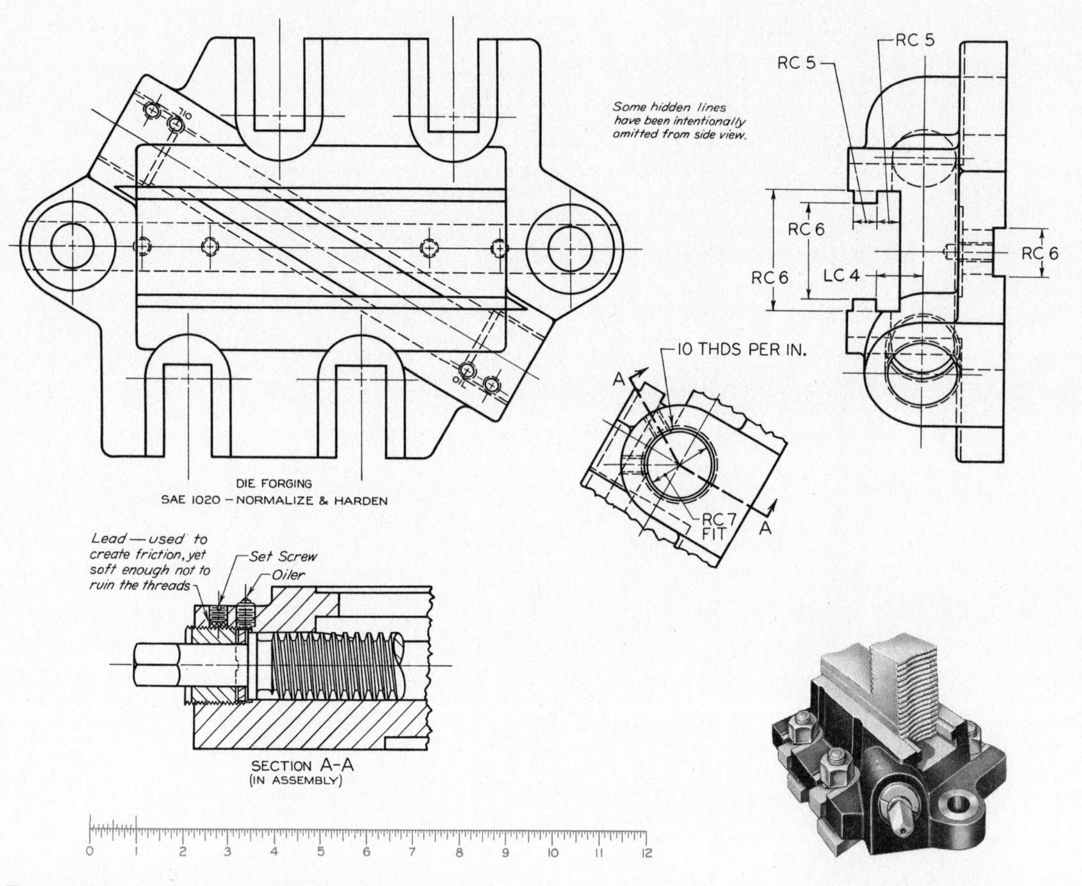


Fig. 14.61 **Jaw Base for Chuck Jaw.**
Given: Top, right-side, and partial auxiliary views.
Required: Top, left-side (beside top), front, and partial auxiliary views complete with dimensions. If assigned, use decimal or metric dimensions. Use American National Standard tables for indicated fits. (See Appendixes 5 to 9.)

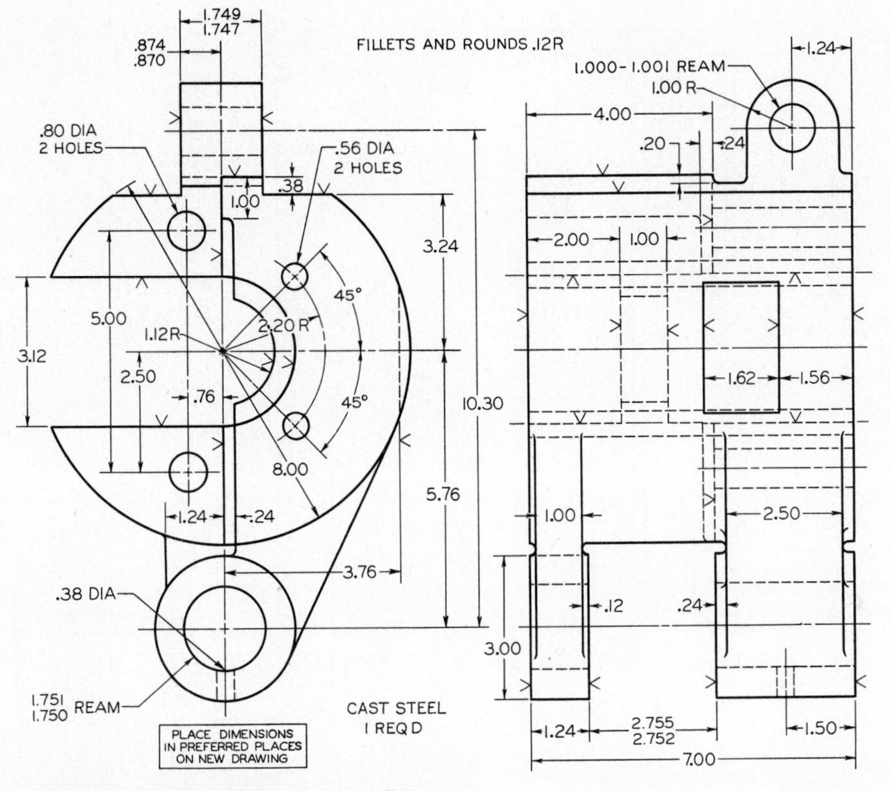


Fig. 14.62 **Fixture Base for 60-Ton Vertical Press.**
Given: Front and right-side views.
Required: Revolve front view 90° clockwise; then add top and left-side views. Draw half size on Size C sheet.

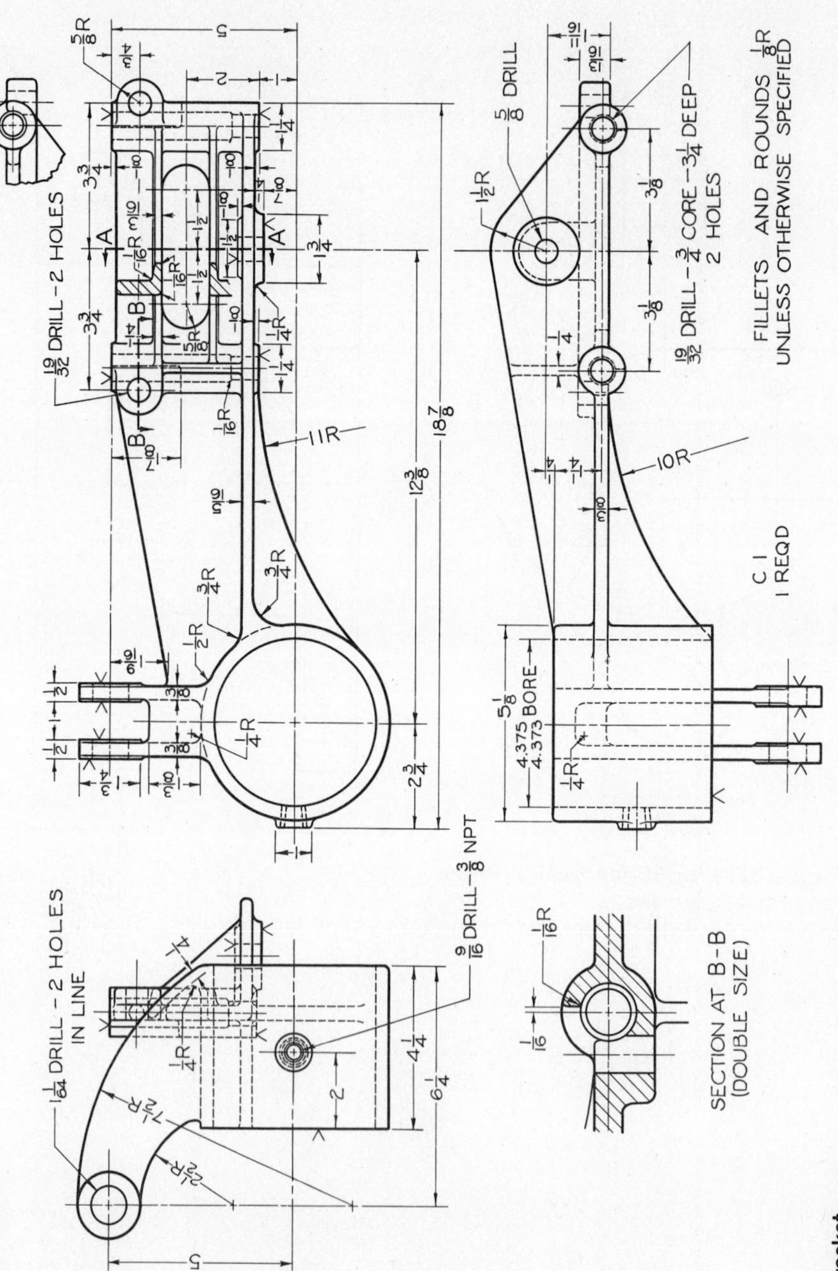


**Fig. 14.63 Bracket.**
Given: Front, left-side, and bottom views, and partial removed section.
Required: Make detail drawing. Draw front, top, and right-side views, and removed sections A–A and B–B. Draw half size on Size C sheet. Draw section B–B full size. If assigned, convert dimensions to decimal or metric system.

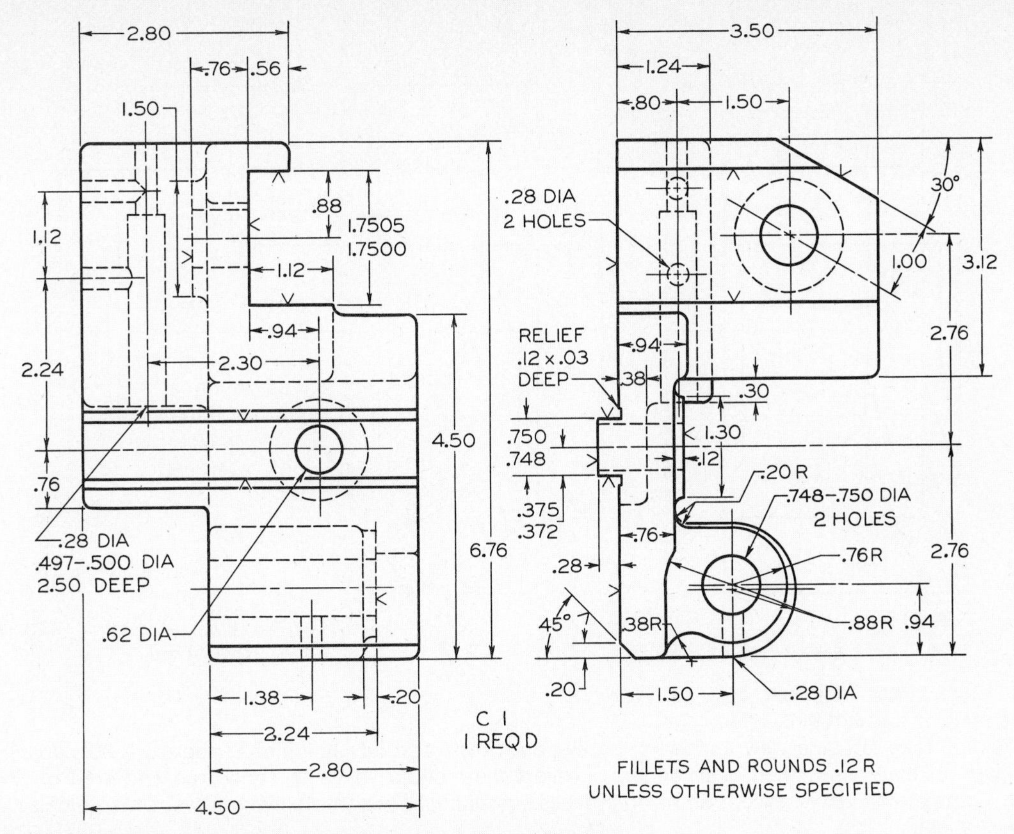


Fig. 14.64 **Roller Rest Bracket for Automatic Screw Machine.**
Given: Front and left-side views.
Required: Revolve front view 90° clockwise; then add top and left-side views. Draw half size on Size C sheet.

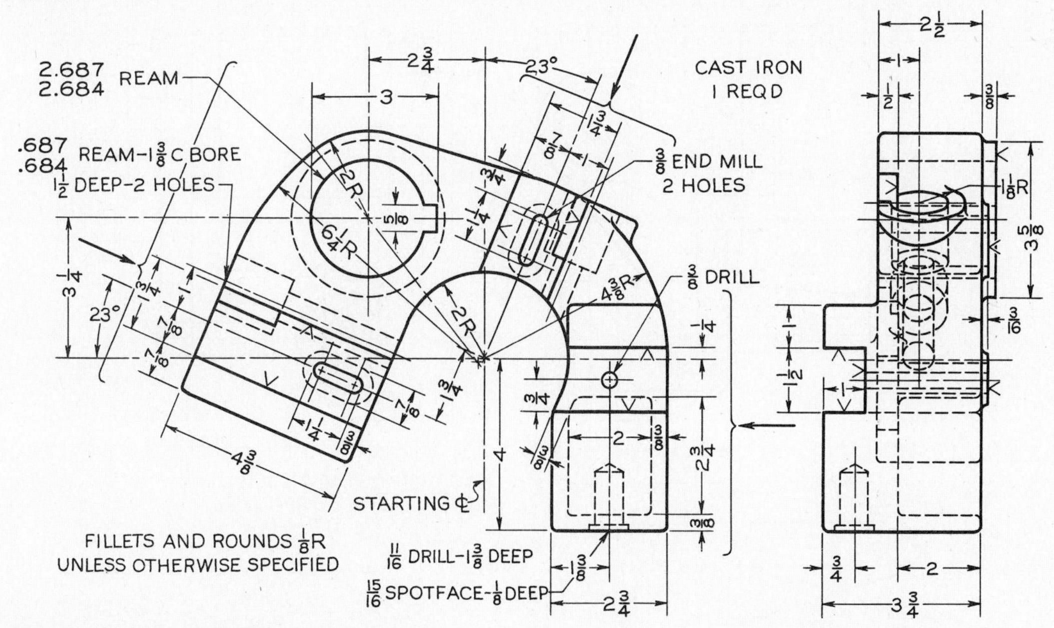


Fig. 14.65 **Guide Bracket for Gear Shaper.**
Given: Front and right-side views.
Required: Front view, a partial right-side view, and two partial auxiliary views taken in direction of arrows. Draw half size on Size C sheet. If assigned, use unidirectional two-place decimals for all fractional dimensions.

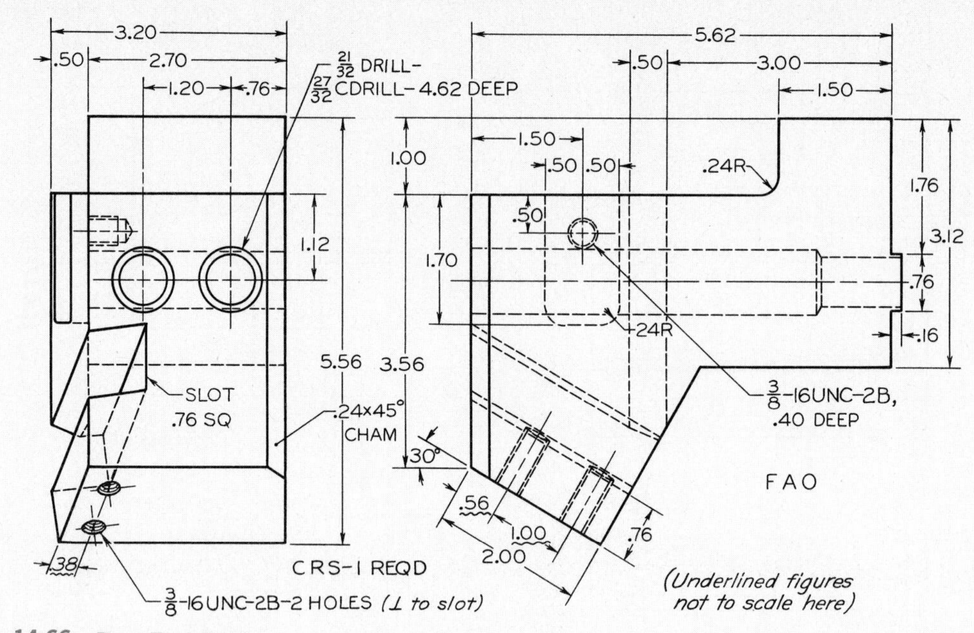


Fig. 14.66 **Rear Tool Post.**
Given: Front and left-side views.
Required: Take left-side view as new top view; add front and left-side views, approx. 8.50″ apart, a primary auxiliary view, then a secondary auxiliary view taken so as to show true end view of .76″ slot. Complete all views, except show only necessary hidden lines in auxiliary views. Draw full size on Size C sheet.

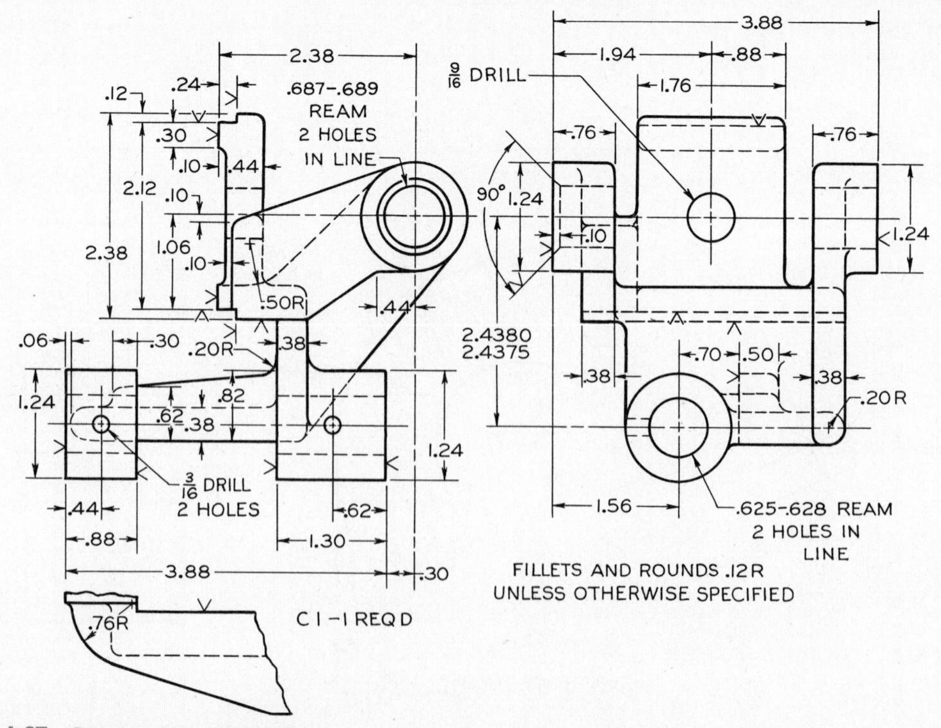


Fig. 14.67 **Bearing for a Worm Gear.**
Given: Front and right-side views.
Required: Front, top, and left-side views. Draw full size on Size C sheet.

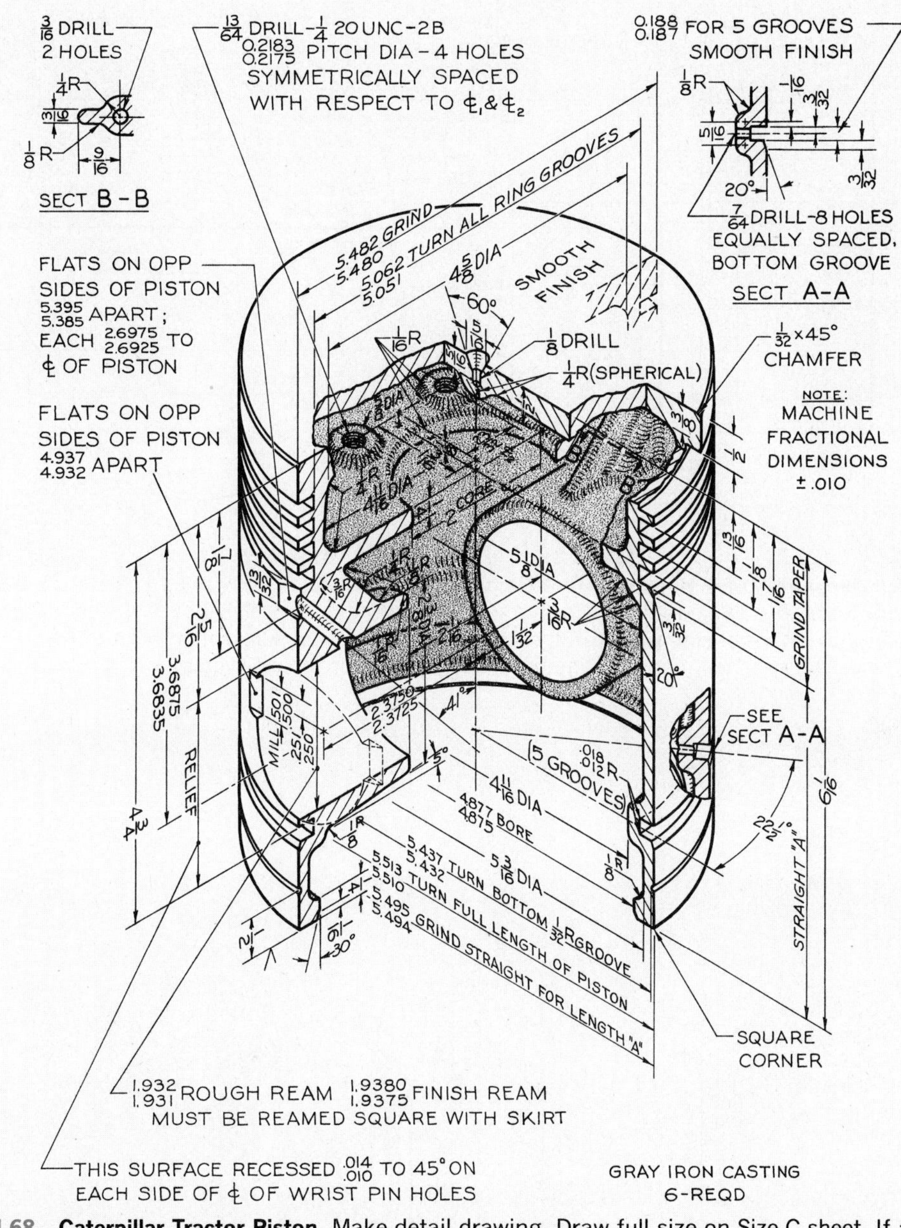


Fig. 14.68 **Caterpillar Tractor Piston.** Make detail drawing. Draw full size on Size C sheet. If assigned, use unidirectional two-place decimals for all fractional dimensions.

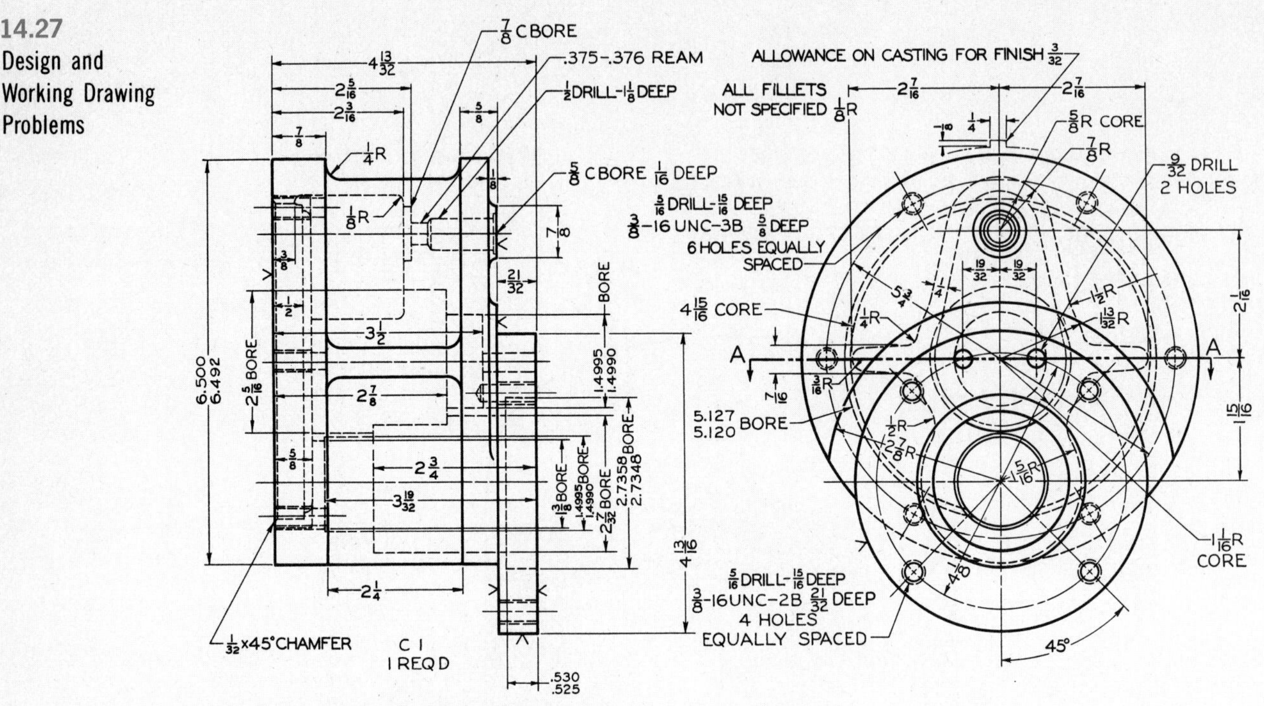


Fig. 14.69 **Generator Drive Housing.**
Given: Front and left-side views.
Required: Front view, right-side view in full section, and top view in full section on A–A. Draw full size on Size C sheet. If assigned, convert dimensions to decimal or metric system.

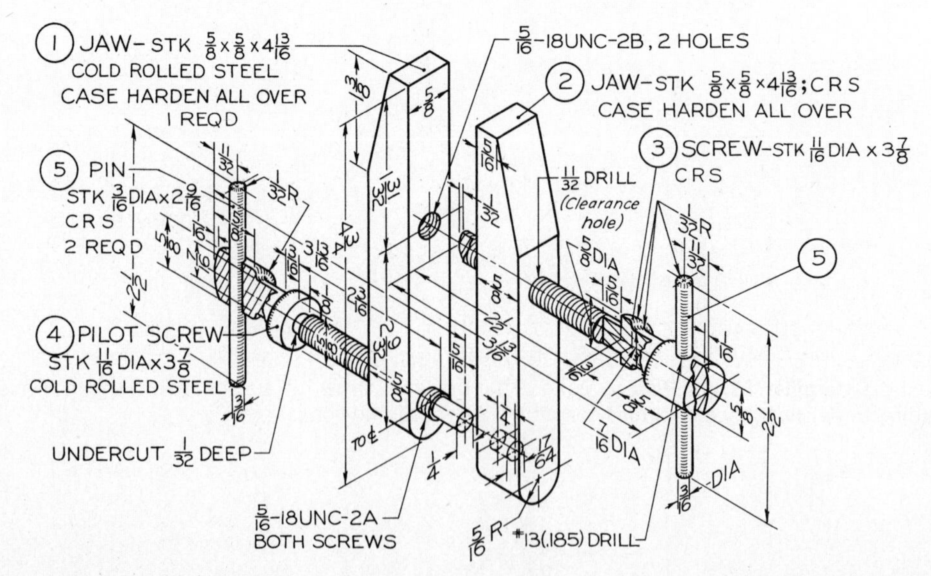


Fig. 14.70 **Machinist's Clamp.** Draw details and assembly. If assigned, use unidirectional two-place decimal dimensions.

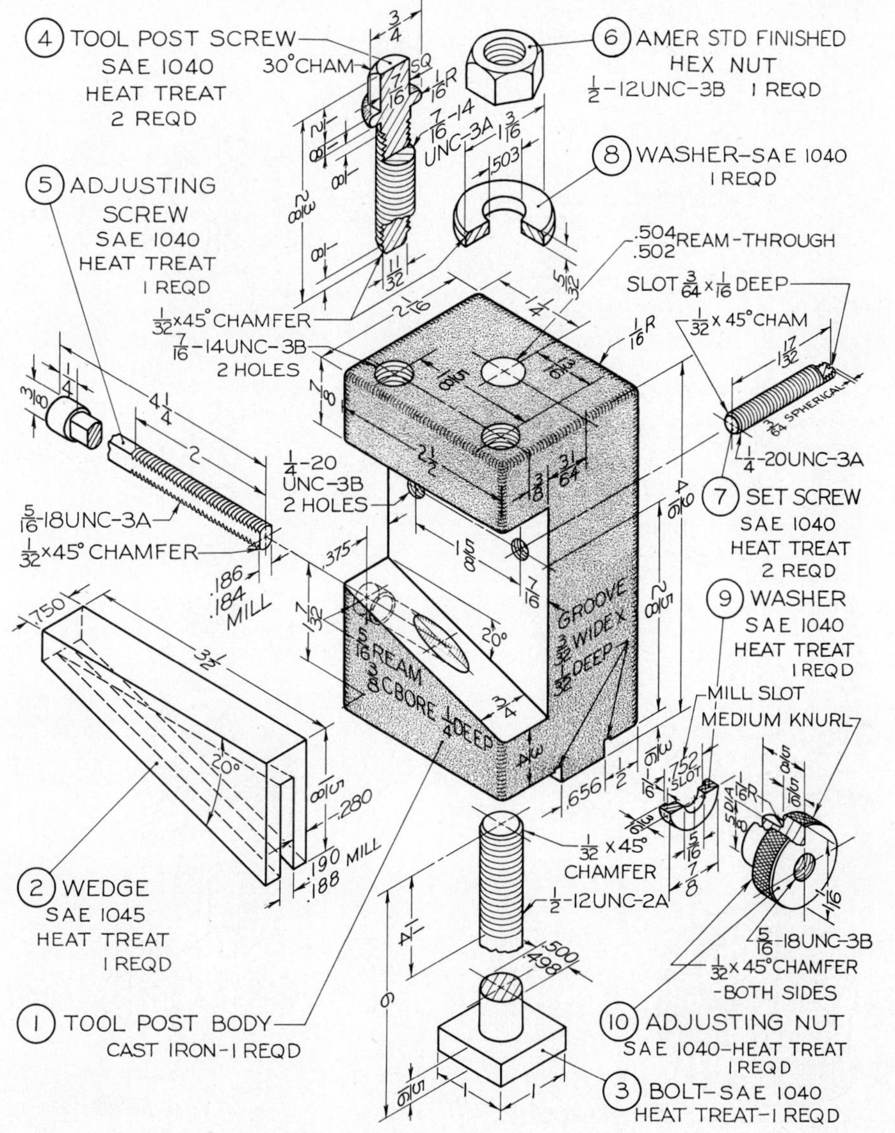


Fig. 14.71 **Tool Post.** (1) Draw details. (2) Draw assembly. If assigned, use unidirectional two-place decimals for all fractional dimensions.

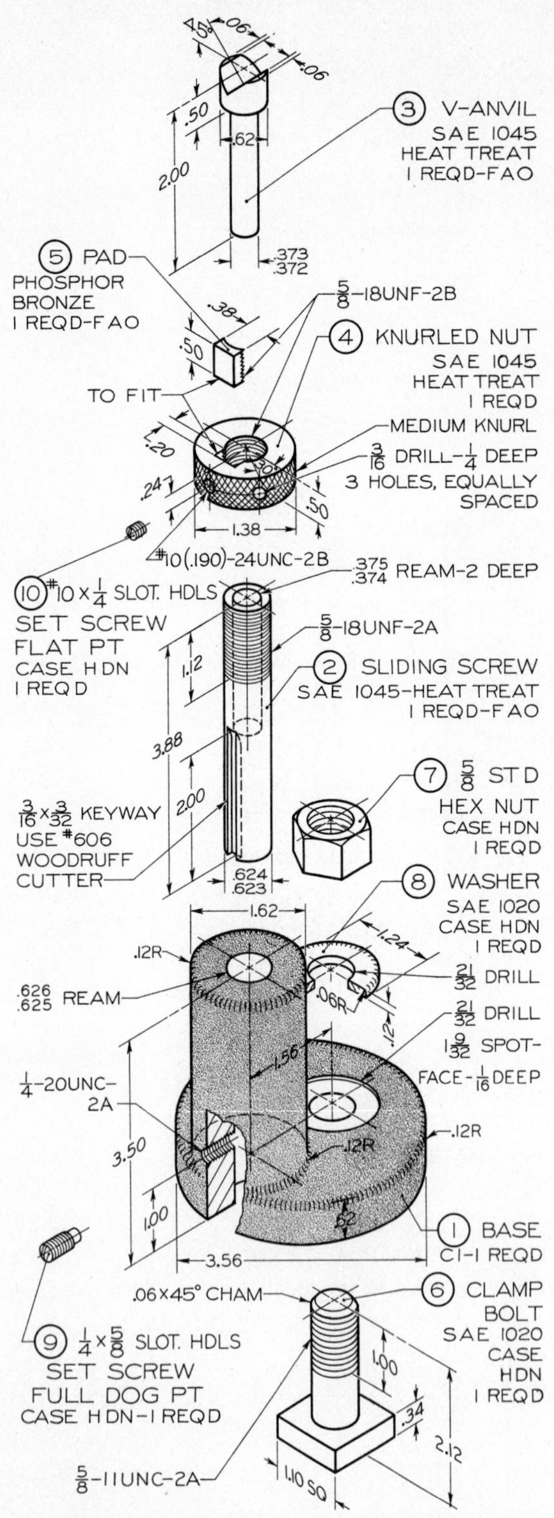

Fig. 14.72 **Milling Jack.** (1) Draw details. (2) Draw assembly.

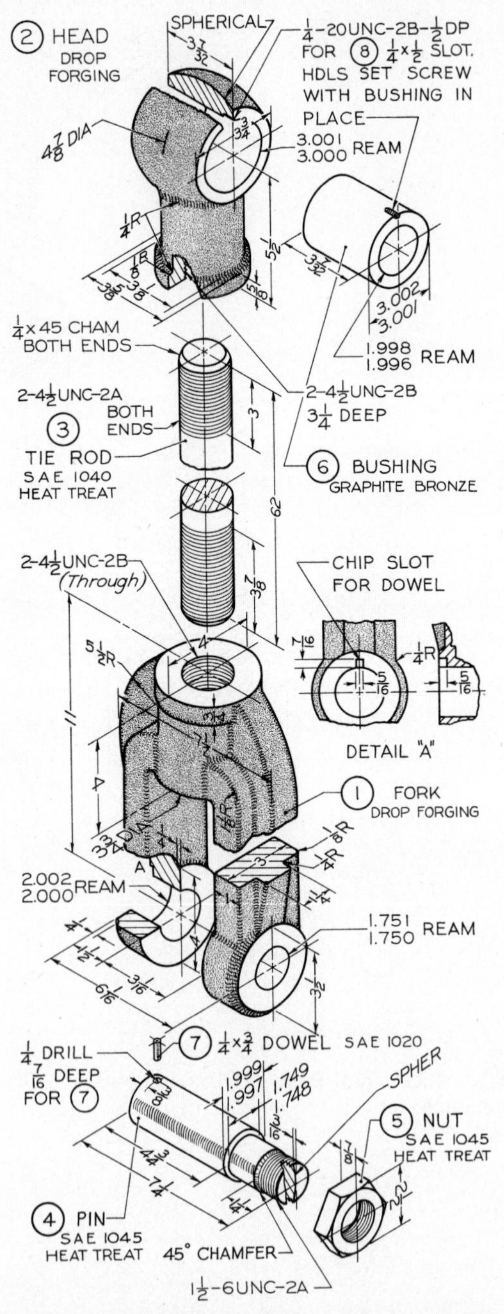

Fig. 14.73 **Connecting Bar.** (1) Draw details. (2) Draw assembly. If assigned, convert dimensions to decimal or metric system.

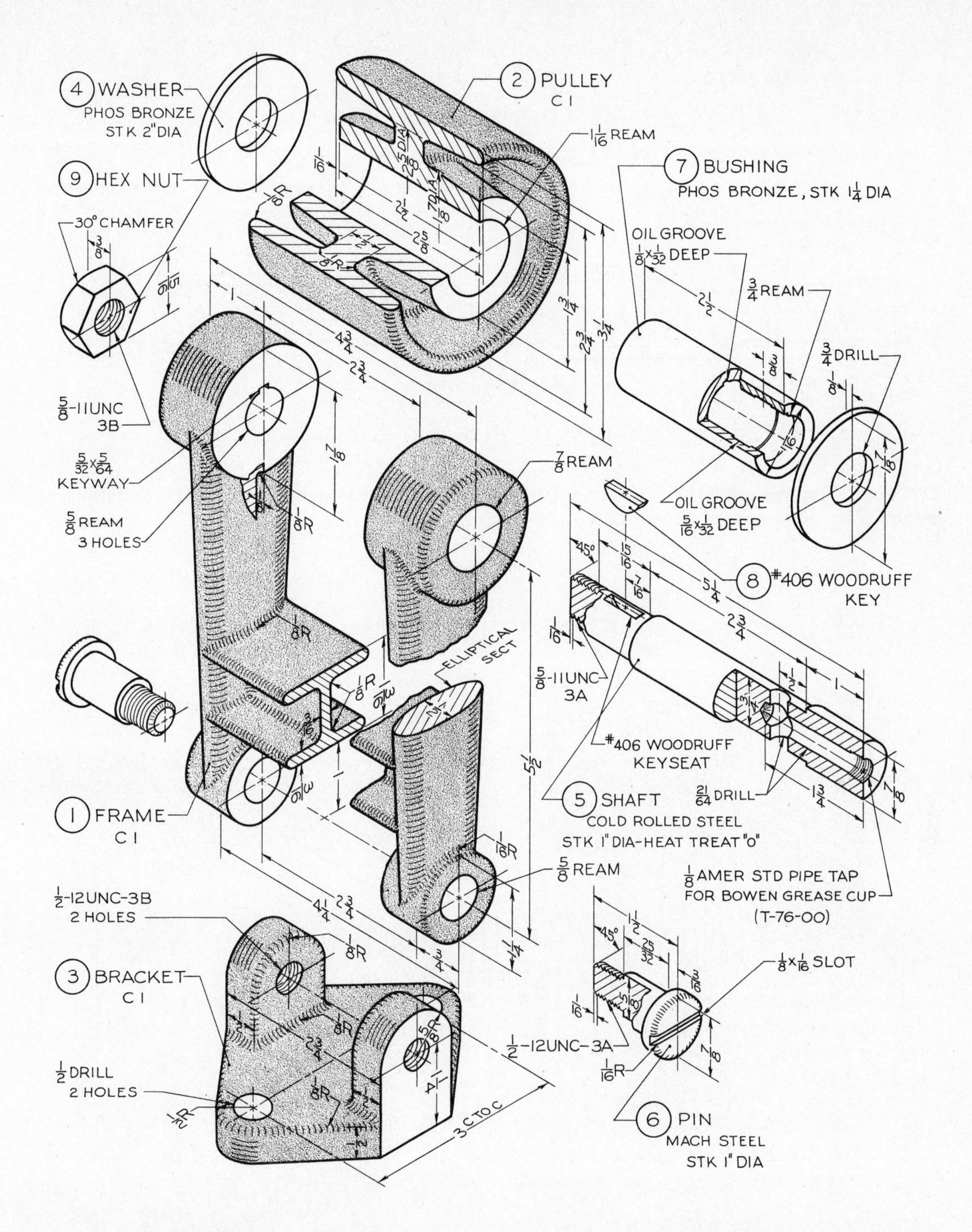


Fig. 14.74 **Belt Tightener.** (1) Draw details. (2) Draw assembly. It is assumed that the parts are to be made in quantity and they are to be dimensioned for interchangeability on the detail drawings. If assigned, convert dimensions to decimal or metric system. Using Tables of Limits in Appendixes 5 to 9, give dimensions as follows:

a. Bushing fit in pulley: Class LN 2 fit.
b. Shaft fit in bushing: Class RC 5 fit.
c. Shaft fits in frame: Class RC 2 fit.
d. Pin fit in frame: Class RC 5 fit.
e. Pulley hub length plus washers fit in frame: Allowance .005 and tolerances .004.
f. Make bushing .010″ shorter than pulley hub.
g. Bracket fit in frame: same as 5 above.

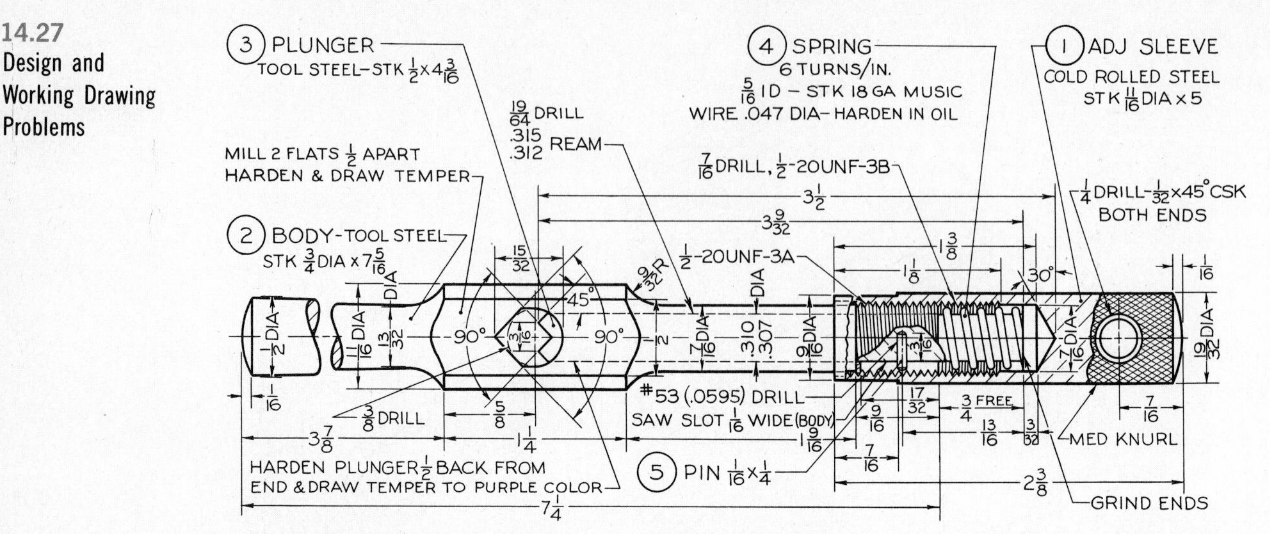

Fig. 14.75 **Tap Wrench.** (1) Draw details. (2) Draw assembly. If assigned, use unidirectional two-place decimals for all fractional dimensions.

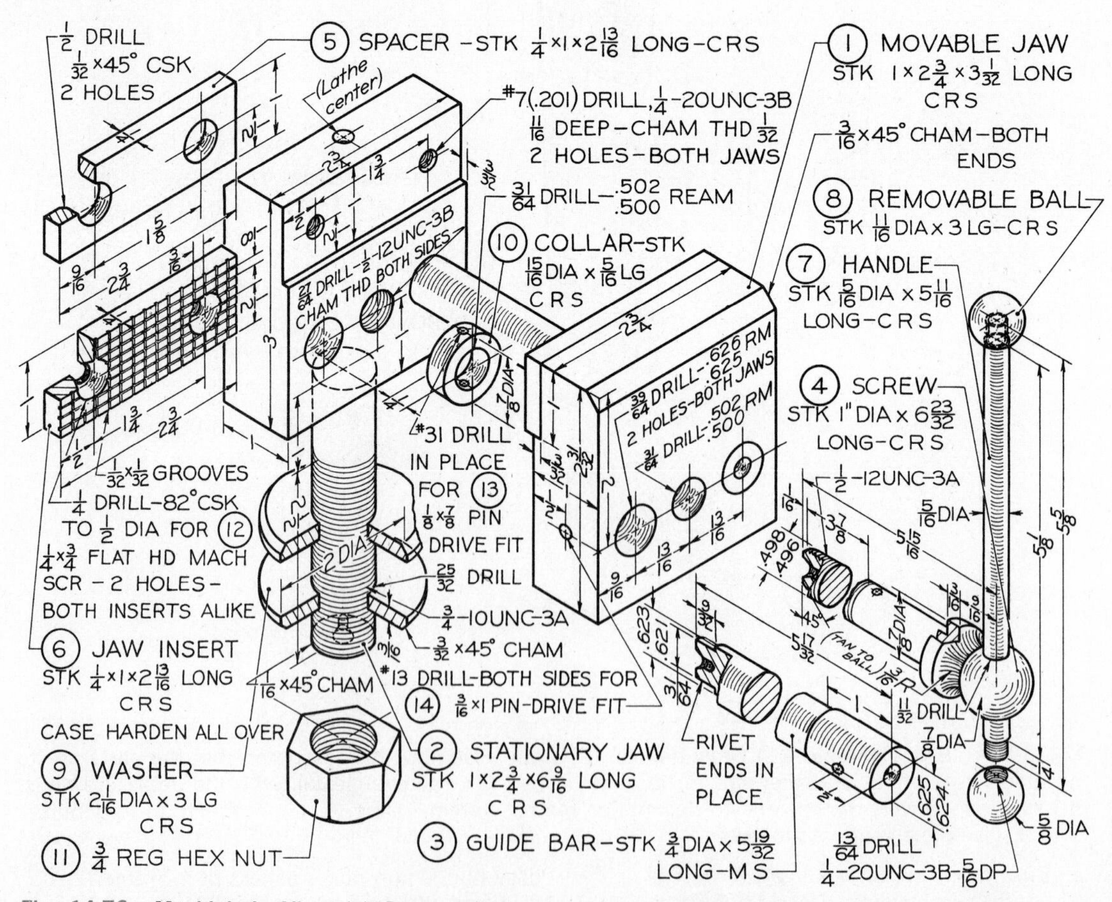

Fig. 14.76 **Machinist's Vise.** (1) Draw details. (2) Draw assembly. If assigned, use unidirectional two-place decimals for all fractional dimensions.

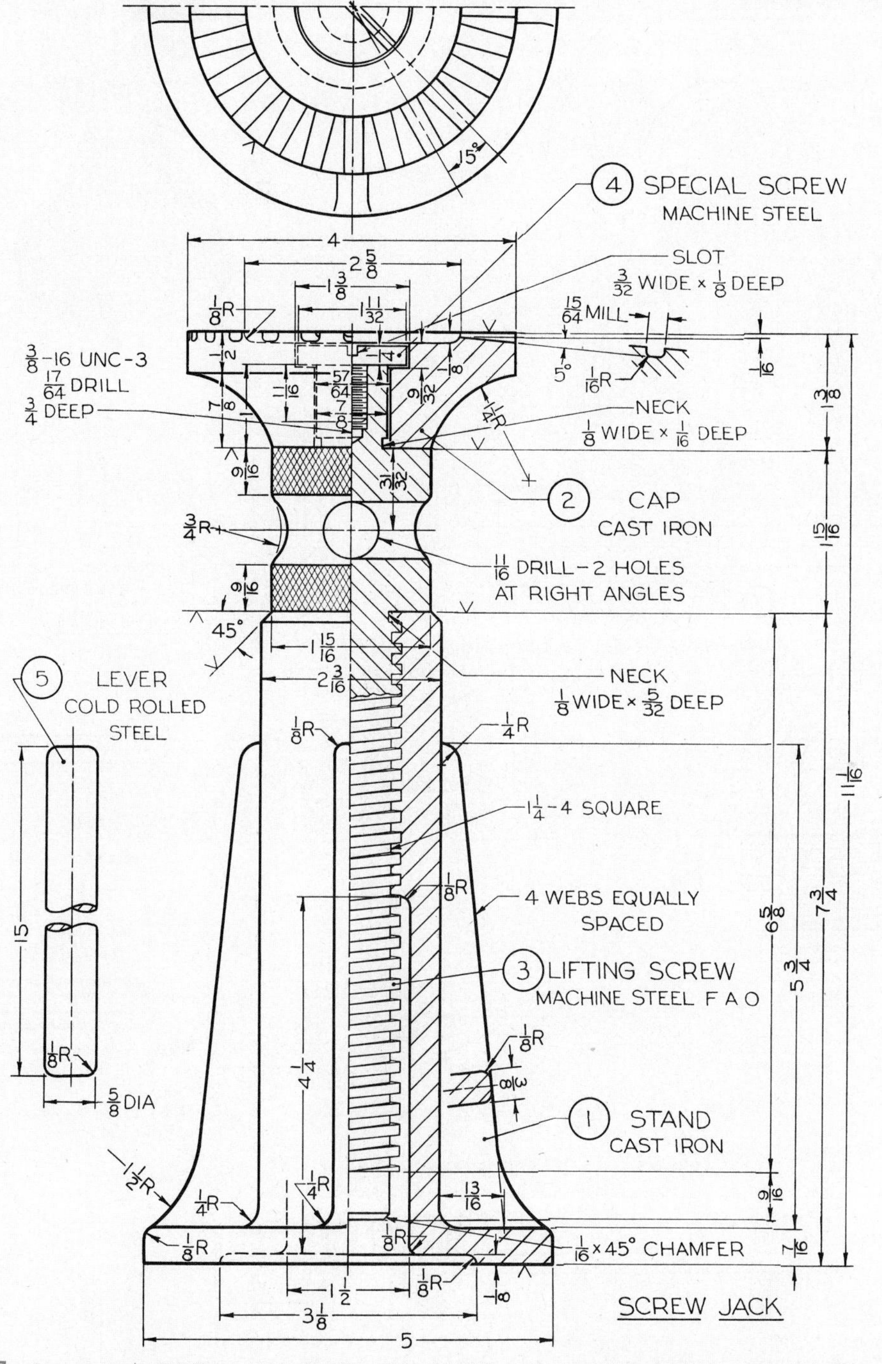


Fig. 14.77 **Screw Jack.** (1) Draw details. (See Fig. 14.23, showing "blocked-in" views on Sheet Layout C-678; see inside back cover.) (2) Draw assembly. If assigned, convert dimensions to decimal or metric system.

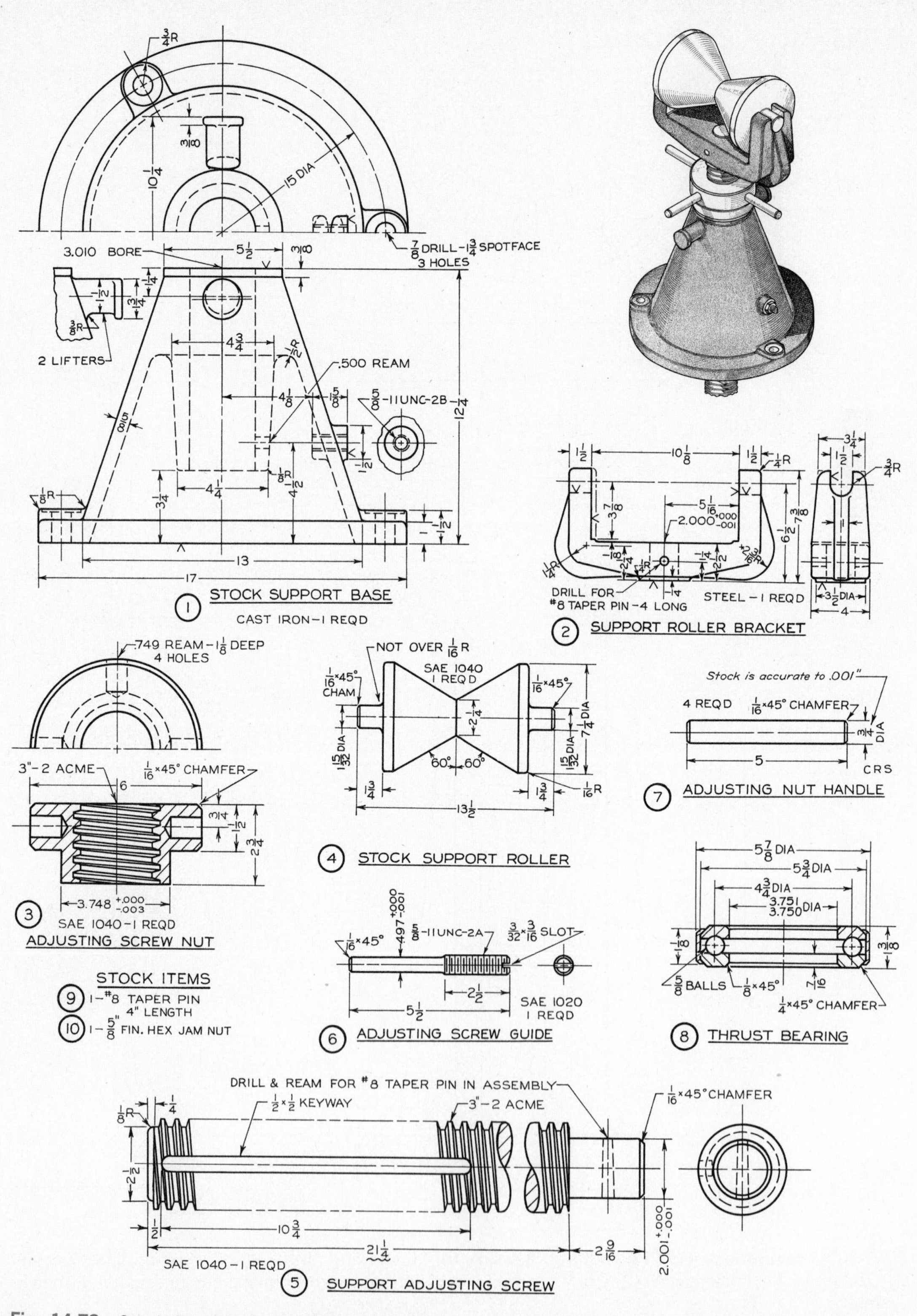


Fig. 14.78 **Stock Bracket for Cold Saw Machine.** (1) Draw details. (2) Draw assembly. If assigned, use unidirectional decimal or metric dimensions.

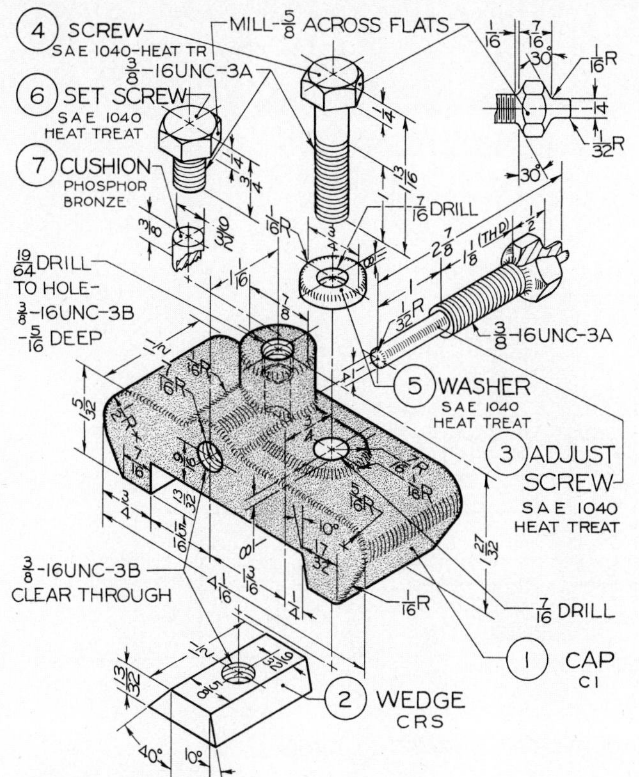


**Fig. 14.79 Clamp Stop.** (1) Draw details. (2) Draw assembly. If assigned, convert dimensions to decimal or metric system.

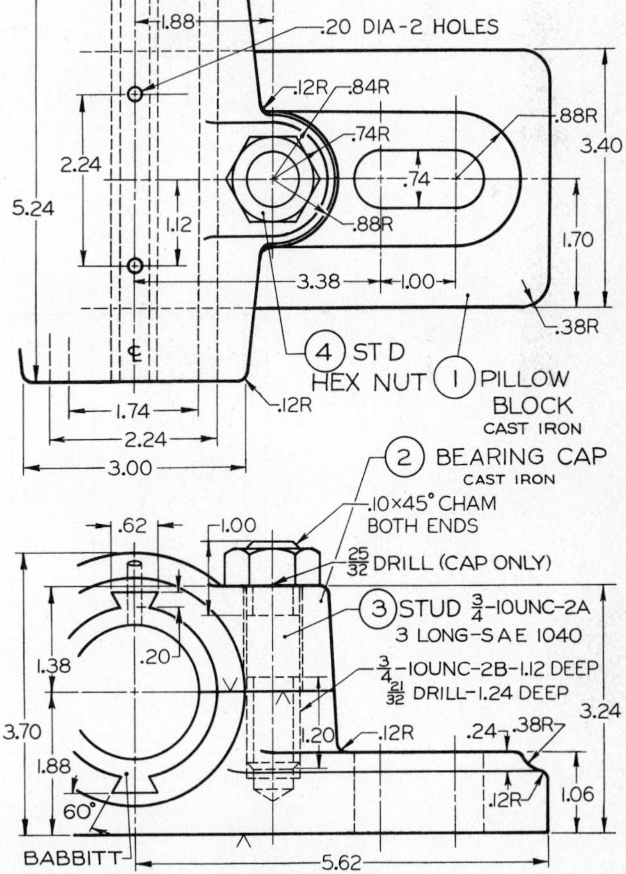


**Fig. 14.80 Pillow Block Bearing.** (1) Draw details. (2) Draw assembly. If assigned, convert dimensions to metric system.

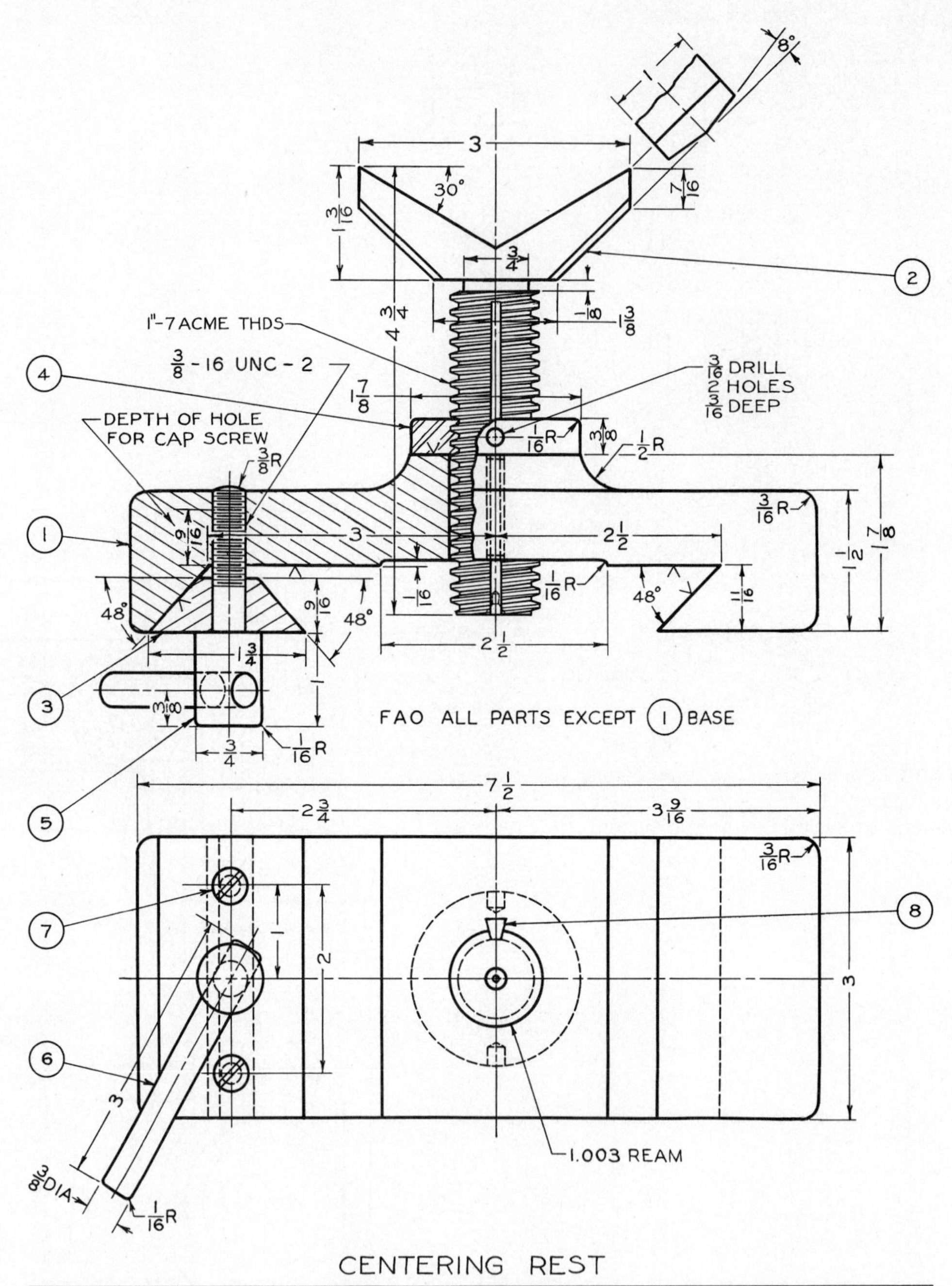


| PARTS LIST | | | | | | | |
|---|---|---|---|---|---|---|---|
| NO. | PART NAME | MATL | REQD | NO. | PART NAME | MATL | REQD |
| 1 | BASE | C I | 1 | 5 | CLAMP SCREW | SAE 1020 | 1 |
| 2 | REST | SAE 1020 | 1 | 6 | CLAMP HANDLE | SAE 1020 | 1 |
| 3 | CLAMP | SAE 1020 | 1 | 7 | $\frac{1}{4}$ X 1 FIL HD CAP SCREW | | 2 |
| 4 | ADJUSTING NUT | SAE 1020 | 1 | 8 | $\frac{7}{32}$ X $\frac{7}{32}$ X $\frac{1}{8}$ - 1 LONG - KEY | SAE 1030 | 1 |

Fig. 14.81 **Centering Rest.** (1) Draw details. (2) Draw assembly. If assigned, convert dimensions to decimal or metric system.

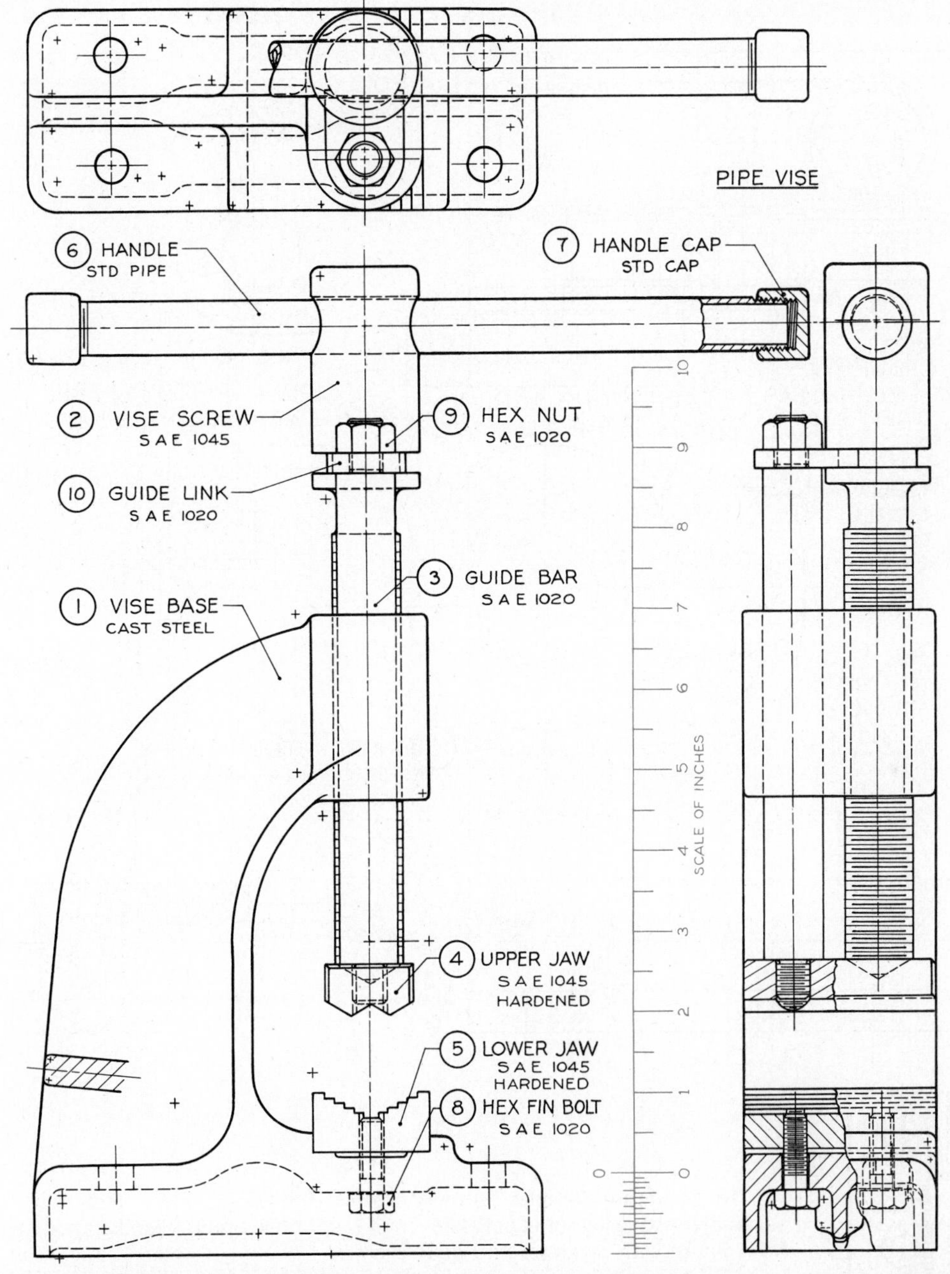


Fig. 14.82 **Pipe Vise.** (1) Draw details. (2) Draw assembly. To obtain dimensions, take distances directly from figure with dividers; then set dividers on printed scale and read measurements in inches. All threads are American National Standard Unified Coarse threads except the American National Standard pipe threads on handle and handle caps.

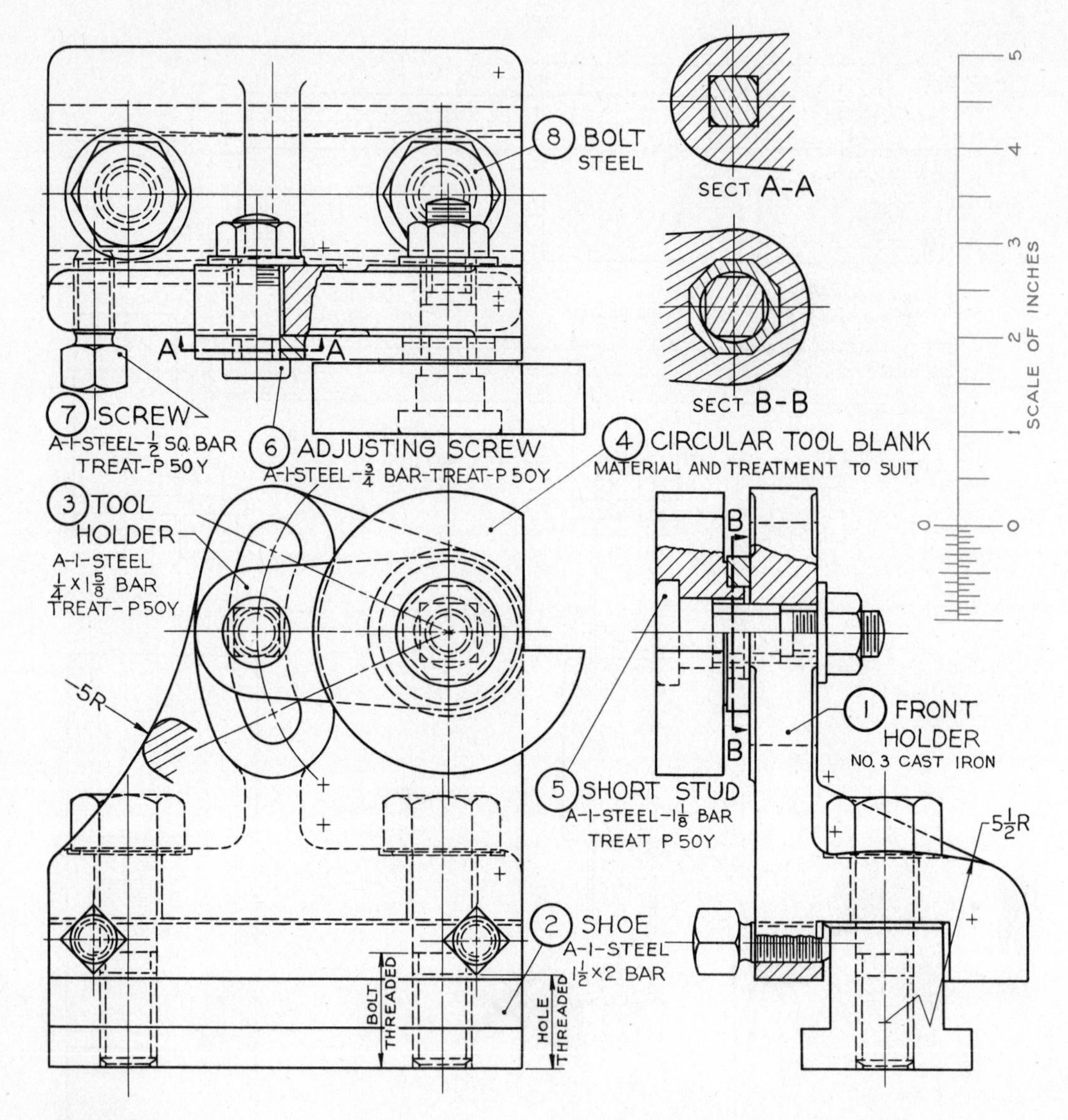


**Fig. 14.83 Front Circular Forming Cutter Holder.** (1) Draw details. (2) Draw assembly. Above layout is half size. To obtain dimensions, take distances directly from figure with dividers, and double them.

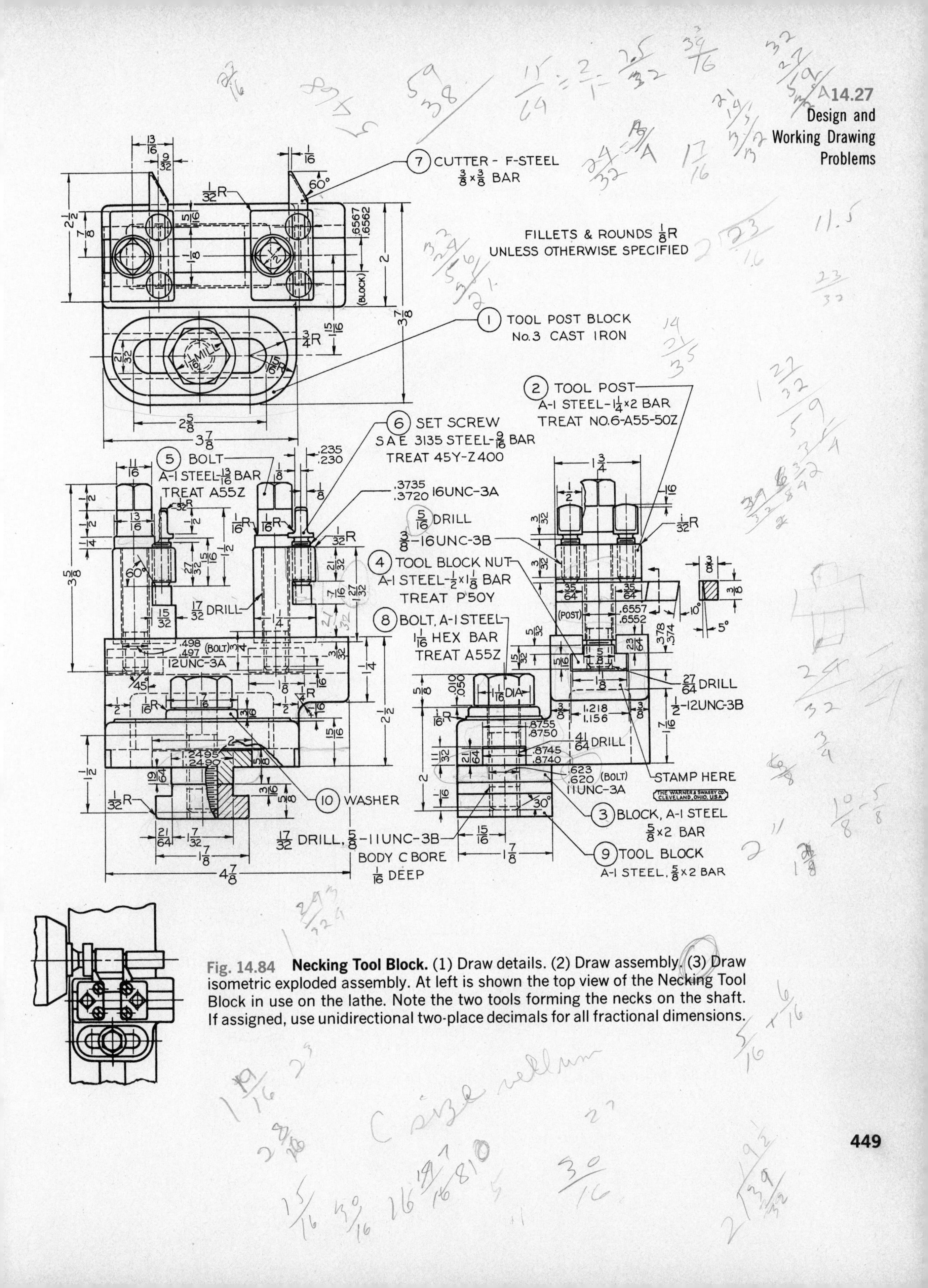


**Fig. 14.84 Necking Tool Block.** (1) Draw details. (2) Draw assembly. (3) Draw isometric exploded assembly. At left is shown the top view of the Necking Tool Block in use on the lathe. Note the two tools forming the necks on the shaft. If assigned, use unidirectional two-place decimals for all fractional dimensions.

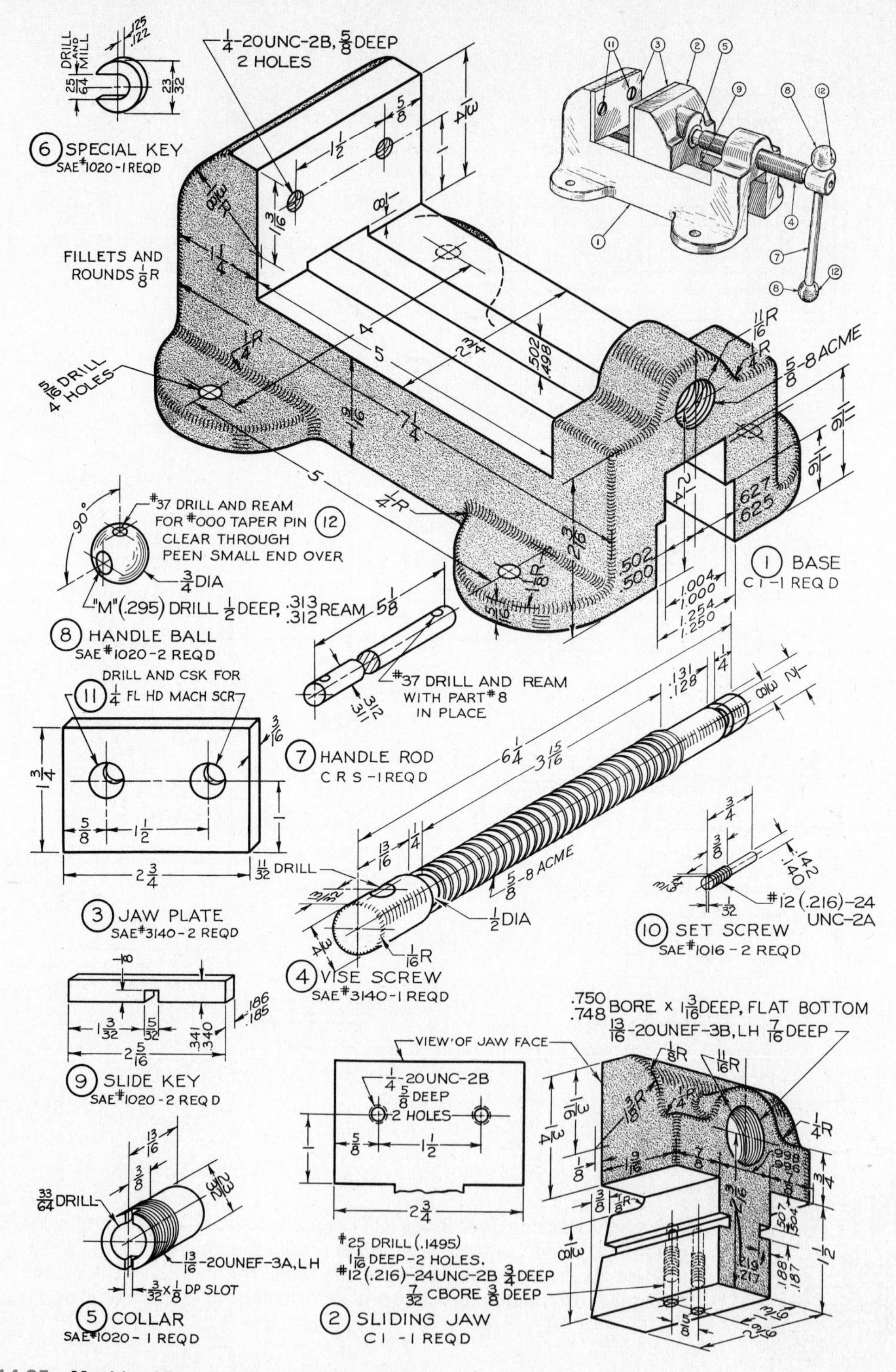


Fig. 14.85 **Machine Vise.** (1) Draw details. (2) Draw assembly. If assigned, convert dimensions to the decimal or metric system.

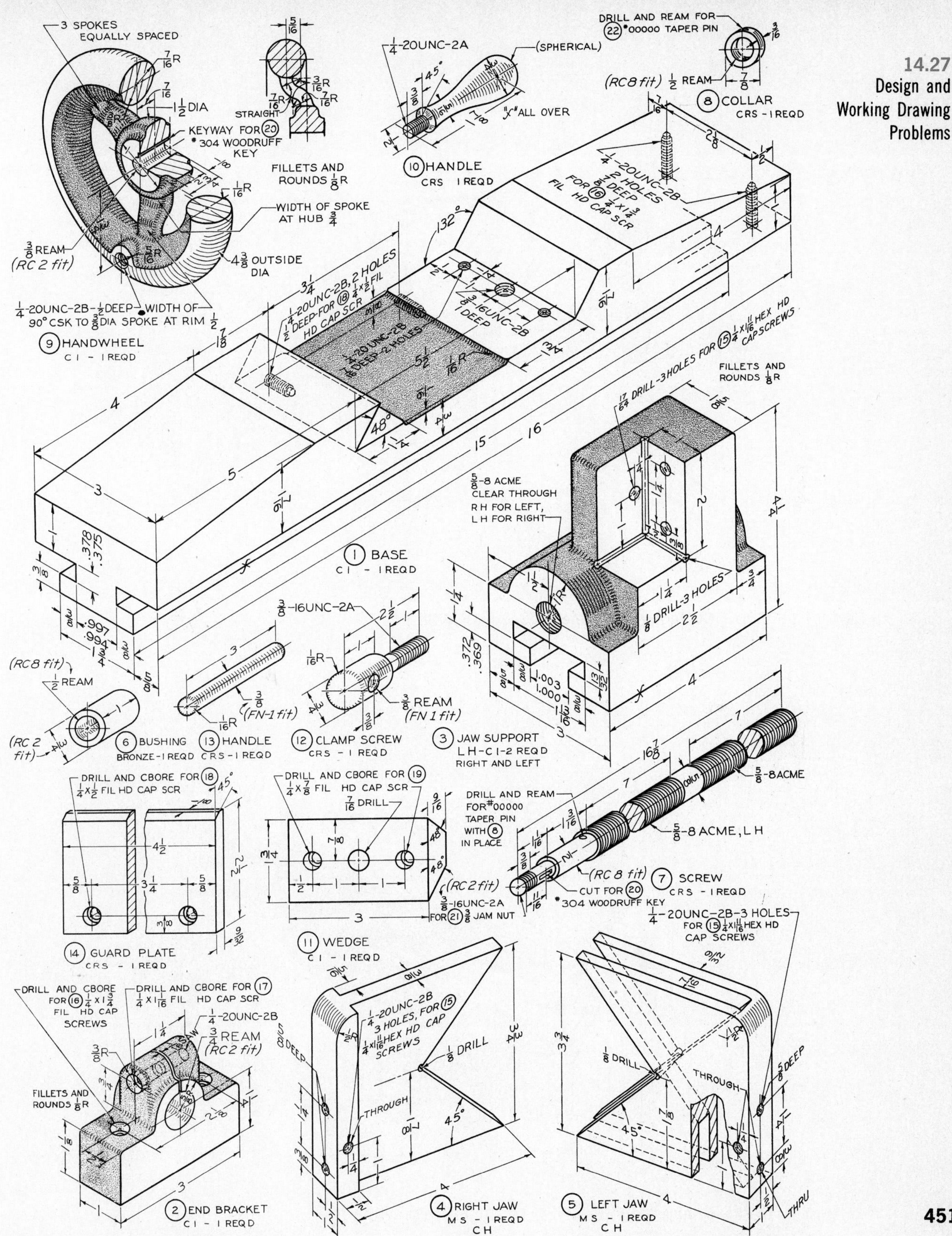




Fig. 14.86 **Centering Attachment.** (1) Draw details. Use American National Standard tables for fits indicated. (See Appendixes 5 to 9.) (2) Draw assembly. (See Fig. 14.87.) If assigned, convert dimensions to decimal or metric system.

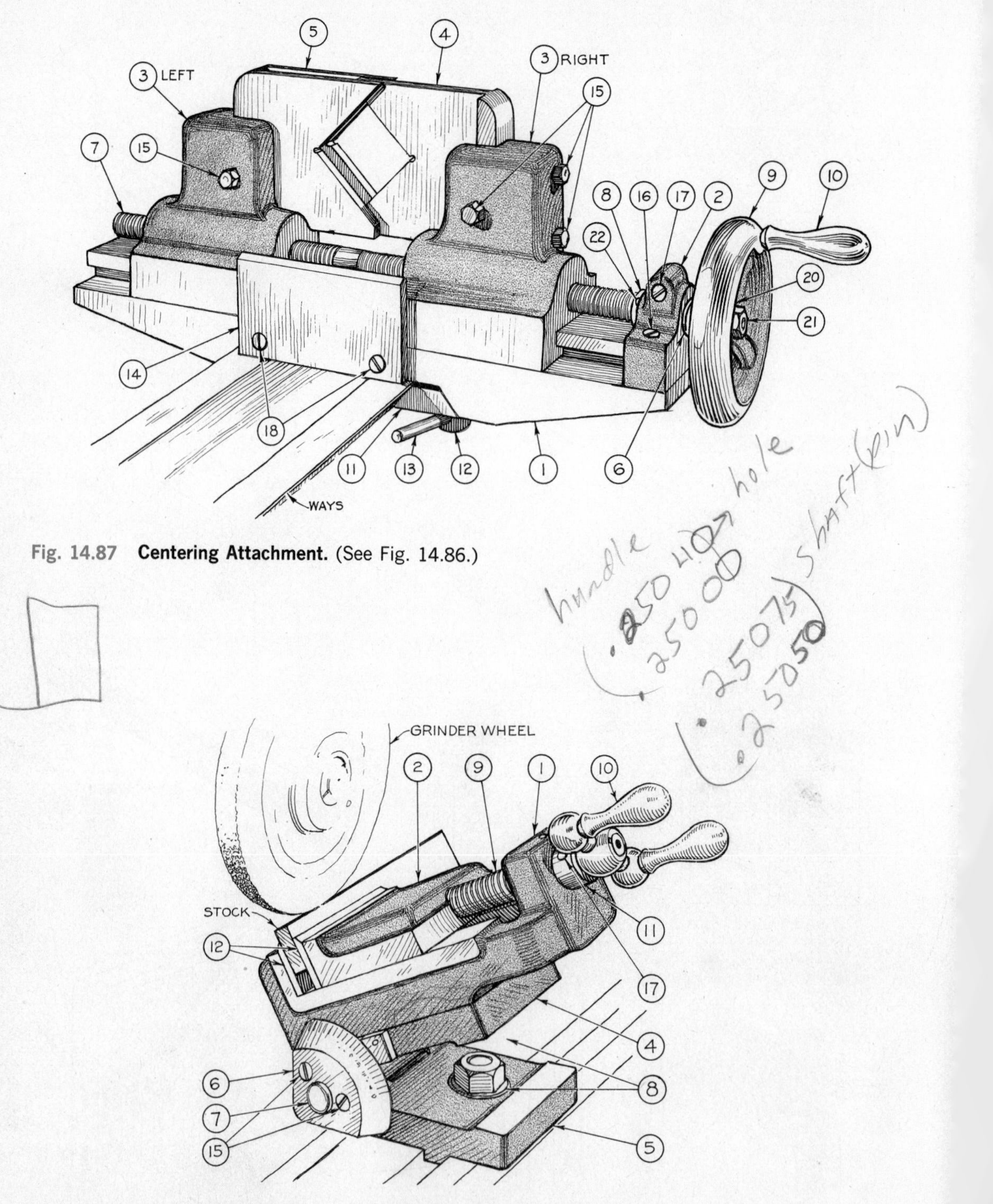


Fig. 14.87 **Centering Attachment.** (See Fig. 14.86.)

Fig. 14.88 **Grinder Vise.** (See Figs. 14.89 and 14.90.)

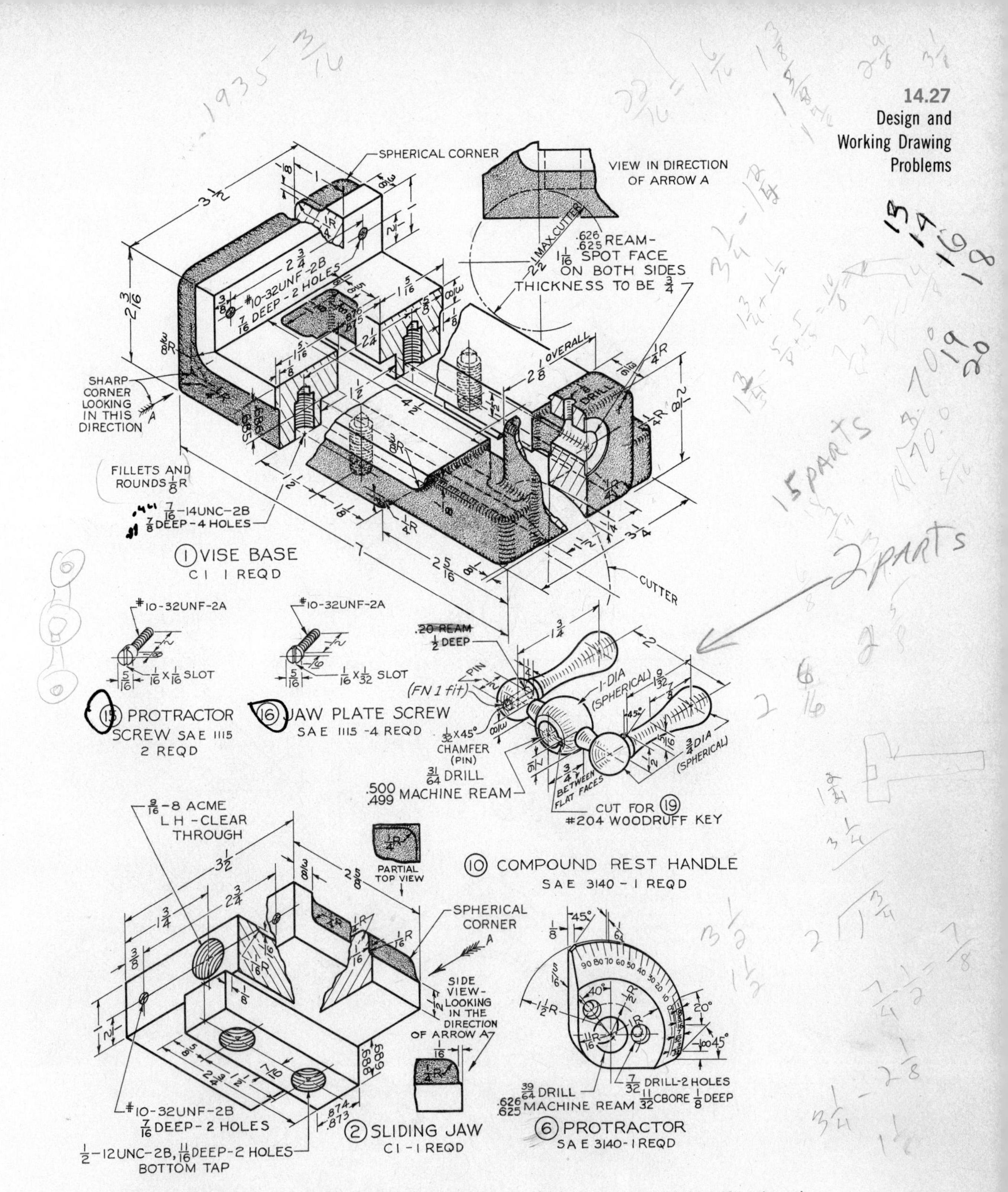


**Fig. 14.89 Grinder Vise.** (1) Draw details. (2) Draw assembly. (See Figs. 14.88 and 14.90.) If assigned, convert dimensions to decimal or metric system.

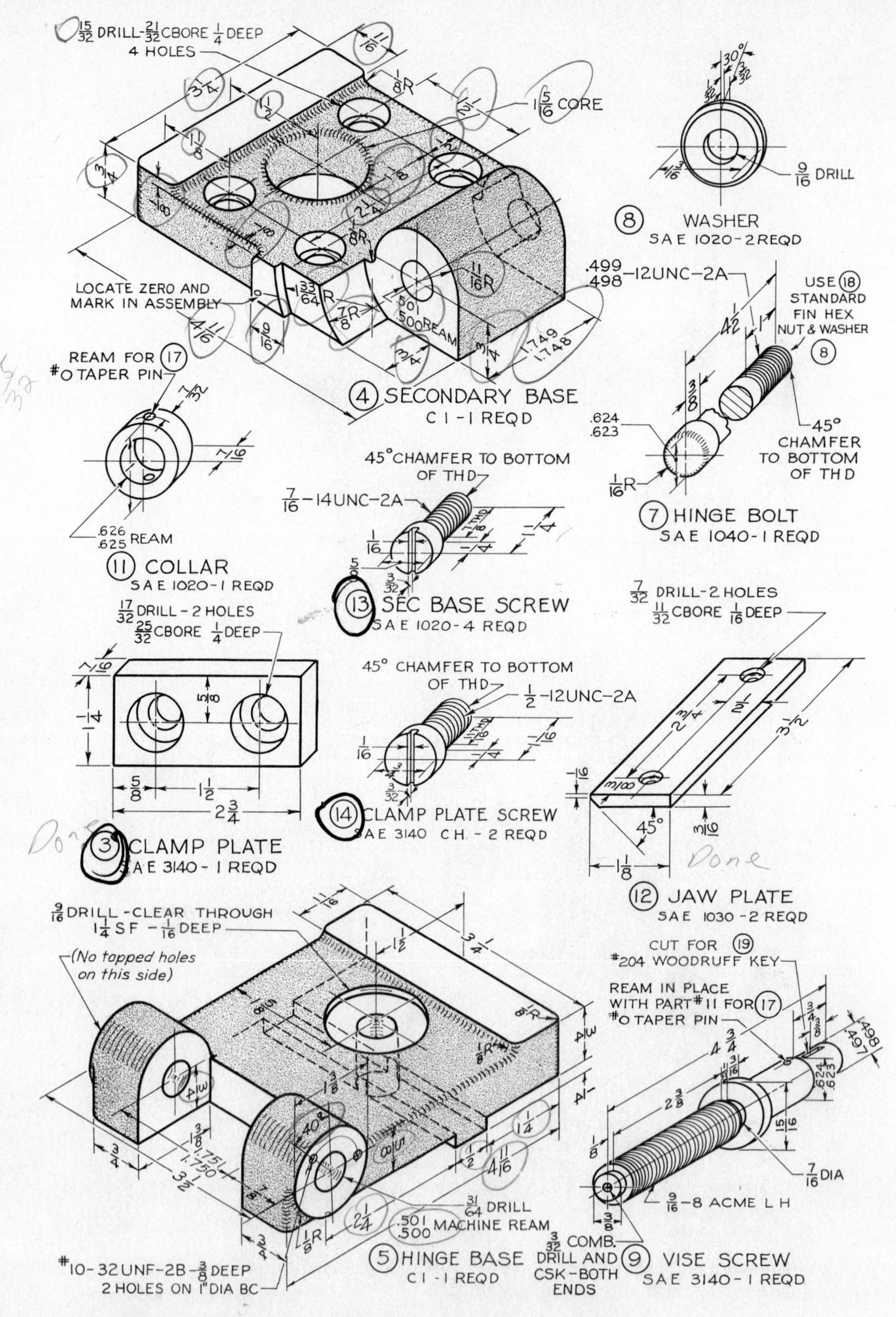


Fig. 14.90 **Grinder Vise** (Continued). See Fig. 14.89 for instructions.

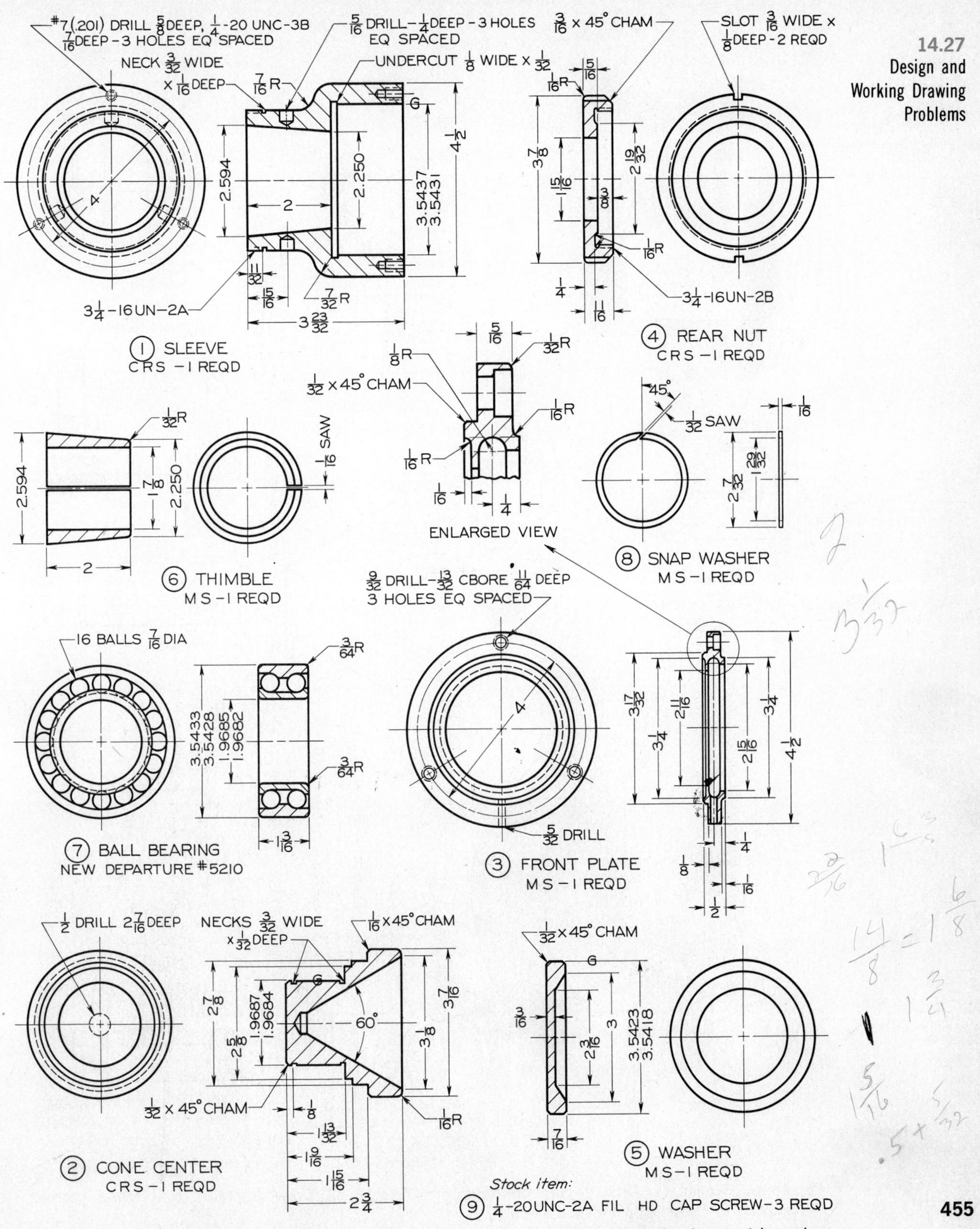




Fig. 14.91 **Cup Center.** Draw assembly. If assigned, convert dimensions to decimal or metric system.

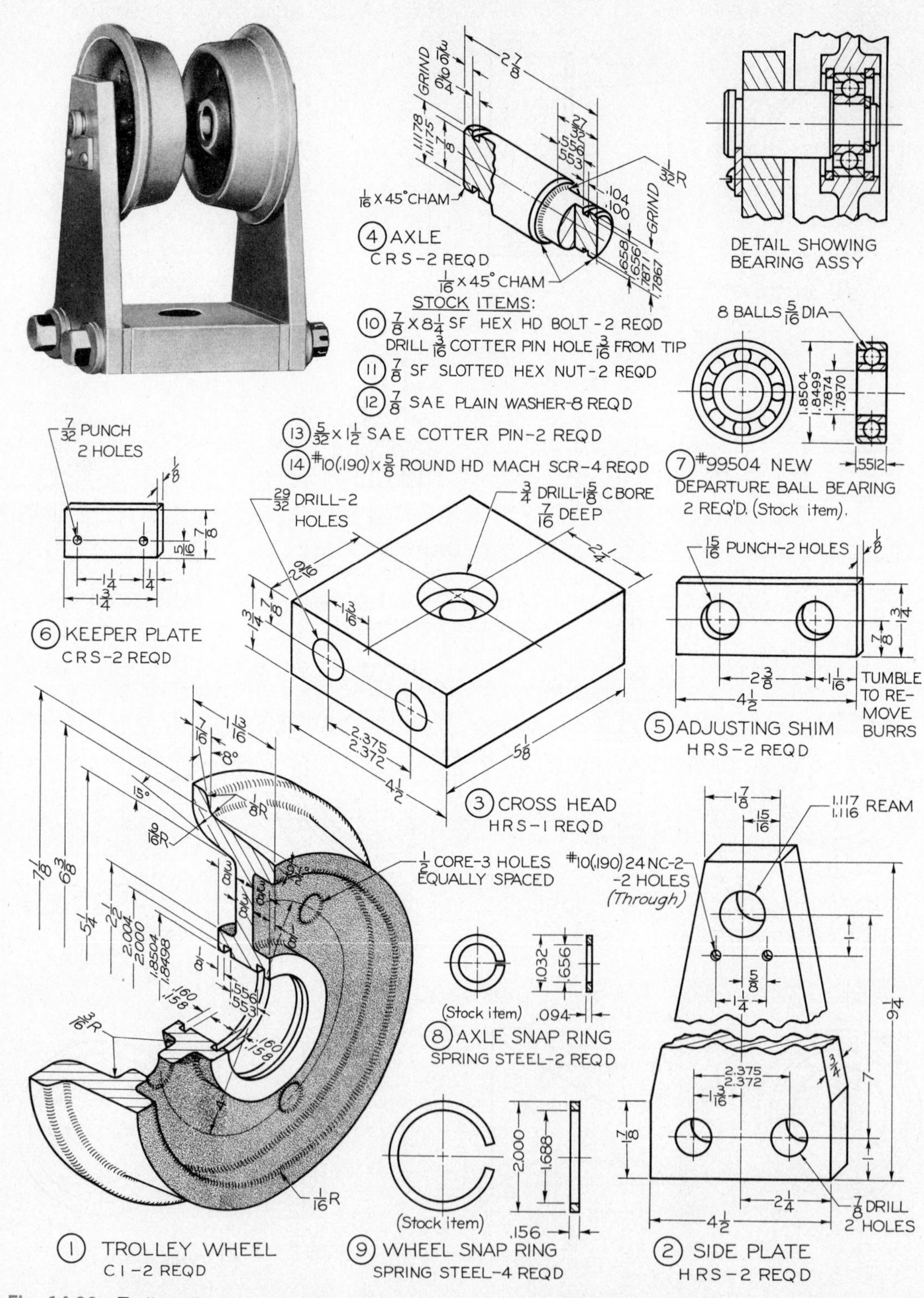


Fig. 14.92 **Trolley.** (1) Draw details, omitting parts 7 to 14. (2) Draw assembly. If assigned, convert dimensions to decimal or metric system.

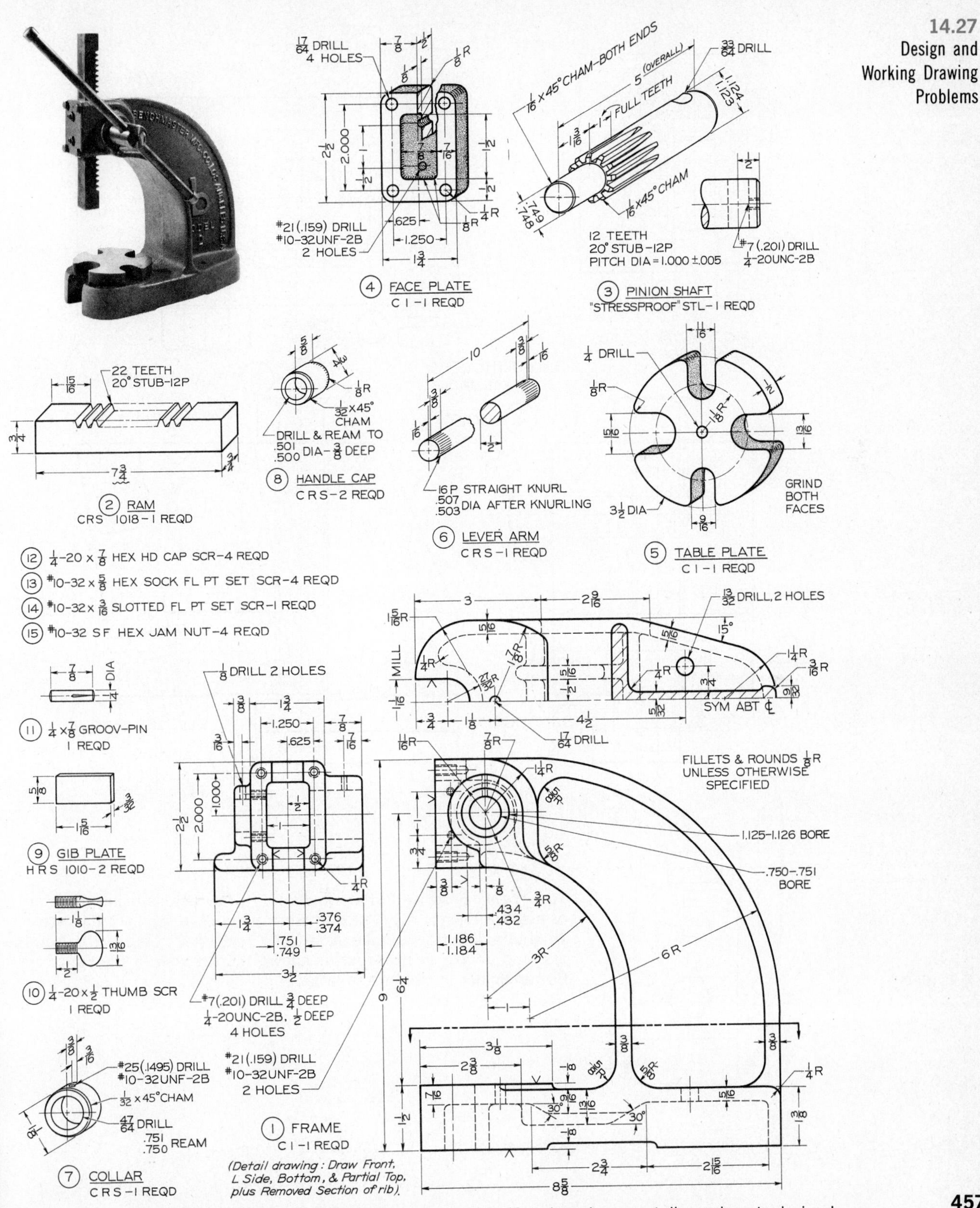


**Fig. 14.93** **Arbor Press.** (1) Draw details. (2) Draw assembly. If assigned, convert dimensions to decimal or metric system.

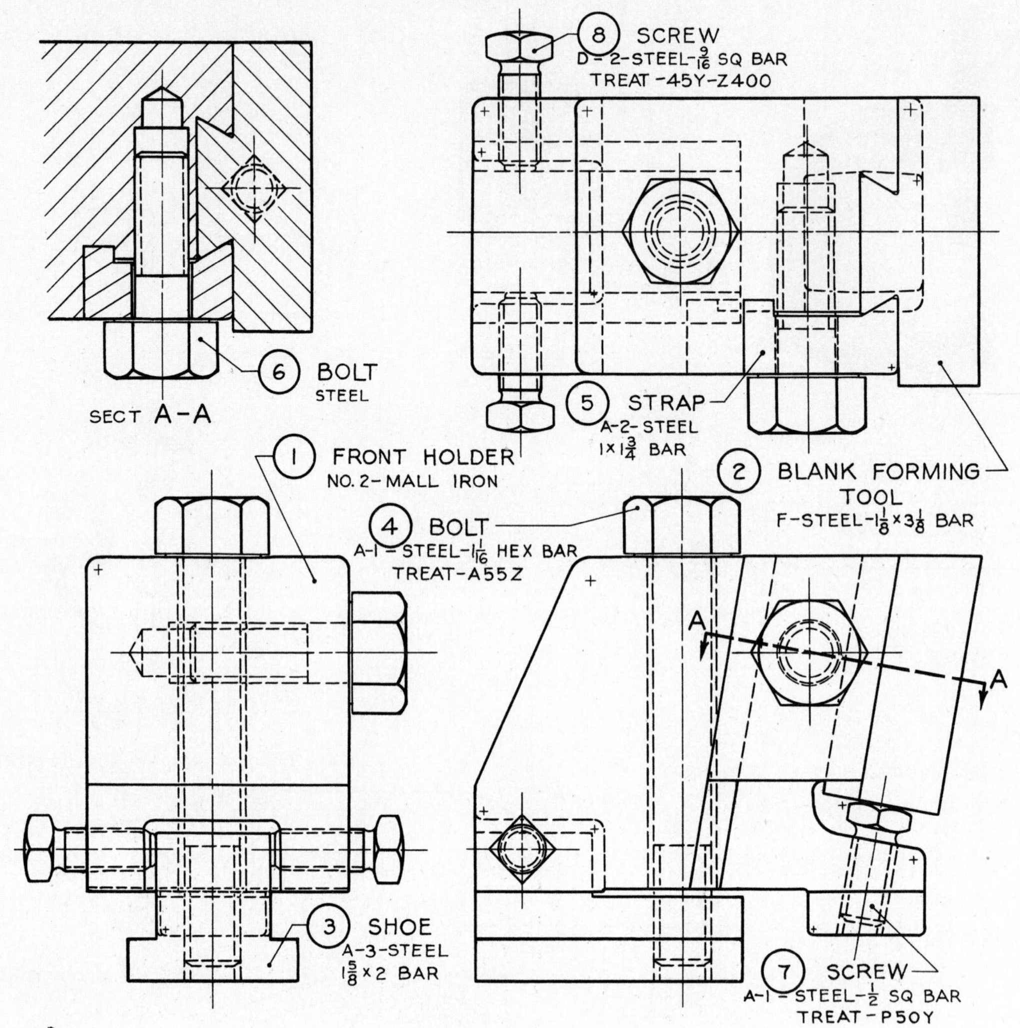


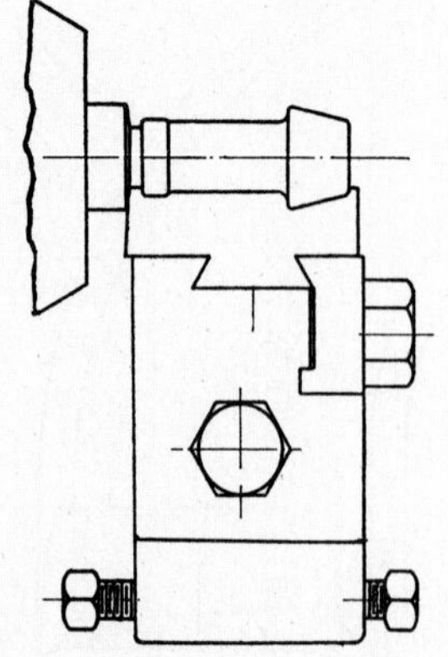

Fig. 14.94 **Forming Cutter Holder.** (1) Draw details using decimal or metric dimensions. (2) Draw assembly. Above layout is half size. To obtain dimensions, take distances directly from figure with dividers, and double them. At left is shown the top view of the Forming Cutter Holder in use on the lathe.

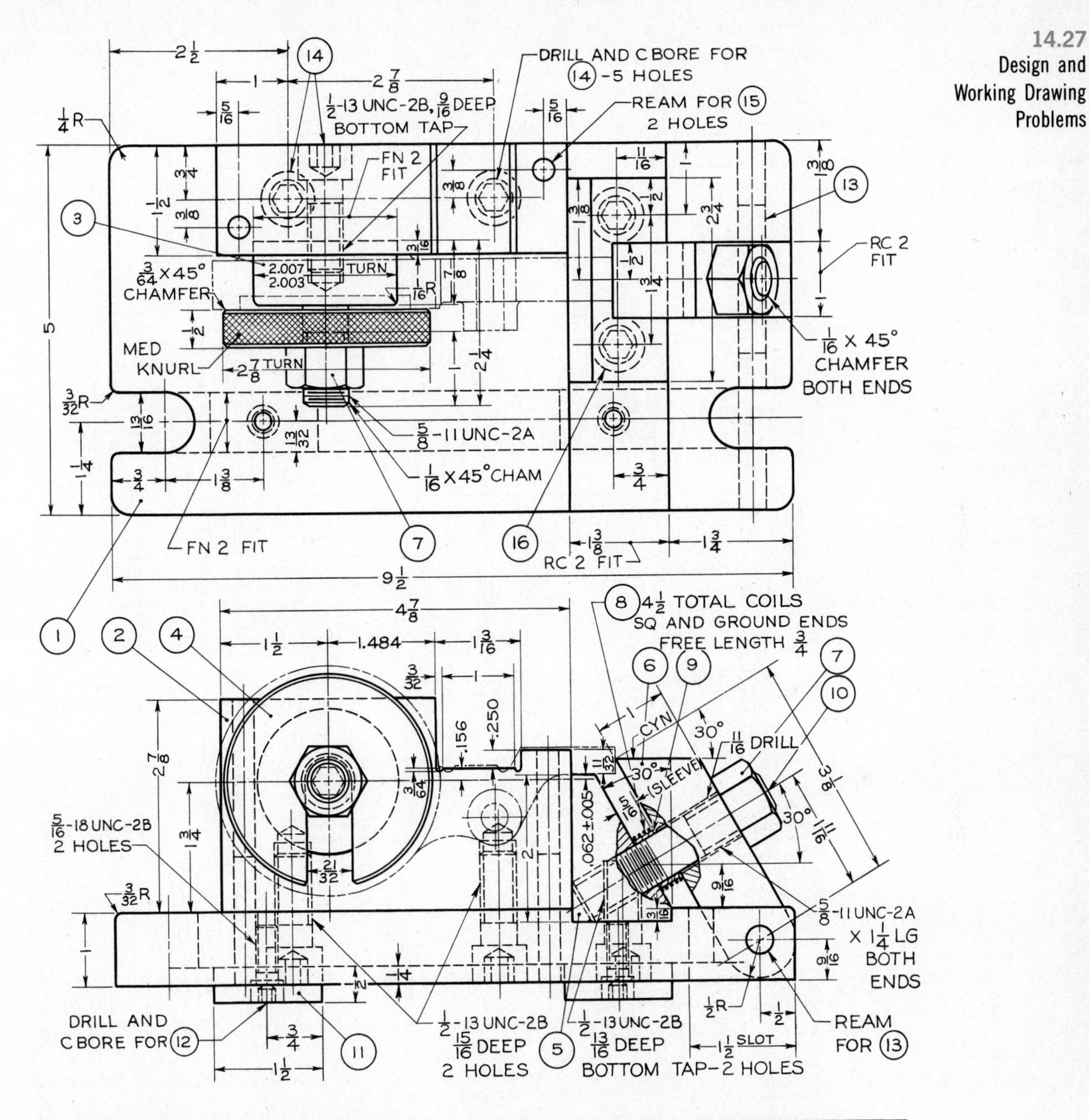


| Item | NAME | Amt | MATL | REMARKS | Item | NAME | Amt | MATL | REMARKS |
|---|---|---|---|---|---|---|---|---|---|
| 1 | BASE PLATE | 1 | CRS | $1\times5\times9\frac{1}{2}$ | 9 | SLEEVE | 1 | BRONZE | O D $\frac{23}{32}$ – I D $\frac{41}{64}$ |
| 2 | GAGE BLOCK | 1 | CRS | $1\frac{1}{2}\times2\frac{7}{8}\times4\frac{7}{8}$ | 10 | STUD | 1 | CRS | $\frac{5}{8}$ DIA × 3 |
| 3 | LOCATING PLUG | 1 | CRS | 2.005 DIA × $2\frac{1}{4}$ | 11 | KEY | 2 | CRS | $\frac{1}{2}\times\frac{13}{16}\times1\frac{1}{2}$ |
| 4 | C-WASHER | 1 | CRS | $2\frac{7}{8}$ DIA × $\frac{1}{2}$ | 12 | SOC HD CAP SCR | 2 | STK | $\frac{5}{16}\times\frac{3}{4}$ |
| 5 | REST BLOCK | 1 | CRS | $1\frac{3}{8}\times2\times2\frac{3}{4}$ | 13 | PIN | 1 | DR | $\frac{3}{8}\times2$ |
| 6 | CLAMP | 1 | CRS | $1\times1\times3\frac{5}{8}$ | 14 | SOC HD CAP SCR | 3 | STK | $\frac{1}{2}\times1\frac{1}{4}$ |
| 7 | $\frac{5}{8}$ STD HEX NUT | 2 | STK | | 15 | DOWEL PIN | 2 | STK | $\frac{5}{16}$ DIA × $1\frac{1}{2}$ |
| 8 | SPRING | 1 | MUSIC WIRE | WIRE .054 – O D $\frac{7}{8}$ | 16 | SOC HD CAP SCR | 2 | STK | $\frac{1}{2}\times1$ |

Fig. 14.95 **Milling Fixture for Clutch Arm.** (1) Draw details using decimal or metric dimensions, if assigned. (2) Draw assembly.

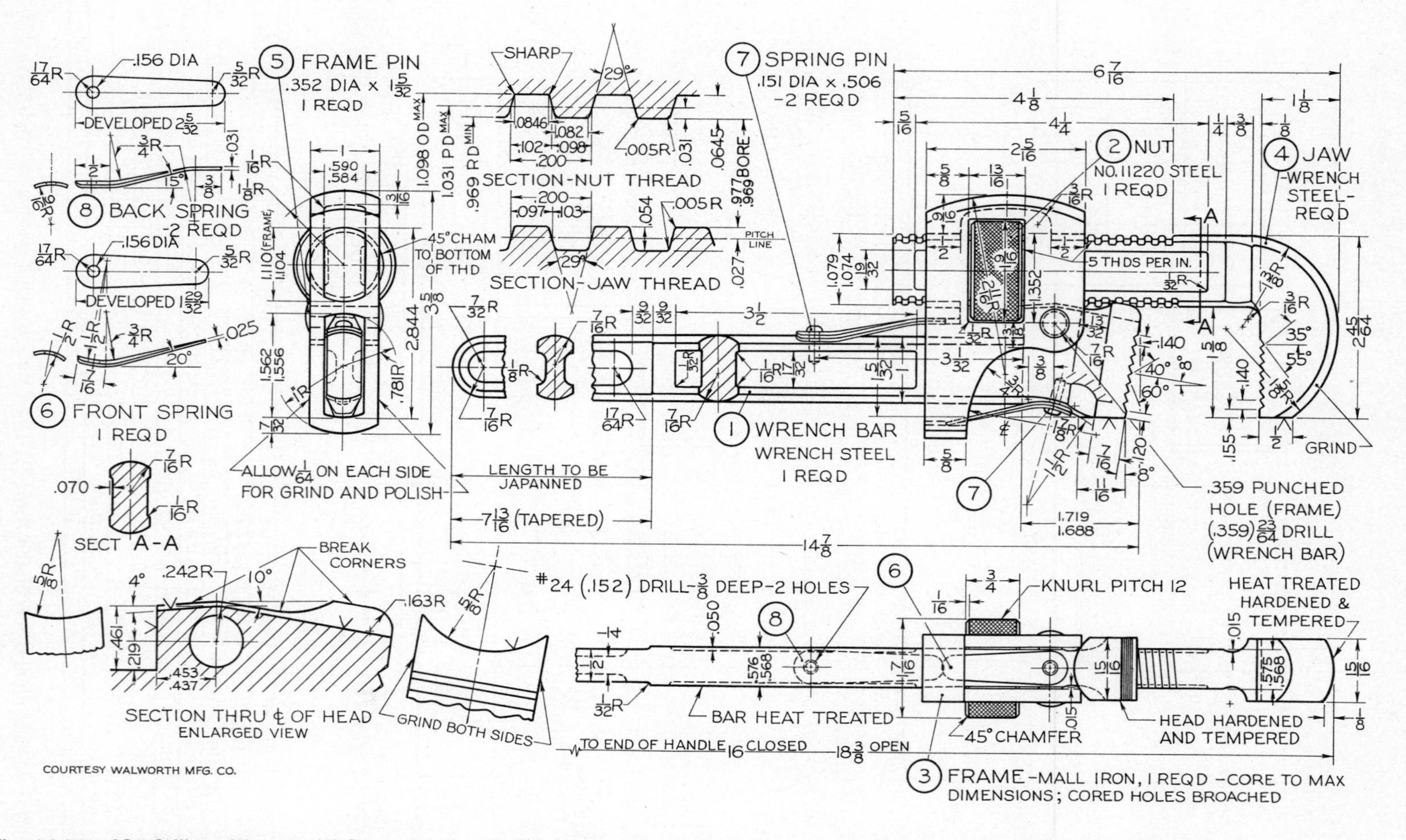


**Fig. 14.96 18″ Stillson Wrench.** (1) Draw details. (2) Draw assembly. If assigned, convert dimensions to decimal or metric system.

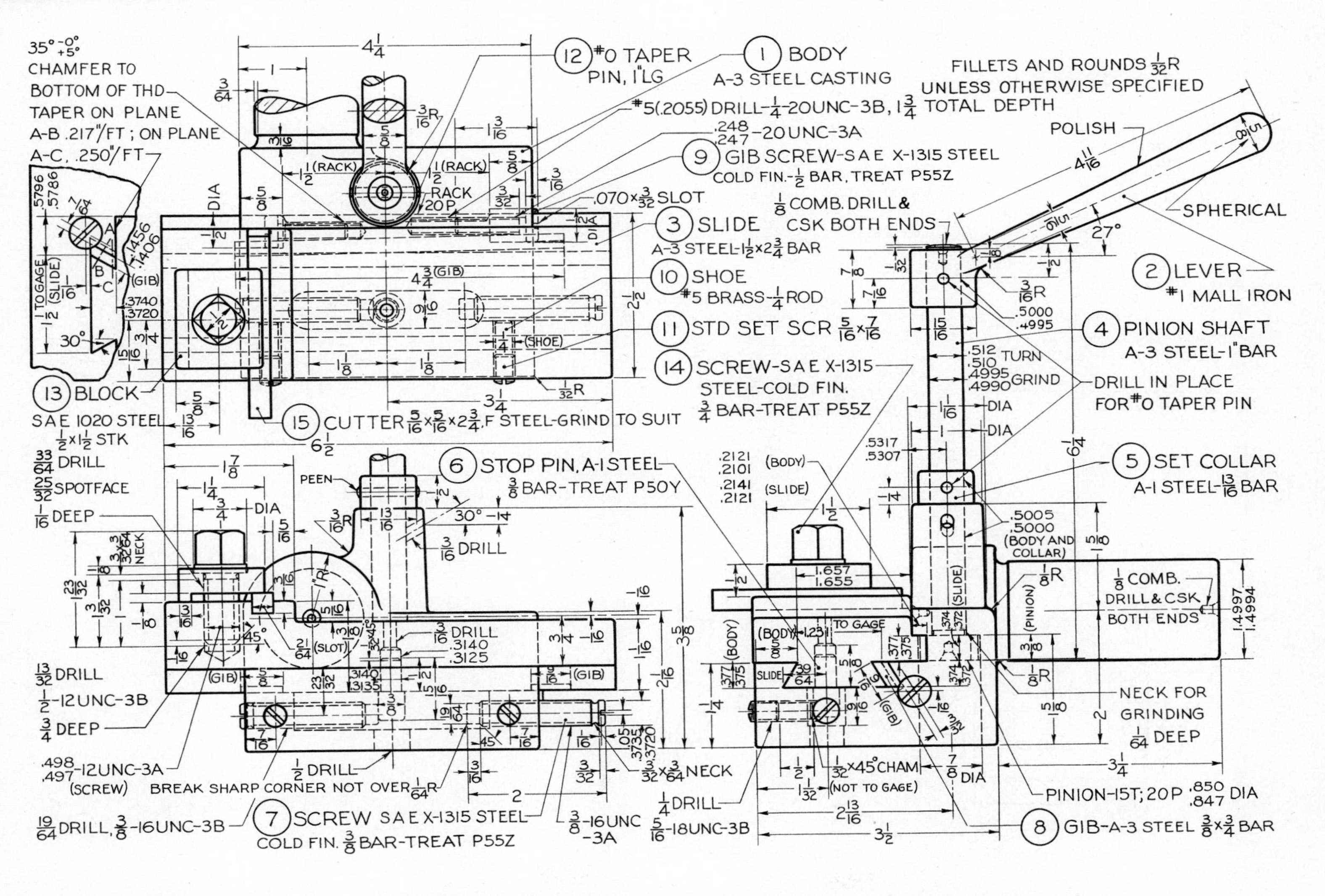


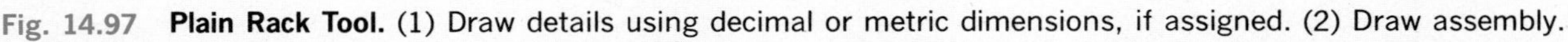
Fig. 14.97 **Plain Rack Tool.** (1) Draw details using decimal or metric dimensions, if assigned. (2) Draw assembly.

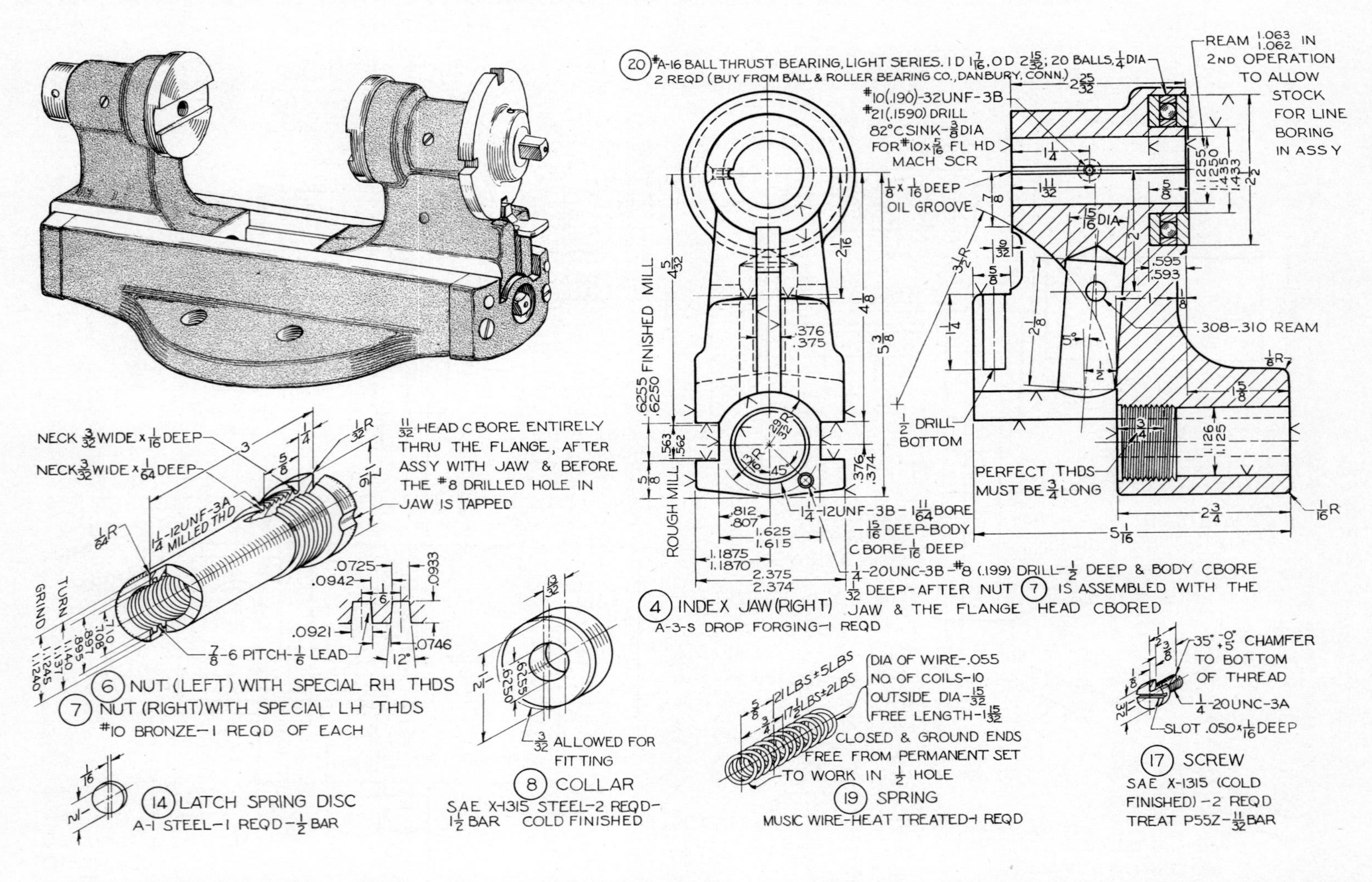


Fig. 14.98 **Revolving Jaw Chuck.** (1) Draw details using decimal or metric dimensions, if assigned, (2) Draw assembly. (See Figs. 14.99 to 14.101.)

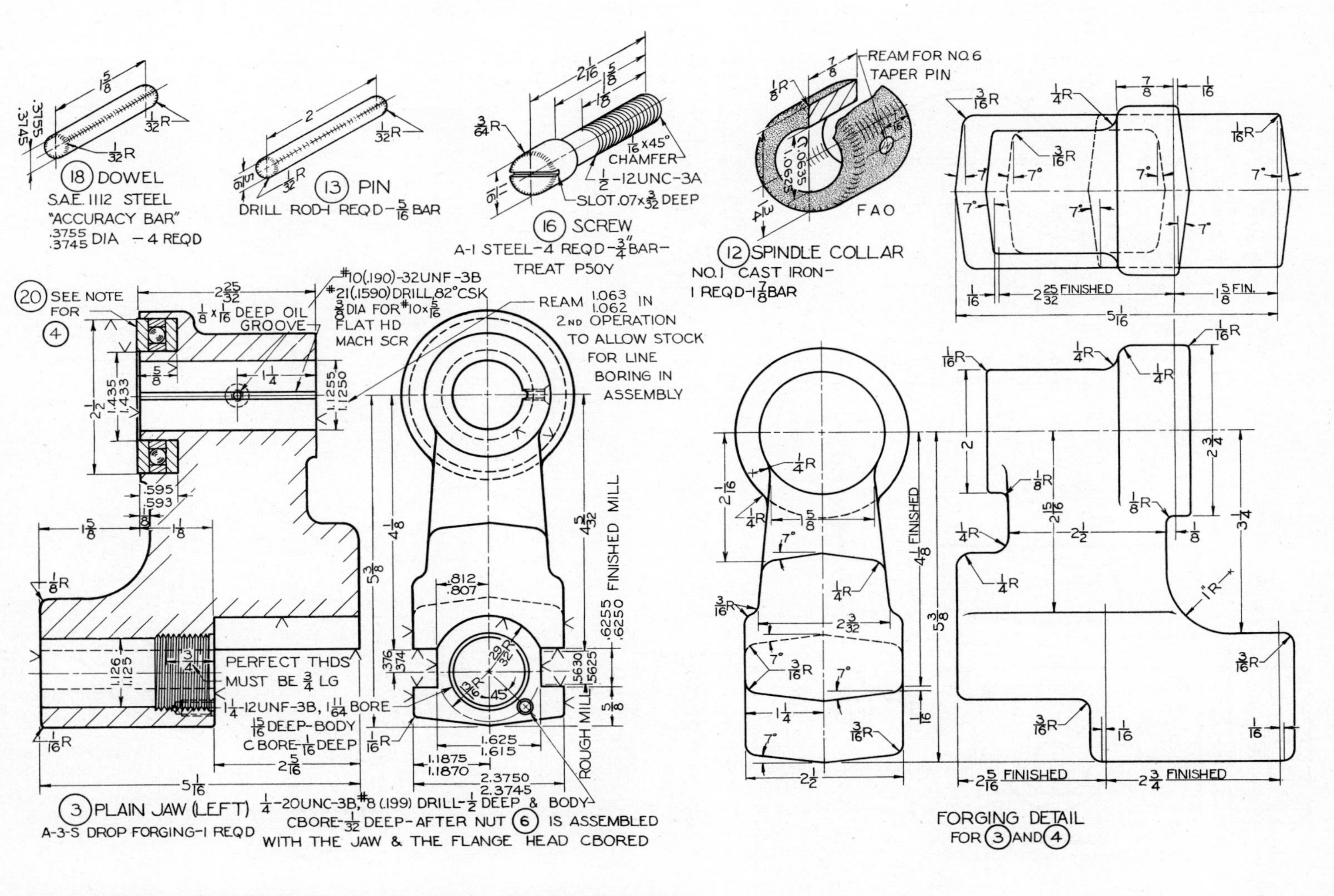


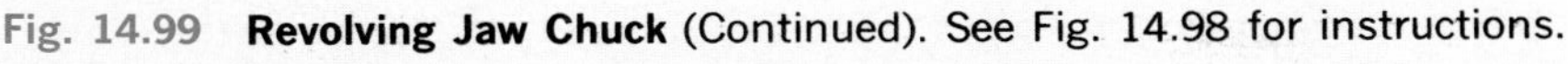

Fig. 14.99 **Revolving Jaw Chuck** (Continued). See Fig. 14.98 for instructions.

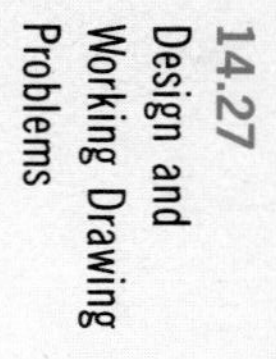


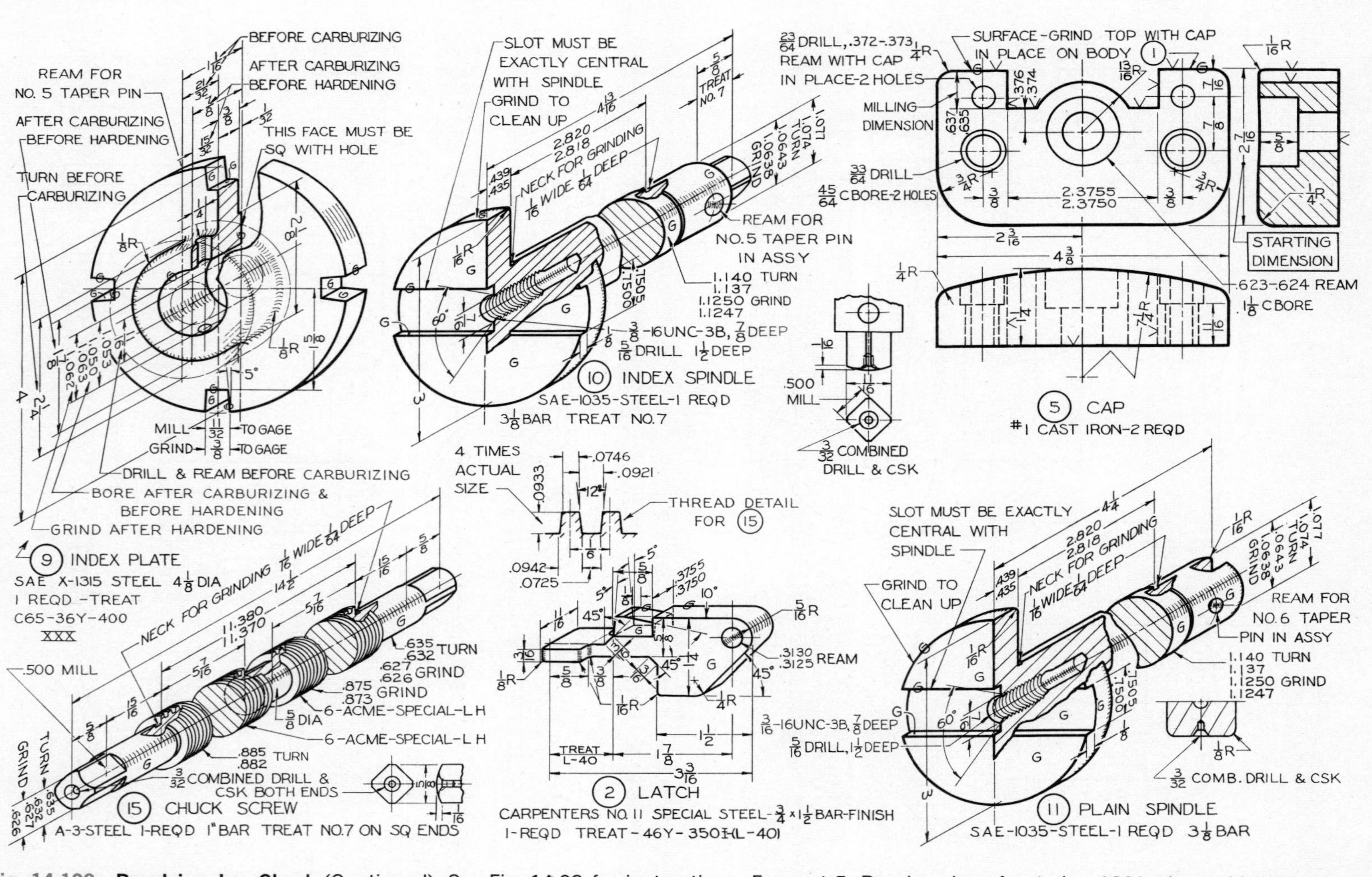


**Fig. 14.100** **Revolving Jaw Chuck** (Continued). See Fig. 14.98 for instructions. For part 5: Revolve given front view 180°; then add right-side and bottom views.

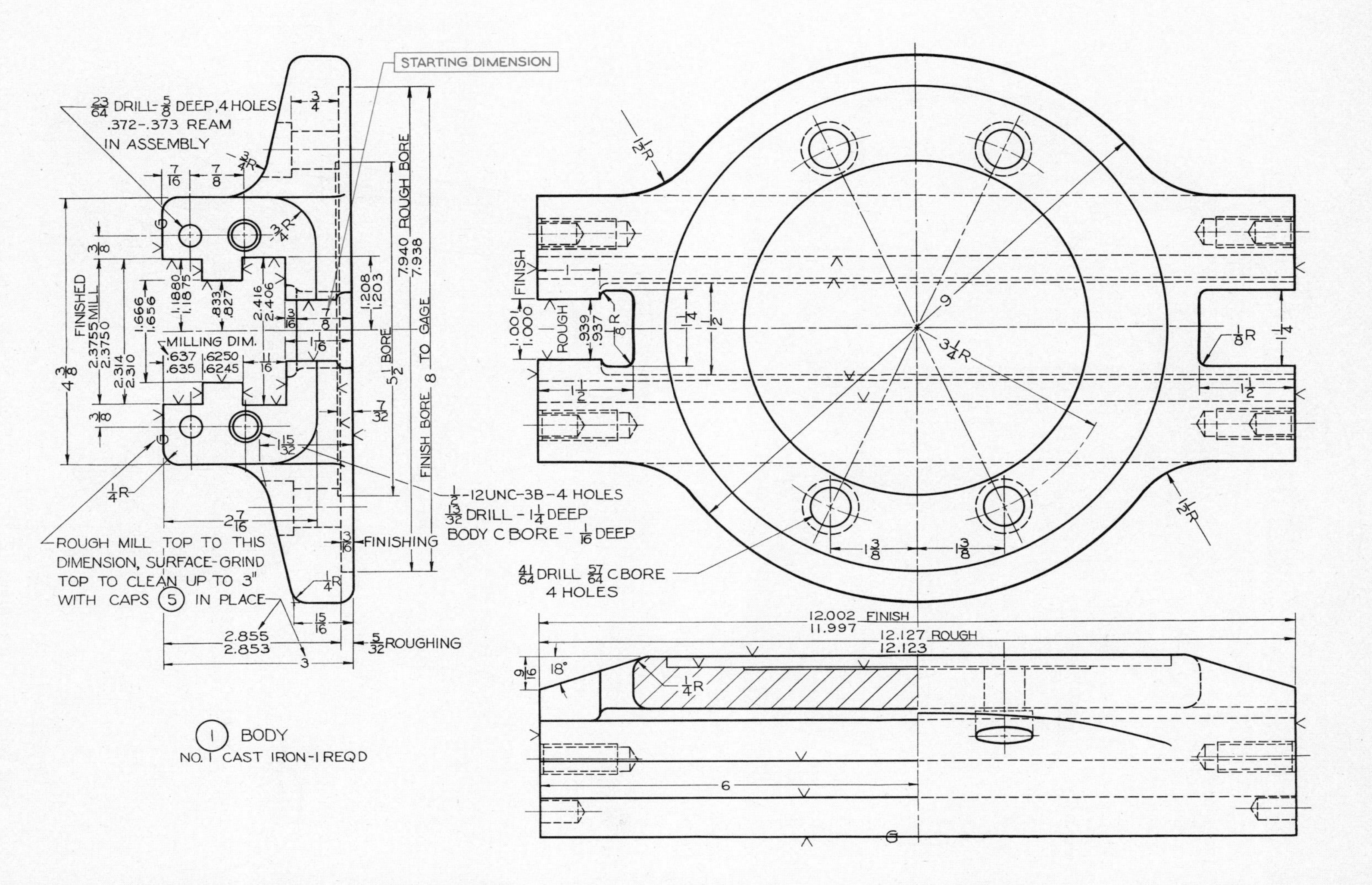


**Fig. 14.101 Revolving Jaw Chuck** (Continued). See Fig. 14.98 for instructions. For part 1: Take given left-side view as the right-side view in the new drawing; then add front view, and bottom view in half section.

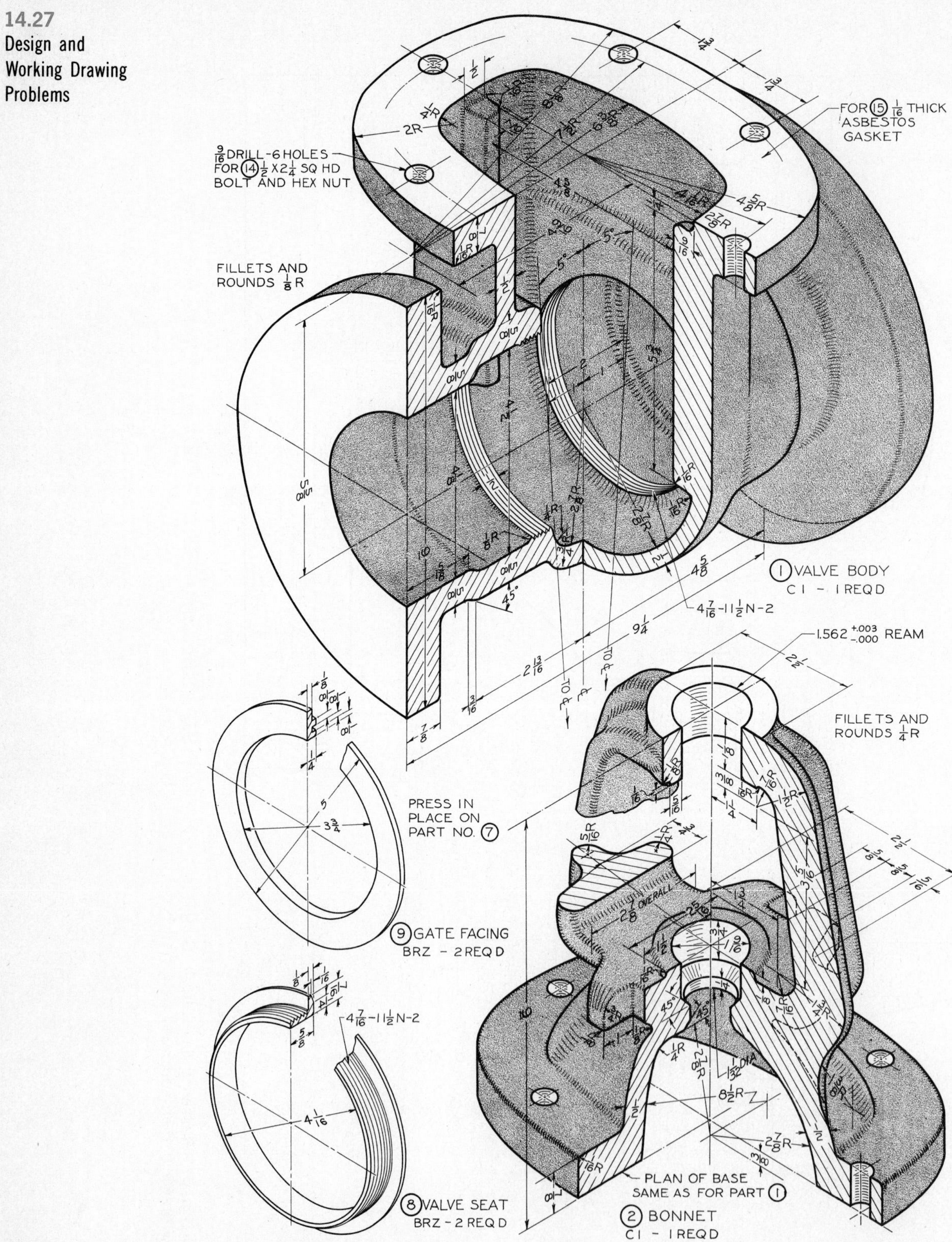


Fig. 14.102 **Gate Valve.** (1) Draw details. (2) Draw assembly. (See also Fig. 14.103.) If assigned, convert dimensions to decimal or metric system.

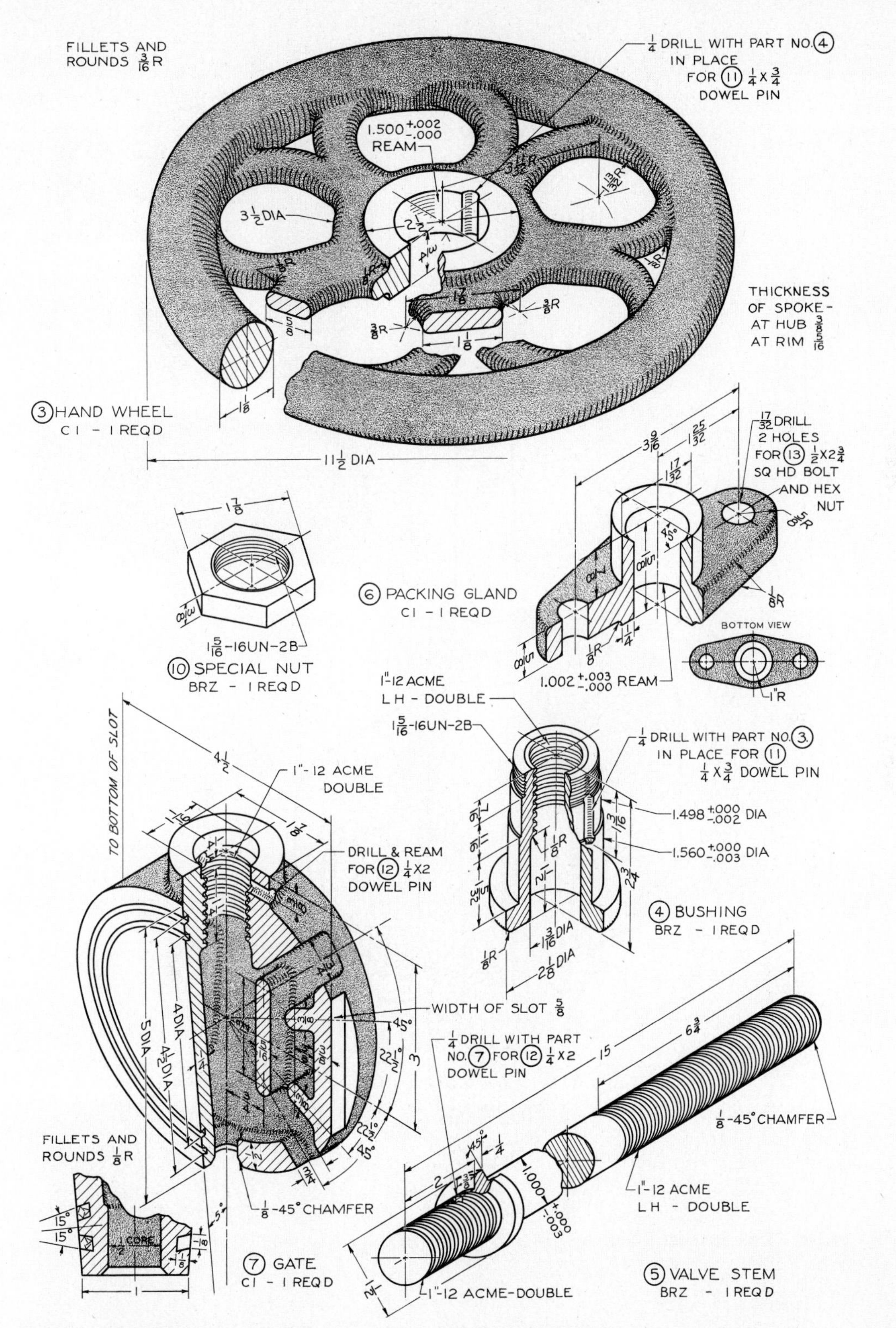


Fig. 14.103 **Gate Valve** (Continued). See Fig. 14.102 for instructions.

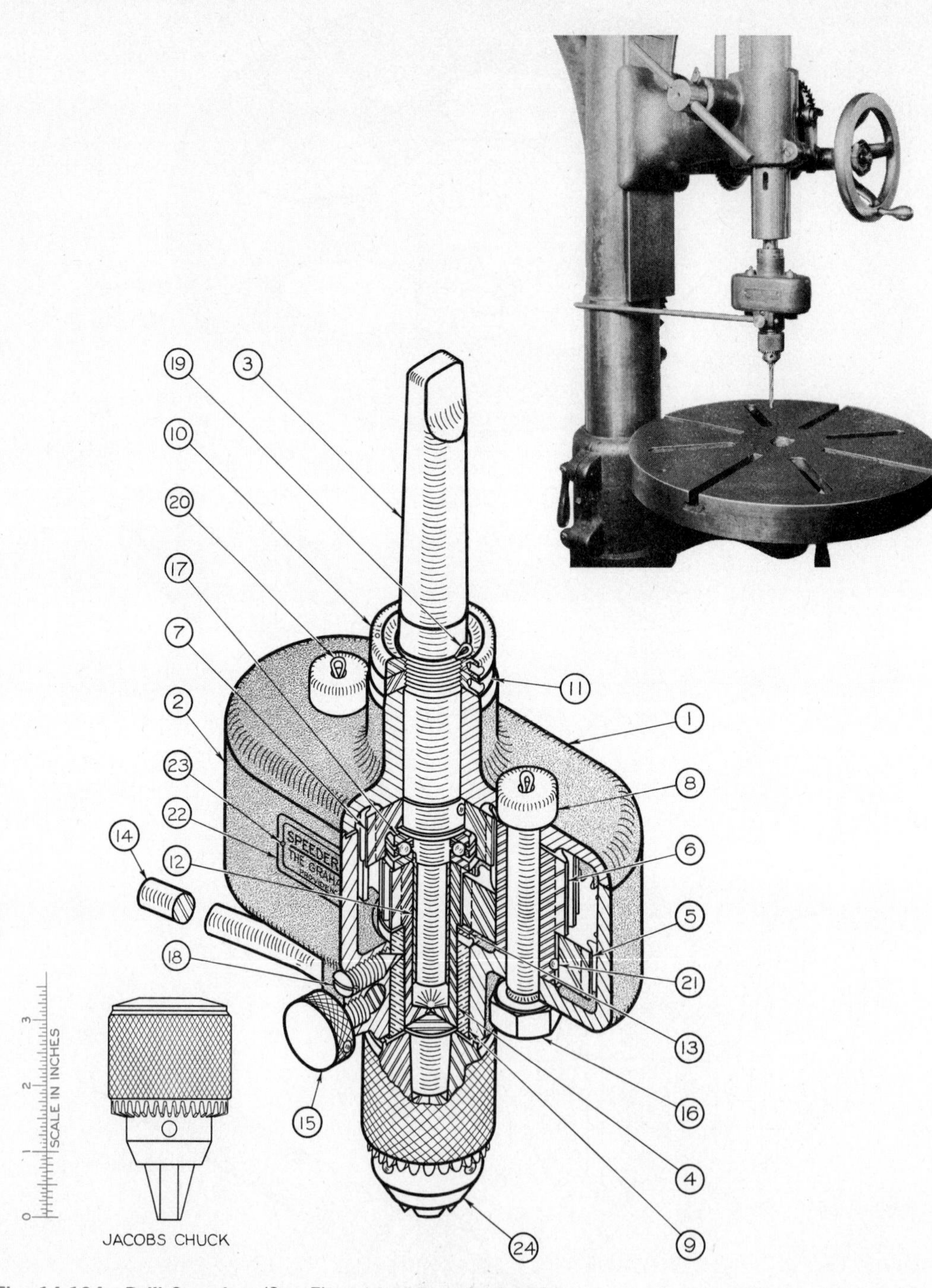


Fig. 14.104 **Drill Speeder.** (See Figs. 14.105 and 14.106.)

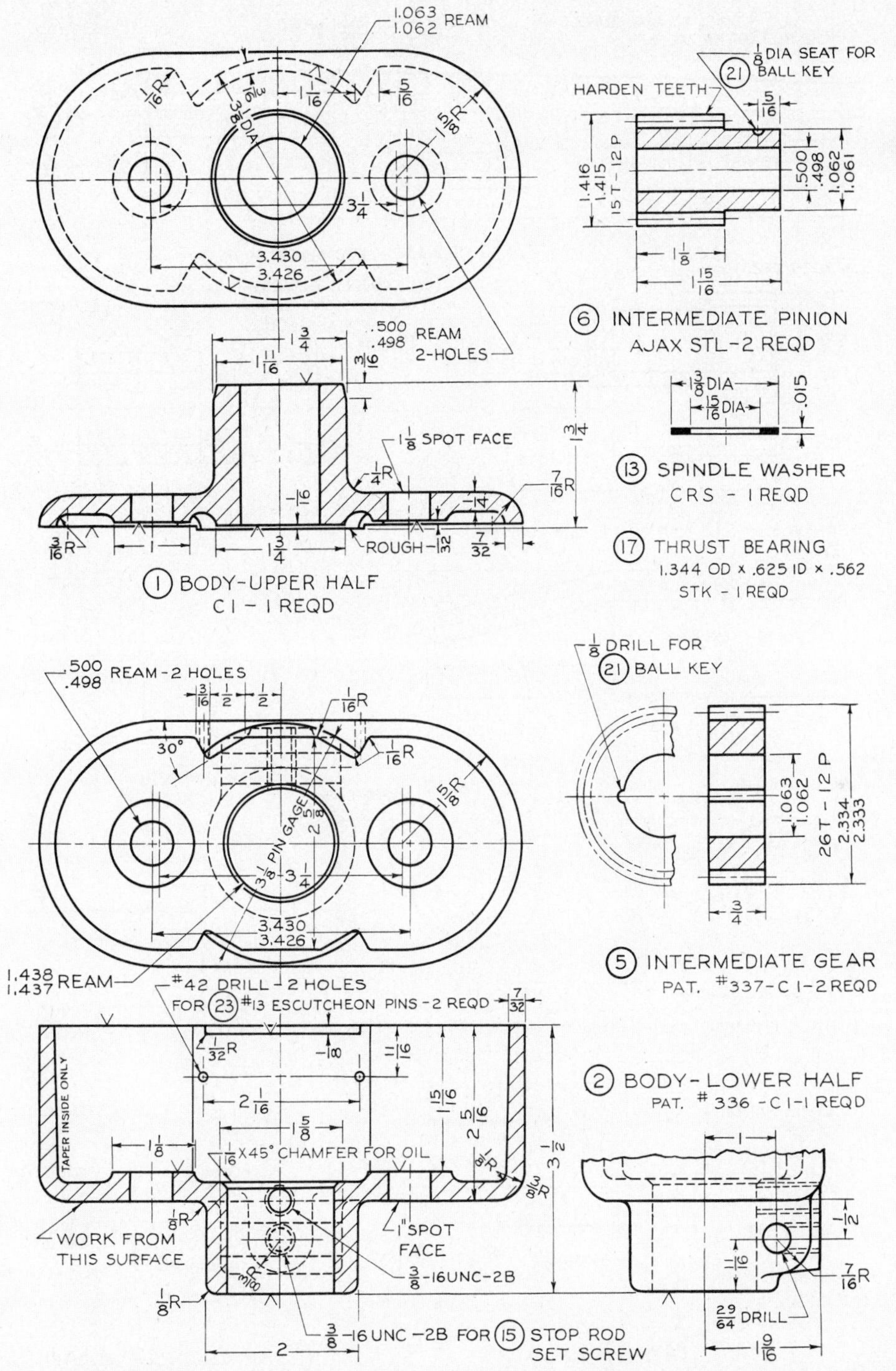


Fig. 14.105 **Drill Speeder** (Continued). (1) Draw details. (2) Draw assembly. (See Fig. 14.104.) If assigned, convert dimensions to decimal or metric system.

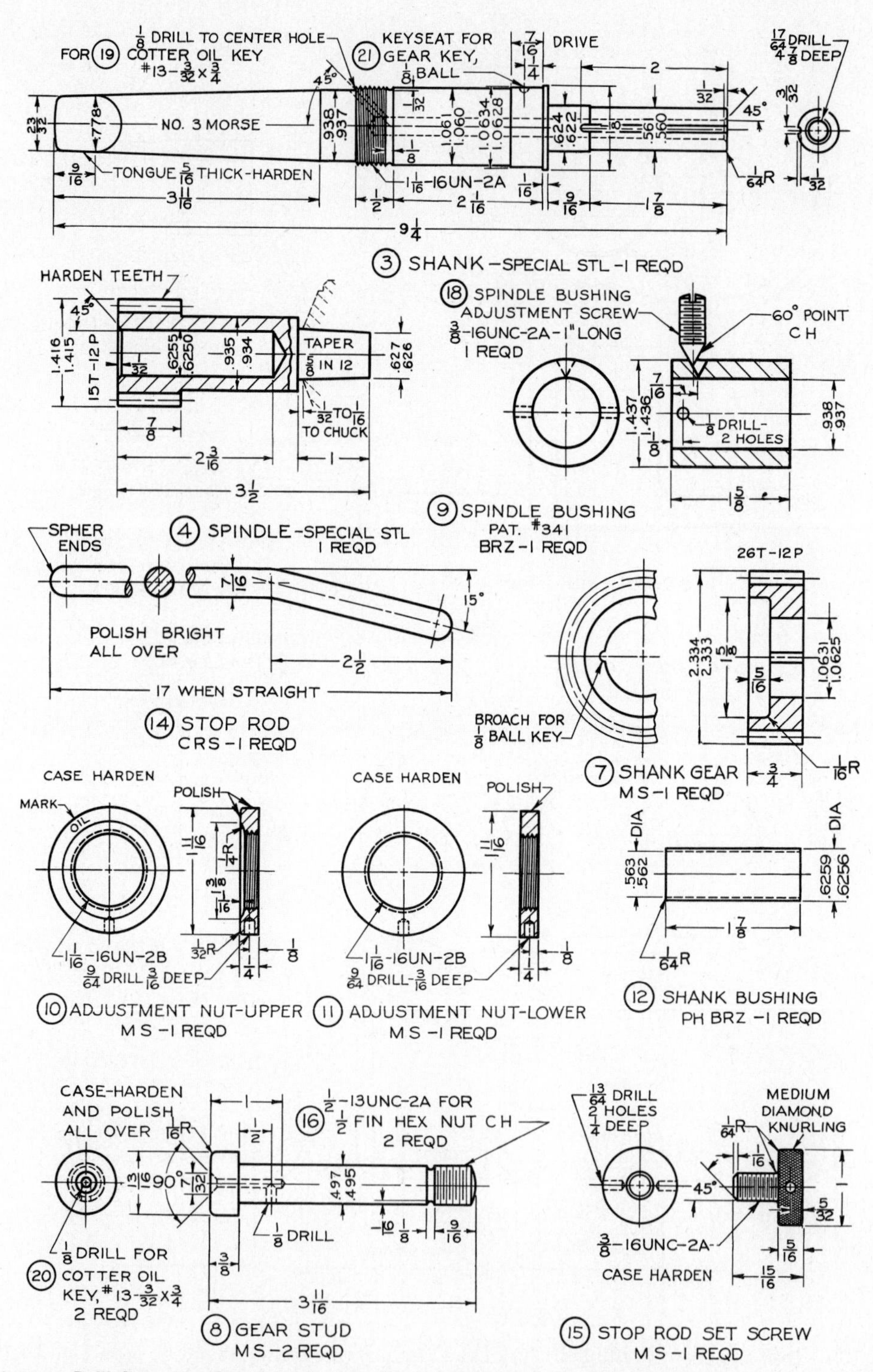


Fig. 14.106 **Drill Speeder** (Continued). See Fig. 14.105 for instructions.

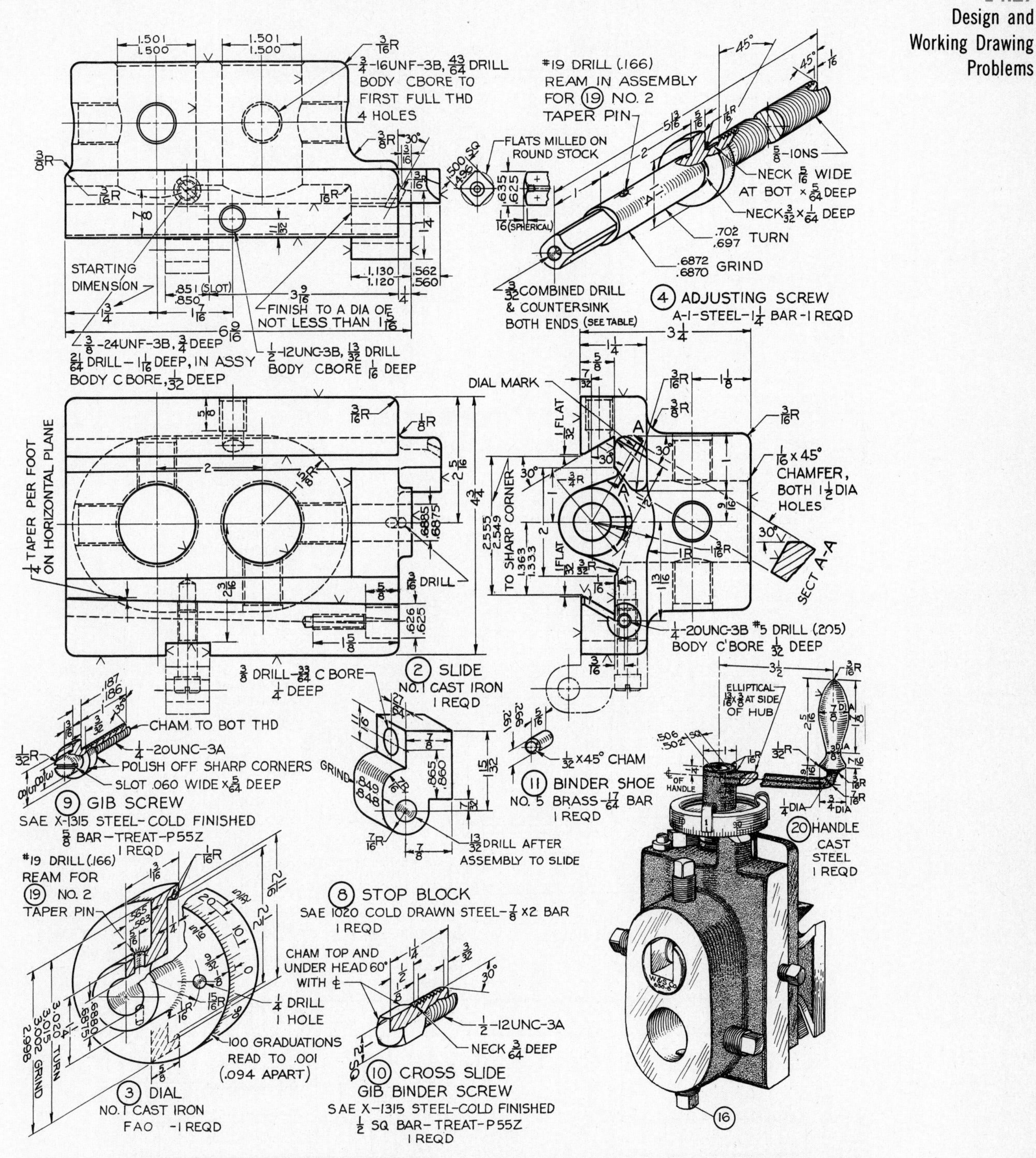


Fig. 14.107 **Vertical Slide Tool.** (1) Draw details. If assigned, convert dimensions to decimal or metric system. (2) Draw assembly. For part 2: Take given top view as front view in the new drawing; then add top and right-side views. (See also Fig. 14.108.) If assigned, use unidirectional dimensions.

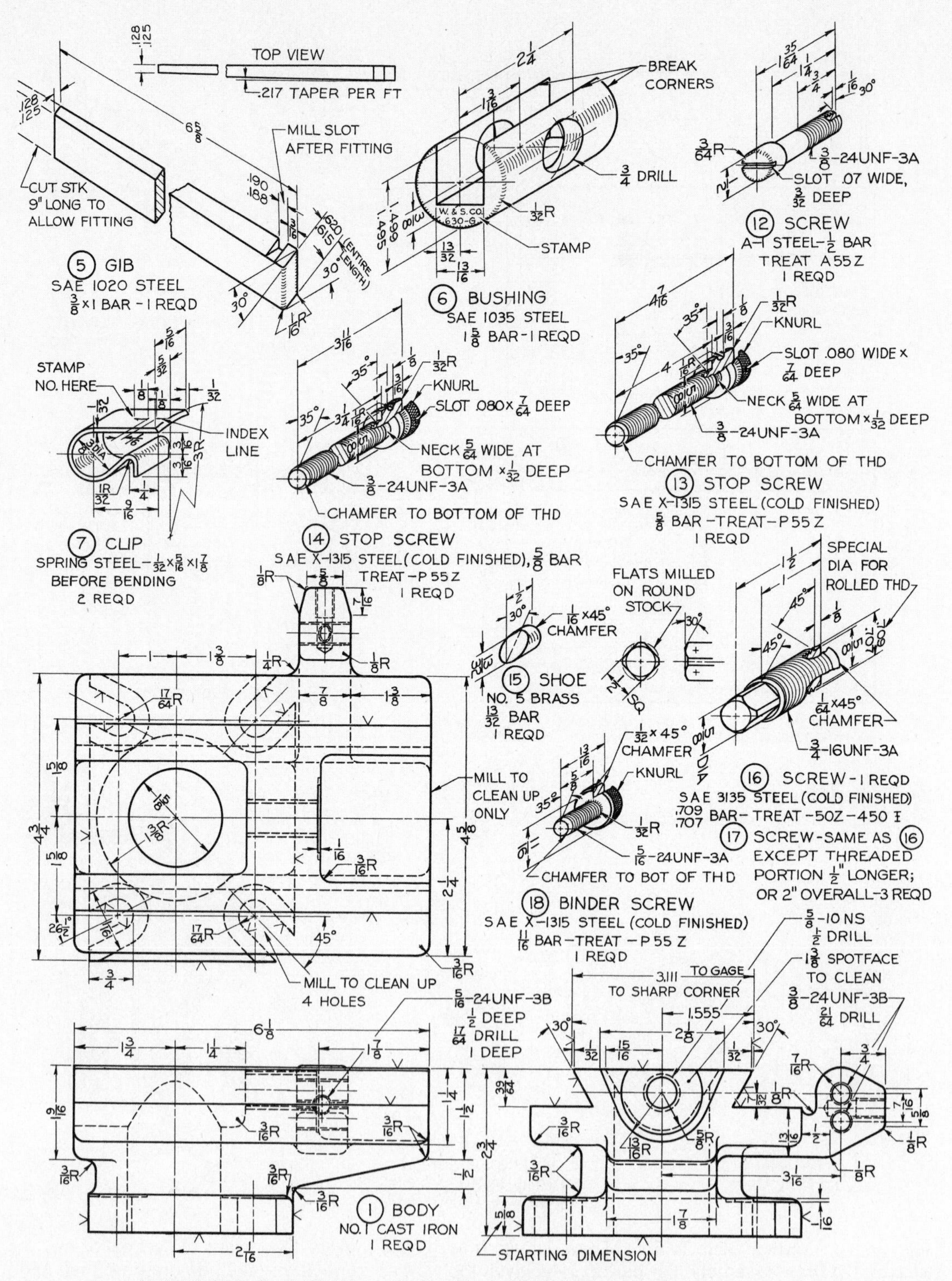


Fig. 14.108 **Vertical Slide Tool** (Continued). See Fig. 14.107 for instructions. For part 1: Take top view as front view in the new drawing; then add top and right-side views.

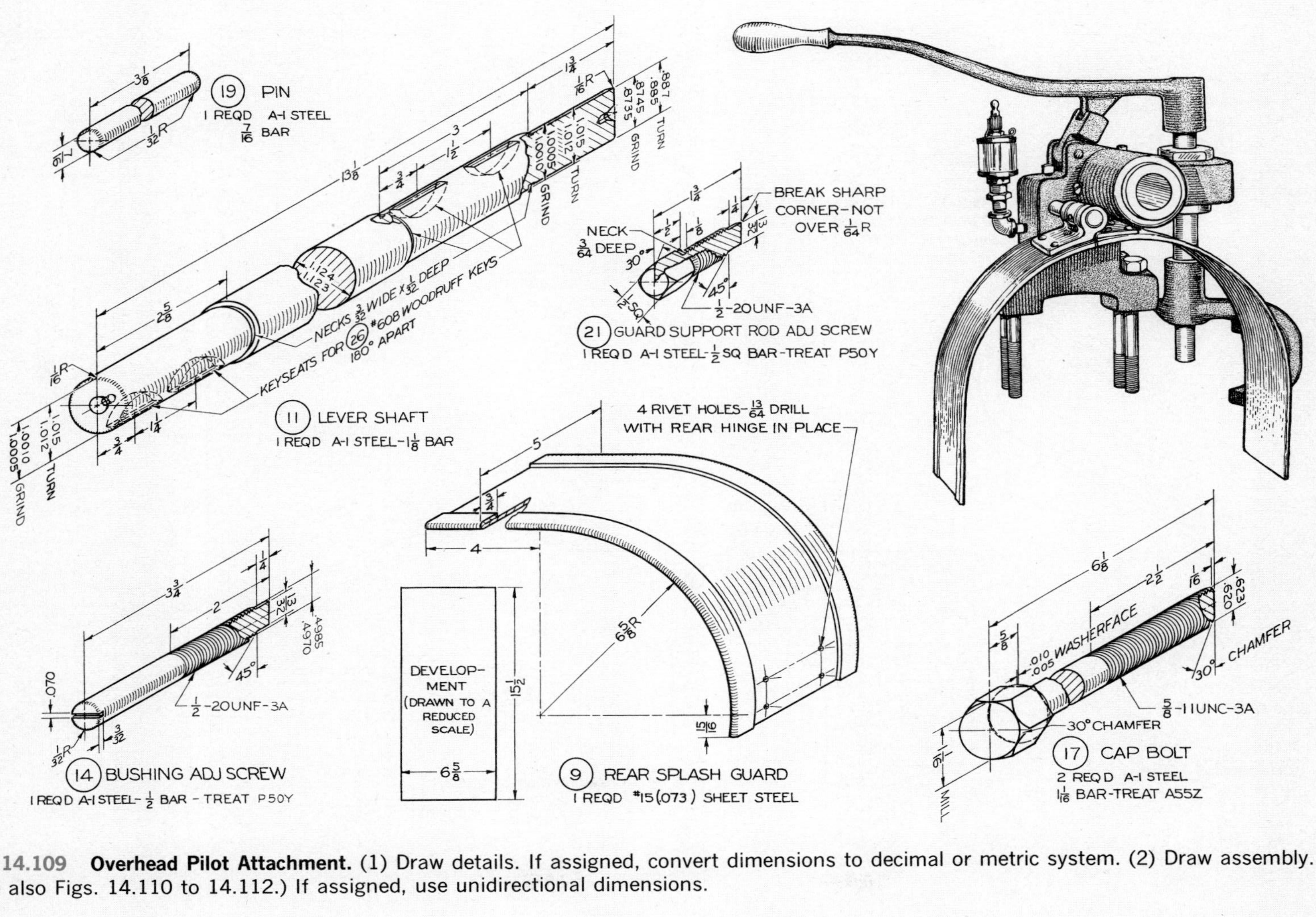


Fig. 14.109 **Overhead Pilot Attachment.** (1) Draw details. If assigned, convert dimensions to decimal or metric system. (2) Draw assembly. (See also Figs. 14.110 to 14.112.) If assigned, use unidirectional dimensions.

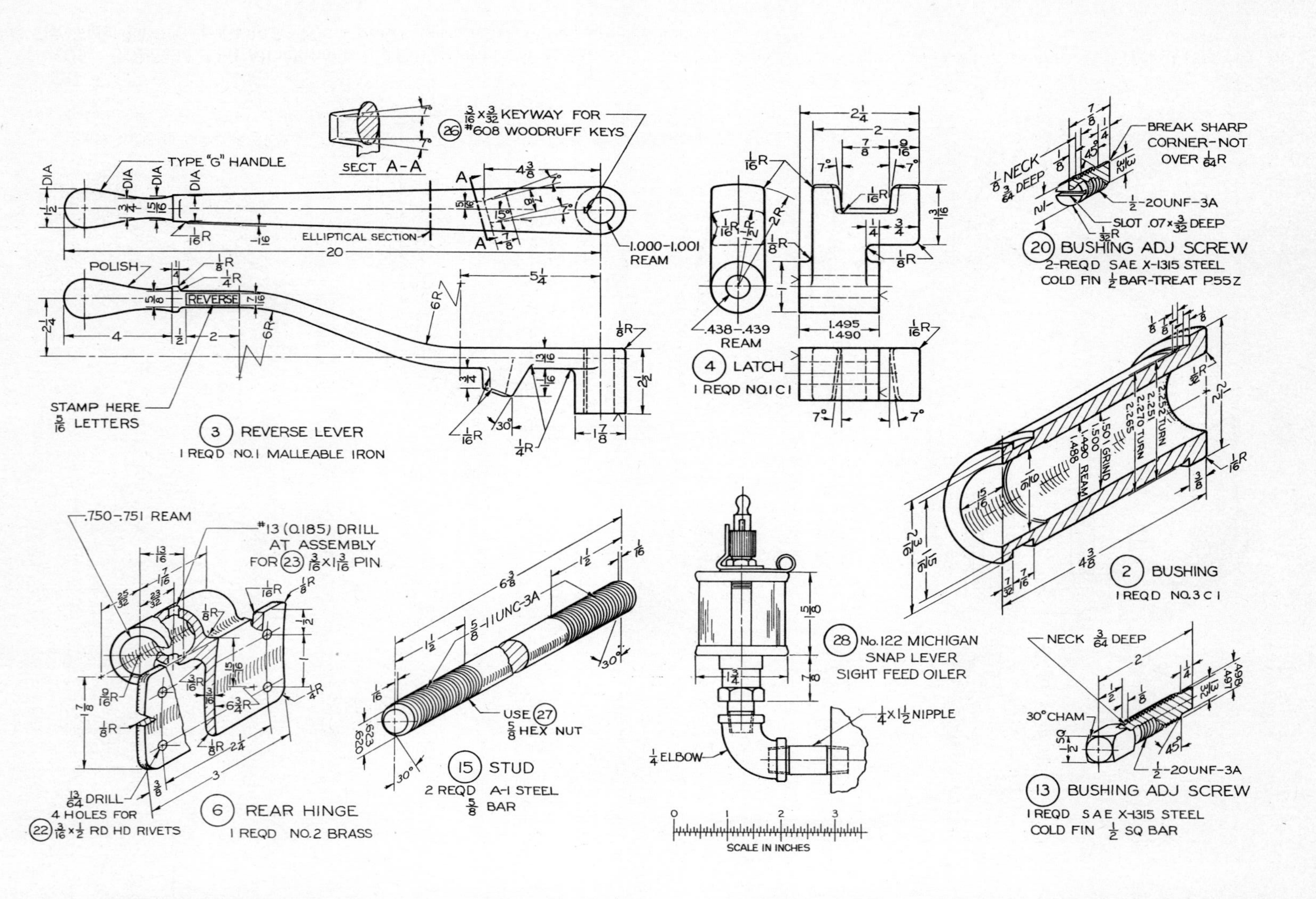


Fig. 14.110 **Overhead Pilot Attachment** (Continued). See Fig. 14.109 for instructions. Draw details. For part 4: Draw front, top, and right-side views.

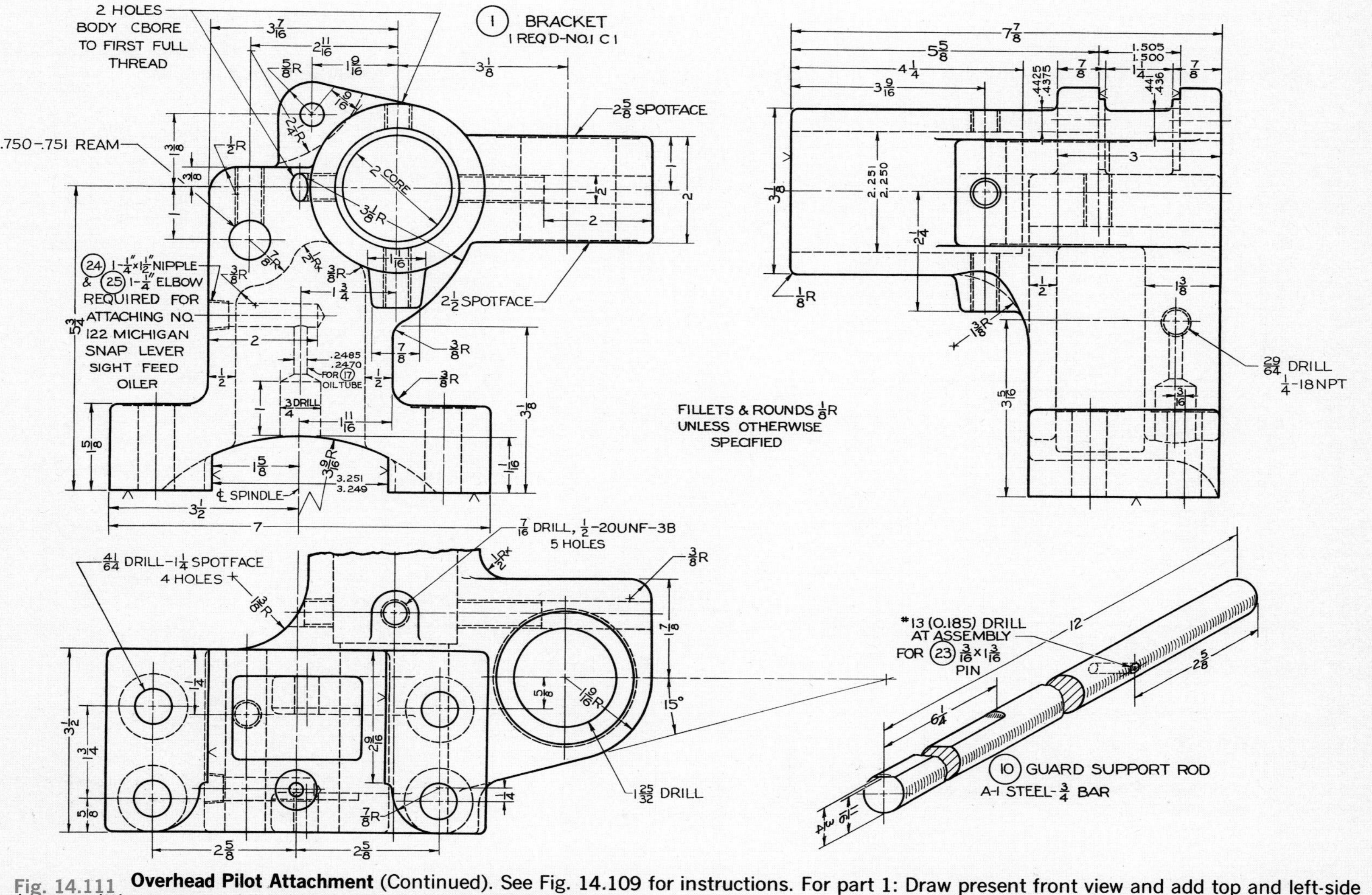


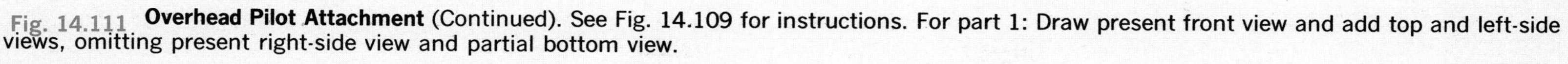

**Fig. 14.111 Overhead Pilot Attachment** (Continued). See Fig. 14.109 for instructions. For part 1: Draw present front view and add top and left-side views, omitting present right-side view and partial bottom view.

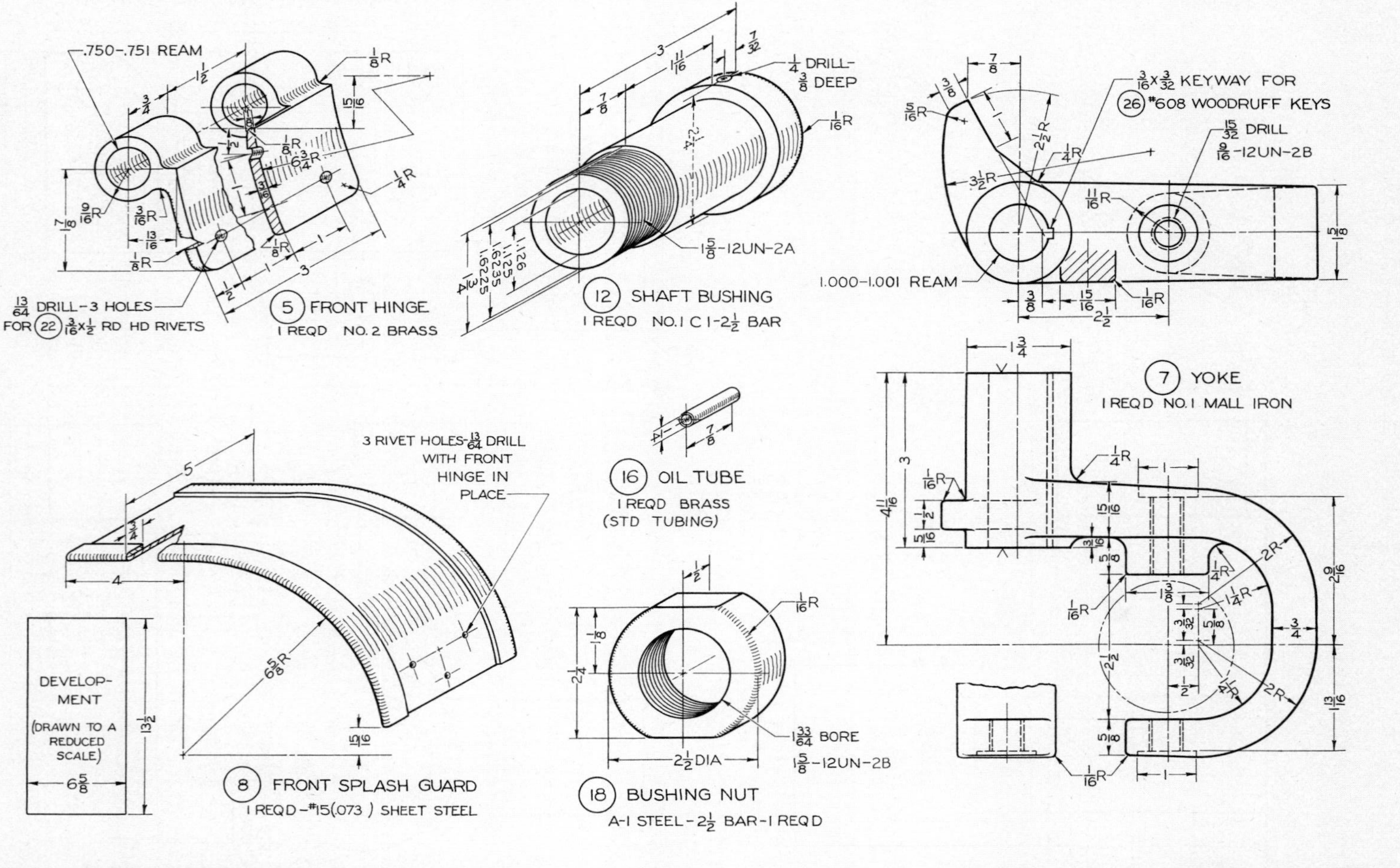


Fig. 14.112 **Overhead Pilot Attachment.** (Continued). See Fig. 14.109 for instructions. For part 7: Revolve given top view 180°; then add front and right-side views.

Fig. 14.113 **Slide Tool.** Make assembly drawing. (See Figs. 14.115 to 14.117.)

**PARTS LIST** NO. OF SHEETS 2 SHEET NO. 1 MACHINE NO. M-219

NAME NO. 4 SLIDE TOOL (SPECIFY SIZE OF SHANK REQ'D.) LOT NUMBER

NO. OF PIECES

| TOTAL ON MACH. | NO. PCS. | NAME OF PART | PART NO. | CAST FROM PART NO. | TRACING NO. | MATERIAL | ROUGH WEIGHT PER PC. | DIA. | LENGTH | MILL | PART USED ON | NO. REQ. FINISH |
|---|---|---|---|---|---|---|---|---|---|---|---|---|
| | 1 | Body | 219-12 | | D-17417 | A-3-S D F | | | | | | |
| | 1 | Slide | 219-6 | | D-19255 | A-3-S D F | | | | | 219-12 | |
| | 1 | Nut | 219-9 | | E-19256 | #10 BZ | | | | | 219-6 | |
| | 1 | Gib | 219-1001 | | C-11129 | S A E 1020 | | | | | 219-6 | |
| | 1 | Slide Screw | 219-1002 | | C-11129 | A-3-S | | | | | 219-12 | |
| | 1 | Dial Bush. | 219-1003 | | C-11129 | A-1-S | | | | | 219-1002 | |
| | 1 | Dial Nut | 219-1004 | | C-11129 | A-1-S | | | | | 219-1002 | |
| | 1 | Handle | 219-1011 | | E-18270 | (Buy from Cincinnati Ball Crank Co.) | | | | | 219-1002 | |
| | 1 | Stop Screw (Short) | 219-1012 | | E-51950 | A-1-S | | | | | 219-6 | |
| | 1 | Stop Screw (Long) | 219-1013 | | E-51951 | A-1-S | | | | | 219-6 | |
| | 1 | Binder Shoe | 219-1015 | | E-51952 | #5 Brass | | | | | 219-6 | |
| | 1 | Handle Screw | 219-1016 | | E-62322 | X-1315 C F | | | | | 219-1011 | |
| | 1 | Binder Screw | 219-1017 | | E-63927 | A-1-S | | | | | 219-6 | |
| | 1 | Dial | 219-1018 | | E-39461 | A-1-S | | | | | 219-1002 | |
| | 2 | Gib Screw | 219-1019 | | E-52777 | A-1-S | | $\frac{1}{4}$-20 | 1 | | 219-6 | |
| | 1 | Binder Screw | 280-1010 | | E-24962 | A-1-S | | | | | 219-1018 | |
| | 2 | Tool Clamp Screws | 683-F-1002 | | E-19110 | D-2-S | | | | | 219-6 | |
| | 1 | Fill Hd Cap Scr | 1-A | | | A-1-S | | $\frac{3}{8}$ | 1 $\frac{3}{8}$ | | 219-6 219-9 | |
| | 1 | Key | No.404 Woodruff | | | | | | | | 219-1002 | |

Fig. 14.114 **Slide Tool Parts List.**

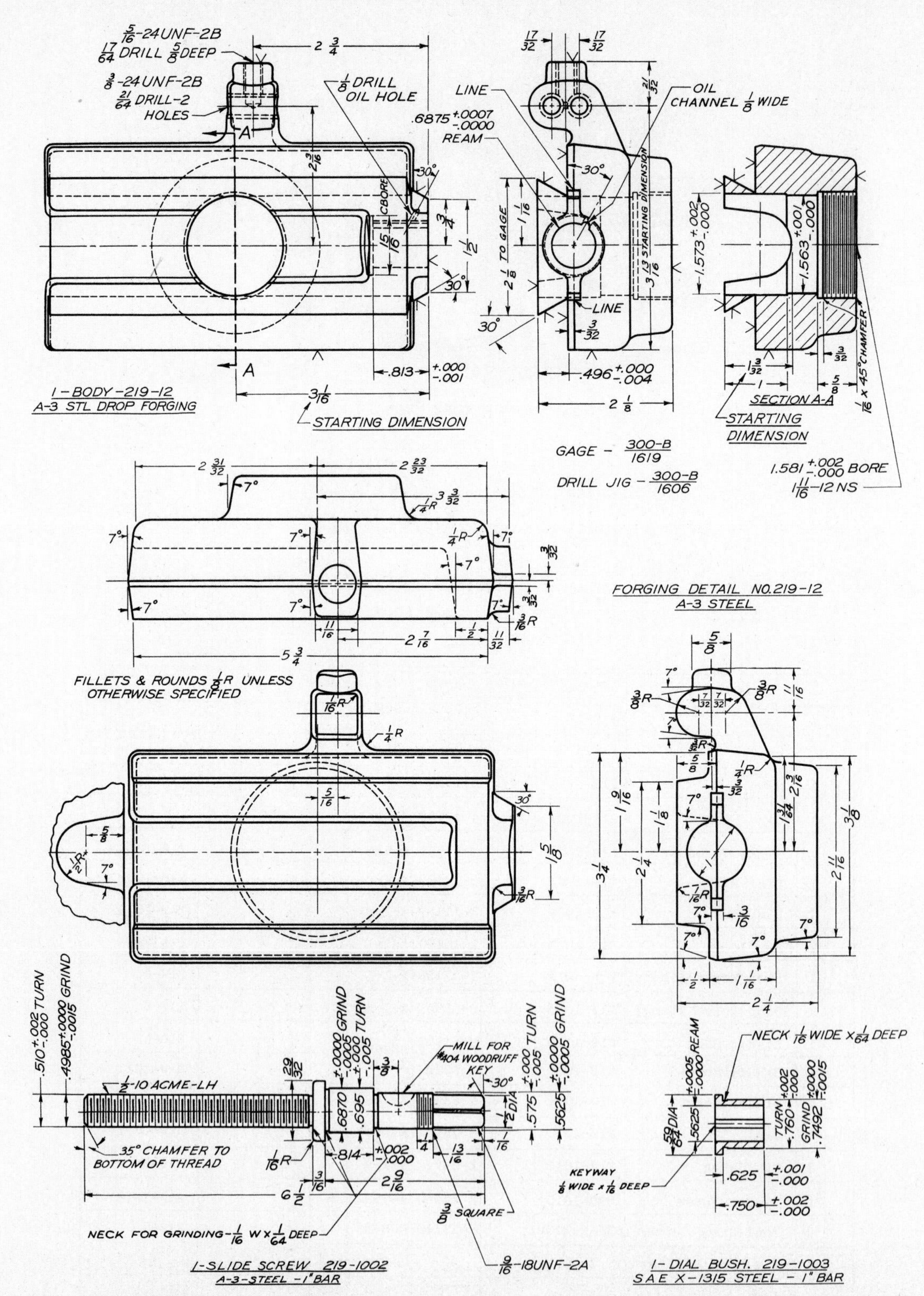


Fig. 14.115 **Slide Tool** (Continued). (1) Draw details using decimal or metric dimensions, if assigned. (2) Draw assembly. (See Fig. 14.113.)

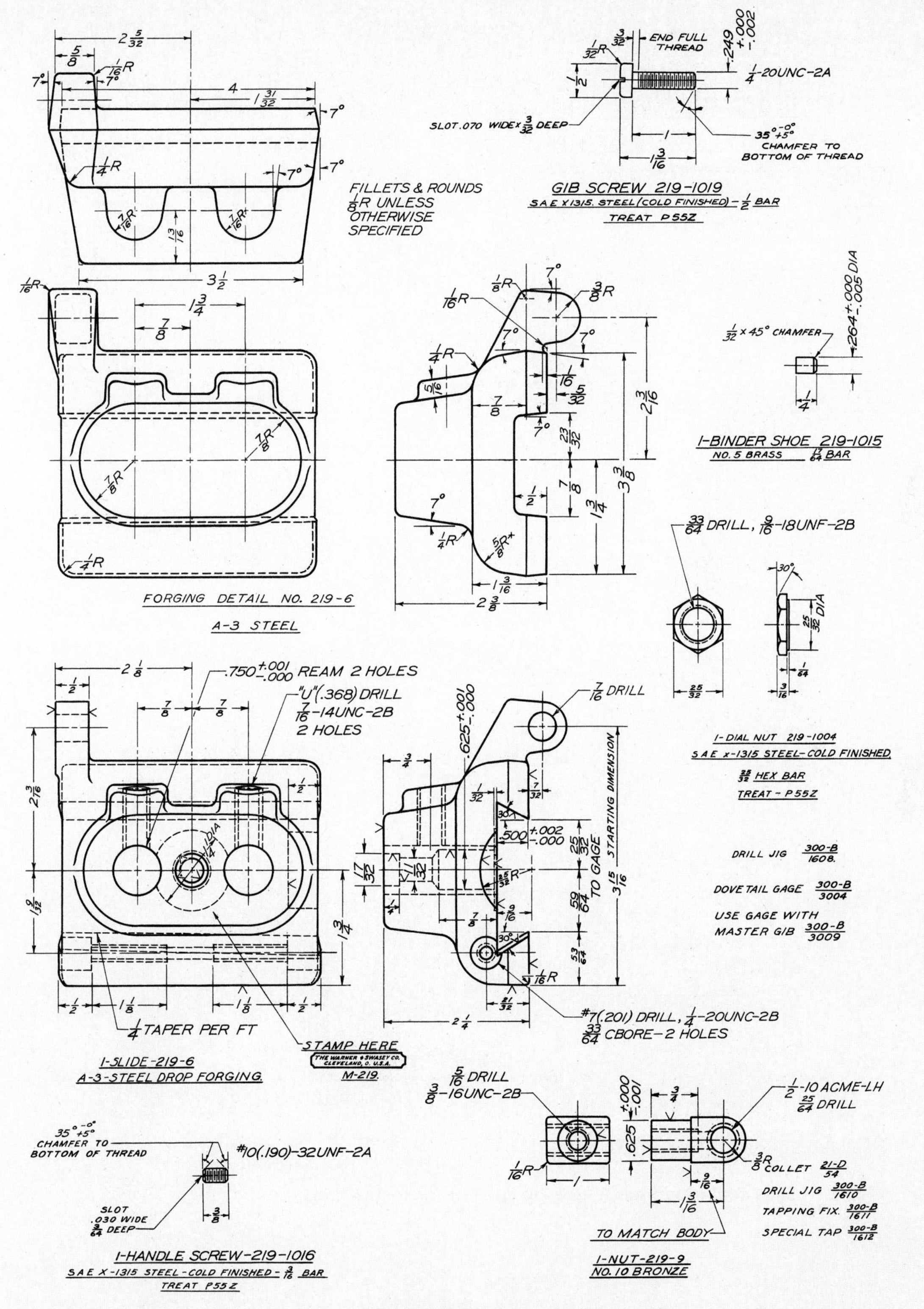
FILLETS & ROUNDS
1/8 R UNLESS
OTHERWISE
SPECIFIED
FORGING DETAIL NO. 219-6
A-3 STEEL
SLOT .070 WIDE x 3/32 DEEP
END FULL THREAD
1/4-20UNC-2A
CHAMFER TO
BOTTOM OF THREAD
GIB SCREW 219-1019
SAE X1315 STEEL (COLD FINISHED) - 1/2 BAR
TREAT P55Z
1/32 x 45° CHAMFER
1-BINDER SHOE 219-1015
NO. 5 BRASS 17/64 BAR
33/64 DRILL, 9/16-18UNF-2B
1-DIAL NUT 219-1004
SAE x-1315 STEEL-COLD FINISHED
25/32 HEX BAR
TREAT - P55Z
.750 +.001 -.000 REAM 2 HOLES
"U" (.368) DRILL
7/16-14UNC-2B
2 HOLES
7/16 DRILL
STARTING DIMENSION
TO GAGE
1/4 TAPER PER FT
STAMP HERE
THE WARNER & SWASEY CO.
CLEVELAND, O. U.S.A.
M-219.
#7(.201) DRILL, 1/4-20UNC-2B
33/64 CBORE-2 HOLES
1-SLIDE-219-6
A-3-STEEL DROP FORGING.
DRILL JIG 300-B/1608.
DOVE TAIL GAGE 300-B/3004
USE GAGE WITH
MASTER GIB 300-B/3009
CHAMFER TO
BOTTOM OF THREAD
#10(.190)-32UNF-2A
SLOT
.030 WIDE
3/64 DEEP
1-HANDLE SCREW-219-1016
SAE X-1315 STEEL-COLD FINISHED - 3/16 BAR
TREAT P55Z
5/16 DRILL
3/8-16UNC-2B
1/2-10 ACME-LH
25/64 DRILL
COLLET 21-D/54
DRILL JIG 300-B/1610
TAPPING FIX. 300-B/1611
SPECIAL TAP 300-B/1612
TO MATCH BODY
1-NUT-219-9
NO. 10 BRONZE

Fig. 14.116 **Slide Tool** (Continued). See Fig. 14.115 for instructions.

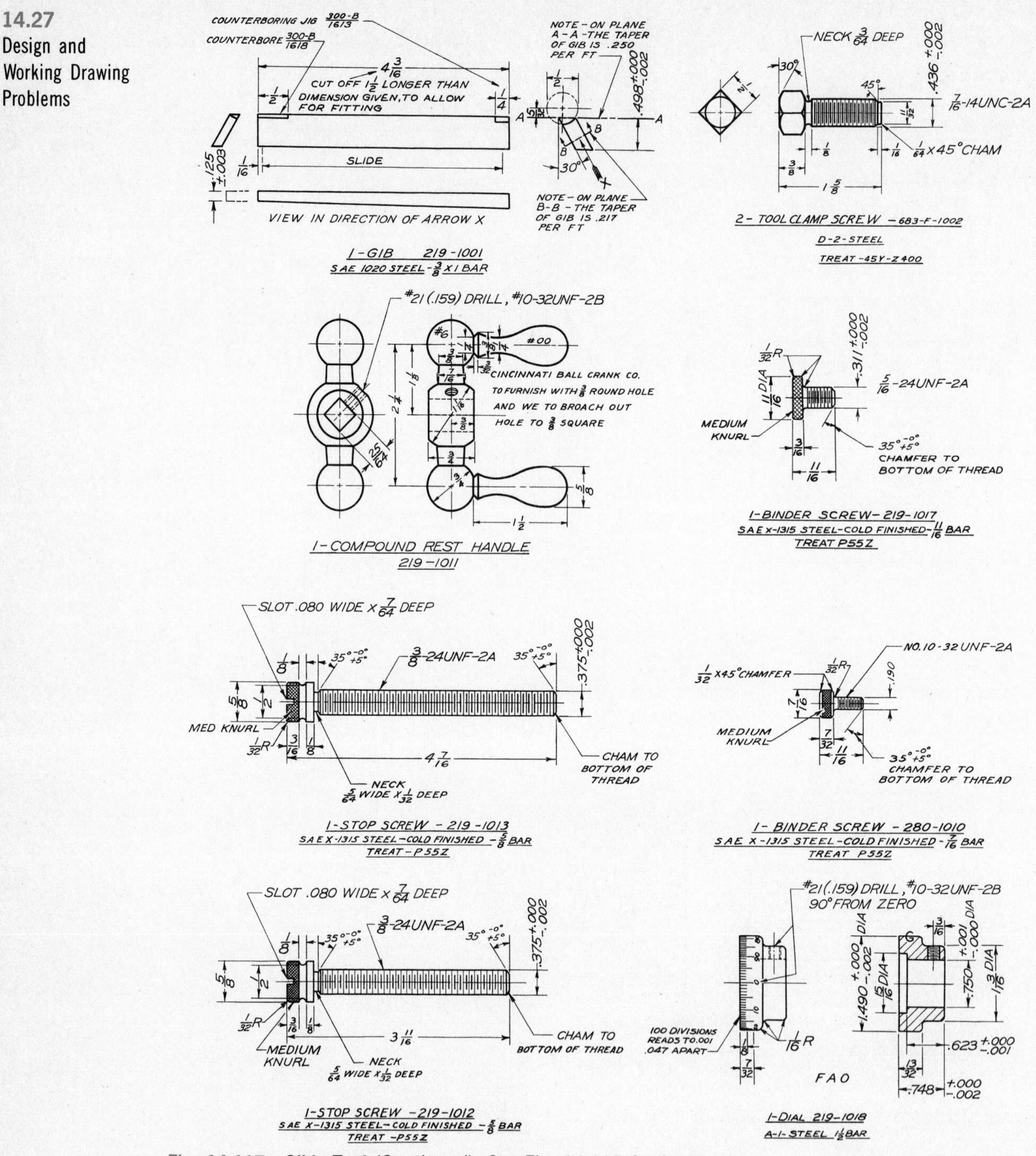


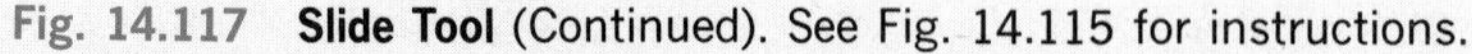

Fig. 14.117 **Slide Tool** (Continued). See Fig. 14.115 for instructions.

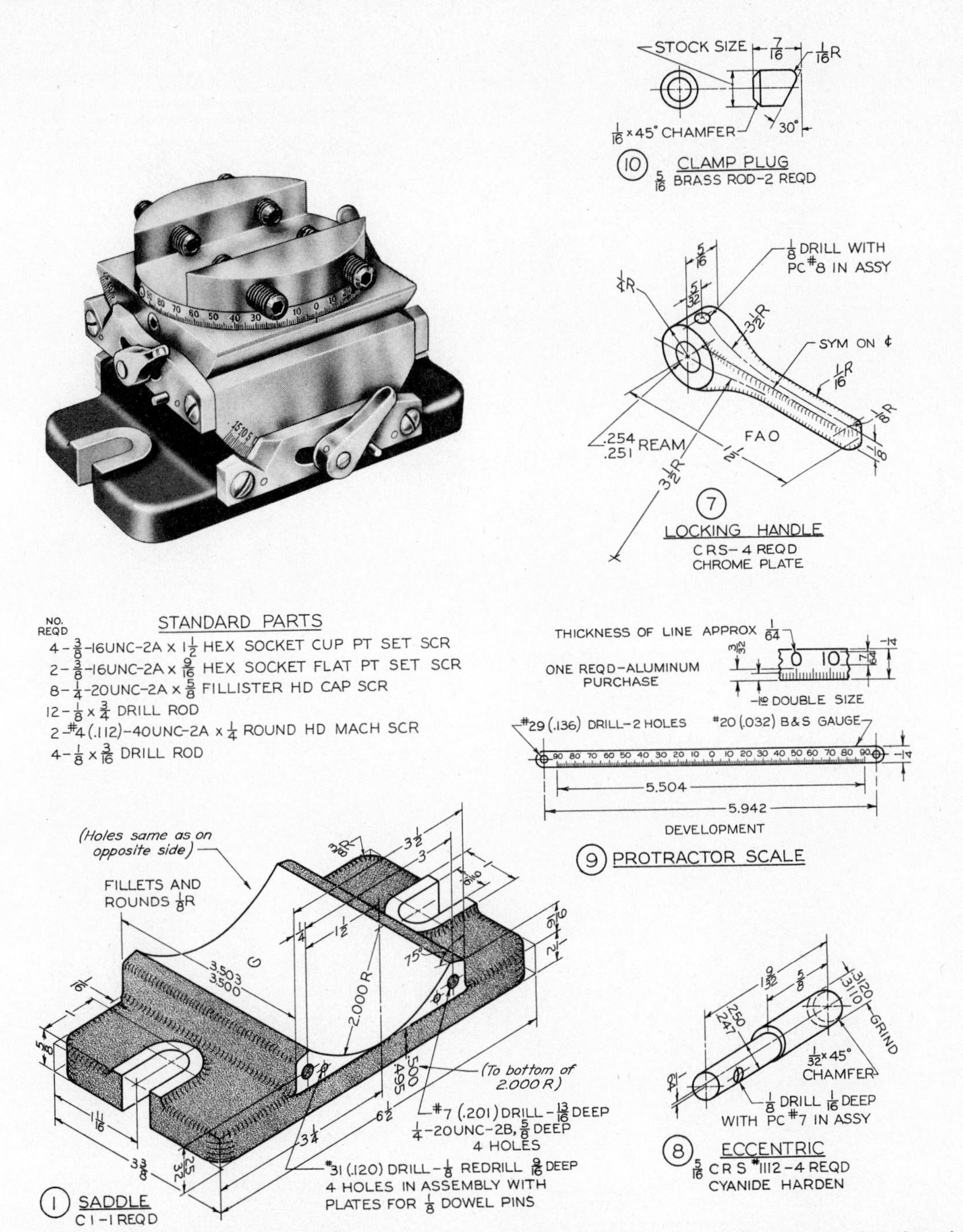


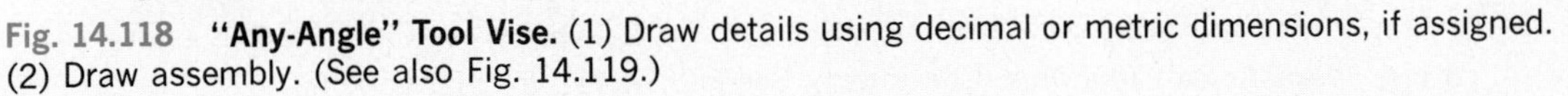
Fig. 14.118 **"Any-Angle" Tool Vise.** (1) Draw details using decimal or metric dimensions, if assigned. (2) Draw assembly. (See also Fig. 14.119.)

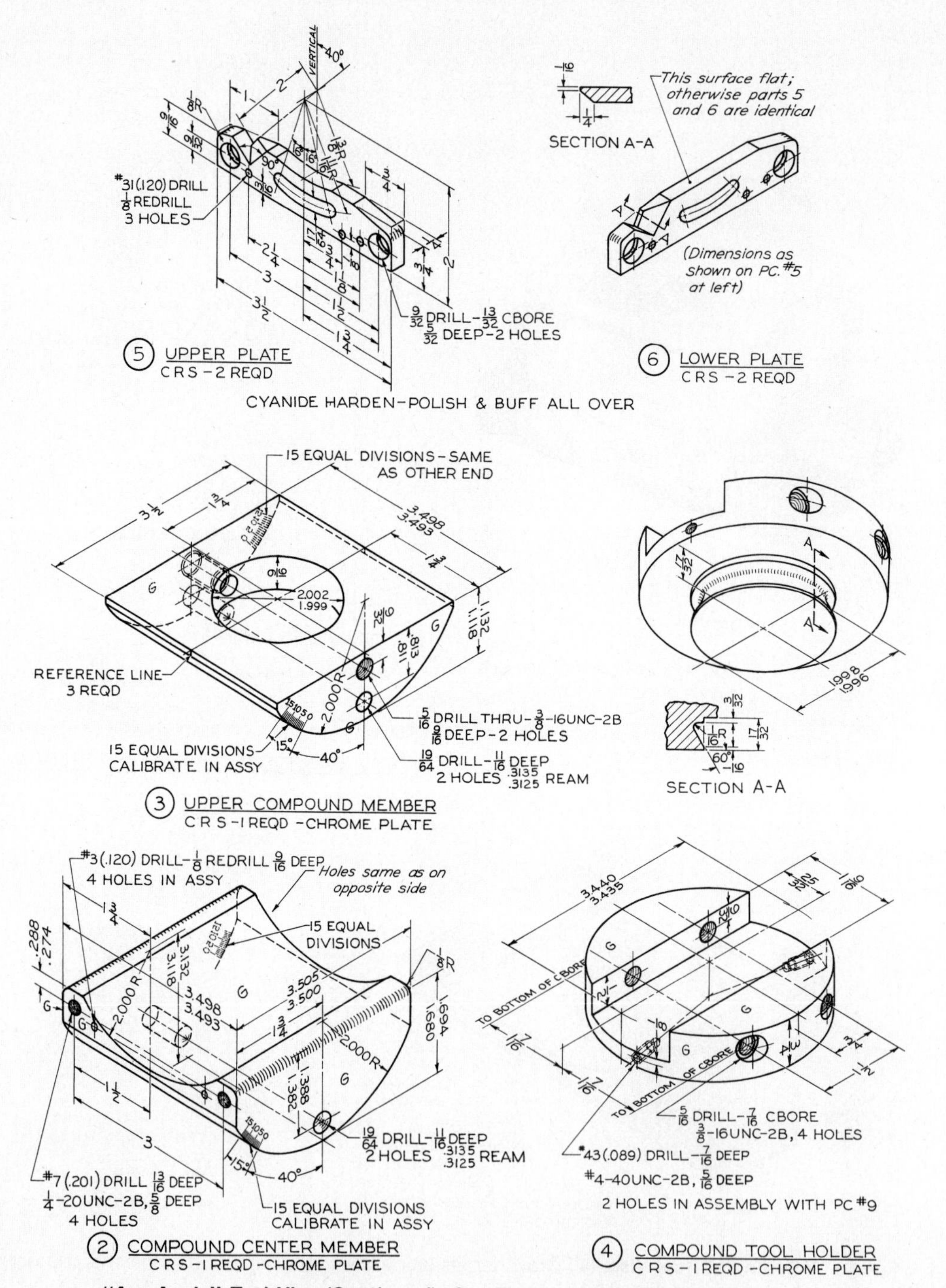


Fig. 14.119 **"Any-Angle" Tool Vise** (Continued). See Fig. 14.118 for instructions.

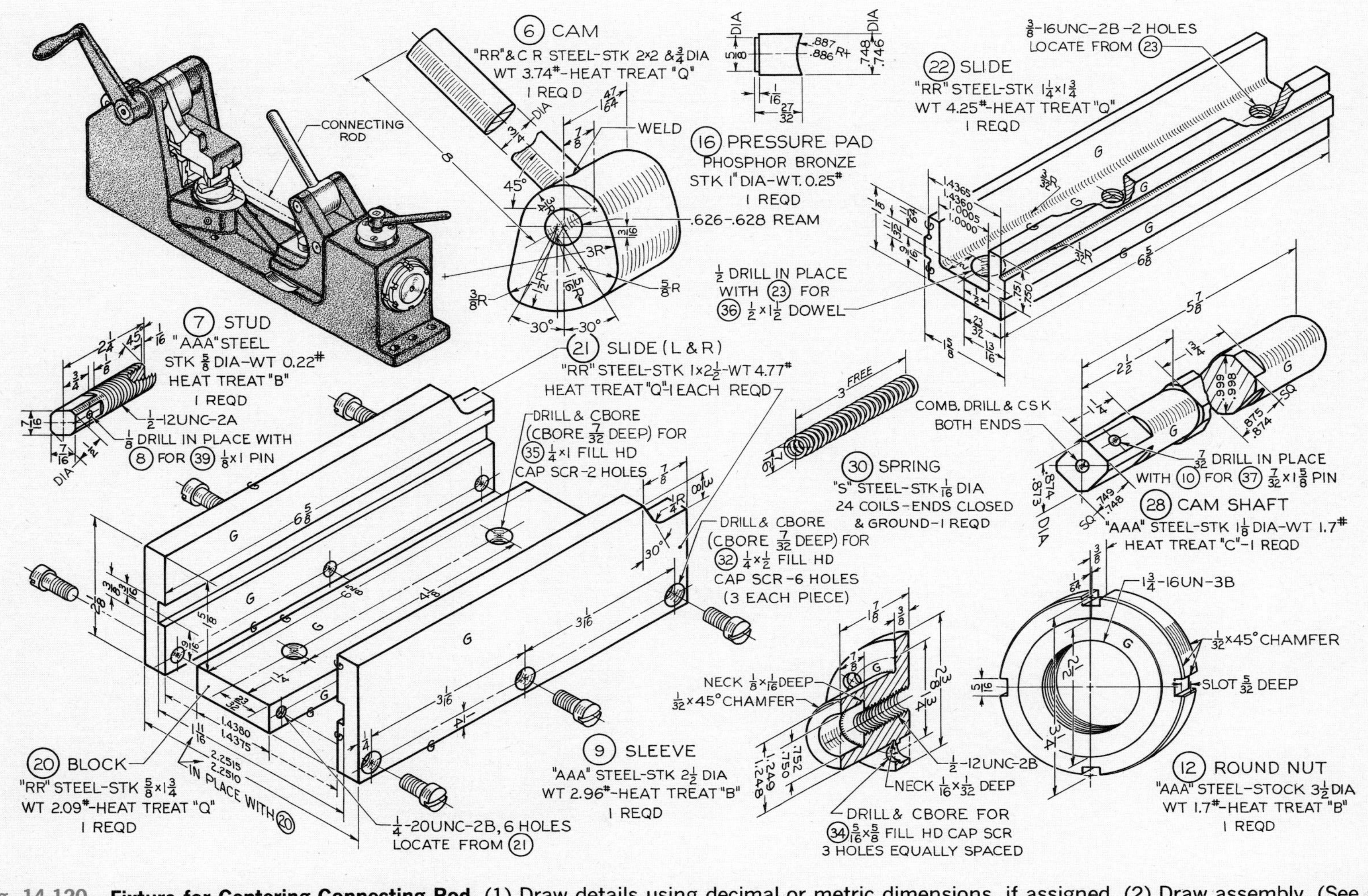


Fig. 14.120 **Fixture for Centering Connecting Rod.** (1) Draw details using decimal or metric dimensions, if assigned. (2) Draw assembly. (See also Figs. 14.121 and 14.122.)

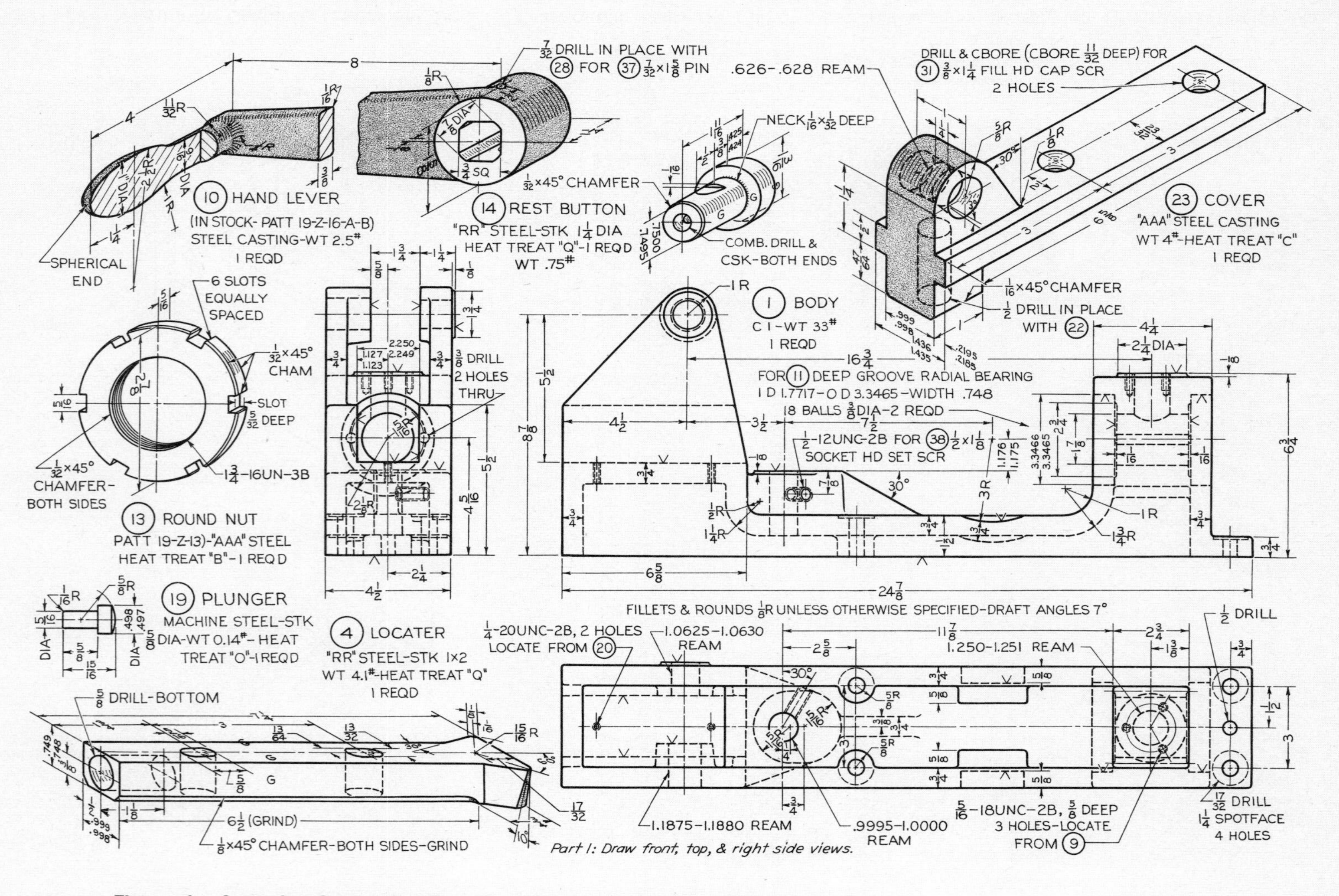


Fig. 14.121 **Fixture for Centering Connecting Rod** (Continued). See Fig. 14.120 for instructions.

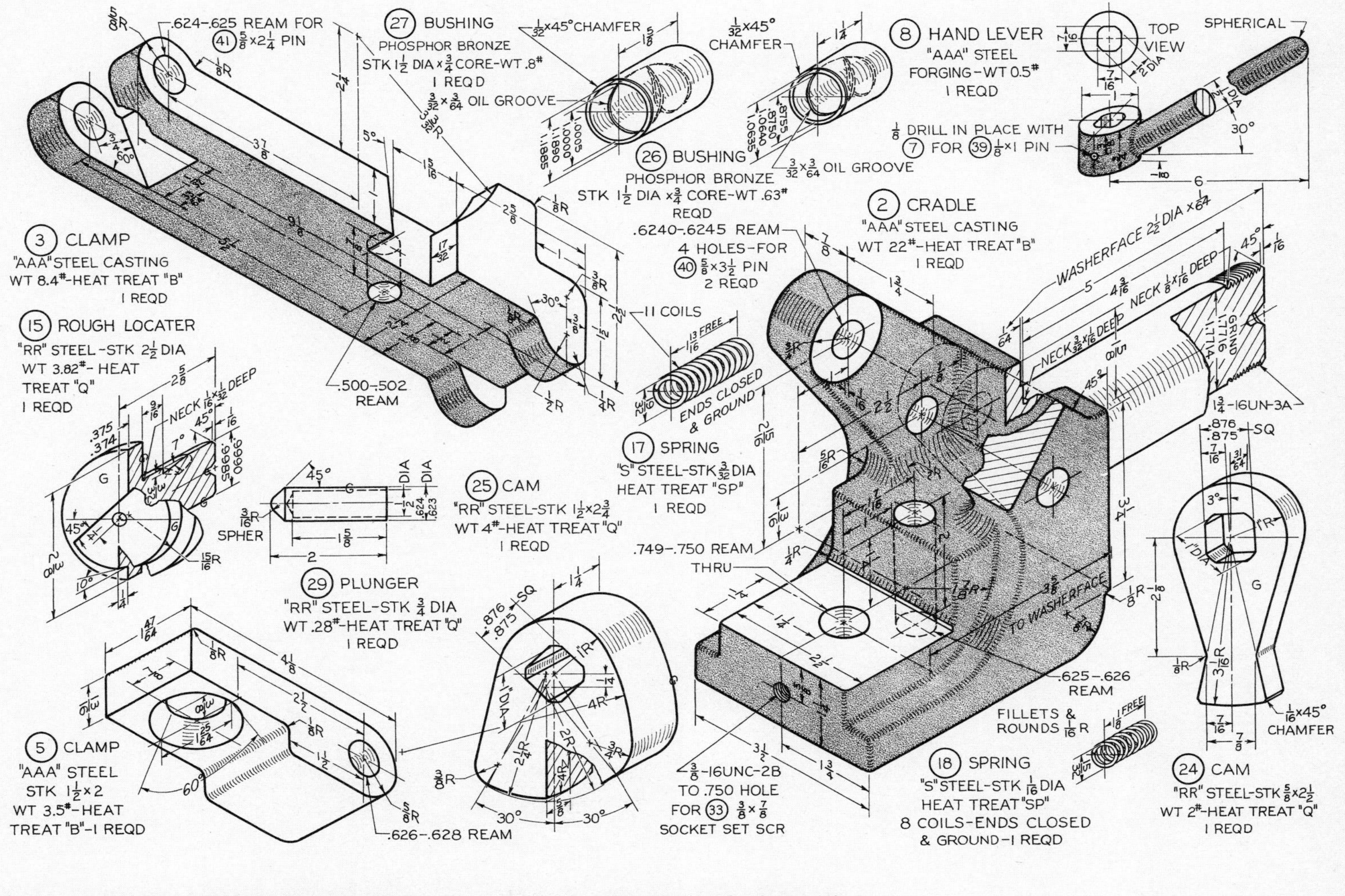


Fig. 14.122 **Fixture for Centering Connecting Rod** (Continued). See Fig. 14.120 for instructions.

# 15

# reproduction and control of drawings

**15.1 Introduction.** Since the average engineering drawing has required a considerable economic investment, adequate control and protection of the original drawing are mandatory. Such items as drawing numbers, §14.13, methods of filing, microfilming, security files, print making and distribution, drawing changes, and retrieval of drawings are all inherent in proper drawing control.

A proper drawing-control system will enable those in charge of drawings (a) to know the location and status of the drawing at all times, (b) to minimize the damage to original drawings from the handling required for revisions, printing, etc., and (c) to provide distribution of prints to proper persons.

**15.2 Drawing Storage.** Drawings may be stored flat in large flat-drawer files or hung vertically in cabinets especially designed for the purpose. Exceptionally large drawings are often rolled and stored in tubes in racks or cabinets. Prints are often folded and stored in standard office file cases. Proper control procedures will enable the user of the drawing to find it in the file, to return it to its proper place, and to know where the drawing is when not in the file.

**15.3 Reproduction of Drawings.** After the drawings of a machine or structure have been completed, it is usually necessary to supply copies to many different persons and firms. Obviously, therefore, some means of exact, rapid, and economical reproduction must be used.

**15.4 Blueprint Process.** Of the several processes in use for reproduction, the *blueprint*

*Courtesy Charles Bruning Company, Division of Addressograph Multigraph Corp.*

Fig. 15.1 **Automatic Blueprinting Machine.**

process is the oldest and is still used for making prints of large architectural and structural drawings. It is essentially a photographic process in which the original drawing is the negative.

Blueprint papers are made by applying a coating of a solution of potassium ferricyanide and ferric ammonium citrate, a chemical preparation sensitive to light. They are available in various speeds and in rolls of various widths, or may be supplied in sheets of specified size. The coated side of fresh paper is a light greenish yellow color. It will gradually turn to a greyish blue color, if not kept away from light, and may eventually be rendered useless.

After the paper has been exposed a sufficient length of time, it is subjected to a developing bath or a fixing bath, or to a fixing bath only, according to the method employed.

Modern blueprint machines are available in *noncontinuous* types in which cut sheets are fed through the machine for exposure only and then washed in a separate washer. The continuous blueprint machine, Fig. 15.1, combines exposure, washing, and drying in one continuous operation.

Fig. 15.2 **Sun Frame.**

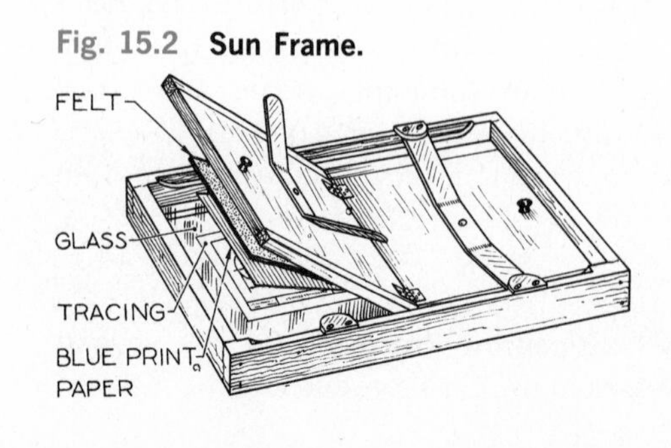


When only a small number of prints are required and better facilities are not available, prints can be made by exposing the tracing and print paper to sunlight while they are held tightly together against a window pane or under a piece of glass. The tracing must be against the glass with its face toward the light and the sensitized surface of the print paper against the back of the tracing. A *sun frame* for this purpose is shown in Fig. 15.2.

Although best results are obtained when the original tracing is drawn in ink on cloth or vellum, excellent prints may be made from penciled drawings or tracings if the tracing paper or pencil tracing cloth is of good quality and if the draftsman has made all required lines and lettering jet black.

Notations and corrections can be made on blueprints with any alkaline solution of sufficient strength to destroy the blue compound; for instance, with a 1.5 percent solution of caustic soda.

**15.5 Vandyke Prints and Blue-Line Blueprints.** A *negative Vandyke print* is composed of white lines on a dark brown background made by printing, in the same manner as for blueprinting, upon a special thin Vandyke paper from an original pencil or ink tracing. This negative Vandyke is then used as an original to make positive blueprints or "blue-line blueprints." In this way the Vandyke print replaces the original tracing as the negative, and the original is not subjected to wear each time a run of prints is made. The blue-line blueprints have blue lines on white backgrounds, and are often preferred because they can be easily marked upon with an ordinary pencil or pen. They have the disadvantage of soiling easily in the shop.

**15.6 Diazo-Moist Prints.** A black-and-white print, composed of nearly black lines on a white background, may be made from ordinary pencil or ink tracings by exposure in the same manner as for blueprints, directly upon special blackprint paper, cloth, or film.

Exposure may be made in a blueprint machine or any machine using light in a similar way. However, the prints are not washed as in blueprinting, but must be fed through a special developer which dampens the coated side of the paper with a developing solution.

Colored-line prints in red, brown, or blue lines on white backgrounds may be made on the same machine simply by using the appropriate paper in each case.

These prints, together with *diazo-dry* prints, §15.7, have largely replaced the more cumbersome blueprint process.

**15.7 The Diazo-Dry Process.** The *diazo-dry* process is based on the sensitivity to light of certain dyestuff intermediates that have the characteristic of decomposing into colorless substances if exposed to actinic light, and of reacting with coupling components to form an azo dyestuff upon exposure to ammonia vapors. It is a contact method of reproduction, and depends upon the transmission of light through the original for the reproduction of positive prints. The subject matter may be pen or pencil lines, typewritten or printed matter, or any opaque image. There is no negative step involved; positives are used to obtain positive prints. Sensitized materials can be handled under normal indoor illumination.

The *diazo* whiteprint method of reproduction consists of two simple steps—exposure and dry development by means of ammonia vapors. Exposure is made in a printer equipped with a source of ultraviolet light, a mercury vapor lamb, or carbon arc, or even by sunlight. The light emitted by these light sources brings about a photochemical decomposition of the light-sensitive yellow coating of the paper except in those places where the surface is protected by the opaque lines of the original. The exposed print is developed dry in a few seconds in a dry-developing machine by the alkaline medium produced by ammonia vapors.

A popular combination printer (exposer) and developer, the *Blu-Ray 747*, is shown in Fig. 15.3. The tracing and the sensitized paper are fed into the machine, and when they emerge, the print is practically dry and ready for use.

*Courtesy Blu-Ray, Inc.*

Fig. 15.3 **Blu-Ray 747.**

Another exposer and developer combined in one machine is the *Ozalid Streamliner 420* shown in Fig. 15.4. Two operations are involved: (1) the tracing and the sensitized paper

Fig. 15.4 **Ozalid Streamliner 420.**

*Courtesy General Aniline & Film Corp.*

are fed into the printer slot for exposure to light, as shown in Fig. 15.5 (a); and (2) the paper is then fed through the developer slot for exposure to ammonia vapors, as shown at (b). If it is desired to remove the ammonia odor completely, the print is then fed through the printer with the back of the sheet next to the warm glass surrounding the light.

Ozalid prints may have black, blue, or red lines on white backgrounds, according to which paper is used. All have the advantage of being easily marked upon with pencil, pen, or crayon.

Ozalid "intermediates" are made in the same manner as regular prints, but upon special translucent paper, cloth, or foil (transparent cellulose acetates). These are used in place of the original to produce regular prints. They may be used to save wear on the original or to permit changes to be made. Changes may be made by painting out parts with correction solution, Fig. 15.6 (a) and (b), and then drawing the new lines or lettering directly on the intermediate in pencil or ink, (c). Special masking and cut-out techniques can also be used.

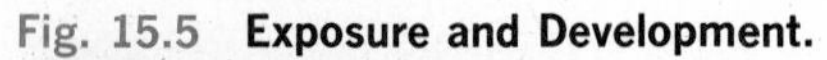

Fig. 15.5 **Exposure and Development.**

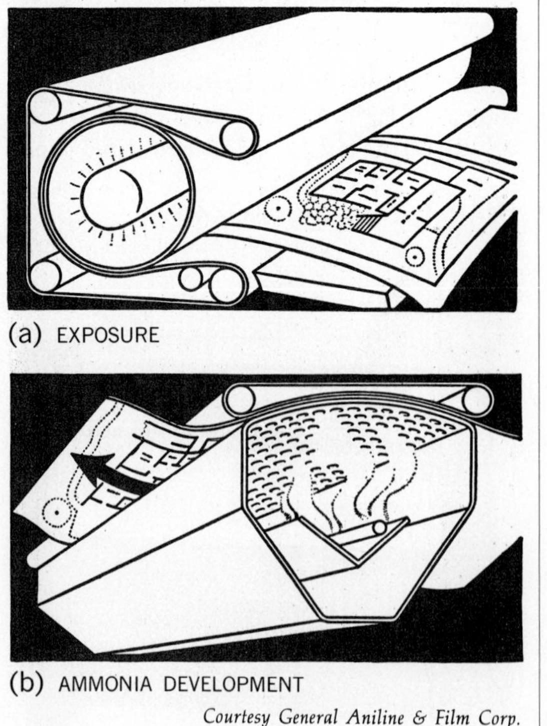

(a) EXPOSURE

(b) AMMONIA DEVELOPMENT

*Courtesy General Aniline & Film Corp.*

Several types of Ozalid foil intermediates are available, including matte surfaces to facilitate pen and pencil additions. By means of foils, many new procedures are possible, such as a composite print in which a wiring system is superimposed over a drawing.

## 15.8 Thermo-Fax Process.

The *Thermo-Fax* copy system is an infrared heat process in which the print paper is sensitive to heat instead of light. Copies are dry and have black lines on a light background. The original drawing need not be on transparent paper. The Thermo-Fax machine is easy to operate, and produces a copy in a few seconds, but the process is rather expensive if several copies of a drawing are required.

## 15.9 Verifax Process.

The *Verifax* process is a dye-transfer process. The original is exposed in the Verifax machine to produce a light-activated matrix, which is then placed in contact with the copy paper. The two are then separated, with the print transferred to the copy paper. Additional copies may be had from the same matrix. The Verifax process is sensitive to all types of copy, but it is a wet process and relatively slow.

## 15.10 Mimeographing and Hectographing.

While *mimeographing* is especially adaptable for reproducing typed material, it can also be very satisfactory in reproducing small and relatively simple drawings. The excellence of the reproduction of such drawings will depend upon the skill of the draftsman in drawing upon the stencil. The A. B. Dick Company, of Chicago, manufacturer of the Mimeograph, has developed a photochemical process by means of which a complicated drawing may be reduced and incorporated into the stencil, which is then used to produce very satisfactory prints.

In the *hectographing* process, an original is

produced by typing on plain paper through a special carbon paper, or drawing with a special pencil or ink. This sheet of paper is then brought into contact with a gelatin pad, which absorbs the coloring from the lines of the original. The original is then removed, and prints are produced by bringing sheets of blank paper in contact with the gelatin. A number of different machines using this basic principle are available.

## 15.11 Xerography.

*Xerox* prints are positive prints with black lines on a white background. A selenium-coated and electrostatically charged plate is used. A special camera is used to project the original onto the plate; hence, reduced or enlarged reproductions are possible. A negatively charged plastic powder is spread across the plate and adheres to the positively charged areas of the image. The powder is then transferred to paper by means of a positive electric charge and baked onto the surface to produce the final print. Prints can be made inexpensively and quickly in the fully automated Xerox copy machines. The process is dry and sensitive to all types of copy. The Xerox process is used also to produce mats for the Multilith offset duplicating method, §15.12.

Recent application of Xerography includes volume print making from original drawings or from microfilms, §15.17. In addition, the portable Xerox 400 Telecopier can receive or send over standard telephone lines in the office or in the field $8\frac{1}{2}'' \times 11''$ documents, Fig. 15.7. After the telephone circuit is established, the telephone receivers are placed in position on the machines and the document is fed into the sending machine. The copy is read and translated into signals for the receiving machine, which reproduces the document in four minutes.

## 15.12 Offset Printing.

The *Photolith*, *Multilith*, and *Planograph* methods are generally known as *offset printing*. A camera is used to reproduce the original, enlarged or reduced if

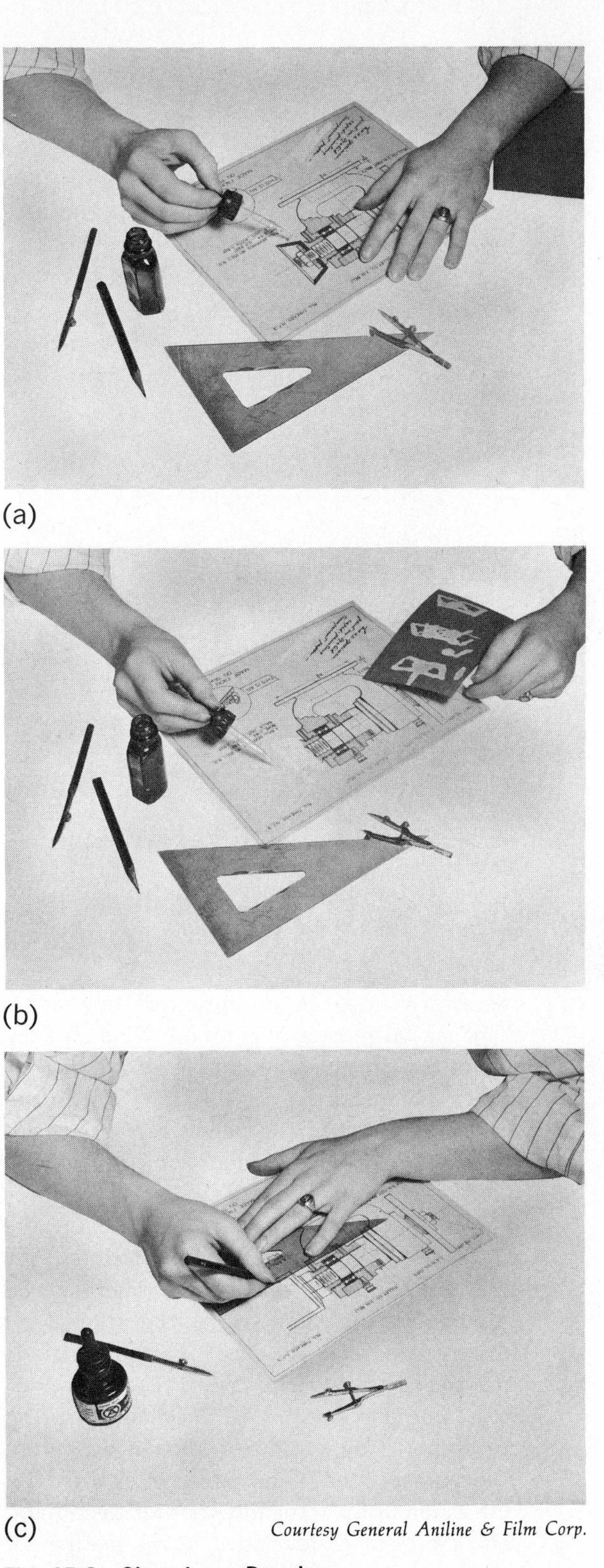

(a) (b) (c)

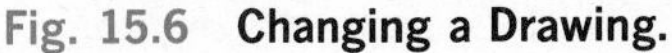

*Courtesy General Aniline & Film Corp.*

Fig. 15.6 **Changing a Drawing.**

(a) OFFICE USAGE

Courtesy Xerox Corp.

(b) FIELD USAGE

Fig. 15.7 **Xerox 400 Telecopier**

necessary, upon an aluminum or zinc sheet. This master plate is then mounted on a rotary drum that revolves in contact with a rubber roller that picks up the ink from the image and transfers it to the paper. The prints are excellent positive reproductions.

## 15.13 Photographic Contact Prints.

Either transparent or opaque drawings may be reproduced the same size as the original by means of *contact printing*. The original is pressed tightly against a sheet of special photographic paper, either by mechanical spring pressure or by means of suction as in the "vacuum printer." The paper is exposed by the action of the transmitted or reflected light, and the print is developed in the manner of a photograph in a dark or semidark room.

The *Portograph*, manufactured by Remington Rand, Inc., is a popular machine for making contact prints of this type. The Portograph is available in various sizes with copying areas from 10″ × 15″ up to 40″ × 60″.

Excellent duplicate tracings can be made on paper, on film, and on either opaque map cloth or transparent tracing cloth through this process. Poor pencil drawings or tracings can be duplicated and improved by intensifying the lines so as to be much better than the original. Pencil drawings can be transformed into "ink-like" tracings. Also, by this process, a reproduction can be made directly from a blueprint.

The Eastman Kodak Company has developed a new *Kodagraph Autopositive* paper by means of which an excellent positive print can be made directly from either a transparent or an opaque original on an ordinary blueprint machine, ammonia-developing machine,

Fig. 15.8 **Stat King II Machine** *Courtesy Visual Graphics Corp.*

Copyflex machine, or any similar machine, without use of a darkroom or costly photographic equipment.

## 15.14 Photostats.

The *Stat King* machine, Fig. 15.8, is essentially a specialized camera. A photostat print may be the same size, or larger or smaller than the original, while photographic contact prints, §15.13, must be the same size.

The original may be transparent or opaque. It is simply fastened in place, the camera is adjusted to obtain the desired size of print, and the print is made, developed, and dried in the machine (no darkroom is required). The result is a negative print with white lines on a near-black background. A positive print having near-black lines on a white background is made by photostating the negative print.

## 15.15 Duplicate Tracings on Cloth.

Specially prepared tracing cloth is available upon which a drawing may be reproduced from a negative Vandyke print of the original. Exposure of the duplicate tracing cloth may be made in a regular blueprint machine or in any machine employing light in the same way.

Excellent duplicate tracings can also be produced with special materials on any of the several types of photographic print machines, as described in §15.13.

## 15.16 Line Etching.

*Line etching* is a photographic method of reproduction. The drawing, in black lines on white paper or on tracing cloth, is placed in a frame behind a glass and photographed. This photographic negative is then mounted on a pane of glass and is printed upon a sheet of planished zinc or copper. After the print has been specially treated to render the lines acid resistant, the plate is washed in a nitric acid solution, which eats away the metal between the lines, leaving them standing above the surface of the plate like type. The plate is then mounted upon a hardwood base, which can be used in any printing press as are blocks of type.

## 15.17 Microfilm.

*Microfilm* has come into considerable use, especially where large numbers of drawings are involved, to furnish complete duplication of all drawings in a small filing space. Microfilm also provides duplicate

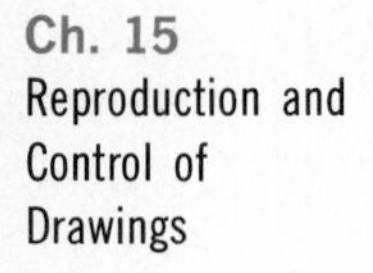


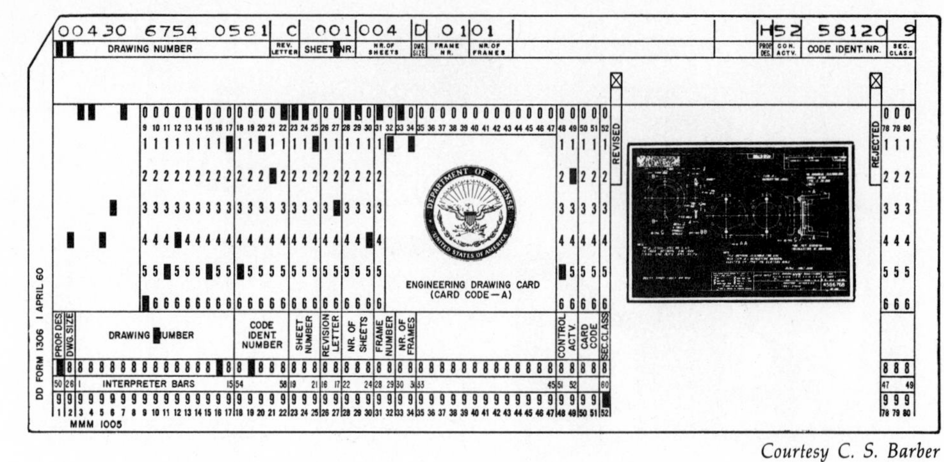


Courtesy C. S. Barber

Fig. 15.9 **A Standard Aperture Card.**

Courtesy 3M Co.

Fig. 15.10 **Microfilm Reader-Printer.**

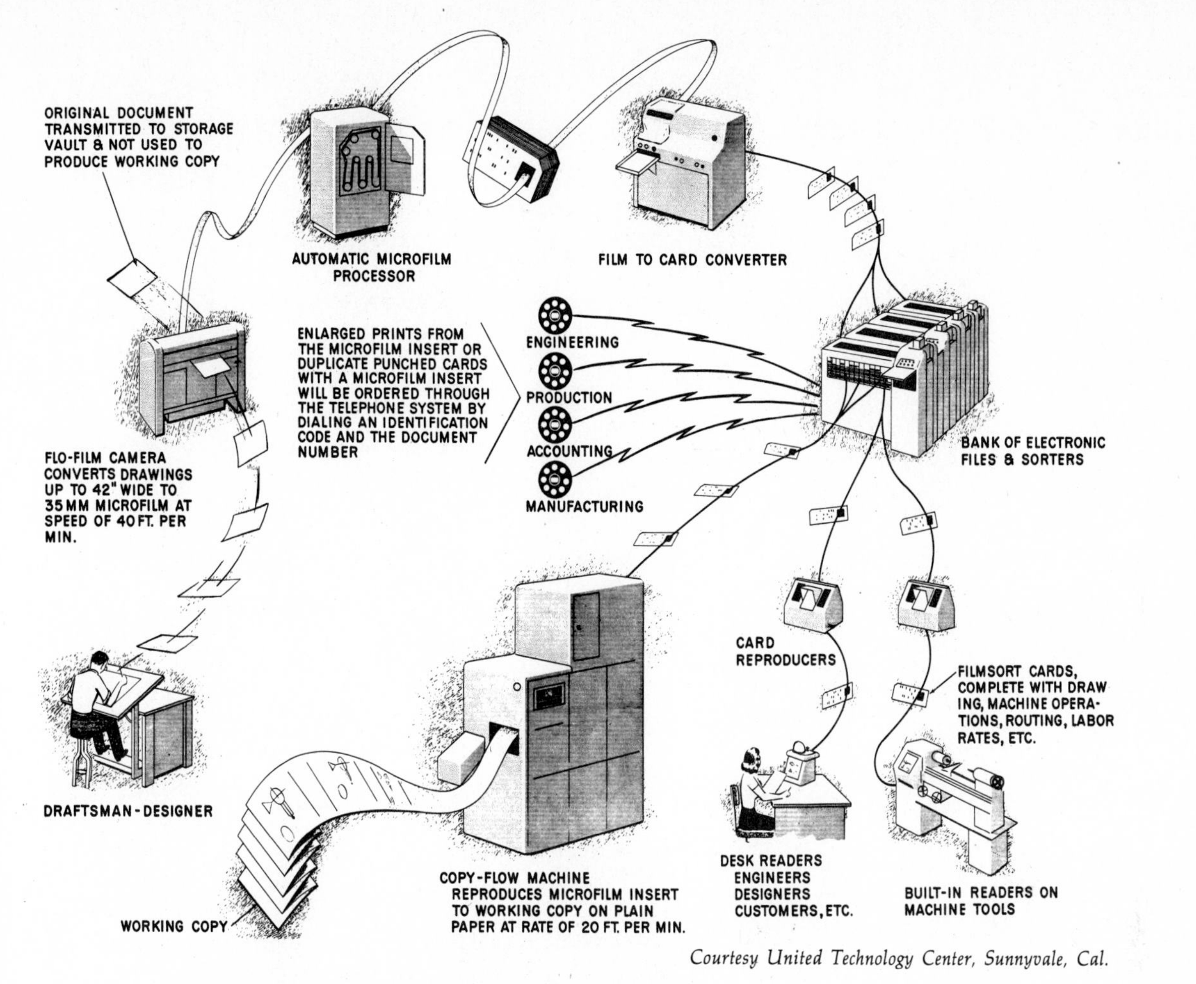
ORIGINAL DOCUMENT TRANSMITTED TO STORAGE VAULT & NOT USED TO PRODUCE WORKING COPY
AUTOMATIC MICROFILM PROCESSOR
FILM TO CARD CONVERTER
ENLARGED PRINTS FROM THE MICROFILM INSERT OR DUPLICATE PUNCHED CARDS WITH A MICROFILM INSERT WILL BE ORDERED THROUGH THE TELEPHONE SYSTEM BY DIALING AN IDENTIFICATION CODE AND THE DOCUMENT NUMBER
ENGINEERING
PRODUCTION
ACCOUNTING
MANUFACTURING
BANK OF ELECTRONIC FILES & SORTERS
FLO-FILM CAMERA CONVERTS DRAWINGS UP TO 42" WIDE TO 35 MM MICROFILM AT SPEED OF 40 FT. PER MIN.
CARD REPRODUCERS
FILMSORT CARDS, COMPLETE WITH DRAWING, MACHINE OPERATIONS, ROUTING, LABOR RATES, ETC.
DRAFTSMAN-DESIGNER
DESK READERS ENGINEERS DESIGNERS CUSTOMERS, ETC.
COPY-FLOW MACHINE REPRODUCES MICROFILM INSERT TO WORKING COPY ON PLAIN PAPER AT RATE OF 20 FT. PER MIN.
BUILT-IN READERS ON MACHINE TOOLS
WORKING COPY

*Courtesy United Technology Center, Sunnyvale, Cal.*

Fig. 15.11 **A Mechanized Engineering Data Handling System.**

copies for security reasons. Enlargement copies can be made from the film through photographic, electrostatic, or photocopy methods. The films may be 16, 35, or 70 mm and may be produced in rolls or individually mounted on *aperture cards,* Fig. 15.9.

The individual cards may be viewed in a *reader* and, if desired, a full-size print may be had from the reader-printer unit in Fig. 15.10.

Since the aperture card is essentially a standard data card used for electronic computers, mechanized equipment is capable of sorting, filing, and retrieving the cards. Duplicate aperture cards and reproductions of the drawings to the size desired may be produced on specialized equipment. A typical mechanized engineering data handling system is shown in Fig. 15.11.

for thur.

162 question # 4, 6, 9, 12, 27

# 16 axonometric projection

**16.1 Pictorial Drawing.*** By means of multiview drawing, it is possible to represent accurately the most complex forms of a design by showing a series of exterior views and sections. This type of representation has, however, two limitations: its execution requires a thorough understanding of the principles of multiview projection, and its reading requires a definite exercise of the constructive imagination.

Frequently, it is necessary to prepare drawings for the presentation of a design idea that are accurate and scientifically correct and can be easily understood by persons without technical training. Such drawings show several faces of an object at once, approximately as they appear to the observer. This type of drawing is called *pictorial drawing,* to distinguish it from multiview drawing discussed in Chapter 6. Since pictorial drawing shows only the appearances of parts or devices, it is not satisfactory for completely describing complex or detailed forms.

Pictorial drawing enables the person without technical training to visualize the design represented. It also enables the designer to visualize the successive stages of the design and to develop it in a satisfactory manner.

Various types of pictorial drawing are used extensively in catalogs, in general sales literature, and also in technical work,* to supplement and amplify multiview drawings. For example, pictorial drawing is used in Patent

*See §1.11. See also ANSI Y14.4-1957.

*Practically all of the pictorial drawings in this book were drawn by the methods described in Chapters 16, 17, and 18. See especially Figs. 6.53 to 6.112 for examples.

Office drawings, in piping diagrams, and in machine, structural, and architectural designs, and in furniture design.

## 16.2 Methods of Projection.

The four principal types of projection are illustrated in Fig. 16.1, and all except the regular multiview projection, (a), are pictorial types since they show several sides of the object in a single view. In all cases the views, or projections, are formed by the piercing points in the plane of projection of an infinite number of visual rays or projectors.

In both multiview projection, (a), and *axonometric projection,* (b), the observer is considered to be at infinity, and the visual rays are parallel to each other and perpendicular to the plane of projection. Therefore, both are classified as *orthographic projections,* §1.11.

In *oblique projection,* (c), the observer is considered to be at infinity, and the visual rays are parallel to each other but oblique to the plane of projection. See Chapter 17.

In *perspective,* (d), the observer is considered to be at a finite distance from the object, and the visual rays extend from the observer's eye, or the Station Point (SP), to all points of the object to form a so-called "cone of rays." See Chapter 18.

## 16.3 Types of Axonometric Projection.

The distinguishing feature of axonometric projection, as compared to multiview projection, is the inclined position of the object with respect to the plane of projection. Since the principal edges and surfaces of the object are inclined to the plane of projection, the lengths of the lines, the sizes of the angles, and the

Fig. 16.1 **Four Types of Projection.**

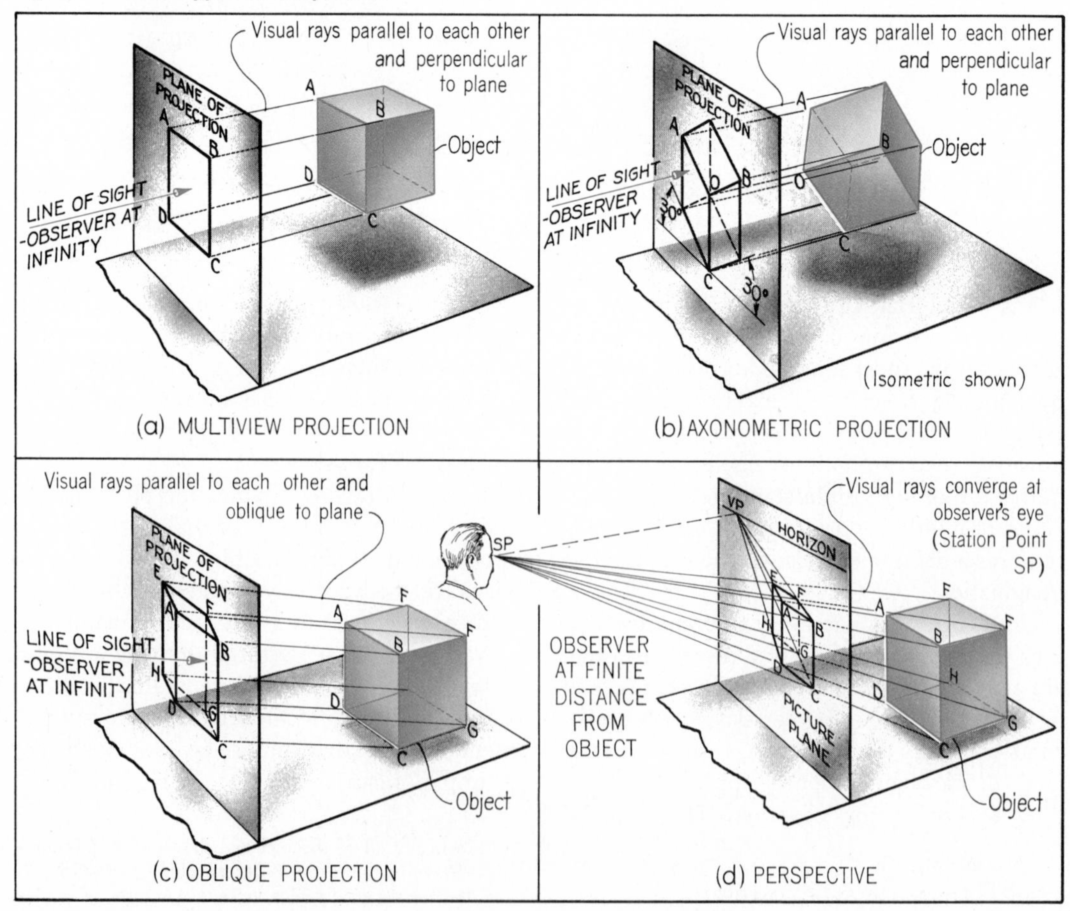

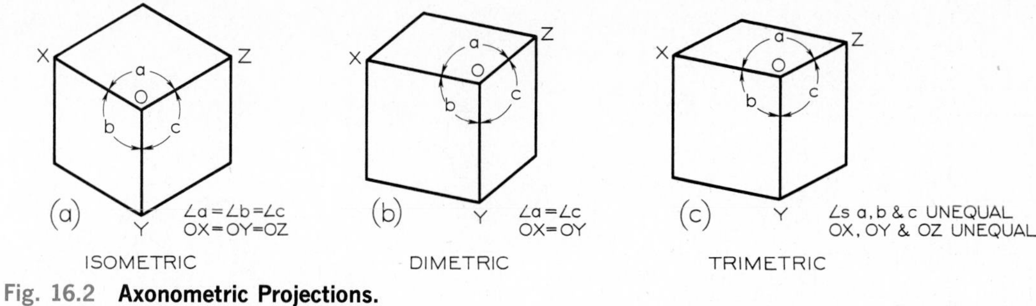


Fig. 16.2 **Axonometric Projections.**

general proportions of the object vary with the infinite number of possible positions in which the object may be placed with respect to the plane of projection. Three of these are shown in Fig. 16.2.

In these cases the edges of the cube are inclined to the plane of projection and are therefore foreshortened. See Fig. 6.20 (c). The degree of foreshortening of any line depends on its angle with the plane of projection; the greater the angle, the greater the foreshortening. If the degree of foreshortening is determined for each of the three edges of the cube that meet at one corner, scales can be easily constructed for measuring along these edges or any other edges parallel to them. See Figs. 16.41 (a) and 16.46 (a).

It is customary to consider the three edges of the cube that meet at the corner nearest the observer as the *axonometric axes.* In Fig. 16.1 (b), the axonometric axes, or simply the *axes,* are OA, OB, and OC. As shown in Fig. 16.2, axonometric projections are classified as (a) *isometric projection,* (b) *dimetric projection,* and (c) *trimetric projection,* depending upon the number of scales of reduction required.

## Isometric Projection

**16.4 Isometric Projection.** To produce an isometric projection (isometric means "equal measure"), it is necessary to place the object so that its principal edges, or axes, make equal angles with the plane of projection and are therefore foreshortened equally. See Fig. 5.21. In this position the edges of a cube would be projected equally and would make equal angles with each other (120°), as shown in Fig. 16.2 (a).

In Fig. 16.3 (a) is shown a multiview drawing of a cube. At (b) the cube is shown revolved through 45° about an imaginary vertical axis. Now an auxiliary view in the direction of the arrow will show the cube diagonal ZW as a point, and the cube appears as a true isometric projection. However, instead of the auxiliary view at (b) being drawn, the cube may be further revolved as shown at (c), this time the cube being tilted forward about an imaginary horizontal axis until the three edges OX, OY, and OZ make equal angles with the frontal plane of projection and are therefore foreshortened equally. Here again, a diagonal of the cube, in this case OT, appears as a point in the isometric view. The front view thus obtained is a true isometric projection. In this projection the twelve edges of the cube make angles of about 35° 16′ with the frontal plane of projection. The lengths of their projections are equal to the lengths of the edges multiplied by $\sqrt{\frac{2}{3}}$, or by 0.816, approximately. Thus the projected lengths are about 80 percent of the true lengths, or still more roughly, about three fourths of the true lengths. The projections of the axes OX, OY, and OZ make angles of 120° with each other and are called the *isometric axes.* Any line parallel to one of these

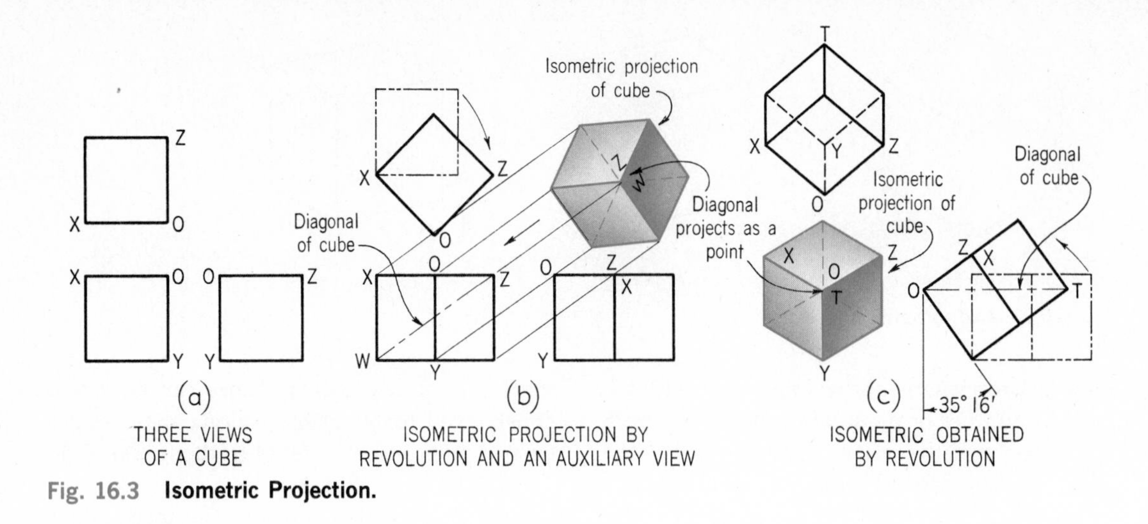


Fig. 16.3 **Isometric Projection.**

is called an *isometric line;* a line that is not parallel is called a *nonisometric line.* It should be noted that the angles in the isometric projection of the cube are either 120° or 60° and that all are projections of 90° angles. In an isometric projection of a cube, the faces of the cube, and any planes parallel to them, are called *isometric planes.*

## 16.5 The Isometric Scale.

A correct isometric projection may be drawn with the use of a special *isometric scale,* prepared on a strip of paper or cardboard, Fig. 16.4. All distances in the isometric scale are $\sqrt{\frac{2}{3}}$ times true size, or approximately 80 percent of true size. The use of the isometric scale is illustrated in Fig. 16.5 (a). A scale of 9″ = 1′-0, or $\frac{3}{4}$-size scale, could be used to approximate the isometric scale.

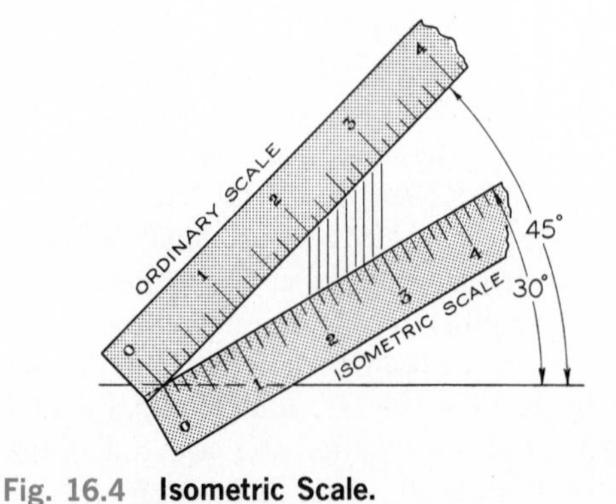


Fig. 16.4 **Isometric Scale.**

## 16.6 Isometric Drawing.

When a drawing is prepared with an isometric scale, or otherwise as the object is actually *projected* on a plane of projection, it is an *isometric projection,* as illustrated in Fig. 16.5 (a). When it is prepared with an ordinary scale, it is an *isometric drawing,* illustrated at (b). The isometric drawing, (b), is about 25 percent larger than the isometric projection (a), but the pictorial value is obviously the same in both.

Since the isometric projection is foreshortened and an isometric drawing is full scale size, it is usually advantageous to make an isometric drawing rather than an isometric projection. The drawing is much easier to execute and, for all practical purposes, is just as satisfactory as the isometric projection.

## 16.7 Steps in Making an Isometric Drawing.

The steps in constructing an isometric drawing of an object composed only of normal surfaces, §6.19, are illustrated in Fig. 16.6. Notice that all measurements are made parallel to the main edges of the enclosing box, that is, parallel to the isometric axes. No measurement along a diagonal (nonisometric line) on any surface or through the object can be set off directly with the scale. The object may be drawn in the same position by beginning at the corner Y, or any other corner, instead of at the corner X.

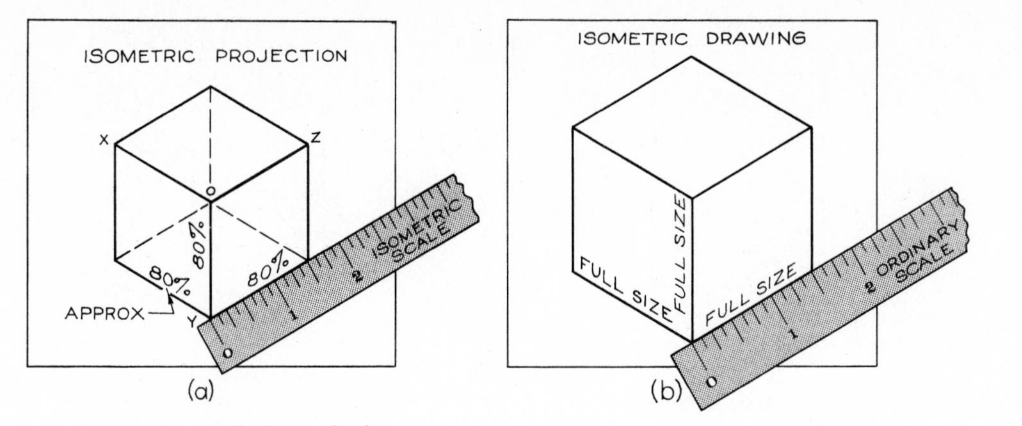


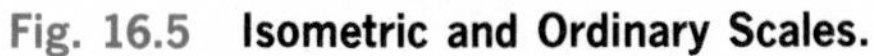

Fig. 16.5 **Isometric and Ordinary Scales.**

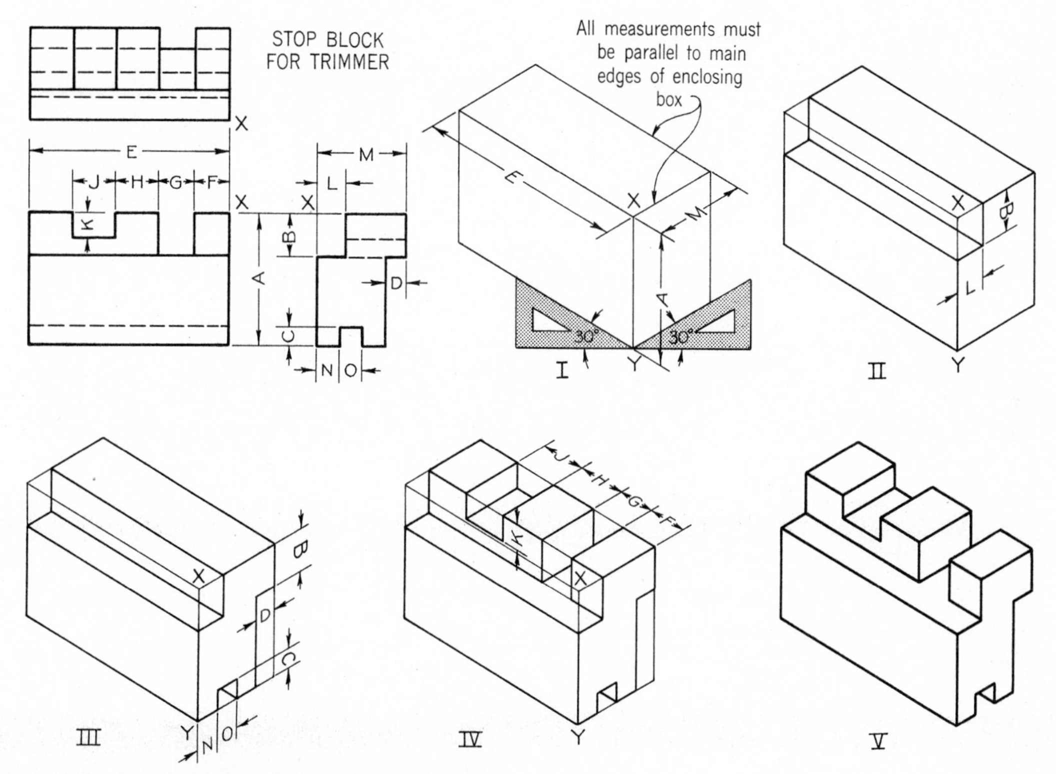


Fig. 16.6 **Isometric Drawing of Normal Surfaces.**

The method of constructing an isometric drawing of an object composed partly of inclined surfaces (and oblique edges) is shown in Fig. 16.7. Notice that inclined surfaces are located by *offset measurements* along isometric lines. For example, dimensions E and F are set off to locate the inclined surface M, and dimensions A and B are used to locate surface N.

For sketching in isometric, see §§5.11 to 5.14.

## 16.8 Oblique Surfaces in Isometric.

Oblique surfaces in isometric may be drawn by establishing the intersections of the oblique surface with the isometric planes. For example,

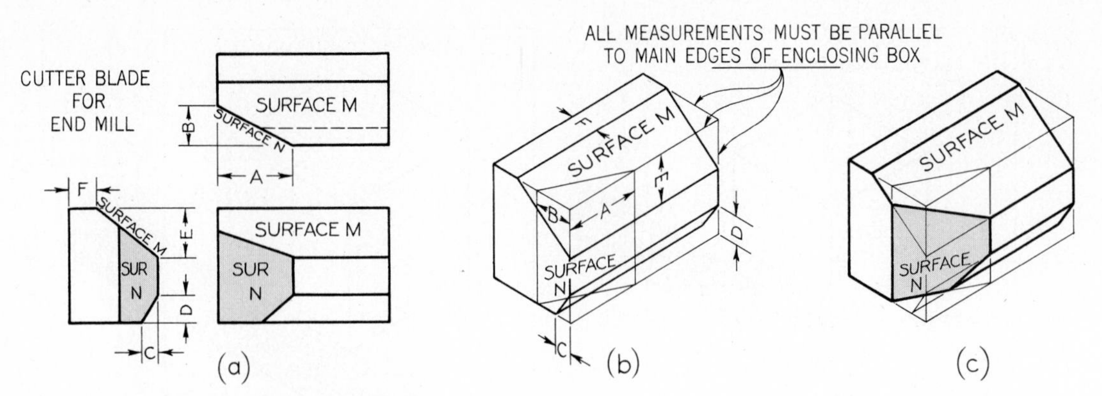


Fig. 16.7 **Inclined Surfaces in Isometric.**

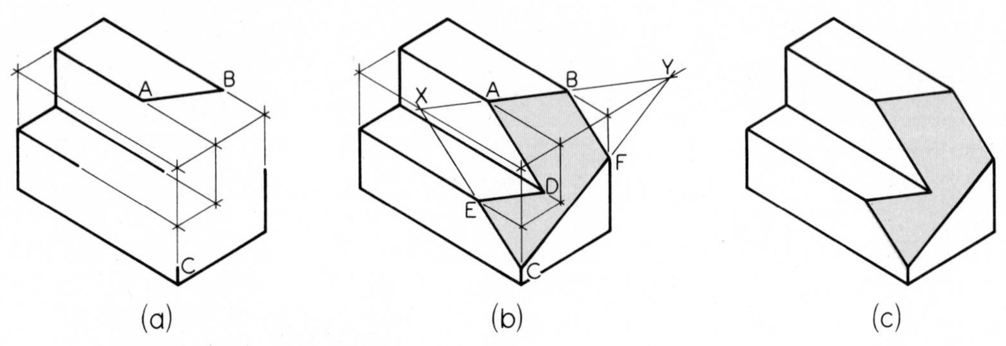


Fig. 16.8 **Oblique Surfaces in Isometric.**

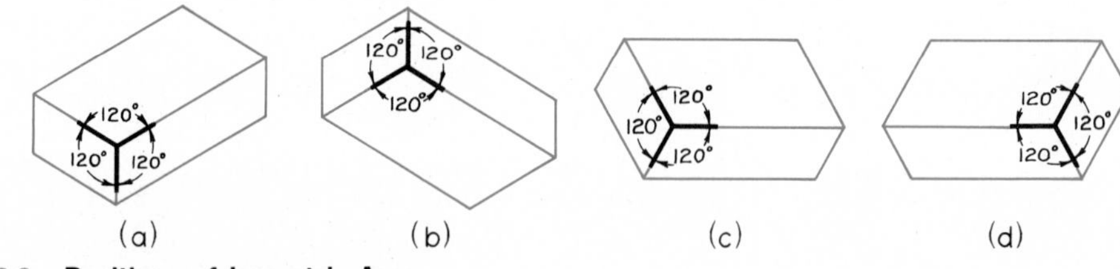


Fig. 16.9 **Positions of Isometric Axes.**

in Fig. 16.8 (a), the oblique plane is known to contain points A, B, and C. To establish the plane, (b), line AB is extended to X and Y, which are in the same isometric planes as C. Lines XC and YC locate points E and F. Finally AD and ED are drawn, using the rule of parallelism of lines. The completed drawing is shown at (c).

**16.9 Other Positions of the Isometric Axes.** The isometric axes may be placed in any desired position according to the requirements of the problem, as shown in Fig. 16.9, but the angle between the axes must remain 120°. The choice of the directions of the axes is determined by the position from which the object is usually viewed, Fig. 16.10, or by the position that best describes the shape of the object. If possible, both requirements should be met.

If the object is characterized by considerable length, the long axis may be placed horizontally for best effect, as shown in Fig. 16.11.

## 16.10 Offset Location Measurements.

The method of locating one part with respect to another is illustrated in Figs. 16.12 and 16.13. In each case, after the main block has been drawn, the offset lines CA and BA in the multiview drawing are drawn full size in the isometric drawing, thus locating corner A of the small block or rectangular recess. These measurements are called *offset measurements*, and since they are parallel to certain edges of the main block in the multiview drawings, they will be parallel respectively to the same edges in the isometric drawings, §6.25.

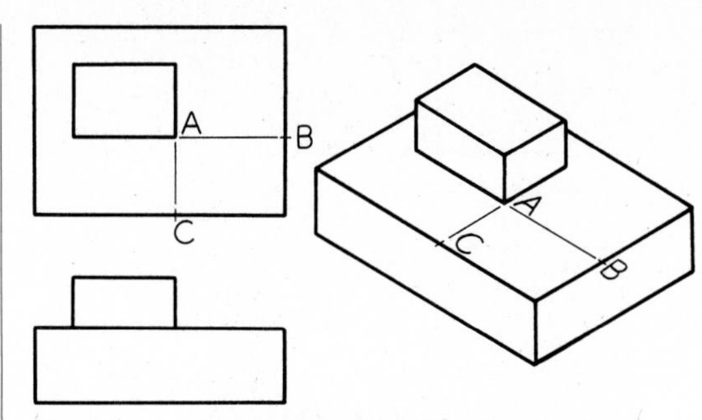

Fig. 16.12 **Offset Location Measurements.**

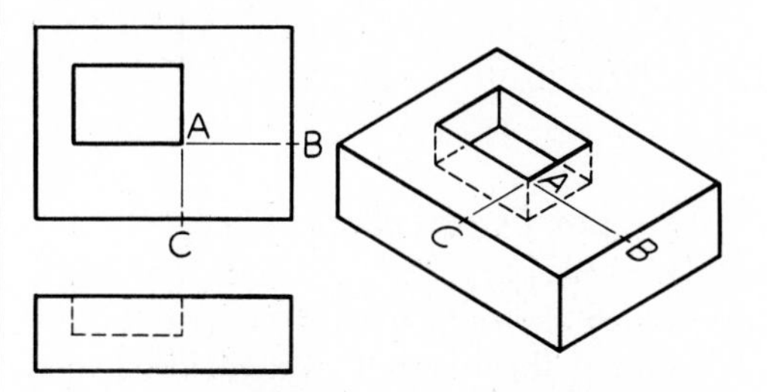


Fig. 16.13 **Offset Location Measurements.**

## 16.11 Hidden Lines.

The use of hidden lines in isometric drawing is governed by the same rules as in all other types of projection: *Hidden lines are omitted unless they are needed to make the drawing clear.* A case in which hidden lines are needed is illustrated in Fig. 16.14, in which a projecting part cannot be clearly shown without the use of hidden lines.

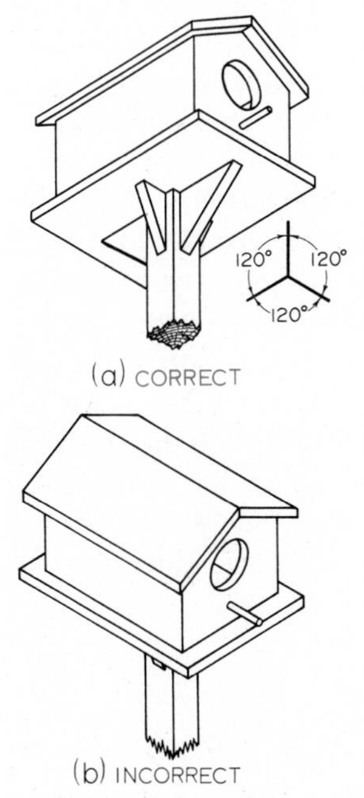


Fig. 16.10 **An Object Naturally Viewed from Below.**

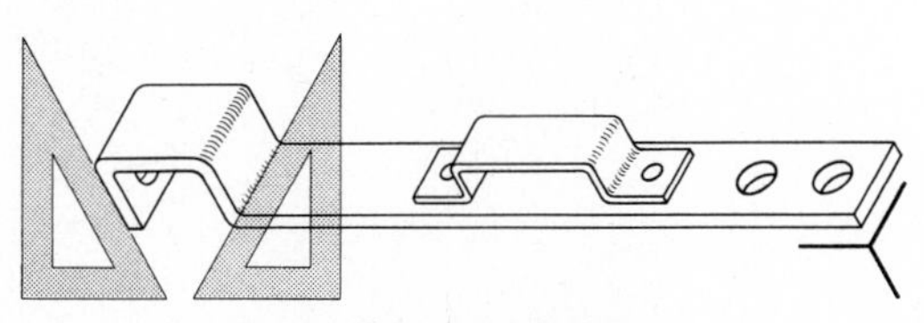

Fig. 16.11 **Long Axis Horizontal.**

## 16.12 Center Lines.

The use of center lines in isometric drawing is governed by the same rules as in multiview drawing: *Center lines are drawn if they are needed to indicate symmetry, or if they are needed for dimensioning,* Fig. 16.14. In general, center lines should be used sparingly and omitted in cases of doubt. The use of too many center lines may produce a confusion of lines, which diminishes the clearness of the drawing. Examples in which center lines are not needed are shown in Fig. 16.10 and

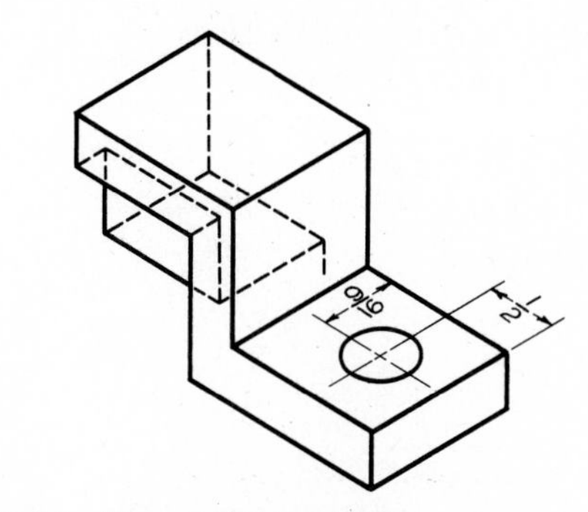

Fig. 16.14 **Use of Hidden Lines.**

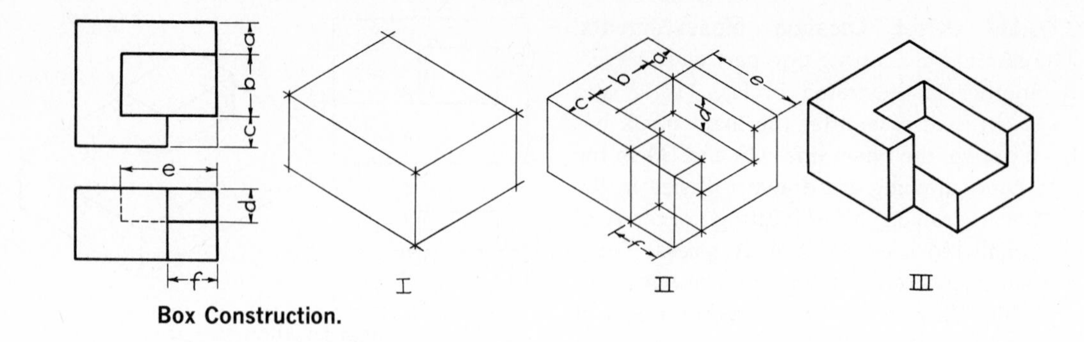


**Box Construction.**

16.11. Examples in which they are needed are seen in Figs. 16.39 (a), 14.30, and 14.31.

## 16.13 Box Construction.

Objects of rectangular shape may be more easily drawn by means of *box construction,* which consists simply in imagining the object to be enclosed in a rectangular box whose sides coincide with the main faces of the object. For example, in Fig. 16.15 the object shown in two views is imagined to be enclosed in a construction box. This box is then drawn lightly with construction lines, as shown at I, the irregular features are then constructed as shown at II, and finally, as shown at III, the required lines are made heavy.

## 16.14 Nonisometric Lines.

Since the only lines of an object that are drawn true length in an isometric drawing are the isometric axes or lines parallel to them, *nonisometric* lines cannot be set off directly with the scale. For example, in Fig. 16.16 (a), the inclined lines BA and CA are shown in their true lengths $2\frac{1}{8}''$ in the top view, but since they are not parallel to the isometric axes, they will not be true length in the isometric. Such lines are drawn in isometric by means of box construction and offset measurements. First, as shown at I, the measurements $1\frac{3}{4}''$, $\frac{3}{4}''$, and $\frac{7}{8}''$ can be set off directly since they are made along isometric lines. The nonisometric $2\frac{1}{8}''$ dimension cannot be set off directly, but if one half of the given top view is constructed full size to scale as shown at (b), the dimension X can be determined. This dimension is parallel to an isometric axis and can be transferred with dividers to the isometric at II. The dimensions $\frac{15}{16}''$ and $\frac{3}{8}''$ are parallel to isometric lines and can be set off directly, as shown at III.

To realize the fact that nonisometric lines will not be true length in the isometric drawing, set your dividers on BA of II and then compare with BA on the given top view at (a). Do the same for line CA. It will be seen that BA is shorter and CA is longer in the isometric than the corresponding lines in the given views.

Fig. 16.16 **Nonisometric Lines.**

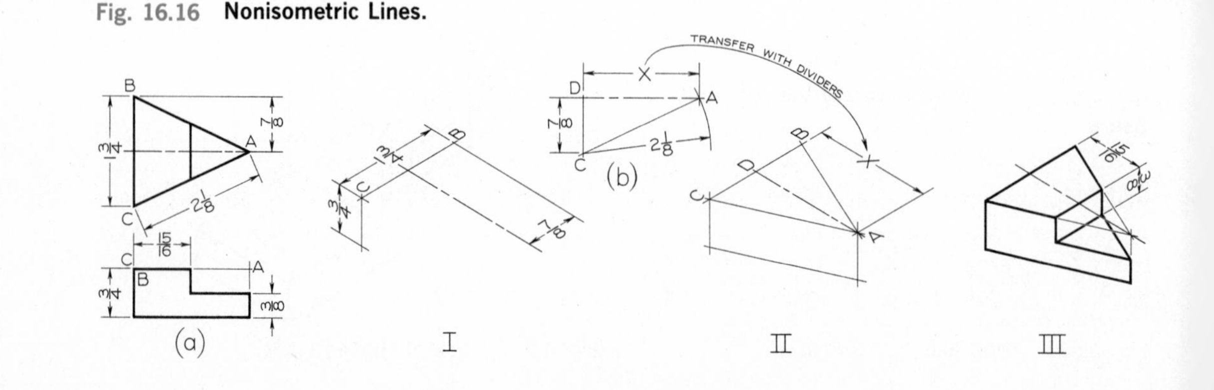

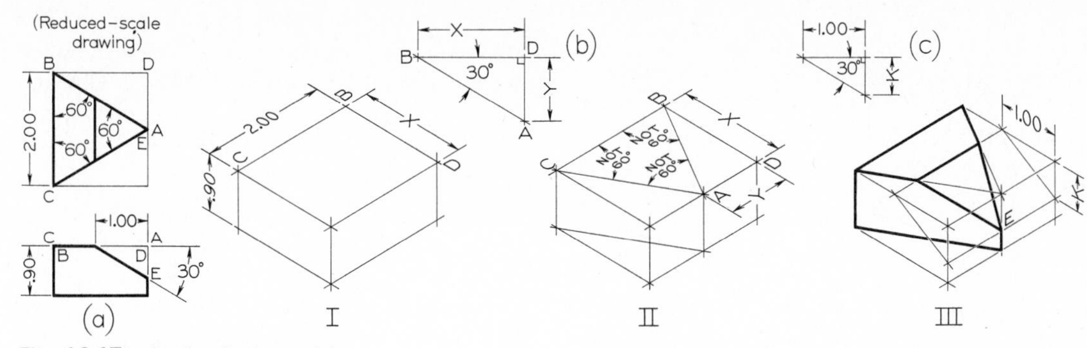


Fig. 16.17 **Angles in Isometric.**

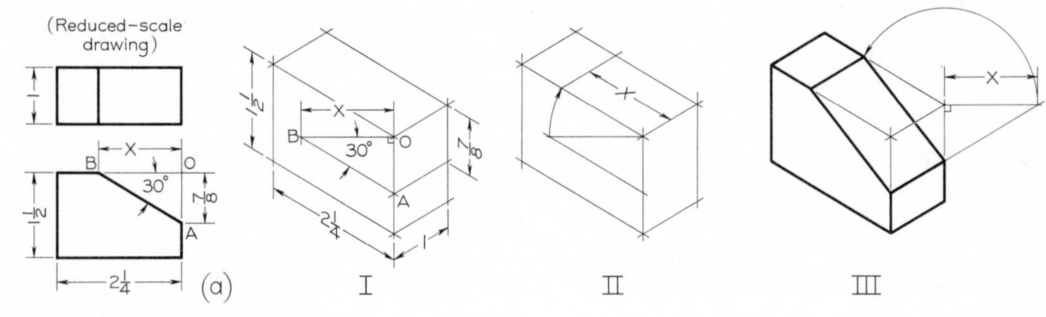


Fig. 16.18 **Angle in Isometric.**

## 16.15 Angles in Isometric.

As shown in §6.26, angles project true size only when the plane of the angle is parallel to the plane of projection. An angle may project larger or smaller than true size, depending upon its position. Since in isometric the various surfaces of the object are usually inclined to the plane of projection, it follows that angles generally will not be projected true size. For example, in the multiview drawing in Fig. 16.17 (a), none of the three 60° angles will be 60° in the isometric drawing. To realize this fact, measure each angle in the isometric of II with the protractor and note the number of degrees compared to the true 60°. No two angles are the same; two are smaller and one larger than 60°.

As shown in I, the enclosing box can be drawn from the given dimensions, except for dimension X, which is not given. To find dimension X, draw triangle BDA from the top view full size, as shown at (b). Transfer dimension X to the isometric in I, to complete the enclosing box.

In order to locate point A in II, dimension Y must be used, but this is not given in the top view at (a). Dimension Y is found by the same construction as at (b) and then transferred to the isometric, as shown. The completed isometric is shown at III where point E is located by using dimension K, as shown.

Thus, in order to set off angles in isometric, the regular protractor cannot be used.* *Angular measurements must be converted to linear measurements along isometric lines.*

In Fig. 16.18 (a) are two views of an object to be drawn in isometric. Point A can easily be located in the isometric, step I, by measuring $\frac{7}{8}''$ down from point O. However, in the given drawing at (a) the location of point B depends upon the 30° angle, and to locate B in the isometric the linear dimension X must be known. This distance can be found graphically by drawing the right triangle BOA attached to the isometric, as shown. The distance X is then transferred to the isometric

*Isometric protractors for setting off angles on isometric surfaces are available from dealers.

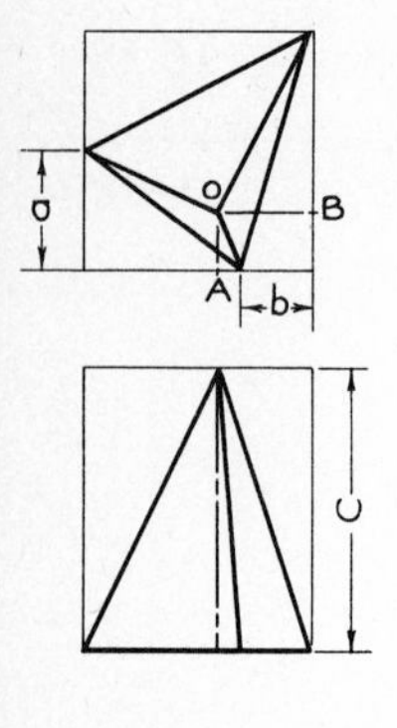


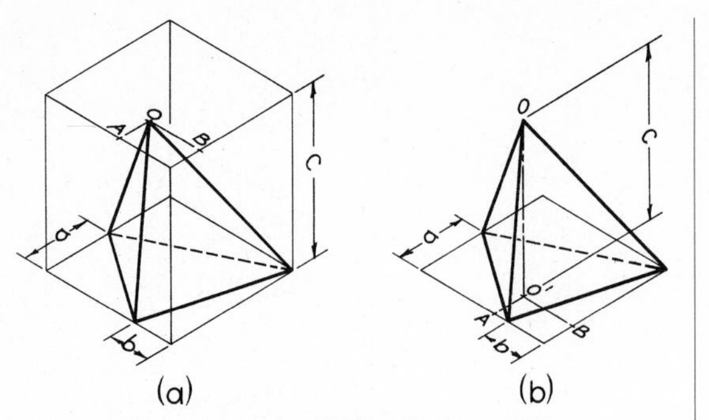


Fig. 16.19 **Irregular Object in Isometric.**

with the compass or dividers, as shown at II. Actually, the triangle could be attached in several different positions. One of these is shown at III.

When angles are given in degrees, it is necessary to convert the angular measurements into linear measurements. This is best done by drawing a right triangle separately, as in Fig. 16.17 (b), or attached to the isometric, as in Fig. 16.18.

## 16.16 Irregular Objects.

If the general shape of an object does not conform somewhat to a rectangular pattern, as shown in Fig. 16.19, it may be drawn as shown at (a) by using the box construction discussed previously. Various points of the triangular base are located by means of offsets a and b along the edges of the bottom of the construction box. The vertex is located by means of offsets OA and OB on the top of the construction box.

However, it is not necessary to draw the complete construction box. If only the bottom of the box is drawn, as shown at (b), the triangular base can be constructed as before. The orthographic projection of the vertex O′ on the base can then be located by offsets O′A and O′B, as shown, and from this point the vertical center line O′O can be erected, using measurement C.

An irregular object may be drawn by means of a series of sections, as illustrated in Fig. 16.20. The edge views of a series of imaginary cutting planes are shown in the top and front views of the multiview drawing at (a). At I the various sections are constructed in isometric, and at II the object is completed by drawing lines through the corners of the sections. In the isometric at I, all height dimensions are taken from the front view at (a), and all depth dimensions from the top view.

## 16.17 Curves in Isometric.

Curves may be drawn in isometric by means of a series of offset measurements similar to those discussed in §16.10. In Fig. 16.21 any desired number of points, such as A, B, and C, are selected at random along the curve in the given top view at (a). Enough points should be chosen to fix accurately the path of the curve; the more points used, the greater the accuracy. Offset grid lines are then drawn from each point parallel to the isometric axes.

As shown at I, offset measurements a and b are laid off in the isometric to locate point A on the curve. Points B, C, and D are located in a similar manner, as shown at II. A light freehand curve is sketched smoothly through the points as shown at III. Points A′, B′, C′, and D′ are located directly under points A, B, C, and D, as shown at IV, by drawing vertical

Fig. 16.20 **Use of Sections in Isometric.**

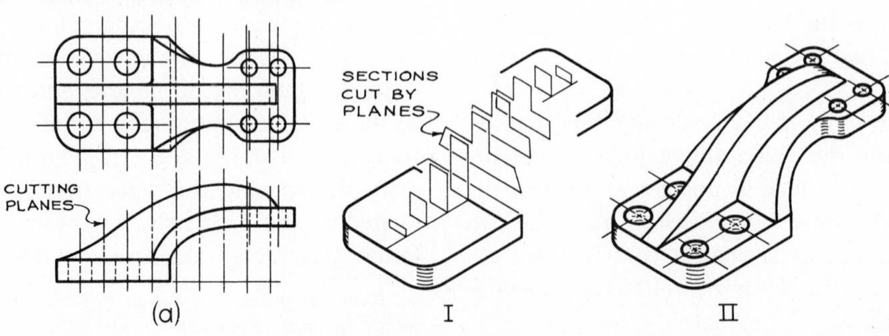

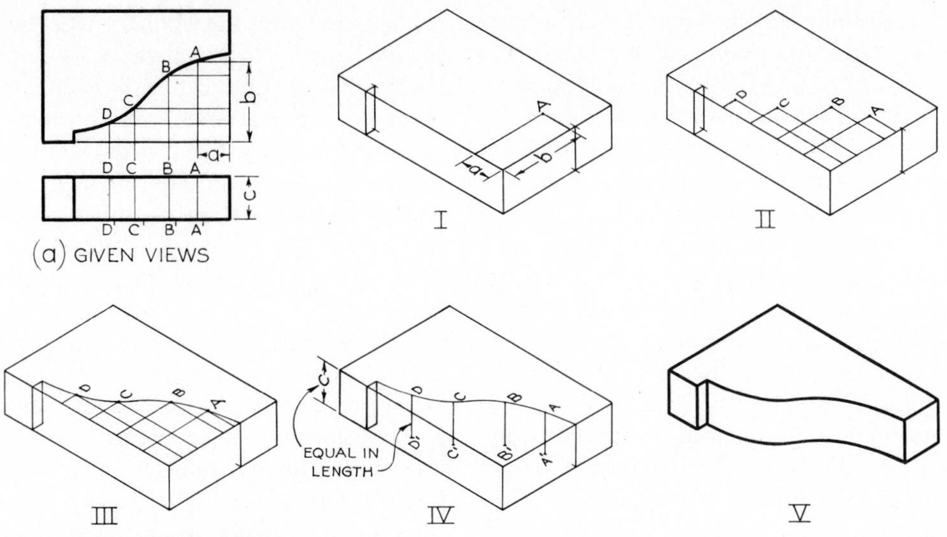


Fig. 16.21 **Curves in Isometric.**

lines downward, making all equal to dimension C, the height of the block. A light freehand curve is then drawn through the points. The final curve is heavied in with the aid of the irregular curve, §2.58, and all straight lines are darkened to complete the isometric at V.

## 16.18 True Ellipses in Isometric.

As shown in §§4.52, 5.13, and 6.30, if a circle lies in a plane that is not parallel to the plane of projection, the circle will be projected as a true ellipse. The ellipse can be constructed by the method of offsets, §16.17. As shown in Fig. 16.22 (a), draw parallel lines, spaced at random, across the circle; then transfer these lines to the isometric as shown at (b), with the aid of the dividers. To locate points in the lower ellipse, transfer points of the upper ellipse down a distance equal to the height d of the block and draw the ellipse, part of which will be hidden, through these points. Draw the final ellipses with the aid of the irregular curve, §2.58.

A variation of the method of offsets, which provides eight points on the ellipse, is illustrated at (c) and (d). If more points are desired, parallel lines, as at (a), can be added. As shown at (c), circumscribe a square around the given circle, and draw diagonals. Through the points of intersection of the diagonals and the circle, draw another square, as shown. Draw this construction in the isometric, as shown at (d), transferring distances a and b with the dividers.

A similar method that provides twelve

Fig. 16.22 **True Isometric Ellipse Construction.**

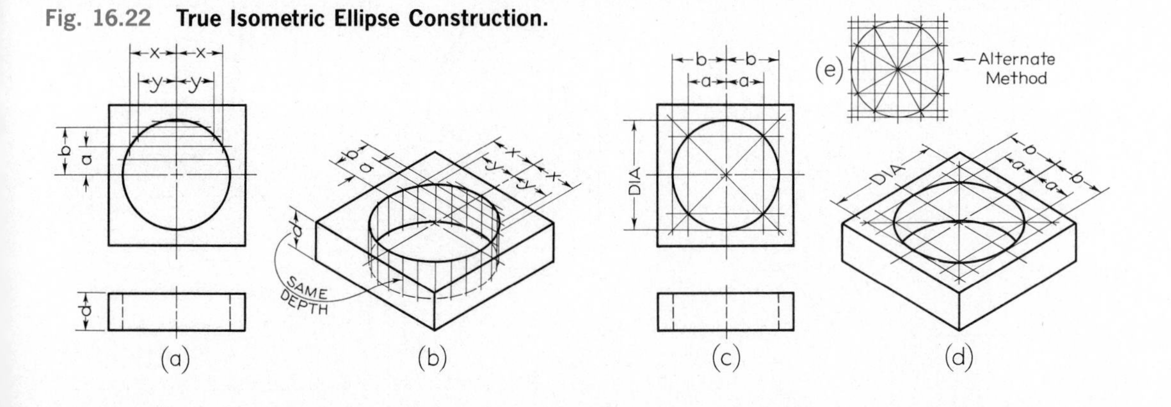

points on the ellipse is shown at (e). The given circle is divided into twelve equal parts, using the 30° × 60° triangle, Fig. 2.26. Lines parallel to the sides of the square are drawn through these points. The entire construction is then drawn in isometric, and the ellipse is drawn through the points of intersection.

When the center lines shown in the top view at (a) are drawn in isometric, (b), they become the *conjugate diameters* of the ellipse. The ellipse can then be constructed on the conjugate diameters by the methods of Figs. 4.51 and 4.52 (b).

When the 45° diagonals at (c) are drawn in isometric at (d), they coincide with the major and minor axes of the ellipse, respectively. Note that the minor axis is equal in length to the sides of the inscribed square at (c). The ellipse can be constructed upon the major and minor axes by any of the methods in §§4.49 to 4.52.

Remember the rule: *The major axis of the ellipse is always at right angles to the center line of the cylinder, and the minor axis is at right angles to the major axis and coincides with the center line.*

Accurate ellipses may be drawn with the aid of ellipse guides, §§4.57 and 16.22, or with a special *ellipsograph.*

If the curve lies in a nonisometric plane, all offset measurements cannot be applied directly. For example, in Fig. 16.23 (a) the elliptical face shown in the auxiliary view lies in an inclined nonisometric plane. The cylinder is enclosed in a construction box, and the box is then drawn in isometric, as shown at I. The base is drawn by the method of offsets, as shown in Fig. 16.22. The inclined ellipse is constructed by locating a number of points on the ellipse in the isometric and drawing the final curve by means of the irregular curve, §2.58.

Measurements a, b, c, etc., are parallel to an isometric axis, and can be set off in the isometric at I on each side of the center line X–X, as shown. Measurements e, f, g, etc., are not parallel to any isometric axis, and cannot be set off directly in isometric. However, when these measurements are projected to the front view and down to the base, as shown at (a), they can then be set off along the lower edge of the construction box, as shown at I. The completed isometric is shown at II.

The ellipse may also be drawn with the aid of an appropriate ellipse template selected to fit the major and minor axes established along X–X and Y–Y, respectively. See Fig. 4.55.

## 16.19 Approximate Four-Center Ellipse.

An approximate ellipse is sufficiently accurate for nearly all isometric drawings. The method commonly used, called the *four-center ellipse,* is

Fig. 16.23 **Ellipse in Inclined Plane.**

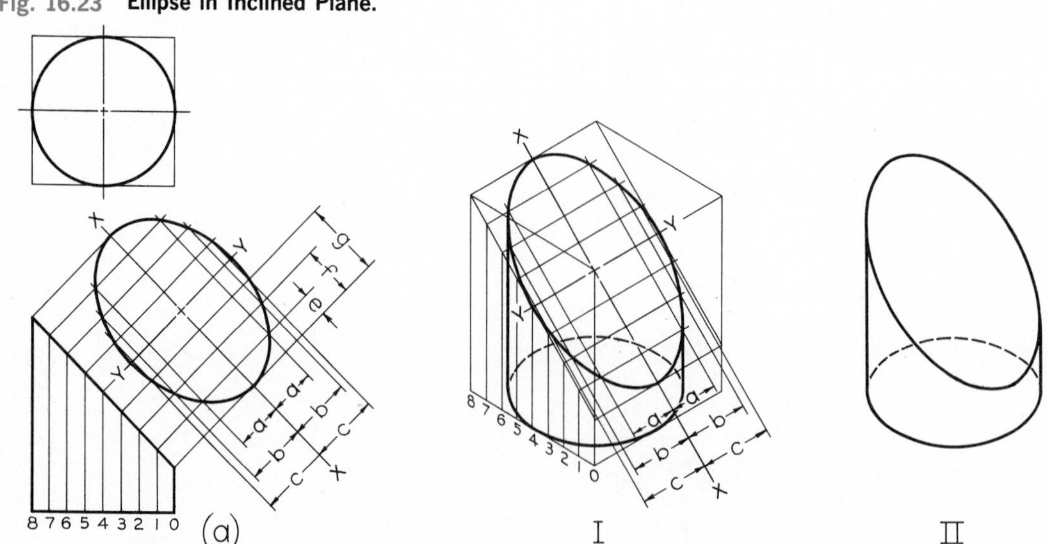

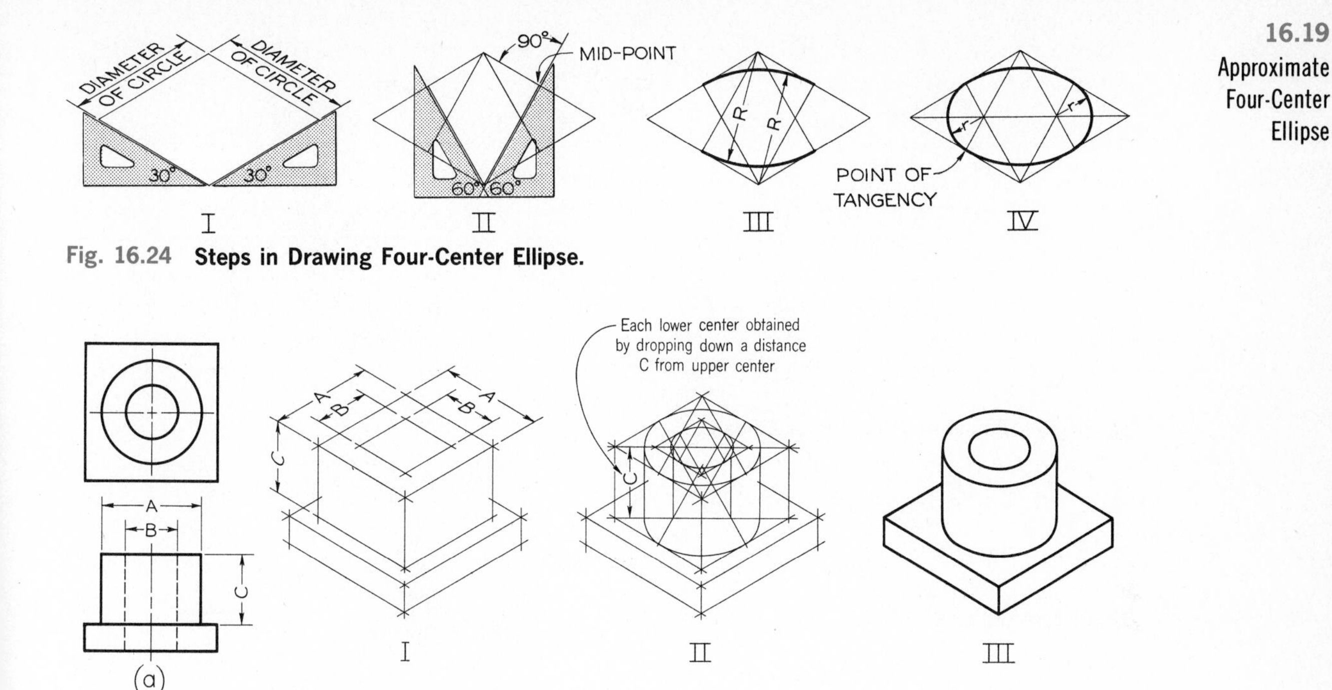


Fig. 16.24 **Steps in Drawing Four-Center Ellipse.**

Fig. 16.25 **Isometric Drawing of a Bearing.**

illustrated in Figs. 16.24, 16.25, and 16.26. It can be used only for ellipses in isometric planes.

To apply this method, Fig. 16.24, draw, or conceive to be drawn, a square around the given circle in the multiview drawing; then:

I. Draw the isometric of the square, which is an equilateral parallelogram whose sides are equal to the diameter of the circle.

II. Erect perpendicular bisectors to each side, using the 30° × 60° triangle as shown. These perpendiculars will intersect at four points, which will be centers for the four circular arcs.

III. Draw the two large arcs, with radius R, from the intersections of the perpendiculars in the two closest corners of the parallelogram, as shown.

IV. Draw the two small arcs, with radius r, from the intersections of the perpendiculars within the parallelogram, to complete the ellipse. As a check on the accurate location of these centers, a long diagonal of the parallelogram may be drawn, as shown. The midpoints of the sides of the parallelogram are points of tangency for the four arcs.

A typical drawing with cylindrical shapes is illustrated in Fig. 16.25. Note that the centers of the larger ellipse cannot be used for the smaller ellipse, though the ellipses represent concentric circles. Each ellipse has its own parallelogram and its own centers. Observe also that the centers of the lower ellipse are obtained by projecting the centers of the upper large ellipse down a distance equal to the height of the cylinder.

The construction of the four-center ellipse upon the three visible faces of a cube is shown in Fig. 16.26, a study of which shows that all diagonals are horizontal or 60° with horizontal; hence the entire construction is made with the T-square and 30° × 60° triangle.

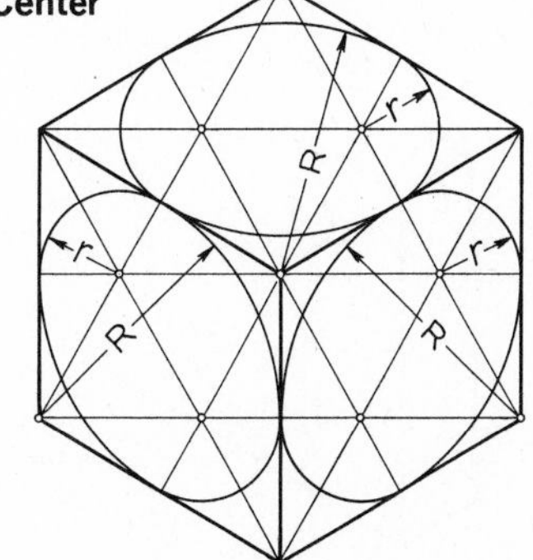


Fig. 16.26 **Four-Center Ellipses.**

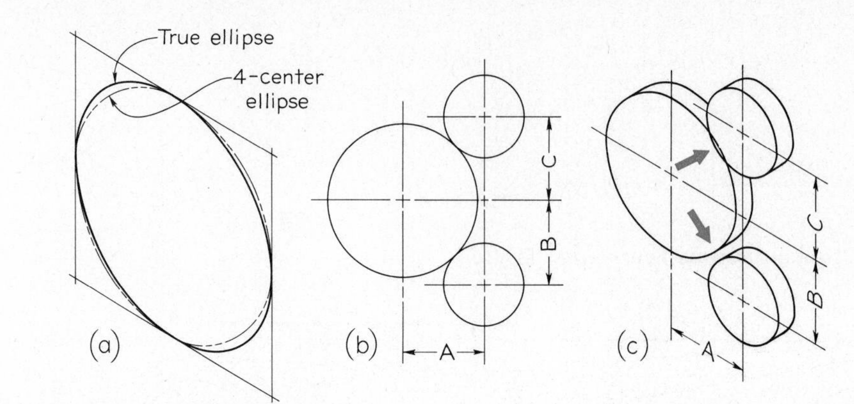


Fig. 16.27 **Faults of Four-Center Ellipse.**

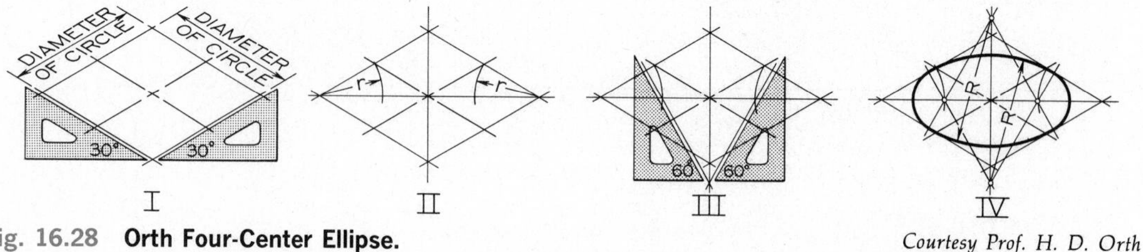


Fig. 16.28 **Orth Four-Center Ellipse.**

*Courtesy Prof. H. D. Orth*

Actually the four-center ellipse deviates considerably from the true ellipse. As shown in Fig. 16.27 (a), the four-center ellipse is somewhat shorter and "fatter" than the true ellipse. In constructions where tangencies or intersections with the four-center ellipse occur in the zones of error, the four-center ellipse is unsatisfactory, as shown at (b) and (c).

For a much closer approximation to the true ellipse, the Orth four-center ellipse, Fig. 16.28, which requires only one more step than the regular four-center ellipse, will be found sufficiently accurate for almost any problem.

When it is more convenient to start with the isometric center lines of a hole or cylinder in drawing the ellipse, rather than the enclosing parallelogram, the *alternate four-center ellipse* is recommended, Fig. 16.29. A completely constructed ellipse is shown at (a), and the steps followed are shown at the right in the figure.

I. Draw the isometric center lines. From the center, draw a construction circle equal to the actual diameter of the hole or cylinder. The circle will intersect the center lines at four points A, B, C, and D.

II. From the two intersection points on one

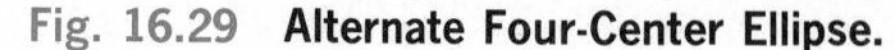

Fig. 16.29 **Alternate Four-Center Ellipse.**

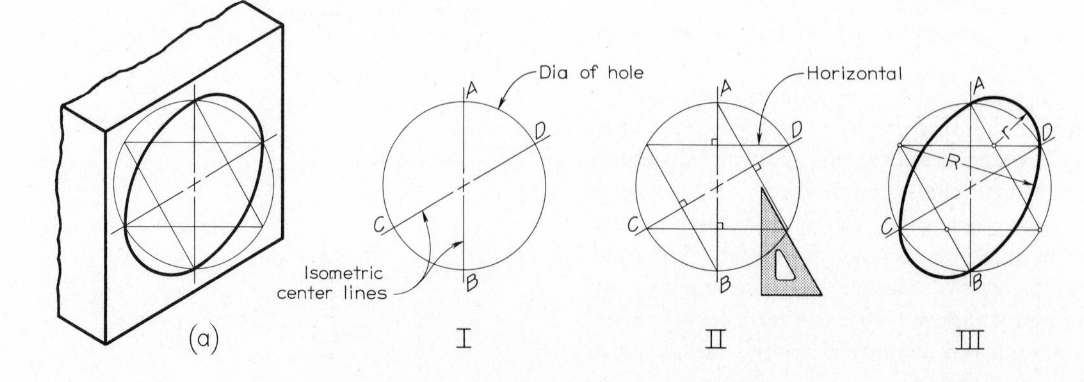

center line, erect perpendiculars to the other center line; then from the two intersection points on the other center line, erect perpendiculars to the first center line.

III. With the intersections of the perpendiculars as centers, draw two small arcs and two large arcs, as shown.

*Note:* The above steps are exactly the same as for the regular four-center ellipse of Fig. 16.24 except for the use of the isometric center lines instead of the enclosing parallelogram.

## 16.20 Screw Threads in Isometric.

Parallel partial ellipses spaced equal to the symbolic thread pitch, Fig. 13.9 (a), are used to represent the crests only of a screw thread in isometric, Fig. 16.30. The ellipses may be drawn by the four-center method of §16.19, or with the ellipse template, which is much more convenient, §§4.57 and 16.22.

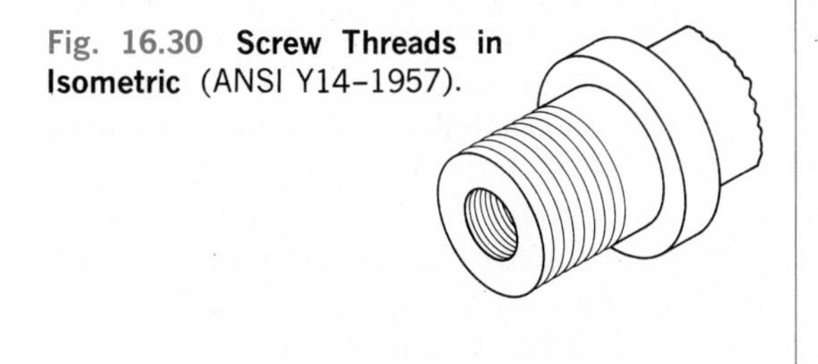

Fig. 16.30 **Screw Threads in Isometric** (ANSI Y14–1957).

## 16.21 Arcs in Isometric.

The four-center ellipse construction is used in drawing circular arcs in isometric, as shown in Fig. 16.31. At (a) the complete construction is shown. However, it is not necessary to draw the complete constructions for arcs, as shown at (b) and (c).

Fig. 16.31 **Arcs in Isometric.**

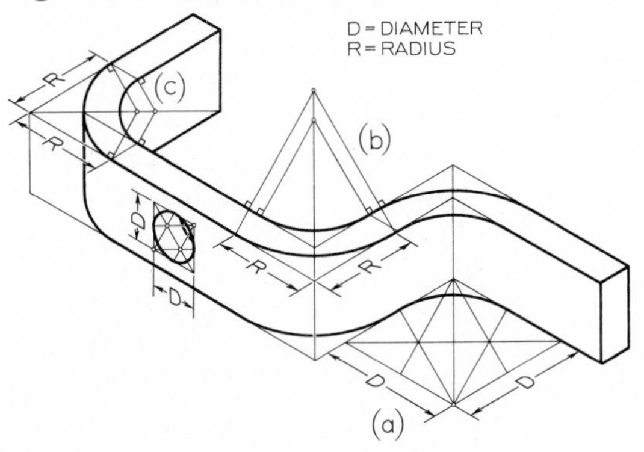


In each case the radius R is set off from the construction corner; then at each point, perpendiculars to the lines are erected, their intersection being the center of the arc. Note that the R distances are equal in both cases, (b) and (c), but that the actual radii used are quite different.

If a truer elliptic arc is required, the Orth construction, Fig. 16.28, can be used. Or a true elliptic arc may be drawn by the method of offsets, §16.18, or with the aid of an ellipse guide, §16.22.

## 16.22 Ellipse Guides.

One of the principal time-consuming elements in pictorial drawing is the construction of ellipses. A wide variety of ellipse guides, or templates, is available for ellipses of various sizes and proportions. See §4.57. They are not available in every possible size, of course, and it may be necessary to "use the fudge factor," such as leaning the pencil or pen when inscribing the ellipse, or shifting the template slightly for drawing each quadrant of the ellipse.

The design of the ellipse template, Fig. 16.32, combines the angles, scales, and ellipses on the same instrument. The ellipses are provided with markings to coincide with the isometric center lines of the holes—a convenient feature in isometric drawing.

Fig. 16.32 **Instrumaster Isometric Template.**

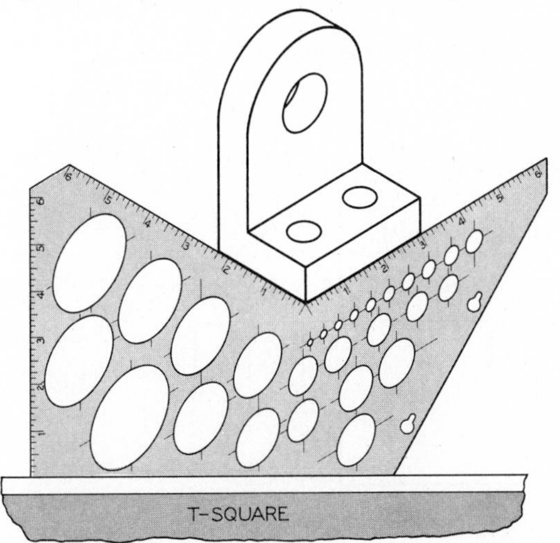

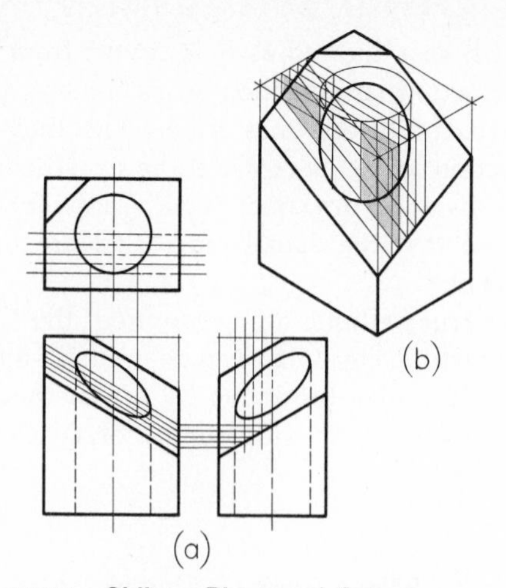


Fig. 16.33 **Oblique Plane and Cylinder.**

## 16.23 Intersections.

To draw the elliptical intersection of a cylindrical hole in an oblique plane in isometric, Fig. 16.33, draw the ellipse in the isometric plane on top of the construction box, (b); then project points down to the oblique plane as shown. It will be seen that the construction for each point forms a trapezoid, which is produced by a slicing plane parallel to a lateral surface of the block.

To draw the curve of intersection between two cylinders, Fig. 16.34, pass a series of imaginary cutting planes through the cylinders parallel to their axes, as shown. Each plane will cut elements on both cylinders that intersect at points on the curve of intersection, as shown at (b). As many points should be plotted as necessary to assure a smooth curve. For most accurate work, the ends of the cylinders should be drawn by the Orth construction, or with ellipse guides, or by one of the true-ellipse constructions.

Fig. 16.34 **Intersection of Cylinders.**

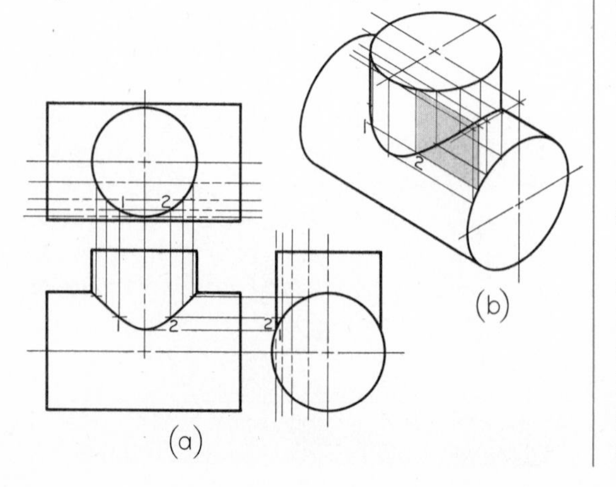


## 16.24 The Sphere in Isometric.

The isometric drawing of any curved surface is evidently the envelope of all lines which can be drawn on that surface. For the sphere, the great circles (circles cut by any plane through the center) may be selected as the lines on the surface. Since all great circles, except those which are perpendicular or parallel to the plane of projection, are shown as ellipses having equal major axes, it follows that their envelope is a circle whose diameter is the major axis of the ellipses.

In Fig. 16.35 (a) two views of a sphere enclosed in a construction cube are shown. The cube is drawn at I, together with the isometric of a great circle that lies in a plane parallel to one face of the cube. Actually, the ellipse need not be drawn, for only the points on the diagonal located by measurements a are needed. These points establish the ends of the major axis from which the radius R of the sphere is determined. The resulting drawing shown at II is an *isometric drawing,* and its diameter is, therefore, $\sqrt{\frac{3}{2}}$ times the actual diameter of the sphere. The *isometric projection* of the sphere is simply a circle whose diameter is equal to the true diameter of the sphere, as shown at III.

## 16.25 Isometric Sectioning.

In drawing objects characterized by open or irregular interior shapes, isometric sectioning is as appropriate as in multiview drawing. An *isometric full section* is shown in Fig. 16.36. In such cases it is usually best to draw the cut surface first, and then to draw the portion of the object that lies behind the cutting plane. Other examples of isometric full sections are shown in Figs. 6.90, 7.11, and 7.12.

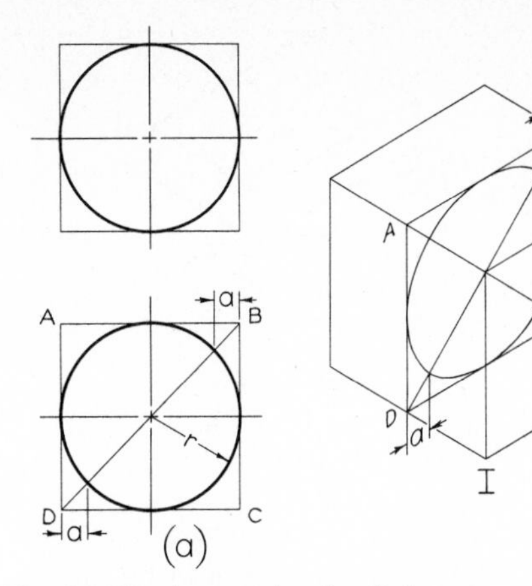


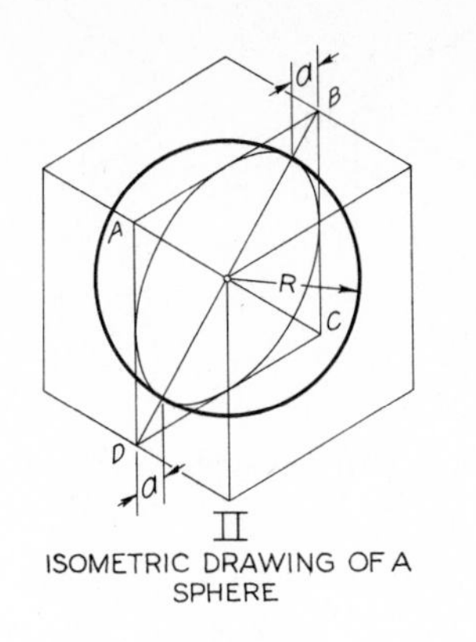


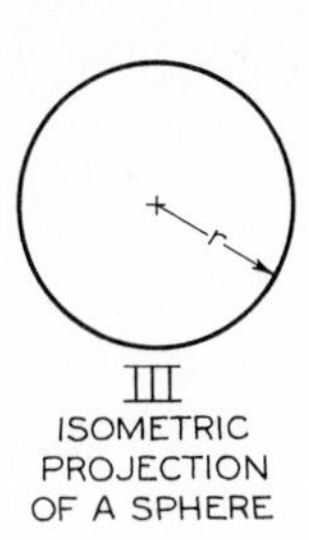


Fig. 16.35 **Isometric of a Sphere.**

An *isometric half section* is shown in Fig. 16.37. The simplest procedure in this case is to make an isometric drawing of the entire object and then the cut surfaces. Since only a quarter of the object is removed in a half section, the resulting pictorial drawing is more useful than full sections in describing both exterior and interior shapes together. Other typical isometric half sections are shown in Figs. 7.13, 7.40, and 7.41.

*Isometric broken-out sections* are also sometimes used. Examples are shown in Figs. 6.97, 7.50, and 14.47.

Section lining in isometric drawing is similar to that in multiview drawing. Section lining at an angle of 60° with horizontal, Figs. 16.36 and 16.37, is recommended, but the direction should be changed if at this angle the lines would be parallel to a prominent visible line bounding the cut surface, or to other adjacent lines of the drawing.

Fig. 16.36 **Isometric Full Section.**

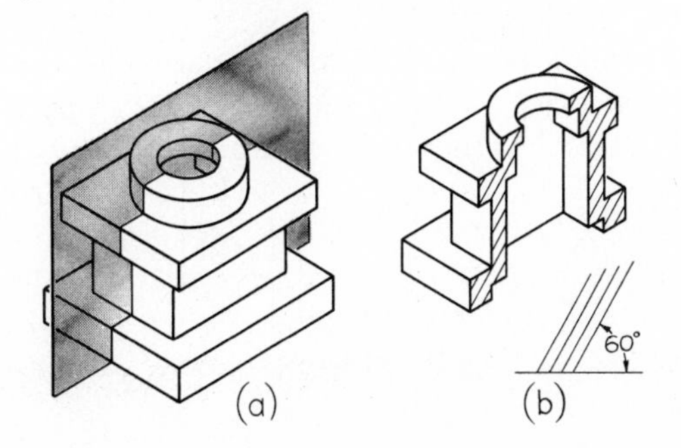


Fig. 16.37 **Isometric Half Section.**

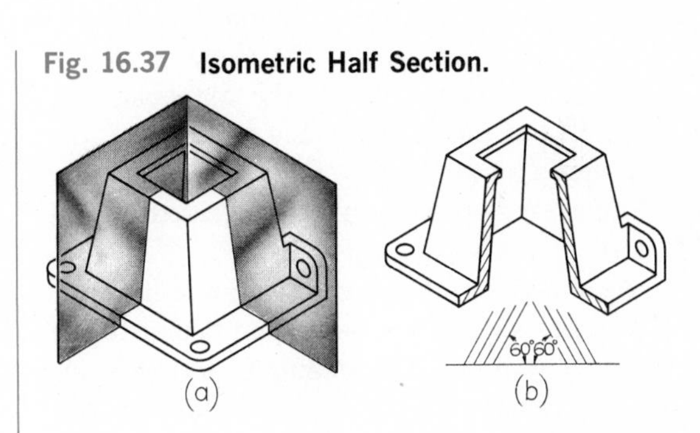


## 16.26 **Isometric Dimensioning.**

Isometric dimensions are similar to ordinary dimensions used on multiview drawings, but are expressed in pictorial form. Two methods of dimensioning are approved by ANSI: namely, the pictorial plane (aligned) system and the unidirectional system, Fig. 16.38. Note that *vertical lettering* is used for either system of dimensioning. Inclined lettering is not recommended for pictorial dimensioning. The method of drawing numerals and arrowheads for the two systems is shown at (a) and (b). For the $2\frac{1}{2}''$ dimension in the aligned system at (a), the extension lines, dimension lines, and lettering are all drawn in the isometric plane of one face of the object. The "horizontal" guide lines for the lettering are drawn parallel to the dimension line, and the "vertical" guide lines are drawn parallel to the extension lines. The barbs of the arrowheads should line up parallel to the extension lines.

For the $2\frac{1}{2}''$ dimension in the unidirectional

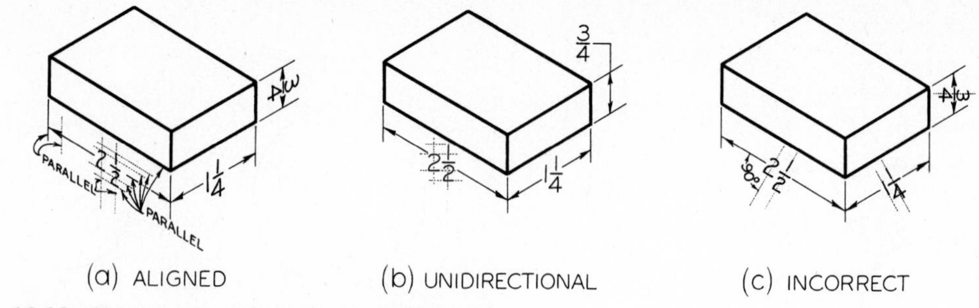


**Fig. 16.38 Numerals and Arrowheads in Isometric.**

system at (b), the extension lines and dimension lines are all drawn in the isometric plane of one face of the object and the barbs of the arrowheads should line up parallel to the extension lines, all exactly the same as at (a). However, the lettering for the dimensions is vertical and reads from the bottom of the drawing. This simpler system of dimensioning is often used on pictorials for production purposes.

As shown at (c), the vertical guide lines for the letters should not be perpendicular to the dimension lines. The example at (c) is incorrect because the $2\frac{1}{2}''$ and $1\frac{1}{4}''$ dimensions are lettered neither in the plane of the corresponding dimension and extension lines nor in a vertical position to read from the bottom of the drawing. The $\frac{3}{4}''$ dimension is awkward to read because of its position.

Correct and incorrect practice in isometric dimensioning using the aligned system of dimensioning is shown in Fig. 16.39. At (b) the $3\frac{1}{8}''$ dimension runs to a wrong extension line at the right, and consequently the dimension does not lie in an isometric plane. Near the left side, a number of lines cross one another unnecessarily, and terminate on the wrong lines. The upper $\frac{1}{2}''$ drill hole is located from the edge of the cylinder when it should be dimensioned from its center line. Study these two drawings carefully to discover additional mistakes at (b).

Many examples of isometric dimensioning are given in the problems at the end of Chapters 6, 7, 8, and 14, and the student should study these to find examples of almost any special case he may encounter. See especially Figs. 6.53 to 6.58.

## 16.27 Exploded Assemblies.

*Exploded assemblies* are often used in design presentations, catalogs, sales literature, and in the shop, to show all of the parts of an assembly and how

**Fig. 16.39 Correct and Incorrect Isometric Dimensioning (Aligned System).**

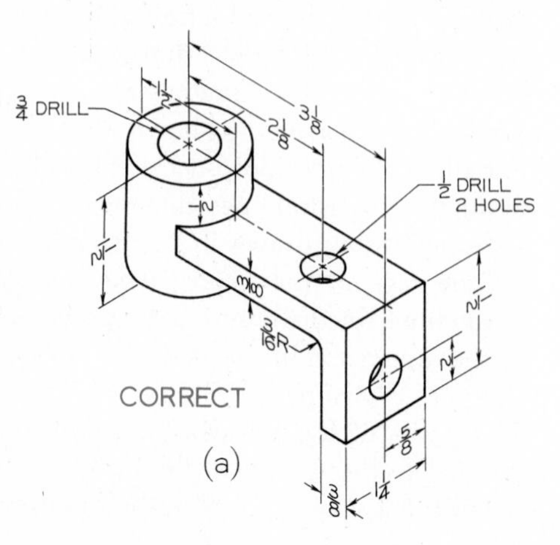


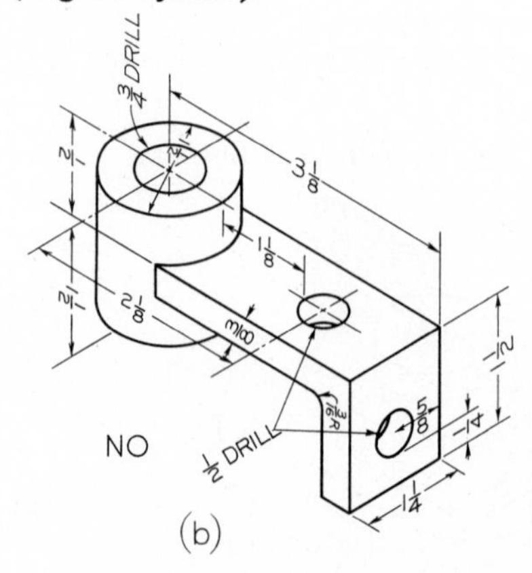

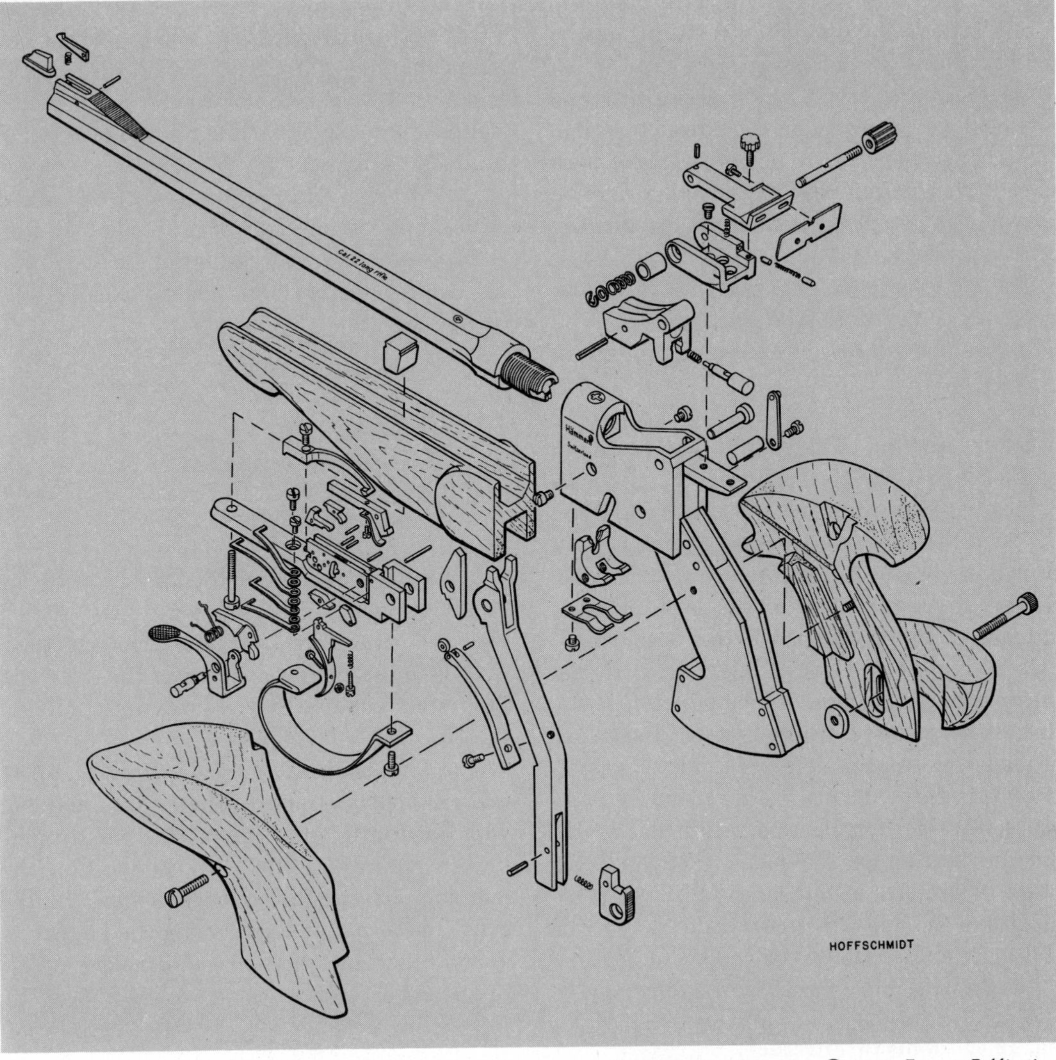


*Courtesy True Magazine © 1955, Fawcett Publications*

Fig. 16.40 **Isometric Exploded Assembly of Hammerli Match Pistol.**

they fit together. They may be drawn by any of the pictorial methods, including isometric, Fig. 16.40. Other isometric exploded assemblies are shown in Figs. 14.71 to 14.74, 14.76, and 14.79.

**16.28 Piping Diagrams.** Isometric and oblique drawings are well suited for representation of piping layouts, as well as for all other structural work to be represented pictorially.

## Dimetric Projection

**16.29 Method of Projection.** A *dimetric projection* is an axonometric projection of an object so placed that two of its axes make equal angles with the plane of projection, and the third axis makes either a smaller or a greater angle. Hence, the two axes making equal angles with the plane of projection are foreshortened equally, while the third axis is foreshortened in a different ratio.

Generally the object is so placed that one axis will be projected in a vertical position. However, if the relative positions of the axes have been determined, the projection may be drawn in any revolved position, as in isometric drawing. See §16.9.

The angles between the *projection of the axes* must not be confused with the angles the *axes themselves* make with the plane of projection.

**16.30 Dimetric Projection.** The positions of the axes may be assumed such that any two angles between the axes are equal and over 90°, and the scales determined graphically, as shown in Fig. 16.41 (a), in which OP, OL, and OS are the projections of the axes or converging edges of a cube. In this case, angle POS = angle LOS. Lines PL, LS, and SP are the lines of intersection of the plane of projection with the three visible faces of the cube. From descriptive geometry we know that since line LO is perpendicular to the plane POS, in space, its projection LO is perpendicular to PS, the intersection of the plane POS and the plane of projection. Similarly, OP is perpendicular to SL, and OS is perpendicular to PL.

If the triangle POS is revolved about the line PS as an axis into the plane of projection, it will be shown in its true size and shape as PO′S. If regular full-size scales are marked along the lines O′P and O′S, and the triangle is counterrevolved to its original position, the dimetric scales may be laid off on the axes OP and OS, as shown.

In order to avoid the preparation of special scales, use can be made of available scales on the architects scale by assuming the scales and calculating the positions of the axes, as follows:

$$\cos a = -\frac{\sqrt{2h^2v^2 - v^4}}{2hv}$$

where $a$ is one of the two equal angles between the projections of the axes, $h$ is one of the two equal scales, and $v$ is the third scale.

Examples are shown in the upper row of Fig. 16.42, in which the assumed scales, shown encircled, are taken from the architects scale. One of these three positions of the axes will be found suitable for almost any practical drawing.

The *Instrumaster Dimetric Template* Fig. 16.41 (b), has angles of approximately 11° and 39° with horizontal, which provides a picture similar to that in Fig. 16.42 (III). In addition, the template has ellipses corresponding to the axes, and accurate scales along the edges.

For other information on drawing of ellipses, see §16.34

Fig. 16.41 **Dimetric Projection.**

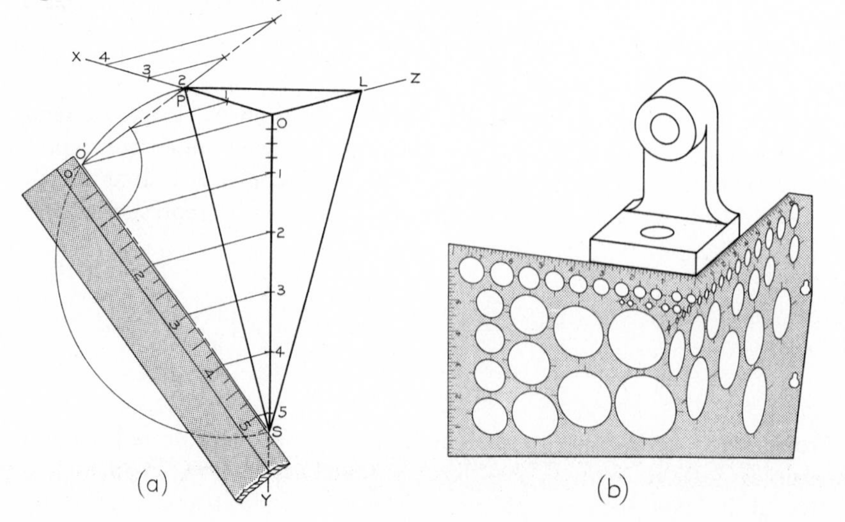


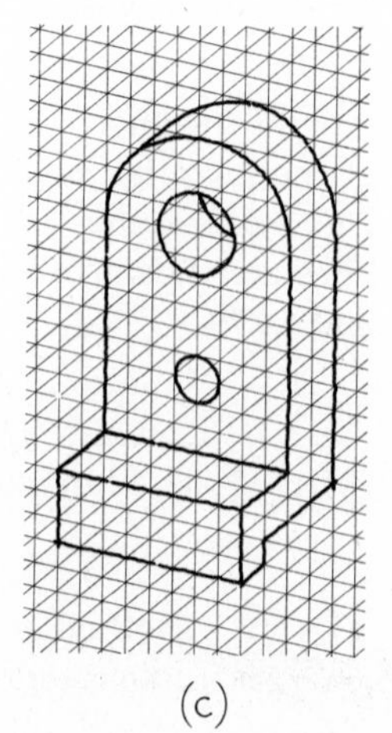

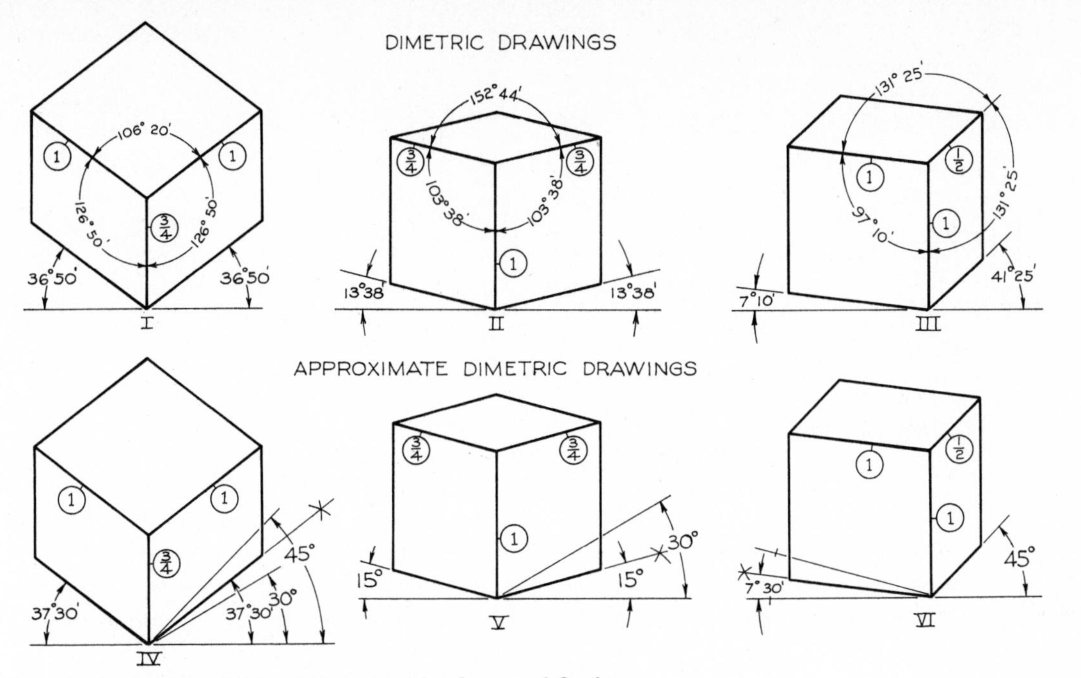


Fig. 16.42 **Angles of Axes Determined by Assumed Scales.**

The *Instrumaster Dimetric Graph* paper, Fig. 16.41 (c), can be used to sketch in dimetric as easily as to sketch isometrics on isometric paper. The grid lines slope in conformity to the angles on the Dimetric Template at (b) and, when printed on vellum, the grid lines do not reproduce on prints.

## 16.31 Approximate Dimetric Drawing.

Approximate dimetric drawings, which closely resemble true dimetrics, can be constructed by substituting for the true angles shown in the upper half of Fig. 16.42, angles that can be obtained with the ordinary triangles and compass, as shown in the lower half of the figure. The resulting drawings will be sufficiently accurate for all practical purposes.

The procedure in preparing an approximate dimetric drawing, using the position of VI in Fig. 16.42, is shown in Fig. 16.43. The offset method of drawing a curve is shown in the figure. Other methods for drawing ellipses are the same as in trimetric drawing, §16.34.

Fig. 16.43 **Steps in Dimetric Drawing.**

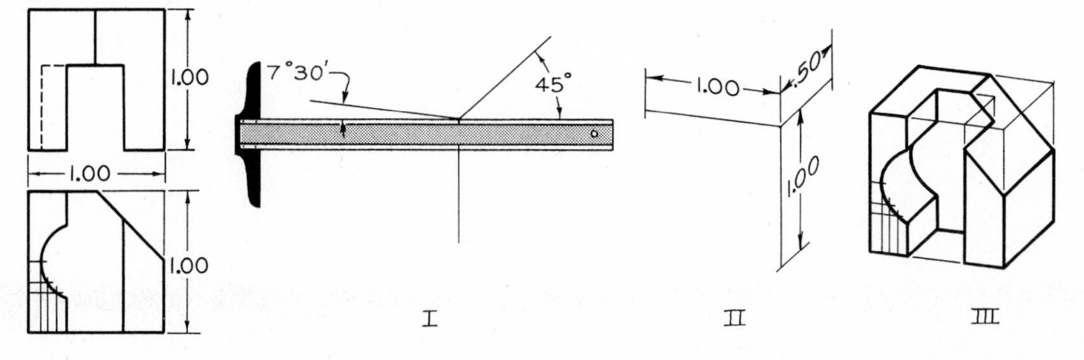

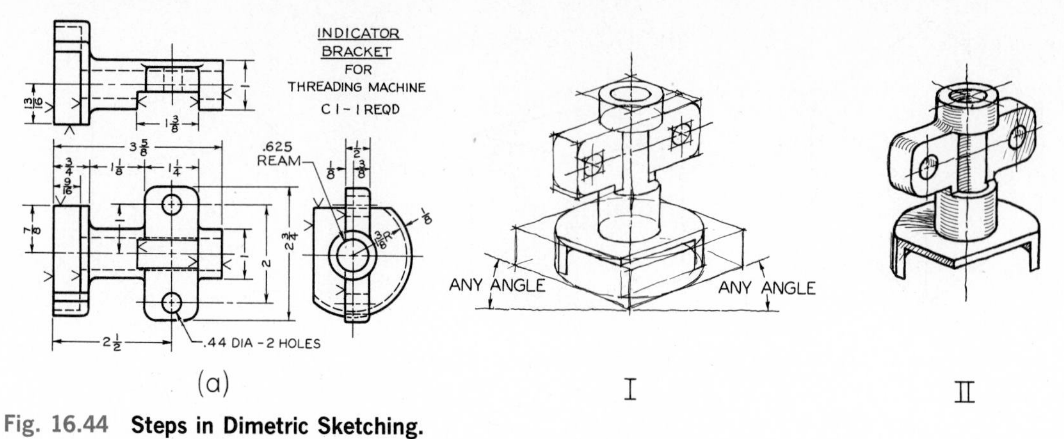


Fig. 16.44 **Steps in Dimetric Sketching.**

The steps in making a dimetric sketch, using a position similar to that in Fig. 16.42 (V), are shown in Fig. 16.44. The two angles are equal and about 20° with horizontal for the most pleasing effect.

An exploded approximate dimetric drawing of an adding machine is shown in Fig. 16.45. The dimetric axes used are those in Fig. 16.42 (IV). Pictorials such as this are often used in service manuals.

Fig. 16.45 **Exploded Dimetric of an Adding Machine.**

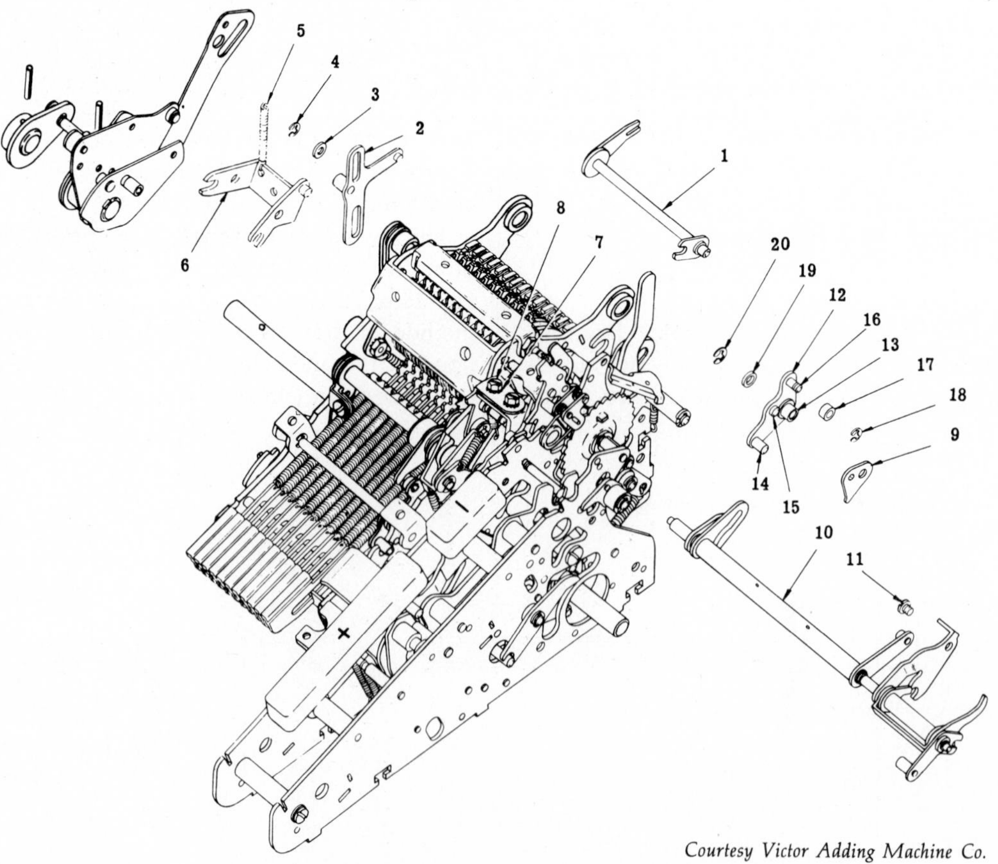


*Courtesy Victor Adding Machine Co.*

## 16.32 Method of Projection.

A *trimetric projection* is an axonometric projection of an object so placed that no two axes make equal angles with the plane of projection. In other words, each of the three axes and the lines parallel to them, respectively, have different ratios of foreshortening when projected to the plane of projection. If the three axes are assumed in any position on paper such that none of the angles is less than 90°, and if neither an isometric nor a dimetric position is deliberately arranged, the result will be a trimetric projection.

## 16.33 Trimetric Scales.

Since the three axes are foreshortened differently, three different trimetric scales must be prepared and used. The scales are determined as shown in Fig. 16.46 (a), the method being the same as explained for the dimetric scales in §16.30. As shown at (a), any two of the three triangular faces can be revolved into the plane of projection to show the true lengths of the three axes. In the revolved position, the regular scale is used to set off inches or fractions thereof. When the axes have been counterrevolved to their original positions, the scales will be correctly foreshortened, as shown. These dimensions should be transferred to the edges of three thin cards and marked OX, OZ, and OY for easy reference.

A special trimetric angle may be prepared from Bristol Board or plastic, as shown at (b). Perhaps six or seven such guides, using angles for a variety of positions of the axes, would be sufficient for all practical requirements.*

## 16.34 Trimetric Ellipses.

The trimetric center lines of a hole, or on the end of a cylinder, become the conjugate diameters of the ellipse when drawn in trimetric. The ellipse may be drawn upon the conjugate diameters by the methods of Fig. 4.51 or 4.52 (b). Or the major and minor axes may be determined from the conjugate diameters, Fig. 4.53 (c), and the ellipse constructed upon them by any of the methods of Figs. 4.48 to 4.50, and 4.52 (a), or with the aid of an ellipse guide, Fig. 4.55.

One of the advantages of trimetric is the infinite number of positions of the object available. The angles and scales can be handled without too much difficulty, as shown in

*Plastic templates of this type are available from Charles Bruning Co., Inc., Mt. Prospect, Ill. 60058.

Fig. 16.46 **Trimetric Scales.**

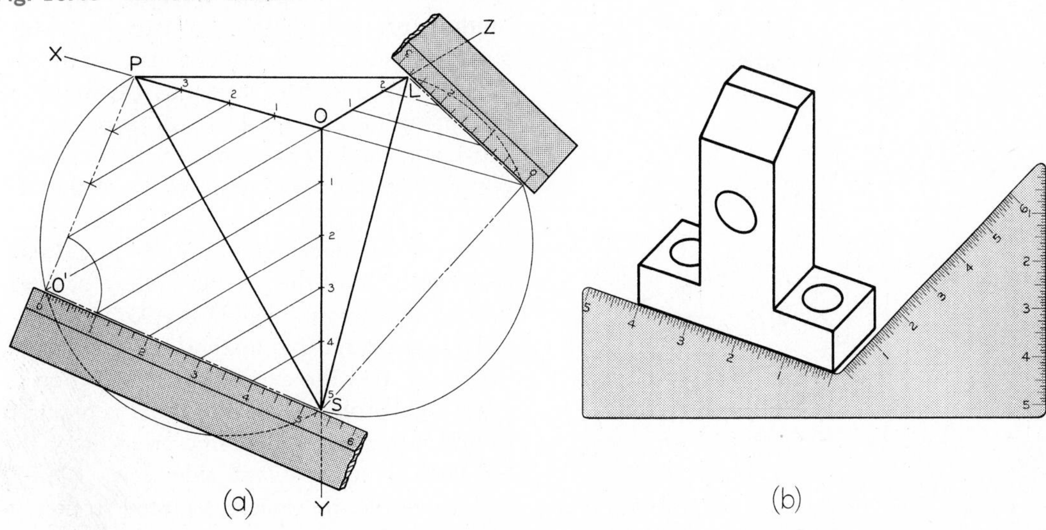

§16.33. However, the infinite variety of ellipses has been a discouraging factor.

In drawing any axonometric ellipse, keep the following in mind:

1. On the drawing, the major axis is always perpendicular to the center line, or axis, of the cylinder.
2. The minor axis is always perpendicular to the major axis; that is, on the paper it coincides with the axis of the cylinder.
3. The length of the major axis is equal to the actual diameter of the cylinder.

Thus we know at once the directions of both the major and minor axes, and the length of the major axis. *We do not know the length of the minor axis.* If we can find it, we can easily construct the ellipse with the aid of an ellipse guide or any of a number of ellipse constructions mentioned above.

In Fig. 16.47 (a), center O is located as desired, and horizontal and vertical construction lines that will contain the major and minor axes are drawn through O. Note that the major axis will be on the horizontal line perpendicular to the axis of the hole, and the minor axis will be perpendicular to it, or vertical.

Set the compass for the actual radius of the hole and draw the semicircle, as shown, to establish the ends A and B of the major axis. Draw AF and BF parallel to the axonometric edges WX and YX, respectively, to locate F

Fig. 16.47 **Ellipses in Trimetric.**

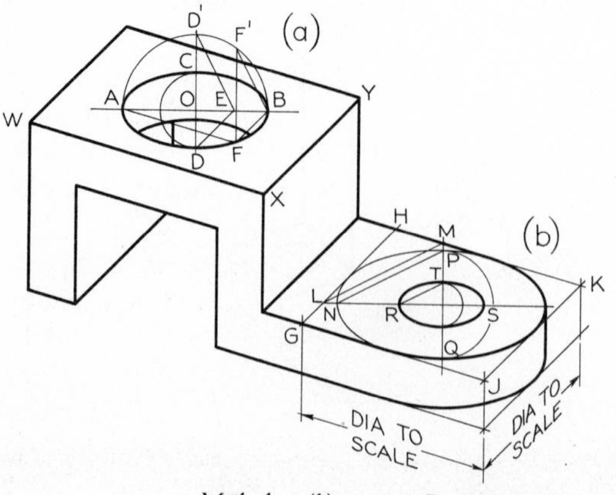


*Method at (b) courtesy Prof. H. E. Grant.*

which lies on the ellipse. Draw a vertical line through F to intersect the semicircle at F′ and join F′ to B as shown. From D′ where the minor axis, extended, intersects the semicircle, draw D′E and ED parallel to F′B and BF, respectively. Point D is one end of the minor axis. From center O, strike arc DC to locate C, the other end of the minor axis. Upon these axes, a true ellipse can be constructed, or drawn with the aid of an ellipse guide. A simple method for finding the "angle" of ellipse guide to use is shown in Fig. 4.55 (c). If an ellipse guide is not available, an approximate four-center ellipse, Fig. 4.56, will be found satisfactory in most cases.

In constructions where the enclosing parallelogram for an ellipse is available or easily constructed, the major and minor axes can be readily determined as shown at (b). The directions of both axes, and the length of the major axis, are known. Extend the axes to intersect the sides of the parallelogram at L and M, and join the points with a straight line. From one end N of the major axis, draw a line NP parallel to LM. The point P is one end of the minor axis. To find one end T of the minor axis of the smaller ellipse, it is only necessary to draw RT parallel to LM or NP.

The method of constructing an ellipse on an oblique plane in trimetric is similar to that shown for isometric in Fig. 16.33.

## 16.35 Axonometric Projection by the Method of Intersections.

Instead of constructing axonometric projections with the aid of specially prepared scales, as explained in the preceding paragraphs, an axonometric projection can be obtained directly by projection from two orthographic views of the object. This method is called the *method of intersections;* it was developed by Profs. L. Eckhart and T. Schmid of the Vienna College of Engineering and was published in 1937.

To understand this method, let us assume, Fig. 16.48, that the axonometric projection of a rectangular object is given, and it is required to find its three orthographic projections: the top view, front view, and side view.

Assume that the object is placed so that its

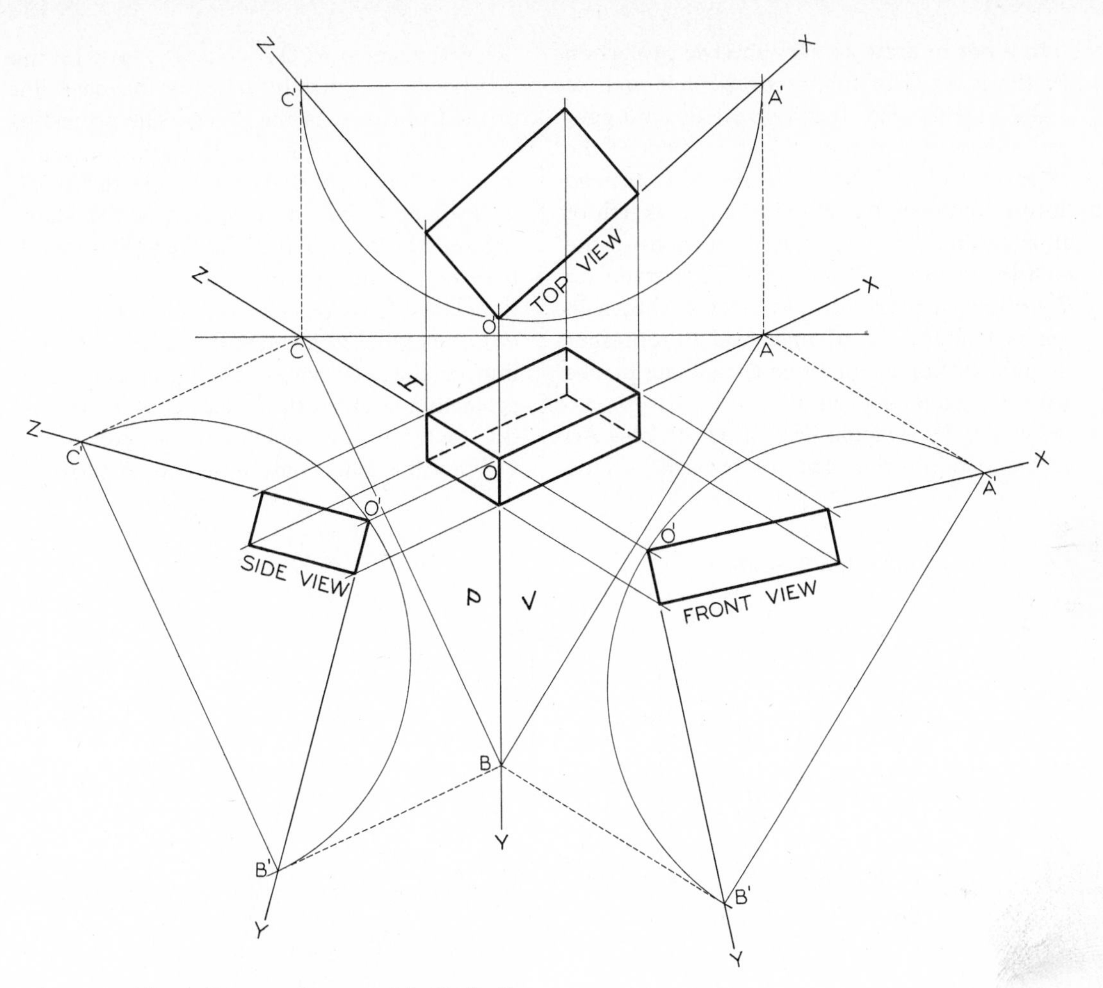


Fig. 16.48 **Views from an Axonometric Projection.**

principal edges coincide with the coordinate axes, and assume that the plane of projection (the plane upon which the axonometric projection is drawn) intersects the three coordinate planes in the triangle ABC. From descriptive geometry we know that lines BC, CA, and AB will be perpendicular, respectively, to axes OX, OY, and OZ. Any one of the three points A, B, or C may be assumed anywhere on one of the axes, and the triangle ABC drawn.

To find the true size and shape of the top view, revolve the triangular portion of the horizontal plane AOC, which is in front of the plane of projection, about its base CA, into the plane of projection. In this case, the triangle is revolved *inward* to the plane of projection through the smallest angle made with it. The triangle will then be shown in its true size and shape, and the top view of the object can be drawn in the triangle by projection from the axonometric projection, as shown, since all width dimensions remain the same. In the figure, the base CA of the triangle has been moved upward to C′A′ so that the revolved position of the triangle will not overlap its projection.

In the same manner, the true sizes and shapes of the front view and side view can be found, as shown.

It is evident that if the three orthographic projections, or in most cases any two of them, are given in their relative positions, as shown in Fig. 16.48, the directions of the projections could be reversed so that the intersections of the projecting lines would determine the required axonometric projection.

In order to draw an axonometric projection by the method of intersections, it is well to make a sketch, Fig. 16.49, of the desired general appearance of the projection. Even if the object is a complicated one, this sketch need not be complete, but may be only a sketch of an enclosing box. Draw the projections of the coordinate axes OX, OY, and OZ, parallel to the principal edges of the object as shown in the sketch, and the triangle ABC to represent the intersection of the three coordinate planes with the plane of projection.

Revolve the triangle ABO about its base AB as the axis into the plane of projection. Line OA will revolve to O′A, and this line, or one parallel to it, must be used as the base line of the front view of the object. The projecting lines from the front view to the axonometric must be drawn parallel to the projection of the unrevolved Z-axis, as indicated in the figure.

Similarly, revolve the triangle COB about its base CB as the axis into the plane of projection. Line CO will revolve to CO″, and this line, or one parallel to it, must be used as the base line of the side view. The direction of the projecting lines must be parallel to the projection of the unrevolved X-axis, as shown.

Draw the front-view base line at pleasure,

Fig. 16.49 **Axonometric Projection.**

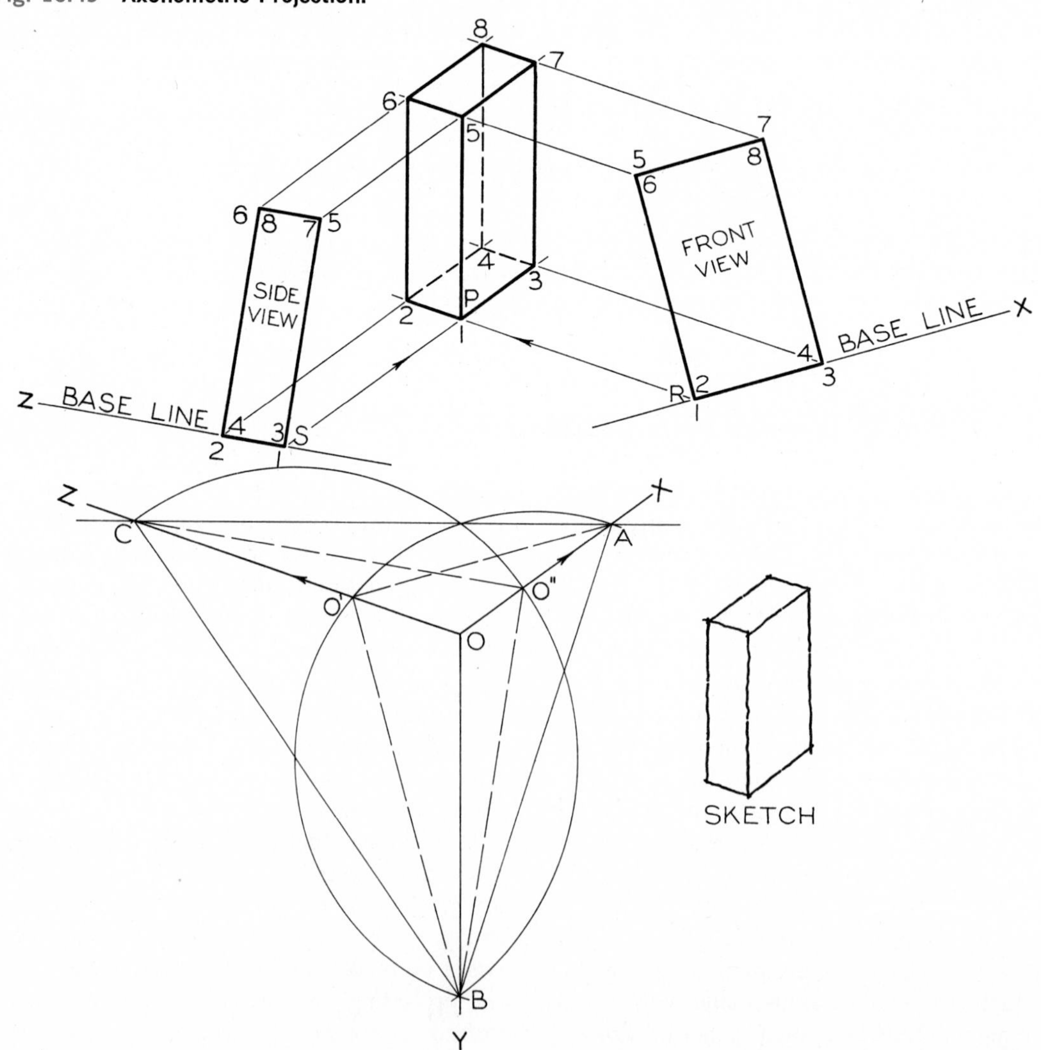

but parallel to O′X, and with it as the base, draw the front view of the object. Draw the side-view base line at pleasure, but parallel to O″C, and with it as the base, draw the side view of the object, as shown. From the corners of the front view, draw projecting lines parallel to OZ, and from the corners of the side view, draw projecting lines parallel to OX. The intersections of these two sets of projecting lines determine the desired axonometric projection. It will be an isometric, dimetric, or a trimetric projection, depending upon the form of the sketch used as the basis for the projections, §16.3. If the sketch is drawn so that the three angles formed by the three coordinate axes are equal, the resulting projection will be an isometric projection; if two of the three angles are equal, the resulting projection will be a dimetric projection; and if no two of the three angles are equal, the resulting projection will be a trimetric projection.

In order to place the desired projection in a specific location on the drawing, Fig. 16.49, select the desired projection P of the point 1, for example, and draw two projecting lines PR and PS to intersect the two base lines and thereby to determine the locations of the two views on their base lines.

Another example of this method of axonometric projection is shown in Fig. 16.50. In this case, it was deemed necessary only to draw a sketch of the plan or base of the object in the desired position, as shown. The axes are then drawn with OX and OZ parallel respectively to the sides of the sketch plan, and the remaining axis OY is assumed in a vertical



Fig. 16.50 **Axonometric Projection.**

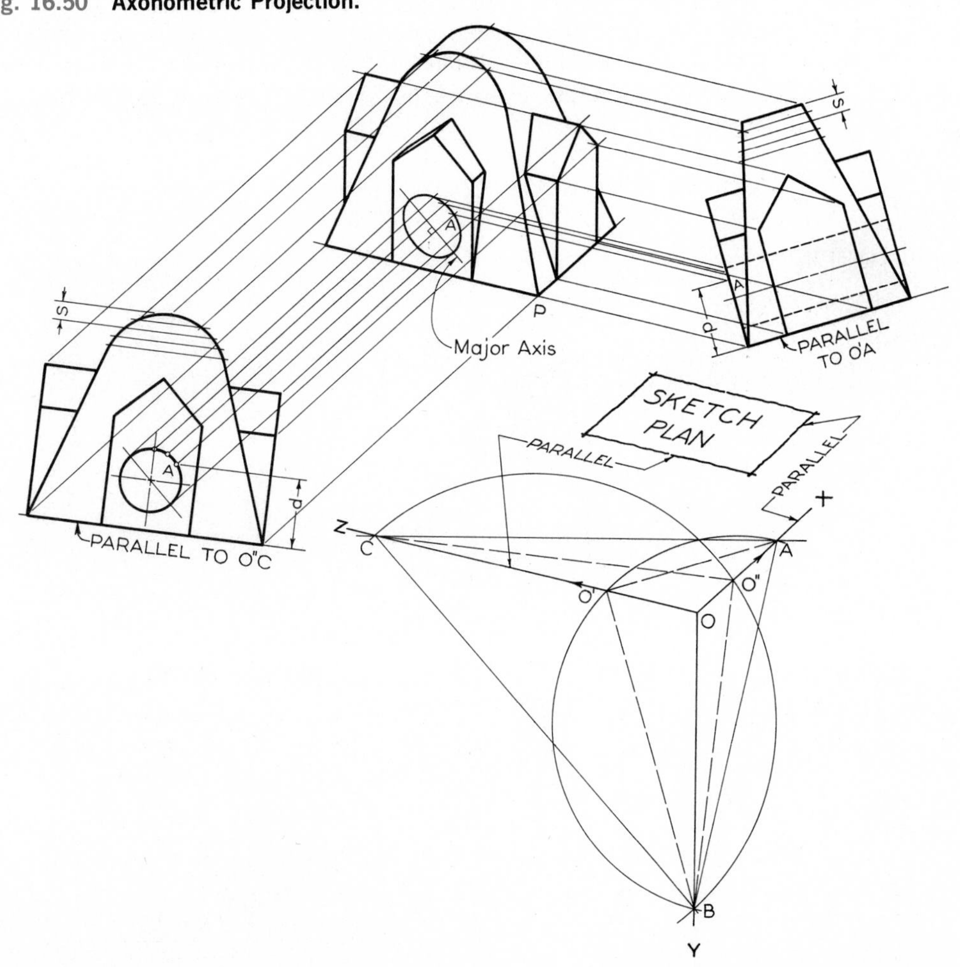

position. The triangles COB and AOB are revolved, and the two base lines drawn parallel to O″C and O′A as shown. Point P, the lower front corner of the axonometric drawing, was then chosen at pleasure, and projecting lines drawn toward the base lines parallel to axes OX and OZ to locate the positions of the views on the base lines. The views are drawn upon the base lines, or cut apart from another drawing and fastened in place with drafting tape or thumbtacks.

To draw the elliptical projection of the circle, assume any points, such as A, on the circle in both front and side views. Note that point A is the same altitude d above the base line in both views. The axonometric projection of point A is found simply by drawing the projecting lines from the two views. The major and minor axes may be easily found by projecting in this manner, or by methods shown in Fig. 16.47, and the true ellipse drawn by any of the methods of Figs. 4.48 to 4.50 and 4.52 (a), or with the aid of an ellipse guide, §§4.57 and 16.22. Or an approximate ellipse, which is satisfactory for most drawings, may be used, Fig. 4.56.

## 16.36 Axonometric Problems.

A large number of problems to be drawn axonometrically are given in Figs. 16.51 to 16.54. The earlier isometric sketches may be drawn on isometric paper, §5.14; later sketches should be made on plain drawing paper. On drawings to be executed with instruments, show all construction lines required in the solutions.

For additional problems, see Figs. 6.50 to 6.52, 17.23 to 17.25, and 18.33.

Axonometric problems in convenient form for solution may be found in *Technical Drawing Problems, Series 1,* by Giesecke, Mitchell, Spencer and Hill; in *Technical Drawing Problems, Series 2,* by Spencer and Hill; and in *Technical Drawing Problems, Series 3,* by Spencer and Hill, all designed to accompany this text and published by Macmillan Publishing Co., Inc.

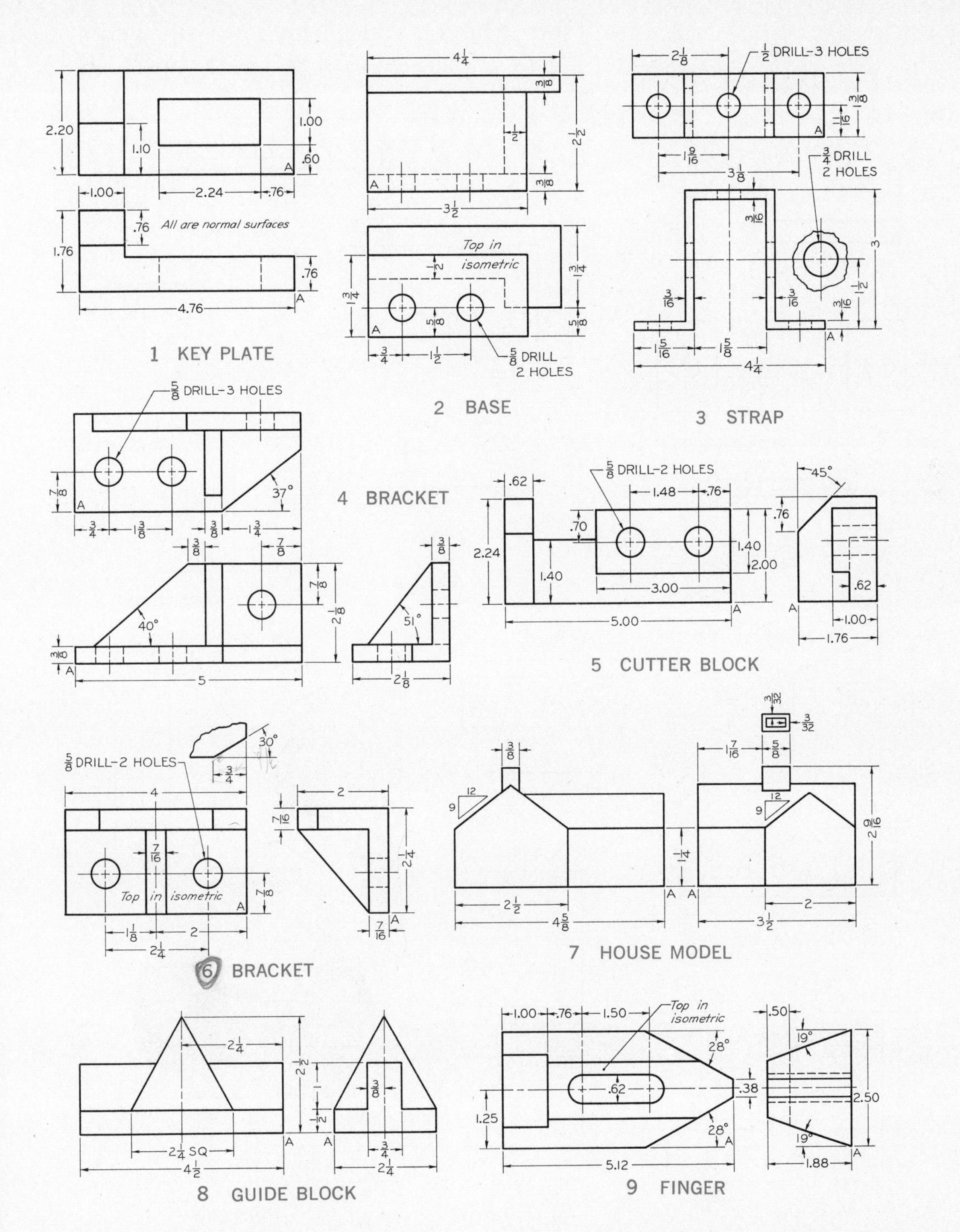


Fig. 16.51 (1) Make freehand isometric sketches. (2) Make isometric drawings with instruments on Layout A-2. (3) Make dimetric drawings with instruments, using Layout A-2, and position assigned from Fig. 16.42. (4) Make trimetric drawings, using instruments, with axes chosen to show the objects to best advantage. If dimensions are required, study §16.26.

## 16.36 Axonometric Problems

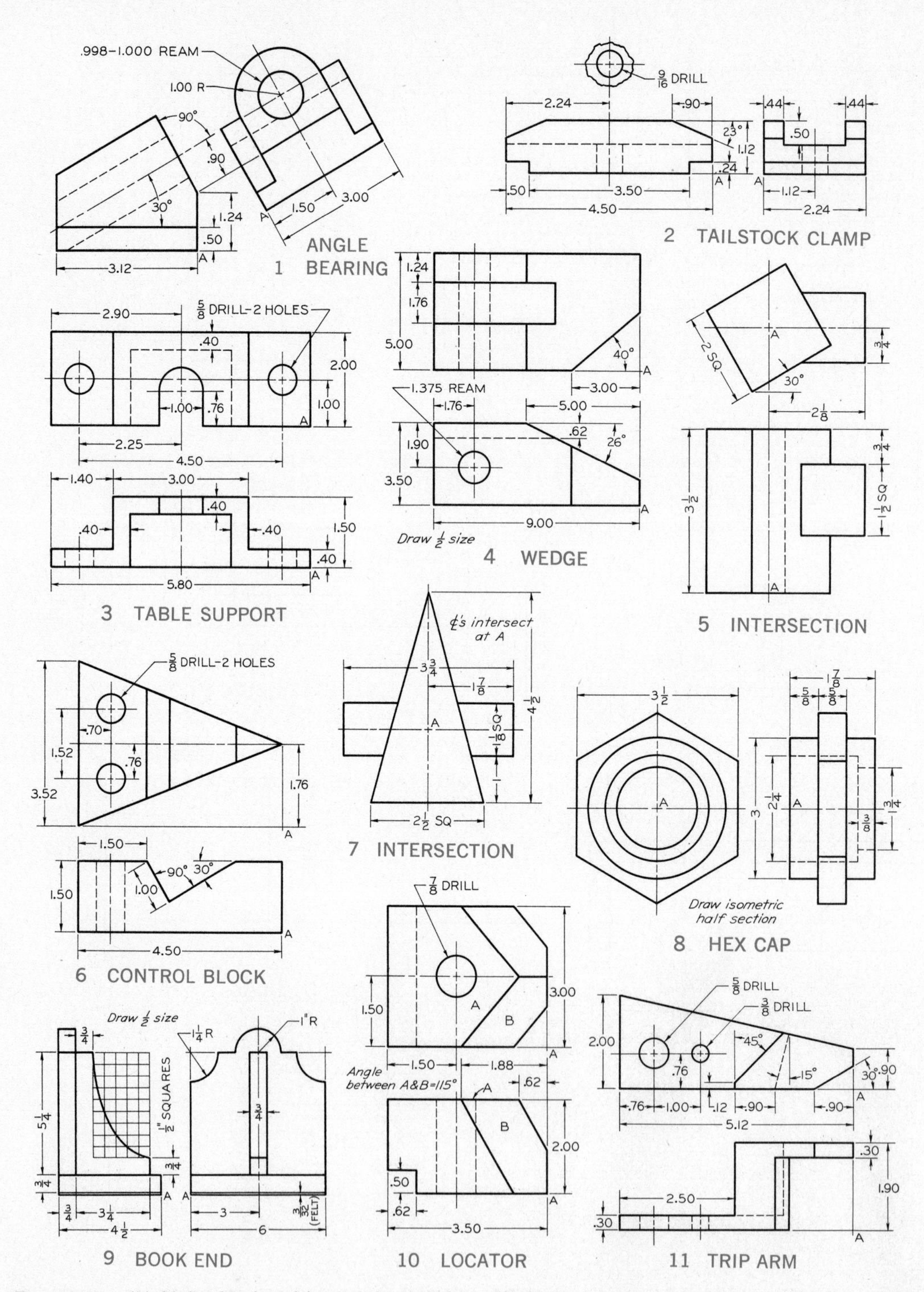


Fig. 16.52 (1) Make freehand isometric sketches. (2) Make isometric drawings with instruments on Layout A–2. (3) Make dimetric drawings with instruments, using Layout A–2, and position assigned from Fig. 16.42. (4) Make trimetric drawings, using instruments, with axes chosen to show the objects to best advantage. If dimensions are required, study §16.26.

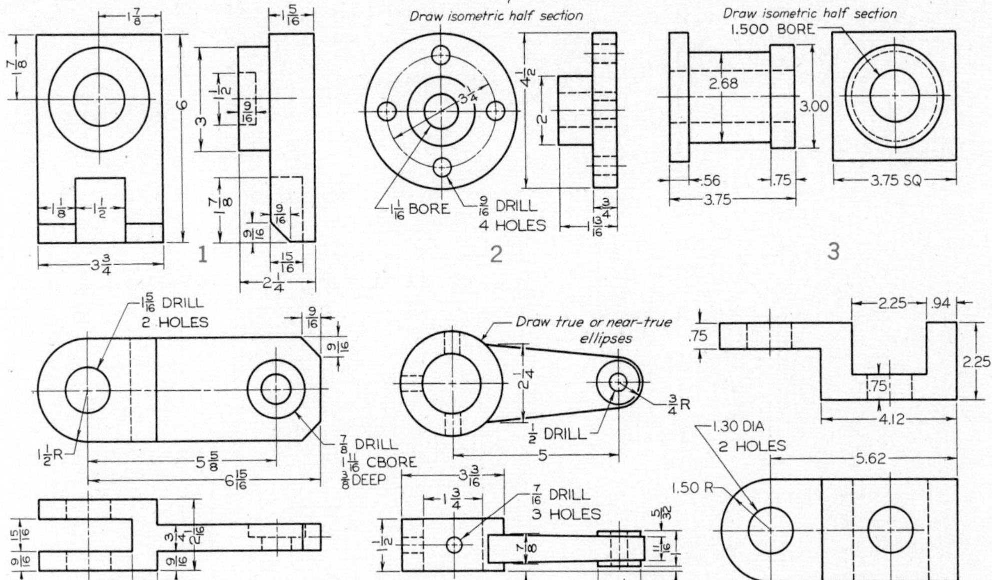


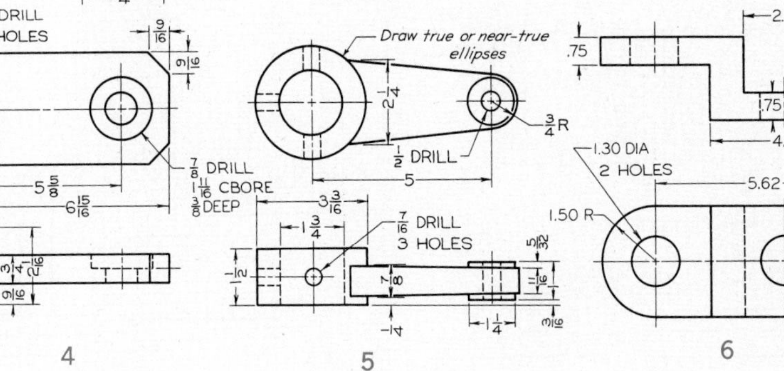


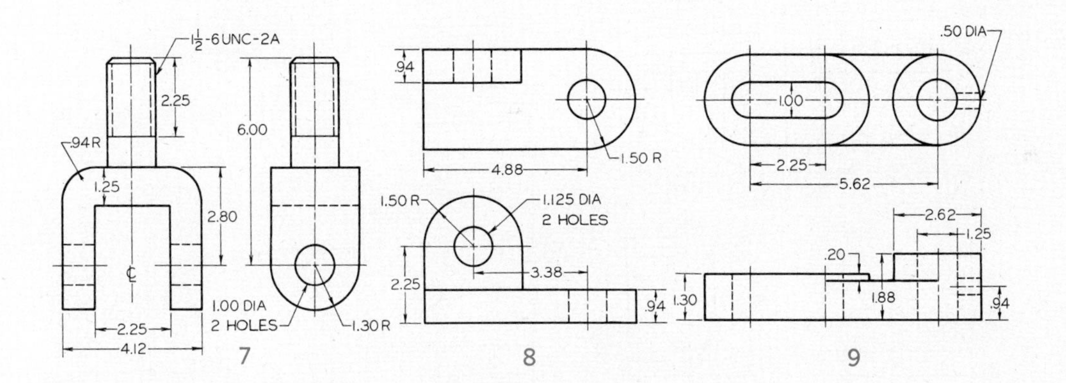


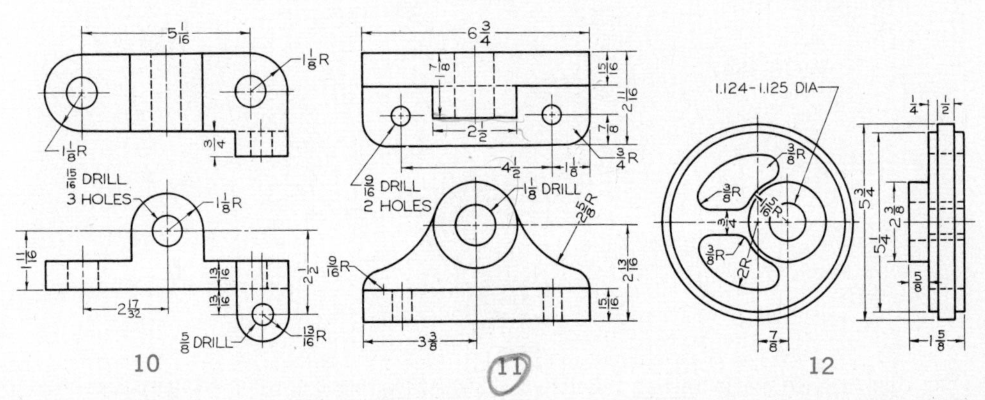


Fig. 16.53 (1) Make isometric freehand sketches. (2) Make isometric drawings with instruments, using Size A or Size B sheet, as assigned. (3) Make dimetric drawings with instruments, using Size A or Size B sheet, as assigned, and position assigned from Fig. 16.42. (4) Make trimetric drawings, using instruments, with axes chosen to show the objects to best advantage. If dimensions are required, study §16.26.

## 16.36 Axonometric Problems

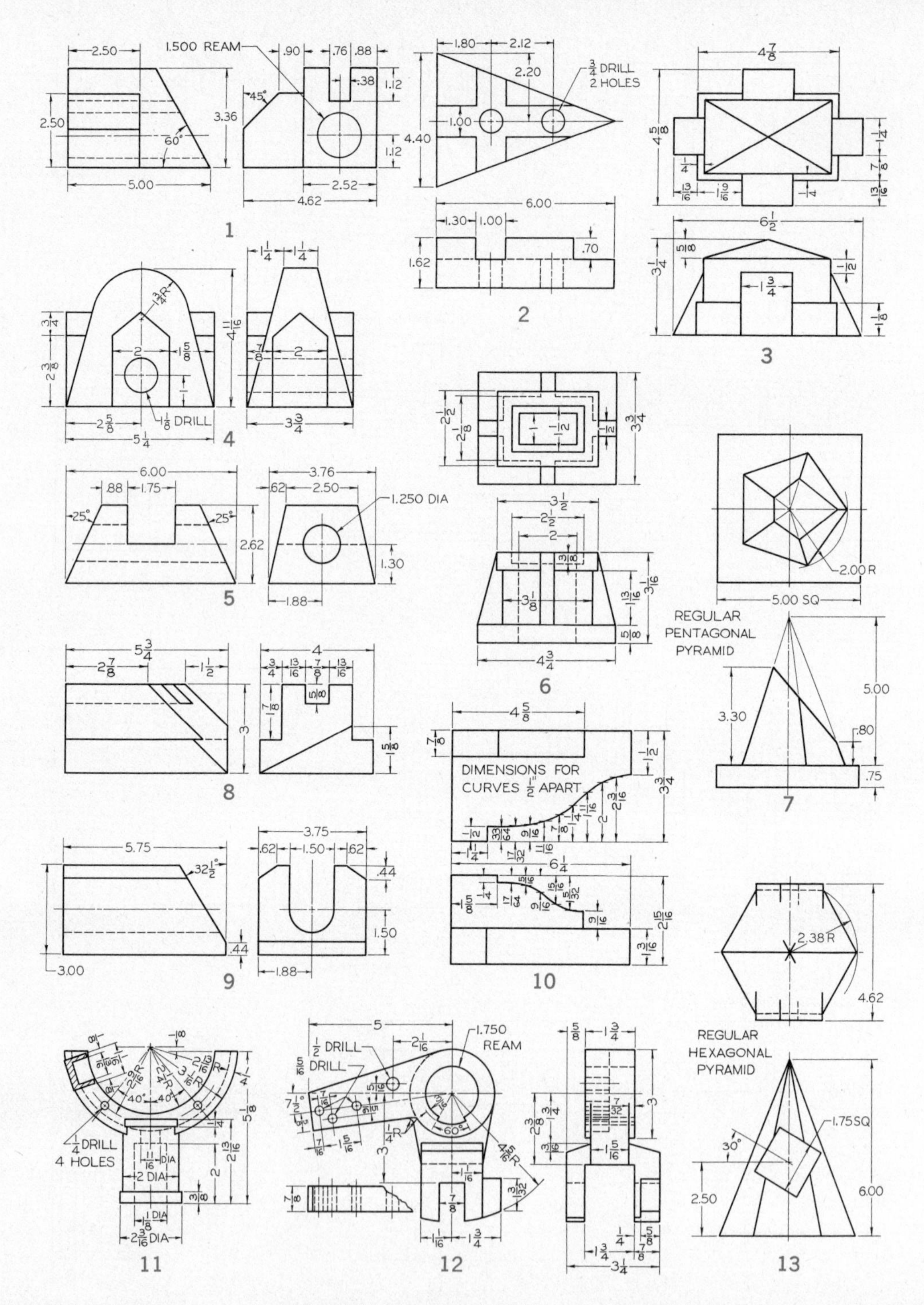


**Fig. 16.54** (1) Make isometric freehand sketches. (2) Make isometric drawings with instruments, using Size A or Size B sheet, as assigned. (3) Make dimetric drawings with instruments, using Size A or Size B sheet, as assigned, and position assigned from Fig. 16.42. (4) Make trimetric drawings, using instruments, with axes chosen to show the objects to best advantage. If dimensions are required, study §16.26. For additional problems, assignments may be made from any of the problems in Figs. 17.23 to 17.25 and 18.32.

# 17
# oblique projection

**17.1 Oblique Projection.*** If the observer is considered to be at an infinite distance from the object, Fig. 16.1 (c), and looking toward the object so that the projectors are parallel to each other and oblique to the plane of projection, the resulting drawing is an *oblique projection.* As a rule, the object is placed with one of its principal faces parallel to the plane of projection. This is equivalent to holding the object in the hand and viewing it approximately as shown in Fig. 5.27.

A comparison of oblique projection and orthographic projection is shown in Fig. 17.1. The front face A′B′C′D′ in the oblique projection is identical with the front view, or orthographic projection, $A^V B^V C^V D^V$. Thus if an object is placed with one of its faces parallel to the plane of projection, that face will be projected true size and shape in oblique projection as well as in orthographic or multiview projection. This is the reason why oblique projection is preferable to axonometric projection in representing certain objects pictorially. Note that surfaces of the object that are not parallel to the plane of projection will not project in true size and shape. For example, surface ABFE on the object (a square) projects as a parallelogram A′B′F′E′ in the oblique projection.

In axonometric projection, circles on the object nearly always lie in surfaces inclined to the plane of projection and project as ellipses. In oblique projection, the object may be positioned so that those surfaces are parallel to the plane of projection, in which case the circles will project as full-size true circles, and can be easily drawn with the compass.

A comparison of the oblique and orthographic projections of a cylindrical object is

*See ANSI Y14.4–1957.

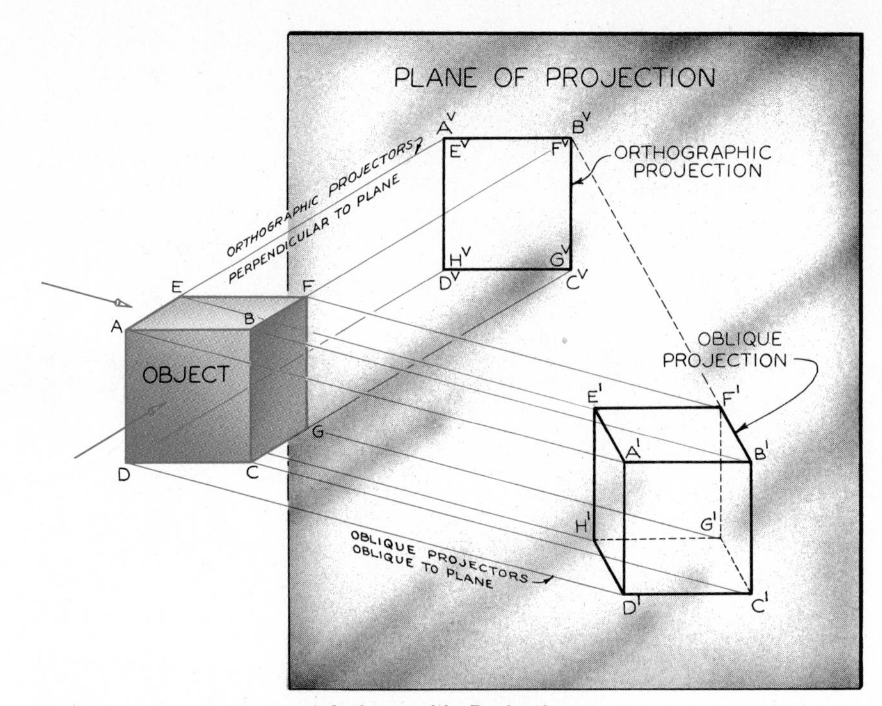


Fig. 17.1 **Comparison of Oblique and Orthographic Projections.**

shown in Fig. 17.2. In both cases, the circular shapes project as true circles. Note that although the observer, looking in the direction of the oblique arrow, does see these shapes as ellipses, the drawing, or projection, represents not what he sees, but what is projected upon the plane of projection. This curious situation is peculiar to oblique projection.

Observe that the axis AB of the cylinder projects as a point $A^V B^V$ in the orthographic projection, since the line of sight is parallel to AB. But in the oblique projection, the axis projects as a line A′B′. The more nearly the direction of sight approaches the perpendicular with respect to the plane of projection —that is, the larger the angle between the projectors and the plane—the closer the oblique projection moves toward the orthographic projection, and the shorter A′B′ becomes.

## 17.2 **Directions of Projectors.**

In Fig. 17.3, the projectors make an angle of 45° with the plane of projection; hence the line CD′, which is perpendicular to the plane, projects true length at C′D′. If the projectors make a greater angle with the plane of projection, the oblique projection is shorter, and if the projectors make a smaller angle with the plane of projection, the oblique projection is longer. Theoretically, CD′ could project in any length from zero to infinity. However, the line AB is parallel to the plane and will project in true length regardless of the angle the projectors make with the plane of projection.

In Fig. 17.1 the lines AE, BF, CG, and DH are perpendicular to the plane of projection, and project as parallel inclined lines A′E′, B′F′, C′G′, and D′H′ in the oblique projection. These lines on the drawing are called the *receding lines.* As we have seen above, they may be any length, from zero to infinity, depending upon the direction of the line of sight. Our next concern is: What angle do these lines make on paper with respect to horizontal?

In Fig. 17.4, the line AO is perpendicular to the plane of projection, and all the projectors make angles of 45° with it; therefore, all of the oblique projections BO, CO, DO, etc., are equal in length to the line AO. It can be seen from the figure that the projectors may be

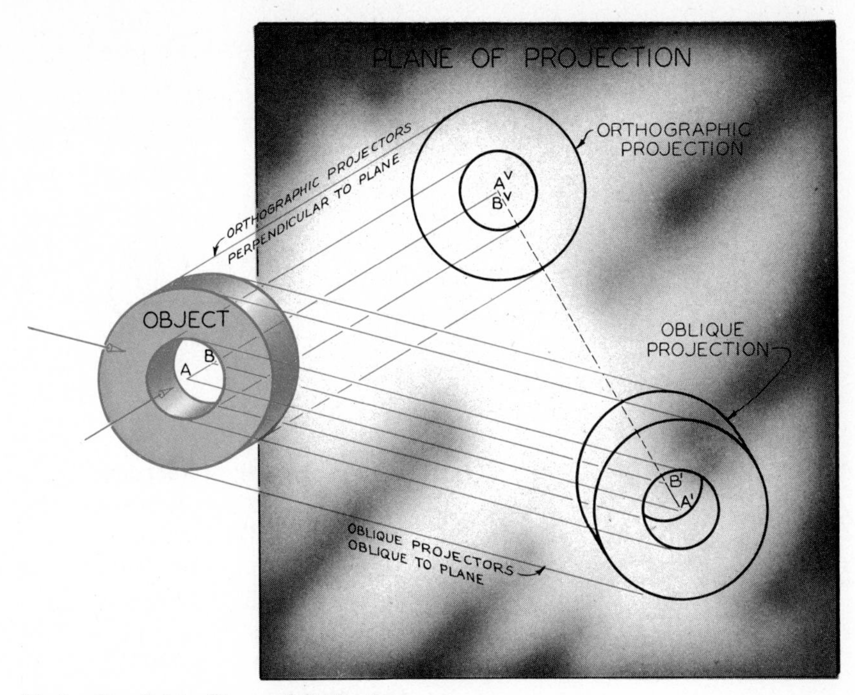


Fig. 17.2 **Circles Parallel to Plane of Projection.**

selected in any one of an infinite number of directions and yet maintain any desired angle with the plane of projection. It is also evident that the directions of the projections BO, CO, DO, etc., are independent of the angles the projectors make with the plane of projection. Ordinarily, this inclination of the projection is 45° (CO in the figure), 30°, or 60° with horizontal, since these angles may be easily drawn with the triangles.

Fig. 17.3 **Lengths of Projections.**

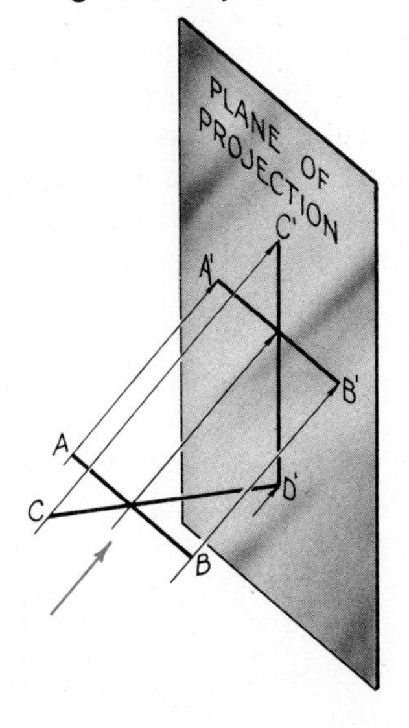


Fig. 17.4 **Directions of Projectors.**

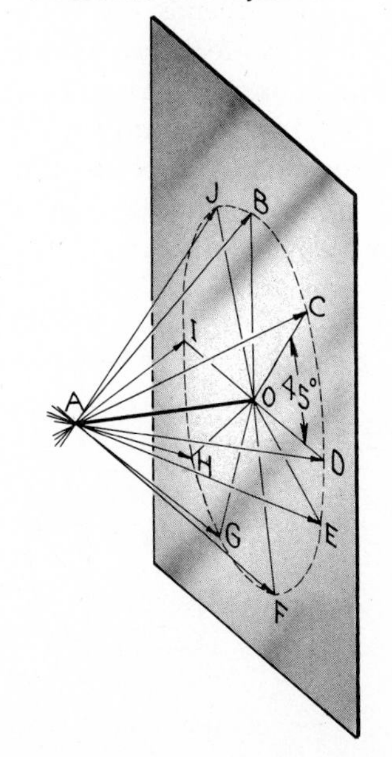

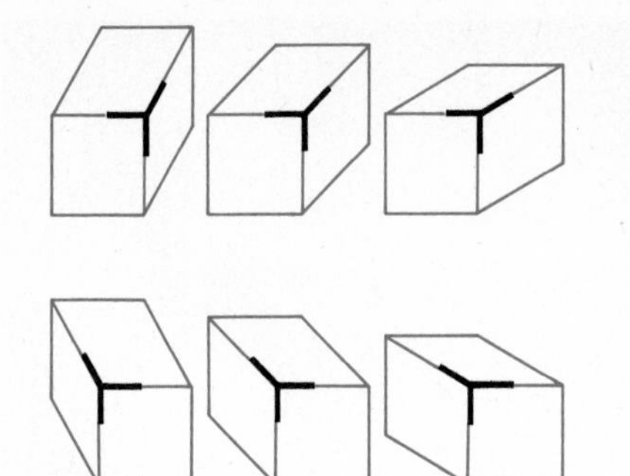
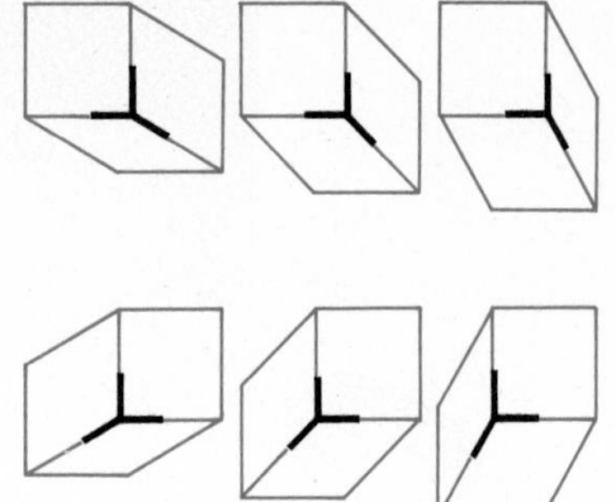

Fig. 17.5 **Variation in Direction of Receding Axis.**

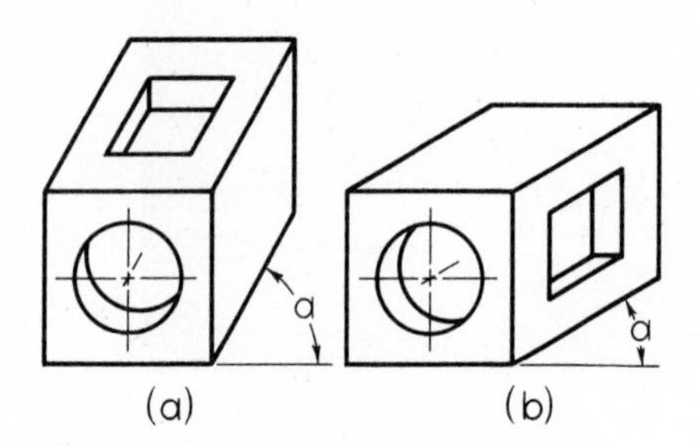


Fig. 17.6 **Angle of Receding Axis.**

**17.3 Angles of Receding Lines.** The receding lines may be drawn at any convenient angle. Some typical drawings with the receding lines in various directions are shown in Fig. 17.5. The angle that should be used in an oblique drawing depends upon the shape of the object and the location of its significant features. For example, in Fig. 17.6 (a) a large angle was used in order to obtain a better view of the rectangular recess on the top, while at (b) a small angle was chosen to show a similar feature on the side.

**17.4 Length of Receding Lines.** Since the eye is accustomed to seeing objects with all receding parallel lines appearing to converge, an oblique projection presents an unnatural appearance, the seriousness of the distortion depending upon the object shown. For example, the object shown in Fig. 17.7 (a) is a cube, the receding lines being full length; but the receding lines appear to be too long and to diverge toward the rear of the block. A striking example of the unnatural appearance of an oblique drawing when compared with the natural appearance of a perspective is shown in Fig. 17.8. This example points up one of the chief limitations of oblique projection: objects characterized by great length should not be drawn in oblique with the long dimension perpendicular to the plane of projection.

The appearance of distortion may be materially lessened by decreasing the length of the receding lines (remember, we established in §17.2 that they could be any length). In Fig. 17.7 a cube is shown in five oblique drawings with varying degrees of foreshortening of the

Fig. 17.7 **Foreshortening of Receding Lines.**

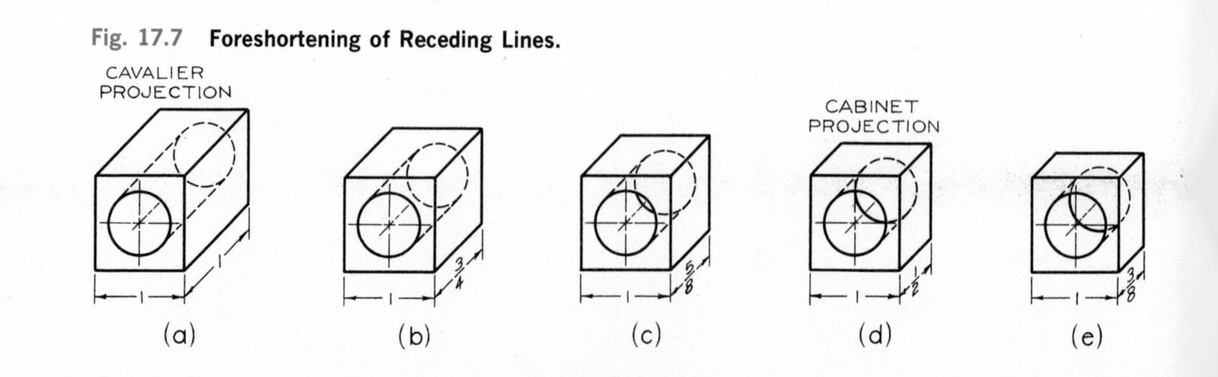

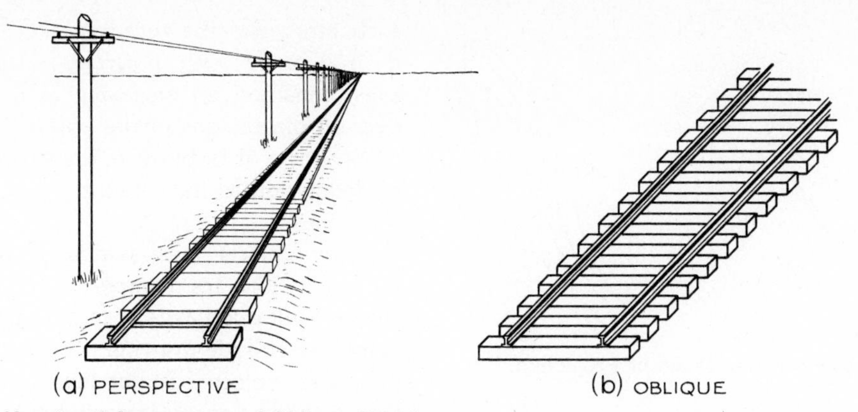

(a) PERSPECTIVE (b) OBLIQUE

Fig. 17.8 **Unnatural Appearance of Oblique Drawing.**

receding lines. The range of scales chosen is sufficient for almost all problems, and most of the scales are available on the architects scale.

When the receding lines are true length—that is, when the projectors make an angle of 45° with the plane of projection—the oblique drawing is called a *cavalier projection,* Fig. 17.7 (a). Cavalier projections originated in the drawing of medieval fortifications and were made upon horizontal planes of projection. On these fortifications the central portion was higher than the rest and it was called *cavalier* because of its dominating and commanding position.

When the receding lines are drawn to half size, as at (d), the drawing is commonly known as a *cabinet projection.* The term is attributed to the early use of this type of oblique drawing in the furniture industries. A comparison of cavalier projection and cabinet projection is shown in Fig. 17.9.

## 17.5 Choice of Position.

The face of an object showing the essential contours should generally be placed parallel to the plane of projection, Fig. 17.10. If this is done, distortion will be kept at a minimum and labor reduced. For example, at (a) and (c) the circles and circular arcs are shown in their true sizes and shapes and may be quickly drawn with the compass, while at (b) and (d) these curves are not shown in their true sizes and shapes and must be plotted as free curves or in the form of ellipses.

The longest dimension of an object should

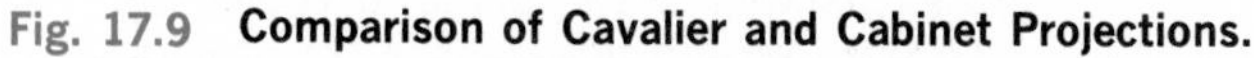

Fig. 17.9 **Comparison of Cavalier and Cabinet Projections.**

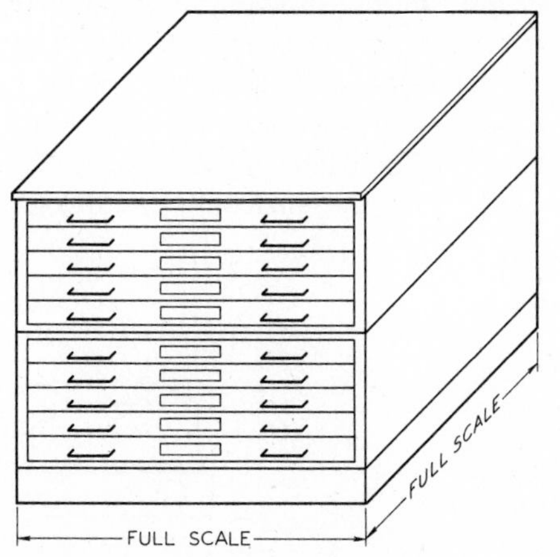


(a) CAVALIER PROJECTION

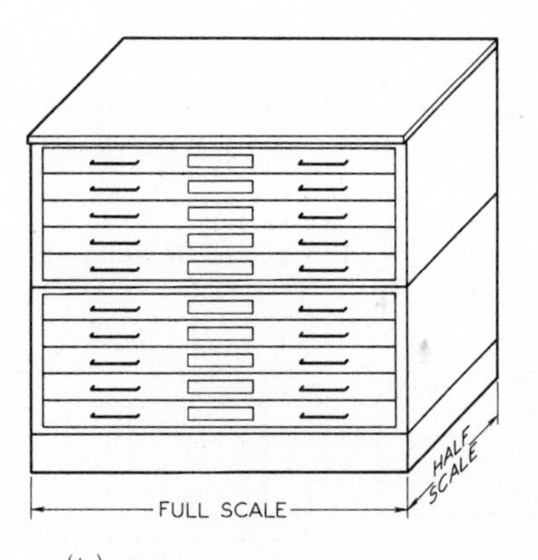


(b) CABINET PROJECTION

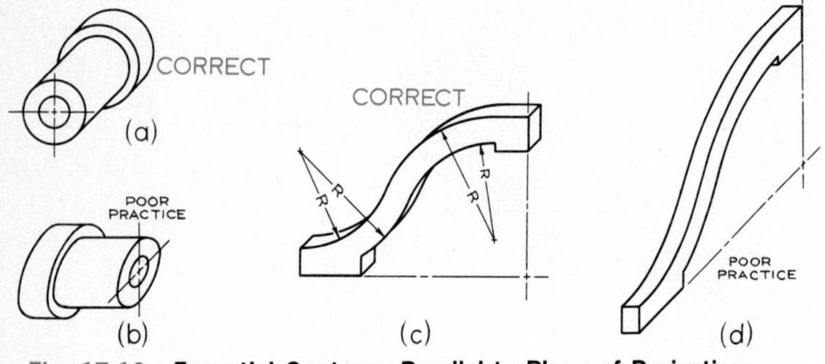

**Fig. 17.10 Essential Contours Parallel to Plane of Projection.**

**Fig. 17.11 Long Axis Parallel to Plane of Projection.**

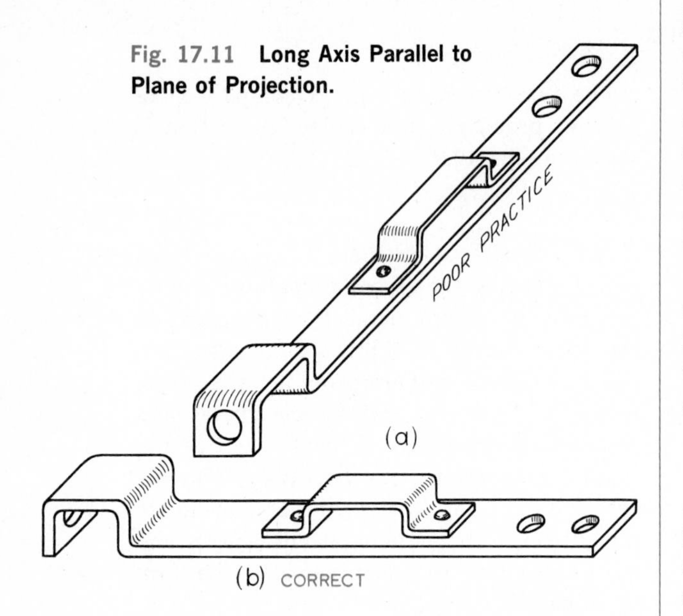


generally be placed parallel to the plane of projection, as shown in Fig. 17.11 (b).

## 17.6 Steps in Oblique Drawing.

The steps in drawing a cavalier drawing of a rectangular object is shown in Fig. 17.12. As shown in step I, draw the axes OX and OY perpendicular to each other and the receding axis OZ at any desired angle with horizontal. Upon these axes, construct an enclosing box, using the over-all dimensions of the object.

As shown at II, block in the various shapes in detail, and as indicated at III, heavy in all final lines.

Many objects most adaptable to oblique representation are composed of cylindrical shapes built upon axes or center lines. In such cases, the oblique drawing is best constructed upon the projected center lines, as shown in Fig. 17.13. The object is positioned so that the circles shown in the given top view are parallel to the plane of projection and hence can be readily drawn with the compass in their true sizes and shapes. The general procedure is to draw the center-line skeleton, as shown in steps I and II, and then to build the drawing upon these center lines.

It is very important to construct all points of tangency, as shown in step IV, especially if the drawing is to be inked. For a review of tangencies, see §§4.34 to 4.42. The final cavalier drawing is shown in step V.

## 17.7 Four-Center Ellipse.

It is not always possible to place an object so that all of its significant contours are parallel to the plane of projection. For example, the object shown in Fig. 17.14 (a) has two sets of circular contours in different planes, and both cannot be placed parallel to the plane of projection.

In the oblique drawing at (b), the regular four-center method of Fig. 16.24 was used to construct ellipses representing circular curves not parallel to the plane of projection. This method can be used only in cavalier drawing in which case the enclosing parallelogram is

**Fig. 17.12 Steps in Oblique Drawing—Box Construction.**

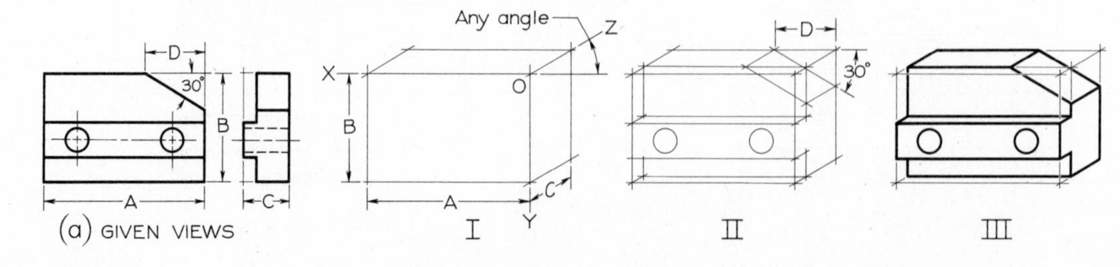

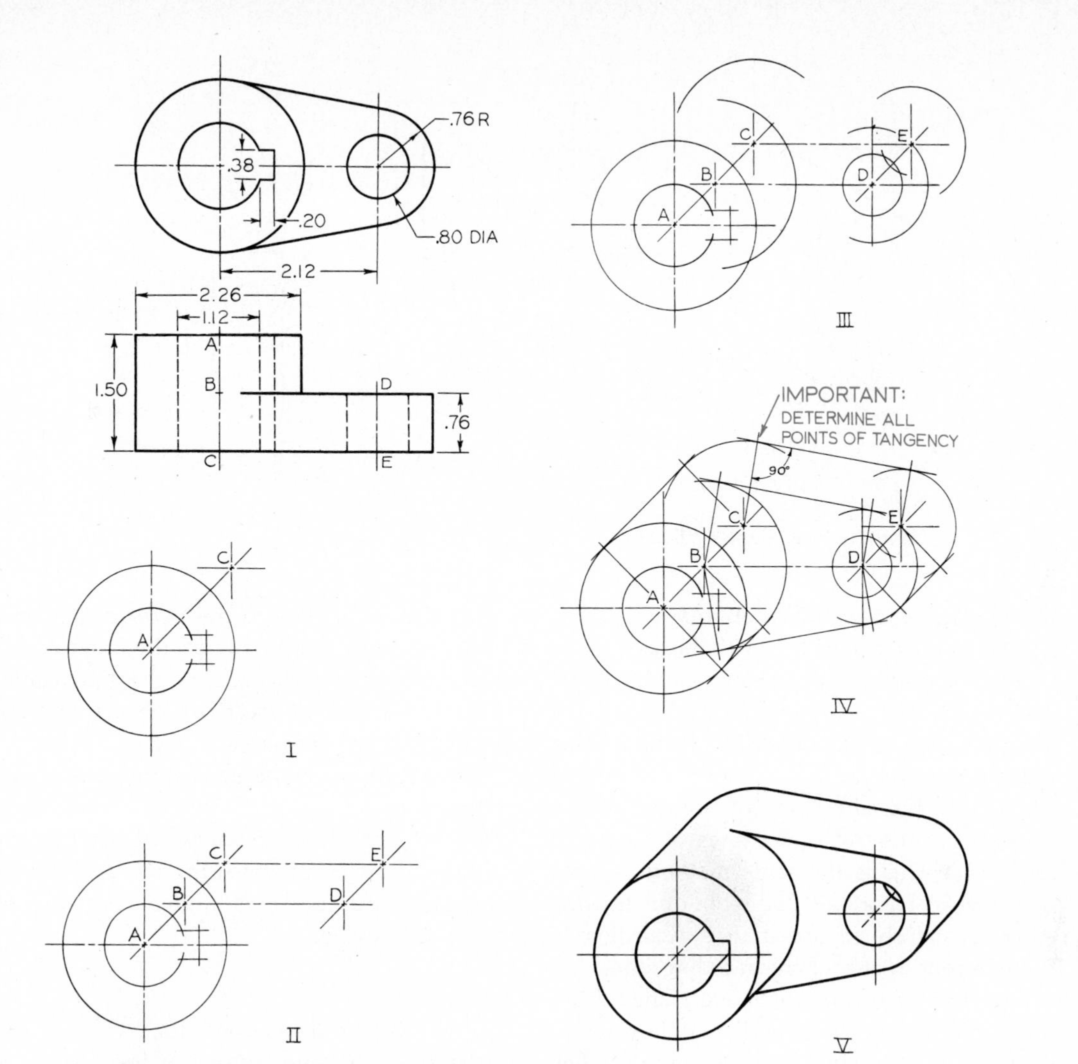


Fig. 17.13 **Steps in Oblique Drawing—Skeleton Construction.**

equilateral—that is, the receding axis is drawn to full scale. The method is the same as in isometric: erect perpendicular bisectors to the four sides of the parallelogram; their intersections will be centers for the four circular arcs. If the angle of the receding lines is other than 30° with horizontal, as in this case, the centers of the two large arcs will not fall in the corners of the parallelogram.

The regular four-center method described

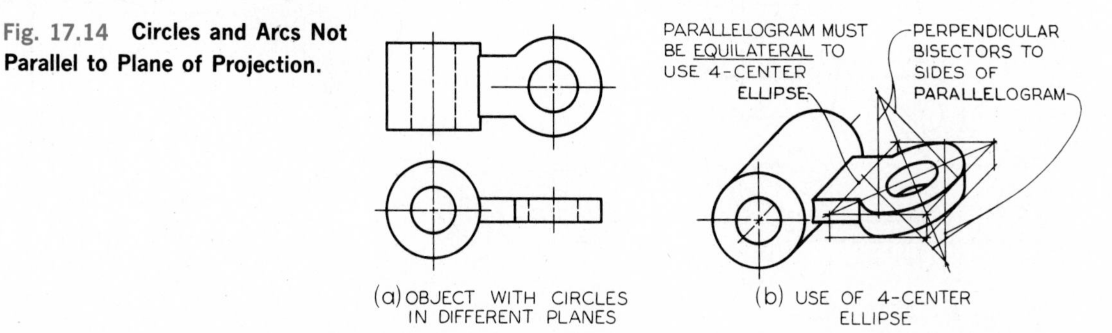


Fig. 17.14 **Circles and Arcs Not Parallel to Plane of Projection.**

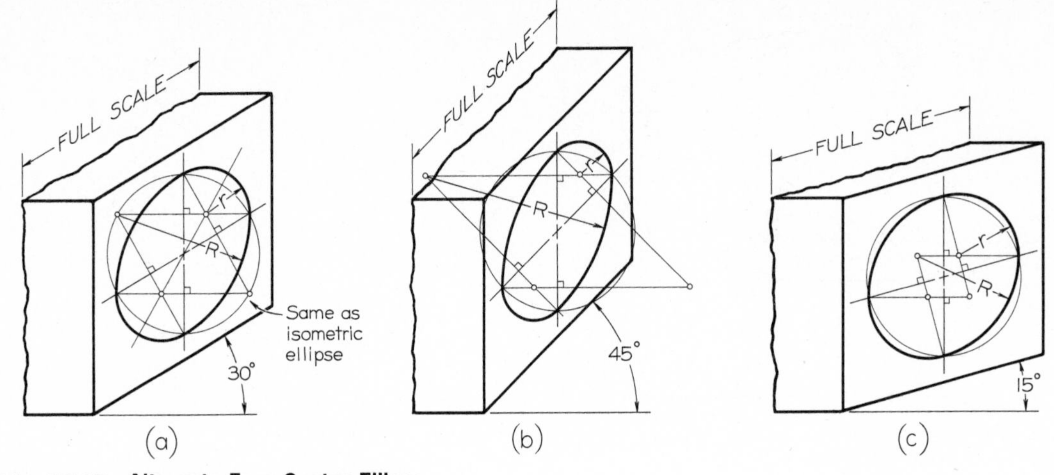


Fig. 17.15 **Alternate Four-Center Ellipse.**

above is not convenient in oblique unless the receding lines make 30° with horizontal so that the perpendicular bisectors may be drawn easily with the 30° × 60° triangle and the T-square without the necessity of first finding the midpoints of the sides. A more convenient method is the alternate four-center ellipse drawn upon the two center lines, as shown in Fig. 17.15. This is the same method as used in isometric, Fig. 16.29, but in oblique drawing it varies slightly in appearance according to the different angles of the receding lines.

First, draw the two center lines. Then, from the center, draw a construction circle equal in diameter to the actual hole or cylinder. The circle will intersect each center line at two points. From the two points on one center line, erect perpendiculars to the other center line; then, from the two points on the other center line, erect perpendiculars to the first center line. From the intersections of the perpendiculars, draw four circular arcs, as shown.

It must be remembered that the four-center ellipse can be inscribed only in an *equilateral* parallelogram; hence it cannot be used in any oblique drawing in which the receding axis is foreshortened. Its use is limited, therefore, to cavalier drawing.

## 17.8 Offset Measurements.

Circles, circular arcs, and other curved or irregular lines may be drawn by means of offset measurements, as shown in Fig. 17.16. The offsets are

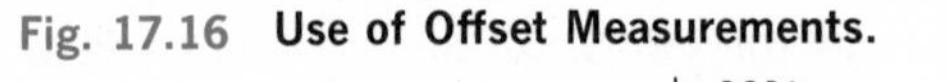

Fig. 17.16 **Use of Offset Measurements.**

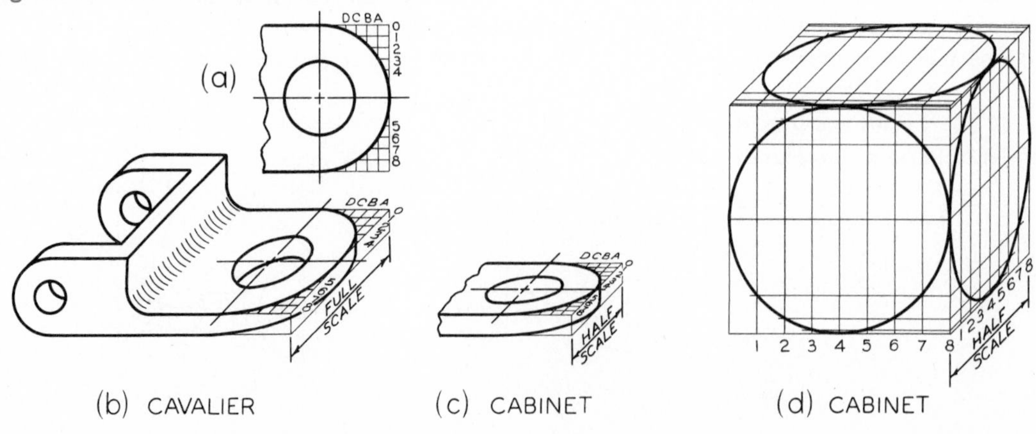

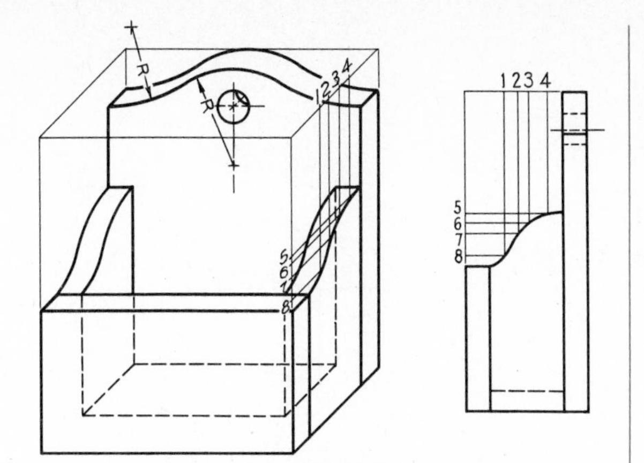


Fig. 17.17 **Use of Offset Measurements.**

first drawn on the multiview drawing of the curve, as shown at (a), and these are transferred to the oblique drawing, as shown at (b). In this case, the receding axis is full scale, and therefore all offsets can be drawn full scale. The four-center ellipse could be used, but the method here is more accurate. The final curve is drawn with the aid of the irregular curve, §2.58.

If the oblique drawing is a cabinet drawing, as shown at (c), or any oblique drawing in which the receding axis is drawn to a reduced scale, the offset measurements parallel to the receding axis must be drawn to the same reduced scale. In this case, there is no choice of methods, since the four-center ellipse could not be used. A method of drawing ellipses in a cabinet drawing of a cube is shown at (d).

As shown in Fig. 17.17, a free curve may be drawn in oblique by means of offset measurements. This figure also illustrates a case in which hidden lines are used to make the drawing clearer.

The use of offset measurements in drawing an ellipse in a plane inclined to the plane of projection is shown in Fig. 17.18. At (a) a number of parallel lines are drawn to represent imaginary cutting planes. Each plane will cut a rectangular surface between the front end of the cylinder and the inclined surface. These rectangles are drawn in oblique, as shown at (b), and the curve is drawn through corner points, as indicated. The final cavalier drawing is shown at (c).

## 17.9 Angles in Oblique Projection.

If an angle that is specified in degrees lies in a receding plane, it is necessary to convert the angle into linear measurements in order to draw the angle in oblique. For example, in Fig. 17.19 (a) an angle of 30° is given. In order to draw the angle in oblique, we need to know dimensions AB and BC. The distance AB is given as $1\frac{1}{4}''$ and can be set off directly in the cavalier drawing, as shown at (b). Distance BC is not known, but can easily be found by constructing the right triangle ABC at (c) from the given dimensions in the top view at (a). The length BC is then transferred with the dividers to the cavalier drawing, as shown.

In cabinet drawing, it must be remembered that *all receding dimensions* must be reduced to half size. Thus, in the cabinet drawing at (d), the distance BC must be half the side BC of the right triangle at (e), as shown.

## 17.10 Oblique Sections.

Sections are often useful in oblique drawing, especially in the representation of interior shapes. An

Fig. 17.18 **Use of Offset Measurements.**

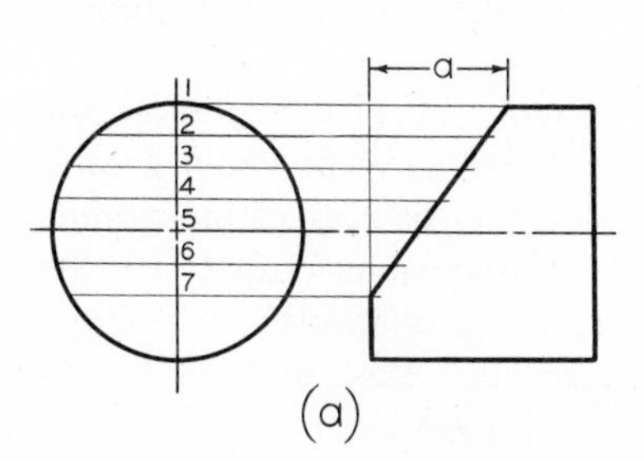


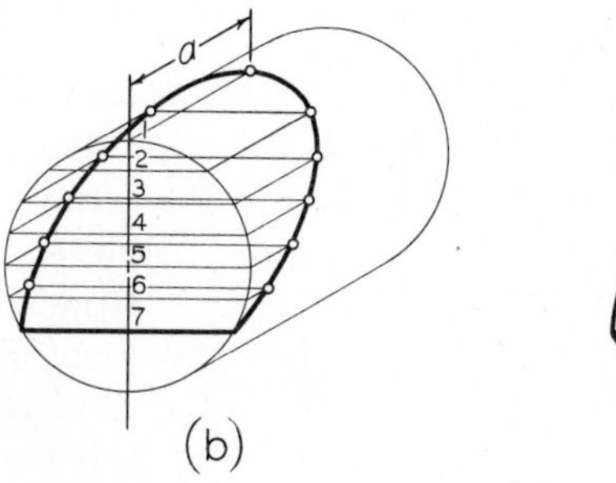


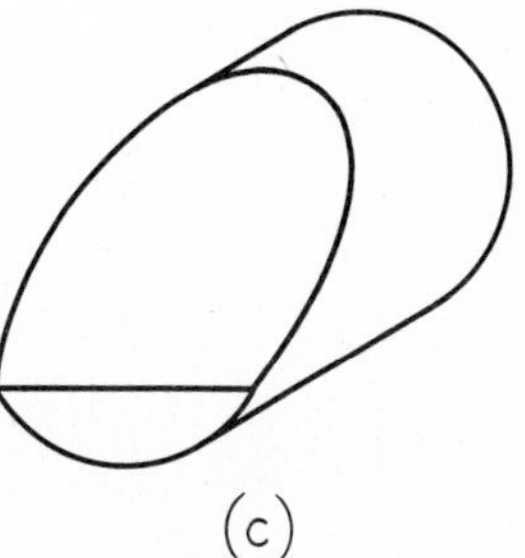

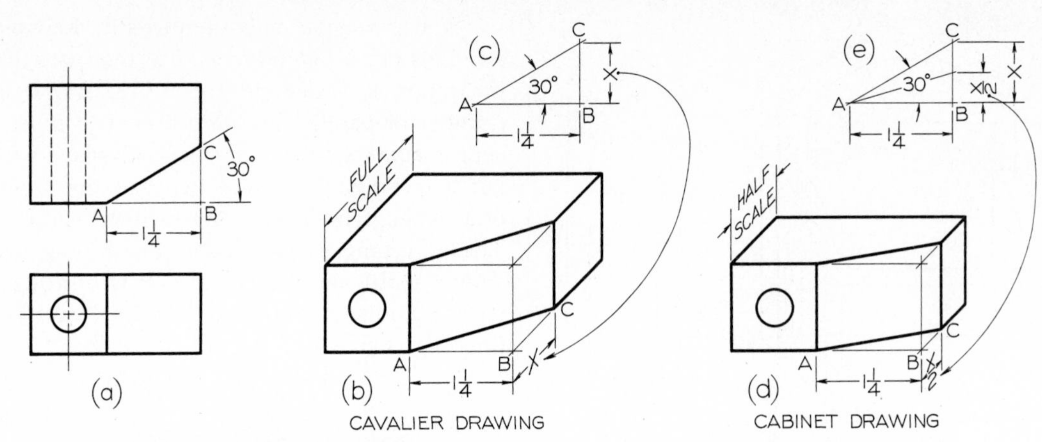


Fig. 17.19 **Angles in Oblique Projection.**

Fig. 17.20 **Oblique Half Section.**

*oblique half section* is shown in Fig. 17.20. Other examples are shown in Figs. 7.40, 7.43, 7.44, and following figures. *Oblique full sections,* in which the plane passes completely through the object, are seldom used because they do not show enough of the exterior shapes. In general, all the types of sections discussed in §16.25 for isometric drawing may be applied equally to oblique drawing.

Fig. 17.21 **Screw Threads in Oblique** (ANSI Y14.4–1957).

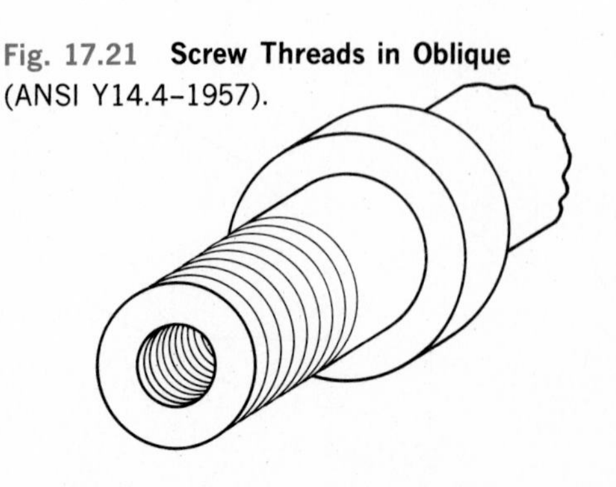

## 17.11 Screw Threads in Oblique.

Parallel partial circles spaced equal to the symbolic thread pitch, Fig. 13.17 (a), are used to represent the crests only of a screw thread in oblique, Fig. 17.21. If the thread is so positioned to require ellipses, they may be drawn by the four-center method of §17.7.

## 17.12 Oblique Dimensioning.

An oblique drawing may be dimensioned in a similar manner to that described in §16.26 for isometric drawing, as shown in Fig. 17.22. The general principles of dimensioning, as outlined in Chapter 11, must be followed. As shown in the figure, all dimension lines, extension lines, and arrowheads must lie in the planes of the object to which they apply. The dimension figures also will lie in the plane when the aligned dimensioning system is used as shown at (a). For the unidirectional system of dimensioning, (b), all dimension figures are set horizontal and read from the bottom of the drawing. This simpler system is often used on pictorials for production purposes. *Vertical lettering* should be used for all pictorial dimensioning.

Dimensions should be placed outside the outlines of the drawing except when greater clearness or directness of application results from placing the dimensions directly on the view. For many other examples of oblique dimensioning, see Figs. 6.59, 6.63, 6.64, and others on following pages.

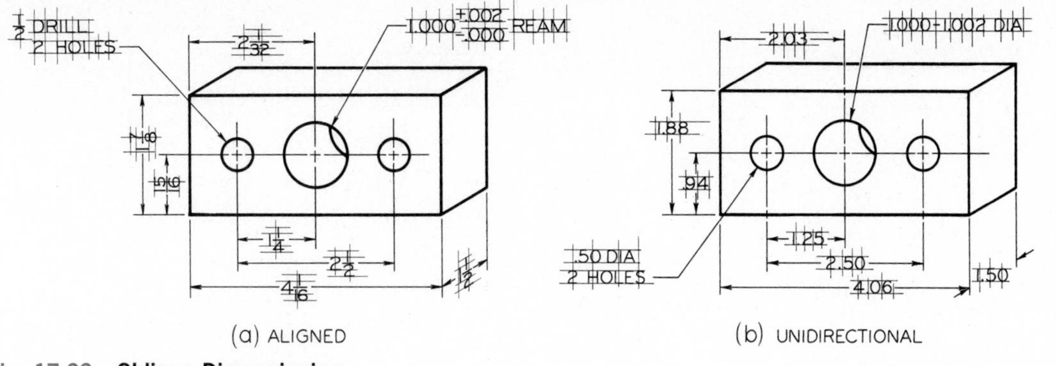


(a) ALIGNED (b) UNIDIRECTIONAL

Fig. 17.22 **Oblique Dimensioning.**

**17.13 Oblique Sketching.** Methods of sketching in oblique on plain paper are illustrated in Fig. 5.27. Ordinary rectangular cross-section paper is very useful in oblique sketching, Fig. 5.28. The height and width proportions can be easily controlled by simply counting the squares. A very pleasing depth proportion can be obtained by sketching the receding lines at 45° diagonally through the squares and through half as many squares as the actual depth would indicate.

**17.14 Oblique Projection Problems.** A large number of problems to be drawn in oblique—either cavalier or cabinet—are given in Figs. 17.23 to 17.25. They may be drawn freehand, §5.15, using cross-section paper or plain drawing paper as assigned by the instructor, or they may be drawn with instruments. In the latter case, all construction lines should be shown on the completed drawing.

Many additional problems suitable for oblique projection will be found in Figs. 6.50 to 6.52, 8.27 to 8.29, 16.51 to 16.54, and 18.33.

Oblique problems in convenient form for solution may be found in *Technical Drawing Problems, Series 1,* by Giesecke, Mitchell, Spencer, and Hill; in *Technical Drawing Problems, Series 2,* by Spencer; and in *Technical Drawing Problems, Series 3,* by Spencer and Hill, all designed to accompany this text and published by Macmillan Publishing Co., Inc.

## 17.14 Oblique Projection Problems

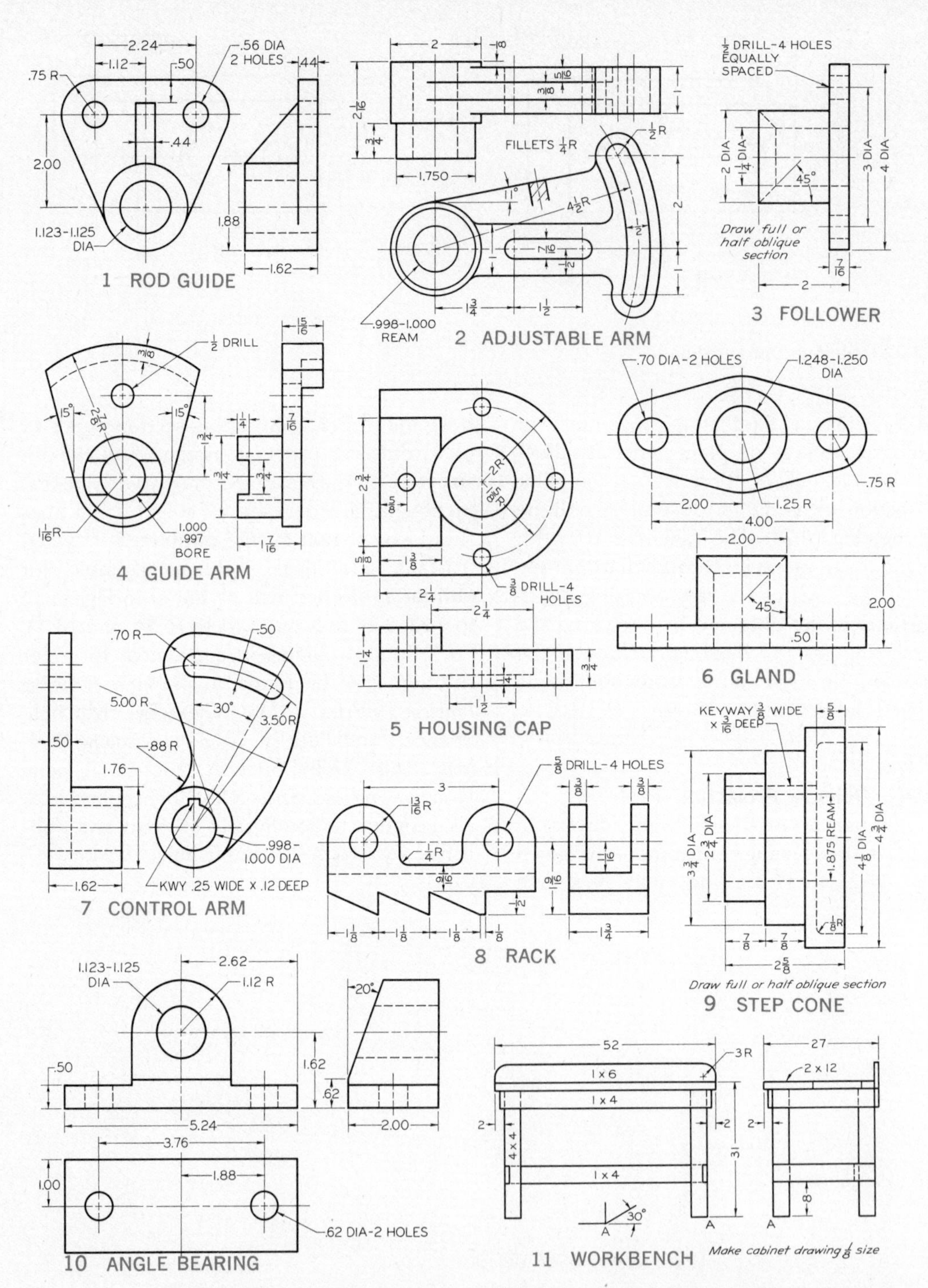


Fig. 17.23 (1) Make freehand oblique sketches. (2) Make oblique drawings with instruments, using Size A or Size B sheet, as assigned. If dimensions are required, study §17.12.

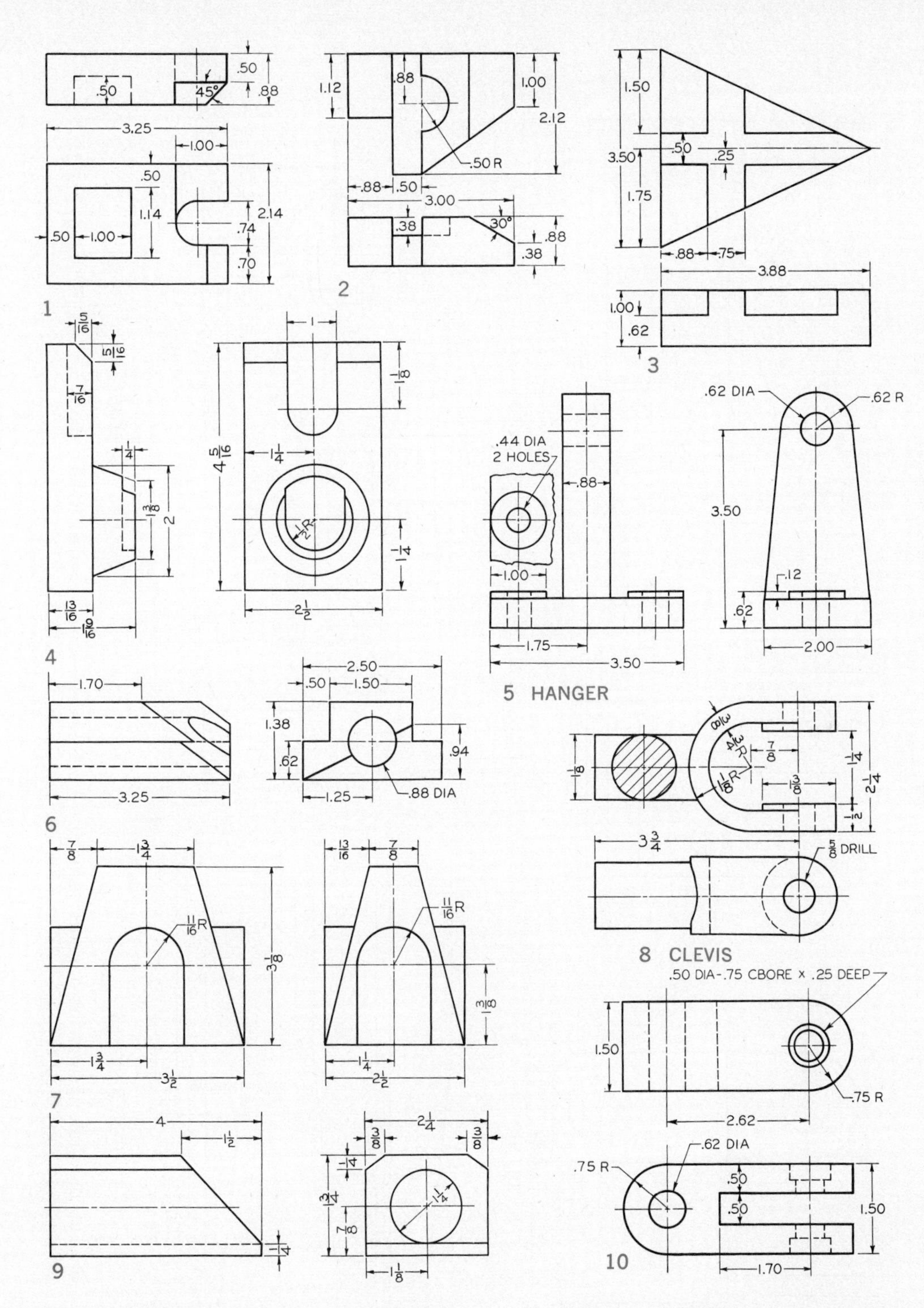


Fig. 17.24 Make oblique drawings with instruments, using Size A or Size B sheet, as assigned. If dimensions are required, study §17.12.

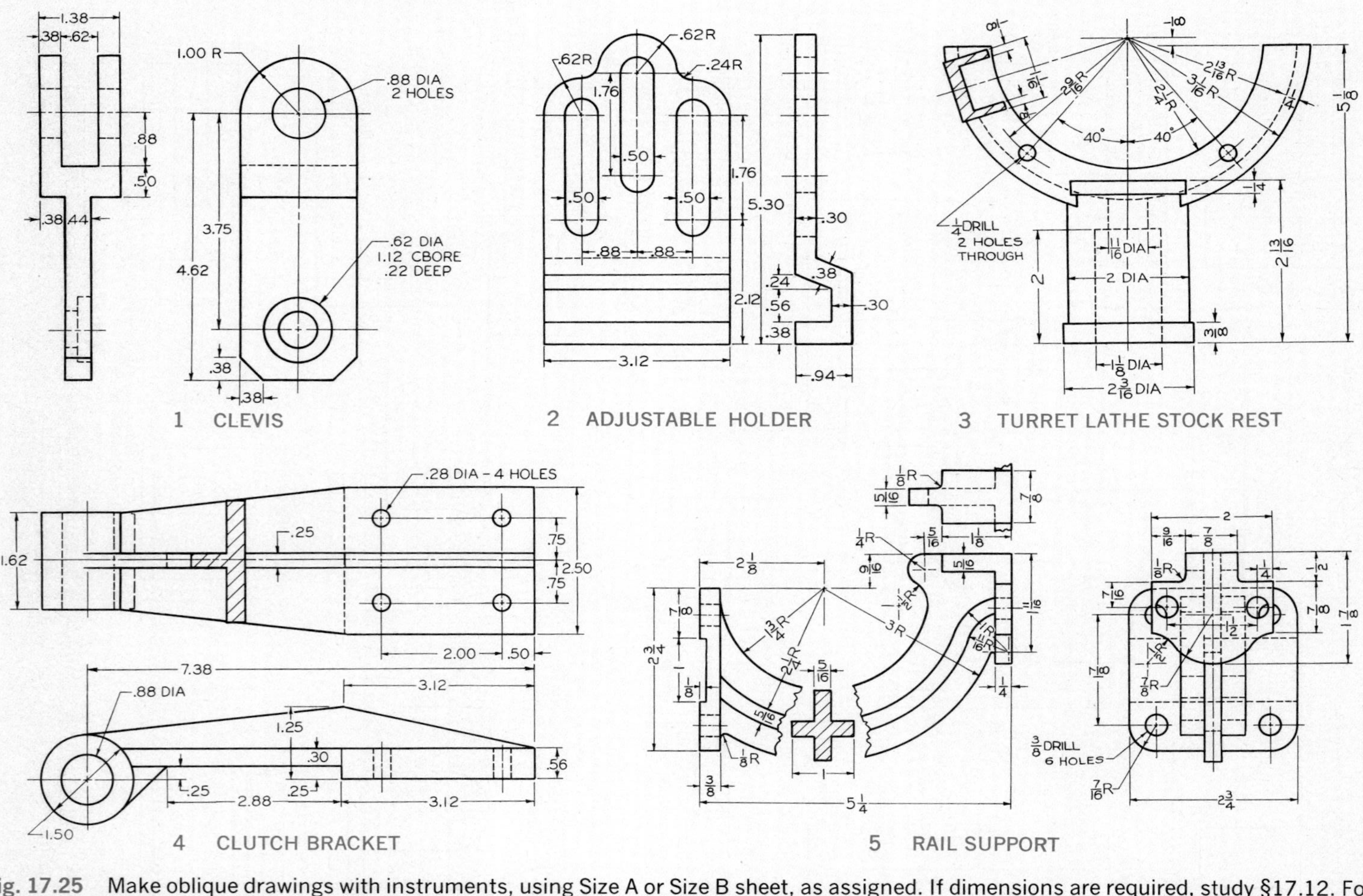


Fig. 17.25 Make oblique drawings with instruments, using Size A or Size B sheet, as assigned. If dimensions are required, study §17.12. For additional problems, see Figs. 6.50 to 6.52, 8.27 to 8.29, 16.51 to 16.54, and 18.33.

# 18 perspective

**18.1 General Principles.** *Perspective,** or *central projection,* excels all other types of projection in the pictorial representation of objects because it more closely approximates the view obtained by the human eye, Fig. 18.1. Geometrically, an ordinary photograph is a perspective. While perspective is of major importance to the architect, industrial designer, or illustrator, the engineer at one time or another is apt to be concerned with the pictorial representation of objects and should understand the basic principles of perspective.

As explained in §1.11, a perspective involves four main elements: (1) the observer's eye, (2) the object being viewed, (3) the plane of projection, and (4) the projectors from the observer's eye to all points on the object. The plane of projection is placed between the observer and the object,* as shown in Fig. 1.10, and the collective piercing points in the plane of projection of all of the projectors produce the perspective.

In Fig. 18.2, the observer is shown looking along a boulevard and through an imaginary plane of projection. This plane is called the *picture plane* or simply PP. The position of the observer's eye is called the *station point* or simply SP. The lines from SP to the various points in the scene are the projectors, or more properly in perspective, *visual rays.* The points where the visual rays pierce PP are the *perspectives* of the respective points. Collectively, these piercing points form the perspective of the object or the scene as viewed by the ob-

*See ANSI Y14.4–1957.

*Except as explained in §18.6.

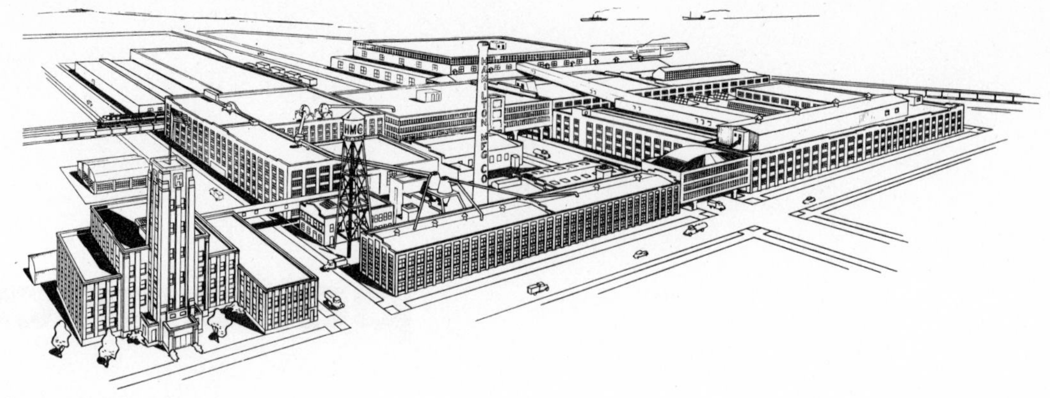

*Courtesy Hamilton Mfg. Co.*

**Fig. 18.1 Perspective of a Factory.**

server. The perspective thus obtained is shown in Fig. 18.3.

In Fig. 18.2, the perspective of lamp post 1–2 is shown at 1′–2′, on the picture plane; the perspective of lamp post 3–4 is shown at 3′–4′, and so on. Each succeeding lamp post, as it is farther from the observer, will be projected smaller than the one preceding. A lamp post at an infinite distance from the observer would appear as a point on the picture plane. A lamp post in front of the picture plane would be projected taller than it is, and a lamp post in the picture plane would be projected in true length. In the perspective, Fig. 18.3, the diminishing heights of the posts are apparent.

In Fig. 18.2, the line representing the *horizon* is the edge view of the *horizon plane,* which is parallel to the ground plane and passes through SP. In the perspective, Fig. 18.3, the horizon is the line of intersection of this plane with the picture plane, and represents the eye level of the observer, or SP. Also, in Fig. 18.2, the *ground plane* is the edge view of the ground upon which the object usually rests. In Fig.

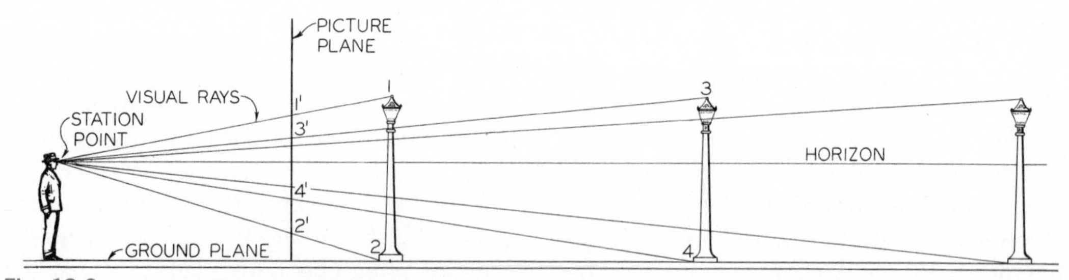

**Fig. 18.2 Looking Through the Picture Plane.**

**Fig. 18.3 A Perspective.**

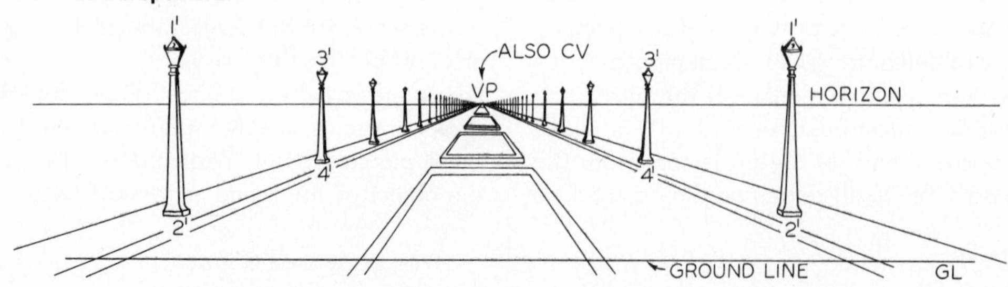

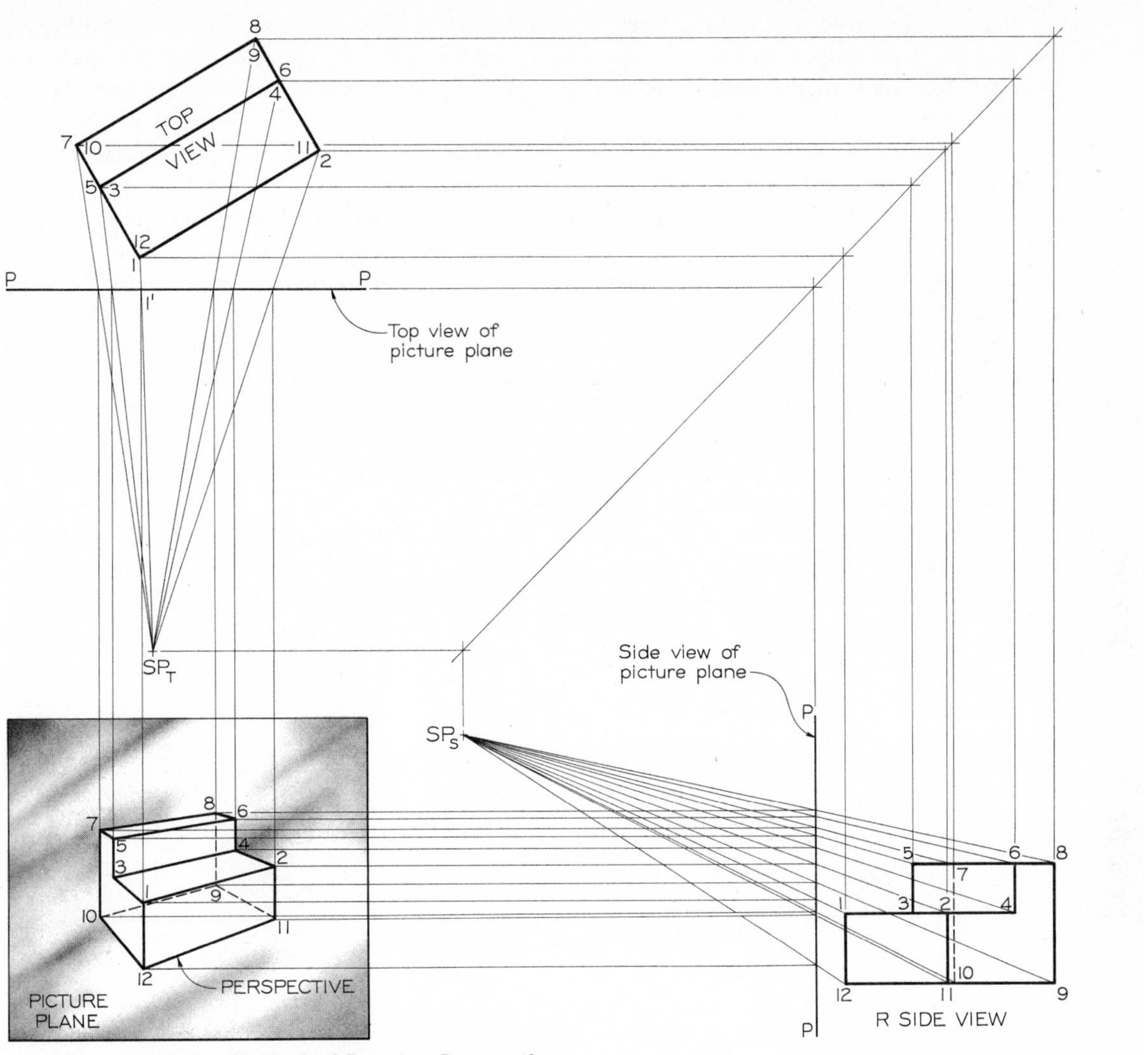


Fig. 18.4 **Multiview Method of Drawing Perspective.**

18.3, the *ground line,* or GL, is the intersection of the ground plane with the picture plane.

In Fig. 18.3, it will be seen that lines that are parallel to each other but not parallel to the picture plane, such as curb lines, sidewalk lines, and lines along the tops and bottoms of the lamp posts, all converge toward a single point on the horizon. This point is called the *vanishing point,* or VP, of the lines. Thus, the first rule to learn in perspective is this: *All parallel lines that are not parallel to* PP *vanish at a single vanishing point, and if these lines are parallel to the ground, the vanishing point will be on the horizon.* Parallel lines that are also parallel PP, such as the lamp posts, remain parallel and do not converge toward a vanishing point.

## 18.2 Multiview Perspective.

A perspective can be drawn by the ordinary methods of multiview projection, as shown in Fig. 18.4. In the upper portion of the drawing are shown the top view of the station point, the picture plane, the object, and the visual rays. In the right-hand portion of the drawing are shown the right-side view of the same station point, picture plane, object, and visual rays. In the front view, the picture plane coincides with the plane of the paper, and the perspective is drawn upon it. Note the method of projecting from the top view to the side view, which conforms to the usual multiview methods shown in Fig. 6.7.

To obtain the perspective of point 1, a visual ray is drawn in the top view from $SP_T$ to point 1 on the object. From the intersection 1′ of this ray with the picture plane, a projection line is drawn downward till it meets a similar projection line from the side view. This intersection is the perspective of point 1, and the perspectives of all other points are found in a similar manner.

Observe that all parallel lines that are also parallel to the picture plane (the vertical lines) remain parallel and do not converge, whereas the other two sets of parallel lines converge toward vanishing points. However, the vanishing points are not needed in the multiview construction of Fig. 18.4, and, therefore, are not shown; but if the converging lines should be extended, it will be found that they meet at two vanishing points (one for each set of parallel lines).

The perspective of any object may be constructed in this way, but if the object is placed at an angle with the picture plane, as is usually the case, the method is a bit cumbersome because of the necessity of constructing the side view in a revolved position. The revolved side view can be dispensed with, as shown in the following section.

## 18.3 The Set-up for a Simple Perspective.

The construction of a perspective of a simple form is shown in Fig. 18.5. The upper

Fig. 18.5 **Perspective of a Prism.**

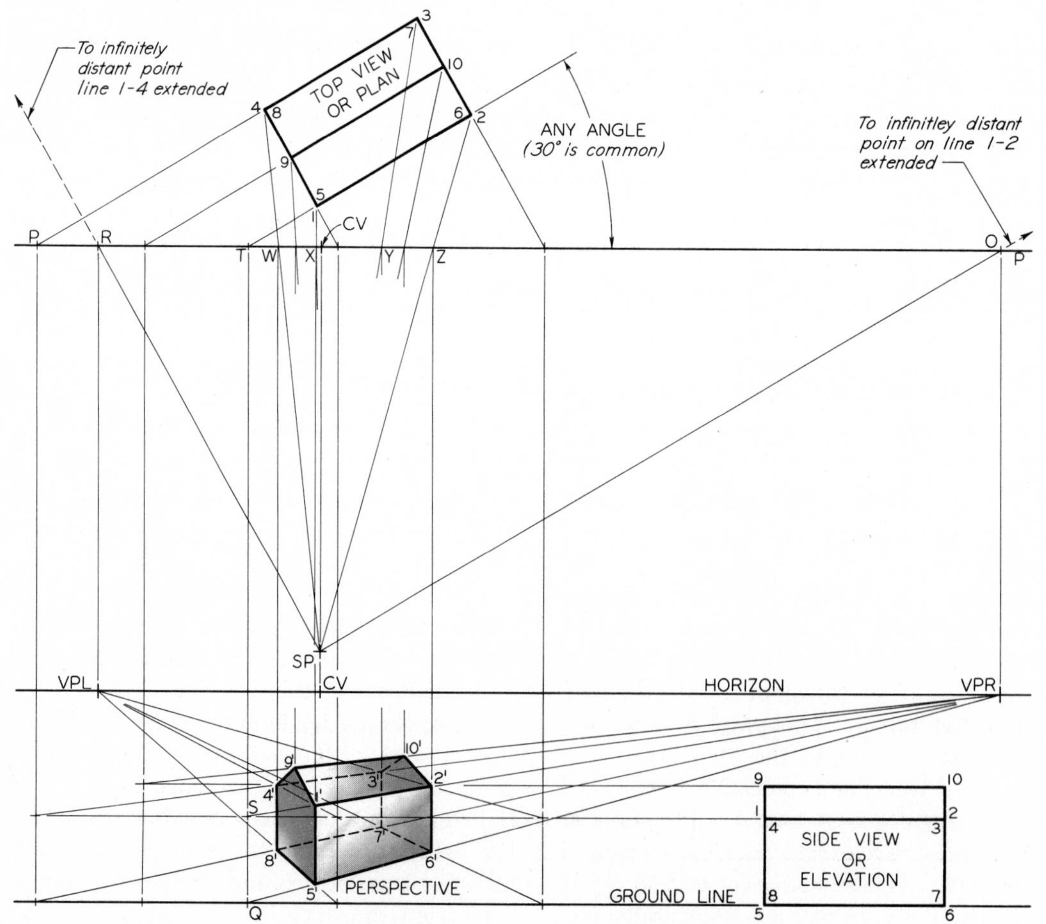

portion of the drawing, as in Fig. 18.4, shows the top views of SP, PP, and of the object. The lines SP-1, SP-2, SP-3, and SP-4 are the top views of the visual rays.

In the side view, a departure from Fig. 18.4 is made, in that a revolved side view is not required. All that is needed is any elevation view that will provide the necessary elevation or height measurements. If these dimensions are known, no view is required.

The perspective itself is drawn in the front-view position, the picture plane being considered as the plane of the paper upon which the perspective is drawn. The ground line is the edge view of the ground plane or the intersection of the ground plane with the picture plane. The horizon is a horizontal line in the picture plane that is the line of intersection of the horizon plane with the picture plane. Since the horizon plane passes through the observer's eye, or SP, the horizon is drawn at the level of the eye; that is, at the distance above the ground line representing, to scale, the altitude of the eye above the ground.

The *center of vision*, or CV, is the orthographic projection (or front view) of SP on the picture plane, and since the horizon is at eye level, CV will always be on the horizon.* In Fig. 18.5, the top view of CV is CV′, found by dropping a perpendicular from SP to PP. The front view CV is found by projecting downward from CV′ to the horizon.

## 18.4 To Draw an Angular Perspective.

Since objects are defined principally by edges that are straight lines, the drawing of a perspective resolves itself into drawing the *perspective of a line.* If a draftsman can draw the perspective of a line, he can draw the perspective of any object, no matter how complex.

To draw the perspective of any horizontal straight line not parallel to PP—for example, the line 1-2 in Fig. 18.5—proceed as follows:

I. *Find the piercing point in* PP *of the line.* In the top view, extend line 1-2 until it pierces PP at T; then project downward to the level of the line 1-2 projected horizontally from the side view. The point S is the piercing point of the line.

II. *Find the vanishing point of the line.* The vanishing point of a line is the piercing point in PP of a line drawn through SP parallel to that line. Hence, the vanishing point VPR of the line 1-2 is found by drawing a line from SP parallel to that line and finding the top view of its piercing point O, and then projecting downward to the horizon. The line SP-O is actually a visual ray drawn toward the infinitely distant point on line 1-2 of the object, extended, and the vanishing point is the intersection of this visual ray with the picture plane. The vanishing point is, then, the perspective of the infinitely distant point on the line extended.

III. *Join the piercing point and the vanishing point with a straight line.* The line S-VPR is the line joining these two points, and it is the perspective of a line of infinite length containing the required perspective of the line 1-2.

IV. *Locate the end points of the perspective of the line.* The end points 1′ and 2′ can be found by projecting down from the piercing points of the visual rays in PP, or by simply drawing the perspectives of the remaining horizontal edges of the object. In practice, it is best to use both methods as a check on the accuracy of the construction. To locate the end points by projecting from the piercing points, draw visual rays from SP to the points 1 and 2 on the object in the top view. The top views of the piercing points are X and Z. Since the perspectives of points 1 and 2 must lie on the line S-VPR, project downward from X and Z to locate points 1′ and 2′.

After the perspectives of the horizontal edges have been drawn, the vertical edges and inclined edges can be drawn, as shown, to complete the perspective of the object. Note that *vertical heights can be measured only in the picture plane.* If the front vertical edge 1-5 of the object was actually in PP—that is, if the object was situated with the front edge in PP—the vertical height could be set off directly full size. If the vertical edge is behind PP, a plane of the object, such as surface 1-2-5-6 can be extended forward until it intersects PP in line TQ. The line TQ is called

*Except in three-point perspective, §18.11.

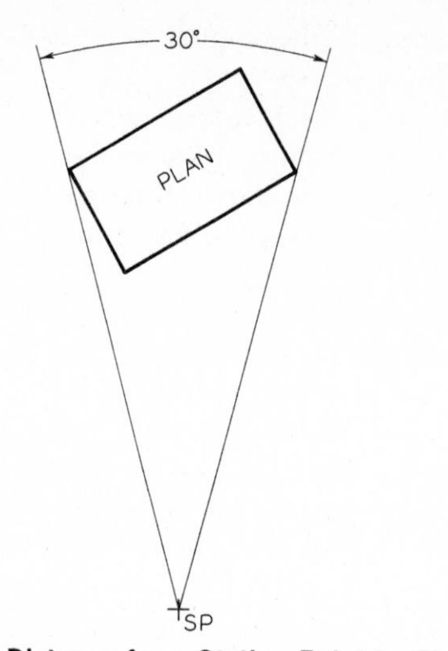


Fig. 18.6 **Distance from Station Point to Object.**

a *measuring line,* and the true height SQ of line 1–5 can be set off with a scale or projected from the side view as shown.

If a large drawing board is not available, one vanishing point, such as VPR, may fall off the board. By using one vanishing point VPL, and projecting down from the piercing points in PP, vanishing point VPR may be eliminated. However, a valuable means of checking the accuracy of the construction will be lost.

## 18.5 Position of the Station Point.

The center line of the cone of visual rays should be directed toward the approximate center, or center of interest, of the object. In a perspective of the type shown in Fig. 18.5,* the location of SP in the plan view should be slightly to the left, not directly in front of the center of the object, and at such a distance from it that the object can be viewed at a glance without turning the head. This is accomplished if a cone of rays with its vertex at SP and a vertical angle of about 30° entirely encloses the object, as shown in Fig. 18.6.

In the perspective portion of Fig. 18.5, SP does not appear, because SP is in front of the picture plane. However, the orthographic projection CV of SP in the picture plane does show the height of SP with respect to the ground plane. Since the horizon is at eye level, it also shows the altitude of SP. Therefore, in the perspective portion of the drawing, the horizon is drawn a distance above the ground line at which it is desired to assume SP. For most small and medium-size objects, such as machine parts or furniture, SP is best assumed slightly above the top of the object. Large objects, such as buildings, are usually viewed from a station point about the altitude of the eye above the ground, or about 5′–6.

## 18.6 Location of the Picture Plane.

In general, the picture plane is placed in front of the object, as in Fig. 18.7 (b) and (c). However, it may be placed behind the object, as shown at (a), and it may even be placed behind SP, as shown at (d), in which event the perspective

*Two-point perspective, §18.10.

Fig. 18.7 **Location of Picture Plane.**

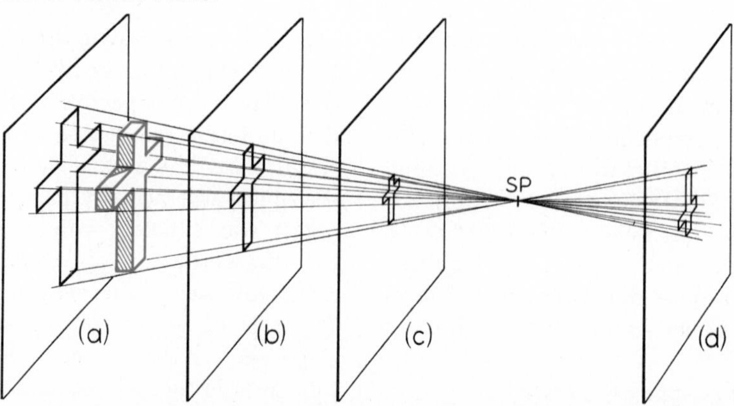

is reversed, as in the case of a camera. Of course the usual position of the picture plane is between SP and the object. The perspectives in Fig. 18.7 differ in size but not in proportion. As shown at (b) and (c), the farther the picture plane is from the object, the smaller the perspective will be. This distance may be assumed, therefore, with the thought of controlling the scale of the perspective. In practice, however, the object is usually assumed with the front corner in the picture plane to facilitate vertical measurements. See Fig. 18.12.

## 18.7 Position of the Object with Respect to the Horizon.

To compare the elevation of the object with that of the horizon is equivalent to referring it to the level of the eye or SP, because the horizon is on a level with the eye.* The differences in effect produced by placing the object on, above, or below the horizon are shown in Fig. 18.8.

If the object is placed above the horizon, it is above the level of the eye, or above SP, and will appear as seen from below. Likewise, if the object is below the horizon, it will appear as seen from above.

## 18.8 The Three Types of Perspectives.

Perspectives are classified according to the number of vanishing points required, which in turn depends upon the position of the object with respect to the picture plane.

If the object is situated with one face parallel to the plane of projection, only one vanishing point is required, and the result is a *one-point perspective* or *parallel perspective,* §18.9.

If the object is situated at an angle with the picture plane but with vertical edges parallel to the picture plane, two vanishing points are required, and the result is a *two-point perspective* or an *angular perspective.* This is the most common type of perspective and is the one described in §18.4. See also §18.10.

If the object is situated so that no system of parallel edges is parallel to the picture plane, three vanishing points are necessary, and the result is a *three-point perspective,* §18.11.

*Except in three-point perspective, §18.11.

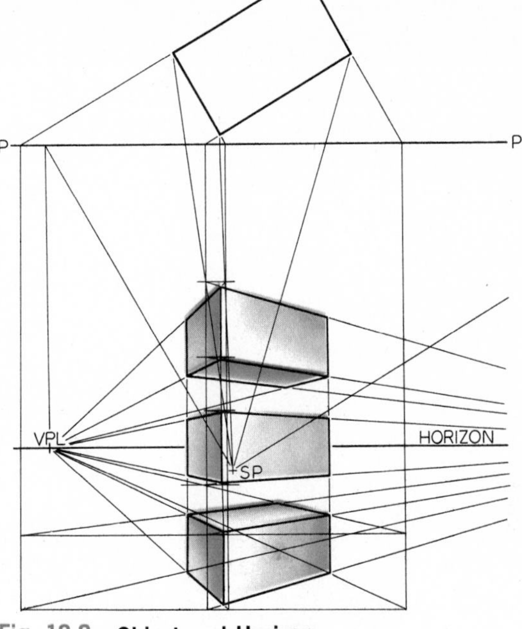


Fig. 18.8 **Object and Horizon.**

## 18.9 One-Point Perspective.

In one-point perspective, Fig. 16.1 (d), the object is placed so that two sets of its principal edges are parallel to PP, and the third set is perpendicular to PP. This third set of parallel lines will converge toward a single vanishing point in perspective, as shown.

In Fig. 18.9, the view shows the object with one face parallel to the picture plane. If desired, this face could be placed *in* the picture plane. The piercing points of the eight edges perpendicular to PP are found by extending them to PP and then projecting downward to the level of the lines as projected across from the elevation view.

To find the VP of these lines, a visual ray is drawn from SP parallel to them (the same as in step II of §18.4), and it is found that the *vanishing point of all lines perpendicular to* PP *is in* CV. By connecting the eight piercing points with the vanishing point CV, the indefinite perspectives of the eight edges are obtained.

To cut off on these lines the definite lengths of the edges of the object, horizontal lines are drawn from the ends of one of the edges in the top view and at any desired angle with PP, 45° for example, as shown. The piercing

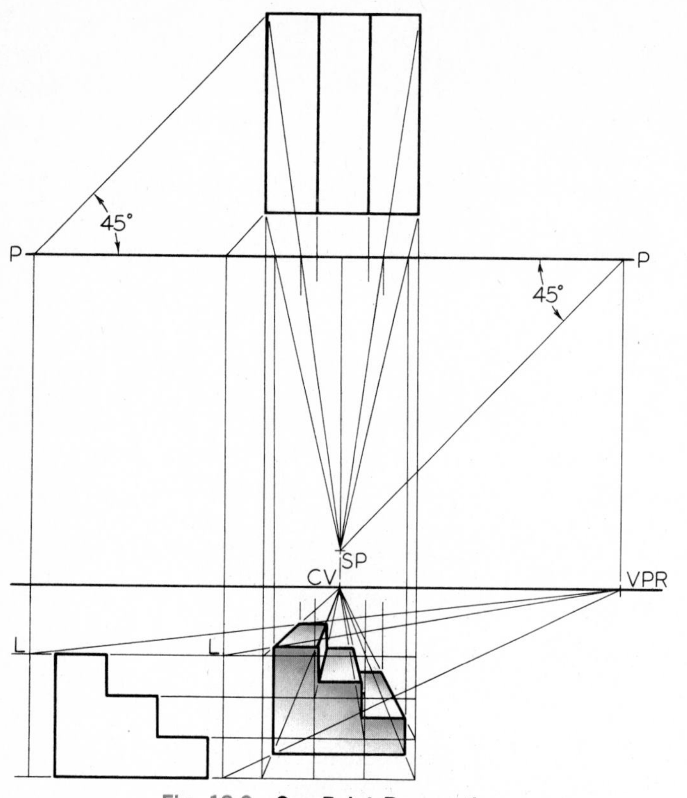

Fig. 18.9 **One-Point Perspective.**

points and the vanishing point VPR of these lines are found, and the perspectives of the lines drawn. The intersections of these with the perspectives of the corresponding edges of the object determine the lengths of the receding edges. The perspective of the object may then be completed as shown.

One of the most common uses of parallel perspective is in the representation of interiors of buildings, as illustrated in Fig. 18.10.

An adaptation of one-point perspective, which is simple and effective in representing machine parts, is shown in Fig. 18.11. The front surface of the cylinder is placed in PP, and all circular shapes are parallel to PP; hence these shapes will be projected as circles and circular arcs in the perspective. SP is located in front and to one side of the object, and the horizon is placed well above the ground line. The single vanishing point is on the horizon in CV.

The two circles and the keyway in the front surface of the object will be drawn true size because they lie in PP. The circles are drawn with the compass on center O′. To locate R′, the perspective center of the large arc, draw visual ray SP–R; then, from its intersection X

Fig. 18.10 **One-Point Perspective.** *Courtesy Eaton Manufacturing Company*

with PP, project down to the center line of the large cylinder, as shown.

To find the radius T′W′ at the right end of the perspective, draw visual rays SP–T and SP–W, and from their intersections with PP, project down to T′ and W′ on the horizontal center line of the hole.

## 18.10 Two-Point Perspective.

In two-point perspective, the object is placed so that one set of parallel edges is vertical and has no vanishing point, while the two other sets each have vanishing points. This is the most common type and is the method discussed in §18.4. It is suitable especially for representing buildings in architectural drawing, or large structures in civil engineering, such as dams or bridges.

The perspective of a small building is shown in Fig. 18.12. It is common practice (1) to assume a vertical edge of an object in PP so that direct measurements may be made on it, and (2) to place the object so that its faces make unequal angles with PP; for example, one angle may be 30° and the other 60°. In practical work, complete multiview drawings are usually available, and the plan and elevation may be fastened in position, used in the construction of the perspective, and later removed.

Since the front corner AB lies in PP, its perspective A′B′ may be drawn full size by projecting downward from the plan and across from the elevation. The lengths of the receding lines from this corner are cut off by vertical lines SC′ and RE′ drawn from the intersections S and R, respectively, of the visual rays to these points of the object. The perspectives of the tops of the windows and the door are determined by the lines A′–VPR and A′–VPL, and their widths and lateral spacings are determined by projecting downward from the intersections with PP of the respective visual rays. The bottom lines of the windows are determined by the lines V′–VPR and V′–VPL.

The perspective of the line containing the ridge of the roof is found by joining N′, the point where the ridge line pierces the picture plane, and VPR. The ridge ends O′ and Q′ are found by projecting downward from the intersections of the visual rays with PP, or by drawing the perspectives of any two lines intersecting at the points. The perspective of the roof is completed by joining the points O′ and Q′ to the ends of the eaves.

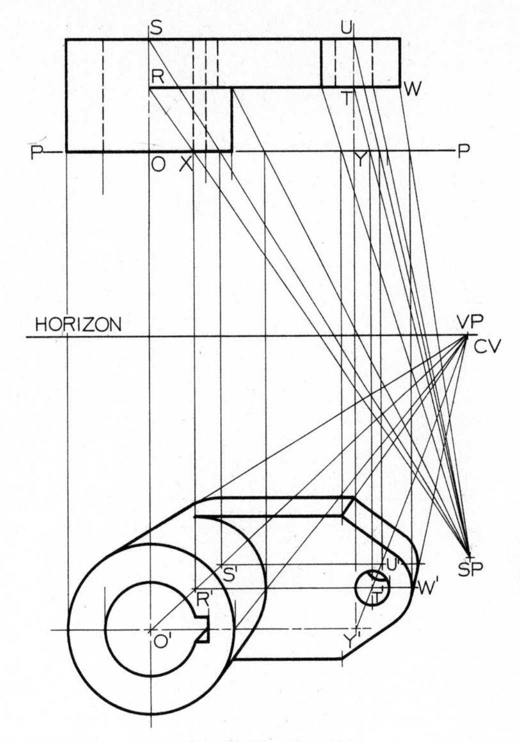


Fig. 18.11 **One-Point Perspective.**

## 18.11 Three-Point Perspective.

In three-point perspective, the object is placed so that none of its principal edges is parallel to PP; therefore, each of the three sets of parallel edges will have a separate VP, Fig. 18.13. The picture plane is assumed approximately perpendicular to the center line of the cone of rays.

In this figure, think of the paper as the picture plane, with the object behind the paper and placed so that all of its edges make an angle with the picture plane. If a point CV is chosen, it will be the orthographic projection of your eye, or SP, on PP. The vanishing points P, Q, and R are found by conceiving lines to be drawn from SP in space parallel to the principal axes of the object, and finding

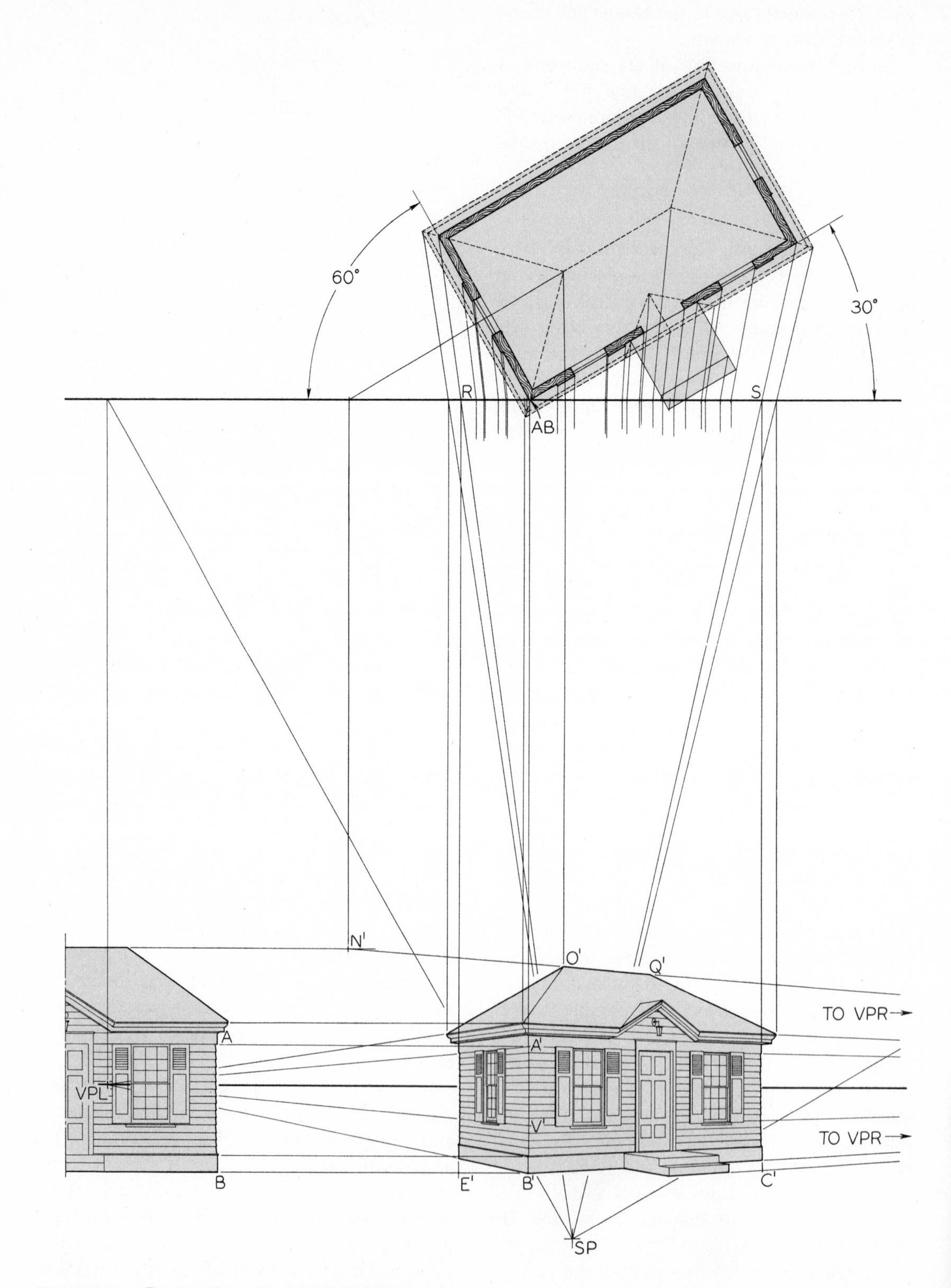




Fig. 18.12 **Perspective of a Small Building.**

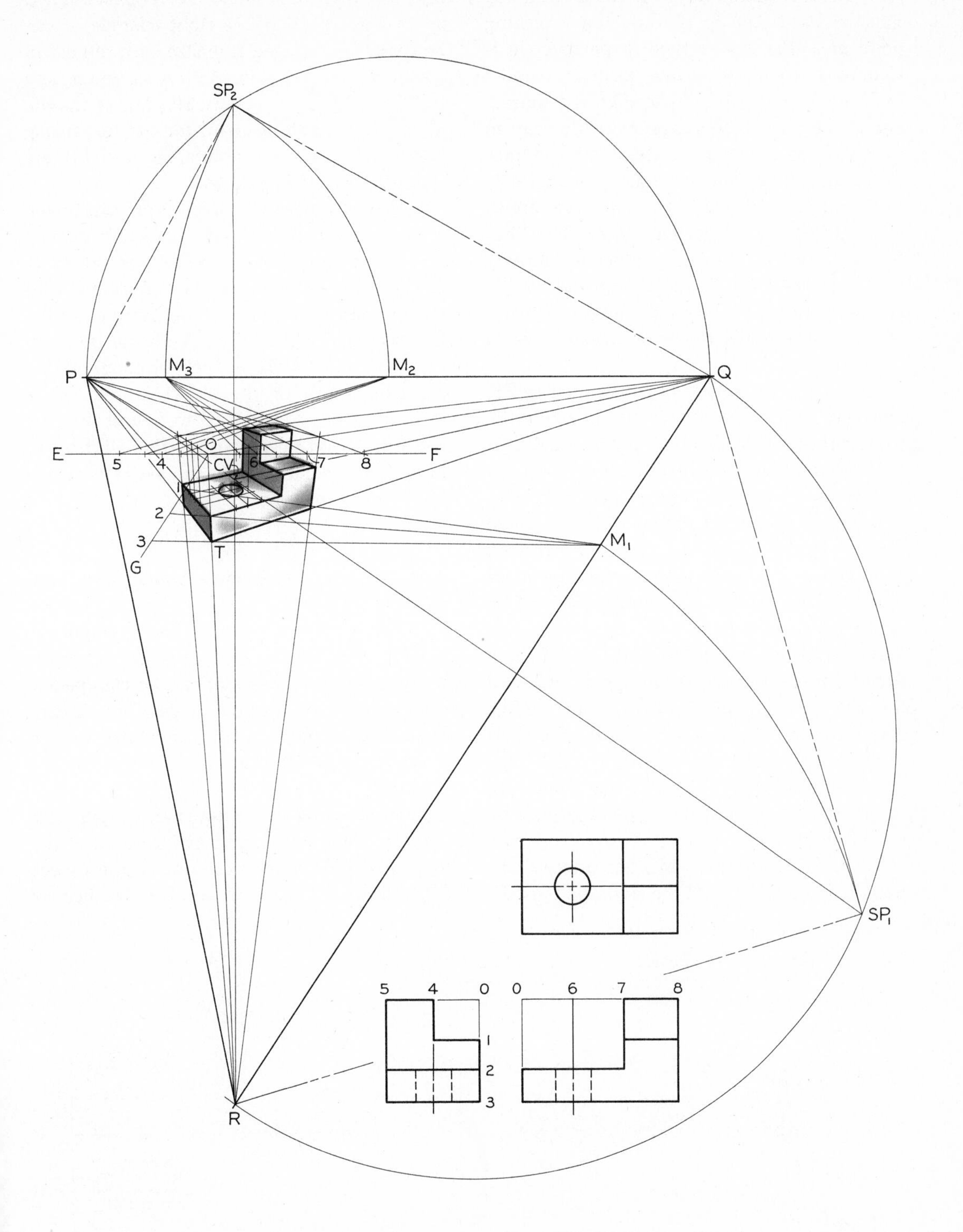


Fig. 18.13 **Three-Point Perspective.**

their piercing points in PP. It will be recalled that the basic rule for finding the vanishing point of a line in any type of perspective is to draw a visual ray, or line, from SP parallel to the edge of the object whose VP is required, and finding the piercing point of this ray in PP. Since the object is rectangular, these lines to the vanishing points are at right angles to each other in space exactly as the axes are in axonometric projection, Fig. 16.46. The lines PQ, QR, and RP are perpendicular, respectively, to CV–R, CV–P, and CV–Q, and are the *vanishing traces*, or horizon lines, of planes through SP parallel to the principal faces of the object.

The imaginary corner O is assumed in PP, and may coincide with CV; but as a rule the front corner is placed at one side near CV, thus determining how nearly the observer is assumed to be directly in front of this corner.

In this method the perspective is drawn directly from measurements and not projected from views. The dimensions of the object are given by the three views, and these will be set off on *measuring lines* GO, EO, and OF. See §18.13. The measuring lines EO and OF are drawn parallel to the vanishing trace PQ, and the measuring line GO is drawn parallel to RQ. These measuring lines are actually the lines of intersection of principal surfaces of the object, extended, with PP. Since these lines are in PP, true measurements of the object can be set off along them.

Three *measuring points* $M_1$, $M_2$, and $M_3$ are used in conjunction with the *measuring lines.* To

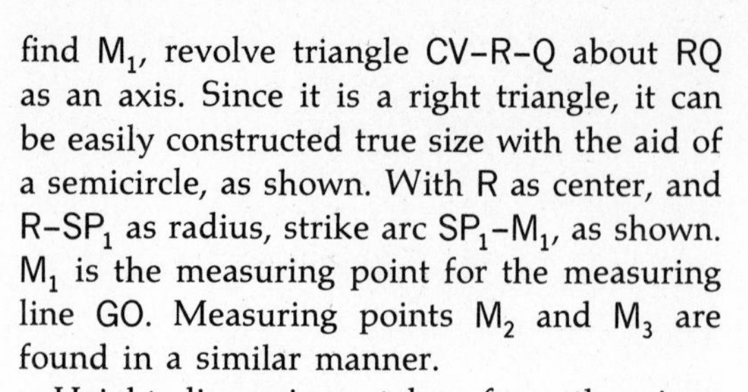

find $M_1$, revolve triangle CV–R–Q about RQ as an axis. Since it is a right triangle, it can be easily constructed true size with the aid of a semicircle, as shown. With R as center, and R–$SP_1$ as radius, strike arc $SP_1$–$M_1$, as shown. $M_1$ is the measuring point for the measuring line GO. Measuring points $M_2$ and $M_3$ are found in a similar manner.

Height dimensions, taken from the given views, are set off full size or to any desired scale, along measuring line GO, at points 3, 2, and 1. From these points, lines are drawn to $M_1$, and heights on the perspective are the intersections of these lines with the perspective front corner OT of the object. Similarly, the true depth of the object is set off on measuring line EO from O to 5, and the true width is set off on measuring line OF from O to 8. Intermediate points can be constructed in a similar manner.

## 18.12 The Perspective Linead and Template.

1. Perspective Linead. The perspective linead, Fig. 18.14 (a), consists of three straight-edged blades that can be clamped to each other at any desired angles. This instrument is convenient in drawing lines toward a vanishing point outside the limits of the drawing.

Before starting such a drawing, a small-scale diagram should be made, as indicated at (b), in which the relative positions of the object, PP, and SP are assumed, and the distances of

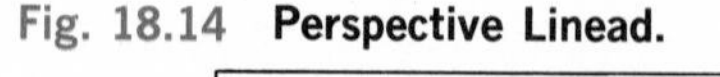

Fig. 18.14 **Perspective Linead.**

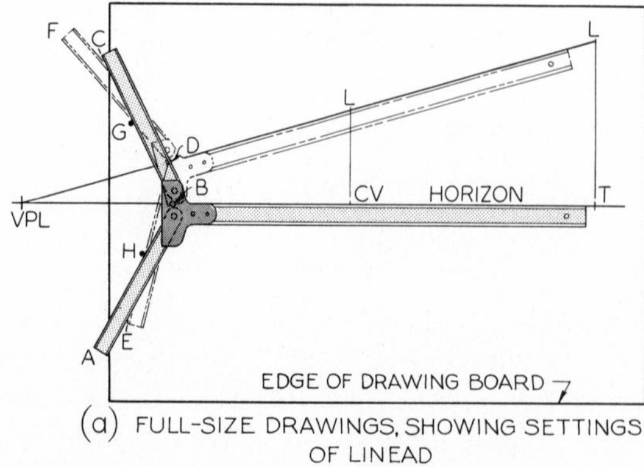


(a) FULL-SIZE DRAWINGS, SHOWING SETTINGS OF LINEAD

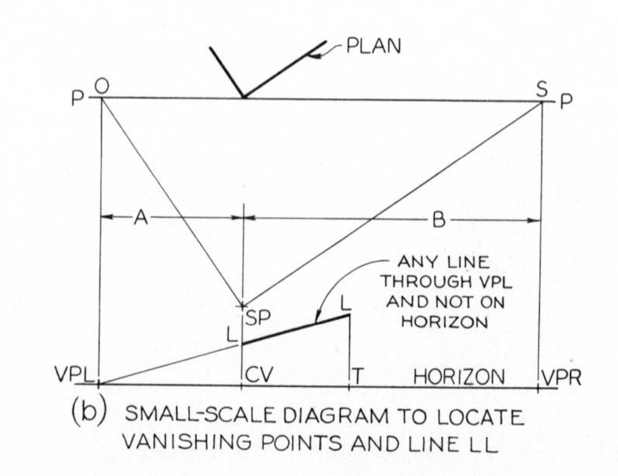


(b) SMALL-SCALE DIAGRAM TO LOCATE VANISHING POINTS AND LINE LL the vanishing points from CV determined. Draw any line LL through a vanishing point as shown; then on the full-size drawing, assume CV and locate LL, as shown at (a).

To set the linead, clamp the blades in any convenient position; set the edge of the long blade along the horizon, and draw the lines BA and BC along the short blades. Then set the edge of the long blade along the line LL, and draw the lines DE and DF to intersect the lines first drawn at points G and H. Set pins at these points. If the linead is moved so that the short blades touch the pins, all lines drawn along the edge of the long blade will pass through VPL. This method is based on the principle that an angle inscribed in a circle is measured by half the arc it subtends.

2. Templates. Fig. 18.15 A template of thin wood or heavy cardboard, cut in the form of a circular arc, may be used instead of a perspective linead. If the template is attached to the drawing board so that the inaccessible VP is at the center of the circular arc, and the T-square is moved so that the head remains in contact with the template, lines drawn along the edge of the blade will, if extended, pass through the inaccessible VP.

If the edge of the blade does not pass through the center of the head, the lines drawn will be tangent to a circle whose center is at VP and whose radius is equal to the distance from the center of the head to the edge of the blade.

3. The Klok Perspective Drawing Board. Fig. 18.16. A specially designed board used for perspective drawing that employs the preceding template idea in a ready-made form. Several useful scales are printed on the board for easy reference.

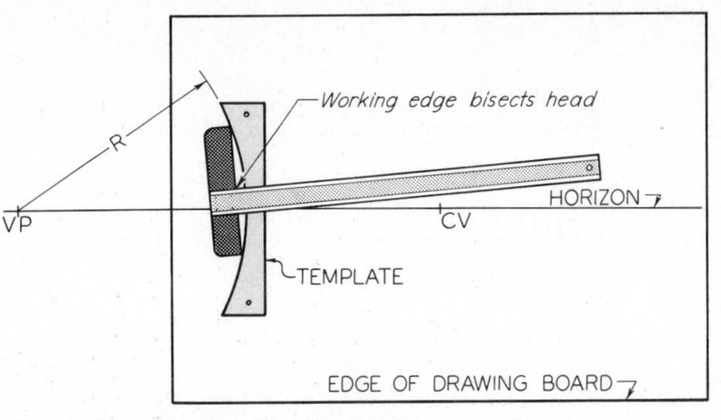


Fig. 18.15 **Perspective Template.**

*Courtesy of Modulux Division, United States Gypsum Co.*

Fig. 18.16 **Klok Perspective Drawing Board.**

## 18.13 Measurements in Perspective.

As explained in §18.1, all lines in PP are shown in their true lengths, and all lines behind PP are foreshortened.

Let it be required to draw the perspective of a line of telephone poles, Fig. 18.17. Let OB be the line of intersection of PP with the vertical plane containing the poles. In this line, the height AB of a pole is set off directly to the scale desired, and the heights of the perspectives of all poles are determined by drawing lines from A and B to VPR.

To locate the bottoms of the poles along the line B–VPR, set off along PP the distances 0–1, 1–2, 2–3, . . . , equal to the distance from pole to pole; draw the lines 1–1, 2–2, 3–3, . . . , forming a series of isosceles triangles 0–1–1, 0–2–2, 0–3–3, . . . . The lines 1–1, 2–2, 3–3, . . . , are parallel to each other, and, therefore, have a common vanishing point MP, which is found in the usual manner by drawing from SP a line SP–T parallel to the lines 1–1, 2–2, 3–3, . . . , and finding its piercing point MP (*measuring point*) in PP.

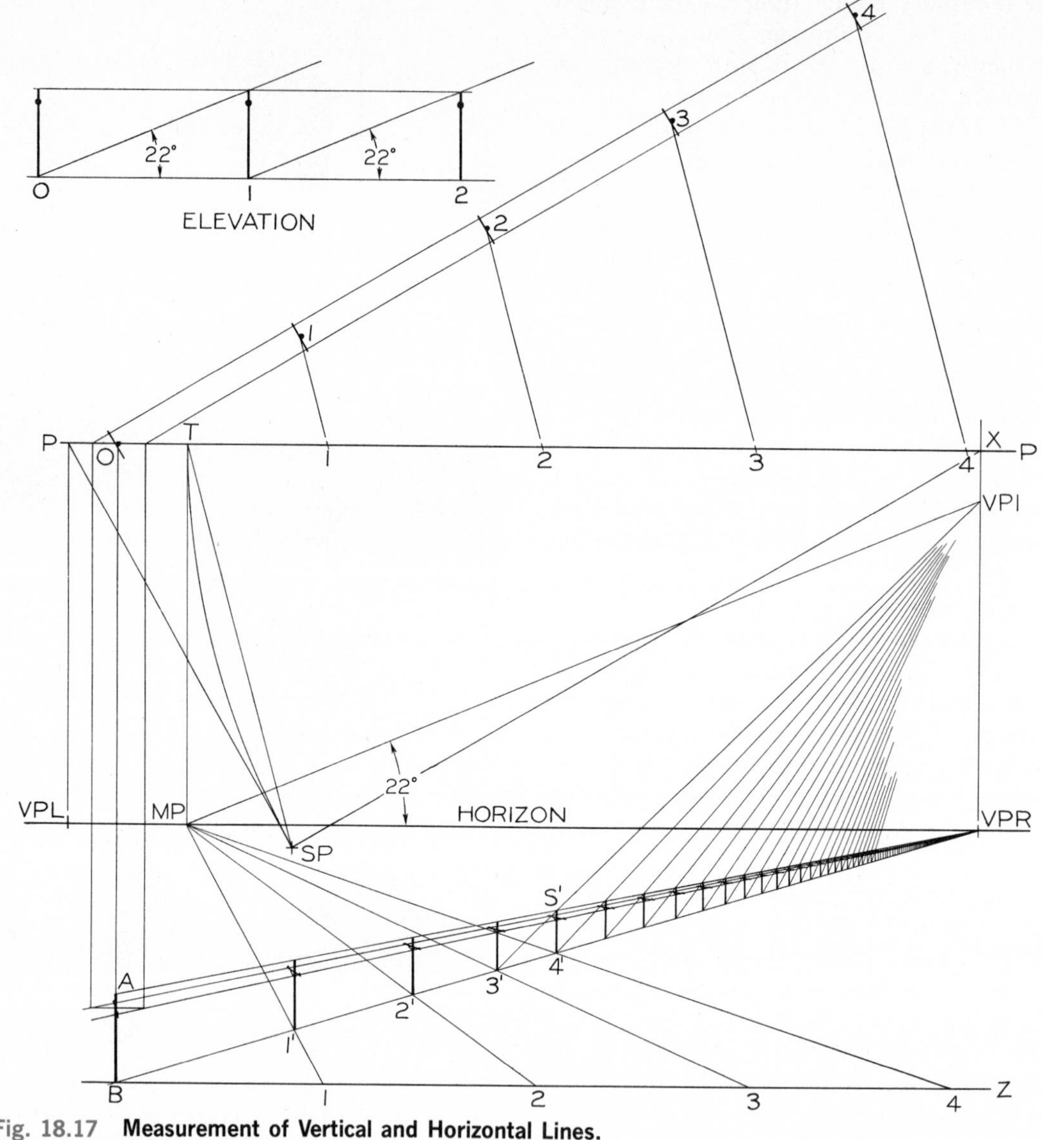


**Fig. 18.17 Measurement of Vertical and Horizontal Lines.**

Since the line SP–X is parallel to the line of poles 1–2–3, . . . , the triangle SP–X–T is an isosceles triangle, and T is the top view of MP. The point T may be determined by setting off the distance X–T equal to SP–X or simply by drawing the arc SP–T with center at X and radius SP–X.

Having the measuring point MP, find the piercing points in PP of the lines 1–1, 2–2, 3–3, . . . , and draw their perspectives as shown. Since these lines are horizontal lines, their piercing points fall in a horizontal line BZ in PP, at the bottom of the drawing. Along BZ the true distances between the poles are set off; hence BZ is called a *measuring line.* The intersections 1′, 2′, 3′, . . . , of the perspectives of the lines 1–1, 2–2, 3–3, . . . , with the line B–VPR determine the spacing of the poles.

It will be seen that only a few measurements may be made along the measuring line BZ within the limits of the drawing. For additional measurements, the *diagonal method* of spacing may be employed, as shown. Since all diagonals from the bottom of each pole to the top

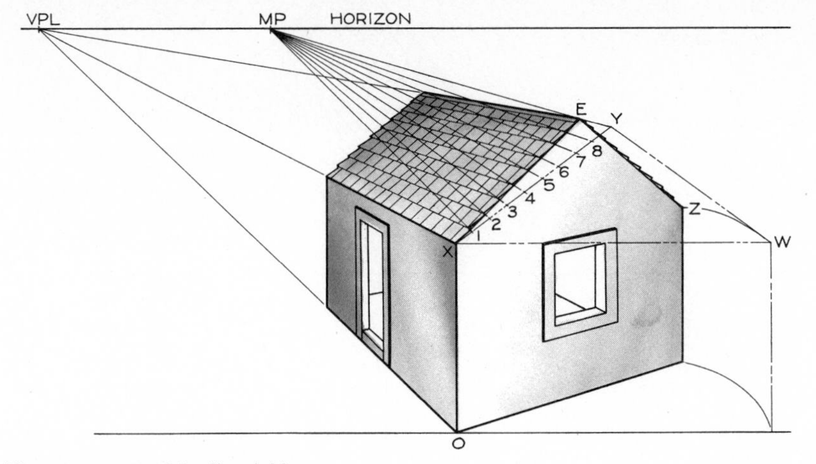


Fig. 18.18 **Measurement of Inclined Lines.**

of the succeeding pole are parallel, they have a common vanishing point VPl, which may be found as explained in §18.14. It is evident that the diagonal method exclusively may be used in the solution of this problem.

The method of direct measurements may be applied also to lines inclined to PP and to the ground plane, as illustrated in Fig. 18.18 for the line XE, which pierces PP at X. If the end of the house is conceived to be revolved about a vertical axis XO into PP, the line XE would be shown in its true length and inclination at XY. This line XY may be used as the measuring line for XE; it remains only to find the corresponding measuring point MP. The line YE is the horizontal base of an isosceles triangle having its vertex at X, and a line drawn parallel to it through SP will determine MP, as described for Fig. 18.17.

## 18.14 Vanishing Points of Inclined Lines.

The vanishing point of an inclined line is determined, as for all other lines, by finding the piercing point in PP of a line drawn from SP parallel to the given line.

In Fig. 18.19 is shown the perspective of a small building. The vanishing point of the inclined roof line C′E′ can be determined as follows: If a plane is conceived to be passed through the station point and parallel to the end of the house (plan view), it would intersect PP in the line XY, through VPL, and perpendicular to the horizon. Since the line drawn from SP parallel to C′E′ (in space) is in the plane SP–X–Y, it will pierce PP at some point T in XY. To find the point T, conceive the plane SP–X–Y revolved about the line XY as an axis into PP. The point SP will then fall on the horizon at a point shown by O in the top view, and by MR in the front view. From the point MR draw the revolved position of the line SP–T (now MR–T) making an angle of 30° with the horizon and thus determining the point T, which is the vanishing point of the line C′E′ and of all lines parallel to that line. The vanishing point S of the line D′E′ is evidently in the line XY, because D′E′ is in the same vertical plane as the line C′E′. The vanishing point S is as far below the horizon as T is above the horizon, because the line E′D′ slopes downward at the same angle at which the line C′E′ slopes upward.

The perspectives of inclined lines can generally be found without finding the vanishing points, by finding the perspectives of the end points and joining them. The *perspective of any point may be determined by finding the perspectives of any two lines intersecting at the point.* Obviously, it would be best to use horizontal lines, parallel respectively to systems of lines whose vanishing points are already available. For example, in Fig. 18.19, to find the perspective of the inclined line EC, the point E′ is the intersection of the horizontal lines R′–VPR and B′–VPL. The point C′ is already established, since it is in PP; but if it were not in PP, it could be easily found in the same manner. The

**Fig. 18.19 Vanishing Points of Inclined Lines.**

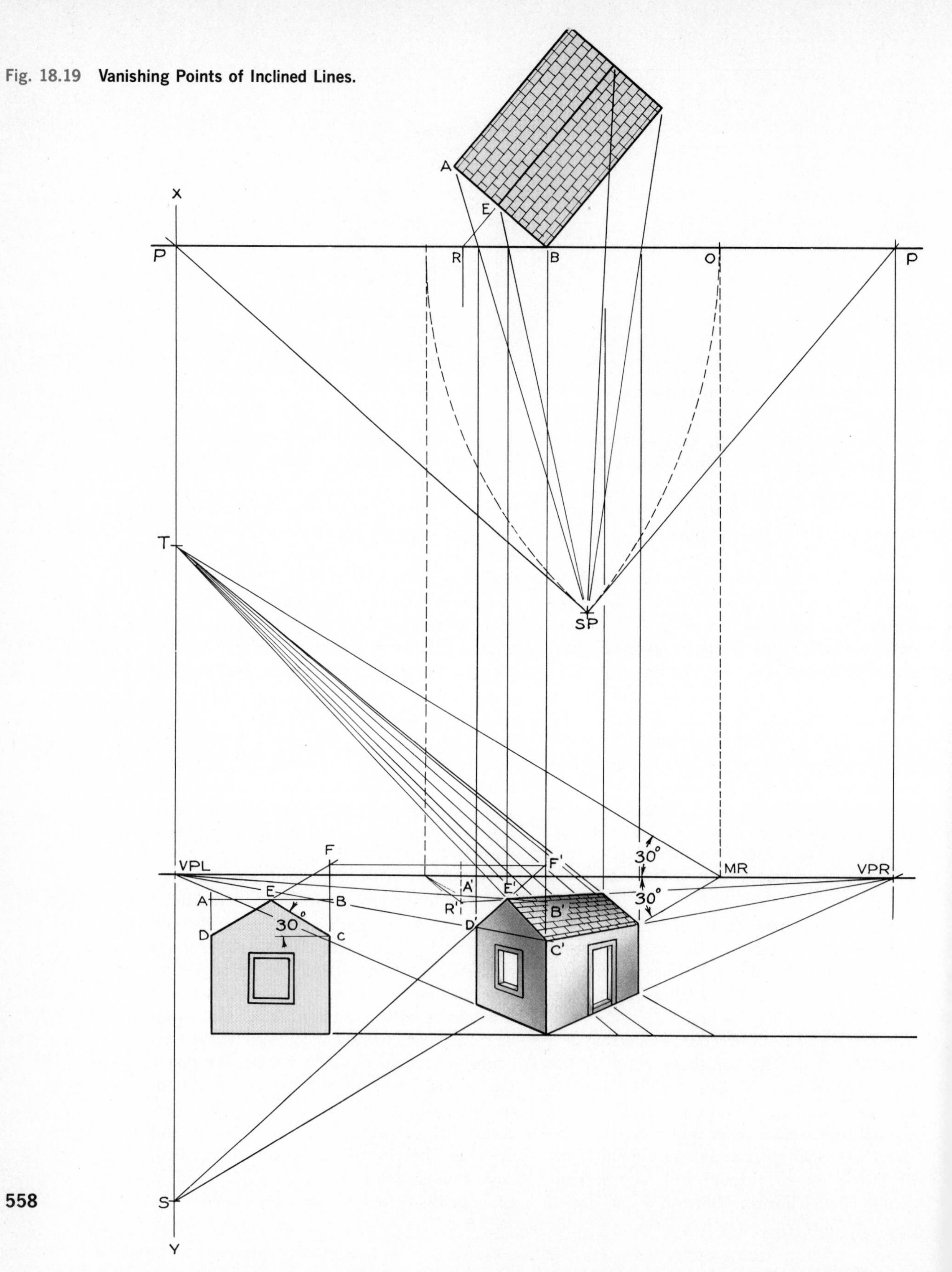

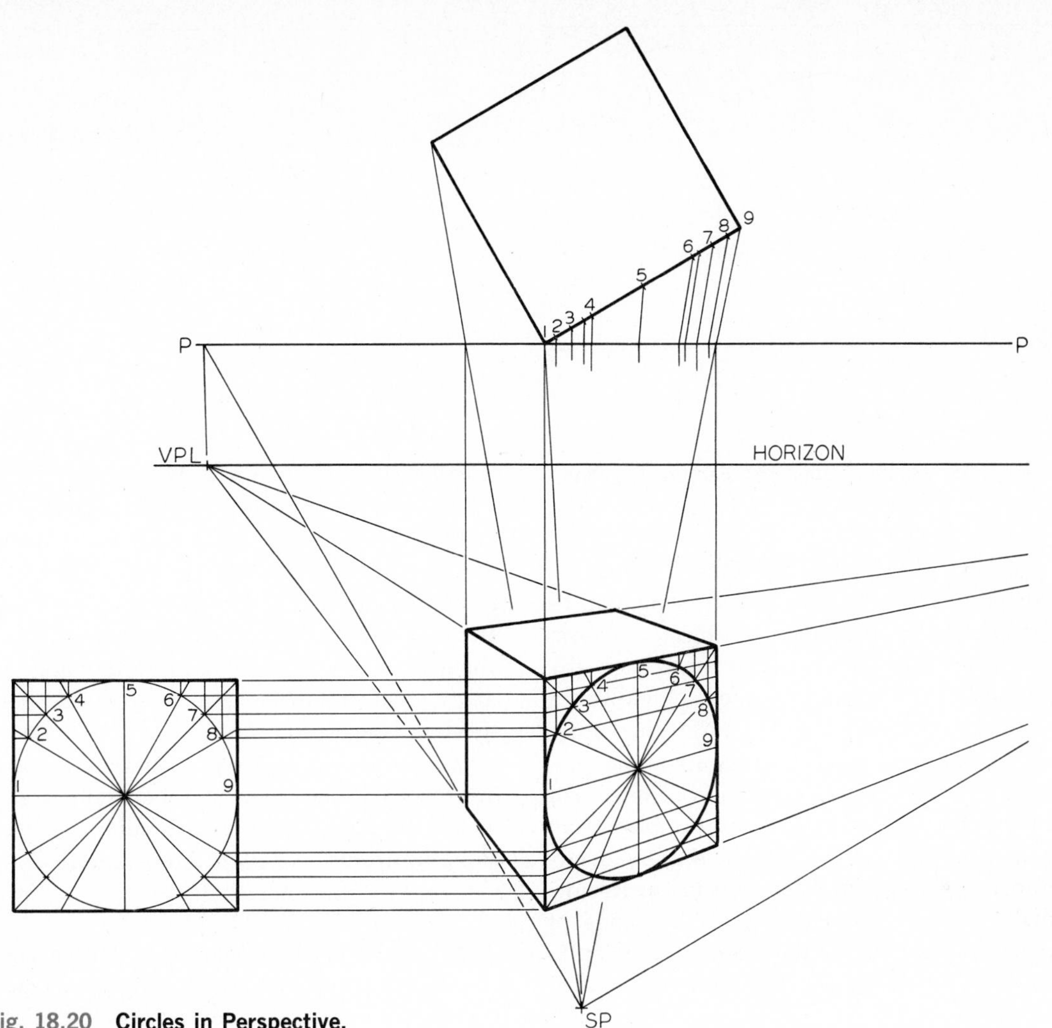


Fig. 18.20 **Circles in Perspective.**

perspective of the inclined line EC is, therefore, the line joining the perspectives of the end points E′ and C′.

## 18.15 Curves and Circles in Perspective.

If a circle is parallel to PP, its perspective is a circle. If the circle is inclined to PP, its perspective may be any one of the conic sections, in which the base of the cone is the given circle, the vertex is SP, and the cutting plane is PP. But since the center line of the cone of rays should be approximately perpendicular to the picture plane, the perspective will generally be an ellipse. The ellipse may be constructed by means of lines intersecting the circle, as shown in Fig. 18.20. The radial lines in the elevation view at the left can be easily drawn with the 45° and 30° × 60° triangles. A convenient method for determining the perspective of any plane curve is shown in Fig. 18.21.

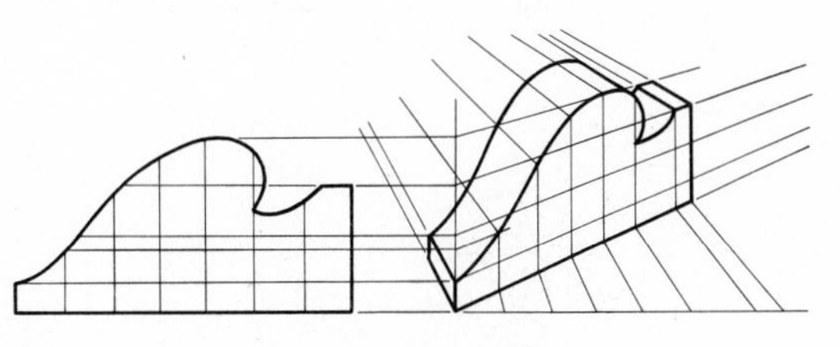

Fig. 18.21 **Curves in Perspective.**

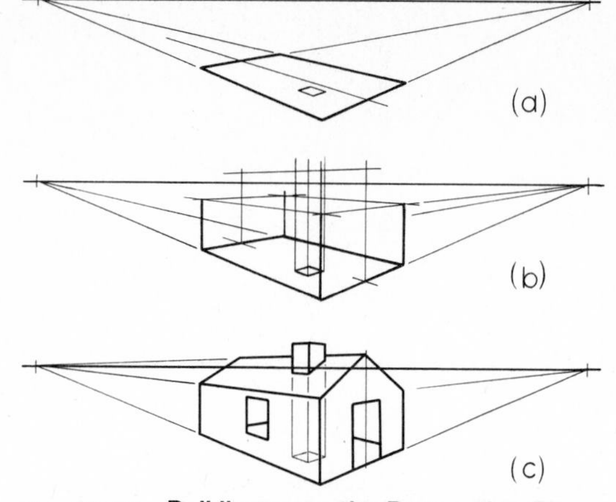


Fig. 18.22 **Building upon the Perspective Plan.**

## 18.16 The Perspective Plan Method.

A perspective may be drawn by drawing first the perspective of the plan of the object, as shown in Fig. 18.22 (a), then the vertical lines, (b), and finally the connecting lines, (c). However, in drawing complicated structures, the superimposition of the perspective upon the perspective plan causes a confusion of lines. For this reason, the perspective of the plan from which the location of vertical lines is determined is drawn either above or below its normal location. A suggestion of the range of possible positions of the perspective plan is given in Fig. 18.23; use of the perspective plan

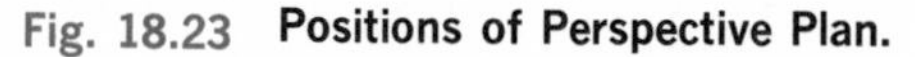
Fig. 18.23 **Positions of Perspective Plan.**

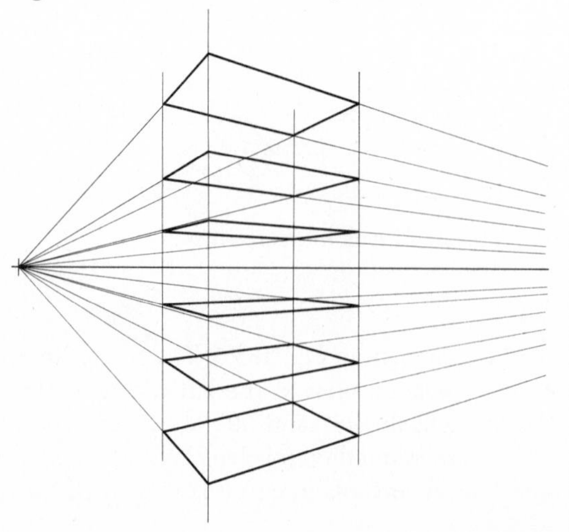

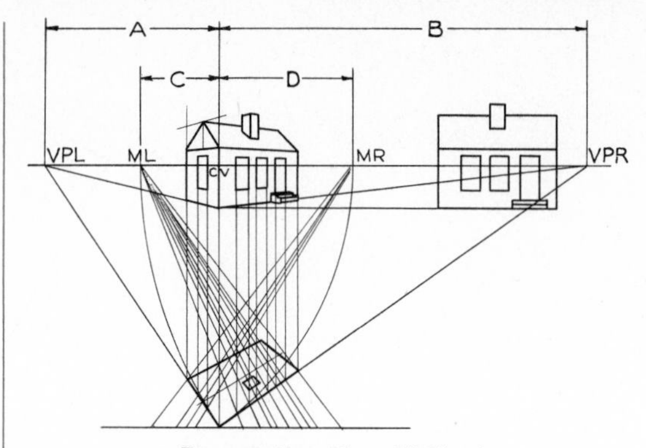


Fig. 18.24 **Perspective Plan Method.**

below the perspective is shown in Fig. 18.24.

The chief advantages of the perspective plan method over the ordinary plan method are that the vertical lines of the perspective can be spaced more accurately and that a considerable portion of the construction can be made above or below the perspective drawing, so that a confusion of lines on the required perspective is avoided.

When the perspective plan method is used, the ordinary plan view can be omitted and measuring points used to determine distances along horizontal edges in the perspective.

## 18.17 Perspective Diagram.

The spacing of vanishing points and measuring points may be determined graphically, or may be calculated. In Fig. 18.25 a simple diagram of the plan layout shows the position of the object, the picture plane, the station point, and the constructions for finding the vanishing points and measuring points for the problem in Fig. 18.24. As indicated in the figure, the complete

Fig. 18.25 **Perspective Diagram.**

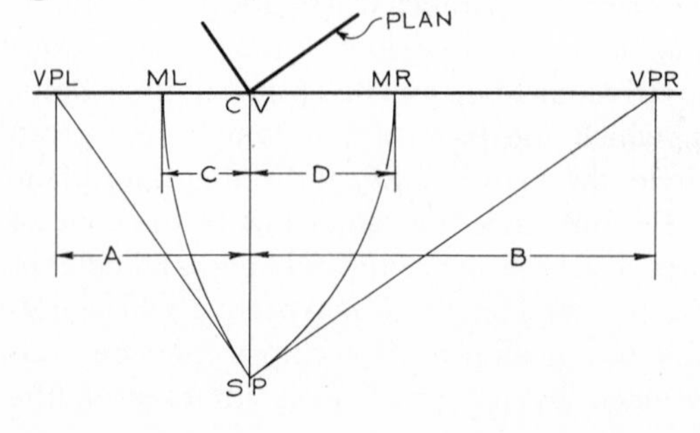

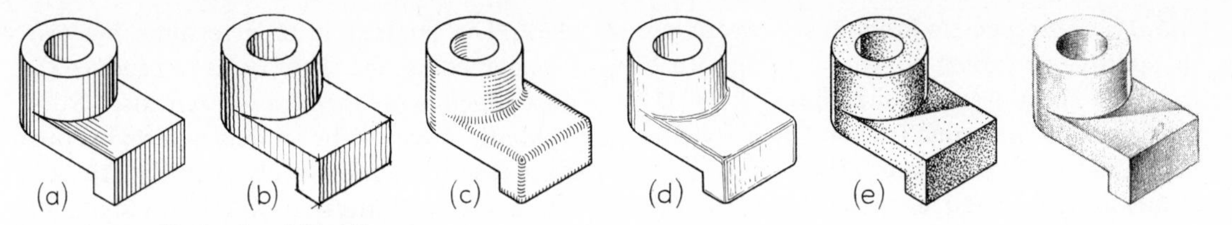


Fig. 18.26 **Methods of Shading.**

plan need not be drawn. The diagram should be drawn to any small convenient scale, and vanishing points and measuring points set off in the perspective to the larger scale desired.

In practice, structures are usually considered in one of a limited number of simple positions with reference to the picture plane, such as 30° × 60°, 45° × 45°, and 20° × 70°. Therefore, a table of measurements for locating vanishing points and measuring points may be easily prepared, to avoid the necessity of a special construction for each drawing.

## 18.18 Shading.

The effect of light can be utilized advantageously in describing the shapes of objects and in the finish and embellishment of such drawings as display drawings, patent drawings, and industrial pictorial drawings. Ordinary working drawings are not shaded.

Since the purpose of an industrial pictorial drawing is to show clearly the shape and not to be artistic, the shading should be simple and limited to producing a clear picture. Some of the common types of shading are shown in Fig. 18.26. Pencil or ink lines are drawn mechanically at (a) or freehand at (b). Two methods of shading fillets and rounds are shown at (c) and (d). Shading produced with pen dots is shown at (e), and pencil "tone" shading is shown at (f). Pencil shading applied to pictorial drawings on tracing paper may be reproduced with good results by making an Ozalid print or a blueprint.

Examples of line shading on pictorial drawings often used in industrial sales literature are shown in Figs. 18.27 and 18.28.

Fig. 18.27 **Surface Shading Applied to Pictorial Drawing of Display Case.**

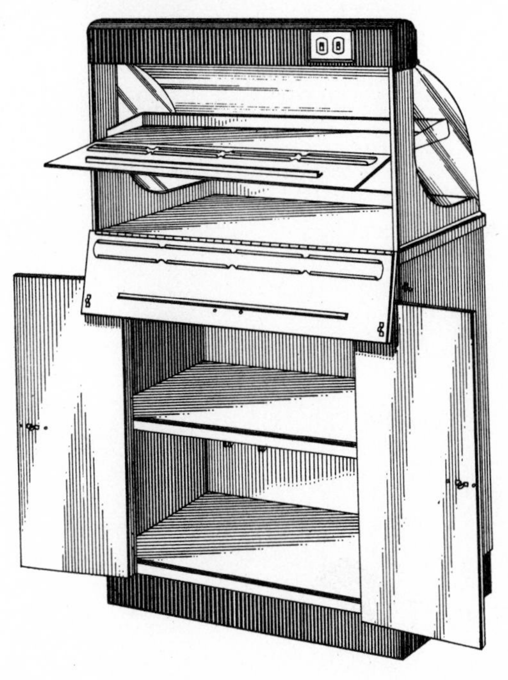

Fig. 18.28 **A Line-Shaded Drawing of an Adjustable Support for Grinding.**

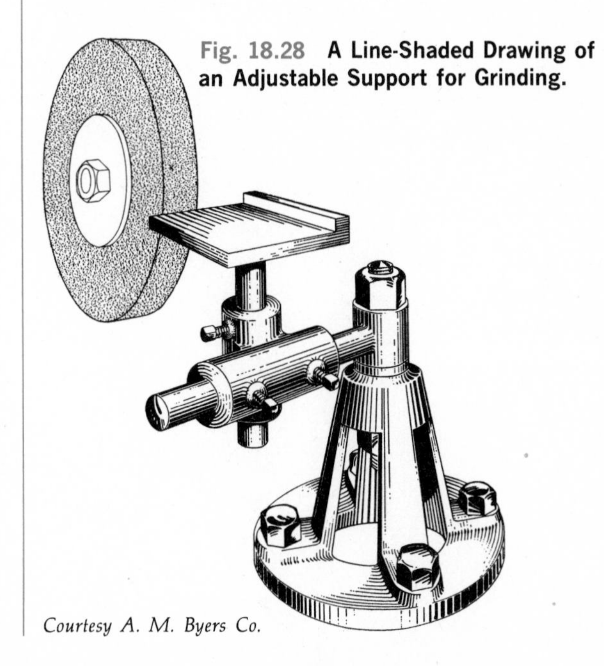

*Courtesy A. M. Byers Co.*

**18.19 Perspective Problems.** Layouts for perspective problems are given in Figs. 18.29 to 18.32. These are to be drawn on 11" × 17" paper, with $\frac{1}{2}$" margins. The student is to letter his name, date, class, and other information $\frac{1}{2}$" below the border as specified by the instructor.

Additional problems for perspective drawings on 11" × 17" paper are given in Fig. 18.33. The student is to determine his own arrangement on the sheet, so as to produce the most effective perspective in each case. Other problems in which the student selects his own sheet size and scale are given in Fig. 18.34.

In addition to these problems, many suitable problems for perspective will be found among the axonometric and oblique problems at the ends of Chapters 16 and 17.

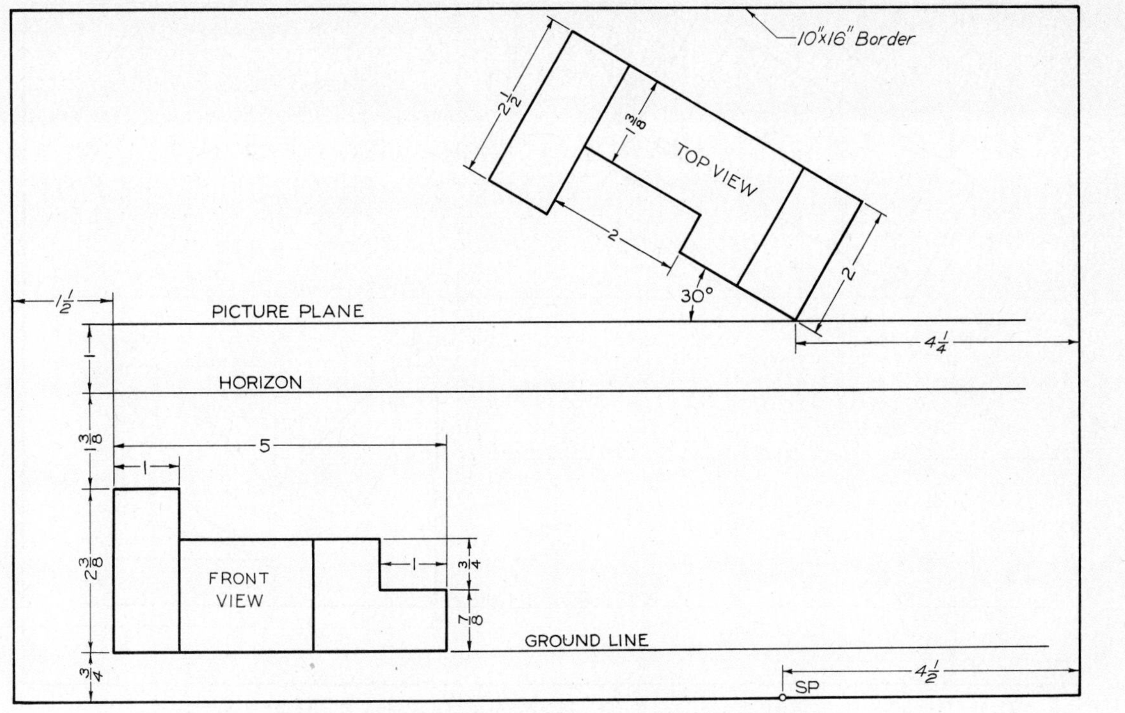


Fig. 18.29 Draw views and perspective. Omit dimensions. Use Size B sheet.

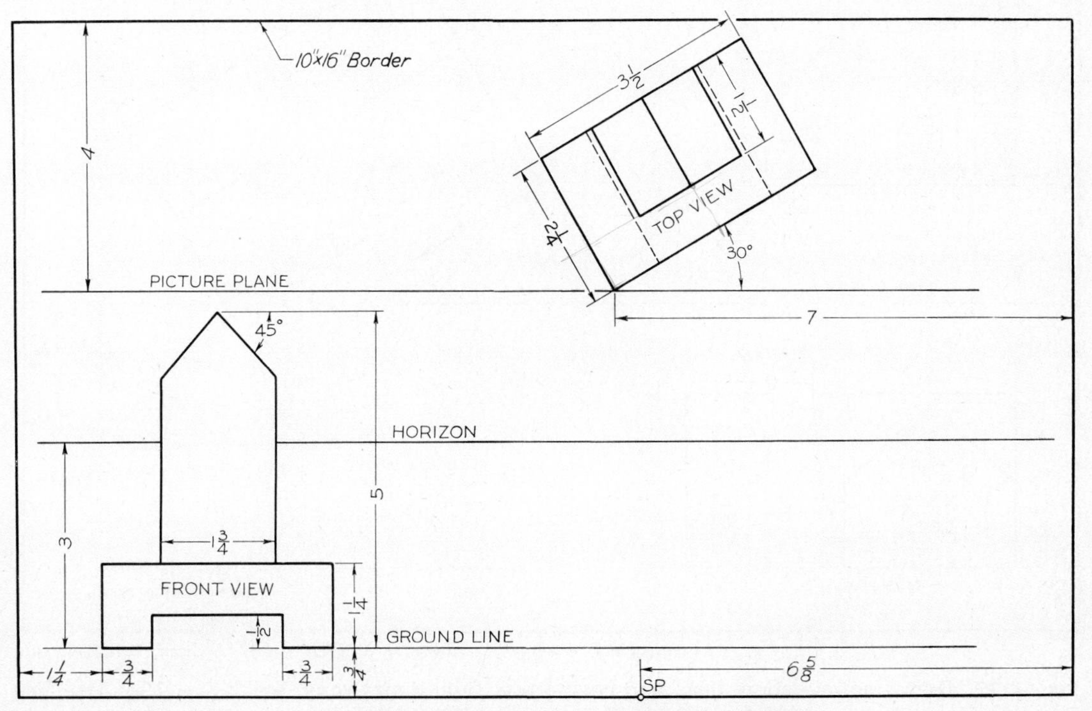


Fig. 18.30 Draw views and perspective. Omit dimensions. Use Size B sheet.

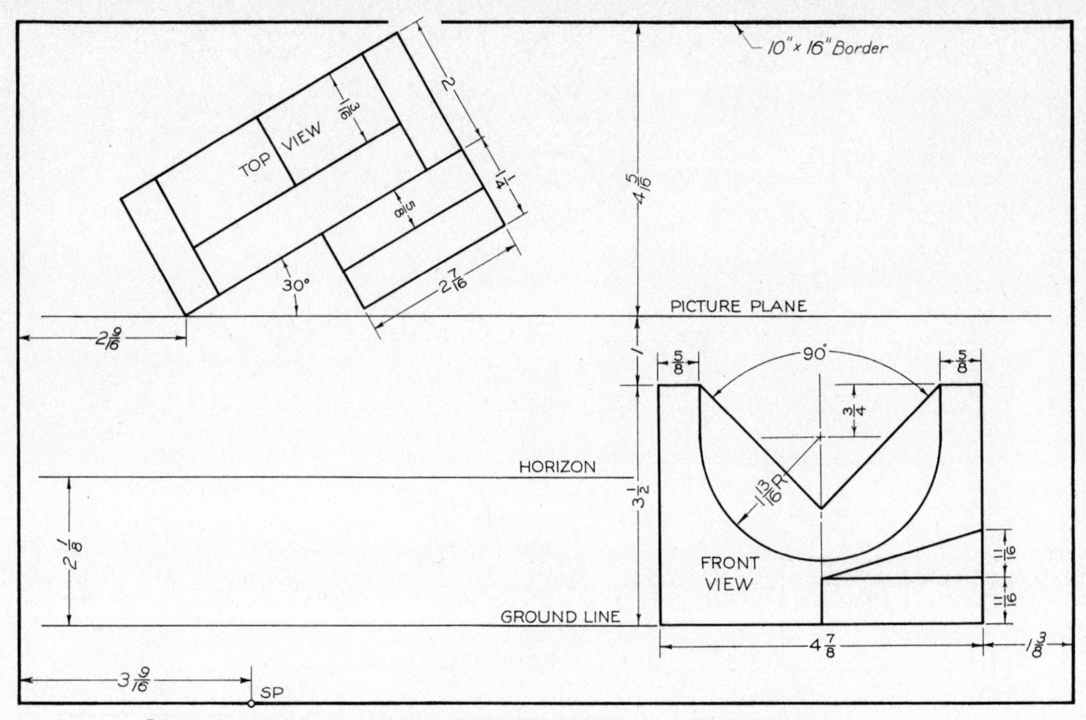


Fig. 18.31 Draw views and perspective. Omit dimensions. Use Size B sheet.

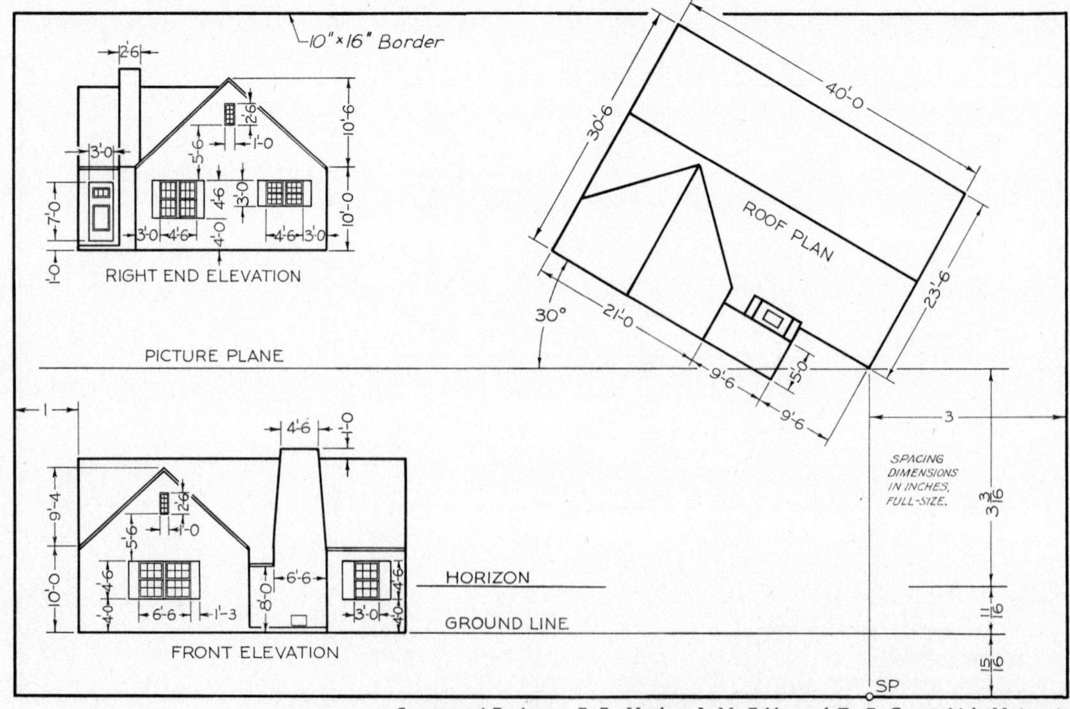


*Courtesy of Professors F. R. Hughes, J. N. Eckle, and D. F. Grant, Yale University*

Fig. 18.32 Draw front elevation, plan, and perspective. Omit dimensions. Scale: $\frac{1}{8}'' = 1'$-0. Use Size B sheet.

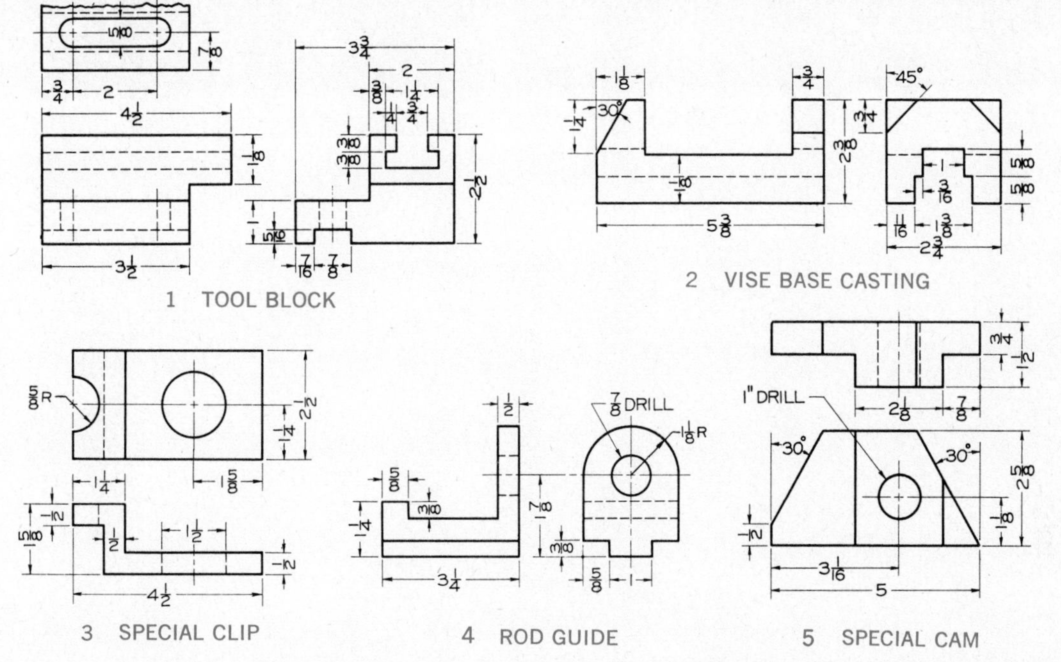


Fig. 18.33 Draw side or front elevation, plan, and perspective of assigned problem. Omit dimensions. Use Size B sheet.

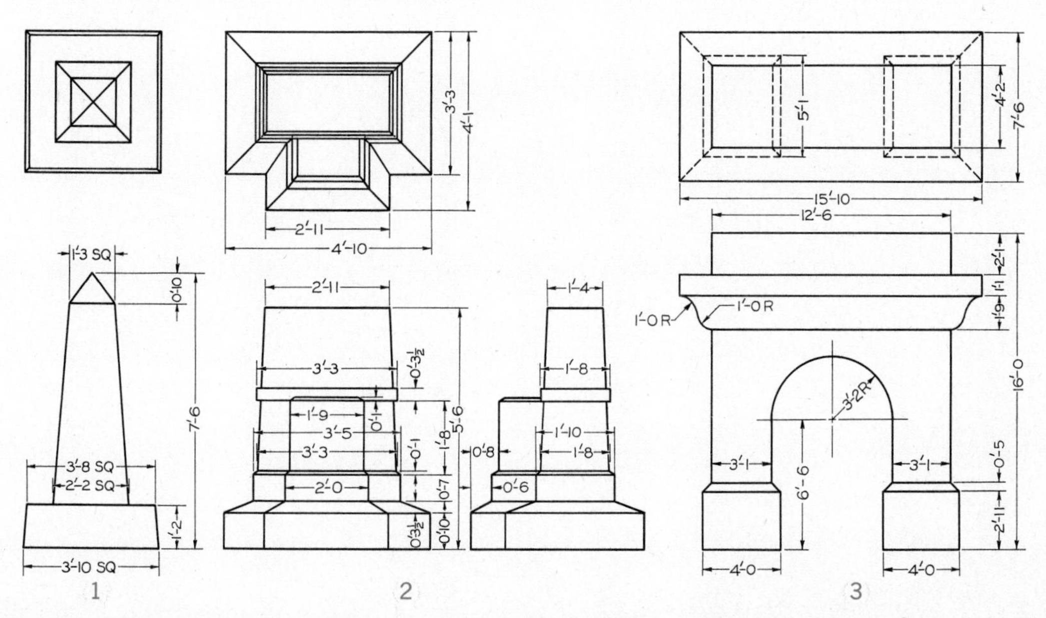


Fig. 18.34 Draw side or front elevation, plan, and perspective of assigned problem. Omit dimensions. Select sheet size and scale.

# 19

# intersections and developments

**19.1 Surfaces.** A *surface* is a geometric magnitude having two dimensions. A surface may be generated by a line, called the *generatrix* of the surface. Any position of the generatrix is an *element* of the surface, Fig. 6.31 (a).

A *ruled surface* is one which may be generated by a straight line and may be a *plane,* a *single-curved surface,* or a *warped surface.*

*A plane* is a ruled surface that may be generated by a straight line one point of which moves along another straight line, while the generatrix remains parallel to its original position. Many of the geometric solids are bounded by plane surfaces, Fig. 4.7.

A *single-curved surface* is a developable ruled surface; that is, it can be unrolled to coincide with a plane. Any two adjacent positions of the generatrix lie in the same plane. Examples are the cylinder and the cone, Fig. 4.7.

A *warped surface* is a ruled surface that is not developable, Fig. 19.1. No two adjacent positions of the generatrix lie in the same plane. Many exterior surfaces on an airplane or automobile are warped surfaces.

A double-curved surface may be generated only by a curved line and has no straight-line elements. Such a surface, generated by revolving a curved line about a straight line in the plane of the curve, is called a *double-curved surface of revolution.* Common examples are the *sphere, torus, ellipsoid,* Fig. 4.7, and the *hyperboloid,* Fig. 19.1 (d).

A *developable surface* is one which may be unfolded or unrolled so as to coincide with a plane, §19.4. Surfaces composed of single-curved surfaces, or of planes, or of combinations of these types, are developable. Warped surfaces and double-curved surfaces are not

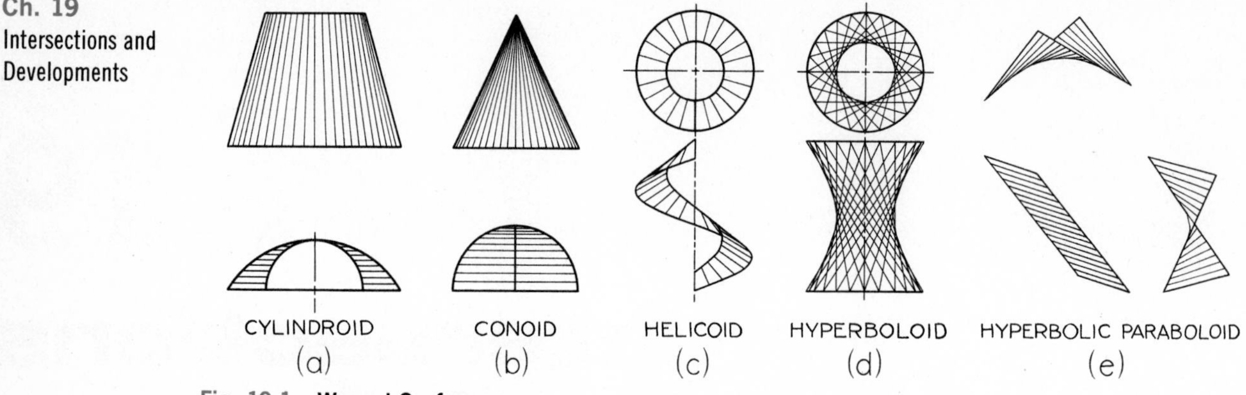


Fig. 19.1 **Warped Surfaces.**

developable. They may be developed approximately by dividing them into sections and substituting for each section a developable surface—that is, a plane or a single-curved surface. If the material used is sufficiently pliable, the flat sheets may be stretched, pressed, stamped, spun, or otherwise forced to assume the desired shape. Nondevelopable surfaces are often produced by a combination of developable surfaces, which are then formed slightly to produce the required shape.

**19.2 Solids.** *Solids* bounded by plane surfaces are *polyhedra,* the most common of which are the pyramid and prism, Fig. 4.7. Convex solids whose faces are all equal regular polygons are *regular polyhedra.* The simple regular polyhedra are the *tetrahedron, cube, octahedron, dodecahedron,* and *icosahedron,* known as the five *Platonic solids.*

Plane surfaces that bound polyhedra are *faces* of the solids. Lines of intersection of faces are *edges* of the solids.

A solid generated by revolving a plane figure about an axis in the plane of the figure is a *solid of revolution.*

Solids bounded by warped surfaces have no group name. The most common example of such solids is the screw thread.

## Intersections of Planes and Solids

**19.3 Principles of Intersections.** The principles involved in intersections of planes and solids have their practical application in the cutting of openings in roof surfaces for flues and stacks and in wall surfaces for pipes and chutes, etc., and in the building of sheet-metal structures (tanks, boilers, etc.).

In such cases, the problem is generally one of determining the true size and shape of the intersection of a plane and one of the more common geometric solids. The intersection of a plane and a solid is the locus of the points of intersection of the elements of the solid with the plane. For solids bounded by plane surfaces, it is necessary only to find the points of intersection of the edges of the solid with the plane, and to join these points, in consecutive order, with straight lines. For solids bounded by curved surfaces, it is necessary to find the points of intersection of several elements of the solid with the plane and to trace a smooth curve through these points. The curve of intersection of a plane and a circular cone is a *conic section.* The various conic sections are defined and illustrated in §4.48 and Fig. 4.46.

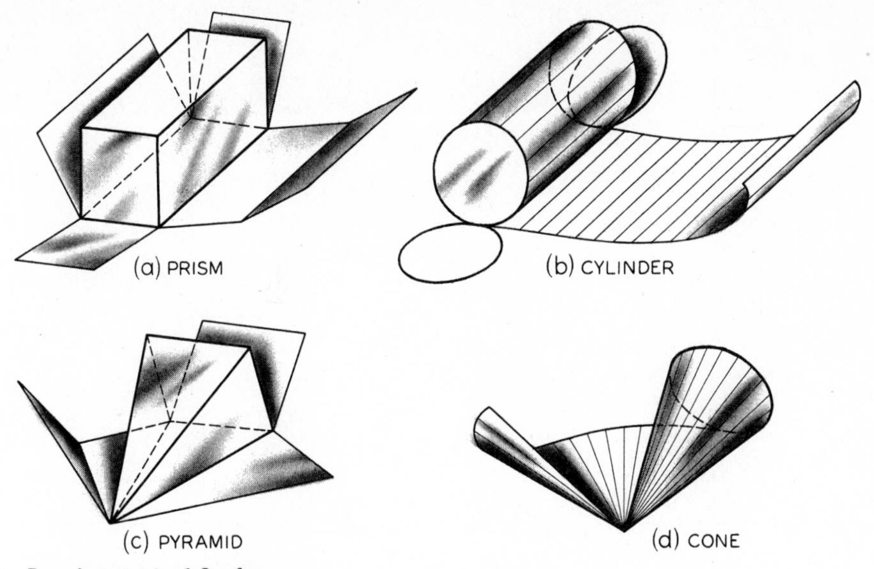


Fig. 19.2 **Development of Surfaces.**

**19.4 Developments.** The *development* of a surface is that surface laid out on a plane, Fig. 19.2. Practical applications of developments occur in sheet-metal work, stone cutting, and pattern making.

Single-curved surfaces and the surfaces of polyhedra can be developed. Warped surfaces and double-curved surfaces can be developed only approximately. See §19.1.

In sheet-metal layout, extra material must be provided for laps or seams. If the material is heavy, the thickness may be a factor, and the crowding of metal in bends must be considered. See §11.39. The draftsman must also take stock sizes into account, and should make his layouts so as to economize in the use of material and of labor. In preparing developments, it is best to put the seam at the shortest edge and to attach the bases at edges where they match, to economize in soldering, welding, or riveting.

It is common practice to draw development layouts widh the *inside surfaces up.* In this way, all fold lines and other markings are related directly to inside measurements, which are the important dimensions in all ducts, pipes, tanks, and other vessels; and in this position they are also convenient for use in the fabricating shop.

**19.5 Hems and Joints for Sheet Metal.** A wide variety of hems and joints are used in the fabrication of sheet metal developments, Fig. 19.3. Hems are used to eliminate the raw edge and also to stiffen the material. Joints and

Fig. 19.3 **Sheet Metal Hems and Joints.**

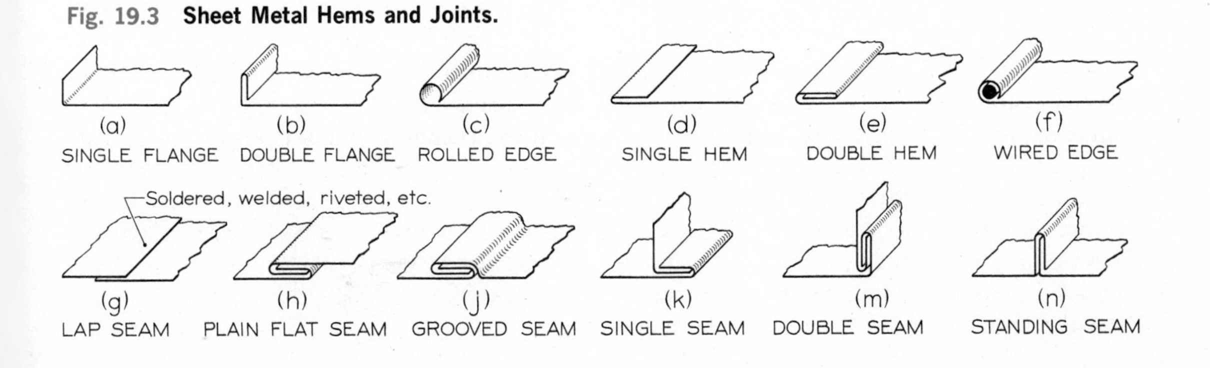

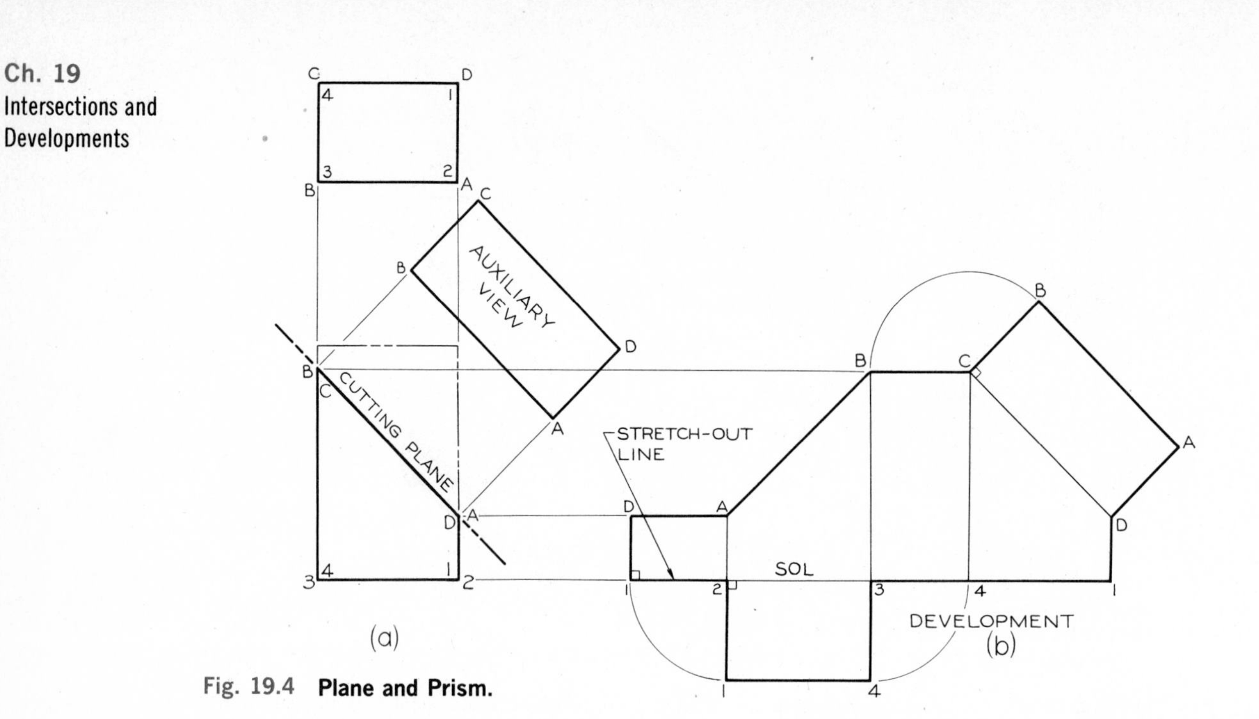


Fig. 19.4 **Plane and Prism.**

seams may be made by bending, welding, riveting, or soldering.

Sufficient material as required for hems and joints must be added to the layout or development. The amount of allowance depends on the thickness of the material and the production equipment; therefore no specific dimensions are given in this chapter. See §11.39.

## 19.6 To Find the Intersection of a Plane and a Prism and the Development of the Prism. Fig. 19.4.

Intersection. Fig. 19.4 (a). The true size and shape of the intersection is shown in the auxiliary view. See Chapter 8. The length AB is the same as AB in the front view, and the width AD is the same as AD in the top view.

Development. Fig. 19.4 (b). On the straight line 1–1, called the *stretch-out line* (SOL), set off the widths of the faces 1–2, 2–3, . . . , taken from the top view. At the division points, erect perpendiculars to 1–1, and set off on each the length of the respective edge, taken from the front view. The lengths can be projected across from the front view, as shown. Join the points thus found by straight lines to complete the development of the lateral surface. Attach to this development the lower base and the upper base, or auxiliary view, to obtain the development of the entire surface of the frustum of the prism.

## 19.7 To Find the Intersection of a Plane and a Cylinder and the Development of the Cylinder. Fig. 19.5.

Intersection. Fig. 19.5 (a). The intersection is an ellipse whose points are the piercing points in the secant plane of the elements of the cylinder. In spacing the elements, it is best, though not necessary, to divide the circumference of the base into *equal* parts, and to draw an element at each division point. In the auxiliary view, the widths BC, DE, . . . , are taken from the top view at 2–16, 3–15, . . . , respectively, and the curve is traced through the points thus determined, with the aid of the irregular curve, §2.53.

The major axis AH and the minor axis JK are shown true length in the front view and the top view, respectively; therefore, the ellipse may also be constructed as explained in §§4.49 to 4.52, or with the aid of an ellipse template, §4.57.

Development. Fig. 19.5 (b). The base of the

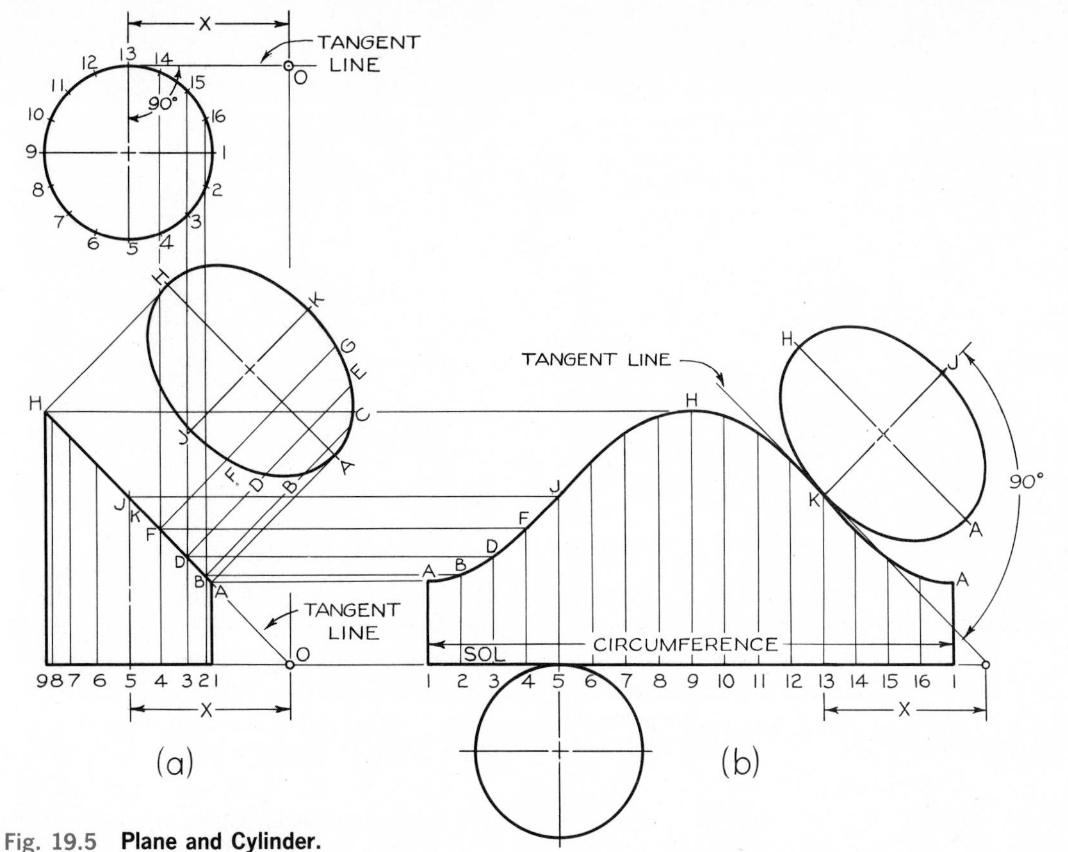


Fig. 19.5 **Plane and Cylinder.**

cylinder develops into a straight line 1–1, the stretch-out line (SOL), equal to the circumference of the base, whose length may be determined by calculation ($\pi d$), by setting off with the bow dividers, or by rectifying the arcs of the base 1–2, 2–3, . . . , §4.46. Divide the stretch-out line into the same number of equal parts as the circumference of the base, and draw an element through each division perpendicular to the line. Set off on each element its length, projected from the front view, as shown; then trace a smooth curve through the points A, B, D, . . . , §2.58, and attach the bases.

## 19.8 To Find the Intersection of a Plane and an Oblique Prism and the Development of the Prism. Fig. 19.6.

Intersection. Fig. 19.6 (a). The right section cut by the plane WX is a regular hexagon, as shown in the auxiliary view; the oblique section, cut by the horizontal plane YZ, is shown in the top view.

Development. Fig. 19.6 (b). The right section develops into the straight line WX, the stretch-out line (SOL). Set off, on the stretch-out line, the widths of the faces 1–2, 2–3, . . . , taken from the auxiliary view, and draw a line through each division perpendicular to the line. Set off, from the stretch-out line, the lengths of the respective edges measured from WX in the front view. Join the points A, B, C, . . . , with straight lines, and attach the bases, which are shown in their true sizes in the top view.

## 19.9 To Find the Intersection of a Plane and an Oblique Cylinder. Fig. 19.7.

Intersection. Fig. 19.7 (a). The right section cut by the plane WX is a circle, shown in the auxiliary view. The intersection of the horizontal plane YZ with the cylinder is an ellipse

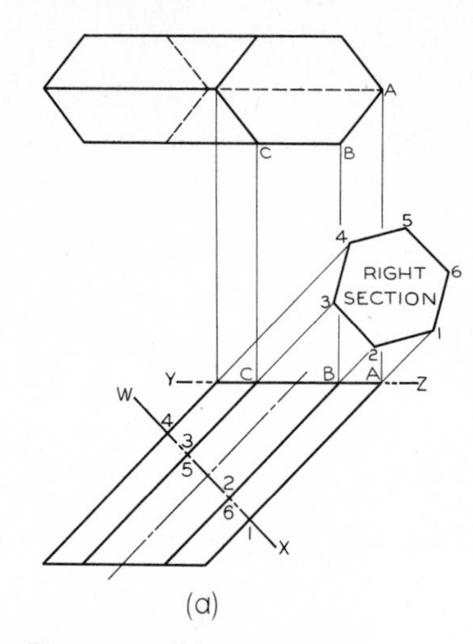


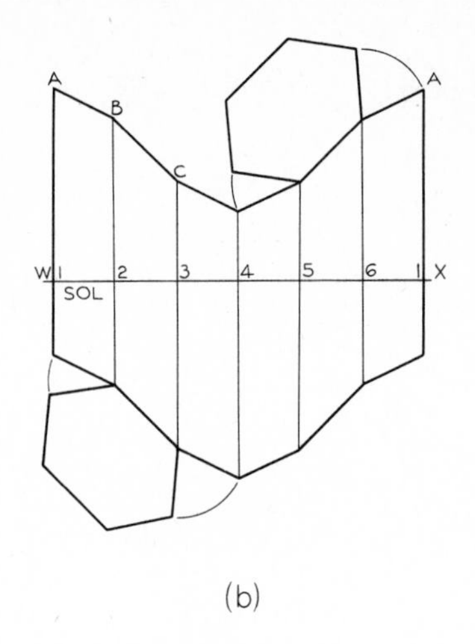


Fig. 19.6 **Plane and Oblique Prism.**

shown in the top view, whose points are found as explained for the auxiliary view in Fig. 19.5 (a), §19.7. The major axis AH is shown true length in the top view, and the minor axis JK is equal to the diameter of the cylinder; therefore, the ellipse may be constructed as explained in §§4.49 to 4.52, or with the aid of an ellipse template, §4.57.

Development. Fig. 19.7 (b). The cylinder may be considered as a prism having an infinite number of edges; therefore, the development is found in a manner similar to that of the oblique prism shown in Fig. 19.6.

The circle of the right section cut by plane WX develops into a straight line 1–1, the stretch-out line (SOL), equal in length to the

Fig. 19.7 **Plane and Oblique Circular Cylinder.**

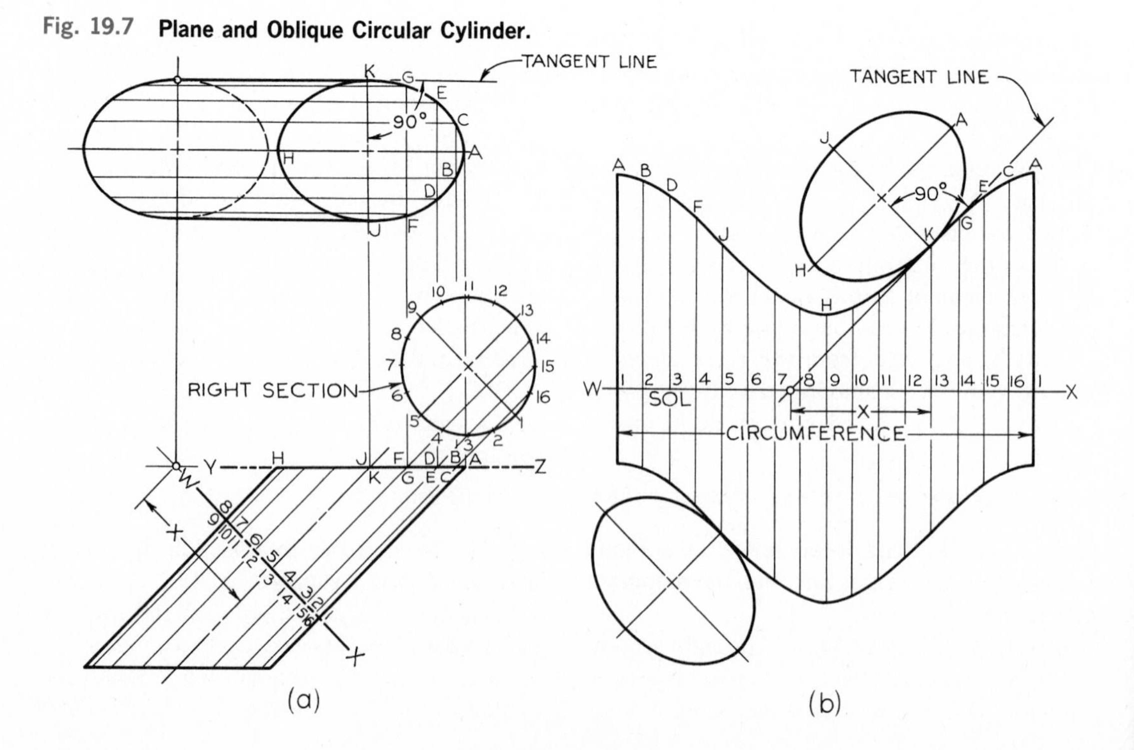

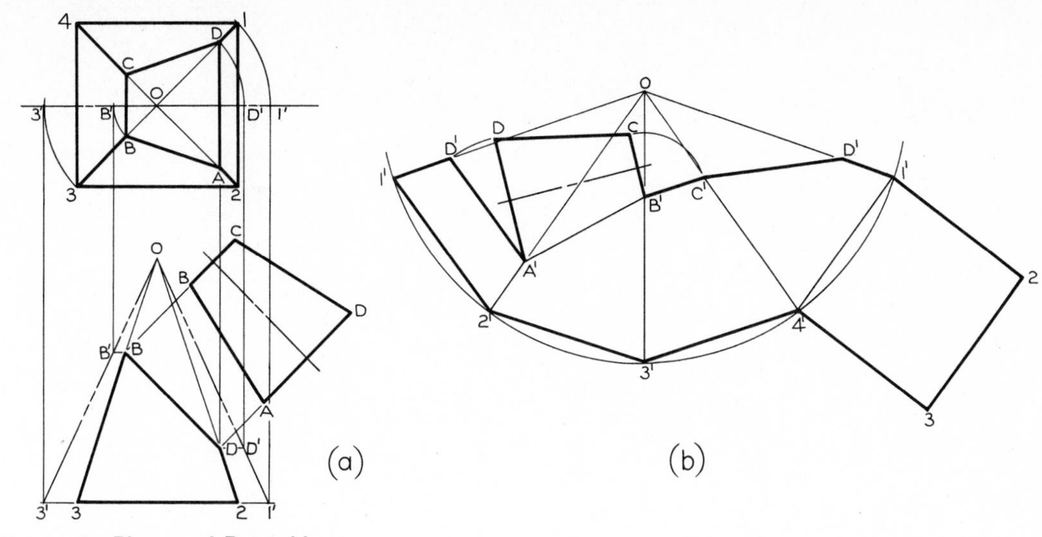


Fig. 19.8 **Plane and Pyramid.**

circumference of the circle ($\pi d$). Divide the stretch-out line into the same number of equal parts as the circumference of the circle as shown in the auxiliary view, and draw elements through these points perpendicular to the line. Set off on each element its length, taken from the front view with dividers, as shown; then trace a smooth curve through the points A, B, D, . . . , §2.58, and attach the bases.

## 19.10 To Find the Intersection of a Plane and a Pyramid and to Develop the Resulting Truncated Pyramid. Fig. 19.8.

Intersection. Fig. 19.8 (a). The intersection is a trapezoid whose vertices are the points in which the edges of the pyramid pierce the secant plane. In the auxiliary view, the altitude of the trapezoid is projected from the front view, and the widths AD and BC are transferred from the top view with dividers.

Development. Fig. 19.8 (b). With O in the development as center and O–1′ in the front view (the true length of one of the edges) as radius, draw the arc 1′–2′–3′ . . . Inscribe the cords 1′–2′, 2′–3′, . . . , equal respectively to the sides of the base, as shown in the top view. Draw the lines 1′–O, 2′–O, . . . , and set off the true lengths of the lines OD′, OA′, OB′, . . . , respectively, taken from the true lengths in the front view, §9.10.

To complete the development, join the points D′, A′, B′, . . . , by straight lines, and attach the bases to their corresponding edges. To transfer an irregular figure, such as the trapezoid shown here, refer to §§4.29 and 4.30.

## 19.11 To Find the Intersection of a Plane and a Cone and to Develop the Lateral Surface of the Cone. Fig. 19.9.

Intersection. Fig. 19.9 (a). The intersection is an ellipse. If a series of horizontal cutting planes is passed perpendicular to the axis, as shown, each plane will cut a circle from the cone that will show in true size and shape in the top view. Points in which these circles intersect the original secant plane are points on the ellipse. Since the secant plane is shown edgewise in the front view, all of these piercing points may be found in that view and projected to the others, as shown.

*Fig. 19.9 (b).* This method is most suitable when a development also is required, since it utilizes elements that are also needed in the development. The piercing points of these elements in the secant plane are points on the intersection. Divide the base into any number of equal parts, and draw an element at each division point. These elements pierce the secant plane in points A, B, C, . . . . The top views of these points are found by projecting

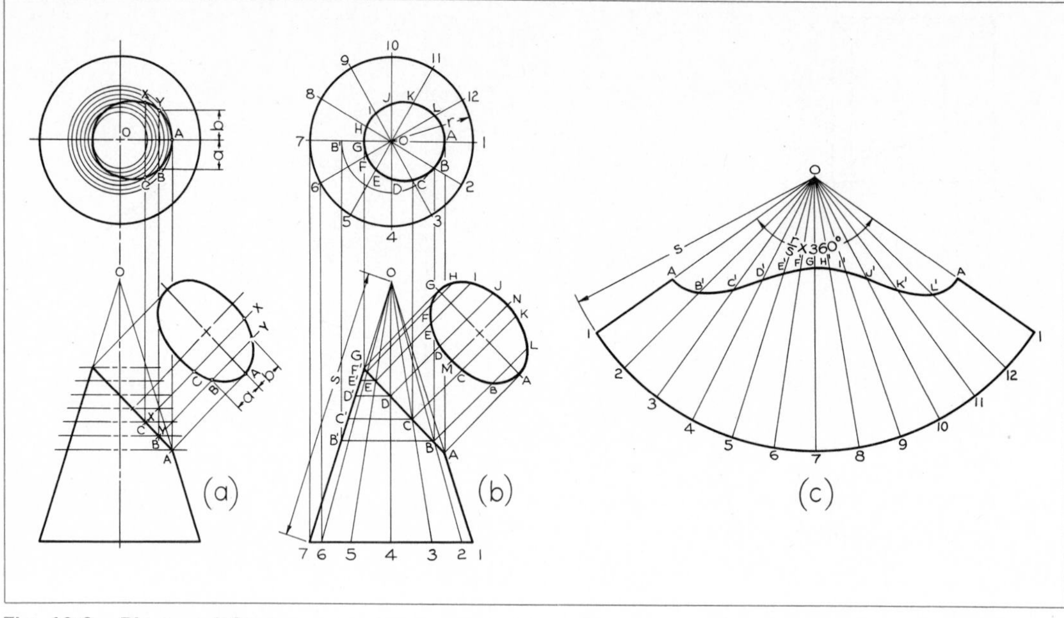


Fig. 19.9 **Plane and Cone.**

upward from the front view, as shown. In the auxiliary view, the widths BL, CK, . . . , are taken from the top view. The ellipse is then drawn with the aid of the irregular curve, §2.58.

The major axis of the ellipse, shown in the auxiliary view, is equal to AG in the front view. The minor axis MN bisects the major axis and is equal to the minor axis of the ellipse in the top view. With these axes, the ellipse may also be constructed, as explained in §§4.49 to 4.52 or with the aid of an ellipse template, §4.57.

DEVELOPMENT. Fig. 19.9 (c). The cone may be considered as a pyramid having an infinite number of edges; hence the development is found in a manner similar to that explained for the pyramid in §19.10. The base of the cone develops into a circular arc, with the slant height of the cone as its radius and the circumference of the base as its length, §4.46. The lengths of the elements in the development are taken from the element O–7 or O–1 in the front view, (b). Instead of our finding the true circumference of the base, the vertical angle 1–O–1 in the development can be set off equal to $\frac{r}{s}$ 360° (where r is the radius of the base and s the slant height of the cone).

## 19.12 To Find the Development of a Hood and Flue.

Fig. 19.10. Since the hood is a conical surface, it may be developed as described in §19.11. The two end sections of the elbow are cylindrical surfaces and may be developed as described in §19.7. The two middle sections of the elbow are cylindrical surfaces, but since their bases are not perpendicular to the axes, they will not develop into straight lines. They will be developed in a manner similar to that for an oblique cylinder, §19.9, Fig. 19.7 (b). If the auxiliary planes AB and DC are passed perpendicular to the axes, they will cut right sections from the cylinders, which will develop into the straight lines AB and CD in the developments.

If the developments are arranged as shown, the elbow can be constructed from a rectangular sheet of metal without wasting material. The patterns are shown separated after cutting. Before cutting, the adjacent curves coincided.

## 19.13 To Find the Development of a Truncated Oblique Rectangular Pyramid.

Fig. 19.11. None of the four lateral surfaces is shown in the multiview drawing in true size

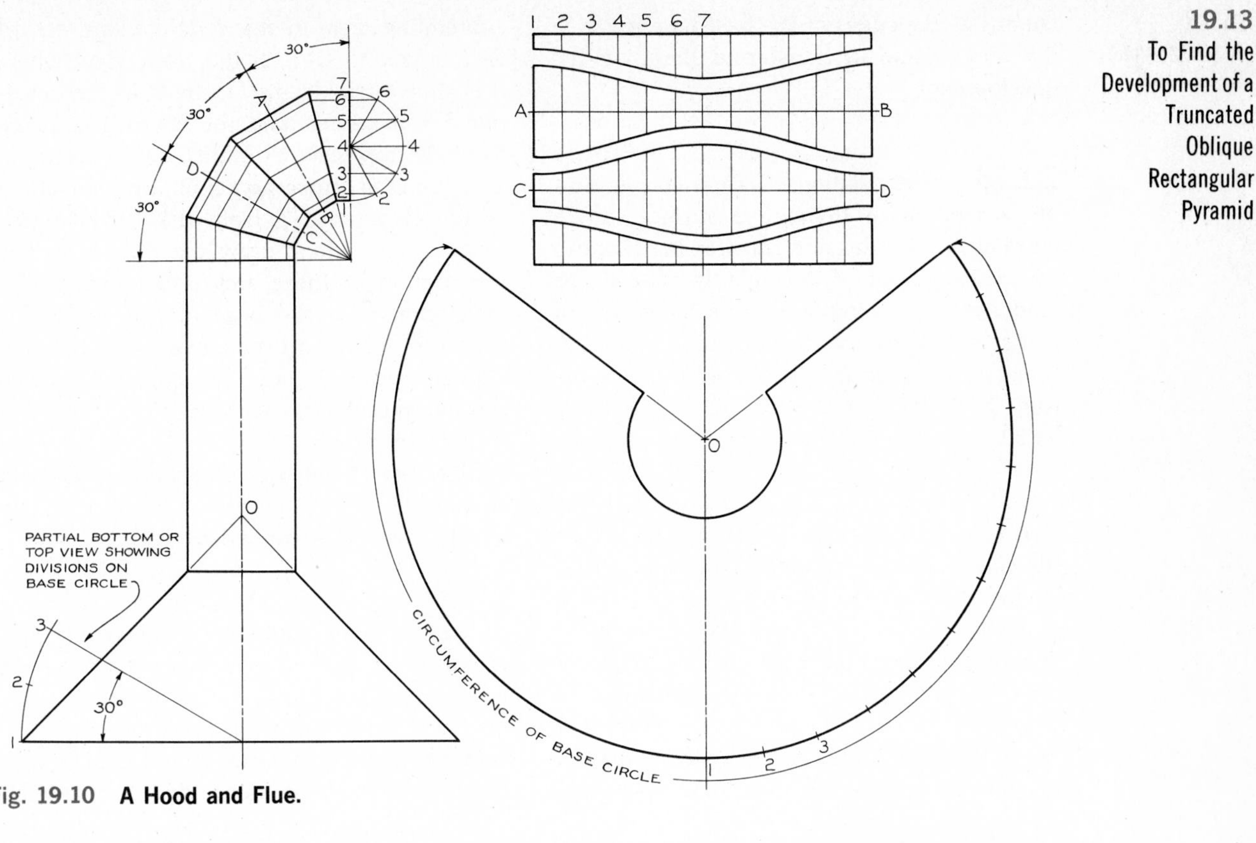


**Fig. 19.10 A Hood and Flue.**

and shape. Using the method of §9.10, revolve each edge until it appears in true length in the front view, as shown. Thus, O–2 revolves to O–2′, O–3 revolves to O–3′, and so on. These true lengths are transferred from the front view to the development with the compass, as shown. Notice that true lengths OD′, OA′, OB′, . . . , are found and transferred. The true

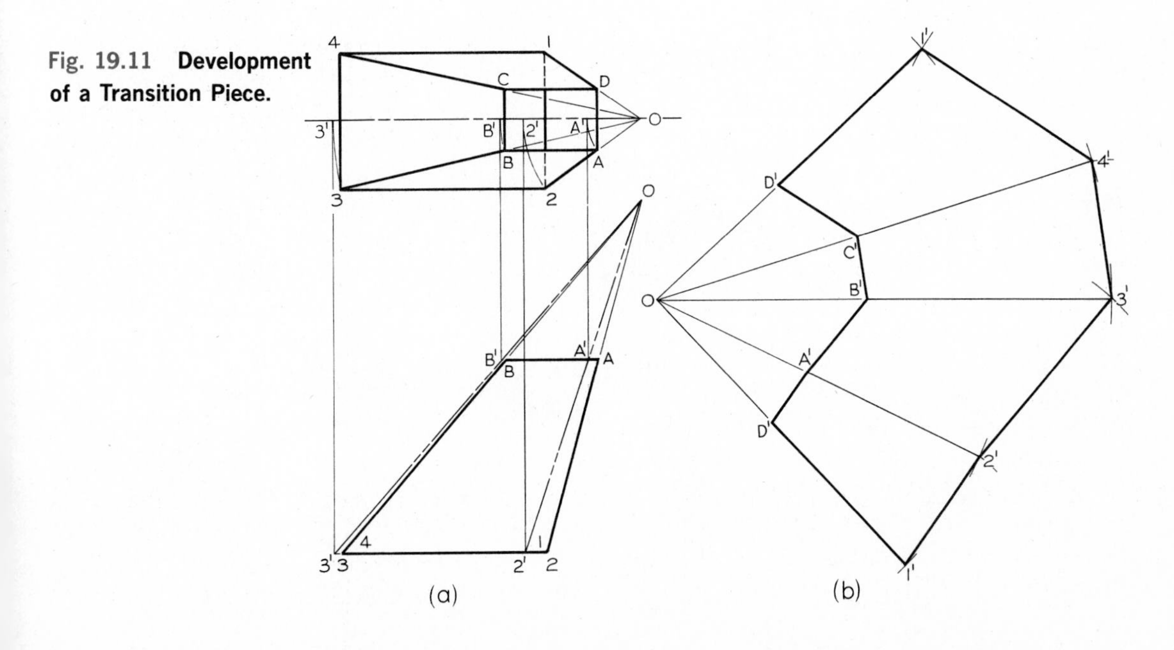


**Fig. 19.11 Development of a Transition Piece.**

lengths of the edges of the bases are given in the top view and are transferred directly to the development.

## 19.14 Triangulation.

*Triangulation* is simply a method of dividing a surface into a number of triangles and transferring them to the development. A triangle is said to be "indestructible," because if its sides are of given lengths, it can be only one shape. A triangle can be easily transferred by transferring the sides with the aid of the compass, §4.29.

## 19.15 To Find the Development of an Oblique Cone by Triangulation.

Fig. 19.12. Divide the base, in the top view, into any number of equal parts, and draw an element at each division point. Find the true length of each element, §9.10. If the divisions of the base are comparatively small, the lengths of the chords may be set off in the development as representing the lengths of the respective subtending arcs. In the development, set off O–1′ equal to O–1 in the front view where it is shown true length. With 1′ in the development as center, and the chord 1–2 taken from the top view as radius, strike an arc at 2′. With O as center, and O–2′, the true length of the element O–2 from the "true-length" diagram, as radius, draw the arc at 2′. The intersection of these arcs is a point on the development of the base of the cone. The points 3′, 4′, . . . , in the curve are found in a similar manner, and the curve is traced through these points with the aid of the irregular curve, §2.58.

Since the development is symmetrical about element O–7′, it is necessary to lay out only half the development, as shown.

## 19.16 Transition Pieces.

A *transition piece* is one that connects two differently shaped, differently sized, or skewed-position openings, Fig. 19.13. In most cases, transition pieces are composed of plane surfaces and conical surfaces, the latter being developed by triangula-

Fig. 19.12 **Development of an Oblique Cone by Triangulation.**

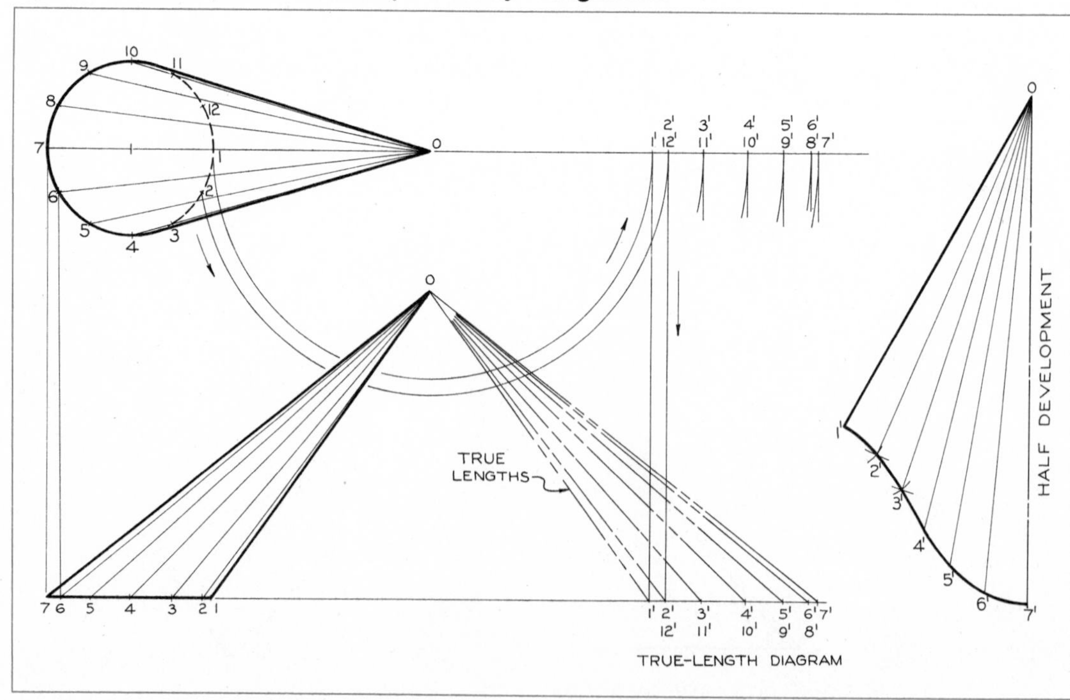

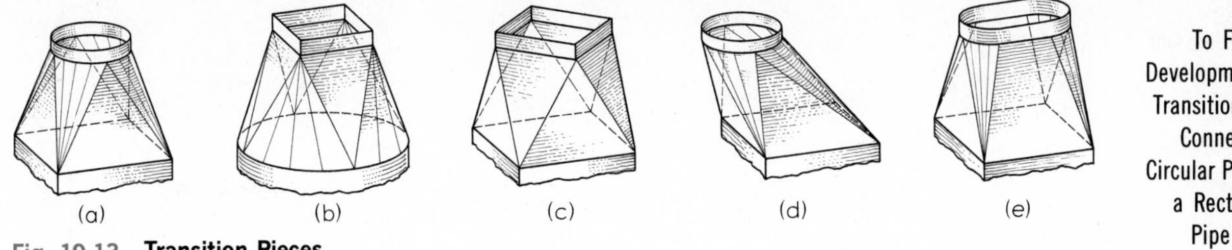


Fig. 19.13 **Transition Pieces.**

tion. Triangulation can also be used to develop, approximately, certain warped surfaces. Transition pieces are extensively used in air conditioning, heating, ventilating, and similar construction.

## 19.17 To Find the Development of a Transition Piece Connecting Rectangular Pipes on the Same Axis.

Fig. 19.14 (a). The transition piece is a frustum of a pyramid. Find the vertex O of the pyramid by extending its edges to their intersection. Find the true lengths of the edges by any one of the methods explained in §9.10. The development can then be found as explained in §19.10.

If the transition piece is not a frustum of a pyramid, as in Fig. 19.14 (b), it can best be developed by triangulation, §19.14, as shown for the faces 1–5–8–4 and 2–6–7–3, or by extending the sides to form triangles, as shown for faces 1–2–6–5 and 3–4–8–7, and then finding the true lengths of the sides of the triangles, §9.10, and setting them off as shown.

As a check on the development, lines parallel on the surface must also be parallel on the development; for example, 8′–5′ must be parallel to 4′–1′ on the development.

## 19.18 To Find the Development of a Transition Piece Connecting a Circular Pipe and a Rectangular Pipe on the Same Axis.

Fig. 19.15. The transition piece is composed of four isosceles triangles and four conical surfaces. The seam is along line S–1. Begin the development on the line 1′–S, and draw the right triangle 1′–S–A, whose base SA is equal to half the side AD and whose hypotenuse A–1′ is equal to the true length of side A–1.

Fig. 19.14 **Development of a Transition Piece.**

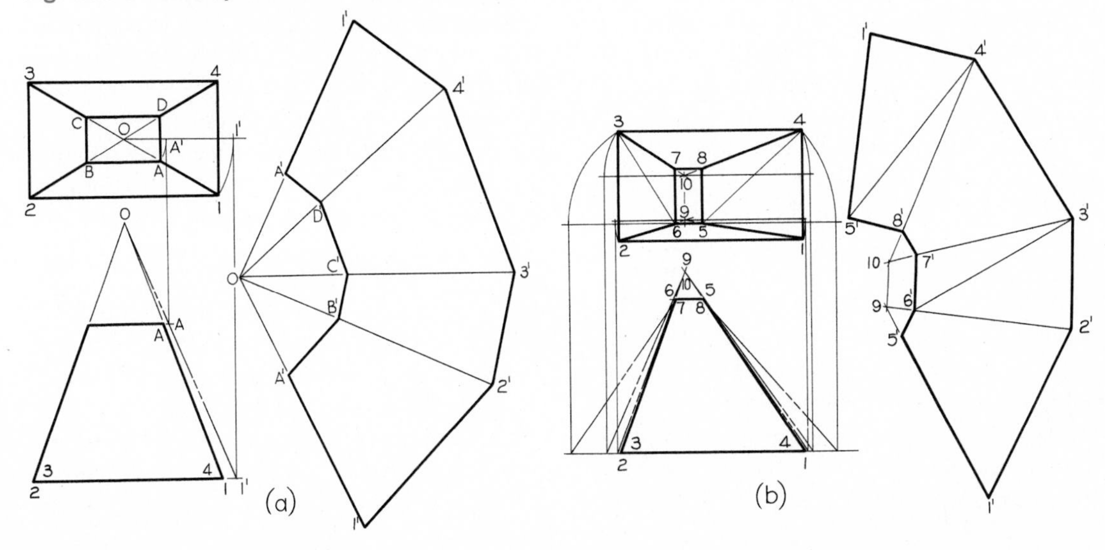

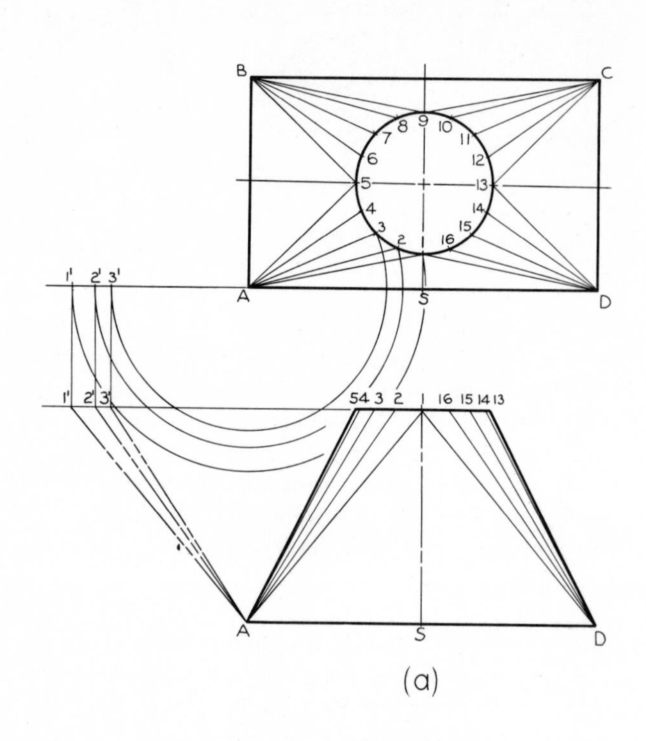


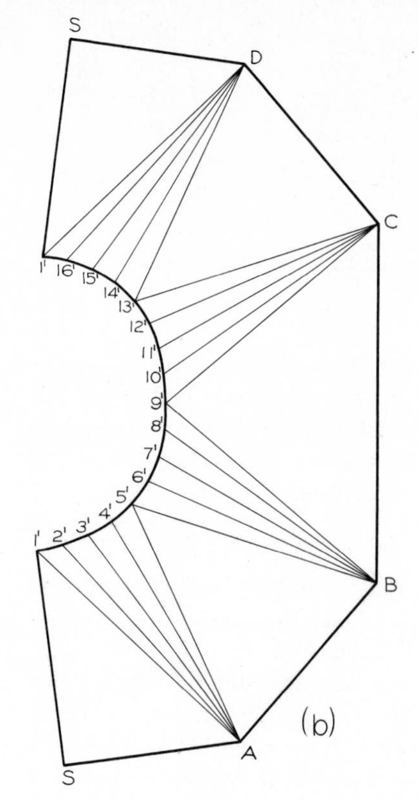


Fig. 19.15 **Development of a Transition Piece.**

The conical surfaces are developed by triangulation as explained in §§19.14 and 19.15.

## 19.19 To Find the Development of a Transition Piece Connecting Two Cylindrical Pipes on Different Axes.

Fig. 19.16. The transition piece is a frustum of a cone, the vertex of which may be found by extending the contour elements to their intersection A.

The development can be found by triangulation, as explained in §§19.14 and 19.15. The sides of each triangle are the true lengths of two adjacent elements of the cone, and the base is the true length of the curve of the base of the cone between the two elements. This curve is not shown in its true length in either view, and the plane of the base of the frustum must therefore be revolved until it is horizontal in order to find the distance from the foot of one element to the foot of the next. When the plane of the base is thus revolved, the foot of any element, such as 7, revolves to 7′, and the curve 6′–7′ (top view) is the true length of the curve of the base between the elements 6 and 7. In practice, the chord distances between these points are generally used to approximate the curved distances.

After the conical surface has been developed, the true lengths of the elements on the truncated section of the cone are set off from the vertex A of the development to secure points on the upper curve of the development.

If the transition piece is not a frustum of a cone, its development is found by another variation of triangulation, as shown in Fig. 19.17. The circular intersection with the large vertical pipe is shown true size in the top view, and the circular intersection with the small inclined pipe is shown true size in the auxiliary view. Since both intersections are true circles, and the planes containing them are not parallel, the lateral surface of the transition piece is a warped surface and not conical

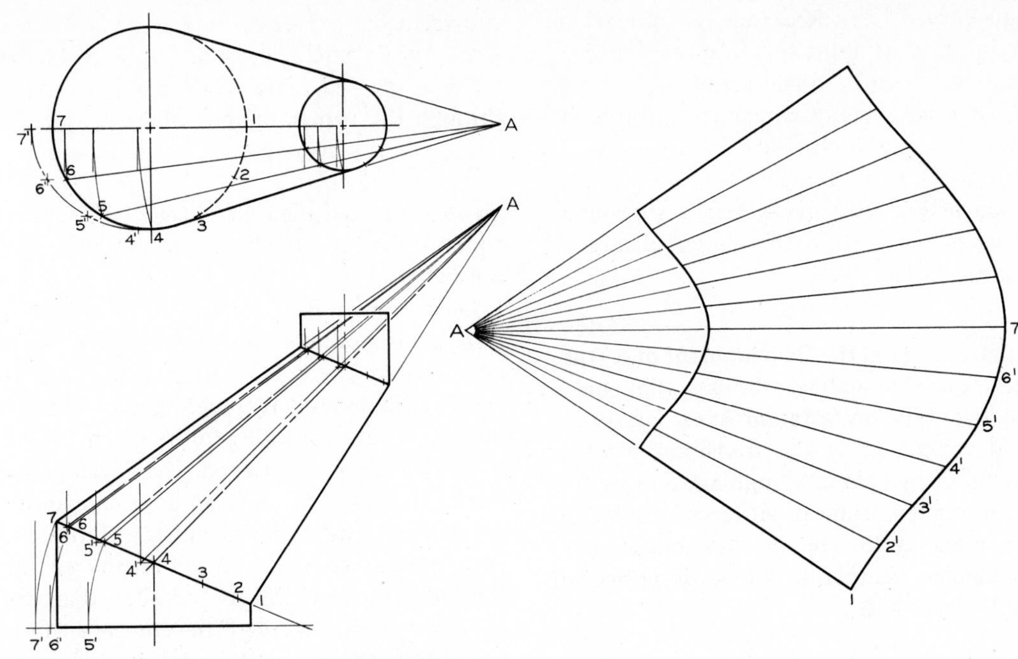

Fig. 19.16 **Development of a Transition Piece.**

Fig. 19.17 **Development of a Transition Piece.**

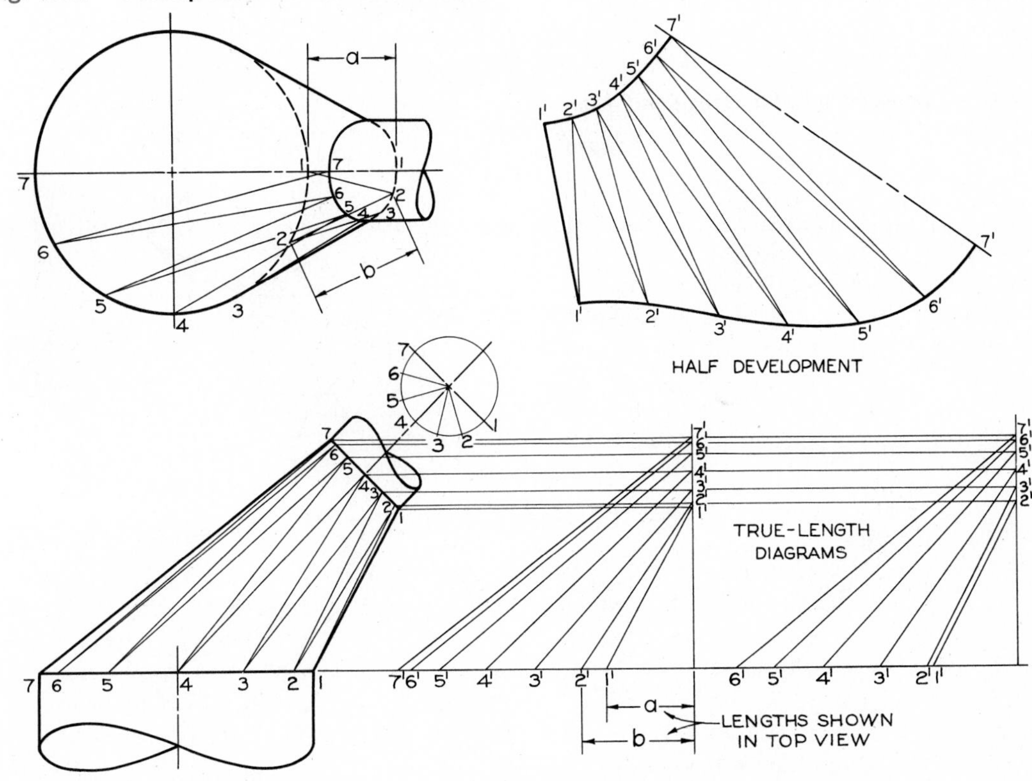

(single-curved). It is theoretically nondevelopable, but may be approximately developed by considering it to be made up of plane triangles, alternate ones of which are inverted, as shown in the development. The true lengths of the sides of the triangles are found by the method of Fig. 9.10 (d), but in a systematic manner so as to form true-length diagrams, as shown in Fig. 19.17.

## 19.20 To Find the Development of a Transition Piece Connecting a Square Pipe and a Cylindrical Pipe on Different Axes. Fig. 19.18.

The development of the transition piece is made up of five plane triangular surfaces and four triangular conical surfaces similar to those in Fig. 19.15. The development is made in a similar manner to those described in §§19.15 and 19.18.

## 19.21 To Find the Intersection of a Plane and a Sphere and to Find the Approximate Development of the Sphere. Fig. 19.19.

INTERSECTION. Fig. 19.19 (a). The intersection of a plane and a sphere is a circle, as shown in the top views in Fig. 19.19, the diameter of the circle depending upon where the plane is passed. Any circle cut by a plane through the center of the sphere is called a *great circle*. If a plane passes through the center and perpendicular to the axis, the resulting great circle is called the *equator*. If a plane contains the axis, it will cut a great circle called a *meridian*.

DEVELOPMENT. Fig. 19.19 (a). The surface of a sphere is a double-curved surface and is not developable, §19.1. The surface may be developed approximately by dividing it into a series of zones and substituting for each zone a frustum of a right-circular cone. The development of the conical surfaces is an approximate development of the spherical surface. If the conical surfaces are inscribed within the sphere, the development will be smaller than the spherical surface, while if the conical surfaces are circumscribed about the sphere, the development will be larger. If the conical surfaces are partly within and partly without the sphere, as indicated in the figure, the resulting development very closely approximates the spherical surface.

Fig. 19.18 **Development of a Transition Piece.**

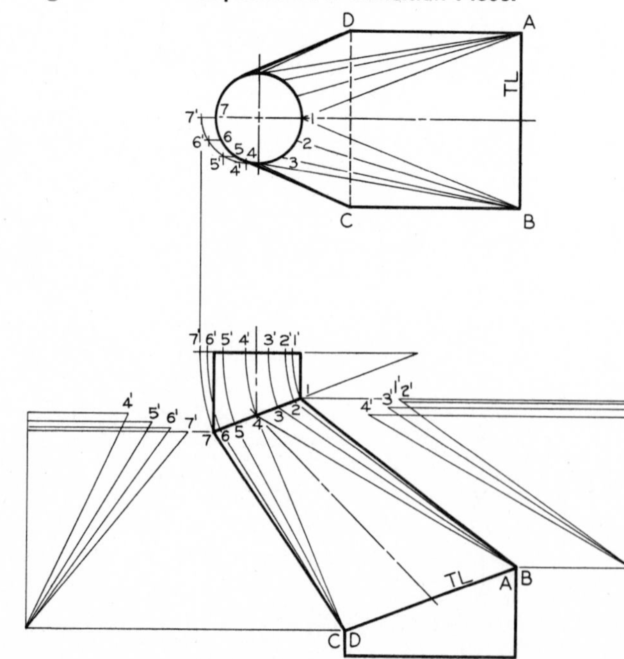


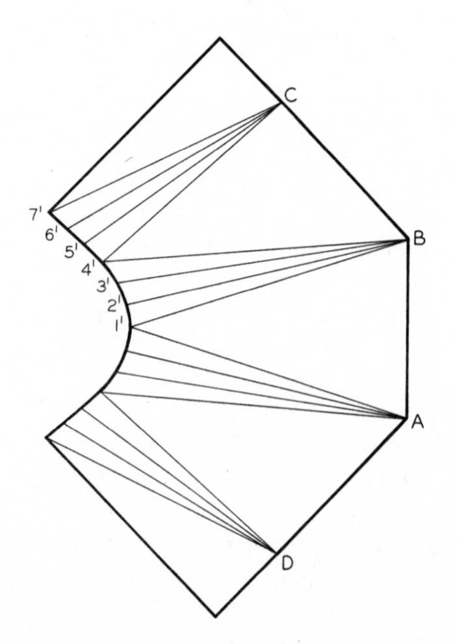

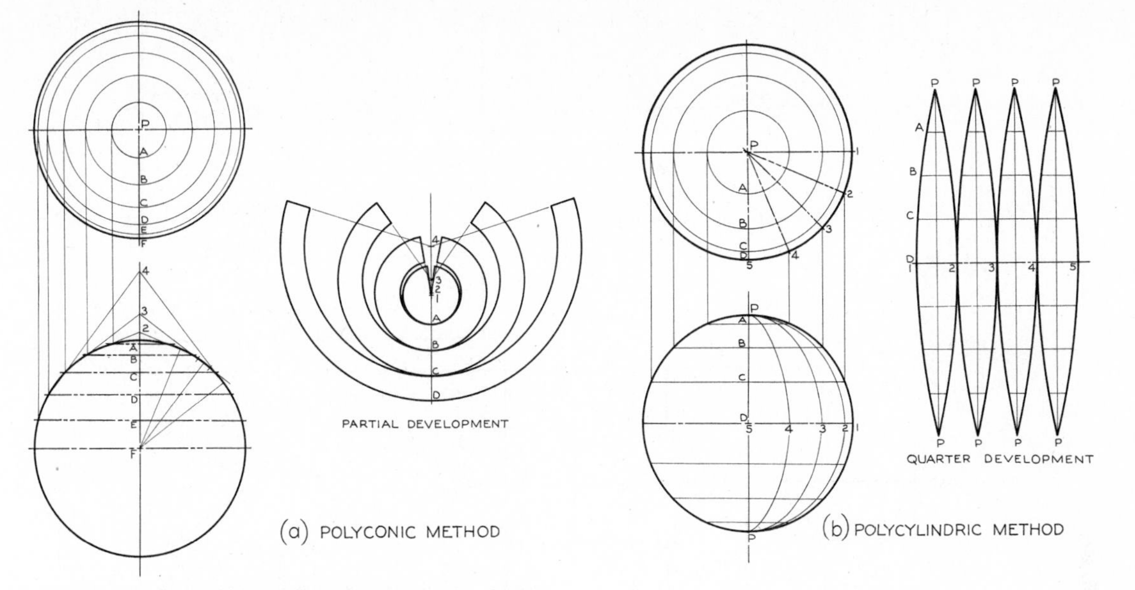


Fig. 19.19 **Approximate Development of a Sphere.**

This method of developing a spherical surface is the *polyconic* method. It is used on all government maps of the United States.

*Fig. 19.19 (b).* Another method of making an approximate development of the double curved surface of a sphere is to divide the surface into equal sections with meridian planes and substitute cylindrical surfaces for the spherical sections. The cylindrical surfaces may be inscribed within the sphere, or circumscribed about it, or located partly within and partly without. The development of the series of cylindrical surfaces is an approximate development of the spherical surface. This method is the *polycylindric* method, sometimes designated as the *gore* method.

## Intersections of Solids

### 19.22 Principles of Intersections.

Intersections of solids are generally regarded as in the province of descriptive geometry, and for information on the more complicated intersections the student is referred to any standard text on that subject. However, most of the intersections encountered in drafting practice do not require a knowledge of descriptive geometry, and some of the more common solutions may be found in the paragraphs that follow.

An intersection of two solids is referred to as a *figure of intersection.* Two plane surfaces intersect in a straight line; hence if two solids which are composed of plane surfaces intersect, the figure of intersection will be composed of straight lines, as shown in Figs. 19.20 to 19.23. The method generally consists in finding the piercing points of the edges of one solid in the surfaces of the other solid and joining these points with straight lines.

If curved surfaces intersect, or if curved surfaces and plane surfaces intersect, the figure of intersection will be composed of curves, as shown in Figs. 19.5, 19.9, and 19.24 to 19.29. The method generally consists in finding the piercing points of *elements* of one solid in the surfaces of the other. A smooth curve is then traced through these points, with the aid of the irregular curve, §2.58.

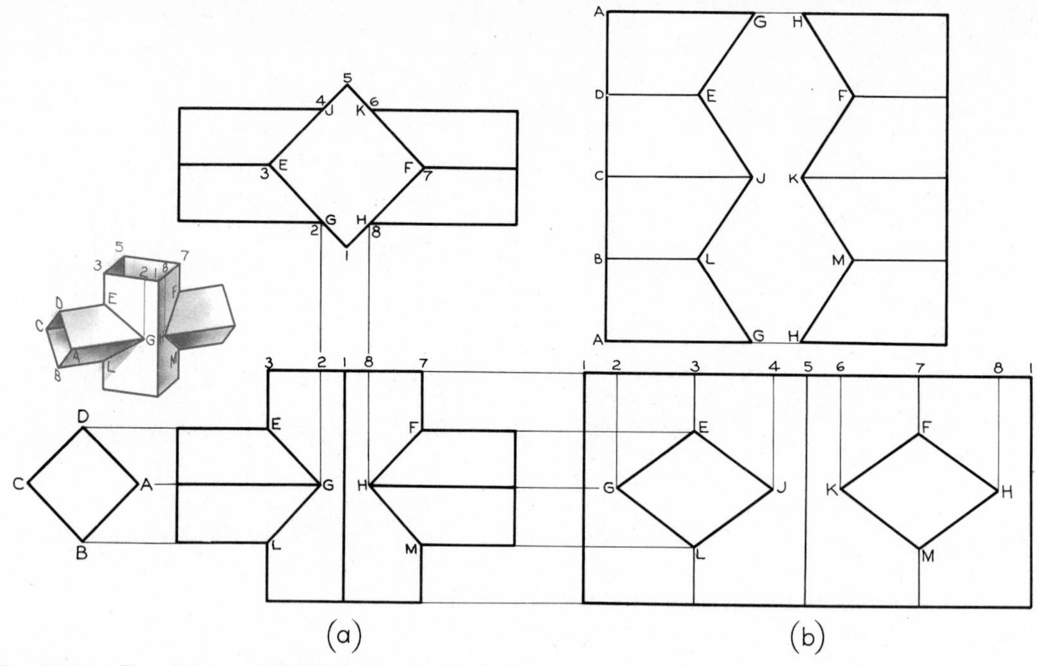


Fig. 19.20 **Two Prisms at Right Angles to Each Other.**

## 19.23 To Find the Intersection and Developments of Two Prisms. Fig. 19.20.

INTERSECTION. Fig. 19.20 (a). The points in which the edges A, B, C, and D of the horizontal prism pierce the vertical prism are vertices of the intersection. The edges D and B of the horizontal prism intersect the edges 3 and 7 of the vertical prism at the points E, F, L, and M. The edges A and C of the horizontal prism intersect the faces of the vertical prism at the points G, H, J, and K. The intersection is completed by joining these points in order by straight lines.

DEVELOPMENTS. Fig. 19.20 (b). To develop the lateral surface of the horizontal prism, set off on the vertical stretch-out line A–A the widths of the faces AB, BC, . . . , taken from the end view, and draw the edges through these points, as shown. Set off, from the stretch-out line, the lengths of the edges AG, BL, . . . , taken from the front view or from the top view, and join the points G, L, J, . . . , by straight lines.

To develop the lateral surface of the vertical prism, set off on the stretch-out line 1–1 the widths of the faces 1–2, 3–5, . . . , taken from the top view, and draw the edges through these points, as shown. Set off on the stretch-out line, the distances 1–2, 5–4, 5–6, and 1–8, taken from the top view, and draw the intermediate elements parallel to the principal edges. Take the lengths of the principal edges and of the intermediate elements from the front view, and join the points E, G, L, . . . , in order with straight lines, to complete the development.

## 19.24 To Find the Intersection and Developments of Two Prisms. Fig. 19.21.

INTERSECTION. Fig. 19.21 (a). The points in which the edges ACEH of the horizontal prism pierce the surfaces of the vertical prism are found in the top view and are projected downward to the corresponding edges ACEH in the front view. The points in which the edges 5 and 11 of the vertical prism pierce the surfaces of the horizontal prism are found in the left-side view at G, D, J, and B, and are projected horizontally to the front view, intersecting the corresponding edges as shown. The intersection is completed by joining these points in order by straight lines.

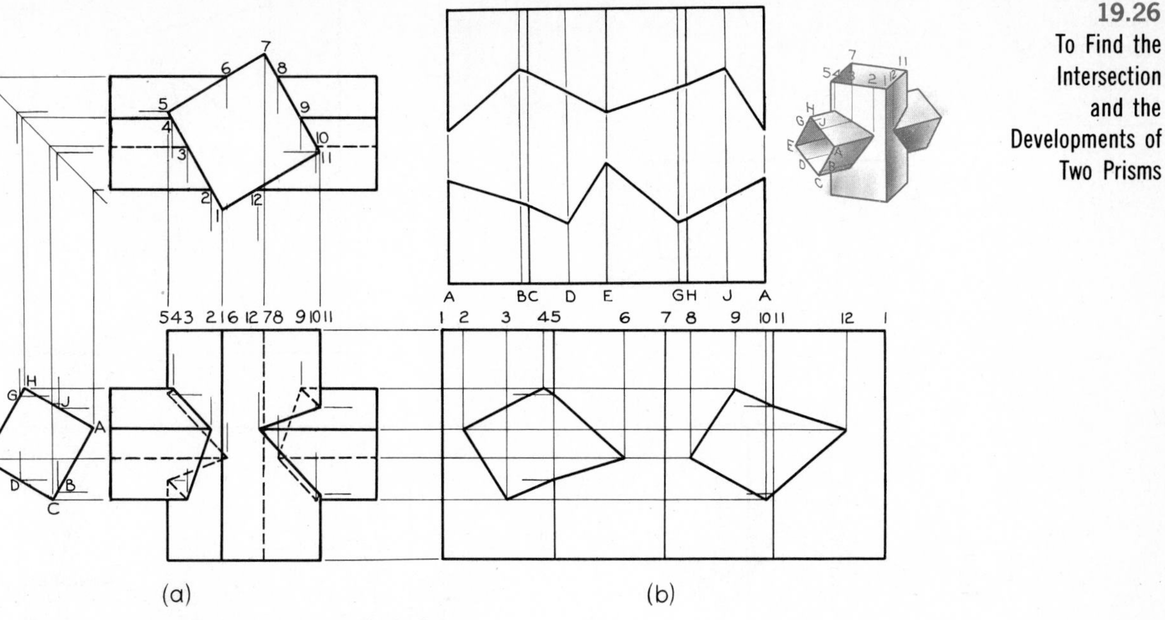


Fig. 19.21 **Two Prisms at Right Angles to Each Other.**

Developments. Fig. 19.21 (b). The lateral surfaces of the two prisms are developed as explained in §19.23. True lengths of all lateral edges and lines parallel to them are shown in the front view of Fig. 19.21 at (a).

## 19.25 To Find the Intersection and Developments of Two Prisms. Fig. 19.22.

Intersection. Fig. 19.22 (a). The points in which edges 1–2–3–4 of the inclined prism pierce the surfaces of the vertical prism are vertices of the intersection. These points, found in the top view, are projected downward to the corresponding edges 1–2–3–4 in the front view, as shown. The intersection is completed by joining these points in order by straight lines.

Developments. Fig. 19.22 (b). The lateral surfaces of the two prisms are developed as explained in §19.23. True lengths of all edges of both prisms are shown in the front view of Fig. 19.22 at (a).

## 19.26 To Find the Intersection and the Developments of Two Prisms. Fig. 19.23.

In this case the edges of the oblique prism are oblique to the planes of projection, and in the front and top views none of the edges is shown true length, §6.24, and none of the faces is shown true size, §6.23. Furthermore, none of the angles, including the angle of inclination, is shown true size, §6.26. Therefore, it is necessary to draw a secondary auxiliary view, §8.19, to obtain the true size and shape of the right section of the oblique prism.

The direction of sight, indicated by arrow A, is assumed perpendicular to the end face 1–2–3, that is, parallel to the principal edges of the prism. The primary auxiliary view, taken in the direction of arrow B, shows the true lengths of the edges, the true inclination of the prism with respect to the horizontal and, incidentally, the true length and inclination of arrow A. In the secondary auxiliary view, arrow A is shown as a point, and the end face 1–2–3 is shown in its true size.

Intersection. Fig. 19.23 (a). The points in which the edges 1–2–3 of the oblique prism pierce the surfaces of the vertical prism are vertices of the intersection, found first in the top view and then projected downward to the front view.

Developments. Fig. 19.23 (b). The lateral surfaces of the two prisms are developed as

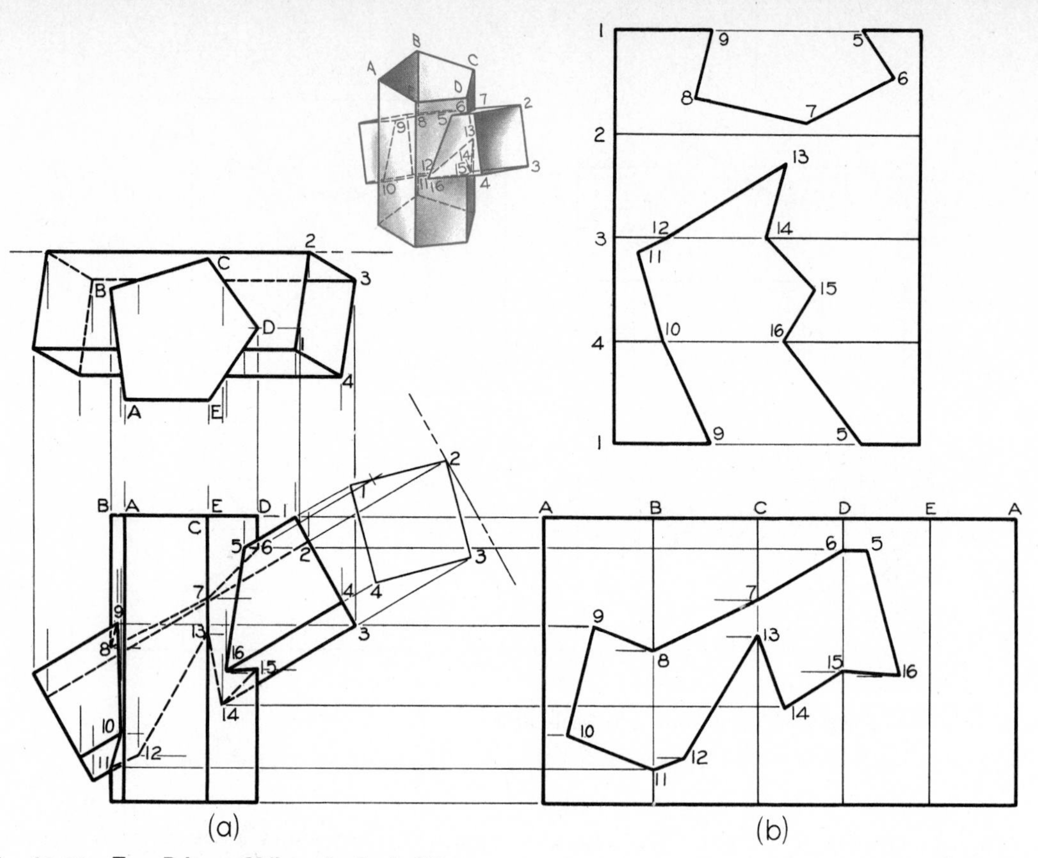


Fig. 19.22 **Two Prisms Oblique to Each Other.**

Fig. 19.23 **Two Prisms Oblique to Each Other.**

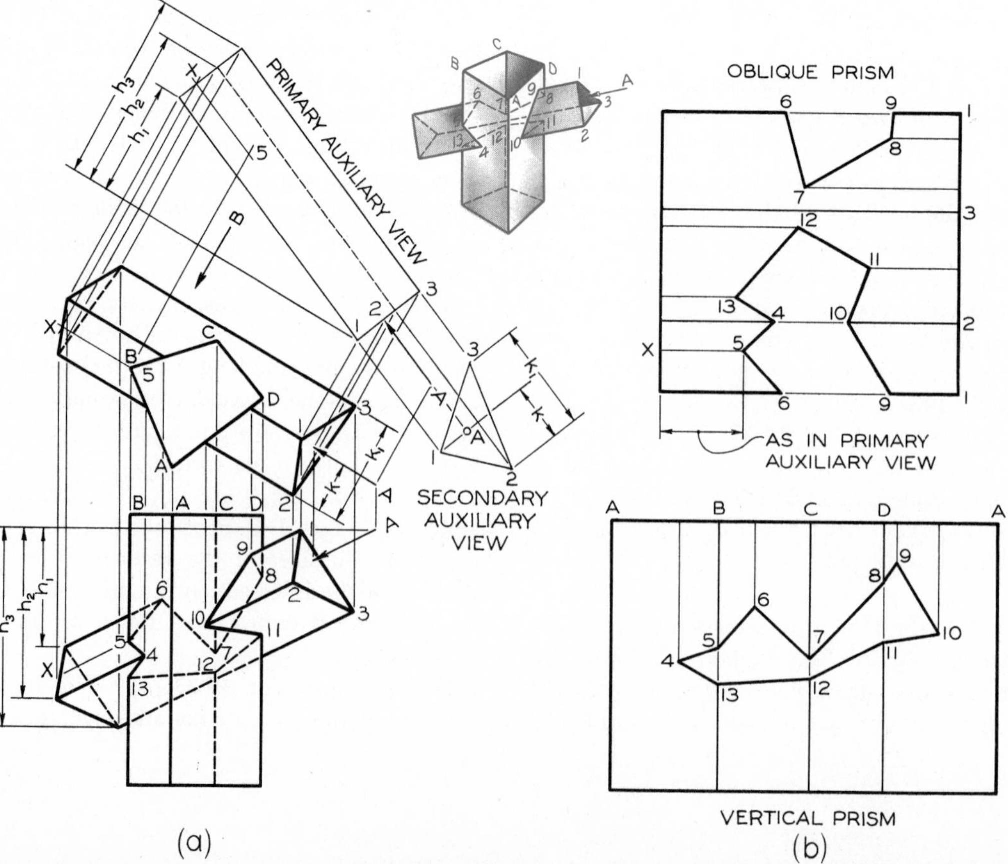

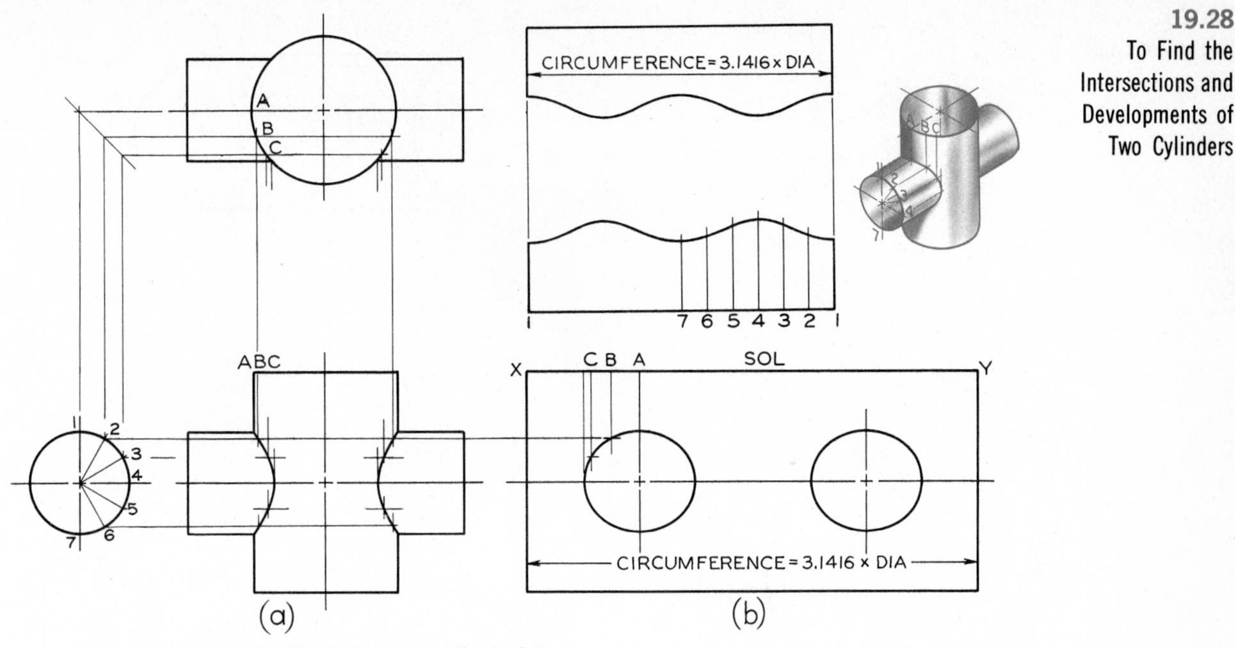


Fig. 19.24 **Two Cylinders at Right Angles to Each Other.**

explained in §19.23. True lengths of the edges of the vertical prism are shown in the front view. True lengths of the edges of the oblique prism can be shown in the primary auxiliary view; true lengths to the vertices of the intersection may be found in this view, as shown for line X–5.

## 19.27 To Find the Intersection and Developments of Two Cylinders. Fig. 19.24.

INTERSECTION. Fig. 19.24 (a). Assume a series of elements (preferably equally spaced) on the horizontal cylinder, numbered 1, 2, 3, . . . , in the side view, and draw their top and front views. Their points of intersection with the surface of the vertical cylinder are shown in the top view at A, B, C, . . . , and may be found in the front view by projecting downward to their intersections with the corresponding elements 1, 2, 3, . . . , in the front view. When a sufficient number of points have been found to determine the intersection, the curve is traced through the points with the aid of the irregular curve, §2.58. See also Fig. 6.38 (c).

DEVELOPMENTS. Fig. 19.24 (b). The lateral surfaces of the two cylinders are developed as explained in §19.7. True lengths of all elements of both cylinders are shown in the front view. Since both cylinders have bases at right angles to the center lines, the circles will develop as straight lines, and the developments will be rectangular, as shown. The length XY of the stretch-out line for the development of the vertical cylinder is equal to the circumference of the cylinder, or $\pi d$, and the length 1–1 of the stretch-out line for the development of the horizontal cylinder is determined in the same way. Those elements of the large cylinder which pierce the small cylinder can be identified in the top view as elements A, B, C, . . . . When these are drawn in the development, the points of intersections are found at their intersections with the corresponding elements of the horizontal cylinder taken from the front view, thus determining one of the figures of intersection, as shown in Fig. 19.24 (b).

## 19.28 To Find the Intersection and Developments of Two Cylinders. Fig. 19.25.

INTERSECTION. Fig. 19.25 (a). A revolved right section of the inclined cylinder is divided

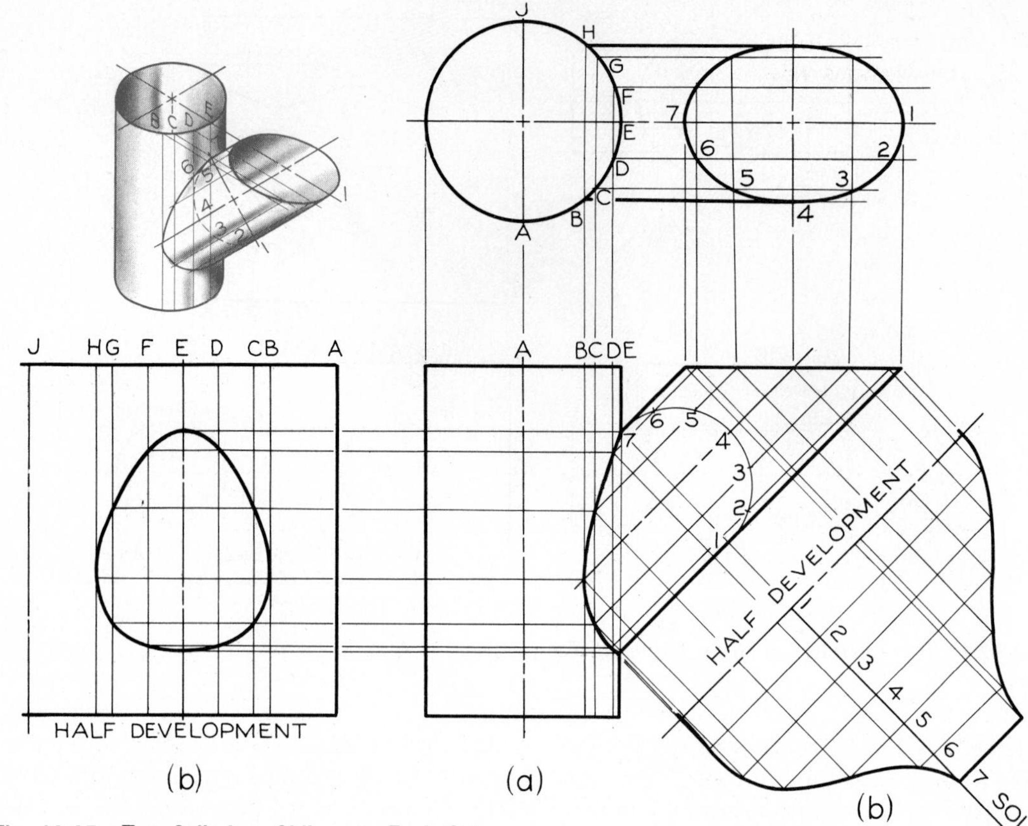


Fig. 19.25 **Two Cylinders Oblique to Each Other.**

into a number of equal parts, and an element is drawn at each of the division points, 1, 2, 3, . . . . The points of intersection of these elements with the surface of the vertical cylinder are shown in the top view at B, C, D, . . . , and are found in the front view by projecting downward to intersect the corresponding elements 1, 2, 3, . . . . The curve is traced through these points with the aid of the irregular curve, §2.58.

Developments. Fig. 19.25 (b). The lateral surfaces of the two cylinders are developed as explained in §§19.7 and 19.9. True lengths of all elements of both cylinders are shown in the front view.

## 19.29 To Find the Intersection and Developments of a Prism and a Cone. Fig. 19.26.

Intersection. Fig. 19.26 (a). Points in which the edges of the prism intersect the surface of the cone are shown in the side view at A, C, and F. Intermediate points such as B, D, E, and G are piercing points of any lines on the lateral surface of the prism parallel to the edges. Through all of the piercing points in the side view, elements of the cone are drawn, and then drawn in the top and front views. The intersections of the elements of the cone with the edges of the prism (and lines along the prism drawn parallel thereto) are points of the intersections. The figures of intersections are traced through these points with the aid of the irregular curve, §2.58.

The elements 6, 5, 4, . . . , in the side view of the cone may be regarded as the edge views of cutting planes which cut these elements on the cone and edges or elements on the prism. The intersection of corresponding edges or elements on the two solids are points on the figure of intersection.

Another method of finding the figure of intersection is to pass a series of horizontal parallel planes through the solids in the manner of Fig. 19.9 (a). The plane will cut circles on the cone and straight lines on the prism, and their intersections will be points on the figure of intersection. See also Fig. 19.27 (b).

Developments. Fig. 19.26 (b). The lateral surface of the prism is developed as explained in §19.23. True lengths of all edges and lines parallel thereto are shown in both the front and top views.

The lateral surface of the cone, Fig. 19.26 (c), is developed as explained in §19.11. True lengths of elements from the vertex to points on the intersections are found as shown in Fig. 9.10 (a).

## 19.30 To Find the Intersection of a Prism and a Cone with Edges of Prism Parallel to Axis of Cone. Fig. 19.27.

*Fig. 19.27 (a).* Since the lateral surfaces of the prism are parallel to the axis of the cone, the figure of intersection will be composed of a series of hyperbolas, §§4.48 and 4.61. If a series of planes is assumed containing the axis of the cone, each plane will contain edges of the prism or will cut lines parallel to them along the prism, and will cut elements on the cone that intersect these at points on the figure of intersection.

*Fig. 19.27 (b).* The intersection is the same as at (a), but found in a different manner. Here a series of parallel planes perpendicular to the axis of the cone cut circles of varying diameters on the cone. These circles are shown true size in the top view, where the piercing points of these circles in the vertical plane surfaces of the prism are also shown. The front views of these piercing points are found by projecting downward to the corresponding cutting-plane lines.

*Fig. 19.27 (c).* The chamfer of an ordinary hexagon bolt head or hexagon nut is actually a conical surface that intersects the six vertical sides of a hexagonal prism to form hyperbolas. At (c) the methods of both (a) and (b) are shown to illustrate how points may be found by either method.

In machine drawings of bolts and nuts, these hyperbolic curves are approximated by means of circular arcs, as shown in Fig. 13.28.

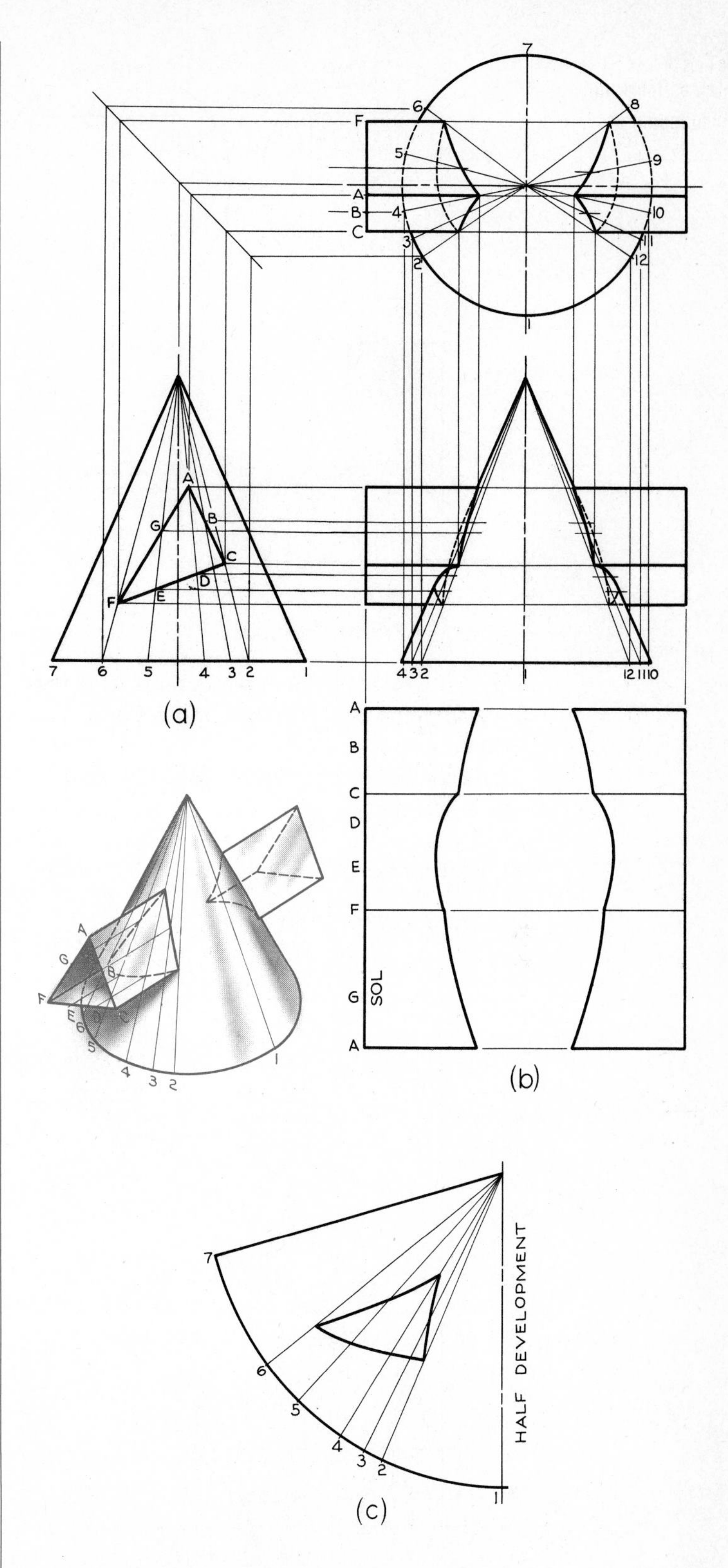


Fig. 19.26 **Prism and Cone.**

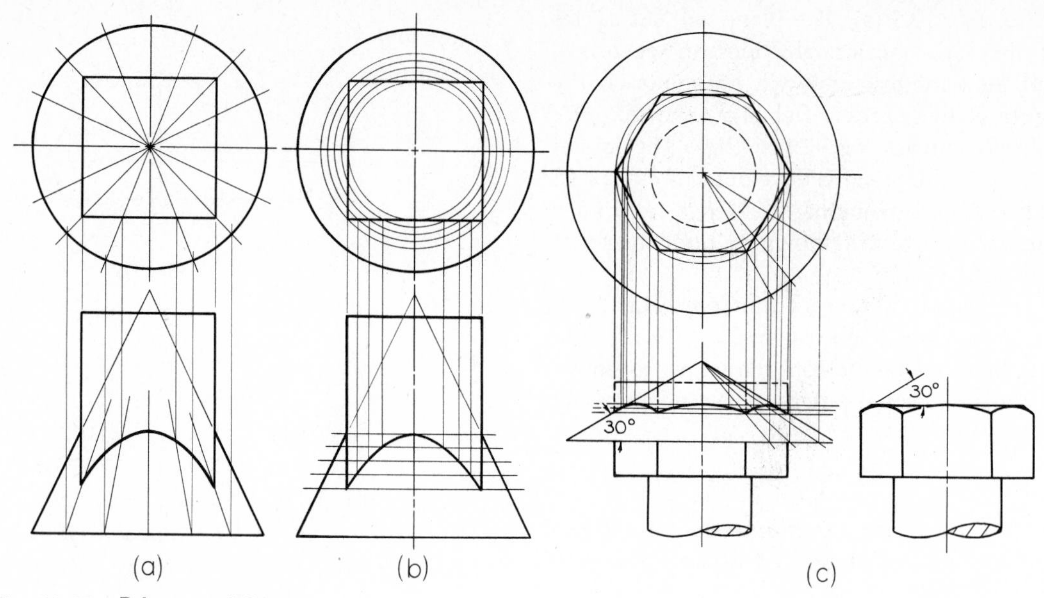


Fig. 19.27 **Prisms and Cones.**

## 19.31 To Find the Intersections and the Developments of a Cylinder and a Cone. Fig. 19.28.

INTERSECTIONS. Fig. 19.28 (a). Points in which elements of the cylinder (preferably equally spaced to facilitate the development) intersect the surface of the cone are shown in the side view at A, B, C, . . . . The elements of the cylinder are here shown as points. Elements of the cone are then drawn from the vertex through each of these points, and then drawn in their correct locations in the top and front views. The intersections of these elements with the elements A, B, C, . . . , of the

Fig. 19.28 **Cone and Cylinder.**

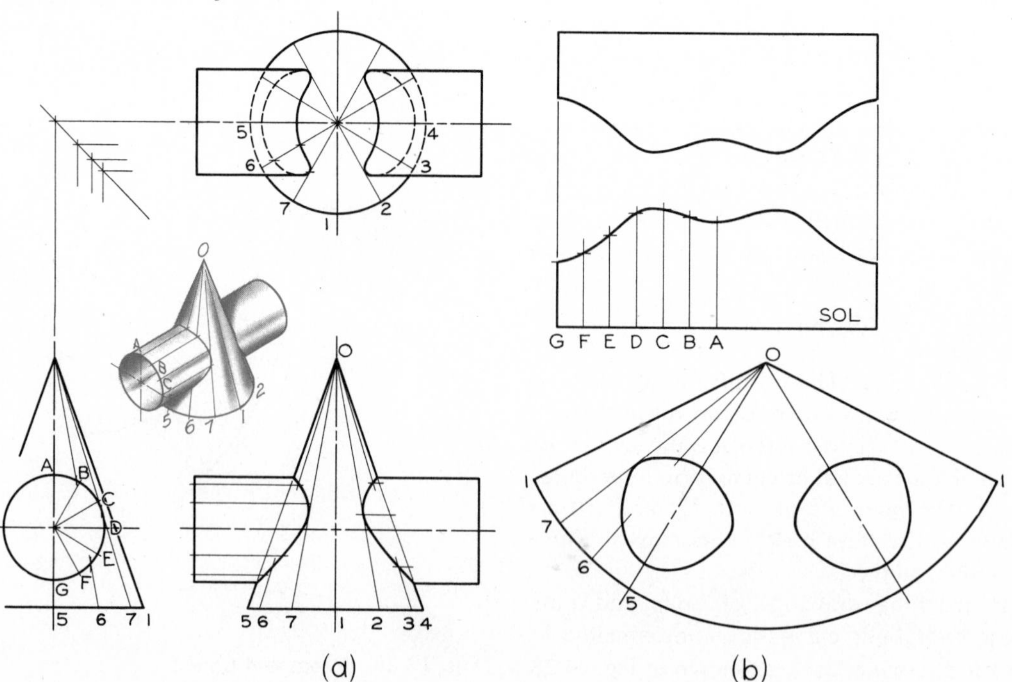

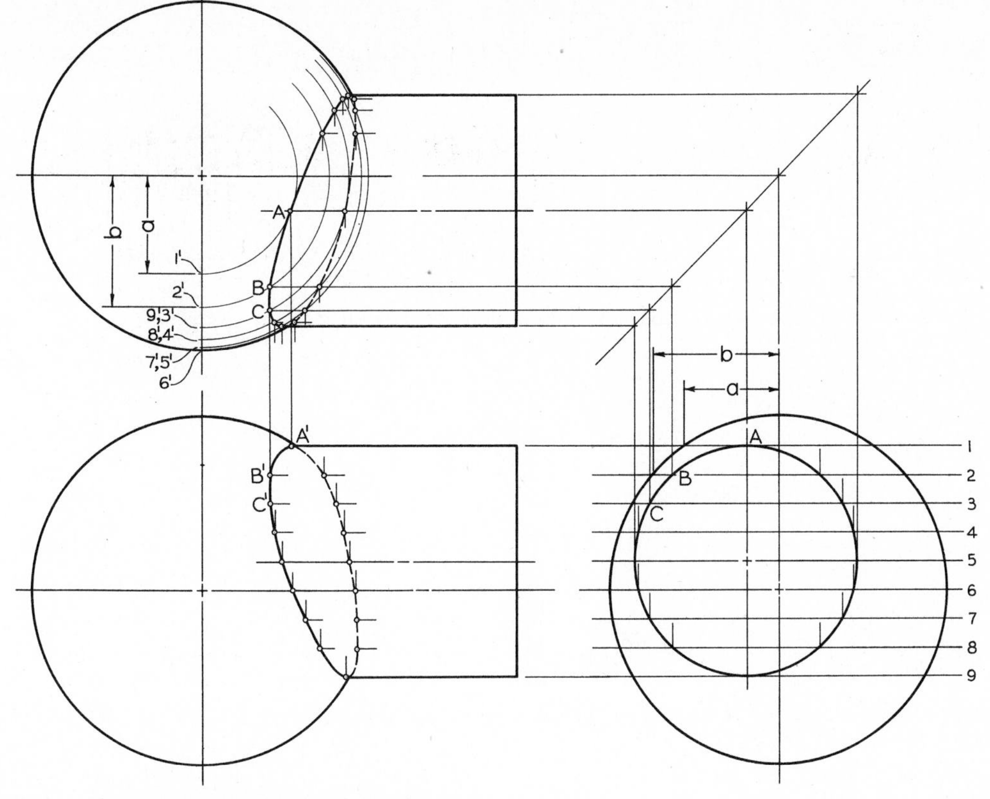


Fig. 19.29 **Intersection of Sphere and Cylinder.**

cylinder are points on the figures of intersection. The curves are then traced through these points with the aid of the irregular curve, §2.58.

As explained in §19.29, the elements 5, 6, 7, . . . , in the side view of Fig. 19.28 (a) could be regarded as edge views of cutting planes that cut elements from both the cone and the cylinder, the elements meeting at points on the figure of intersection. Or a series of horizontal parallel planes can be passed through the solids that will cut circles from the cone and elements from the cylinder that intersect at points on the figure of intersection.

Developments. Fig. 19.28 (b). The lateral surface of the cylinder is developed as explained in §19.27. True lengths of all elements are shown in both the front and top views. The lateral surface of the cone is developed as explained in §19.11. True lengths of elements from the vertex to points on the intersections are found as shown in Fig. 9.10 (a).

## 19.32 To Find the Intersection of a Cylinder and a Sphere.

Fig. 19.29. Horizontal planes 1, 2, 3, . . . , which appear edgewise in the front and side views, cut elements A, B, C, . . . , from the cylinder and circular arcs 1′, 2′, 3′, . . . , from the sphere. The intersections of the elements with the arcs produced by the corresponding planes are points on the figure of intersection. Join the points with a smooth curve, §2.58.

## 19.33 Intersection and Development Problems.

A wide selection of intersection and development problems is provided in Figs. 19.30 to 19.38. These problems are designed to fit on 11″ × 17″ sheets. Dimensions should be included on the given views. The student is cautioned to take special pains to obtain accuracy in these drawings, and to draw smooth curves as required.

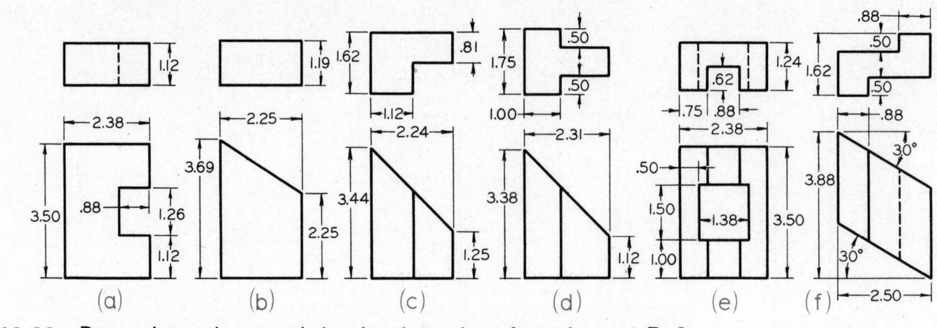


Fig. 19.30 Draw given views and develop lateral surface. Layout B-3.

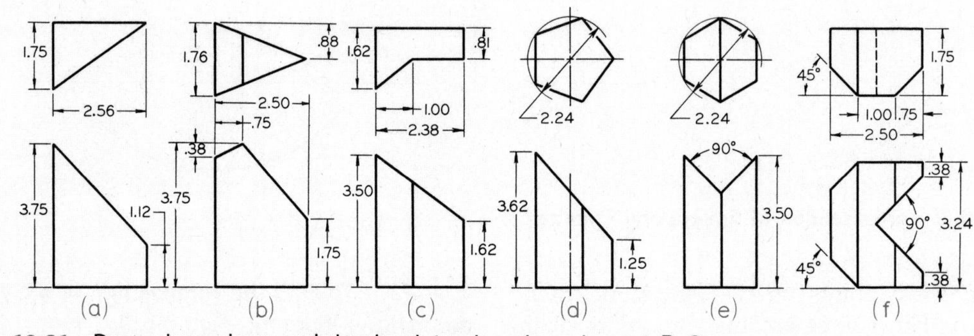


Fig. 19.31 Draw given views and develop lateral surface. Layout B-3.

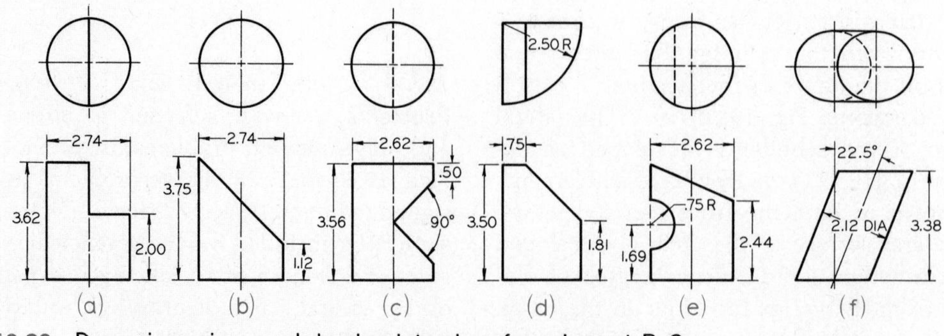


Fig. 19.32 Draw given views and develop lateral surface. Layout B-3.

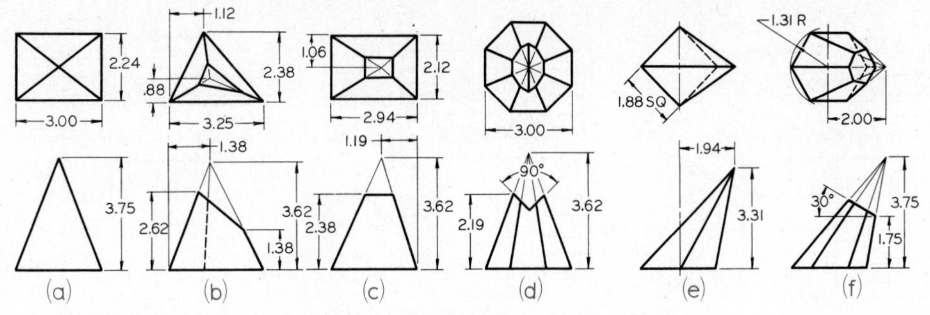


Fig. 19.33 Draw given views and develop lateral surface. Layout B-3.

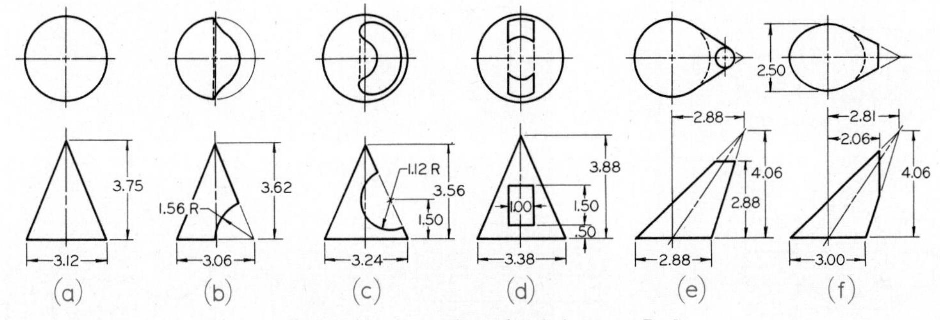


Fig. 19.34 Draw given views and develop lateral surface. Layout B-3.

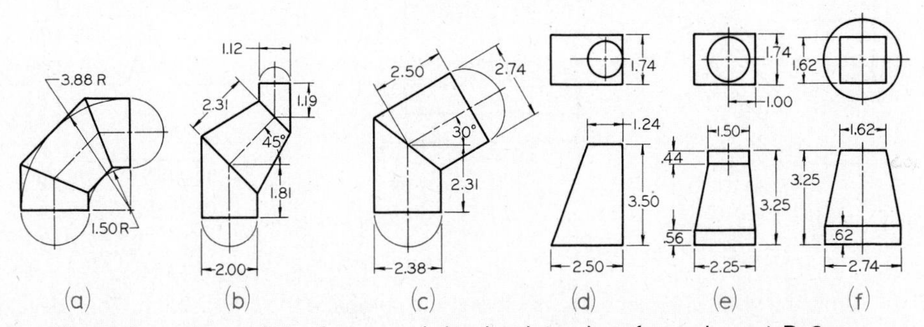


Fig. 19.35 Draw given views of the forms and develop lateral surfaces. Layout B-3.

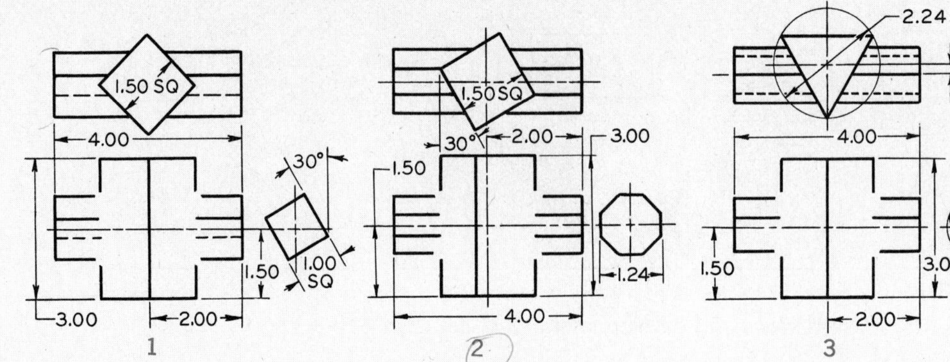


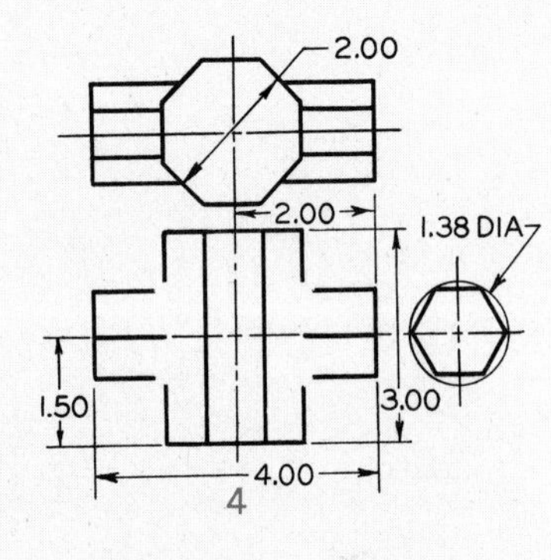


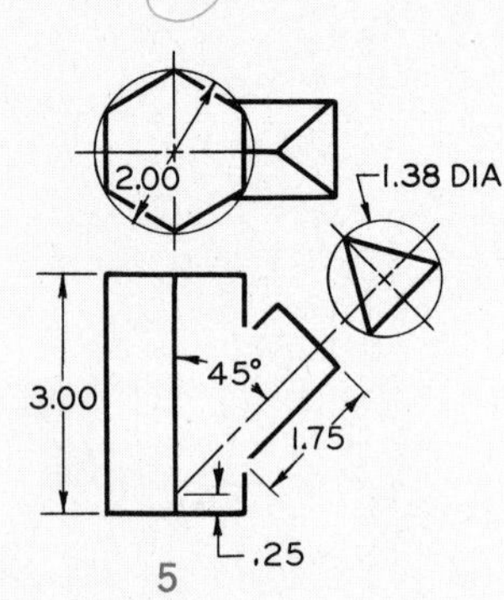


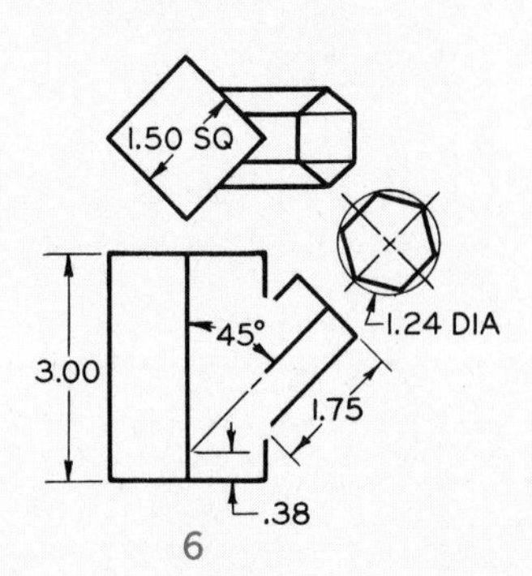


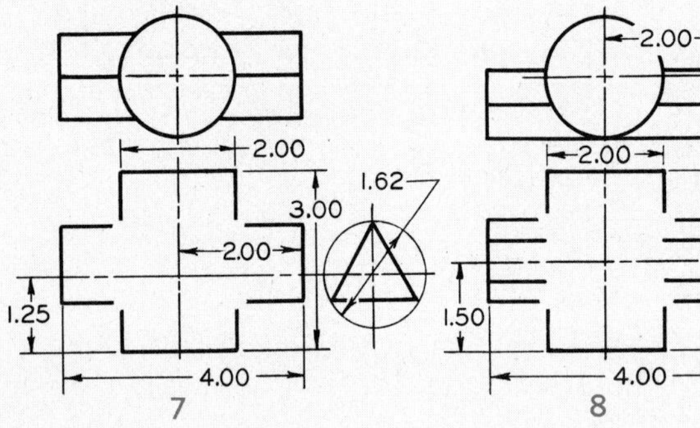


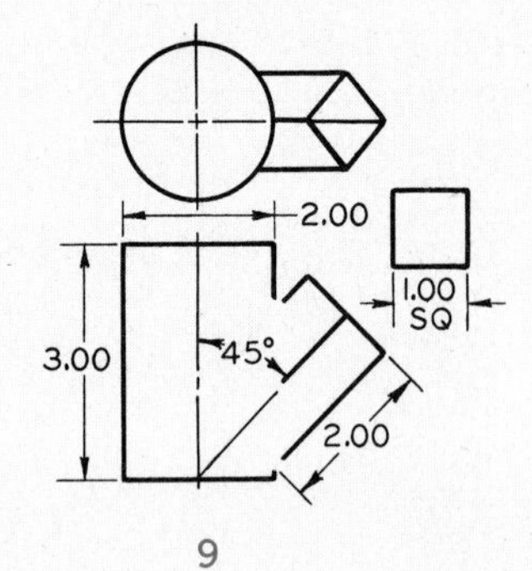


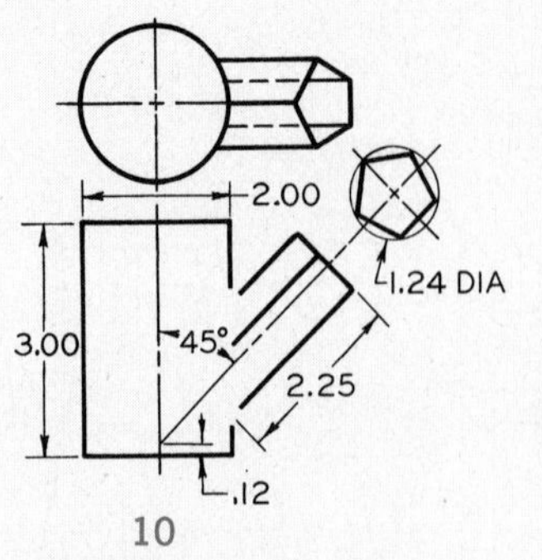


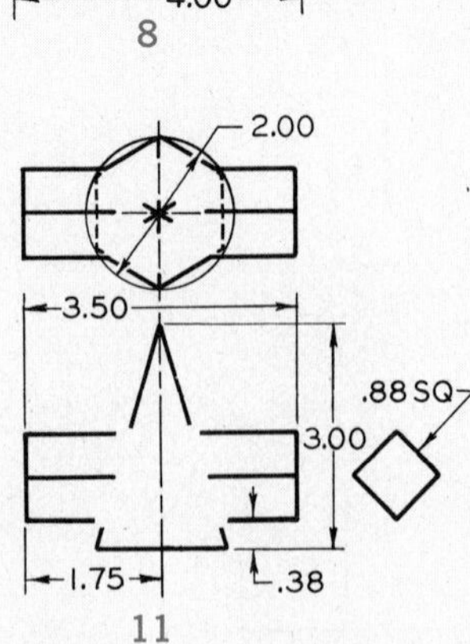


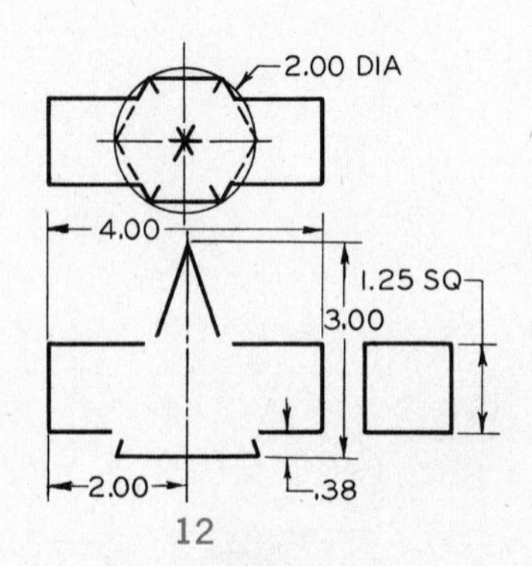


Fig. 19.36 Draw the given views of assigned form, and complete the intersection. Then develop lateral surfaces. Layout B-3.

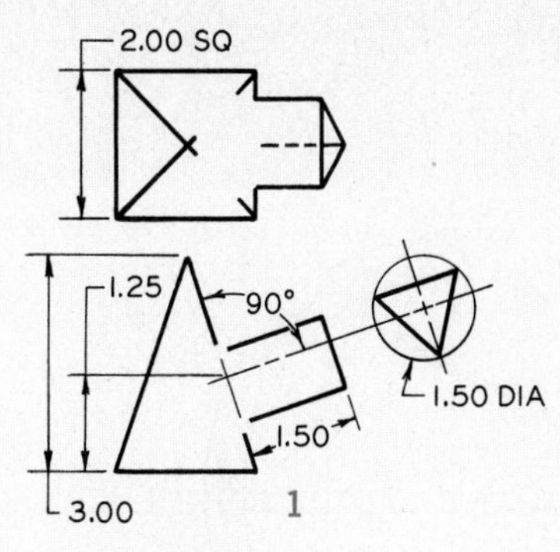


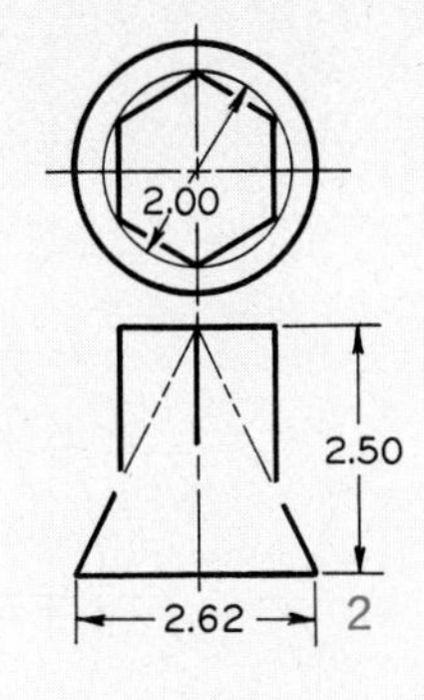


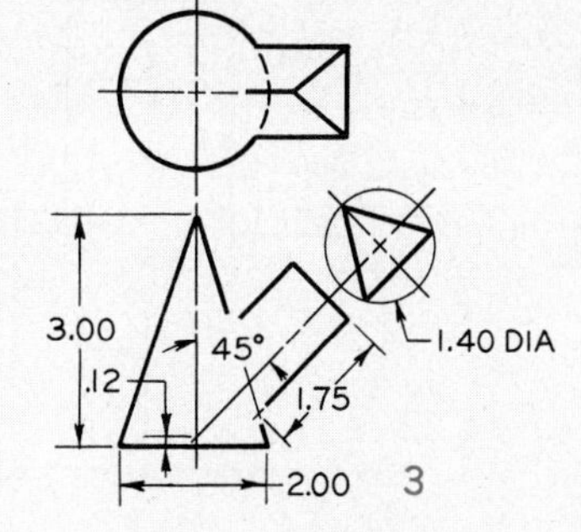


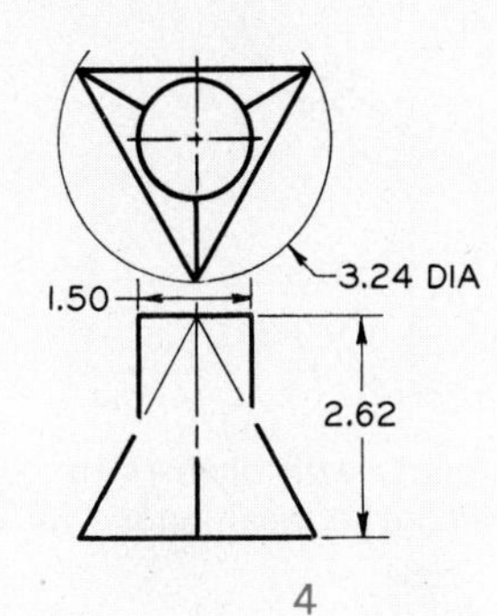


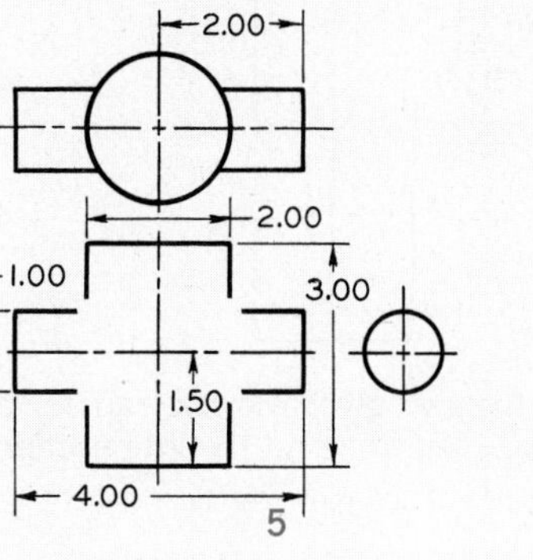


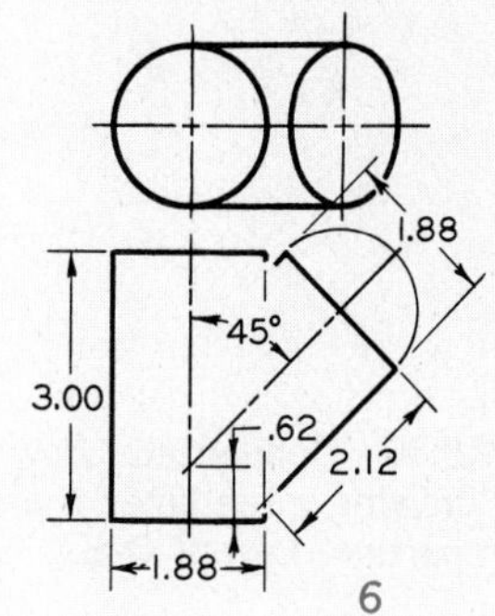


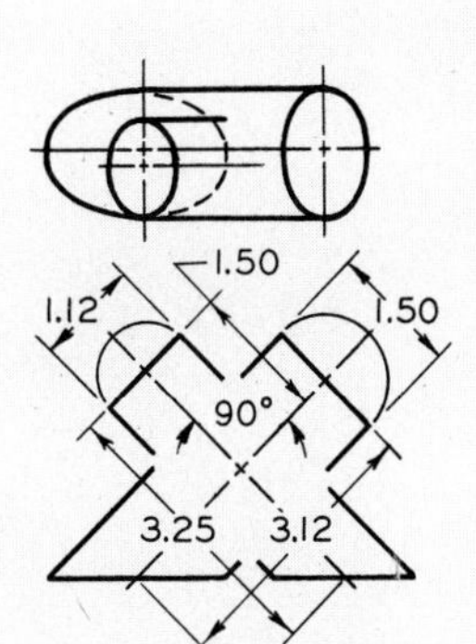


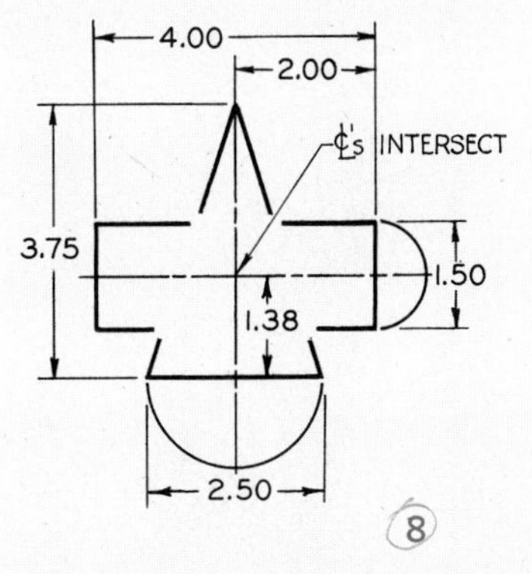


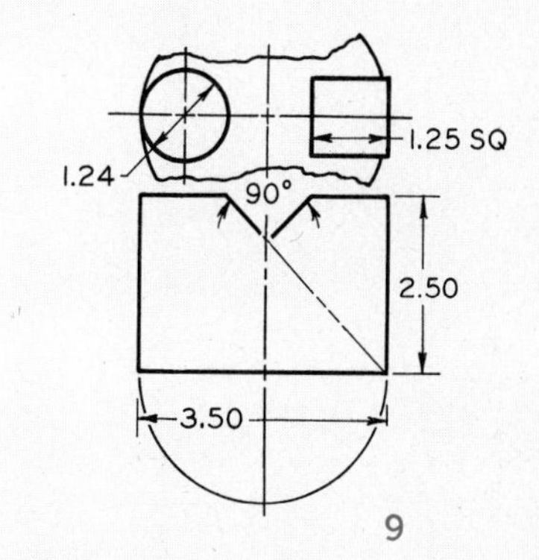


1.00
2.50
2.50
10

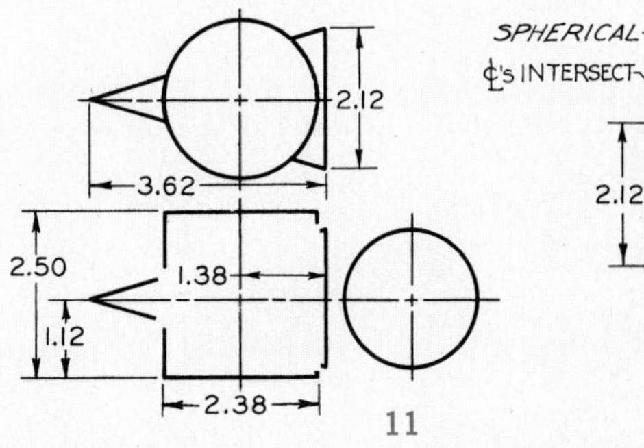


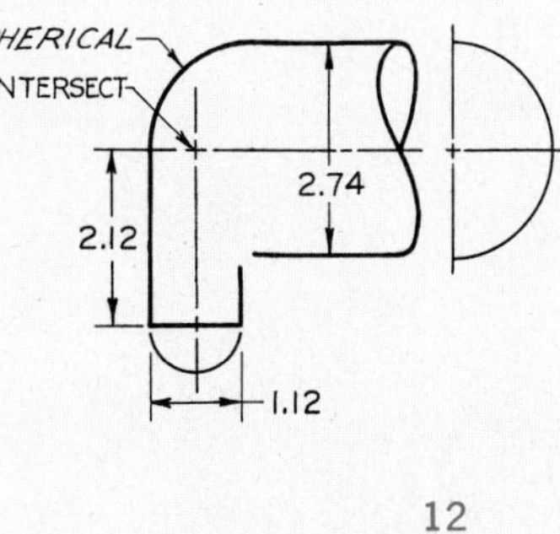


Fig. 19.37 Draw the given views of assigned form, and complete the intersection. Then develop lateral surfaces. Layout B–3.

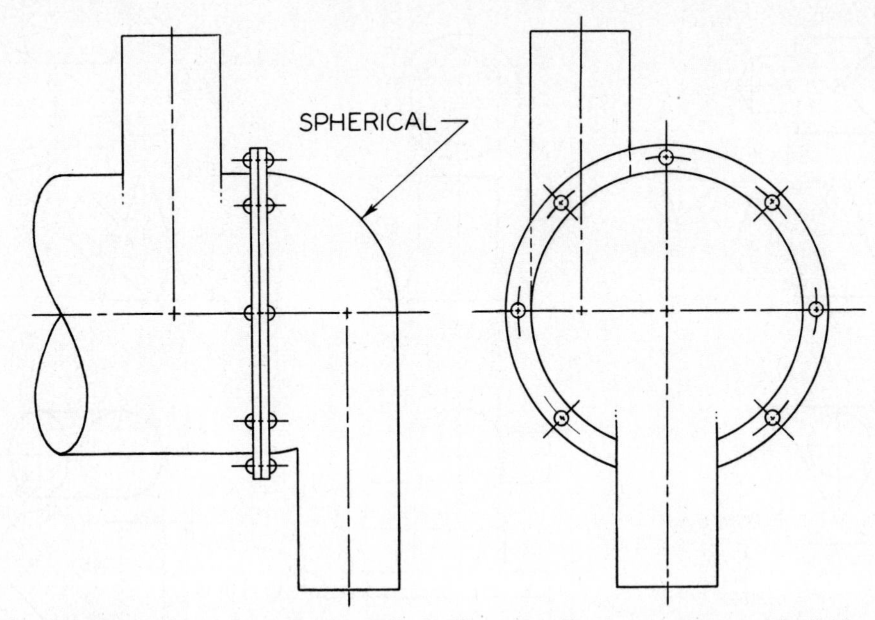


Fig. 19.38 **Condenser.** Draw the two given views. Transfer all measurements with dividers, making your drawing three times the size shown. Find the intersections of the small cylindrical pipes with the main portion. Layout C–3.

# 20

# gearing and cams

by B. Leighton Wellman*

**20.1 Gears.** *Gears,* Fig. 20.1, are used to transmit motion, rotating or reciprocating, from one machine part to another. They may be classified according to the position of the shafts that they connect. Parallel shafts, for example, may be connected by *spur gears, helical gears,* or *herringbone gears.* Intersecting shafts may be connected by *bevel gears* having either straight, skew, or spiral teeth. Nonparallel, nonintersecting shafts may be connected by *crossed helical gears, hypoid gears,* or a *worm* and *worm gear.* A spur gear meshed with a *rack* will convert rotary motion to reciprocating motion.

**20.2 Spur Gears.** The *friction wheels* shown in Fig. 20.2 (a) will transmit motion and power from one shaft to another parallel shaft. However, friction gears are subject to slipping, and excessive pressure is required between the wheels to obtain the necessary frictional force. If teeth of the proper shape are provided on the cylindrical surfaces, the resulting spur gears, Fig. 20.2 (b), will transmit the same motion and power without slipping and with greatly reduced bearing pressures.

If a friction wheel of diameter D turns at n rpm, the linear velocity of a point on its periphery will be $\pi$Dn. But the pitch circles of a pair of mating spur gears correspond exactly to the outside diameters of the friction wheels, and since the gears turn in contact

*Professor Emeritus of Mechanical Engineering, Worcester Polytechnic Institute.

Fig. 20.1 **An Assortment of Gears.** *Courtesy Charles Bond Co.*

without slipping, they must have the same linear velocity at the pitch line. Therefore,

$$\pi D_G n_G = \pi D_P n_P$$

$$\text{or} \quad \frac{D_G}{D_P} = \frac{n_P}{n_G} = m_G$$

where $D_G$ and $D_P$ are the pitch diameters of the larger gear (called the *gear*) and the smaller gear (called the *pinion*); $n_G$ and $n_P$ are the rpm of the gears; $m_G$ is the gear ratio, expressed as the ratio of larger gear to smaller.

The teeth on mating gears must be of equal width and spacing; hence the number of teeth on each gear, N, is directly proportional to its pitch diameter, or

$$\frac{N_G}{N_P} = \frac{D_G}{D_P} = \frac{n_P}{n_G} = m_G$$

## 20.3 Spur Gear Definitions and Formulas.

Proportions and shapes of gear teeth are well standardized, and the terms defined below and in Fig. 20.3 are common to all spur gears. The dimensions relating to tooth height are for full-depth $14\frac{1}{2}°$ or $20°$ involute teeth.

*Pitch circle.*—An imaginary circle that corresponds to the circumference of the friction gear from which the spur gear is derived.

*Pitch diameter* ($D_G$ or $D_P$).—The diameter of the pitch circle of gear or pinion.

*Number of teeth* ($N_G$ or $N_P$).—The number of teeth on the gear or pinion.

*Diametral pitch* (P).—A ratio equal to the number of teeth on the gear per inch of pitch diameter. $P = N/D$.

*Circular pitch* (p).—The distance measured along the pitch circle from a point on one tooth to the corresponding point on the adjacent tooth. It thus includes one tooth and one space. $p = \pi D/N$. It is useful to note that $p \times P = \pi$.

*Addendum* (a).—The radial distance from the pitch circle to the top of the tooth. $a = 1/P$.

*Dedendum* (b).—The radial distance from the pitch circle to the bottom of the tooth space. $b = 1.157/P$.

*Outside diameter* ($D_O$).—The diameter of the addendum circle. It is equal to the pitch diameter plus twice the addendum. $D_O = D + 2a = (N + 2)/P$.

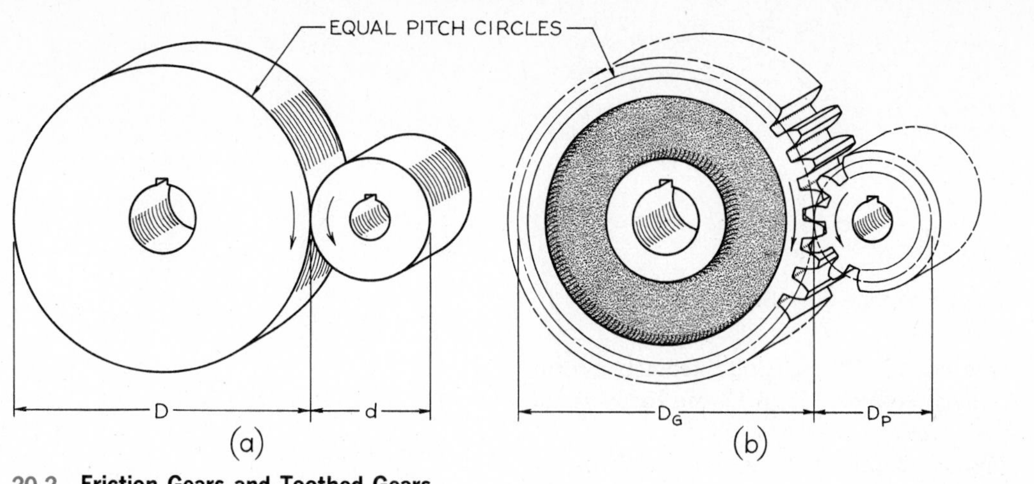

Fig. 20.2 **Friction Gears and Toothed Gears.**

*Root diameter* ($D_R$).—The diameter of the root circle. It is equal to the pitch diameter minus twice the dedendum. $D_R = D - 2b = (N - 2.314)/P$.

*Whole depth* ($h_t$).—The total height of the tooth. It is equal to the addendum plus the dedendum. $h_t = a + b = 2.157/P$.

*Working depth* ($h_k$).—The distance that a tooth projects into the mating space. It is equal to twice the addendum. $h_k = 2a = 2/P$.

*Clearance* (c).—The distance between the top of a tooth and bottom of the mating space. It is equal to the dedendum minus the addendum. $c = b - a = 0.157/P$.

*Circular thickness* (t).—The thickness of a tooth measured along the pitch circle. It is equal to one-half the circular pitch. $t = p/2 = \pi/2P$.

*Chordal thickness* ($t_c$).—The thickness of a tooth measured along a chord of the pitch circle. $t_c = D \sin(90°/N)$.

*Chordal addendum* ($a_c$).—The radial distance from the top of a tooth to the chord of the pitch circle. $a_c = a + \frac{1}{2}D\,[1 - \cos(90°/N)]$.

*Pressure angle* ($\phi$).—the angle that determines the direction of pressure between contacting teeth, and that designates the shape of involute teeth—for example, $14\frac{1}{2}°$ involute. It also determines the size of the base circle.

*Base Circle.*—The circle from which the involute profile is generated.

## 20.4 The Shape of the Tooth.

If gears are to operate smoothly with a minimum of noise and vibration, the curved surface of the tooth

Fig. 20.3 **Gear Tooth Nomenclature.**

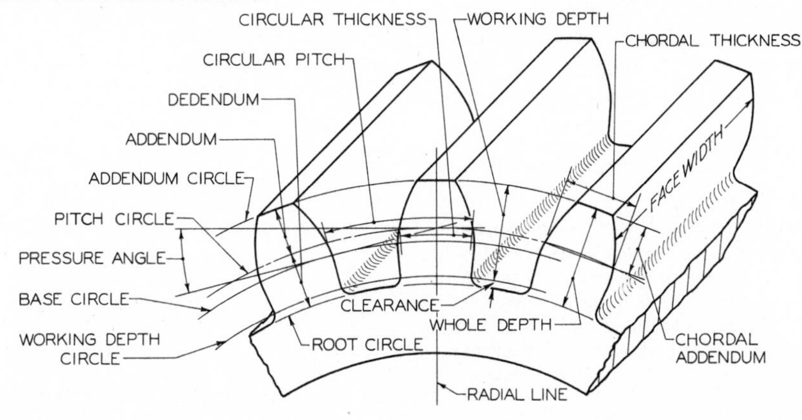

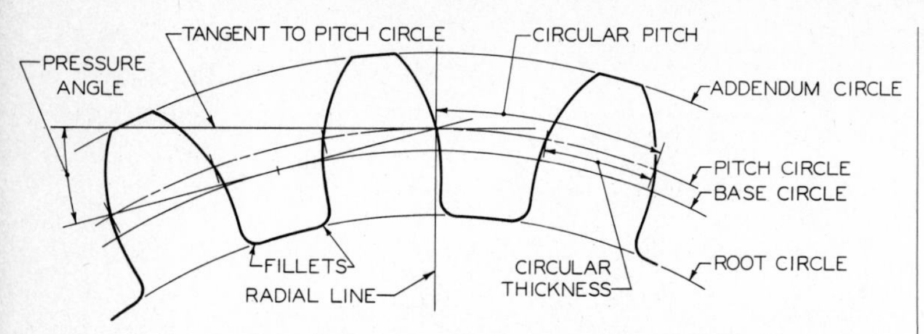


**Fig. 20.4 Representation of Involute Spur Gear Teeth.**

profile must be of a definite geometric form. The most common form in use today is the *involute profile.*

In the involute system, the shape of the tooth depends basically upon the *pressure angle,* which is ordinarily $14\frac{1}{2}°$ or 20°. This pressure angle determines the size of the *base circle;* from this the involute curve is generated in the following manner. At any point on the pitch circle, Fig. 20.4, a line is drawn tangent to it; a second line is drawn through the point of tangency at the required pressure angle ($14\frac{1}{2}°$ is frequently approximated at 15° on the drawing); the base circle is then drawn tangent to the pressure-angle line.

If the exact shape of the tooth is desired, the portion of the profile from the base circle to the addendum circle can be drawn as the involute of the base circle. The method of construction is shown in Fig. 4.64 (d) and (e). That part of the profile which is below the base circle is drawn as a radial line that terminates in the fillet at the root circle. The fillet should be equal in radius to one and one-half times the clearance.

For display drawings and more rapid construction, the involute curve can be closely approximated with two circular arcs, as shown in Fig. 20.5. This method, originally devised by Grant,* employs a table of arc radii, an *Involute Odontograph,* for gears having various numbers of teeth. The base circle is drawn as

*G. B. Grant, "Teeth of Gears," 1891. Grant's tabulated radii, which have been used for many years, were based on a pressure angle of 15° and were empirically shortened to correct for interference at the tooth tips. The new radii given in Fig. 20.5 have been computed for both $14\frac{1}{2}°$ and 20° to more closely approximate the true involute profile. To distinguish clearly the two tables, the new values are called *Wellman's Involute Odontograph.*

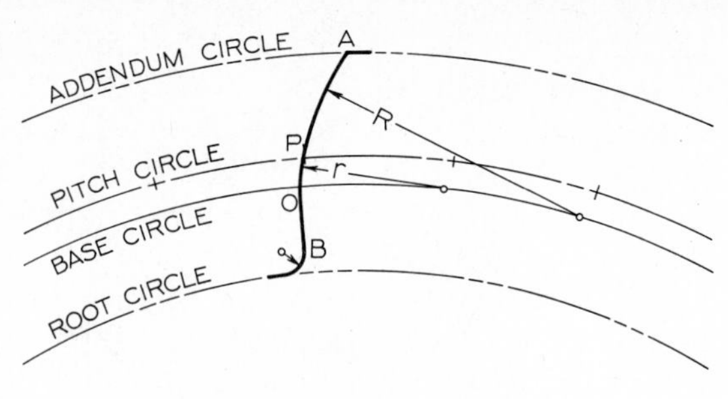


Divide radii given in table by Diametral Pitch.

| No. of Teeth (N) | $14\frac{1}{2}°$ | | 20° | |
|---|---|---|---|---|
| | R | r | R | r |
| 12 | 2.87 | 0.79 | 3.21 | 1.31 |
| 13 | 3.02 | 0.88 | 3.40 | 1.45 |
| 14 | 3.17 | 0.97 | 3.58 | 1.60 |
| 15 | 3.31 | 1.06 | 3.76 | 1.75 |
| 16 | 3.46 | 1.16 | 3.94 | 1.90 |
| 17 | 3.60 | 1.26 | 4.12 | 2.05 |
| 18 | 3.74 | 1.36 | 4.30 | 2.20 |
| 19 | 3.88 | 1.46 | 4.48 | 2.35 |
| 20 | 4.02 | 1.56 | 4.66 | 2.51 |
| 21 | 4.16 | 1.66 | 4.84 | 2.66 |
| 22 | 4.29 | 1.77 | 5.02 | 2.82 |
| 23 | 4.43 | 1.87 | 5.20 | 2.98 |
| 24 | 4.57 | 1.98 | 5.37 | 3.14 |
| 25 | 4.70 | 2.08 | 5.55 | 3.29 |
| 26 | 4.84 | 2.19 | 5.73 | 3.45 |
| 27 | 4.97 | 2.30 | 5.90 | 3.61 |
| 28 | 5.11 | 2.41 | 6.08 | 3.77 |
| 29 | 5.24 | 2.52 | 6.25 | 3.93 |
| 30 | 5.37 | 2.63 | 6.43 | 4.10 |
| 31 | 5.51 | 2.74 | 6.60 | 4.26 |
| 32 | 5.64 | 2.85 | 6.78 | 4.42 |
| 33 | 5.77 | 2.96 | 6.95 | 4.58 |
| 34 | 5.90 | 3.07 | 7.13 | 4.74 |
| 35 | 6.03 | 3.18 | 7.30 | 4.91 |
| 36 | 6.17 | 3.29 | 7.47 | 5.07 |
| 37–39 | 6.36 | 3.46 | 7.82 | 5.32 |
| 40–44 | 6.82 | 3.86 | 8.52 | 5.90 |
| 45–50 | 7.50 | 4.46 | 9.48 | 6.76 |
| 51–60 | 8.40 | 5.28 | 10.84 | 7.92 |
| 61–72 | 9.76 | 6.54 | 12.76 | 9.68 |
| 73–90 | 11.42 | 8.14 | 15.32 | 11.96 |
| 91–120 | 0.118N | | 0.156N | |
| 121–180 | 0.122N | | 0.165N | |
| Over 180 | 0.125N | | 0.171N | |

**Fig. 20.5 Wellman's Involute Odontograph.**

described above, and the spacing of the teeth is set off along the pitch circle. Then the face of the tooth from P to A is drawn with the *face radius* R, and the portion of the flank from P to O is drawn with the *flank radius* r. Both arcs are drawn from centers located on the base circle. The table gives the correct face and flank radii for gears of *one* diametral pitch. For other pitches, the values in the table must be divided by the diametral pitch. For gears with more than 90 teeth, use a single radius (let $R = r$) computed from the formula, and then divide by diametral pitch. Below the base circle, the flank of the tooth is completed with a radial line OB and a fillet.

When the pressure angle is 20° and the height of the tooth is reduced, the teeth are called *stub* teeth. Stub teeth are drawn in the same manner except that $a = 0.8/P$, $b = 1/P$, and the pressure angle is made 20°. The major advantage of stub teeth is that they are stronger than the $14\frac{1}{2}°$ standard full-depth teeth.

When the gear teeth are formed on a flat surface, the result is a *rack*, Fig. 20.6. In the involute system, the sides of rack teeth are straight, and are inclined at an angle equal to the pressure angle. To mesh with a gear, it is obvious that the linear pitch of the rack must be the same as the circular pitch of the gear, and the rack teeth must have the same height proportions as the gear teeth.

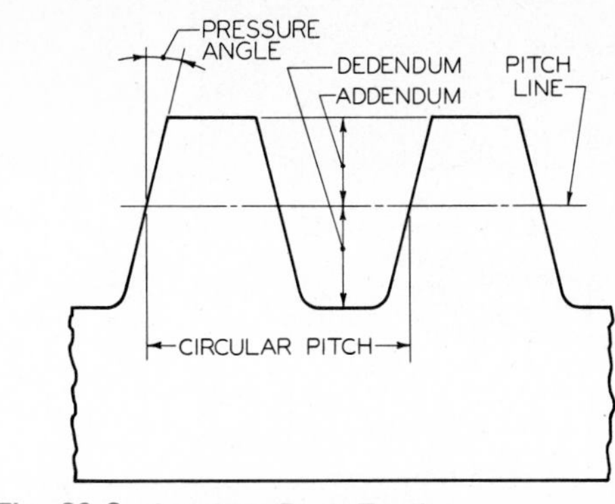


Fig. 20.6 **Involute Rack Teeth.**

## 20.5 Working Drawings of Spur Gears.

A typical working drawing of a spur gear is shown in Fig. 20.7. Since the teeth are cut to standard shape with special cutters, it is not necessary to show individual teeth on the drawing. Instead, the addendum and root circles are drawn as phantom lines, Fig. 2.15, and the pitch circle as a center line. Thus, the drawing actually shows only a gear blank—a gear complete except for teeth. Since the machining of the blank and the cutting of the teeth are separate operations in the shop, the necessary dimensions are arranged in two groups: the blank dimensions are shown on the views, and the cutting data are given in a note or table.

Before laying out the working drawing, the draftsman must calculate the gear dimensions.

Fig. 20.7 **Working Drawing of a Spur Gear.**

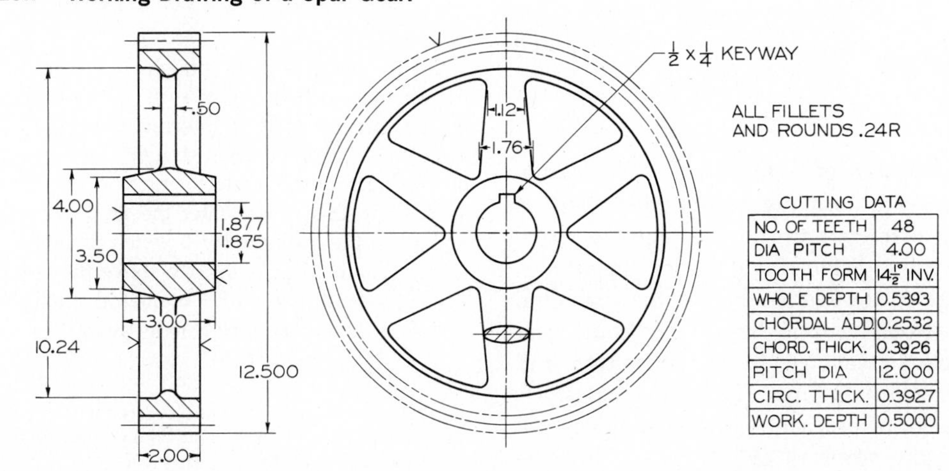


| CUTTING DATA | |
|---|---|
| NO. OF TEETH | 48 |
| DIA PITCH | 4.00 |
| TOOTH FORM | $14\frac{1}{2}°$ INV. |
| WHOLE DEPTH | 0.5393 |
| CHORDAL ADD | 0.2532 |
| CHORD. THICK. | 0.3926 |
| PITCH DIA | 12.000 |
| CIRC. THICK. | 0.3927 |
| WORK. DEPTH | 0.5000 |

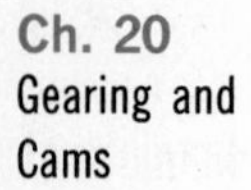

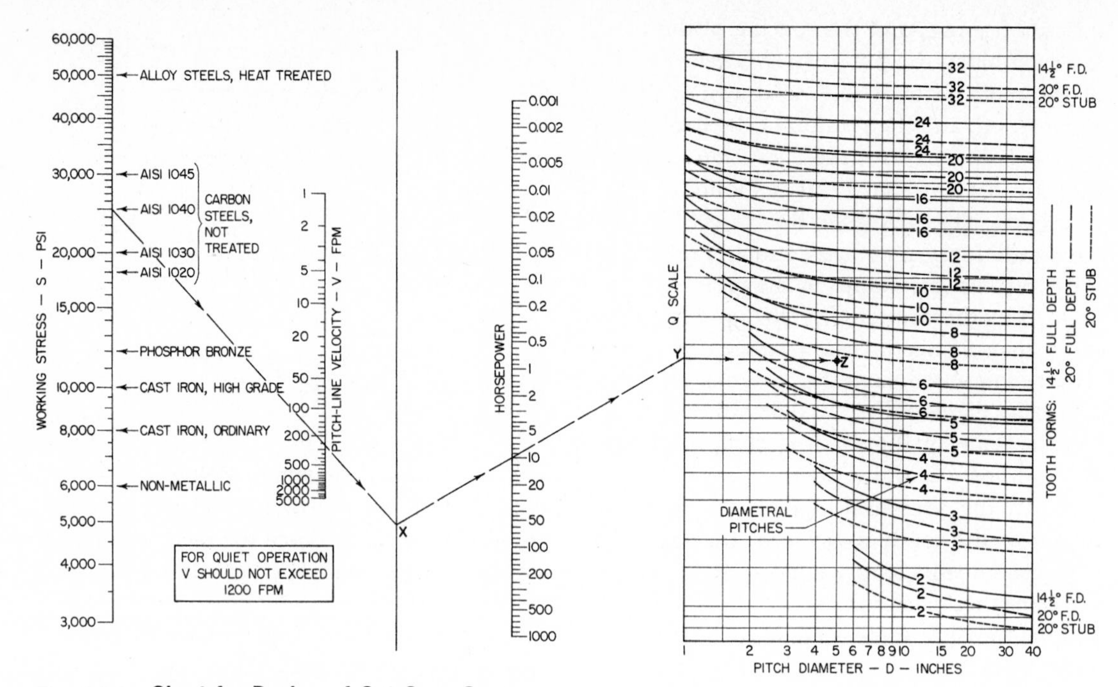


Fig. 20.8 **Chart for Design of Cut Spur Gears.**

For example, if the gear must have 48 teeth of 4 diametral pitch, with $14\frac{1}{2}°$ full-depth involute profile, as in Fig. 20.7, then the following items should be calculated in this order: pitch diameter, addendum, dedendum, outside diameter, root diameter, whole depth, chordal thickness, and chordal addendum. The dimensions shown in the figure are the minimum requirements for a spur gear. The chordal addendum and chordal thickness are given to aid in checking the finished gear. Other special data may be given in the table, according to the degree of precision required and the method of manufacture.

## 20.6 Design of Spur Gears.

The design of spur gears normally begins with selection of pitch diameters to suit the required speed ratio, center distance, and space limitations. The size of the teeth—that is, the diametral pitch—depends upon the gear speeds, gear materials, horsepower to be transmitted, and the selected tooth form. The complete analysis and design of precision gears is complex and beyond the scope of this textbook, but the chart in Fig. 20.8 will quickly yield a suitable diametral pitch for ordinary cut spur gears.*

*Example:* Determine the diametral pitch and face width for a 5″ pitch diameter steel (AISI 1040) pinion having $14\frac{1}{2}°$ full-depth teeth that must transmit 10 hp at 200 rpm.

Calculate the pitch-line velocity V as follows:

$$V = \frac{\pi}{12} D_p n_p$$

$$V = 0.262 \times 5 \times 200 = 262 \text{ fpm}$$

On the chart, draw a straight line from 25,000 psi on the working stress scale through 262 fpm on the velocity scale to intersect the pivot line at X. From X, draw a second straight line through 10 on the horsepower scale to intersect the Q scale at Y. From this point, enter the graph of pitches and go to the ordinate for 5″ pitch diameter. The junction point Z falls slightly above the curve for $14\frac{1}{2}°$ full-depth teeth of 6 diametral pitch, hence choose 6 diametral pitch for the gears.

*The chart is based on the Lewis equation, with the Barth velocity modification, and assumes gear-face width equal to three times the circular pitch.

For good proportions the face width of a spur gear should be about three times the circular pitch, and this proportion is incorporated in the chart. Therefore, for 6 diametral pitch the circular pitch is $\pi/6$ or .5236″, and the face width is 3 × .5236 or 1.5708″. This width should be rounded off to $1\frac{5}{8}$″.

When pinion and gear are of the same material, the smaller gear is the weaker, and the design should be based on the pinion. When the materials are different, the chart should be used to determine the horsepower capacity of each gear.

## 20.7 Bevel Gears.

*Bevel gears* are used to transmit power between shafts whose axes intersect. The analogous friction drive would consist of a pair of cones having a common apex at the point of intersection of the axes. The axes may intersect at any angle, but axes at right angles occur most frequently, and this is the only case that will be considered here. Bevel gear teeth have the same involute shape as teeth on spur gears, but the bevel gear teeth are tapered toward the cone apex; hence the height and width of a bevel gear tooth vary as the distance from the cone apex. Whereas spur gears are interchangeable—a spur gear of given pitch will run properly with any other spur gear of the same pitch and tooth form—this is not true of bevel gears, which must be designed in pairs and will run only with each other.

The speed ratio of bevel gears can be calculated from the same formulas given for spur gears in §20.2.

## 20.8 Bevel Gear Definitions and Formulas.

The design of bevel gears is very similar to that of spur gears;* therefore, many of the spur gear terms are applied with slight modification to bevel gears. The *pitch diameter*

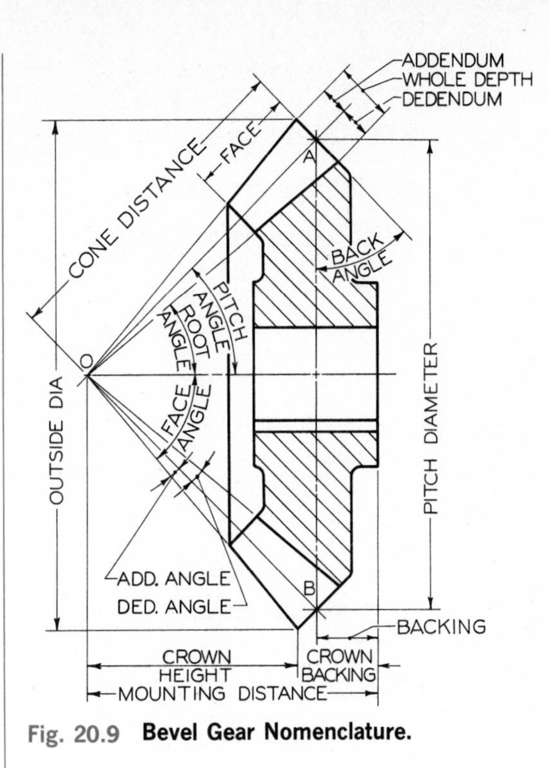


Fig. 20.9 **Bevel Gear Nomenclature.**

D of a bevel gear is the diameter of the base of the pitch cone, and the *circular pitch* p of the teeth is measured along this circle. The *diametral pitch* P is also based on this circle; hence the relationship of these three items is the same as for spur gears.

The important dimensions and angles of a bevel gear are illustrated in Fig. 20.9. The *pitch cone* is shown as the triangle OAB. Examination of Fig. 20.10 will reveal the pitch cones for the mating gear and pinion shown there. Evidently the pitch angle of each gear depends upon the relative diameters of the gears. Therefore, the pitch angles, Γ (*gamma*), are determined from the following equations:

$$\tan \Gamma_G = \frac{D_G}{D_P} = \frac{N_G}{N_P}$$

and $$\tan \Gamma_P = \frac{D_P}{D_G} = \frac{N_P}{N_G}$$

Other terms are defined as follows:

*Pitch diameter* ($D_G$ or $D_P$).—The diameter of the base of the pitch cone. $D_G = N_G/P$ and $D_P = N_P/P$.

*In the simplified formulas of this article, tooth proportions are assumed equal on both gears, but in modern practice bevel gears are often designed with unequal addendums and unequal tooth thicknesses to balance the strength of gear and pinion. See ANSI B6.13–1965, AGMA 208.02, and the publications of the Gleason Works, Rochester, N.Y.

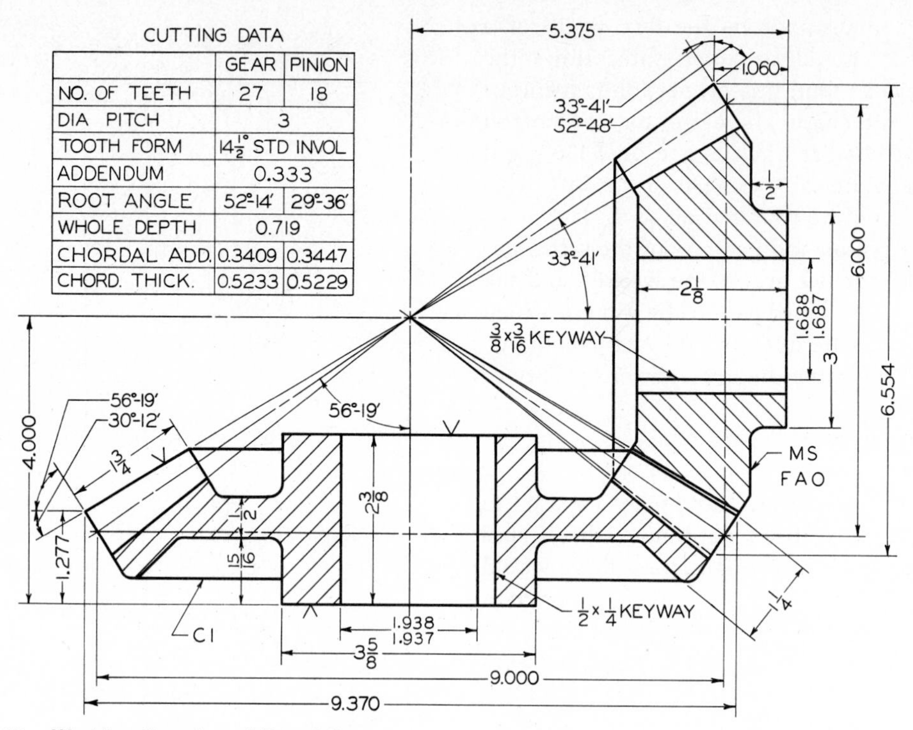


Fig. 20.10 **Working Drawing of Bevel Gears.**

*Cone distance* (A).—The slant height of the pitch cone; hence the same for both gear and pinion. $A = D/2 \sin \Gamma$.

*Addendum* (a).—The same as for a spur gear of the same diametral pitch. It is measured at the large end of the tooth. $a = 1/P$.

*Dedendum* (b).—The same as for a spur gear of the same diametral pitch. It is measured at the large end of the tooth. $b = 1.157/P$.

*Addendum angle* ($\alpha$).—The angle subtended by the addendum. It is the same for both gear and pinion. $\tan \alpha = a/A$.

*Dedendum angle* ($\delta$).—The angle subtended by the dedendum. It is the same for both gear and pinion. $\tan \delta = b/A$.

*Face angle* ($\Gamma_0$).—The angle between the top of the teeth and the gear axis. $\Gamma_0 = \Gamma + \alpha$.

*Root angle* ($\Gamma_R$).—The angle between the root of the teeth and the gear axis. $\Gamma_R = \Gamma - \delta$.

*Back angle.*—This angle is usually equal to the pitch angle.

*Outside diameter* ($D_0$).—The diameter of the outside or crown circle of the gear. $D_0 = D + 2a \cos \Gamma$.

*Crown height* (X).—The distance parallel to the gear axis from the cone apex to the crown of the gear. $X = \frac{1}{2}D_0/\tan \Gamma_0$.

*Backing* (Y).—The distance from the base of the pitch cone to the rear of the hub.

*Crown backing* (Z).—For shop use, the crown backing is more practical than the backing; hence dimension Z is given on drawings instead of Y. $Z = Y + a \sin \Gamma$.

*Mounting distance* (M).—This dimension is used primarily for inspection and assembly purposes. $M = Y + \frac{1}{2}D/\tan \Gamma$.

*Face width* (F).—The face width should not exceed $\frac{1}{3}$A.

*Chordal addendum* ($a_c$) and *Chordal thickness* ($t_c$).—For bevel gears the formulas given for spur gears, §20.3, can be used if D is replaced by $D/\cos \Gamma$ and N is replaced by $N/\cos \Gamma$.

## 20.9 Working Drawings of Bevel Gears.

As in the case of spur gears, a working drawing of a bevel gear gives only the dimensions of the gear blank. The necessary data for cutting the teeth are given in a note or table. A single sectional view, Fig. 20.10, usually will

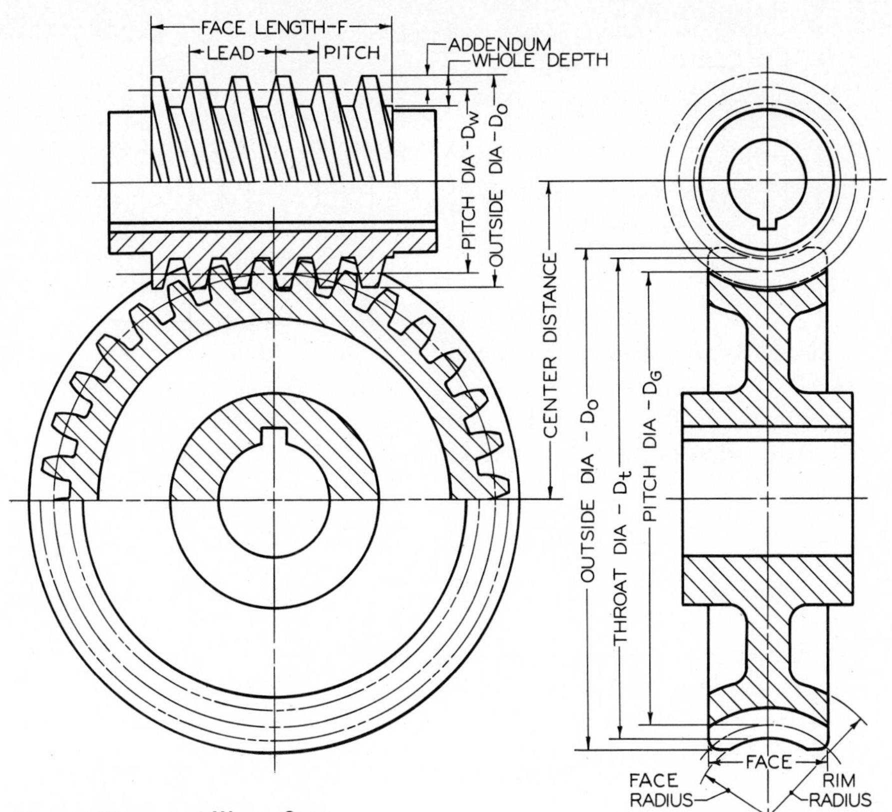


Fig. 20.11 **Double Thread Worm and Worm Gear.**

provide all necessary information. If a second view is required, only the gear blank is drawn, and the tooth profiles are omitted. Two gears are shown in their operating relationship. On detail drawings, each gear is usually drawn separately, Fig. 20.9, and fully dimensioned. Proper placing of the gear-blank dimensions is largely dependent upon the shop methods used in producing the gear, but the scheme shown is commonly followed.

**20.10 Worm Gears.** Worm gears are used to transmit power between nonintersecting shafts that are at right angles to each other. The *worm* is a screw having a thread of the same shape as a rack tooth. The *worm wheel* is similar to a spur gear except that the teeth have been twisted and curved to conform to the shape of the worm. A large speed ratio is obtainable with worm gearing, since a *single-thread worm* in one revolution advances the worm wheel only one tooth and a space.

In Fig. 20.11, a worm and worm wheel are shown engaged. The section taken through the center of the worm and perpendicular to the axis of the worm wheel shows that the worm section is identical with a rack, and that the wheel section is identical with a spur gear. Consequently, in this plane the height proportions of thread and gear teeth are the same as for a spur gear of corresponding pitch.

*Pitch* (p).—The axial pitch of the worm is the distance from a point on one thread to the corresponding point on the next thread measured parallel to the worm axis. The pitch of the worm must exactly equal the circular pitch of the gear.

*Lead* (L).—The lead is the distance that the thread advances axially in one turn. The lead is always a multiple of the pitch. Thus for a single-thread worm, the lead equals the pitch; for a double-thread worm, the lead is twice the pitch, etc.

*Lead angle* ($\lambda$).—The angle between a tangent to the helix at the pitch diameter and a plane

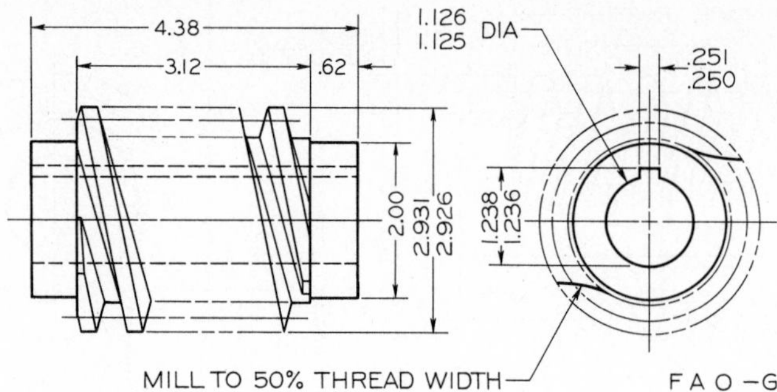


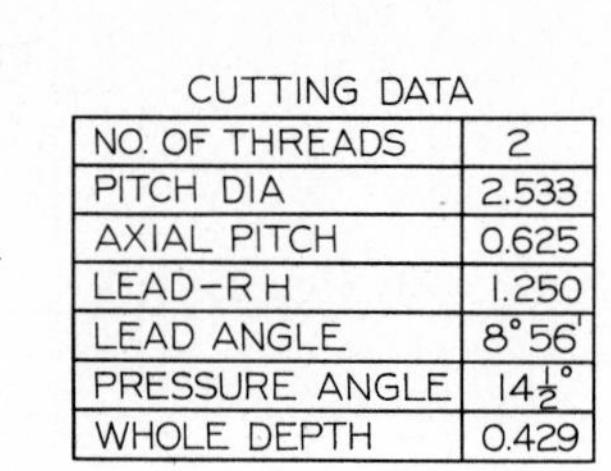

CUTTING DATA

| NO. OF THREADS | 2 |
|---|---|
| PITCH DIA | 2.533 |
| AXIAL PITCH | 0.625 |
| LEAD–R H | 1.250 |
| LEAD ANGLE | 8°56' |
| PRESSURE ANGLE | $14\frac{1}{2}$° |
| WHOLE DEPTH | 0.429 |

Fig. 20.12 **Working Drawing of a Worm.**

perpendicular to the axis of the worm. The lead angle can be calculated from the formula

$$\tan \lambda = \frac{L}{\pi D_W}$$

where $D_W$ = pitch diameter of the worm.

The speed ratio of worm gears depends only upon the *number of threads* on the worm and the *number of teeth* on the gear. Therefore,

$$m_G = \frac{N_G}{N_W}$$

where $N_G$ = number of teeth on the gear, and $N_W$ = number of threads on the worm.

For $14\frac{1}{2}$° standard involute teeth and single-thread or double-thread worms, the following proportions are the recommended practice of the AGMA (American Gear Manufacturers' Association). All formulas are expressed in terms of circular pitch p instead of diametral pitch. It is easier to machine the worm and the hob used to cut the gear if the circular pitch has an even rational value such as $\frac{5}{8}$''.

*For the worm:*

| | |
|---|---|
| Pitch diameter | $D_W = 2.4p + 1.1^*$ |
| Whole depth | $h_t = 0.686p$ |
| Outside diameter | $D_O = D_W + 0.636p$ |
| Face length | $F = p(4.5 + N_G/50)$ |

*Recommended value, but may be varied.

*For the gear:*

| | |
|---|---|
| Pitch diameter | $D_G = p(N_G/\pi)$ |
| Throat diameter | $D_t = D_G + 0.636p$ |
| Outside diameter | $D_O = D_t + 0.4775p$ |
| Face radius | $R_f = \frac{1}{2}D_W - 0.318p$ |
| Rim radius | $R_r = \frac{1}{2}D_W + p$ |
| Face width | $F = 2.38p + 0.25$ |
| Center distance | $C = \frac{1}{2}(D_G + D_W)$ |

## 20.11 Working Drawings of Worm Gears.

In an assembly drawing, the engaged worm and gear can be shown as in Fig. 20.11, but it is customary to omit the gear teeth and represent the gear blank conventionally as shown in the lower half of the circular view. On detail drawings, the worm and gear are usually drawn separately as shown in Figs. 20.12 and 20.13. Although dimensioning of these parts is again largely dependent upon the method of production, it is standard practice to dimension the blanks on the views and to give the cutting data in tabular form, as shown. Note that those dimensions that closely affect the engagement of the gear and worm have been given as three-place decimal or limit dimensions; other dimensions, such as rim radius, face lengths, and gear outside diameter have been rounded to convenient two-place decimal values.

## 20.12 Cams.

*Cams* provide a simple means for obtaining unusual and irregular motions that would be difficult to produce

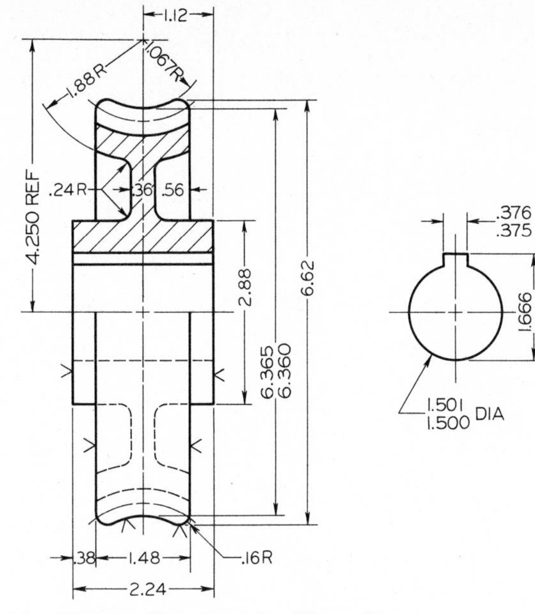


CUTTING DATA

| | |
|---|---|
| NO. OF TEETH | 30 |
| PITCH DIA | 5.967 |
| ADDENDUM | 0.199 |
| WHOLE DEPTH | 0.429 |
| NO. OF THREADS | 2 |
| AXIAL PITCH | 0.625 |
| LEAD—R H | 1.250 |
| LEAD ANGLE | 8°56' |
| PRESSURE ANGLE | $14\frac{1}{2}$° |

Fig. 20.13 **Working Drawing of a Worm Gear.**

otherwise. Figure 20.14 (a) illustrates the basic principle of the cam. A shaft rotating at uniform speed carries an irregularly shaped disk called the *cam;* a reciprocating plunger, called the *follower,* presses a small roller against the curved surface of the cam. Rotation of the cam thus causes the follower to reciprocate with a definite cyclic motion according to the shape of the cam profile. The roller is held in contact with the cam by gravity or a spring. The problem of the draftsman is to construct the cam profile necessary to obtain the desired motion of the follower.

An automobile valve cam that operates a flat-faced follower is shown at (b). The profile

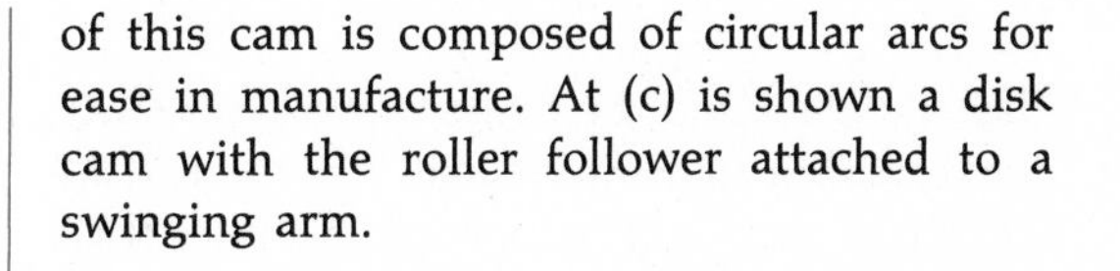
of this cam is composed of circular arcs for ease in manufacture. At (c) is shown a disk cam with the roller follower attached to a swinging arm.

## 20.13 Displacement Diagrams.

Since the motion of the follower is of primary importance, its rate of speed and its various positions should be carefully planned in a *displacement diagram* before the cam profile is constructed. A displacement diagram, Fig. 20.15, is a curve showing the displacement of the follower as ordinates erected on a base line that represents one revolution of the cam. The

Fig. 20.14 **Disk Cams.**

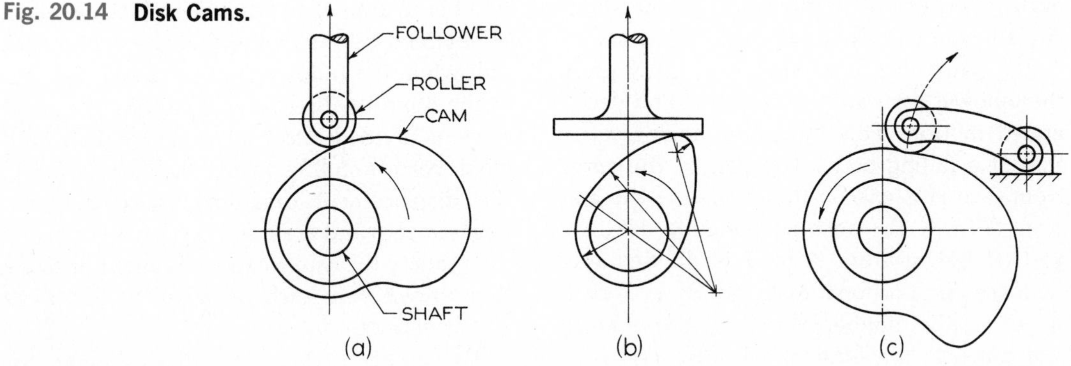

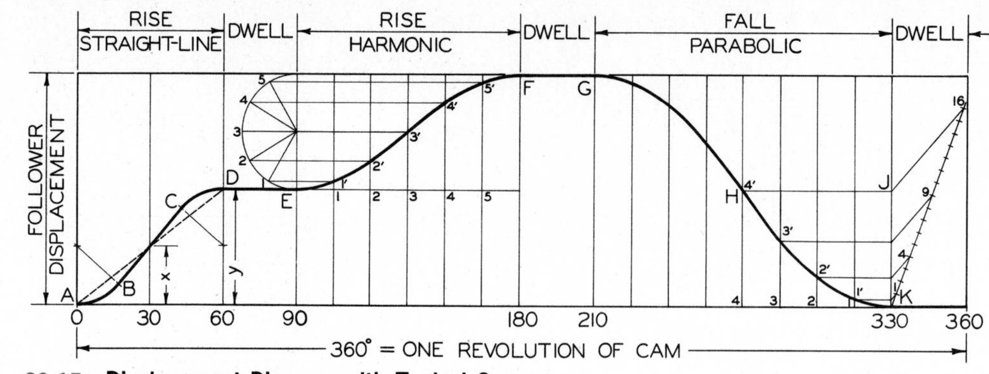


Fig. 20.15 **Displacement Diagram with Typical Curves.**

follower displacement should be drawn to scale, but any convenient length can be used to represent the 360° of cam rotation.

The motion of the follower as it rises or falls depends upon the shape of the curves in the displacement diagram. In this diagram, four commonly employed types of curves are shown. If a straight line is used, such as the dashed line AD in the figure, the follower will move with a uniform velocity, but it will be forced to start and stop very abruptly. This straight-line motion can be modified as shown in the curve ABCD, where arcs have been introduced at the beginning and at the end of the period.

The curve shown at EF is one which gives harmonic motion to the follower. To construct this curve, a semicircle is drawn whose diameter is equal to the desired rise. The circumference of the semicircle is divided into equal arcs, the number of divisions being the same as the number of horizontal divisions. Points on the curve are then found by projecting horizontally from the divisions on the semicircle to the corresponding ordinates.

The parabolic curve shown at GHK gives the follower constantly accelerated and decelerated motion. This motion is analogous to that of a falling body. The half of the curve from G to H is exactly the reverse of the half from H to K. To construct the curve HK, the vertical height from K to J is divided into distances proportional to $1^2$, $2^2$, $3^2$, . . . , or 1, 4, 9, . . . , the number of such divisions being the same as the number of horizontal divisions. See §§4.17 and 4.59. Points on the curve are found by projecting horizontally from the divisions on the line JK to the corresponding ordinates.

## 20.14 **The Cam Profile.**

The general procedure in constructing a cam profile is shown in Fig. 20.16. The disk cam rotating counterclockwise on its shaft raises and lowers the roller follower along the straight line AB. The axis of the follower is *offset* from the center line of the cam. The displacement diagram at the bottom of the figure shows the desired follower motion.

With the follower in its lowest or initial position, the center of the roller is at A, and OA is the radius of the *base circle*. With the same center O, an *offset circle* is drawn with radius equal to the offset. As the cam turns, the extended center line of the follower will always be tangent to this offset circle. But since the cam must remain stationary while it is being drawn, an equivalent rotative effect is obtained by imagining that the cam stands still while the follower rotates about the cam in the opposite direction. Therefore, the offset circle is divided into twelve equiangular divisions corresponding to the divisions used in the displacement diagram. These divisions begin at zero and are numbered in an opposite direction to the cam rotation. Tangent lines are then drawn from each point on the offset circle, as shown.

The points 1, 2, . . . , on the follower axis

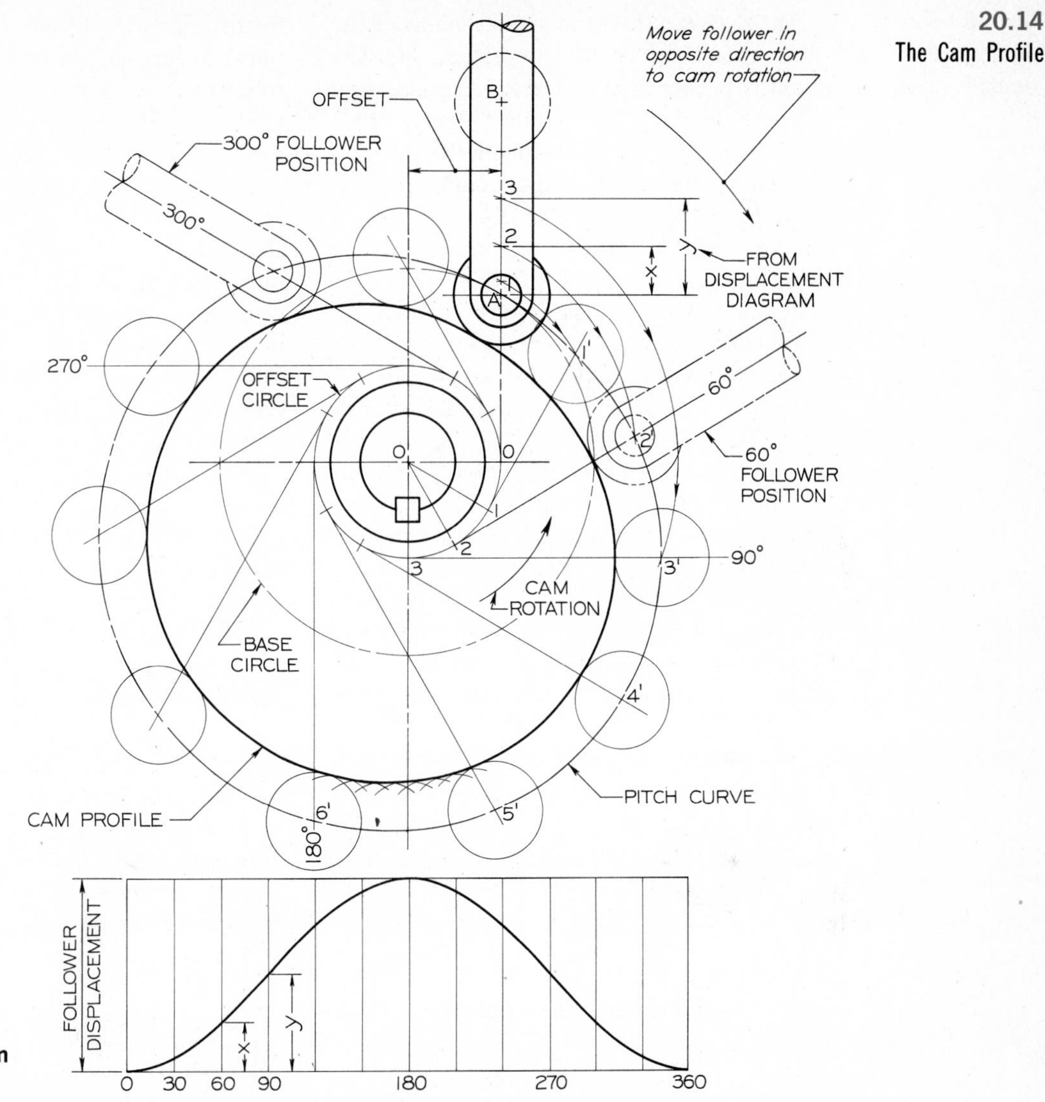


Fig. 20.16 **Disk Cam Profile Construction.**

AB indicate successive positions of the center of the roller, and are located by transferring ordinates such as x and y from the displacement diagram. Thus, when the cam has rotated 60°, the follower roller must rise a distance x to position 2, and after 90° of rotation, a distance y to position 3, and so on.

It should now be observed that while the center of the roller moved from its initial position A to position 2, for example, the cam rotated 60° *counterclockwise*. Therefore, point 2 must be revolved *clockwise* about the cam center O to the corresponding 60° tangent line to establish point 2′. In this position, the complete follower would appear as shown by the phantom outline. Points 1′, 2′, 3′, . . . , represent consecutive positions of the roller center, and a smooth curve drawn through these points is called the *pitch curve.* To obtain the actual cam profile, the roller must be drawn in a number of positions, and the cam profile drawn tangent to the roller circles, as shown. The best results are obtained by first drawing the pitch curve very carefully, and then drawing a great many closely spaced roller circles with centers on the pitch curve as shown between points 5′ and 6′.

If the roller is on a pivoted arm, as shown

in Fig. 20.17 (a), then the displacement of the roller center is along the circular arc AB. The height of the displacement diagram (not shown) should be made equal to the rectified length of arc AB. Ordinates from the diagram are then transferred to arc AB to locate the roller positions 1, 2, 3,.... As the follower is revolved about the cam, pivot point C moves in a circular path of radius OC to the consecutive positions $C_1$, $C_2$, .... Length AC is constant for all follower positions; hence, from each new position of point C the follower arc of radius R is drawn as shown at the 90° position. The roller centers 1, 2, 3,..., are now revolved about the cam center O to intersect the follower arcs at 1′, 2′, 3′,.... After the pitch curve is completed, the cam profile is drawn tangent to the roller circles.

The construction for a flat-faced follower is shown at (b). The initial point of contact is at A, and points 1, 2, 3,..., represent consecutive positions of the follower face. Then for the 90° position, point 3 must be revolved 90°, as shown, to position 3′, and the flat face of the follower is drawn through point 3′ at right angles to the cam radius. When this procedure has been repeated for each position, the cam profile will be enveloped by a series of straight lines, and the cam profile is drawn inside and tangent to these lines. Note that the point of contact, initially at A, changes as the follower rises: at 90°, for example, contact is at D, a distance X to the right of the follower axis.

## 20.15 **Cylindrical Cams.**

When the follower movement is in a plane parallel to the cam shaft, some form of cylindrical cam must be employed. In Fig. 20.18, for example, the follower rod moves vertically parallel to the cam axis as the attached roller follows the groove in the rotating cam cylinder.

A cylindrical cam of diameter D is required to lift the follower rod a distance AB with harmonic motion in 180° of cam motion and to return the rod in the remaining 180° with the same motion. The displacement diagram is drawn first and conveniently placed directly opposite the front view. The 360° length of the diagram must be made equal to $\pi D$ so that the resulting curves will be a true development of the outer surface of the cam cylinder. The pitch curve is drawn to represent the required motion, and a series of roller circles are then drawn to establish the sides of the groove tangent to these circles. This completes the development of the outer cylinder and actually

**Fig. 20.17 Pivoted and Flat-Faced Followers.**

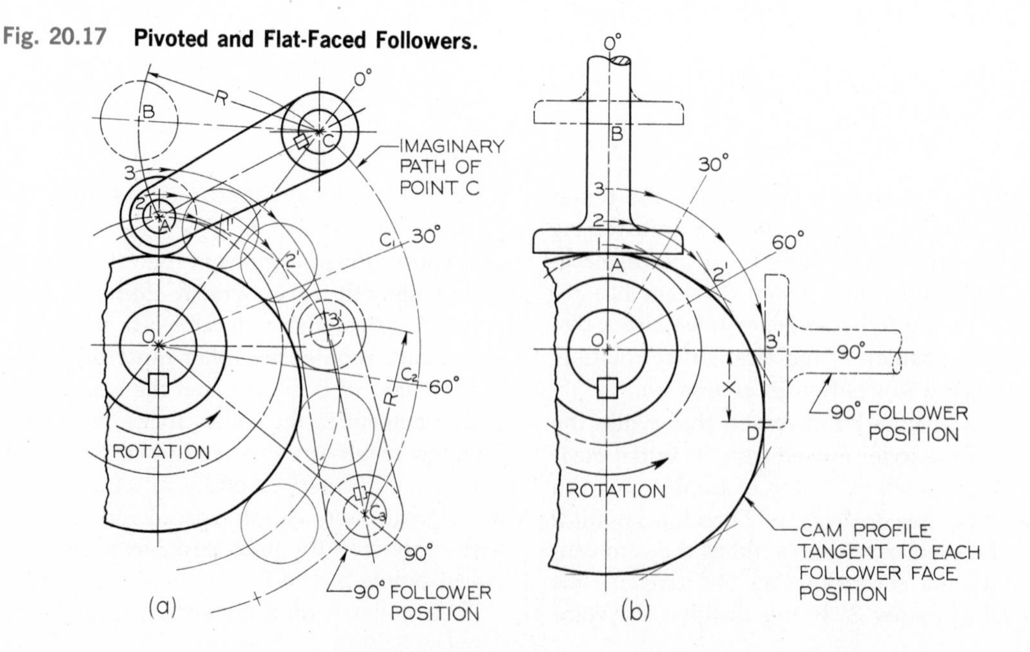

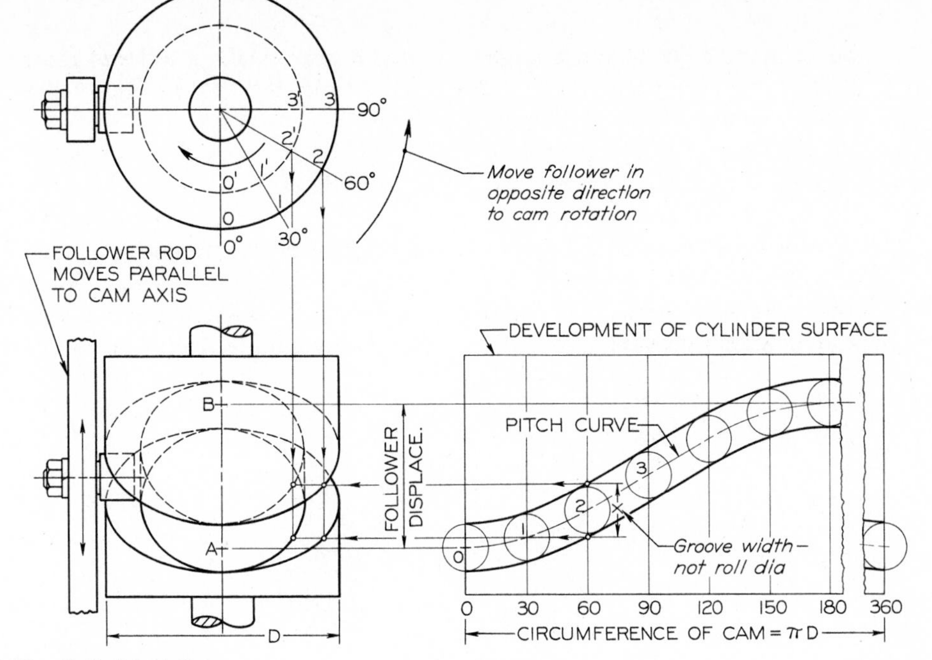


Fig. 20.18 **Cylindrical Cam.**

provides all information needed for making the cam; hence it is not uncommon to omit the curves in the front view.

To complete the front view, points on the curves are projected horizontally from the development. For example, at 60° in the development, the width of the groove measured parallel to the cam axis is X, a distance slightly greater than the actual roller diameter. This width X is projected to the front view directly below point 2, the corresponding 60° position in the top view, to establish two points on the outer curves. The inner curves for the bottom of the groove can be established in the same manner, except that the groove width X is located below point 2′, which lies on the inner diameter. The inner curves are only approximate because the width at the bottom of the groove is actually slightly greater than X, but the exact bottom width can only be determined by drawing a second development for the inner cylinder.

## 20.16 Problems on Gearing and Cams.

The following problems are given to provide practice in laying out and making working drawings of the common types of gears and cams. Where paper sizes are not given, the student is to select his own scale and sheet layout.

## 20.16 Problems on Gearing and Cams

**Prob. 20.1** A 12-tooth 1 DP pinion engages a 15-tooth gear. Make a full-size drawing of a segment of each gear showing how the teeth mesh. Construct the $14\frac{1}{2}°$ involute teeth exactly, noting any points where the teeth appear to interfere.

**Prob. 20.2** The same as Prob. 20.1, but use 20° stub teeth.

**Prob. 20.3** The same as Prob. 20.1, but use a rack in place of the 12-tooth pinion.

**Prob. 20.4** Make a display drawing of the pinion in Fig. 20.19. Show two views, drawing the teeth by the Odontograph method shown in Fig. 20.5. Draw double size.

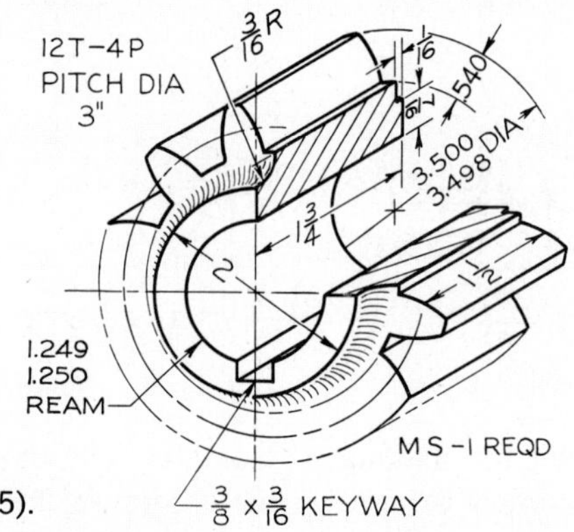

Fig. 20.19 **Pinion** (Prob. 20.5).

**Prob. 20.5** Make a pictorial display drawing of the intermediate pinion shown in Fig. 14.104 (part 6). Show the gear as an oblique half section, §17.10, similar to Fig. 20.19. Draw four-times size, reducing the 30° receding lines by one-half as in cabinet projection, §17.4. Draw the teeth by the Odontograph method.

**Prob. 20.6** A spur gear has 60 teeth of 5 diametral pitch. The face width is $1\frac{1}{2}''$. The shaft is $1\frac{3}{16}''$ in diameter. Make the hub 2″ long, and $2\frac{1}{4}''$ in diameter. Calculate accurately all dimensions, and make a working drawing of the gear. Show six spokes, each $1\frac{1}{8}''$ wide at the hub, tapering to $\frac{3}{4}''$ wide at the rim, and $\frac{1}{2}''$ thick. Use your own judgment for any dimensions not given. Draw half size on Layout A–3 or full size on Layout B–3.

**Prob. 20.7** Make a working drawing of the intermediate pinion shown in Fig. 14.104 (part 6). Check the gear dimensions by calculation.

**Prob. 20.8** The same as Prob. 20.7, but use the intermediate gear shown in Fig. 14.104 (part 5).

**Prob. 20.9** The same as Prob. 20.7, but use the pinion shown in Fig. 20.19.

**Prob. 20.10** The same as Prob. 20.7, but use the spur gear shown in Fig. 20.20.

**Prob. 20.11** A pair of bevel gears have teeth of 4 diametral pitch. The pinion has 13 teeth, the gear 25 teeth. The face width is $1\frac{1}{8}''$. The pinion shaft is $\frac{15}{16}''$ in diameter, and the gear shaft is $1\frac{3}{16}''$ in diameter. Calculate accurately all dimensions, and make a working drawing showing

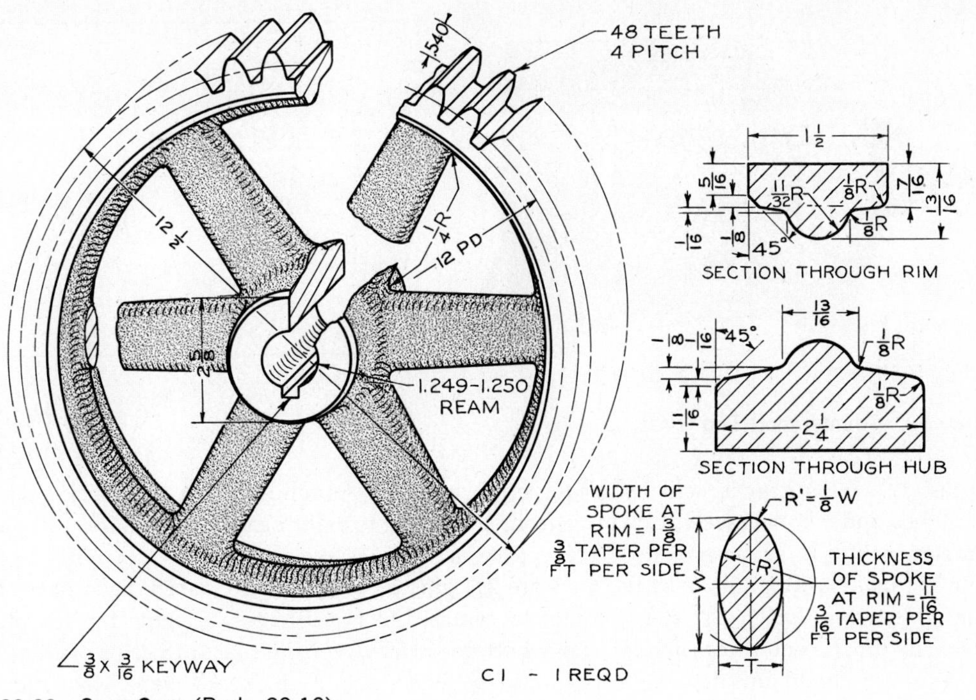


Fig. 20.20 **Spur Gear** (Prob. 20.10).

the gears engaged as in Fig. 20.10. Make the hub diameters approximately twice the shaft diameters. Select key sizes from Appendix Table 16. The backing for the pinion must be $\frac{5}{8}''$; for the gear $1\frac{1}{4}''$. Use your own judgment for any dimensions not given.

**Prob. 20.12** Make a working drawing of the pinion in Prob. 20.11.

**Prob. 20.13** Make a working drawing of the gear in Prob. 20.11.

**Prob. 20.14** The same as Prob. 20.12, but show two views of the pinion.

**Prob. 20.15** The same as Prob. 20.11, but use 5 diametral pitch, 30 and 15 teeth. The face is 1″. Shafts: pinion, 1″ dia.; gear, $1\frac{1}{2}''$ dia. Backing: pinion, $\frac{1}{2}''$; gear, 1″.

**Prob. 20.16** The same as Prob. 20.11, but use 4 diametral pitch, both gears 20 teeth. Select the correct face width. Shafts: $1\frac{3}{8}''$ dia. Backing: $\frac{3}{4}''$.

**Prob. 20.17** Fig. 20.21 (a) shows a countershaft end used on a trough conveyor, and (b) shows the layout for an assembly drawing of a similar unit. Using a scale of half size, make an assembly drawing of the complete unit (given layout dimensions are full size). The following full-size dimensions are sufficient to establish the position and general outline of the parts; the minor detail dimensions, web and rib shapes, and fillets and rounds should be designed by the student with the help of the photograph.

The gears are identical in size and are similar in shape to the one in Fig. 20.9. There are 24 teeth of 3 diametral pitch in each gear. Face width, 2″. Shafts, $1\frac{1}{2}''$ dia., extending 7″ beyond left bearing, 4″ beyond rear (break shafts as shown). Hub diameters, $2\frac{3}{4}''$. Backing, 1″. Hub

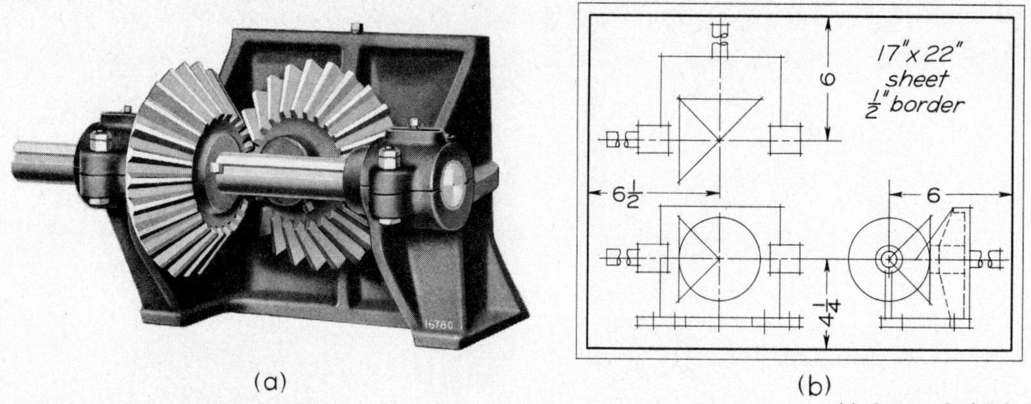


Fig. 20.21 **Countershaft End** (Prob. 20.17).

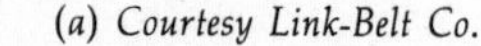
(*a*) *Courtesy Link-Belt Co.*

lengths, $3\frac{1}{4}''$. Front gear is held by a square gib key (see Appendix Table 16); back gear by a square key and a $\frac{3}{8}''$ set screw. Collar on front shaft next to right bearing is $\frac{3}{4}''$ thick, $2\frac{1}{4}''$ outside diameter, with a $\frac{3}{8}''$ set screw.

On the main casting, the split bearings are $2\frac{3}{4}''$ dia., 3″ long, and 10″ apart. Each bearing cap is held by two $\frac{1}{2}''$ bolts, $3\frac{1}{2}''$ apart, center to center. Oil holes have a $\frac{1}{4}''$ pipe tap for plug or grease cup. Shaft center lines are $6\frac{1}{2}''$ above bottom surface, 8″ from rear surface of casting. Main casting is $11\frac{1}{2}''$ high, and its base is 16″ long, $8\frac{3}{4}''$ wide, and $\frac{3}{4}''$ thick. All webs, ribs, and walls are uniformly $\frac{1}{2}''$ thick. In the base are eight holes (not shown in the photograph) for $\frac{1}{2}''$ bolts, two outside each end, and two inside each end.

Proceed by first blocking in the gears in each view, and then block in the principal main casting dimensions. Fill in details only after principal dimensions are clearly established.

**Prob. 20.18** The worm and worm gear shown in Fig. 20.11 have a circular pitch of $\frac{5}{8}''$, and the gear has 32 teeth of $14\frac{1}{2}°$ involute form. The worm is double threaded. Make an assembly drawing similar to Fig. 20.11. Draw the teeth on the gear by the Odontograph method of Fig. 20.5. Calculate dimensions accurately, and use AGMA proportions. Shafts: worm, $1\frac{1}{8}''$ dia.; gear, $1\frac{5}{8}''$ dia.

**Prob. 20.19** Make a working drawing of the worm in Prob. 20.18.

**Prob. 20.20** Make a working drawing of the gear in Prob. 20.18.

**Prob. 20.21** The same as Prob. 20.19, but the worm is single thread.

**Prob. 20.22** A single-thread worm has a lead of $\frac{3}{4}''$. The worm gear has 28 teeth of standard form. Make a working drawing of the worm. The shaft is $1\frac{1}{4}''$ in diameter.

**Prob. 20.23** Make a working drawing of the gear in Prob. 20.22. The shaft is $1\frac{1}{2}''$ in diameter.

## CAMS

**Prob. 20.24** Fig. 20.22 (a). Draw the displacement diagram, and determine the cam profile that will give the radial roller follower this motion: up $1\frac{1}{2}''$ in 120°, dwell 60°, down in 90°, dwell 90°. Motions are to be unmodified straight line and of uniform velocity. The roller is $\frac{3}{4}''$ in diameter, and the base circle is 3″ in diameter. Note that the follower has zero offset. The cam rotates clockwise.

**Prob. 20.25** The same as Prob. 20.24, except that the straight-line motions are to be modified by arcs whose radii are equal to one-half the rise of the follower.

**Prob. 20.26** The same as Prob. 20.24, except that the upward motion is to be harmonic, and the downward motion parabolic.

**Prob. 20.27** The same as Prob. 20.26, except that the follower is offset 1″ to the left of the cam center line.

**Prob. 20.28** Fig. 20.22 (b). Draw the displacement diagram, and determine the cam profile that will give the flat-faced follower this motion: dwell 30°, up $1\frac{1}{2}$″ on a parabolic curve in 180°, dwell 30°, down with harmonic motion in 120°. The base circle is 3″ in diameter. After completing the cam profile, determine the necessary width of face of the follower by finding the position of the follower where the point of contact with the cam is farthest from the follower axis. The cam rotates counterclockwise.

**Prob. 20.29** Fig. 20.22 (c). Draw the displacement diagram, and determine the cam profile that will swing the pivoted follower through an angle of 30° with the same motion as prescribed in Prob. 20.28. The radius of the follower arm is $2\frac{3}{4}$″, and in its lowest position the center of the roller is directly over the center of the cam. The base circle is $2\frac{1}{2}$″ in diameter, and the roller is $\frac{3}{4}$″ in diameter. The cam rotates counterclockwise.

**Prob. 20.30** The same as Prob. 20.29, except that the motion is to be that given in Probs. 20.24 and 20.25.

**Prob. 20.31** The same as Prob. 20.29, except that the motion is to be that given in Probs. 20.24 and 20.26.

**Prob. 20.32** Using an arrangement like Fig. 20.18, construct a *half* development, and complete the front view for the following cylindrical cam. Cam, $2\frac{1}{2}$″ dia., $2\frac{1}{2}$″ high. Roller, $\frac{1}{2}$″ dia.; cam groove, $\frac{3}{8}$″ deep. Cam shaft, $\frac{5}{8}$″ dia.; follower rod, $\frac{5}{8}$″ wide, $\frac{5}{16}$″ thick. Motion: up $1\frac{1}{2}$″ with harmonic motion in 180°, down with same motion in remaining 180°. Cam rotates counterclockwise. Assume lowest position of the follower at the front center of cam. Use Layout A–3.

**Prob. 20.33** The same as Prob. 20.32, but construct a *full* development for the following motion: up $1\frac{1}{2}$″ with parabolic motion in 120°, dwell 90°, down with harmonic motion in 150°. Use Layout B–3.

Fig. 20.22 **Cam Problems** (Probs. 20.24 to 20.31; Layout A–1).

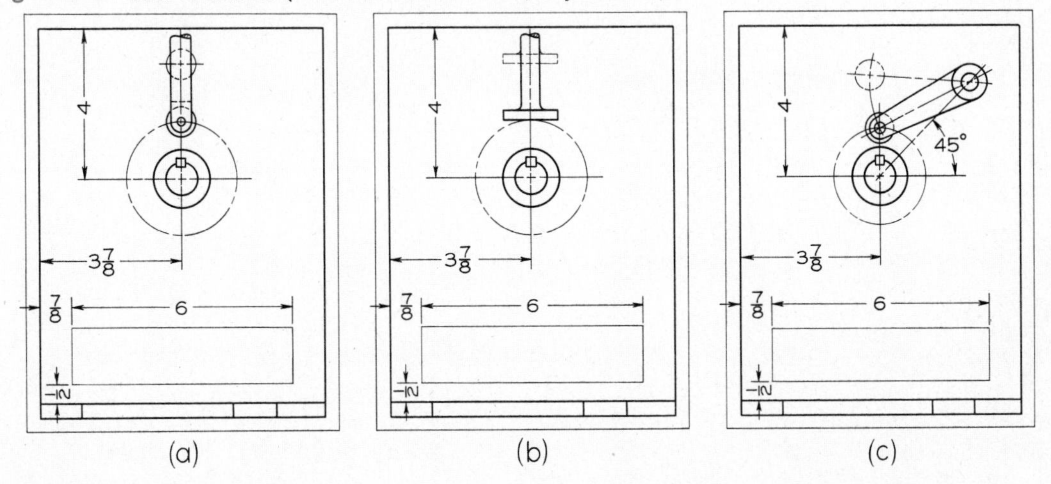

# 21

# electronic diagrams

by R. O. Loving*

**21.1 Introduction.** Working drawings for the fabrication of electrical machinery, switching devices, chassis for electronic equipment, cabinets, and other "mechanical" elements associated with electrical equipment are based on the same principles as given in Chapter 14, "Design and Working Drawings." Figure 21.1 shows a working drawing of a chassis for an electronic device. Because of the complexity of modern electrical circuitry, however, a considerable number of graphic symbols has been developed for describing the connections and functions of such circuits. Those approved by the American National Standards Institute (ANSI) are published in "Graphic Symbols for Electrical and Electronic Diagrams," ANSI Y32.2. For selected portions of this standard, see Appendix 31.

Recommendations for the preparation of electrical diagrams used by the electronics and communications industries, employing the American National Standard symbols, appear in "Electrical and Electronics Diagrams," ANSI Y14.15. Figure 21.2 is an example of such a diagram. Much of the following material is extracted or adapted from that standard.

**21.2 Definitions.** The following definitions are quoted from ANSI Y14.15–1966:

*Single-Line or One-Line Diagram.* A diagram which shows, by means of single lines and graphic symbols, the course of an electric circuit or system of circuits, and the component devices or parts used therein.

*Professor of Engineering Graphics, Illinois Institute of Technology.

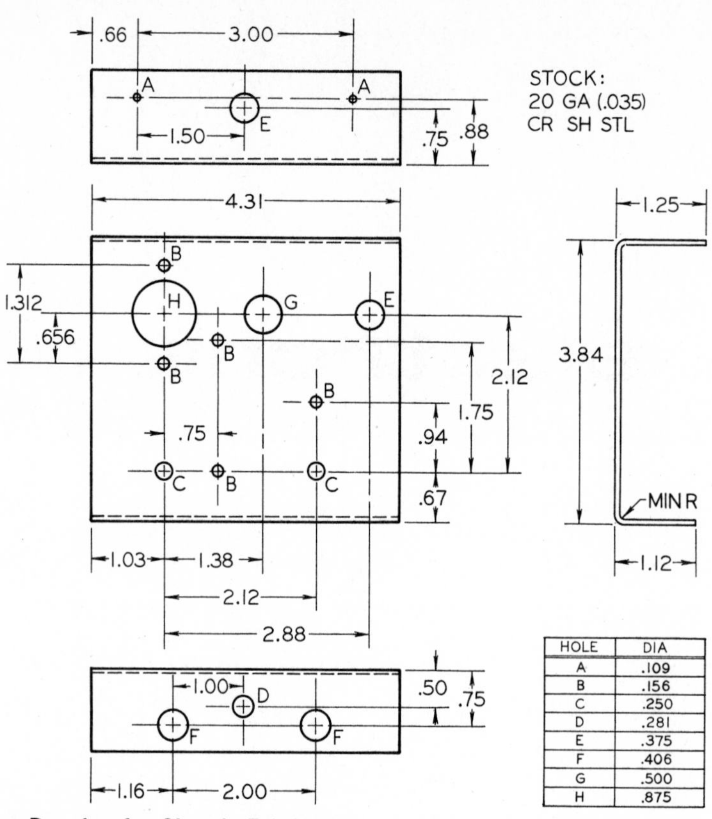


| HOLE | DIA |
|---|---|
| A | .109 |
| B | .156 |
| C | .250 |
| D | .281 |
| E | .375 |
| F | .406 |
| G | .500 |
| H | .875 |

Fig. 21.1 **Working Drawing for Chassis Fabrication.**

*Schematic or Elementary Diagram.* A diagram which shows, by means of graphic symbols, the electrical connections and functions of a specific circuit arrangement. The schematic diagram facilitates tracing the circuit and its functions without regard to the actual physical size, shape, or location of the component device or parts.

*Connection or Wiring Diagram.* A diagram which shows the connections of an installation or its component devices or parts. It may cover internal or external connections, or both, and contains such detail as is needed to make or trace connections that are involved. The connection diagram usually shows general physical arrangement of the component devices or parts.

*Interconnection Diagram.* A form of connection or wiring diagram which shows only external connections between unit assemblies or equipment. The internal connections of the unit assemblies or equipment are usually omitted.

## 21.3 Drawing Size, Format, and Title.

The sizes and formats for drawings of electrical diagrams should conform to American National Standard Y14.1-1957. See §2.66.

The title of a drawing of one of the types defined in §21.2 should include the standardized wording in the definition, plus modifying phrases pointing out the application.

*Examples:*

SINGLE-LINE DIAGRAM—AM-FM RECEIVER
SCHEMATIC DIAGRAM—250 WATT AMATEUR TRANSMITTER.

Frequently it becomes expedient to include some fragmentary wiring information on schematic diagrams, such as connection details of coils and switches. The title of such a *combined form* is not altered, however, but describes the primary function of the diagram.

## 21.4 Line Conventions and Lettering.

As in any drawings intended for reproduction, line thicknesses and letter sizes should be

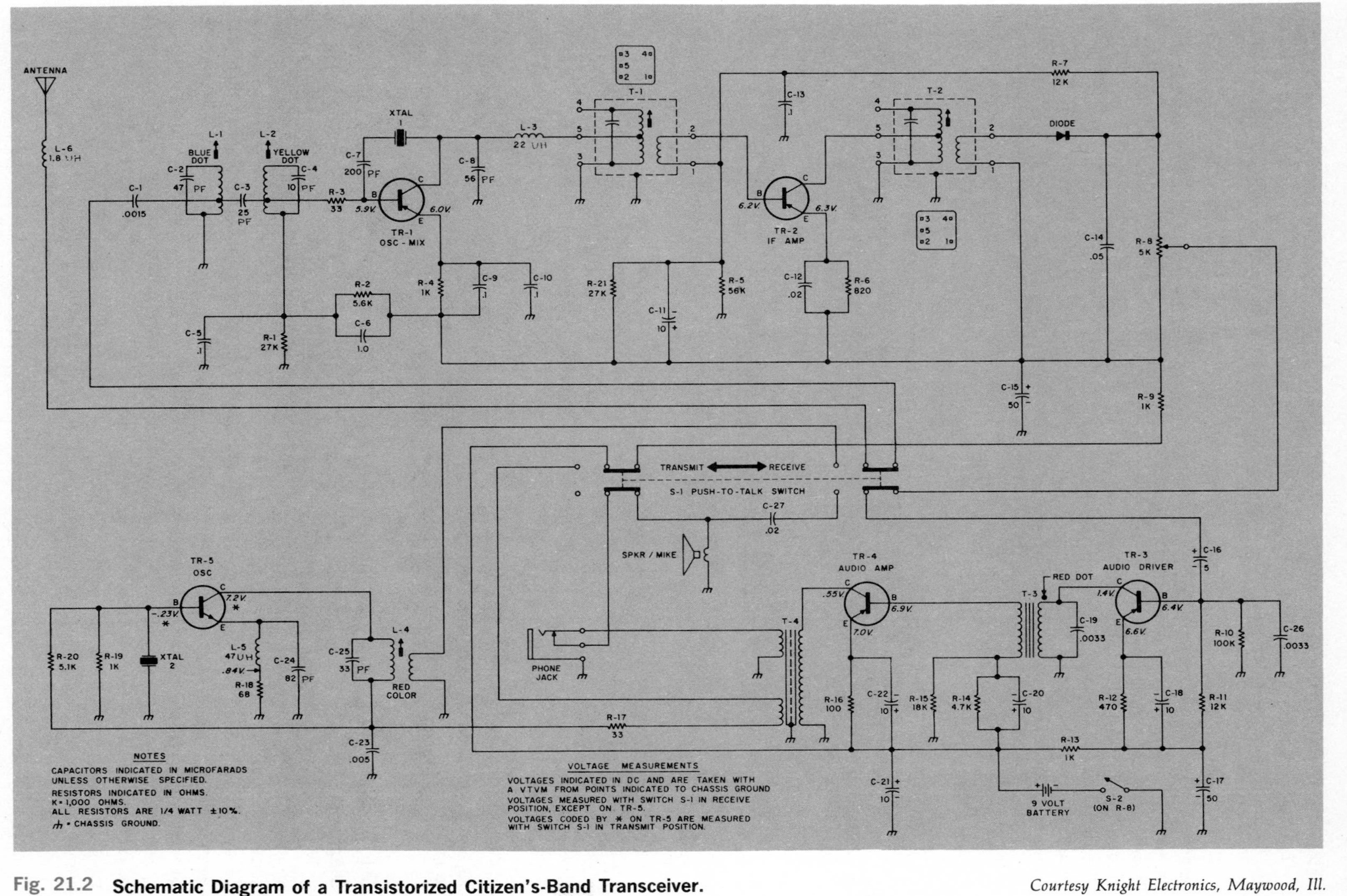


Fig. 21.2 **Schematic Diagram of a Transistorized Citizen's-Band Transceiver.**

*Courtesy Knight Electronics, Maywood, Ill.*

selected according to the amount of reduction or enlargement involved, to the extent dictated by legibility requirements.

> A line of medium thickness is recommended for general use on electrical diagrams. A thin line may be used for brackets, leader lines, etc. When emphasis of special features such as main or transmission paths is essential, a line thickness sufficient to provide the desired contrast may be used. Line thickness and lettering . . . shall, in general, conform to American National Standard Y14.2.*

Line conventions for electrical diagrams are shown in Fig. 21.3.

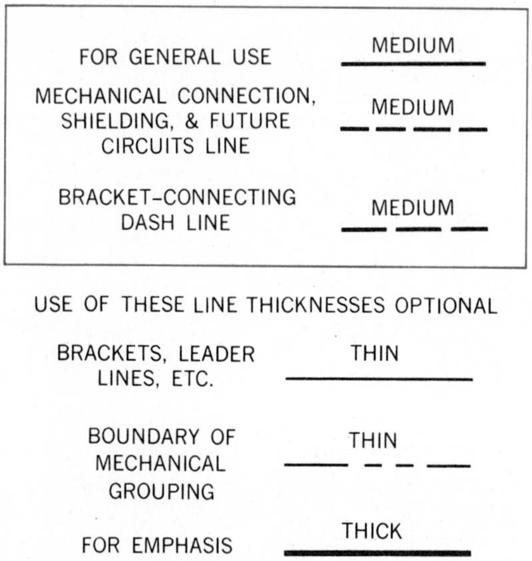

| | |
|---|---|
| FOR GENERAL USE | MEDIUM |
| MECHANICAL CONNECTION, SHIELDING, & FUTURE CIRCUITS LINE | MEDIUM |
| BRACKET-CONNECTING DASH LINE | MEDIUM |

USE OF THESE LINE THICKNESSES OPTIONAL

| | |
|---|---|
| BRACKETS, LEADER LINES, ETC. | THIN |
| BOUNDARY OF MECHANICAL GROUPING | THIN |
| FOR EMPHASIS | THICK |

Fig. 21.3 **Line Conventions for Electronic Diagrams** (ANSI Y14.15–1966).

## 21.5 Conventional Practices in Electrical Diagrams.

Symbols. As stated earlier, symbols used should conform to ANSI Y32.2–1970 or to some other nationally approved standard. If no symbol is available, a special symbol may be devised (or a standard symbol augmented) provided an explanatory note is included.

Size of Symbols. It is recommended in ANSI Y14.15–1966 that symbols be drawn roughly 1.5 times the size of the samples shown in ANSI Y32.2 (Appendix 31), provided anticipated reduction is no greater than 2.5 to 1. Following this recommendation, circular tube envelopes will be about .88″ to 1.00″ in diameter and circular envelopes for semiconductors will be about .62″ to .75″ in diameter, although the envelope for semiconductors may be omitted if no confusion results. Some textbooks on this subject contain much more detailed recommendations about symbol sizes, as well as spacing and arrangement.*

Switches and Relays. These in general should be shown in the "normal" position—with no operating force or applied energy. If exceptions are necessary, as for switches that may operate in several positions with no applied force, an explanatory note should describe the conditions shown.

Abbreviations. As in other fields of technical drawing, abbreviations on electrical diagrams should conform to ANSI recommendations.†

Grouping of Parts. When certain parts or components are naturally grouped as in separately obtained subassemblies or assembled components, such as relays, tuned circuit transformers, hermetically sealed units, and printed circuit boards, this grouping may be indicated by means of an enclosing dashed-line "box" as in Fig. 21.6, or by extra space from adjacent circuitry.

## 21.6 Single-Line Diagrams.

The single-line diagram is intended to describe the basic functions of a circuit or system. As such, it normally includes little of the detailed information and individual-parts identification of the schematic diagram. The individual lines connecting the symbols may represent single conductors or multiple conductors. The emphasis is on the function of each stage of a device rather than on the composition of the stage. A single-line diagram as used in electronics and communications is shown in Fig. 21.4.

Layout of a single-line diagram involves

*ANSI Y14.15–1966.

*Bibliography, Appendix 1. See, for instance, *Electronic Drafting* by Shiers.
†ANSI Z32.13–1950. See Appendix 4.

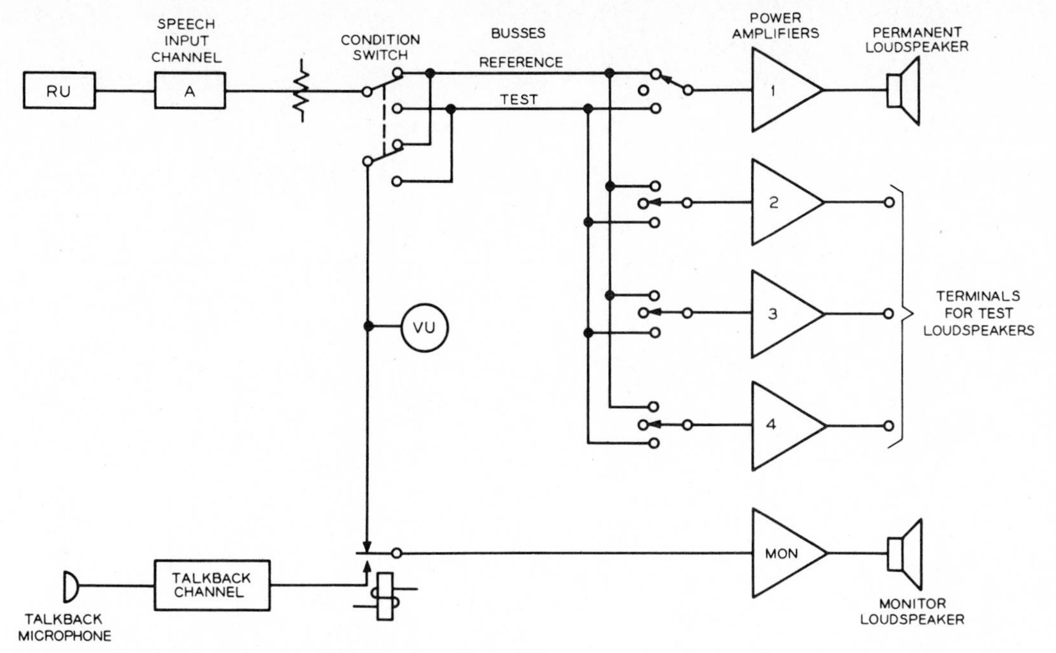


Fig. 21.4 **Single-Line Diagram** (ANSI Y14-15-1966).

essentially the same principles and procedures (except for lesser detail) suggested for schematic diagrams, §21.7.

## 21.7 Schematic Diagrams.

A convenient procedure for making a schematic (or single-line) diagram is to begin with a freehand sketch on cross-section paper, Fig. 21.5. It is frequently desirable to make the final drawing, Fig. 21.6, on tracing paper or film placed over similar cross-section paper. Note that some details are rearranged in Fig. 21.6 to improve the final diagram. The use of templates, such as that shown in Fig. 21.7, is a convenient time-saver in drawing symbols.

The various parts or symbols should be arranged so as to provide a pleasing balance between blank areas and lines. It is very important that sufficient blank spaces be provided adjacent to symbols for insertion of reference designations and notes. Exceptionally large spaces will, of course, give an unbalanced effect and should be avoided unless it is necessary to provide for possible later circuit additions.

SIGNAL PATH. It is standard practice to arrange schematic and single-line diagrams so that the signal or transmission path from input to output proceeds from left to right, and from

Fig. 21.5 **A Portion of a Preliminary Freehand Sketch of a Schematic Diagram.**

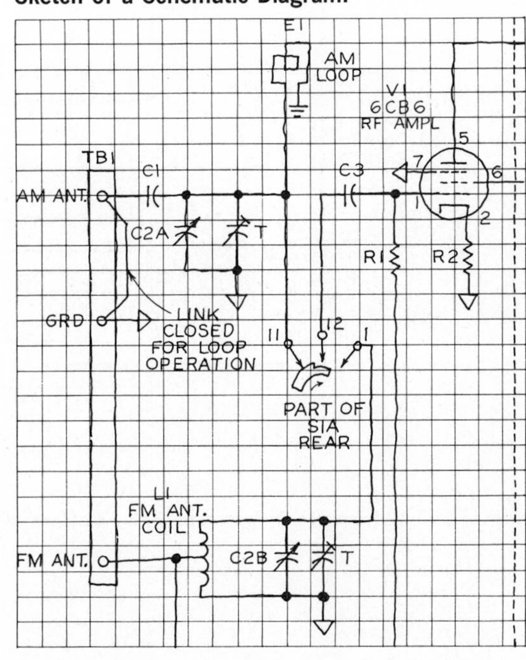

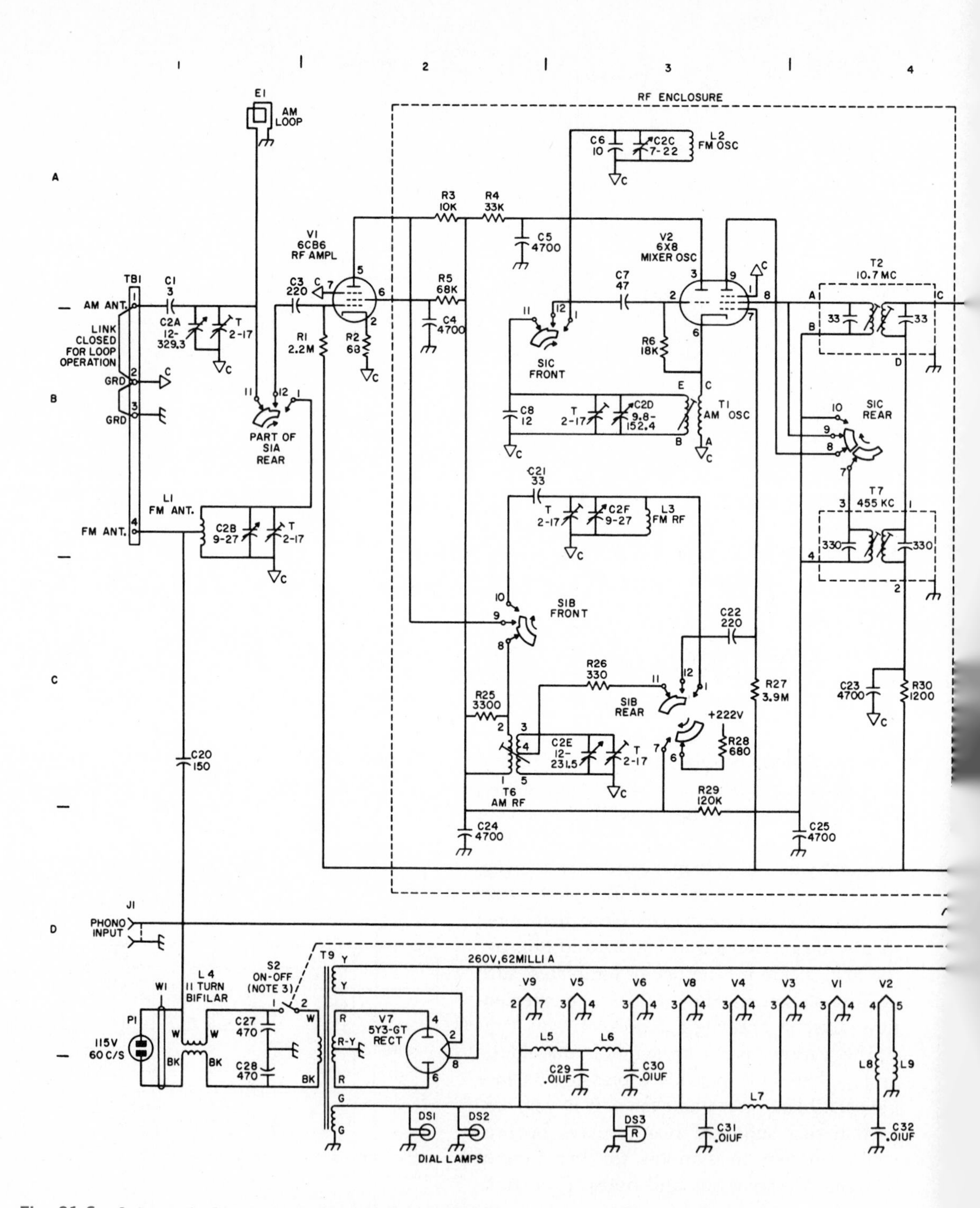

Fig. 21.6 **Schematic Diagram of AM–FM Radio Receiver, LH Half** (ANSI Y14.15–1966).

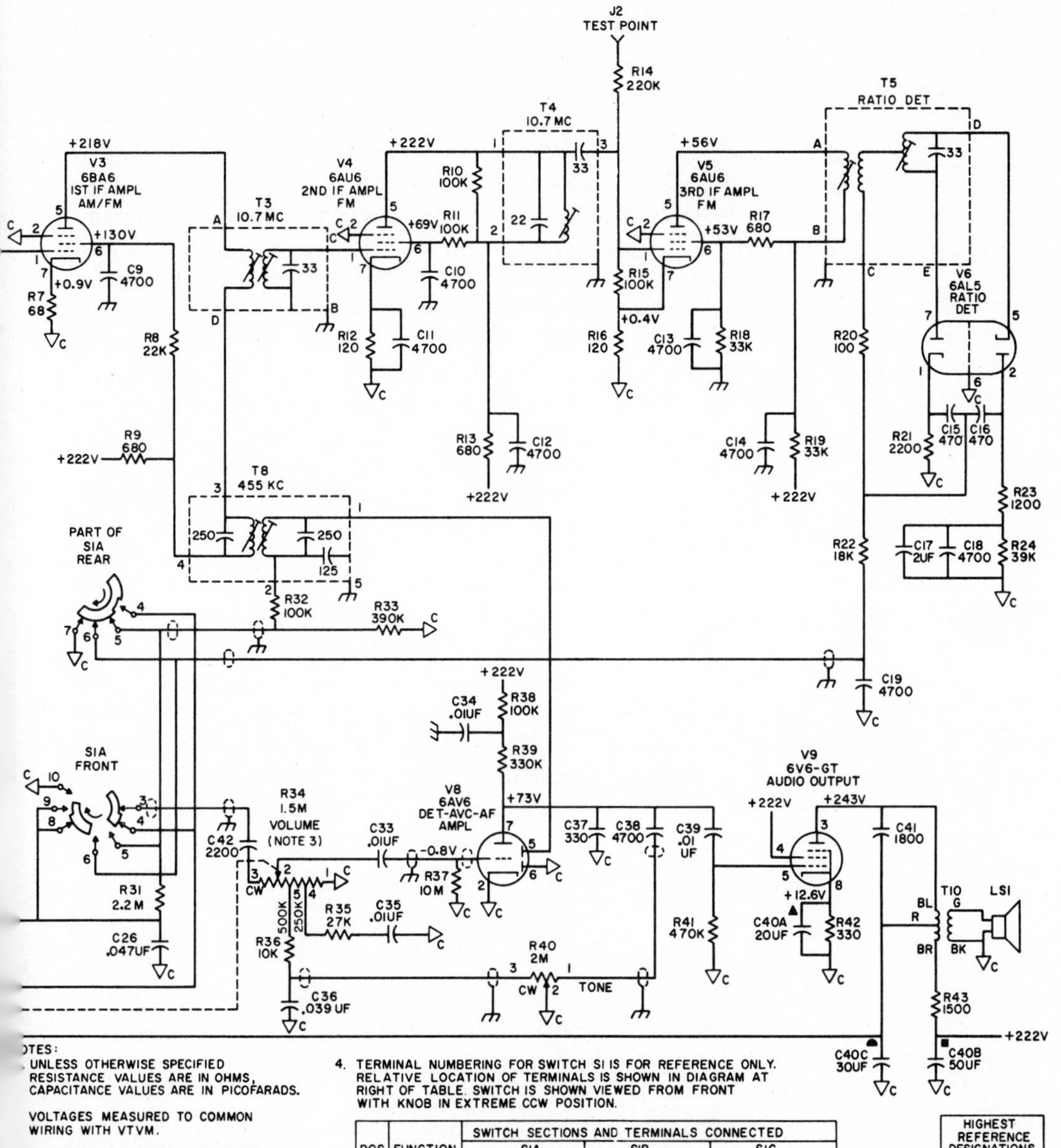


)TES:
. UNLESS OTHERWISE SPECIFIED RESISTANCE VALUES ARE IN OHMS, CAPACITANCE VALUES ARE IN PICOFARADS.

VOLTAGES MEASURED TO COMMON WIRING WITH VTVM.

SWITCH S2 IS PART OF POTENTIOMETER R34. TERMINAL NUMBERING IS FOR REFERENCE ONLY. RELATIVE LOCATION OF TERMINALS IS SHOWN IN DIAGRAM AT RIGHT.

R34 TERMINAL SIDE

4. TERMINAL NUMBERING FOR SWITCH S1 IS FOR REFERENCE ONLY. RELATIVE LOCATION OF TERMINALS IS SHOWN IN DIAGRAM AT RIGHT OF TABLE. SWITCH IS SHOWN VIEWED FROM FRONT WITH KNOB IN EXTREME CCW POSITION.

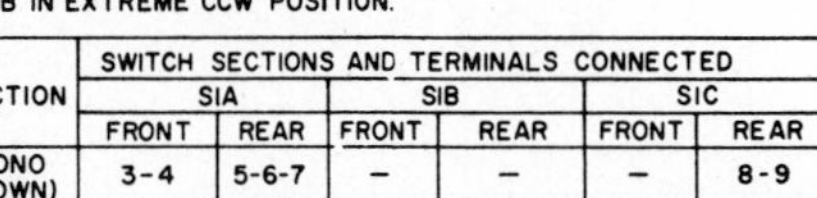

| POS | FUNCTION | SWITCH SECTIONS AND TERMINALS CONNECTED | | | | | |
|---|---|---|---|---|---|---|---|
| | | S1A | | S1B | | S1C | |
| | | FRONT | REAR | FRONT | REAR | FRONT | REAR |
| 1 | PHONO (SHOWN) | 3-4 | 5-6-7 | — | — | — | 8-9 |
| 2 | AM RADIO | 3-5, 8-9 | 4-6-7, 11-12 | 8-9 | 6-7, 11-12 | 11-12 | 7-8, 9-10 |
| 3 | FM RADIO | 3-6, 9-10 | 4-5-7, 12-1 | 9-10 | 6-7, 12-1 | 12-1 | 8-9 |

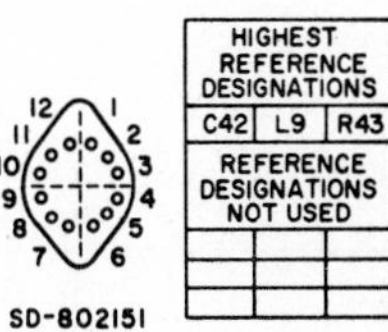

SD-802151

| HIGHEST REFERENCE DESIGNATIONS | | |
|---|---|---|
| C42 | L9 | R43 |
| REFERENCE DESIGNATIONS NOT USED | | |
| | | |
| | | |
| | | |

Fig. 21.6 **(Continued) RH Half.**

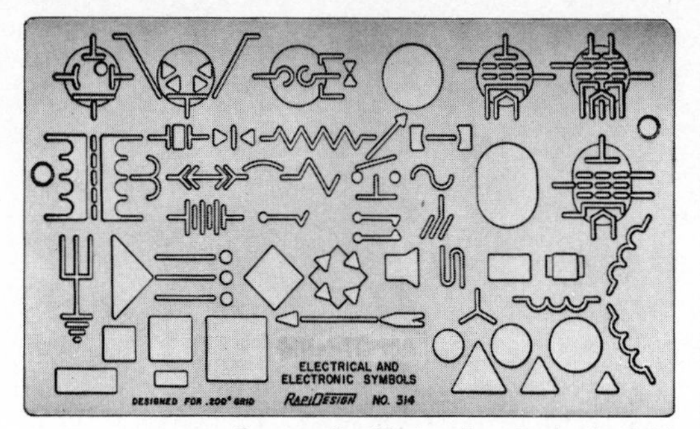


*Courtesy RapiDesign, Inc., Burbank, Calif.*

Fig. 21.7 **Electronic Symbol Template.**

top to bottom if it is necessary to draw the diagram in successive "layers." Supplementary circuits, such as a power supply and an oscillator circuit, are usually drawn below the main circuit.

*Stages* of an electronic device are groups of components, usually associated with an electronic tube or transistor or other semiconductor, which together perform one function of the device. For example, the sketch of Fig. 21.5 shows the RF (Radio Frequency) amplifier stage of the AM–FM Receiver whose schematic is shown in Fig. 21.6. Some other stages that may be observed in Fig. 21.6 are the Mixer-Oscillator stage, associated with electron tube V2, several IF (Intermediate Frequency Amplifier) stages, and a Ratio Detector stage (for the Frequency Modulation portion). These finally lead to the output stage (V9) and speaker. The power supply, which rectifies alternating house current through the use of V7 to produce direct-current voltage for the tube plate circuits, is shown at the lower left.

The experienced electronic draftsman is able to visualize fairly accurately the space required by the circuitry involved in each stage. He generally keeps the tube envelope ovals in horizontal lines and groups the associated circuitry in reasonably symmetrical arrangement about them. Thus, each stage is fairly well confined to an assigned area of the drawing. Note again that *the signal path from input to output follows the general path from left to right and top to bottom.* The input is at the upper left and the output at the lower right in Fig. 21.6. Supplementary circuitry (the power supply) is in the lower portion of the diagram.

*Connecting lines* (for conductors) are drawn horizontally or vertically, for the most part, minimizing bends and crossovers. Long interconnecting lines should be avoided. *Interrupted paths* may be used in place of long, awkward interconnecting lines or where a diagram occupies more than one sheet. When parallel connecting lines must be drawn close together, the spacing between lines should not be less than .06″ after reduction. As a further visual aid, parallel lines should be grouped with consideration of function, and with at least double spacing between groups.

*Crossovers* are not usually completely avoidable. When they are necessary, they should be handled with care. The looped crossovers of Fig. 21.8 (a) are not American National Standard but are still employed when there is possibility of confusion. A simpler practice recognized by ANSI is shown at (b). Connection of more than three lines at one point, shown at X, is not recommended and can usually be avoided by moving or "staggering" one or more lines as at Y.

The ANSI recommended practice is shown at (c).* In this system it is understood that

* ANSI Y14.15–1966.

Fig. 21.8 **Crossovers.**

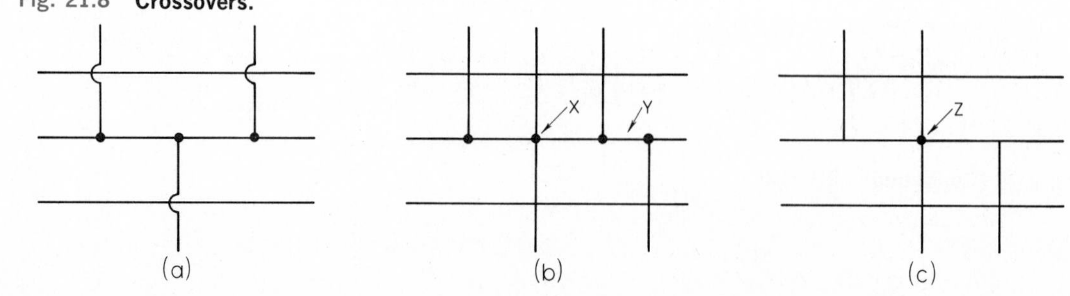

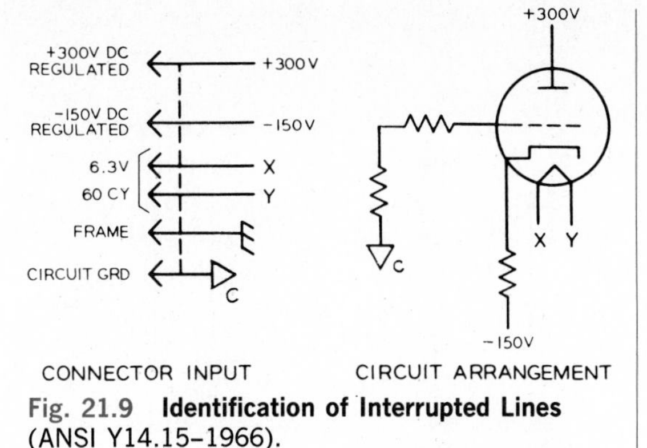


Fig. 21.9 **Identification of Interrupted Lines** (ANSI Y14.15–1966).

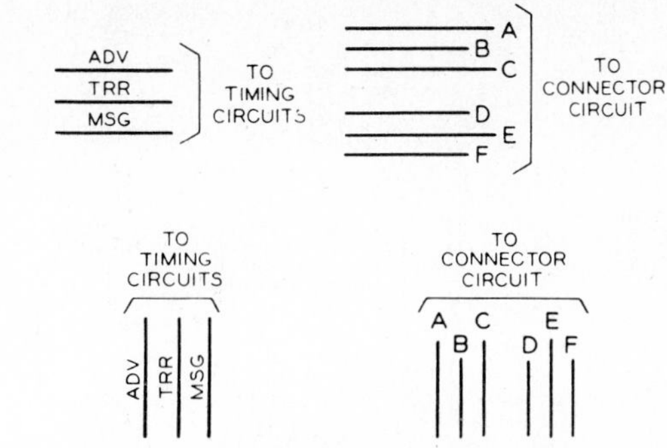


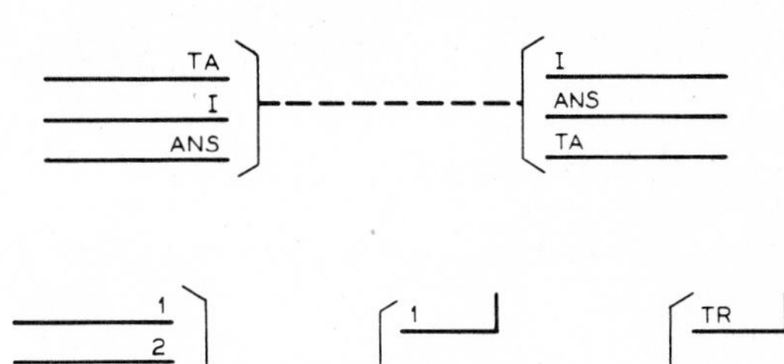


Fig. 21.10 **Typical Arrangement of Line Identifications and Destinations** (ANSI Y14.15–1966).

termination of a line signifies a connection. If more than three lines come together, as at Z, the dot symbol becomes necessary. This is avoidable, however, by staggering the connecting lines.

*Interrupted paths,* either for a single line or groups of lines, may be used where desirable for over-all simplification of a diagram, Figs. 21.9 and 21.10. These must be carefully labeled with letters, numbers, abbreviations, or other identification so that their destinations are unmistakable. Grouped lines are also bracketed as well as labeled, Fig. 21.10. When convenient, interrupted grouped lines may be connected by dash lines, as in Fig. 21.11.

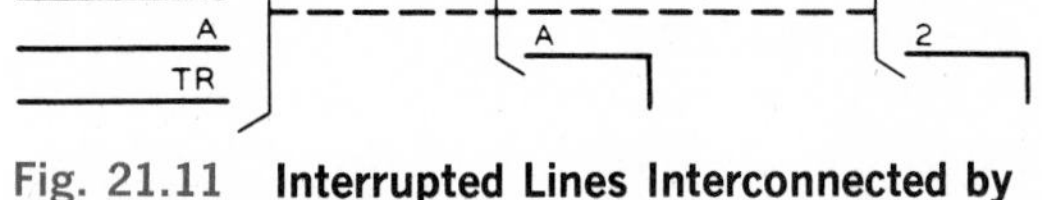


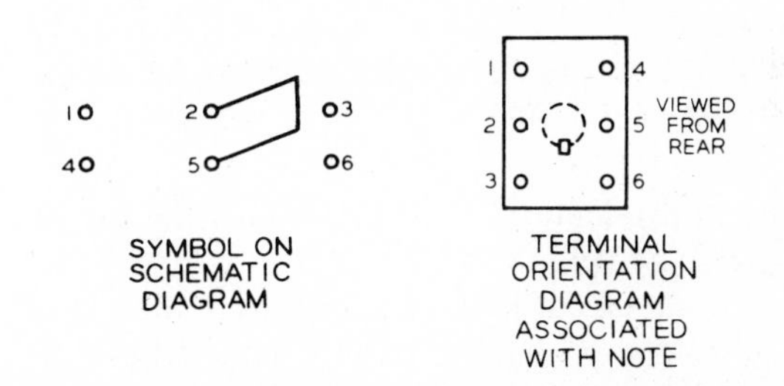


Fig. 21.11 **Interrupted Lines Interconnected by Dash Lines** (ANSI Y14.15–1966).

SYMBOL ON SCHEMATIC DIAGRAM

TERMINAL ORIENTATION DIAGRAM ASSOCIATED WITH NOTE

VIEWED FROM REAR

Fig. 21.12 **Terminal Identification—Toggle Switch** (ANSI Y14.15–1966).

Fig. 21.13 **Terminal Identification—Rotary Switch** (ANSI Y14.15–1966).

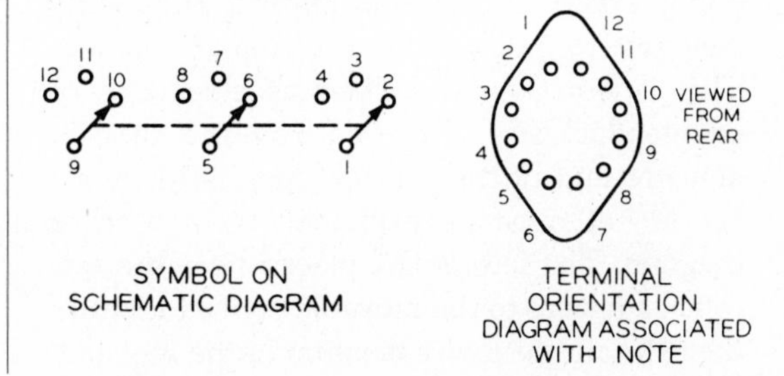


## 21.8 Terminals.

Terminal circles need not be shown unless they are needed for clarity and identification. Switch S1 (note its several parts) of Fig. 21.6 illustrates the use of terminal symbols and their identification. In this same figure, transformers T2, T3, and some others are shown without the terminal circles; instead, the terminal identifications are placed just outside the enclosures.

When actual physical markings appear on or near terminals of a component, these may be shown on the electrical diagram. Otherwise, for convenient reference it is well to assign arbitrary numbers or letters to the terminals. These designations must then be explained by a simple diagram which associates the arbitrarily assigned numbers or letters with the actual physical arrangement of terminals on the component, Figs. 21.12, 21.13, and

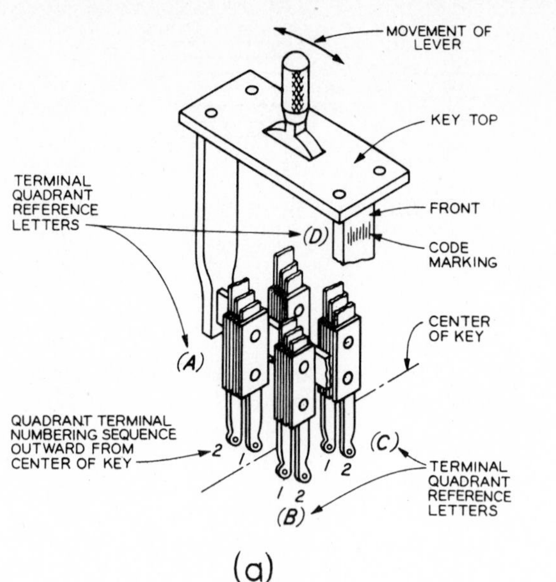


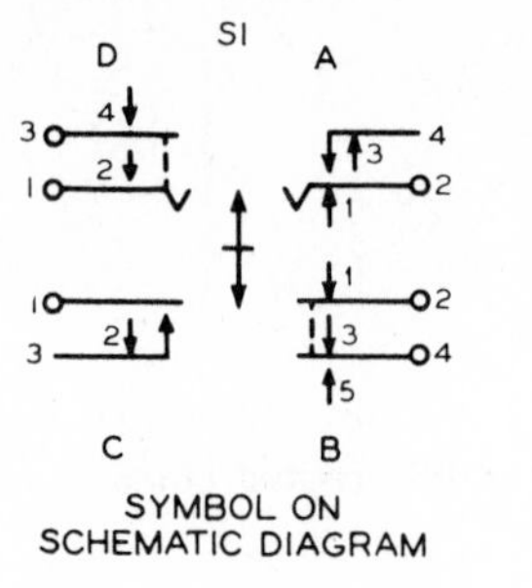


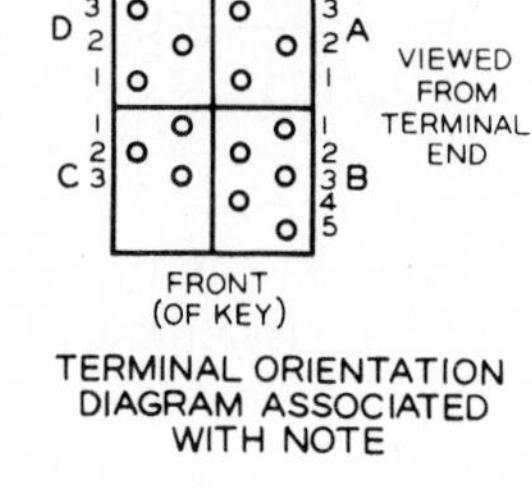


Fig. 21.14 **Terminal Identification—Lever-Type Key** (ANSI Y14.15–1966).

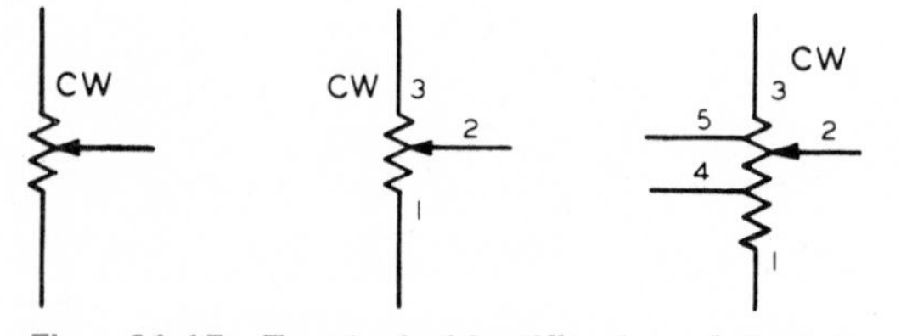


Fig. 21.15 **Terminal Identification—Adjustable Resistor** (ANSI Y14.15–1966).

21.14 (b). Figure 21.14 (a) is a pictorial explanation of (b).

Terminals or leads are frequently identified by colors or symbols, which should be indicated on the diagram. In Fig. 21.6, color coding is shown for transformers T9 and T10, while geometric symbol designation is shown for capacitor C40.

Terminal identification for rotary adjustable resistors follows practices similar to the foregoing, except that in addition it is frequently desirable to indicate the direction of rotation. This is normally described as clockwise or counterclockwise as viewed from the knob or actuator end of the control. Any contrary arrangement must be explained by a note or diagram. The letters CW placed near the terminal adjacent to the movable contact identify the extreme clockwise position of the movable terminal, Fig. 21.15. When terminals are numbered, Fig. 21.15, number 2 is assigned to the movable contact. Additional fixed taps are assigned sequential numbers as at the right.

Switch position, as it affects function, should also be shown on a schematic diagram. Depending on complexity, the designation may range from the simple ON–OFF of switch S2 in Fig. 21.6, to the functional labeling of the terminals in Fig. 21.16 (left), to the tabular form of function listing in Fig. 21.17 (right). The tabular forms indicate by dashes which terminals are connected for each position. For example, in Fig. 21.17 (right), position 2 connects terminal 1 to terminal 3, terminal 5 to terminal 7, and terminal 9 to terminal 11.

## 21.9 Division of Parts.

For clarity, sections of multielement parts may be drawn separately in a schematic diagram, with the subdivisions indicated by suffix letters. Thus, for the rotary switch of Fig. 21.18, the first portion S1A of switch S1 is identified as S1A FRONT and S1A REAR, viewed as usual from the actuator end. A similar symbol is used for switch S1 in Fig. 21.6. Note in Fig. 21.6 that sections S1B FRONT and S1B REAR and S1C

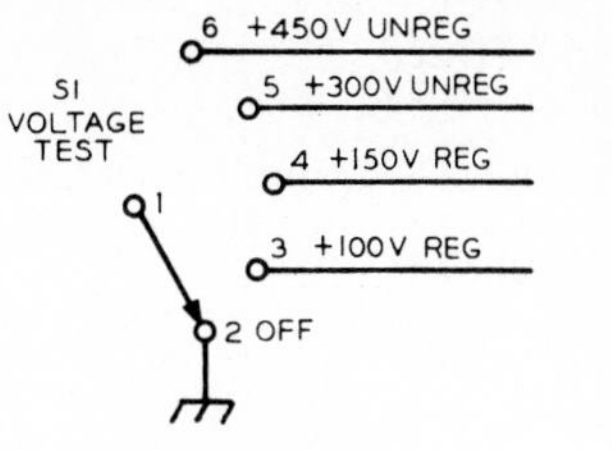

| S1 VOLTAGE TEST | |
|---|---|
| FUNCTION | TERM. |
| OFF | 1-2 |
| + 100 V REG | 1-3 |
| + 150 V REG | 1-4 |
| +300 V UNREG | 1-5 |
| +450 V UNREG | 1-6 |
| | |

FUNCTIONS SHOWN
IN TABULAR FORM

Fig. 21.16 **Position-Function Relationships for Rotary Switches—Optional Methods** (ANSI Y14.15-1966).

S1
REAR

(SWITCH VIEWED
FROM REAR)

SYMBOL ON
SCHEMATIC DIAGRAM

| S1 | | |
|---|---|---|
| POS | FUNCTION | TERM. |
| 1 | OFF(SHOWN) | 1-2,5-6,9-10 |
| 2 | STANDBY | 1-3,5-7,9-11 |
| 3 | OPERATE | 1-4,5-8,9-12 |
| | | |

FUNCTIONS SHOWN
IN TABULAR FORM

Fig. 21.17 **Position-Function Relationships for Rotary Switches—Tabular Method Only** (ANSI Y14.15-1966).

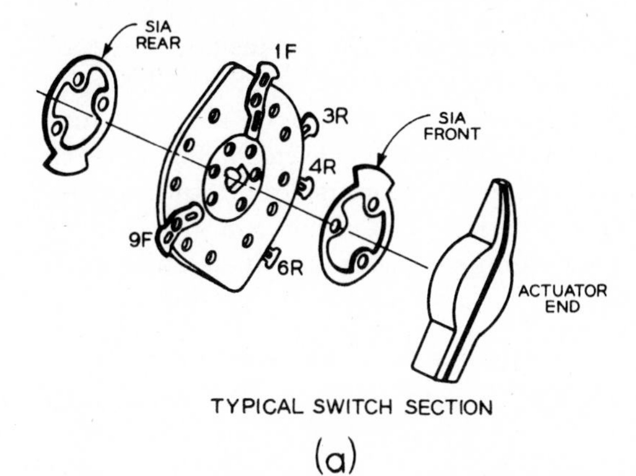


(a)

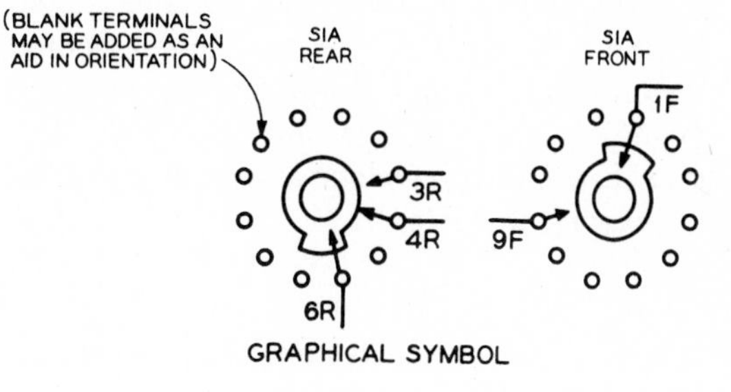


(b)

Fig. 21.18 **Development of a Graphical Symbol—Rotary Switch** (ANSI Y14.15-1966).

FRONT and S1C REAR are quite widely separated from section S1A, as is one portion of S1A REAR. Note also that the situation is further explained in NOTE 3.

Another example of suffixing is shown in Fig. 21.19. Here the two piezoelectric crystals A and B are enclosed in one unit, Y1. They are then referred to as Y1A and Y1B. A very common example is shown in Fig. 21.6 for capacitor C40. This is a triple capacitor (constructed as one unit) of which the sections C40A, C40B, and C40C are electrically independent.

Another method of identification of parts of components that are functionally separated on the diagram is with the PART OF prefix, Fig. 21.20 (left). This method has also been noted in connection with switch S1A REAR of Fig. 21.6. If the portion is indicated incomplete by a broken line, Fig. 21.20 (right), the words PART OF may be omitted.

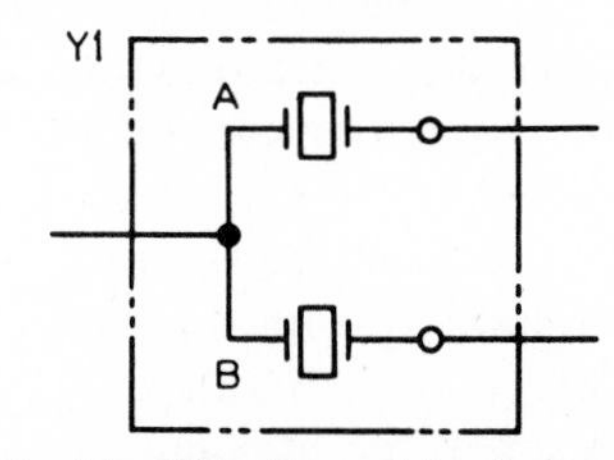


Fig. 21.19 **Identification of Parts by Suffix Letters** (ANSI Y14.15-1966).

Fig. 21.20 **Identification of Portions of Items** (ANSI Y14.15-1966).

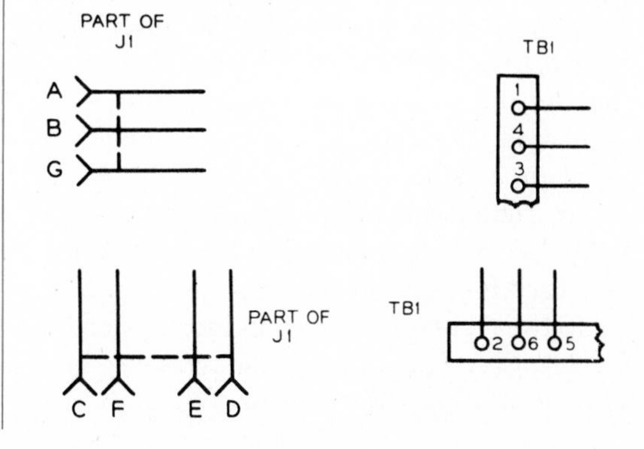

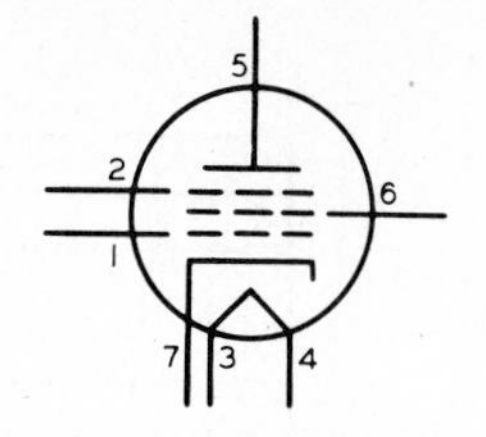


Fig. 21.21 **Terminal Identification for Electron Tube Pins** (ANSI Y14.15–1966).

## 21.10 Electron Tube Pin Identification.

Electron tube pins are conventionally numbered clockwise from the tube base key or other point of reference, with the tube viewed from the bottom. In the recommended method of pin identification on a diagram, the corresponding numbers are shown immediately outside the tube envelope and adjacent to the connecting line, Fig. 21.21.

## 21.11 Reference Designations.

An essential part of any electrical diagram is the identification of each separately replaceable part, shown with appropriate combinations of letters and numbers. In addition, portions not separately replaceable are frequently identified, as in previous examples. It should be noted that in electronics and communications, the recognized form is the appropriate letter followed by a number of the same size and on the same line, with no separating hyphen or space. This is turn is followed by any suffix needed, again with no hyphen.

The assignment of numbers for each category (resistances, capacitances, inductances, etc.) should start in the upper left corner of the diagram and proceed from left to right, then top to bottom, ending at the lower right corner. If in revision some parts are deleted, the remaining components should not be renumbered. The numbers not used should be listed in a table, such as Fig. 21.22. If the circuit contains many parts, it may also be helpful to include, as shown, the highest numbers used in each category. See also Fig. 21.6.

| HIGHEST REFERENCE DESIGNATIONS | |
|---|---|
| R65 | C35 |
| REFERENCE DESIGNATIONS NOT USED | |
| R7, R9<br>R60, R62 | C11, C14<br>C19, C23 |

Fig. 21.22 **Table Indicating Omitted and Highest Reference Designations** (ANSI Y14.15–1966).

Electron tubes and semiconductors such as transistors are identified by the reference designation and by the type number. It is frequently desirable also to include the tube function below the type number. All this information should be placed near the tube symbol—above it, if convenient, Fig. 21.23. In Fig. 21.2 are some examples of transistor reference designations, although in this case the transistor types were identified on another sheet.

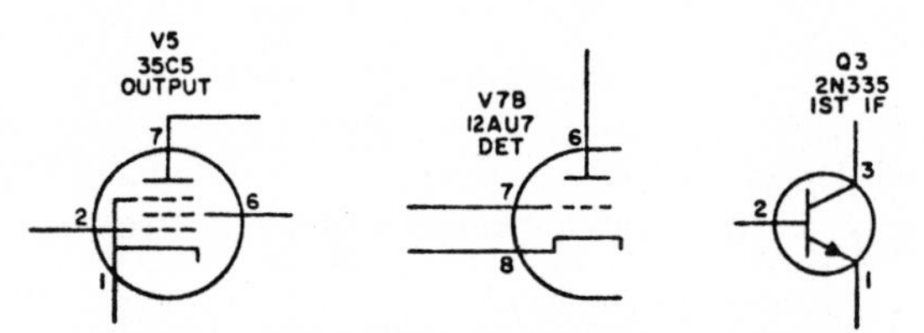


Fig. 21.23 **Reference Designation, Type Number, and Function for Electron Tubes and Semiconductors** (ANSI Y14.15–1966).

## 21.12 Numerical Values.

Along with reference designations, numerical values for resistance, capacitances, and inductances should be shown, preferably in the form employing the fewest ciphers. This may be done by a judicious combination of the multipliers shown in Fig. 21.24 (a) and the symbols in (b) and (c). Note that commas are not used in four-digit numbers: 4700 is recommended, not 4,700.

INDUCTANCE. According to magnitude, inductance may be expressed in *henries* (H), *millihenries* (mH or MILLI H), or *microhenries* ($\mu$H or UH). Following the principle of minimizing the number of ciphers, 2 $\mu$H is preferable to .002 mH, and 5 mH is better than either .005 H or 5000 UH.

| Multiplier | Prefix | Symbol | |
|---|---|---|---|
| | | Method 1 | Method 2 |
| $10^{12}$ | tera | T | T |
| $10^{9}$ | giga | G | G |
| $10^{6}$ (1,000,000) | mega | M | M |
| $10^{3}$ (1000) | kilo | k | K |
| $10^{-3}$ (.001) | milli | m | MILLI |
| $10^{-6}$ (.000001) | micro | $\mu$ | U |
| $10^{-9}$ | nano | n | N |
| $10^{-12}$ | pico | p | P |
| $10^{-15}$ | femto | f | F |
| $10^{-18}$ | atto | a | A |

(a) MULTIPLIERS

| Range in Ohms | Express as | Example |
|---|---|---|
| Less than 1000 | ohms | .031<br>470 |
| 1000 to 99,999 | ohms or kilohms | 1800<br>15,853<br>10 k<br>22 K |
| 100,000 to 999,999 | kilohms or megohms | 380 k<br>.38 M |
| 1,000,000 or more | megohms | 3.3 M |

(b) RESISTANCE

| Range in Picofarads | Express as | Example |
|---|---|---|
| Less than 10,000 | picofarads | 152.4 pF<br>4700 PF |
| 10,000 or more | microfarads | .015 $\mu$F<br>30 UF |

(c) CAPACITANCE

Fig. 21.24 **Numerical Values for Components.**

*Note:* Much repetition of units of measurement may be avoided by notes such as the following: UNLESS OTHERWISE SPECIFIED, RESISTANCE VALUES ARE IN OHMS AND CAPACITANCE VALUES ARE IN MICROFARADS.

PART VALUE PLACEMENT. For clarity, numerical values should be immediately adjacent to the symbol. Suggested forms are shown in Fig. 21.25.

## 21.13 Functional Identification and Other Information.

It sometimes contributes to readability of a diagram to include special functional identification of certain parts or stages. In particular, where functional designations (TUNER, OUTPUT, etc.) are to be shown on a panel or chassis surface, they should also be shown on the electrical diagram in an appropriate place.

*Test points* may be identified by the words TEST POINT on the drawing, Fig. 21.6 (upper right, zone A7). If several are to be shown, the abbreviated form TP1, TP2, and so on may be used.

Still other information that may be shown on schematic diagrams, if desired, includes the following:

DC resistance of winding and coils.
Critical input or output impedance values.
Voltage or current wave shapes at selected points.
Wiring requirements for critical ground points, shielding pairing, etc.
Power or voltage ratings of parts.
Indication of operational controls or circuit functions.

## 21.14 Printed Circuits.

In the interests of miniaturization and mass production, *printed circuit boards* have come into wide usage comparatively recently. A variety of procedures is used in the fabrication of such boards. A typi-

Fig. 21.25 **Methods of Reference Designation and Part Value Placement** (ANSI Y14.15–1966).

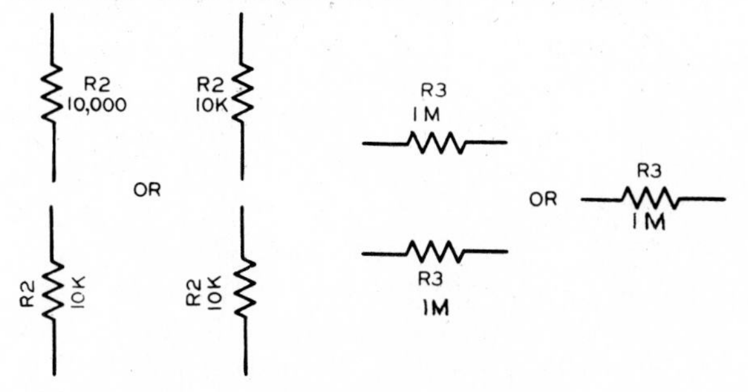

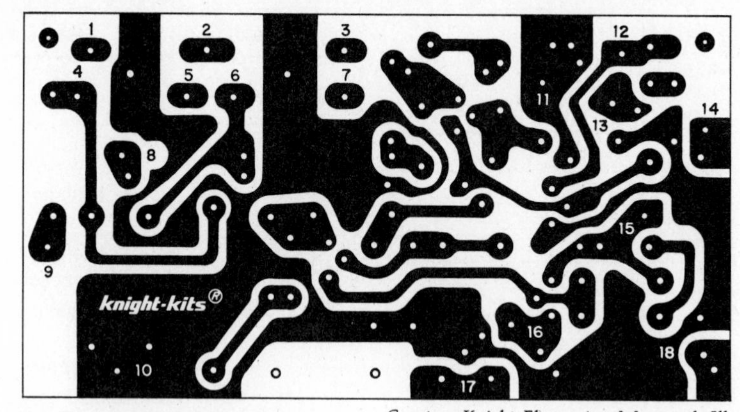


Fig. 21.26 **Printed Circuit Master Drawing.** *Courtesy Knight Electronics, Maywood, Ill.*

cal board is made from a commercially available laminated phenolic plastic sheet with a metallic foil, such as copper, bonded to one or both sides.

The master drawing for a printed circuit is usually made at a scale of 2 to 1 or 4 to 1 on a stable drawing material such as polyester film. A master drawing for a one-sided printed-circuit board is shown in Fig. 21.26. (It is also possible to "print" certain components.) The circuit is then reproduced photographically on the foil surface of the board and is protected by a coating or "resist." The unprotected metal is etched away, leaving the circuits as originally designed. Mounting holes for components and hardware are drilled or punched, and the board is then ready for mounting and soldering the circuit components.

**21.15 Electronic Diagram Problems.** The following exercises in the construction of schematic diagrams are designed for Layout A–3. If freehand sketches are assigned, $8\frac{1}{2}'' \times 11''$ cross-section paper with $\frac{1}{4}''$ grid squares is suggested. For a mechanical drawing, tracing paper or film is convenient, the drawing being preceded by a sketch on cross-section paper to work out the details of arrangement and spacing. Refer to Appendix 31 for standard graphical symbols.

Fig. 21.27 **Crystal Calibrator** (Prob. 21.1).

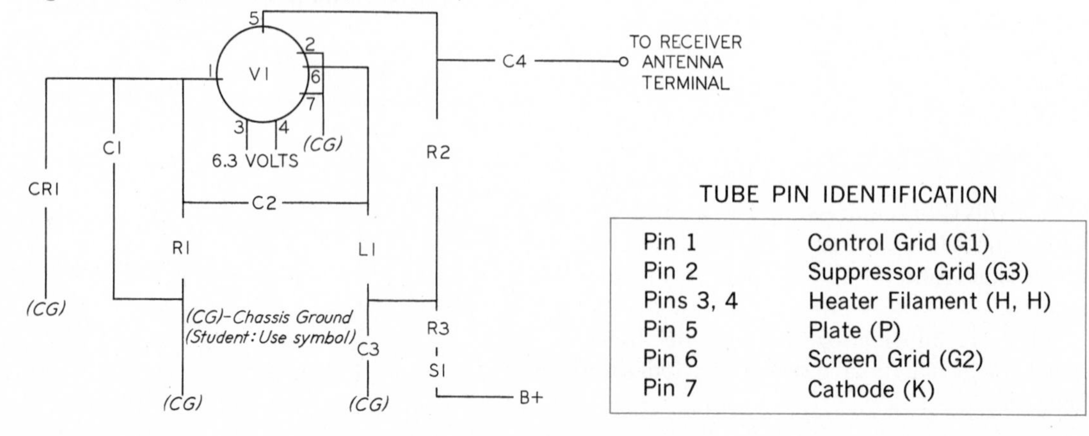


TUBE PIN IDENTIFICATION

| | |
|---|---|
| Pin 1 | Control Grid (G1) |
| Pin 2 | Suppressor Grid (G3) |
| Pins 3, 4 | Heater Filament (H, H) |
| Pin 5 | Plate (P) |
| Pin 6 | Screen Grid (G2) |
| Pin 7 | Cathode (K) |

**Prob. 21.1** The unit shown in Fig. 21.27 is used to calibrate the tuning dials of communications receivers by providing a check signal every 100 kilohertz throughout the commonly used communication bands.

Reproduce the schematic diagrams approximately double size, freehand or mechanically, as assigned. Add standard symbols and include reference designations, component values, and other data following recommended practices (Layout A–3).

COMPONENTS LIST

| Capacitors | | Resistors (ohms) | |
|---|---|---|---|
| C1 | 7-45 PF Trimmer | R1 | 4.7M |
| C2 | 3.3 PF | R2 | 0.47M |
| C3 | .01 UF | R3 | 22K |
| C4 | 10 PF | | |
| **Crystal** | | **Tube** | |
| CR1 | 100 KC | V1 | 6AK6 (Pentode) |
| **Inductor** | | **Switch** | |
| L1 | RF Choke, 5 MILLI H, Iron Core | S1 | SPST (Single Pole, Single Throw) |

**Prob. 21.2** The circuit shown in Fig. 21.28 is suitable for classroom or laboratory demonstrations. After a local radio station is tuned with a regular 9-volt battery in place, other devices such as photoelectric cells and homemade batteries are substituted.

Make a freehand or mechanical schematic diagram of the circuit, approximately twice the size of Fig. 21.28. Use standard symbols and show component values and reference designations, following recommended practices (Layout A–3).

COMPONENTS LIST

| | | | |
|---|---|---|---|
| C1 | 365 PF Variable Capacitor | T1 | Output Transformer, 3–4 ohm |
| D1 | Germanium Diode | TR1 | 2N170 Type NPN Transistor (*B* = *Base*, *C* = *Collector*, *E* = *Emitter*) |
| L1 | Loopstick Antenna Coil with Center Tap | | |
| | | TR2 | 2N107 Type PNP Transistor |

Fig. 21.28 **Circuit for Demonstrating the Generation of Electrical Energy** (Prob. 21.2).

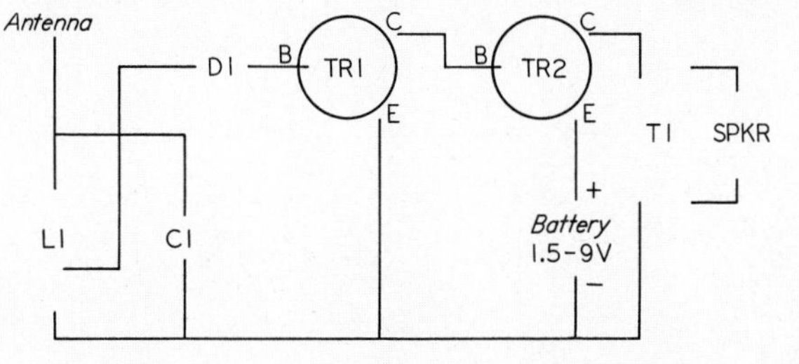

**Prob. 21.3** The circuit of Fig. 21.29 provides fundamental frequencies for transmitters operating in the 2-meter and 6-meter amateur radio ("ham") bands.

Make a freehand or mechanical schematic diagram, approximately twice the size of Fig. 21.29. Use standard symbols and include reference designations and component values, following recommended practices (Layout A–3).

COMPONENTS LIST

Capacitors

| | |
|---|---|
| C1 | 220 PF |
| C2 | 5.5-20 PF *Variable* |
| C3 | 4.5-25 PF *Variable* |
| C4, C5 | 390 PF |
| C6, C7 | .001 UF |
| C8 | 2.3-14.2 PF *Variable* |
| C9 | 100 PF |

Resistors (ohms)

| | |
|---|---|
| R1 | 68 |
| R2 | 47K |
| R3 | 5K Wire-wound |

Inductors

| | |
|---|---|
| L1 | 32 turns, slug-turned |
| L2 | RF choke .75 MILLI H |
| L3 | 26 turns, slug-turned |

Voltage Regulator Tube and Pin Identification
VR 1 Type OA2 Cold cathode, gas-filled

| | |
|---|---|
| Pins 1, 5 | Plate (P) |
| Pins 2, 4, 7 | Cathode (K) |

Electron Tube and Pin Identification
V 1 Type 6417 Pentode

| | |
|---|---|
| Pin 1 | Plate (P) |
| Pin 2 | Not connected |
| Pin 3 | Suppressor Grid (G3) |
| Pins 4, 5 | Heater (H, H) |
| Pin 6 | Screen Grid (G2) |
| Pin 7 | Cathode (K) |
| Pins 8, 9 | Control Grid (G1) |

Fig. 21.29 **Variable-Frequency Oscillator** (Prob. 21.3).

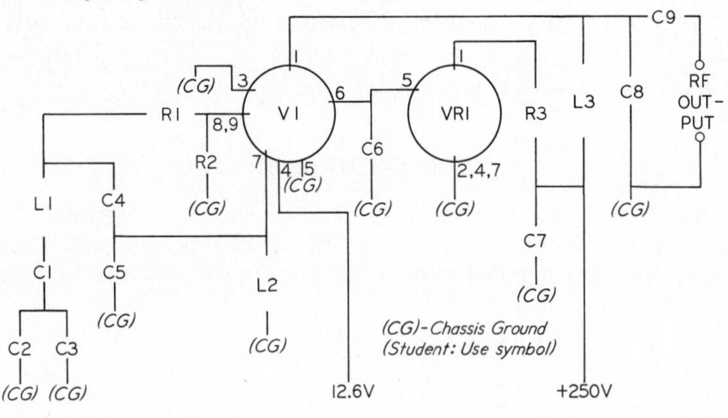

**Prob. 21.4** Make a freehand or mechanical schematic diagram of the circuit shown in Fig. 21.30, approximately double size. Use standard symbols and show reference designations and component values, following recommended practices (Layout A-3).

*Note:* By rearranging slightly a few connecting lines, you can eliminate the three dot symbols.

COMPONENTS LIST

**Capacitors**

| | |
|---|---|
| C1 | 10 PF |
| C2 | 10-365 PF *Variable* |
| C3 | 90 UF |
| C4 | .005 UF |
| C5 | 5 UF |
| C6 | .001 UF |
| C7 | 5 UF |
| C8 | .01 UF |
| C9 | .005 UF |
| C10 | .005 UF |
| C11 | 90 UF |
| C12 | .005 UF |

**Resistors (ohms)**

| | |
|---|---|
| R1 | 47K |
| R2 | 82K |
| R3 | 2200 |
| R4 | 1000 |
| R5 | 10K |
| R6 | 82 |
| R7 | 10K *Variable (Potentiometer)* |
| R8 | 82K |
| R9 | 5600 |
| R10 | 3900 |

**Miscellaneous**

| | |
|---|---|
| S1 | On-Off Switch *(mounted on R-7)* |
| CR 1 | Diode Detector |

**Coils**

| | |
|---|---|
| L1 | Loopstick Antenna *(same symbol as magnetic core transformer)* |
| L2 | Magnetic core, 0.5 MILLI H |
| L3 | Magnetic core, 1.0 MILLI H |

**Transistors**

*(E = Emitter, B = Base, C = Collector)*

| | |
|---|---|
| TR1 | PNP Type, RF-AF Amplifier |
| TR2 | PNP Type, Audio Amplifier |

Fig. 21.30 **Transistor Pocket Radio** (Prob. 21.4).

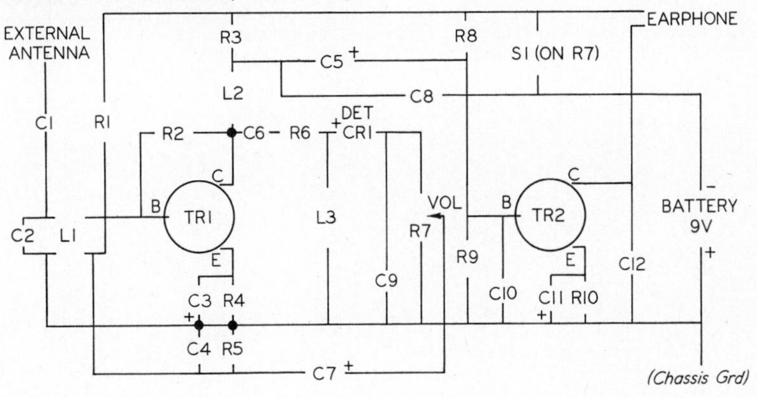

**Prob. 21.5** Figures 21.31 (a) and (b) are pictorial wiring diagrams of a transistorized code practice oscillator. The unit provides for practicing International Morse Code either by listening to an audible tone over headphones or speaker or by watching a flashing light.

Study the illustrations and the components list caredully, making preliminary sketches of connections. Then make a complete schematic diagram of the circuit following recommended practices and including component values and reference designations (Layout A–3).

*Note:* Terminal Strips TS–1 and TS–2 and the Battery Holder are for mechanical function and convenience in wiring. They should *not* be shown in a schematic diagram.

COMPONENTS LIST

Capacitors

C1 .075 UF

Resistors (ohms)

R1 100K
R2 100
R3 220
R4 50K *Variable (Potentiometer)*
R5 10
R6 10

Transistors

(*E = Emitter, B = Base, C = Collector*)

TR1 Oscillator, type NPN
TR2 Amplifier, type PNP

Miscellaneous

Battery "C" Type, $1\frac{1}{2}$ Volt
Battery Holder
Jack 2-conductor (*for headphones*)
Key Telegraph Key (*not shown—connected between terminals 1 and 2 of TS2*)
Lamp
Lamp Socket
Speaker
Switch 3-position Slide Switch
Connections:
Position One (OFF-LIGHT)—Terminals 1–5, 3–4
Position Two (SPEAKER)—Terminals 1–6, 2–4
Position Three (PHONES)—Terminals 2–5
Terminal Strips TS-1 and TS-2

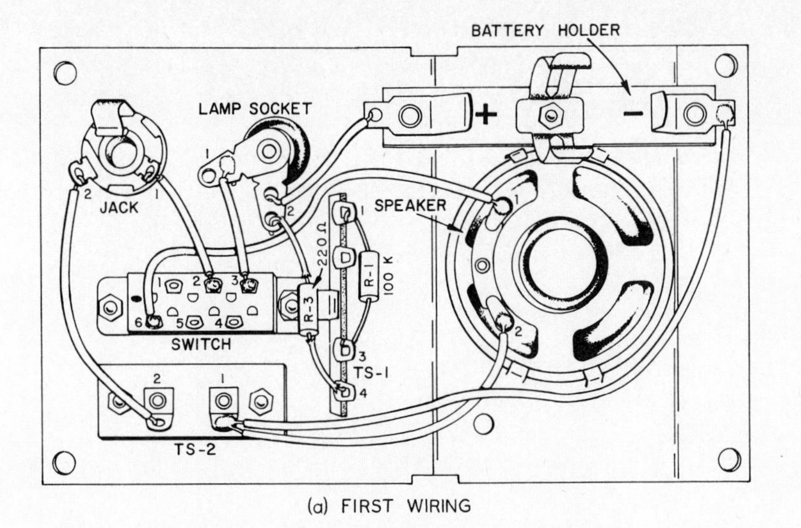


(a) FIRST WIRING

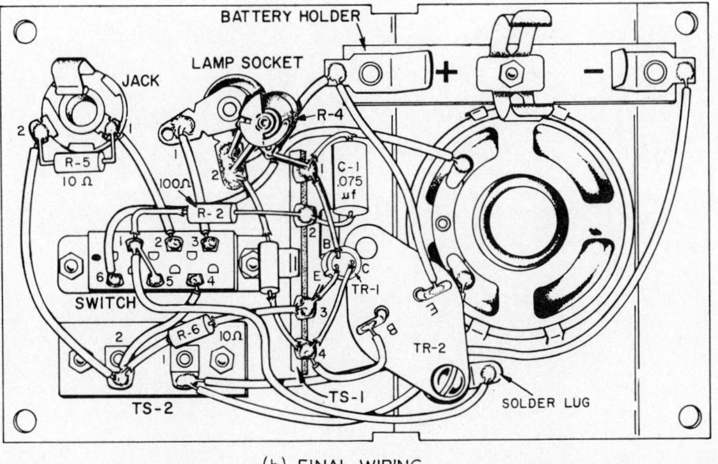


(b) FINAL WIRING

*Courtesy Knight Electronics, Maywood, Ill.*

Fig. 21.31 **Code Practice Oscillator** (Prob. 21.5).

# 22 structural drawing

by E. I. Fiesenheiser*

**22.1 Structural Drawing.** *Structural drawing* consists of the preparation of design and working drawings for buildings, bridges, tanks, towers, and other structures, and comprises a very large field for the draftsman, as distinguished from the making of drawings for machines and machine parts. Although the basic principles used are the same for both types of work, certain of the methods of representation are different in structural drafting. The structural draftsman should be somewhat familiar with structural design principles. He must know a great deal about materials and the methods of fastening various members together in structures. It is also important that he be able to design connections of adequate strength to transmit the forces in a member to the other members to which it is joined.

Ordinarily the designing civil engineer determines the form and shape of a structure as well as the sizes of the main members to be used. The draftsman then makes the detail drawings, frequently under the engineer's supervision. Thus, the draftsman is the engineer's first assistant and has the opportunity to learn something about design. In fact, in many cases, the draftsman's position is regarded as a stepping stone to one of greater responsibility.

The materials most commonly used in construction are wood, mild steel, concrete (plain, reinforced, and prestressed), structural clay products, and stone masonry. This chapter

*Professor Emeritus of Civil Engineering, Illinois Institute of Technology.

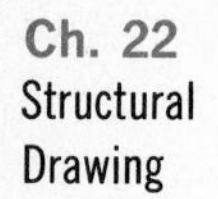

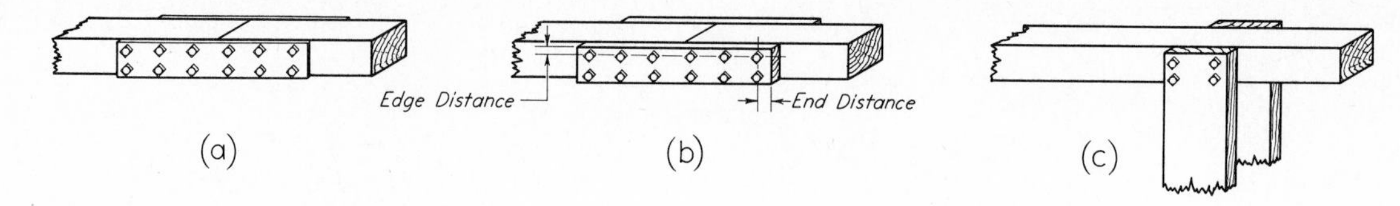


Fig. 22.1 **Typical Bolted Joints.**

will be limited to discussions of the uses of these materials.

For structural drafting standards applied to structures used in connection with mechanical or electrical equipment, see ANSI Y14.14-1961, "Mechanical Assemblies."

## 22.2 Wood Construction.

Many different types of wood are used as structural timber, among which are ash, birch, cedar, cypress, Douglas fir, elm, oak, pine, poplar, redwood, and spruce. Authentic information concerning the strength properties of the various types and grades can be obtained from the *Wood Handbook No. 72* prepared by the U.S. Forest Products Laboratory in 1955.* For allowable stresses and design of connections, the student should refer to *National Design Specification for Stress-Grade Lumber and Its Fastenings*, published by the National Forest Products Association, Washington, D.C.

Wood is in common use in the construction of homes and other buildings in the form of sills, columns, studs, floor joists, roof rafters, purlins, trusses, and roof sheathing. Typical drawings of wood structures and construction details may be obtained from the various publications of the National Forest Products Association. Common methods of fastening timber members together involve the use of nails, screws, lag screws, drift bolts, bolts, steel plates, and various special timber connectors. Ordinarily a structural timber is cut so that the wood fibers, or grain, run parallel to the length. The strength resistance of wood is not the same in a direction perpendicular to the grain as it is parallel to the grain. Therefore, in designing connections, the direction of the force to be transmitted must be taken into account. Also, a proper spacing, edge distance, and end distance must be maintained for screws, bolts, and other connectors.

Typical bolted joints are shown in Fig. 22.1. To transmit the forces, either steel or wood splice plates may be used, as shown at (a) and (b). A detail without splice plates is shown at (c). It should be realized that each type of connection requires a different design.

The use of *split-ring metal connectors*, Fig. 22.3, is now common. In Fig. 22.2 is shown a drawing of the left half of a timber roof truss in which this type of connector is used. The left half only is drawn, since the truss is symmetrical about the center line. By referring to the views of the top and bottom chords, the student may visualize the relative positions of the connecting members of the structure. In drawings of trusses, the view of the top chord is simply an auxiliary view. However, it is customary to show the lower chord by means of a section taken just above the lower chord with the *line of sight downward.* The lower chord, therefore, is shown in first-angle projection, §6.38.

Ordinarily, wood is surfaced or dressed. Because of the wood loss in surfacing, a nominal 1″ thickness is reduced to a minimum of $\frac{3}{4}$″ for dressed thickness if dry, and $\frac{25}{32}$″ if the wood is green. For nominal dimensions from 2″ to 4″, the dressed sizes, dry, are less in each case by $\frac{1}{2}$″, or by $\frac{7}{16}$″ if green. For 5″, 6″, and 7″ dimensions, the dressed-size reduction is $\frac{1}{2}$″ for dry wood and $\frac{3}{8}$″ for green wood. For 8″ and greater sizes, the reduction is $\frac{3}{4}$″ for dry and $\frac{1}{2}$″ for green wood. Lumber is considered "green" if its moisture content exceeds 19 percent. The symbols S2S and S2E indicate, respectively, surfacing of two sides and two edges. If all four faces are to be surfaced, the symbol S4S is used. The working drawings must show whether standard dressed, standard rough, or special sizes are to be used.

*Superintendent of Documents, U.S. Government Printing Office, Washington, D.C. 20402.

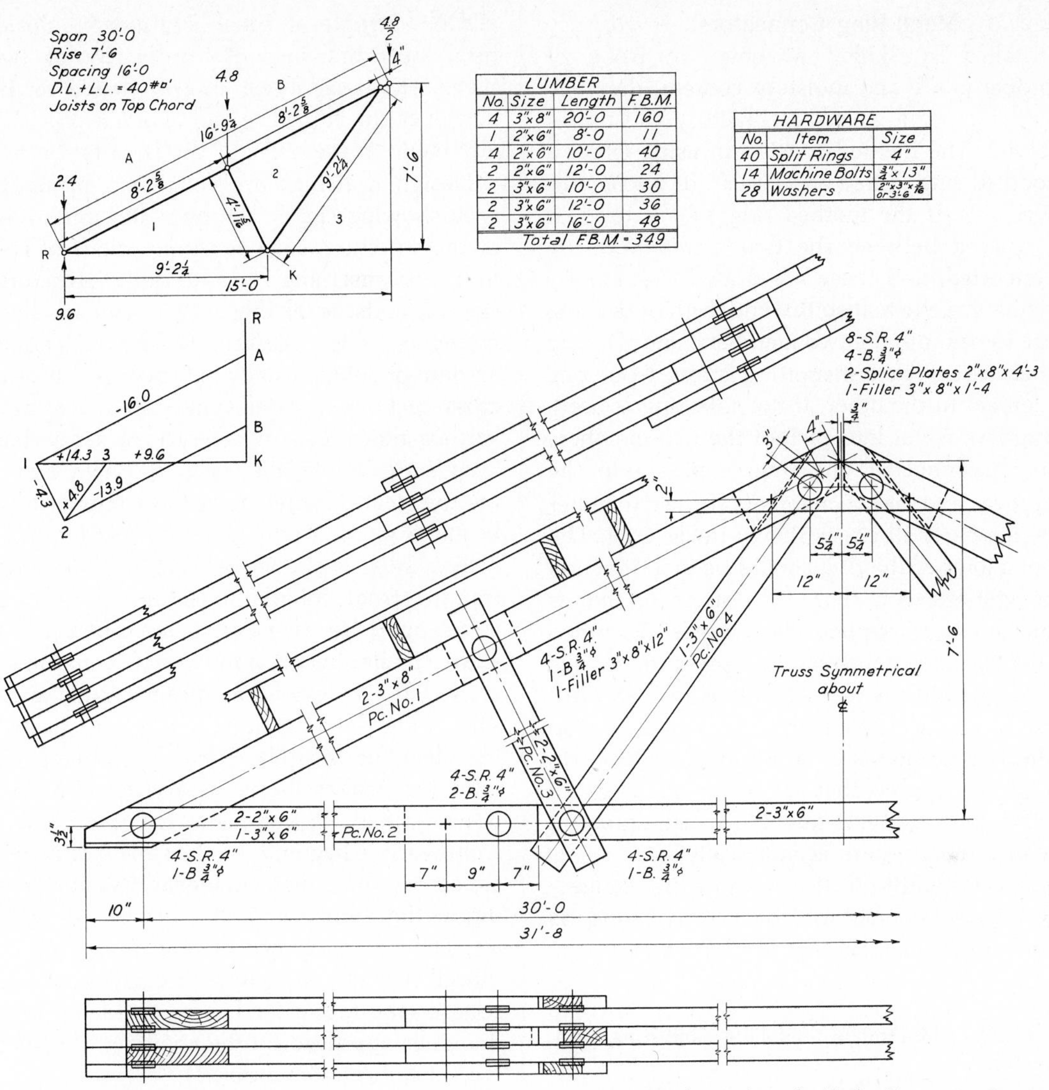


| LUMBER | | | |
|---|---|---|---|
| No. | Size | Length | F.B.M. |
| 4 | 3"x8" | 20'-0 | 160 |
| 1 | 2"x6" | 8'-0 | 11 |
| 4 | 2"x6" | 10'-0 | 40 |
| 2 | 2"x6" | 12'-0 | 24 |
| 2 | 3"x6" | 10'-0 | 30 |
| 2 | 3"x6" | 12'-0 | 36 |
| 2 | 3"x6" | 16'-0 | 48 |
| Total F.B.M. = 349 | | | |

| HARDWARE | | |
|---|---|---|
| No. | Item | Size |
| 40 | Split Rings | 4" |
| 14 | Machine Bolts | 3/4"x 13" |
| 28 | Washers | 2"x3"x 3/16 or 3-6 |

*Based on a design by Timber Engineering Company*

Fig. 22.2 **A Roof Truss.**

Fig. 22.3 **Metal Connectors for Timber Structures.**

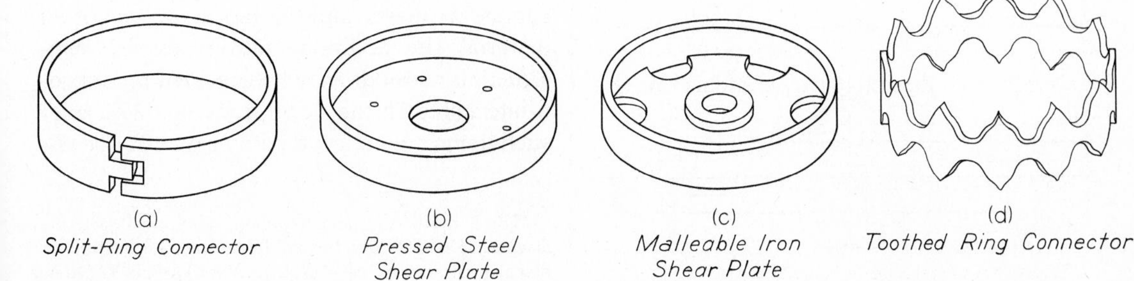

## 22.3 Metal Ring Connectors.

If properly installed by skilled workmen in wood of proper grade and moisture content, the *metal ring connector* is a very satisfactory and useful device. The method consists in using either a toothed ring, called an *alligator,* or a *split ring,* Fig. 22.3. If the toothed ring, (d), is used, it is placed between the two members to be connected, and these are drawn together by tightening the bolt so that the teeth of the ring are forced into the two members, and the ring thus assists in transmitting stress from one member to the other. If the split ring is used, a groove is cut into each of the two members to be connected, the ring is placed in the grooves, and the two members are held together by means of a bolt, as shown in Fig. 22.4. The open joint of the ring should be in a direction at right angles to that of the stress, so that as the stress is applied the ring is deformed slightly and transmits the pressure to the wood within the ring as well as to that without. With this connection, the tensile and shearing strengths of wood are developed to a higher degree than by other methods of connection, and it is possible to use timber in tension much more economically.

It is standard practice to show ring connectors by solid lines in order to save time, as shown in Fig. 22.2.

Fig. 22.4 **Method of Installing Split-Ring Connectors.**

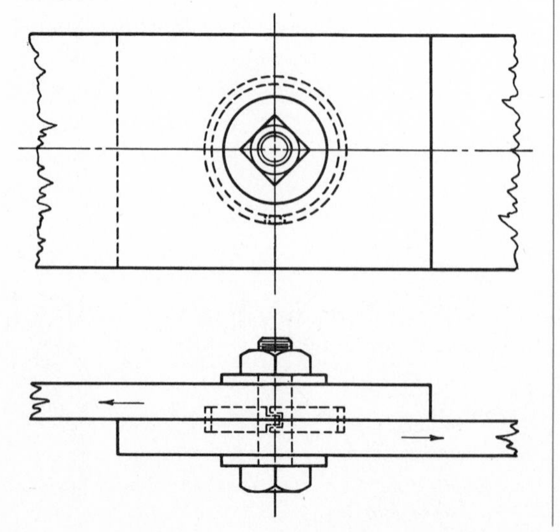

## 22.4 Structural Steel Drafting.*

Structural steel drawings are ordinarily of two types: *engineering design drawings* made in the design engineer's office and *shop drawings* usually made in the office of the steel fabricator.

Design drawings are concerned primarily with showing clearly the over-all dimensions of the structure, such as the locations of columns, beams, angles, and other structural shapes, and the listing of the sizes of these members. It is necessary to show a certain amount of detail, usually in the form of typical cross sections, special connections required, various notes, etc. In the case of a building floor, for example, a *floor plan* is drawn, showing the steel columns in cross section and the beam or girder framing by the use of single heavy lines, as shown in Fig. 22.5. Members framing from column to column, providing end support for other beams, are called *girders,* while smaller beams framing between girders are called *filler beams.* The designer's plans are sent to the steel fabricator who is to furnish the steel for the job. From these plans the fabricator makes the necessary detailed shop drawings and erection plans. However, before shop work is begun, the fabricator's drawings are sent to the design engineer for final checking, as the engineer has the authority to make any changes necessary to conform to the required strength and safety of connections. As soon as the fabricator has received the shop drawings approved by the engineer, the shop work may be carried out.

Shop drawings consist of detail drawings of all parts of the entire structure, showing exactly how the parts are to be made. See Figs. 22.8 to 22.12 and Fig. 22.14. Essentially, such drawings show all dimensions necessary for fabrication calculated to the nearest $\frac{1}{16}''$, the location of all holes for connections, details of connection parts, and the required sizes of all material. In addition, fabrication or construction methods may be specified by appropriate notes on the detail drawings whenever such items are not covered by separate specifi-

*For a more complete treatment of this subject, see *Structural Steel Detailing,* Second Ed., American Institute of Steel Construction, 101 Park Ave., New York, N.Y. 10017.

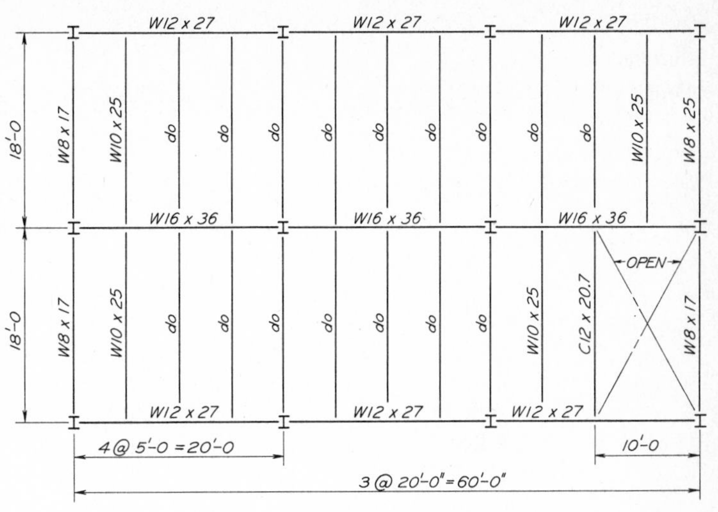


Fig. 22.5 **Typical Steel Floor Design Plan.**

cations. Obviously, fabrication and shop methods, as well as suitable field construction methods, must be fully understood by the detailer. The design of details and connections is an important part of the engineering of the structure and should not be neglected, as the connections of the various members must be adequate to transmit the forces in these members. Connection details should be drawn to a scale sufficiently large to show them clearly without crowding, although over-all lengths of members need not be drawn to scale. All dimensions should be shown, since detail drawings should never be scaled in the shop or in the field by the workmen making the piece. An adequate system of piece marking should be employed. Each piece that is separately handled should have its own piece mark, and this piece mark should be shown wherever the member appears on the drawings. This mark also is painted on the member in the shop, and later serves as a shipping mark and erection mark in the field for final assembly of the member in the structure.

Erection plans, ordinarily made by the steel fabricator, are essentially skeleton assembly drawings showing the relationship of the various members or parts to be fitted together in the final structure in the field. In all cases, the piece marks of the individual members are shown on the erection plans. Only sufficient detail to enable the complete assembly of the various members by skilled workmen is required, because the detail drawings, already made, fully describe each member and its connections. In most cases, line diagrams in which members are represented simply by heavy straight lines are adequate, although when complex assemblies are required, these should be shown in greater detail. Assembly views should be drawn to scale but, like the detail shop drawings, are never scaled by workmen to obtain dimensions. Appropriate notes on these drawings may be used to indicate how the structure is to be assembled. An erection plan of roof steel framing, which is an addition to an existing building, is shown in Fig. 22.6. This industrial structure houses a pulp mill for the manufacture of paper roofing products. New steel is shown by full lines, whereas existing roof members are shown by dashed lines. Connections to existing steel members are shown in sectional views, and timber framing for the support of large roof ventilators is also shown.

## 22.5 Structural Steel Shapes.*

Although aluminum and magnesium shapes are available for use in construction, only steel shapes are within the scope of this chapter. Structural

*See *Manual of Steel Construction,* American Institute of Steel Construction, 101 Park Ave., New York, N.Y. 10017.

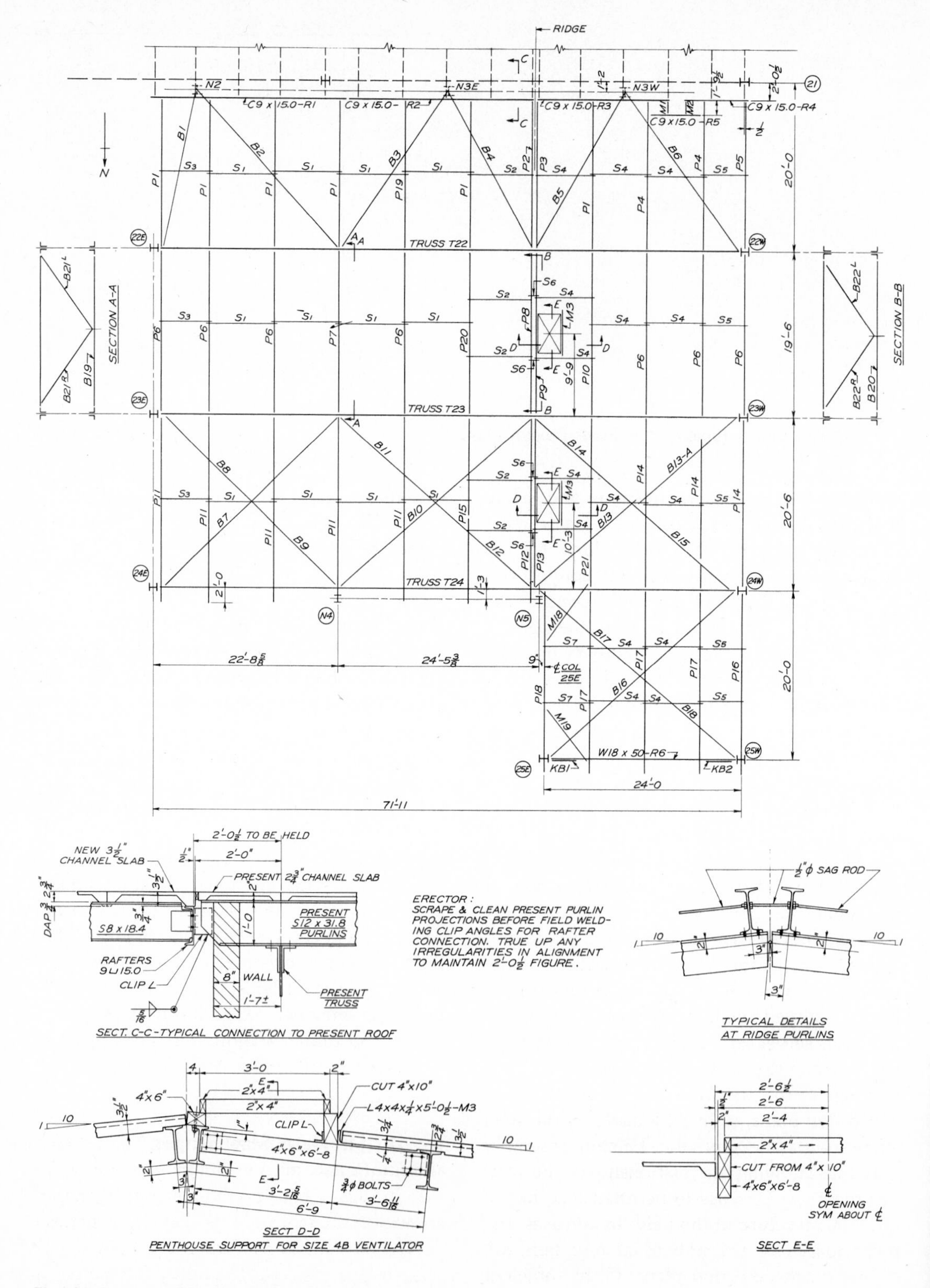




Fig. 22.6 **Roof Steel Erection Plan.**

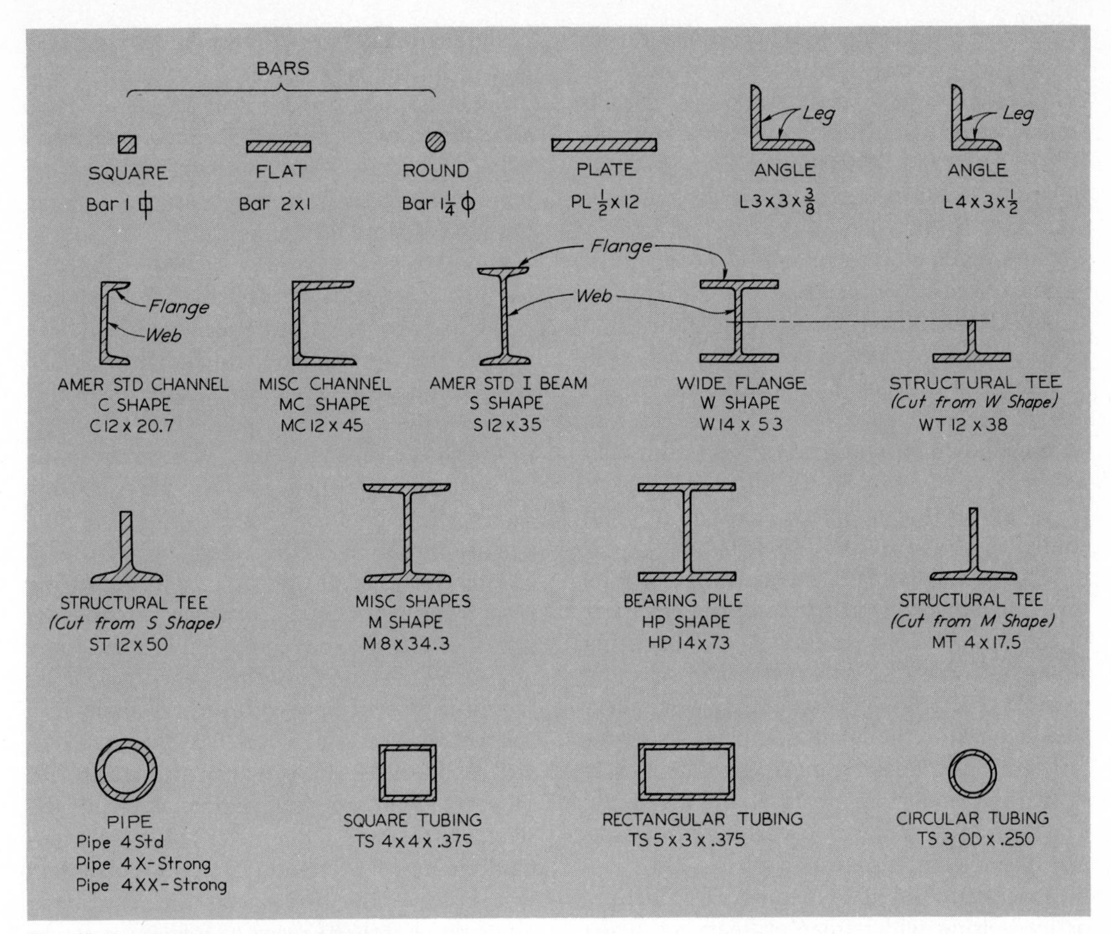


Fig. 22.7 **Structural Steel Shapes.**

steel is available in many standard shapes that are formed at the mill by rolling steel billets under high temperatures. The shapes available are square, flat, or round bars; plates; equal-leg and unequal-leg angles; American Standard and miscellaneous channels; S, W, M, and HP shapes for use as beams, columns, and bearing piles; structural tees, mill cut from W, S, or M shapes; pipe, in standard, strong, and extra-strong weights; and tubing, in square, rectangular, or circular cross sections. Figure 22.7 shows some typical cross sections of these shapes with the conventional manner of designation or billing in each case. The correct designations must be used on both the design and the detailed drawings in every case where the structural member appears. As a typical example, a wide flange section of 14″ nominal depth, weighing 53 lb per ft, and having an over-all length of 26′-2⅛″, would be designated as W14 × 53 × 26′-2⅛.

## 22.6 **Scales for Detailing.**

Details should be drawn to a scale of ¾″ = 1′-0, or 1″ = 1′-0, using the architects scale, §2.26. Over-all lengths of members, however, need not be drawn to scale.

## 22.7 **Specifications.**

The *Manual of Steel Construction* gives detailed information concerning dimensions, weights, and properties of rolled steel structural shapes, rolling mill practice, and miscellaneous data for designing and estimating. The information contained in this manual is essential to structural steel drafting.

Currently, thirteen different types or grades of steel are available. These differ in chemical composition and physical properties. Each is controlled in its manufacture by a separate ASTM (American Society for Testing Materials) specification, and there are considerable variations in the costs of the various grades. The designing engineer must be aware of the various physical properties of steels, such as strengths, ductilities, corrosion resistance, and costs, if he is to make an economical selection of the grade of steel. The types used, in any case, must be specified by ASTM designation on the drawings. The grade most commonly used now is ASTM A36, the number 36 specifying that the guaranteed minimum yield strength is 36 ksi (36,000 lb/in.$^2$).

Main members of a steel structure must be joined together to form the complete structure in the field. In the modern steel fabricating shop the most common method of forming suitable connections is by welding of connection material to the main members, with provision of open holes for insertion of field bolts to join with other members of the assembly. The connection material, consisting, for example, of angles or connection plates, may also be attached to the main members by riveting, or by bolting with either ordinary or high-strength steel bolts in either bearing-type or friction-type connections. Formerly, shop riveting and field riveting were used exclusively in structural work. Although still used to some extent, the riveted connection tends to be replaced by welding or high-strength bolting. High-strength bolts must be field-tightened by a definite amount of torque applied to the wrench.

**22.8 Riveting.** Structural rivets are made of soft carbon steel and are available in diameters ranging from $\frac{1}{2}''$ to $1\frac{1}{4}''$. Rivets driven in the shop are called *shop rivets,* and those driven in the field (at the construction site) are called *field rivets.* Rivets are usually of the button-head type, Fig. 13.41 (a), and are driven hot, into holes $\frac{1}{16}''$ larger than the rivet diameter. The length of a rivet is the thickness (grip) of the parts being connected, plus the length of the shank necessary to form the driven head and to fill the hole. Excess shank length will produce capped heads, whereas lengths too short will not permit the formation of a full head. Shop rivets are ordinarily driven by large riveting machines that are part of the permanent shop equipment.

Field rivets are usually heated in a coal-burning forge with the use of a hand bellows, and when properly heated the shank will have a light cherry-red color and the head a dull red color. Riveting crews consist of four workmen: (1) the *heater,* who passes the hot rivets to the *sticker;* (2) the sticker, who receives the rivets and enters them into the holes; (3) the *bucker,* who holds the rivet firmly in the hole against the force of the rivet gun by use of a dolly bar; and (4) the *riveter,* who forms the head with a pneumatic hammer, forcing the shank to fill the hole completely. If the forge is at some distance from the work, rivets are tossed to the sticker, who catches them in a special bucket. After catching, the sticker strikes the head sharply against metal to remove cinders and scale before entering the rivet into the hole. The riveter holds the pneumatic hammer or rivet gun against the rivet with considerable force, and during driving rotates it slightly to assist in forming a round smooth head.

A shop drawing for a riveted roof truss is shown in Fig. 22.8. Only the left half is drawn, since the truss is symmetrical about the center line. The use of the gage lines of the members should be noted. Gage lines should be located as closely as possible to the centroidal axes of the members; and at the joints where members intersect, these gage lines should intersect at a single point to avoid unnecessary moment stresses due to eccentricities. It is noted that the locations of the open holes indicate where the field splices are to be, namely at the center line of the truss (℄) for the top chord, and at $5'$-$2\frac{5}{16}$ from the center line for the bottom chord.

**22.9 Framed Beam Connections.** Because of their common usage, the American Institute of Steel Construction recommends certain

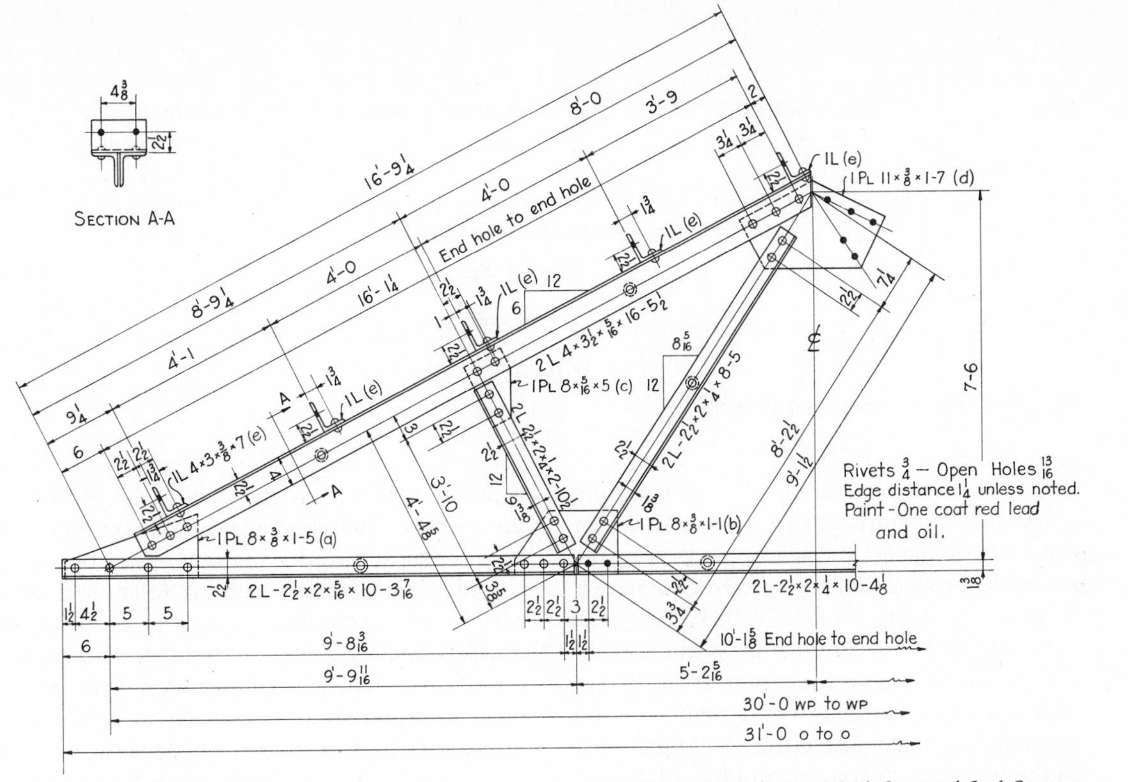

**Fig. 22.8 A Riveted Truss.**

*Based on a design by Fort Worth Structural Steel Company*

standard connections for attaching beams to other members. These connections are ordinarily adequate to transmit the end forces that beams carry. However, the draftsman should know the strength of these connections and use them only when they are sufficient. For complete information, see the AISC *Manual of Steel Construction.* This manual shows details of standard framed beam connections using $\frac{3}{4}''$, $\frac{7}{8}''$, and 1″ diameter bolted or riveted connections, and allowable loads for various sizes of structural beams up to 36″ W with 1 to 10 vertical rows of fasteners. For heavy beam loading requiring heavy connections, a section has been included in the *Manual* covering heavy framed beam connections ranging from 2 to 10 vertical rows of fasteners.

The use of angles for standard two-angle framed beam connections in a typical detail drawing of a floor beam is shown in Fig. 22.9. This drawing illustrates several important features: shop rivets are shown as open circles on shop drawings, whereas holes for field rivets or bolts are blacked in solid, Fig. 13.43, *gage lines* (lines passing through rivets or holes like center lines) are always shown, and it is desirable to line up holes or rivets on these lines where possible, rather than to "break the gage." It is necessary to locate the gage line of an angle for each leg in all cases, unless it has already been shown for the identical angle elsewhere on the drawing. The edge distance, from end rivet or hole to the end of the angle, must be given at one end, the billed length of the piece being worked out to provide the necessary edge distance at the other end. It is not necessary that the beam extend the full length of the distance back-to-back of end angles. In this case, as is customary, it is shown "set back" at both ends, the length of beam called for being 1″ less than the $13'$–$7\frac{3}{4}$ distance. Below the sketch, the mark B25 is the piece or shipping mark which appears on the erection plan, and is to be painted on the member in the shop for identification. The end connection angles are fully detailed at the left

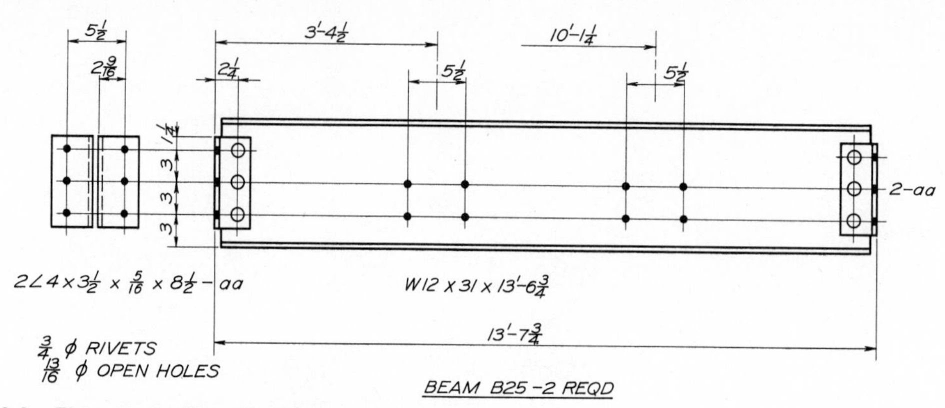


Fig. 22.9 **Floor Beam Shop Drawing.**

end of the beam and are given the assembly mark aa. Therefore at the right end where these same angles are again used, only the assembly mark 2–aa, to indicate two angles, is given. The figure 10′–1¼ is called an *extension figure,* as this is the distance from the back of the left-end angles to the center of the group of four holes. Note that this dimension is on the same horizontal line as the 3′–4½ figure just to the left. It is customary, in giving feet and inches, to give the foot mark (′) after dimensions in feet, but not to give the (″) mark, designating inches.

## 22.10 Welding.

The use of welding as a method of connecting steel members of buildings and bridges is common. Most steel fabricators have riveting, bolting, and welding equipment available, although a few are equipped to handle only welded fabrication. The metal-arc process is used, energy being supplied through an electrode to unite both the metal of the electrode and the parent or base metal. Electrodes may be either bare or coated, although most welding today is done with coated electrodes. Of all types of welds, the *fillet weld* is most common in structural steel fabrication. For additional information, the student should refer to Chapter 25 and to the specifications of the AISC *Manual of Steel Construction*. When a structure is to be welded, it should be designed throughout for this method of fabrication, as it is not possible to obtain the maximum economy of construction by merely substituting welding for riveting. The designations of welds by the use of standard symbols, Fig. 25.3, have greatly simplified the making of shop drawings.

A beam with end connection angles shop welded to the beam web is shown in Fig. 22.10. The outstanding legs of the angles are to be welded to the connecting columns in the field, as shown in the end view. This view

Fig. 22.10 **Detail of Welded Beam.**

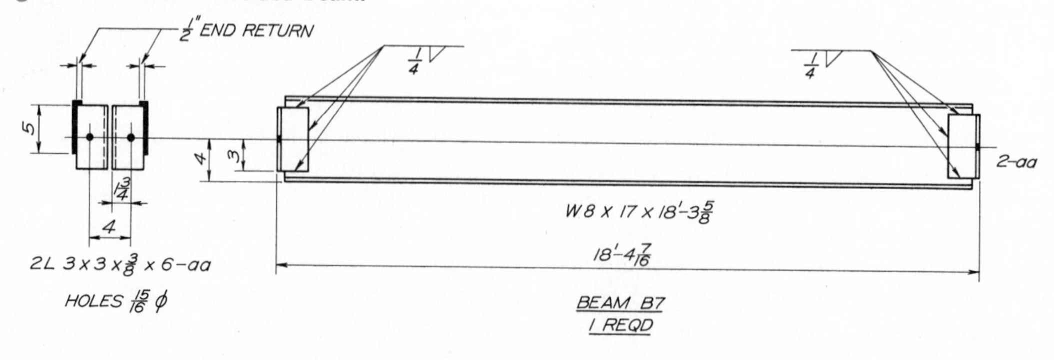

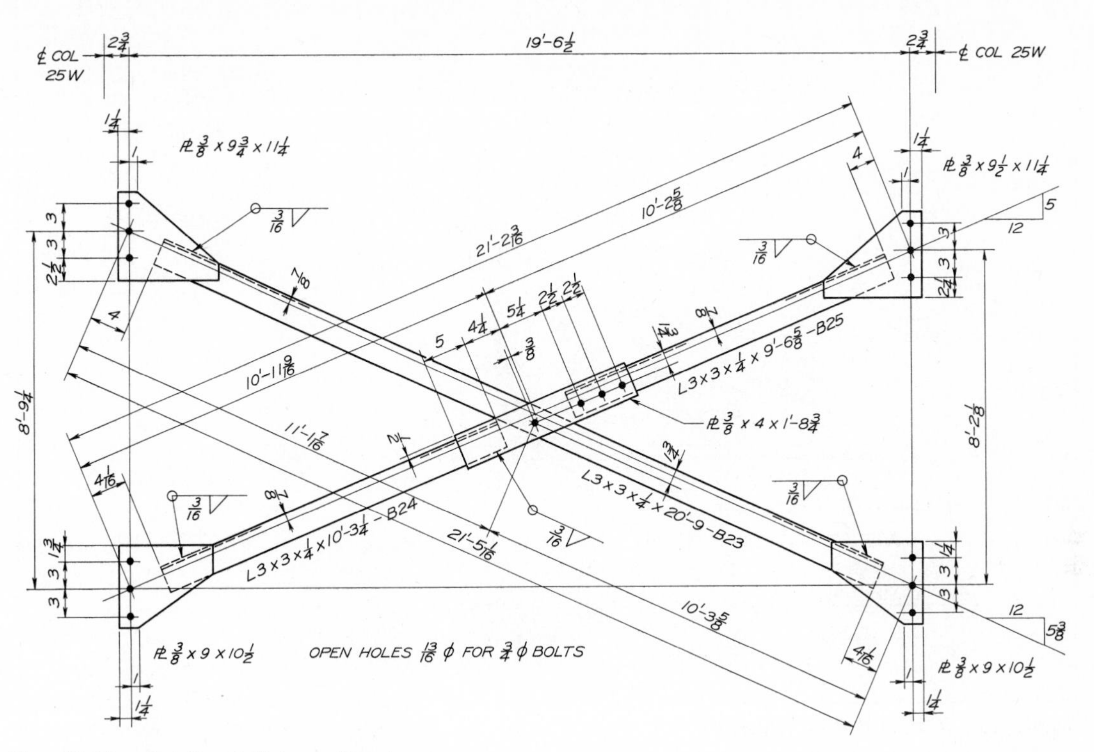


Fig. 22.11 **Details of Column Bracing.**

pertains only to field erection. Open holes in the outstanding legs are for bolts, to facilitate positioning.

A shop drawing of diagonal bracing between two columns is shown in Fig. 22.11. Here the diagonal angle members are shop welded to gusset plates that are to be bolted to the column flanges as a permanent installation in the field.

Figure 22.12 is the complete shop drawing for a symmetrical welded roof truss. Because it is symmetrical, it is only necessary to draw the left half of the structure; symmetry is shown on the drawing by the note Sym. abt. ℄. The clip angles marked aa are to be used for the attachment of roof purlins to the truss. In this structure the only plate material needed is the small gusset plate marked pb, most of the connections of web members to chords being made by simply fillet welding the angles against the webs of the chords. The chords are of structural tee (WT) type. The top chords are joined at the ridge by butt welding, which is also used at the end of the truss where the chords join. Note the use of welding symbols, the marking of members, and the listing of all sizes and quantities in the Bill of Material. The drawing is uncluttered and clear, with considerable important information conveyed under the heading General Notes.

## 22.11 High Strength Bolting for Structural Joints.

There are two basic types of high-strength steel bolts in common use, known as ASTM A325 and A490. The type A449 is similar in physical properties to A325, except that ordinary rather than special nuts may be used with this type. These bolts are heat treated by quenching and tempering. The A325 bolt is made of medium carbon steel, whereas the A490 is of alloy steel. At the time of installation, such bolts are tightened a prescribed amount either by the "turn of the nut method" or by use of a calibrated torque wrench.

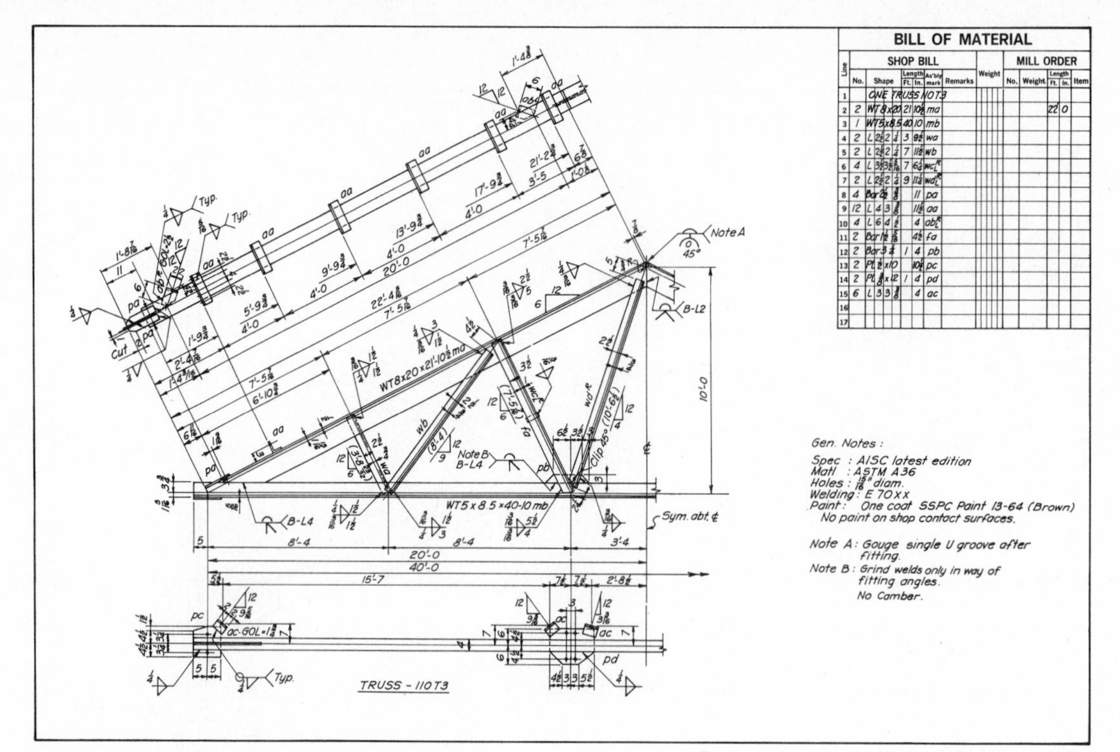


Fig. 22.12 **A Welded Roof Truss.** *Courtesy American Institute of Steel Construction*

The use of a high-strength steel bolt to transmit a force from the center plate into the two outside plates is shown in Fig. 22.13. When fully tightened, the connecting parts are held together by friction, thus preventing joint slip. Fatigue failures due to impact or stress reversals are then minimized. The slip resistance depends not only on the amount of clamping torque but also upon the nature of the contact surfaces. Figure 22.14 shows hardened washers under both head and nut. Whether one or two washers are needed, or none, depends upon the method of tightening used, the yield stress of the material being joined, and whether the joint is to be of "friction" or of "bearing" type. In the bearing-type connection, no allowance is made for friction due to clamping action, and the strength of the bolt shank bearing against the material is relied upon. Numerous specifications govern accepted practice. These are well covered, together with installation and inspection procedures, design examples, and reference tables in a publication entitled *High Strength Bolting for Structural Joints.**

Figure 22.14 represents the complete shop drawing of a steel column of two-floor height, the floors being indicated by reference lines Ⓑ, ①, and ② of the drawing. Here, high-strength bolts are used to attach connection material to the W12 × 99 column shaft. The top views at levels ① and ② show the outstanding legs of the angles in sectional plan

Fig. 22.13 **High-Tensile Steel Bolt.**

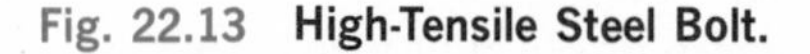

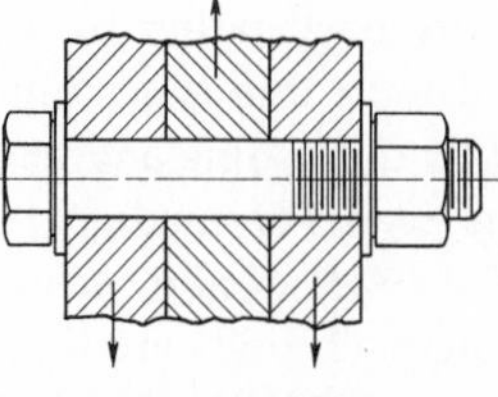

HEAVY HEX STRUCTURAL BOLT
HEAVY HEX NUT
HEAVY PLAIN HARDENED
WASHERS

*Published by Bethlehem Steel Corporation, Bethlehem, Pa. 18016.

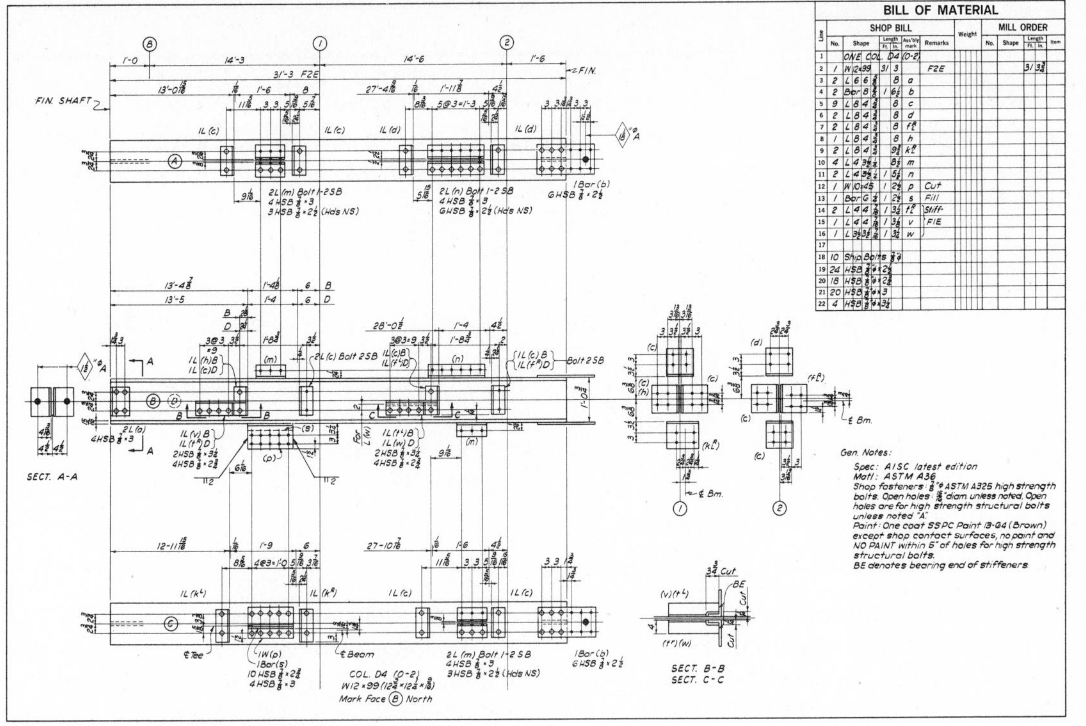


Fig. 22.14 **Steel Column Detail—Bolted Connections.** *Courtesy American Institute of Steel Construction*

views, with open holes shown by the round dots located by exact dimensions taken from the column center lines. The open holes are for the insertion of field bolts in beam connections when the entire assembly is erected.

Faces Ⓐ, Ⓑ, Ⓒ, and Ⓓ are designated. Viewing a column from above, the faces are designated in that order counterclockwise, with faces Ⓐ and Ⓒ designating the flanges. With regard to web faces Ⓑ and Ⓓ, only face Ⓑ needs to be drawn in the usual case. However, when the connections on face Ⓓ are exceptionally complicated this face also should be drawn as an additional elevation view.

Note the manner of showing the shop bolts and their designation "HSB" (high-strength bolts).

At the top of this column, splice connection material is provided by Bars (b), for attachment of the shaft above. All material for this column is listed in the Bill of Material which is a part of the drawing.

## 22.12 The Calculation of Dimensions.

Perhaps the most important part of the structural draftsman's work is the accurate calculation of dimensions. If there are incorrect dimensions on the drawings, they will result in serious errors and misfits when members are assembled in the field. Correction of such errors not only entails considerable expense but often results in delaying the completion of the work. Since in ordinary steel work, dimensions are given to the closest $\frac{1}{16}''$, considerable precision is demanded. For skewed members, such as those in a truss, some trigonometry is involved. To facilitate such calculations, tables of logarithms and squares are extremely useful. Since dimensions are given in feet, inches, and fractions of inches, it is necessary to convert fractions of inches to inch decimals and then to convert inches to decimals of feet so that all dimensions are in foot units, when using ordinary logarithmic tables. Fortunately, tables are available in which these conversions have already been made. Such

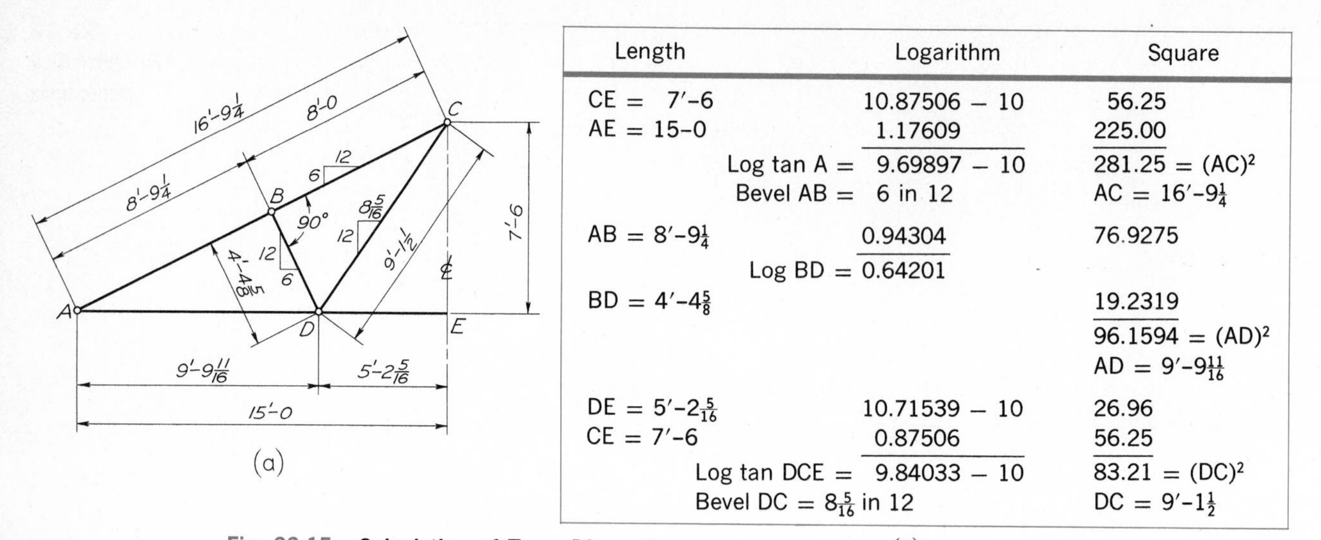


| Length | | Logarithm | Square |
|---|---|---|---|
| CE = 7′-6 | | 10.87506 − 10 | 56.25 |
| AE = 15-0 | | 1.17609 | 225.00 |
| | Log tan A = | 9.69897 − 10 | 281.25 = $(AC)^2$ |
| | Bevel AB = | 6 in 12 | AC = 16′-9¼ |
| AB = 8′-9¼ | | 0.94304 | 76.9275 |
| | Log BD = | 0.64201 | |
| BD = 4′-4⅝ | | | 19.2319 |
| | | | 96.1594 = $(AD)^2$ |
| | | | AD = 9′-9 11/16 |
| DE = 5′-2 5/16 | | 10.71539 − 10 | 26.96 |
| CE = 7′-6 | | 0.87506 | 56.25 |
| | Log tan DCE = | 9.84033 − 10 | 83.21 = $(DC)^2$ |
| | Bevel DC = | 8 5/16 in 12 | DC = 9′-1½ |

Fig. 22.15 **Calculation of Truss Dimensions.** (b)

tables* include logarithms and squares for dimensions to 1/16″ for distances up to 100 ft, and to ⅛″ for distances to 200 ft. Both natural and logarithmic functions of angles are also included, as well as bevels, slopes, and rises. See §11.14. Ordinarily angles on shop drawings are not given in degrees, minutes, and seconds, but rather in terms of a *bevel,* which is the rise, or height of a right triangle whose base is 12. The slope is the hypotenuse of the triangle. The use of such tables implies an adequate knowledge of the trigonometry involved.

To illustrate the use of the above tables, the main dimensions for the truss of Fig. 22.8 will be calculated. For this purpose, the line diagram of Fig. 22.15 (a), which shows the gage lines of the truss members, will be used. Length AE, one-half the truss span, and CE, the height of truss, are the known dimensions. Point B is located 8′-9¼ from A, as shown. From these dimensions, the other lengths and the bevels are to be determined. In a Fink truss such as this, the web member BD is perpendicular to the top chord AC. An arrangement of computations in tabular form is desirable, as in Fig. 22.15 (b).

*For example, *Smoley's New Combined Tables,* C. K. Smoley & Sons, Inc., Chatauqua, N.Y.

## 22.13 Concrete Construction.

Concrete is a building material made by mixing sand and gravel or other fine and coarse aggregate with *Portland cement* and water. The strength of concrete varies with the quality and relative quantities of the materials, with the manner of mixing, placing, and curing, and with the age of the concrete. The compressive strength of concrete depends on the mix design, but has been manufactured to develop an ultimate strength at 28 days as high as 7000 psi. The tensile strength of the material is limited to about one-tenth the compressive strength. Portland cement is a controlled, manufactured product as compared to natural cements found in some localities. It derived its name from its color, which resembles that of a famous building stone found on the island of Portland in southern England.

Since the tensile strength of concrete is very limited, the usefulness of concrete as a building material can be materially improved by embedding steel reinforcing bars in it in such a way that the steel resists the tension, and the concrete mainly the compression. In this way the two materials act together in resisting forces and flexure. Concrete, combined in this way with steel, is called *reinforced concrete;* without the addition of steel rods or wires, it is called *plain concrete.* When the steel is pretensioned before the application of the superimposed load, thus producing an interior force within the member, the material is called *prestressed concrete.*

## 22.14 Reinforced Conrete Drawings.

The design of the reinforcing for a reinforced concrete structure and the preparation of the corresponding drawings are complicated. In order to simplify this work and to secure uniformity in the many engineering offices, the American Concrete Institute has prepared a *Manual of Standard Practice for Detailing Reinforced Concrete Structures,* which has been approved as an ACI Standard.*

It is recommended in the manuals that two sets of drawings be prepared, an *engineering drawing* and a *placing drawing.* The engineering drawing is prepared by the engineer who designs the structure, and the placing drawing is prepared by the manufacturer who fabricates the reinforcing steel. The engineering drawing is to show the general arrangement of the structure, the sizes and reinforcements of the several members, and such other information as may be necessary for the correct interpretation of the designer's ideas. The placing drawing is to show the sizes and shapes of the several rods, stirrups, hoops, ties, and so forth, and to arrange them in tabular forms for ready reference by the building contractor. The method of preparing an engineering drawing for a two-way slab and beam floor of a multistory building is illustrated in Fig. 22.16. For methods of preparing placing drawings, the student should consult the manual referred to above.

The design drawing for a reinforced concrete pier, one of the supporting members for a highway bridge, is shown in Fig. 22.17. Note that the steel bars, even though embedded, are shown by full lines, and that concrete is always stippled in cross-section. Unlike shop drawings for structural steel, concrete drawings are ordinarily made to scale in both directions. Usually a scale of $\frac{1}{4}''$ to the foot is adequate, although when the structure is complicated, scales of $\frac{3}{8}''$ or $\frac{1}{2}''$ to the foot may be used. An effort should be made to avoid a cluttered appearance, usually the result of crowding the drawing with many notes. Cluttering can largely be avoided by the use of tables and schedules for listing of bar sizes and other necessary data and by covering many important points in a single set of notes, as shown in Fig. 22.17.

*American Concrete Institute, P.O. Box 4754, Redford Station, Detroit, Mich. 48219.

## 22.15 Structural Clay Products.

Brick and tile, which are manufactured clay products, have been in use for centuries and comprise some of the best-known forms of building construction. Brick and tile units are made from many different types of clay and in many different shapes, forms, and colors. Ordinarily they are built into masonry forms by the skilled brick or tile mason, who places the units one at a time in a soft mortar. After the mortar hardens, it becomes an integral part of the structure. Typical mortars contain sand, lime, Portland cement, and water. Although the compressive and tensile strengths of the clay units themselves are considerable, the over-all strength of the structure is limited by the strength of the mortar joints. As with concrete, therefore, the result is a structure of high compressive strength and relatively low tensile strength. Similarly, it is possible to reinforce brick and tile masonry by embedding steel rods within the members, thus adding greatly to their tension resistance and strength. When this is done, the material is called *reinforced brick* or *tile masonry* (RBM).

Information concerning the manufacture, the weight and strength properties, and the various uses and applications of structural clay products can be obtained from the handbooks *Principles of Brick Engineering* and *Principles of Tile Engineering.** These references will be found invaluable to both the designer and the draftsman concerned with designs in this material.

Bricks are made of various sizes, the $2\frac{1}{4}'' \times 3\frac{3}{4}'' \times 8''$ building brick being the most common. Thickness of mortar joints varies

*Published by the Brick Institute of America, 1750 Old Meadow Rd., McLean, Va. 22101.

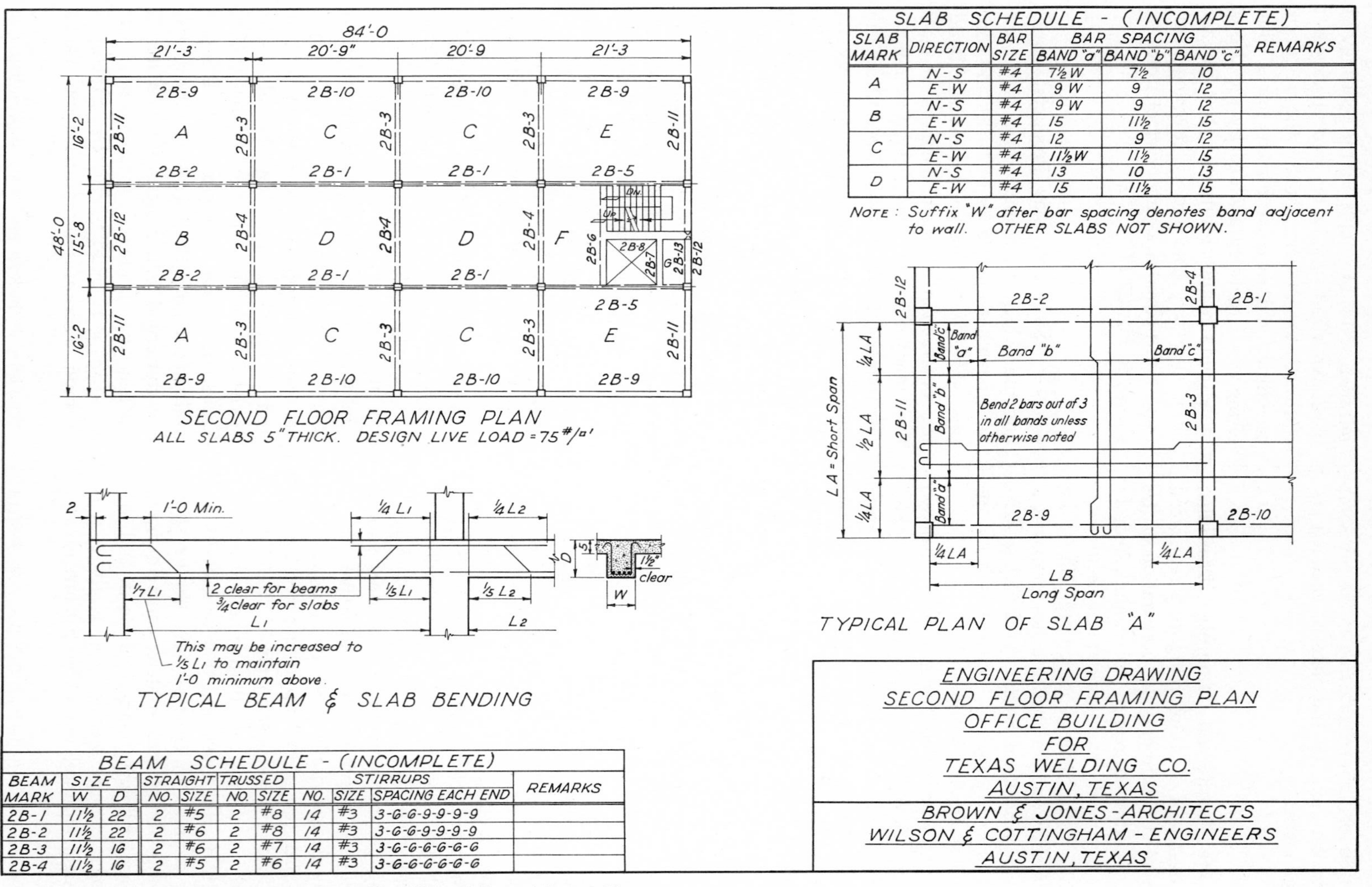


SLAB SCHEDULE - (INCOMPLETE)

| SLAB MARK | DIRECTION | BAR SIZE | BAR SPACING BAND "a" | BAND "b" | BAND "c" | REMARKS |
|---|---|---|---|---|---|---|
| A | N-S | #4 | 7½ W | 7½ | 10 | |
| | E-W | #4 | 9 W | 9 | 12 | |
| B | N-S | #4 | 9 W | 9 | 12 | |
| | E-W | #4 | 15 | 11½ | 15 | |
| C | N-S | #4 | 12 | 9 | 12 | |
| | E-W | #4 | 11½ W | 11½ | 15 | |
| D | N-S | #4 | 13 | 10 | 13 | |
| | E-W | #4 | 15 | 11½ | 15 | |

NOTE: Suffix "W" after bar spacing denotes band adjacent to wall. OTHER SLABS NOT SHOWN.

BEAM SCHEDULE - (INCOMPLETE)

| BEAM MARK | SIZE W | D | STRAIGHT NO. | SIZE | TRUSSED NO. | SIZE | STIRRUPS NO. | SIZE | SPACING EACH END | REMARKS |
|---|---|---|---|---|---|---|---|---|---|---|
| 2B-1 | 11½ | 22 | 2 | #5 | 2 | #8 | 14 | #3 | 3-6-6-9-9-9-9 | |
| 2B-2 | 11½ | 22 | 2 | #6 | 2 | #8 | 14 | #3 | 3-6-6-9-9-9-9 | |
| 2B-3 | 11½ | 16 | 2 | #6 | 2 | #7 | 14 | #3 | 3-6-6-6-6-6-6 | |
| 2B-4 | 11½ | 16 | 2 | #5 | 2 | #6 | 14 | #3 | 3-6-6-6-6-6-6 | |

**Fig. 22.16 Engineering Drawing for a Two-Way Slab and Beam Floor.**

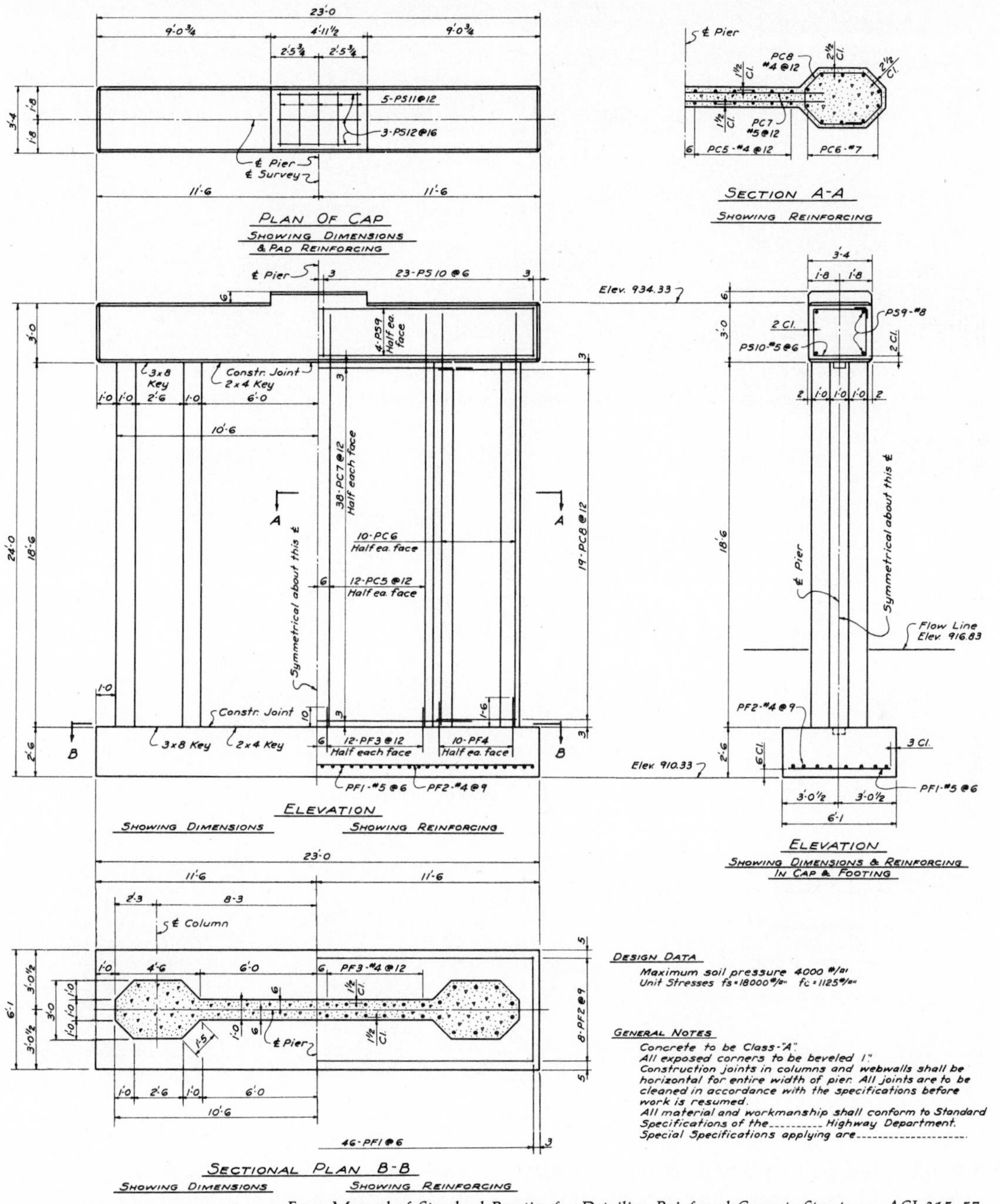


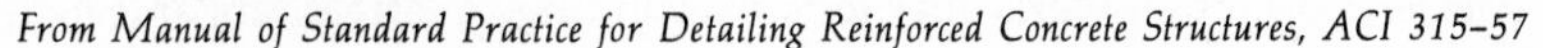
*From Manual of Standard Practice for Detailing Reinforced Concrete Structures, ACI 315–57*

Fig. 22.17 **Reinforced Concrete Pier for Deck Girder.**

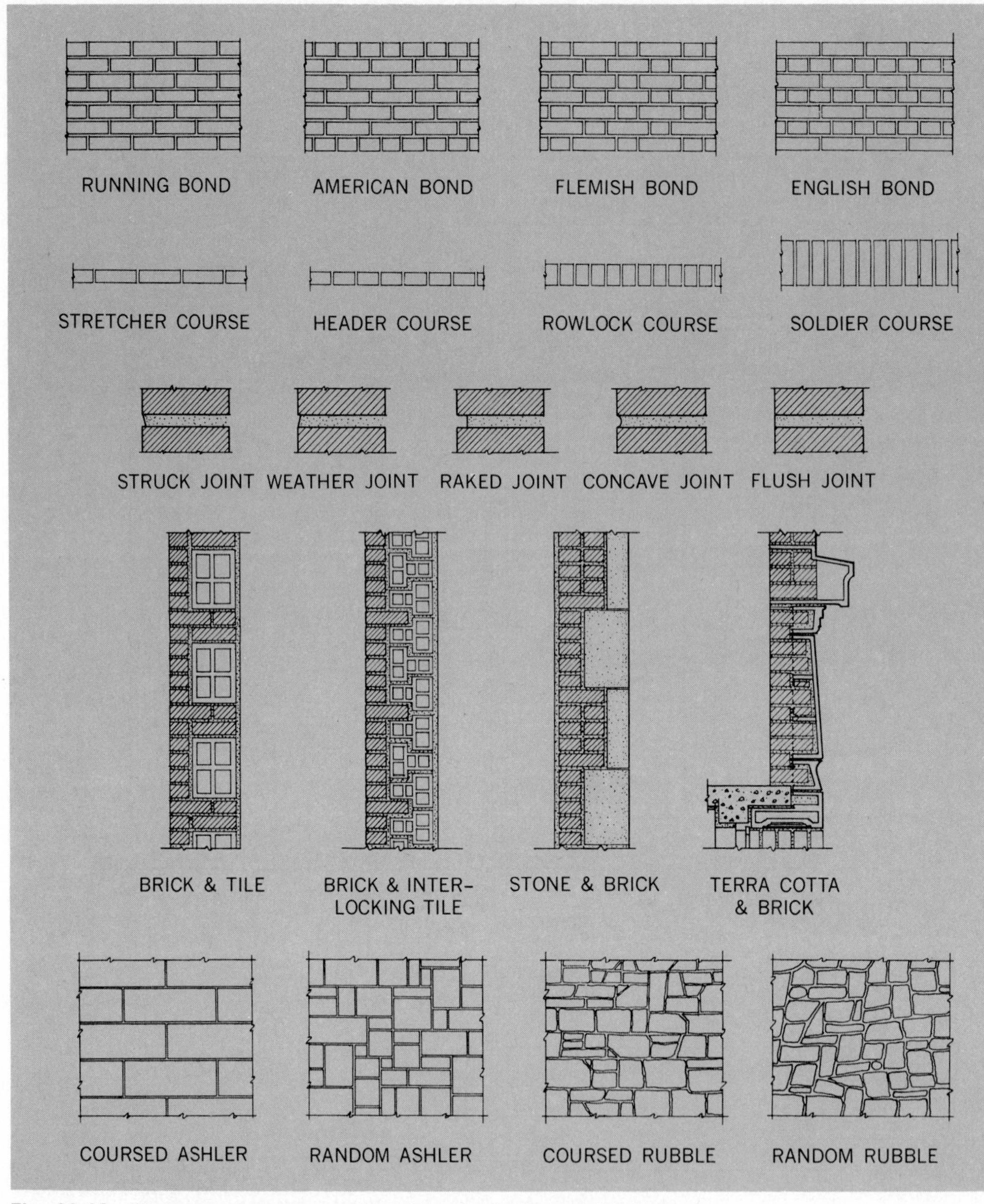


Fig. 22.18 **Methods of Laying and Bonding Brick, Tile, and Stone.**

usually from $\frac{1}{4}''$ to $\frac{3}{4}''$, with $\frac{3}{8}''$ and $\frac{1}{2}''$ most common. Of the several methods of bonding brick, Fig. 22.18, the following are the most common: *running bond*—all face brick are stretchers and are generally bonded to the backing by metal ties; *American bond*—the face brick are laid alternately, five courses of stretchers and one course of headers; *Flemish bond*—the face brick are laid with alternate stretchers and headers in every course; *English bond*—the face brick are laid alternately, one course of stretchers and one course of headers. In modern work, these standard methods of bonding are frequently modified to produce various artistic effects. Typical brick lintel arches are shown in Fig. 22.19.

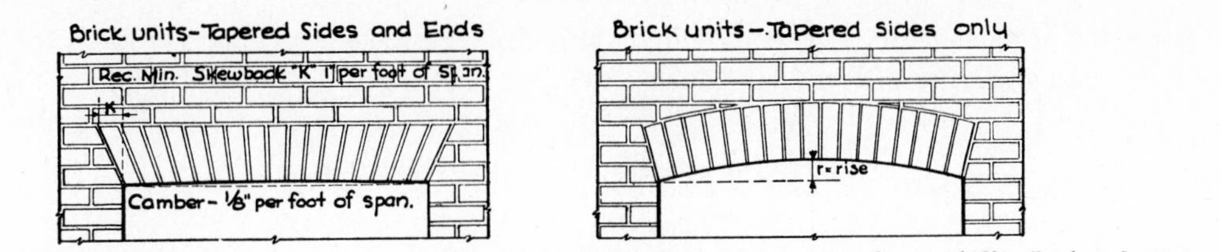


Fig. 22.19 **Typical Lintels.** *From Brick and Tile Engineering, courtesy Structural Clay Products Institute*

In addition to its use as a basic building material, tile is extensively used in fireproofing of structural steel members. Most building codes require that the steel members be enclosed in concrete or masonry so that fire will not cause collapse of the structure. Hollow tile units, being light and relatively inexpensive, are well adapted to this usage.

## 22.16 Stone Construction.

*Natural stone,* used in masonry construction—most commonly today for ornamental facing—is generally limestone, marble, sandstone, or granite, Fig. 22.18.

*Ashlar* (or *ashler*) masonry is formed of stones cut accurately to rectangular faces and laid in regular courses or at random with thin mortar joints.

*Rubble* masonry is formed of stones of irregular shapes and laid in courses or at random with mortar joints of varying thickness.

*Manufactured stone* is concrete made of fine aggregate for the facing, and coarse aggregate for the backing. The fine aggregate consists of screenings of limestone, marble, sandstone, or granite, so that the manufactured stone presents an appearance similar to that of natural stone. Manufactured stone is made of any desired shape, with or without architectural ornament.

*Architectural terra cotta* is a hard-burned clay product and is used primarily for architectural decoration and for wall facing and wall coping.

Brick, stone, tile, and terra cotta are combined in many different ways in masonry construction. A few examples are shown in Fig. 22.18.

## 22.17 Structural Drawing Problems.

The following problems are intended to afford practice in drawing and dimensioning simple structures and in illustrating methods of construction.

## 22.17 Structural Drawing Problems

**Prob. 22.1** Calculate point-to-point lengths (the distances between centers of joints) of the web members of the truss of Fig. 22.2. Make a detailed drawing of web member piece No. 4.

**Prob. 22.2** Make a complete detail of the top chord member of the truss shown in Fig. 22.2, for a 30° angle of inclination.

**Prob. 22.3** Assuming riveted construction, with rivets of $\frac{3}{4}$" dia, make a complete shop drawing for a typical filler beam of Fig. 22.5. Detail the same beam for welded construction. Consult AISC Manual.

**Prob. 22.4** Assuming the column size to be W8 × 31, detail the W16 × 36 girder at the center of the drawing of Fig. 22.5.

**Prob. 22.5** Referring to Fig. 22.12, make a complete detail for a truss of the same length, but change the height from 10′-0 to 6′-8. Use angle members of the same cross section sizes as those shown, but of different lengths, as needed.

**Prob. 22.6** Referring to Fig. 22.11, make a similar bracing detail, changing the distance between column centers from 20′-0 to 18′-6.

**Prob. 22.7** Draw cross sections through panel D of Fig. 22.16, in both directions. Include the supporting beams in each cross section, and show all dimensions, size and spacing of reinforcing steel, and dimensions to locate the ends of bars and the points of bend for the bent bars. Also show and locate the stirrups in these views.

**Prob. 22.8** Detail the brickwork surrounding a window frame for an opening 4′-0$\frac{7}{8}$ wide by 6′-9 high. Use type of curved arch lintel similar to that of Fig. 22.19. Assume standard-size building brick with $\frac{1}{2}$" mortar joints.

**Prob. 22.9** Consult *Principles of Tile Engineering,* §22.15, and draw a cross-section view through a 12" wall of composite brick and tile construction.

# 23

# topographic drawing and mapping

by E. I. Fiesenheiser*

**23.1 Introduction.** Up to this point, we have been concerned with the methods and techniques used in drawing man-made objects. *Topographic drawing* and *mapping* have to do with the representation of portions of the earth's surface—mainly its natural features—to a convenient scale. On such drawings the relative positions of natural features, with respect to certain definitely located points, are shown. Since the shape of the earth is spherical, any representation on a plane, such as a piece of paper, is necessarily somewhat distorted. See §19.21. In drawing large areas, therefore, some method of projection must be used that results in a minimum of distortion. In such work, the positions of the control or reference points are usually defined by spherical coordinates of latitude and longitude (meridians and parallels), which are shown as reference lines on the drawing. In drawing small areas to a relatively large scale, the distortion due to earth curvature is so slight as to be unnoticeable and it may therefore be entirely neglected. Orthographic projection, as used in technical drawings, in which the line of sight is assumed to be perpendicular to the plane of the map, is the method most commonly used in topographic representation.

**23.2 Definitions.** The purpose for which a map is to be used determines what features should be represented, what scale should be used, and what detail should be included. Certain commonly used symbols are employed in representing natural features and man-made objects. Also, certain terms and various types of maps are common. Therefore, the following definitions are essential to an understanding of the subject.

*Professor Emeritus of Civil Engineering, Illinois Institute of Technology.

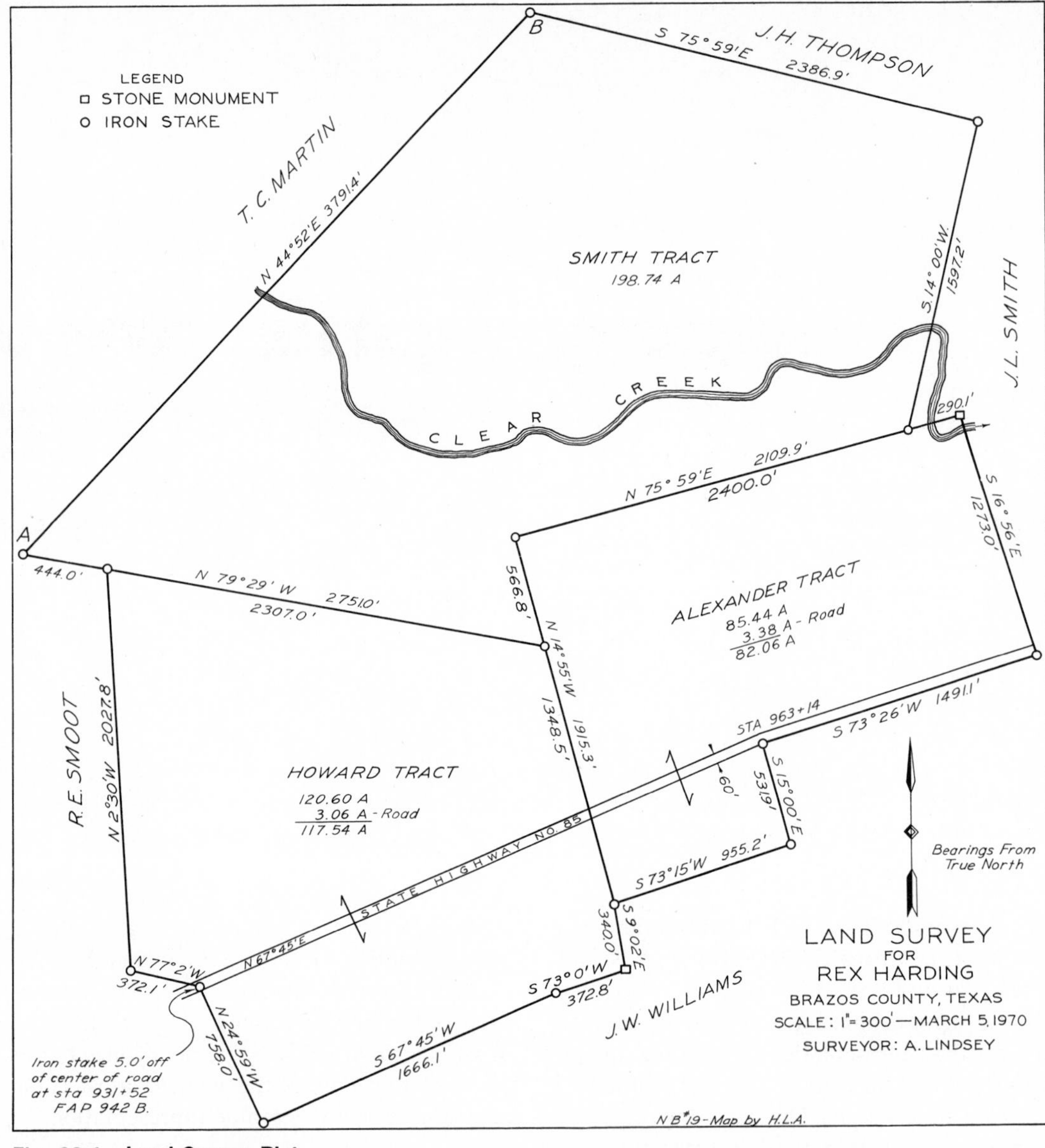


**Fig. 23.1 Land Survey Plat.**

A *plat* is a map, usually of a small area, plotted directly from a land survey. It does not ordinarily show relative ground elevations and is drawn for some specific purpose, such as the calculation of areas, the location of property lines, or the location of a building project. Usually it contains a *traverse.* See Fig. 23.1.

A *traverse* consists of a series of intersecting straight lines of accurately measured lengths. At the points of intersection, the deflection angles between adjacent lines are measured and recorded. Starting at one point, therefore, rectangular coordinates of the other intersection points can be calculated by trigonometry. A *closed traverse* thus becomes a closed polygon and provides a method for checking the accuracy of the work. The land survey plat of Fig. 23.1 illustrates a closed traverse.

*Elevations* are vertical distances above a common *datum* or reference plane. The elevation of a point on the surface of the ground is usually determined by differential leveling from some other point of known elevation. Commonly, elevations are referenced to mean sea-level datum.

A *profile* is a line contained in a vertical plane, and it depicts the relative elevations of various points along the line. Thus, if a vertical section were to be cut into the earth, the top line of this section would represent the ground profile.

*Contours* are lines drawn on a map to locate, in the plan view, points of equal ground elevation. On a single contour line, therefore, all points have the same elevation.

*Hatchures* are short, parallel, or slightly divergent lines drawn in the direction of the slope. They are closely spaced on steep slopes and converge toward the tops of ridges and hills. Hatchures are shade lines to show relief. See Appendix 28.

*Monuments* are special installations of stone or concrete to mark the locations of points accurately determined by precise surveying. It is intended that monuments be permanent or nearly so, and they are usually tied in by references to nearby natural features, such as trees and large boulders.

*Cartography* is the science or art of map making.

*Topographic* maps depict (1) water, including seas, lakes, ponds, rivers, streams, canals, swamps, and other features; (2) *relief* or elevations of mountains, hills, valleys, cliffs, and the like; (3) *culture,* or the works of man, such as towns, cities, roads, railroads, airfields, and boundaries. See Figs. 23.2 and 23.12.

*Hydrographic* maps convey information concerning bodies of water, such as shoreline locations; relative elevations of points of lake, stream, or ocean beds; and sounding depths.

*Cadastral* maps are accurately drawn maps of cities and towns, showing property lines and other features that control property ownership.

*Military* maps contain information of military importance in the area represented.

*Nautical* maps and charts show navigation features and aids, such as locations of buoys, shoals, lighthouses and beacons, and sounding depths.

*Aeronautical* maps and charts show prominent landmarks, towers, beacons, and elevations for the use of air navigators.

*Engineering* maps are made for special projects as an aid to locations and construction. See Figs. 23.5 and 23.8 to 23.10.

*Landscape* maps are used in planning installations of trees, shrubbery, drives, and other garden features in the artistic design of area improvements. See Fig. 23.7.

Fig. 23.2 **A Topographic Map.**

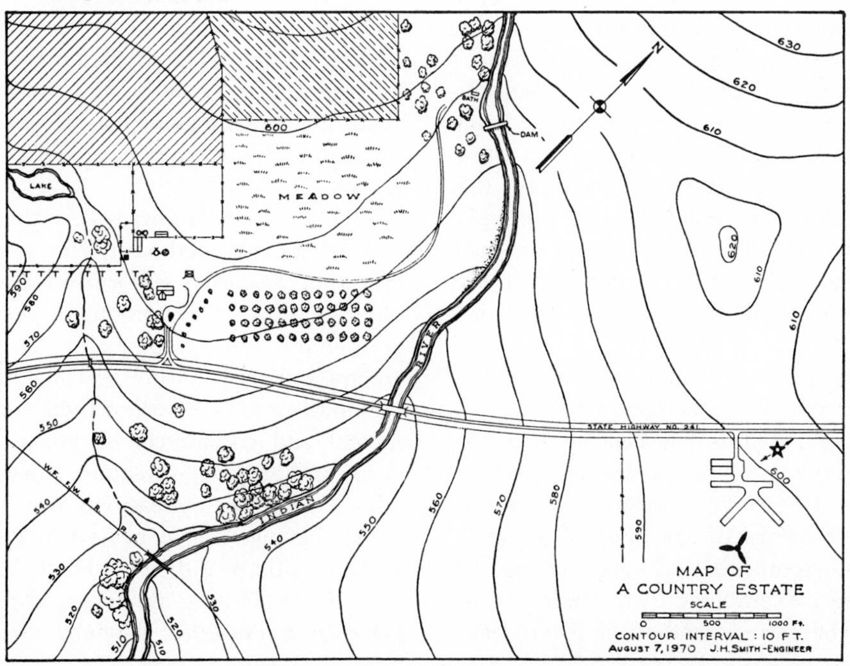

## 23.3 Sources of Information.

The basis of all maps and topographic drawing is the *survey.* Surveying is the actual measurement of distances and elevations on the earth's surface. Hence, all maps are plotted from field data provided by the surveyor. It will be appropriate to discuss briefly the surveying methods upon which map information is based.

Short distances are ordinarily measured by steel tape, with stakes being driven to mark the points between field measurements. Distances may also be determined by measuring aerial photographs, when the scale of the photograph is known. An instrumental method, known as the *stadia* method, is also widely used in map making. The *stadia transit* is an optical instrument used in conjunction with a special *stadia rod.* By sighting on the rod and using the necessary conversion factor, the instrument reading can easily be converted to distance.

Recent developments in the art of surveying have revolutionized distance measurement by the use of electronic instruments. To measure to a distant point, the surveyor directs the instrument's aiming head toward the point, where a passive reflector or prism has been set. The instrument generates either a modulated infrared light signal, focused into a narrow beam, or a laser beam, aimed directly at the reflector. When the reflector bounces the beam back to the aiming head, the beam's travel time is electronically measured and directly converted into the distance to the point. A major advantage of the electronic measurement over the distance-taping method is that moving traffic does not have to be stopped while measurements are being taken. Such time-saving instruments are capable of distance measurements up to four miles, depending upon the quality of the instrument. Accuracy varies from .01 to .03 ft, more than sufficient for topographic surveying.

By use of a *compass,* the bearing of a line, which is the angle between the line and the *magnetic north,* may be read on the compass. When all bearings of the lines of a *traverse* have been determined, the angles between the lines are easily computed by addition or subtraction. The correct method of listing bearings requires that they be referenced either to the north or to the south. A bearing, therefore, is either north, and east or west; or it is south, and east or west. See N 44°52′E for traverse line AB of Fig. 23.1. Compass readings are not to be regarded as of high accuracy, since magnetic north and true north are not quite the same; also local magnetism may affect the position of the compass needle.

When accurate measurement of angles is desired, the *transit* is the instrument usually employed. This optical instrument may be set up directly over a point, then sighted successively on two other points, after which the deflection angle in a horizontal plane can be read on the transit. This instrument is also used for the measurement of vertical angles.

For accurate orientation of lines on a map, the angle of a line with the *true north* is often desired. For this purpose the transit may again be used for sighting on a star (usually *Polaris,* the North Star), or on the sun. From instrument readings, the true angle of a line may then be calculated.

The *level,* also an optical instrument, is equipped with a telescope for sighting long distances. This instrument is commonly used to determine differences in elevation in the field, which is called the process of *differential leveling.* When this accurate instrument is leveled, the line of sight of its telescope is horizontal. A level rod, graduated in feet and decimals of feet, may be held on various points. Instrument readings of the rod then serve to determine the differences in elevations of the points.

*Photogrammetry* is now widely used for map surveying. This method utilizes actual photographs of the earth's surface and of man-made objects on the earth. Originally, aerial photogrammetry was used mainly in mapping enemy territory during wartime. Now this method is used for such activities as governmental and commercial surveying, explorations, and property valuation. It has the great advantage of being easy to use in difficult terrain having steep slopes, where ground surveying would be difficult or nearly impossible. The utilization of aerial photographs is called *aerial photogrammetry,* whereas that utilizing photographs taken from ground stations with the axis of the camera lens nearly horizontal, is known as *terrestrial photogrammetry.* By combining the results of both types of observations, it is possible not only to determine the relative positions of objects in a horizontal plane, but it is also possible to determine relative elevations. Thus the science of photogrammetry can be the basis of contour mapping as well as plan mapping. Generally, aërial photographs are used in matched groups to form a *mosaic.* To form a satisfactory mosaic map, the group photographs must overlap slightly. A distinct advantage of photogrammetry is that a large area can be mapped from a single clear photograph. The method may be used in connection with ground surveying by photographing *control points* already located on the ground by precise surveying.

For large land developments and large construction projects, more modern methods for producing topographic maps have evolved. Some state highway departments and most engineering firms who specialize in surveying and mapping now make use of aerial photography with computers, terrain digitizers, stereoplotters, and various new photolaboratory techniques. Investment in this expensive equipment can be justified when the volume of work is great and sufficient man-hours can be saved through its use, as is frequently the case.

For additional information, which is beyond the scope of this chapter, the student may refer to the following texts: *Advanced Surveying and Mapping and Elements of Photogrammetry* by George D. Whitmore (International Textbook Company); *Elementary Topography and Map Reading* by Samuel L. Greitzer (McGraw-Hill Book Company); and *Manual of Surveying Instructions for the Survey of the Public Lands of the United States,* prepared and published by the Bureau of Land Management (U.S. Government Printing Office, Washington, D.C.). Also see Appendix 1.

## 23.4 **Contours.**

Although contours have already been defined, their uses and characteristics and the methods of plotting them require further discussion.

A *contour interval* is the vertical distance between horizontal planes passing through successive contours. For example, in Fig. 23.3 the contour interval is 10′. The contour interval should not change on any one map. It is customary to show every fifth contour by a line heavier than those representing intermediate contours.

If extended far enough, every contour line will close. At streams, contours form V's pointing upstream. Should successive contours be evenly spaced, this means that the ground slopes uniformly, whereas closely spaced contours indicate steep slopes.

Locations of points on contour lines are determined by *interpolation.* In Fig. 23.3, the locations and elevations of seven control points are determined, and contour lines will be drawn on the assumption that the slope of the surface of the ground is uniform between station A and the six adjacent stations. To draw the contour lines, a contour interval of 10′ was adopted, and the locations of the points of intersection of the contour lines with the straight lines, joining the point A and the six adjacent points, was calculated as follows:

The horizontal distance between stations A and B is 740′. The difference in elevation of those stations is 61′. The difference in eleva-

**Fig. 23.3 Contours Determined from Control Points.**

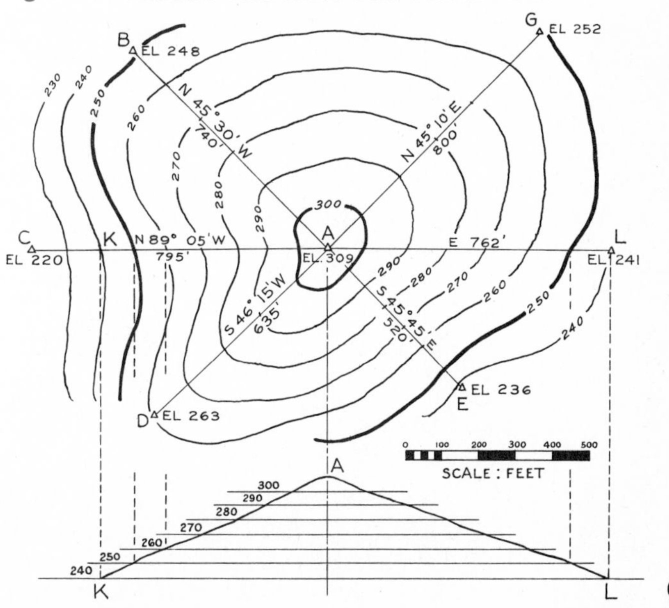

tion of station A and contour 300 is 9′; therefore, contour 300 crosses the line AB at a distance from station A of $\frac{9}{61}$ of 740, or 109.1′. Contour 290 crosses the line AB at a distance from contour 300 of $\frac{10}{61}$ of 740, or 121.3′. This 121.3′ distance between contour lines is constant along the line AB and can be set off without further calculation.

In the same way, points in which the contours cross the other lines of the survey can be interpolated. After this process is finished, the several contour lines can be drawn through points of equal elevation, as shown.

After contours have been plotted, it is easy to construct a profile of the ground line in any direction. In Fig. 23.3, the profile of line KAL is shown in the lower or front view. It is customary, as shown here, to draw the profile to a larger vertical scale than that of the plan in order to emphasize the varying slopes.

Contour lines may also be plotted by use of the recorded elevations of points on the ground surface, as in Fig. 23.4. This figure illustrates a *checkerboard survey,* in which lines are drawn at right angles to each other, dividing the survey into 100 ft. squares, and where elevations have been determined at the corners of the squares. The contour interval is taken as 2′, and the slope of the ground between adjacent stations is assumed to be uniform.

The points where the contour lines cross the

Fig. 23.4 **Contours Determined from Readings at Regular Intervals.**

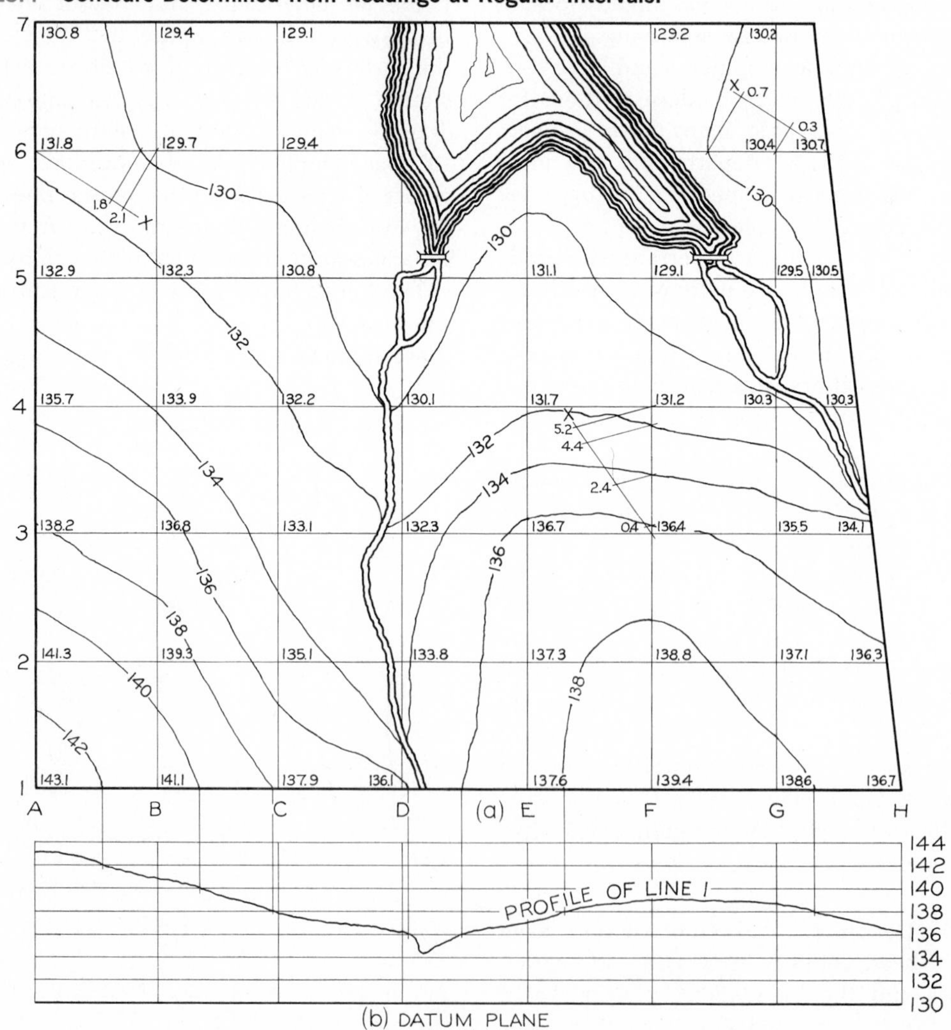

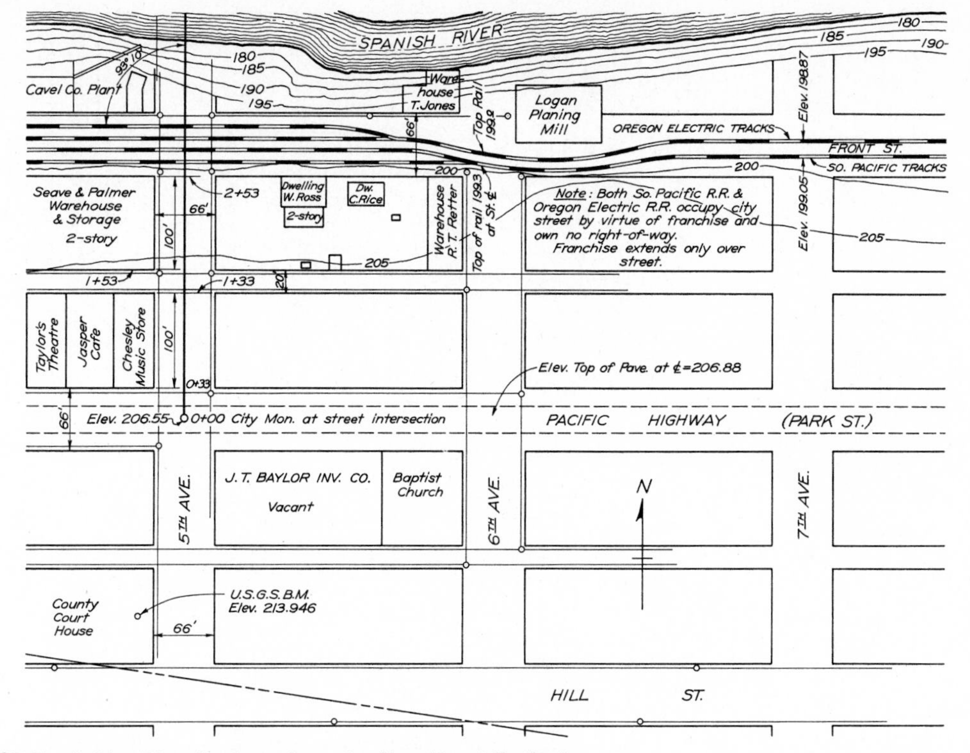


Fig. 23.5 **A City Plan for Location of a New Road Project.**

survey lines can be located approximately by inspection, accurately by the graphical method shown in Fig. 4.14, or by the numerical method explained above for Fig. 23.3.

The points of intersection of contour lines with survey lines may also be found by constructing a profile of each line of the survey, as shown for line 1 in Fig. 23.4 (b). Horizontal lines are drawn at elevations at which it is desired to show contours. The points in which the profile line intersects these horizontal lines indicate the elevations of points in which corresponding contour lines cross the survey line 1 and, therefore, can be projected upward, as shown, to locate these points.

It is obvious that the profile of any line can be constructed from the contour map by the converse of the process just described.

## 23.5 Symbols.

Various natural and man-made features are designated by special symbols. A reference list of the most commonly used map symbols is given in Appendix 28, and the student should refer to this list for identification of the symbols used in the figures of this chapter.

## 23.6 City Maps.

The special use of a map determines what features are to be shown. Maps of city areas may be put to many uses, some of which will be described. Figure 23.5, a city plan for location of a new road construction, shows only those features of importance to the location and construction of the road. The transit line starts at the center-line intersection of Park St. and 5th Ave., and it is marked as station 0 + 00. From here it extends north over the railroad yard to cross the river. Features near the transit line, such as buildings, are shown and identified by name. Street widths are important and are shown. Contour lines between the railroad yard and the river indicate the steeply sloping terrain.

Maps perform an important function for those who plan the layout of lots and streets.

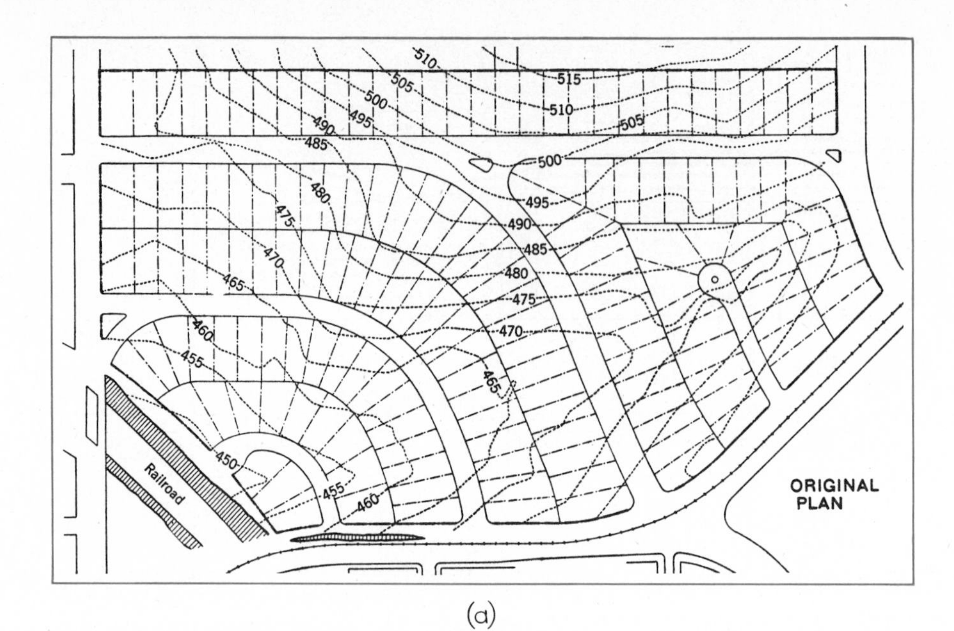


(a)

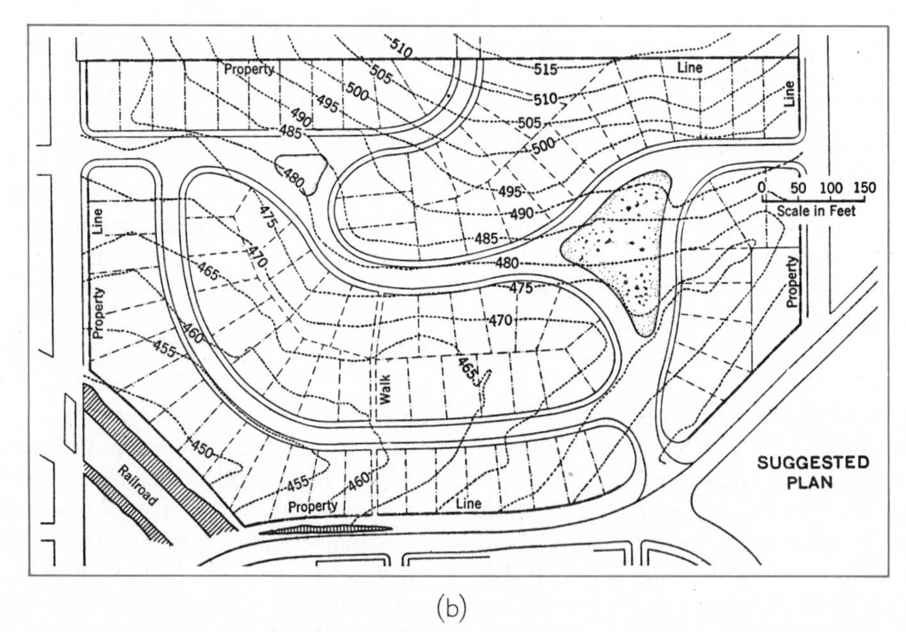


(b)

*From Land Subdivision, ASCE Manual No. 16 of Engineering Practice*

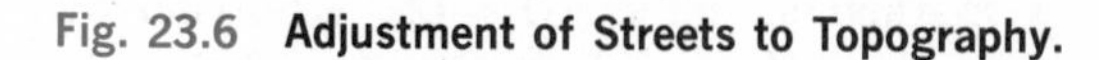

Fig. 23.6 **Adjustment of Streets to Topography.**

For example, Fig. 23.6 (a) shows an original layout of these features for a new residential area. An examination of the contours will show that this layout is not satisfactory, since the directions of the streets do not fit the natural ground slopes. Streets should be arranged so that the subdivision can be entered from a low point and so that a maximum number of lots will be above street grade. The layout at (b) is a decided improvement, in that the streets curve to fit the topography and the entrance is located at a low point.

Maps have a definite use in landscape planning. Figure 23.7 is a landscape map showing a proposed layout of lots and trees for beautification of a new subdivision.

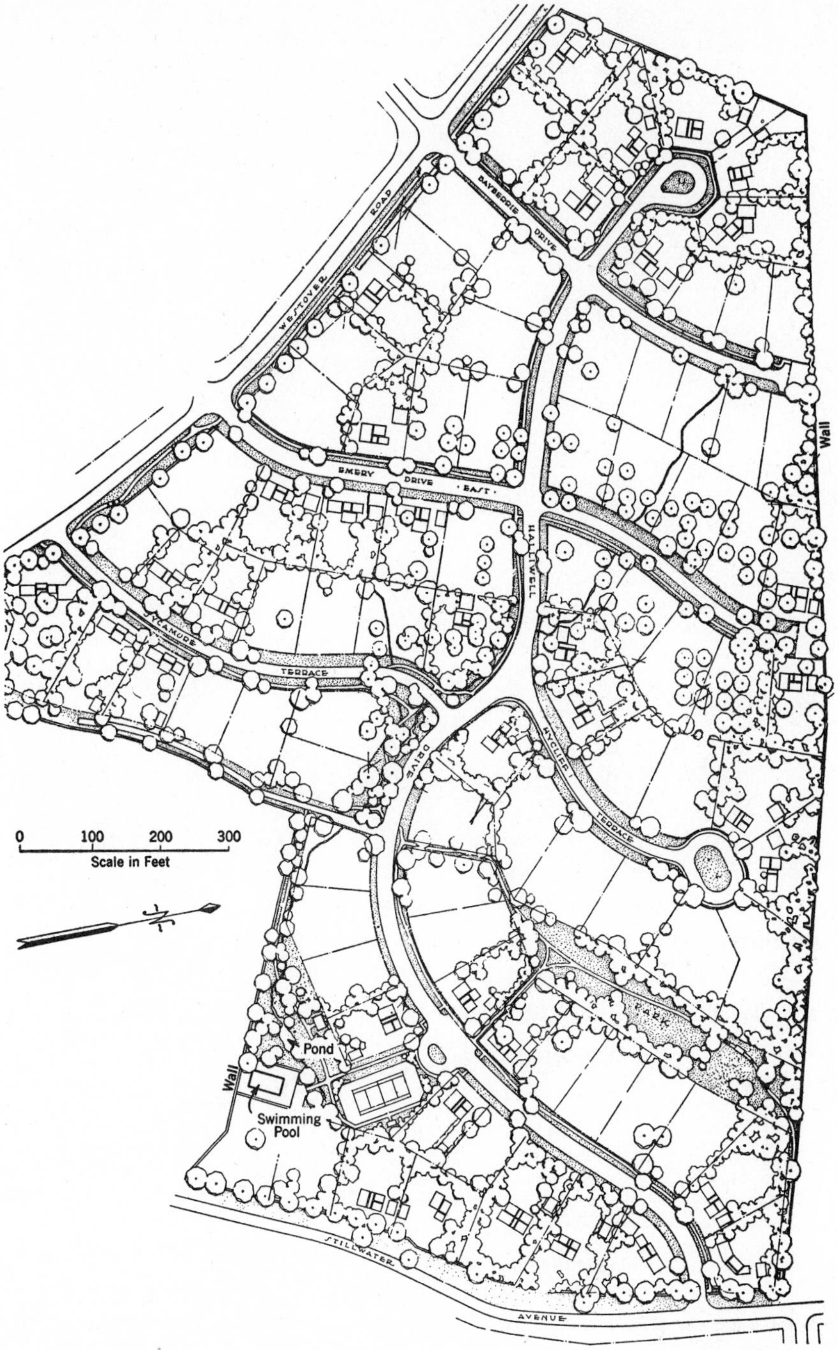

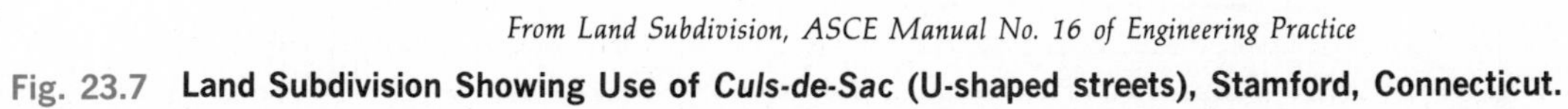


*From Land Subdivision, ASCE Manual No. 16 of Engineering Practice*

Fig. 23.7 **Land Subdivision Showing Use of *Culs-de-Sac* (U-shaped streets), Stamford, Connecticut.**

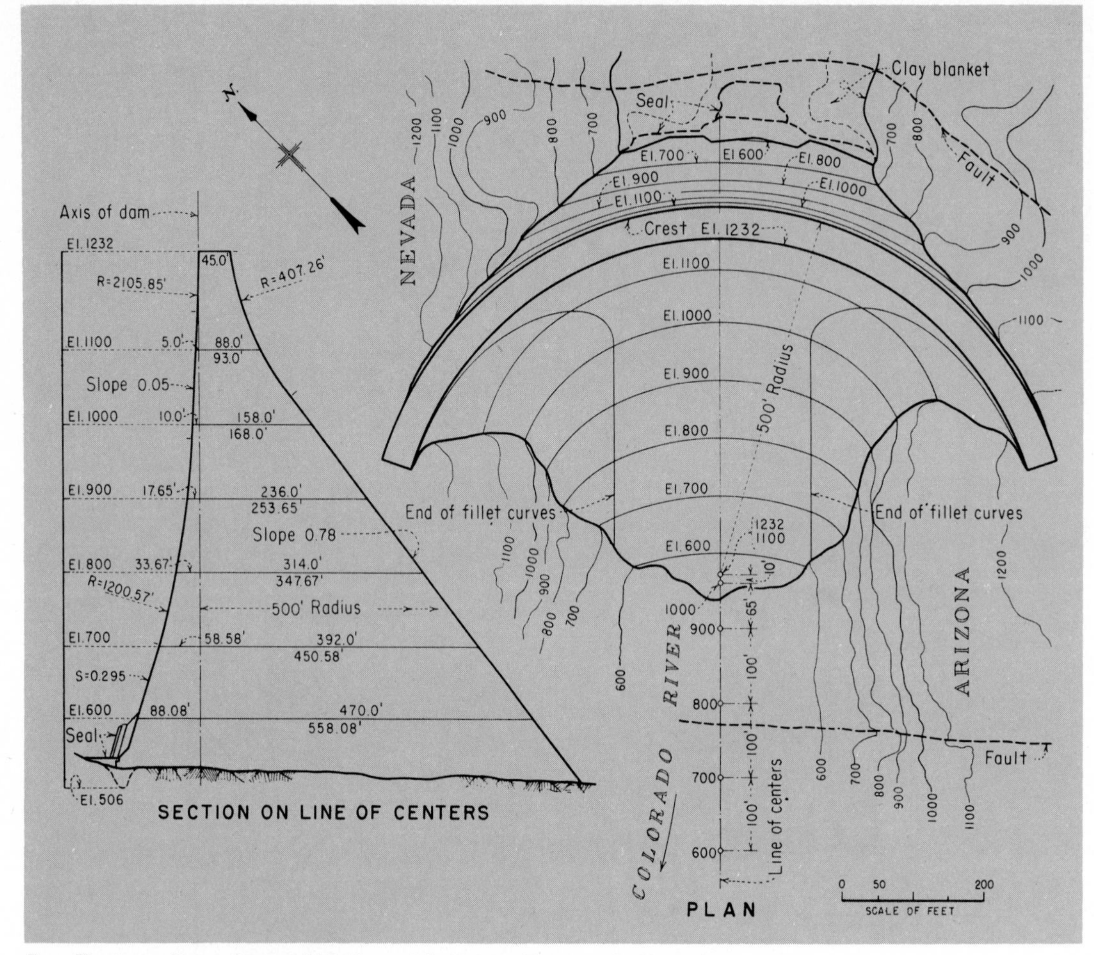


From Treatise on Dams, Courtesy U.S. Dept. of the Interior, Bureau of Reclamation

Fig. 23.8 **Plan for Hoover Dam.**

Fig. 23.9 **Plan and Elevation of a Bridge Structure.**

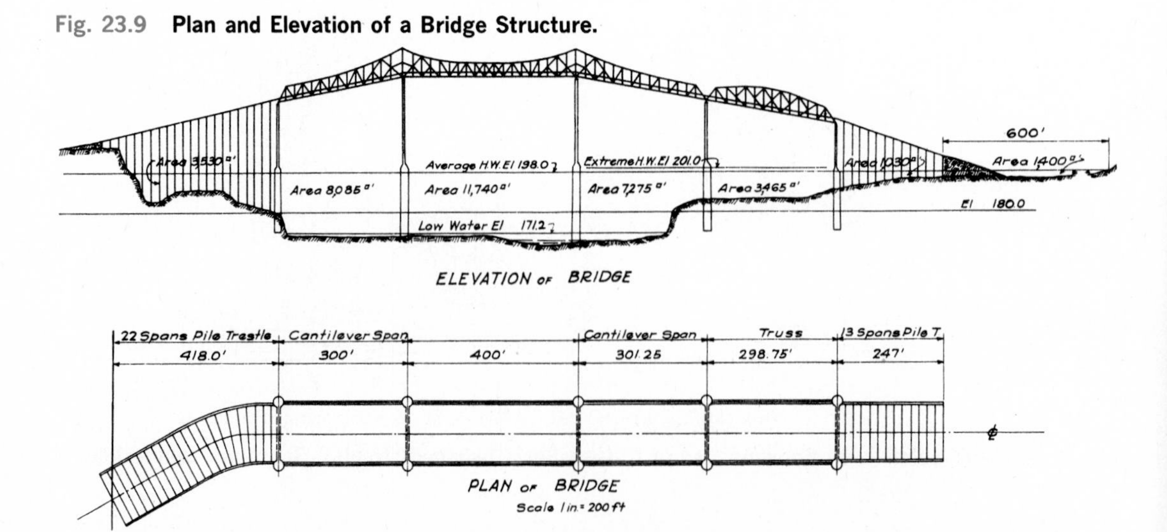

## 23.7 Structure Location Plans.

When a construction project is contemplated in a particular area, the construction should be located most advantageously to fit the topography of the area. Thus, a location plan on paper is of great advantage. A project location plan for a dam is shown in Fig. 23.8. This map shows the important natural features, contours, a plan view of the structure, and a cross section through Hoover Dam.

Although many drawings, perhaps hundreds, comprise the complete detail drawings for a large bridge, one of the most important early drawings is a general arrangement plan and elevation in the form of a line diagram. As an example, Fig. 23.9 shows a plan and elevation of a large bridge structure.

## 23.8 Highway Plans.

Before highway construction starts, it is necessary first to plan the location and to arrange both the horizontal and vertical alignment most advantageously. Commonly, both plan and profile are drawn on the same sheet, as in Fig. 23.10. In the plan at the top of the figure, the topography, consisting of such features as trees, fences, and farmhouses along the right of way, is shown. The transit line, locating the center line of the new road, is drawn with stations spotted at 100′ distances apart. Data for laying in the horizontal curves in the field have been calculated and are listed. The point of curve (P.C.) at station 17 + 00 is the point at which the line begins to curve with a 600′ radius, for a curve length of 400′. The central angle is 38°12′, and the degree of curve (angle subtended by a 100′ chord) is shown as 9°33′. The reverse curve of 800′ radius begins at station 21 + 00. Note also the north point and the bearing N73°E of the transit line.

The vertical alignment is shown in profile below the plan. Note the listing of station numbers below the profile. The scale of this

Fig. 23.10 **A Highway Plan and Profile.**

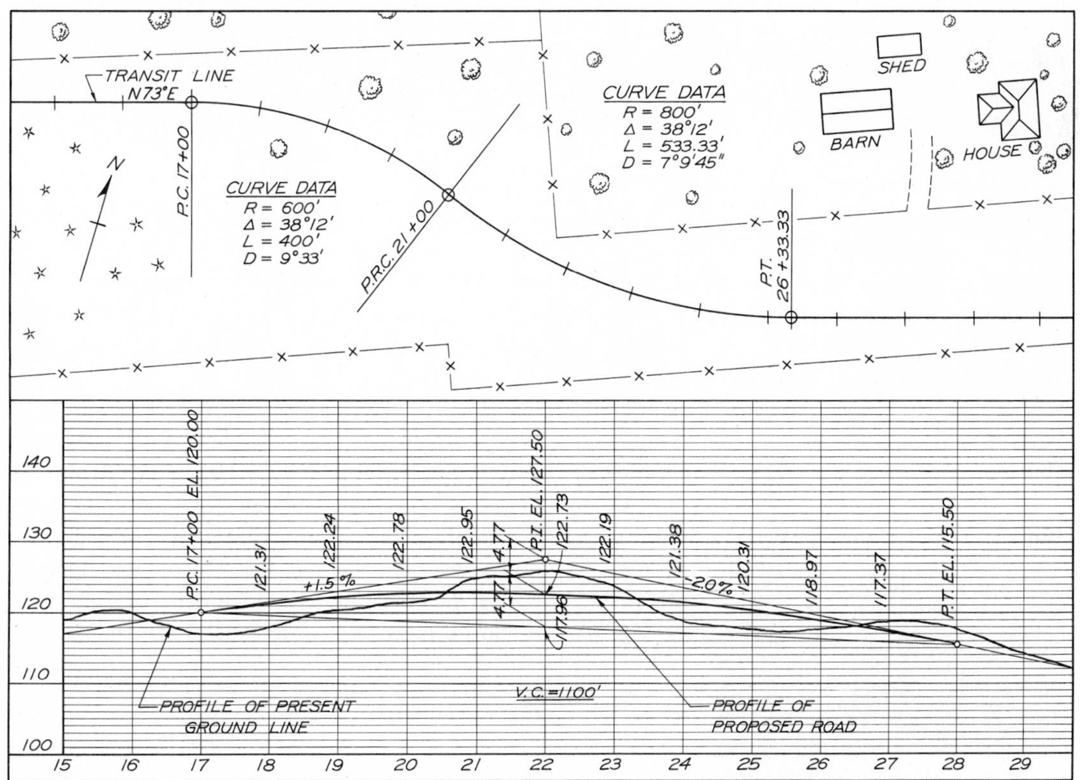

| Station | Tangent Elevations | Ordinate | Curve Elevations |
|---|---|---|---|
| 18 | 121.50 | 0.19* | 121.31 |
| 19 | 123.00 | 0.76 | 122.24 |
| 20 | 124.50 | 1.72 | 122.78 |
| 21 | 126.00 | 3.05 | 122.95 |
| 22 | 127.50 | 4.77 | 122.73 |
| 23 | 125.50 | 3.31 | 122.19 |
| 24 | 123.50 | 2.12 | 121.38 |
| 25 | 121.50 | 1.19 | 120.31 |
| 26 | 119.50 | 0.53 | 118.97 |
| 27 | 117.50 | 0.13 | 117.37 |

Fig. 23.11 **Calculation of Vertical Curve Elevations.**

view is larger in a vertical than in a horizontal direction in order to show the elevations clearly. The existing ground profile along the center line of the road is shown, as well as the profile of the proposed vertical alignment. The symbol P.I. denotes the point of intersection of the grade lines, and grade slopes are given in percentages. A 1 percent grade would rise vertically 1′ in each 100′ of horizontal distance. Station 22 + 00, therefore, is the intersection point of an upgrade of 1.5 percent and a downgrade of −2.0 percent.

To provide a smooth transition between these grades, a vertical curve (V.C.) of 1100′ length is used. This curve is parabolic and tangent to grade at stations 17 + 00 and 28 + 00. The straight line joining these points has an elevation 117.96′ directly below the P.I. At this point the parabolic curve must pass through the midpoint of the vertical distance, at a height of 4.77′ below the P.I. Ordinates to parabolas, measured from tangents, are proportional to the squares of the horizontal distances from the points of tangency.* Therefore, it is possible to calculate the elevations of points along the curve by first determining the grade elevations, then subtracting the parabolic curve ordinates. The final profile elevations are given in the figure. Calculations are given in Fig. 23.11.

$$\frac{{}^{*}(100)^2\ 4.77'}{(500)^2} = 0.19'.$$

## 23.9 United States Maps.

United States maps, prepared by the U.S. Coast and Geodetic Survey and the U.S. Geological Survey, are excellent examples of topographic mapping. A small section of a typical U.S. Geological Survey map is shown in Fig. 23.12. Such maps cover large areas, the largest scale used being 1:62,500, or very nearly one mile to the inch. The contour interval in this example is 20′. To so small a scale, it would be impossible to show clearly any small features—vegetation and fences, for example. Therefore, these maps can show only the main features of the terrain, such as contours, roads, railroads, rivers, lakes, and streams. These maps are very reliable, as they are based upon precise surveying.

## 23.10 Topographic Drawing Problems

The following problems are given to afford practice in topographic drawing. The drawings are designed for a 11″ × 17″ or Size B sheet. The position and arrangement of the titles should conform approximately to that of Fig. 23.1.

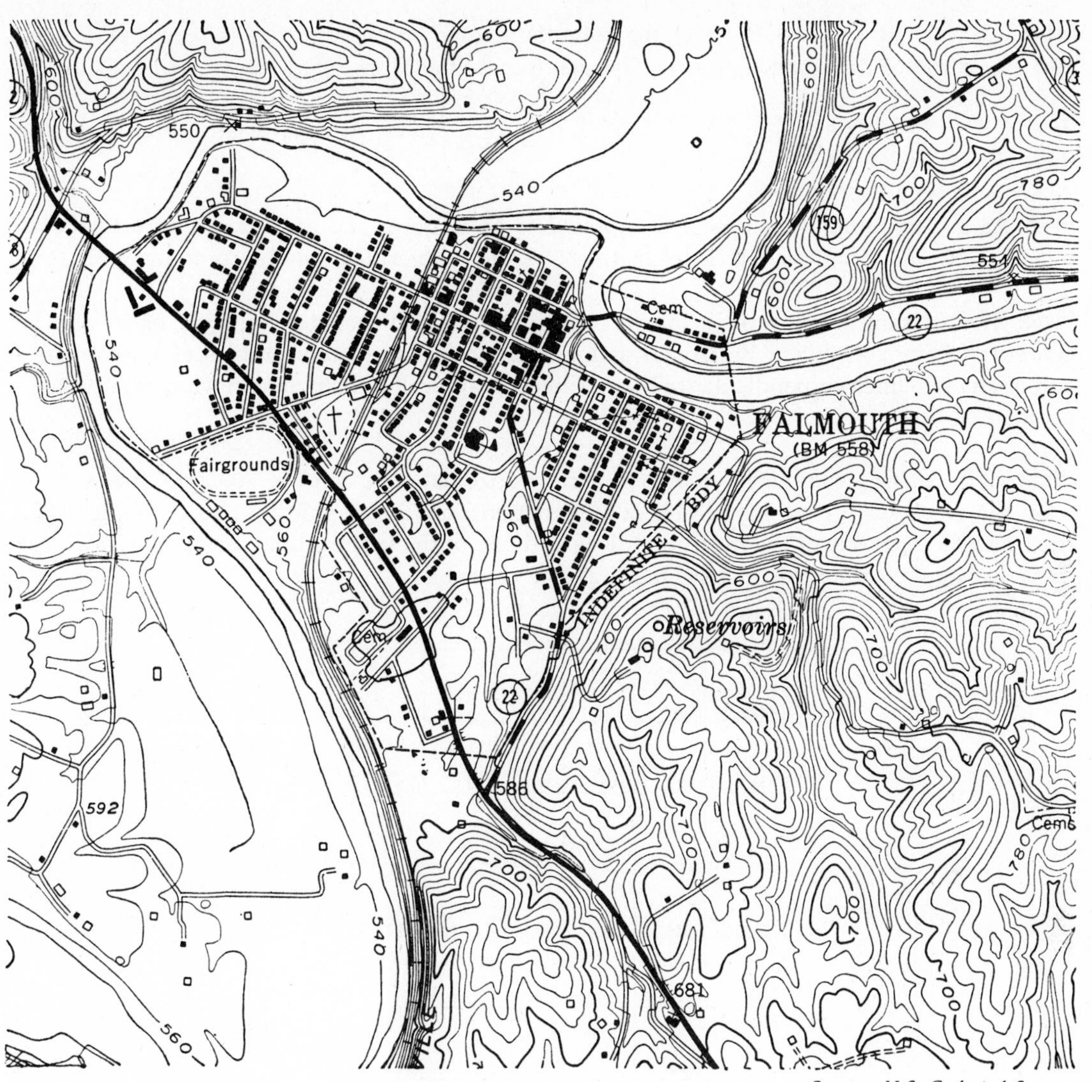


Courtesy U.S. Geological Survey

Fig. 23.12 **A Typical Map of the U.S. Geological Survey.**

## 23.10 Topographic Drawing Problems

**Prob. 23.1** Draw symbols of six of the common natural surface features (streams, lakes, etc.) and six of the common development features (roads, buildings, etc.) shown in Appendix 28.

**Prob. 23.2** Draw, to assigned horizontal and vertical scales, profiles of any three of the six lines shown in Fig. 23.3.

**Prob. 23.3** Assuming the slope of the ground to be uniform and assuming a horizontal scale of 1″ = 200′ and a contour interval, §23.4, or 5′, plot, by interpolation, the contours of Fig. 23.3.

**Prob. 23.4** Using the elevations shown in Fig. 23.4 (a) and a contour interval of 1′, plot the contours to any convenient horizontal and vertical scales, and draw profiles of lines 3 and 5 and of any two lines perpendicular to them. Check, graphically, the points in which the contours cross these lines.

**Prob. 23.5** Using a contour interval of 1′ and a horizontal scale of 1″ = 100′, plot the contours from the elevations given, Fig. 23.13, at 100′ stations; check, graphically, the points in which the contours cross one of the horizontal lines and one of the vertical lines, using a vertical scale of 1″ = 10′; sketch, approximately, the drainage channels.

| | | | | | | |
|---|---|---|---|---|---|---|
| 144.7 | 139.2 | 143.1 | 144.6 | 144.3 | 143.5 | 142.2 |
| 142.5 | 138.0 | 139.0 | 141.3 | 142.7 | 139.3 | 139.1 |
| 140.7 | 137.5 | 136.1 | 138.6 | 138.0 | 136.1 | 137.2 |
| 138.8 | 136.5 | 135.0 | 136.2 | 135.7 | 135.9 | 136.1 |
| 139.1 | 136.4 | 134.6 | 133.5 | 133.7 | 134.1 | 135.8 |
| 135.3 | 134.5 | 133.0 | 132.7 | 132.0 | 131.9 | 132.3 |
| 135.9 | 134.0 | 132.7 | 131.3 | 130.8 | 129.6 | 131.5 |

Fig. 23.13 **To Draw Contours** (Prob. 23.5).

**Prob. 23.6** Draw a plat of the survey shown in Fig. 23.1 to as large a scale as practicable. Use an engineers scale to set off distances and a protractor to set off bearings; if the drawing is accurate, the plat will close.

**Prob. 23.7** Draw a topographic map of a country estate similar to that shown in Fig. 23.2.

**Prob. 23.8** Calculate profile elevations for a vertical curve 800′ long to join grades of +3.00 percent and −3.00 percent. Assume grade elevations at points of tangency to be 100.00′.

# 24

# piping drawing

**24.1 Introduction.** Pipe is used for transporting liquids and gases and for structural elements such as columns and handrails. The choice of the type of pipe is determined by the purpose for which it is to be used.

**24.2 Kinds of Pipe.** Pipe is made of aluminum, brass, clay, concrete (concrete made with ordinary aggregates and in combination with asbestos, etc.), copper, glass, iron, lead, plastics, rubber, wood, and other materials or combinations of them. Cast iron, steel, wrought iron, brass, copper, and lead pipes are most commonly used for transporting water, steam, oil, or gases.

**Steel and Wrought-Iron Pipe.** Steel or wrought-iron pipe is in common use for water, steam, oil, and gas. Up to the early 1930's it was available in three weights known as "standard," "extra strong," and "double extra strong" only. At that time increasing pressures and temperatures, particularly for steam service, made the availability of more diversified wall thicknesses desirable. The American National Standards Institute Sectional Committee B36 has developed dimensions for ten different Schedules of pipe. See Appendix 32. It will be observed, in this table, that dimensions are established for nominal sizes from $\frac{1}{8}''$ to 24", and that dimensions have not been established for all Schedules. In the different Schedules, the outside diameters (OD) are maintained for each nominal size to facilitate threading and the uniform use of fittings and valves.

Certain of the Schedule dimensions correspond to the dimensions of "standard" and "extra strong" pipe. These are shown in **bold-**

**face** type in the appendix. The Schedule dimensions so shown for Schedules 30 and 40 correspond to standard pipe, and those for Schedule 80 correspond to extra strong pipe. There are no Schedule dimensions corresponding to "double extra strong" pipe. Pipe corresponding to all of the established Schedule dimensions is not always commercially available, and should be investigated before specifying pipe on drawings. Generally Schedules 40, 80, and 160 are readily available; others may or may not be.

Note that the actual outside diameter of pipe in nominal sizes $\frac{1}{8}''$ to 12″ inclusive is larger than the nominal size, whereas the outside diameter of pipe in nominal sizes 14″ and larger corresponds to the nominal size. This pipe in nominal sizes 14″ and larger is commonly referred to as OD pipe.

Pipe is available as welded or as seamless pipe. Welded pipe is available in Schedules 40 and 80 in the smaller sizes. Lap-welded pipe is made in sizes up to and including 2″. Butt-welded pipe is available as furnace-welded material, where a formed length is heated in a furnace and then welded, in sizes up to and including 3″. Butt-welded pipe is also available as continuous welded pipe, where the finished pipe is continuously heated, formed, and welded from a roll of strip steel, in sizes up to and including 4″. Seamless pipe is made in both small and large sizes.

Many of today's applications require the use of alloys to withstand the pressure-temperature conditions without having to be excessively thick. Numerous alloys in both ferritic and austenitic material are available. Reference should be made to the various specifications of the American Society for Testing Materials (ASTM)* for these alloys and for dimensional tolerances.

Steel pipe can be obtained as *black pipe* or as *galvanized pipe*. Galvanized pipe is not always available from the original producer, due to lack of facilities at the pipe mill.

Steel or wrought-iron pipe is available in lengths up to about 40′ in the small sizes, the length decreasing with increasing size and wall thickness.

*1916 Race St., Philadelphia, Pa. 19103.

## 24.4 Cast-Iron Pipe.

Cast-iron pipe is generally used for water or gas service and as soil pipe. For water and gas pipe, it is available in sizes from 3″ to 60″ inclusive and in standard lengths of 12′. Various wall thicknesses, depending on the internal pressure, can be secured. The dimensions and the pressure ratings of the various Classes are shown in Appendix 33.

Generally speaking, water and gas pipes are connected with *bell and spigot* joints, Fig. 24.1 (a), or *flanged* joints, (b), although other types of joints, (c), are also used.

As soil pipe, cast-iron pipe is available in sizes 2″ to 15″ inclusive, in standard lengths of 5′, and in *service* and *extra heavy* weights. Soil pipe is generally connected with bell and spigot joints, but soil pipe with threaded ends is available in sizes up to 12″.

In using cast-iron pipe, the designer must consider both the internal pressure and the external loading due to fill and other loadings, such as roads and tracks. Cast-iron pipe is brittle, and settlement can cause fracture unless sufficient flexibility is provided in the joints. For this reason, flange joints are not usually employed for buried pipes unless adequately supported.

## 24.5 Seamless Brass and Copper Pipe.

Pipe made of brass and copper, and having approximately the same dimensions as "standard" and "extra strong" steel pipe, is available. Such pipe is suitable for plumbing, including supply, soil, waste drain, and vent lines. It is also particularly suitable for process work where scale and oxidation of steel pipe are objectionable. Brass and copper pipe is available in straight lengths up to 12 ft.

Brass pipe is generally known as *red brass pipe*, which is an alloy of approximately 85 percent copper and 15 percent zinc. Copper pipe is practically pure copper with less than 0.1 percent of alloying elements.

Brass and copper pipe should be joined with fittings of copper-base alloy in order to avoid galvanic action resulting in corrosion. Where screwed joints are used, fittings similar to cast or malleable iron fittings, Fig. 24.5, are avail-

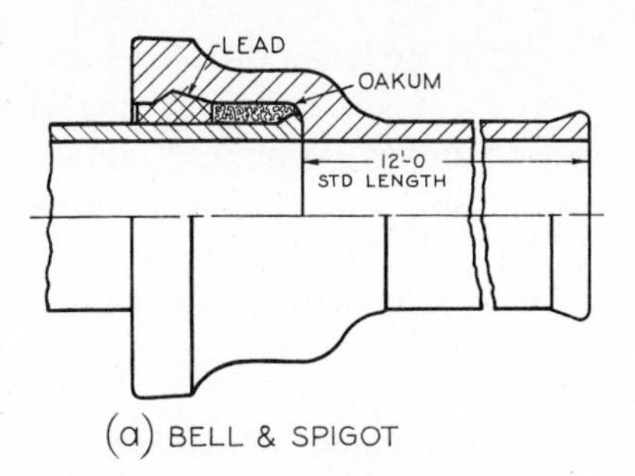

(a) BELL & SPIGOT

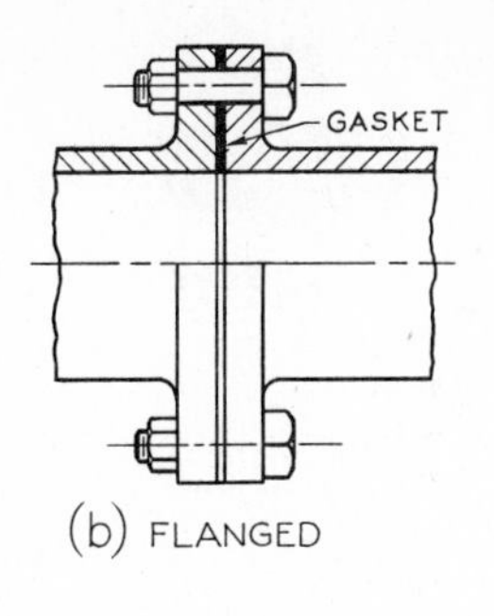

(b) FLANGED

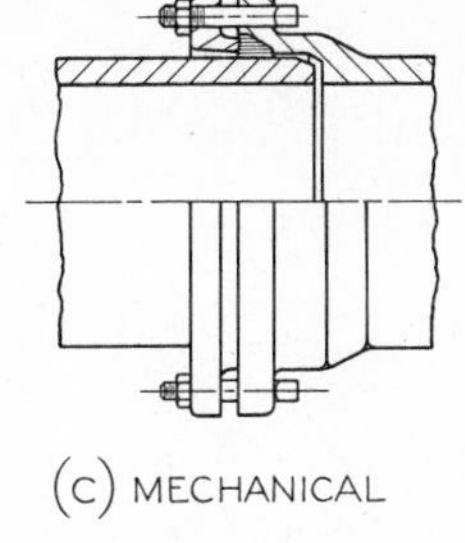
(c) MECHANICAL

Fig. 24.1 **Cast-Iron Pipe Joints.**

able. Flanged fittings of brass and copper are of different dimensions than those made of ferrous material. Reference should be made to dimensional standards published by the American National Standards Institute (ANSI B16.24) for dimensions of such fittings.

## 24.6 Copper Tubing.

Where nonferrous construction in sizes below 2" is used, copper tubing is frequently employed. Such tubing is suitable for process work as mentioned in §24.5, for plumbing (particularly supply lines for hot and cold water), and for heating systems (particularly radiant heating).

Copper tubing is made as *hard temper* and as *soft tubing*. Hard temper tubing is much stiffer than soft tubing and should be used where rigidity is desired. Soft tubing can be easily bent and, therefore, is generally used where bending during assembly is required. Neither hard nor soft temper tubing has the rigidity of iron or steel pipe, and consequently must be supported at much more frequent intervals than the latter. Where multiple runs of parallel tubes are employed for long distances (20' or more), it is common practice to use soft tubing and to lay the parallel runs in a trough construction, thus obtaining continuous support.

Copper tubing joints are usually made with *flared* joints, Fig. 24.2 (a), or with *solder* joints, (b). There are several types of flared joints, but the basic design of making a metal-to-metal joint is common to all. Fittings, such as tees, elbows, and couplings, are available for flared joints.

Solder joints are also known as *capillary* joints because the annular space between the tube and the fitting is so small that the molten solder is drawn into the space by capillary action. The solder may be introduced through a hole in the fitting, Fig. 24.3, through the outer end of the annular space, Fig. 24.2 (b), or fittings having a factory-assembled ring of solder in the fitting can be purchased. Solder joints may be made with soft solder (usually 50–50 or 60–40 tin and lead) or with silver solder. This latter material has a higher melting point than soft solder, makes a stronger joint, and is suitable for higher operating temperatures.

Fig. 24.2 **Copper Pipe Fittings.**

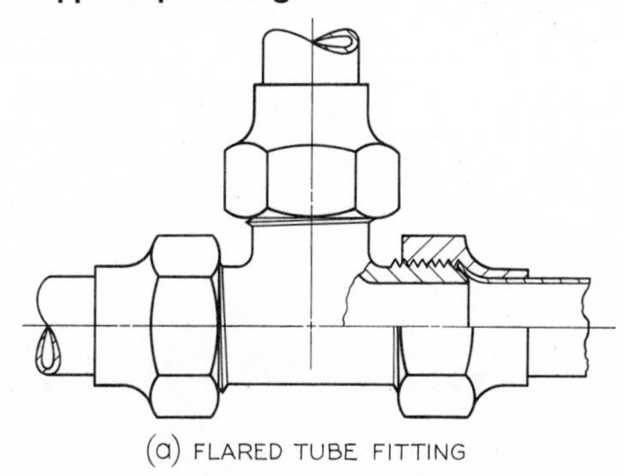
(a) FLARED TUBE FITTING

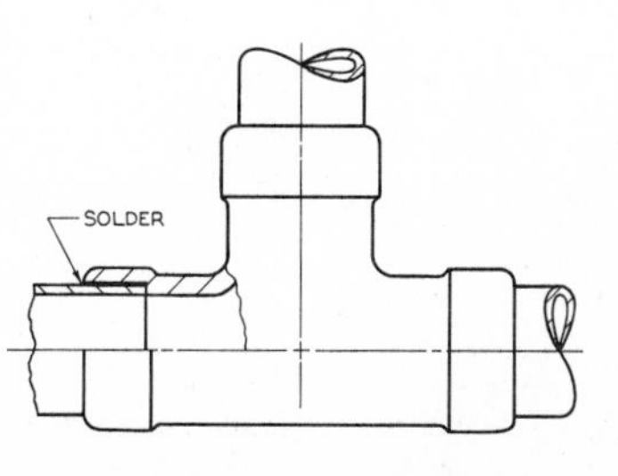

(b) SOLDER TUBE FITTING

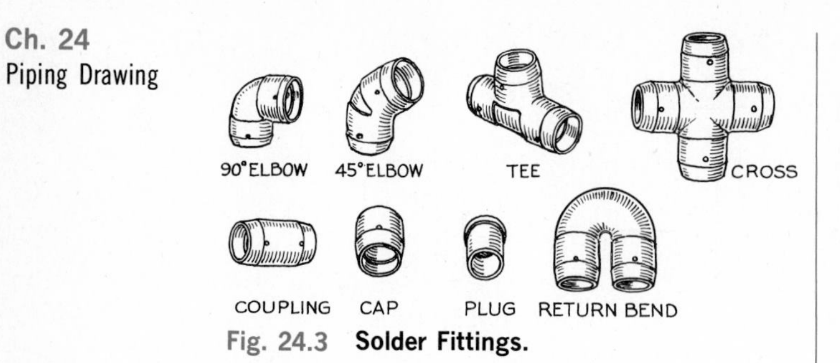

Fig. 24.3 **Solder Fittings.**

Copper pipe or tubing has an upper operating temperature limit of 406°F. If solder fittings are used, the upper temperature limit will be dependent on the softening point of the solder rather than the limit of temperature of the base material.

Copper tubing can be connected to threaded pipe or fittings by means of *adapters.* These adapters are available with either male or female pipe threads, and with either flared or solder connections for the tubing. Two types of such adapters are shown in Fig. 24.4.

Copper tubing is available in straight lengths up to 20′ or in coils of 60′ for soft temper material. Hard temper material is available in straight lengths only, since it cannot successfully be coiled. Installation costs of coiled material are lower than for straight material, because of the fewer number of joints to be made. Copper tubing is available in both OD and nominal sizes. American Society for Testing Materials Specifications B88 and B251 give details of dimensions of such tubing.

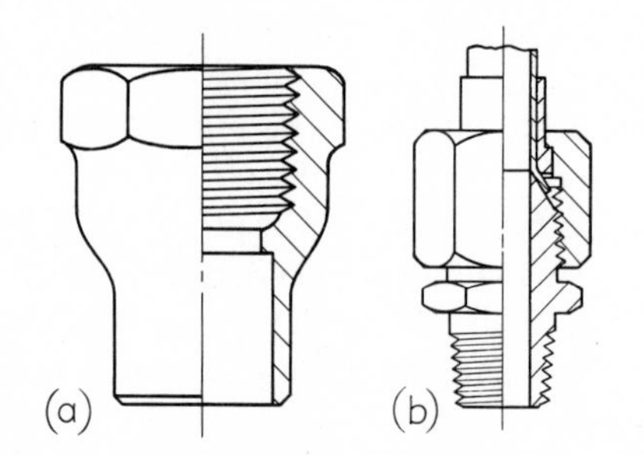

Fig. 24.4 **Adapters—Copper Tube to Threaded Pipe.**

**24.7 Special Pipes.** Pipe and tubing of other materials, such as aluminum and stainless steel, are also available. Pipe and tubing of plastics are being increasingly used as these materials undergo development. Service for which these miscellaneous materials are suitable varies widely because of the physical properties and temperature-pressure limitations of the materials.

**24.8 Pipe Fittings.** Pipe fittings are used to join adjacent lengths of pipe and frequently also to provide changes of direction, to provide branch connections at different angles, or to effect a change in size. They are made of cast iron, malleable iron, cast or forged steel, nonferrous alloys, and other materials for special applications. They can be obtained in various weights that should be matched to the pipe with which they are to be used. Ferrous fittings are made for threaded, welded, or flanged joints. Nonferrous fittings are made for threaded, solder, flared, or flanged joints. The common types of fittings for threaded joints are shown in Fig. 24.5, for welded joints in Fig. 24.6, for flanged joints in Fig. 24.7, and for solder joints in Fig. 24.3.

Where both or all ends of a fitting are the same nominal size, the fitting is designated by the nominal size and the description—for ex-

Fig. 24.5 **Screwed Fittings.**

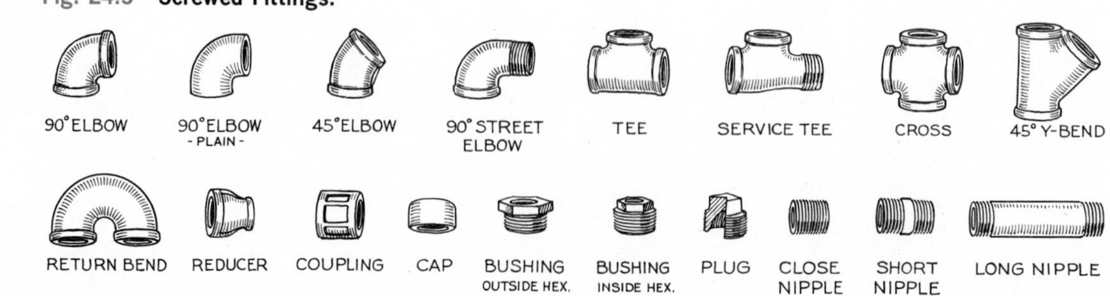

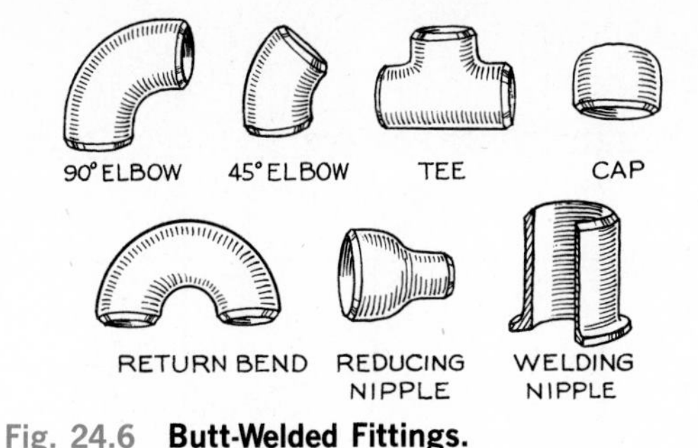


Fig. 24.6 **Butt-Welded Fittings.**

ample, a *2″ screwed tee.* Where two or more ends of a fitting are not the same nominal size, the fitting is designated as a *reducing fitting,* and the dimensions of the run precede those of the branches, and the dimension of the larger opening precedes that of the smaller opening—for example, a *2″* × *1½″* × *1″ screwed reducing tee.* See Fig. 24.8 for typical designations.

The threads of screwed fittings conform to the pipe thread with which they are to be used, either male or female as the case may be. See §24.9.

Dimensions of 125 lb cast-iron screwed fittings, 250 lb cast-iron screwed fittings, 125 lb cast-iron flanged fittings, and 250 lb cast-iron flanged fittings are shown in Appendixes 34, 35, 36, and 39. See §24.9 for reference to steel fittings.

## 24.9 Pipe Joints.

The joints between pipes, fittings, and valves may be *screwed, flanged, welded,* or for nonferrous materials, joints may be soldered.

The American National Standard pipe threads are illustrated in Figs. 13.21 to 13.23, and tabular dimensions are shown in Appendix 32. The threads of the American Petroleum Institute (API) differ somewhat from the American National Standard pipe threads. Refer to the API Standards for these differences.

Threaded joints can be made up tightly by simply screwing the cleaned threads together. However, it is common practice to use *pipe compound* in making such joints, as this provides lubrication for the threads and enables them to be screwed together more tightly. It also serves to seal irregularities, thus providing a tighter joint. Such material should be

Fig. 24.7 **Flanged Fittings.**

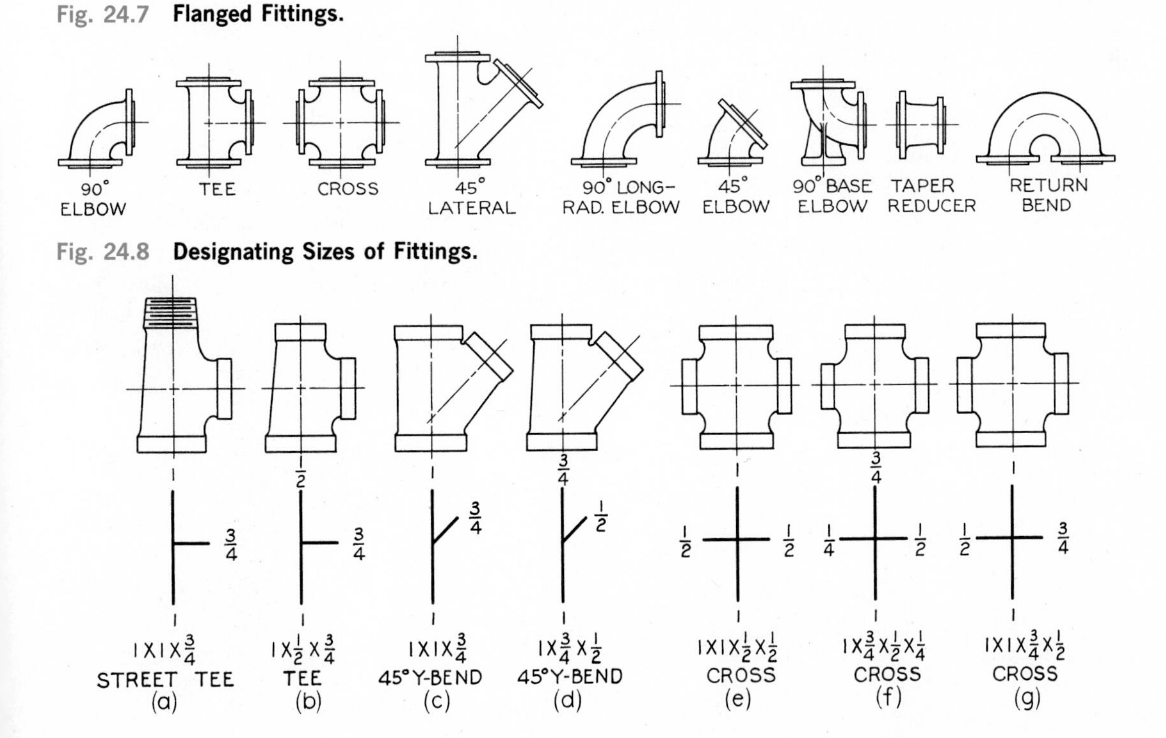


Fig. 24.8 **Designating Sizes of Fittings.**

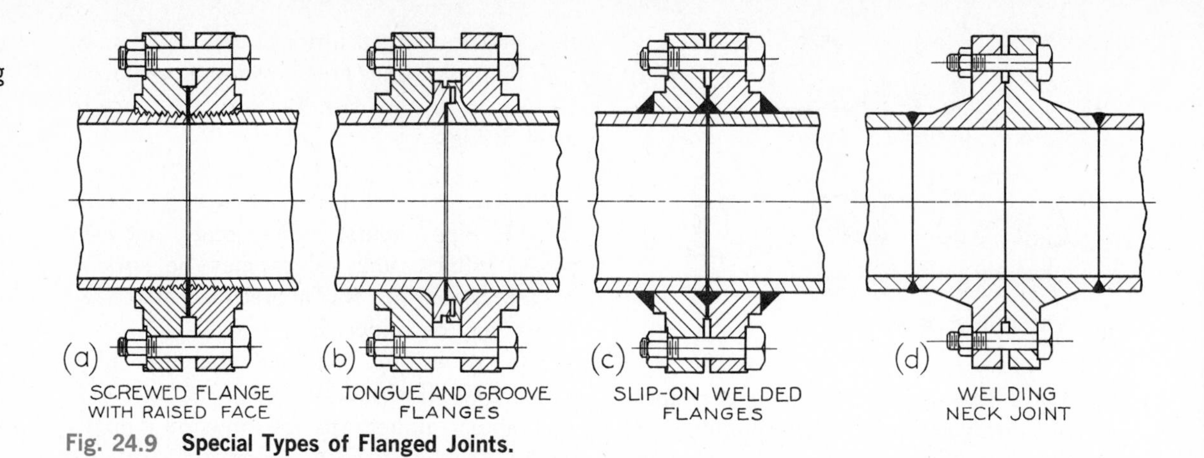


Fig. 24.9 **Special Types of Flanged Joints.**

applied to the male thread only, to avoid forcing it into the pipe where contamination or obstruction may result.

Flanged joints are made by bolting two flanges together with a resilient *gasket* between the flange faces. Flanges may be attached to the pipe, fitting, or appliance by means of a screwed joint, by welding, by lapping the pipe, or by being cast integrally with the pipe, fitting, or appliance.

The faces of the flanges between which the gasket is placed have different standard *facings*, such as *flat face*, $\frac{1}{16}''$ *raised face*, $\frac{1}{4}''$ *raised face*, *male and female*, *tongue and groove*, and *ring joints*. Flat face and $\frac{1}{16}''$ raised face are standard for cast-iron flanges in the 125 lb and 250 lb classes, respectively. The other types of facing are standard for steel flanges.

The number and size of the bolts joining these flanges varies with the size and the working pressure of the joint. Bolting for Class 125 cast-iron and Class 250 cast-iron flanges is shown in Appendixes 37 and 40, respectively.

For dimensions of the various flange facings and for flange and bolting dimensions of the various sizes and pressure standards of steel flanges, refer to the American National Standard for Steel Pipe Flanges and Flanged Fittings (ANSI B16.5), which is too voluminous to be included here. Some typical types of flanged joints are shown in Fig. 24.9.

Piping construction employing welded joints is in almost universal use today, particularly for higher pressure and temperature conditions. Such joints may either be *socket welded* or *butt welded*, Fig. 24.10. Socket-welded joints are limited to use in small sizes. The contours of the butt-welded joints shown at (b) and (c) are those shown in American National Standard B16.25.

## 24.10 Valves.

Valves are used to stop or to regulate the flow of fluids in a pipe line. The more common types are *gate valves*, *globe valves*, and *check valves*. Other types of valves, such as *pressure reducing valves* and *safety valves*,

Fig. 24.10 **Welded Joints.**

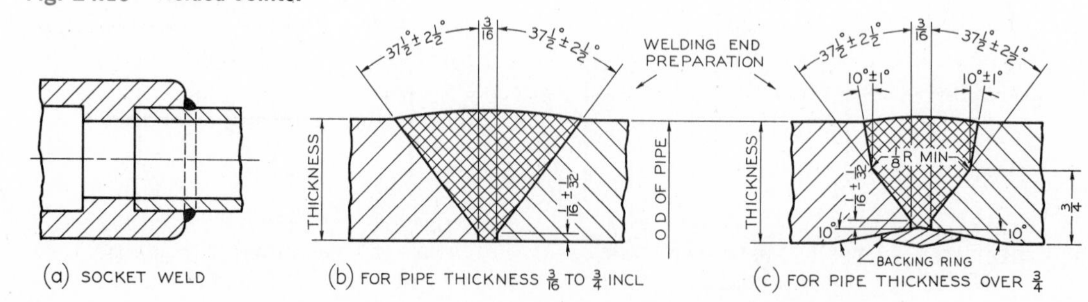

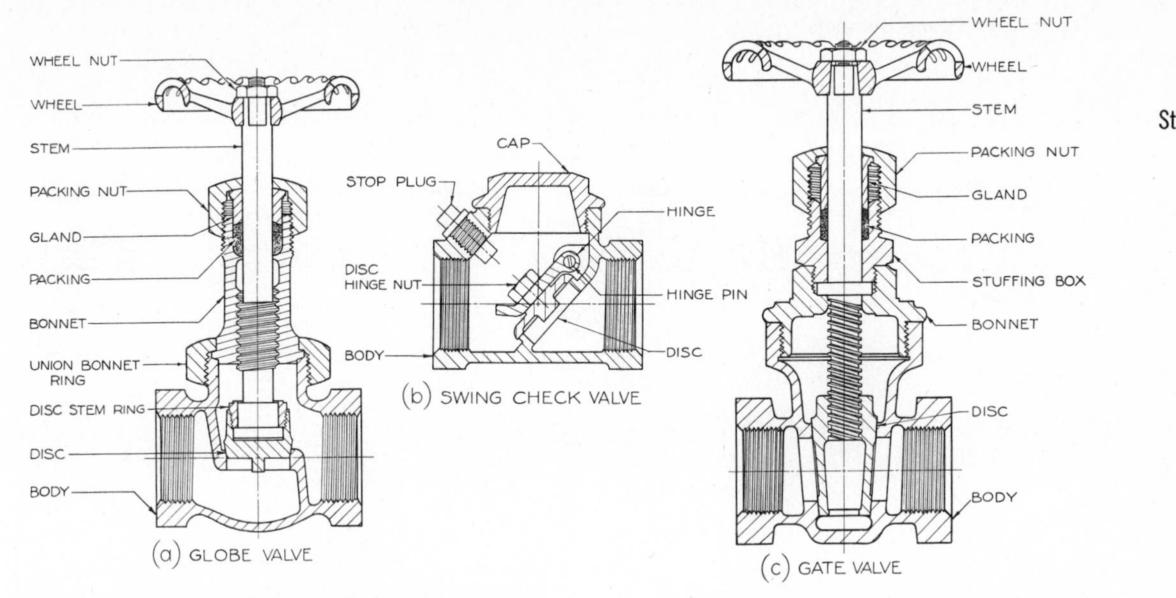


Fig. 24.11 **Gate, Globe, and Check Valves.**

are special devices used to maintain automatically a desired lower pressure on the downstream side of the valve or to prevent automatically undesirable overpressure, respectively.

## 24.11 Globe Valves.

Globe valves have approximately spherical bodies with the seating surface at either a right or an acute angle to the center line of the pipe, Fig. 24.11 (a). In such a valve the flowing fluid must make abrupt turns in the body, thus resulting in considerably higher pressure loss than for a gate valve.

Globe valves are commonly used where close regulation of flow is desired, because they lend themselves better to this type of regulation and are less subject to cutting action in throttling service than gate valves.

Valves of the *inside screw* and *outside screw and yoke* (O S & Y) types are available, §24.13. Angle valves and needle valves are special designs of the general class of globe valves.

## 24.12 Check Valves.

Check valves are used to limit the flow of fluids to one direction only. The disc may be hinged so as to swing partially out of the stream, Fig. 24.11 (b), or it may be guided in such a manner that it can rise vertically from its seat. The two types are called *swing checks* and *lift checks*, respectively.

## 24.13 Gate Valves.

Gate valves have full-sized straighway openings which offer small resistance to the flow of fluids. The gate, or disc, may rise on the stem (*inside screw* type), Fig. 24.11 (c), or the gate may rise with the stem, which in turn rises out of the body (*rising stem*, or *outside screw* and *yoke—O S & Y*—type). Inside screw type valves are employed in the smaller sizes and lower pressures.

Seating may be on nonparallel seats, in which case the disc is solid and wedge shaped. There is also a type of gate valve employing parallel seats. In this type two discs are hung loosely on the stem and are free of the seats until an adjusting wedge reaches a lug at the closed position of the valve, when further movement of the stem causes the wedge to spread the discs and form a tight joint on the parallel seats. Such valves are used only on low pressure and temperature services.

## 24.14 American National Standard Code for Pressure Piping.

The American National Standards Institute has adopted an American

Fig. 24.12 **Piping Symbols.**

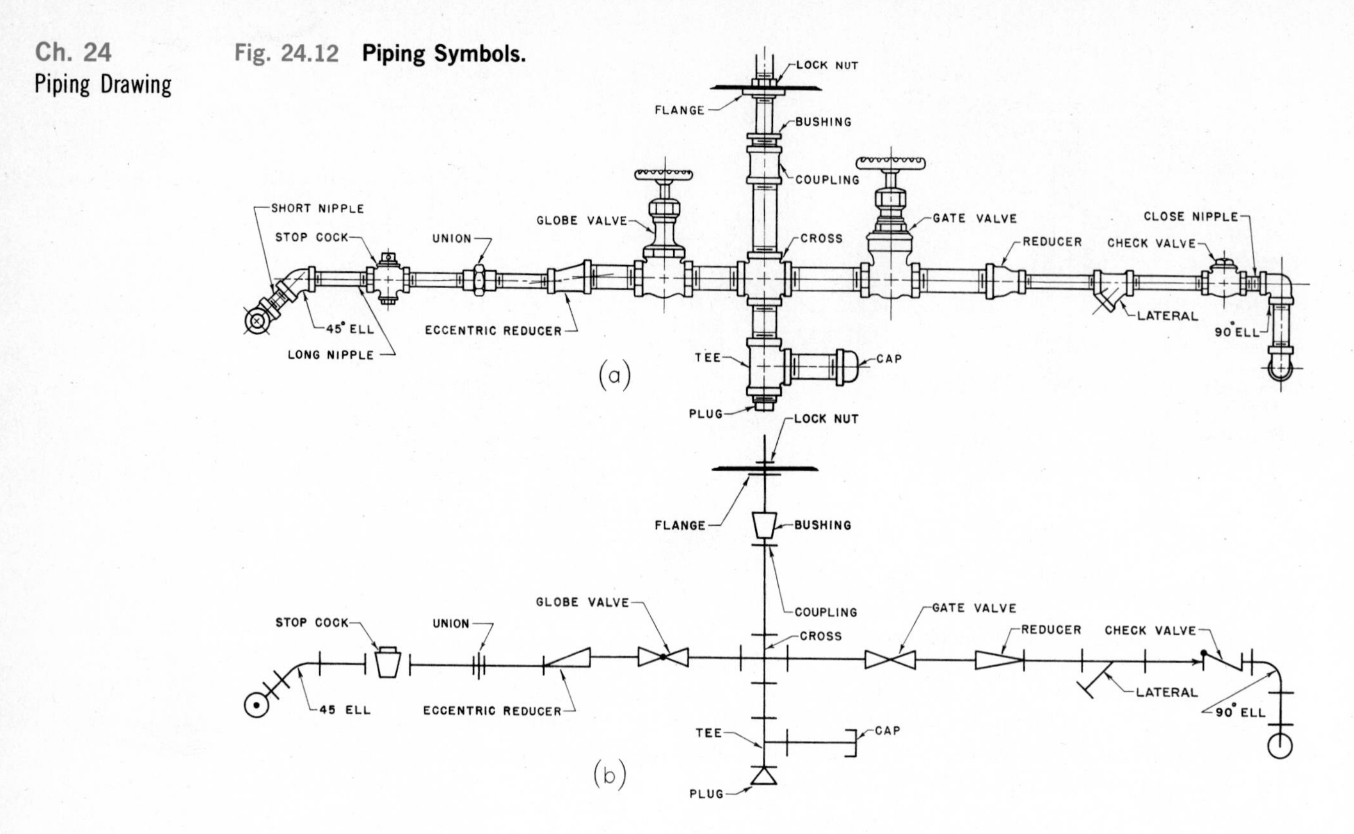


Fig. 24.13 **Representations of Pipe Expansion Joint.**

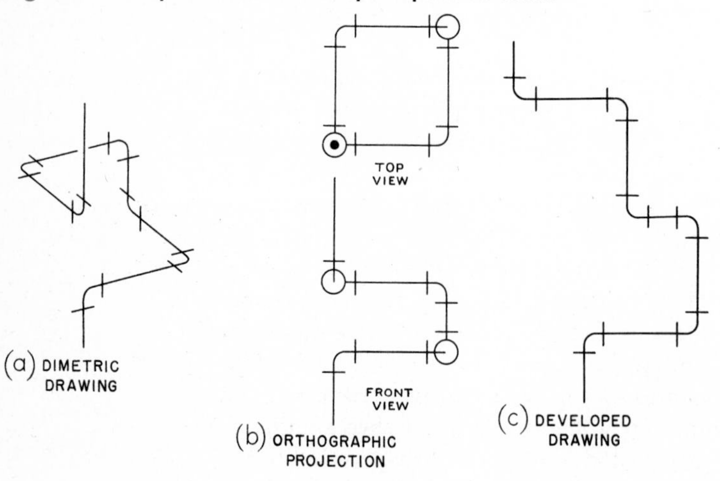


National Standard Code for Pressure Piping (ANSI B31.1). This is a compilation of recommended practices and minimum safety standards covering various types of piping, such as power piping, industrial gas and air piping, oil refinery piping, oil transportation piping, refrigerating piping, chemical industry process piping, and gas transmission and distribution piping.

## 24.15 Piping Drawings.

To simplify the preparation of working drawings of piping systems, the set of symbols shown in Appen-

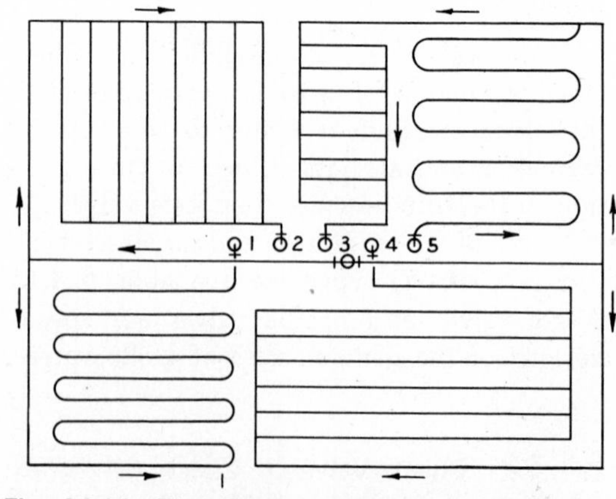


Fig. 24.14 **Pipe Grids and Serpentine Coils for a Panel Radiant Heating System.**

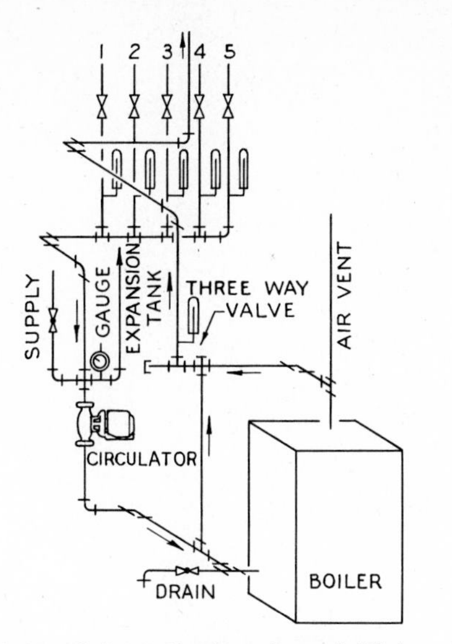


Fig. 24.15 **Schematic Drawing of Piping Connecting Boiler to Heating Coils.**

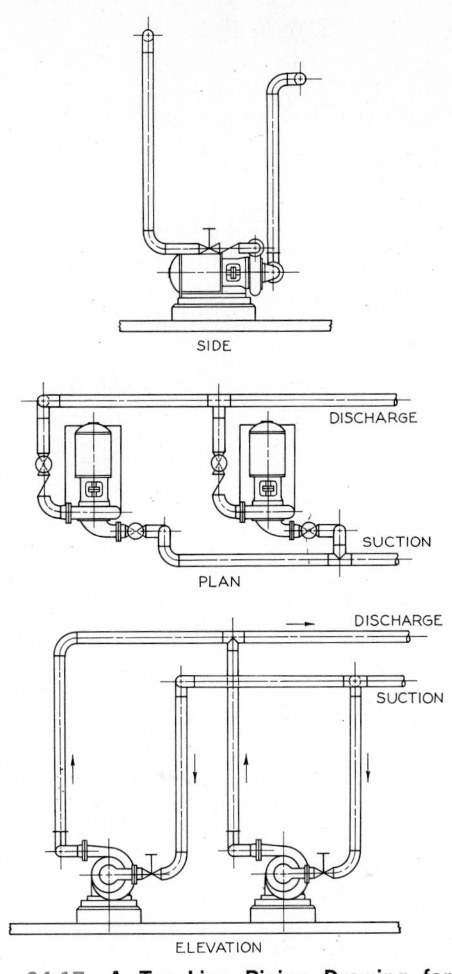


Fig. 24.17 **A Two-Line Piping Drawing for a Pumping Plant.**

dix 29 has been developed to represent the various pipe fittings and valves in common use. An application of these symbols in a piping drawing is shown in Fig. 24.12 (b).

Drawings of piping systems may be made as *single-line* drawings, Figs. 24.12 (b) and 24.13 to 24.15, or as *double-line* drawings, Figs. 24.12 (a), 24.17, and 24.18. Either type of drawing may be made as multiview projections, Figs. 24.13 (b), 24.17, and 24.19; as axonometric projections, Figs. 24.13 (a) and 24.20; or as oblique projections, Figs. 24.15 and 24.16. The oblique projection in Fig. 24.16 is a modified form of oblique projection generally used in representing the piping arrangement for heating systems. In these cases, the pipe mains are shown in plan and the risers in oblique projection in various directions so as to make the representation as clear as possible.

Fig. 24.16 **A One-Pipe Steam Heating System.**

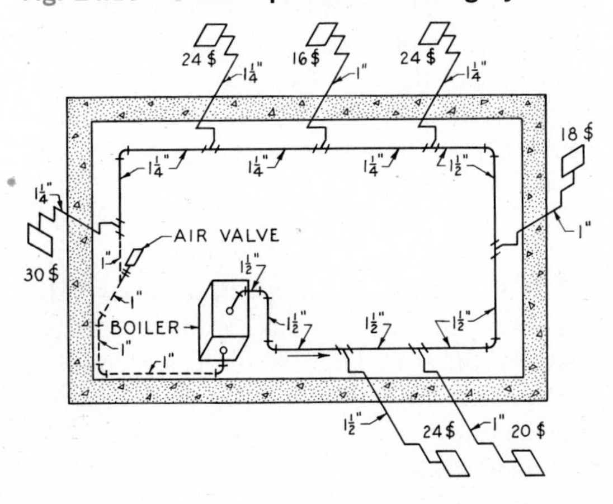


In most installations, some pipes are vertical and some are horizontal. If the vertical pipes are assumed to be revolved into the horizontal plane or the horizontal pipes revolved into the vertical plane by turning some of the fittings, Fig. 24.13 (c), the entire installation can be shown in one plane. Such a drawing is a *developed piping drawing*.

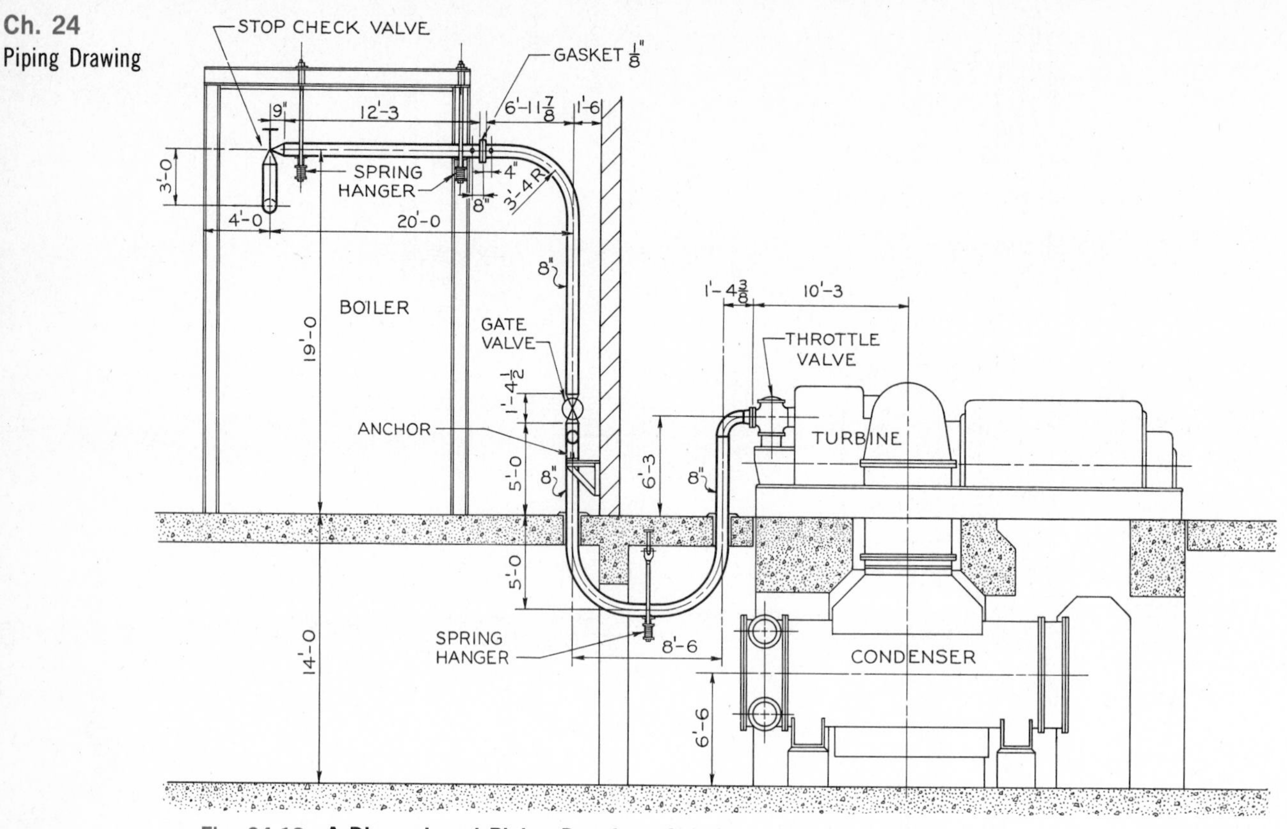


Fig. 24.18 **A Dimensioned Piping Drawing—Side View of Steam Piping.**

In complicated systems where a large amount of piping of various sizes is run in close proximity, and where clearances are important, the use of double-line multiview drawings, made accurately to scale, is desirable. The use of such drawings, which show the relative positions of component parts in all views, greatly reduces the probability of interferences when the piping is erected and is almost a necessity where piping components are prefabricated in a shop and sent to the job in finished dimensions. Such method of fabrication is universal in large systems using large piping, most piping $2\frac{1}{2}''$ and larger being shop fabricated. See Figs. 24.17 and 24.18.

## 24.16 Dimensioning.

In dimensioning a piping drawing, distances from center to center (c to c), center to end (c to e), or end to end (e to e) of fittings or valves and the lengths of all straight runs of pipe should be given, Fig. 24.18. Fully dimensioned single-line drawings need not be drawn to scale. Allowances in pipe lengths for make-up in fittings and valves must be made in preparing a bill of materials. All double-line drawings should have center lines if they are to be dimensioned. The size of the pipe for each run is shown by a numeral or by a note at the side of the pipe, with a leader when necessary.

## 24.17 Piping Drawing Problems.

The drawings for the first six problems are to be three times as large as the corresponding illustrations in the book, unless otherwise specified.

**Prob. 24.1** Make a double-line drawing, similar to Fig. 24.12 (a) showing the following fittings: a union, a 45° Y-bend, an eccentric reducer, a globe valve, a tee, a stopcock, and a 45° ell. Use ½" and 1" wrought-steel pipe and 125 lb C.I. screwed fittings.

**Prob. 24.2** Make a single-line drawing, similar to Fig. 24.12 (b), showing the following fittings: a 45° ell, a union, a 45° Y-bend, an eccentric reducer, a tee, a reducer, a gate valve, a plug, a cap, and a cross.

**Prob. 24.3** Make a single-line drawing of the system of pipe coils and grids shown in Fig. 24.14; show, by their respective standard symbols, the elbows and tees that must be used to connect pipes meeting at right angles if welding is not used to make the joints.

**Prob. 24.4** Make an oblique projection, similar to that shown in Fig. 24.15, of the one-pipe steam heating system shown in Fig. 24.16. Show the pipes by single lines, the fittings by their standard symbols, and the boiler and radiators as parallelepipeds.

Fig. 24.19 **Pipe Layout for Battery of Air Receivers** (Probs. 24.5 and 24.7).

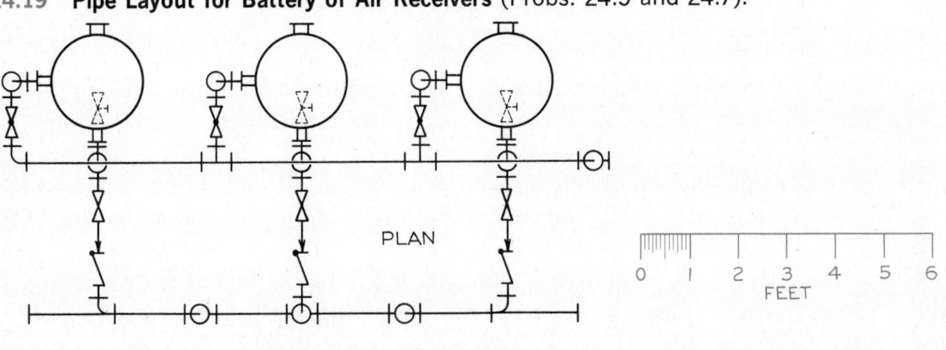


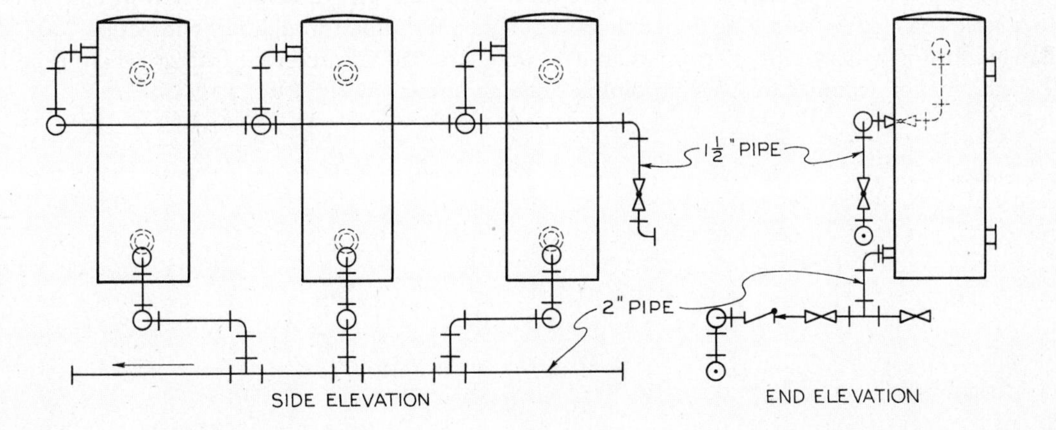

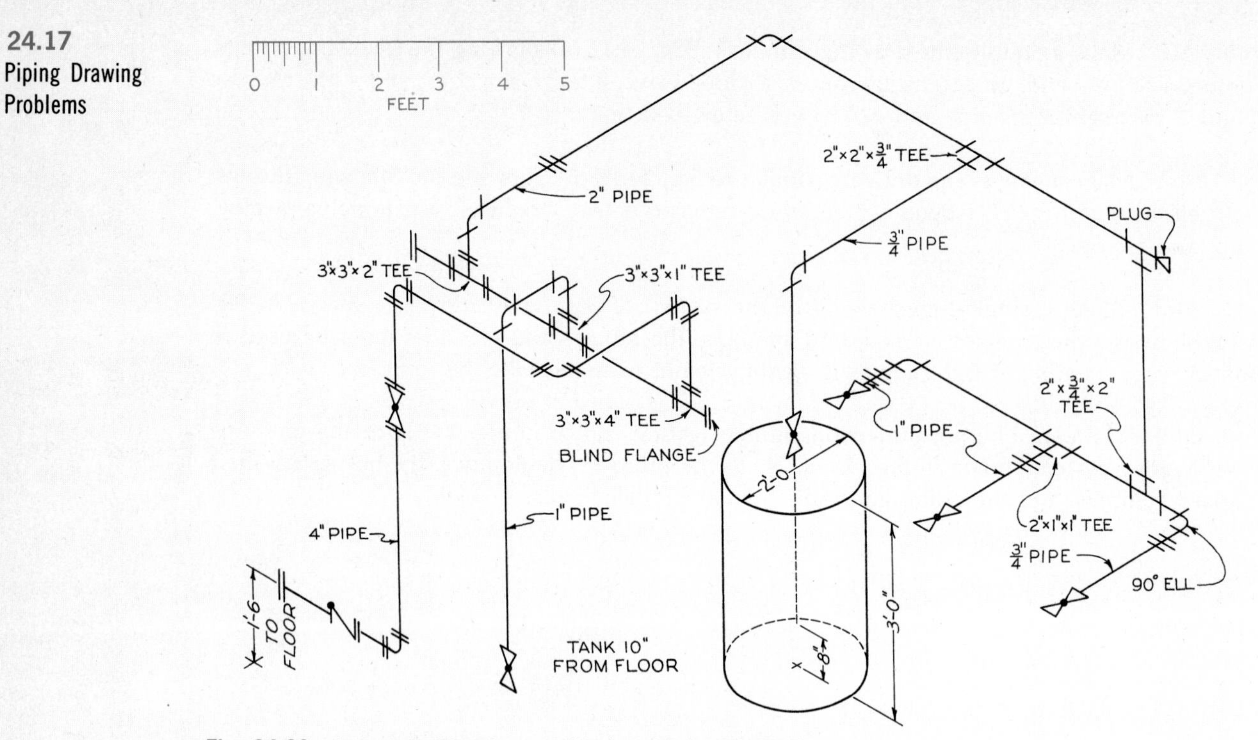


Fig. 24.20 **Isometric Pipe Layout** (Probs. 24.6 and 24.8).

Prob. 24.5 Make a single-line isometric drawing of the piping layout shown in Fig. 24.19. Use a scale of ¾″ = 1′-0, and a 17″ × 22″ sheet. (Similar to isometric layout shown in Fig. 24.20.)

Prob. 24.6 Make a single-line multiview drawing of the piping layout shown in Fig. 24.20. Use a scale of 1″ = 1′-0 and a 17″ × 22″ sheet. (Similar to piping layout shown in Fig. 24.19.)

Prob. 24.7 Make a double-line multiview drawing of the piping layout shown in Fig. 24.19, to a scale selected by the student. (Similar to two-line piping drawing shown Fig. 24.17.)

Prob. 24.8 Make a double-line multiview drawing of the piping layout shown in Fig. 24.20, to a scale selected by you. Use Schedule 80 wrought-steel pipe throughout, with Class 250 C.I. flanged fittings where pipe is larger than 2″, and Class 250 C.I. screwed fittings where pipe is 2″ and smaller. (Similar to two-line piping drawing shown in Fig. 24.17.)

# 25
# welding representation

**25.1 Welding Drawings.*** In recent years, welding has been increasingly used for fastening parts together permanently in place of bolts, screws, rivets, or other fasteners. Welding is also being used extensively in fabricating machine parts or other structures that formerly would have been formed by casting or forging, and it is used to a considerable extent in the erection of structural steel frames for buildings, ships, and other structures.

Since welding is used so extensively, and for so large a variety of purposes, it is essential to have an accurate method of showing on the working drawings of machines or structures, the exact types, sizes, and locations of welds desired by the designer. The old practice of simply lettering a note on the drawing, "To be welded throughout," or "To be completely welded," which actually shifted responsibility for welding control to the welding shop, is not only dangerous, but may be unnecessarily expensive because shops will usually "play safe" by welding more than necessary.

To provide a means for placing complete welding information on the drawing in a simple manner, a system of welding symbols was developed by the American Welding Society and published in 1947 under the title "Standard Welding Symbols."* These symbols were adopted by the American National Standards Institute and are currently published as ANSI Y32.3–1969.

A typical welding drawing is shown in Fig. 25.1. It is an assembly drawing in the sense that it is composed of a number of separate pieces fastened together as a unit. The welds

*The text matter and most of the illustrations in this chapter are based on ANSI Y32.3–1969.

*American Welding Society, 2501 NW 7th Street, Miami, Fla. 33125.

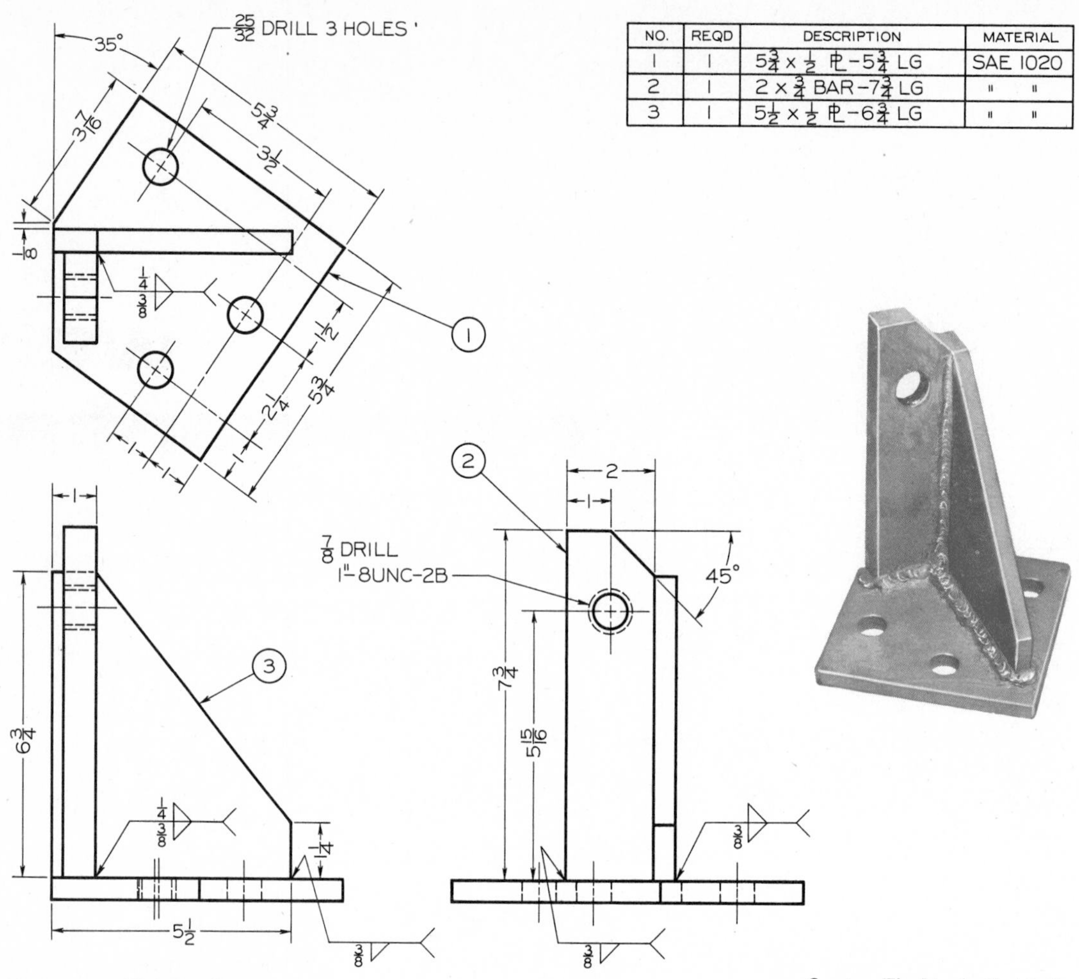


| NO. | REQD | DESCRIPTION | MATERIAL |
| --- | --- | --- | --- |
| 1 | 1 | 5¾ x ½ PL-5¾ LG | SAE 1020 |
| 2 | 1 | 2 x ¾ BAR-7¾ LG | " " |
| 3 | 1 | 5½ x ½ PL-6¾ LG | " " |

Fig. 25.1 **Welding Drawing.**

*Courtesy The Lincoln Electric Co.*

themselves are not drawn, but are clearly and completely indicated by the welding symbols. The joints are all shown as they would appear before welding. Dimensions are given to show the sizes of the individual pieces to be cut from stock. Each component piece is identified by encircled numbers and by specifications in the parts list, as shown.

## 25.2 Welding Processes.

Three of the principal methods of welding are the oxy-acetylene method, generally known as *gas welding,* the electric-arc method, generally known as *arc welding,* and electric-resistance welding, generally called *resistance welding.*

Gas welding was originated in 1895, when the French chemist Le Châtelier discovered that the combustion of acetylene gas with oxygen produced a flame of very high temperature—high enough to melt metals. This discovery was soon followed by the development of practical methods of producing and transporting oxygen and acetylene, and by the construction of suitable torches and welding rods.

In arc welding, the heat of an electric arc is used to fuse the metals that are to be welded or cut. The first arc welding was done in 1881 by De Meritens in France, and for a number of years the development of arc welding was very slow. During World War I, the U.S. Navy used welding to a limited extent for repairing machinery. Since that time, as the result of extensive research, basic improvements have been made in the manufacture of electrodes

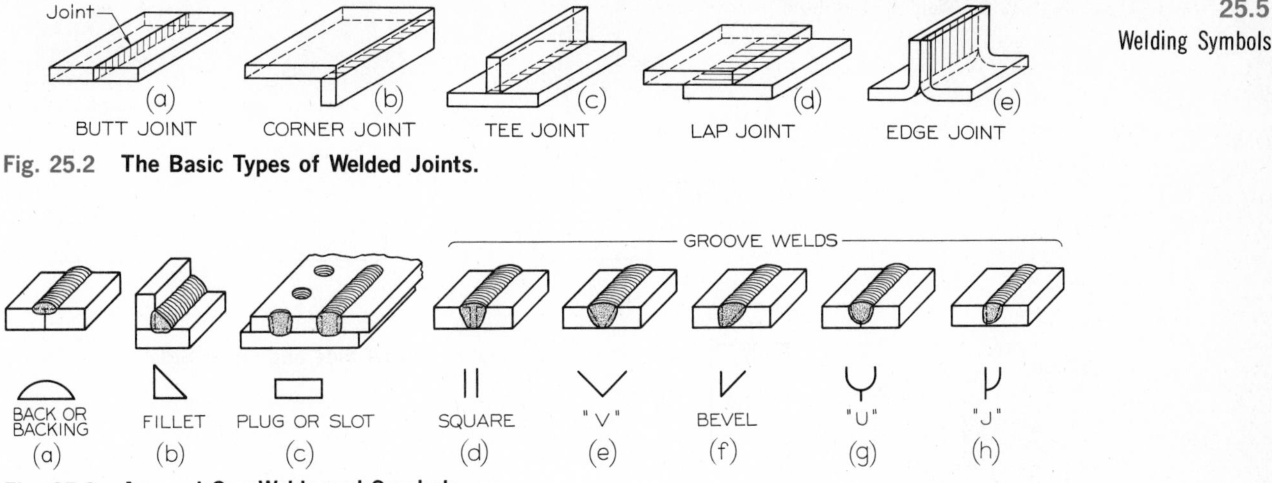


**Fig. 25.2 The Basic Types of Welded Joints.**

**Fig. 25.3 Arc and Gas Welds and Symbols.**

and in the mechanical equipment used for welding, so that now arc and gas welding are important construction processes in industry.

In resistance welding, two pieces of metal are held together under some pressure, and a large amount of electric current is passed through the parts. The resistance of the metals to the passage of the current causes great heating at the junction of the two pieces, resulting in the welding of the metals.

**25.3 Types of Welded Joints.** There are five basic types of welded joints, classified according to the positions of the parts being joined, as shown in Fig. 25.2. A number of different types of welds are applicable to each type of joint, depending upon the thickness of metal, the strength of joint required, and other considerations.

**25.4 Types of Welds.** The four types of arc and gas welds, Fig. 25.3, are the *back* or *backing,* (a), the *fillet,* (b), the *plug* or *slot,* (c), and *groove.* The groove welds are further classified as square, V, bevel, U, and J, shown in (d) to (h). More than one type of weld may be applied to a single joint. For example, a V weld may be on one side and a back weld on the other side. Frequently, the same type of weld is used on opposite sides, forming such welds as a double-V, a double-U, or a double-J.

The four basic resistance welds are the *spot weld, projection weld, seam weld,* and *flash* or *upset weld.* The corresponding weld symbols are shown in Fig. 25.4. Supplementary symbols are shown in Fig. 25.5.

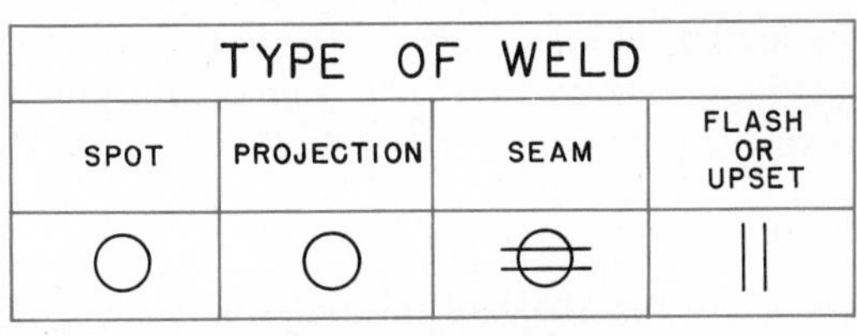


**Fig. 25.4 Resistance Weld Symbols.**

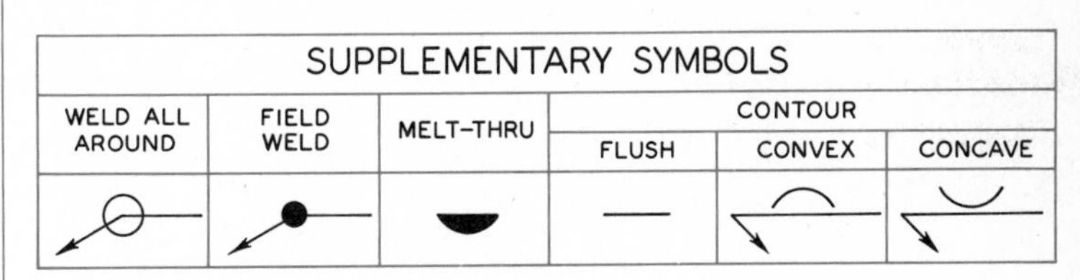


**Fig. 25.5 Supplementary Symbols.**

**25.5 Welding Symbols.** The basic element of the symbol is the "bent" arrow, Fig. 25.6 (a). The arrow points to the joint where the weld is to be made, (b), and attached to the reference line, or shank, of the arrow is the weld symbol for the desired weld. The symbol would be one of those shown in Figs. 25.3 or

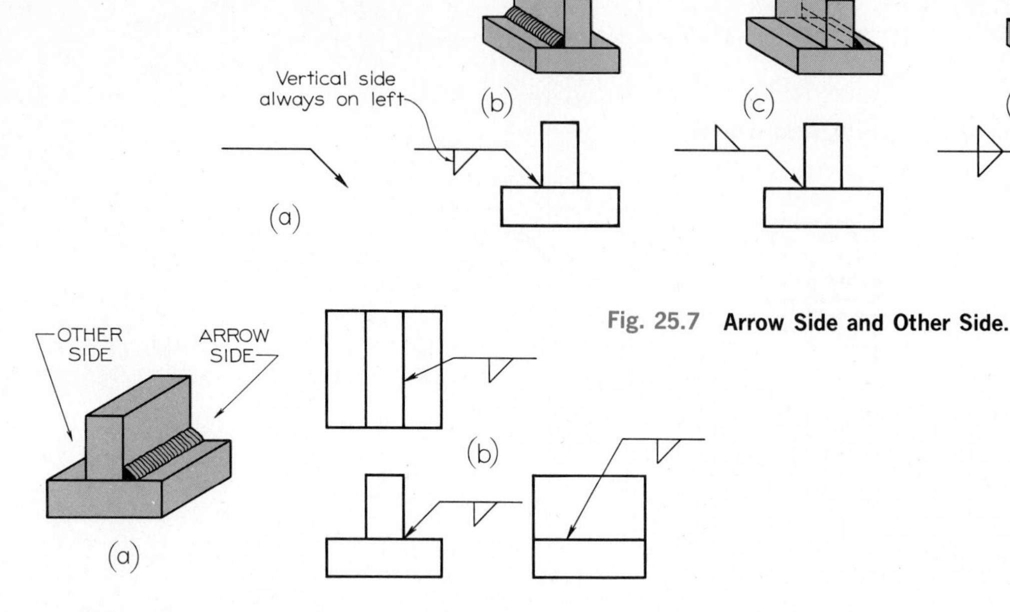


**Fig. 25.6 Welding Symbols.**

**Fig. 25.7 Arrow Side and Other Side.**

25.4. In this case a fillet weld symbol is shown.

The weld symbol is placed below the reference line if the weld is to be on the *arrow side* of the joint as at (b), or above the reference line if the weld is to be on the *other side* of the joint as at (c). If the weld is to be on both the *arrow side* and the *other side* of the joint, weld symbols are placed on both sides of the reference line, (d). This rule for placement of the weld symbol is followed for all the arc or gas weld symbols.

When a joint is shown by a single line on a drawing, as in the top and side views of Fig. 25.7 (b), the *arrow side* of the joint is regarded as the "near" side to the reader of the drawing, according to the usual conventions of drafting.

**Fig. 25.8 The Standard Locations of the Elements of a Welding Symbol.**

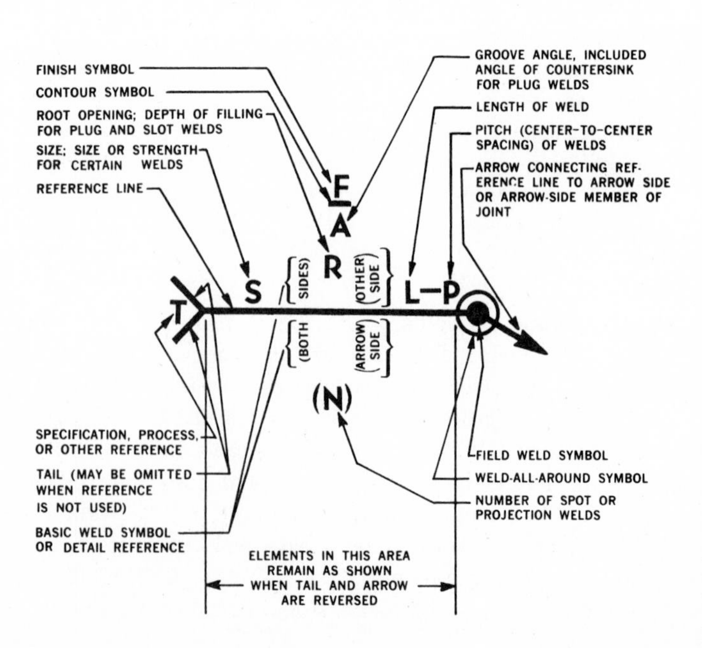

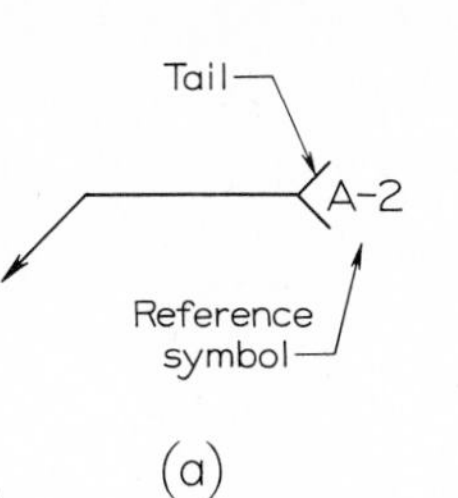


(b) (c)

**Fig. 25.9 Welding Symbols.**

In plug, slot, seam, and projection welding symbols, the arrow points to the outer surface of one of the members at the center line of the weld. In such cases, the arrow side of the joint is the one to which the arrow points, or the "near" side to the reader. See Figs. 25.20 and 25.24.

Note that for all fillet or groove symbols, the *perpendicular side of the symbol is always drawn on the left,* Fig. 25.6 (b).

For best results, the welding symbols should be drawn mechanically, but in certain cases where necessary, the symbols may be drawn freehand.

The complete welding symbol, enlarged, is shown in Fig 25.8. See Appendix 27 for additional welding symbol details. In practice, some companies will need to use only a simple symbol composed of the minimum elements, the arrow and the weld symbol; others will make use of additional components.

Reference to a specification, process, or other supplementary information is indicated by any desired symbol in the tail of the arrow, Fig. 25.9 (a). Otherwise a general note may be placed on the drawing, such as UNLESS OTHERWISE INDICATED, MAKE ALL WELDS PER SPECIFICATION NO. XXX. If no reference is indicated in the symbol, the tail may be omitted.

To avoid repeating the same information on many welding symbols on a drawing, general notes may be used, such as FILLET WELDS $\frac{5}{16}$'' UNLESS OTHERWISE INDICATED, or ROOT OPENINGS FOR ALL GROOVE WELDS $\frac{3}{16}$'' UNLESS OTHERWISE INDICATED.

Welds extending completely around a joint are indicated by an open circle around the elbow of the arrow, as shown at (b). *When the weld-all-around symbol is not used, the welding symbol is understood to apply between abrupt changes in direction of the weld, unless otherwise shown.* A solid round dot at the elbow of the arrow indicates a weld to be made "in the field" (on the site) rather than in the fabrication shop, as shown at (c).

Spot, seam, flash, or upset symbols usually do not have *arrow-side* or *other-side* significance, and are simply centered on the reference line of the arrow, as shown in Fig. 25.10 (a) to (c). Spot and seam symbols are shown on the drawing as indicated at (d) and (e). Note the required process must be specified in the tail of the symbol (RSW—resistance spot weld, EBW—electron beam weld).

For bevel- or J-groove welds, the arrow

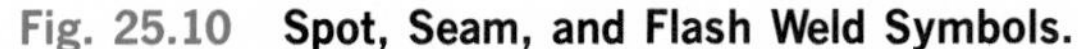

**Fig. 25.10 Spot, Seam, and Flash Weld Symbols.**

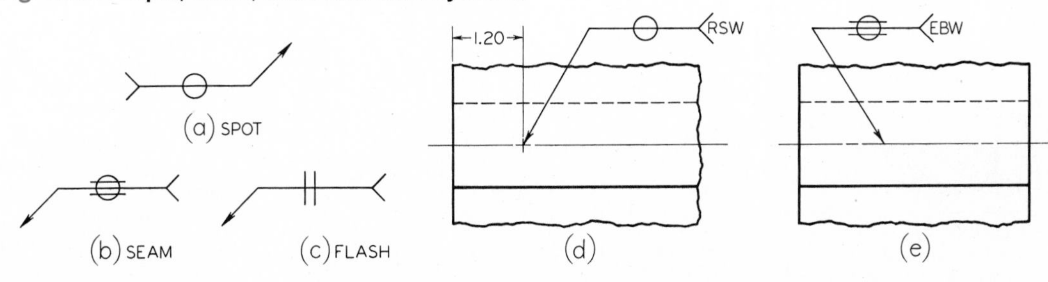

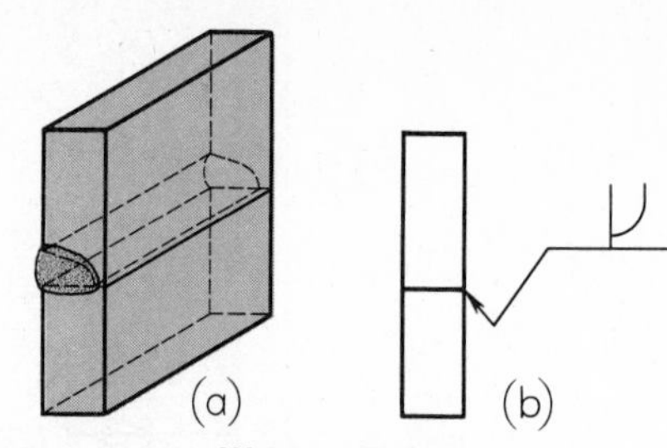


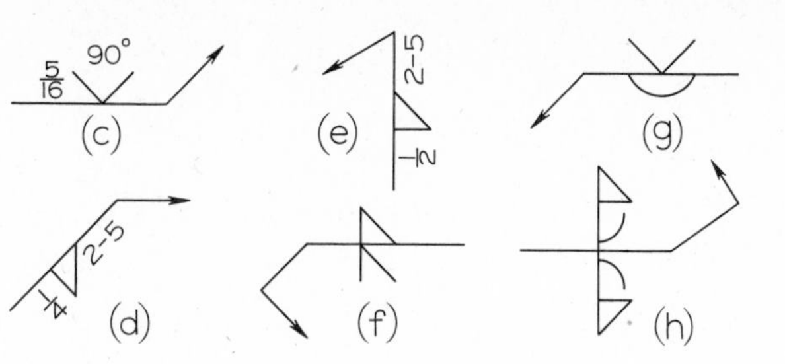


Fig. 25.11 **Welding Symbols.**

should point with a definite change of direction, or break, *toward the member* that is to be beveled or grooved, as shown in Fig. 25.11 (a) and (b). In this case, the upper member is grooved. The break is omitted if the location of the bevel or groove is obvious.

Lettering in symbols should be placed to read from the bottom or from the right side of the drawing in accordance with the aligned system, as shown at (c) to (e).

When a joint has more than one weld, the combined symbols are shown, as at (f) to (h).

## 25.6 **Fillet Welds.**

The usual fillet weld has equal legs, Fig. 25.12 (a). The size of the weld is the length of one leg, and is indicated, (b), by a dimension figure at the left of the weld symbol. For fillet welds on both sides of a joint, the dimensions may be indicated on one side or both sides of the reference line, (c). The lengths of the welds, and the pitch (c-to-c spacing of welds) are indicated as shown. When the welds on opposite sides are different in size, the sizes are given as shown at (d). If a fillet weld has unequal legs, the weld orientation is shown on the drawing, if necessary, and the lengths of the legs given in parentheses to the left of the weld symbol, as at (e). If a general note is given on the drawing, such as ALL FILLET WELDS $\frac{5}{16}$'' UNLESS OTHERWISE NOTED, the size dimensions are omitted from the symbols.

No length dimension is needed for a weld that extends the full distance between abrupt changes of direction. For each abrupt change in direction, an additional arrow is added to the symbol, except when the weld-all-around symbol is used.

Lengths of fillet welds may be indicated by symbols in conjunction with dimension lines, Fig. 25.13 (a). The extent of fillet welding may be shown graphically by means of section lining, (b), if desired.

Chain intermittent fillet welding is indicated as shown in Fig. 25.14 (a). If the welds are staggered, the weld symbols are staggered as shown at (b).

Unfinished flat-faced fillet welds are indicated by adding the flush symbol, Fig. 25.5, to the weld symbol, as shown in Fig. 25.15 (a). If fillet welds are to be made flat faced by mechanical means, the flush-contour symbol and the user's standard finish symbol are added to the weld symbol, as shown at (b) to (d). These finish symbols indicate the method of finishing C = chipping, G = grinding, M = machining, R = rolling, H = hammering), and not the degree of finish. If fillet welds are to be finished to a convex contour, the

Fig. 25.12 **Dimensioning of Fillet Welds.**

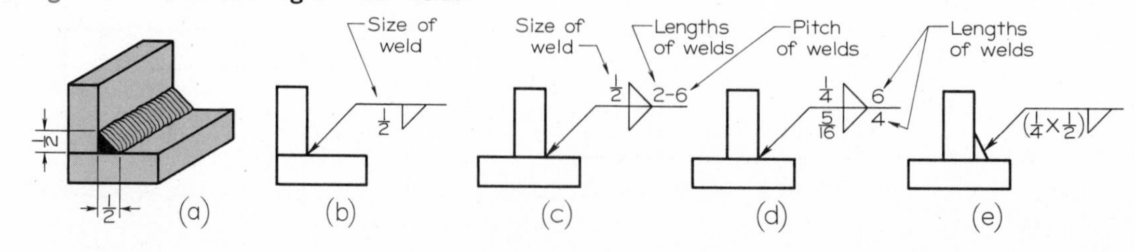

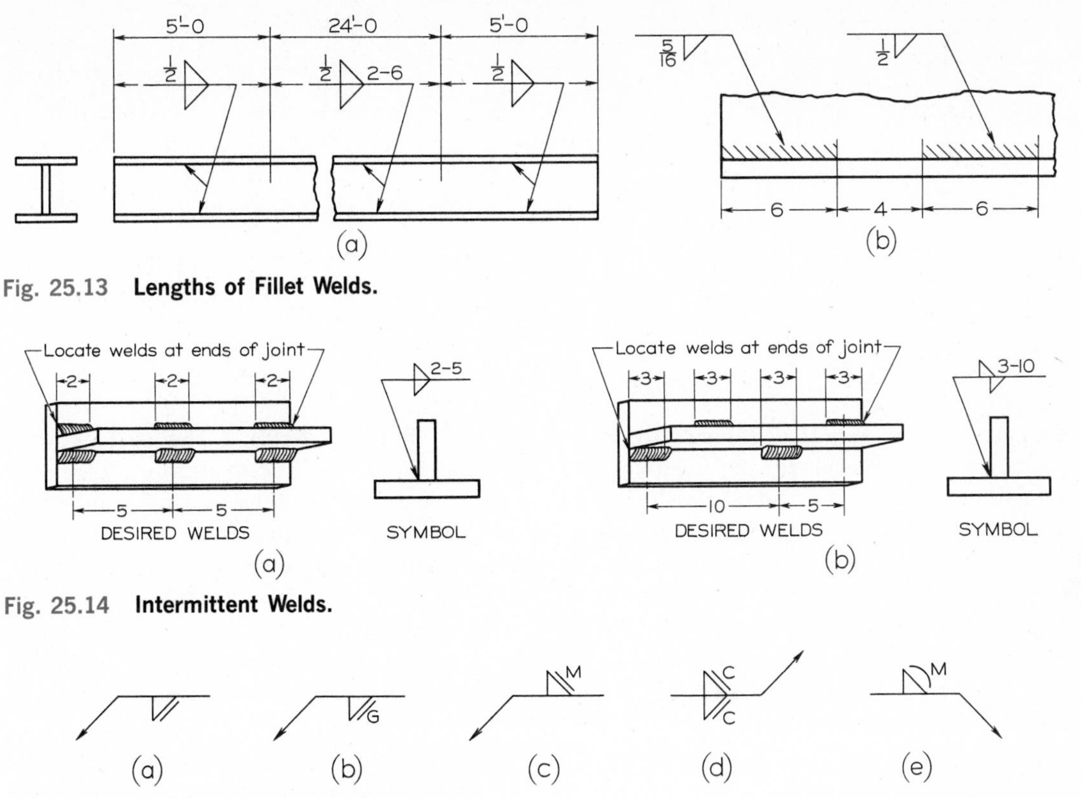


Fig. 25.13 **Lengths of Fillet Welds.**

Fig. 25.14 **Intermittent Welds.**

Fig. 25.15 **Surface Contour of Fillet Welds.**

convex-contour symbol is added, together with the finish symbol, as shown at (e).

## 25.7 Groove Welds.

In Fig. 25.16 various groove welds are shown above, and the corresponding symbolic representations below. The sizes of the groove welds (depth of the V, bevel, U, or J) are indicated at the left of the weld symbol. For example, at (a) the size of the V-weld is $\frac{1}{2}''$, at (b) the sizes are $\frac{1}{4}''$ and $\frac{7}{8}''$, at (c) the size is $\frac{3}{4}''$, and at (d) the size is $\frac{1}{4}''$. In this symbol at (d), the size is followed by $\frac{1}{8}''$, which is the additional "root penetration" of the weld. At (e), the root penetration is $\frac{5}{32}''$ from zero, or from the outside of the members. Note the overlap of the root penetration in this case.

The root opening or space between members, when not covered by a company standard, is shown inside the weld symbol. At (a) and (b) the root openings are $\frac{1}{8}''$. At (c) to (e) the openings are zero.

The groove angles, when not covered by a company standard, are shown just outside the openings of the weld symbols, as shown at (a) to (d).

A general note may be used on the drawing to avoid repetition on the symbols, such as ALL V-GROOVE WELDS TO HAVE 60° GROOVE ANGLE UNLESS OTHERWISE SHOWN. However, when the dimensions of one or both of two opposite welds differ from the general note, both welds should be completely dimensioned.

When single-groove or symmetrical double-groove welds extend completely through, the size need not be given in the welding symbol. For example, in Fig. 25.16 (a), if the V-groove extended entirely through the joint, the depth or size would be simply the thickness of the stock and would not need to be given in the welding symbol.

When groove welds are to be approximately flush without finishing, the flush-contour symbol, Fig. 25.5, is added to the weld sym-

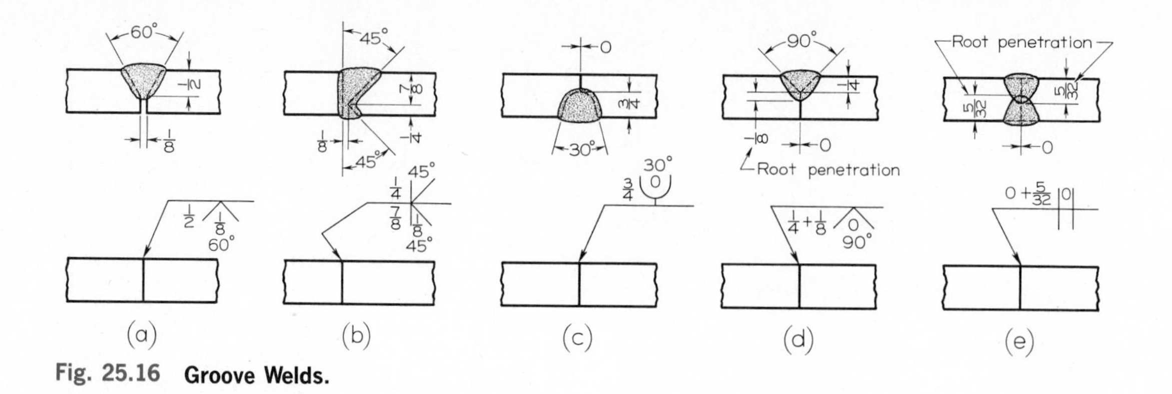


Fig. 25.16 **Groove Welds.**

bols as shown in Fig. 25.17 (a) and (b). If the welds are to be machined, the flush-contour symbol and the user's standard finish symbol are added to the weld symbol as shown at (c) and (d). These finish symbols indicate the method of finishing (C = chipping, G = grinding, M = machining), and not the degree of finish. If a groove weld is to be finished with a convex contour, the convex-contour and finish symbols are added, as at (e).

## 25.8 Back or Backing Welds.

The bead-type welds used as back or backing welds on single-groove welds are indicated, Fig. 25.18 (a), by a back or backing symbol opposite the groove weld symbol. Dimensions of such back or backing welds are not shown on the symbol, but may be shown, if necessary, directly on the drawing.

When back or backing welds are to be approximately flush without machining, the flush-contour symbol is added to the weld symbols, as shown at (b). If they are to be machined, the user's finish symbol is added, (c) and (d). If the welds are to be finished with a convex contour, the convex-contour symbol and the finish symbol are added to the weld symbol, as at (e).

## 25.9 Surface Welds.

The surface weld symbol is used to indicate a surface to be built up by welding, whether by single- or multiple-pass bead-type welds, as shown in Fig. 25.19. Since this symbol does not indicate a welded joint, there is no arrow-side or other-side significance; hence the symbol is always drawn below the reference line. The minimum height of the weld deposit is indicated at the left of the weld symbol, as shown, except where no specific height is required. When a specific area of a surface is to be built up, the dimensions of the area are given directly on the drawing.

## 25.10 Plug and Slot Welds.

As shown in Fig. 25.20 (c), the same symbol is used for either plug welds or slot welds. A hole or a slot is made in one member to receive the weld, as shown in Fig. 25.20 (a) and (d). If the hole or slot is in the arrow-side member, the weld symbol is placed below the reference line, (b) and (c); if in the other-side member, the weld symbol is placed above the line, (c) and (f).

The size of a plug weld, which is the smallest diameter of the hole if countersunk, is placed at the left of the weld symbol, as shown

Fig. 25.17 **Surface Contour of Groove Welds.**

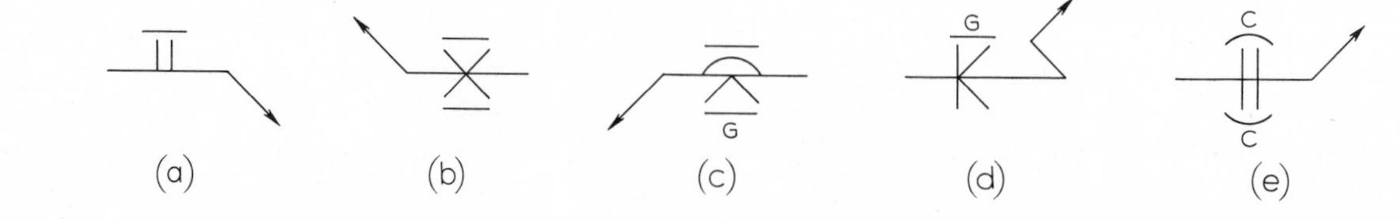

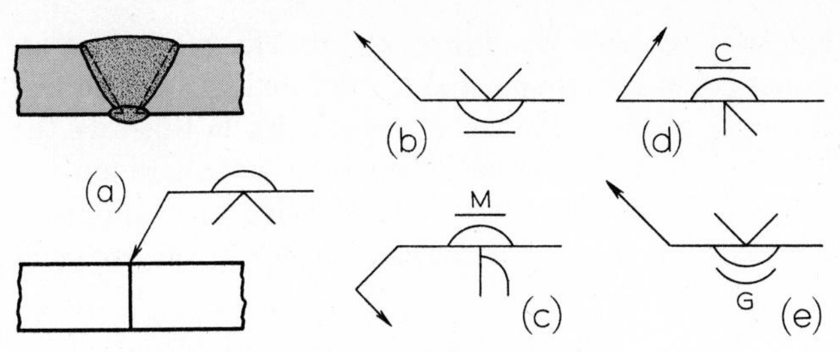


Fig. 25.18 **Back or Backing Weld Symbols.**

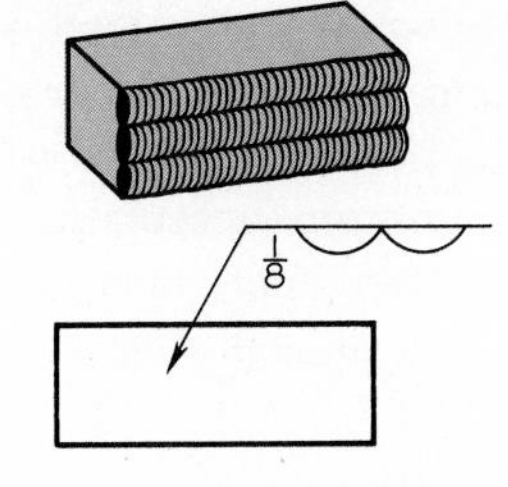

Fig. 25.19 **Surface Weld Symbol.**

at (b). If the included angle of countersink of plug welds is in accordance with the user's standard, it is omitted from the welding symbol; otherwise, it is indicated adjacent to the weld symbol, as shown at (b).

A plug weld is understood to fill the depth of the hole unless the depth of plug is indicated by a number, in inches, inside the weld symbol, Fig. 25.21 (a). The pitch (center-to-center spacing) of plug welds is shown in inches at the right of the weld symbol, as shown at (b). If the weld is to be approximately flush without finishing, the flush-contour symbol is added, (c). If the weld is to be made flush by mechanical means, a finish symbol is added, as at (d). The flush-contour and finish symbols are used in the same manner for slot welds and for plug welds.

The depth of filling of slot welds is indicated in the same manner as for plug welds, (a). The size and location dimensions of slot welds cannot be shown on the welding symbol, and must be shown directly on the drawing, Fig. 25.20 (f), or by a detail with a reference to it on the welding symbol, as shown in Fig. 25.21 (e).

## 25.11 Spot Welds.

The spot weld symbol, with the required welding process indicated in the tail, Fig. 25.22, may or may not have arrow-side or other-side significance. Dimensions shall be shown on the same side of the reference line as the symbol, or on either side when the symbol is centered on the reference line and no arrow-side or other-side significance is intended.

The size of a spot weld is its diameter. This value, expressed in hundredths of an inch,

Fig. 25.20 **Plug and Slot Welds.**

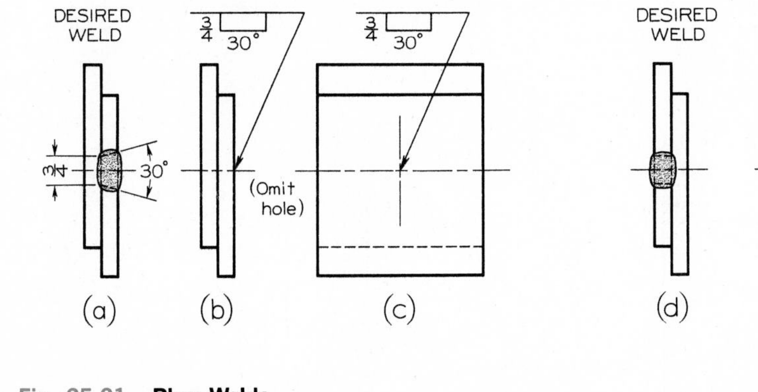


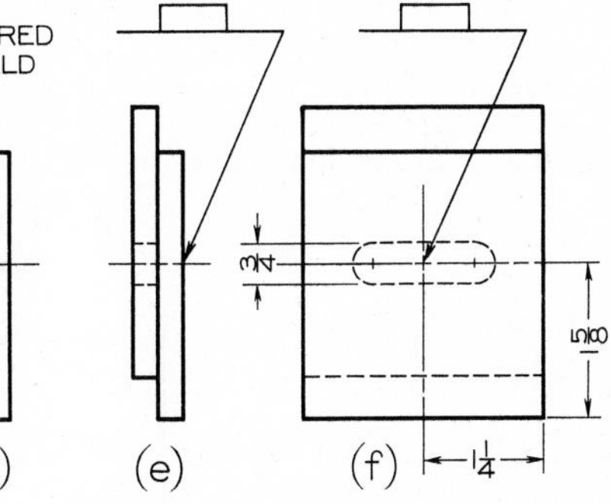


Fig. 25.21 **Plug Welds.**

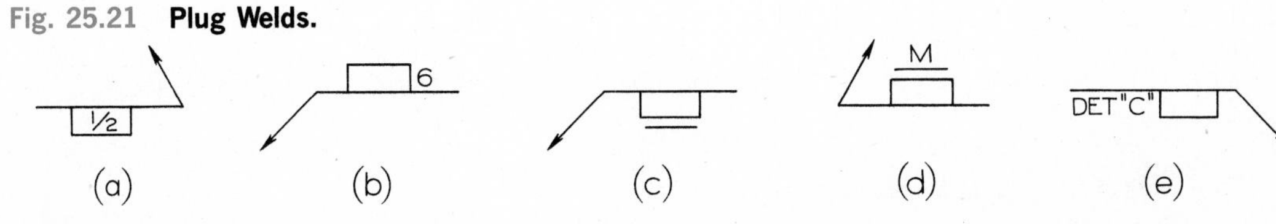

may be shown with inch marks at the left of the weld symbol on either side of the reference line, as shown at (a). If it is desired to indicate the minimum acceptable shear strength in pounds per spot, instead of the size of the weld, this value is placed at the left of the weld symbol, as shown at (b). The pitch (center-to-center spacing) is indicated in inches at the right of the weld symbol, (b). In this case the spot welds are 3″ apart. If a joint requires a certain number of spot welds, the number is given in parenthesis above or below the sym-

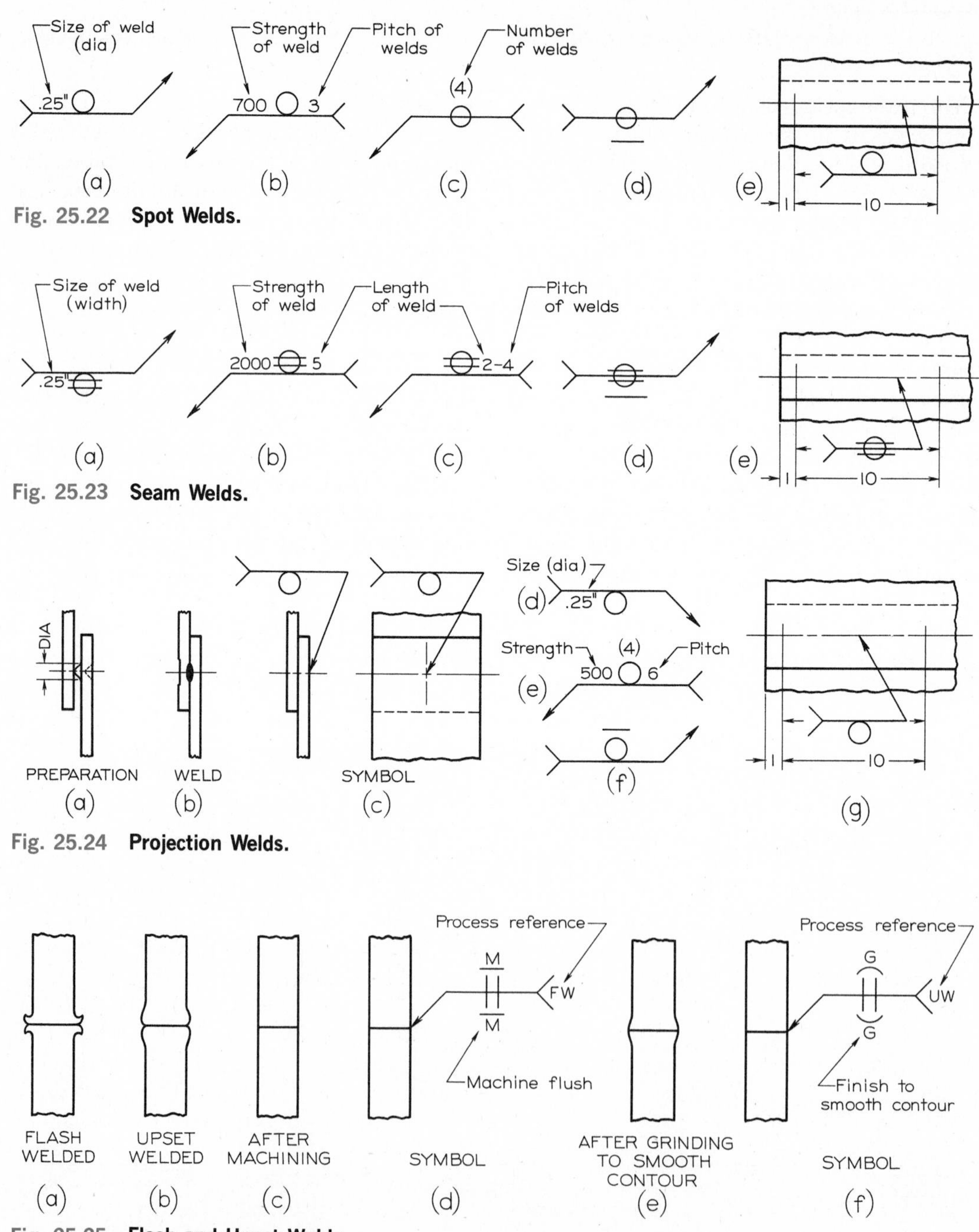


Fig. 25.22 **Spot Welds.**

Fig. 25.23 **Seam Welds.**

Fig. 25.24 **Projection Welds.**

Fig. 25.25 **Flash and Upset Welds.**

bol, as at (c). If the exposed surface of one member is to be flush, the flush-contour symbol is added, above the symbol if it is the other-side member, and below if it is the arrow-side member, as shown at (d). The use of the welding symbol in conjunction with ordinary dimensions is shown at (e).

## 25.12 Seam Welds.

The seam weld symbol, with the welding process indicated in the tail, Fig. 25.23, may or may not have arrow-side or other-side significance. Dimensions shall be shown on the same side of the reference line as the symbol, or on either side when the symbol is centered on the reference line and no arrow-side or other-side significance is intended.

The size of a seam weld is its width. This value, expressed in hundredths of an inch, may be shown with inch marks at the left of the weld symbol, on either side of the reference line, as shown at (a). If it is desired to indicate the minimum acceptable shear strength in pounds per linear inch, instead of the size of the weld, this value is placed at the left of the weld symbol, as shown at (b). The length of a seam weld may be shown in inches at the right of the weld symbol, (b). In this case, the seam weld is 5″ long. If the weld extends the full distance between abrupt changes of direction, no length dimension in the symbol is given.

The pitch (center-to-center spacing) of intermittent seam welding is the distance between centers of lengths of welding. The pitch, in inches, is shown at the right of the length figure, as shown at (c). In this case, the welds are 2″ long and spaced 4″ center-to-center.

When the exposed surface of one member is to be flush, the flush-contour symbol is added, above the symbol if it is the other-side member and below if it is the arrow-side member, as shown at (d). The use of the welding symbol in conjunction with ordinary dimensions is shown at (e).

## 25.13 Projection Welds.

In projection welding, one member is embossed in preparation for the weld, as shown in Fig. 25.24 (a). When welded, the joint appears in section as at (b). The weld symbols, in this case, are placed below the reference lines, as shown at (c), to indicate that the arrow-side member is the one that is embossed. The weld symbols would be placed above the lines if the other member were embossed.

Projection welds are dimensioned by either size or strength. The size is the diameter of the weld in hundredths of an inch. This value is shown, with inch marks, to the left of the weld symbol, as shown at (d). If it is desired to indicate the minimum acceptable shear strength in pounds per weld, the value is placed at the left of the weld symbol, as at (e). The pitch (center-to-center spacing) is indicated in inches at the right of the weld symbol, as at (e). In this case, the welds are spaced 6″ apart. If the joint requires a definite number of welds, the number is given in parentheses, as shown at (e). If the exposed surface of one member is to be flush, the flush-contour symbol is added, as shown at (f). The use of the welding symbol in conjunction with ordinary dimensions is shown at (g). The welding process reference is required in the tail of the symbol.

## 25.14 Flash and Upset Welds.

Flash and upset weld symbols have no arrow-side or other-side significance, but the supplementary symbols do. A flash-welded joint is shown in Fig. 25.25 (a), and an upset welded joint at (b). The joint after machining flush is shown at (c). The complete symbol at (d) includes the weld symbol together with the flush-contour and machining symbols.

If the joint is ground to smooth contours, (e), the resulting welding drawing and symbol would be constructed as shown at (f), which includes convex-contour and grind symbols. At either (d) or (f), the joint may be finished on only one side, if desired, by indicating the contour and machining symbols on the appropriate side of the reference line. The dimensions of flash and upset welds are not shown on the welding symbol. Note that the process reference for flash welding (FW) or upset

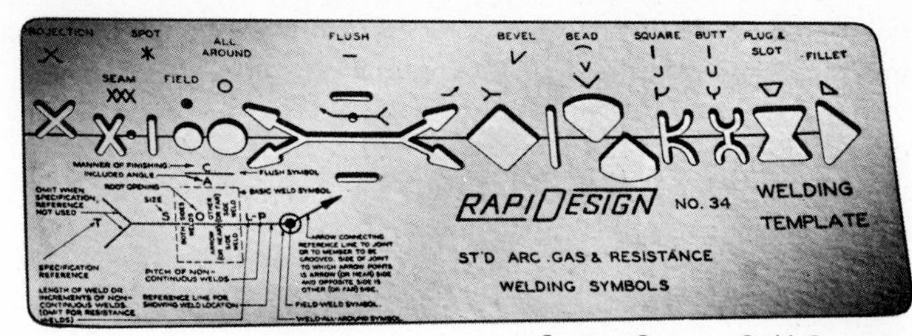


Fig. 25.26 **Welding Template.**

*Courtesy Gramercy Guild Group, Inc.*

welding (UW) must be placed in the tail of the symbol.

## 25.15 Welding Templates.

To simplify the drawing of welding symbols, and to speed up the drafting, welding templates are available. The template shown in Fig. 25.26 has all the forms needed for drawing the arrow, the weld symbols, and the supplementary symbols, as well as an illustration of the complete composite welding symbol for quick reference. The symbols may be drawn in pencil or ink. For the latter, the technical fountain pen is recommended, Fig. 2.60.

## 25.16 Welding Applications.

A typical example of welding fabrication for machine parts is shown in Fig. 25.1. In many cases, especially when only one or only a few identical parts are required, it is cheaper to produce by welding than to make patterns, sand castings, and do the necessary machining. Thus, welding is particularly adaptable to custom-built constructions.

Welding is also suitable for large structures that are difficult or impossible to fabricate entirely in the shop and is coming into greater use for steel structures, such as building frames, bridges, and ships. A welded beam is shown in Fig. 22.10, and a welded assembly of diagonal bracing between two columns is shown in Fig. 22.11. A welded truss is shown in Fig. 25.27. It is easier to place members in such a welded truss so that their center-of-gravity axes coincide with the working lines of the truss than is the case in a riveted truss. The student should compare this welded truss with the riveted truss of Fig. 22.8.

## 25.17 Welding Drawing Problems.

The following problems are given to familiarize the student with some applications of welding symbols to machine construction and to steel structures.

**Prob. 25.1** Fig. 6.59. Change to a welded part. Make working drawing, using appropriate welding symbols.

**Prob. 25.2** Fig. 6.62. Same instructions as for Prob. 25.1.

**Prob. 25.3** Fig. 6.67. Same instructions as for Prob. 25.1.

**Prob. 25.4** Fig. 6.78. Same instructions as for Prob. 25.1.

**Prob. 25.5** Fig. 6.85. Same instructions as for Prob. 25.1.

**Prob. 25.6** Fig. 6.97. Same instructions as for Prob. 25.1.

**Prob. 25.7** Fig. 6.106. Same instructions as for Prob. 25.1.

**Prob. 25.8** Fig. 6.110. Same instructions as for Prob. 25.1.

**Prob. 25.9** Fig. 6.115. Same instructions as for Prob. 25.1.

**Prob. 25.10** Make a half-size drawing of the joint at the center of the lower chord of the truss in Fig. 25.27 (page 694) where the chord is supported by two vertical angles. The chord is a structural tee, cut from an $8 \times 5\frac{1}{4}$, 17 lb. wide-flange shape. Draw the front and side views, and show the working lines, the two angles, the structural tee, and all welding symbols.

**Prob. 25.11** Make a half-size front view, showing the welding symbols, of any joint of the truss in Fig. 25.27 (page 694) in which three or four members meet.

**Prob. 25.12** Draw half-size front, top, and left-side views of the end joint of the truss in Fig. 25.27 (page 694), showing the welding symbols.

**Fig. 25.27 A Welded Truss** (Probs. 25.10 to 25.12).

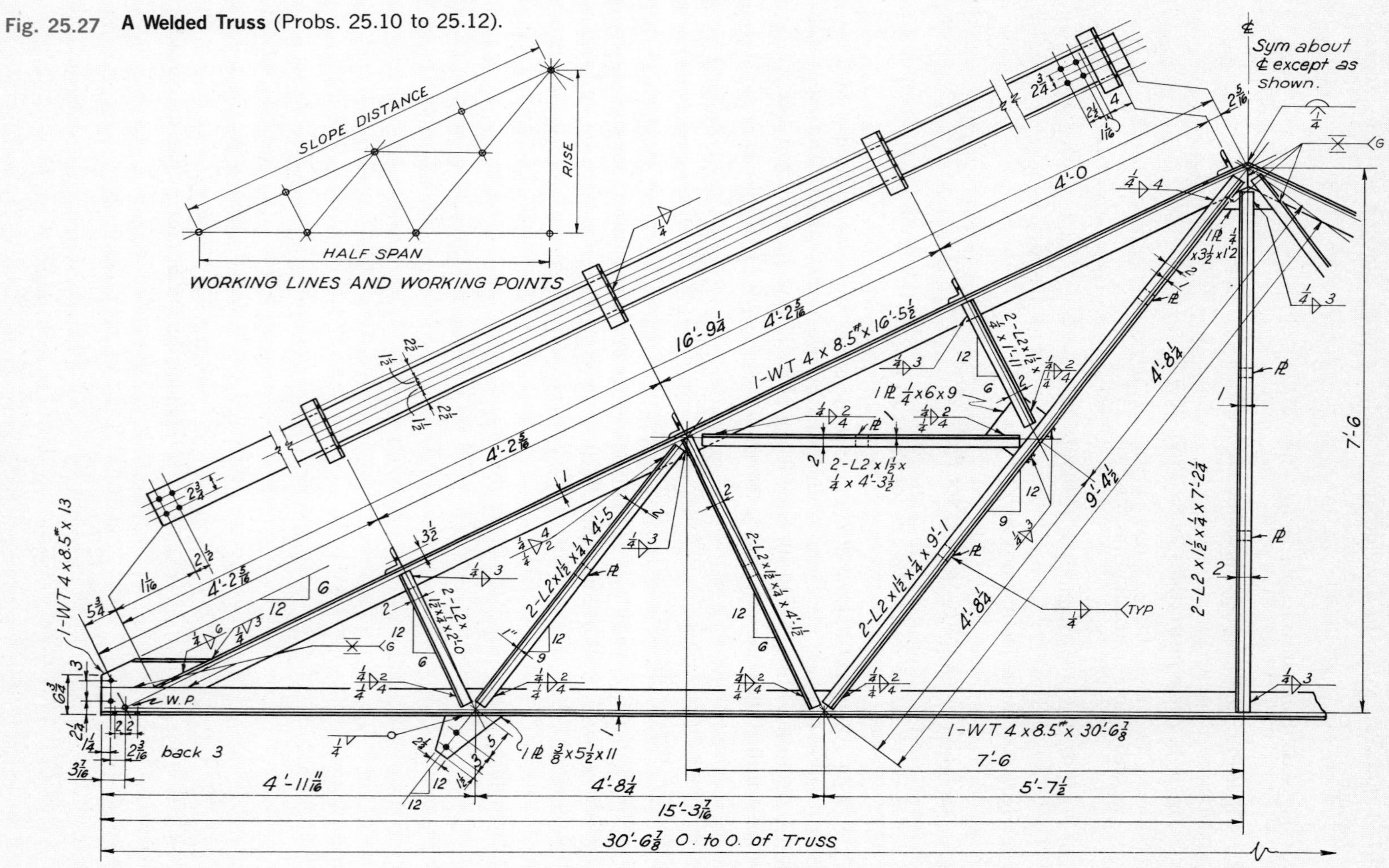

# 26

# graphs

by E. J. Mysiak*

**26.1 Graphical Representation.** In previous chapters we have seen how graphical representation is used instead of word descriptions to describe the size, shape, material, and fabrication methods for the manufacture of actual objects. Graphical representation is also used extensively to represent engineering facts, statistics, and laws of phenomena. A pictorial or graphical description is much more impressive and easier to understand than a numerical tabulation or a verbal description. These graphical descriptions are synonymously termed *graphs, charts,* or *diagrams.*

Tabulated data in Fig. 26.1 (a), showing the growing market for engineers for the years 1900 to 1978 inclusive, are presented as a *line graph* at (b) and a *bar graph* at (c). The greater effectiveness of graphical representation is evident.

Graphical constructions are used in three general ways: (1) for graphical presentation of data, (2) for graphical analysis, and (3) for graphical computation. Graphic presentations and analysis are discussed in this chapter, and graphical computations are discussed in Chapters 27 to 29.

The type of graphical presentation used depends upon any of the following factors:

1. The type of reader to be reached.
2. The most efficient type of graph to help the reader visualize the significant features.
3. General purpose of graph.
4. Features of a relationship that are considered significant.
5. Occasion for its use.
6. Nature or amount of data.

*Mechanical Engineer, International Electro-Magnetics, Inc.

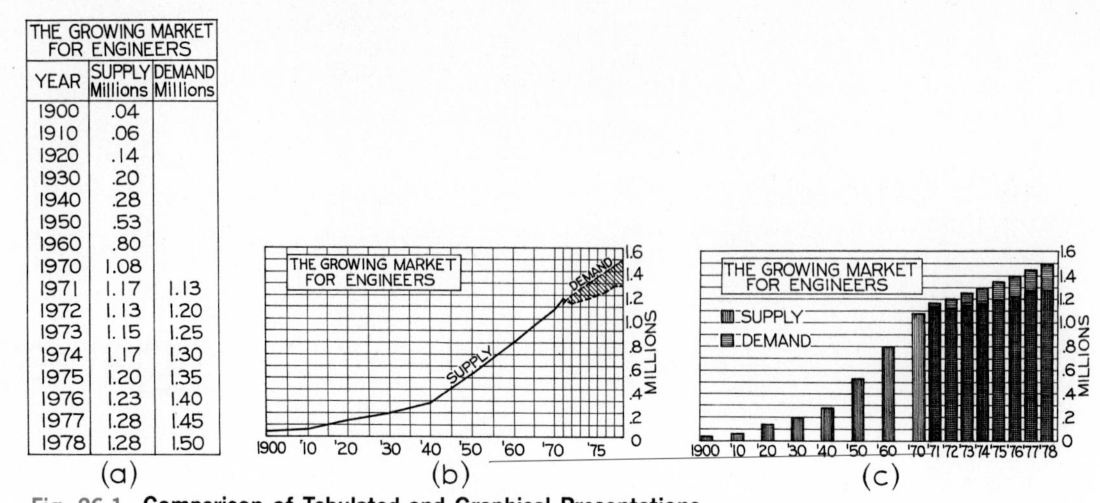

| THE GROWING MARKET FOR ENGINEERS | | |
|---|---|---|
| YEAR | SUPPLY Millions | DEMAND Millions |
| 1900 | .04 | |
| 1910 | .06 | |
| 1920 | .14 | |
| 1930 | .20 | |
| 1940 | .28 | |
| 1950 | .53 | |
| 1960 | .80 | |
| 1970 | 1.08 | |
| 1971 | 1.17 | 1.13 |
| 1972 | 1.13 | 1.20 |
| 1973 | 1.15 | 1.25 |
| 1974 | 1.17 | 1.30 |
| 1975 | 1.20 | 1.35 |
| 1976 | 1.23 | 1.40 |
| 1977 | 1.28 | 1.45 |
| 1978 | 1.28 | 1.50 |



Fig. 26.1 **Comparison of Tabulated and Graphical Presentations.**

## 26.2 Rectangular Coordinate Line Graphs.

The *rectangular coordinate line graph* is the type in which values of two related variables are plotted on coordinate paper, and the points, joined together successively, form a continuous line or "curve."

The following are some of the purposes for which a line graph can be used to advantage:

1. Comparison of a large number of plotted values in a compact space.
2. Comparison of the relative movements (trends) of several curves on the same graph. There should not be more than two or three curves on the same graph, and there should be some definite relationship between them.
3. Interpolation of intermediate values.
4. Representation of movement or over-all trend (relative change) of a series of values, rather than the difference between values (absolute amounts).

Line graphs are *not* particularly suited for: (1) presenting relatively few plotted values in a series, (2) emphasizing changes or difference in absolute amounts, or (3) showing extreme or irregular movement of data.

Rectangular line graphs may be classified as (1) *mathematical graphs,* (2) *time series charts,* or (3) *engineering graphs.* Any of these may have one or more curves on the same graph. If the values plotted along the axes are pure numbers (positive and negative), showing the relationship of an equation, the plot is commonly called a mathematical graph, Fig. 26.2 (a). When one of the variables is any unit of time, the chart is known as a time-series chart, (b). This is one of the most common forms, since time is frequently one of the variables. Line, bar, or surface chart forms may be used for time-series charts, line charts being the most widely used in engineering practice. The plotting of values of any two related physical variables on a rectangular coordinate grid is referred to as an engineering chart, graph, or diagram, (c).

Line curves are generally presented for any of three types of relationships:

1. Observed relationships, usually plotted with observed data points connected by straight, irregular lines, Fig. 26.3 (a).
2. Empirical relationships, (b), normally reflecting the author's interpretation of his series of observations, represented as smooth curves or straight lines fitted to the data by eye or by formulas chosen empirically.

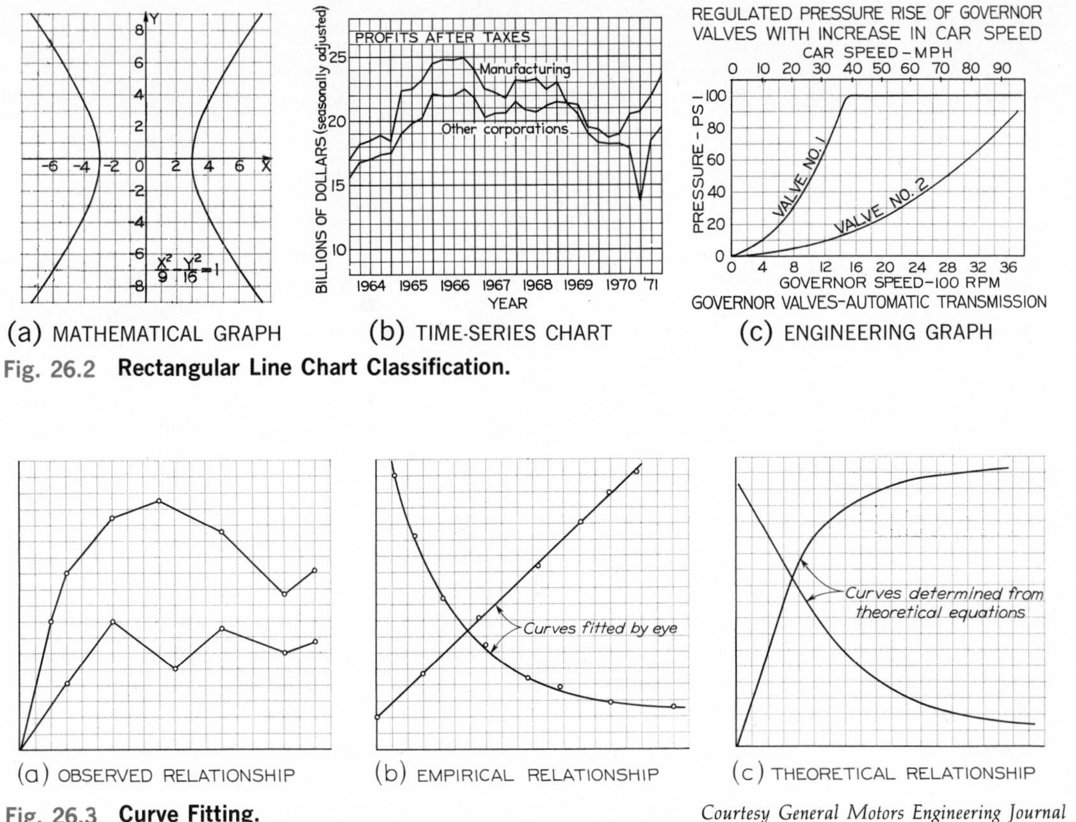


(a) MATHEMATICAL GRAPH (b) TIME-SERIES CHART (c) ENGINEERING GRAPH

**Fig. 26.2 Rectangular Line Chart Classification.**

(a) OBSERVED RELATIONSHIP (b) EMPIRICAL RELATIONSHIP (c) THEORETICAL RELATIONSHIP

**Fig. 26.3 Curve Fitting.** *Courtesy General Motors Engineering Journal*

3. Theoretical relationships, (c), in which the curves are smooth and without point designations, though observed values may be plotted to compare them with a theoretical curve if desired. The curve thus drawn is based on theoretical considerations only, in which a theoretical function (equation) is used to compute values for the curve.

## 26.3 Design and Layout of Rectangular Coordinate Line Graphs.

The steps in drawing a typical coordinate line graph are shown in Fig. 26.4:

I. a. Compute and/or assemble data in a convenient arrangement.

b. Select the type of graph and coordinate paper most suitable, §26.4.

c. Determine the size of the paper and locate the axes.

d. Determine the variable for each axis and choose the appropriate scales, §26.5. Letter the unit values along the axes, §26.5.

II. Plot the points representing the data and draw the curve or curves, §26.6.

III. Identify the curves by lettering names or symbols, §26.6. Letter the title, §26.7. Ink in the graph, if desired.

The completed graph, which includes the curves, captions, and designations, should result in a balanced arrangement relative to the axes. Much of the above procedure is also applicable to the other forms of charts, graphs, and diagrams to be discussed in subsequent sections of this chapter.

## 26.4 Grids and Composition.

To simplify the plotting of values along the perpendicular axes and to eliminate the use of a special scale to locate them, *coordinate paper,* or "cross-section paper," ruled with grids, is generally

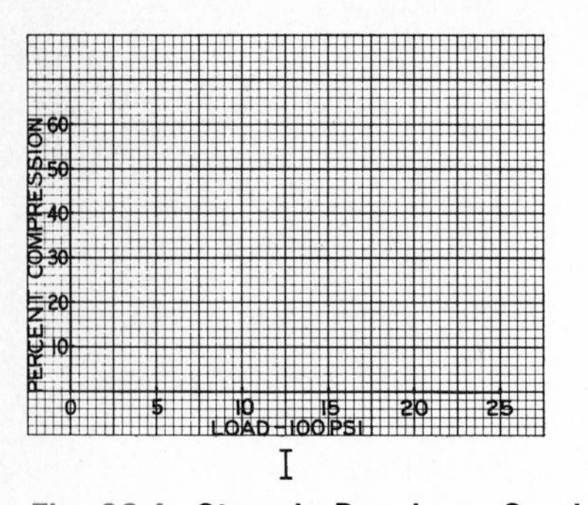


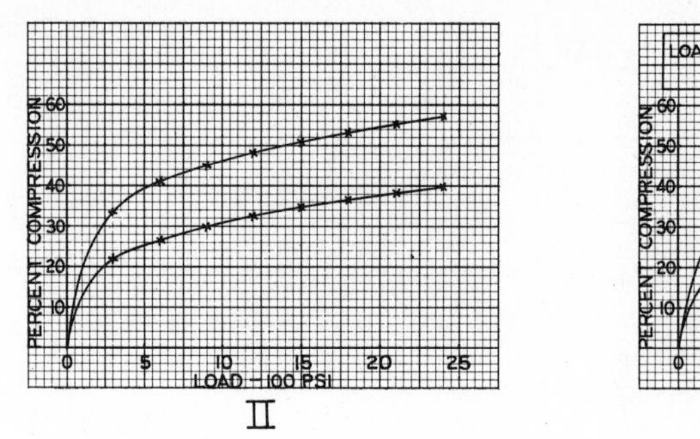


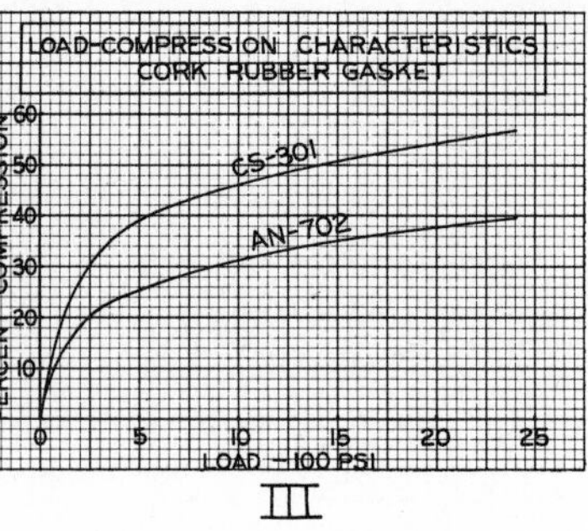


Fig. 26.4 **Steps in Drawing a Graph.**

used and can be purchased already printed; or the grids can be drawn on blank paper, if desired.

Printed coordinate papers are available in various sizes and spacings of grids, $8\frac{1}{2}'' \times 11''$ being the most common paper size. The spacing of the grid lines may be $\frac{1}{10}''$, $\frac{1}{20}''$, or multiples of $\frac{1}{16}''$. A spacing of $\frac{1}{8}''$ to $\frac{1}{4}''$ is preferred. Closely spaced coordinate ruling is generally avoided for publications, charts reduced in size, and charts used for lantern slides. Much of engineering graphical analysis, however, requires (1) close study, (2) interpolation, and (3) only one copy, with possibly a few prints that can be readily prepared with little effort. Therefore, engineering graphs are usually plotted on the closely spaced, printed coordinate paper. Printed papers can be obtained in several colors of lines and in various weights and grades. A thin translucent paper is used when prints are required. A special nonreproducible-grid coordinate paper is available in which reproductions will not show the grids on the prints.

Scale values and captions should be placed outside the grid axes, if possible. Since printed papers do not have sufficient margins to accommodate the axes and nomenclature, the axes should be placed far enough inside the grid area to permit sufficient space for axes and lettering, as shown in Fig. 26.5 (a) and (b). As much of the remaining grid space as possible should be used for the curve—that is, the scale should be such as to spread the curve out over the available space. A title block (and tabular data, if any) should be placed in an open space on the chart, as shown. If only one copy of the chart is required, tabular data may be placed on the back of the graph or on a separate sheet.

Charts prepared for printed publications, conferences, or projection (lantern slides) generally do not require accurate or detailed interpolation and should emphasize the major facts presented. For such graphs, coordinate grids drawn or traced on blank paper, cloth, or film have definite advantages when compared to printed paper. The charts in Fig. 26.5 (b) and (c) show the same information plotted on printed coordinate paper and plain paper, respectively. The specially prepared sheet should have as few grid rulings as necessary—or none, as at (c)—to allow a clear interpretation of the curve. For ease of reading, lettering is not placed upon grid lines. The title and other data can be lettered in open areas, completely free of grid lines.

The layout of specially prepared grids is restricted by the over-all paper size required, or space limitations for lantern slides and other considerations. Space is first provided for margins and for axes nomenclature; the remaining space is then divided into a sufficient number of grid spaces needed for the range of values to be plotted. Another important consideration of composition that affects the spacing of grids is the slope or trend of the curve, as discussed in §26.5.

Since independent variable values are generally placed along the horizontal axes, especially in time-series charts, vertical rulings can be made for each value plotted, if uniformly spaced, Fig. 26.6 (a). If there are many values to plot, intermediate values can be designated

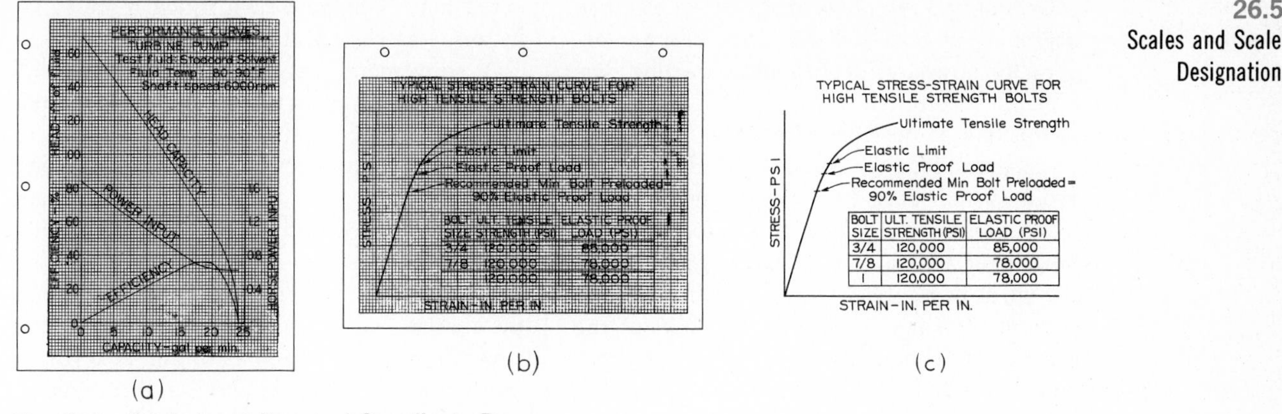


**Fig. 26.5 Printed and Prepared Coordinate Paper.**

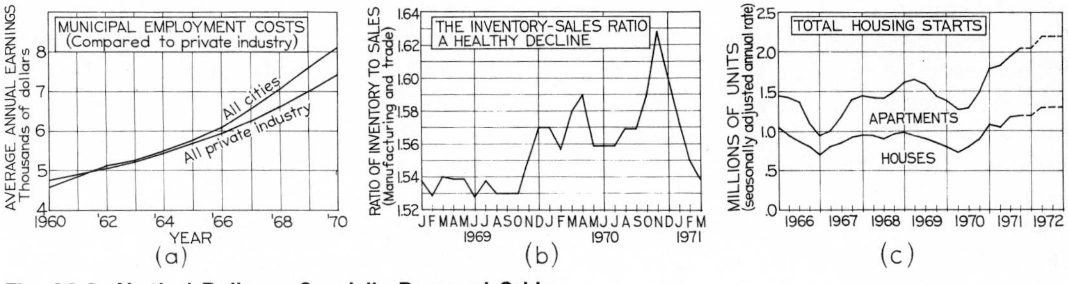


**Fig. 26.6 Vertical Rulings—Specially Prepared Grids.**

by *ticks* on the curves or along the axis, (b) and (c), with the grid rulings omitted. The horizontal lines are generally spaced according to the available space and the range of values.

The weight of the grid rulings should be thick enough to guide the eye in reading the values, but thin enough to provide contrast and emphasize the curve. The thickness of the lines generally should decrease as the number of rulings increase. As a general rule, as few rulings should be used as possible, but if a large number of rulings is necessary, major divisions should be drawn heavier than the subdivision rulings, for ease of reading.

## 26.5 Scales and Scale Designation.

The choice of scale is the most important factor of composition and curve significance. Rectangular coordinate line graphs have values of the two related variables plotted with reference to two mutually perpendicular coordinate axes, meeting at a zero point or origin, Fig. 26.7 (a). The horizontal axis, normally designated as an X-axis, is called the *abscissa.* The vertical axis is denoted a Y-axis and is called the *ordinate.* It is common practice to place independent values along the abscissa and the dependent values along the ordinate. For example, if in an experiment at certain time intervals, related values are observed, recorded, or determined, the amount of these values is dependent upon the time intervals (independent or controlled) chosen. The values increase from the point of origin toward the right on the X-axis and upward on the Y-axis.

Mathematical graphs, (b), quite often contain positive and negative values, which necessitates the division of the coordinate field into four quadrants, numbered counterclockwise as shown. Positive values increase toward the right on the X-axis and upward on the Y-axis, from the origin. Negative values increase (negatively) to the left on the X-axis and downward on the Y-axis.

Generally, a full range of values is desirable, beginning at zero and extending slightly beyond the largest value, to avoid crowding. The available coordinate area should be used as completely as possible. However, certain circumstances require special consideration to avoid wasted space. For example, if the values to be plotted along one of the axes do not range near zero, a "break" in the grid may be shown, as in Fig. 26.8. However, when relative amount of change is required, Fig. 26.5 (a), the axes or grid should not be broken or the zero line omitted. If the absolute amount is the important consideration, the zero line may be omitted, as in Fig. 26.7 (b) to (d). Time designations of years naturally are fixed and have no relation to zero.

If a few given values to be plotted are widely separated in amount from the others, the total range may be very great, and when this is compressed to fit on the sheet, the resulting curve will tend to be "flat," as shown in Fig. 26.9 (a). In such cases it is best to arrange for such values to fall off the sheet and to indicate them as "freak" values. The curve may then be drawn much more satisfactorily, as shown at (b).

A convenient manner in which to show related curves having the same units along the abscissa, but different ordinate units, is to place one or more sets of ordinate units along the left margin and another set of ordinate values along the right margin, as shown in Fig. 26.5 (a), using the same rulings. Multiple scales are also sometimes established along the abscissa, such as for time units of months covering multiple years, as in Fig. 26.6 (b). A more compact arrangement for the curve is shown in Fig. 26.9 (c), where the purpose is to compare the inventory/sales ratios for 1969 and 1970 on a monthly basis.

The choice of scales deserves careful consideration, since it has a controlling influence on the depicted rate of change of the dependent variable. The *slope* of the curve (trend) should be chosen to represent a true picture of the data or a correct impression of the trend.

The slope of a curve is affected by the spacing of the rulings and its designations. A slope

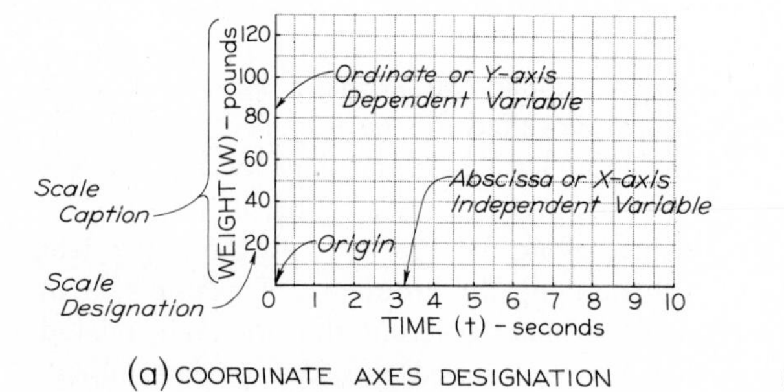


(a) COORDINATE AXES DESIGNATION

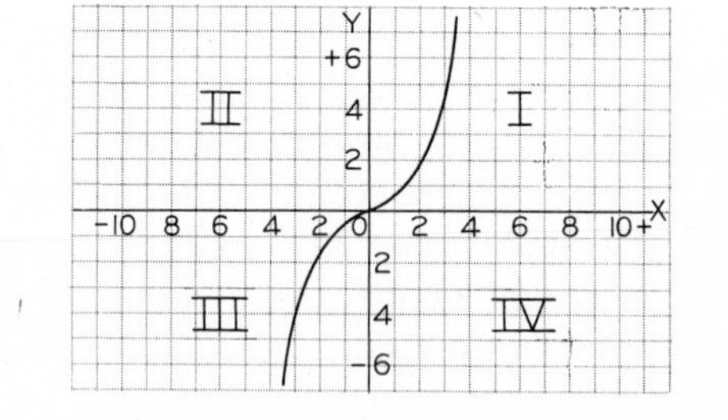

(b) MATHEMATICAL GRAPH AXES DESIGNATION

**Fig. 26.7 Axes Designation.**

**Fig. 26.8 Axes Scale "Breaks."**

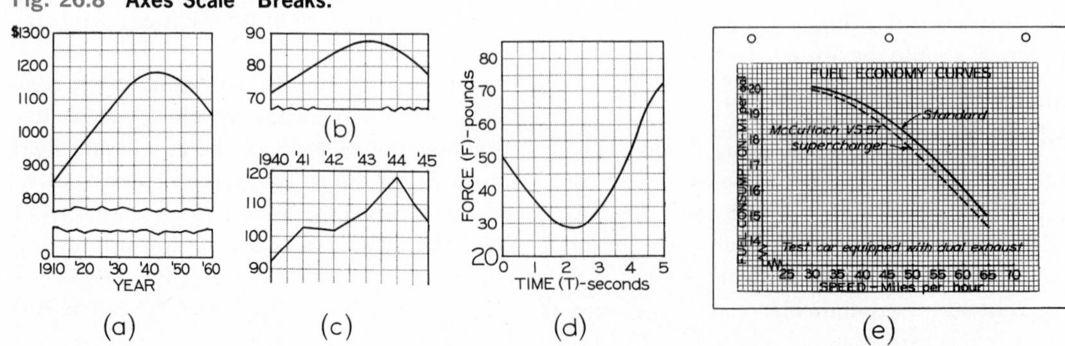


(a) (b) (c) (d) (e)

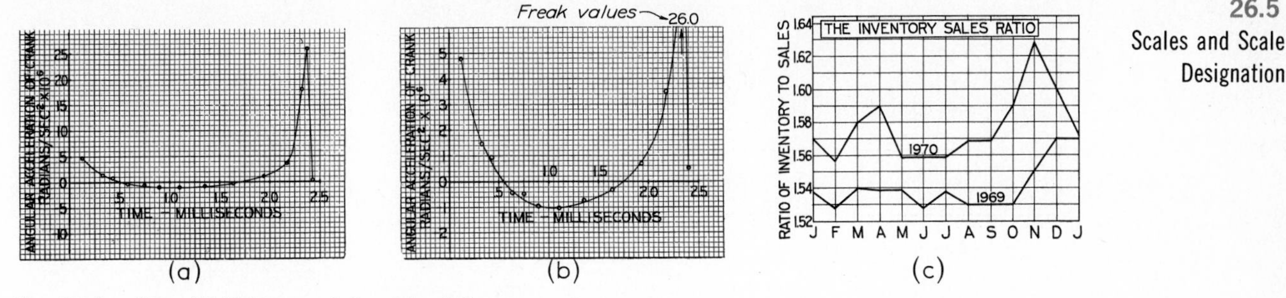


Fig. 26.9 **"Freak" Values and Combined Curves.**

or trend can be made to appear "steeper" by increasing the ordinate scale or decreasing the abscissa scale, Fig. 26.10 (a), and "flatter" by increasing the abscissa scale or decreasing the ordinate scale, (b). As shown in Fig. 26.11, a variety of slopes or shapes can be obtained by expanding or contracting the scales. A deciding factor is the impression desired to be conveyed graphically.

Normally, an angle greater than 40° with the horizontal gives an impression of a significant rise or increase of ordinate values, while an angle of 10° or less suggests an insignificant trend, Fig. 26.12 (a) and (b). *The slope chosen should emphasize the significance of the data plotted.* Some relationships are customarily presented in a conventional shape, as shown at (c) and (d). In this case, an expanded abscissa scale, as shown at (d), should be avoided.

Scale designations should be placed outside the axes, where they can be shown clearly. Abscissa nomenclature is placed along the axis so that it can be read from the bottom of the graph. Ordinate values are generally lettered so that they can also be read from the bottom; but ordinate captions, if lengthy, are lettered to be read from the right. The values can be shown on both the right and left sides of the graph, or along the top and bottom, if necessary for clearness, as when the graph is exceptionally wide or tall or when the rulings are closely spaced and hard to follow. When the major interest (e.g., maximum or minimum values) is situated at the right, the ordinate designations may be placed along the right, Fig. 26.1 (b) and (c). This arrangement

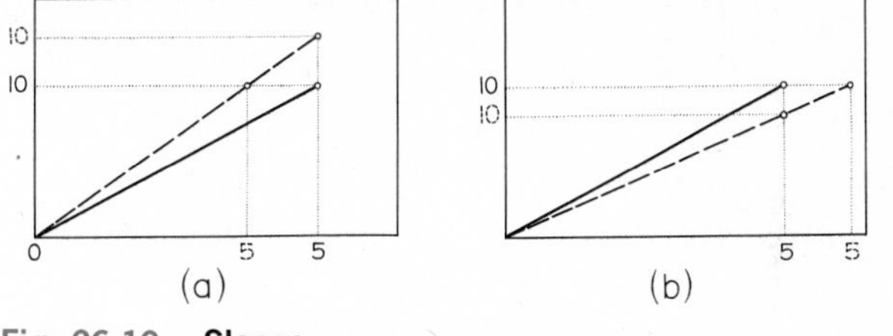

Fig. 26.10 **Slopes.**

also encourages reading the chart first and then the scale magnitudes.

When grid rulings are specially prepared on blank paper, every major division ruling should have its value designated, Fig. 26.13 (a). The labeled divisions should not be closer than .25″ and rarely more than 1″ apart. Intermediate values (rulings or ticks) should not be identified and should be spaced no closer than .05″. If the rulings are numerous and close together, as on printed graph paper, only the major values are noted, as at (b) and

Fig. 26.11 **Effects of Scale Designation.**

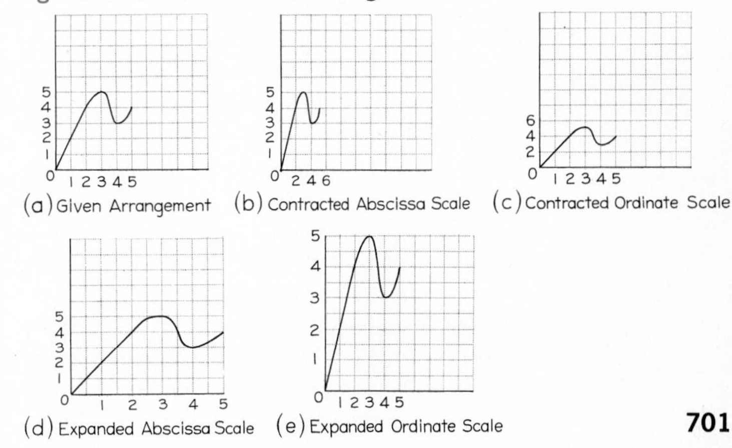

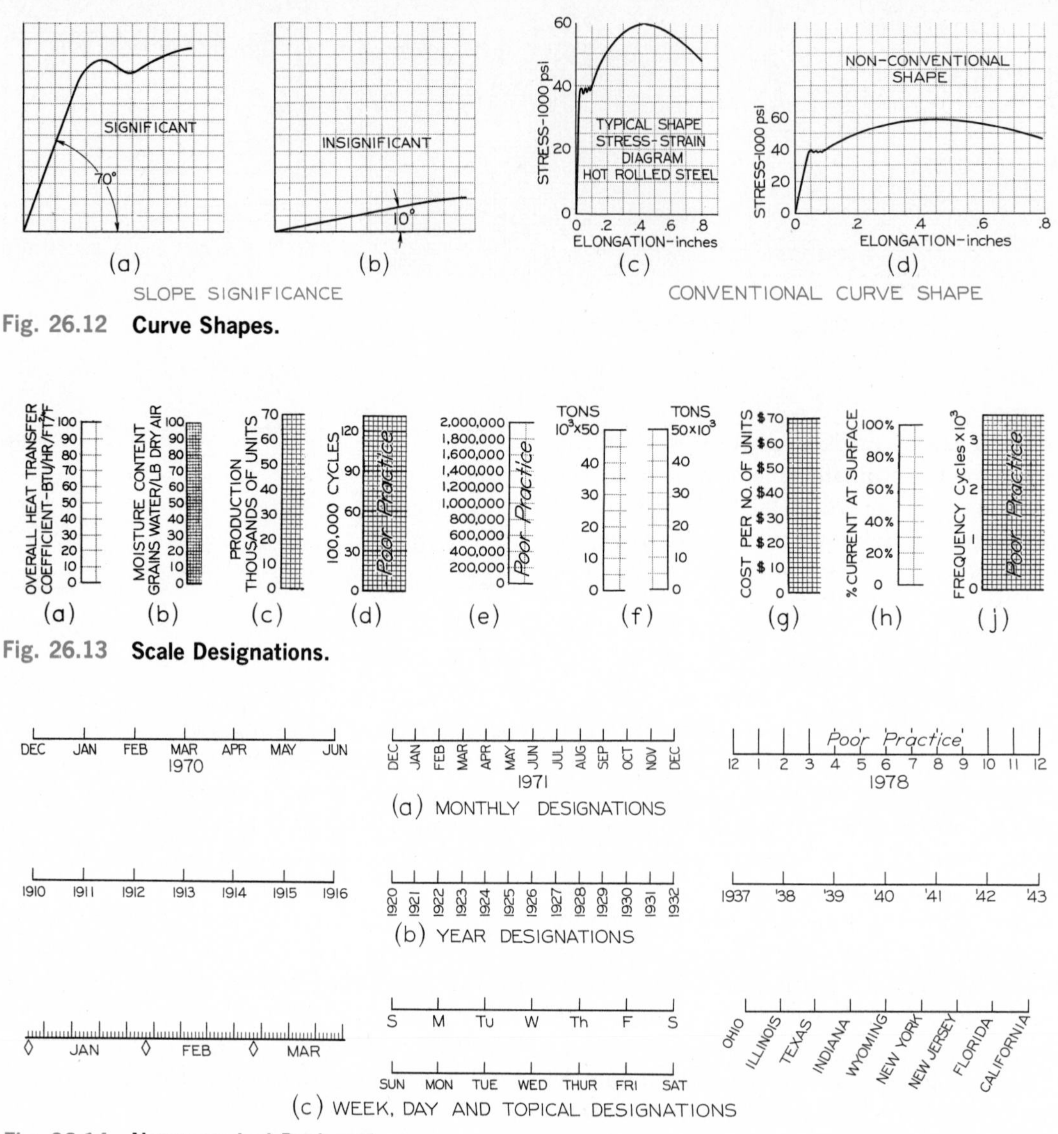


**Fig. 26.12 Curve Shapes.**

**Fig. 26.13 Scale Designations.**

**Fig. 26.14 Nonnumerical Designations.**

(c). The assigned values should be consistent with the minor divisions. For example, major divisions designated as 0, 5, 10, . . . , should not have 2 or 4 minor intervals, since resulting values of 1.25, 2.5, 3.75, . . . , are undesirable. Similarly, odd-numbered major divisions of 3, 5, 7, . . . , or multiples of odd numbers with an even number of minor divisions, should be avoided as shown at (d). The numbers, if three digits or smaller, can be fully given. If the numbers are larger than three digits, (e), dropping the ciphers is recommended, if the omission is indicated in the scale caption, as at (c). Values are shortened to even hundreds, thousands, or millions, in preference to tens of thousands, for example. Graphs for technical use can have the values shortened by indicating the shortened number times some power of 10, as at (f). In special cases, such as when giving values in dollars or percent, the symbols may be given adjacent to the numbers, as at (g) and (h).

Designations other than numbers usually require additional space; therefore, standard abbreviations should be used, Fig. 26.14. Abscissa values for these may be lettered vertically as in the center at (a) and (b), or inclined, as at the right in (c), to fit the designations along the axes.

Scale captions (or titles) should be placed along the scales so that they can be read from the bottom for the abscissa, and from the right for the ordinate. Captions include the name of the variable, symbol (if any), units of measurement, and any explanation of digit omission in the values. If space permits, the designations are lettered completely, but if necessary, standard abbreviations may be used for the units of measurement. Notations such as shown in Fig. 26.13 (j) should be avoided, since it is not clear whether the values shown are to be multiplied by the power of ten or already have been. Short captions may be placed above the values, Fig. 26.13 (f), especially when graphs are prepared for projection slides, since reading from the right is difficult.

## 26.6 Points, Curves, and Curve Designations.

In mathematical and popular-appeal graphs, curves without designated points are commonly used, since the purpose is to emphasize the general significance of the curves. On graphs prepared from observed data, as in laboratory experiments, points are usually designated by various symbols, Fig. 26.15. If more than one curve is plotted on the same grid, a combination of these symbols may be used, one type for each curve, although labels are preferable if clear. The use of open-point symbols is recommended, except in cases of

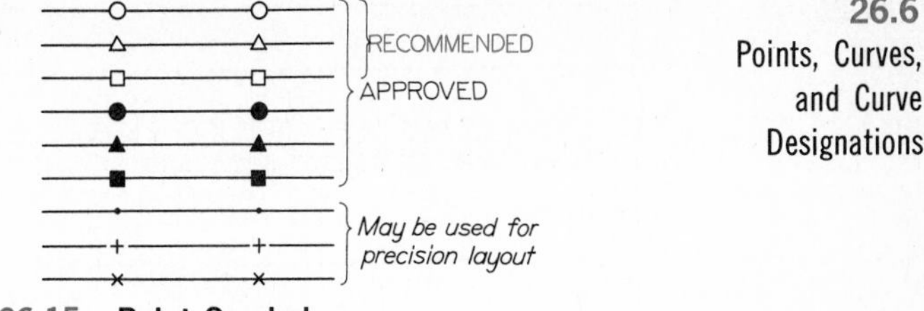


Fig. 26.15 **Point Symbols.**

"scatter" diagrams, Fig. 26.37, where the filled-in points are more visible. In general, filled-in symbols should be used only when more than three curves are plotted on the same graph, and a different identification is required for each curve. The curve should not be drawn through the point symbols as they may be needed for reference later for additional information.

When several curves are to be plotted on the same grid, they can be distinguished by the use of various types of lines, Fig. 26.16 (a). However, solid lines are used for the curves wherever possible, while the dashed line is commonly used for *projections* (estimated values, such as future expectations), as shown at (b). The curve should be heavier in weight than the grid rulings, but a difference in weight can also be made between various curves to emphasize a preferred curve or a total value curve (sum of two or more curves), as shown at (c). A *key*, or *legend*, should be placed in an isolated portion of the grid, preferably enclosed by a border, to denote point symbols or line types that are used for the curves. If the grid lines are drawn on blank paper, a space should be left vacant for this information, Fig. 26.17. Keys may be placed

Fig. 26.16 **Curve Lines.**

(a)
CURVE LINE TYPES

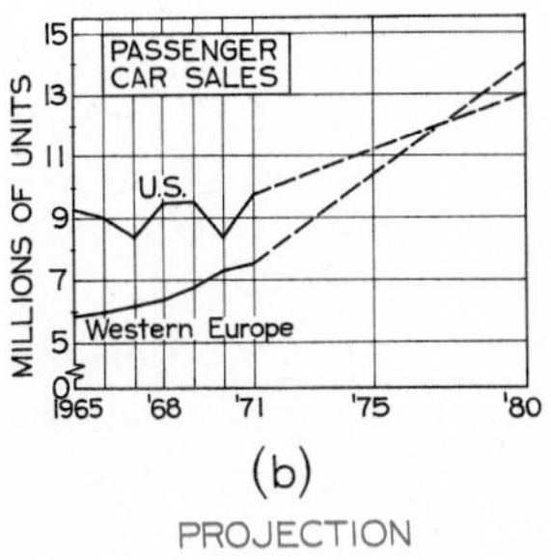


(b)
PROJECTION

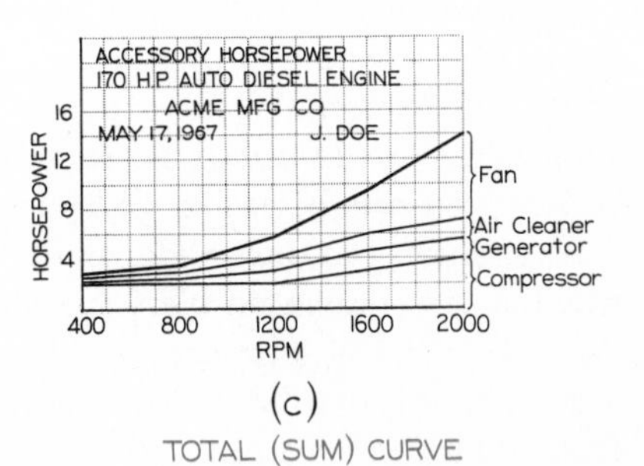


(c)
TOTAL (SUM) CURVE

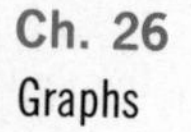


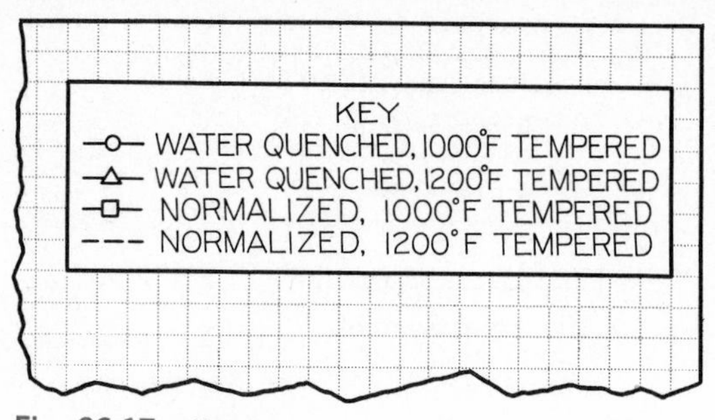


Fig. 26.17 **Keys.**

off the grids below the title, if space permits. However, it is preferable to designate curves with labels, rather than letters, numbers, or keys, if possible, Fig. 26.2 (c). Colored lines are very effective for distinguishing the various curves on a grid, but they may not be suitable for multiple copies.

## 26.7 Titles.

Titles for a graph may be placed on or off the grid surface. If placed on the grid, white space should be left for the title block, but if printed coordinate paper is used, a heavy border should enclose the title block. If further emphasis is desired, the title may be underlined. The contents of title blocks varies according to method of presentation. Typical title blocks include title, subtitle, institution or company, date of preparation, and name of the author, Fig. 26.16 (c). Some relationships may be given an appropriate name. For example, a number of curves showing the performance of an engine are commonly entitled "Performance Characteristics." If two variables plotted do not have a suitable title, "Dependent variable (name) vs. independent variable (name)" will suffice. For example: GOVERNOR PRESSURE vs SPEED.

Notes, when required, may be placed under the title for general information, Fig. 26.18 (a); labeled adjacent to the curve, (b), or along the curve, Fig. 26.4 (a); or referred to by means of reference symbols, Fig. 26.18 (c) and (d).

Any chart can be made more effective, whether it is drawn on blank paper or upon printed paper, if it is inked. For reproduction purposes, as for slides or for publications, inking is necessary.

## 26.8 Semilogarithmic Coordinate Line Charts.

A *semilog chart,* also known as a *rate-of-change* or *ratio chart,* is a type in which two viables are plotted on semilogarithmic coordinate paper to form a continuous straight line or curve. Semilog paper contains uniformly spaced vertical rulings and logarithmically spaced horizontal rulings.

Semilog charts have the same advantages as rectangular coordinate line charts (arithmetic charts), §26.2. When rectangular coordinate line charts give a false impression of the trend of a curve, the semilog charts would be more effective in revealing whether the rate of change is increasing, decreasing, or constant. Semilog charts are also useful in the derivation of empirical equations, Chapter 28.

Semilog charts, like rectangular coordinate line graphs, are not recommended for presenting only a few plotted values in a series, for emphasizing change in absolute amounts, or for showing extreme or irregular movement or trend of data.

In Fig. 26.19 (a) and (b), data are plotted on rectangular coordinate grids (arithmetic) and on semilogarithmic coordinate grids, respectively. The same data, which produce curves on the arithmetic graph, produce straight lines on the semilog grid. The straight lines permit an easier analysis of the trend or movements of the variables. If the logarithms of the ordinate values are plotted on a rectangular coordinate grid, instead of the actual values, straight lines will result on the arithmetic graph, as shown at (c). The straight lines produced on semilog grid provide a simple means of deriving empirical equations, as discussed in Chapter 28.

Straight lines are not necessarily obtained on a semilog grid, but if they do occur, it means that the rate of change is constant, Fig. 26.20 (a). Irregular curves can be compared to constant-rate scales individually or between a series of curves, as shown at (b).

## 26.9 Design and Layout of Semilog Charts.

Semilog graphs are usually prepared on printed semilog coordinate paper. Graphs for publication, however, requiring fewer grid

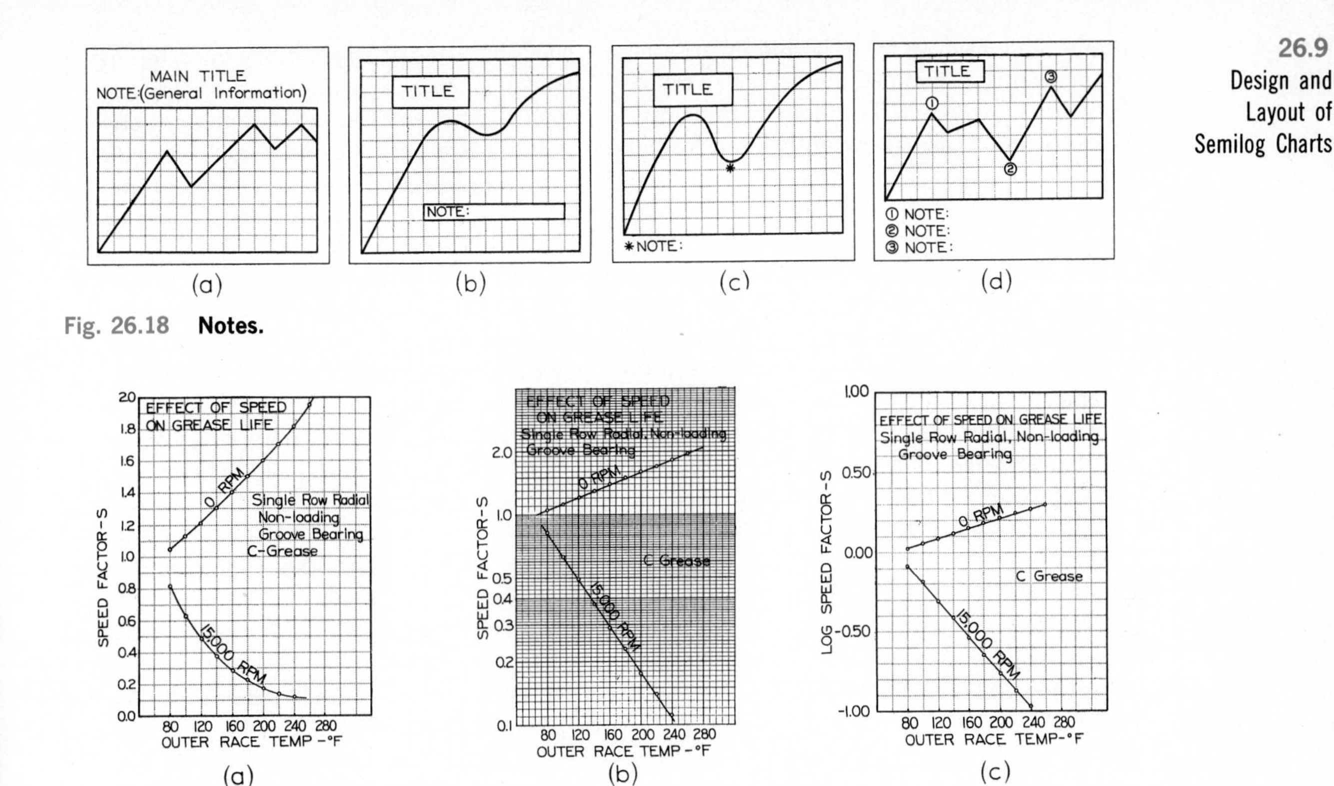


Fig. 26.18 **Notes.**

Fig. 26.19 **Arithmetic and Semilogarithmic Plottings.**

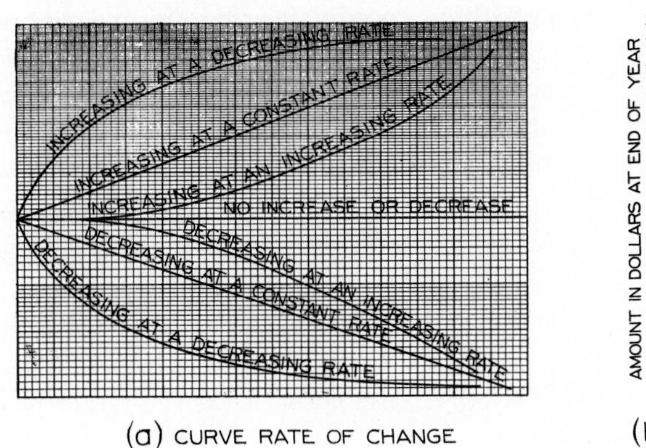


(a) CURVE RATE OF CHANGE

THE AMOUNT OF ONE DOLLAR AT COMPOUND INTEREST

AMOUNT IN DOLLARS AT END OF YEAR

TIME - YEARS

(b) CURVE COMPARISON TO CONSTANT RATES OF CHANGE

Fig. 26.20 **Rates of Change.**

rulings, are plotted on specially prepared grid scales or traced from the original plot. As illustrated in Fig. 26.21 (a), the logarithmically divided scale is generally placed along the ordinate. The abscissa scale is uniformly divided into the required number of rulings for the independent variable. Logarithmic scales are nonuniform, beginning with the largest space at the bottom and decreasing in size for ten divisions, which is known as one *log cycle.* The locations of major divisions and subdivisions for the semilog grid are determined by taking the logarithm to the base 10 (common logarithm) of the value to be plotted. As shown at (b), the major division of one log cycle, from 1 to 10, ranges in value (logarithmically) from 0 to 1. If this logarithmic proportion is maintained, the log cycle can be laid out to any height, but the remaining divisions depend upon the height chosen for one unit of log cycle. For example, if a length of 10″ for one log cycle is suitable, the number two division is at 3.01″ = (.301 × 10″) from the origin of the cycle (number one division), and the number three is at 4.77″ from the origin. A convenient device for laying out logarithmic scales is a *log cycle sector* prepared on a separate sheet of paper, as shown at (c). The sector is folded along the length of log cycle desired, and divisions are marked on the scale being prepared adjacent to the folded edge.

Printed semilog paper is available with as many as five log cycles on an $8\frac{1}{2}'' \times 11''$ sheet. Nonreproducible grids for prints are also available.

The composition and layout techniques of the semilog graphs are similar to those of the rectangular coordinate graphs, discussed in §26.5.

As can be observed from the layout of a logarithmic cycle scale, the first designation (origin of cycle) is at log 1 = 0, and the last designation is at log 10 = 1. Log cycle designation must start at some power of 10 and end at the next power of 10 (e.g., 1 to 10, 10 to 100, .1 to 1.0, .01 to .1, or correspondingly $10^0$ to $10^1$, $10^1$ to $10^2$, $10^{-1}$ to $10^0$, $10^{-2}$ to $10^{-1}$), Fig. 26.22. As can also be observed from a log scale, the values plotted can be decreasingly small in quantity, approaching zero, but never reaching zero. Logarithms of numbers greater than one are positive, and of numbers between 0 and 1 are negative. As a positive variable quantity approaches zero, its logarithm becomes negatively infinite. *Semilog charts, therefore, never have a zero value on the logarithmic scale.* The cycle designation determines the number of cycles required as a result of observing the range of values to be plotted. Cycle designation also permits the plotting of an extensive range of values, since each cycle accommodates an entire range of one power of 10.

The techniques of scale and chart designation and captions are similar to those discussed for rectangular coordinate line graphs. Values shown along the logarithmic axis, however, can be designated in a number of ways, as shown in Fig. 26.23. The scale shown at (a) is preferred, since it is the simplest and will be understood by any person who makes frequent use of semilog charts.

## 26.10 Logarithmic Coordinate Line Charts.

*Logarithmic charts* have two variables plotted on a logarithmic coordinate grid to form a continuous line or a "curve." Printed logarithmic paper contains logarithmically spaced horizontal and vertical rulings. As in the case of semilog charts, paper containing as many as five cycles on an axis can be purchased. See §26.9.

Fig. 26.21 **Semilog Chart Design.**

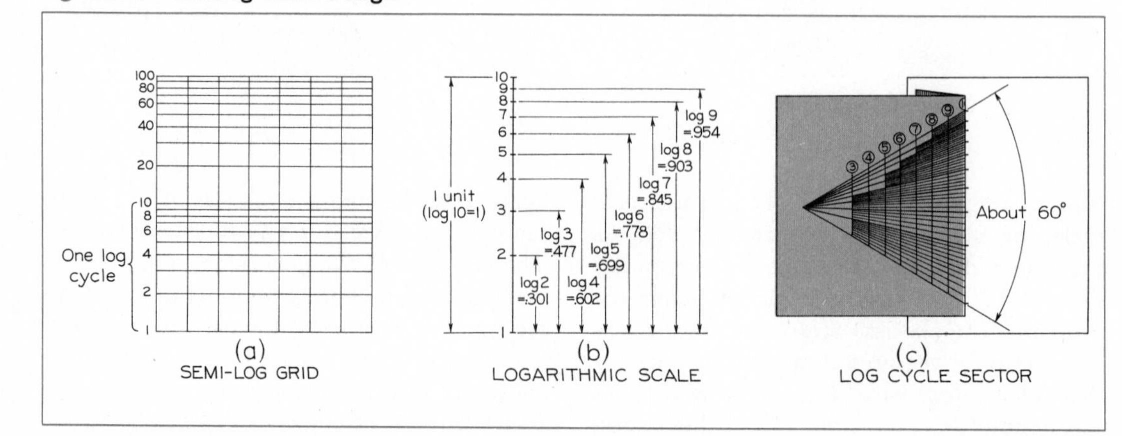

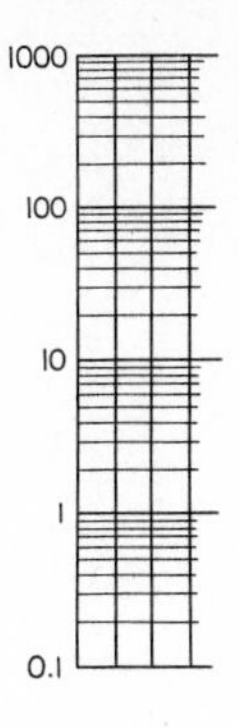


Fig. 26.22 **Cycle Designation.**

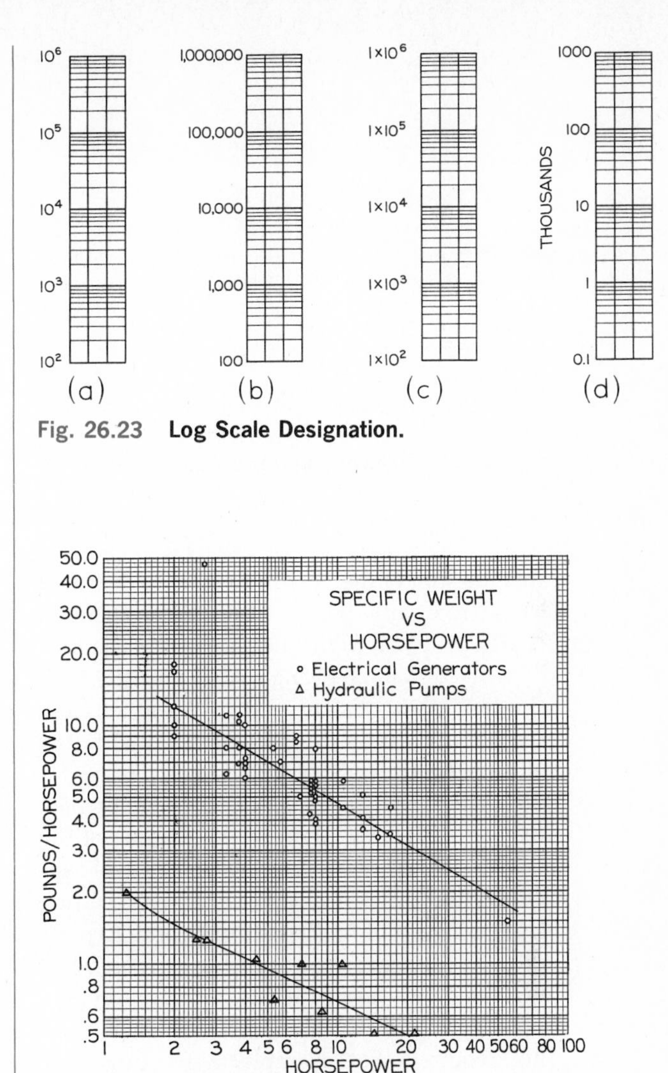


Fig. 26.23 **Log Scale Designation.**

Fig. 26.24 **Logarithmic Chart.**

Log charts are applicable for the comparison of a large number of plotted values in a compact space, and for the comparison of the relative trends of several curves on the same chart. This form of graph is not the best form for presentation of relatively few plotted values in a series or for emphasizing change in absolute amounts. The designation of log cycles, however, permits the plotting of very extensive ranges of values.

Logarithmic charts are primarily used to determine empirical equations by fitting a single straight line to a series of plotted points, Chapter 28. They are also used to obtain straight-line relationships when the data are suitable, as in Fig. 26.24. The design and layout of log charts is the same for semilog charts, §26.9, and similar in many respects to that for rectangular coordinate charts, §§26.3 to 26.7.

## 26.11 Trilinear Coordinate Line Charts.

*Trilinear charts* have three related variables plotted on a coordinate paper in the form of an equilateral triangle. The points joined together successively form a continuous straight line or "curve."

Trilinear charts are particularly suited for the following uses:

1. Comparison of three related variables relative to their total composition (100 percent).
2. Analysis of the composition structure by a combination of curves—for example, metallic microstructure of trinary alloys, Fig. 26.26 (a).
3. Emphasizing change in amount or differences between values.

The trilinear chart is *not* recommended for (1) emphasizing movement or trend of data or for (2) comparison of three related but dissimilar physical quantities—for example, force, acceleration, and time.

Trilinear charts are most frequently applied in the metallurgical and chemical fields, because of the frequency of three variables in metallurgical and chemical composition. The basis of application is the geometric principle

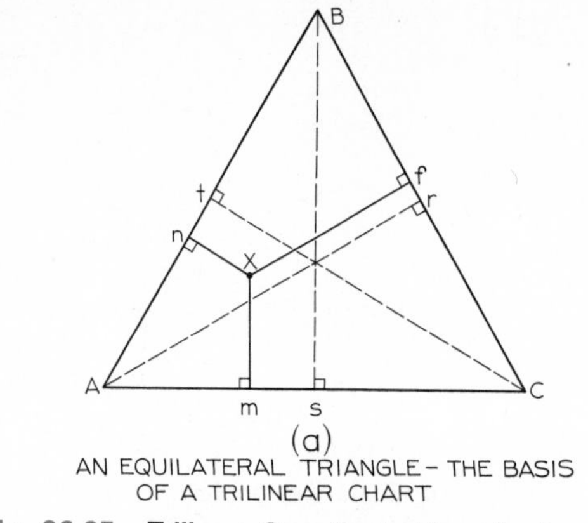


(a)
AN EQUILATERAL TRIANGLE – THE BASIS OF A TRILINEAR CHART

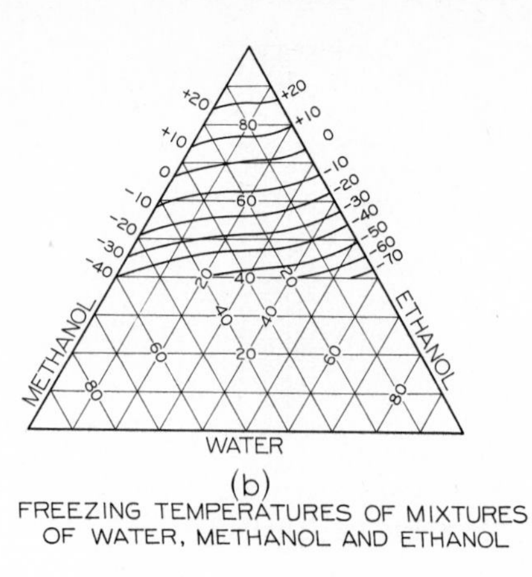


(b)
FREEZING TEMPERATURES OF MIXTURES OF WATER, METHANOL AND ETHANOL

Fig. 26.25 **Trilinear Coordinate Line Charts.**

that the sum of the perpendiculars to the three sides from any point within an equilateral triangle is equal to the altitude of the triangle.

In an equilateral triangle ABC, Fig. 26.25 (a), the sum of the distances xf, xm, and xn from the point x within the triangle is equal to the altitude Ar, Bs, or Ct of the triangle. For example, if the distances xf, xm, and xn are respectively 50, 30, and 20 units, the altitude of the triangle is 100 units, and the point x will represent the function composed of, or resulting from, 50 parts of A, 30 parts of B, and 20 parts of C. At (b) is shown a chart for various freezing temperatures, with the mixture proportions by volume of water, methanol, and ethanol required. For example, a freezing temperature of −40°F can be established by mixing 50 parts of water with 10 parts of ethanol and 40 parts of methanol.

**26.12 Design and Layout of Trilinear Charts.** Trilinear charts are also usually drawn on printed grid paper. Charts for publications, requiring maximum clarity, can be plotted on grids drawn on plain paper or traced from an original plotting on a printed sheet. The trilinear coordinate grid is an equilateral triangle, which must have the same space between rulings along each axis. Normally, each side of the triangle is divided into ten divisions for percent plots, Fig. 26.26, with

Fig. 26.26 **Metallurgical Trilinear Charts.**

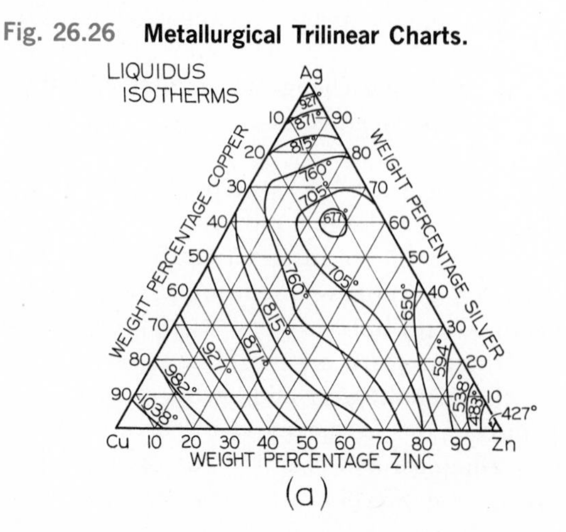


(a)

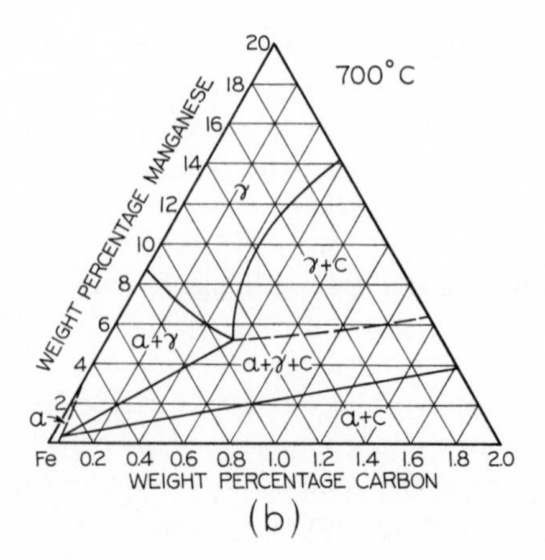


(b)

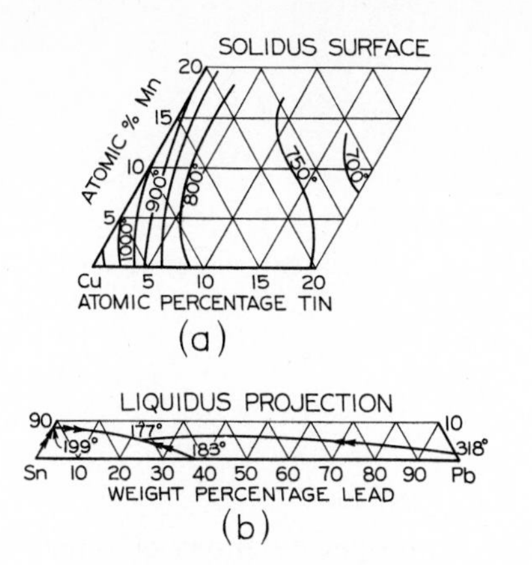


Fig. 26.27 **Partial Trilinear Charts.**

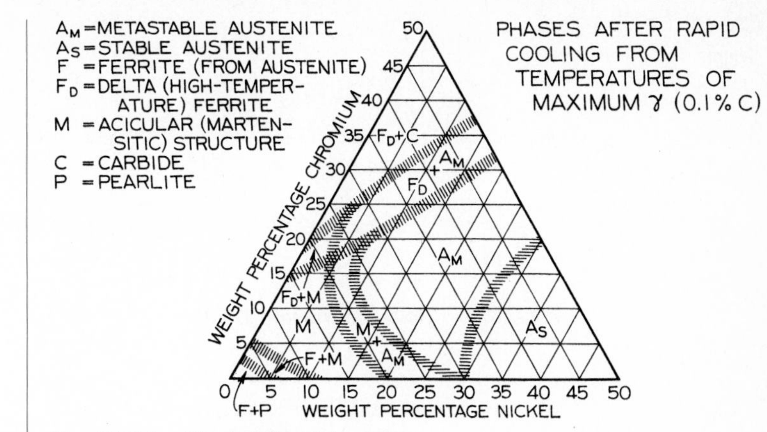


Fig. 26.28 **Shaded-Band Curves.**

as many subdivisions as required. If only a portion of a coordinate field is to be used, only a portion of the grid need be shown, Fig. 26.27.

The scale designations can be placed to be read from the bottom of the sheet or can be tilted as shown in Fig. 26.25 (b), but for ease of reading, all scale notations should be placed outside the grid area. Other techniques of scale designations are the same as for rectangular coordinate line graphs, §26.5.

Since individual points are generally not shown on trilinear charts, the curves are drawn as continuous lines.

Curves are frequently single-weight solid lines, but dashed lines are used for projected (anticipated) information. Fig. 26.26 (b). When the curve-fitting information varies within a range, shaded bands are used, Fig. 26.28.

When the chart indicates the composition of various parts or different parts, the symbols are placed within the appropriate areas—for example, $\alpha$, $\beta$, $\alpha + \gamma$, which are microstructure symbols, Fig. 26.26 (b). Curve designations are placed along the curves or in breaks in the curve—for example, the temperature values shown in Fig. 26.27 (a).

Since most of the grid area is utilized by the curves, the title, keys, and notes are placed off the grid, as shown in Fig. 26.28.

## 26.13 Polar Coordinate Line Charts.

*Polar charts* have two variables, one a *linear* magnitude and the other an *angular* quantity, plotted on a polar coordinate grid with respect to a pole (origin) to form a continuous line or "curve."

Polar charts are particularly adaptable to the following applications:

1. Comparison of two related variables, one being a linear magnitude (called a *radius vector*), and the second an angular value.
2. Indicating movement or trend or location with respect to a pole point.

Polar charts are *not* suited for (1) emphasizing changes in amounts or differences between values, or for (2) interpolating intermediate values.

As shown in Fig. 26.29 (a), the zero degree line is the horizontal right axis. To locate a point P, it is necessary to know the radius vector r, and an angle $\theta$ (e.g., 5, 70°). The point P could also be denoted as 5, 430°), (5, −290°), and (−5, 250°), (−5, −110°), for example. If we plot the equation $r = a$ (no angular designation), we will obtain a circle with the center at the pole, as shown at (b). The value a is a constant value, which determines the relative size of the radius vector and the curve. The equation $r = 2a \cos \theta$ produces a circle going through the pole point, with its center on the polar axis, (c). The plot of $r = a \sin 3\theta$ produces a "three-leaved rose," as shown at (d). The above are charts of mathematical equations; however, many practical applications are not concerned with mathematical relationships, but with the magnitude

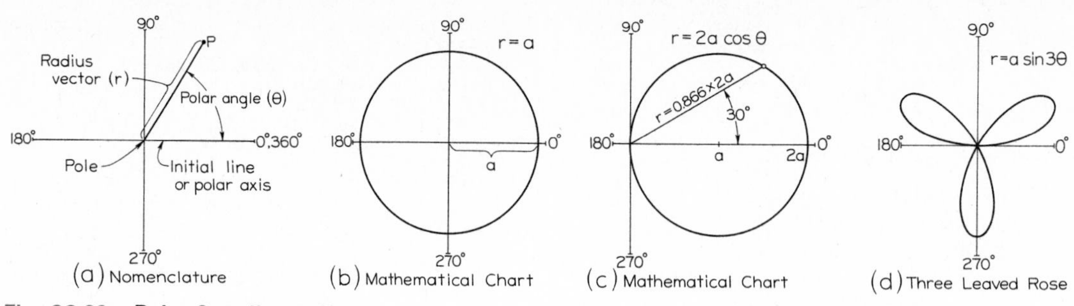


Fig. 26.29 **Polar Coordinate Charts.**

of some values and their location with respect to a pole point. For example, Fig. 26.30 (a) and (b) illustrate stress charts from experimentation and for stress visualization, respectively. At (c) is shown a polar chart of a bearing load diagram, and the graph at (d) indicates the noise distribution from a jet engine.

## 26.14 Design and Layout of Polar Charts.

Polar charts are usually prepared on printed polar grids on $8\frac{1}{2}'' \times 11''$ paper, Fig. 26.30 (c). Charts for publication may be plotted on a grid drawn upon blank paper, which involves the required number of subdivisions of 360° and a sufficient number of concentric circles, equally spaced, to denote the values of radius vectors. See Fig. 26.30 (d).

The initial line, or polar axis, is normally the horizontal line extending from the pole toward the right. Upon many polar charts and printed grids, however, the upper vertical line is used as a polar axis. The sense of direction from 0° to 360° can be either clockwise or counterclockwise. A proper sense of direction may at times be determined by a physical sense of rotation, as for the load diagram in Fig. 26.30 (c). Mathematical diagrams, in which the standard four quadrants are used, follow a counterclockwise sense of direction and a horizontal polar axis.

Angular designations are placed along the outskirts of the polar grid, while radius vector values are lettered along either the horizontal or vertical axis, wherever space permits. As in the case of rectangular coordinate graphs, numbers no larger than three digits should be used, with proper notations of actual values, Fig. 26.30 (a). All captions should be placed to be read from the bottom of the graph.

Points are normally not designated on polar charts. As in other forms of charts, curves may be distinguished by various types of lines, Fig. 26.30 (b), when more than one curve is required. Curves may be designated with a leader and arrow to a note, or with a legend.

Since in many polar charts the grid area is more or less filled with the curve, the scale and curve designations, titles, keys, and additional notes are placed off the grid. If a printed grid is used, the title information may be placed on white paper and fastened onto the grid to facilitate ease of reading, as at (c).

## 26.15 Nomographs or Alignment Charts.

*Nomographs* or *alignment charts* consist of straight or curved scales, arranged in various configurations so that a straight line drawn across the scales intersects them at values satisfying the equation represented.

Alignment charts can be used for analysis, but the predominant application is for computation. The design and layout of some common forms of nomographs are discussed in Chapter 27.

## 26.16 Bar or Column Charts.

*Bar charts* are graphic representations of numerical values by lengths of bars, beginning at a base line, indicating the relationship between two or more related variables.

Bar charts are particularly suited for the following purposes:

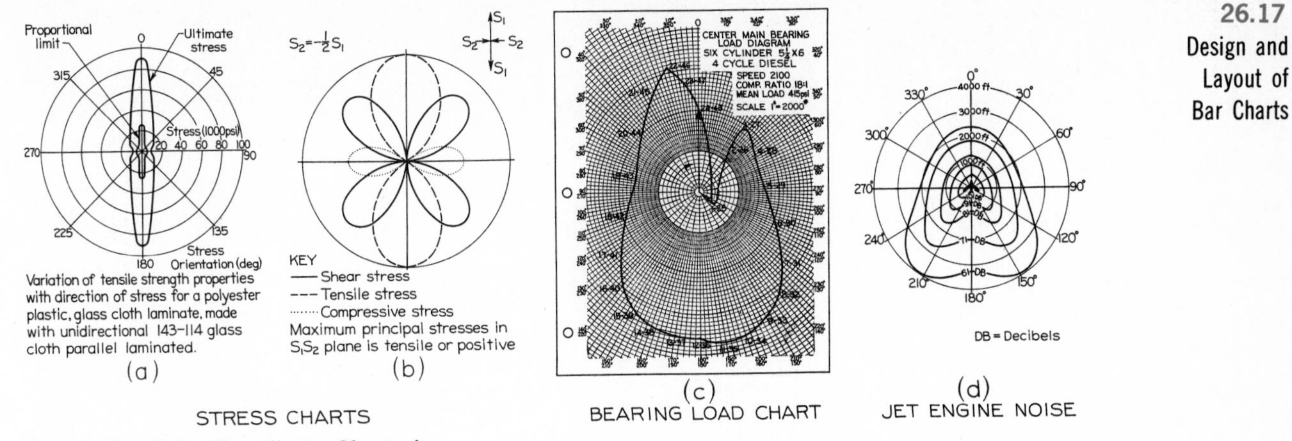


**Fig. 26.30 Polar Coordinate Charts.***

1. Presentation for the nontechnical reader.
2. A simple comparison of two values along two axes.
3. Illustration of relatively few plotted values.
4. Representation of data for a total period of time in comparison to point data.

Bar or column charts are *not* recommended for (1) comparing several series of data or for (2) plotting a comparatively large number of values.

The bar chart is effective for nontechnical use because it is most easily read and understood. Therefore, it is used extensively by newspapers, magazines, and similar publications. The bars may be placed horizontally or vertically; when they are placed vertically, the presentation is sometimes called a *column chart.*

Bar charts are plotted with reference to two mutually perpendicular coordinate axes, similar to those in rectangular coordinate line charts. The charts may be a simple bar chart with two related variables, Fig. 26.31 (a); a grouped bar chart (three or more related variables), (b); or a combined bar chart (three or more related variables), (c). Another example of bar graphs for three or more related variables employs pictorial symbols, composed to form bars, as shown at (d). Bar charts can also effectively indicate a "deviation" or difference between values, (e).

*(a) and (b) adapted from charts by Robert L. Stedfeld and F. W. Kinsman, respectively, with permission of *Machine Design.* (c) Adapted from R. R. Slaymaker, *Bearing Lubrication Analysis,* copyright 1955 by John Wiley and Sons, with permission of the publisher and Clevite Corporation. (d) Adapted from chart by G. S. Schairer, with permission of author and Society of Automotive Engineers.

## 26.17 Design and Layout of Bar Charts.

If only a few values are to be represented by vertical bars, the chart should be higher than wide, Fig. 26.31 (a). When a relatively large number of values are plotted, a chart wider than high is preferred, (b) and (c). Composition is dictated by the number of bars used, whether they are to be vertical or horizontal, and the available space.

A convenient method of spacing bars is to divide the available space into twice as many equal spaces as bars are required, Fig. 26.32 (a). Center the bars on every other division mark, beginning with the first division at each end, as shown at (b). When the series of data is incomplete, the missing bars should be indicated by the use of ticks, (c), indicating the lack of data. The bars should be spaced uniformly when the data used are distributed. When irregularities in data exist, the bars should be spaced accordingly, as shown.

Bar charts may be drawn on printed coordinate paper; however, clarity for popular use is promoted by the use of blank paper and the designation of only the major rulings perpendicular to the bars. If the values of bars are individually noted, the perpendicular rulings may be omitted completely.

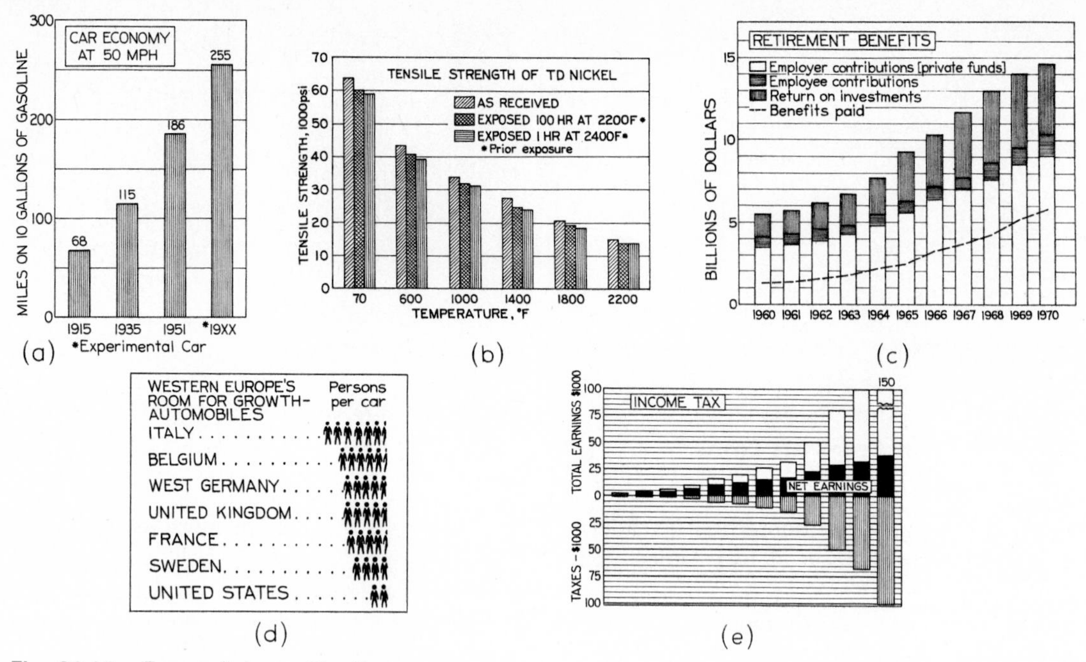


Fig. 26.31 **Bar or Column Charts.**

Since bar charts are used extensively to show differences in values for given periods of time, the values or amounts should be proportionate to the heights or lengths of the bars. The zero or principal line of reference should never be omitted. Normally, the full length of the bars should be shown to the scale chosen. When a few exceptionally large values exist, the columns may be broken as shown in Fig. 26.31 (e), with a notation included indicating the full value of the bar.

Scale designations are normally placed along the base line of the bars and adjacent to the rulings.

Standard abbreviations may be used for bar designations. The techniques of scale designations used for line curves, §26.5, are also applicable to bar charts. The values of the bars may be designated above the bars for simple and grouped bar charts, Fig. 26.31 (a). Spaces between bars should be wider than the bars when there are relatively few bars. Spaces between bars should be narrower than the bars when there are many bars, as at (b) and (c).

Bars are normally emphasized by shading or by filling in solid. Some of the common forms of shading are shown in the figure. Weight and spacing of shading depend upon the amount of area to be shaded and the final size of the chart and are therefore a matter of judgment.

Bar designations may be placed across several bars when possible, as shown at (e). Other

Fig. 26.32 **Bar Composition.**

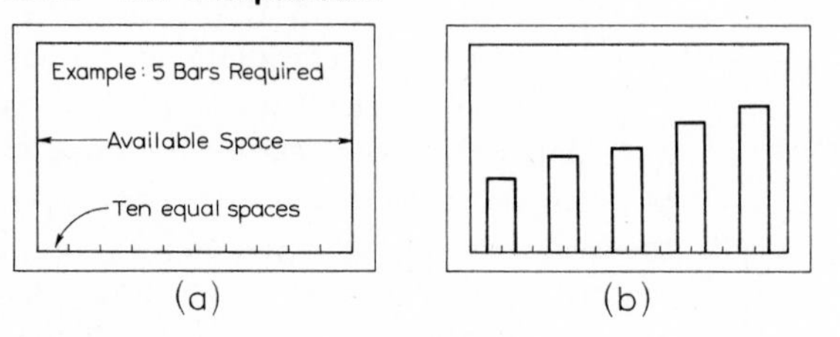


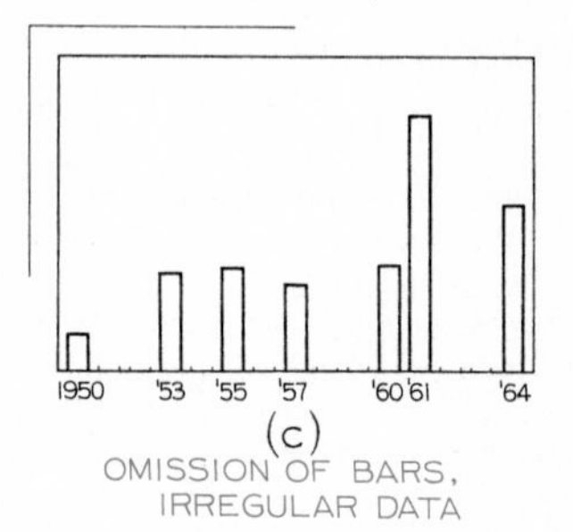

methods include the use of notes with leaders and arrowheads and the use of keys.

As in the case of line graphs, §26.7, titles for bar charts are placed where space permits and contain similar information.

## 26.18 Rectangular Coordinate Surface or Area Charts.

A *surface chart,* or *area chart,* is a line or bar chart with ordinate values accentuated by shading the entire area between the outlining curve and the abscissa axis, Fig. 26.33 (a).

A surface or area chart can be a simple chart (called a staircase chart), (b), a multiple surface or "strata" chart with one surface or layer on top of another, (c), or a combined surface chart indicating the distribution of components in relation to their total, (d).

Surface charts are used effectively to:

1. Accentuate or emphasize data that appear weak as a line chart.
2. Emphasize amount or ordinate values.
3. Represent the components of a total, usually expressed as a percent of a total, or compared to 100 percent.
4. Present a general picture.

Surface charts are *not* recommended for (1) emphasizing accurate reading of values of charts containing more than one curve or for (2) showing line curves that intersect or cross one another.

A map of terrestrial or geographic features is a form of area or surface chart. It represents a graphic picture of areas, surfaces, or the relationship of their component parts to the total configuration.



## 26.19 Design and Layout of Surface Charts.

Since surface charts are used as a general picture, and accurate reading of values is difficult, only major rulings need be shown. Techniques of grids and composition are the same as for line curves, §26.4. Printed grids generally should not be used, since the printed grid lines would interfere with the shading of areas.

Similarly to bar charts, surface charts represent a comparison of values from a zero line or base line; therefore, the ordinate scale should not be broken, nor should the zero (base) line be omitted. Other procedures and techniques of scale and scale designations are the same as for line curves.

The shading of surfaces is accomplished by the use of black (solid) and white areas, lines, and dots, Fig. 26.33. The darker shading tones should be used at the bottom of the chart, and progressively lighter tones should be used as each strata is shaded proceeding upward, as shown at (c). The weight and spacing of lines and dots are dependent upon the final size of the chart, and are a matter of judgment on the part of the draftsman. Projections, or extensions of a curve beyond present available data, can be distinguished by dashed outlines and lighter line shading.

Surfaces should be designated by placing the labels entirely on the surface, where possible, as at (c). Small surfaces can be denoted by a label with a leader and arrowhead, (a). The area of the labels should be clear of shading for ease in reading. Legends should not be used as a means of designation if direct labeling is possible.

The methods of chart designations are the same as for rectangular coordinate line graphs.

**Fig. 26.33 Surface or Area Charts.**

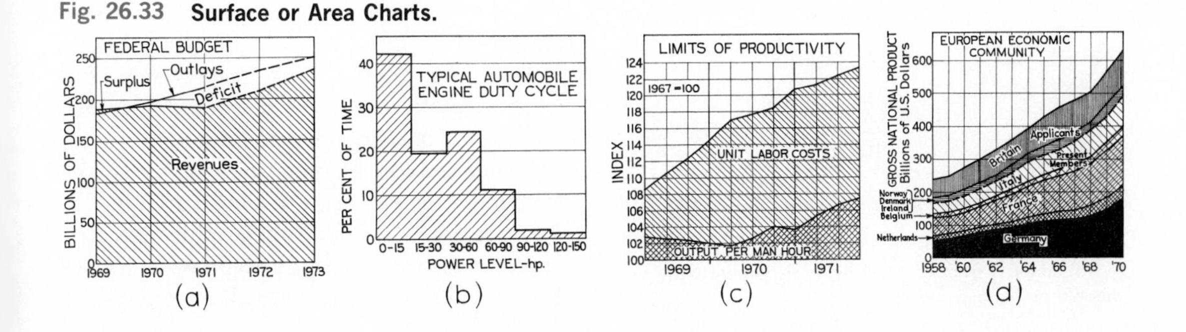

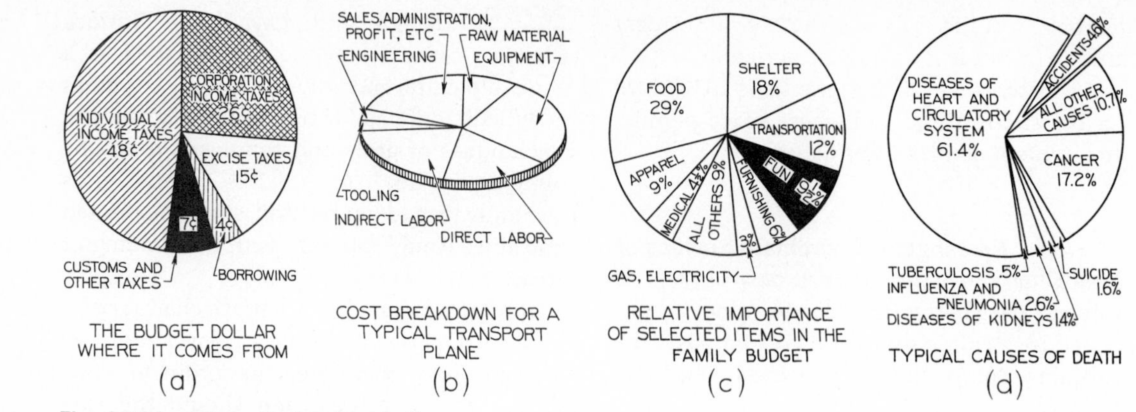


**Fig. 26.34 Pie or Sector Charts.***

## 26.20 Pie Charts.

*Pie charts,* or *sector charts,* are used to compare component parts in relation to their total by the use of circular areas.

Pie charts are an effective method for:

1. Representing data on a percent basis.
2. Popular presentation of a general picture.
3. Showing relatively few plotted values.
4. Emphasizing amounts rather than the trend of data.

A pie chart is normally presented as a true circular area, Fig. 26.34 (a), or in pictorial form, (b). Since most applications are concerned with monetary values, a disk or "coin" is commonly used for the circular area.

## 26.21 Design and Layout of Pie Charts.

Grids are not used for pie charts. The circular area is drawn to a desired size, within a permissible space. The determination of the various sizes of the sectors is based upon 360° being equivalent to 100 percent. Therefore, the following relationship exists:

$$\frac{100\%}{360^\circ} = \frac{x\%}{\theta^\circ}$$

or

$$\theta^\circ = \frac{x\%\ (360^\circ)}{100\%} = 3.6^\circ\ (x\%)$$

and 1 percent is represented by 3.6°, 2 percent by 7.2°, etc.

*(b) adapted from drawing by E. A. Green, with permission of *Machine Design.*

Sector values (percent and amount) are placed within the sectors where possible. If a sector is small, a note with a leader and arrowhead will suffice. Labels should be clear of any shading and lettered to read from the bottom of the chart, where possible, Fig. 26.34 (c) and (d). Another technique of sector designation is to shade the areas, (a). If one of the parts is to be emphasized and compared with the other parts, it can be shaded a different tone from the others, (c), or separated from the circular area, as shown at (d).

Titles should be placed above or below the figure.

## 26.22 Volume Charts.

A *volume chart* is the graphic representation of three related variables with respect to three mutually perpendicular axes in space. Volume charts are generally difficult to prepare, so they are not often used. The method of construction is not discussed in detail, but some of the forms will be considered in the following paragraphs.

Figure 26.35 illustrates line volume charts plotted with respect to three axes (Cartesian coordinates), shown in isometric and oblique projection at (a) and (b), respectively. Bar graphs can be similarly presented in three dimensions on an isometric grid.

A combination of bars and maps may be used to represent three related variables, as shown in Fig. 26.36 (a). Topographic map construction also is a graphic representation of three related variables (two dimensions in a

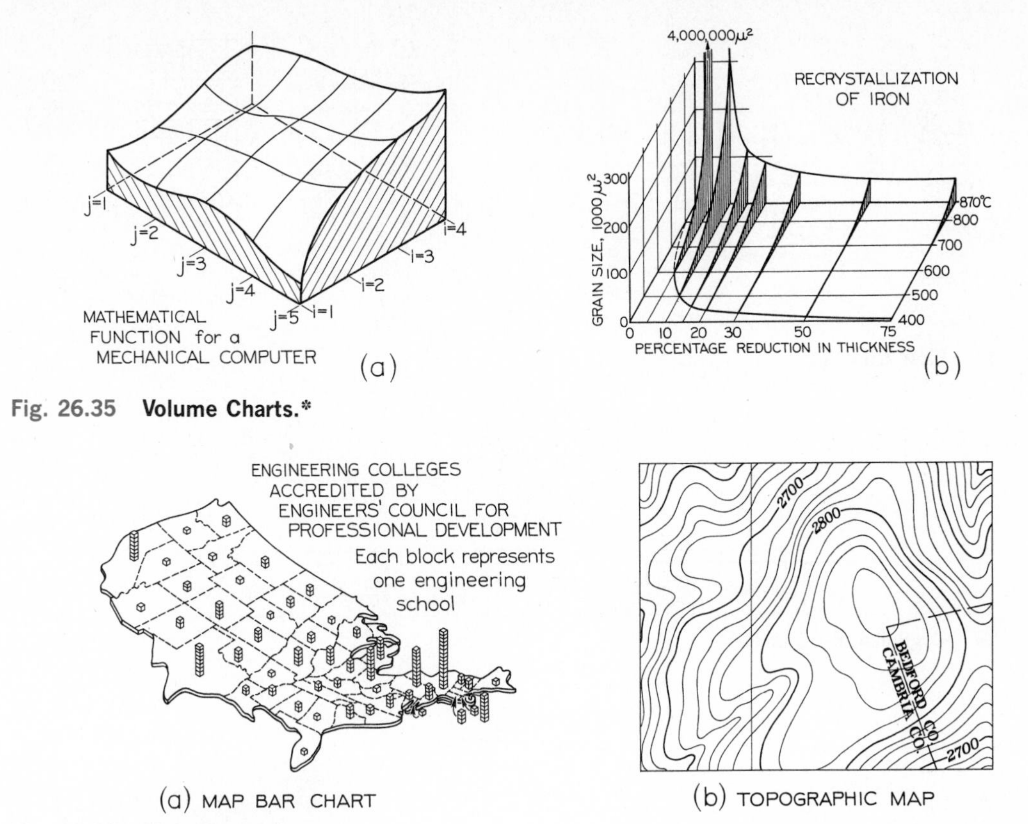


Fig. 26.35 **Volume Charts.***

Fig. 26.36 **Map Charts.†**

horizontal plane and one dimension in a vertical direction), all drawn in one plane of the drawing paper, as shown at (b).

## 26.23 Rectangular Coordinate Distribution Charts.

When the data observed or obtained vary greatly, the data can be plotted on a rectangular coordinate grid for the purpose of observing the distribution or areas of major concentration, with no attempt to fit a curve, Fig. 26.37. Charts of this nature are commonly called *scatter diagrams.*

## 26.24 Flow Charts.

*Flow charts* are predominantly schematic representations of the flow of a process—for example, manufacturing production processes and electric or hydraulic circuits, Fig. 26.38. Pictorial forms may be used as shown at (a). Schematic symbols are also used, if applicable, (b) and (c). Blocks with captions are used in the simplest form of flow chart, as illustrated at (d).

*Organization charts* are similar to flow charts, except that they are usually representations of the arrangement of personnel and physical items of a definite organization, Fig. 26.39.

Fig. 26.37 **"Scatter" Diagram.**

*Courtesy of Product Engineering*

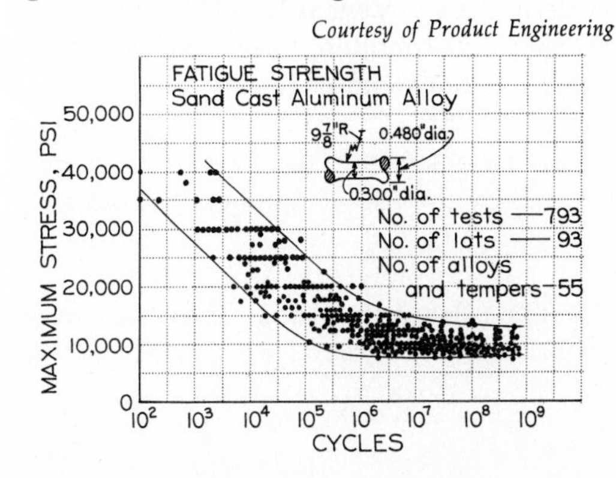


*(a) adapted from drawing by Eugene W. Pike and Thomas R. Silverberg, with permission of *Machine Design.*

†(a) adapted from *Engineering: A Creative Profession,* 3rd Ed., copyright by Engineers' Council for Professional Development, with permission of the publisher.

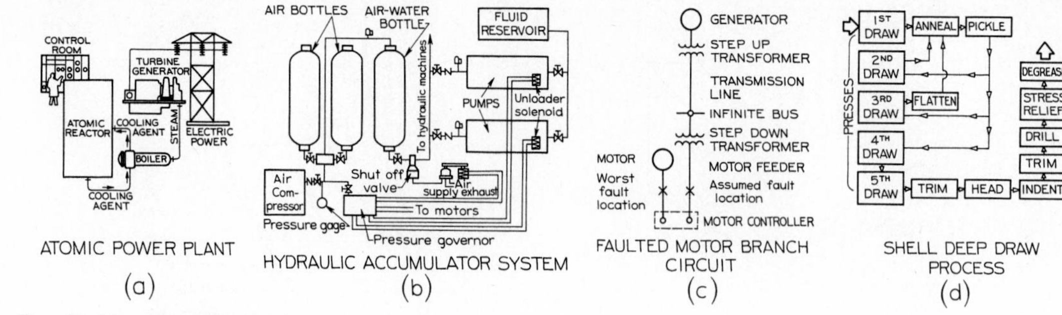


Fig. 26.38 **Flow Charts.***

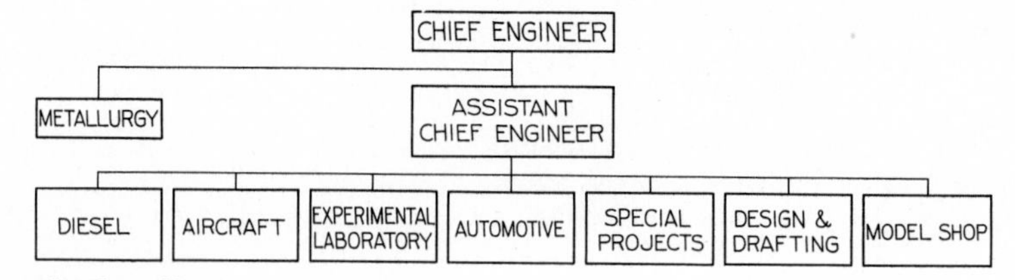


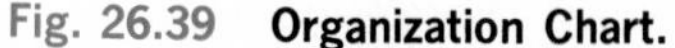

Fig. 26.39 **Organization Chart.**

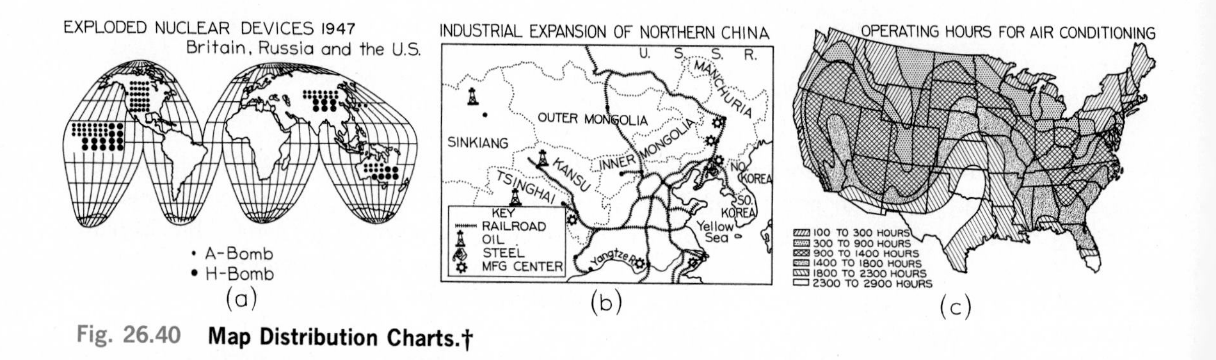


Fig. 26.40 **Map Distribution Charts.†**

**26.25 Map Distribution Charts.** When it is desired to present data according to geographical distribution, maps are commonly used. Locations and emphasis of data can be shown by dots of various sizes, Fig. 26.40 (a), by the use of symbols, (b), by shading of areas, (c), and by the use of numbers or colors.

**26.26 Graph Problems.** Construct an appropriate form of graph for the data listed. The determination of graph form (line, bar, surface, etc.) is left to the discretion of the instructor or the student and should be based on the nature of the data or the form of presentation desired. In some cases, more than one curve or more than one series of bars are required.

*(b) adapted from chart by A. F. Welsh, with permission of *Machine Design*.

†(a), (b), and (c) courtesy of *Look* Magazine, *Life* Magazine, and *Heating, Piping and Air Conditioning*, respectively.

**Prob. 26.1** United States Population.

| Year | Population |
|---|---|
| 1800 | 5,308,483 |
| 1810 | 7,239,881 |
| 1820 | 9,638,453 |
| 1830 | 12,866,020 |
| 1840 | 17,069,453 |
| 1850 | 23,191,876 |
| 1860 | 31,443,321 |
| 1870 | 38,558,371 |
| 1880 | 56,155,783 |
| 1890 | 62,947,714 |
| 1900 | 75,994,575 |
| 1910 | 96,977,266 |
| 1920 | 105,710,620 |
| 1930 | 122,775,046 |
| 1940 | 131,669,275 |
| 1950 | 151,132,000 |
| 1960 | 179,325,175 |
| 1970 | 211,400,000 |
| 1980 | 250,000,000[a] |
| 1990 | 300,000,000[a] |
| 2000 | 360,000,000[a] |

[a] Estimated

**Prob. 26.2** Chevy Pickup Trucks (Based on 1971 Data).

| Model | Percent Still Working |
|---|---|
| 1956 | 55.7 |
| 1957 | 63.9 |
| 1958 | 65.1 |
| 1959 | 71.1 |
| 1960 | 78.9 |
| 1961 | 79.4 |
| 1962 | 87.4 |
| 1963 | 90.3 |
| 1964 | 93.1 |
| 1965 | 96.6 |
| 1966 | 97.9 |
| 1967 | 97.7 |

**Prob. 26.3** Effect of Accuracy of Gear Manufacture on Available Strength Horsepower (60 teeth gear and 30 teeth pinion of 6 diametral pitch, 1.5″ face width, $14\frac{1}{2}°$ pressure angle; 500 Brinell Case hardness).

| Pitch Line Velocity, ft/min | Horsepower | | | |
|---|---|---|---|---|
| | Perfect Gear | Aircraft Quality Gear | Accurate Quality Gear | Commercial Quality Gear |
| 0 | 0 | 0 | 0 | 0 |
| 1000 | 140 | 100 | 75 | 50 |
| 2000 | 290 | 180 | 105 | 60 |
| 3000 | Straight line | 250 | 120 | 70 |
| 4000 | curve through | 320 | 130 | 72 |
| 5000 | two points | 380 | 140 | 72 |

Noise limit: 2750 ft/min—accurate quality gear; 1400 ft/min—commercial quality gear.

**Prob. 26.4** Lumber Prices—Softwood Plywood.

| Year | Index 1957–59 = 100 |
|---|---|
| 1968 | |
| Jan | 84 |
| Mar | 90 |
| June | 96 |
| Sept | 110 |
| Dec | 133 |
| 1969 | |
| Mar | 178 |
| June | 90 |
| Sept | 86 |
| Dec | 95 |
| 1970 | |
| Mar | 91 |
| June | 95 |
| Sept | 95 |
| Dec | 90 |
| 1971 | |
| Mar | 110 |
| June | 105 |

**Prob. 26.5** Comparison Curves of Horsepower at the Rear Wheels (as shown by dynamometer tests).

| Engine rpm | Horsepower at Rear Wheels | | | |
|---|---|---|---|---|
| | McCulloch Supercharged with Dual Exhausts | McCulloch Supercharged | Unsupercharged —Dual Exhausts | Unsupercharged |
| 2000 | 77.0 | 73.0 | 64.5 | 59.5 |
| 2200 | 82.0 | 77.5 | 70.5 | 65.0 |
| 2400 | 88.0 | 83.5 | 75.0 | 69.0 |
| 2600 | 95.5 | 91.5 | 79.0 | 73.5 |
| 2800 | 105.0 | 99.0 | 82.5 | 76.5 |
| 3000 | 112.5 | 105.5 | 84.5 | 78.5 |
| 3200 | 117.0 | 109.5 | 85.5 | 79.5 |
| 3400 | 119.0 | 111.5 | 83.5 | 77.0 |
| 3600 | 118.5 | 111.5 | | |

**Prob. 26.6** Psychological Analysis of Work Efficiency and Fatigue.

| Hours of Work | Relative Production Index | |
|---|---|---|
| | Heavy Work | Light Work |
| 9–10 A.M. | 100 | 96 |
| 10–11 A.M. | 108 | 104 |
| 11–12 A.M. | 104 | 104 |
| 12–1 P.M. | 98 | 103 |
| Lunch | | |
| 2–3 P.M. | 103 | 100 |
| 3–4 P.M. | 99 | 102 |
| 4–5 P.M. | 98 | 101 |
| After 8-hour day | | |
| 5–6 P.M. | 91 | 94 |
| 6–7 P.M. | 86 | 93 |
| 7–8 P.M. | 68 | 83 |

**Prob. 26.7** Metal Hardness Comparison (Mohs' hardness scale).

| Metal | Comparative Degree of Hardness (Diamond = 10) |
|---|---|
| Lead | 1.5 |
| Tin | 1.8 |
| Cadmium | 2.0 |
| Zinc | 2.5 |
| Gold | 2.5 |
| Silver | 2.7 |
| Aluminum | 2.9 |
| Copper | 3.0 |
| Nickel | 3.5 |
| Platinum | 4.3 |
| Iron | 4.5 |
| Cobalt | 5.5 |
| Tungsten | 7.5 |
| Chromium | 9.0 |
| Diamond | 10.0 |

**Prob. 26.8** U.S. Personal Consumption Expenditures.

| Year | Billions of Dollars | | |
|---|---|---|---|
| | Durables | Nondurables | Services |
| 1960 | 45 | 150 | 130 |
| 1965 | 70 | 190 | 175 |
| 1970 | 87 | 265 | 265 |
| 1975 | 115[a] | 365[a] | 400[a] |
| 1980 | 140[a] | 480[a] | 575[a] |

[a] Projections

**Prob. 26.9** U.S. Defense Dept. (Pentagon) Use of Computers.

| Year | Computers (thousands) |
|---|---|
| 1967 | 2.35 |
| 1968 | 2.75 |
| 1969 | 2.90 |
| 1970 | 3.25 |

**Prob. 26.10** Growth Rate—Publications in Science of Mechanisms.

| Year | Number of Publications |
|---|---|
| 1900–1910 | 450 |
| 1911–1920 | 650 |
| 1921–1930 | 850 |
| 1931–1940 | 1350 |
| 1941–1950 | 2000 |
| 1951–1960 | 3000 |
| 1961–1970 | 4600 |

**Prob. 26.11** Profits and Presidents: How Manufacturers Have Fared.

| President/ Year | Net Income as Percent of Sales |
|---|---|
| Truman | |
| 1947 | 6.65 |
| 1948 | 6.95 |
| 1949 | 5.85 |
| 1950 | 7.00 |
| 1951 | 4.80 |
| 1952 | 4.25 |
| Eisenhower | |
| 1953 | 4.25 |
| 1954 | 4.45 |
| 1955 | 5.35 |
| 1956 | 5.25 |
| 1957 | 4.75 |
| 1958 | 4.25 |
| 1959 | 4.75 |
| 1960 | 4.45 |
| Kennedy–Johnson | |
| 1961 | 4.25 |
| 1962 | 4.50 |
| 1963 | 4.70 |
| 1964 | 5.20 |
| 1965 | 5.60 |
| 1966 | 5.60 |
| 1967 | 5.05 |
| 1968 | 5.10 |
| Nixon | |
| 1969 | 4.75 |
| 1970 | 4.00 |

**Prob. 26.12** The World Economy—Industrial Production.

| Year | Index 1958 = 100 | | | | | |
|---|---|---|---|---|---|---|
| | Britain | France | U.S. | West Germany | Italy | Japan |
| 1963 | 120 | 130 | 135 | 138 | 170 | 215 |
| 1964 | 130 | 140 | 143 | 149 | 171 | 247 |
| 1965 | 132 | 145 | 154 | 158 | 181 | 256 |
| 1966 | 135 | 155 | 167 | 160 | 201 | 290 |
| 1967 | 135 | 160 | 170 | 156 | 215 | 345 |
| 1968 | 143 | 165 | 177 | 175 | 230 | 407 |
| 1969 | 146 | 183 | 183 | 196 | 237 | 470 |
| 1970 | 149 | 195 | 180 | 210 | 255 | 545 |
| 1971 | 148 | 202 | 178 | 215 | 252 | 575 |

Plot on semilog shows rate of change of index.

**Prob. 26.13** Essential Qualities of a Successful Engineer (average estimate based on 1500 inquiries from practicing engineers).

| Quality | Percent |
|---|---|
| Character | 41 |
| Judgment | $17\frac{1}{2}$ |
| Efficiency | $14\frac{1}{2}$ |
| Understanding human nature | 14 |
| Technical knowledge | 13 |
| Total | 100 |

**Prob. 26.14** Automobile Accident Analysis.

| Type of Accident | Percent |
|---|---|
| Cross traffic (grade crossing, highway, railway) | 21 |
| Same Direction | 30 |
| Head-on | 21 |
| Fixed Object | 11 |
| Pedestrian | 10 |
| Misc. | 7 |
| Total | 100 |

**Prob. 26.15** U.S. Population Movement—1970.

| Area | Millions of People | Percent Growth 1960–1970 |
|---|---|---|
| Total U.S. | 204 | 13.7 |
| Total metropolitan | 136 | 15.2 |
| Central cities | 62 | 4.5 |
| Suburbs | 74 | 25.3 |
| Nonmetroplitan | 65 | 6.0 |

**Prob. 26.16** How the World Uses Its Manpower.

| Country | Percent of Workers | | | | |
|---|---|---|---|---|---|
| | Agriculture | Utilities and transport | Services[a] | Commerce | Manufacturing, Mining, Construction |
| U.S.A. | 13 | 8 | 21 | 23 | 35 |
| Canada | 19 | 10 | 20 | 16 | 35 |
| France | 28 | 7 | 15 | 15 | 35 |
| Japan | 41 | 5 | 13 | 17 | 24 |
| Italy | 42 | 6 | 12 | 9 | 31 |
| U.S.S.R. | 43 | 7 | 15 | 5 | 30 |
| Mexico | 60 | 3 | 11 | 9 | 17 |
| India | 70 | 3 | 11 | 4 | 12 |
| Honduras | 84 | 2 | 5 | 2 | 7 |

[a] Shoe repair, laundry, etc.

**26.26 Graph Problems**

Plot the data for the following problems on rectangular coordinate paper and on semilog paper.

**Prob. 26.17** Rupture Strength of T.D. Nickel—High-Temperature Alloy.

| Temperature (T), °F. | 100 hr Rupture Stress ($s_r$), psi × 1000 (*log scale*) |
|---|---|
| 1200 | 24 |
| 1400 | 17 |
| 1600 | 12.5 |
| 1800 | 9 |
| 2000 | 6.5 |
| 2200 | 4.75 |
| 2400 | 3.5 |

**Prob. 26.18** Nuisance Noise.

| Frequency (f), cps (*log scale*) | Octave Band Level, db | | | |
|---|---|---|---|---|
| | Hearing Loss Risk Region | | Power Lawn Mower at 3 ft | 5 hp Chainsaw at 3 ft |
| | Negligible | Serious | | |
| 53 | 104 | 122 | 84 | 93 |
| 106 | 93 | 113 | 93 | 103 |
| 220 | 87 | 107 | 94 | 103 |
| 425 | 85 | 105 | 90 | 111 |
| 850 | 85 | 105 | 84 | 112 |
| 1700 | 85 | 105 | 84 | 107 |
| 3400 | 85 | 105 | 82 | 104 |
| 6800 | 85 | 105 | 75 | 98 |

Plot the data for the following problems on rectangular coordinate paper and on log-log paper.

**Prob. 26.19** Loss of Head for Water Flowing in Iron Pipes.

| Velocity (v), ft/sec | Loss of Head, ft/1000 ft | | | |
|---|---|---|---|---|
| | 1″ Pipe | 2″ Pipe | 4″ Pipe | 6″ Pipe |
| 0 | 0 | 0 | 0 | 0 |
| 1 | 6.0 | 2.9 | 1.6 | 0.7 |
| 2 | 23.5 | 9.5 | 4.2 | 2.4 |
| 3 | 50 | 20 | 8.2 | 4.7 |
| 4 | | 34 | 13.5 | 7.7 |
| 5 | | 51 | 20 | 11.4 |
| 6 | | | 28 | 15.5 |
| 7 | | | 37 | 20 |

**Prob. 26.20** Material and Process Economics.

| Weight of Steel Forging (W), lb | Unit Cost (C) for Forging, $ | | |
|---|---|---|---|
| | Simple | Average | Complex |
| 0.1 | 0.22 | | |
| 0.2 | 0.38 | | |
| 0.4 | 0.67 | | |
| 1.0 | 1.4 | | |
| 2.0 | 2.5 | 2.75 | 4.5 |
| 4.0 | 4.4 | 5.0 | 8.5 |
| 10.0 | 9.2 | 11.0 | 19.5 |
| 20.0 | 16.0 | 20.0 | 37.0 |
| 40.0 | 28.5 | 36.5 | 70.0 |
| 100 | 60.0 | 80.0 | 165.0 |

Plot the data for the following problems on trilinear coordinate paper.

**Prob. 26.21** Phases in the Temperature Range for Maximum $\gamma$ (900° to 1300°C).

Top apex, right scale: Cr (weight percentage chromium) increases in value from bottom to top, 0 to 100%.

Left apex, left scale: Fe (weight percentage iron) increases in value from right toward left apex, 0 to 100%.

Right apex, bottom scale: Ni (weight percentage nickel) increases in value from left down toward right apex, 0 to 100%.

| $\alpha$-Ferrite (area above curve) | | |
|---|---|---|
| Ni, % wt | Cr, % wt | Fe, % wt |
| 36 | 64 | 0 |
| 32 | 58 | 10 |
| 27 | 53 | 20 |
| 22 | 48 | 30 |
| 17 | 43 | 40 |
| 13 | 37 | 50 |
| 8 | 32 | 60 |
| 5 | 25 | 70 |
| 2 | 18 | 80 |
| 0 | 12 | 88 |

Projected

| $\gamma$-Austenite (below curve) | | |
|---|---|---|
| Fe, % wt | Ni, % wt | Cr, % wt |
| 0 | 54 | 46 |
| 10 | 47 | 43 |
| 20 | 41 | 39 |
| 30 | 33 | 37 |
| 40 | 26 | 34 |
| 50 | 21 | 29 |
| 60 | 14 | 26 |
| 70 | 8 | 22 |
| 80 | 3 | 17 |
| 88 | 0 | 12 |

$\alpha + \gamma$ between two curves.

**Prob. 26.22** Carbon-Iron-Manganese at 800°C.

Top apex, left scale: Mn (weight percentage manganese) increases in value from bottom to top, 0 to 20% (ten divisions—equilateral triangle).

Left apex, no scale: Fe (iron)

Right apex, bottom scale: C (weight percentage carbon) increases in value from left toward right, 0 to 2.0% (ten divisions—equilateral triangle).

Plot the following data on polar coordinate paper. Use the upper vertical axis as the zero axis. The data for 0° to 180° are identical for both sides of a 360° plot.

| $\alpha$ = Solid Solution of Carbon and Manganese in $\alpha$-Iron (area to left of curve) | |
|---|---|
| Mn, % wt | C, % wt |
| 0.00 | 0.04 |
| 2.0 | 0.0 |

| $\alpha + \gamma$ (area to left of curve) | |
|---|---|
| Mn, % wt | C, % wt |
| 0.0 | 0.5 |
| 2.00 | 0.26 |
| 4.0 | 0.0 |

| $\gamma$ = Solid Solution of Carbon and Manganese in $\gamma$-Iron | |
|---|---|
| Mn, % wt | C, % wt |
| 0.00 | 1.02 |
| 2.00 | 0.9 |
| 4.00 | 0.82 |
| 6.00 | 0.74 |
| 8.00 | 0.68 |
| 10.00 | 0.65 |
| 12.00 | 0.68 |
| 13.00 | 0.71 |

$\gamma$ + C area to right of curve; C is carbide.

**Prob. 26.23** Light Distribution in a Vertical Plane for a Bulb Suspended from the Ceiling with the Filament at the Origin of the Polar Chart.

| Orientation, degrees | Candle Power |
|---|---|
| 0 | 140 |
| 10 | 210 |
| 20 | 310 |
| 30 | 320 |
| 40 | 310 |
| 50 | 310 |
| 60 | 300 |
| 70 | 290 |
| 80 | 280 |
| 90 | 250 |
| 100 | 270 |
| 110 | 290 |
| 120 | 295 |
| 130 | 300 |
| 140 | 315 |
| 150 | 330 |
| 160 | 340 |
| 170 | 350 |
| 180 | 340 |

**Prob. 26.24** Variation of Modulus of Elasticity (E) with Direction in Copper.

| Orientation, degrees | E × $10^6$ psi | |
|---|---|---|
| | As Rolled | Annealed |
| 0 | 20.0 | 9.5 |
| 15 | 18.5 | 11.0 |
| 30 | 16.5 | 14.5 |
| 45 | 15.0 | 17.5 |
| 60 | 16.5 | 14.5 |
| 75 | 18.5 | 11.0 |
| 90 | 20.0 | 9.5 |
| 105 | 18.5 | 11.0 |
| 120 | 16.5 | 14.5 |
| 135 | 15.0 | 17.5 |
| 150 | 16.5 | 14.5 |
| 165 | 18.5 | 11.0 |
| 180 | 20.0 | 9.5 |

Direction of rolling is from 0° to 180°.

**Prob. 26.25** Pressurized Square Tubing, Fig. 26.41. Data are given for one quadrant; quadrants are identical.

| Orientation, degrees | Stress Ratio ($\sigma_1/P$) | |
|---|---|---|
| | D/a = 0.80 | D/a = 0.86 |
| 0 | 2.0 | 2.6 |
| 15 | 3.2 | 4.3 |
| 30 | 4.2 | 6.3 |
| 45 | 4.3 | 5.0 |
| 60 | 4.2 | 6.3 |
| 75 | 3.2 | 4.3 |
| 90 | 2.0 | 2.6 |

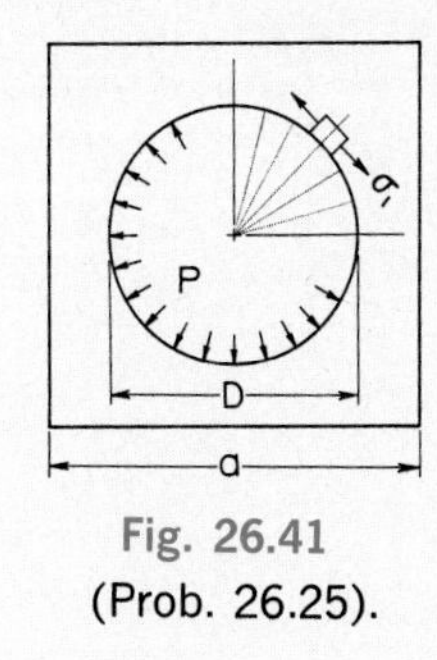


Fig. 26.41
(Prob. 26.25).

**Prob. 26.26** Main Bearing Load Diagram (4000 rpm, no counterweight).

| Orientation, degrees | Load, 1000 lb | Orientation, degrees | Load, 1000 lb |
|---|---|---|---|
| 0 | 6.6 | 190 | 5.3 |
| 10 | 4.8 | 200 | 5.2 |
| 20 | 4.2 | 210 | 4.9 |
| 30 | 3.7 | 220 | 4.5 |
| 40 | 3.2 | 230 | 4.2 |
| 50 | 3.0 | 240 | 3.7 |
| 60 | 2.8 | 250 | 3.3 |
| 70 | 2.7 | 260 | 3.1 |
| 80 | 2.7 | 270 | 2.9 |
| 90 | 2.8 | 280 | 2.8 |
| 100 | 2.9 | 290 | 2.7 |
| 110 | 3.3 | 300 | 2.8 |
| 120 | 3.6 | 310 | 3.0 |
| 130 | 4.1 | 320 | 3.2 |
| 140 | 4.6 | 330 | 3.7 |
| 150 | 4.9 | 340 | 4.4 |
| 160 | 5.0 | 350 | 5.3 |
| 170 | 5.2 | 360 | 6.6 |
| 180 | 5.3 | | |

Max. load = 6600 lb; mean load = 4530 lb

# 27

# alignment charts

by E. J. Mysiak*

**27.1 Introduction.** The engineer, in performing his design work, uses mathematics for calculations with equations, for deriving equations, and when solving particular solutions of equations. Graphical solutions of equations are performed with the use of alignment charts and coordinate axes graphs. Alignment charts are explained in the subsequent sections of this chapter. The derivation of equations is covered in Chapter 28, "Empirical Equations." Particular solutions of equations are explained in Chapter 29 under the topics of Rectangular Coordinate Algebraic Solution Graphs and the Graphical Calculus, §§29.2 and 29.3.

**27.2 Nomographs or Alignment Charts.** A nomograph is a diagram or a combination of diagrams representing a mathematical law or equation. The Greek roots *nomos* (law) and *graphein* (to write) suggest this definition. The rectangular coordinate graphs discussed in the preceding chapter are the most common examples, showing graphically the relationship of two or more variables and their function. The term "nomograph," however, is more popularly applied to a combination of scales, arranged properly to represent mathematical laws (equations) for computational purposes. Some of the more common forms of nomographs are shown in Fig. 27.1.

Basically, a nomograph is used to solve a three-variable equation. A straight line (*isopleth*) joining known or given values of two of the variables intersects the scale of the third variable at a value that satisfies the equation represented, as at (a), (c), (f), (g), and (h). For this reason they are also called *alignment charts.* Alignment charts are nomographs; however, nomographs are not necessarily alignment charts. A rectangular coordinate graph can be

* Mechanical Engineer, International Electro-Magnetics, Inc.

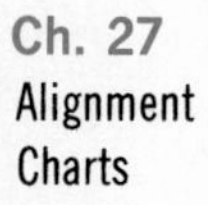

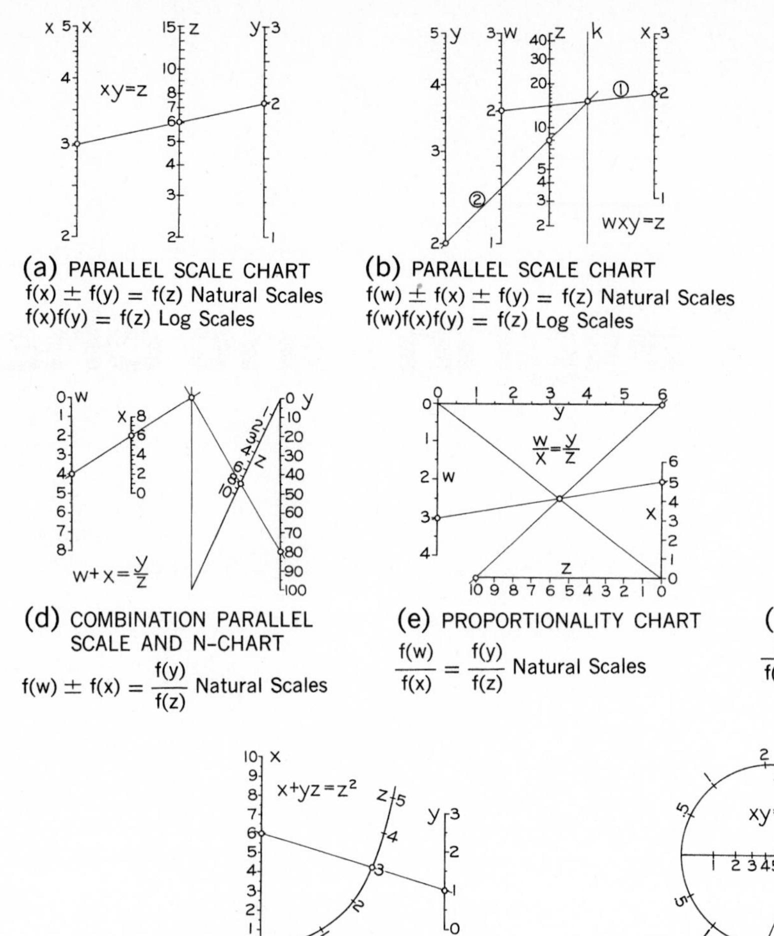


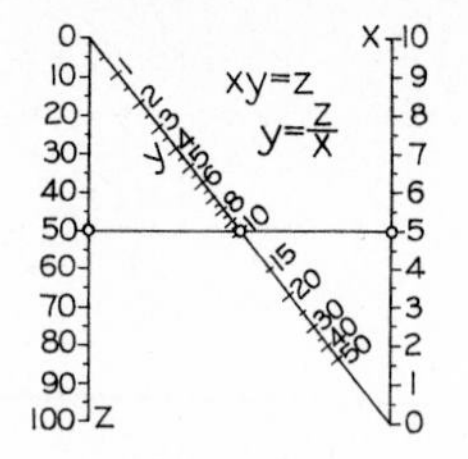


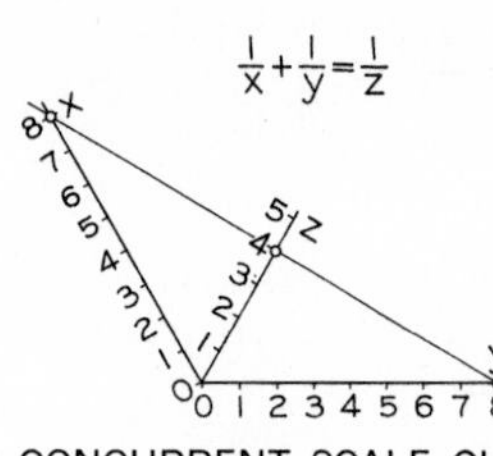

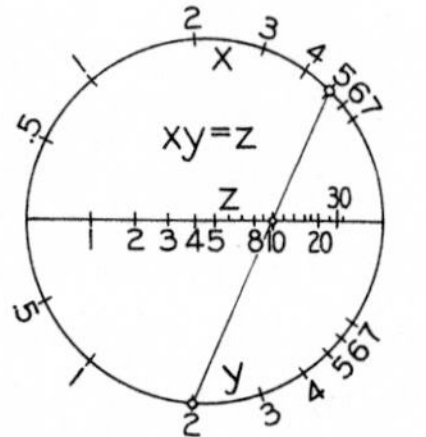


(a) PARALLEL SCALE CHART
$f(x) \pm f(y) = f(z)$ Natural Scales
$f(x)f(y) = f(z)$ Log Scales

(b) PARALLEL SCALE CHART
$f(w) \pm f(x) \pm f(y) = f(z)$ Natural Scales
$f(w)f(x)f(y) = f(z)$ Log Scales

(c) N-CHART
$f(x)f(y) = f(z)$ Natural Scales
$f(x) = \frac{f(z)}{f(y)}$ Natural Scales

(d) COMBINATION PARALLEL SCALE AND N-CHART
$f(w) \pm f(x) = \frac{f(y)}{f(z)}$ Natural Scales

(e) PROPORTIONALITY CHART
$\frac{f(w)}{f(x)} = \frac{f(y)}{f(z)}$ Natural Scales

(f) CONCURRENT SCALE CHART
$\frac{1}{f(x)} + \frac{1}{f(y)} = \frac{1}{f(z)}$ Natural Scales

(g) CHART WITH ONE CURVED SCALE
$f(x) \pm f(y)f_1(z) = f_2(z)$ Natural Scales

(h) CIRCULAR NOMOGRAM
$f(x)f(y) = f(z)$

Fig. 27.1 **Common Forms of Nomographs or Alignment Charts.**

classified as a nomograph. Two or more such charts can sometimes be combined to solve an equation containing more than three variables, as at (b), (d), and (e). The forms shown have fixed scales and a movable alignment line. However, movable-scale nomographs can be designed with a fixed direction alignment line, the slide rule being an example of this form.

Although alignment charts require time to construct, they have considerable popular appeal for the following reasons:

1. They save time when it is necessary to make repeated calculations of certain numerical relationships (equations).
2. They enable one unskilled or lacking a background in mathematics to handle analytical solutions.
3. When constructed properly, they are limited to the scale values for which the equation is valid. This prevents the use of the equation with values which are not applicable.

The design of alignment charts as described in the subsequent articles is accomplished by the use of plane geometry. For the design of alignment charts by means of determinants (matrix algebra), more advanced textbooks should be consulted.

## 27.3 Functional Scales.

A mathematical equation expresses the relationship of a group of *variables* and *constants*. For example, the equation

$$y = x^2 + 2x + 3$$

contains two variables, x and y, which are letter symbols designating quantities that may have several values in the equation. The variable x is expressed as the *function* $x^2 + 2x$; that is, the function of x is $f(x) = x^2 + 2x$. The function of y is $f(y) = y$. Other typical expressions of functions of a variable are $\frac{1}{x}$, log x, sin x, $x^2 + \frac{3}{x^3}$, and $x - 1$.

A *constant* is any quantity that always has the same value, such as the number 3 in the equation. The number 2 of 2x is a *constant coefficient.*

The common forms of nomographs are composed of scales, each representing only one variable; hence, scales are designed for each individual variable. To draw a scale to a certain length L, representing the function of a variable between definite limits, the difference between the extreme values is multiplied by a proportionality factor called the *functional modulus,* m.

$$L = m[f(x)_{max} - f(x)_{min}]$$
$$= m[f(x_2) - f(x_1)]$$

where $x_2$ and $x_1$ represent the values of the variable x corresponding respectively to the maximum and minimum values of the function, Fig. 27.2.

In this relationship a convenient modulus could be chosen to simplify construction of the scale and consequently determine the length, which is a preferred procedure by those experienced in scale layout.

It is sometimes difficult to choose a functional modulus that will result in a scale length long enough or short enough to fit the paper size to be used. Rearranging the equation to

$$m = \frac{L}{f(x_2) - f(x_1)}$$

the functional modulus can be defined as the length on the scale for a unit value of the function variable. The equation in this form with the determined range of values, $f(x_2)$ and $f(x_1)$, and an assumed scale length, L, may result in an inconvenient functional modulus, m. Since a certain convenient scale length is seldom necessary, the resulting functional modulus, m, can then be rounded off to a convenient value for scale layout, as explained in §27.6. The rounded-off functional modulus will result in a new scale length, which should be close enough to the desired length.

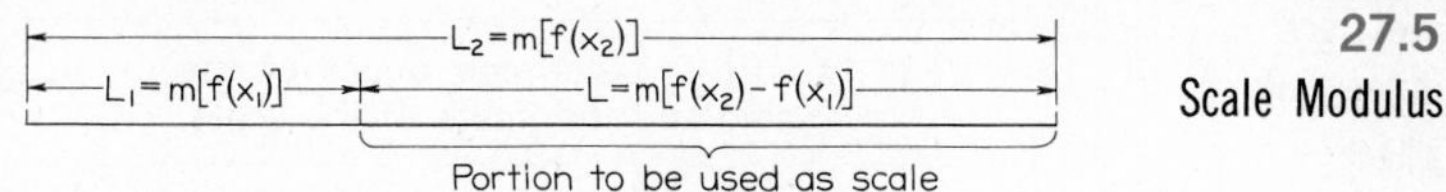


Fig. 27.2 **Functional Scale Relationship.**

If any of the values for the function are negative, they must be treated algebraically as negative values. For example, if $f(x_1)$ is negative,

$$L = m[f(x_2) - f(-x_1)]$$
$$= m[f(x_2) + f(x_1)]$$

## 27.4 Scale Layout.

As an example of functional scale construction, the equation of §27.3 is used, and the typical procedure for scale layout is illustrated in Fig. 27.3 for the variable x, $f(x) = x^2 + 2x + 3$. The consequence of including the constant in the function of the variable is to add $\frac{3}{8}''$ to the scale, which is not used. Therefore, the resulting scale in effect is for the terms with variables only, $f(x) = x^2 + 2x$.

## 27.5 Scale Modulus.

When each term of a variable in a function contains a constant coefficient, it may be more convenient to plot values from the function after combining the constant coefficient with the functional modulus. For example, if $f(x) = 2x^2 + 2x$, with x ranging in value from .5 to 5.0,

$$L = m[f(x_2) - f(x_1)]$$
$$= m[2(25) + 2(5)] - [2(.25) + 2(.5)]$$
$$= m[60 - 1.5] = 58.5m$$

If the length $L = 10''$ is chosen, then $10'' = 58.5m$ and $m = .171$. The scale could

| | | | | | | | |
|---|---|---|---|---|---|---|---|
| STEP 1. Assume or determine values of the variable and the limits of the scale. | x | 0 | 1 | 2 | 3 | 4 | 5 |
| STEP 2. Compute the values of the function, substituting the values of step 1. | $f(x) = x^2 + 2x + 3$ | 3 | 6 | 11 | 18 | 27 | 38 |
| STEP 3. Choose a convenient functional modulus and determine total length of the scale. | $L = m[f(x_2) - f(x_1)]$<br>If $m = \frac{1}{8}$, $L = \frac{1}{8}[(x_2^2 + 2x_2 + 3) - (x_1^2 + 2x_1 + 3)]$<br>$= \frac{1}{8}[(5^2 + 2(5) + 3) - (3)] = \frac{35}{8} = 4.375''$ | | | | | | |
| STEP 4. If scale length is suitable, multiply functional modulus by the function values. These are the measuring lengths for the graduations on the scale for the values of the function. | $\left(m = \frac{1}{8}\right)$<br>$mf(x) =$ | $\frac{3}{8} =$ .375 | $\frac{6}{8} =$ .750 | $\frac{11}{8} =$ 1.375 | $\frac{18}{8} =$ 2.250 | $\frac{27}{8} =$ 3.375 | $\frac{38}{8} =$ 4.750 |
| STEP 5. Lay off the distances obtained along a chosen line. | 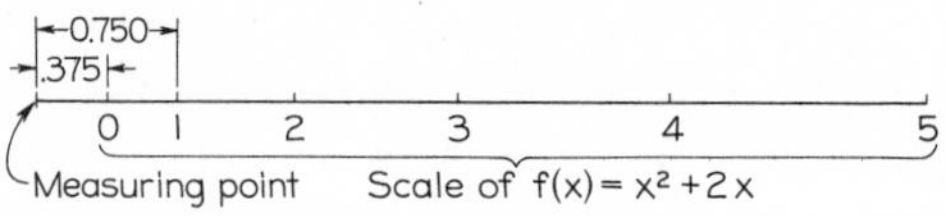 0.750<br>.375 | | | | | | |
| STEP 6. Denote the values of the variable (not the function) at the corresponding points determined. | 0 1 2 3 4 5<br>Measuring point<br>Scale of $f(x) = x^2 + 2x$ | | | | | | |

Fig. 27.3 **Functional Scale Layout and Design.**

be prepared from tabulated values for $2x^2 + 2x$. A more convenient method is to use the equation as

$$L = 2m[(x_2^2 + x_2) - (x_1^2 + x_1)].$$

A common factor has been withdrawn from the coefficients and combined with the functional modulus. The product of the functional modulus and the constant coefficient is called a *scale modulus* and is commonly designated by a capital letter M. The scale modulus would be $M = 2m = .342$, and the function of the variable used would be $x^2 + x$.

A further advantage in preparing this scale would be to choose a more convenient scale modulus, resulting in a length close to the length of 10″ desired. If $M = .333 = \frac{1}{3}$ is chosen, then

$$L = .333[(25 + 5) - (.25 + .5)] = 9.74''$$

Since $M = \frac{1}{3} = \frac{10}{30}$, the 30 scale on an engineers scale could be used, §27.6.

## 27.6 Engineers Scale.

The use of an engineers scale, where appropriate, eliminates the tedious operation of multiplying each functional value by the scale modulus to obtain the measurement of the graduations. The scale graduations are laid out directly, using the function values on the engineers scale. The functional or scale moduli are determined only to denote which scale to use and to determine the total scale length. If a definite scale length is not required, it is best to choose a scale modulus of 1, $\frac{1}{2}$, $\frac{1}{3}$, $\frac{1}{4}$, $\frac{1}{5}$, or $\frac{1}{6}$, or a multiple of these such as $\frac{10}{3}$, $\frac{1}{30}$, and $\frac{10}{30}$, that would permit the use of the 10, 20, 30, 40, 50, or 60 scales on the engineers scale, §2.27.

The numbers 10, 20, 30, 40, and so on represent the number of subdivisions per inch on the scale. For example, the 30 scale, Fig. 27.4 (a), has one inch divided into 30 parts. The number 3 *on the engineers scale* can be taken to represent .3, 3, 30, or 300, depending upon the scale modulus used: $\frac{10}{3}$, $\frac{1}{3}$, $\frac{1}{30}$, or $\frac{1}{300}$, respectively. At (b) is shown the equation of §27.5. Since $M = \frac{1}{3}$, the figure at 1″ represents 3 as a value of the function $(x^2 + x)$. Number 5 (a value of the variable x) would be at 30 on the engineers scale, number 2 would be at 6, and so on. Notice that the function values are measured directly from the engineers scale onto the scale line. When an odd scale modulus or length is necessary (occurs frequently

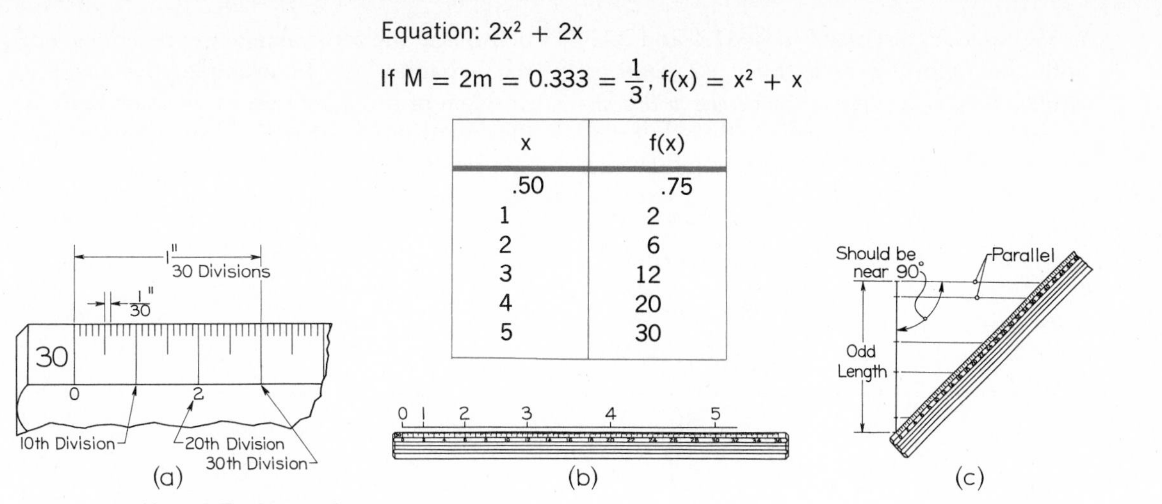


| x | f(x) |
|---|---|
| .50 | .75 |
| 1 | 2 |
| 2 | 6 |
| 3 | 12 |
| 4 | 20 |
| 5 | 30 |

Fig. 27.4 **Use of Engineers Scale.**

for the middle scale of a three-scale alignment chart), the odd length can be laid off by the parallel-line method, (c). In this case the functional or scale moduli are not used for scale calibration. The engineers scale is used to calibrate a line at an angle to the scale line, and it is then projected onto the scale line within the determined scale length. For accuracy in projection, the parallel lines should be as near 90° as possible to the scale line. In the example of §27.5, $f(x) = x^2 + x$, $M = 2m = .342$, $L = 10''$; the scale can be laid out for a length of 10″. The scale modulus for the final scale can still be .342. On a line at an angle to the final scale line, the 30 engineers scale ($M = .333$), or any other appropriate scale, can be used to calibrate a scale with the functional values of .75, 2, 6, 12, 20, and 30 units, which is then projected onto the final 10″ long scale line.

## 27.7 Approximate Subdivision of Nonuniform Spaces.

After the major graduations are located along the scale, the scale should be subdivided into smaller divisions. If the spaces between the major divisions are nonuniform, an approximate method of subdivision is to use a *sector*, constructed on a separate sheet of translucent paper, tracing cloth, or film, as shown in Fig. 27.5. In use, the sector is fitted to three previously located points on the scale, such as 3, 3.5, and 4 in the example. The

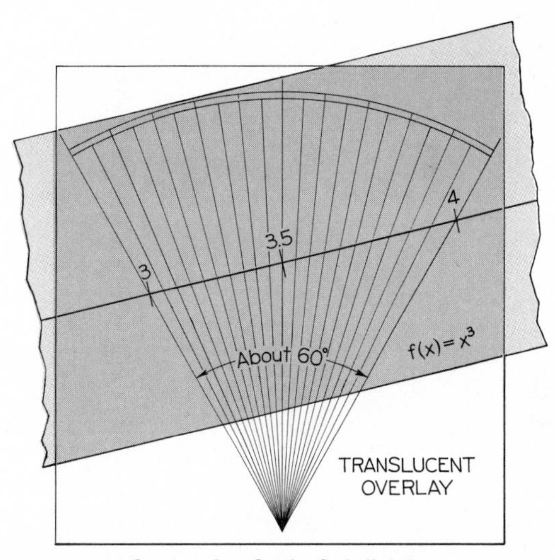


Fig. 27.5 **Sector for Scale Subdivision.**

subdivisions are located on the scale by piercing through the sector paper with a compass or divider needle point. If the scale is drawn on a transparent medium, the scale may be placed over the sector chart.

## 27.8 Conversion Scales.

An equation with two variables can be arranged to equate the function of one variable to the function of the second variable. If functional scales are designed for each function of the two related variables and placed adjacent to one another, the resulting chart is known as a *conversion scale*. The procedure for designing each scale is the same as discussed in §§27.4 and 27.5, with the following exceptions: *the functional modulus must be the same for both scales; both scales are laid out from the same origin,* although it may not appear on the scale. The origin (measuring point) is at the zero value of the functions, not the variables. As an example, the classical problem of a temperature conversion scale is illustrated in Fig. 27.6. The equation for converting °C to °F is

**Fig. 27.6 Design and Layout of Temperature Conversion Chart.**

| Variable °C | Function $1.8°C + 32$ | M × Function $\frac{1}{50}(1.8°C + 32)$ |
|---|---|---|
| −40 | −40 | $-\frac{40}{50}$ |
| −30 | −22 | $-\frac{22}{50}$ |
| −20 | − 4 | $-\frac{4}{50}$ |
| −10 | 14 | $\frac{14}{50}$ |
| 0 | 32 | $\frac{32}{50}$ |
| 10 | 50 | $\frac{50}{50}$ |
| 20 | 68 | $\frac{68}{50}$ |
| 30 | 86 | $\frac{86}{50}$ |
| 40 | 104 | $\frac{104}{50}$ |
| 50 | 122 | $\frac{122}{50}$ |
| 60 | 140 | $\frac{140}{50}$ |
| 70 | 158 | $\frac{158}{50}$ |
| 80 | 176 | $\frac{176}{50}$ |
| 90 | 194 | $\frac{194}{50}$ |
| 100 | 212 | $\frac{212}{50}$ |

| Variable °F | Function °F | M × Function $\frac{1}{50}(°F)$ |
|---|---|---|
| −40 | −40 | $-\frac{40}{50}$ |
| −20 | −20 | $-\frac{20}{50}$ |
| 0 | 0 | $-\frac{0}{50}$ |
| 20 | 20 | $\frac{20}{50}$ |
| 40 | 40 | $\frac{40}{50}$ |
| 60 | 60 | $\frac{60}{50}$ |
| 80 | 80 | $\frac{80}{50}$ |
| 100 | 100 | $\frac{100}{50}$ |
| 120 | 120 | $\frac{120}{50}$ |
| 140 | 140 | $\frac{140}{50}$ |
| 160 | 160 | $\frac{160}{50}$ |
| 180 | 180 | $\frac{180}{50}$ |
| 200 | 200 | $\frac{200}{50}$ |
| 212 | 212 | $\frac{212}{50}$ |

If $M = m = \frac{1}{50}$, $L_{°C} = \frac{1}{50}[212 - (-40)] = \frac{1}{50}(252) = 5.04''$

When °C = −40°, °F = 1.8(−40) + 32 = −40°F; when °C = 100°, °F = 1.8(100) + 32 = 212°F

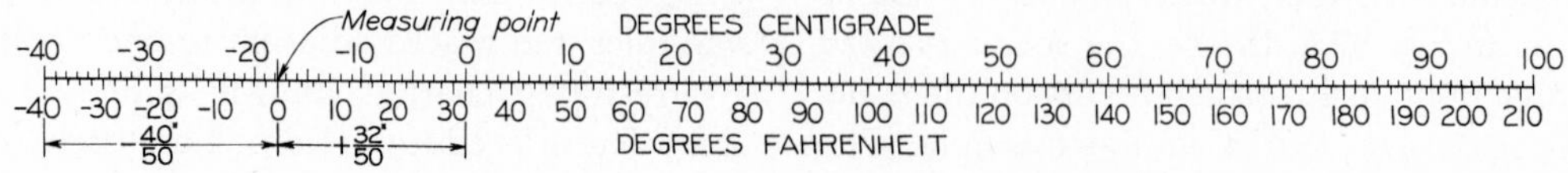

Measuring point
DEGREES CENTIGRADE
-40 -30 -20 -10 0 10 20 30 40 50 60 70 80 90 100
-40 -30 -20 -10 0 10 20 30 40 50 60 70 80 90 100 110 120 130 140 150 160 170 180 190 200 210
$-\frac{40''}{50}$ $+\frac{32''}{50}$
DEGREES FAHRENHEIT

$$°F = \tfrac{9}{5}°C + 32 = 1.8°C + 32$$

The first step is to separate the two variables, expressing their functions individually. Thus, $f(°F) = °F$; $f(°C) = 1.8°C + 32$. If °C is to range from −40°C to 100°C, the range of °F must be determined from the equation.

After tabulation of an appropriate distribution of intermediate values and the corresponding computed function values, the scale modulus multiplied by the function need not be determined if a convenient modulus is chosen for the use of the engineers scale. If the centigrade scale is the first scale laid out, the measuring point can be located $\frac{40}{50}''$ from the left end. This point can be used to locate the remaining calibrations; negative values to the left, positive values to the right. Notice the measuring point (function value origin) is at −17.8°C and 0°F after the variable values have been added. For this particular conversion scale, the origin (zero value) for the function and the °F variable value are the same point.

Another convenient form of conversion scale is the series illustrated in Fig. 27.7. The scales are parallel, but not adjacent, and a horizontal line must be drawn at right angles to the scales to determine the various functions of x. Thus, if from 6 on the scale x, a

Fig. 27.7 **Conversion Scales.**

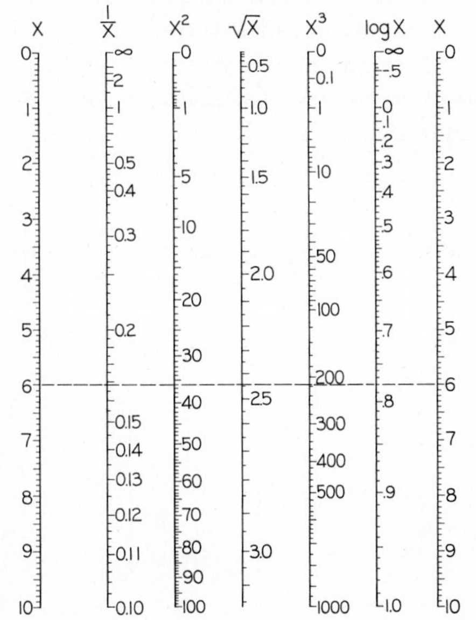

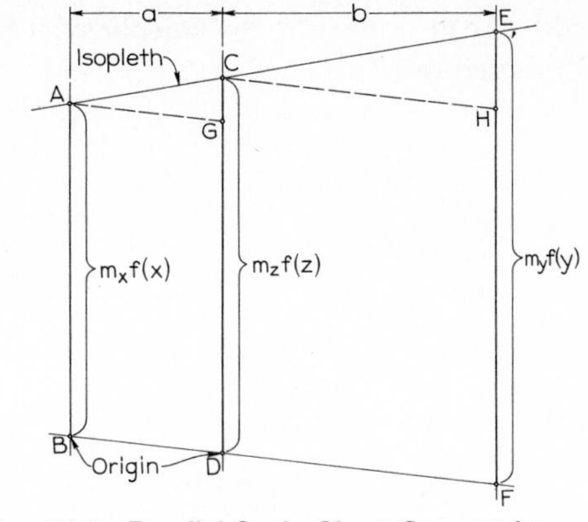


Fig. 27.8 **Parallel-Scale Chart Geometric Derivation.**

horizontal line is drawn across to scale $x^2$, the value 36 is found. These scales are called *natural (functional) scales.* If the function is the log of a variable, it is a natural scale, but if it is a function laid out logarithmically, it is called a log scale, §27.11. In Fig. 27.7 only the first and the last scale are uniformly divided; the intermediate scales are nonuniform.

## 27.9 Natural Parallel-Scale Nomographs, $f(x) \pm f(y) = f(z)$.

The construction of nomographs for equations with three variables, involving the sum or difference of two variables, is based upon the following geometric derivation.

In Fig. 27.8, AB, CD, and EF represent three parallel functional scales. The origins are aligned at any angle, and an *isopleth* (index line) crosses the three scales in any direction. Lines AG and CH are drawn parallel to BF.

From similar triangles,

$$\frac{CD - GD}{a} = \frac{EF - HF}{b}$$

Correspondingly,

$$\frac{m_zf(z) - m_xf(x)}{a} = \frac{m_yf(y) - m_zf(z)}{b}$$

Collecting terms,

$$m_xf(x) + \frac{a}{b}m_yf(y) = \left(1 + \frac{a}{b}\right)m_zf(z)$$

If f(x) ± f(y) = f(z), the coefficients must be equal; therefore,

$$m_x = \frac{a}{b}m_y = \left(1 + \frac{a}{b}\right)m_z$$

or
$$\frac{m_x}{m_y} = \frac{a}{b} \qquad (1)$$

Substituting equation (1) into the coefficient of f(z),

$$m_x = \left(1 + \frac{m_x}{m_y}\right)m_z$$

or
$$m_z = \frac{m_x m_y}{m_x + m_y} \qquad (2)$$

Equation (1) is the relationship for spacing the outside scales after choosing moduli for these scales. Equation (2) is used to determine the modulus of the middle scale. These two relationships are used on all parallel-scale nomographs. Note that the two variables on one side of the equality sign, f(x) and f(y), are the outside scales.

A few principles applicable to all parallel-scale charts are illustrated in Fig. 27.9. When positive values are laid out in one direction, negative values must be laid out in the opposite direction, (b). Note that a function with a negative sign has an inverted scale and the origins are in alignment. Another approach would be to rearrange this relationship to be x = z + y, x being the middle scale. In this case, no scale would be inverted, which is preferred.

As mentioned previously, changing the alignment of origins to any angle, including horizontal, does not affect the equation.

## 27.10 Design and Layout of Three-Variable Parallel-Scale Nomographs.

The example to be used is the equation for the outside diameter of a 20° stub-tooth gear,

$$d_0 = d + 2a$$

where $d_0$ = outside diameter of pinion, inches; d = pitch diameter, 3.000″ to .250″ (for diametral pitches 4 to 54); and a = addendum, .250″ to .0393″ (for diametral pitches 4 to 54, respectively).

The first step is to determine or assume the range of values (maximum and minimum) for two of the functions of the equation. The maximum and the minimum values of the third function can then be determined. The computation must be performed, since the layout of the alignment chart is based on the alignment of the three scales within these specific maximum and minimum values.

The next step is to separate the three functions and determine an appropriate length (for $8\frac{1}{2}$″ × 11″ paper size) and functional moduli for the two outside scales. See upper portion of Fig. 27.10 (a). The variables d and a were selected for the outside scales, since they are positive value scales. If $d_0$ was selected as an outside scale, one of the other variables would become negative and inverted on the alignment chart ($d_0 - d = 2a$ or $d_0 - 2a = d$). Remember: the two variables on one side of the equality sign are the outside scales.

**Fig. 27.9 Parallel-Scale Variations.**

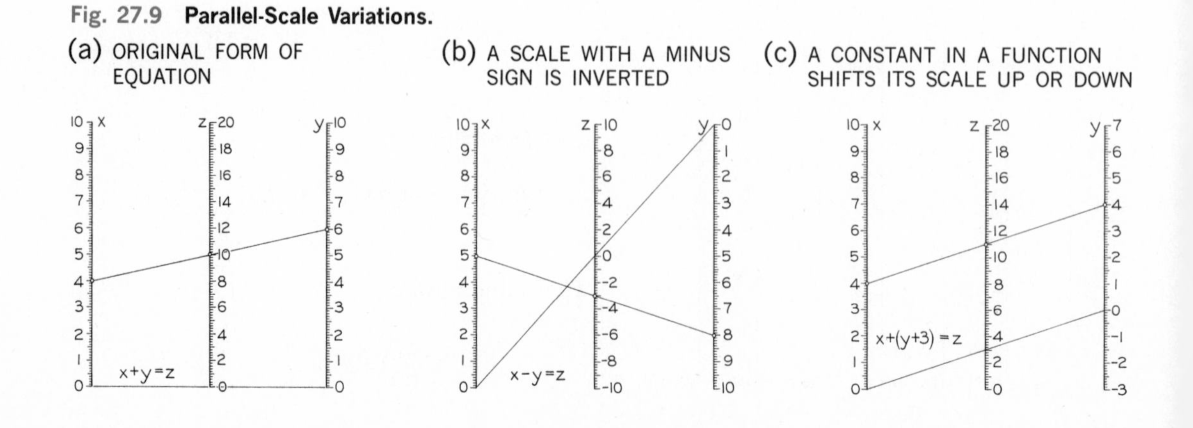

| Equation | Var. | Funct. | Variable Range | Funct. Modulus (m) | Scale Modulus (M) | Scale Length | Distance from Center Scale | Dir. | Scale Used |
|---|---|---|---|---|---|---|---|---|---|
| $d_0 = d + 2a$ | d | d | 3.00″–.250″ | $3 = \frac{90}{30}$ | $1 \times \frac{90}{30} = \frac{90}{30}$ | 8.250″ | .60″ | ↑ | 30 |
| | a | 2a | .250″–.0393″ | $20 = \frac{1000}{50}$ | $2 \times \frac{1000}{50} = \frac{2000}{50}$ | 8.428″ | 4.00″ | ↑ | 50 |
| | $d_0$ | $d_0$ | 3.50″–.329″ | 2.61 | $1 \times 2.61 = 2.61$ | 8.277″ | — | ↑ | 40 |

(a) DESIGN

| d | f(d) = d | a | f(a) = a | $d_0$ | $f(d_0) = d_0$ |
|---|---|---|---|---|---|
| .25 | .25 | .039 | .039 | .329 | .329 |
| .50 | .50 | .05 | .05 | .50 | .50 |
| 1.00 | 1.00 | .10 | .10 | 1.0 | 1.0 |
| 1.50 | 1.50 | .15 | .15 | 1.5 | 1.5 |
| 2.00 | 2.00 | .20 | .20 | 2.0 | 2.0 |
| 2.50 | 2.50 | .25 | .25 | 2.5 | 2.5 |
| 3.00 | 3.00 | | | 3.0 | 3.0 |
| | | | | 3.5 | 3.5 |

(b) DATA

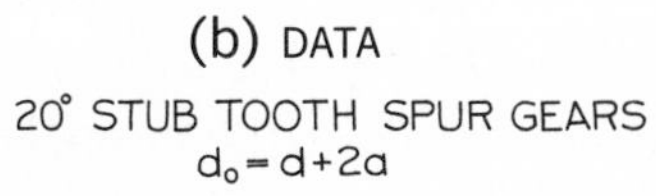


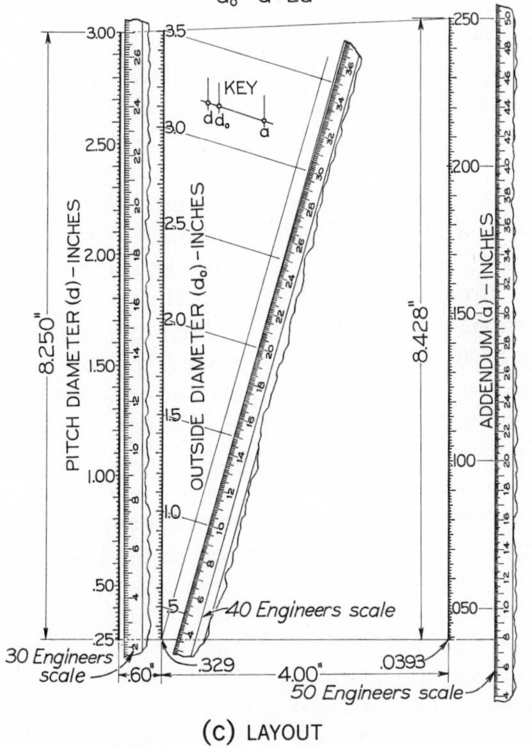


(c) LAYOUT

**Fig. 27.10 Three-Variable Parallel-Scale Nomograph.**

If $m_d = 3$,

$$L_d = m_d(d_2 - d_1)$$
$$= 3(3.00 - .250) = 8.250''$$

If $m_a = 20$,

$$L_a = m_a(a_2 - a_1)$$
$$= 20(2)(.250 - .0393) = 8.428''$$

Correspondingly, the spacing of the outside scales is

$$\frac{a}{b} = \frac{m_d}{m_a} = \frac{3}{20} = \frac{.60}{4.00}$$

The functional modulus of the middle scale is

$$m_{d_0} = \frac{m_d m_a}{m_d + m_a} = \frac{(3)(20)}{3 + 20} = 2.61$$

The range of $d_0$ is found by substituting the maximum and minimum values into the equation.

$$L_{d_0} = m_{d_0}(d_{0_2} - d_{0_1})$$
$$= 2.61(3.500 - .329) = 8.277''$$

The scale modulus is found by multiplying the functional modulus by the constant coefficient of each variable function.

Note that *functional moduli are used to determine spacing and length of scales.* A convenient scale modulus, however, aids in the layout of the scale.

It is not necessary to compute the length of the middle scale, $v_f$, but its computation is desirable, since it serves as a check on the preceding work.

To construct the chart, three parallel lines are drawn, spaced proportionate to .60 :: 4.00, about 9″ to 10″ long, Fig. 27.10. *Caution:* One of the most frequently made mistakes is erroneous placing of the outside scales because of interchanging the proportionate distances. Since all the terms in the equation are positive, the three scales increase in value in the same direction, normally upward, as shown here. In order to align the scale origins, draw a base line across the three scales. Usually the base line is drawn in a horizontal direction for convenience. With the 30 engineers scale, lay out the d scale, with 1″ representing 30. The major values of .50, 1.00, 1.50, and 2.00 are located by multiplying these values by 9 (since the scale modulus is 90/30) and measuring them at 4.50, 9.00, 13.50, and 18.00, respectively. The use of the engineers scale eliminates the need for multiplying each functional value by the functional modulus. The a scale is laid out in a similar manner by using the 50 scale. Since the constant coefficient was accounted for in the functional modulus, it is not necessary to multiply each value by 2 or the functional modulus.

After the two outside scales are laid out for their entire range, a line joining their top ends determines the correct height of the middle scale. The calculated length of 8.277″ should check with the layout, if outside scales were laid out correctly. The major scale divisions of the middle scale could be located by multiplying each functional value by the functional modulus and measuring this distance with a regular scale from the origin point. Since the modulus is odd, however, a more convenient method is to use a slightly larger engineers scale (40 or 50) and locate the major divisions by the parallel-line method, as shown. From the equation, the values of $d_0$ range from .329 to 3.500, and their functional values are the same, since $f(d_0) = d_0$. With the 40 scale placed with 3.29 at the origin (1″ = 4), the major values are set off with the engineers scale. The limit values of the scale designations need not be exactly .329 and 3.500. The scale can be extended to round figures of .250″ and 3.750″, but the values of .329 and 3.500 must lie on the base line and top isopleth, respectively.

After the scales have been properly arranged, the subdivisions can be located with the aid of a sector chart, Fig. 27.5. Addition of title, key, and scale captions completes the chart.

As a final check, two or three arithmetic computations should be compared to the same data on the alignment chart.

For a second illustration, the example to be used is the equation for the velocity of a free-falling body,

$$v_f^2 = v_0^2 + 2gs$$

where $v_f$ = final velocity, ft/sec;
$v_0$ = original velocity, 10 to 50 ft/sec;

$g$ = constant acceleration due to gravity, 32.2 ft/sec$^2$; and $s$ = displacement of body, 5′ to 50′.

If $m_{v_0} = \frac{1}{300}$,

$$L_{v_0} = m_{v_0}(v_{0_2}{}^2 - v_{0_1}{}^2)$$
$$= \tfrac{1}{300}(2500 - 100) = 8''$$

If $m_s = \frac{1}{322}$,

$$L_s = m_s(2g)(s_2 - s_1)$$
$$= \tfrac{1}{322}(64.4)(50 - 5) = 9''$$

$m_s = \frac{1}{322}$ was chosen since 322 is a multiple divisor of 64.4 and conveniently results in a scale modulus of $M_s = \frac{10}{50}$. The use of the 50 engineers scale is anticipated.

Correspondingly, the spacing of the outside scales is

$$\frac{a}{b} = \frac{m_{v_0}}{m_s} = \frac{\frac{1}{300}}{\frac{1}{322}} = \frac{322}{300} = \frac{3.22}{3.00}$$

Note the algebraic treatment of the reciprocal functional moduli. The functional modulus of the middle scale is

$$m_{v_f} = \frac{m_{v_0} m_s}{m_{v_0} + m_s} = \frac{(\frac{1}{300})(\frac{1}{322})}{(\frac{1}{300}) + (\frac{1}{322})} = \frac{1}{622}$$

The range of $v_f$ is found by substituting the

| Equation | Var. | Funct. | Variable Range | Funct. Modulus (m) | Scale Modulus (M) | Scale Length | Distance from Center Scale | Dir. | Scale Used |
|---|---|---|---|---|---|---|---|---|---|
| | $V_0$ | $V_0^2$ | 10–50 ft/sec | $\frac{1}{300}$ | $1 \times \frac{1}{300}$ | 8″ | 3.22″ | ↑ | 30 |
| $V_f^2 = V_0^2 + 2gs$ | s | 2gs | 5–50 ft | $\frac{1}{322}$ | $\frac{64.4}{322} = \frac{2}{10} = \frac{10}{50}$ | 9″ | 3.00″ | ↑ | 50 |
| | $V_f$ | $V_f^2$ | 20.5–75.7 ft/sec | $\frac{1}{622}$ | $1 \times \frac{1}{622}$ | 8.518″ | 0 | ↑ | 40 |

(a) DESIGN

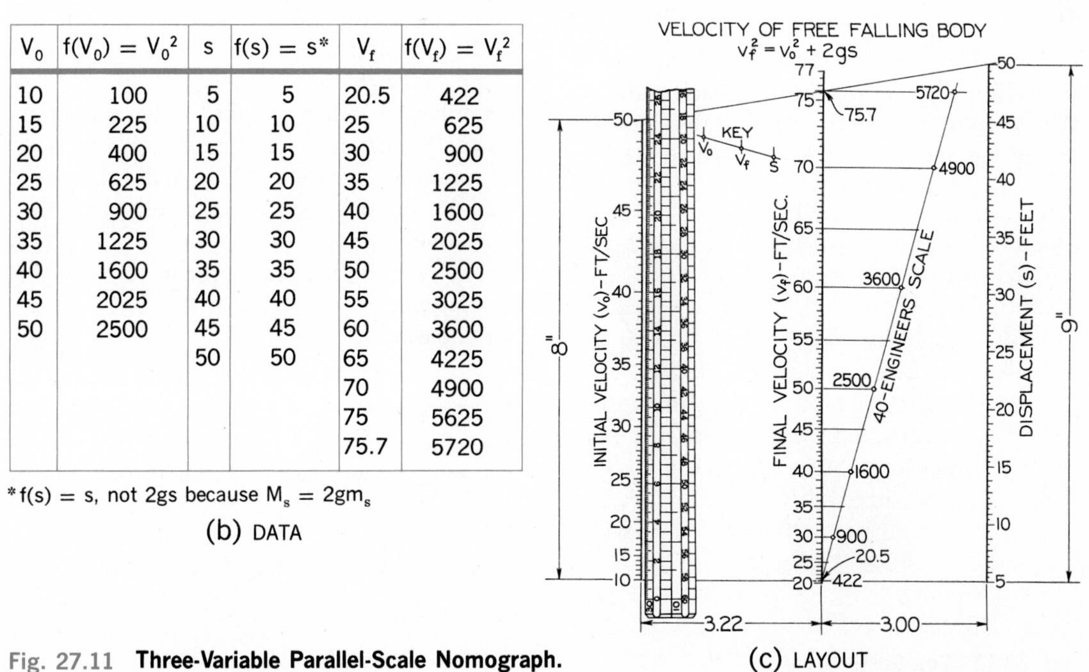

| $V_0$ | $f(V_0) = V_0^2$ | s | f(s) = s* | $V_f$ | $f(V_f) = V_f^2$ |
|---|---|---|---|---|---|
| 10 | 100 | 5 | 5 | 20.5 | 422 |
| 15 | 225 | 10 | 10 | 25 | 625 |
| 20 | 400 | 15 | 15 | 30 | 900 |
| 25 | 625 | 20 | 20 | 35 | 1225 |
| 30 | 900 | 25 | 25 | 40 | 1600 |
| 35 | 1225 | 30 | 30 | 45 | 2025 |
| 40 | 1600 | 35 | 35 | 50 | 2500 |
| 45 | 2025 | 40 | 40 | 55 | 3025 |
| 50 | 2500 | 45 | 45 | 60 | 3600 |
| | | 50 | 50 | 65 | 4225 |
| | | | | 70 | 4900 |
| | | | | 75 | 5625 |
| | | | | 75.7 | 5720 |

*f(s) = s, not 2gs because $M_s = 2gm_s$

(b) DATA



(c) LAYOUT

**Fig. 27.11 Three-Variable Parallel-Scale Nomograph.**

maximum and minimum values into the equation.

$$L_{v_f} = m_{v_f}(v_{f_2}{}^2 - v_{f_1}{}^2)$$
$$= \frac{1}{622}(5720 - 422) = 8.518''$$

The design, data, and layout for the alignment chart are shown in Fig. 27.11.

## 27.11 Logarithmic Parallel-Scale Nomographs, f(x) f(y) = f(z) or f(x) = f(z)/f(y).

Many engineering equations appear in the form of the product of variables. This form can be transformed into the sum or difference of variables by taking logarithms of both sides of the equation:

$$f(x)f(y) = f(z)$$

or $$\log f(x) + \log f(y) = \log f(z)$$

$$f(x) = f(z)/f(y)$$

or $$\log f(x) = \log f(z) - \log f(y)$$

Parallel-scale charts can be constructed for this form of equation by the use of logarithmic scales instead of natural scales. A convenient device for the construction of logarithmic scales is a log sector chart, Fig. 27.12. The use of a logarithmic sector eliminates the necessity of multiplying the logarithm of each value by the modulus to construct the scale. The use of the log sector chart is detailed in §27.12.

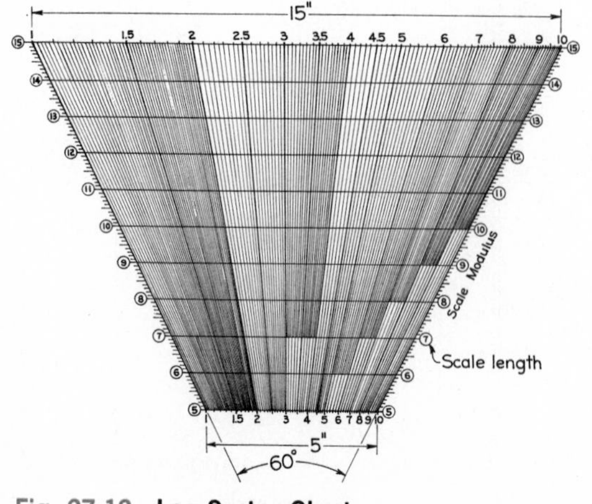


Fig. 27.12 **Log Sector Chart.**

## 27.12 Design and Layout of Logarithmic Parallel-Scale Nomographs.

The primary design of log scale nomographs is similar to natural scale nomographs. The majority of the calculations are tabulated in Fig. 27.13 (a), for the example equation:

$$P = VI$$

The first step is to compute the range of values for P, using the maximum and minimum values for V and I. Since P = VI,

$$\log P = \log V + \log I$$

V and I will be on the outside scales. The function of the variables for scale layout is the log of the original function. The determination of scale and functional modulus, scale lengths, and spacing is similar to the form discussed in §27.10.

In the chart layout, (b), notice that the scale modulus is the length of one complete log cycle used. A portion of a cycle, or more than one cycle, may be used. Cycles longer than those on the sector can be constructed by doubling or tripling a smaller cycle length. Odd-length cycles can be conveniently determined by folding the sector at the appropriate subdivision between two even-length cycles—for example, the 1.875″ cycle for P is below the 2″ cycle and is extended lower until a cycle length of 1.875″ is obtained.

Similar to other parallel-scale alignment charts, the maximum and minimum values of the variables, as determined and calculated, must be on the top isopleth and base line, respectively.

For a second illustration, the majority of the calculations are tabulated in Fig. 27.14 (a) for the equation:

$$f = 4.42\sqrt{F_0/W}$$

Since $f = 4.42\sqrt{F_0/W}$,

$$f^2 = 19.5\ F_0/W$$

and

$$2 \log f = \log 19.5 + \log F_0 - \log W \qquad (1)$$

or

$$f^2W = 19.5\ F_0$$

and

$$2 \log f + \log W = \log 19.5 + \log F_0 \quad (2)$$

or

$$W = 19.5\, F_0/f^2$$

and

$$\log W = \log 19.5 + \log F_0 - 2 \log f \quad (3)$$

Equation (2) was selected as the best arrangement, since all the scales for the variables will be positive. Equations (1) and (3) would have a negative or inverted scale ($-\log W$ or $-2 \log f$, respectively). W and f will be on the outside scales.

The scale for the variable W is one complete log cycle, since the range of values is from .1 to 1.0, and will be 8″ long. The 20″ cycle for f can be constructed by doubling lengths measured on the 10″ cycle, only a portion of a cycle being used (from 19.75 to 44.2). The 4.44″ cycle for $F_0$ is between 4″ and 5″ cycles, and is laid out almost twice, since $F_0$ ranges from 2 to 10 and from 10 to 100. Notice that tabulated values of the variables for scale layout are not necessary, since any variable to a power merely increases the log scale length required—for example, $f^2$ becomes $2 \log f$. The

**Fig. 27.13 Logarithmic Parallel-Scale Nomograph.**

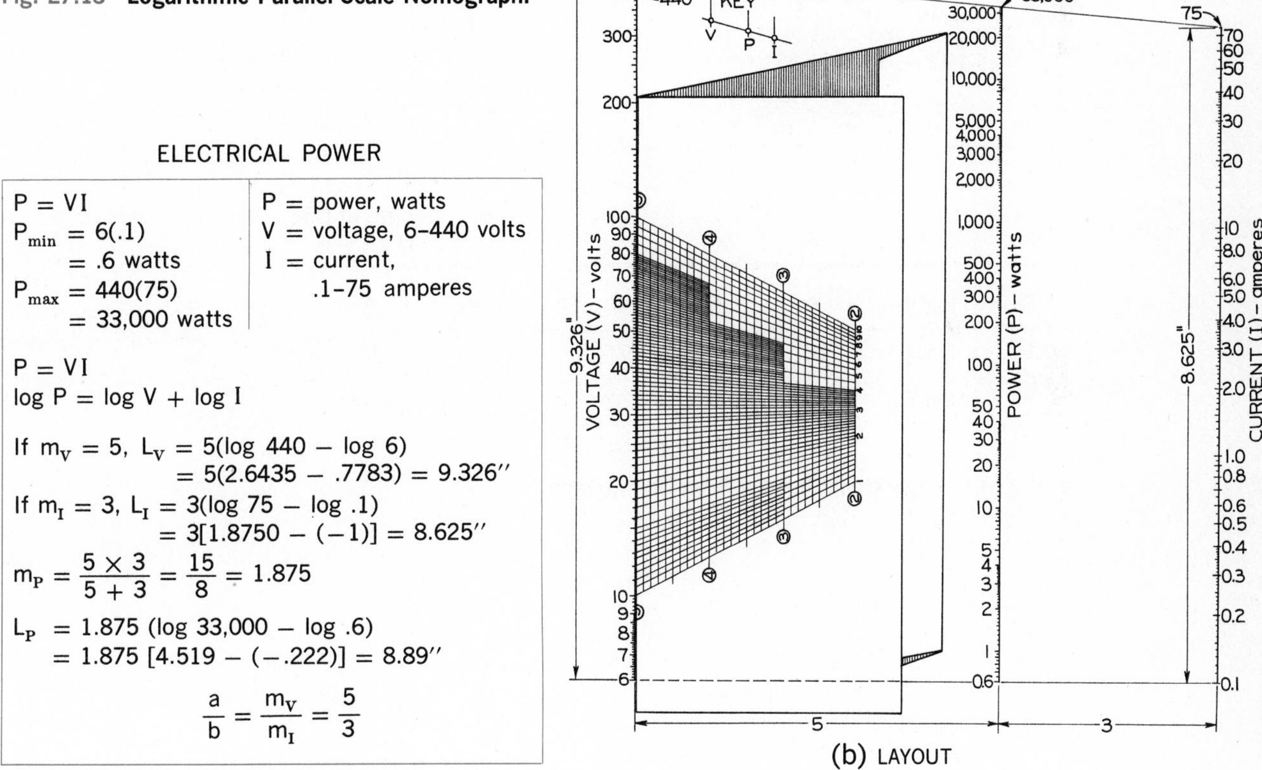


| Var. | Funct. | m | M | Scale Length | Scale Dir. | Length of One Log Cycle |
|---|---|---|---|---|---|---|
| V | log V | 5 | 5 | 9.326″ | ↑ | 5″ |
| I | log I | 3 | 3 | 8.625″ | ↑ | 3″ |
| P | log P | 1.875 | 1.875 | 8.890″ | ↑ | 1.875″ |

(a) DESIGN

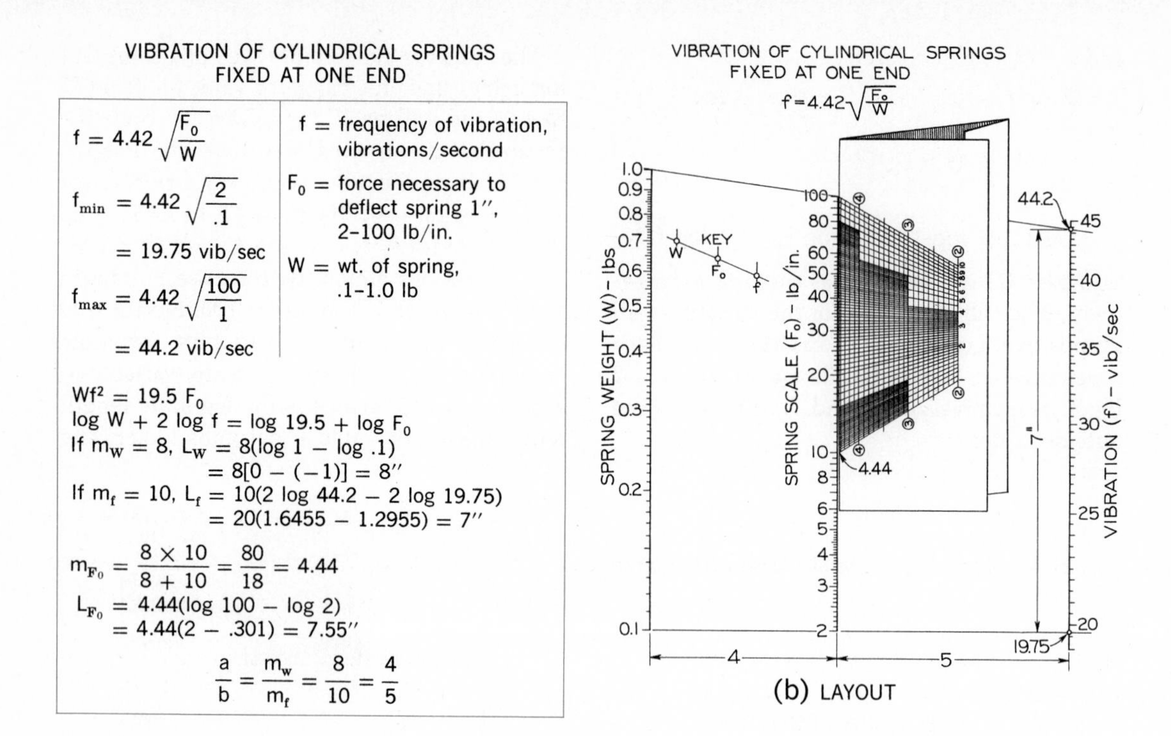

VIBRATION OF CYLINDRICAL SPRINGS
FIXED AT ONE END
$f = 4.42\sqrt{\frac{F_0}{W}}$
$f_{min} = 4.42\sqrt{\frac{2}{.1}}$
$= 19.75$ vib/sec
$f_{max} = 4.42\sqrt{\frac{100}{1}}$
$= 44.2$ vib/sec
$f =$ frequency of vibration, vibrations/second
$F_0 =$ force necessary to deflect spring 1″, 2–100 lb/in.
$W =$ wt. of spring, .1–1.0 lb
$Wf^2 = 19.5\ F_0$
$\log W + 2\log f = \log 19.5 + \log F_0$
If $m_W = 8$, $L_W = 8(\log 1 - \log .1)$
$= 8[0 - (-1)] = 8''$
If $m_f = 10$, $L_f = 10(2\log 44.2 - 2\log 19.75)$
$= 20(1.6455 - 1.2955) = 7''$
$m_{F_0} = \frac{8 \times 10}{8 + 10} = \frac{80}{18} = 4.44$
$L_{F_0} = 4.44(\log 100 - \log 2)$
$= 4.44(2 - .301) = 7.55''$
$\frac{a}{b} = \frac{m_w}{m_f} = \frac{8}{10} = \frac{4}{5}$
VIBRATION OF CYLINDRICAL SPRINGS
FIXED AT ONE END
$f = 4.42\sqrt{\frac{F_0}{W}}$
SPRING WEIGHT (W) - lbs
SPRING SCALE ($F_0$) - lb/in.
VIBRATION (f) - vib/sec
KEY
4.44
44.2
19.75
7″
4
5
(b) LAYOUT

| Var. | Funct. | m | M | Scale Length | Scale Dir. | Length of One Log Cycle |
|---|---|---|---|---|---|---|
| W | log W | 8 | 8 | 8″ | ↑ | 8″ |
| f | 2 log f | 10 | 20 | 7″ | ↑ | 20″ |
| $F_0$ | log $F_0$ | 4.44 | 4.44 | 7.55″ | ↑ | 4.44″ |

(a) DESIGN

Fig. 27.14 **Logarithmic Parallel-Scale Nomograph.**

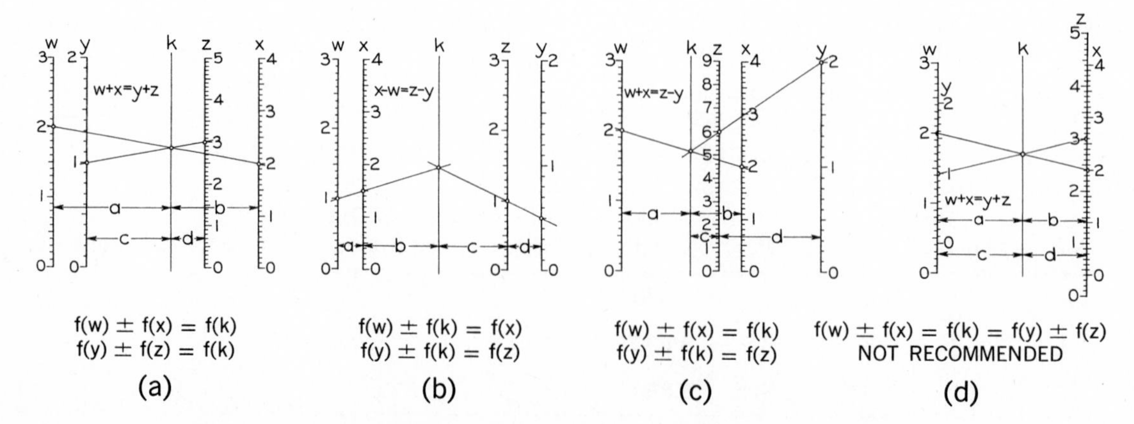


Fig. 27.15 **Arrangements of Four-Variable Parallel-Scale Nomographs.**

calibrations are laid out directly with the log sector chart.

Accounting for the term with a constant C (log 19.5) may be questioned. Whether the form of the equation is $x + y = z + C$ or $xy = zC$, the constant C is accounted for by determining the range of values for all variables from the equation with the constant. Then the primary concern is to place the last scale in the proper position to accommodate the constant. This is automatically accomplished by placing the maximum and minimum values chosen or determined on the upper isopleth and base line, respectively.

## 27.13 Four-Variable Parallel-Scale Nomographs, $f(w) \pm f(x) \pm f(y) = f(z)$ or $f(w)\,f(x)\,f(y) = f(z)$ or $f(w)\,f(x) = f(z)\,f(y)$.

Parallel-scale nomographs may be constructed for any of the four-variable equation forms shown.

The design of four-variable nomographs requires transforming the equation into two groups of two variables each and equating these two groups to a third term, f(k).

$$f(w) \pm f(x) = f(z) \pm f(y) = f(k)$$

and

$$\log f(w) \pm \log f(x) = \log f(z) \pm \log f(y) = \log f(k)$$

From the transformed equation, two separate three-variable parallel-scale nomographs are constructed. Both nomographs include the new term f(k). The f(k) is used to determine a scale and functional modulus, which *must be the same for both nomographs.* The scale for f(k) is never calibrated, but is used as a *pivot scale* and must have the same direction as the other scales.

Several scale arrangements for four-variable nomographs are shown in Fig. 27.15.

## 27.14 Design and Layout of Four-Variable Parallel-Scale Nomographs.

Since the method of design and layout is similar to the other forms discussed, the information for an example problem is tabulated and illustrated in Fig. 27.16 (a) and (b). As shown in the upper portion of the figure at (a), equation (1) could be used, but would require an inverted scale; therefore, equation (2) is preferred. One three-variable nomograph will contain scales for H, k, and I. Scales I and k will be outside scales. The second three-variable nomograph will have scales for k, R, and t. Variables R and t will be outside scales. The k term will be an outside scale for one nomograph and the middle scale for the second nomograph. The functional and scale moduli for k, $m_k$, and $M_k$, calculated for one nomograph, *must be the same moduli values for the second nomograph.* The use of a log sector chart to lay out the logarithmic scales is the same as explained in §27.12. The scale moduli are the lengths of the log cycles used for direct calibration of complete, partial, or multiple log cycles, depending on the range of values for the variables. The constant value (log 2.39) is disregarded, since it was used to determine the range of values for the variables and is automatically accounted for by placing maximum and minimum values on the upper isopleth and base line, respectively. Note that the functional modulus of a middle scale does not have to be the last modulus determined. A common error, to be especially avoided, is incorrect association of the scale spacings a, b, c, and d, with their respective scales.

## 27.15 Natural-Scale N-charts, $f(z)\,f(y) = f(x)$ or $f(z) = f(x)/f(y)$.

Equations that are in the form of the product or division of two variables equaling a third variable may be constructed with natural (nonlogarithmic) scales in the form of an *N-chart* (also called a *Z-chart*).

In Fig. 27.17, AB and CD represent two parallel scales for the variables that are the dividend and divisor in tl.e division equation form. The term that represents the quotient appears on the diagonal scale AE.

The functional moduli and lengths of the two vertical scales are arbitrarily chosen, limited only by the paper size. *These scales must begin at zero* and extend in opposite directions. The diagonal scale connects the two zero

JOULE'S LAW OF HEATING OF ELECTRICAL CONDUCTORS

$H = .239\,RI^2t$

$\frac{H}{I^2} = .239\,Rt$

(1) $\log H - 2\log I = \log k$
$= \log 2.39 + \log R + \log t$

(2) $\log H = \log k + 2\log I$

$\log k = \log 2.39 + \log R + \log t$

$H$ = heat, calories
$R$ = resistance, 1-10 ohms
$I$ = current, 5-25 amperes
$t$ = time, 1-60 seconds

$H = 0.239(1)(25)(1)$
$= 5.975$ cal

$H = 0.239(10)(625)(60)$
$= 89{,}600$ cal

If $m_H = 2$, $L_H = 2(\log 89{,}600 - \log 5.975)$
$= 2(4.9525 - .7765) = 8.352''$

If $m_I = 6$, $L_I = 6(2\log 25 - 2\log 5)$
$= 6(2)(1.398 - .699) = 8.388''$

Since f(H) is the middle scale, $m_H = \frac{m_I m_k}{m_I + m_k}$; $2 = \frac{6m_k}{6 + m_k}$; $m_k = 3$

$$\frac{a}{b} = \frac{m_I}{m_k} = \frac{6}{3} = \frac{2}{1}$$

If $m_R = 8$, $L_R = 8(\log 10 - \log 1) = 8(1 - 0) = 8''$

Since f(k) is the middle scale, $m_k = \frac{m_R m_t}{m_R + m_t}$; $3 = \frac{8m_t}{8 + m_t}$; $m_t = 4.8$

$L_t = 4.8(\log 60 - \log 1) = 4.8(1.778 - 0) = 8.53''$

$$\frac{c}{d} = \frac{m_R}{m_t} = \frac{8}{4.8} = \frac{2}{1.2}$$

(a) DESIGN

JOULE'S LAW OF HEATING OF ELECTRICAL CONDUCTORS

H=0.239 RI²t

(b) LAYOUT

| Equation | Var. | Funct. | Variable Range | m | M | Scale Length | Dir. | Distance from Center Scale | Length of One Log Cycle |
|---|---|---|---|---|---|---|---|---|---|
| | H | log H | 5.975–89,600 | 2 | 2 | 8.352″ | ↑ | 0 | 2″ |
| log H = log k + 2 log I | I | 2 log I | 5–25 | 6 | 12 | 8.388″ | ↑ | 1.500 | 12″ |
| | k | log k | — | 3 | 3 | — | ↑ | .750 | — |
| | k | log k | — | 3 | 3 | — | ↑ | 0 | — |
| log k = log 2.39 + log R + log t | R | log R | 1–10 | 8 | 8 | 8.000″ | ↑ | 4.000 | 8″ |
| | t | log t | 1–60 | 4.8 | 4.8 | 8.530″ | ↑ | 2.400 | 4.8″ |

Fig. 27.16 **Four-Variable Parallel-Scale Nomograph.**

points from A to C and is graduated from A to the maximum value desired. The distance between the vertical scales is any convenient distance.

The outside scales are graduated in the same manner as previously described in §§27.4 to 27.6. The diagonal scale may be calibrated analytically or graphically. The analytical method may be found in the texts listed in the bibliography. A simpler graphical method is discussed in §27.16.

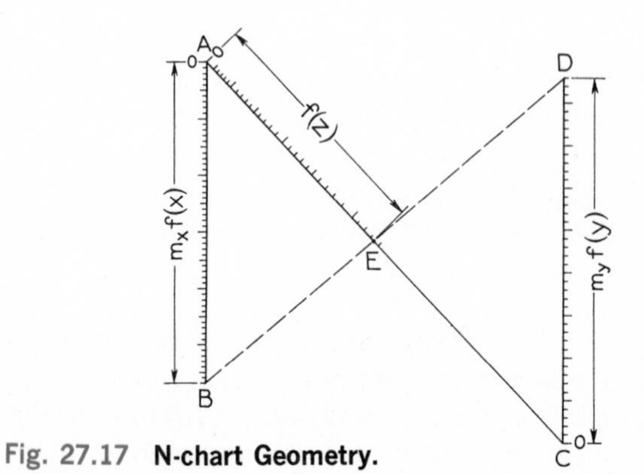


Fig. 27.17 **N-chart Geometry.**

## 27.16 Design and Layout of an N-chart.

The calculations and layout of an example problem are shown in Fig. 27.18. The equation may be arranged in any of the following forms:

$$\frac{I}{b} = \frac{h^3}{12}; \quad \frac{I}{h^3} = \frac{b}{12}; \quad \text{or} \quad \frac{12\,I}{b} = h^3.$$

If the variable to a power were one of the vertical scales, it would be a non-uniform functional scale, requiring calibration and computation. Since the diagonal scale is always nonuniform and calibrated, $f(h) = h^3$ was chosen for this scale. This corresponds to the equation form $\frac{12\,I}{b} = h^3$. The zero values of the function 12 I and $h^3$ will coincide.

After reasonable scale lengths have been determined for f(I) and f(b), construct these scales parallel to each other and any convenient distance apart. Usually, a roughly square chart is satisfactory. Proper selection of moduli permits direct use of the engineers scale. A base line for alignment is not required.

Draw a line connecting the zero values of the two vertical scales. To calibrate the diagonal scale for f(h), select one convenient value of b (or I). This value should be so selected as to simplify the form of the equation for computation of the other values, when substituted into the equation. A value of $b = 6$ simplifies the equation to $I = \frac{h^3}{2}$. With this expression, determine the values of I for various values of h. Draw lines from 6 on the b scale to the values of I on the I scale to locate the corresponding values of h on the diagonal scale.

## 27.17 Large Value N-charts.

When the range of values for an N-chart are large numbers and do not extend to zero, the chart may be constructed with an additional computation. As shown in Fig. 27.19, the scales of the chart are nonintersecting within the area of the paper. The conditions for an N-chart, however, are met, since the scales extended to their zero values would intersect. The lengths of the two vertical scales are determined for their large-number range of values. The two lengths (L) must be the same.

After computing new scale lengths from the zeros to the maximum values, the necessary horizontal distances b and c can be easily determined, since by similar triangles,

$$\frac{a}{d_1} = \frac{b}{d_2} = \frac{c}{d_3}$$

## 27.18 Combination Parallel-Scale and N-chart, $f(w) \pm f(x) = f(y)/f(z)$ [or $= f(y)\,f(z)$].

The arrangement of variables within an equation for a nomograph makes it necessary that all the scales have the same type of graduations, either logarithmic or natural. An equation with addition or subtraction of variables has natural scales, while an equation with a multiplication or division of variables suggests logarithmic scales. If an equation of four variables contains any combination of addition or subtraction with multiplication or division, the use of an N-chart (natural scales) for the multiplication or division terms in combination with a natural parallel-scale for the addition or subtraction terms, can properly represent the equation, as shown in Fig. 27.20. A fifth term, f(k), equated to the equation, is used as an uncalibrated vertical outside scale for the parallel-scale nomograph and as a vertical scale for the N-chart, since

$$f(w) \pm f(x) = f(k) = \frac{f(y)}{f(z)} \quad [\text{or} = f(y)\,f(z)]$$

## 27.19 Design and Layout of a Four-Variable Combination Nomograph.

The method of design and layout is similar to the other forms discussed; the given data for an example problem are tabulated and illustrated in Fig. 27.21. The equations are arranged as shown in step (2), since the f(k) scale is to be an outside scale for the parallel-scale chart and a vertical scale for the N-chart. If the equation contains a product of two functions, either variable can be placed on the diagonal scale of the N-chart. If the equation contains a division of two functions, the variable in the de-

MOMENT OF INERTIA—RECTANGULAR CROSS SECTION

$$I = \frac{bh^3}{12}$$

$$h^3 = \frac{12\,I}{b}$$

$I$ = moment of inertia, in.$^4$

b = width, 0–8″

h = height, 0–12″

$$I = \frac{8(12)^3}{12} = 1152 \text{ in.}^4$$

If $m_b = 1$, $L_b = m_b(b_2 - b_1) = 1(8 - 0) = 8''$

If $m_I = \frac{1}{2400}$, $L_I = m_I\, 12(I_2 - I_1) = \frac{1}{2400}\, 12(1152 - 0)$

$$= \frac{1}{200}(1152) = 5.76''$$

| Equation | Var. | Funct. | Variable Range | m | M | Scale Length | Dir. | Scale Used |
|---|---|---|---|---|---|---|---|---|
| | b | b | 0–8″ | 1 | 1 | 8″ | ↑ | 10 |
| $h^3 = \frac{12\,I}{b}$ | $I$ | $12\,I$ | 0–1152 in.$^4$ | $\frac{1}{2400}$ | $12m = \frac{12}{2400} = \frac{1}{200}$ | 5.76″ | ↓ | 20 |
| | h | $h^3$ | 0–12″ | — | — | — | ↘ | — |

(a) DESIGN

(b) DATA

| h | Values of $I$ with b = 6″ $I = h^3/2$ |
|---|---|
| 1 | .5 |
| 2 | 4.0 |
| 3 | 13.5 |
| 4 | 32.0 |
| 5 | 62.5 |
| 6 | 108.0 |
| 7 | 171.5 |
| 8 | 256.0 |
| 9 | 364.5 |
| 10 | 500.0 |
| 11 | 665.5 |
| 12 | 864.0 |

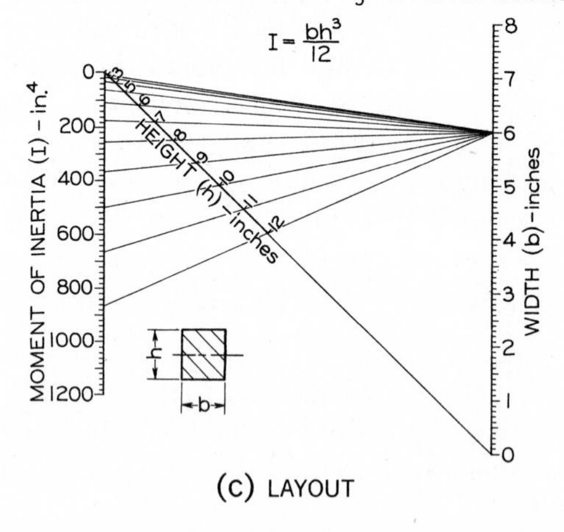


(c) LAYOUT

Fig. 27.18 **N-chart.**

nominator must be placed on the diagonal scale. The lengths of only the three vertical scales must be determined. The resulting functional moduli for the parallel-scale chart are used for spacing the scales, while the N-chart spacing is not definite and is arranged simply for convenience.

The 10 engineers scale is used to calibrate the $v_0$ scale and the 20 scale is used to calibrate the $v_f$ scale. The $v_0$ scale was designed for a shorter length in order to keep the pivot point of the k scale on the paper (within reason) when using low $v_0$ values and large $v_f$ values.

To calibrate the diagonal scale of the N-chart, a value must be located on the uncalibrated k scale as a pivot point (k = 4000).

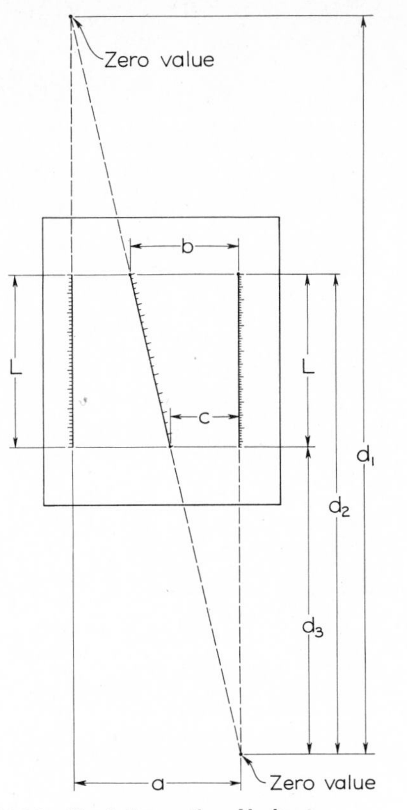


Fig. 27.19 **Nonintersecting N-chart.**

A second point (k = 2500) was found to locate the value of a = .25 ft/sec$^2$ only.

## 27.20 Proportional Charts, f(w)/f(x) = f(y)/f(z).

Equations with four variables in the form of a proportion, or in the form f(w)f(z) = f(y)f(x), can be represented by a nomograph with natural scales, called a *proportional chart.* The geometry of its construction is shown in Fig. 27.22 (a).

From similar triangles,

$$\frac{AB}{EF} = \frac{a}{b} = \frac{AC}{DF}$$

If AB = $L_w = m_w f(w)$ and EF = $L_x = m_x f(x)$, etc.,

$$\frac{m_w f(w)}{m_x f(x)} = \frac{m_y f(y)}{m_z f(z)}$$

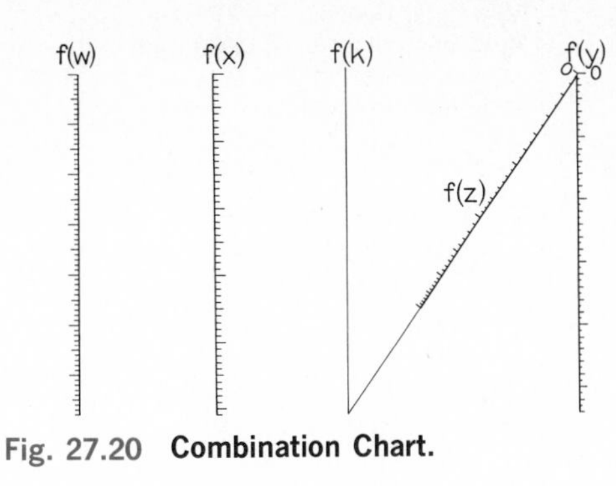


Fig. 27.20 **Combination Chart.**

and since

$$\frac{f(w)}{f(x)} = \frac{f(y)}{f(z)}$$

then

$$\frac{m_w}{m_x} = \frac{m_y}{m_z}$$

The proportionality of the functional moduli is the only relationship necessary for the construction of the chart. After three moduli have been determined for the scale lengths desired, the fourth modulus must be in proportion to the other values. Many arrangements of scale are possible; the most convenient form is shown at (b). The uncalibrated diagonal line connects the zero values of the four scales and is used as an index or pivot line. The relationship of the reversed directions for the vertical and horizontal scales, as based on the geometry employed, should be noted. The f(y) scale must be opposite the f(z) scale, and in the reverse direction. The same is true of the f(x) and f(w) scales.

The design calculations and layout of an illustrative example of a proportional chart are shown in Fig. 27.23.

## 27.21 Alignment Chart Problems.

The problems on pages 748–52 provide laboratory experience in applying the methods described in this chapter and, in addition, introduce many typical engineering problems to the student. For convenience in assignment, the problems are classified according to type.

LINEAR MOTION—CONSTANT ACCELERATION

$V_f^2 - V_0^2 = 2as$

(1) $f(V_f) - f(V_0) = f(k)$

$= f(a)f(s)$

(2) $V_f^2 = k + V_0^2$; $2a = \dfrac{k}{s}$

$V_f$ = final velocity, 0–120 mph
$V_0$ = original velocity, 0–60 mph
$a$ = acceleration, 0–2.2 ft/sec²
$s$ = displacement, 0–1 mile

$$L_{V_f} = m_{V_f}(V_{f_2}{}^2 - V_{f_1}{}^2) = m_{V_f}(14400 - 0)$$

If $m_{V_f} = .0005$, $L_{V_f} = 7.20''$

$$L_{V_0} = m_{V_0}(V_{0_2}{}^2 - V_{0_1}{}^2) = m_{V_0}(3600 - 0)$$

If $m_{V_0} = .001$, $L_{V_0} = 3.60''$

$$m_{V_f} = \frac{m_{V_0}m_k}{m_{V_0} + m_k};\ .0005 = \frac{.001\ m_k}{.001 + m_k};\ m_k = .001;$$

$$\frac{a}{b} = \frac{m_{V_0}}{m_k} = \frac{1}{1}$$

$L_s = m_s(5280 - 0)$; if $m_s = .001$, $L_s = 5.28''$

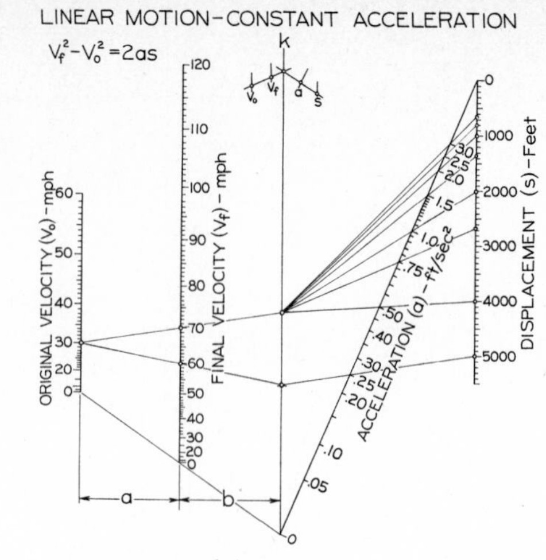


(c) LAYOUT

| Equation | Variable | Function | m | M | Scale Length | Distance from Center Scale | Direction | Scale Used |
|---|---|---|---|---|---|---|---|---|
| $V_f^2 = k + V_0^2$ | $V_f$ | $V_f^2$ | $.0005 = \frac{1}{2000}$ | $\frac{1}{2000}$ | 7.20″ | 0 | ↑ | 20 |
| | $V_0$ | $V_0^2$ | $.001 = \frac{1}{1000}$ | $\frac{1}{1000}$ | 3.60″ | 1 | ↑ | 10 |
| | k | k | $.001 = \frac{1}{1000}$ | $\frac{1}{1000}$ | — | 1 | ↑ | — |
| $2a = \frac{k}{s}$ | k | k | $.001 = \frac{1}{1000}$ | $\frac{1}{1000}$ | — | — | ↑ | — |
| | s | s | $.001 = \frac{1}{1000}$ | $\frac{1}{1000}$ | 5.28″ | — | ↓ | 10 |
| | a | 2a | — | — | — | — | ↗ | — |

(a) DESIGN

| $V_f$ | $f(V_f) = V_f^2$ | $V_0$ | $f(V_0) = V_0^2$ |
|---|---|---|---|
| 0 | 0 | 0 | 0 |
| 10 | 100 | 10 | 100 |
| 20 | 400 | 20 | 400 |
| 30 | 900 | 30 | 900 |
| 40 | 1,600 | 40 | 1600 |
| 50 | 2,500 | 50 | 2500 |
| 60 | 3,600 | 60 | 3600 |
| 70 | 4,900 | | |
| 80 | 6,400 | | |
| 90 | 8,100 | | |
| 100 | 10,000 | | |
| 110 | 12,100 | | |
| 120 | 14,400 | | |

DATA FOR N-CHART DIAGONAL CALIBRATION

| a | s |
|---|---|
| .50 | 4000 |
| .75 | 2670 |
| 1.00 | 2000 |
| 1.50 | 1333 |
| 2.00 | 1000 |
| 2.50 | 800 |
| 3.00 | 666 |

(b) DATA

$k = V_f^2 - V_0^2$

when $V_f = 70$, $V_0 = 30$, $k = 70^2 - 30^2 = 4000$

when $k = 4000$, $s = \dfrac{4000}{2a} = \dfrac{2000}{a}$

when $V_f = 60$, $V_0 = 30$, $k = 2500$

when $k = 2500$, $a = .25$, $s = 5000$

**Fig. 27.21 Combination Parallel Scale and N-chart.**

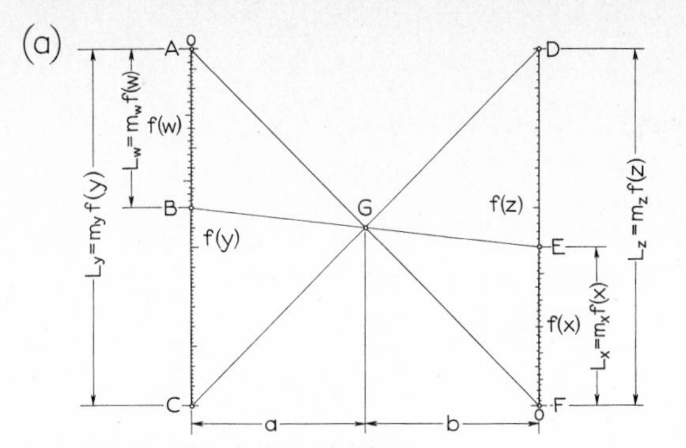


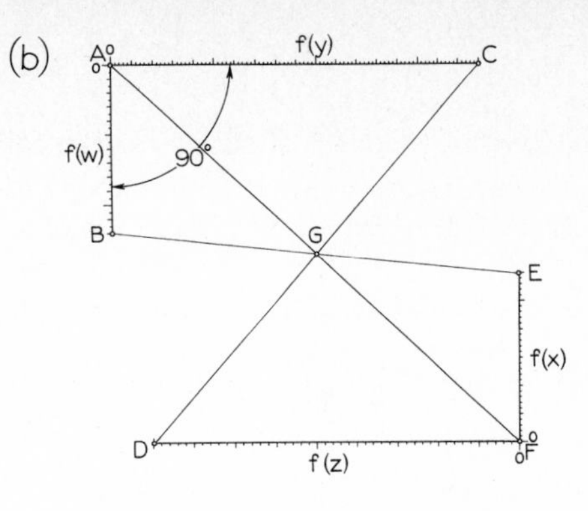

Fig. 27.22 **Proportional Chart.**

Fig. 27.23 **Proportional Chart.**

FIBER STRESS—HELICAL COMPRESSION AND EXTENSION SPRINGS

| | |
|---|---|
| $s = \dfrac{8PD}{\pi d^3}$ | s = fiber stress in torsion, 0–63,600 psi<br>P = axial load, 0–20,000 lb |
| $\dfrac{s}{P} = \dfrac{8D}{\pi d^3}$ | D = mean diameter of spring, 0–10″<br>d = wire diameter, 0–2.0″ |

$$L_s = m_s[f(s_2) - f(s_1)] = m_s(63{,}600 - 0)$$

If $m_s = .0001$, $L_s = 6.36''$

$$L_P = m_P[f(P_2) - f(P_1)] = m_P(20{,}000 - 0)$$

If $m_P = .0004$, $L_P = 8.000''$

$$L_D = m_D[f(D_2) - f(D_1)] = m_D[8(10) - 8(0)] = 80\ m_D$$

If $m_D = .05$, $L_D = 4.00''$

$$\frac{m_s}{m_P} = \frac{m_D}{m_d};\ \frac{.0001}{.0004} = \frac{.05}{m_d};\ m_d = .20$$

$$L_d = m_d[f(d_2) - f(d_1)] = .2[\pi(2)^3 - \pi(0)^3] = 5.026$$

(a) DESIGN

| Variable | Function | m | M | Scale Length | Dir. | Scale Used |
|---|---|---|---|---|---|---|
| s | s | $.0001 = \dfrac{1}{10{,}000}$ | $\dfrac{1}{10{,}000}$ | 6.36″ | ↓ | 10 |
| P | P | $.0004 = \dfrac{4}{10000} = \dfrac{2}{5000}$ | $\dfrac{2}{5000}$ | 8.00″ | ↑ | 50* |
| D | 8D | $.05 = \dfrac{5}{100} = \dfrac{1}{20}$ | $8m = \dfrac{8}{20} = \dfrac{2}{5}$ | 4.00″ | → | 50* |
| d | $\pi d^3$ | $.2 = \dfrac{2}{10} = \dfrac{1}{5}$ | $\dfrac{1}{5}$ | 5.026″ | ← | 50 |

*Halved. (Actually $M = \frac{1}{25}$)

| f(s) = s | f(P) = P | f(D) = D | d | $f(d) = \pi d^3$ |
|---|---|---|---|---|
| 0 | 0 | 0 | 0 | 0 |
| 10,000 | 5,000 | 1 | 1 | 3.14 |
| 20,000 | 10,000 | 2 | 1.5 | 10.60 |
| 30,000 | 15,000 | 3 | 2.0 | 25.12 |
| 40,000 | 20,000 | 4 | | |
| 50,000 | | 5 | | |
| 60,000 | | 6 | | |
| 63,600 | | 7 | | |
| | | 8 | | |
| | | 9 | | |
| | | 10 | | |

(b) DATA

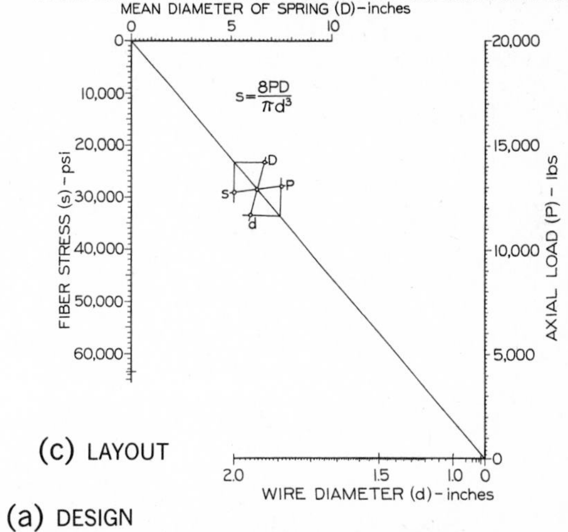


(c) LAYOUT

**27.21 Alignment Chart Problems**

SCALE LAYOUT. Construct scales for the following:

**Prob. 27.1** $f(d) = \pi d$ — $d$ = diameter of circle, 0″ to 6″

**Prob. 27.2** $f(x) = 2x - 3$ — $x = 0$ to 15

**Prob. 27.3** $f(x) = 2 \log x$ — $x = 1$ to 10

**Prob. 27.4** $f(x) = 3x$ — $x = 1$ to 10

**Prob. 27.5** $f(x) = \frac{1}{1 + x}$ — $x = 1$ to 10

TWO VARIABLES—Conversion Scales. Construct two-variable conversion nomographs for the following:

**Prob. 27.6** Area of a Circle.

$$A = \frac{\pi d^2}{4}$$

$A$ = area of a circle, in.$^2$
$d$ = diameter of circle, 0″ to 6″

**Prob. 27.7** Radian-Degrees Conversion.

$$r = \frac{N°}{57.3}$$

$r$ = radians
$N°$ = number of degrees, 0° to 360°
(1 revolution, 360° = $2\pi$ radians; 1 radian = 57.3°)

**Prob. 27.8** Velocity of Efflux (liquid flow through an orifice).

$$v = \sqrt{2gh}$$

$v$ = velocity, ft/sec
$h$ = height of liquid surface above orifice, 0′ to 5′
$g$ = acceleration due to gravity, constant = 32.2 ft/sec$^2$

**Prob. 27.9** Major Diameter of Numbered Thread Sizes.

$$D = .013'' N + .060''$$

$D$ = major diameter of thread, in.
$N$ = thread numbers, 0, 1, 2, 3, 4, 5, 6, 8, 10, 12

**Prob. 27.10** Hardness Conversion.

$$R_C = 88\, B^{.162} - 192$$

$R_C$ = Rockwell-C hardness number
$B$ = Brinell hardness number, 153 to 767

**Prob. 27.11** Hardness Conversion.

$$R_B = \frac{(B - 47)}{.0074B + .154}$$

$R_B$ = Rockwell-B hardness number
$B$ = Brinell hardness number, 100 to 352

**Prob. 27.12** Coefficient of Friction between Steel Wheels and Cast Iron Blocks—Rubbing Velocity.

$$f = \frac{.6}{\sqrt[3]{v}}$$

$f$ = coefficient of friction
$v$ = rubbing velocity, 0 to 88 ft/sec

THREE VARIABLES. Construct parallel-scale nomographs or N-charts for the following:

**Prob. 27.13** Velocity of Free-Falling Body.

$$v_f = v_0 + gt$$

$v_f$ = final velocity, ft/sec
$v_0$ = initial velocity, 0 to 50 ft/sec
$t$ = time, 0 to 20 sec
$g$ = acceleration due to gravity, constant = 32.2 ft/sec$^2$

**Prob. 27.14** Rectangular Moment of Inertia for Circular Hollow Cylinder.

$$I = \frac{\pi(D^4 - d^4)}{64}$$

$I$ = rectangular moment of inertia, in.$^4$
$D$ = outside diameter, .5″ to 5.0″
$d$ = inside diameter, .375″ to 4.75″

**Prob. 27.15** Weight of Steel Tube per Foot.

$$W = 2.65\,(d_o^2 - d_i^2)$$

$W$ = weight per foot, lb
$d_o$ = outside diameter, 1″ to 12″
$d_i$ = inside diameter, $\frac{7}{8}$″ to $11\frac{3}{4}$″

**Prob. 27.16** Tube—Polar Radius of Gyration.

$$k_0 = .354\,\sqrt{D_1^2 + D_2^2}$$

$k_0$ = polar radius of gyration, in.
$D_1$ = outside diameter of tube, 1″ to 12″
$D_2$ = inside diameter of tube, $\frac{3}{4}$″ to $11\frac{1}{2}$″

**Prob. 27.17** Differential Chain-Block Rise, Fig. 27.24.

$$H = \frac{\pi}{2}(D_a - D_b)$$

$H$ = rise, in. per one revolution of sheaves in upper block
$D_a$ = large sheave diameter, upper block, 4″ to 12″
$D_b$ = small sheave diameter, upper block, 3″ to 11″

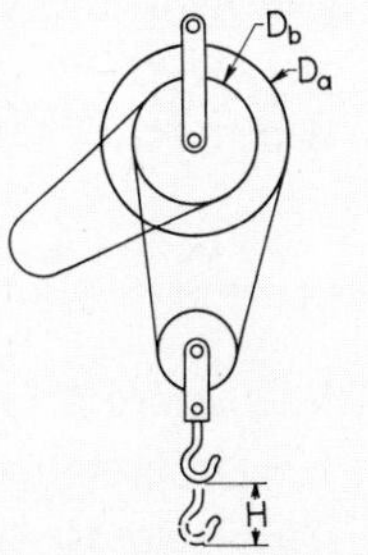

Fig. 27.24 (Prob. 27.17).

**Prob. 27.18** Valley Slopes for Bin Hoppers, Fig. 27.25.

$$\cot^2 C = \cot^2 A + \cot^2 B$$

$A$ = side angle, 30° to 75°
$B$ = side angle, 30° to 75°
$C$ = valley angle, degrees

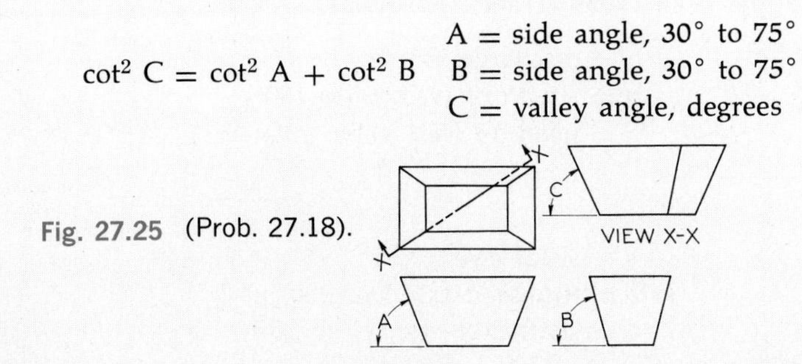


Fig. 27.25 (Prob. 27.18).

**Prob. 27.19** Ohm's Law.

$$V = IR$$

V = potential (voltage), volts
I = electric current, .1 to 10 amperes
R = resistance, 1 to 100 ohms

**Prob. 27.20** Coefficient of Friction.

$$\mu = \frac{f}{N}$$

$\mu$ = coefficient of friction, .15 to .6
f = frictional force, lb
N = normal force (to plane of movement), 5 to 150 lb

**Prob. 27.21** Kinetic Energy.

$$E_k = \frac{Wv^2}{2g}$$

$E_k$ = kinetic energy, ft-lb
W = weight of body in motion, 10 to 200 lb
v = velocity of body, 5 to 100 ft/sec
g = acceleration due to gravity = 32.2 ft/sec$^2$

**Prob. 27.22** Torque–rpm–hp Relationship.

$$T = \frac{63025 \text{ hp}}{N}$$

T = torque, 1000 to 100,000 in.-lb
hp = horsepower, .02 to 3000 hp
N = shaft speed, rpm

**Prob. 27.23** Pipe Flow Loss.

$$H = 6.2\frac{V^{1.9}}{D^{1.16}}$$

H = head loss, ft per 1000′ of pipe length
V = flow velocity, 1.0 to 50 ft/sec
D = internal pipe diameter, 1″ to 36″

**Prob. 27.24** Internal Stress—Thick Cylinders (Lame's equation for brittle materials), Closed-End Cylinders.

$$s_t = p\left(\frac{1 + R^2}{R^2 - 1}\right)$$

$s_t$ = working stress, tension at inner radius, psi
p = working pressure in cylinder, 10 to 150 psi
R = $r_0/r_i$ = ratio of outside radius to inside radius of cylinder, 1.2 to 5.0

**Prob. 27.25** Energy of a Magnetic Field.

$$W = \tfrac{1}{2}LI^2$$

W = energy, joules
L = inductance, .01 to 4.0 henries
I = current, 1 to 30 amperes

**Prob. 27.26** Discharge of Suppressed Weirs (Francis Formula).

$$q = 3.33bh^{3/2}$$

q = discharge, ft$^3$/sec
b = length of weir, 1′ to 150′
h = head on weir, .1′ to 15′

**Prob. 27.27** Shaft Design.

$$s = 5.09\frac{T}{d^3}$$

s = shear stress, psi
T = torque, 2500 to 90,000 in.-lb
d = shaft diameter, 1″ to 10″

FOUR VARIABLES. Construct the appropriate four-variable nomograph (parallel-scale, combination N-chart and parallel-scale nomographs, or proportional chart) for the following:

**Prob. 27.28** Automotive Gear Ratios.

$$R\ S_c = D\ S_e$$

R = reduction ratios, 1 to 16
$S_c$ = car speed, 0 to 120 mph
D = wheel diameter, in.
$S_e$ = engine speed, 0 to 3500 rpm

**Prob. 27.29** Electromagnetic Field Intensity.

$$H = \frac{2\pi N\ I}{10r}$$

H = field intensity, Oersteds
N = coil turns, 5 to 1000 turns
I = current, 1 to 20 amperes
r = coil radius, 1 to 10 cm

**Prob. 27.30** Vibration Frequency of Spring.

$$n = \frac{761{,}500\ d}{ND^2}$$

n = vibration per minute of spring
d = wire diameter, 0″ to .1875″
D = mean diameter of spring, 0″ to 1.0″
N = number of active coils, 4 to 50

**Prob. 27.31** Centrifugal Force.

$$F = .00034\ N^2 rW$$

F = centrifugal force, lb
N = rotational speed, 10 to 10,000 rpm
r = radius of center of gravity, .1′ to 5.0′
W = weight of rotating body, .5 to 500 lb

**Prob. 27.32** Section Moduli of Floor Beams.

$$s = \frac{12\ WBL^2}{8(16{,}000)}$$

s = section modulus of beam or girder, in.$^2$ (for maximum fiber stress in beam of 16,000 psi)
W = uniform load, 15 to 700 lb/ft$^2$
B = mean spacing of beams, 4′ to 30′
L = span of beam, length of girder, 5′ to 50′

**Prob. 27.33** Orifice Flow.

$$R = 19.64\ Cd^2 \sqrt{h}$$

R = rate of flow, gal/min
C = orifice constant, .5 to 1.6
d = diameter of orifice, .06″ to 1.0″
h = head of water, 3′ to 100′

**Prob. 27.34** Electrodeposit Design.

$$A = .167\ CBL$$

A = plating current, amperes
C = current density, 1 to 1000 amperes/ft$^2$
B = width of strip, 1″ to 100″
L = length of immersed strip, 1′ to 100′

**Prob. 27.35** Cylinder Bending Stress.

$$s = \frac{M}{\pi R^2 T}$$

s = maximum allowable stress, tensile or compressive, psi
M = bending moment, 100 to 100,000,000 in.-lb
R = cylinder radius, 1″ to 10″
T = cylinder wall thickness, .01″ to 1.0″

**Prob. 27.36** Critical Speed of End-Supported Bare Steel Shafts.

$$N_c = \frac{46.886(10)^5 \sqrt{D_1^2 + D_2^2}}{L^2}$$

$N_c$ = critical speed, rpm
L = shaft length, 20″ to 60″
$D_1$ = outside diameter, 1″ to 4″
$D_2$ = inside diameter, .50″ to 3.0″

For aluminum, multiply $N_c$ by 1.0026.
For magnesium, multiply $N_c$ by .9879.

**Prob. 27.37** Moment of Inertia of Rectangular Bar.

$$I = \frac{M}{12}(b^2 + l^2)$$

I = moment of inertia of rectangular bar about axis through its center and at right angles to dimensions b and l, lb-ft²
M = mass of bar, 2 to 10 lb
b = width of bar, .042′ to .25′
l = length of bar, .50′ to 2.0′

**Prob. 27.38** Steel Tape—Temperature Correction.

$$C_t = .0000065\ s(T - T_0)$$

$C_t$ = temperature correction to measured length, ft
s = measured length, ft
T = temperature at which measurements are made, 0° to 120° Fahrenheit
$T_0$ = temperature at which tape is standardized, 65° to 75° Fahrenheit (68°F = Std)

**Prob. 27.39** Maximum Deflection of Simple Steel Beam Under Uniform Load.

$$\Delta_{max} = \frac{5}{384}\left(\frac{1728\ WL^4}{EI}\right)$$

$\Delta_{max}$ = maximum deflection, in.
W = total load per ft, 100 to 30,000 lb/ft
L = span in ft, 5′ to 60′
= modulus of elasticity = $29 \times 10^6$ lb/in.² (constant value)
I = moment of inertia, 100 to 30,000 in.⁴

**Prob. 27.40** Reynolds Number—Hydraulics.

$$R = \frac{7750\ VD}{v}$$

R = Reynolds number
V = velocity, 1 to 100 ft/sec
D = pipe inside diameter, .10″ to 2.0″
v = kinematic viscosity, 5 to 500 centistokes

# 28

# empirical equations

by E. J. Mysiak*

**28.1 Introduction.** Empirical equations by definition are equations derived from experimental data or experience as distinguished from equations derived from logical reasoning or hypothesis (rational equations). At times, tabulated data or the analysis of a graph are inadequate, and an equation for the data is required. A graphical plot shows the trend of the data and the value of one variable relative to the corresponding second variable. The derivation of empirical equations is a more comprehensive method of analysis and can be used to calculate additional data not obtained in experimentation.

The derivations of equations are varied in methods. A basic procedure is to plot the data on rectangular, semilogarithmic, or logarithmic coordinate graph paper, in an attempt to obtain a straight line. If the plot results in a reasonably straight line on one of these papers, an approximate (empirical) equation can be derived by geometric and algebraic methods. Three of the more common methods of deriving the equation of a straight line are described in the following sections. The reader is referred to the references in the bibliography, Appendix 1, for additional methods of derivation and forms of empirical equations.

* Mechanical Engineer, International Electro-Magnetics, Inc.

## 28.2 Data Plotting.

The plot and graph-paper size should be large enough to suit the accuracy of the data. In most cases an $8\frac{1}{2}'' \times 11''$ sheet will suffice. However, if the data contain three or four significant figures that cannot be accurately located, larger graph paper may be used. The major divisions of the rectangular grid should be subdivided into ten parts for convenient plotting. See §26.5.

The standard practice of plotting the independent variable on the X-axis and the dependent variable on the Y-axis is followed. The one exception to this practice may be when attempting to obtain a straight line on semilogarithmic paper. If a straight line cannot be obtained when the Y-axis is on the logarithmic grid, the data can be plotted on reversed axes. If a straight line is obtained, the notation of the equation is correspondingly reversed.

If the derived equation is to be of any value, the initial plotting must be done with care and accuracy. The points should be denoted carefully with small, sharp, light crosses or dots, which are better than circles, Fig. 28.1. Circles can be used later when preparing the graph for presentation.

## 28.3 Curve Fitting.

After all the points of data are plotted, the straight line can be fitted by eye, Fig. 28.1. As many points as possible should be on the line. It is normal to have some points above and below the line, owing to deviations in reading instruments when obtaining the data. The straight line should be fitted with as many points above the line as below the line to "balance out" the errors. The properly drawn straight line minimizes and accounts for the error of the data resulting in an average line. Occasionally a "wild" point will be plotted, which is at an abnormal distance from the general path of the curve. A point of this type may be due to a misread instrument or a mistake in recording the data and should be disregarded.

Fig. 28.1 **Data Plotting and Curve Fitting.**

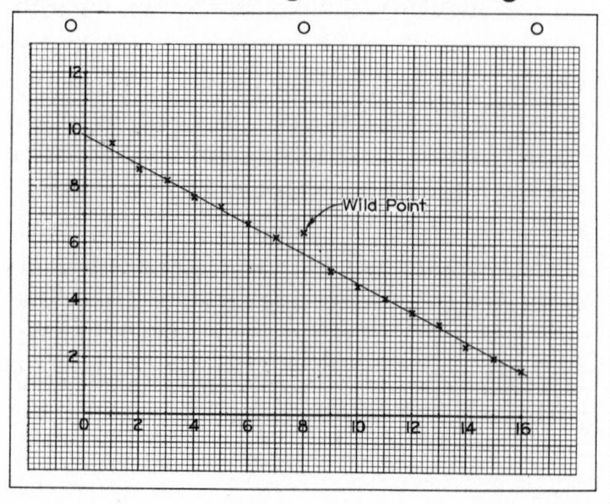


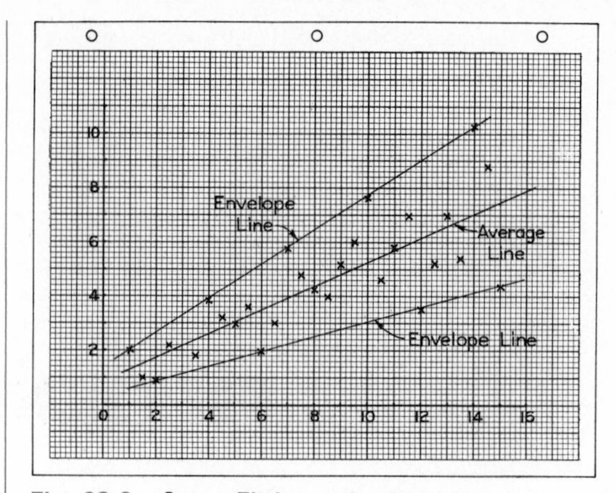


Fig. 28.2 **Curve Fitting—"Scatter" Data.**

When the data are numerous and the deviations of instrument readings are greater, a "scatter" plotting will result, Fig. 28.2. The straight line is fitted by enveloping the maximum and minimum points with light straight lines. An average line is determined, which will be near the mid-points within the envelopment, depending upon the density of the points plotted.

## 28.4 Limitations and Accuracy.

The derived empirical equation is at best an approximate representation of the data. The equation is valid only within the range of the experiment or observed data. Extrapolation of the curve (extension beyond the known range) or calculations from the equation for extrapolated values should be treated with caution.

The equation derived cannot be more accurate than the data used. If the data are accurate to three significant figures, the equation must also be expressed to the same number of significant figures.

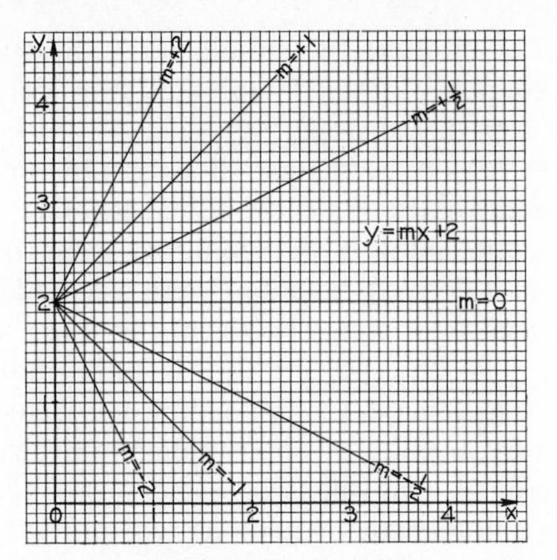


Fig. 28.3 **Equation of Straight Lines on Rectangular Grid.**

| Torque (T), oz-in. | Current Input to Motor (I), amperes |
|---|---|
| 10 | 6.6 |
| 20 | 8.1 |
| 30 | 9.6 |
| 40 | 11.2 |
| 50 | 12.6 |
| 60 | 14.3 |

(a) TABULATED DATA

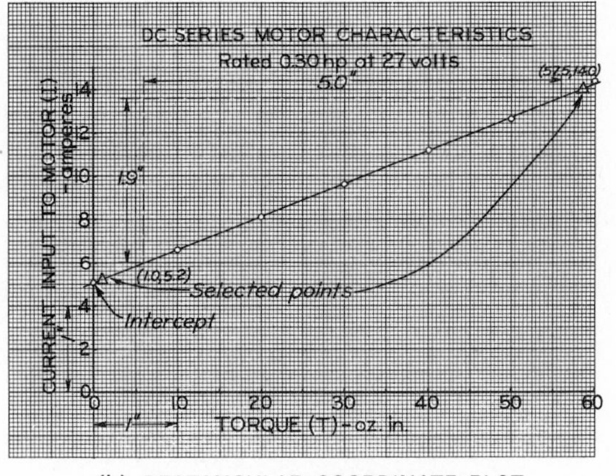


(b) RECTANGULAR COORDINATE PLOT

Fig. 28.4 **Rectangular Coordinate Solution.**

Errors are possible for a number of reasons. Variations of instrument accuracy (sensitivity) or the readings taken, which result in deviations of the points plotted from a straight line, are self-compensating in the graphical solution. This is accomplished by fitting the average line to the plotting and does not affect the validity of the data. "Wild" points, which occur infrequently due to erroneous readings or data recording, are noticeable in a graphical method and can be disregarded. Errors due to faulty instruments, calibrations, or settings will result in all the data being high or low. Graphical solutions will not compensate for this situation. The instrument must be corrected.

## 28.5 Empirical Equations—Solution by Rectangular Coordinates.

The equation for a straight line on a rectangular coordinate grid is $y = mx + b$. As shown in Fig. 28.3, m represents the slope of the line (the tangent of the angle between the line and the X-axis) and b is the intercept on the Y-axis (when $x = 0$). A negative slope is inclined downward to the right. A positive slope has an upward trend to the right. The intercept may be positive or negative.

In Fig. 28.4, the data presented at (a), plotted on rectangular coordinate paper, allow a straight line to be fitted to the points, as shown at (b). The procedure for derivation of the equation requires the determination of the slope of the line m and the I-intercept b when $T = 0$.

Slope-Intercept Method. The slope-intercept method requires the use of a graphical plot. Since in a majority of cases the scale modulus on the Y-axis does not equal the scale modulus on the X-axis, the trigonometric slope of the line s must be divided by the ratio of the axes scale moduli r to determine the slope of the line.

$$m = \frac{s}{r} = \frac{\text{trig. slope} = \dfrac{\text{y distance}}{\text{x distance}}}{\dfrac{\text{in. per unit Y-axis}}{\text{in. per unit X-axis}}}$$

$$= \frac{\dfrac{1.9}{5.0}}{\dfrac{.250}{.10}} = \frac{.380}{2.50} = .152$$

The slope of the line may also be obtained as follows:

$$m = \frac{I_2 - I_1}{T_2 - T_1}$$

$$= \frac{14.0 - 5.2}{57.5 - 1.0} = \frac{8.8}{56.5} = .156$$

The use of coordinates for slope determination will automatically result in the appropriate slope sign if treated algebraically. The difference in the slope values obtained from the two methods of calculation, although small, exists because the data used were different in each case. The first method required measurements of a y distance and the corresponding x distance. Selected coordinates on the straight line were used for the second method.

By observation, the intercept value (I value, when T = 0) is 5.05. The derived equation is

$$I = .152T + 5.05.$$

SELECTED POINTS METHOD. Since two unknowns must be determined, m and b, two equations solved simultaneously provides another method of equation derivation. Two points are selected; then data values on the line or points are established on the line. The points selected should be widely spread. The T and I values for these two points, substituted into the equation $I = mT + b$, will provide a solution.

Selected points: (57.5, 14.0) and (1.0, 5.2)

$I = mT + b$:

$$14.0 = 57.5m + b \quad (1)$$

$$5.2 = 1.0m + b \quad (2)$$

Subtracting:

$$8.8 = 56.5m$$

$$m = .156$$

Substituting in (1):

$$14.0 = 57.5(.156) + b$$

$$b = 14.0 - 8.97 = 5.03$$

Equation is

$$I = .156T + 5.03$$

The solution of the simultaneous equations, algebraically treated, automatically denotes the sign of the slope.

METHOD OF AVERAGES. If a definite number of data values of y were added together, and an equal number of computed values of y (from a derived equation) were also added, a difference of zero between the two sums ($\Sigma y_{data} - \Sigma y_{computed} = 0$) would indicate close agreement. On this premise, if the equation $\Sigma y = m\,\Sigma x + nb$ is applied to two equal groups of data values of x and y, solution of two simultaneous equations will provide the solution for the slope and intercept values. The term n in the equation is the number of data values of x and y summed.

In the example used, six items of data are available; therefore, two groups of three are added together.

$$\Sigma I = m\Sigma T + nb$$

Sum of second 3 data items: $38.1 = 150m + 3b$ (1)

Sum of first 3 data items: $24.3 = 60m + 3b$ (2)

Subtracting: $13.8 = 90m$

$$m = .153$$

Substituting in (1):

$$38.1 = 150(.153) + 3b$$

$$b = \frac{15.15}{3} = 5.05$$

Equation is

$$I = .153T + 5.05$$

The method of averages is based on a numerical derivation from the data only. The graphical plot is only required to verify the form of the curve and need not be accurate. This method requires additional time and care must be taken to prevent errors in calculation. Equation derivation by the slope-intercept method or selected points method is dependent on the accuracy of the representative straight line and not the data.

| Observed Data | | Slope-Intercept $I = .152\ T + 5.05$ | | Selected Points $I = .156\ T + 5.03$ | | Method of Averages $I = .153\ T + 5.05$ | |
|---|---|---|---|---|---|---|---|
| T | I | $I_{c1}$ | Residual | $I_{c2}$ | Residual | $I_{c3}$ | Residual |
| 10 | 6.6 | 6.6 | .0 | 6.6 | .0 | 6.6 | .0 |
| 20 | 8.1 | 8.1 | .0 | 8.2 | −.1 | 8.1 | .0 |
| 30 | 9.6 | 9.6 | .0 | 9.7 | −.1 | 9.6 | .0 |
| 40 | 11.2 | 11.1 | +.1 | 11.3 | −.1 | 11.2 | .0 |
| 50 | 12.6 | 12.7 | −.1 | 12.8 | −.2 | 12.7 | −.1 |
| 60 | 14.3 | 14.2 | +.1 | 14.4 | −.1 | 14.2 | +.1 |
| | | Sum = +.1 | | Sum = −.6 | | Sum = .0 | |

Fig. 28.5 **Computation of Residuals for Equations in §28.5.**

## 28.6 Computation of Residuals.

In order to determine the best approximation of the three derived equations, *residuals* should be computed. With the use of the derived empirical equations, the *dependent variable* is computed, using the independent variable data values. A comparison of computed values with observed data values reveals a plus or minus difference (residual), Fig. 28.5. If the observed value is smaller than the computed value, the residual is commonly denoted as minus. The algebraic sum of the residuals for each equation suggests which is the best equation (the smallest sum, plus or minus). For the example problem used in §28.5, the equation derived by the method of averages seems to be the best approximation.

If the data are accurate and fit a straight line closely, any of the three methods will produce good results. If the data are approximate and still fit a straight line well, the selected points method will produce the best results. When the straight line must be extended an appreciable distance to obtain the intercept value, inaccuracy can be expected, and the slope-intercept method is not a reliable method.

As a verification of the fact that the accuracy of derivation is dependent on the data accuracy in comparison to the method (graphical or numerical), the following correlation was performed.

In the method of averages, the first three and second three data values were added to obtain the equation $I = .153T + 5.05$. If the 1st, 3rd, and 5th data values, and the 2nd, 4th, and 6th data values were added instead, the solution of the simultaneous equations results in the equation $I = .160T + 4.80$. A comparison of the two equations indicates a great difference and a poor correlation because of the data. The numerical method cannot be assumed to be automatically a more accurate method in comparison to the graphical or semigraphical methods of slope-intercept or selected points, respectively.

## 28.7 Empirical Equation—Semilog Coordinates, $y = b(10)^{mx}$ or $y = be^{mx}$.

Data plotted on rectangular coordinate paper, which do not result in a straight line, may rectify to an approximate straight line graph on semilogarithmic coordinate grid, if the rectangular coordinate curve resembles an exponential curve, as shown in Fig. 28.6 (a) or (b). The base of the exponent may be either e (e = 2.718) or 10. An exponential curve intersects one of the

Fig. 28.6 **Exponential Curves.**

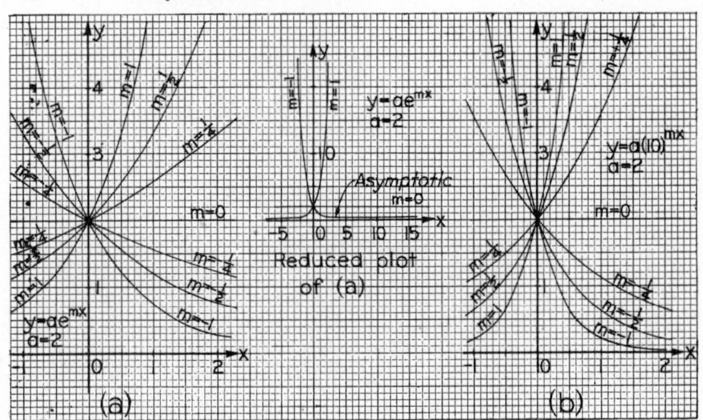

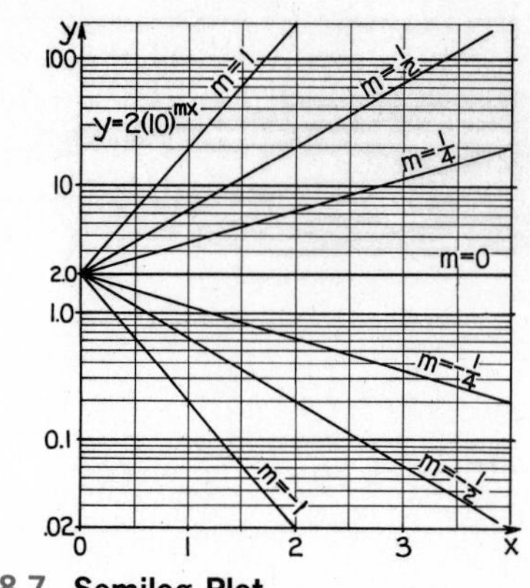


Fig. 28.7 **Semilog Plot.**

axes at a steep angle and is asymptotic to the other axis. The same data used to plot some of the curves in Fig. 28.6 (b), when plotted on semilog coordinate paper, rectify to straight lines as shown in Fig. 28.7.

Since semilog paper has logarithmic divisions along one of the axes (normally designated the Y-axis), the equation for a straight line on this type of graph paper is

$$\log y = mx + \log b$$

or $$\ln y = mx + \ln b$$

therefore,

$$y = b(10)^{mx} \qquad \text{or} \qquad y = be^{mx}$$

The derivation of the empirical equation requires the solution for the values b (Y-axis intercept, when $x = 0$) and m (the slope of the straight line to the X-axis). The same three methods used for rectangular coordinate solutions, §28.5, are applicable.

An alternate method is to plot log y values

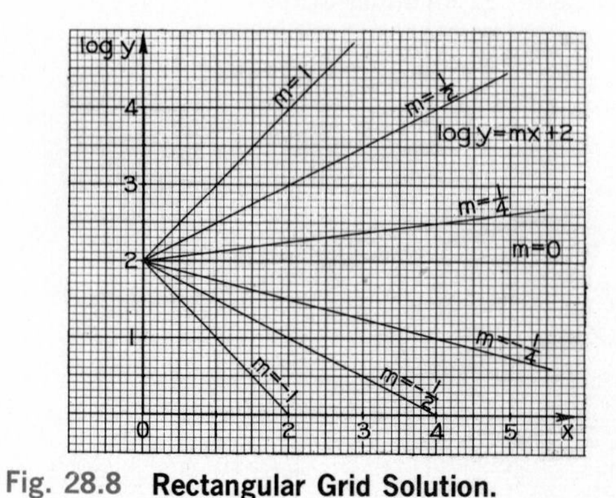


Fig. 28.8 **Rectangular Grid Solution.**

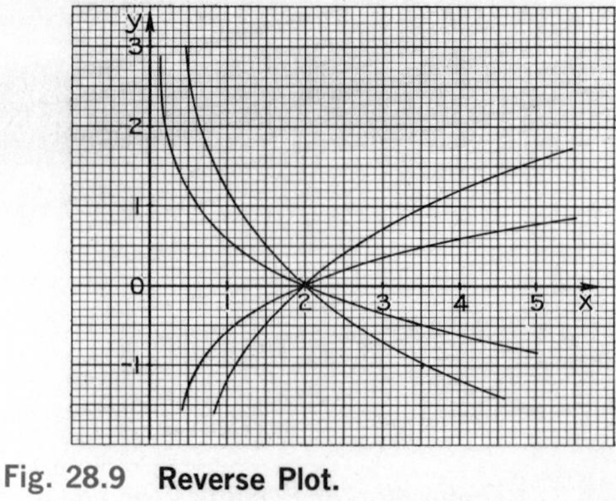


Fig. 28.9 **Reverse Plot.**

on rectangular coordinate graph paper, if semilog coordinate paper is not readily available, Fig. 28.8.

When a rectangular coordinate plot results in an X-axis intercept and a curve appearing asymptotic to the Y-axis, Fig. 28.9, the semilog plotting may rectify to a straight line if the logarithmic scale is placed on the X-axis. The equation becomes

$$x = a(10)^{my} \qquad \text{or} \qquad x = ae^{my}$$

The data for friction in thrust bearings, Fig. 28.10 (a), when plotted on rectangular coordinate paper, result in the curve shown at (b). The Y-axis intercept and appearance of being asymptotic to the X-axis suggest the possibility of a straight line plot on semilog coordinate paper, (c).

Slope-Intercept Method

Intercept = .980

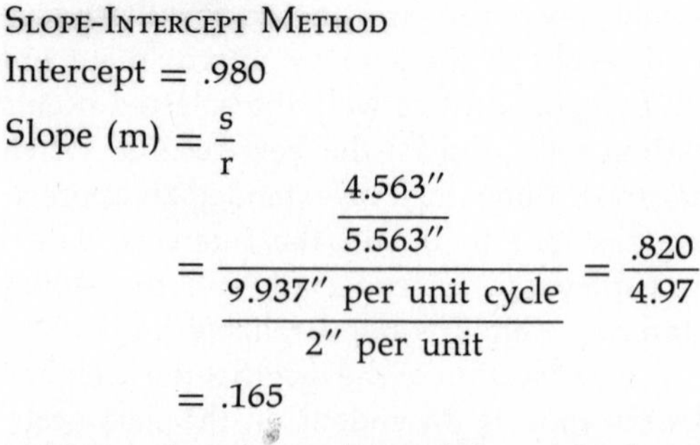

$$\text{Slope (m)} = \frac{s}{r}$$

$$= \frac{\dfrac{4.563''}{5.563''}}{\dfrac{9.937'' \text{ per unit cycle}}{2'' \text{ per unit}}} = \frac{.820}{4.97}$$

$$= .165$$

Notice that the scale moduli are the length of one log cycle on the log axis, and the length per unit (one) on the rectangular coordinate axis. The log cycle should be measured, since printing variations occur. The 10″ log cycle measured 9.937″ on the semilog graph paper used in this example. The intercept in this example is .980.

| Friction coefficient factor (k) | Log k | Ratio (l/a) |
|---|---|---|
| 1.00 | — | .0 |
| 1.05 | .021 | .25 |
| 1.15 | .061 | .50 |
| 1.30 | .114 | .75 |
| 1.45 | .161 | 1.00 |
| 1.65 | .218 | 1.25 |
| 1.80 | .255 | 1.50 |
| 2.00 | .301 | 1.75 |
| 2.15 | .332 | 2.00 |
| 2.35 | .371 | 2.25 |
| 2.60 | .415 | 2.50 |
| 2.80 | .447 | 2.75 |
| 3.00 | .477 | 3.00 |

(a) DATA

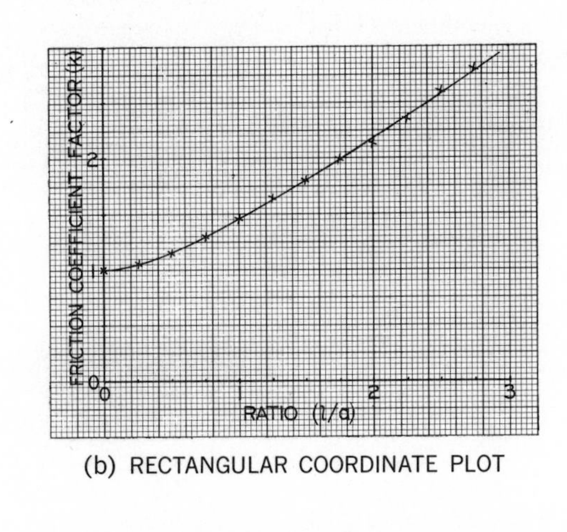


(b) RECTANGULAR COORDINATE PLOT

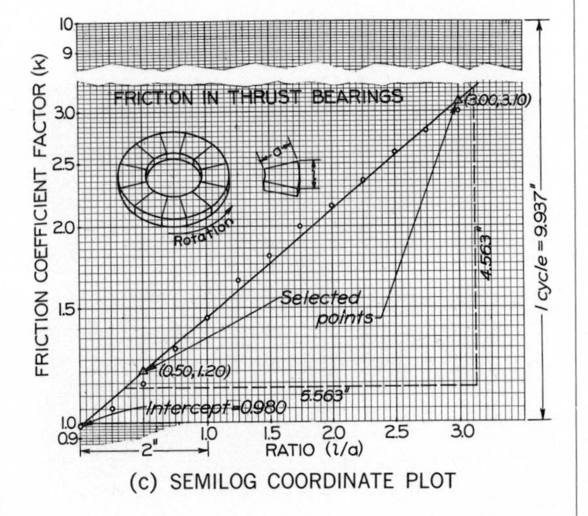


(c) SEMILOG COORDINATE PLOT

Fig. 28.10 **Semilog Empirical Equation Derivation.**

Alternate Slope Determination (Coordinate Method)

$$\text{Slope (m)} = \frac{\log y_2 - \log y_1}{x_2 - x_1}$$

$$= \frac{\log 3.1 - \log 1.2}{3.00 - .5}$$

$$= \frac{.4125}{2.50} = .165$$

$y = .980(10)^{.165x}$ $\quad$ $k = .980(10)^{.165(l/a)}$

If logarithms to the base e are preferred,

$$\text{Slope (m)} = \frac{\ln y_2 - \ln y_1}{x_2 - x_1}$$

$$= \frac{\ln 3.1 - \ln 1.2}{3.00 - .5}$$

$$= \frac{1.132 - .183}{2.50}$$

$$= \frac{.949}{2.50} = .380$$

$y = .980\ e^{.380x}$ $\quad$ $k = .980\ e^{.380(l/a)}$

Selected Points Method

Selected points: (.50, 1.2) and (3.00, 3.10)

$y = b(10)^{mx}$:

$$\log y = \log b + mx \log 10$$

$$\log 3.10 = \log b + m(3.00)$$

$$.492 = \log b + 3.00m \qquad (1)$$

$$\log 1.20 = \log b + m(.50)$$

$$.079 = \log b + .50m \qquad (2)$$

Subtracting (2) from (1) gives

$$.413 = 2.50m$$

$$m = .165$$

Substituting into equation (1):

$$.492 = \log b + 3.00(.165)$$

$$\log b = .492 - .495$$

$$= -.003 = 9.997 - 10$$

$b = .993$ $\quad$ $k = .993(10)^{.165(l/a)}$

| Observed Data | | Slope-Intercept $k = .980(10)^{.165(l/a)}$ | | Selected Points $k = .993(10)^{.165(l/a)}$ | | Method of Averages $k = .980(10)^{.168(l/a)}$ | |
|---|---|---|---|---|---|---|---|
| l/a | k | $k_{c_1}$ | Residual | $k_{c_2}$ | Residual | $k_{c_3}$ | Residual |
| .00 | 1.00 | .98 | +.02 | .99 | +.01 | .98 | +.02 |
| .25 | 1.05 | 1.08 | −.03 | 1.09 | −.04 | 1.08 | −.03 |
| .50 | 1.15 | 1.19 | −.04 | 1.20 | −.05 | 1.19 | −.04 |
| .75 | 1.30 | 1.30 | .00 | 1.32 | −.02 | 1.31 | −.01 |
| 1.00 | 1.45 | 1.42 | +.03 | 1.45 | .00 | 1.44 | +.01 |
| 1.25 | 1.65 | 1.58 | +.07 | 1.60 | +.05 | 1.59 | +.06 |
| 1.50 | 1.80 | 1.73 | +.07 | 1.76 | +.04 | 1.75 | +.05 |
| 1.75 | 2.00 | 1.91 | +.09 | 1.93 | +.07 | 1.93 | +.07 |
| 2.00 | 2.15 | 2.09 | +.06 | 2.12 | +.03 | 2.12 | +.03 |
| 2.25 | 2.35 | 2.31 | +.04 | 2.33 | +.02 | 2.34 | +.01 |
| 2.50 | 2.60 | 2.53 | +.07 | 2.57 | +.03 | 2.58 | +.02 |
| 2.75 | 2.80 | 2.79 | +.01 | 2.83 | −.03 | 2.83 | −.03 |
| 3.00 | 3.00 | 3.07 | −.07 | 3.11 | −.11 | 3.13 | −.13 |
| | | Sum = +.32 | | Sum = .00 | | Sum = +.03 | |

Fig. 28.11 **Computation of Residuals for Equations in §28.7.**

The solution for logarithms to the base e is as follows:

$y = be^{mx}$:

$$\ln y = \ln b + mx \ln e$$
$$\ln 3.10 = \ln b + m\,(3.00) \ln e$$
$$1.132 = \ln b + 3.00m \qquad (1)$$
$$\ln 1.2 = \ln b + m\,(.50) \ln e$$
$$0.183 = \ln b + .50m \qquad (2)$$

Subtracting (2) from (1) gives

$$.949 = 2.50m$$
$$m = .380$$

Substituting into equation (1):

$$1.132 = \ln b + 3.00(.380)$$
$$\ln b = 1.132 - 1.140$$
$$= -.008$$
$$b = .992$$

$y = .992\, e^{.380x}$ $\qquad k = .992\, e^{.380(l/a)}$

Fig. 28.12
**Power Curves.**

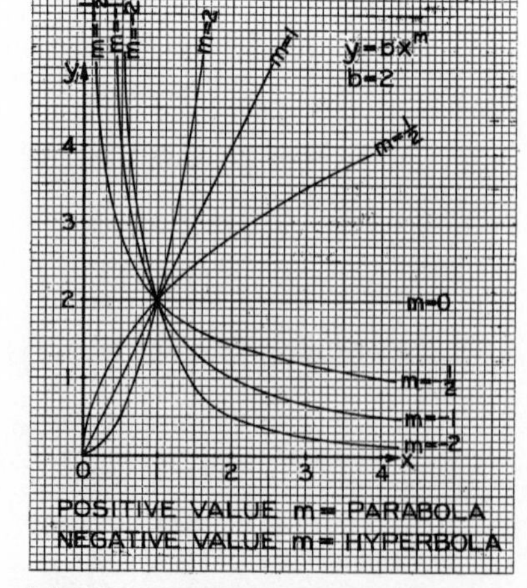


(a) RECTANGULAR COORDINATE PLOT

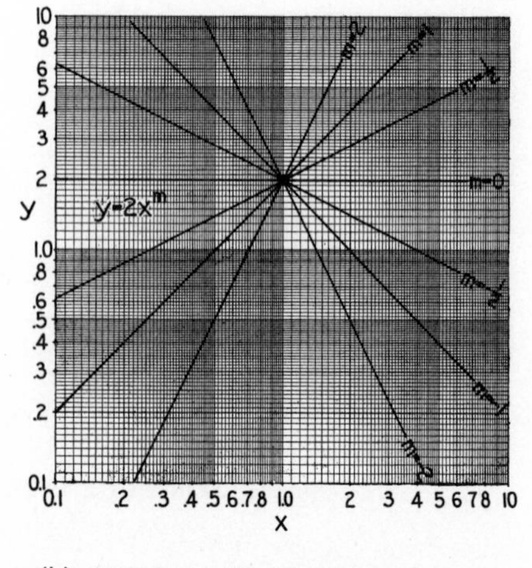


(b) LOGARITHMIC COORDINATE PLOT

$$\Sigma \log y = n \log b + m\Sigma x \log 10$$

Sum of last 6 terms:

$$2.343 = 6 \log b + m(14.25) \quad (1)$$

Sum of first 6 terms:

$$.830 = 6 \log b + m(\ 5.25) \quad (2)$$

Subtracting (2) from (1) gives

$$1.513 = \qquad m(\ 9.00)$$
$$m = .168$$

Substituting into equation (2):

$$.830 = 6 \log b + 5.25(.168)$$

$$\log b = \frac{.830 - .883}{6}$$

$$= -.009 = 9.991 - 10$$

$b = .980$ $\qquad$ $k = .980(10)^{.168(1/a)}$

The solution for logarithms to the base e is as follows:

$$\Sigma \ln y = n \ln b + m\Sigma x \ln e$$

Sum of last 6 terms:

$$5.397 = 6 \ln b + m(14.25) \quad (1)$$

Sum of first 6 terms:

$$1.911 = 6 \ln b + m(\ 5.25) \quad (2)$$

Subtracting (2) from (1) gives

$$3.486 = \qquad m(\ 9.00)$$
$$m = .387$$

Substituting into equation (2):

$$1.911 = 6 \ln b + .387(5.25)$$

$$\ln b = \frac{1.911 - 2.032}{6}$$

$$= -.0202$$
$$b = .980$$

$y = .980\ e^{.387x}$ $\qquad$ $k = .980\ e^{.387(1/a)}$

Residuals for the equation form $y = b(10)^{mx}$ were computed for the best approximation and are tabulated in Fig. 28.11.

**28.8 Empirical Equations—Logarithmic Coordinates, $y = bx^m$.** Curves that plot as a parabola through the origin or a hyperbola asymptotic to the X- and Y-axes, are known as power curves, Fig. 28.12 (a), and can be rectified to a straight line by plotting the same data on logarithmic coordinate paper, as shown at (b). The equation for a straight line on logarithmic paper is

$$\log y = m \log x + \log b$$

or

$$y = bx^m$$

The derivation of an empirical equation requires the determination of the values for the slope m of the line and the intercept value b (y intercept when $x = 1$). The intercept is at the axis value of 1.0 since $\log 1.0 = 0$. The three methods of equation derivation used for rectangular and semilog coordinate solutions, §§28.5 and 28.7, are applicable to logarithmic coordinate plottings.

As an example of equation derivation, the data for head discharge for a 1.00″ diameter circular orifice, Fig. 28.13 (a), were plotted on rectangular coordinate grid paper, (b). The resulting curve, a parabola through the origin, suggests a power curve. Plotting the same data on logarithmic graph paper yields a straight line graph, as shown in (c).

SLOPE-INTERCEPT METHOD. In order to obtain the intercept value b on the Y-axis, when $x =$ 1.0, the straight line must be extended downward and to the left, since the Y-axis values are only to .1. Three-cycle log paper was used; however, the intercept is still below the bottom cycle. The intercept can be measured, as shown. The use of four-cycle log paper would result in a less accurate plotting. This procedure is necessary only for the slope-intercept method.

Intercept $b = .077$, when $D = 1.00$

If the lengths of the log cycles are the same along both axes, the slope m is the tangent of the angle between the straight line and the X-axis.

$$\text{Slope } m = \frac{19.55}{10} = 1.955$$

**Fig. 28.13 Logarithmic Empirical Equation Derivation.**

(a) DATA

| Discharge (D), ft³/sec | Head (H), ft³ |
|---|---|
| 4 | 1.15 |
| 5 | 1.75 |
| 6 | 2.5 |
| 7 | 3.4 |
| 8 | 4.5 |
| 9 | 5.5 |
| 10 | 6.9 |
| 15 | 15.5 |
| 20 | 27 |
| 25 | 43 |
| 30 | 63 |

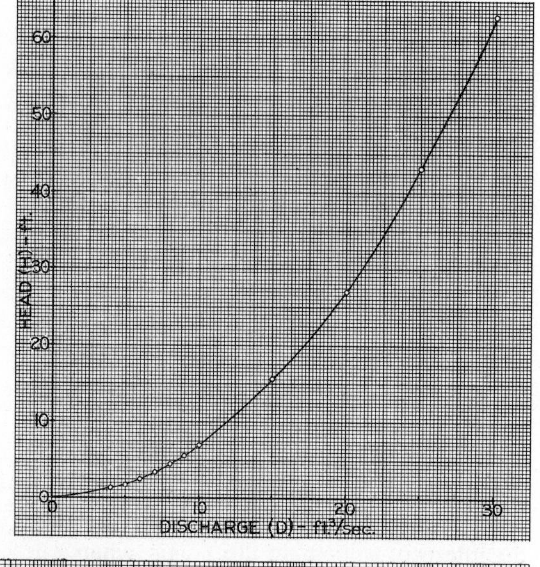


(b) RECTANGULAR COORDINATE PLOT

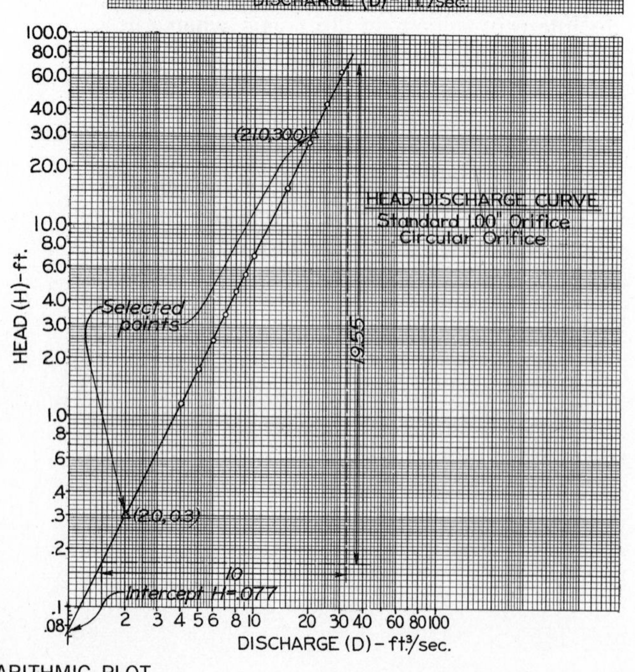


(c) LOGARITHMIC PLOT

Equation is

$$H = .077\ D^{1.955}$$

Selected Points Method

Selected points: (2.0, .3) and (21.0, 30.0)

$y = bx^m$:

$$\log y = \log b + m \log x$$
$$1.477 = \log b + m(1.322) \quad (1)$$
$$-.523 = \log b + m(.301) \quad (2)$$

Subtracting: $2.000 = m(1.021)$

$$m = 1.960$$

Substituting in equation (1):

$$1.477 = \log b + 1.960(1.322)$$
$$\log b = 1.477 - 2.590$$
$$= -1.113 = 8.887 - 10$$
$$b = .0771 \qquad H = .0771\ D^{1.960}$$

Method of Averages

$y = bx^m$:

$$\log y = \log b + m \log x$$
$$\Sigma \log y = n \log b + m\Sigma \log x$$

Sum of second 5:

$$5.836 = 5 \log b + m(5.829) \quad (1)$$

Sum of first 5:

$$1.887 = 5 \log b + m(3.827) \quad (2)$$

Subtracting: $3.949 = m(2.002)$

$$m = 1.971$$

Substituting in equation (1):

$$5.836 = 5 \log b + (1.971)(5.829)$$
$$\log b = \frac{5.836 - 11.49}{5}$$
$$= -1.131 = 8.869 - 10$$
$$b = .0739 \qquad H = .0739\ D^{1.971}$$

To test for the best approximation, residuals were computed and tabulated in Fig. 28.14.

As a second example of equation derivation, the data for influence of tolerance on cost of machining, Fig. 28.15 (a), were plotted on rectangular coordinate grid paper, (b). The re-

| Observed Data | | Slope-Intercept $H = .077\ D^{1.955}$ | | Selected Points $H = .0771\ D^{1.960}$ | | Method of Averages $H = .0739\ D^{1.971}$ | |
|---|---|---|---|---|---|---|---|
| D | H | $H_{c_1}$ | Residual | $H_{c_2}$ | Residual | $H_{c_3}$ | Residual |
| 4 | 1.15 | 1.16 | − .01 | 1.17 | − .02 | 1.14 | + .01 |
| 5 | 1.75 | 1.80 | − .05 | 1.81 | − .06 | 1.77 | − .02 |
| 6 | 2.50 | 2.56 | − .06 | 2.59 | − .09 | 2.53 | − .03 |
| 7 | 3.40 | 3.47 | − .07 | 3.51 | − .11 | 3.44 | − .04 |
| 8 | 4.50 | 4.52 | − .02 | 4.57 | − .07 | 4.47 | + .03 |
| 9 | 5.50 | 5.67 | − .17 | 5.74 | − .24 | 5.67 | − .17 |
| 10 | 6.90 | 6.98 | − .08 | 7.05 | − .15 | 6.95 | − .05 |
| 15 | 15.50 | 15.5 | .00 | 15.7 | − .20 | 15.5 | .00 |
| 20 | 27 | 27.1 | − .10 | 27.5 | − .50 | 27.4 | − .40 |
| 25 | 43 | 42 | +1.0 | 42.8 | + .20 | 42.7 | + .30 |
| 30 | 63 | 60 | +3.0 | 60.9 | +2.1 | 60.6 | +2.4 |
| | | Sum = +3.44 | | Sum = + .86 | | Sum = +2.03 | |

Fig. 28.14 **Computation of Residuals for Equation in §28.8.**

Fig. 28.15 **Logarithmic Empirical Equation Derivation.**

INFLUENCE OF TOLERANCE ON COST OF MACHINING

| Tolerance (T), in. | Relative Cost of Machining (C) | Log T | Log C |
|---|---|---|---|
| .001 | 5.00 | −3.000 | .699 |
| .002 | 2.60 | −2.699 | .415 |
| .003 | 1.70 | −2.523 | .231 |
| .004 | 1.25 | −2.398 | .097 |
| .005 | 1.00 | −2.301 | .000 |
| .006 | .90 | −2.222 | −.046 |
| .007 | .80 | −2.155 | −.097 |

(a) DATA

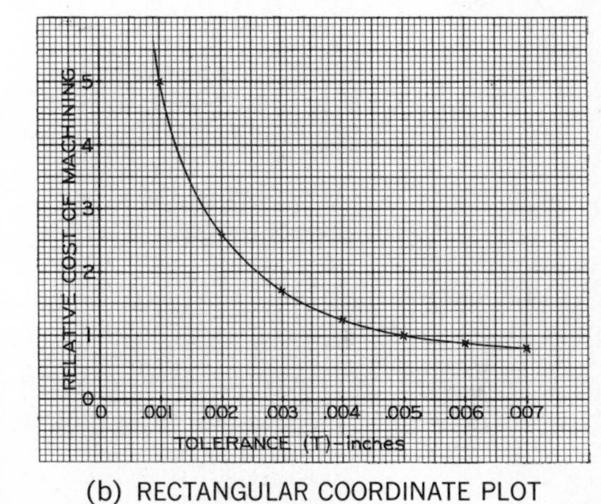


(b) RECTANGULAR COORDINATE PLOT

sulting curve, asymptotic to both axes, suggests a power curve. The same data plotted on logarithmic graph paper rectifies to a straight line graph, as shown in (c).

Slope-Intercept Method. In order to obtain the intercept value b on the Y-axis, when x = 1.0, the straight line must be extended downward and to the right, since the X-axis values are only to .10. Point A was shifted to point B (corresponding value) on the upper portion of the graph paper to enable the extension of the curve on the same graph paper. This procedure is necessary only for the Slope-Intercept Method.

Intercept b = .0062, when T = 1.00

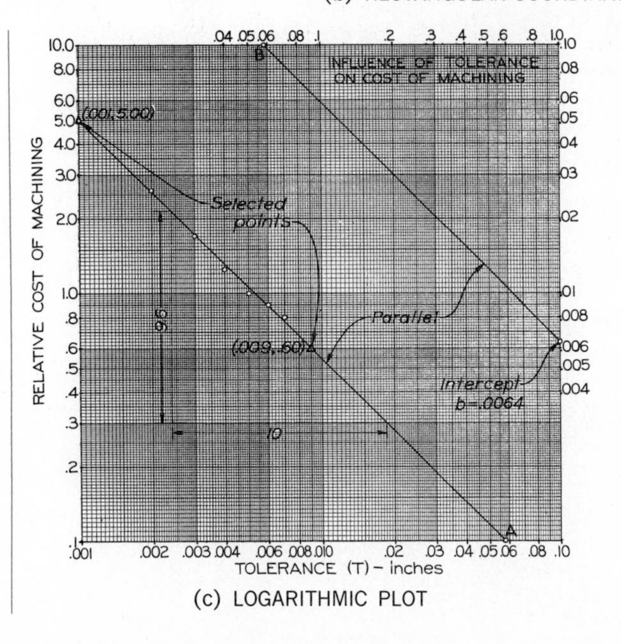


(c) LOGARITHMIC PLOT

| Observed Data | | Slope-Intercept $C = .0062T^{-.960}$ | | Selected Points $C = .00637T^{-.965}$ | | Method of Averages $C = .00528T^{-.994}$ | |
|---|---|---|---|---|---|---|---|
| T | C | $C_{C_1}$ | Residual | $C_{C_2}$ | Residual | $C_{C_3}$ | Residual |
| .001 | 5.00 | 4.77 | +.23 | 4.97 | +.03 | 5.02 | −.02 |
| .002 | 2.60 | 2.43 | +.17 | 2.55 | +.05 | 2.54 | +.06 |
| .003 | 1.70 | 1.64 | +.06 | 1.73 | −.03 | 1.69 | +.01 |
| .004 | 1.25 | 1.25 | .00 | 1.31 | −.06 | 1.28 | −.03 |
| .005 | 1.00 | 1.01 | −.01 | 1.06 | −.06 | 1.02 | −.02 |
| .006 | .90 | .85 | +.05 | .89 | +.01 | .86 | +.04 |
| .007 | .80 | .73 | +.07 | .77 | +.03 | .73 | +.07 |
| | | | Sum = +.57 | | Sum = −.03 | | Sum = +.11 |

Fig. 28.16 **Computation of Residuals for Equations in §28.8.**

The lengths of the log cycles are usually the same along both axes; the slope m is the tangent of the angle between the straight line and the X-axis.

$$\text{Slope } m = \frac{-9.6}{10.0} = -.960$$

The sign of the slope must be included in the computation.

$$C = .0062T^{-.960}$$

SELECTED POINTS METHOD. Selected points: (.001, 5.00) and (.009, .60)

$y = bx^m$:

$$\log y = \log b + m \log x$$

$$.699 = \log b + m(-3.00) \quad (1)$$

$$9.778 - 10 = \log b + m(7.954 - 10)$$

or

$$-.222 = \log b + m(-2.046) \quad (2)$$

$$\text{Subtracting: } .921 = m(-.954)$$

$$m = -.965$$

Substituting in equation (1):

$$.699 = \log b + (-.965)(-3.000)$$

$$\log b = .699 - 2.895$$

$$= -2.196 = 7.804 - 10$$

$$b = .00637 \qquad C = .0064T^{-.965}$$

METHOD OF AVERAGES

$y = bx^m$:

$$\log y = \log b + m \log x;$$

$$\Sigma \log y = n \log b + m\Sigma \log x$$

Sum of first 3:

$$1.345 = 3 \log b + m(-8.222) \quad (1)$$

Sum of second 3:

$$.051 = 3 \log b + m(-6.921) \quad (2)$$

Subtracting (2) from (1) gives

$$1.294 = m(-1.301)$$

$$m = -.994$$

Substituting in equation (1):

$$1.345 = 3 \log b + (-.994)(-8.222)$$

$$\log b = \frac{1.345 - 8.175}{3}$$

$$= -2.277 = 7.723 - 10$$

$$b = .00528 \qquad C = .0053\ T^{-.994}$$

To test for the best approximation, residuals were computed and tabulated in Fig. 28.16.

**28.9 Empirical Equation Problems.** Plot the data for the following problems on the appropriate form of graph paper. Determine the equation by the three methods included in the text, and analyze with residuals the best equation.

**Prob. 28.1** Thermal Conductivity of Insulating Materials.

| Mean Temperature (T), °F | Thermal Conductivity (C), Btu/ft²/hr/°F/in. | |
|---|---|---|
| | Celotex | Rockwool Blanket |
| 40 | .35 | .24 |
| 80 | .355 | .27 |
| 120 | .36 | .30 |
| 160 | .365 | .32 |
| 200 | .372 | .35 |
| 240 | .380 | .38 |

**Prob. 28.2** Power at Different Altitudes for Compound Compression of Air.

| Altitude (A), ft above sea level | Power (P), indicated hp/100 ft³/min | |
|---|---|---|
| | 80 lb gage | 125 lb gage |
| 1000 | 13.55 | 16.83 |
| 2000 | 13.30 | 16.50 |
| 3000 | 13.05 | 16.17 |
| 4000 | 12.8 | 15.83 |
| 5000 | 12.55 | 15.50 |
| 6000 | 12.30 | 15.17 |
| 7000 | 12.05 | 14.85 |
| 8000 | 11.80 | 14.50 |
| 9000 | 11.55 | 14.17 |
| 10,000 | 11.30 | 13.82 |

**Prob. 28.3** Impact Breaking Strength of Gray Iron (specimen: cantilever beam $\frac{9}{16}$″ long, $\frac{1}{2}$″ wide; Class 20 gray cast iron).

| Thickness of $\frac{1}{2}$″ Wide Specimen (t), in. | Fracture Energy (E), ft-lb (scatter data) |
|---|---|
| 0 | 0 |
| $\frac{1}{8}$ | 1–2 |
| $\frac{1}{4}$ | $2\frac{3}{8}$–$3\frac{7}{8}$ |
| $\frac{3}{8}$ | $3\frac{3}{4}$–$5\frac{1}{2}$ |
| $\frac{1}{2}$ | $5\frac{1}{2}$–7 |
| $\frac{5}{8}$ | 7–$8\frac{5}{8}$ |
| $\frac{3}{4}$ | $8\frac{1}{2}$–$10\frac{1}{4}$ |

**Prob. 28.4** Thermal Aging of Adhesives.

| Aging Time (T), yr | Shear Strength ($S_s$), 1000 psi | | | |
|---|---|---|---|---|
| | Nitrile Phenolic | | Epoxy Phenolic | |
| | At 250°F | At 350°F | At 250°F | At 350°F |
| 0 | 2.20 | 1.60 | 2.08 | 1.95 |
| $\frac{1}{4}$ | 2.18 | 1.50 | 1.93 | 1.70 |
| $\frac{1}{2}$ | 2.13 | 1.40 | 1.80 | 1.45 |
| $\frac{3}{4}$ | 2.10 | 1.30 | 1.65 | 1.20 |
| 1 | 2.05 | 1.20 | 1.52 | .90 |
| $1\frac{1}{4}$ | 2.00 | 1.10 | 1.38 | .65 |
| $1\frac{1}{2}$ | 1.97 | 1.00 | 1.25 | .40 |
| $1\frac{3}{4}$ | 1.93 | .90 | 1.10 | — |
| 2 | 1.87 | .80 | .95 | — |
| $2\frac{1}{4}$ | 1.83 | .70 | .80 | — |
| $2\frac{1}{2}$ | 1.80 | .60 | .68 | — |

**Prob. 28.5** Warmup Heat.

| Temperature Rise ($T_{final} - T_{initial} = \Delta t$), °F | Warmup Heat (H), Btu/lb | | | |
|---|---|---|---|---|
| | Water | Paraffin Wax | Aluminum, Air, Nitrogen | Copper, Steel, Stainless |
| 0 | 0 | 0 | 0 | 0 |
| 50 | 50 | 35 | 10 | 5 |
| 100 | 100 | 70 | 25 | 10 |
| 150 | 150 | 105 | 35 | 15 |
| 200 | 200 | 140 | 47 | 20 |
| 250 | — | 175 | 58 | 25 |
| 300 | — | 210 | 70 | 30 |
| 350 | — | 245 | 83 | 35 |
| 400 | — | 280 | 95 | 40 |

**Prob. 28.6** Coefficient of Friction vs Feed Rate (material: SAE 4340 steel, 200 BHN; cutting speed: 542 rpm; tool: sintered carbide, +0° rake angle).

| Feed Rate (t), in./revolution | Coefficient of Friction ($\mu$) |
|---|---|
| .00055 | 1.175 |
| .0011 | 1.075 |
| .00235 | 1.050 |
| .00295 | 1.025 |
| .00370 | .975 |
| .00620 | .850 |

**Prob. 28.7** Gear Pump Efficiency (internal-external gear pump at 600 rpm).

| Pressure (P), psi | Volumetric Efficiency (V), % |
|---|---|
| 100 | 82.0 |
| 200 | 75.0 |
| 300 | 68.5 |
| 400 | 61.5 |
| 500 | 54.5 |
| 600 | 47.5 |
| 700 | 40.0 |
| 800 | 33.0 |

## SEMILOG COORDINATE SOLUTION

**Prob. 28.8** Allowable Working Stresses for Compression Springs (chrome-vanadium wire, ASTM-A231, SAE 6150).

| Wire Diameter (D), in. | Torsional Stress ($S_t$), 1000 psi | |
|---|---|---|
| | Severe Service | Average Service |
| .03 | 91 | 121 |
| .05 | 84 | 112 |
| .07 | 79 | 105 |
| .09 | 76 | 101 |
| .11 | 73 | 97 |
| .13 | 71 | 94 |
| .15 | 69 | 92 |
| .17 | 67 | 90 |
| .19 | 66 | 87 |
| .21 | 64 | 86 |
| .23 | 63 | 84 |
| .25 | 62 | 83 |

**Prob. 28.9** Electrical Resistance of 95% Alumina Ceramic.

| Temperature (T), °C | Electrical Resistance (R), megohms |
|---|---|
| 400 | 240 |
| 450 | 120 |
| 500 | 60 |
| 550 | 31 |
| 600 | 15 |
| 650 | 7.5 |
| 700 | 3.7 |
| 750 | 2.0 |
| 800 | .9 |
| 850 | .47 |

**Prob. 28.10** High-Temperature Alloy (René 41 strength).

| Rupture Life at 1800°F (L), hr | Stress (S), 1000 psi |
|---|---|
| 20 | 14.3 |
| 40 | 12.9 |
| 60 | 12.0 |
| 80 | 11.5 |
| 100 | 11.0 |
| 200 | 9.7 |
| 400 | 8.3 |
| 600 | 7.5 |
| 800 | 6.9 |
| 1000 | 6.5 |

**Prob. 28.11** K Factors for Preliminary Estimates of Helical and Spur Gear Size.

| Brinell Hardness Number (BHN) | K Factor (scatter data) |
|---|---|
| 200 | 200–420 |
| 250 | 240–560 |
| 300 | 350–680 |
| 350 | 460–840 |
| 400 | 575–1050 |
| 450 | 750–1380 |
| 500 | 900–1700 |
| 550 | 1050–2100 |
| 600 | 1250–2500 |

**Prob. 28.12** Velocity of Sound in Air at 70°F.

| Pressure (p), 1000 psi | Velocity (V), ft/sec |
|---|---|
| .5 | 1150 |
| 1.0 | 1175 |
| 1.5 | 1215 |
| 2.0 | 1260 |
| 2.5 | 1310 |
| 3.0 | 1365 |
| 3.5 | 1420 |
| 4.0 | 1480 |
| 5.0 | 1610 |

**Prob. 28.13** Bearing Lubrication Test (single row, radial, nonloading groove bearings, operating at 3000 rpm with negligible load and one-fourth standard pack of New Departure Code C grease).

| Bearing Outer Race Temperature (T), °F | Base Grease Life (B), hr |
|---|---|
| 80 | 40,000 |
| 100 | 28,500 |
| 120 | 21,000 |
| 140 | 15,000 |
| 160 | 10,750 |
| 180 | 7,600 |
| 200 | 5,500 |
| 220 | 4,000 |
| 240 | 2,850 |

## LOG–LOG COORDINATE SOLUTION

**Prob. 28.14** Power Requirements for Rotary Drilling Operations (300 hydraulic hp required to pump 80 lb mud through a $4\frac{1}{2}''$ full hole string).

| Mud Circulation (G), gal/min | Standpipe Pressure (P), 1000 psi |
|---|---|
| 300 | 17.2 |
| 400 | 12.8 |
| 500 | 10.2 |
| 600 | 8.5 |
| 700 | 7.3 |
| 800 | 6.35 |
| 900 | 5.7 |
| 1000 | 5.1 |

**Prob. 28.15** Gas Carburizing Heat-Treatment of Low Carbon Steel at 1700°F.

| Carburizing Time (T), hr | Case Depth (D), in. |
|---|---|
| 4 | .047 |
| 8 | .067 |
| 12 | .084 |
| 16 | .096 |
| 20 | .110 |
| 24 | .124 |
| 28 | .136 |

**Prob. 28.16** Thermalizing Heat-Treatment Process of Molybdenum Alloy (Mo–.5% Ti).

| Distance from Hardened Surface (d), in. | Microhardness (H) (20-gram load) |
|---|---|
| 0.0003 | 1800 |
| 0.001 | 950 |
| 0.002 | 660 |
| 0.003 | 540 |
| 0.004 | 460 |
| 0.005 | 410 |
| 0.006 | 380 |
| 0.007 | 350 |
| 0.008 | 330 |
| 0.009 | 310 |

**Prob. 28.17** Welded Stainless Honeycomb Sandwich Panels (unsupported core made from 17–7 PH alloy; shear is in longitudinal direction, parallel to plane of nodal spotwelds).

| Cell Size (d), in. | Core Shear Modulus (G), psi | | | |
|---|---|---|---|---|
| | t = .001 | t = .002 | t = .003 | t = .004 |
| .1 | 165 | — | — | — |
| .15 | — | 190 | — | — |
| .2 | 70 | 139 | 205 | — |
| .25 | — | — | — | 215 |
| .3 | 45 | 89 | 128 | 175 |
| .4 | 31 | 62 | 93 | 126 |
| .5 | 24 | 49 | 74 | 99 |
| .6 | 19 | 40 | 61 | 81 |
| .7 | 16 | 34 | 52 | 69 |
| .8 | 13 | 30 | 44 | 60 |

**Prob. 28.18** Glass Bead Filter Efficiency.

| Rate of Flow (Q), gal/min | Flow Loss (p), in. of Hg |
|---|---|
| 5 | .14 |
| 10 | .5 |
| 15 | 1.1 |
| 20 | 1.9 |
| 25 | 2.9 |
| 30 | 4.0 |
| 35 | 5.4 |

**Prob. 28.19** Fluorothene Dielectric Strength.

| Thickness (t), mils | Dielectric Strength (E), volts/mil |
|---|---|
| 13 | 1400 |
| 19 | 1200 |
| 25 | 1000 |
| 40 | 800 |
| 50 | 700 |
| 70 | 600 |
| 90 | 500 |
| 120 | 450 |

# 29
# graphical mathematics

by E. J. Mysiak*

**29.1 Introduction.** In general, mathematical problems may be solved *algebraically* (using numerical and mathematical symbols) or *graphically* (using drawing techniques). The algebraic method is predominantly a verbal approach in comparison to the visual methods of graphics; therefore, errors and discrepancies are more evident and subject to detection in a graphical presentation. The advantages of graphics are quite evident in any mathematics text, since most writers in the field of mathematics supplement their algebraic notations with graphical illustrations to illuminate their writings and improve the visualization of the solutions.

Naturally, the question of accuracy arises, but it must be remembered that measuring instruments are themselves graphical devices. The data obtained from an instrument reading, or from measurements of physical quantities, in many cases are recorded to three significant figures. Such a degree of accuracy is practical and readily substantiated by graphical methods of computation.

The graphical methods cannot be used exclusively, but neither can the algebraic methods be used to the fullest degree of effectiveness without the use of graphics. The engineer should be familiar with both methods in order to convey a clearer understanding of his analyses and designs.

Since equations are mathematical expressions, mathematical operations can be: (1) computations with equations, (2) derivation of equations, and (3) solutions of particular equations. Graphic computations with the use of coordinate paper are included in this chap-

*Mechanical Engineer, International Electro-Magnetics, Inc.

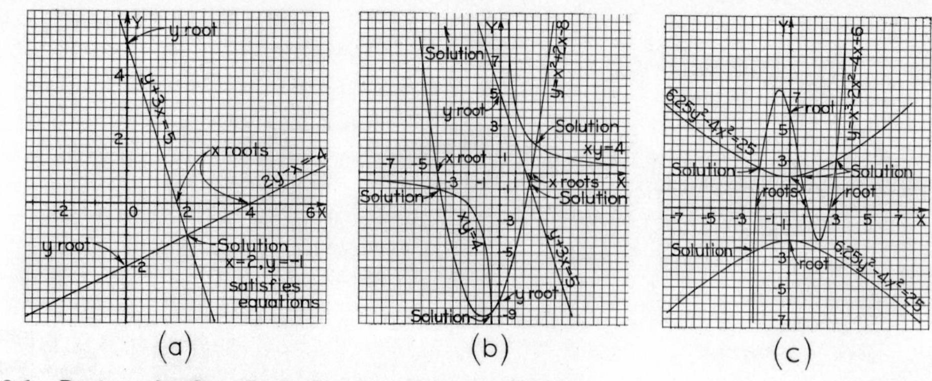


Fig. 29.1 **Rectangular Coordinate Graphs—Algebraic Solutions.**

ter as well as solutions of particular equations, algebra and the calculus. The use of alignment charts for graphic computations and the specific topic of graphic empirical equation derivation are discussed in Chapters 27 and 28, respectively.

## 29.2 Rectangular Coordinate Algebraic Solution Graphs.

Rectangular coordinate paper can be effectively used to visualize algebraic solutions. A common application is to solve two simultaneous polynomial equations with two unknowns. A graph of two linear equations is shown in Fig. 29.1 (a), indicating the solution as the intersection of the two lines. Linear equations are simple to solve algebraically; however, a graphical approach is advantageous in the solution of a quadratic equation for a parabola ($y = x^2 + 2x - 8$) with a linear equation for a straight line ($y + 3x = 5$) or an equation for an equilateral hyperbola ($xy = 4$), Fig. 29.1 (b). The graphical plot can also be used for further analysis. Note the linear equation and the equation for the equilateral hyperbola do not have a real simultaneous equation solution. The ease of visualization and solution is evident as the equations become more complex. Figure 29.1 (c) is a graphical solution for the cubic equation $y = x^3 - 2x^2 - 4x + 6$ and the equation for a hyperbola $6.25y^2 - 4x^2 = 25$. If the data were empirical, without an equation, the graphical solution would be necessary.

Another graphic application to algebraic problems is the determination of the roots for the equations. The roots of the equation are determined by the intersection of the curve and the corresponding axis. The linear equation of Fig. 29.1 (a) has one root per variable, as denoted. The highest power of the quadratic equation for the x term of Fig. 29.1 (b) being 2, there are two roots on the X-axis. The y term has only one root. The equilateral hyperbola, in the same graph, is asymptotic to both axes and therefore has no real roots. The symmetrical hyperbola of Fig. 29.1 (c) has one plus and one minus root. A cubic equation can have one or three real roots, depending on the location of the curve relative to the axes. The cubic equation of Fig. 29.1 (c) has three roots.

A more convenient procedure to determine roots of an equation is to separate the equation and plot two separate curves. The quadratic equation of Fig. 29.1 (b), $y = x^2 + 2x - 8$, was separated to $y = x^2 = 8 - 2x$. The equation $y = x^2$ results in a symmetrical parabola, Fig. 29.2. The remaining portion of the quadratic equation, $y = 8 - 2x$, is a linear equation. The intersection of the parabola and the straight line provides the same root solution for the quadratic equation. The parabolic curve can be used with other linear portions of similar quadratic equations to provide the respective root solutions on the same plot.

## 29.3 The Graphical Calculus.

If two variables are so related that the value of one of them depends on the value assigned the other, then the first variable is said to be a function

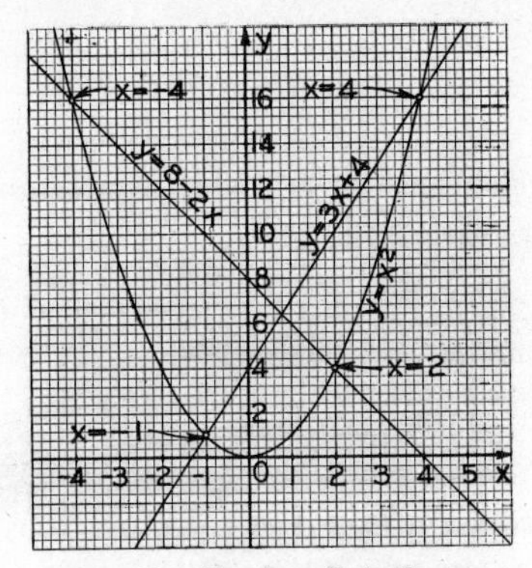


Fig. 29.2 **Graphic Algebra Solutions.**

of the second. For example, the area of a circle is a function of the radius. *The calculus* is that branch of mathematics pertaining to the change of values of functions due to finite changes in the variables involved. It is a method of analysis called the *differential calculus* when concerned with the determination of the *rate of change* of one variable of a function with respect to a related variable of the same function. A second principal operation, called *the integral calculus,* is the inverse of the differential calculus and is defined as a *process of summation* (finding the total change).

## 29.4 The Differential Calculus.

Fundamentally, the differential calculus is a means of determining for a given function the limit of the ratio of change in the dependent variable to the corresponding change in the independent variable, as this change approaches zero. This limit is the *derivative* of the function with respect to the independent variable. The derivative of a function $y = f(x)$ may be denoted by $dy/dx$, $f'(x)$, $y'$, or $D_x y$.

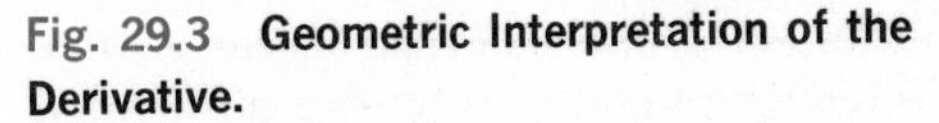

Fig. 29.3 **Geometric Interpretation of the Derivative.**

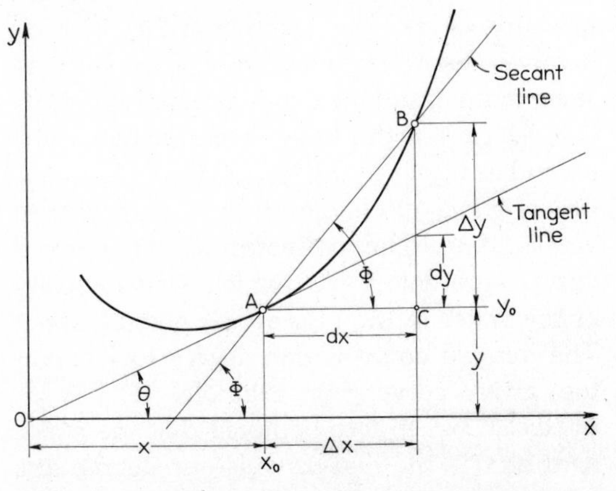


## 29.5 Geometric Interpretation of the Derivative.

As illustrated in Fig. 29.3:

1. The symbol $f(x)$ is used to denote a function of a single variable $x$ and is read "f of x."
2. The value of the function when $x$ has the value $x_0$ may be denoted by $f(x_0)$ or $F(x_0)$.
3. The increment $\Delta y$ (read: "delta y") is the change in y produced by increasing x from $x_0$ to $x_0 + \Delta x$; therefore, $\Delta y = f(x_0 + \Delta x) - f(x_0)$.
4. The differential, $dy$, of y is the value that $\Delta y$ would have if the curve coincided with its tangent. The differential, $dx$, of x is the same as $\Delta x$, when x is the independent variable.
5. The slope of the secant line is represented by the ratio $\Delta y/\Delta x$.
6. Ratio $dy/dx$ represents the slope of the tangent line.

Draw a secant through A (coordinates x, y) and a neighboring point B on the curve (coordinates $x + \Delta x$, $y + \Delta y$). The slope of the secant line through A and B is

$$\frac{BC}{AC} = \frac{(y + \Delta y) - y}{(x + \Delta x) - x} = \frac{\Delta y}{\Delta x}$$

$$= \text{tangent of angle BAC} = \text{tangent } \Phi$$

Let point B move along the curve and approach point A indefinitely. The secant line will revolve about point A, and its limiting position is the tangent line at A, when $\Delta x$, varying, approaches zero as a limit. The angle of inclination, $\theta$, of the tangent line at A, becomes the limit of the angle $\Phi$ when $\Delta x$ approaches zero. Therefore,

$$\tan \Phi = \tan \theta = \frac{dy}{dx}$$

$$= \text{slope of tangent line at A}$$

The result of this analysis is the following theorem:

*The value of the derivative at any point of a curve is equal to the slope of the tangent line to the curve at that point.*

Accordingly, graphical differentiation is a process of drawing tangents or the equivalent (chords parallel to tangents) to a curve at any point and determining the value of a differential (value of the slope of the tangent line to the curve or to a parallel chord, at the point selected). The value of the slope is the corresponding ordinate value on the derived curve.

## 29.6 Graphical Differentiation—The Slope Law.

The slope of the curve at any point is the tangent of the angle ($\theta$) between the X-axis of the graph and a tangent line to the curve at the selected point, Fig. 29.4. The slope is also the rise or fall of the tangent line in the y direction per one unit of travel in the x direction, the value of the slope being calculated in terms of the scale units of the X- and Y-axes.

Fig. 29.4 **Graphical Differentiation.**

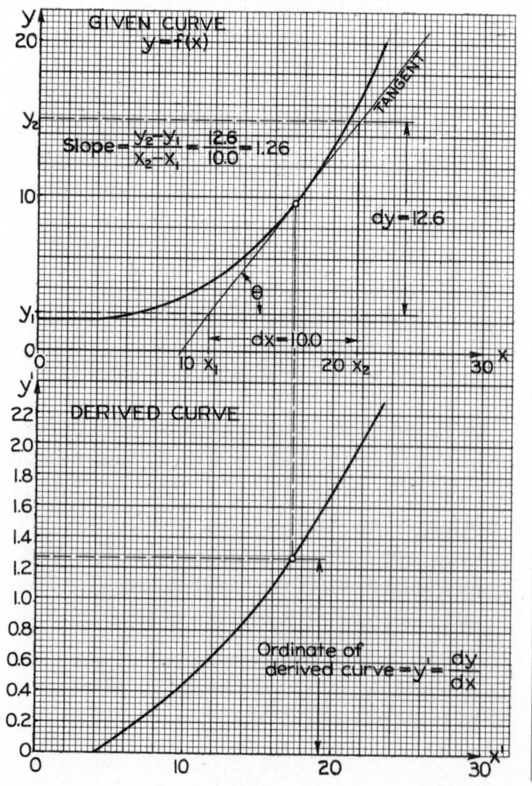


The *Slope Law* as applied to differentiation may be stated as follows:

*The slope at any point on a given curve is equal to the ordinate of the corresponding point on the next lower derived curve.*

The Slope Law is the graphical equivalent of differentiation of the calculus. The applications of the principles of the Slope Law are illustrated in the following examples.

In Fig. 29.5 (a), the slopes of OA and CD are constant and positive, indicated by the straight lines increasing in value upward (positively). Curve AB has a constant zero slope; curve BC has a constant negative slope as shown by the straight line decreasing in value downward (negatively). The constant slopes produce level derivative curves, which are derived in a definite order and graphed in the same order in projection below the given curve. Units assigned to the axes are at any convenient scale.

At (b), any point on the curve has a slope equal to dy/dx. The corresponding point on the derived curve has an ordinate y′ equal to the numerical value of dy/dx. Negative slopes have negative ordinate values on the derived curve. Zero slopes at B and D have zero values at B′ and D′ and are on the X-axis on the derived curve.

Since the differential calculus is concerned with the determination of a rate of change, it is applicable to problems involving the change of a physical quantity as a second related quantity varies. The example at Fig. 29.5 (a) involves the displacement of a body with respect to time, and the derived curve represents the velocity of the body at any instant. The second example, at (b), is a graph of the quantity of water discharged from a pipe plotted versus time. The ordinates of the derived curve represent $\frac{\text{quantity}}{\text{time}}$, which is the velocity of the water in the pipe at any given instant. The derived curve is one degree lower than the given curve. For example, if $y = x^3$, $dy/dx = 3x^2$. It is to be noted that the exponent of x in the derivative is one degree less

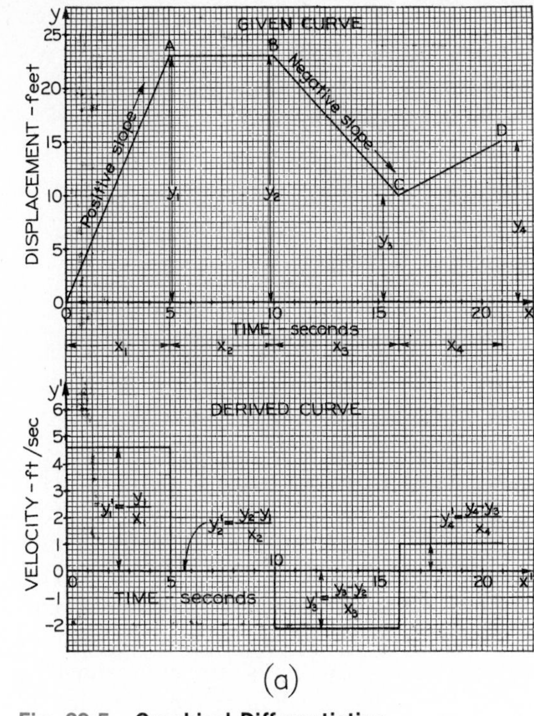
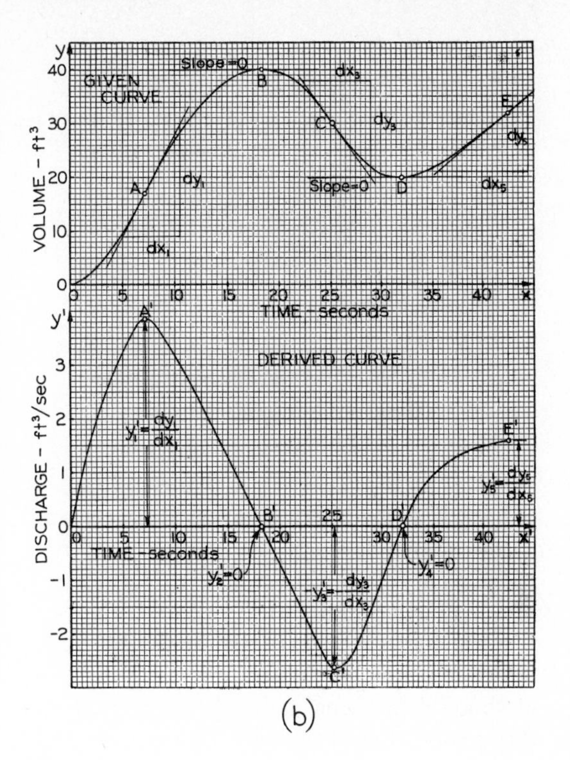

Fig. 29.5 **Graphical Differentiation.**

or lower. If the derived curve $\frac{\text{velocity}}{\text{time}}$ were differentiated, an $\frac{\text{acceleration}}{\text{time}}$ curve would be derived, completing a family of curves, which includes the $\frac{\text{displacement}}{\text{time}}$, $\frac{\text{velocity}}{\text{time}}$, and $\frac{\text{acceleration}}{\text{time}}$ curves.

## 29.7 Tangent-Line Constructions.

The principal manipulation involved in graphical differentiation is the drawing of lines tangent to the given curve. Generally, it is more convenient to draw a tangent line in a given direction and then determine the point of tangency than it is to select a point of tangency and then determine a tangent line at this point.

Under certain conditions, the tangent line can be constructed geometrically. If a portion of the curve is assumed to be approximately circular, the point of tangency will lie on the perpendicular bisector of any chord, and the tangent line will be parallel to the chord, Fig. 29.6 (a).

If the curve approximates a parabola, ellipse, hyperbola, or circle, the tangent point

Fig. 29.6 **Tangent Lines.**

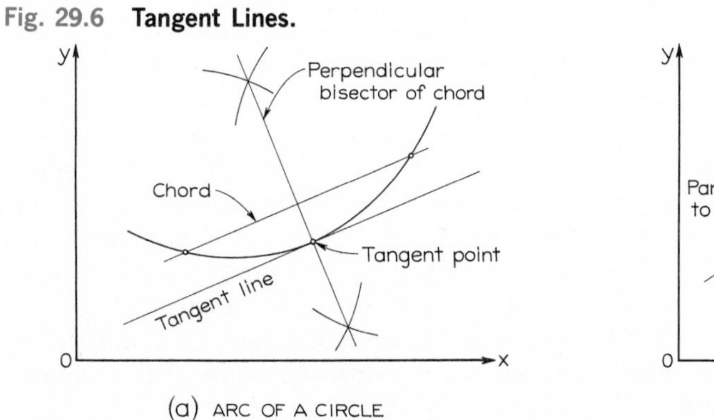


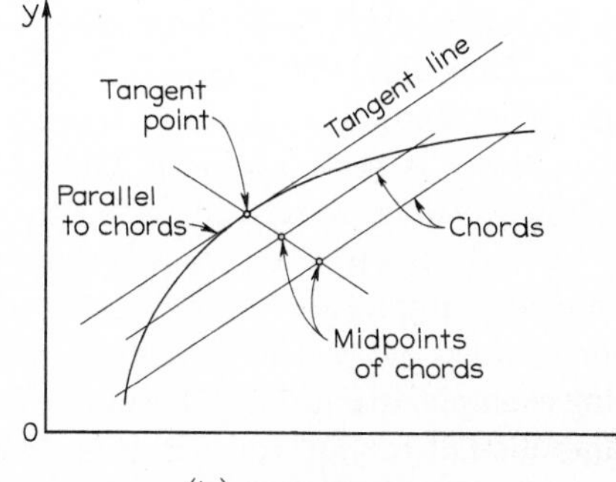

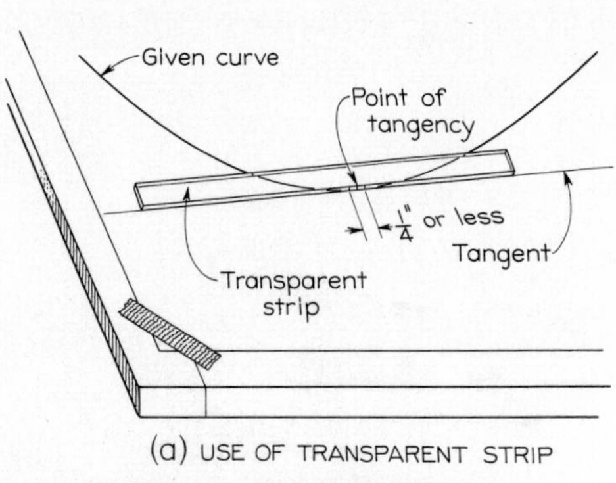


(a) USE OF TRANSPARENT STRIP

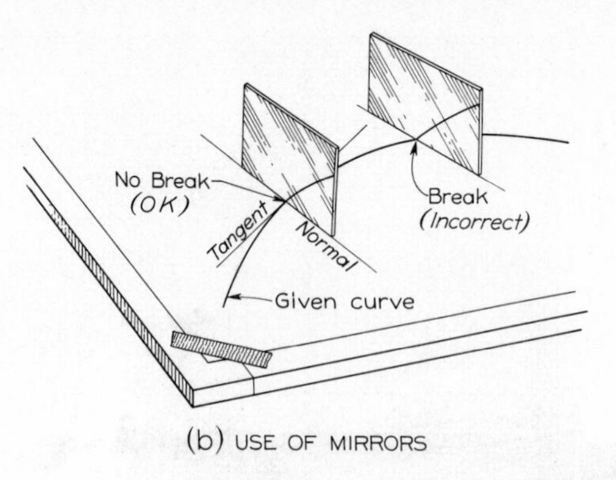


(b) USE OF MIRRORS

Fig. 29.7 **Tangent Line Determination.**

is located at the intersection of the curve and a line drawn through the midpoint of two parallel chords, as shown at (b). The tangent line is drawn parallel to the chords.

If the curve is of unknown analytical form, difficulties arise in drawing the tangent. One method of drawing tangents is to mark a transparent strip at the edge with two dots no more than $\frac{1}{4}''$ apart. If the dots lie on the curve, the edge of the strip approximately coincides with the tangent—the point of tangency is located midway between the dots, Fig. 29.7 (a).

A second method requires the use of a stainless steel or glass mirror, as shown at (b). The mirror is placed perpendicular to the graph surface and across the curve. The mirror is then moved until the curve and image coincide to form an unbroken line, as shown at the left at (b). A line drawn along the front edge of the stainless steel or the back edge of the glass mirror determines a normal to the tangent, as shown.

## 29.8 Graphical Differentiation Using Ray, String, or Funicular Polygons—The Chord Method.

The method of differentiating described in §29.6 is semigraphical. The computed tangent values were transferred to the derived curve ordinate scale. A more convenient, entirely graphical, method using *ray, string,* or *funicular polygons* is described in the following example. In Fig. 29.8 a harmonic-rise and parabolic-fall motion for a cam is determined in the displacement diagram (given curve), and a derived curve is required. The given curve is divided into a number of short arcs, with equally spaced ordinates. Draw chords from O to A, A to B, and so on. The distance between ordinates along the X-axis is chosen small enough so that the chords between the two points are approximately parallel to a tangent line at the midpoint of the chosen arc.

Locate in projection below the given curve the origin O′ of the derived curve and a *pole point* P at a convenient distance d from the origin point O′ on the X-axis. To avoid "flat" derived curves, d is generally chosen at some multiple distance of the scale along the X-axis. The slope values of the tangent points determined on the given curve, and plotted on the ordinate scale of the derived curve, are computed as the y value per *unit* x value. Slope $= \dfrac{y_2 - y_1}{x_2 - x_1}$ accounts for the axes scale designations, whatever they may be. If the pole distance, which is an x value, is any multiple of 1 (unit value), the ordinate values of the derived curve are reduced by a proportional amount, in order to transfer the same slope value from one graph to the other. If the pole distance is twice the unit x value, the ordinate scale of the derived curve is $\frac{1}{2}$ the scale unit of the given curve ordinates. The pole distance of the example problem is 3 units; therefore the same length for one unit on the given curve ordinate scale equals $\frac{1}{3}$ of a unit on the ordinate scale of the derived curve.

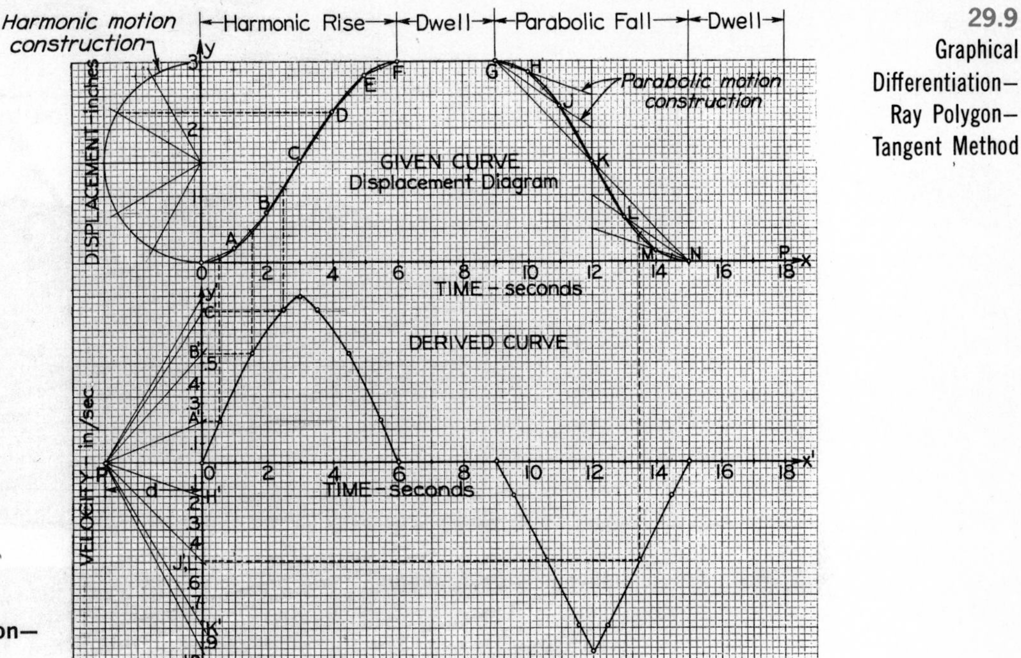


Fig. 29.8 **Differentiation—Ray Polygon Method.**

From the pole point P, draw *rays* parallel to the chords in the given curve. For example, PA′ is parallel to OA and PB′ is parallel to AB.

Locate the tangent points on the given curve for each subtended arc, as described in §29.7. Project each tangent point from the given curve downward to the derived curve until it intersects a horizontal line drawn from the appropriate points on the Y′-axis of the derived curve. The process is repeated for all the tangent points and their corresponding points on the Y′-axis. Construct the derived (derivative) curve by drawing a smooth curve through the located points. The computation of the slope values is not involved.

## 29.9 Graphical Differentiation—Ray Polygon—Tangent Method.

Another method similar to the chord method consists in the determination of the tangent points and lines, without concern for the chords. In Fig. 29.9, a derived curve is obtained from a given curve $y = 2x^2$ by the tangent method. A ray is drawn from the pole point parallel to each tangent line, and the derived curve points are located by projecting the tangent points vertically from the given curve until the lines intersect the appropriate horizontal projection lines from the Y′-axis of the derived curve.

The pole point P in this case is at a distance equal to $1\frac{1}{2}$ times the unit scale; therefore the ordinate scale for the first derived curve (y′) is $\frac{2}{3}$ of the ordinate scale of the given curve.

The differentiating of the derived curve (successive differentiation) results in a second derived curve. The pole point selected, P′, is at a distance equal to the unit scale of the X′-axis of the derived curve; therefore the ordinate scale for the second derived curve (y″) is equal to the ordinate scale of the first derived curve (y′).

If clarity permits, the successively derived curves can be conveniently placed one on top of the other, since the abscissa scale is the same for all the curves. Each curve, however, requires a different ordinate scale, since a "flat" curve is to be avoided.

In the example used, the first derived curve is an inclined straight line, which is expected, since if $y = 2x^2$,

1st derivative: $\dfrac{dy}{dx} = 4x$

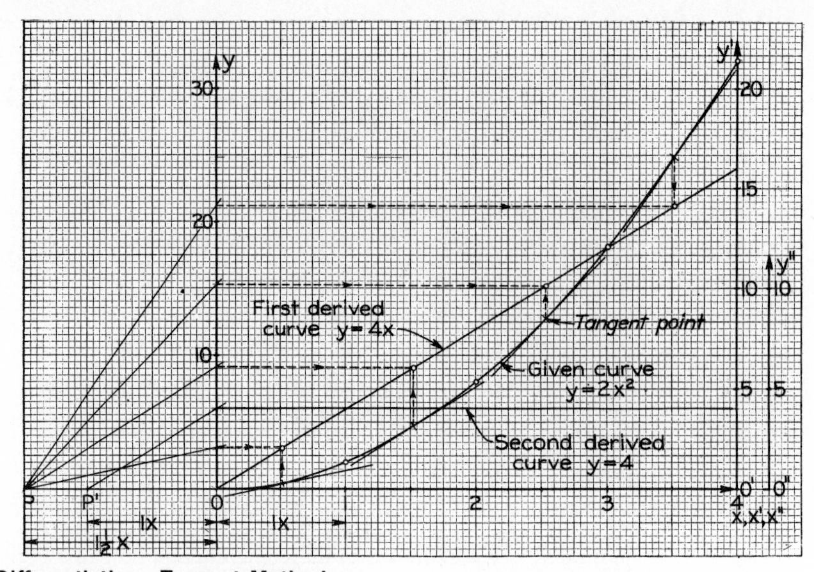


Fig. 29.9 **Differentiation—Tangent Method.**

The second derived curve appears as a line parallel to the X-axis, with a value of $y = 4$, which is also expected, since if $y = 4x$

$$\text{2nd differential:} \quad \frac{d^2y}{dx^2} = 4$$

**29.10 Area Law.** Given a curve, Fig. 29.10 (upper portion), a tangent to the arc is to be constructed at T. A length of arc AB is chosen so that T is at the midpoint of the length of arc. A chord is constructed through A and B, and the tangent is drawn through T parallel to the chord. The coordinates of A are $x_1$, $y_1$, and of B are $x_2$, $y_2$. Since the chord and tangent are parallel, their slopes are equal. The slope of the tangent line at T is equal to the mean ordinate $y'_m$ of T′ in the derived curve

$$y'_m = \frac{y_2 - y_1}{x_2 - x_1}$$

$$\text{and} \quad y_2 - y_1 = y'_m\,(x_2 - x_1)$$

but

$$y'_m\,(x_2 - x_1) = \text{area of rectangle } CDx_2x_1$$

Therefore

$$y_2 - y_1 = \text{area of rectangle } CDx_2x_1$$

The law derived from this analysis, the *Area Law,* may be stated as follows:

*The difference in the length of any two ordinates to a continuous curve equals the total net area between the corresponding ordinates of the next lower curve.*

Fig. 29.10 **Area Law.**

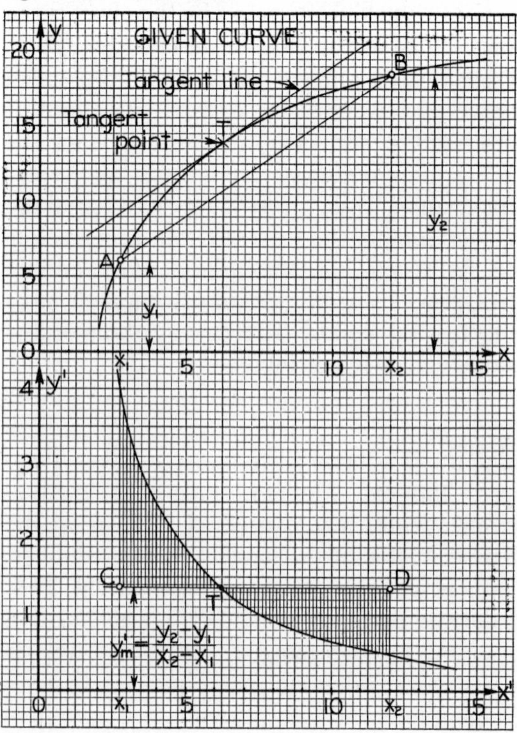

The application of the Area Law as stated, which provides for dividing the given curve into short arcs, permits the determination of the derivative curve for the given curve. The law is also applicable to the process of integration, as discussed in §29.14.

## 29.11 Graphical Differentiation—Area Law.

**29.11 Graphical Differentiation—Area Law.** As an example of the application of the Area Law, a practical problem in rate of interest may be used, as follows: An office building being erected costs $60,000 for the first story, $62,500 for the second, $65,000 for the third, and so forth. A basic cost of $400,000 is for the lot, plans, basement, and other preliminaries. The net annual income is $4000 per story. What number of stories will give the greatest rate of interest on the investment?

Initial computations are the following:

| Stories | Cost (1) | Income (2) | Percent Interest $\frac{(2)}{(1)} \times 100$ |
|---|---|---|---|
| 1 | $460,000 | $ 4,000 | .87% |
| 2 | 522,500 | 8,000 | 1.53% |
| 3 | 587,500 | 12,000 | 2.04% |
| Etc. | | | |

In Fig. 29.11, the given curve is plotted using the computed coordinate values. The derived (derivative) curve is determined from the given curve by applying the Area Law.

Divide the given curve into a number of short arcs, with equally spaced ordinates $y_1$, $y_2$, . . . . The intervals between the ordinates along the X-axis should be chosen so that the chord intercepting the two points on the arcs OA, AB, BC, . . . , will be approximately parallel to the tangent lines at the midpoints of the subtended arcs.

Determine the length of each mean ordinate for the appropriate interval by dividing the difference in length of the ordinates by the x distance between them. For example,

$$y'_m = \frac{y_2 - y_1}{x_2 - x_1}$$

The solution by this procedure is semigraphical.

If the intervals between ordinates are made equal, the length of the mean ordinate for each interval of the derived curve will be proportional to SA, TB, UC, . . . , which eliminates the calculations involved in the semigraphical method.

The ray-polygon method of transferring ordinate values from one graph to the other also can be used.

Many of the derived curves determined are apt to appear "flat." To avoid a flat curve, the scale for the ordinates of the derived curve may be twice the scale of the ordinates of the given curve,

$$y'_m = 2\left(\frac{y_2 - y_1}{x_2 - x_1}\right)$$

The distance on the ordinate of the derived curve can be rapidly determined by applying the dividers to the given curve ordinate differences SA, TB, UC, . . ., and placing twice these distances on the corresponding ordinates of the derived curve.

Fig. 29.11 **Differentiation—Area Law.**

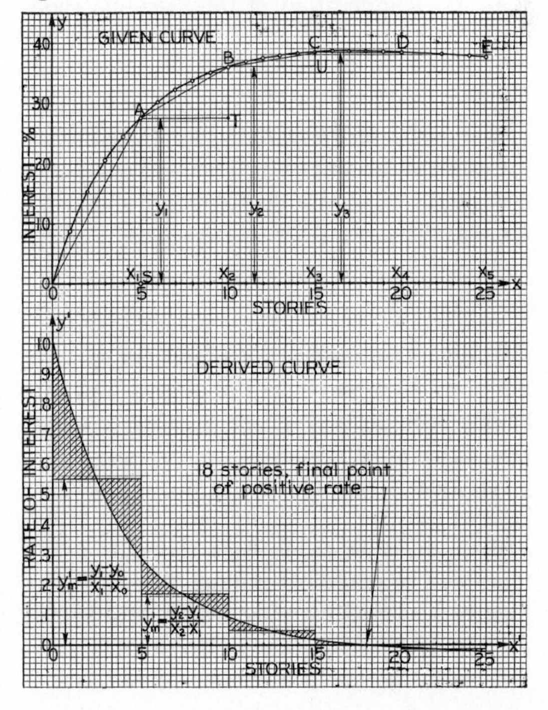

| Plotted Data | | Derivative of Derived Curve |
|---|---|---|
| Independent | Dependent | |
| Time | Displacement | Velocity |
| Time | Velocity | Acceleration |
| Time | Amount of:<br>Population<br>Inches<br>Volume<br>Temperature | Rate of:<br>Growth<br>Growth<br>Flow<br>Cooling, Heating |
| Time | Energy or Work | Power |
| Displacement | Energy or Work | Force |
| Quantity | Cost | Rate of Cost |
| Variable #2 | Related variable #1 | Rate of change of variable #1 as variable #2 changes |

Fig. 29.12 **Differentiation Applications.**

Horizontal lines for each interval and a smooth curve through the centers of horizontal lines are then drawn so that the triangular areas above and below the curve appear by eye to be approximately equal in area.

## 29.12 Practical Applications—Differentiation.

The application of the calculus, especially by graphical methods, is based on the practicality and knowledge of the subject matter. The tabulation of Fig. 29.12 is a summary of but a few of the practical applications for the differential calculus. The derived curve also shows maximum and minimum values for the derivative and if the rate is constant, variable, or zero.

## 29.13 The Integral Calculus.

*Integration* is a process of summation, the integral calculus having been devised for the purpose of calculating the areas bounded by curves. If the given area is assumed to be divided into an infinite number of infinitesimal parts called *elements,* the sum of all these elements is the total area required. The integral sign $\int$, the long S, was used by early writers and is still used in calculations to indicate "sum."

One of the most important applications of the integral calculus is the determination of a function from a given derivative, the inverse of differentiation. The process of determining such a function is *integration,* and the resulting function is called the *integral* of the given derivative or *integrand.* In many cases, however, graphical integration is used to determine the area bounded by curves, in which case the expression of the function is not necessary.

Graphical solutions may be classified into three general groups: graphical, semigraphical, and mechanical methods. Each of these methods will be discussed.

## 29.14 Graphical Solution—Area Law.

Since integration is the inverse of differentiation, the Area Law as analyzed for differentiation, §29.10, is applicable to integration, but in reverse. The Area Law as applied to integration may be stated as follows:

*The area between any two ordinates of a given curve is equal to the difference in length between the corresponding ordinates of the next higher curve.*

If an $\frac{\text{acceleration}}{\text{time}}$ curve were given, the integral of this derivative curve would be the $\frac{\text{velocity}}{\text{time}}$ curve. If the $\frac{\text{velocity}}{\text{time}}$ curve is integrated as shown in Fig. 29.13, the $\frac{\text{displacement}}{\text{time}}$ curve is obtained (the next higher curve).

In Fig. 29.13, the integral curve is drawn directly above the given curve, using the same scale along the X-axis as that on the X′-axis. The scale for the Y-axis must provide the maximum value expected, and its length can be estimated. The value is equal to the total area under the given curve. For this example, the mean ordinate of the given curve appears to be about 17; therefore, the total area is approximately 102 square units (17 × 6). The ordinate scale for the integral curve is then selected to accommodate slightly more than this value.

The procedure for locating points on the integral curve is as follows: On the given derivative curve, uniformly (equally) spaced ordinates are established, and a mean ordinate is approximated between two consecutive ordinates by drawing a horizontal line that cuts the arc between the two ordinates, forming a pair of approximately equal "triangles," as shown in the lower part of the figure. It must be remembered that the area between any two ordinates on the given curve is calculated. In the example used, the distance between ordinates is one-second unit value; therefore, the area is equal to the mean ordinate value multiplied by 1. If the distance between ordinates is greater or less than unit value, this value must be considered in calculating the areas.

The points on the integral curve in this example may be obtained directly from observation of the given curve. The first ordinate value ($y_1$) is at the end of the first one-second interval and is equal to the mean ordinate ($y'_1$) obtained from the given curve since $y'_1 \times 1 = y'_1$. The second ordinate value ($y_2$) is for the end of the second one-second interval and is equal to the first mean ordinate value plus the second mean ordinate value ($y_2 = y'_1 + y'_2$), etc. If the scales of the ordinates for the given and integral curves are equal or proportional, these additions can be made graphically with the dividers.

## 29.15 Graphical Solution—Funicular Polygon Method.

Another graphical method of integration is the *funicular* or *string polygon* method. If the lower-degree curve is given, Fig. 29.14, and the area between the curve ABCDEF and the X-axis is required (integration), the area obviously equals the sum of the rectangles ABKG, KLDC, and EFML.

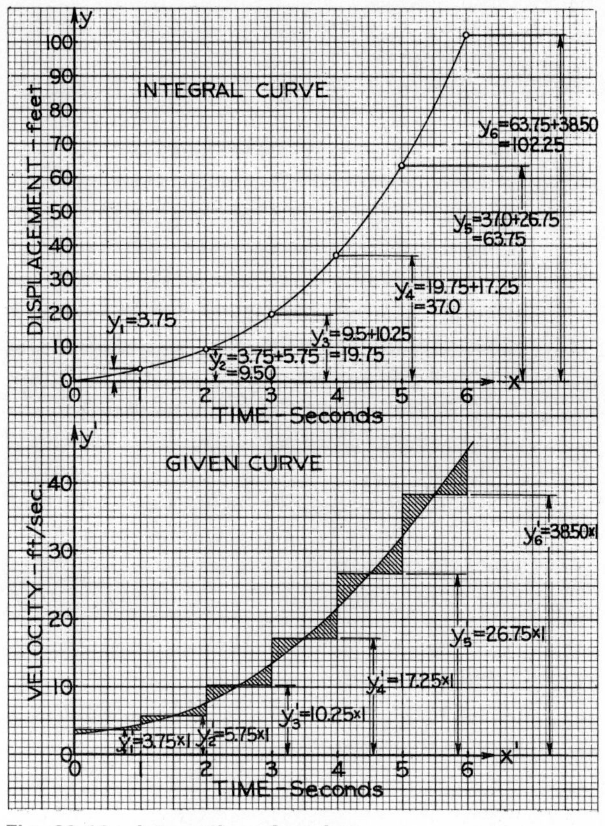


Fig. 29.13 **Integration—Area Law.**

Fig. 29.14 **Integration—Funicular Polygon.**

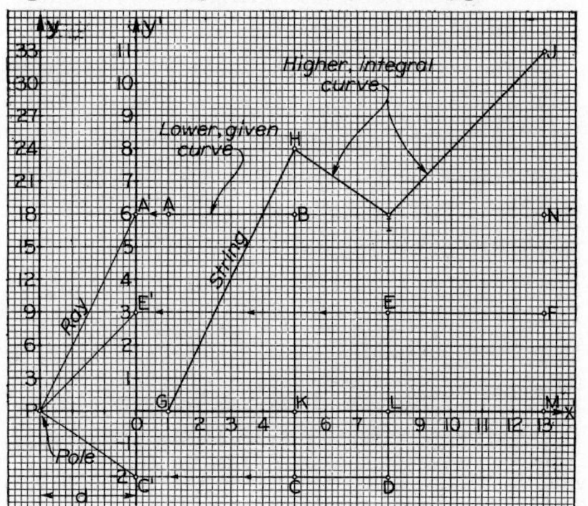

Locate a pole point P at a distance d to the left of the origin O on the extended X-axis. See §29.16. Project the lines AB, CD, and EF to the Y-axis to determine the points A′, C′, and E′, respectively. Draw rays from the pole point to points A′, C′, and E′. From point G draw a straight line (string) parallel to ray PA′ until it intersects the ordinate CKB, extended to the point of intersection at H. From point H draw a string parallel to ray PC′ until it intersects the ordinate DLE, extended to the point of intersection at I. From point I draw a string parallel to ray PE′ until it intersects the ordinate MF, extended to the intersection at J.

The curve GHIJ is the higher-degree integral curve. Ordinate KH represents the area under line AB, ordinate IL represents the area under line ABCD, and ordinate JM represents the area under line ABCDEF.

From Fig. 29.14,

$$\text{Slope of ray PA}' = \frac{OA'}{PO} = \frac{OA'}{d}$$

$$\text{Slope of string GH} = \frac{KH}{GK}$$

since these slopes were constructed equal,

$$\frac{OA'}{d} = \frac{KH}{GK}$$

$$\text{or} \quad OA' = d\left(\frac{KH}{GK}\right)$$

The ordinate OA′ of the lower curve therefore equals the slope of the string or upper curve multiplied by a constant d, the pole distance.

It will be seen that this analysis is a verification of the Slope Law, §29.6. A rearrangement of the slope relationship verifies the Area Law, since $OA' = d(KH/GK)$, $OA'(GK) = d(KH)$ = area of rectangle ABGK. KH is the ordinate of the higher degree integral curve.

## 29.16 Ordinate Scale of Integral Curve.

The ordinate values placed along the Y-axis of the next higher degree (integral) curve are dependent upon the location of the pole point used to obtain strings and rays. The abscissa scale along the X-axis is kept the same as that of the original curve.

It is desirable to choose the unit scale of the integral curve ordinates a proper length so as to avoid "flat" integral curves. Larger units along the Y-axis, Fig. 29.15, in comparison to the units along the Y′-axis, will produce a steeper curve. The factor to consider is the choice of the pole distance (H). An even whole number multiple of the unit scale length along the abscissa will simplify the layout of the Y-axis (ordinate) scale units.

In the figure, H was chosen equal to four times the unit length of the abscissa scale, to the left of the origin. The vertical (ordinate) scale units of the integral curve are, therefore, four times the unit scale of the ordinate of the given curve (based on areas under the curves).

The problem in Fig. 29.15 was arranged to obtain the best use of the available space, plus increased accuracy for the required integral. After the given curve is plotted within the available space, a line 0′–m is drawn from the origin to the desired end point of the integral curve. An average ordinate, line n–n, is estimated so that the two approximately triangular areas appear to be equal. An auxiliary ordinate axis for the integral curve is erected at the intersection of line n–n and 0′–m. Divide the given curve into small increments (Area Law method) and establish an average ordinate for each increment, a, b, c, etc. Project these ordinates to the auxiliary ordinate axis, a to a′, b to b′, and so on. From the pole point, a ray is drawn to each of these points, Pa′, Pb′, Pc′, etc. Draw a segment 0′–a″ across the first increment from the origin parallel to the corresponding ray Pa′. This segment coincides with 0′–a′. From the end of this segment, a second segment a″b″ is drawn across the second increment parallel to the corresponding second ray Pb′. This process is continued for the remaining increments. A smooth curve is drawn through the end points of the segments.

## 29.17 Constants of Integration.

The integral curves in many cases begin at zero, but

it cannot be assumed that this is always the case.

In Fig. 29.16, since the area under the given curve between any pair of ordinates gives only the difference between the lengths of the corresponding ordinates of the integral curve, the integration of the given curve gives only the shape of the integral curve. Its position with respect to the X-axis is determined by initial conditions that must be known or assumed. Therefore, when a curve is integrated by the calculus and limits are not stated, a quantity C, called the *constant of integration,* must be added to the resulting equation to provide for initial conditions—that is, to fix the curve with respect to the coordinate axes.

The actual constant value can be determined by knowing one point on the integral curve or the relationship between the given and integral values of the variable. In the example, the integral curve determines the fuel consumed. If the tank capacity is known and if the tank was full at the start—that is, it held 385 gallons—the integral curve should be shifted upward until the largest integral value equals the tank capacity. The constant value is the amount of fuel remaining in the tank.

Whether the solution is algebraic or graphic, a constant of integration must be considered. The location of the integral curve relative to the X-axis is the major concern when considering the constant of integration.

## 29.18 Semigraphical Integration.

When it becomes necessary to determine only an area and not an integral curve, many semigraphical methods are applicable, some of which are the Rectangular Rule, Trapezoid Rule, Simpson's Rule, Durand's Rule, Weddle's Rule, Method of Gauss, and the Method of Parabolas. All of these rules and methods may be found in the references listed in the bibliography of this text. Simpson's Rule, which is one of the more accurate methods, is described below. The solutions from these methods become partly graphical, since the data used are measured from the graphical plot.

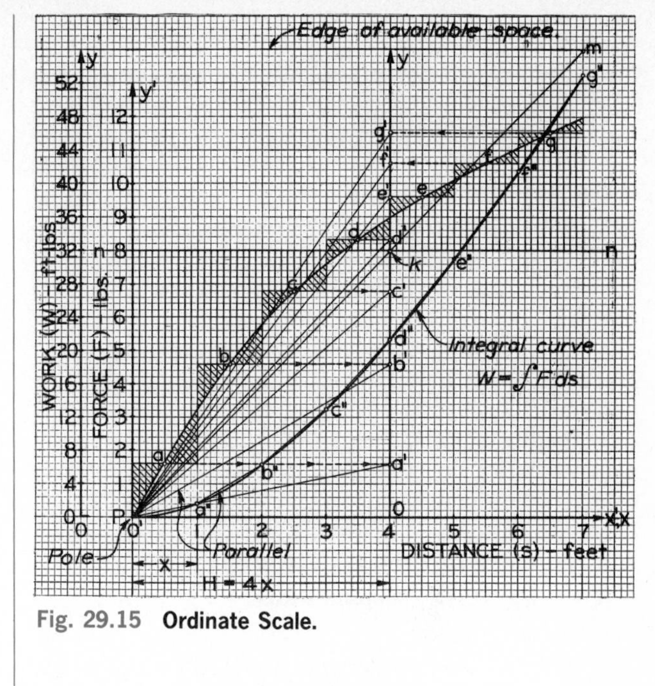


Fig. 29.15 **Ordinate Scale.**

Fig. 29.16 **Constants of Integration.**

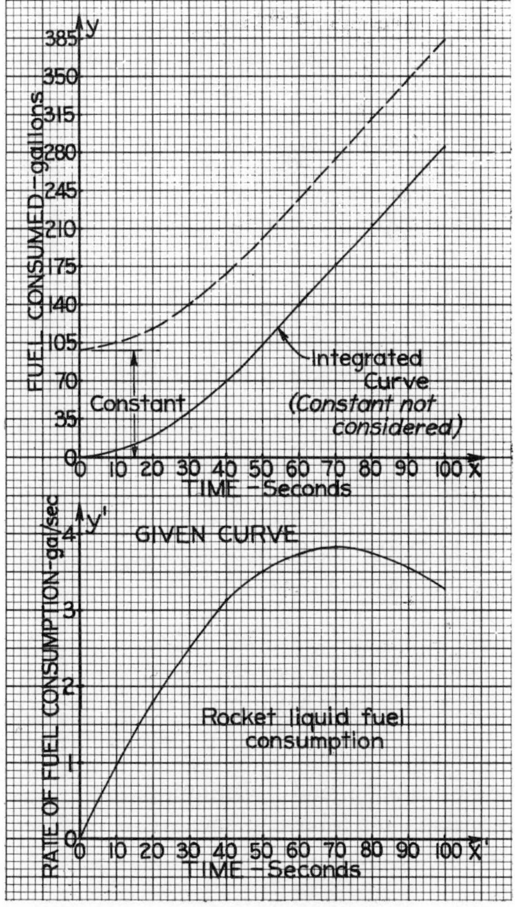

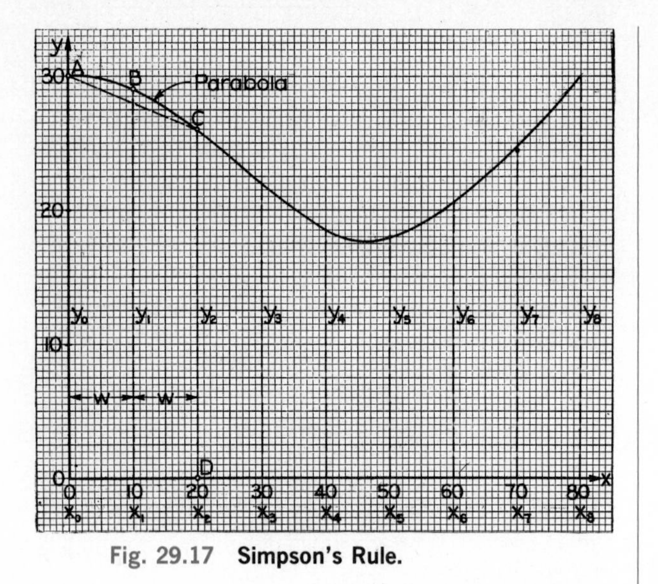


Fig. 29.17 **Simpson's Rule.**

If we consider the curve ABC, Fig. 29.17, as part of a parabola, and analyze the area 0ABCD on the same premise, then the configuration bounded by the curve ($y = ax^2 + bx + c$), the X-axis, and the ordinates $x = x_0$ and $x = x_2$, has an area as follows:

$$A = \int_{x_0}^{x_2} (ax^2 + bx + c)dx$$

$$= \left| \frac{ax^3}{3} + \frac{bx^2}{2} + cx \right|_{x_0}^{x_2}$$

$$= \frac{a}{3}(x_2^3 - x_0^3) + \frac{b}{2}(x_2^2 - x_0^2) + c(x_2 - x_0)$$

$$= \frac{x_2 - x_0}{6}[2a(x_2^2 + x_2x_0 + x_0^2) + 3b(x_2 + x_0) + 6c]$$

$$y_0 = ax_0^2 + bx_0 + c;$$

$$y_2 = ax_2^2 + bx_2 + c;$$

$$w = \frac{x_2 - x_0}{2}$$

$$y_1 = ax_1^2 + bx_1 + c$$

$$= a\left(\frac{x_2 + x_0}{2}\right) + b\left(\frac{x_2 + x_0}{2}\right) + c$$

Rearranging and substituting these relationships, the following equation is obtained:

$$A = \tfrac{1}{3}w(y_0 + 4y_1 + y_2)$$

If we divide the given curve into an *even* number of intervals and apply the equation for the area to the successive areas under the parabolic arcs, we obtain

$$\begin{aligned} A_s &= \tfrac{1}{3}w(y_0 + 4y_1 + y_2) \\ &\quad + \tfrac{1}{3}w(y_2 + 4y_3 + y_4) \\ &\quad + \cdots + \tfrac{1}{3}w(y_{n-2} + 4y_{n-1} + y_n) \\ &= \tfrac{1}{3}w(y_0 + 4y_1 + 2y_2 + 4y_3 + 2y_4 \\ &\quad + \cdots + 2y_{n-2} + 4y_{n-1} + y_n) \\ &= \tfrac{1}{3}w[(y_0 + y_n) \\ &\quad + 4(y_1 + y_3 + y_5 + \cdots + y_{n-1})] \\ &\quad + 2(y_2 + y_4 + y_6 \cdots + y_{n-2})] \end{aligned}$$

To apply Simpson's Rule, the given curve *must be divided into an even number of intervals,* and the required area is approximately equal to the sum of the extreme ordinates, plus four times the sum of the ordinates with odd subscripts, plus twice the sum of the ordinates with even subscripts, all multiplied by one-third the common distance between the ordinates. In the example shown in Fig. 29.17,

$$\begin{aligned} A_s &= \tfrac{1}{3}w[(y_0 + y_n) \\ &\quad + 4(y_1 + y_3 + y_5 + y_7) \\ &\quad + 2(y_2 + y_4 + y_6)] \\ &= \tfrac{1}{3}(10)\ [(30 + 30) \\ &\quad + 4(29 + 22 + 18 + 24.8) \\ &\quad + 2(26 + 18.5 + 20.5)] \\ &= 3.33\ [60 + 375 + 130] \\ &= 1883 \text{ square units per axis scale, not actual area.} \end{aligned}$$

## 29.19 Integration—Mechanical Methods.

When the requirement for integration is the determination of the area, mechanical integrators called *planimeters* may be used. The operator manually traces the outline of the area, and the instrument automatically records on a dial the area circumnavigated.

A common type of planimeter is the polar planimeter, Fig. 29.18 (b). As illustrated at (a), by means of the polar arm OM, one end M of the tracer arm is caused to move in a circle, and the other end N is guided around a closed curve bounding the area measured. The area $M_1N_1NN_2M_2MM_1$ is "swept out" twice, but

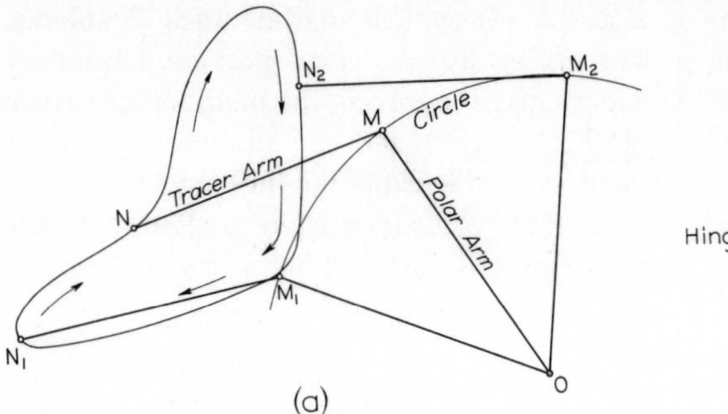


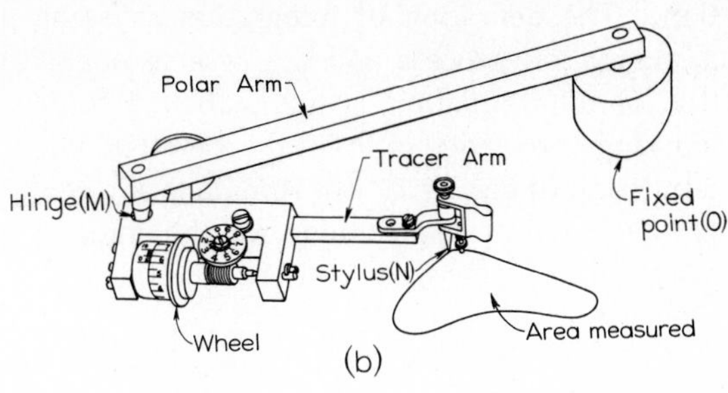


Fig. 29.18 **Polar Planimeter.**

in opposite directions. The resulting displacement reading (difference in the two sweeps) on the integrating wheel indicates the amount of the area. The circumference of the wheel is graduated so that one revolution corresponds to a definite number of square units of area.

The ordinary planimeter used to measure indicator diagrams has a length $L = 4''$, and a wheel circumference of 2.5″, so that one revolution of the wheel is $4 \times 2.5'' = 10$ sq in. The wheel is graduated into ten parts, each part being further subdivided into ten parts, and a vernier scale facilitates a further subdivision into ten parts, enabling a reading to the nearest hundredth of a square inch.

Fig. 29.19 **Practical Applications—Integration.**

| Plotted Data | | Integral of Integration Curve |
|---|---|---|
| Independent | Dependent | |
| Time | Acceleration | Velocity |
| Time | Velocity | Displacement |
| Y values | X values | Area of X–Y plot enclosed by X-axis and/or Y-axis and the curve |
| Area | Pressure | Force |
| Displacement | Force | Work or Energy |
| Time | Power | Work or Energy |
| Volume | Pressure | Work or Energy |
| Time | Force | Momentum |
| Velocity | Momentum | Impulse |
| Time | Rate based on time | Total change or cumulative change |
| Quantity #2 | Rate of change of quantity #1 based on change of quantity #2 | Max. or min. quantity #1 for values of quantity #2 |

**29.20 Practical Applications of Integration.** The definition of integration as being a summation process and the reverse operation of differentiation is not sufficient for a working knowledge of the subject matter. The tabulation of Fig. 29.19 is a summary of a few of the practical applications for the integral calculus.

**29.21 Graphical Mathematics Problems.** The following problems provide laboratory experience in applying the methods described in this chapter and, in addition, introduce many typical engineering problems to the student. For convenience in assignment, the problems are classified according to type.

## ALGEBRAIC SOLUTIONS

ROOTS AND SIMULTANEOUS EQUATIONS. Plot the following equations and determine the x and y roots (if any). Plot any two of the following equations and determine their solution (simultaneous), if any.

**Prob. 29.1** $y = 4 + 3x$
**Prob. 29.2** $y = 6 - 5x$
**Prob. 29.3** $2y = 4 + 8x$
**Prob. 29.4** $3y = 6 - 2x$
**Prob. 29.5** $4y = 3x$
**Prob. 29.6** $-y = 2x + 2$
**Prob. 29.7** $y = 2x - 3$
**Prob. 29.8** $y = x^2 + 3x - 4$
**Prob. 29.9** $y = x^2 - 2x - 3$
**Prob. 29.10** $y = x^2 + 4x + 3$
**Prob. 29.11** $y = x^2 - 3x + 2$
**Prob. 29.12** $y = x^3 + 6x^2 - 30x - 40$
**Prob. 29.13** $y = x^3 + 4x^2 - 9x - 36$
**Prob. 29.14** $x = y^3 + 2y^2 - 3y - 12$
**Prob. 29.15** $x^2y = 16$
**Prob. 29.16** $xy^2 = 15$
**Prob. 29.17** $xy = 9$
**Prob. 29.18** $2y^2 - 3x = 5$
**Prob. 29.19** $\log y = 2x^3$
**Prob. 29.20** $\log y = 5x^2$
**Prob. 29.21** $x^2 + y^2 = 9$
**Prob. 29.22** $4y^2 - 9x^2 = 16$

## THE GRAPHICAL CALCULUS

In the following problems, plot the given data and perform the graphical differentiation or integration by any of the methods included in the text.

### DIFFERENTIAL CALCULUS

**Prob. 29.23** Determine the acceleration (mph/sec) versus time curve for the following tests performed on an 8-cylinder automobile engine.

| Time (T), sec | Velocity (V), mph | |
|---|---|---|
| | Without Supercharger | With Supercharger |
| 0 | 0 | 0 |
| 5 | 28 | 35 |
| 10 | 45 | 55 |
| 15 | 54 | 67 |
| 20 | 61 | 75 |
| 25 | 67 | 81 |
| 30 | 71 | 86 |
| 35 | 75 | 90 |
| 47 | — | 97.1 |
| 57 | 86.7 | — |

**Prob. 29.24** From the data of tank content versus time for a rocket, determine the rate of fuel consumption, gal/sec. Plot fuel consumed versus time.

| Time (T), sec | Liquid Fuel Tank Content (C), gal |
|---|---|
| 0 | 940 |
| 5 | 900 |
| 10 | 810 |
| 15 | 660 |
| 20 | 430 |
| 25 | 290 |
| 30 | 195 |
| 35 | 130 |
| 40 | 75 |
| 45 | 40 |
| 50 | 20 |
| 55 | 5 |
| 60 | 0 |

**Prob. 29.25** Determine the power curve for the following data. Power is the rate of doing work. 1 hp = 33,000 ft-lb/min.

| Time (T), min | Work (W), ft-lb |
|---|---|
| 1 | 100,000 |
| 2 | 70,000 |
| 3 | 50,000 |
| 4 | 37,000 |
| 5 | 29,500 |
| 6 | 24,500 |
| 7 | 21,500 |
| 8 | 19,500 |

**Prob. 29.26** Plot the velocity data for the ram of a shaper machine and determine the acceleration by graphical differentiation.

| Time (T), sec | Velocity (V) ft/sec |
|---|---|
| 0 | 0 |
| .25 | .333 |
| .50 | .533 |
| .75 | .650 |
| 1.00 | .733 |
| 1.25 | .783 |
| 1.50 | .800 |
| 1.75 | .817 |
| 2.00 | .800 |
| 2.25 | .783 |
| 2.50 | .733 |
| 2.75 | .650 |
| 3.00 | .533 |
| 3.25 | .400 |
| 3.50 | .200 |
| 3.75 | −.133 |
| 4.00 | −.533 |
| 4.25 | −1.017 |
| 4.50 | −1.417 |
| 4.75 | −1.625 |
| 4.85 | −1.650 |
| 5.00 | −1.608 |
| 5.25 | −1.350 |
| 5.50 | −.917 |
| 5.75 | −.417 |
| 6.00 | 0 |

**Prob. 29.27** Plot the data for the recovery of hydrogen-embrittled mild steel, .505″ diameter rods and determine the rate of cooling.

| Time (t), hr | Temperature (T), °C |
|---|---|
| $\frac{1}{2}$ | 700 |
| 1 | 500 |
| 2 | 425 |
| 4 | 340 |
| 6 | 275 |
| 8 | 225 |
| 10 | 200 |
| 12 | 170 |
| 14 | 150 |
| 16 | 125 |
| 18 | 120 |
| 20 | 110 |
| 22 | 100 |
| 24 | 95 |

**Prob. 29.28** A manufacturer of machine parts can sell x parts per week at a price $p = 200 - 1.5x$ = (price per part). The cost of production $c = .5x^2 + 20x + 600$. Determine the total selling price and production cost for increments of five parts, for a total of 50 parts. Plot each curve, total selling price versus number of parts and production cost versus number of parts on the same sheet of graph paper.

Determine the "break-even" point, where selling price equals production costs. Determine the number of parts to be manufactured with the largest profit (profit = selling price minus production cost).

Determine the rate of production cost and the rate of the selling price.

## INTEGRAL CALCULUS

**Prob. 29.29** For the data of Prob. 29.23, determine the displacement versus time curves.

**Prob. 29.30** Determine the work curve for the following data. Plot volume as abscissa and pressure as ordinate. The integral curve is the work curve, inch-pounds versus cubic inches, and is equivalent to the area under the given curve. The units for work are inch-pounds.

| Volume (V), in.$^3$ | Pressure (p), lb/in.$^2$ (psi) |
|---|---|
| 0 | 100 |
| 10 | 57 |
| 20 | 41 |
| 30 | 33 |
| 40 | 27.5 |
| 50 | 24 |
| 60 | 22 |
| 70 | 21 |
| 80 | 20 |
| 90 | 20 |

**Prob. 29.31** Plot the X and Y coordinates given. The resulting shape is one-fourth the cross-sectional area of an aircraft fuel tank. After integrating the given curve to determine the area, compute the volume of the tank, if the length of the tank is 3 ft. Lay out a line EF to an appropriate scale representing the vertical height of the tank and calibrate it as a measuring stick to the nearest 10 gal (1 ft$^3$ = 7.48 gal).

| X, ft. | Y, ft. |
|---|---|
| 2.0 | 0 |
| 1.975 | .25 |
| 1.890 | .50 |
| 1.740 | .75 |
| 1.500 | 1.00 |
| 1.120 | 1.25 |
| .750 | 1.40 |
| 0 | 1.50 |

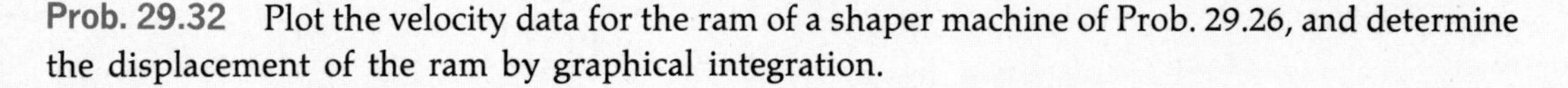

**Prob. 29.32** Plot the velocity data for the ram of a shaper machine of Prob. 29.26, and determine the displacement of the ram by graphical integration.

**Prob. 29.33** Plot the extrusion force data, with force as the ordinate axis. Integrate graphically the curve. The resultant integral is the work done by the punch, since $W = \int F ds$.

IMPACT EXTRUSION OF SOFT COPPER

| Punch Stroke (s), mm | Extruding Force (F), tons |
|---|---|
| 0 | 0 |
| .35 | .75 |
| .60 | 1.25 |
| .70 | 1.50 |
| 1.00 | 2.75 |
| 1.30 | 5.25 |
| 1.40 | 6.50 |
| 1.45 | 7.25 |
| 1.50 | 8.25 |
| 1.70 | 10.25 |
| 1.80 | 10.75 |
| 2.10 | 11.75 |
| 2.25 | 11.90 |
| 2.50 | 12.10 |
| 2.65 | 11.95 |
| 3.05 | 11.75 |
| 3.60 | 11.25 |
| 3.80 | 10.75 |
| 4.25 | 10.50 |
| 4.35 | 10.25 |
| 4.50 | 10.50 |
| 4.65 | 10.75 |
| 4.80 | 11.50 |
| 4.85 | 11.00 |
| 4.90 | 0 |

**Prob. 29.34** Plot the acceleration data for the explosion stroke of a piston of a 6-cylinder gas automobile engine (engine rpm: 2800; cylinder diameter: $3\frac{3}{8}''$; stroke: $5\frac{1}{2}''$; connecting rod length: 11″). Determine the velocity of the piston by graphical integration.

| Crank Angle Position ($\theta$), degrees | Acceleration (a), ft/sec$^2$ |
|---|---|
| 0 | 24,630 |
| 15 | 23,500 |
| 30 | 19,600 |
| 45 | 13,980 |
| 60 | 7,420 |
| 75 | 753 |
| 90 | 5,160 |
| 105 | 9,500 |
| 120 | 12,300 |
| 135 | 13,830 |
| 150 | 14,460 |
| 165 | 14,730 |
| 180 | 14,770 |

**Prob. 29.35** Plot the piston force data (explosion stroke), with the force as the ordinate axis; data for the engine specified in Prob. 29.34. Integrate graphically the curve. The resultant integral is the work done by the piston, since $W = \int F ds$.

| Piston Stroke (s), in. | Piston Force (F), lb |
|---|---|
| 0 | 834 |
| $\frac{1}{8}$ | 2,550 |
| $\frac{7}{16}$ | 2,005 |
| $\frac{31}{32}$ | 1,370 |
| $1\frac{5}{8}$ | 950 |
| $2\frac{3}{8}$ | 682 |
| $3\frac{3}{32}$ | 518 |
| $3\frac{25}{32}$ | 406 |
| $4\frac{3}{8}$ | 327 |
| $4\frac{7}{8}$ | 288 |
| $5\frac{7}{32}$ | 251 |
| $5\frac{7}{16}$ | 187 |
| $5\frac{1}{2}$ | 101 |

# appendix

CONTENTS OF APPENDIX

# 1 Bibliography of Technical Drawing and Related Subjects


### Aeronautical Drafting

Anderson, Newton. *Aircraft Layout and Detail Design.* McGraw-Hill
Katz, Hyman H. *Aircraft Drafting.* Macmillan
Le Master, C. A. *Aircraft Sheet Metal Work.* American Technical Society
Liming, Roy. *Practical Analytic Geometry with Applications of Aircraft.* Macmillan
Meadowcroft, Norman. *Aircraft Detail Drafting.* McGraw-Hill
*SAE Aeronautical Drafting Manual.* Society of Automotive Engineers, Inc., 485 Lexington Ave., New York, N.Y. 10017.
Svensen, C. L. *A Manual of Aircraft Drafting.* D. Van Nostrand

### American National Standards

American National Standards Institute, 1430 Broadway, New York, N.Y. 10018. For complete listing of standards, see ANSI *Price List and Index.*

*Abbreviations*
Abbreviations for Use on Drawings, Z32.13–1950
Abbreviations for Scientific and Engineering Terms, Z10.1–1941

*Bolts and Screws*
Hexagon Head Cap Screws, Slotted Head Cap Screws, Square Head Set Screws, and Slotted Headless Set Screws, B18.6.2–1956
Plow Bolts, B18.9–1958
Round Head Bolts, B18.5–1952 (R 1959)
Slotted and Recessed Head Machine Screws and Machine Screw Nuts, B18.6.3–1962
Slotted and Recessed Head Wood Screws, B18.6.1–1961
Socket Cap, Shoulder, and Set Screws, B18.3–1969
Square and Hex Bolts and Screws, B18.2.1–1965
Square and Hex Nuts, B18.2.2–1965
Track Bolts and Nuts, B18.10–1963

*Charts and Graphs*
Illustrations for Publication and Projection, Y15.1–1959
Time-Series Charts, Y15.2–1960

*Dimensioning and Surface Finish*
Preferred Limits and Fits for Cylindrical Parts, B4.1–1967
Rules for Rounding Off Numerical Values, Z25.1–1940 (R 1961)
Scales to Use with Decimal-Inch Dimensioning, Z75.1–1955
Surface Texture, B46.1–1962

*Drafting Manual*
Sect. 1 Size and Format, Y14.1–1957
Sect. 2 Line Conventions and Lettering, Y14.2-1973
Sect. 3 Projections, Y14.3–1957
Sect. 4 Pictorial Drawing, Y14.4–1957
Sect. 5 Dimensioning and Tolerancing for Engineering Drawings, Y14.5–1966
Sect. 6 Screw Threads, Y14.6–1957
Sect. 7 Gears, Splines and Serrations, Y14.7–1958

Sect. 8 Castings, Y14.8—In preparation
Sect. 9 Forging, Y14.9-1958
Sect. 10 Metal Stampings, Y14.10-1959
Sect. 11 Plastics, Y14.11-1958
Sect. 12 Die Castings, Y14.12—In preparation
Sect. 13 Springs, Helical and Flat, Y14.13—In preparation
Sect. 14 Mechanical Assemblies, Y14.14-1961
Sect. 15 Electrical Diagrams, Y14.15-1966
Sect. 16 Tools, Dies and Gages, Y14.16—In preparation
Sect. 17 Fluid Power Diagrams, Y14.17-1966

*Gears*
System for Straight Bevel Gears, B6.13-1965
Tooth Proportions for Coarse-Pitch Involute Spur Gears, B6.1-1968
Tooth Proportions for Fine-Pitch Involute Spur and Helical Gears, B6.7-1967

*Graphic Symbols*
Graphic Electrical Wiring Symbols for Architectural and Electrical Layout Drawings, Y32.9-1962
Graphic Symbols for Electrical and Electronics Diagrams, Y32.2-1970
Graphic Symbols for Heat-Power Apparatus, Z32.2.6-1950 (R 1956)
Graphic Symbols for Heating, Ventilating, and Air Conditioning, Z32.2.4-1949 (R 1953)
Graphic Symbols for Pipe Fittings, Valves, and Piping, Z32.2.3-1949 (R 1953)
Graphic Symbols for Plumbing, Y32.4-1955
Graphic Symbols for Use on Railroad Maps and Profiles, Y32.7-1957
Graphic Symbols for Welding, Y32.3-1969
Symbols for Engineering Mathematics, Y10.17-1961

*Keys and Pins*
Machine Pins, B5.20-1958
Woodruff Keys and Keyseats, B17.2-1967

*Piping*
Cast-Iron Pipe Centrifugally Cast in Sand-Lined Molds, A21.8-1962
Cast-Iron Pipe Flanges and Flanged Fittings, 25, 125, 250, and 800 lb, B16.1-1967
Cast-Iron Screwed Fittings, 125 and 250 lb, B16.4-1963
Ferrous Plugs, Bushings, and Locknuts with Pipe Threads, B16.14-1965
Malleable-Iron Screwed Fittings, 150 and 300 lb, B16.3-1963
Steel Butt-Welding Fittings, B16.9-1964
Steel Pipe Flanges and Flanged Fittings, B16.5-1968
Wrought-Steel and Wrought-Iron Pipe, B36.10-1959

*Rivets*
Large Rivets ($\frac{1}{2}$ Inch Nominal Dia. and Larger), B18.4-1960
Small Solid Rivets, B18.1-1965

*Small Tools and Machine Tool Elements*
Jig Bushings, B5.6-1962
Machine Tapers, B5.10-1963
Milling Cutters and End Mills, B94.19-1968
Reamers, B94.2-1964

T-Slots—Their Bolts, Nuts, Tongues, and Cutters, B5.1-1949
Taps, Cut and Ground Threads, B94.9-1967
Twist Drills, Straight Shank and Taper Shank, B94.11-1967

*Threads*
Dry Seal Pipe Threads, B2.2-1968
Nomenclature, Definitions, and Letter Symbols for Screw Threads, B1.7-1965
Pipe Threads, B2.1-1968
Unified and American National Screw Threads, B1.1-1960

*Washers*
Lock Washers, B27.1-1965
Plain Washers, B27.2-1965

*Miscellaneous*
Knurling, B94.6-1966
Preferred Thicknesses for Uncoated Thin Flat Metals, B32.1-1952 (R1968)


### Architectural Drawing

Field, W. B. *Architectural Drawing.* McGraw-Hill
Kenny, J. E., and McGrail, J. P. *Architectural Drawing for the Building Trades.* McGraw-Hill
Martin, C. L. *Architectural Graphics.* Macmillan
Morgan, S. W. *Architectural Drawing.* McGraw-Hill
Ramsey, C. G., and Sleeper, H. R. *Architectural Graphic Standards.* John Wiley
Sleeper, H. R. *Architectural Specifications.* John Wiley

### Blueprint Reading

DeVette, W. A., and Kellogg, D. E. *Blueprint Reading for the Metal Trades.* Bruce Pub. Co., Milwaukee
Heine, G. M., and Dunlap, C. H. *How to Read Electrical Blueprints.* American Technical Society, Chicago
Ihne, R. W., and Streeter, W. E. *Machine Trades Blueprint Reading.* American Technical Society, Chicago
Kenny, J. E. *Blueprint Reading for the Building Trades.* McGraw-Hill
Lincoln Electric Co. *Simple Blueprint Reading* (Welding). Cleveland, O.
Svensen, C. L., and Street, W. E. *A Manual of Blueprint Reading.* D. Van Nostrand
Wright, William N. *A Simple Guide to Blueprint Reading* (Aircraft). McGraw-Hill

### Cams

Furman, F. Der. *Cams, Elementary and Advanced.* John Wiley
Jensen, P. W. *Cam Design and Manufacture.* Industrial Press.
Rothbert, H. A. *Cams.* John Wiley

### Charts and Graphs

Haskell, A. C. *How to Make and Use Graphic Charts.* Codex Book Co., New York, N.Y.
Karsten, K. G. *Charts and Graphs.* Prentice-Hall
Leicey, N. W. *Graphic Charts,* Lefax order No. 11-248; Lefax, Philadelphia, Pa.
Lutz, R. R. *Graphic Presentation Simplified.* Funk and Wagnalls
Schmid, C. *Handbook of Graphic Presentation.* Ronald Press

### Descriptive Geometry

Grant, H. E. *Practical Descriptive Geometry.* McGraw-Hill
Hood, G. J., and Palmerlee, A. S. *Geometry of Engineering Drawing.* McGraw-Hill
Howe, H. B. *Descriptive Geometry.* Ronald Press
Johnson, L. O., and Wladaver, I. *Elements of Descriptive Geometry.* Prentice-Hall
Levens, A. S., and Eggers, H. *Descriptive Geometry.* Harper
Paré, E. G., Hill, I. L., and Loving, R. O. *Descriptive Geometry.* Macmillan
Rowe, C. E., and McFarland, J. D. *Engineering Descriptive Geometry.* D. Van Nostrand
Slaby, S. M. *Engineering Descriptive Geometry.* Barnes & Noble
Street, W. E. *Technical Descriptive Geometry.* D. Van Nostrand
Warner, F. M., and McNeary, M. *Applied Descriptive Geometry.* McGraw-Hill
Wellman, B. L. *Technical Descriptive Geometry.* McGraw-Hill

### Descriptive Geometry Workbooks

Earle, J. H., et al. *Design and Descriptive Geometry Problems, Series 1, 2, and 3.* Addison-Wesley
Howe, H. B. *Problems for Descriptive Geometry.* Ronald Press
Johnson, L. O., and Wladaver, I. *Elements of Descriptive Geometry* (Problems). Prentice-Hall
Paré, E. G., Loving, R. O., and Hill, I. L. *Descriptive Geometry Worksheets, Series A, B, and C.* Macmillan
Rowe, C. E., and McFarland, J. D. *Engineering Descriptive Geometry Problems.* D. Van Nostrand
Street, W. E., Perryman, C. C., and McGuire, J. G. *Technical Descriptive Geometry Problems.* D. Van Nostrand
Wellman, B. L. *Problem Layouts for Technical Descriptive Geometry.* McGraw-Hill

### Drawing Instruments and Supplies

Charles Bruning Co., Chicago, Ill.
Eugene Dietzgen Co., Chicago, Ill.
Frederick Post Co., Chicago, Ill.
Gramercy Guild Group, Inc., Denver, Colo.
Keuffel & Esser Co., Hoboken, N.J.
Theo. Altender & Sons, Philadelphia, Pa.
V & E Manufacturing Co., Pasadena, Calif.

### Electrical Drawing

Baer, C. J. *Electrical and Electronic Drawing.* McGraw-Hill
Bishop, C. C. *Electrical Drafting and Design.* McGraw-Hill
Carini, L. F. D. *Drafting for Electronics.* McGraw-Hill
Kocher, S. E. *Electrical Drafting.* International Textbook
Shiers, G. *Electronic Drafting.* Prentice-Hall
Van Gieson, D. W. *Electrical Drafting.* McGraw-Hill

### Engineering as a Vocation

Beakley, G. C., and Leach, H. W. *Careers in Engineering and Technology.* Macmillan
Carlisle, N. D. *Your Career in Engineering.* E. P. Dutton
Grinter, L. E., Spencer, H. C., et al. *Engineering Preview.* Macmillan
McGuire, J. G., and Barlow, H. W. *An Introduction to the Engineering Profession.* Addison-Wesley
Smith, R. J. *Engineering as a Career.* McGraw-Hill
Williams, C. C. *Building an Engineering Career.* McGraw-Hill

ENGINEERING DESIGN

Edel, D. H. Jr. *Introduction to Creative Design.* Prentice-Hall
Beakley, G. C., and Leach, H. W. *Engineering—An Introduction to a Creative Profession.* Macmillan
Hill, P. H. *The Science of Engineering Design.* Holt, Rinehart, and Winston
Krick, E. V. *An Introduction to Engineering and Engineering Design.* John Wiley
Spotts, M. F. *Design Engineering Projects.* Prentice-Hall

ENGINEERING DRAWING

French, T. E., and Svensen, C. L. *Mechanical Drawing.* McGraw-Hill
French, T. E., and Vierck, C. J. *Engineering Drawing.* McGraw-Hill
Giesecke, F. E., Mitchell, A., Spencer, H. C., and Hill, I. L. *Technical Drawing.* Macmillan
Katz, H. H. *Handbook of Layout and Dimensioning for Production.* Macmillan
Lent, Deane. *Machine Drawing.* Prentice-Hall
Luzadder, W. J. *Fundamentals of Engineering Drawing.* Prentice-Hall
Orth, H. D., Worsencroft, R. R., and Doke, H. B. *Theory and Practice of Engineering Drawing.* Wm. C. Brown Co.
Spencer, H. C. and Dygdon, J. T. *Basic Technical Drawing.* Macmillan
Zozzora, F. *Engineering Drawing.* McGraw-Hill

ENGINEERING DRAWING WORKBOOKS

Dygdon, J. T., Loving, R. O., and Halicki, J. E. *Basic Problems in Engineering Drawing, Vols. I and II.* Holt, Rinehart, and Winston
Giesecke, F. E., Mitchell, A., Spencer, H. C., and Hill, I. L. *Technical Drawing Problems, Series 1.* Macmillan
Johnson, L. O., and Wladaver, I. *Engineering Drawing Problems.* Prentice-Hall
Levens, A. S., and Edstrom, A. E. *Problems in Engineering Drawing.* McGraw-Hill
Luzadder, W. J., et al. *Problems in Engineering Drawing.* Prentice-Hall
McNeary, M., Weidhaas, E. R., and Kelso, E. A. *Creative Problems for Basic Engineering Drawing.* McGraw-Hill
Orth, H. D., Worsencroft, R. R., and Doke, H. B. *Problems in Engineering Drawing.* Wm. C. Brown
Paré, E. G., and Hrachovsky, F. M. *Graphic Representation.* Macmillan
Paré, E. G., and Tozer, E. F. *Engineering Drawing Problems.* D. Van Nostrand
Spencer, H. C., and Hill, I. L. *Technical Drawing Problems, Series 2 and 3.* Macmillan.
Turner, W. W., Buck, C. P., and Ackert, H. P. *Basic Engineering Drawing.* Ronald Press
Vierck, C. J., Cooper, C. D., and Machovina, P. E. *Engineering Drawing Problems.* McGraw-Hill
Zozzora, F. *Engineering Drawing Problems.* McGraw-Hill

ENGINEERING DRAWING AND GRAPHICS TEXTS

(*With or without descriptive geometry and design. In most cases, workbooks are also available.*)

Arnold, J. N. *Introductory Graphics.* McGraw-Hill
Earle, J. H. *Engineering Design Graphics.* Addison-Wesley
French, T. E., and Vierck, C. J. *Graphic Science and Design.* McGraw-Hill
Giesecke, F. E., Mitchell, A., Spencer, H. C., Hill, I. L., and Loving, R. O. *Engineering Graphics.* Macmillan
Hammond, R., et al. *Engineering Graphics for Design and Analysis.* Ronald Press
Hoelscher, R. P., and Springer, C. H. *Engineering Drawing and Geometry.* John Wiley
Levens, A. S. *Graphics in Engineering and Science.* John Wiley
Luzadder, W. J. *Basic Graphics.* Prentice-Hall
Paré, E. G. *Engineering Drawing.* Holt, Rinehart, and Winston
Rising, J. S., Almfeldt, M. W., and DeJong, P. S. *Engineering Graphics.* William C. Brown

Rule, J. T., and Watts, E. F. *Engineering Graphics.* McGraw-Hill
Shupe, H. W., and Machovina, P. E. *Engineering Geometry and Graphics.* McGraw-Hill
Svensen, C. L., and Street, W. E. *Engineering Graphics.* D. Van Nostrand
Wellman, B. L. *Introduction to Graphical Analysis and Design.* McGraw-Hill

### Graphical Computation

Adams, D. P. *An Index of Nomograms.* John Wiley
Allcock, H. J., and Jones, J. P. *The Nomogram.* Pitman
Davis, D. S. *Chemical Engineering Nomographs.* McGraw-Hill
Davis, D. S. *Empirical Equations and Nomography.* McGraw-Hill
Douglass, R. D., and Adams, D. P. *Elements of Nomography.* McGraw-Hill
Heacock, F. A. *Graphic Methods for Solving Problems.* Edwards Bros., Inc.
Hoelscher, R. P., Arnold, J. N., and Pierce, S. H. *Graphic Aids in Engineering Computation.* McGraw-Hill
Johnson, L. H. *Nomography and Empirical Equations.* John Wiley
Kulmann, C. A. *Nomographic Charts.* McGraw-Hill
Levens, A. S. *Nomography.* John Wiley
Lipka, J. *Graphical and Mechanical Computation.* John Wiley
Mackey, C. O. *Graphical Solutions.* John Wiley
Mavis, F. T. *The Construction of Nomographic Charts.* International Textbook
Runge, C. *Graphical Methods.* Columbia Univ. Press
Running, T. R. *Graphical Mathematics.* John Wiley

### Handbooks

*ASHRAE Guide and Data Book.* American Society of Heating, Refrigerating and Air-Conditioning Engineers, Inc., 51 Madison Ave., New York, N.Y. 10010
*ASME Handbook* (4 vols.). McGraw-Hill
Baumeister, T., and Marks, L. S. *Standard Handbook for Mechanical Engineers.* McGraw-Hill
Colvin, F. H., and Stanley, F. A. *American Machinists Handbook.* McGraw-Hill
Dudley, D. W. *Gear Handbook.* McGraw-Hill
Huntington, W. C. *Building Construction.* John Wiley
Kent, W. *Mechanical Engineers Handbook.* John Wiley
Kidder, F. E., and Parker, H. *Architects and Builders Handbook.* John Wiley
Knowlton, A. E. *Standard Handbook of Electrical Engineers.* McGraw-Hill
Oberg, E., and Jones, F. D. *Machinery's Handbook.* Industrial Press
O'Rourke, C. E. *General Engineering Handbook.* McGraw-Hill
Perry, J. H. *Chemical Engineers Handbook.* McGraw-Hill
Raskhodoff, N. M., *Electronic Drafting Handbook.* Macmillan
*SAE Automotive Drafting Standards.* Society of Automotive Engineers, 485 Lexington Ave., New York, N.Y. 10017
*SAE Handbook.* Society of Automotive Engineers, 485 Lexington Ave., New York, N.Y. 10017
*Smoley's New Combined Tables.* C. K. Smoley & Sons, Scranton, Pa.
*Tool Engineers Handbook.* McGraw-Hill
Tweney, C. F., and Hughes, L. E. C. *Chambers Technical Dictionary.* Macmillan
Urquhart, L. C. *Civil Engineers Handbook.* McGraw-Hill

### Lettering

French, T. E., and Meiklejohn, R. *Essentials of Lettering.* McGraw-Hill
French, T. E., and Turnbull, W. D. *Lessons in Lettering, Books 1 and 2.* McGraw-Hill
George, R. F. *Modern Lettering for Pen and Brush Poster Design.* Hunt Pen Co., Camden, N.J.

Hornung, C. P. *Lettering from A to Z.* Ziff-Davis Pub. Co., N.Y.
Ogg, Oscar. *An Alphabet Source Book.* Harper
Svensen, C. L. *The Art of Lettering.* D. Van Nostrand

#### Machine Design

Albert, C. D. *Machine Design and Drawing Room Problems.* John Wiley
Berard, S. J., Watters, E. O., and Phelps, C. W. *Principles of Machine Design.* Ronald Press
Faires, V. M. *Design of Machine Elements.* Macmillan
Jefferson, T. B., and Brooking, W. J. *Introduction to Mechanical Design.* Ronald Press
Norman, C. A., Ault, S., and Zabrosky, I. *Fundamentals of Machine Design.* Macmillan
Spotts, M. F. *Elements of Machine Design.* Prentice-Hall
Vallance, A., and Doughtie, V. L. *Design of Machine Members.* McGraw-Hill

#### Mechanism

Ham, C. W., Crane, E. J., and Rogers, W. L. *Mechanics of Machinery.* McGraw-Hill
Keon, R. M., and Faires, V. M. *Mechanism.* McGraw-Hill
Schwamb, P., Merrill, A. L., and James, W. H. *Elements of Mechanism.* John Wiley

#### Map Drawing

Deetz, C. H. *Elements of Map Projection.* U.S. Government Printing Office
Greitzer, S. L. *Elementary Topography and Map Reading.* McGraw-Hill
Hinks, A. R. *Maps and Surveys.* Macmillan
*Manual of Surveying Instructions for the Survey of the Public Lands of the United States.* U.S. Government Printing Office
Robinson, A. H. *Elements of Cartography.* John Wiley
Sloane, R. C., and Montz, J. M. *Elements of Topographic Drawing.* McGraw-Hill
Whitmore, G. D. *Advanced Surveying and Mapping.* International Textbook
Whitmore, G. D. *Elements of Photogrammetry.* International Textbook

#### Patent Drawings

*Guide for Patent Draftsmen.* U.S. Government Printing Office
Radzinsky, H. *Making Patent Drawings.* Macmillan

#### Perspective

Freese, E. I. *Perspective Projection.* Reinhold
Lawson, P. J. *Practical Perspective Drawing.* McGraw-Hill
Lubchez, B. *Perspective.* D. Van Nostrand
Morehead, J. C., and Morehead, J. C. Jr. *A Handbook of Perspective Drawing.* D. Van Nostrand
Norling, E. *Perspective Made Easy.* Macmillan
Turner, W. W. *Simplified Perspective.* Ronald Press

#### Piping Drawing and Design

Babbitt, H. E. *Plumbing.* McGraw-Hill
*Catalog of Valves, Fittings and Piping.* Crane Co.
Crocker, S. *Piping Handbook.* McGraw-Hill
Day, L. J. *Standard Plumbing Details.* John Wiley
*Fabricated Piping Data.* Pittsburg Piping & Equipment Co.
*Handbook of Cast Iron Pipe.* Cast Iron Pipe Research Association
Littleton, C. T. *Industrial Piping.* McGraw-Hill
*NAVCO Piping Catalog.* National Valve and Manufacturing Co.

Nielsen, L. S. *Standard Plumbing Engineering Design.* McGraw-Hill
*Piping Design and Engineering.* Grinnell Co.
*Primer in Power Plant Piping.* Grinnell Co.
Rase, H. F. *Piping Design for Process Plants.* John Wiley
Thompson, C. H. *Fundamentals of Pipe Drafting.* John Wiley.

### Production Illustration

Farmer, J. H., Hoecker, A. J., and Vavrin, F. F. *Illustrating for Tomorrow's production.* Macmillan
Hoelscher, R. P., Springer, C. H., and Pohle, R. F. *Industrial Production Illustration.* McGraw-Hill
Pyeatt, A. D. *Technical Illustration.* Higgins Ink Co., 41 Dickerson St., Newark, N.J. 07103
Thomas, T. A. *Technical Illustration.* McGraw-Hill
Treacy, J. *Production Illustration.* John Wiley

### Sheet Metal Drafting

Betterley, M. L. *Sheet Metal Drafting.* McGraw-Hill
Dougherty, J. S. *Sheet Metal Pattern Drafting and Shop Problems.* Manual Arts Press
Giachino, J. W. *Basic Sheet Metal Practice.* International Textbook
Jenkins, Rolland, *Sheet Metal Pattern Layout.* Prentice-Hall
Kidder, F. S. *Triangulation Applied to Sheet Metal Pattern Cutting.* Sheet Metal Pub. Co., New York, N.Y.
Neubecker, William. *Sheet Metal Work.* American Technical Society, Chicago, Ill.
O'Rourke, F. J. *Sheet Metal Pattern Drafting.* McGraw-Hill
Paull, J. H. *Industrial Sheet Metal Drawing.* D. Van Nostrand

### Shop Processes and Materials

*Arc Welding in Design, Manufacturing and Construction.* Lincoln Arc Welding Foundation, Cleveland, O.
Begeman, M. L. *Manufacturing Processes.* John Wiley
Boston, O. W. *Metal Processing.* John Wiley
Campbell, H. L. *Metal Castings.* John Wiley
Clapp, W. H., and Clark, D. S. *Engineering Materials and Processes.* International Textbook
Colvin, F. H., and Haas, L. L. *Jigs and Fixtures.* McGraw-Hill
Dubois, J. H. *Plastics.* American Technical Society, Chicago, Ill.
Hesse, H. C. *Engineering Tools and Processes.* D. Van Nostrand
Hinman, C. W. *Die Engineering Layouts and Formulas.* McGraw-Hill
Johnson, C. G. *Forging Practice.* American Technical Society, Chicago, Ill.
Wendt, R. E. *Foundry Work.* McGraw-Hill
Young, J. F. *Materials and Processes.* John Wiley

### Sketching

Guptill, A. L. *Drawing with Pen and Ink.* Reinhold
Guptill, A. L. *Sketching and Rendering in Pencil.* Reinhold
Jones, F. D. *How to Sketch Mechanisms.* Industrial Press
Katz, H. H. *Technical Sketching.* Macmillan
Kautsky, T. *Pencil Broadsides.* Reinhold
Turner, W. W. *Freehand Sketching for Engineers.* Ronald Press
Zipprich, A. E. *Freehand Drafting.* D. Van Nostrand

Structural Drafting and Design

Bishop, C. T. *Structural Drafting.* John Wiley
Bresler, B., and Lin, T. Y. *Design of Steel Structures.* John Wiley
Ketchum, M. S. *Handbook of Standard Structural Details for Buildings.* Prentice-Hall
Lucy, T. A. *Practical Design of Structural Members.* McGraw-Hill
*Manual of Standard Practice for Detailing Reinforced Concrete Highway Structures.* American Concrete Institute
*Manual of Standard Practice for Detailing Reinforced Concrete Structures.* American Concrete Institute
*Manual of Steel Construction.* American Institute of Steel Construction
Parker, H. *Simplified Design of Reinforced Concrete.* John Wiley
Parker, H. *Simplified Design of Structural Steel.* John Wiley
Parker, H. *Simplified Design of Structural Timber.* John Wiley
*Structural Steel Detailing.* American Institute of Steel Construction

Tool Design

Bloom, R. R. *Principles of Tool Engineering.* McGraw-Hill
Cole, C. B. *Tool Design.* American Technical Society, Chicago, Ill.

Welding

*Procedure Handbook of Arc Welding Design and Practice.* Lincoln Electric Co., Cleveland, O.
Rossi, B. E. *Welding Engineering.* McGraw-Hill.

## 2 Technical Terms

*"The beginning of wisdom is to call things by their right names."*
—Chinese Proverb

*n* means a *noun;* *v* means a *verb.*

**Acme** (*n*) Screw thread form, §§13.3 and 13.13.

**Addendum** (*n*) Radial distance from pitch circle to top of gear tooth, §§20.3 and 20.8.

**Allen Screw** (*n*) Special set screw or cap screw with hexagon socket in head, §13.31.

**Allowance** (*n*) Minimum clearance between mating parts, §12.3.

**Alloy** (*n*) Two or more metals in combination, usually a fine metal with a baser metal.

**Aluminum** (*n*) A lightweight but relatively strong metal. Often alloyed with copper to increase hardness and strength.

**Angle Iron** (*n*) A structural shape whose section is a right angle, §10.25.

**Anneal** (*v*) To heat and cool gradually, to reduce brittleness and increase ductility, §10.30.

**Arc-weld** (*v*) To weld by electric arc. The work is usually the positive terminal.

**Babbitt** (*n*) A soft alloy for bearings, mostly of tin with small amounts of copper and antimony.

**Bearing** (*n*) A supporting member for a rotating shaft.

**Bevel** (*n*) An inclined edge, not at right angle to joining surface.

**Bolt Circle** (*n*) A circular center line on a drawing, containing the centers of holes about a common center, §11.25.

**Bore** (*v*) To enlarge a hole with a boring bar or tool in a lathe, drill press, or boring mill, Figs. 10.20 (b) and 10.24 (b).

**Boss** (*n*) A cylindrical projection on a casting or a forging.

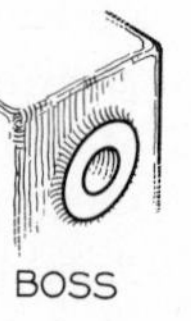
BOSS

**Brass** (*n*) An alloy of copper and zinc.

**Braze** (*v*) To join with hard solder of brass or zinc.

**Brinell** (*n*) A method of testing hardness of metal.

**Broach** (*n*) A long cutting tool with a series of teeth that gradually increase in size which is forced through a hole or over a surface to produce a desired shape, §10.19.

**Bronze** (*n*) An alloy of eight or nine parts of copper and one part of tin.

**Buff** (*v*) To finish or polish on a buffing wheel composed of fabric with abrasive powders.

**Burnish** (*v*) To finish or polish by pressure upon a smooth rolling or sliding tool.

**Burr** (*n*) A jagged edge on metal resulting from punching or cutting.

**Bushing** (*n*) A replaceable lining or sleeve for a bearing.

**Calipers** (*n*) Instrument (of several types) for measuring diameters, §10.21.

**Cam** (*n*) A rotating member for changing circular motion to reciprocating motion.

**Carburize** (*v*) To heat a low-carbon steel to approximately 2000°F in contact with material which adds carbon to the surface of the steel, and to cool slowly in preparation for heat treatment, §10.30.

**Caseharden** (*v*) To harden the outer surface of a carburized steel by heating and then quenching.

**Castellate** (*v*) To form like a castle, as a castellated shaft or nut.

**Casting** (*n*) A metal object produced by pouring molten metal into a mold, §10.4.

**Cast Iron** (*n*) Iron melted and poured into molds, §10.4.

**Center Drill** (*n*) A special drill to produce bearing holes in the ends of a workpiece to be mounted between centers. Also called a "combined drill and countersink," §11.35.

COMBINED DRILL & C SINK

**Chamfer** (*n*) A narrow inclined surface along the intersection of two surfaces.

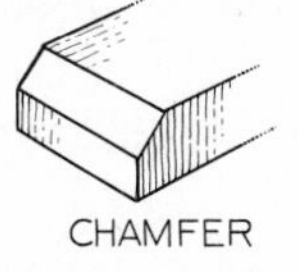

CHAMFER

**Chase** (*v*) To cut threads with an external cutting tool.

**Cheek** (*n*) The middle portion of a three-piece flask used in molding, §10.4.

**Chill** (*v*) To harden the outer surface of cast iron by quick cooling, as in a metal mold.

**Chip** (*v*) To cut away metal with a cold chisel.

**Chuck** (*n*) A mechanism for holding a rotating tool or workpiece.

**Coin** (*v*) To form a part in one stamping operation.

**Cold Rolled Steel** (CRS) (*n*) Open hearth or Bessemer steel containing .12% to .20% carbon that has been rolled while cold to produce a smooth, quite accurate stock.

**Collar** (*n*) A round flange or ring fitted on a shaft to prevent sliding.

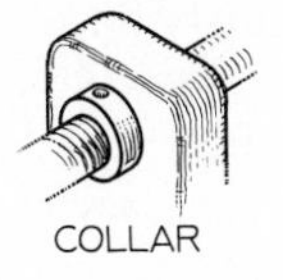

COLLAR

**Colorharden** (*v*) Same as *caseharden,* except that it is done to a shallower depth, usually for appearance only.

**Cope** (*n*) The upper portion of a flask used in molding, §10.4.

**Core** (*v*) To form a hollow portion in a casting by using a dry-sand core or a green-sand core in a mold, §10.4.

**Coreprint** (*n*) A projection on a pattern which forms an opening in the sand to hold the end of a core, §10.4.

**Cotter Pin** (*n*) A split pin used as a fastener, usually to prevent a nut from unscrewing, Fig. 13.32 (g) and (h) and Appendix 25.

**Counterbore** (*v*) To enlarge an end of a hole cylindrically with a *counterbore,* §10.20.

COUNTERBORE

**Countersink** (*v*) To enlarge an end of a hole conically, usually with a *countersink,* §10.20.

COUNTERSINK

**Crown** (*n*) A raised contour, as on the surface of a pulley.

**Cyanide** (*v*) To surface-harden steel by heating in contact with a cyanide salt, followed by quenching.

**Dedendum** (*n*) Distance from pitch circle to bottom of tooth space.

**Development** (*n*) Drawing of the surface of an object unfolded or rolled out on a plane.

**Diametral Pitch** (*n*) Number of gear teeth per inch of pitch diameter.

**Die** (*n*) 1. Hardened metal piece shaped to cut or form a required shape in a sheet of metal by pressing it against a mating die. 2. Also used for cutting small threads. In a sense is opposite to a tap.

**Die Casting** (*n*) Process of forcing molten metal under pressure into metal dies or molds, producing a very accurate and smooth casting, §10.8.

**Die Stamping** (*n*) Process of cutting or forming a piece of sheet metal with a die.

**Dog** (*n*) A small auxiliary clamp for preventing work from rotating in relation to the face plate of a lathe.

**Dowel** (*n*) A cylindrical pin, commonly used to prevent sliding between two contacting flat surfaces.

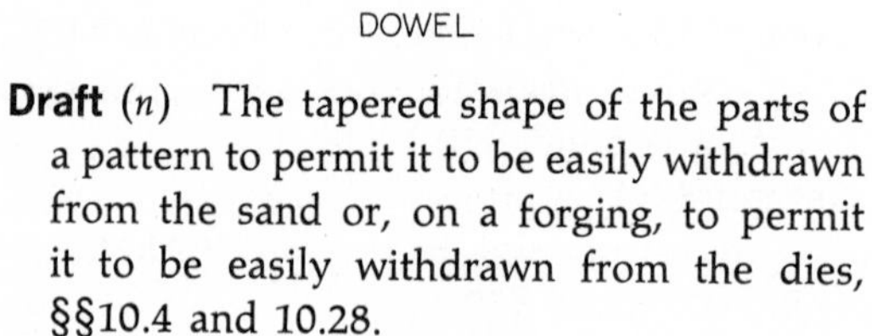


**Draft** (*n*) The tapered shape of the parts of a pattern to permit it to be easily withdrawn from the sand or, on a forging, to permit it to be easily withdrawn from the dies, §§10.4 and 10.28.

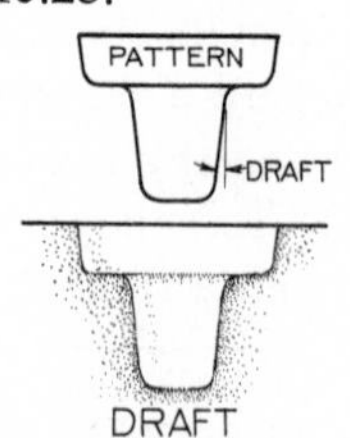


**Drag** (*n*) Lower portion of a flask used in molding, §10.4.

**Draw** (*v*) To stretch or otherwise to deform metal. Also to temper steel.

**Drill** (*v*) To cut a cylindrical hole with a drill. A *blind hole* does not go through the piece, §10.20.

**Drill Press** (*n*) A machine for drilling and other hole-forming operations, §10.13.

**Drop Forge** (*v*) To form a piece while hot between dies in a drop hammer or with great pressure, §10.29.

**Face** (*v*) To finish a surface at right angles, or nearly so, to the center line of rotation on a lathe, Fig. 10.24 (a).

**FAO** Finish all over, §11.17.

**Feather Key** (*n*) A flat key, which is partly sunk in a shaft and partly in a hub, permitting the hub to slide lengthwise of the shaft, §13.34.

**File** (*v*) To finish or smooth with a file.

**Fillet** (*n*) An interior rounded intersection between two surfaces, §§6.34 and 10.6.

**Fin** (*n*) A thin extrusion of metal at the intersection of dies or sand molds.

**Fit** (*n*) Degree of tightness or looseness between two mating parts, as a *loose fit*, a *snug fit*, or a *tight fit*, §§11.9 and 12.1 to 12.4.

**Fixture** (*n*) A special device for holding the work in a machine tool, *but not for guiding the cutting tool*, §10.27.

**Flange** (*n*) A relatively thin rim around a piece.

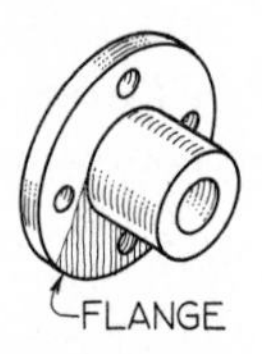


**Flash** (*n*) Same as *fin*.

**Flask** (*n*) A box made of two or more parts for holding the sand in sand molding, §10.4.

**Flute** (*n*) Groove, as on twist drills, reamers, and taps.

**Forge** (*v*) To force metal while it is hot to take on a desired shape by hammering or pressing, §10.29.

**Galvanize** (*v*) To cover a surface with a thin layer of molten alloy, composed mainly of zinc, to prevent rusting.

**Gasket** (*n*) A thin piece of rubber, metal, or some other material, placed between surfaces to make a tight joint.

**Gate** (*n*) The opening in a sand mold at the bottom of the *sprue* through which the molten metal passes to enter the cavity or mold, §10.4.

**Graduate** (*v*) To set off accurate divisions on a scale or dial.

**Grind** (*v*) To remove metal by means of an abrasive wheel, often made of carborundum. Use chiefly where accuracy is required, §10.17.

**Harden** (*v*) To heat steel above a critical temperature and then quench in water or oil, §10.30.

**Heat-treat** (*v*) To change the properties of metals by heating and then cooling, §10.30.

**Interchangeable** (*adj.*) Refers to a part made to limit dimensions so that it will fit any mating part similarly manufactured, §12.1.

**Jig** (*n*) A device *for guiding a tool* in cutting a piece. Usually it holds the work in position, §10.27.

**Journal** (*n*) Portion of a rotating shaft supported by a bearing.

**Kerf** (*n*) Groove or cut made by a saw.

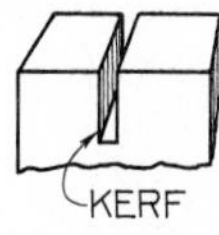


**Key** (*n*) A small piece of metal sunk partly into both shaft and hub to prevent rotation, §13.34.

**Keyseat** (*n*) A slot or recess in a shaft to hold a key, §13.34.

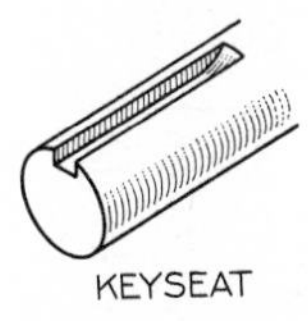


**Keyway** (*n*) A slot in a hub or portion surrounding a shaft to receive a key, §13.34.

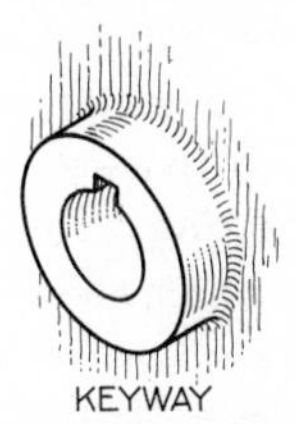


**Knurl** (*v*) To impress a pattern of dents in a turned surface with a knurling tool to produce a better hand grip, §11.37.

**Lap** (*v*) To produce a very accurate finish by sliding contact with a *lap,* or piece of wood, leather, or soft metal impregnated with abrasive powder.

**Lathe** (*n*) A machine used to shape metal or other materials by rotating against a tool, §10.12.

**Lug** (*n*) An irregular projection of metal, but not round as in the case of a *boss,* usually with a hole in it for a bolt or screw.

**Malleable Casting** (*n*) A casting that has been made less brittle and tougher by annealing.

**Mill** (*v*) To remove material by means of a rotating cutter on a milling machine, §10.14.

**Mold** (*n*) The mass of sand or other material that forms the cavity into which molten metal is poured, §10.4.

**MS** (*n*) Machinery steel, sometimes called *mild steel* with a small percentage of carbon. Cannot be hardened.

**Neck** (*v*) To cut a groove called a *neck* around a cylindrical piece.

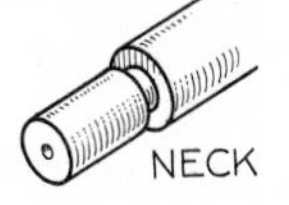


**Normalize** (*v*) To heat steel above its critical temperature and then to cool it in air, §10.30.

**Pack-harden** (*v*) To *carburize,* then to *case-harden,* §10.30.

**Pad** (*n*) A slight projection, usually to provide a bearing surface around one or more holes.

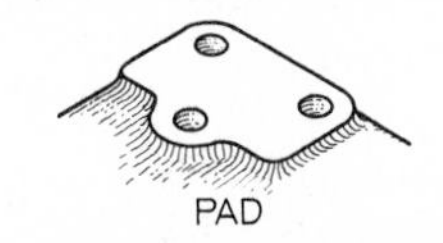


**Pattern** (*n*) A model, usually of wood, used in forming a mold for a casting. In sheet metal work a pattern is called a *development.*

**Peen** (*v*) To hammer into shape with a ball-peen hammer.

**Pickle** (*v*) To clean forgings or castings in dilute sulphuric acid.

**Pinion** (*n*) The smaller of two mating gears.

**Pitch Circle** (*n*) An imaginary circle corresponding to the circumference of the friction gear from which the spur gear was derived.

**Plane** (*v*) To remove material by means of the *planer,* §10.16.

**Planish** (*v*) To impart a planished surface to sheet metal by hammering with a smooth-surfaced hammer.

**Plate** (*v*) To coat a metal piece with another metal, such as chrome or nickel, by electrochemical methods.

**Polish** (*v*) To produce a highly finished or polished surface by friction, using a very fine abrasive.

**Profile** (*v*) To cut any desired outline by moving a small rotating cutter, usually with a master template as a guide.

**Punch** (*v*) To cut an opening of a desired shape with a rigid tool having the same shape, by pressing the tool through the work.

**Quench** (*v*) To immerse a heated piece of metal in water or oil in order to harden it.

**Rack** (*n*) A flat bar with gear teeth in a straight line to engage with teeth in a gear.

**Ream** (*v*) To enlarge a finished hole slightly to give it greater accuracy, with a *reamer,* §10.20.

**Relief** (*n*) An offset of surfaces to provide clearance for machining.

RELIEF

**Rib** (*n*) A relatively thin flat member acting as a brace or support.

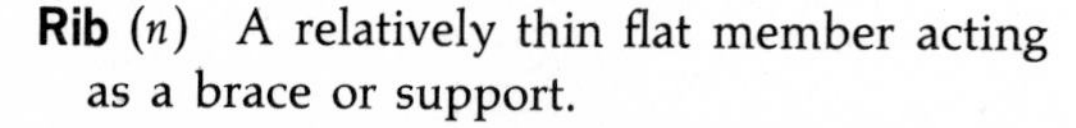

RIB

**Rivet** (*v*) To connect with rivets or to clench over the end of a pin by hammering, §13.36.

**Round** (*n*) An exterior rounded intersection of two surfaces, §§6.34 and 10.6.

**SAE** Society of Automotive Engineers.

**Sandblast** (*v*) To blow sand at high velocity with compressed air against castings or forgings to clean them.

**Scleroscope** (*n*) An instrument for measuring hardness of metals.

**Scrape** (*v*) To remove metal by scraping with a hand scraper, usually to fit a bearing.

**Shape** (*v*) To remove metal from a piece with a *shaper,* §10.15.

**Shear** (*v*) To cut metal by means of shearing with two blades in sliding contact.

**Sherardize** (*v*) To galvanize a piece with a coating of zinc by heating it in a drum with zinc powder, to a temperature of 575° to 850°F.

**Shim** (*n*) A thin piece of metal or other material used as a spacer in adjusting two parts.

**Solder** (*v*) To join with solder, usually composed of lead and tin.

**Spin** (*v*) To form a rotating piece of sheet metal into a desired shape by pressing it with a smooth tool against a rotating form.

**Spline** (*n*) A keyway, usually one of a series cut around a shaft or hole.

SPLINED HOLE

**Spotface** (*v*) To produce a round *spot* or bearing surface around a hole, usually with a *spotfacer.* The spotface may be on top of a boss or it may be sunk into the surface, §§6.33 and 10.20.

SPOTFACE

**Sprue** (*n*) A hole in the sand leading to the *gate* which leads to the mold, through which the metal enters, §10.4.

**Steel Casting** (*n*) Like cast-iron casting except that in the furnace scrap steel has been added to the casting.

**Swage** (*v*) To hammer metal into shape while it is held over a *swage,* or die, which fits in a hole in the *swage block,* or anvil.

**Sweat** (*v*) To fasten metal together by the use of solder between the pieces and by the application of heat and pressure.

**Tap** (*v*) To cut relatively small internal threads with a *tap,* §10.20.

**Taper** (*n*) Conical form given to a shaft or a hole. Also refers to the slope of a plane surface, §11.33.

**Taper Pin** (*n*) A small tapered pin for fastening, usually to prevent a collar or hub from rotating on a shaft.

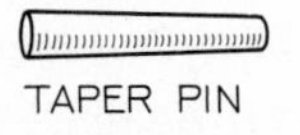

TAPER PIN

**Taper Reamer** (*n*) A tapered reamer for producing accurate tapered holes, as for a taper pin, §§11.33 and 13.37.

**Temper** (*v*) To reheat hardened steel to bring it to a desired degree of hardness, §10.30.

**Template** or **Templet** (*n*) A guide or pattern used to mark out the work, guide the tool in cutting it, or check the finished product.

**Tin** (*n*) A silvery metal used in alloys and for coating other metals, such as tin plate.

**Tolerance** (*n*) Total amount of variation permitted in limit dimension of a part, §12.3.

**Trepan** (*v*) To cut a circular groove in the flat surface at one end of a hole.

**Tumble** (*v*) To clean rough castings or forgings in a revolving drum filled with scrap metal.

**Turn** (*v*) To produce, on a lathe, a cylindrical surface parallel to the center line, §10.12.

**Twist Drill** (*n*) A drill for use in a drill press, §10.20.

**Undercut** (*n*) A recessed cut or a cut with inwardly sloping sides.

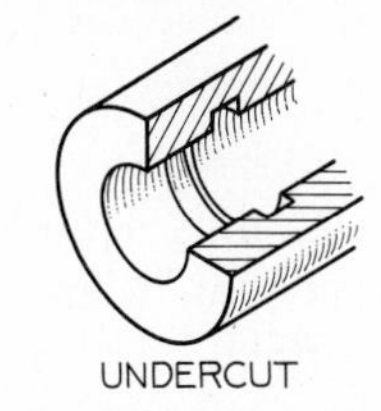

UNDERCUT

**Upset** (*v*) To form a head or enlarged end on a bar or rod by pressure or by hammering between dies.

**Web** (*n*) A thin flat part joining larger parts. Also known as a *rib*.

**Weld** (*v*) Uniting metal pieces by pressure or fusion welding processes, §10.26.

**Woodruff Key** (*n*) A semicircular flat key, §13.34.

WOODRUFF KEYS

**Wrought Iron** (*n*) Iron of low carbon content useful because of its toughness, ductility, and malleability.

## 3 Visual Aids for Technical Drawing

The following visual aids are suggested. However, the instructor should preview each visual aid before using it in class, in order to determine its suitability for his purposes. All motion pictures listed are on 16 mm. film. An asterisk indicates that a follow-up filmstrip is also available. The sources for the aids listed below are as follows:

DA Du Art Film Laboratories, U.S. Government Film Services, 245 W. 55th St., New York, N.Y. 10019

GM General Motors Corp., Public Relations Staff Film Library, GM Bldg., Detroit, Mich. 48202

JH Jam Handy Organization, 2821 E. Grand Blvd., Detroit, Mich. 48211

McG McGraw-Hill Book Co., Text Film Division, 330 W. 42nd St., New York, N.Y. 10036

P Purdue University, Audio Visual Center, Lafayette, Ind. 47907

PSU Pennsylvania State University, Audio Visual Aids Library, University Park, Pa. 16802

UC University of California, Extension Media Center, 2223 Fulton St., Berkeley, Calif. 94720

### The Graphic Language

According to Plan*—9 min. sound movie—McG.

The Draftsman—10 min. sound movie—P.

The Language of Drawing—10 min. sound movie—McG.

### Freehand Sketching

Freehand Drafting—13 min. silent movie—P.

Pictorial Sketching*—11 min. sound movie—McG.

MECHANICAL DRAWING

T-squares and Triangles, Parts 1 & 2—film strips—JH.

Use of T-squares and Triangles—20 min. silent movie—P.

LETTERING

Capital Letters—21 min. sound movie—P.

Lettering Instructional Materials—22 min. sound movie—PSU.

Lower Case Letters—17 min. sound movie—P.

Technical Lettering—filmstrips (a series of five)—JH.

GEOMETRY OF TECHNICAL DRAWING

Applied Geometry—16 min. silent movie—P.

VIEWS OF OBJECTS

Behind the Shop Drawing—18 min. sound movie—P.

Multiview Drawing—24 min. silent movie—P.

Orthographic Projection*—18 min. sound movie—McG.

Reading a Three-View Drawing—9 min. sound movie—DA.

Shape Description—31 min. sound movie—P.

Shape Description, Part I—11 min. sound movie—McG.

Shape Description, Part II—8 min. sound movie—McG.

Visualizing an Object—10 min. sound movie—DA.

TECHNIQUES AND APPLICATIONS

Basic Reproduction Processes in the Graphic Arts—25 min. sound movie—PSU.

Drafting Tips—28 min. sound movie—PSU.

INKING

Tracing with Ink—32 min. silent movie—P.

DIMENSIONING

Principal Dimensions, References, Surfaces and Tolerances—12 min. sound movie—DA.

Selection of Dimensions*—18 min. sound movie—McG.

Size Description—13 min. sound movie—McG.

COMPUTER GRAPHICS

Design Augmented by Computer—13 min. sound movie—GM

SHOP PROCESSES

Cutting a Keyway on the End of a Finished Shaft—13 min. sound movie—PSU.

Drawings and the Shop*—15 min. sound movie—McG.

The Drill Press—10 min. sound movie—DA.

Fixed Gages—16 min. sound movie—P.

The Lathe—15 min. sound movie—DA.

Laying Out Small Castings—16 min. sound movie—PSU.

Machining a Cast Iron Rectangular Block—25 min. sound movie—PSU.

Machining a Tool Steel V-Block—21 min. sound movie—PSU.

The Milling Machine—15 min. sound movie—DA.

Precision Layout and Measuring—11 min. sound movie—PSU.

The Shaper—15 min. sound movie—DA.

Shop Drawing—22 min. sound movie—PSU.

Shop Procedures—17 min. sound movie—McG.

Shop Work—27 min. silent movie—P.

SECTIONAL VIEWS

Sections and Conventions*—15 min. sound movie—McG.

Sectional Views—15 min. silent movie—P.

Sectional Views and Projections, Finish Marks—15 min. sound movie—DA.

AUXILIARY VIEWS

Auxiliary Views—17 min. silent movie—P.

Auxiliary Views, Parts I and II—11 and 10 min. sound movies—McG.

Auxiliary Views: Single Auxiliaries*—23 min. sound movie—McG.

Auxiliary Views: Double Auxiliaries—13 min. sound movie—McG.

THREADS AND FASTENERS

Cutting an Internal Acme Thread—22 min. sound movie—PSU.

Cutting an External Acme Thread—12 min. sound movie—PSU.

Cutting an External National Fine Thread—11 min. sound movie—P.

Cutting Threads with Taps and Dies—20 min. sound movie—P.
Screw Threads—22 min. silent movie—P.

WORKING DRAWINGS

Concepts and Principles of Functional Drafting—20 min. sound movie—PSU.
Design for Manufacture—29 min. sound movie—PSU.

PICTORIAL DRAWINGS

Discovering Perspective—14 min. sound movie—UC.
Perspective Drawing—10 min. sound movie—UC.
Pictorial Drawing (Isometric)—21 min. silent movie—P.
Pictorial Sketching*—11 min. sound movie—McG.

DEVELOPMENTS AND INTERSECTIONS

Development of Surfaces—22 min. silent movie—P.
Finding the Line of Intersection of Two Solids—22 min. sound movie—PSU.
Intersection of Surfaces—9 min. silent movie—P.
Oblique Cones and Transition Developments*—11 min. sound movie—McG.

## 4 American National Standard Abbreviations for Use on Drawings

(From ANSI Z32.13-1950)

**A**

| | |
|---|---|
| Absolute | ABS |
| Accelerate | ACCEL |
| Accessory | ACCESS. |
| Account | ACCT |
| Accumulate | ACCUM |
| Actual | ACT. |
| Adapter | ADPT |
| Addendum | ADD. |
| Addition | ADD. |
| Adjust | ADJ |
| Advance | ADV |
| After | AFT. |
| Aggregate | AGGR |
| Air Condition | AIR COND |
| Airplane | APL |
| Allowance | ALLOW |
| Alloy | ALY |
| Alteration | ALT |
| Alternate | ALT |
| Alternating Current | AC |
| Altitude | ALT |
| Aluminum | AL |
| American National Standard | AMER NATL STD |
| American Wire Gage | AWG |
| Amount | AMT |
| Ampere | AMP |
| Amplifier | AMPL |
| Anneal | ANL |
| Antenna | ANT. |
| Apartment | APT. |
| Apparatus | APP |
| Appendix | APPX |
| Approved | APPD |
| Approximate | APPROX |
| Arc Weld | ARC/W |
| Area | A |
| Armature | ARM. |
| Armor Plate | ARM-PL |
| Army Navy | AN |
| Arrange | ARR. |
| Artificial | ART. |
| Asbestos | ASB |
| Asphalt | ASPH |
| Assemble | ASSEM |
| Assembly | ASSY |
| Assistant | ASST |
| Associate | ASSOC |
| Association | ASSN |
| Atomic | AT |
| Audible | AUD |
| Audio Frequency | AF |
| Authorized | AUTH |
| Automatic | AUTO |
| Auto-Transformer | AUTO TR |
| Auxiliary | AUX |
| Avenue | AVE |
| Average | AVG |
| Aviation | AVI |
| Azimuth | AZ |

**B**

| | |
|---|---|
| Babbitt | BAB |
| Back Feed | BF |
| Back Pressure | BP |
| Back to Back | B to B |
| Backface | BF |
| Balance | BAL |
| Ball Bearing | BB |
| Barometer | BAR |
| Base Line | BL |
| Base Plate | BP |
| Bearing | BRG |
| Bench Mark | BM |
| Bending Moment | M |
| Bent | BT |
| Bessemer | BESS |
| Between | BET. |
| Between Centers | BC |
| Between Perpendiculars | BP |
| Bevel | BEV |
| Bill of Material | B/M |
| Birmingham Wire Gage | BWG |
| Blank | BLK |

| | |
|---|---|
| Block | BLK |
| Blueprint | BP |
| Board | BD |
| Boiler | BLR |
| Boiler Feed | BF |
| Boiler Horsepower | BHP |
| Boiling Point | BP |
| Bolt Circle | BC |
| Both Faces | BF |
| Both Sides | BS |
| Both Ways | BW |
| Bottom | BOT |
| Bottom Chord | BC |
| Bottom Face | BF |
| Bracket | BRKT |
| Brake | BK |
| Brake Horsepower | BHP |
| Brass | BRS |
| Brazing | BRZG |
| Break | BRK |
| Brinell Hardness | BH |
| British Standard | BR STD |
| British Thermal Units | BTU |
| Broach | BRO |
| Bronze | BRZ |
| Brown & Sharpe (Wire Gage, same as AWG) | B&S |
| Building | BLDG |
| Bulkhead | BHD |
| Burnish | BNH |
| Bushing | BUSH. |
| Button | BUT. |

## C

| | |
|---|---|
| Cabinet | CAB. |
| Calculate | CALC |
| Calibrate | CAL |
| Cap Screw | CAP SCR |
| Capacity | CAP |
| Carburetor | CARB |
| Carburize | CARB |
| Carriage | CRG |
| Case Harden | CH |
| Cast Iron | CI |
| Cast Steel | CS |
| Casting | CSTG |
| Castle Nut | CAS NUT |
| Catalogue | CAT. |
| Cement | CEM |
| Center | CTR |
| Center Line | CL |
| Center of Gravity | CG |
| Center of Pressure | CP |
| Center to Center | C to C |
| Centering | CTR |
| Chamfer | CHAM |
| Change | CHG |
| Channel | CHAN |
| Check | CHK |
| Check Valve | CV |
| Chord | CHD |
| Circle | CIR |
| Circular | CIR |
| Circular Pitch | CP |
| Circumference | CIRC |
| Clear | CLR |
| Clearance | CL |
| Clockwise | CW |
| Coated | CTD |
| Cold Drawn | CD |
| Cold Drawn Steel | CDS |
| Cold Finish | CF |
| Cold Punched | CP |
| Cold Rolled | CR |
| Cold Rolled Steel | CRS |
| Combination | COMB. |
| Combustion | COMB |
| Commercial | COML |
| Company | CO |
| Complete | COMPL |
| Compress | COMP |
| Concentric | CONC |
| Concrete | CONC |
| Condition | COND |
| Connect | CONN |
| Constant | CONST |
| Construction | CONST |
| Contact | CONT |
| Continue | CONT |
| Corner | COR |
| Corporation | CORP |
| Correct | CORR |
| Corrugate | CORR |
| Cotter | COT |
| Counter | CTR |
| Counter Clockwise | CCW |
| Counterbore | CBORE |
| Counterdrill | CDRILL |
| Counterpunch | CPUNCH |
| Countersink | CSK |
| Coupling | CPLG |
| Cover | COV |
| Cross Section | XSECT |
| Cubic | CU |
| Cubic Foot | CU FT |
| Cubic Inch | CU IN. |
| Current | CUR |
| Customer | CUST |
| Cyanide | CYN |

## D

| | |
|---|---|
| Decimal | DEC |
| Dedendum | DED |
| Deflect | DEFL |
| Degree | (°) DEG |
| Density | D |
| Department | DEPT |
| Design | DSGN |
| Detail | DET |
| Develop | DEV |
| Diagonal | DIAG |
| Diagram | DIAG |
| Diameter | DIA |
| Diametral Pitch | DP |
| Dimension | DIM. |
| Discharge | DISCH |
| Distance | DIST |
| Division | DIV |
| Double | DBL |
| Dovetail | DVTL |
| Dowel | DWL |
| Down | DN |
| Dozen | DOZ |
| Drafting | DFTG |
| Draftsman | DFTSMN |
| Drawing | DWG |
| Drill or Drill Rod | DR |
| Drive | DR |
| Drive Fit | DF |
| Drop | D |
| Drop Forge | DF |
| Duplicate | DUP |

## E

| | |
|---|---|
| Each | EA |
| East | E |
| Eccentric | ECC |
| Effective | EFF |
| Elbow | ELL |

| | |
|---|---|
| Electric | ELEC |
| Elementary | ELEM |
| Elevate | ELEV |
| Elevation | EL |
| Engine | ENG |
| Engineer | ENGR |
| Engineering | ENGRG |
| Entrance | ENT |
| Equal | EQ |
| Equation | EQ |
| Equipment | EQUIP |
| Equivalent | EQUIV |
| Estimate | EST |
| Exchange | EXCH |
| Exhaust | EXH |
| Existing | EXIST. |
| Exterior | EXT |
| Extra Heavy | X HVY |
| Extra Strong | X STR |
| Extrude | EXTR |

### F

| | |
|---|---|
| Fabricate | FAB |
| Face to Face | F to F |
| Fahrenheit | F |
| Far Side | FS |
| Federal | FED. |
| Feed | FD |
| Feet | (′) FT |
| Figure | FIG. |
| Fillet | FIL |
| Fillister | FIL |
| Finish | FIN. |
| Finish All Over | FAO |
| Flange | FLG |
| Flat | F |
| Flat Head | FH |
| Floor | FL |
| Fluid | FL |
| Focus | FOC |
| Foot | (′) FT |
| Force | F |
| Forged Steel | FST |
| Forging | FORG |
| Forward | FWD |
| Foundry | FDRY |
| Frequency | FREQ |
| Front | FR |
| Furnish | FURN |

### G

| | |
|---|---|
| Gage or Gauge | GA |
| Gallon | GAL |
| Galvanize | GALV |
| Galvanized Iron | GI |
| Galvanized Steel | GS |
| Gasket | GSKT |
| General | GEN |
| Glass | GL |
| Government | GOVT |
| Governor | GOV |
| Grade | GR |
| Graduation | GRAD |
| Graphite | GPH |
| Grind | GRD |
| Groove | GRV |
| Ground | GRD |

### H

| | |
|---|---|
| Half-Round | $\frac{1}{2}$RD |
| Handle | HDL |
| Hanger | HGR |
| Hard | H |
| Harden | HDN |
| Hardware | HDW |
| Head | HD |
| Headless | HDLS |
| Heat | HT |
| Heat Treat | HT TR |
| Heavy | HVY |
| Hexagon | HEX |
| High-Pressure | HP |
| High-Speed | HS |
| Horizontal | HOR |
| Horsepower | HP |
| Hot Rolled | HR |
| Hot Rolled Steel | HRS |
| Hour | HR |
| Housing | HSG |
| Hydraulic | HYD |

### I

| | |
|---|---|
| Illustrate | ILLUS |
| Inboard | INBD |
| Inch | (″) IN. |
| Inches per Second | IPS |
| Inclosure | INCL |
| Include | INCL |
| Inside Diameter | ID |
| Instrument | INST |
| Interior | INT |
| Internal | INT |
| Intersect | INT |
| Iron | I |
| Irregular | IRREG |

### J

| | |
|---|---|
| Joint | JT |
| Joint Army-Navy | JAN |
| Journal | JNL |
| Junction | JCT |

### K

| | |
|---|---|
| Key | K |
| Keyseat | KST |
| Keyway | KWY |

### L

| | |
|---|---|
| Laboratory | LAB |
| Laminate | LAM |
| Lateral | LAT |
| Left | L |
| Left Hand | LH |
| Length | LG |
| Length Over All | LOA |
| Letter | LTR |
| Light | LT |
| Line | L |
| Locate | LOC |
| Logarithm | LOG. |
| Long | LG |
| Lubricate | LUB |
| Lumber | LBR |

### M

| | |
|---|---|
| Machine | MACH |
| Machine Steel | MS |
| Maintenance | MAINT |
| Malleable | MALL |
| Malleable Iron | MI |
| Manual | MAN. |
| Manufacture | MFR |
| Manufactured | MFD |
| Manufacturing | MFG |
| Material | MATL |
| Maximum | MAX |
| Mechanical | MECH |
| Mechanism | MECH |
| Median | MED |
| Metal | MET. |

| | |
|---|---|
| Meter | M |
| Miles | MI |
| Miles per Hour | MPH |
| Millimeter | MM |
| Minimum | MIN |
| Minute | (′) MIN |
| Miscellaneous | MISC |
| Month | MO |
| Morse Taper | MOR T |
| Motor | MOT |
| Mounted | MTD |
| Mounting | MTG |
| Multiple | MULT |
| Music Wire Gage | MWG |

N

| | |
|---|---|
| National | NATL |
| Natural | NAT |
| Near Face | NF |
| Near Side | NS |
| Negative | NEG |
| Neutral | NEUT |
| Nominal | NOM |
| Normal | NOR |
| North | N |
| Not to Scale | NTS |
| Number | NO. |

O

| | |
|---|---|
| Obsolete | OBS |
| Octagon | OCT |
| Office | OFF. |
| On Center | OC |
| Opposite | OPP |
| Optical | OPT |
| Original | ORIG |
| Outlet | OUT. |
| Outside Diameter | OD |
| Outside Face | OF |
| Outside Radius | OR |
| Overall | OA |

P

| | |
|---|---|
| Pack | PK |
| Packing | PKG |
| Page | P |
| Paragraph | PAR. |
| Part | PT |
| Patent | PAT. |
| Pattern | PATT |
| Permanent | PERM |
| Perpendicular | PERP |
| Piece | PC |
| Piece Mark | PC MK |
| Pint | PT |
| Pitch | P |
| Pitch Circle | PC |
| Pitch Diameter | PD |
| Plastic | PLSTC |
| Plate | PL |
| Plumbing | PLMB |
| Point | PT |
| Point of Curve | PC |
| Point of Intersection | PI |
| Point of Tangent | PT |
| Polish | POL |
| Position | POS |
| Potential | POT. |
| Pound | LB |
| Pounds per Square Inch | PSI |
| Power | PWR |
| Prefabricated | PREFAB |
| Preferred | PFD |
| Prepare | PREP |
| Pressure | PRESS. |
| Process | PROC |
| Production | PROD |
| Profile | PF |
| Propeller | PROP |
| Publication | PUB |
| Push Button | PB |

Q

| | |
|---|---|
| Quadrant | QUAD |
| Quality | QUAL |
| Quarter | QTR |

R

| | |
|---|---|
| Radial | RAD |
| Radius | R |
| Railroad | RR |
| Ream | RM |
| Received | RECD |
| Record | REC |
| Rectangle | RECT |
| Reduce | RED. |
| Reference Line | REF L |
| Reinforce | REINF |
| Release | REL |
| Relief | REL |
| Remove | REM |
| Require | REQ |
| Required | REQD |
| Return | RET. |
| Reverse | REV |
| Revolution | REV |
| Revolutions per Minute | RPM |
| Right | R |
| Right Hand | RH |
| Rivet | RIV |
| Rockwell Hardness | RH |
| Roller Bearing | RB |
| Room | RM |
| Root Diameter | RD |
| Root Mean Square | RMS |
| Rough | RGH |
| Round | RD |

S

| | |
|---|---|
| Schedule | SCH |
| Schematic | SCHEM |
| Scleroscope Hardness | SH |
| Screw | SCR |
| Second | SEC |
| Section | SECT |
| Semi-Steel | SS |
| Separate | SEP |
| Set Screw | SS |
| Shaft | SFT |
| Sheet | SH |
| Shoulder | SHLD |
| Side | S |
| Single | S |
| Sketch | SK |
| Sleeve | SLV |
| Slide | SL |
| Slotted | SLOT. |
| Small | SM |
| Socket | SOC |
| Space | SP |
| Special | SPL |
| Specific | SP |
| Spot Faced | SF |
| Spring | SPG |
| Square | SQ |
| Standard | STD |
| Station | STA |
| Stationary | STA |
| Steel | STL |

| | |
|---|---|
| Stock | STK |
| Straight | STR |
| Street | ST |
| Structural | STR |
| Substitute | SUB |
| Summary | SUM. |
| Support | SUP. |
| Surface | SUR |
| Symbol | SYM |
| System | SYS |

**T**

| | |
|---|---|
| Tangent | TAN. |
| Taper | TPR |
| Technical | TECH |
| Template | TEMP |
| Tension | TENS. |
| Terminal | TERM. |
| Thick | THK |
| Thousand | M |
| Thread | THD |
| Threads per Inch | TPI |
| Through | THRU |
| Time | T |
| Tolerance | TOL |
| Tongue & Groove | T & G |
| Tool Steel | TS |
| Tooth | T |
| Total | TOT |
| Transfer | TRANS |
| Typical | TYP |

**U**

| | |
|---|---|
| Ultimate | ULT |
| Unit | U |
| Universal | UNIV |

**V**

| | |
|---|---|
| Vacuum | VAC |
| Valve | V |
| Variable | VAR |
| Versus | VS |
| Vertical | VERT |
| Volt | V |
| Volume | VOL |

**W**

| | |
|---|---|
| Wall | W |
| Washer | WASH. |
| Watt | W |
| Week | WK |
| Weight | WT |
| West | W |
| Width | W |
| Wood | WD |
| Woodruff | WDF |
| Working Point | WP |
| Working Pressure | WP |
| Wrought | WRT |
| Wrought Iron | WI |

**X, Y, Z**

| | |
|---|---|
| Yard | YD |
| Year | YR |

## 5 American National Standard Running and Sliding Fits[a]

RC 1 *Close sliding fits* are intended for the accurate location of parts which must assemble without perceptible play.

RC 2 *Sliding fits* are intended for accurate location, but with greater maximum clearance than class RC 1. Parts made to this fit move and turn easily but are not intended to run freely, and in the larger sizes may seize with small temperature changes.

RC 3 *Precision running fits* are about the closest fits which can be expected to run freely, and are intended for precision work at slow speeds and light journal pressures, but are not suitable where appreciable temperature differences are likely to be encountered.

RC 4 *Close running fits* are intended chiefly for running fits on accurate machinery with moderate surface speeds and journal pressures, where accurate location and minimum play are desired.

Basic hole system. Limits are in thousandths of an inch. See §12.9.
Limits for hole and shaft are applied algebraically to the basic size to obtain the limits of size for the parts.
Data in **boldface** are in accordance with ABC agreements.
Symbols H5, g5, etc., are hole and shaft designations used in ABC System.

| Nominal Size Range, inches | Class RC 1 | | | Class RC 2 | | | Class RC 3 | | | Class RC 4 | | |
|---|---|---|---|---|---|---|---|---|---|---|---|---|
| | Limits of Clearance | Standard Limits | | Limits of Clearance | Standard Limits | | Limits of Clearance | Standard Limits | | Limits of Clearance | Standard Limits | |
| Over To | | Hole H5 | Shaft g4 | | Hole H6 | Shaft g5 | | Hole H7 | Shaft f6 | | Hole H8 | Shaft f7 |
| 0 - 0.12 | 0.1 | +0.2 | −0.1 | 0.1 | +0.25 | −0.1 | 0.3 | +0.4 | −0.3 | 0.3 | +0.6 | −0.3 |
| | 0.45 | −0 | −0.25 | 0.55 | −0 | −0.3 | 0.95 | −0 | −0.55 | 1.3 | −0 | −0.7 |
| 0.12- 0.24 | 0.15 | +0.2 | −0.15 | 0.15 | +0.3 | −0.15 | 0.4 | +0.5 | −0.4 | 0.4 | +0.7 | −0.4 |
| | 0.5 | −0 | −0.3 | 0.65 | −0 | −0.35 | 1.12 | −0 | −0.7 | 1.6 | −0 | −0.9 |
| 0.24- 0.40 | 0.2 | +0.25 | −0.2 | 0.2 | +0.4 | −0.2 | 0.5 | +0.6 | −0.5 | 0.5 | +0.9 | −0.5 |
| | 0.6 | −0 | −0.35 | 0.85 | −0 | −0.45 | 1.5 | −0 | −0.9 | 2.0 | −0 | −1.1 |
| 0.40- 0.71 | 0.25 | +0.3 | −0.25 | 0.25 | +0.4 | −0.25 | 0.6 | +0.7 | −0.6 | 0.6 | +1.0 | −0.6 |
| | 0.75 | −0 | −0.45 | 0.95 | −0 | −0.55 | 1.7 | −0 | −1.0 | 2.3 | −0 | −1.3 |
| 0.71- 1.19 | 0.3 | +0.4 | −0.3 | 0.3 | +0.5 | −0.3 | 0.8 | +0.8 | −0.8 | 0.8 | +1.2 | −0.8 |
| | 0.95 | −0 | −0.55 | 1.2 | −0 | −0.7 | 2.1 | −0 | −1.3 | 2.8 | −0 | −1.6 |
| 1.19- 1.97 | 0.4 | +0.4 | −0.4 | 0.4 | +0.6 | −0.4 | 1.0 | +1.0 | −1.0 | 1.0 | +1.6 | −1.0 |
| | 1.1 | −0 | −0.7 | 1.4 | −0 | −0.8 | 2.6 | −0 | −1.6 | 3.6 | −0 | −2.0 |
| 1.97- 3.15 | 0.4 | +0.5 | −0.4 | 0.4 | +0.7 | −0.4 | 1.2 | +1.2 | −1.2 | 1.2 | +1.8 | −1.2 |
| | 1.2 | −0 | −0.7 | 1.6 | −0 | −0.9 | 3.1 | −0 | −1.9 | 4.2 | −0 | −2.4 |
| 3.15- 4.73 | 0.5 | +0.6 | −0.5 | 0.5 | +0.9 | −0.5 | 1.4 | +1.4 | −1.4 | 1.4 | +2.2 | −1.4 |
| | 1.5 | −0 | −0.9 | 2.0 | −0 | −1.1 | 3.7 | −0 | −2.3 | 5.0 | −0 | −2.8 |
| 4.73- 7.09 | 0.6 | +0.7 | −0.6 | 0.6 | +1.0 | −0.6 | 1.6 | +1.6 | −1.6 | 1.6 | +2.5 | −1.6 |
| | 1.8 | −0 | −1.1 | 2.3 | −0 | −1.3 | 4.2 | −0 | −2.6 | 5.7 | −0 | −3.2 |
| 7.09- 9.85 | 0.6 | +0.8 | −0.6 | 0.6 | +1.2 | −0.6 | 2.0 | +1.8 | −2.0 | 2.0 | +2.8 | −2.0 |
| | 2.0 | −0 | −1.2 | 2.6 | −0 | −1.4 | 5.0 | −0 | −3.2 | 6.6 | −0 | −3.8 |
| 9.85-12.41 | 0.8 | +0.9 | −0.8 | 0.8 | +1.2 | −0.8 | 2.5 | +2.0 | −2.5 | 2.5 | +3.0 | −2.5 |
| | 2.3 | −0 | −1.4 | 2.9 | −0 | −1.7 | 5.7 | −0 | −3.7 | 7.5 | −0 | −4.5 |
| 12.41-15.75 | 1.0 | +1.0 | −1.0 | 1.0 | +1.4 | −1.0 | 3.0 | +2.2 | −3.0 | 3.0 | +3.5 | −3.0 |
| | 2.7 | −0 | −1.7 | 3.4 | −0 | −2.0 | 6.6 | −0 | −4.4 | 8.7 | −0 | −5.2 |

[a] From ANSI B4.1-1967. For larger diameters, see the standard.

## 5 American National Standard Running and Sliding Fits (continued)[a]

RC 5, RC 6 — *Medium running fits* are intended for higher running speeds, or heavy journal pressures, or both.

RC 7 — *Free running fits* are intended for use where accuracy is not essential, or where large temperature variations are likely to be encountered, or under both these conditions.

RC 8, RC 9 — *Loose running fits* are intended for use where wide commercial tolerances may be necessary, together with an allowance, on the external member.

| Nominal Size Range, inches | Class RC 5 | | | Class RC 6 | | | Class RC 7 | | | Class RC 8 | | | Class RC 9 | | |
|---|---|---|---|---|---|---|---|---|---|---|---|---|---|---|---|
| | Limits of Clearance | Standard Limits | | Limits of Clearance | Standard Limits | | Limits of Clearance | Standard Limits | | Limits of Clearance | Standard Limits | | Limits of Clearance | Standard Limits | |
| Over To | | Hole H8 | Shaft e7 | | Hole H9 | Shaft e8 | | Hole H9 | Shaft d8 | | Hole H10 | Shaft c9 | | Hole H11 | Shaft |
| 0 - 0.12 | 0.6<br>1.6 | +0.6<br>−0 | −0.6<br>−1.0 | 0.6<br>2.2 | +1.0<br>−0 | −0.6<br>−1.2 | 1.0<br>2.6 | +1.0<br>−0 | − 1.0<br>− 1.6 | 2.5<br>5.1 | +1.6<br>−0 | − 2.5<br>− 3.5 | 4.0<br>8.1 | + 2.5<br>− 0 | − 4.0<br>− 5.6 |
| 0.12- 0.24 | 0.8<br>2.0 | +0.7<br>−0 | −0.8<br>−1.3 | 0.8<br>2.7 | +1.2<br>−0 | −0.8<br>−1.5 | 1.2<br>3.1 | +1.2<br>−0 | − 1.2<br>− 1.9 | 2.8<br>5.8 | +1.8<br>−0 | − 2.8<br>− 4.0 | 4.5<br>9.0 | + 3.0<br>− 0 | − 4.5<br>− 6.0 |
| 0.24- 0.40 | 1.0<br>2.5 | +0.9<br>−0 | −1.0<br>−1.6 | 1.0<br>3.3 | +1.4<br>−0 | −1.0<br>−1.9 | 1.6<br>3.9 | +1.4<br>−0 | − 1.6<br>− 2.5 | 3.0<br>6.6 | +2.2<br>−0 | − 3.0<br>− 4.4 | 5.0<br>10.7 | + 3.5<br>− 0 | − 5.0<br>− 7.2 |
| 0.40- 0.71 | 1.2<br>2.9 | +1.0<br>−0 | −1.2<br>−1.9 | 1.2<br>3.8 | +1.6<br>−0 | −1.2<br>−2.2 | 2.0<br>4.6 | +1.6<br>−0 | − 2.0<br>− 3.0 | 3.5<br>7.9 | +2.8<br>−0 | − 3.5<br>− 5.1 | 6.0<br>12.8 | + 4.0<br>− 0 | − 6.0<br>− 8.8 |
| 0.71- 1.19 | 1.6<br>3.6 | +1.2<br>−0 | −1.6<br>−2.4 | 1.6<br>4.8 | +2.0<br>−0 | −1.6<br>−2.8 | 2.5<br>5.7 | +2.0<br>−0 | − 2.5<br>− 3.7 | 4.5<br>10.0 | +3.5<br>−0 | − 4.5<br>− 6.5 | 7.0<br>15.5 | + 5.0<br>− 0 | − 7.0<br>−10.5 |
| 1.19- 1.97 | 2.0<br>4.6 | +1.6<br>−0 | −2.0<br>−3.0 | 2.0<br>6.1 | +2.5<br>−0 | −2.0<br>−3.6 | 3.0<br>7.1 | +2.5<br>−0 | − 3.0<br>− 4.6 | 5.0<br>11.5 | +4.0<br>−0 | − 5.0<br>− 7.5 | 8.0<br>18.0 | + 6.0<br>− 0 | − 8.0<br>−12.0 |
| 1.97- 3.15 | 2.5<br>5.5 | +1.8<br>−0 | −2.5<br>−3.7 | 2.5<br>7.3 | +3.0<br>−0 | −2.5<br>−4.3 | 4.0<br>8.8 | +3.0<br>−0 | − 4.0<br>− 5.8 | 6.0<br>13.5 | +4.5<br>−0 | − 6.0<br>− 9.0 | 9.0<br>20.5 | + 7.0<br>− 0 | − 9.0<br>−13.5 |
| 3.15- 4.73 | 3.0<br>6.6 | +2.2<br>−0 | −3.0<br>−4.4 | 3.0<br>8.7 | +3.5<br>−0 | −3.0<br>−5.2 | 5.0<br>10.7 | +3.5<br>−0 | − 5.0<br>− 7.2 | 7.0<br>15.5 | +5.0<br>−0 | − 7.0<br>−10.5 | 10.0<br>24.0 | + 9.0<br>− 0 | −10.0<br>−15.0 |
| 4.73- 7.09 | 3.5<br>7.6 | +2.5<br>−0 | −3.5<br>−5.1 | 3.5<br>10.0 | +4.0<br>−0 | −3.5<br>−6.0 | 6.0<br>12.5 | +4.0<br>−0 | − 6.0<br>− 8.5 | 8.0<br>18.0 | +6.0<br>−0 | − 8.0<br>−12.0 | 12.0<br>28.0 | +10.0<br>− 0 | −12.0<br>−18.0 |
| 7.09- 9.85 | 4.0<br>8.6 | +2.8<br>−0 | −4.0<br>−5.8 | 4.0<br>11.3 | +4.5<br>−0 | −4.0<br>−6.8 | 7.0<br>14.3 | +4.5<br>−0 | − 7.0<br>− 9.8 | 10.0<br>21.5 | +7.0<br>−0 | −10.0<br>−14.5 | 15.0<br>34.0 | +12.0<br>− 0 | −15.0<br>−22.0 |
| 9.85-12.41 | 5.0<br>10.0 | +3.0<br>−0 | −5.0<br>−7.0 | 5.0<br>13.0 | +5.0<br>−0 | −5.0<br>−8.0 | 8.0<br>16.0 | +5.0<br>−0 | − 8.0<br>−11.0 | 12.0<br>25.0 | +8.0<br>−0 | −12.0<br>−17.0 | 18.0<br>38.0 | +12.0<br>− 0 | −18.0<br>−26.0 |
| 12.41-15.75 | 6.0<br>11.7 | +3.5<br>−0 | −6.0<br>−8.2 | 6.0<br>15.5 | +6.0<br>−0 | −6.0<br>−9.5 | 10.0<br>19.5 | +6.0<br>−0 | −10.0<br>13.5 | 14.0<br>29.0 | +9.0<br>−0 | −14.0<br>−20.0 | 22.0<br>45.0 | +14.0<br>− 0 | −22.0<br>−31.0 |

[a] From ANSI B4.1-1967. For larger diameters, see the standard.

## 6 American National Standard Clearance Locational Fits[a]

LC *Locational clearance fits* are intended for parts which are normally stationary, but which can be freely assembled or disassembled. They run from snug fits for parts requiring accuracy of location, through the medium clearance fits for parts such as spigots, to the looser fastener fits where freedom of assembly is of prime importance.

Basic hole system. Limits are in thousandths of an inch. See §12.9.
Limits for hole and shaft are applied algebraically to the basic size to obtain the limits of size for the parts.
Data in **boldface** are in accordance with ABC agreements.
Symbols H6, h5, etc., are hole and shaft designations used in ABC System.

| Nominal Size Range, inches | Class LC 1 | | | Class LC 2 | | | Class LC 3 | | | Class LC 4 | | | Class LC 5 | | |
|---|---|---|---|---|---|---|---|---|---|---|---|---|---|---|---|
| | Limits of Clearance | Standard Limits | | Limits of Clearance | Standard Limits | | Limits of Clearance | Standard Limits | | Limits of Clearance | Standard Limits | | Limits of Clearance | Standard Limits | |
| Over To | | Hole H6 | Shaft h5 | | Hole H7 | Shaft h6 | | Hole H8 | Shaft h7 | | Hole H10 | Shaft h9 | | Hole H7 | Shaft g6 |
| 0 - 0.12 | 0<br>0.45 | +0.25<br>−0 | +0<br>−0.2 | 0<br>0.65 | +0.4<br>−0 | +0<br>−0.25 | 0<br>1 | +0.6<br>−0 | +0<br>−0.4 | 0<br>2.6 | +1.6<br>−0 | +0<br>−1.0 | 0.1<br>0.75 | +0.4<br>−0 | −0.1<br>−0.35 |
| 0.12– 0.24 | 0<br>0.5 | +0.3<br>−0 | +0<br>−0.2 | 0<br>0.8 | +0.5<br>−0 | +0<br>−0.3 | 0<br>1.2 | +0.7<br>−0 | +0<br>−0.5 | 0<br>3.0 | +1.8<br>−0 | +0<br>−1.2 | 0.15<br>0.95 | +0.5<br>−0 | −0.15<br>−0.45 |
| 0.24– 0.40 | 0<br>0.65 | +0.4<br>−0 | +0<br>−0.25 | 0<br>1.0 | +0.6<br>−0 | +0<br>−0.4 | 0<br>1.5 | +0.9<br>−0 | +0<br>−0.6 | 0<br>3.6 | +2.2<br>−0 | +0<br>−1.4 | 0.2<br>1.2 | +0.6<br>−0 | −0.2<br>−0.6 |
| 0.40– 0.71 | 0<br>0.7 | +0.4<br>−0 | +0<br>−0.3 | 0<br>1.1 | +0.7<br>−0 | +0<br>−0.4 | 0<br>1.7 | +1.0<br>−0 | +0<br>−0.7 | 0<br>4.4 | +2.8<br>−0 | +0<br>−1.6 | 0.25<br>1.35 | +0.7<br>−0 | −0.25<br>−0.65 |
| 0.71– 1.19 | 0<br>0.9 | +0.5<br>−0 | +0<br>−0.4 | 0<br>1.3 | +0.8<br>−0 | +0<br>−0.5 | 0<br>2 | +1.2<br>−0 | +0<br>−0.8 | 0<br>5.5 | +3.5<br>−0 | +0<br>−2.0 | 0.3<br>1.6 | +0.8<br>−0 | −0.3<br>−0.8 |
| 1.19– 1.97 | 0<br>1.0 | +0.6<br>−0 | +0<br>−0.4 | 0<br>1.6 | +1.0<br>−0 | +0<br>−0.6 | 0<br>2.6 | +1.6<br>−0 | +0<br>−1 | 0<br>6.5 | +4.0<br>−0 | +0<br>−2.5 | 0.4<br>2.0 | +1.0<br>−0 | −0.4<br>−1.0 |
| 1.97– 3.15 | 0<br>1.2 | +0.7<br>−0 | +0<br>−0.5 | 0<br>1.9 | +1.2<br>−0 | +0<br>−0.7 | 0<br>3 | +1.8<br>−0 | +0<br>−1.2 | 0<br>7.5 | +4.5<br>−0 | +0<br>−3 | 0.4<br>2.3 | +1.2<br>−0 | −0.4<br>−1.1 |
| 3.15– 4.73 | 0<br>1.5 | +0.9<br>−0 | +0<br>−0.6 | 0<br>2.3 | +1.4<br>−0 | +0<br>−0.9 | 0<br>3.6 | +2.2<br>−0 | +0<br>−1.4 | 0<br>8.5 | +5.0<br>−0 | +0<br>−3.5 | 0.5<br>2.8 | +1.4<br>−0 | −0.5<br>−1.4 |
| 4.73– 7.09 | 0<br>1.7 | +1.0<br>−0 | +0<br>−0.7 | 0<br>2.6 | +1.6<br>−0 | +0<br>−1.0 | 0<br>4.1 | +2.5<br>−0 | +0<br>−1.6 | 0<br>10 | +6.0<br>−0 | +0<br>−4 | 0.6<br>3.2 | +1.6<br>−0 | −0.6<br>−1.6 |
| 7.09– 9.85 | 0<br>2.0 | +1.2<br>−0 | +0<br>−0.8 | 0<br>3.0 | +1.8<br>−0 | +0<br>−1.2 | 0<br>4.6 | +2.8<br>−0 | +0<br>−1.8 | 0<br>11.5 | +7.0<br>−0 | +0<br>−4.5 | 0.6<br>3.6 | +1.8<br>−0 | −0.6<br>−1.8 |
| 9.85–12.41 | 0<br>2.1 | +1.2<br>−0 | +0<br>−0.9 | 0<br>3.2 | +2.0<br>−0 | +0<br>−1.2 | 0<br>5 | +3.0<br>−0 | +0<br>−2.0 | 0<br>13 | +8.0<br>−0 | +0<br>−5 | 0.7<br>3.9 | +2.0<br>−0 | −0.7<br>−1.9 |
| 12.41–15.75 | 0<br>2.4 | +1.4<br>−0 | +0<br>−1.0 | 0<br>3.6 | +2.2<br>−0 | +0<br>−1.4 | 0<br>5.7 | +3.5<br>−0 | +0<br>−2.2 | 0<br>15 | +9.0<br>−0 | +0<br>−6 | 0.7<br>4.3 | +2.2<br>−0 | −0.7<br>−2.1 |

[a] From ANSI B4.1–1967. For larger diameters, see the standard.

## 6 American National Standard Clearance Locational Fits (continued)[a]

| Nominal Size Range, inches | Class LC 6 | | | Class LC 7 | | | Class LC 8 | | | Class LC 9 | | | Class LC 10 | | | Class LC 11 | | |
|---|---|---|---|---|---|---|---|---|---|---|---|---|---|---|---|---|---|---|
| | Limits of Clearance | Standard Limits | | Limits of Clearance | Standard Limits | | Limits of Clearance | Standard Limits | | Limits of Clearance | Standard Limits | | Limits of Clearance | Standard Limits | | Limits of Clearance | Standard Limits | |
| Over To | | Hole H9 | Shaft f8 | | Hole H10 | Shaft e9 | | Hole H10 | Shaft d9 | | Hole H11 | Shaft c10 | | Hole H12 | Shaft | | Hole H13 | Shaft |
| 0 - 0.12 | 0.3<br>1.9 | +1.0<br>−0 | −0.3<br>−0.9 | 0.6<br>3.2 | +1.6<br>−0 | − 0.6<br>− 1.6 | 1.0<br>3.6 | +1.6<br>−0 | − 1.0<br>− 2.0 | 2.5<br>6.6 | + 2.5<br>− 0 | − 2.5<br>− 4.1 | 4<br>12 | + 4<br>− 0 | − 4<br>− 8 | 5<br>17 | + 6<br>− 0 | − 5<br>−11 |
| 0.12- 0.24 | 0.4<br>2.3 | +1.2<br>−0 | −0.4<br>−1.1 | 0.8<br>3.8 | +1.8<br>−0 | − 0.8<br>− 2.0 | 1.2<br>4.2 | +1.8<br>−0 | − 1.2<br>− 2.4 | 2.8<br>7.6 | + 3.0<br>− 0 | − 2.8<br>− 4.6 | 4.5<br>14.5 | + 5<br>− 0 | − 4.5<br>− 9.5 | 6<br>20 | + 7<br>− 0 | − 6<br>−13 |
| 0.24- 0.40 | 0.5<br>2.8 | +1.4<br>−0 | −0.5<br>−1.4 | 1.0<br>4.6 | +2.2<br>−0 | − 1.0<br>− 2.4 | 1.6<br>5.2 | +2.2<br>−0 | − 1.6<br>− 3.0 | 3.0<br>8.7 | + 3.5<br>− 0 | − 3.0<br>− 5.2 | 5<br>17 | + 6<br>− 0 | − 5<br>−11 | 7<br>25 | + 9<br>− 0 | − 7<br>−16 |
| 0.40- 0.71 | 0.6<br>3.2 | +1.6<br>−0 | −0.6<br>−1.6 | 1.2<br>5.6 | +2.8<br>−0 | − 1.2<br>− 2.8 | 2.0<br>6.4 | +2.8<br>−0 | − 2.0<br>− 3.6 | 3.5<br>10.3 | + 4.0<br>− 0 | − 3.5<br>− 6.3 | 6<br>20 | + 7<br>− 0 | − 6<br>−13 | 8<br>28 | +10<br>− 0 | − 8<br>−18 |
| 0.71- 1.19 | 0.8<br>4.0 | +2.0<br>−0 | −0.8<br>−2.0 | 1.6<br>7.1 | +3.5<br>−0 | − 1.6<br>− 3.6 | 2.5<br>8.0 | +3.5<br>−0 | − 2.5<br>− 4.5 | 4.5<br>13.0 | + 5.0<br>− 0 | − 4.5<br>− 8.0 | 7<br>23 | + 8<br>− 0 | − 7<br>−15 | 10<br>34 | +12<br>− 0 | −10<br>−22 |
| 1.19- 1.97 | 1.0<br>5.1 | +2.5<br>−0 | −1.0<br>−2.6 | 2.0<br>8.5 | +4.0<br>−0 | − 2.0<br>− 4.5 | 3.0<br>9.5 | +4.0<br>−0 | − 3.0<br>− 5.5 | 5<br>15 | + 6<br>− 0 | − 5<br>− 9 | 8<br>28 | +10<br>− 0 | − 8<br>−18 | 12<br>44 | +16<br>− 0 | −12<br>−28 |
| 1.97- 3.15 | 1.2<br>6.0 | +3.0<br>−0 | −1.2<br>−3.0 | 2.5<br>10.0 | +4.5<br>−0 | − 2.5<br>− 5.5 | 4.0<br>11.5 | +4.5<br>−0 | − 4.0<br>− 7.0 | 6<br>17.5 | + 7<br>− 0 | − 6<br>−10.5 | 10<br>34 | +12<br>− 0 | −10<br>−22 | 14<br>50 | +18<br>− 0 | −14<br>−32 |
| 3.15- 4.73 | 1.4<br>7.1 | +3.5<br>−0 | −1.4<br>−3.6 | 3.0<br>11.5 | +5.0<br>−0 | − 3.0<br>− 6.5 | 5.0<br>13.5 | +5.0<br>−0 | − 5.0<br>− 8.5 | 7<br>21 | + 9<br>− 0 | − 7<br>−12 | 11<br>39 | +14<br>− 0 | −11<br>−25 | 16<br>60 | +22<br>− 0 | −16<br>−38 |
| 4.73- 7.09 | 1.6<br>8.1 | +4.0<br>−0 | −1.6<br>−4.1 | 3.5<br>13.5 | +6.0<br>−0 | − 3.5<br>− 7.5 | 6<br>16 | +6<br>−0 | − 6<br>−10 | 8<br>24 | +10<br>− 0 | − 8<br>−14 | 12<br>44 | +16<br>− 0 | −12<br>−28 | 18<br>68 | +25<br>− 0 | −18<br>−43 |
| 7.09- 9.85 | 2.0<br>9.3 | +4.5<br>−0 | −2.0<br>−4.8 | 4.0<br>15.5 | +7.0<br>−0 | − 4.0<br>− 8.5 | 7<br>18.5 | +7<br>−0 | − 7<br>−11.5 | 10<br>29 | +12<br>− 0 | −10<br>−17 | 16<br>52 | +18<br>− 0 | −16<br>−34 | 22<br>78 | +28<br>− 0 | −22<br>−50 |
| 9.85-12.41 | 2.2<br>10.2 | +5.0<br>−0 | −2.2<br>−5.2 | 4.5<br>17.5 | +8.0<br>−0 | − 4.5<br>− 9.5 | 7<br>20 | +8<br>−0 | − 7<br>−12 | 12<br>32 | +12<br>− 0 | −12<br>−20 | 20<br>60 | +20<br>− 0 | −20<br>−40 | 28<br>88 | +30<br>− 0 | −28<br>−58 |
| 12.41-15.75 | 2.5<br>12.0 | +6.0<br>−0 | −2.5<br>−6.0 | 5.0<br>20.0 | +9.0<br>−0 | − 5<br>−11 | 8<br>23 | +9<br>−0 | − 8<br>−14 | 14<br>37 | +14<br>− 0 | −14<br>−23 | 22<br>66 | +22<br>− 0 | −22<br>−44 | 30<br>100 | +35<br>− 0 | −30<br>−65 |

[a] From ANSI B4.1-1967. For larger diameters, see the standard.

## 7 American National Standard Transition Locational Fits[a]

LT *Transition fits* are a compromise between clearance and interference fits, for application where accuracy of location is important, but either a small amount of clearance or interference is permissible.

Basic hole system. Limits are in thousandths of an inch. See §12.9.

Limits for hole and shaft are applied algebraically to the basic size to obtain the limits of size for the mating parts.

Data in **boldface** are in accordance with ABC agreements.

"Fit" represents the maximum interference (minus values) and the maximum clearance (plus values).

Symbols H7, js6, etc., are hole and shaft designations used in ABC System.

**−0.65**
**+0.25**

| Nominal Size Range, inches | Class LT 1 | | | Class LT 2 | | | Class LT 3 | | | Class LT 4 | | | Class LT 5 | | | Class LT 6 | | |
|---|---|---|---|---|---|---|---|---|---|---|---|---|---|---|---|---|---|---|
| | Fit | Standard Limits | | Fit | Standard Limits | | Fit | Standard Limits | | Fit | Standard Limits | | Fit | Standard Limits | | Fit | Standard Limits | |
| Over To | | Hole H7 | Shaft js6 | | Hole H8 | Shaft js7 | | Hole H7 | Shaft k6 | | Hole H8 | Shaft k7 | | Hole H7 | Shaft n6 | | Hole H7 | Shaft n7 |
| 0 - 0.12 | −0.10<br>+0.50 | +0.4<br>−0 | +0.10<br>−0.10 | −0.2<br>+0.8 | +0.6<br>−0 | +0.2<br>−0.2 | | | | | | | −0.5<br>+0.15 | +0.4<br>−0 | +0.5<br>+0.25 | −0.65<br>+0.15 | +0.4<br>−0 | +0.65<br>+0.25 |
| 0.12- 0.24 | −0.15<br>+0.65 | +0.5<br>−0 | +0.15<br>−0.15 | −0.25<br>+0.95 | +0.7<br>−0 | +0.25<br>−0.25 | | | | | | | −0.6<br>+0.2 | +0.5<br>−0 | +0.6<br>+0.3 | −0.8<br>+0.2 | +0.5<br>−0 | +0.8<br>+0.3 |
| 0.24- 0.40 | −0.2<br>+0.8 | +0.6<br>−0 | +0.2<br>−0.2 | −0.3<br>+1.2 | +0.9<br>−0 | +0.3<br>−0.3 | −0.5<br>+0.5 | +0.6<br>−0 | +0.5<br>+0.1 | −0.7<br>+0.8 | +0.9<br>−0 | +0.7<br>+0.1 | −0.8<br>+0.2 | +0.6<br>−0 | +0.8<br>+0.4 | −1.0<br>+0.2 | +0.6<br>−0 | +1.0<br>+0.4 |
| 0.40- 0.71 | −0.2<br>+0.9 | +0.7<br>−0 | +0.2<br>−0.2 | −0.35<br>+1.35 | +1.0<br>−0 | +0.35<br>−0.35 | −0.5<br>+0.6 | +0.7<br>−0 | +0.5<br>+0.1 | −0.8<br>+0.9 | +1.0<br>−0 | +0.8<br>+0.1 | −0.9<br>+0.2 | +0.7<br>−0 | +0.9<br>+0.5 | −1.2<br>+0.2 | +0.7<br>−0 | +1.2<br>+0.5 |
| 0.71- 1.19 | −0.25<br>+1.05 | +0.8<br>−0 | +0.25<br>−0.25 | −0.4<br>+1.6 | +1.2<br>−0 | +0.4<br>−0.4 | −0.6<br>+0.7 | +0.8<br>−0 | +0.6<br>+0.1 | −0.9<br>+1.1 | +1.2<br>−0 | +0.9<br>+0.1 | −1.1<br>+0.2 | +0.8<br>−0 | +1.1<br>+0.6 | −1.4<br>+0.2 | +0.8<br>−0 | +1.4<br>+0.6 |
| 1.19- 1.97 | −0.3<br>+1.3 | +1.0<br>−0 | +0.3<br>−0.3 | −0.5<br>+2.1 | +1.6<br>−0 | +0.5<br>−0.5 | −0.7<br>+0.9 | +1.0<br>−0 | +0.7<br>+0.1 | −1.1<br>+1.5 | +1.6<br>−0 | +1.1<br>+0.1 | −1.3<br>+0.3 | +1.0<br>−0 | +1.3<br>+0.7 | −1.7<br>+0.3 | +1.0<br>−0 | +1.7<br>+0.7 |
| 1.97- 3.15 | −0.3<br>+1.5 | +1.2<br>−0 | +0.3<br>−0.3 | −0.6<br>+2.4 | +1.8<br>−0 | +0.6<br>−0.6 | −0.8<br>+1.1 | +1.2<br>−0 | +0.8<br>+0.1 | −1.3<br>+1.7 | +1.8<br>−0 | +1.3<br>+0.1 | −1.5<br>+0.4 | +1.2<br>−0 | +1.5<br>+0.8 | −2.0<br>+0.4 | +1.2<br>−0 | +2.0<br>+0.8 |
| 3.15- 4.73 | −0.4<br>+1.8 | +1.4<br>−0 | +0.4<br>−0.4 | −0.7<br>+2.9 | +2.2<br>−0 | +0.7<br>−0.7 | −1.0<br>+1.3 | +1.4<br>−0 | +1.0<br>+0.1 | −1.5<br>+2.1 | +2.2<br>−0 | +1.5<br>+0.1 | −1.9<br>+0.4 | +1.4<br>−0 | +1.9<br>+1.0 | −2.4<br>+0.4 | +1.4<br>−0 | +2.4<br>+1.0 |
| 4.73- 7.09 | −0.5<br>+2.1 | +1.6<br>−0 | +0.5<br>−0.5 | −0.8<br>+3.3 | +2.5<br>−0 | +0.8<br>−0.8 | −1.1<br>+1.5 | +1.6<br>−0 | +1.1<br>+0.1 | −1.7<br>+2.4 | +2.5<br>−0 | +1.7<br>+0.1 | −2.2<br>+0.4 | +1.6<br>−0 | +2.2<br>+1.2 | −2.8<br>+0.4 | +1.6<br>−0 | +2.8<br>+1.2 |
| 7.09- 9.85 | −0.6<br>+2.4 | +1.8<br>−0 | +0.6<br>−0.6 | −0.9<br>+3.7 | +2.8<br>−0 | +0.9<br>−0.9 | −1.4<br>+1.6 | +1.8<br>−0 | +1.4<br>+0.2 | −2.0<br>+2.6 | +2.8<br>−0 | +2.0<br>+0.2 | −2.6<br>+0.4 | +1.8<br>−0 | +2.6<br>+1.4 | −3.2<br>+0.4 | +1.8<br>−0 | +3.2<br>+1.4 |
| 9.85-12.41 | −0.6<br>+2.6 | +2.0<br>−0 | +0.6<br>−0.6 | −1.0<br>+4.0 | +3.0<br>−0 | +1.0<br>−1.0 | −1.4<br>+1.8 | +2.0<br>−0 | +1.4<br>+0.2 | −2.2<br>+2.8 | +3.0<br>−0 | +2.2<br>+0.2 | −2.6<br>+0.6 | +2.0<br>−0 | +2.6<br>+1.4 | −3.4<br>+0.6 | +2.0<br>−0 | +3.4<br>+1.4 |
| 12.41-15.75 | −0.7<br>+2.9 | +2.2<br>−0 | +0.7<br>−0.7 | −1.0<br>+4.5 | +3.5<br>−0 | +1.0<br>−1.0 | −1.6<br>+2.0 | +2.2<br>−0 | +1.6<br>+0.2 | −2.4<br>+3.3 | +3.5<br>−0 | +2.4<br>+0.2 | −3.0<br>+0.6 | +2.2<br>−0 | +3.0<br>+1.6 | −3.8<br>+0.6 | +2.2<br>−0 | +3.8<br>+1.6 |

[a] From ANSI B4.1-1967. For larger diameters, see the standard.

## 8 American National Standard Interference Locational Fits[a]

LN *Locational interference fits* are used where accuracy of location is of prime importance, and for parts requiring rigidity and alignment with no special requirements for bore pressure. Such fits are not intended for parts designed to transmit frictional loads from one part to another by virtue of the tightness of fit, as these conditions are covered by force fits.

Basic hole system. Limits are in thousandths of an inch. See §12.9.
Limits for hole and shaft are applied algebraically to the basic size to obtain the limits of size for the parts.
Data in **boldface** are in accordance with ABC agreements.
Symbols H7, p6, etc., are hole and shaft designations used in ABC System.

| Nominal Size Range, inches | Class LN 1 | | | Class LN 2 | | | Class LN 3 | | |
|---|---|---|---|---|---|---|---|---|---|
| | Limits of Interference | Standard Limits | | Limits of Interference | Standard Limits | | Limits of Interference | Standard Limits | |
| Over To | | Hole H6 | Shaft n5 | | Hole H7 | Shaft p6 | | Hole H7 | Shaft r6 |
| 0 - 0.12 | **0**<br>**0.45** | **+0.25**<br>**−0** | **+0.45**<br>**+0.25** | **0**<br>**0.65** | **+0.4**<br>**−0** | **+0.65**<br>**+0.4** | **0.1**<br>**0.75** | **+0.4**<br>**−0** | **+0.75**<br>**+0.5** |
| 0.12– 0.24 | **0**<br>**0.5** | **+0.3**<br>**−0** | **+0.5**<br>**+0.3** | **0**<br>**0.8** | **+0.5**<br>**−0** | **+0.8**<br>**+0.5** | **0.1**<br>**0.9** | **+0.5**<br>**0** | **+0.9**<br>**+0.6** |
| 0.24– 0.40 | **0**<br>**0.65** | **+0.4**<br>**−0** | **+0.65**<br>**+0.4** | **0**<br>**1.0** | **+0.6**<br>**−0** | **+1.0**<br>**+0.6** | **0.2**<br>**1.2** | **+0.6**<br>**−0** | **+1.2**<br>**+0.8** |
| 0.40– 0.71 | **0**<br>**0.8** | **+0.4**<br>**−0** | **+0.8**<br>**+0.4** | **0**<br>**1.1** | **+0.7**<br>**−0** | **+1.1**<br>**+0.7** | **0.3**<br>**1.4** | **+0.7**<br>**−0** | **+1.4**<br>**+1.0** |
| 0.71– 1.19 | **0**<br>**1.0** | **+0.5**<br>**−0** | **+1.0**<br>**+0.5** | **0**<br>**1.3** | **+0.8**<br>**−0** | **+1.3**<br>**+0.8** | **0.4**<br>**1.7** | **+0.8**<br>**−0** | **+1.7**<br>**+1.2** |
| 1.19– 1.97 | **0**<br>**1.1** | **+0.6**<br>**−0** | **+1.1**<br>**+0.6** | **0**<br>**1.6** | **+1.0**<br>**−0** | **+1.6**<br>**+1.0** | **0.4**<br>**2.0** | **+1.0**<br>**−0** | **+2.0**<br>**+1.4** |
| 1.97– 3.15 | **0.1**<br>**1.3** | **+0.7**<br>**−0** | **+1.3**<br>**+0.7** | **0.2**<br>**2.1** | **+1.2**<br>**−0** | **+2.1**<br>**+1.4** | **0.4**<br>**2.3** | **+1.2**<br>**−0** | **+2.3**<br>**+1.6** |
| 3.15– 4.73 | **0.1**<br>**1.6** | **+0.9**<br>**−0** | **+1.6**<br>**+1.0** | **0.2**<br>**2.5** | **+1.4**<br>**−0** | **+2.5**<br>**+1.6** | **0.6**<br>**2.9** | **+1.4**<br>**−0** | **+2.9**<br>**+2.0** |
| 4.73– 7.09 | **0.2**<br>**1.9** | **+1.0**<br>**−0** | **+1.9**<br>**+1.2** | **0.2**<br>**2.8** | **+1.6**<br>**−0** | **+2.8**<br>**+1.8** | **0.9**<br>**3.5** | **+1.6**<br>**−0** | **+3.5**<br>**+2.5** |
| 7.09– 9.85 | **0.2**<br>**2.2** | **+1.2**<br>**−0** | **+2.2**<br>**+1.4** | **0.2**<br>**3.2** | **+1.8**<br>**−0** | **+3.2**<br>**+2.0** | **1.2**<br>**4.2** | **+1.8**<br>**−0** | **+4.2**<br>**+3.0** |
| 9.85–12.41 | **0.2**<br>**2.3** | **+1.2**<br>**−0** | **+2.3**<br>**+1.4** | **0.2**<br>**3.4** | **+2.0**<br>**−0** | **+3.4**<br>**+2.2** | **1.5**<br>**4.7** | **+2.0**<br>**−0** | **+4.7**<br>**+3.5** |

[a] From ANSI B4.1–1967. For larger diameters, see the standard.

## 9 American National Standard Force and Shrink Fits[a]

FN 1 *Light drive fits* are those requiring light assembly pressures, and produce more or less permanent assemblies. They are suitable for thin sections or long fits, or in cast-iron external members.

FN 2 *Medium drive fits* are suitable for ordinary steel parts, or for shrink fits on light sections. They are about the tightest fits that can be used with high-grade cast-iron external members.

FN 3 *Heavy drive fits* are suitable for heavier steel parts or for shrink fits in medium sections.

FN 4, FN 5 *Force fits* are suitable for parts which can be highly stressed, or for shrink fits where the heavy pressing forces required are impractical.

Basic hole system. Limits are in thousandths of an inch. See §12.9.

Limits for hole and shaft are applied algebraically to the basic size to obtain the limits of size for the parts.
Data in **boldface** are in accordance with ABC agreements.
Symbols H7, s6, etc., are hole and shaft designations used in ABC System.

| Nominal Size Range, inches | Class FN 1 | | | Class FN 2 | | | Class FN 3 | | | Class FN 4 | | | Class FN 5 | | |
|---|---|---|---|---|---|---|---|---|---|---|---|---|---|---|---|
| | Limits of Interference | Standard Limits | | Limits of Interference | Standard Limits | | Limits of Interference | Standard Limits | | Limits of Interference | Standard Limits | | Limits of Interference | Standard Limits | |
| Over To | | Hole H6 | Shaft | | Hole H7 | Shaft s6 | | Hole H7 | Shaft t6 | | Hole H7 | Shaft u6 | | Hole H8 | Shaft x7 |
| 0 - 0.12 | 0.05<br>0.5 | **+0.25**<br>**−0** | +0.5<br>+0.3 | **0.2**<br>**0.85** | **+0.4**<br>**−0** | **+0.85**<br>**+0.6** | | | | **0.3**<br>**0.95** | **+0.4**<br>**−0** | **+ 0.95**<br>**+ 0.7** | **0.3**<br>**1.3** | **+0.6**<br>**−0** | **+ 1.3**<br>**+ 0.9** |
| 0.12- 0.24 | 0.1<br>0.6 | **+0.3**<br>**−0** | +0.6<br>+0.4 | **0.2**<br>**1.0** | **+0.5**<br>**−0** | **+1.0**<br>**+0.7** | | | | **0.4**<br>**1.2** | **+0.5**<br>**−0** | **+ 1.2**<br>**+ 0.9** | **0.5**<br>**1.7** | **+0.7**<br>**−0** | **+ 1.7**<br>**+ 1.2** |
| 0.24- 0.40 | 0.1<br>0.75 | **+0.4**<br>**−0** | +0.75<br>+0.5 | **0.4**<br>**1.4** | **+0.6**<br>**−0** | **+1.4**<br>**+1.0** | | | | **0.6**<br>**1.6** | **+0.6**<br>**−0** | **+ 1.6**<br>**+ 1.2** | **0.5**<br>**2.0** | **+0.9**<br>**−0** | **+ 2.0**<br>**+ 1.4** |
| 0.40- 0.56 | 0.1<br>0.8 | **−0.4**<br>**−0** | +0.8<br>+0.5 | **0.5**<br>**1.6** | **+0.7**<br>**−0** | **+1.6**<br>**+1.2** | | | | **0.7**<br>**1.8** | **+0.7**<br>**−0** | **+ 1.8**<br>**+ 1.4** | **0.6**<br>**2.3** | **+1.0**<br>**−0** | **+ 2.3**<br>**+ 1.6** |
| 0.56- 0.71 | 0.2<br>0.9 | **+0.4**<br>**−0** | +0.9<br>+0.6 | **0.5**<br>**1.6** | **+0.7**<br>**−0** | **+1.6**<br>**+1.2** | | | | **0.7**<br>**1.8** | **+0.7**<br>**−0** | **+ 1.8**<br>**+ 1.4** | **0.8**<br>**2.5** | **+1.0**<br>**−0** | **+ 2.5**<br>**+ 1.8** |
| 0.71- 0.95 | 0.2<br>1.1 | **+0.5**<br>**−0** | +1.1<br>+0.7 | **0.6**<br>**1.9** | **+0.8**<br>**−0** | **+1.9**<br>**+1.4** | | | | **0.8**<br>**2.1** | **+0.8**<br>**−0** | **+ 2.1**<br>**+ 1.6** | **1.0**<br>**3.0** | **+1.2**<br>**−0** | **+ 3.0**<br>**+ 2.2** |
| 0.95- 1.19 | 0.3<br>1.2 | **+0.5**<br>**−0** | +1.2<br>+0.8 | **0.6**<br>**1.9** | **+0.8**<br>**−0** | **+1.9**<br>**+1.4** | **0.8**<br>**2.1** | **+0.8**<br>**−0** | **+ 2.1**<br>**+ 1.6** | **1.0**<br>**2.3** | **+0.8**<br>**−0** | **+ 2.3**<br>**+ 1.8** | **1.3**<br>**3.3** | **+1.2**<br>**−0** | **+ 3.3**<br>**+ 2.5** |
| 1.19- 1.58 | 0.3<br>1.3 | **+0.6**<br>**−0** | +1.3<br>+0.9 | **0.8**<br>**2.4** | **+1.0**<br>**−0** | **+2.4**<br>**+1.8** | **1.0**<br>**2.6** | **+1.0**<br>**−0** | **+ 2.6**<br>**+ 2.0** | **1.5**<br>**3.1** | **+1.0**<br>**−0** | **+ 3.1**<br>**+ 2.5** | **1.4**<br>**4.0** | **+1.6**<br>**−0** | **+ 4.0**<br>**+ 3.0** |

## 9 American National Standard Force and Shrink Fits (continued)[a]

| | | | | | | | | | | | | | | | |
|---|---|---|---|---|---|---|---|---|---|---|---|---|---|---|---|
| 1.58– 1.97 | 0.4<br>1.4 | +0.6<br>−0 | +1.4<br>+1.0 | 0.8<br>2.4 | +1.0<br>−0 | +2.4<br>+1.8 | 1.2<br>2.8 | +1.0<br>−0 | + 2.8<br>+ 2.2 | 1.8<br>3.4 | +1.0<br>−0 | + 3.4<br>+ 2.8 | 2.4<br>5.0 | +1.6<br>−0 | + 5.0<br>+ 4.0 |
| 1.97– 2.56 | 0.6<br>1.8 | +0.7<br>−0 | +1.8<br>+1.3 | 0.8<br>2.7 | +1.2<br>−0 | +2.7<br>+2.0 | 1.3<br>3.2 | +1.2<br>−0 | + 3.2<br>+ 2.5 | 2.3<br>4.2 | +1.2<br>−0 | + 4.2<br>+ 3.5 | 3.2<br>6.2 | +1.8<br>−0 | + 6.2<br>+ 5.0 |
| 2.56– 3.15 | 0.7<br>1.9 | +0.7<br>−0 | +1.9<br>+1.4 | 1.0<br>2.9 | +1.2<br>−0 | +2.9<br>+2.2 | 1.8<br>3.7 | +1.2<br>−0 | + 3.7<br>+ 3.0 | 2.8<br>4.7 | +1.2<br>−0 | + 4.7<br>+ 4.0 | 4.2<br>7.2 | +1.8<br>−0 | + 7.2<br>+ 6.0 |
| 3.15– 3.94 | 0.9<br>2.4 | +0.9<br>−0 | +2.4<br>+1.8 | 1.4<br>3.7 | +1.4<br>−0 | +3.7<br>+2.8 | 2.1<br>4.4 | +1.4<br>−0 | + 4.4<br>+ 3.5 | 3.6<br>5.9 | +1.4<br>−0 | + 5.9<br>+ 5.0 | 4.8<br>8.4 | +2.2<br>−0 | + 8.4<br>+ 7.0 |
| 3.94– 4.73 | 1.1<br>2.6 | +0.9<br>−0 | +2.6<br>+2.0 | 1.6<br>3.9 | +1.4<br>−0 | +3.9<br>+3.0 | 2.6<br>4.9 | +1.4<br>−0 | + 4.9<br>+ 4.0 | 4.6<br>6.9 | +1.4<br>−0 | + 6.9<br>+ 6.0 | 5.8<br>9.4 | +2.2<br>− 0 | + 9.4<br>+ 8.0 |
| 4.73– 5.52 | 1.2<br>2.9 | +1.0<br>−0 | +2.9<br>+2.2 | 1.9<br>4.5 | +1.6<br>−0 | +4.5<br>+3.5 | 3.4<br>6.0 | +1.6<br>−0 | + 6.0<br>+ 5.0 | 5.4<br>8.0 | +1.6<br>−0 | + 8.0<br>+ 7.0 | 7.5<br>11.6 | +2.5<br>−0 | +11.6<br>+10.0 |
| 5.52– 6.30 | 1.5<br>3.2 | +1.0<br>−0 | +3.2<br>+2.5 | 2.4<br>5.0 | +1.6<br>−0 | +5.0<br>+4.0 | 3.4<br>6.0 | +1.6<br>−0 | + 6.0<br>+ 5.0 | 5.4<br>8.0 | +1.6<br>−0 | + 8.0<br>+ 7.0 | 9.5<br>13.6 | +2.5<br>−0 | +13.6<br>+12.0 |
| 6.30– 7.09 | 1.8<br>3.5 | +1.0<br>−0 | +3.5<br>+2.8 | 2.9<br>5.5 | +1.6<br>−0 | +5.5<br>+4.5 | 4.4<br>7.0 | +1.6<br>−0 | + 7.0<br>+ 6.0 | 6.4<br>9.0 | +1.6<br>−0 | + 9.0<br>+ 8.0 | 9.5<br>13.6 | +2.5<br>−0 | +13.6<br>+12.0 |
| 7.09– 7.88 | 1.8<br>3.8 | +1.2<br>−0 | +3.8<br>+3.0 | 3.2<br>6.2 | +1.8<br>−0 | +6.2<br>+5.0 | 5.2<br>8.2 | +1.8<br>−0 | + 8.2<br>+ 7.0 | 7.2<br>10.2 | +1.8<br>−0 | +10.2<br>+ 9.0 | 11.2<br>15.8 | +2.8<br>−0 | +15.8<br>+14.0 |
| 7.88– 8.86 | 2.3<br>4.3 | +1.2<br>−0 | +4.3<br>+3.5 | 3.2<br>6.2 | +1.8<br>−0 | +6.2<br>+5.0 | 5.2<br>8.2 | +1.8<br>−0 | + 8.2<br>+ 7.0 | 8.2<br>11.2 | +1.8<br>−0 | +11.2<br>+10.0 | 13.2<br>17.8 | +2.8<br>−0 | +17.8<br>+16.0 |
| 8.86– 9.85 | 2.3<br>4.3 | +1.2<br>−0 | +4.3<br>+3.5 | 4.2<br>7.2 | +1.8<br>−0 | +7.2<br>+6.0 | 6.2<br>9.2 | +1.8<br>−0 | + 9.2<br>+ 8.0 | 10.2<br>13.2 | +1.8<br>−0 | +13.2<br>+12.0 | 13.2<br>17.8 | +2.8<br>−0 | +17.8<br>+16.0 |
| 9.85–11.03 | 2.8<br>4.9 | +1.2<br>−0 | +4.9<br>+4.0 | 4.0<br>7.2 | +2.0<br>−0 | +7.2<br>+6.0 | 7.0<br>10.2 | +2.0<br>−0 | +10.2<br>+ 9.0 | 10.0<br>13.2 | +2.0<br>−0 | +13.2<br>+12.0 | 15.0<br>20.0 | +3.0<br>−0 | +20.0<br>+18.0 |
| 11.03–12.41 | 2.8<br>4.9 | +1.2<br>−0 | +4.9<br>+4.0 | 5.0<br>8.2 | +2.0<br>−0 | +8.2<br>+7.0 | 7.0<br>10.2 | +2.0<br>−0 | +10.2<br>+ 9.0 | 12.0<br>15.2 | +2.0<br>−0 | +15.2<br>+14.0 | 17.0<br>22.0 | +3.0<br>−0 | +22.0<br>+20.0 |
| 12.41–13.98 | 3.1<br>5.5 | +1.4<br>−0 | +5.5<br>+4.5 | 5.8<br>9.4 | +2.2<br>−0 | +9.4<br>+8.0 | 7.8<br>11.4 | +2.2<br>−0 | +11.4<br>+10.0 | 13.8<br>17.4 | +2.2<br>−0 | +17.4<br>+16.0 | 18.5<br>24.2 | +3.5<br>+0 | +24.2<br>+22.0 |

[a] From ANSI B4.1-1967. For larger diameters, see the standard.

# 10 American National Standard Unified and American National Threads[a]

| Nominal Diameter | Coarse[b] NC UNC | | Fine[b] NF UNF | | Extra Fine[c] NEF UNEF | | Nominal Diameter | Coarse[b] NC UNC | | Fine[b] NF UNF | | Extra Fine[c] NEF UNEF | |
|---|---|---|---|---|---|---|---|---|---|---|---|---|---|
| | Thds. per Inch | Tap Drill[d] | Thds. per Inch | Tap Drill[d] | Thds. per Inch | Tap Drill[d] | | Thds. per Inch | Tap Drill[d] | Thds. per Inch | Tap Drill[d] | Thds. per Inch | Tap Drill[d] |
| 0 (.060) | | | 80 | 3/64 | | | 1 | 8 | 7/8 | 12 | 59/64 | 20 | 61/64 |
| 1 (.073) | 64 | No. 53 | 72 | No. 53 | .... | .... | 1 1/16 | .... | .... | .... | .... | 18 | 1 |
| 2 (.086) | 56 | No. 50 | 64 | No. 50 | .... | .... | 1 1/8 | 7 | 63/64 | 12 | 1 3/64 | 18 | 1 5/64 |
| 3 (.099) | 48 | No. 47 | 56 | No. 45 | .... | .... | 1 3/16 | .... | .... | .... | .... | 18 | 1 9/64 |
| 4 (.112) | 40 | No. 43 | 48 | No. 42 | .... | .... | 1 1/4 | 7 | 1 7/64 | 12 | 1 11/64 | 18 | 1 3/16 |
| 5 (.125) | 40 | No. 38 | 44 | No. 37 | .... | .... | 1 5/16 | .... | .... | .... | .... | 18 | 1 17/64 |
| 6 (.138) | 32 | No. 36 | 40 | No. 33 | .... | .... | 1 3/8 | 6 | 1 7/32 | 12 | 1 19/64 | 18 | 1 5/16 |
| 8 (.164) | 32 | No. 29 | 36 | No. 29 | .... | .... | 1 7/16 | .... | .... | .... | .... | 18 | 1 3/8 |
| 10 (.190) | 24 | No. 25 | 32 | No. 21 | .... | .... | 1 1/2 | 6 | 1 11/32 | 12 | 1 27/64 | 18 | 1 7/16 |
| 12 (.216) | 24 | No. 16 | 28 | No. 14 | 32 | No. 13 | 1 9/16 | .... | .... | .... | .... | 18 | 1 1/2 |
| 1/4 | 20 | No. 7 | 28 | No. 3 | 32 | 7/32 | 1 5/8 | .... | .... | .... | .... | 18 | 1 9/16 |
| 5/16 | 18 | F | 24 | I | 32 | 9/32 | 1 11/16 | .... | .... | .... | .... | 18 | 1 5/8 |
| 3/8 | 16 | 5/16 | 24 | Q | 32 | 11/32 | 1 3/4 | 5 | 1 9/16 | .... | .... | .... | .... |
| 7/16 | 14 | U | 20 | 25/64 | 28 | 13/32 | 2 | 4 1/2 | 1 25/32 | .... | .... | .... | .... |
| 1/2 | 13 | 27/64 | 20 | 29/64 | 28 | 15/32 | 2 1/4 | 4 1/2 | 2 1/32 | .... | .... | .... | .... |
| 9/16 | 12 | 31/64 | 18 | 33/64 | 24 | 33/64 | 2 1/2 | 4 | 2 1/4 | .... | .... | .... | .... |
| 5/8 | 11 | 17/32 | 18 | 37/64 | 24 | 37/64 | 2 3/4 | 4 | 2 1/2 | .... | .... | .... | .... |
| 11/16 | .... | .... | .... | .... | 24 | 41/64 | 3 | 4 | 2 3/4 | .... | .... | .... | .... |
| 3/4 | 10 | 21/32 | 16 | 11/16 | 20 | 45/64 | 3 1/4 | 4 | .... | .... | .... | .... | .... |
| 13/16 | .... | .... | .... | .... | 20 | 49/64 | 3 1/2 | 4 | .... | .... | .... | .... | .... |
| 7/8 | 9 | 49/64 | 14 | 13/16 | 20 | 53/64 | 3 3/4 | 4 | .... | .... | .... | .... | .... |
| 15/16 | .... | .... | .... | .... | 20 | 57/64 | 4 | 4 | .... | .... | .... | .... | .... |

[a] ANSI B1.1–1960. For 8-, 12-, and 16-pitch thread series, see next page.
[b] Classes 1A, 2A, 3A, 1B, 2B, 3B, 2, and 3.
[c] Classes 2A, 2B, 2, and 3.
[d] For approximate 75% full depth of thread. For decimal sizes of numbered and lettered drills, see Appendix 11.

## 10 American National Standard Unified and American National Threads (continued)[a]

| Nominal Diameter | 8-Pitch[b] Series 8N and 8UN | | 12-Pitch[b] Series 12N and 12UN | | 16-Pitch[b] Series 16N and 16UN | | Nominal Diameter | 8-Pitch[b] Series 8N and 8UN | | 12-Pitch[b] Series 12N and 12UN | | 16-Pitch[b] Series 16N and 16UN | |
|---|---|---|---|---|---|---|---|---|---|---|---|---|---|
| | Thds. per Inch | Tap Drill[c] | Thds. per Inch | Tap Drill[c] | Thds. per Inch | Tap Drill[c] | | Thds. per Inch | Tap Drill[c] | Thds. per Inch | Tap Drill[c] | Thds. per Inch | Tap Drill[c] |
| 1/2 | .... | .... | 12 | 27/64 | .... | .... | 2 1/16 | .... | .... | .... | .... | **16** | 2 |
| 9/16 | .... | .... | 12[e] | 31/64 | .... | .... | 2 1/8 | .... | .... | 12 | 2 3/64 | 16 | 2 1/16 |
| 5/8 | .... | .... | 12 | 35/64 | .... | .... | 2 3/16 | .... | .... | .... | .... | **16** | 2 1/8 |
| 11/16 | .... | .... | 12 | 39/64 | .... | .... | 2 1/4 | 8 | 2 1/8 | 12 | 2 11/64 | 16 | 2 3/16. |
| 3/4 | .... | .... | 12 | 43/64 | 16[e] | 11/16 | 2 5/16 | .... | .... | .... | .... | **16** | 2 1/4 |
| 13/16 | .... | .... | 12 | 47/64 | 16 | 3/4 | 2 3/8 | .... | .... | 12 | 2 19/64 | 16 | 2 5/16 |
| 7/8 | .... | .... | 12 | 51/64 | 16 | 13/16 | 2 7/16 | .... | .... | .... | .... | **16** | 2 3/8 |
| 5/16 | .... | .... | 12 | 55/64 | 16 | 7/8 | 2 1/2 | 8 | 2 3/8 | 12 | 2 27/64 | 16 | 2 7/16 |
| 1 | 8[e] | 7/8 | 12 | 59/64 | 16 | 15/16 | 2 5/8 | .... | .... | 12 | 2 35/64 | 16 | 2 9/16 |
| 1 1/16 | .... | .... | 12 | 63/64 | 16 | 1 | 2 3/4 | 8 | 2 5/8 | 12 | 2 43/64 | 16 | 2 11/16 |
| 1 1/8 | 8 | 1 | 12[e] | 1 3/64 | 16 | 1 1/16 | 2 7/8 | .... | .... | 12 | .... | 16 | .... |
| 1 3/16 | .... | .... | 12 | 1 7/64 | 16 | 1 1/8 | 3 | 8 | 2 7/8 | 12 | .... | 16 | .... |
| 1 1/4 | 8 | 1 1/8 | 12 | 1 11/64 | 16 | 1 3/16 | 3 1/8 | .... | .... | 12 | .... | 16 | .... |
| 1 5/16 | .... | .... | 12 | 1 15/64 | 16 | 1 1/4 | 3 1/4 | 8 | .... | 12 | .... | 16 | .... |
| 1 3/8 | 8 | 1 1/4 | 12[e] | 1 19/64 | 16 | 1 5/16 | 3 3/8 | .... | .... | 12 | .... | 16 | .... |
| 1 7/16 | .... | .... | 12 | 1 23/64 | 16 | 1 3/8 | 3 1/2 | 8 | .... | 12 | .... | 16 | .... |
| 1 1/2 | 8 | 1 3/8 | 12[e] | 1 27/64 | 16 | 1 7/16 | 3 5/8 | .... | .... | 12 | .... | 16 | .... |
| 1 9/16 | .... | .... | .... | .... | 16 | 1 1/2 | 3 3/4 | 8 | .... | 12 | .... | 16 | .... |
| 1 5/8 | 8 | 1 1/2 | 12 | 1 35/64 | 16 | 1 9/16 | 3 7/8 | .... | .... | 12 | .... | 16 | .... |
| 1 11/16 | .... | .... | .... | .... | 16 | 1 5/8 | 4 | 8 | .... | 12 | .... | 16 | .... |
| 1 3/4 | 8 | 1 5/8 | 12 | 1 43/64 | 16[e] | 1 11/16 | 4 1/4 | 8 | .... | 12 | .... | 16 | .... |
| 1 13/16 | .... | .... | .... | .... | 16 | 1 3/4 | 4 1/2 | 8 | .... | 12 | .... | 16 | .... |
| 1 7/8 | 8 | 1 3/4 | 12 | 1 51/64 | 16 | 1 13/16 | 4 3/4 | 8 | .... | 12 | .... | 16 | .... |
| 1 15/16 | .... | .... | .... | .... | 16 | 1 7/8 | 5 | 8 | .... | 12 | .... | 16 | .... |
| 2 | 8 | 1 7/8 | 12 | 1 59/64 | 16[e] | 1 15/16 | 5 1/4 | 8 | .... | 12 | .... | 16 | .... |

[a] ANSI B1.1–1960.
[b] Classes 2A, 3A, 2B, 3B, 2, and 3.
[c] For approximate 75% full depth of thread.
[d] Boldface type indicates American National threads only.
[e] This is a standard size of the Unified or American National Threads of the coarse, fine, or extra fine series. See preceding page.

## 11 Twist Drill Sizes

All dimensions are in inches.
Drills designated in common fractions are available in diameters 1/64″ to 1¾″ in 1/64″ increments, 1¾″ to 2¼″ in 1/32″ increments, and 2¼″ to 3½″ in 1/16″ increments. Drills larger than 3½″ are seldom used, and are regarded as special drills.

| Size | Drill Diameter | Size | Drill Diameter | Size | Drill Diameter | Size | Drill Diameter | Size | Drill Diameter |
|---|---|---|---|---|---|---|---|---|---|
| 1 | .2280 | 17 | .1730 | 33 | .1130 | 49 | .0730 | 65 | .0350 |
| 2 | .2210 | 18 | .1695 | 34 | .1110 | 50 | .0700 | 66 | .0330 |
| 3 | .2130 | 19 | .1660 | 35 | .1100 | 51 | .0670 | 67 | .0320 |
| 4 | .2090 | 20 | .1610 | 36 | .1065 | 52 | .0635 | 68 | .0310 |
| 5 | .2055 | 21 | .1590 | 37 | .1040 | 53 | .0595 | 69 | .0292 |
| 6 | .2040 | 22 | .1570 | 38 | .1015 | 54 | .0550 | 70 | .0280 |
| 7 | .2010 | 23 | .1540 | 39 | .0995 | 55 | .0520 | 71 | .0260 |
| 8 | .1990 | 24 | .1520 | 40 | .0980 | 56 | .0465 | 72 | .0250 |
| 9 | .1960 | 25 | .1495 | 41 | .0960 | 57 | .0430 | 73 | .0240 |
| 10 | .1935 | 26 | .1470 | 42 | .0935 | 58 | .0420 | 74 | .0225 |
| 11 | .1910 | 27 | .1440 | 43 | .0890 | 59 | .0410 | 75 | .0210 |
| 12 | .1890 | 28 | .1405 | 44 | .0860 | 60 | .0400 | 76 | .0200 |
| 13 | .1850 | 29 | .1360 | 45 | .0820 | 61 | .0390 | 77 | .0180 |
| 14 | .1820 | 30 | .1285 | 46 | .0810 | 62 | .0380 | 78 | .0160 |
| 15 | .1800 | 31 | .1200 | 47 | .0785 | 63 | .0370 | 79 | .0145 |
| 16 | .1770 | 32 | .1160 | 48 | .0760 | 64 | .0360 | 80 | .0135 |

LETTER SIZES

| Size | Drill Diameter | Size | Drill Diameter | Size | Drill Diameter | Size | Drill Diameter | Size | Drill Diameter |
|---|---|---|---|---|---|---|---|---|---|
| A | .234 | G | .261 | L | .290 | Q | .332 | V | .377 |
| B | .238 | H | .266 | M | .295 | R | .339 | W | .386 |
| C | .242 | I | .272 | N | .302 | S | .348 | X | .397 |
| D | .246 | J | .277 | O | .316 | T | .358 | Y | .404 |
| E | .250 | K | .281 | P | .323 | U | .368 | Z | .413 |
| F | .257 | | | | | | | | |

## 12 General Purpose Acme Threads

| Size | Threads per Inch | Size | Threads per Inch | Size | Threads per Inch | Size | Threads per Inch |
|---|---|---|---|---|---|---|---|
| ¼ | 16 | ¾ | 6 | 1½ | 4 | 3 | 2 |
| 5/16 | 14 | ⅞ | 6 | 1¾ | 4 | 3½ | 2 |
| ⅜ | 12 | 1 | 5 | 2 | 4 | 4 | 2 |
| 7/16 | 12 | 1⅛ | 5 | 2¼ | 3 | 4½ | 2 |
| ½ | 10 | 1¼ | 5 | 2½ | 3 | 5 | 2 |
| ⅝ | 8 | 1⅜ | 4 | 2¾ | 3 | . . . | . . |

# 13 American National Standard Square and Hexagon Bolts[a] and Nuts[b] and Hexagon-Head Cap Screws[c]

**Boldface type** indicates product features unified dimensionally with British and Canadian standards.
All dimensions are in inches.
For thread series, minimum thread lengths, and bolt lengths, see §13.25.

| Nominal Size D Body Diameter of Bolt | Regular Bolts | | | | | Heavy Bolts | | |
|---|---|---|---|---|---|---|---|---|
| | Width Across Flats W | | Height H | | | Width Across Flats W | Height H | |
| | Sq. | Hex. | Sq. (Unfin.) | Hex. (Unfin.) | Hex. Cap Scr.[c] (Fin.) | | Hex. (Unfin.) | Hex. Screw (Fin.) |
| **1/4 0.2500** | **3/8** | **7/16** | **11/64** | **11/64** | **5/32** | .... | .... | .... |
| **5/16 0.3125** | **1/2** | **1/2** | **13/64** | **7/32** | **13/64** | .... | .... | .... |
| **3/8 0.3750** | **9/16** | **9/16** | **1/4** | **1/4** | **15/64** | .... | .... | .... |
| **7/16 0.4375** | **5/8** | **5/8** | **19/64** | **19/64** | **9/32** | .... | .... | .... |
| **1/2 0.5000** | **3/4** | **3/4** | **21/64** | **11/32** | **5/16** | **7/8** | 11/32 | 5/16 |
| **9/16 0.5625** | .... | **13/16** | .... | .... | **23/64** | .... | .... | .... |
| **5/8 0.6250** | **15/16** | **15/16** | **27/64** | **27/64** | **25/64** | **1 1/16** | 27/64 | 25/64 |
| **3/4 0.7500** | **1 1/8** | **1 1/8** | **1/2** | **1/2** | **15/32** | **1 1/4** | 1/2 | 15/32 |
| **7/8 0.8750** | **1 5/16** | **1 5/16** | **19/32** | **37/64** | **35/64** | **1 7/16** | 37/64 | 35/64 |
| **1 1.000** | **1 1/2** | **1 1/2** | **21/32** | **43/64** | **39/64** | **1 5/8** | 43/64 | 39/64 |
| **1 1/8 1.1250** | **1 11/16** | **1 11/16** | **3/4** | **3/4** | **11/16** | **1 13/16** | 3/4 | 11/16 |
| **1 1/4 1.2500** | **1 7/8** | **1 7/8** | **27/32** | **27/32** | **25/32** | **2** | 27/32 | 25/32 |
| **1 3/8 1.3750** | **2 1/16** | **2 1/16** | **29/32** | **29/32** | **27/32** | **2 3/16** | 29/32 | 27/32 |
| **1 1/2 1.5000** | **2 1/4** | **2 1/4** | **1** | **1** | **15/16** | **2 3/8** | 1 | 15/16 |
| **1 3/4 1.7500** | .... | **2 5/8** | .... | **1 5/32** | **1 3/32** | **2 3/4** | 1 5/32 | 1 3/32 |
| **2 2.0000** | .... | **3** | .... | **1 11/32** | **1 7/32** | **3 1/8** | 1 11/32 | 1 7/32 |
| 2 1/4 2.2500 | .... | 3 3/8 | .... | 1 1/2 | 1 3/8 | 3 1/2 | 1 1/2 | 1 3/8 |
| 2 1/2 2.5000 | .... | 3 3/4 | .... | 1 21/32 | 1 17/32 | 3 7/8 | 1 21/32 | 1 17/32 |
| 2 3/4 2.7500 | .... | 4 1/8 | .... | 1 13/16 | 1 11/16 | 4 1/4 | 1 13/16 | 1 11/16 |
| 3 3.0000 | .... | 4 1/2 | .... | 2 | 1 7/8 | 4 5/8 | 2 | 1 7/8 |
| 3 1/4 3.2500 | .... | 4 7/8 | .... | 2 3/16 | .... | .... | .... | .... |
| 3 1/2 3.5000 | .... | 5 1/4 | .... | 2 5/16 | .... | .... | .... | .... |
| 3 3/4 3.7500 | .... | 5 5/8 | .... | 2 1/2 | .... | .... | .... | .... |
| 4 4.0000 | .... | 6 | .... | 2 11/16 | .... | .... | .... | .... |

[a] ANSI B18.2.1-1965.
[b] ANSI B18.2.2-1965.
[c] Hexagon cap screws and finished hexagon bolts are combined as a single product.

## 13 American National Standard Square and Hexagon Head Bolts and Nuts and Hexagon-Head Cap Screws (continued)

See ANSI B18.2.2–1965 for jam nuts, slotted nuts, thick nuts, thick slotted nuts, and castle nuts. For methods of drawing bolts and nuts and hexagon-head cap screws, see Figs. 13.30, 13.31, and 13.33.

| Nominal Size D Body Diameter of Bolt | Regular Nuts | | | | | Heavy Nuts | | | |
|---|---|---|---|---|---|---|---|---|---|
| | Width Across Flats W | | Thickness T | | | Width Across Flats W | Thickness T | | |
| | Sq. | Hex. | Sq. (Unfin.) | Hex. Flat (Unfin.) | Hex. (Fin.) | | Sq. (Unfin.) | Hex. Flat (Unfin.) | Hex. (Fin.) |
| 1/4 0.2500 | 7/16 | 7/16 | 7/32 | 7/32 | 7/32 | 1/2 | 1/4 | 15/64 | 15/64 |
| 5/16 0.3125 | 9/16 | 1/2 | 17/64 | 17/64 | 17/64 | 9/16 | 5/16 | 19/64 | 19/64 |
| 3/8 0.3750 | 5/8 | 9/16 | 21/64 | 21/64 | 21/64 | 11/16 | 3/8 | 23/64 | 23/64 |
| 7/16 0.4375 | 3/4 | 11/16 | 3/8 | 3/8 | 3/8 | 3/4 | 7/16 | 27/64 | 27/64 |
| 1/2 0.5000 | 13/16 | 3/4 | 7/16 | 7/16 | 7/16 | 7/8[a] | 1/2 | 31/64 | 31/64 |
| 9/16 0.5625 | .... | 7/8 | .... | 31/64 | 31/64 | 15/16 | .... | 35/64 | 35/64 |
| 5/8 0.6250 | 1 | 15/16 | 35/64 | 35/64 | 35/64 | 1 1/16[a] | 5/8 | 39/64 | 39/64 |
| 3/4 0.7500 | 1 1/8 | 1 1/8 | 21/32 | 41/64 | 41/64 | 1 1/4[a] | 3/4 | 47/64 | 47/64 |
| 7/8 0.8750 | 1 5/16 | 1 5/16 | 49/64 | 3/4 | 3/4 | 1 7/16[a] | 7/8 | 55/64 | 55/64 |
| 1 1.0000 | 1 1/2 | 1 1/2 | 7/8 | 55/64 | 55/64 | 1 5/8[a] | 1 | 63/64 | 63/64 |
| 1 1/8 1.1250 | 1 11/16 | 1 11/16 | 1 | 1 | 31/32 | 1 13/16[a] | 1 1/8 | 1 1/8 | 1 7/64 |
| 1 1/4 1.2500 | 1 7/8 | 1 7/8 | 1 3/32 | 1 3/32 | 1 1/16 | 2[a] | 1 1/4 | 1 1/4 | 1 7/32 |
| 1 3/8 1.3750 | 2 1/16 | 2 1/16 | 1 13/64 | 1 13/64 | 1 11/64 | 2 3/16[a] | 1 3/8 | 1 3/8 | 1 11/32 |
| 1 1/2 1.5000 | 2 1/4 | 2 1/4 | 1 5/16 | 1 5/16 | 1 9/32 | 2 3/8[a] | 1 1/2 | 1 1/2 | 1 15/32 |
| 1 5/8 1.6250 | .... | .... | .... | .... | .... | 2 9/16 | .... | .... | 1 19/32 |
| 1 3/4 1.7500 | .... | .... | .... | .... | .... | 2 3/4 | .... | 1 3/4 | 1 23/32 |
| 1 7/8 1.8750 | .... | .... | .... | .... | .... | 2 15/16 | .... | .... | 1 27/32 |
| 2 2.0000 | .... | .... | .... | .... | .... | 3 1/8 | .... | 2 | 1 31/32 |
| 2 1/4 2.2500 | .... | .... | .... | .... | .... | 3 1/2 | .... | 2 1/4 | 2 13/64 |
| 2 1/2 2.5000 | .... | .... | .... | .... | .... | 3 7/8 | .... | 2 1/2 | 2 29/64 |
| 2 3/4 2.7500 | .... | .... | .... | .... | .... | 4 1/4 | .... | 2 3/4 | 2 45/64 |
| 3 3.0000 | .... | .... | .... | .... | .... | 4 5/8 | .... | 3 | 2 61/64 |
| 3 1/4 3.2500 | .... | .... | .... | .... | .... | 5 | .... | 3 1/4 | 3 3/16 |
| 3 1/2 3.5000 | .... | .... | .... | .... | .... | 5 3/8 | .... | 3 1/2 | 3 7/16 |
| 3 3/4 3.7500 | .... | .... | .... | .... | .... | 5 3/4 | .... | 3 3/4 | 3 11/16 |
| 4 4.0000 | .... | .... | .... | .... | .... | 6 1/8 | .... | 4 | 3 15/16 |

[a] Product feature not unified for heavy square nut.

## 14 American National Slotted[a] and Socket Head[b] Cap Screws

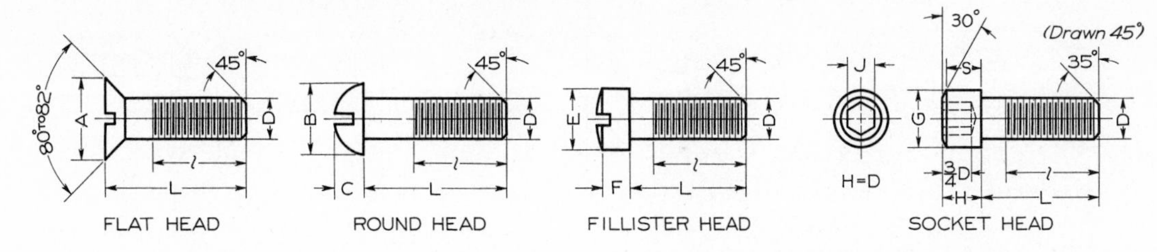


For methods of drawing cap screws, screw lengths, and thread data, see Fig. 13.33.

| Nominal Size D | Flat Head[a] | Round Head[a] | | Fillister Head[a] | | Socket Head[b] | | |
|---|---|---|---|---|---|---|---|---|
| | A | B | C | E | F | G | J | S |
| 0 (.060) | .... | .... | .... | .... | .... | .096 | .05 | .054 |
| 1 (.073) | .... | .... | .... | .... | .... | .118 | 1/16 | .066 |
| 2 (.086) | .... | .... | .... | .... | .... | .140 | 5/64 | .077 |
| 3 (.099) | .... | .... | .... | .... | .... | .161 | 5/64 | .089 |
| 4 (.112) | .... | .... | .... | .... | .... | .183 | 3/32 | .101 |
| 5 (.125) | .... | .... | .... | .... | .... | .205 | 3/32 | .112 |
| 6 (.138) | .... | .... | .... | .... | .... | .226 | 7/64 | .124 |
| 8 (.164) | .... | .... | .... | .... | .... | .270 | 9/64 | .148 |
| 10 (.190) | .... | .... | .... | .... | .... | 5/16 | 5/32 | .171 |
| 1/4 | 1/2 | 7/16 | .191 | 3/8 | 11/64 | 3/8 | 3/16 | .225 |
| 5/16 | 5/8 | 9/16 | .245 | 7/16 | 13/64 | 15/32 | 1/4 | .281 |
| 3/8 | 3/4 | 5/8 | .273 | 9/16 | 1/4 | 9/16 | 5/16 | .337 |
| 7/16 | 13/16 | 3/4 | 21/64 | 5/8 | 19/64 | 21/32 | 3/8 | .394 |
| 1/2 | 7/8 | 13/16 | .355 | 3/4 | 21/64 | 3/4 | 3/8 | .450 |
| 9/16 | 1 | 15/16 | .409 | 13/16 | 3/8 | .... | .... | .... |
| 5/8 | 1 1/8 | 1 | 7/16 | 7/8 | 27/64 | 15/16 | 1/2 | .562 |
| 3/4 | 1 3/8 | 1 1/4 | 35/64 | 1 | 1/2 | 1 1/8 | 5/8 | .675 |
| 7/8 | 1 5/8 | .... | .... | 1 1/8 | 19/32 | 1 5/16 | 3/4 | .787 |
| 1 | 1 7/8 | .... | .... | 1 5/16 | 21/32 | 1 1/2 | 3/4 | .900 |
| 1 1/8 | 2 1/16 | .... | .... | .... | .... | 1 11/16 | 7/8 | 1.012 |
| 1 1/4 | 2 5/16 | .... | .... | .... | .... | 1 7/8 | 7/8 | 1.125 |
| 1 3/8 | 2 9/16 | .... | .... | .... | .... | 2 1/16 | 1 | 1.237 |
| 1 1/2 | 2 13/16 | .... | .... | .... | .... | 2 1/4 | 1 | 1.350 |

[a] ANSI B18.6.2–1956.
[b] ANSI B18.3–1969. For hexagon-head screws, see §13.29 and Appendix 13.

# 15 American National Standard Machine Screws[a]

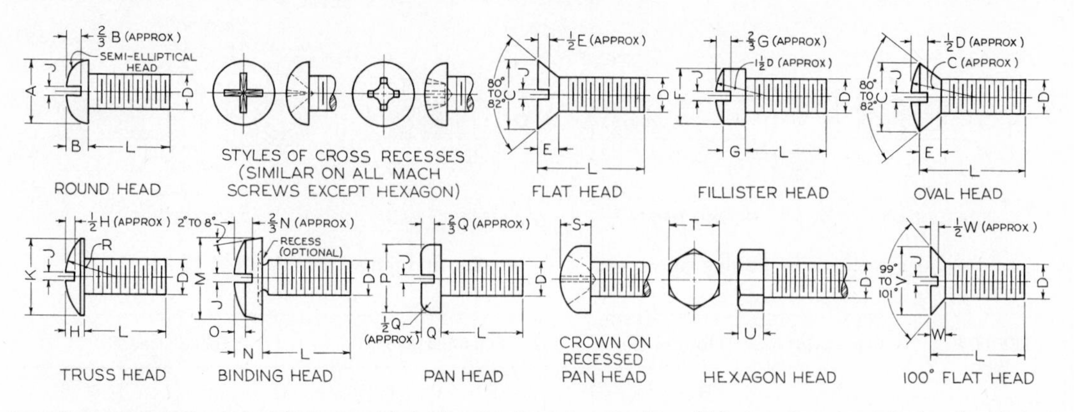


*Length of Thread:* On screws 2″ long and shorter, the threads extend to within two threads of the head and closer if practicable; longer screws have minimum thread length of 1¾″.
*Points:* Machine screws are regularly made with plain sheared ends, not chamfered.
*Threads:* Either Coarse or Fine Thread Series, Class 2 fit.
*Recessed Heads:* Two styles of cross recesses are available on all screws except hexagon head.

| Nominal Size | Max. Diameter D | Round Head | | Flat Heads & Oval Head | | Fillister Head | | Truss Head | | | Slot Width |
|---|---|---|---|---|---|---|---|---|---|---|---|
| | | A | B | C | E | F | G | K | H | R | J |
| 0 | 0.060 | 0.113 | 0.053 | 0.119 | 0.035 | 0.096 | 0.045 | 0.131 | 0.037 | 0.087 | 0.023 |
| 1 | 0.073 | 0.138 | 0.061 | 0.146 | 0.043 | 0.118 | 0.053 | 0.164 | 0.045 | 0.107 | 0.026 |
| 2 | 0.086 | 0.162 | 0.069 | 0.172 | 0.051 | 0.140 | 0.062 | 0.194 | 0.053 | 0.129 | 0.031 |
| 3 | 0.099 | 0.187 | 0.078 | 0.199 | 0.059 | 0.161 | 0.070 | 0.226 | 0.061 | 0.151 | 0.035 |
| 4 | 0.112 | 0.211 | 0.086 | 0.225 | 0.067 | 0.183 | 0.079 | 0.257 | 0.069 | 0.169 | 0.039 |
| 5 | 0.125 | 0.236 | 0.095 | 0.252 | 0.075 | 0.205 | 0.088 | 0.289 | 0.078 | 0.191 | 0.043 |
| 6 | 0.138 | 0.260 | 0.103 | 0.279 | 0.083 | 0.226 | 0.096 | 0.321 | 0.086 | 0.211 | 0.048 |
| 8 | 0.164 | 0.309 | 0.120 | 0.332 | 0.100 | 0.270 | 0.113 | 0.384 | 0.102 | 0.254 | 0.054 |
| 10 | 0.190 | 0.359 | 0.137 | 0.385 | 0.116 | 0.313 | 0.130 | 0.448 | 0.118 | 0.283 | 0.060 |
| 12 | 0.216 | 0.408 | 0.153 | 0.438 | 0.132 | 0.357 | 0.148 | 0.511 | 0.134 | 0.336 | 0.067 |
| ¼ | 0.250 | 0.472 | 0.175 | 0.507 | 0.153 | 0.414 | 0.170 | 0.573 | 0.150 | 0.375 | 0.075 |
| 5/16 | 0.3125 | 0.590 | 0.216 | 0.635 | 0.191 | 0.518 | 0.211 | 0.698 | 0.183 | 0.457 | 0.084 |
| ⅜ | 0.375 | 0.708 | 0.256 | 0.762 | 0.230 | 0.622 | 0.253 | 0.823 | 0.215 | 0.538 | 0.094 |
| 7/16 | 0.4375 | 0.750 | 0.328 | 0.812 | 0.223 | 0.625 | 0.265 | 0.948 | 0.248 | 0.619 | 0.094 |
| ½ | 0.500 | 0.813 | 0.355 | 0.875 | 0.223 | 0.750 | 0.297 | 1.073 | 0.280 | 0.701 | 0.106 |
| 9/16 | 0.5625 | 0.938 | 0.410 | 1.000 | 0.260 | 0.812 | 0.336 | 1.198 | 0.312 | 0.783 | 0.118 |
| ⅝ | 0.625 | 1.000 | 0.438 | 1.125 | 0.298 | 0.875 | 0.375 | 1.323 | 0.345 | 0.863 | 0.133 |
| ¾ | 0.750 | 1.250 | 0.547 | 1.375 | 0.372 | 1.000 | 0.441 | 1.573 | 0.410 | 1.024 | 0.149 |

| Nominal Size | Max. Diameter D | Binding Head | | | Pan Head | | | Hexagon Head | | 100° Flat Head | | Slot Width |
|---|---|---|---|---|---|---|---|---|---|---|---|---|
| | | M | N | O | P | Q | S | T | U | V | W | J |
| 2 | 0.086 | 0.181 | 0.050 | 0.018 | 0.167 | 0.053 | 0.062 | 0.125 | 0.050 | .... | .... | 0.031 |
| 3 | 0.099 | 0.208 | 0.059 | 0.022 | 0.193 | 0.060 | 0.071 | 0.187 | 0.055 | .... | .... | 0.035 |
| 4 | 0.112 | 0.235 | 0.068 | 0.025 | 0.219 | 0.068 | 0.080 | 0.187 | 0.060 | 0.225 | 0.049 | 0.039 |
| 5 | 0.125 | 0.263 | 0.078 | 0.029 | 0.245 | 0.075 | 0.089 | 0.187 | 0.070 | .... | .... | 0.043 |
| 6 | 0.138 | 0.290 | 0.087 | 0.032 | 0.270 | 0.082 | 0.097 | 0.250 | 0.080 | 0.279 | 0.060 | 0.048 |
| 8 | 0.164 | 0.344 | 0.105 | 0.039 | 0.322 | 0.096 | 0.115 | 0.250 | 0.110 | 0.332 | 0.072 | 0.054 |
| 10 | 0.190 | 0.399 | 0.123 | 0.045 | 0.373 | 0.110 | 0.133 | 0.312 | 0.120 | 0.385 | 0.083 | 0.060 |
| 12 | 0.216 | 0.454 | 0.141 | 0.052 | 0.425 | 0.125 | 0.151 | 0.312 | 0.155 | .... | .... | 0.067 |
| ¼ | 0.250 | 0.513 | 0.165 | 0.061 | 0.492 | 0.144 | 0.175 | 0.375 | 0.190 | 0.507 | 0.110 | 0.075 |
| 5/16 | 0.3125 | 0.641 | 0.209 | 0.077 | 0.615 | 0.178 | 0.218 | 0.500 | 0.230 | 0.635 | 0.138 | 0.084 |
| ⅜ | 0.375 | 0.769 | 0.253 | 0.094 | 0.740 | 0.212 | 0.261 | 0.562 | 0.295 | 0.762 | 0.165 | 0.094 |

[a] ANSI B18.3-1969.

## 16 Square and Flat Keys, Plain Taper Keys,[a] and Gib Head Keys

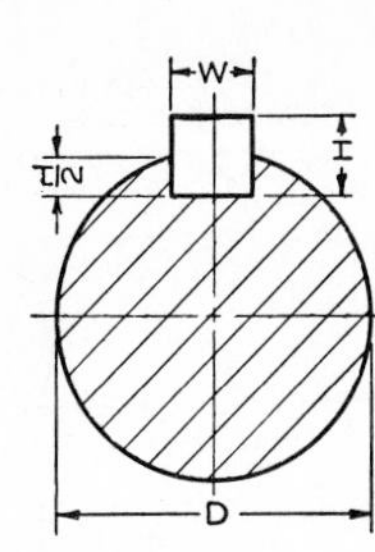


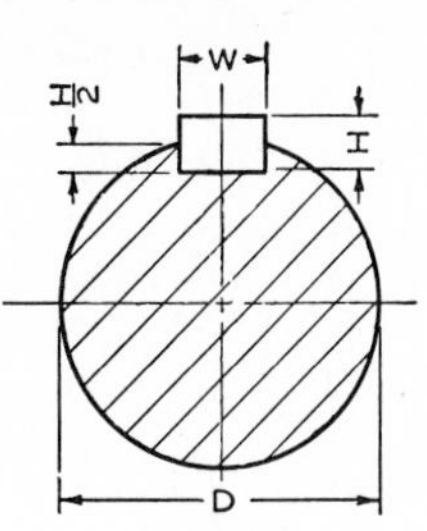


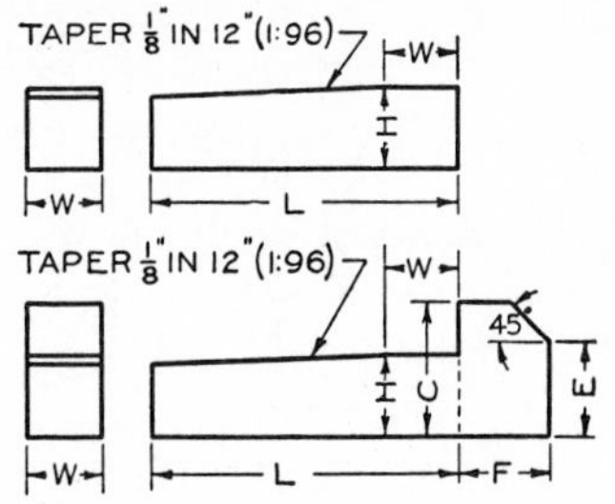


| Shaft Diameters | Square Stock Key | Flat Stock Key | Gib Head Taper Stock Key | | | | | |
|---|---|---|---|---|---|---|---|---|
| | | | *Square* | | | *Flat* | | |
| | | | Height | Length | Height to Chamfer | Height | Length | Height to Chamfer |
| D | W = H | W × H | C | F | E | C | F | E |
| 1/2 to 9/16 | 1/8 | 1/8 × 3/32 | 1/4 | 7/32 | 5/32 | 3/16 | 1/8 | 1/8 |
| 5/8 to 7/8 | 3/16 | 3/16 × 1/8 | 5/16 | 9/32 | 7/32 | 1/4 | 3/16 | 5/32 |
| 15/16 to 1 1/4 | 1/4 | 1/4 × 3/16 | 7/16 | 11/32 | 11/32 | 5/16 | 1/4 | 3/16 |
| 1 5/16 to 1 3/8 | 5/16 | 5/16 × 1/4 | 9/16 | 13/32 | 13/32 | 3/8 | 5/16 | 1/4 |
| 1 7/16 to 1 3/4 | 3/8 | 3/8 × 1/4 | 11/16 | 15/32 | 15/32 | 7/16 | 3/8 | 5/16 |
| 1 13/16 to 2 1/4 | 1/2 | 1/2 × 3/8 | 7/8 | 19/32 | 5/8 | 5/8 | 1/2 | 7/16 |
| 2 5/16 to 2 3/4 | 5/8 | 5/8 × 7/16 | 1 1/16 | 23/32 | 3/4 | 3/4 | 5/8 | 1/2 |
| 2 7/8 to 3 1/4 | 3/4 | 3/4 × 1/2 | 1 1/4 | 7/8 | 7/8 | 7/8 | 3/4 | 5/8 |
| 3 3/8 to 3 3/4 | 7/8 | 7/8 × 5/8 | 1 1/2 | 1 | 1 | 1 1/16 | 7/8 | 3/4 |
| 3 7/8 to 4 1/2 | 1 | 1 × 3/4 | 1 3/4 | 1 3/16 | 1 3/16 | 1 1/4 | 1 | 13/16 |
| 4 3/4 to 5 1/2 | 1 1/4 | 1 1/4 × 7/8 | 2 | 1 7/16 | 1 7/16 | 1 1/2 | 1 1/4 | 1 |
| 5 3/4 to 6 | 1 1/2 | 1 1/2 × 1 | 2 1/2 | 1 3/4 | 1 3/4 | 1 3/4 | 1 1/2 | 1 1/4 |

[a] Plain taper square and flat keys have the same dimensions as the plain parallel stock keys, with the addition of the taper on top. Gib head taper square and flat keys have the same dimensions as the plain taper keys, with the addition of the gib head.

*Stock lengths for plain taper and gib head taper keys:* The minimum stock length equals 4W, and the maximum equals 16W. The increments of increase of length equal 2W.

## 17 Square and Acme Threads[a]

| Size | Threads per Inch | Size | Threads per Inch | Size | Threads per Inch | Size | Threads per Inch |
|---|---|---|---|---|---|---|---|
| 3/8 | 12 | 7/8 | 5 | 2 | 2 1/2 | 3 1/2 | 1 1/3 |
| 7/16 | 10 | 1 | 5 | 2 1/4 | 2 | 3 3/4 | 1 1/3 |
| 1/2 | 10 | 1 1/8 | 4 | 2 1/2 | 2 | 4 | 1 1/3 |
| 9/16 | 8 | 1 1/4 | 4 | 2 3/4 | 2 | 4 1/4 | 1 1/3 |
| 5/8 | 8 | 1 1/2 | 3 | 3 | 1 1/2 | 4 1/2 | 1 |
| 3/4 | 6 | 1 3/4 | 2 1/2 | 3 1/4 | 1 1/2 | over 4 1/2 | 1 |

[a] See Appendix 12 for General Purpose Acme Threads.

## 18 American National Standard Woodruff Keys[a]

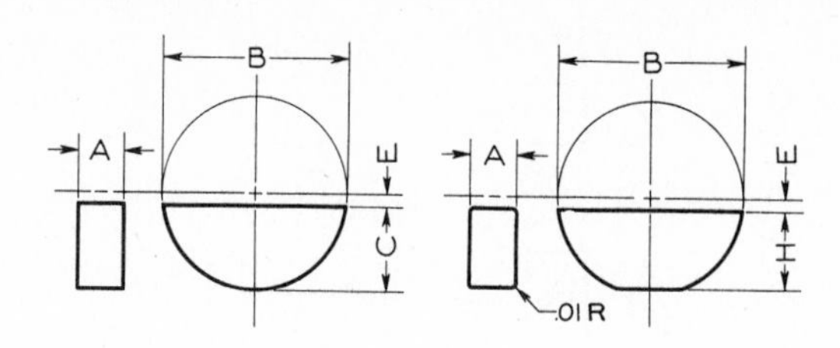


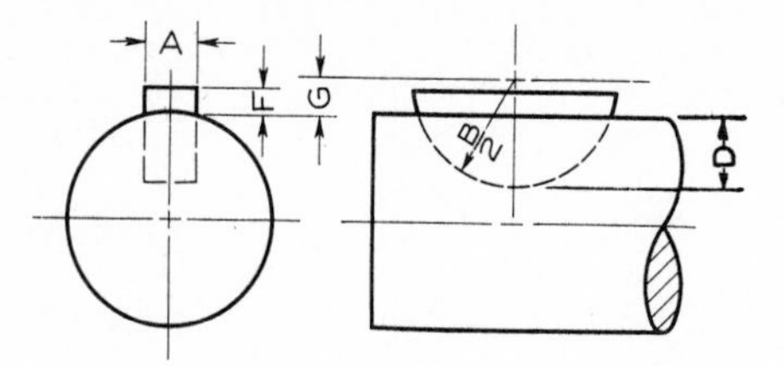


| Key No.[b] | Nominal Sizes | | | | Maximum Sizes | | | Key No.[b] | Nominal Sizes | | | | Maximum Sizes | | |
|---|---|---|---|---|---|---|---|---|---|---|---|---|---|---|---|
| | A × B | E | F | G | H | D | C | | A × B | E | F | G | H | D | C |
| 204 | 1/16 × 1/2 | 3/64 | 1/32 | 5/64 | .194 | .1718 | .203 | 808 | 1/4 × 1 | 1/16 | 1/8 | 3/16 | .428 | .3130 | .438 |
| 304 | 3/32 × 1/2 | 3/64 | 3/64 | 3/32 | .194 | .1561 | .203 | 809 | 1/4 × 1 1/8 | 5/64 | 1/8 | 13/64 | .475 | .3590 | .484 |
| 305 | 3/32 × 5/8 | 1/16 | 3/64 | 7/64 | .240 | .2031 | .250 | 810 | 1/4 × 1 1/4 | 5/64 | 1/8 | 13/64 | .537 | .4220 | .547 |
| 404 | 1/8 × 1/2 | 3/64 | 1/16 | 7/64 | .194 | .1405 | .203 | 811 | 1/4 × 1 3/8 | 3/32 | 1/8 | 7/32 | .584 | .4690 | .594 |
| 405 | 1/8 × 5/8 | 1/16 | 1/16 | 1/8 | .240 | .1875 | .250 | 812 | 1/4 × 1 1/2 | 7/64 | 1/8 | 15/64 | .631 | .5160 | .641 |
| 406 | 1/8 × 3/4 | 1/16 | 1/16 | 1/8 | .303 | .2505 | .313 | 1008 | 5/16 × 1 | 1/16 | 5/32 | 7/32 | .428 | .2818 | .438 |
| 505 | 5/32 × 5/8 | 1/16 | 5/64 | 9/64 | .240 | .1719 | .250 | 1009 | 5/16 × 1 1/8 | 5/64 | 5/32 | 15/64 | .475 | .3278 | .484 |
| 506 | 5/32 × 3/4 | 1/16 | 5/64 | 9/64 | .303 | .2349 | .313 | 1010 | 5/16 × 1 1/4 | 5/64 | 5/32 | 15/64 | .537 | .3908 | .547 |
| 507 | 5/32 × 7/8 | 1/16 | 5/64 | 9/64 | .365 | .2969 | .375 | 1011 | 5/16 × 1 3/8 | 3/32 | 5/32 | 8/32 | .584 | .4378 | .594 |
| 606 | 3/16 × 3/4 | 1/16 | 3/32 | 5/32 | .303 | .2193 | .313 | 1012 | 5/16 × 1 1/2 | 7/64 | 5/32 | 17/64 | .631 | .4848 | .641 |
| 607 | 3/16 × 7/8 | 1/16 | 3/32 | 5/32 | .365 | .2813 | .375 | 1210 | 3/8 × 1 1/4 | 5/64 | 3/16 | 17/64 | .537 | .3595 | .547 |
| 608 | 3/16 × 1 | 1/16 | 3/32 | 5/32 | .428 | .3443 | .438 | 1211 | 3/8 × 1 3/8 | 3/32 | 3/16 | 9/32 | .584 | .4065 | .594 |
| 609 | 3/16 × 1 1/8 | 5/64 | 3/32 | 11/64 | .475 | .3903 | .484 | 1212 | 3/8 × 1 1/2 | 7/64 | 3/16 | 19/64 | .631 | .4535 | .641 |
| 807 | 1/4 × 7/8 | 1/16 | 1/8 | 3/16 | .365 | .2500 | .375 | .... | ......... | .. | .. | .. | .... | .... | .... |

[a] ANSI B17.2-1967.

[b] Key numbers indicate nominal key dimensions. The last two digits give the nominal diameter B in eighths of an inch, and the digits before the last two give the nominal width A in thirty-seconds of an inch.

## 19 Woodruff Key Sizes for Different Shaft Diameters[a]

| Shaft Diameter | 5/16 to 3/8 | 7/16 to 1/2 | 9/16 to 3/4 | 13/16 to 15/16 | 1 to 1 3/16 | 1 1/4 to 1 7/16 | 1 1/2 to 1 3/4 | 1 13/16 to 2 1/8 | 2 3/16 to 2 1/2 |
|---|---|---|---|---|---|---|---|---|---|
| Key Numbers | 204 | 304<br>305 | 404<br>405<br>406 | 505<br>506<br>507 | 606<br>607<br>608<br>609 | 807<br>808<br>809 | 810<br>811<br>812 | 1011<br>1012 | 1211<br>1212 |

[a] Suggested sizes; not standard.

## 20 Pratt and Whitney Round-End Keys

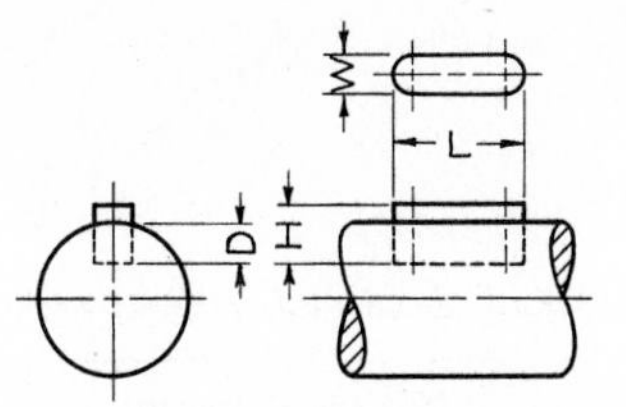


KEYS MADE WITH ROUND ENDS AND KEYWAYS CUT IN SPLINE MILLER

Maximum length of slot is 4″ + W. Note that key is sunk two-thirds into shaft in all cases.

| Key No. | L[a] | W or D | H | Key No. | L[a] | W or D | H |
|---|---|---|---|---|---|---|---|
| 1 | 1/2 | 1/16 | 3/32 | 22 | 1 3/8 | 1/4 | 3/8 |
| 2 | 1/2 | 3/32 | 9/64 | 23 | 1 1/38 | 5/16 | 15/32 |
| 3 | 1/2 | 1/8 | 3/16 | F | 1 3/8 | 3/8 | 9/16 |
| 4 | 5/8 | 3/32 | 9/64 | 24 | 1 1/2 | 1/4 | 3/8 |
| 5 | 5/8 | 1/8 | 3/16 | 25 | 1 1/2 | 5/16 | 15/32 |
| 6 | 5/8 | 5/32 | 15/64 | G | 1 1/2 | 3/8 | 9/16 |
| 7 | 3/4 | 1/8 | 3/16 | 51 | 1 3/4 | 1/4 | 3/8 |
| 8 | 3/4 | 5/32 | 15/64 | 52 | 1 3/4 | 5/16 | 15/32 |
| 9 | 3/4 | 3/16 | 9/32 | 53 | 1 3/4 | 3/8 | 9/16 |
| 10 | 7/8 | 5/32 | 15/64 | 26 | 2 | 3/16 | 9/32 |
| 11 | 7/8 | 3/16 | 9/32 | 27 | 2 | 1/4 | 3/8 |
| 12 | 7/8 | 7/32 | 21/64 | 28 | 2 | 5/16 | 15/32 |
| A | 7/8 | 1/4 | 3/8 | 29 | 2 | 3/8 | 9/16 |
| 13 | 1 | 3/16 | 9/32 | 54 | 2 1/4 | 1/4 | 3/8 |
| 14 | 1 | 7/32 | 21/64 | 55 | 2 1/4 | 5/16 | 15/32 |
| 15 | 1 | 1/4 | 3/8 | 56 | 2 1/4 | 3/8 | 9/16 |
| B | 1 | 5/16 | 15/32 | 57 | 2 1/4 | 7/16 | 21/32 |
| 16 | 1 1/8 | 3/16 | 9/32 | 58 | 2 1/2 | 5/16 | 15/32 |
| 17 | 1 1/8 | 7/32 | 21/64 | 59 | 2 1/2 | 3/8 | 9/16 |
| 18 | 1 1/8 | 1/4 | 3/8 | 60 | 2 1/2 | 7/16 | 21/32 |
| C | 1 1/8 | 5/16 | 15/32 | 61 | 2 1/2 | 1/2 | 3/4 |
| 19 | 1 1/4 | 3/16 | 9/32 | 30 | 3 | 3/8 | 9/16 |
| 20 | 1 1/4 | 7/32 | 21/64 | 31 | 3 | 7/16 | 21/32 |
| 21 | 1 1/4 | 1/4 | 3/8 | 32 | 3 | 1/2 | 3/4 |
| D | 1 1/4 | 5/16 | 15/32 | 33 | 3 | 9/16 | 27/32 |
| E | 1 1/4 | 3/8 | 9/16 | 34 | 3 | 5/8 | 15/16 |

[a] The length L may vary from the table, but equals at least 2W.

# 21 American National Standard Plain Washers[a]

For parts lists, etc., give inside diameter, outside diameter, and the thickness; for example, .344 × .688 × .065 TYPE A PLAIN WASHER.

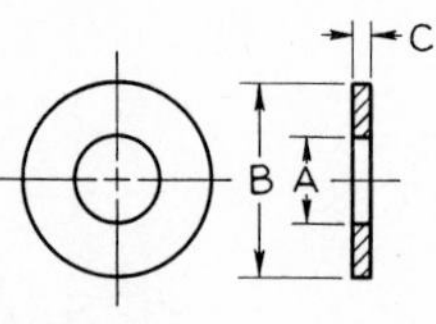


PREFERRED SIZES OF TYPE A PLAIN WASHERS[b]

| Nominal Washer Size[c] | | | Inside Diameter A | Outside Diameter B | Nominal Thickness C |
|---|---|---|---|---|---|
| .... | .... | | 0.078 | 0.188 | 0.020 |
| .... | .... | | 0.094 | 0.250 | 0.020 |
| .... | .... | | 0.125 | 0.312 | 0.032 |
| No. 6 | 0.138 | | 0.156 | 0.375 | 0.049 |
| No. 8 | 0.164 | | 0.188 | 0.438 | 0.049 |
| No. 10 | 0.190 | | 0.219 | 0.500 | 0.049 |
| 3/16 | 0.188 | | 0.250 | 0.562 | 0.049 |
| No. 12 | 0.216 | | 0.250 | 0.562 | 0.065 |
| 1/4 | 0.250 | N | 0.281 | 0.625 | 0.065 |
| 1/4 | 0.250 | W | 0.312 | 0.734 | 0.065 |
| 5/16 | 0.312 | N | 0.344 | 0.688 | 0.065 |
| 5/16 | 0.312 | W | 0.375 | 0.875 | 0.083 |
| 3/8 | 0.375 | N | 0.406 | 0.812 | 0.065 |
| 3/8 | 0.375 | W | 0.438 | 1.000 | 0.083 |
| 7/16 | 0.438 | N | 0.469 | 0.922 | 0.065 |
| 7/16 | 0.438 | W | 0.500 | 1.250 | 0.083 |
| 1/2 | 0.500 | N | 0.531 | 1.062 | 0.095 |
| 1/2 | 0.500 | W | 0.562 | 1.375 | 0.109 |
| 9/16 | 0.562 | N | 0.594 | 1.156 | 0.095 |
| 9/16 | 0.562 | W | 0.625 | 1.469 | 0.109 |
| 5/8 | 0.625 | N | 0.656 | 1.312 | 0.095 |
| 5/8 | 0.625 | W | 0.688 | 1.750 | 0.134 |
| 3/4 | 0.750 | N | 0.812 | 1.469 | 0.134 |
| 3/4 | 0.750 | W | 0.812 | 2.000 | 0.148 |
| 7/8 | 0.875 | N | 0.938 | 1.750 | 0.134 |
| 7/8 | 0.875 | W | 0.938 | 2.250 | 0.165 |
| 1 | 1.000 | N | 1.062 | 2.000 | 0.134 |
| 1 | 1.000 | W | 1.062 | 2.500 | 0.165 |
| 1 1/8 | 1.125 | N | 1.250 | 2.250 | 0.134 |
| 1 1/8 | 1.125 | W | 1.250 | 2.750 | 0.165 |
| 1 1/4 | 1.250 | N | 1.375 | 2.500 | 0.165 |
| 1 1/4 | 1.250 | W | 1.375 | 3.000 | 0.165 |
| 1 3/8 | 1.375 | N | 1.500 | 2.750 | 0.165 |
| 1 3/8 | 1.375 | W | 1.500 | 3.250 | 0.180 |
| 1 1/2 | 1.500 | N | 1.625 | 3.000 | 0.165 |
| 1 1/2 | 1.500 | W | 1.625 | 3.500 | 0.180 |
| 1 5/8 | 1.625 | | 1.750 | 3.750 | 0.180 |
| 1 3/4 | 1.750 | | 1.875 | 4.000 | 0.180 |
| 1 7/8 | 1.875 | | 2.000 | 4.250 | 0.180 |
| 2 | 2.000 | | 2.125 | 4.500 | 0.180 |
| 2 1/4 | 2.250 | | 2.375 | 4.750 | 0.220 |
| 2 1/2 | 2.500 | | 2.625 | 5.000 | 0.238 |
| 2 3/4 | 2.750 | | 2.875 | 5.250 | 0.259 |
| 3 | 3.000 | | 3.125 | 5.500 | 0.284 |

[a] From ANSI B27.2-1965. For complete listings, see the standard.
[b] Preferred sizes are for the most part from series previously designated "Standard Plate" and "SAE." Where common sizes existed in the two series, the SAE size is designated "N" (narrow) and the Standard Plate "W" (wide).
[c] Nominal washer sizes are intended for use with comparable nominal screw or bolt sizes.

## 22 American National Standard Lock Washers[a]

For parts lists, etc., give nominal size and series; for example, 1/4 REGULAR LOCK WASHER

PREFERRED SERIES

| Nominal Washer Size[b] | | Inside Diameter, Min. | Regular | | Extra Duty | | Hi-Collar | |
|---|---|---|---|---|---|---|---|---|
| | | | Outside Diameter, Max. | Thickness, Min. | Outside Diameter, Max. | Thickness, Min. | Outside Diameter, Max. | Thickness, Min. |
| No. 2 | 0.086 | 0.088 | 0.172 | 0.020 | 0.208 | 0.027 | .... | .... |
| No. 3 | 0.099 | 0.101 | 0.195 | 0.025 | 0.239 | 0.034 | .... | .... |
| No. 4 | 0.112 | 0.115 | 0.209 | 0.025 | 0.253 | 0.034 | 0.173 | 0.022 |
| No. 5 | 0.125 | 0.128 | 0.236 | 0.031 | 0.300 | 0.045 | 0.202 | 0.030 |
| No. 6 | 0.138 | 0.141 | 0.250 | 0.031 | 0.314 | 0.045 | 0.216 | 0.030 |
| No. 8 | 0.164 | 0.168 | 0.293 | 0.040 | 0.375 | 0.057 | 0.267 | 0.047 |
| No. 10 | 0.190 | 0.194 | 0.334 | 0.047 | 0.434 | 0.068 | 0.294 | 0.047 |
| No. 12 | 0.216 | 0.221 | 0.377 | 0.056 | 0.497 | 0.080 | .... | .... |
| 1/4 | 0.250 | 0.255 | 0.489 | 0.062 | 0.535 | 0.084 | 0.365 | 0.078 |
| 5/16 | 0.312 | 0.318 | 0.586 | 0.078 | 0.622 | 0.108 | 0.460 | 0.093 |
| 3/8 | 0.375 | 0.382 | 0.683 | 0.094 | 0.741 | 0.123 | 0.553 | 0.125 |
| 7/16 | 0.438 | 0.446 | 0.779 | 0.109 | 0.839 | 0.143 | 0.647 | 0.140 |
| 1/2 | 0.500 | 0.509 | 0.873 | 0.125 | 0.939 | 0.162 | 0.737 | 0.172 |
| 9/16 | 0.562 | 0.572 | 0.971 | 0.141 | 1.041 | 0.182 | .... | .... |
| 5/8 | 0.625 | 0.636 | 1.079 | 0.156 | 1.157 | 0.202 | 0.923 | 0.203 |
| 11/16 | 0.688 | 0.700 | 1.176 | 0.172 | 1.258 | 0.221 | ... | .... |
| 3/4 | 0.750 | 0.763 | 1.271 | 0.188 | 1.361 | 0.241 | 1.111 | 0.218 |
| 13/16 | 0.812 | 0.826 | 1.367 | 0.203 | 1.463 | 0.261 | .... | .... |
| 7/8 | 0.875 | 0.890 | 1.464 | 0.219 | 1.576 | 0.285 | 1.296 | 0.234 |
| 15/16 | 0.938 | 0.954 | 1.560 | 0.234 | 1.688 | 0.308 | .... | .... |
| 1 | 1.000 | 1.017 | 1.661 | 0.250 | 1.799 | 0.330 | 1.483 | 0.250 |
| 1 1/16 | 1.062 | 1.080 | 1.756 | 0.266 | 1.910 | 0.352 | .... | .... |
| 1 1/8 | 1.125 | 1.144 | 1.853 | 0.281 | 2.019 | 0.375 | 1.669 | 0.313 |
| 1 3/16 | 1.188 | 1.208 | 1.950 | 0.297 | 2.124 | 0.396 | .... | .... |
| 1 1/4 | 1.250 | 1.271 | 2.045 | 0.312 | 2.231 | 0.417 | 1.799 | 0.313 |
| 1 5/16 | 1.312 | 1.334 | 2.141 | 0.328 | 2.335 | 0.438 | .... | .... |
| 1 3/8 | 1.375 | 1.398 | 2.239 | 0.344 | 2.439 | 0.458 | 2.041 | 0.375 |
| 1 7/16 | 1.438 | 1.462 | 2.334 | 0.359 | 2.540 | 0.478 | .... | .... |
| 1 1/2 | 1.500 | 1.525 | 2.430 | 0.375 | 2.638 | 0.496 | 2.170 | 0.375 |

[a] From ANSI B27.1-1965. For complete listing, see the standard.
[b] Nominal washer sizes are intended for use with comparable nominal screw or bolt sizes.

# 23 Standards for Wire Gages[a]

Dimensions of sizes in decimal parts of an inch.[b]

| No. of Wire | American or Brown & Sharpe for Non-ferrous Metals | Birming-ham, or Stubs' Iron Wire[c] | American S. & W. Co.'s (Washburn & Moen) Std. Steel Wire | American S. & W. Co.'s Music Wire | Imperial Wire | Stubs' Steel Wire[c] | Steel Manu-facturers' Sheet Gage[b] | No. of Wire |
|---|---|---|---|---|---|---|---|---|
| 7-0's | .651354 | .... | .4900 | .... | .500 | .... | ..... | 7-0's |
| 6-0's | .580049 | .... | .4615 | .004 | .464 | .... | ..... | 6-0's |
| 5-0's | .516549 | .500 | .4305 | .005 | .432 | .... | ..... | 5-0's |
| 4-0's | .460 | .454 | .3938 | .006 | .400 | .... | ..... | 4-0's |
| 000 | .40964 | .425 | .3625 | .007 | .372 | .... | ..... | 000 |
| 00 | .3648 | .380 | .3310 | .008 | .348 | .... | ..... | 00 |
| 0 | .32486 | .340 | .3065 | .009 | .324 | .... | ..... | 0 |
| 1 | .2893 | .300 | .2830 | .010 | .300 | .227 | ..... | 1 |
| 2 | .25763 | .284 | .2625 | .011 | .276 | .219 | ..... | 2 |
| 3 | .22942 | .259 | .2437 | .012 | .252 | .212 | .2391 | 3 |
| 4 | .20431 | .238 | .2253 | .013 | .232 | .207 | .2242 | 4 |
| 5 | .18194 | .220 | .2070 | .014 | .212 | .204 | .2092 | 5 |
| 6 | .16202 | .203 | .1920 | .016 | .192 | .201 | .1943 | 6 |
| 7 | .14428 | .180 | .1770 | .018 | .176 | .199 | .1793 | 7 |
| 8 | .12849 | .165 | .1620 | .020 | .160 | .197 | .1644 | 8 |
| 9 | .11443 | .148 | .1483 | .022 | .144 | .194 | .1495 | 9 |
| 10 | .10189 | .134 | .1350 | .024 | .128 | .191 | .1345 | 10 |
| 11 | .090742 | .120 | .1205 | .026 | .116 | .188 | .1196 | 11 |
| 12 | .080808 | .109 | .1055 | .029 | .104 | .185 | .1046 | 12 |
| 13 | .071961 | .095 | .0915 | .031 | .092 | .182 | .0897 | 13 |
| 14 | .064084 | .083 | .0800 | .033 | .080 | .180 | .0747 | 14 |
| 15 | .057068 | .072 | .0720 | .035 | .072 | .178 | .0763 | 15 |
| 16 | .05082 | .065 | .0625 | .037 | .064 | .175 | .0598 | 16 |
| 17 | .045257 | .058 | .0540 | .039 | .056 | .172 | .0538 | 17 |
| 18 | .040303 | .049 | .0475 | .041 | .048 | .168 | .0478 | 18 |
| 19 | .03589 | .042 | .0410 | .043 | .040 | .164 | .0418 | 19 |
| 20 | .031961 | .035 | .0348 | .045 | .036 | .161 | .0359 | 20 |
| 21 | .028462 | .032 | .0317 | .047 | .032 | .157 | .0329 | 21 |
| 22 | .025347 | .028 | .0286 | .049 | .028 | .155 | .0299 | 22 |
| 23 | .022571 | .025 | .0258 | .051 | .024 | .153 | .0269 | 23 |
| 24 | .0201 | .022 | .0230 | .055 | .022 | .151 | .0239 | 24 |
| 25 | .0179 | .020 | .0204 | .059 | .020 | .148 | .0209 | 25 |
| 26 | .01594 | .018 | .0181 | .063 | .018 | .146 | .0179 | 26 |
| 27 | .014195 | .016 | .0173 | .067 | .0164 | .143 | .0164 | 27 |
| 28 | .012641 | .014 | .0162 | .071 | .0149 | .139 | .0149 | 28 |
| 29 | .011257 | .013 | .0150 | .075 | .0136 | .134 | .0135 | 29 |
| 30 | .010025 | .012 | .0140 | .080 | .0124 | .127 | .0120 | 30 |
| 31 | .008928 | .010 | .0132 | .085 | .0116 | .120 | .0105 | 31 |
| 32 | .00795 | .009 | .0128 | .090 | .0108 | .115 | .0097 | 32 |
| 33 | .00708 | .008 | .0118 | .095 | .0100 | .112 | .0090 | 33 |
| 34 | .006304 | .007 | .0104 | .... | .0092 | .110 | .0082 | 34 |
| 35 | .005614 | .005 | .0095 | .... | .0084 | .108 | .0075 | 35 |
| 36 | .005 | .004 | .0090 | .... | .0076 | .106 | .0067 | 36 |
| 37 | .004453 | .... | .0085 | .... | .0068 | .103 | .0064 | 37 |
| 38 | .003965 | .... | .0080 | .... | .0060 | .101 | .0060 | 38 |
| 39 | .003531 | .... | .0075 | .... | .0052 | .099 | ..... | 39 |
| 40 | .003144 | .... | .0070 | .... | .0048 | .097 | ..... | 40 |

[a] Courtesy Brown & Sharpe Mfg. Co.
[b] Now used by steel manufacturers in place of old U.S. Standard Gage.
[c] The difference between the Stubs' Iron Wire Gage and the Stubs' Steel Wire Gage should be noted, the first being commonly known as the English Standard Wire, or Birmingham Gage, which designates the Stubs' soft wire sizes, and the second being used in measuring drawn steel wire or drill rods of Stubs' make.

## 24 American National Standard Taper Pins[a]

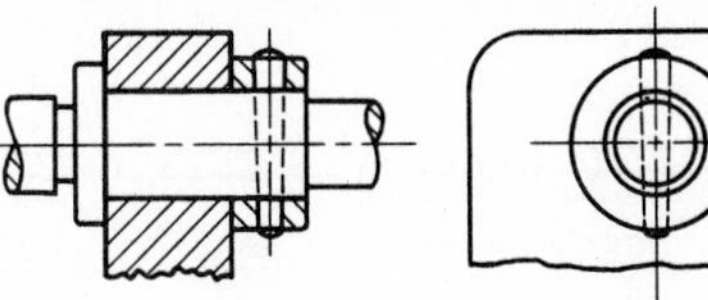

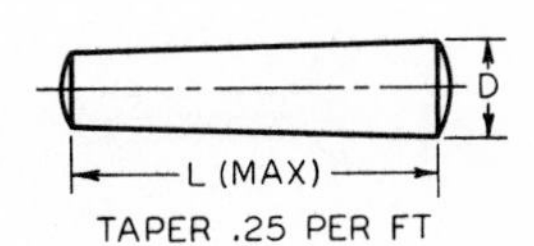


To find small diameter of pin, multiply the length by .02083 and subtract the result from the larger diameter.
All dimensions are given in inches.
Standard reamers are available for pins given above the heavy line.

| Number | 7/0 | 6/0 | 5/0 | 4/0 | 3/0 | 2/0 | 0 | 1 | 2 | 3 | 4 | 5 | 6 | 7 | 8 |
|---|---|---|---|---|---|---|---|---|---|---|---|---|---|---|---|
| Size (Large End) | .0625 | .0780 | .0940 | .1090 | .1250 | .1410 | .1560 | .1720 | .1930 | .2190 | .2500 | .2890 | .3410 | .4090 | .4920 |
| Shaft Diameter (Approx)[b] | | 7/32 | 1/4 | 5/16 | 3/8 | 7/16 | 1/2 | 9/16 | 5/8 | 3/4 | 13/16 | 7/8 | 1 | 1 1/4 | 1 1/2 |
| Drill Size (Before Reamer)[b] | | .0595 | .0785 | .0935 | .104 | .120 | .1405 | .1495 | .166 | .189 | .213 | 1/4 | 9/32 | 11/32 | 13/32 |
| Length, L | | | | | | | | | | | | | | | |
| .375 | X | X | | | | | | | | | | | | | |
| .500 | X | X | X | X | X | X | X | | | | | | | | |
| .625 | X | X | X | X | X | X | X | | | | | | | | |
| .750 | .... | X | X | X | X | X | X | X | X | X | | | | | |
| .875 | .... | .... | .... | .... | X | X | X | X | X | X | | | | | |
| 1.000 | .... | .... | X | X | X | X | X | X | X | X | X | X | | | |
| 1.250 | .... | .... | .... | .... | .... | X | X | X | X | X | X | X | X | | |
| 1.500 | .... | .... | .... | .... | .... | .... | X | X | X | X | X | X | X | | |
| 1.750 | .... | .... | .... | .... | .... | .... | .... | X | X | X | X | X | X | | |
| 2.000 | .... | .... | .... | .... | .... | .... | .... | X | X | X | X | X | X | X | X |
| 2.250 | .... | .... | .... | .... | .... | .... | .... | .... | X | X | X | X | X | X | X |
| 2.500 | .... | .... | .... | .... | .... | .... | .... | .... | X | X | X | X | X | X | X |
| 2.750 | .... | .... | .... | .... | .... | .... | .... | .... | .... | X | X | X | X | X | X |
| 3.000 | .... | .... | .... | .... | .... | .... | .... | .... | .... | X | X | X | X | X | X |
| 3.250 | .... | .... | .... | .... | .... | .... | .... | .... | .... | .... | .... | .... | X | X | X |
| 3.500 | .... | .... | .... | .... | .... | .... | .... | .... | .... | .... | .... | .... | X | X | X |
| 3.750 | .... | .... | .... | .... | .... | .... | .... | .... | .... | .... | .... | .... | X | X | X |
| 4.000 | .... | .... | .... | .... | .... | .... | .... | .... | .... | .... | .... | .... | X | X | X |
| 4.250 | .... | .... | .... | .... | .... | .... | .... | .... | .... | .... | .... | .... | .... | .... | X |
| 4.500 | .... | .... | .... | .... | .... | .... | .... | .... | .... | .... | .... | .... | .... | .... | X |

[a] ANSI B5.20–1958. For Nos. 9 and 10, see the standard. Pins Nos. 11 (size .8600), 12 (size 1.032), 13 (size 1.241), and 14 (size 1.523) are special sizes; hence their lengths are special.
[b] Suggested sizes; not American National Standard.

## 25 American National Standard Cotter Pins[a]

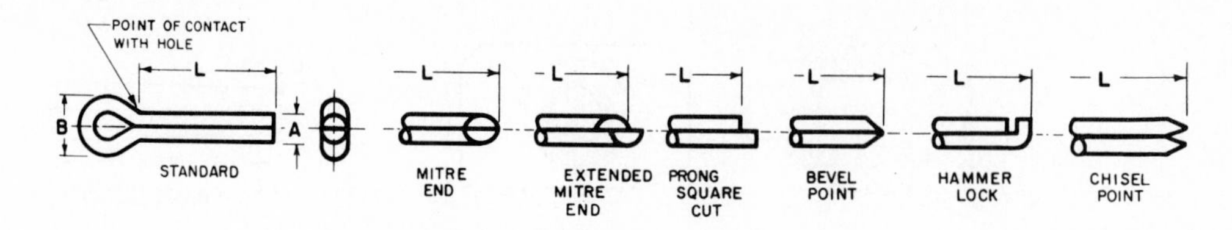


All dimensions are given in inches.

| Diameter, Nominal | Diameter A | | Outside Eye Diameter B, Min. | Hole Sizes Recom-mended | Diameter, Nominal | Diameter A | | Outside Eye Diameter B, Min. | Hole Sizes Recom-mended |
|---|---|---|---|---|---|---|---|---|---|
| | Max. | Min. | | | | Max. | Min. | | |
| 0.031 | 0.032 | 0.028 | 1/16 | 3/64 | 0.188 | 0.176 | 0.172 | 3/8 | 13/64 |
| 0.047 | 0.048 | 0.044 | 3/32 | 1/16 | 0.219 | 0.207 | 0.202 | 7/16 | 15/64 |
| 0.062 | 0.060 | 0.056 | 1/8 | 5/64 | 0.250 | 0.225 | 0.220 | 1/2 | 17/64 |
| 0.078 | 0.076 | 0.072 | 5/32 | 3/32 | 0.312 | 0.280 | 0.275 | 5/8 | 5/16 |
| 0.094 | 0.090 | 0.086 | 3/16 | 7/64 | 0.375 | 0.335 | 0.329 | 3/4 | 3/8 |
| 0.109 | 0.104 | 0.100 | 7/32 | 1/8 | 0.438 | 0.406 | 0.400 | 7/8 | 7/16 |
| 0.125 | 0.120 | 0.116 | 1/4 | 9/64 | 0.500 | 0.473 | 0.467 | 1 | 1/2 |
| 0.141 | 0.134 | 0.130 | 9/32 | 5/32 | 0.625 | 0.598 | 0.590 | 1 1/4 | 5/8 |
| 0.156 | 0.150 | 0.146 | 5/16 | 11/64 | 0.750 | 0.723 | 0.715 | 1 1/2 | 3/4 |

[a] ANSI B5.20–1958.

# 26 Metric Equivalents

## Length

| *U.S. to Metric* | *Metric to U.S.* |
| --- | --- |
| 1 inch = 2.540 centimeters | 1 millimeter = .039 inch |
| 1 foot = .304 meter | 1 centimeter = .394 inch |
| 1 yard = .914 meter | 1 meter = 3.281 feet or 1.094 yards |
| 1 mile = 1.609 kilometers | 1 kilometer = .621 mile |

## Area

| U.S. to Metric | Metric to U.S. |
| --- | --- |
| 1 $inch^2$ = 6.451 $centimeter^2$ | 1 $millimeter^2$ = .00155 $inch^2$ |
| 1 $foot^2$ = .093 $meter^2$ | 1 $centimeter^2$ = .155 $inch^2$ |
| 1 $yard^2$ = .836 $meter^2$ | 1 $meter^2$ = 10.764 $foot^2$ or 1.196 $yard^2$ |
| 1 $acre^2$ = 4,046.873 $meter^2$ | 1 $kilometer^2$ = .386 $mile^2$ or 247.04 $acre^2$ |

## Volume

| U.S. to Metric | Metric to U.S. |
| --- | --- |
| 1 $inch^3$ = 16.387 $centimeter^3$ | 1 $centimeter^3$ = .061 $inch^3$ |
| 1 $foot^3$ = .028 $meter^3$ | 1 $meter^3$ = 35.314 $foot^3$ or 1.308 $yard^3$ |
| 1 $yard^3$ = .764 $meter^3$ | 1 liter = .2642 gallons |
| 1 quart = .946 liter | 1 liter = 1.057 quarts |
| 1 gallon = .003785 $meter^3$ | 1 $meter^3$ = 264.02 gallons |

## Weight

| U.S. to Metric | Metric to U.S. |
| --- | --- |
| 1 ounce = 28.349 grams | 1 gram = .035 ounce |
| 1 pound = .454 kilogram | 1 kilogram = 2.205 pounds |
| 1 ton = .907 metric ton | 1 metric ton = 1.102 tons |

## Velocity

| U.S. to Metric | Metric to U.S. |
| --- | --- |
| 1 foot/second = .305 meter/second | 1 meter/second = 3.282 feet/second |
| 1 mile/hour = .447 meter/second | 1 kilometer/hour = .621 mile/second |

## Acceleration

| U.S. to Metric | Metric to U.S. |
| --- | --- |
| 1 inch/$second^2$ = .0254 meter/$second^2$ | 1 meter/$second^2$ = 3.278 feet/$second^2$ |
| 1 foot/$second^2$ = .305 meter/$second^2$ | |

# 27 American Welding Society Standard Welding Symbols

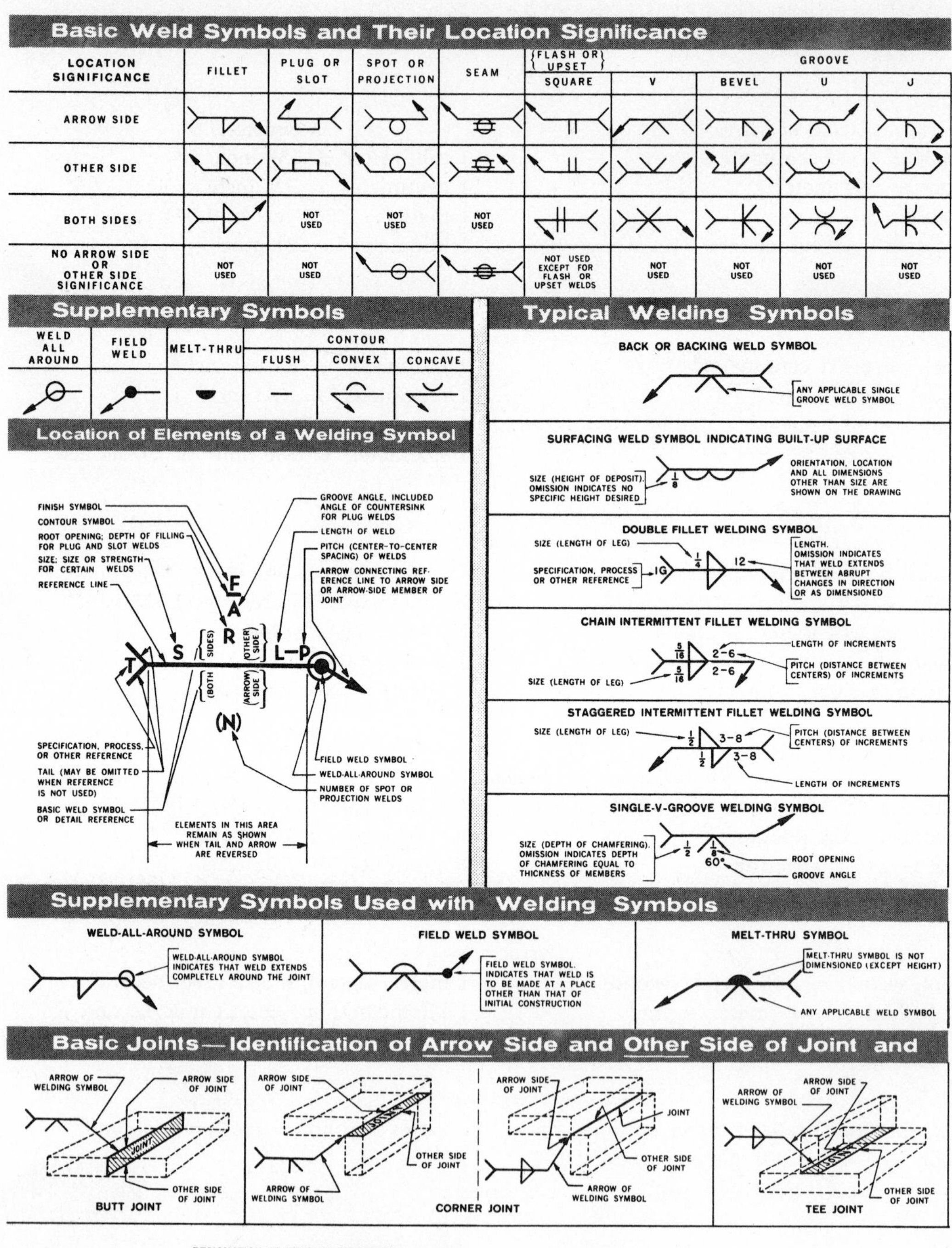


### DESIGNATION OF WELDING PROCESSES BY LETTERS

CAW .............. Carbon-Arc Welding
CW .............. Cold Welding
DB .............. Dip Brazing
DFW .............. Diffusion Welding
EBW .............. Electron Beam Welding
EW .............. Electroslag Welding
EXW .............. Explosion Welding
FB .............. Furnace Brazing
FCAW .............. Flux Cored Arc Welding
FOW .............. Forge Welding
FRW .............. Friction Welding

FW .............. Flash Welding
GMAW .............. Gas Metal-Arc Welding
GTAW .............. Gas Tungsten-Arc Welding
IB .............. Induction Brazing
IRB .............. Infrared Brazing
IW .............. Induction Welding
LBW .............. Laser Beam Welding
OAW .............. Oxyacetylene Welding
OHW .............. Oxyhydrogen Welding
PAW .............. Plasma-Arc Welding
PEW .............. Percussion Welding
PGW .............. Pressure Gas Welding

RB .............. Resistance Brazing
RPW .............. Projection Welding
RSEW .............. Resistance-Seam Welding
RSW .............. Resistance-Spot Welding
SAW .............. Submerged Arc Welding
SMAW .............. Shielded Metal-Arc Welding
SW .............. Stud Welding
TB .............. Torch Brazing
TW .............. Thermit Welding
USW .............. Ultrasonic Welding
UW .............. Upset Welding

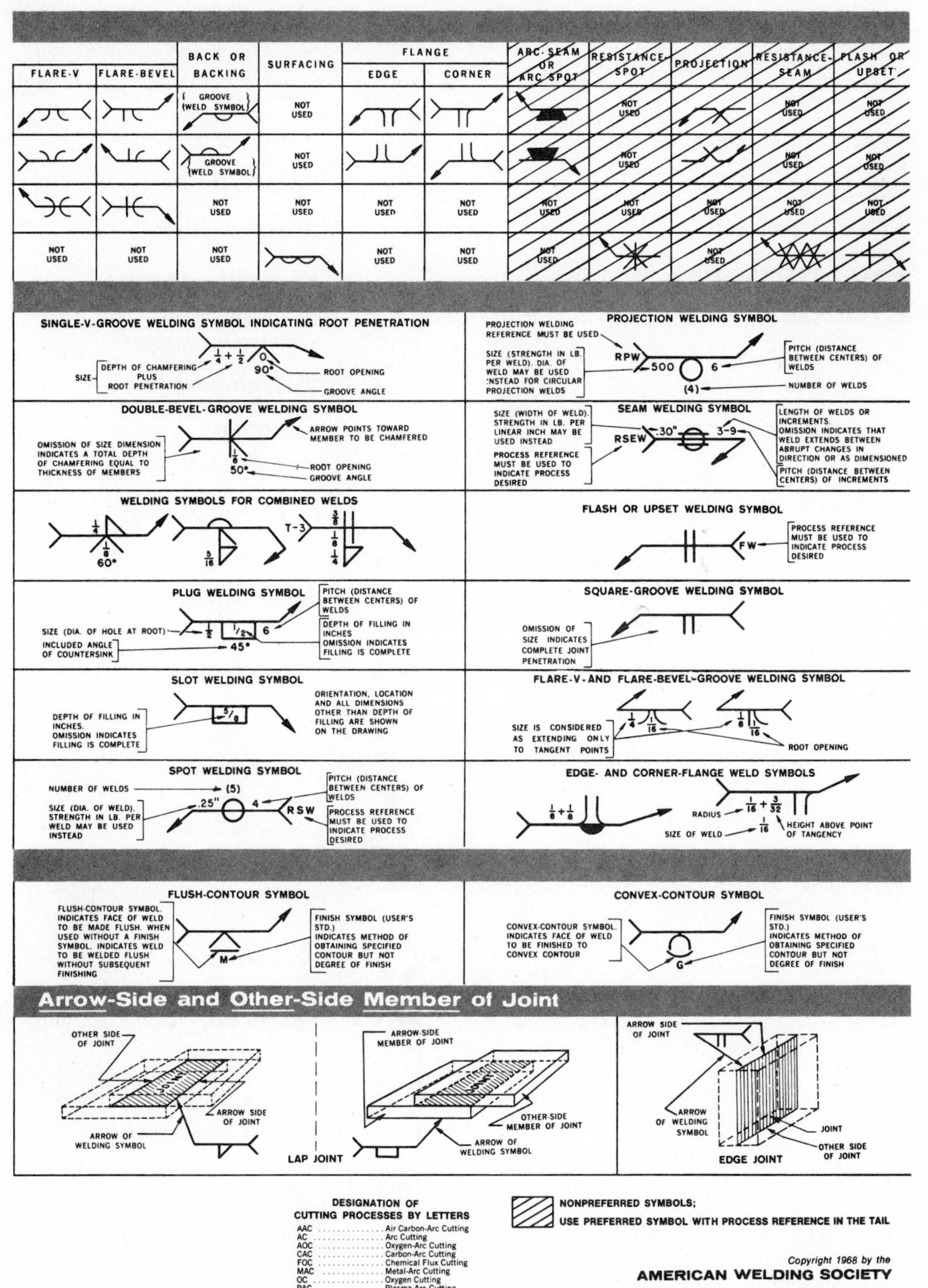


**DESIGNATION OF CUTTING PROCESSES BY LETTERS**

| | |
|---|---|
| AAC | Air Carbon-Arc Cutting |
| AC | Arc Cutting |
| AOC | Oxygen-Arc Cutting |
| CAC | Carbon-Arc Cutting |
| FOC | Chemical Flux Cutting |
| MAC | Metal-Arc Cutting |
| OC | Oxygen Cutting |
| PAC | Plasma-Arc Cutting |
| POC | Metal Powder Cutting |

NONPREFERRED SYMBOLS;
USE PREFERRED SYMBOL WITH PROCESS REFERENCE IN THE TAIL



*AWS A2.1-68*

# 28 Topographic Symbols

Highway

Railroad

Highway Bridge

Railroad Bridge

Drawbridges

Suspension Bridge

Dam

Telegraph or Telephone Line

Power-Transmission Line

Buildings in General

Capital

County Seat

Other Towns

Barbed Wire Fence

Smooth Wire Fence

Hedge

Oil or Gas Wells

Windmill

Tanks

Canal or Ditch

Canal Lock

Canal Lock (POINT UPSTREAM)

Aqueduct or Water Pipe

National or State Line

County Line

Township or District Line

City or Village Line

Triangulation Station

BM ✕ 1232 Bench Mark and Elevation

Any Location Station (WITH EXPLANATORY NOTE)

Streams in General

Lake or Pond

Falls and Rapids

Contours

Hachures

Sand and Sand Dunes

Marsh

Woodland of Any Kind

Orchard

Grassland in General

Cultivated Fields

Commercial or Municipal Field

Airplane Landing Field Marked or Emergency

Mooring Mast

Airway Light Beacon (ARROWS INDICATE COURSE LIGHTS)

Auxiliary Airway Light Beacon, Flashing

## 29 American National Standard Piping Symbols[a]

| | FLANGED | SCREWED | BELL & SPIGOT | WELDED | SOLDERED |
|---|---|---|---|---|---|
| 1. Joint | | | | | |
| 2. Elbow—90° | | | | | |
| 3. Elbow—45° | | | | | |
| 4. Elbow—Turned Up | | | | | |
| 5. Elbow—Turned Down | | | | | |
| 6. Elbow—Long Radius | LR | LR | | | |
| 7. Reducing Elbow | 2 4 | 2 4 | | | 2 4 |
| 8. Tee | | | | | |
| 9. Tee—Outlet Up | | | | | |
| 10. Tee—Outlet Down | | | | | |
| 11. Side Outlet Tee—Outlet Up | | | | | |
| 12. Cross | | | | | |
| 13. Reducer—Concentric | | | | | |
| 14. Reducer—Eccentric | | | | | |
| 15. Lateral | | | | | |
| 16. Gate Valve—Elev. | | | | | |
| 17. Globe Valve—Elev. | | | | | |
| 18. Check Valve | | | | | |
| 19. Stop Cock | | | | | |
| 20. Safety Valve | | | | | |
| 21. Expansion Joint | | | | | |
| 22. Union | | | | | |
| 23. Sleeve | | | | | |
| 24. Bushing | | | 6 4 | | |

[a] ANSI Z32.2.3–1949 (R 1953).

## 30 American National Standard Heating, Ventilating, and Ductwork Symbols[a]

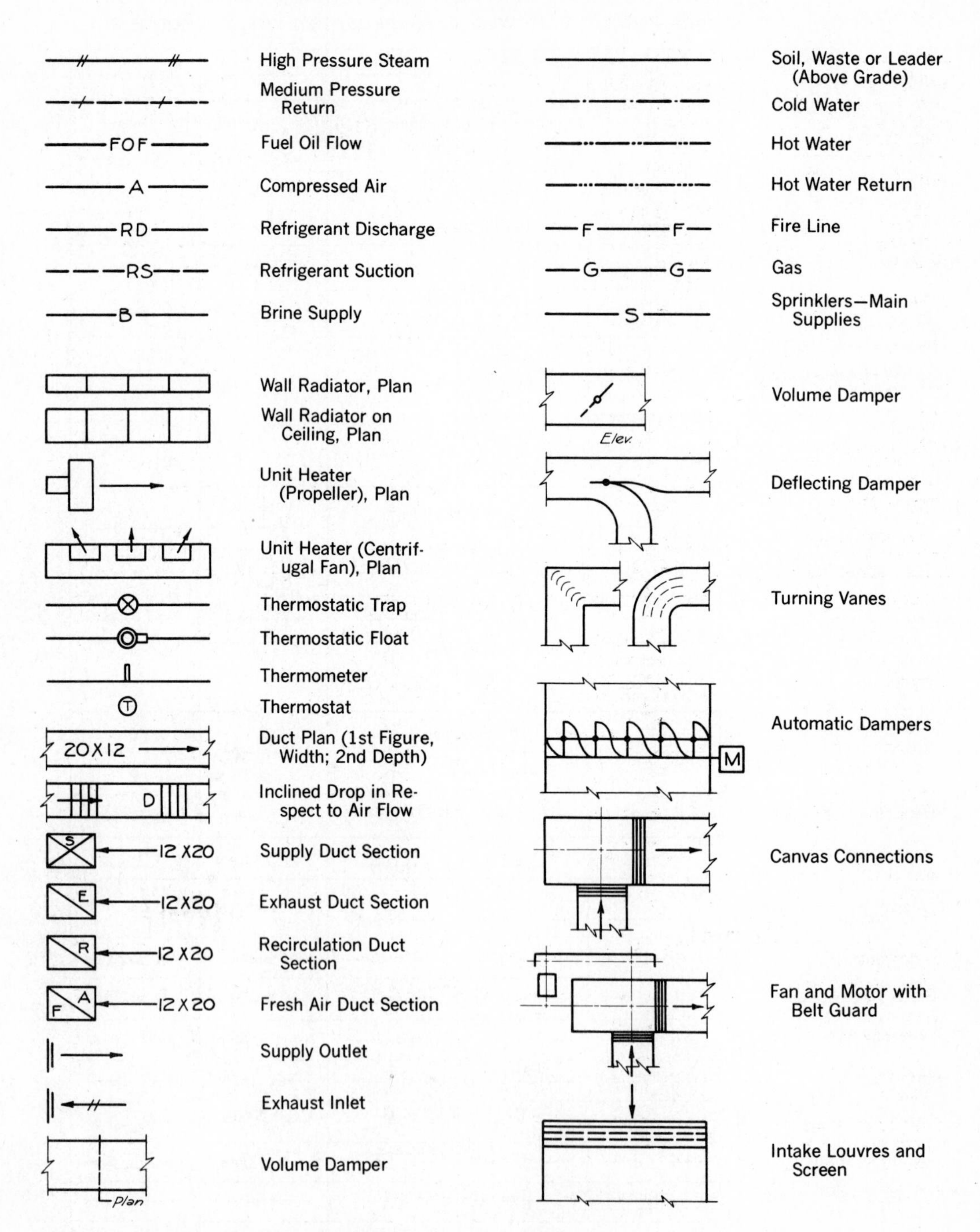


[a] ANSI Z32.2.3–1949 (R 1953) and ANSI Z32.2.4–1949 (R 1953).

# 31 American National Standard Graphical Symbols for Electronic Diagrams[a]

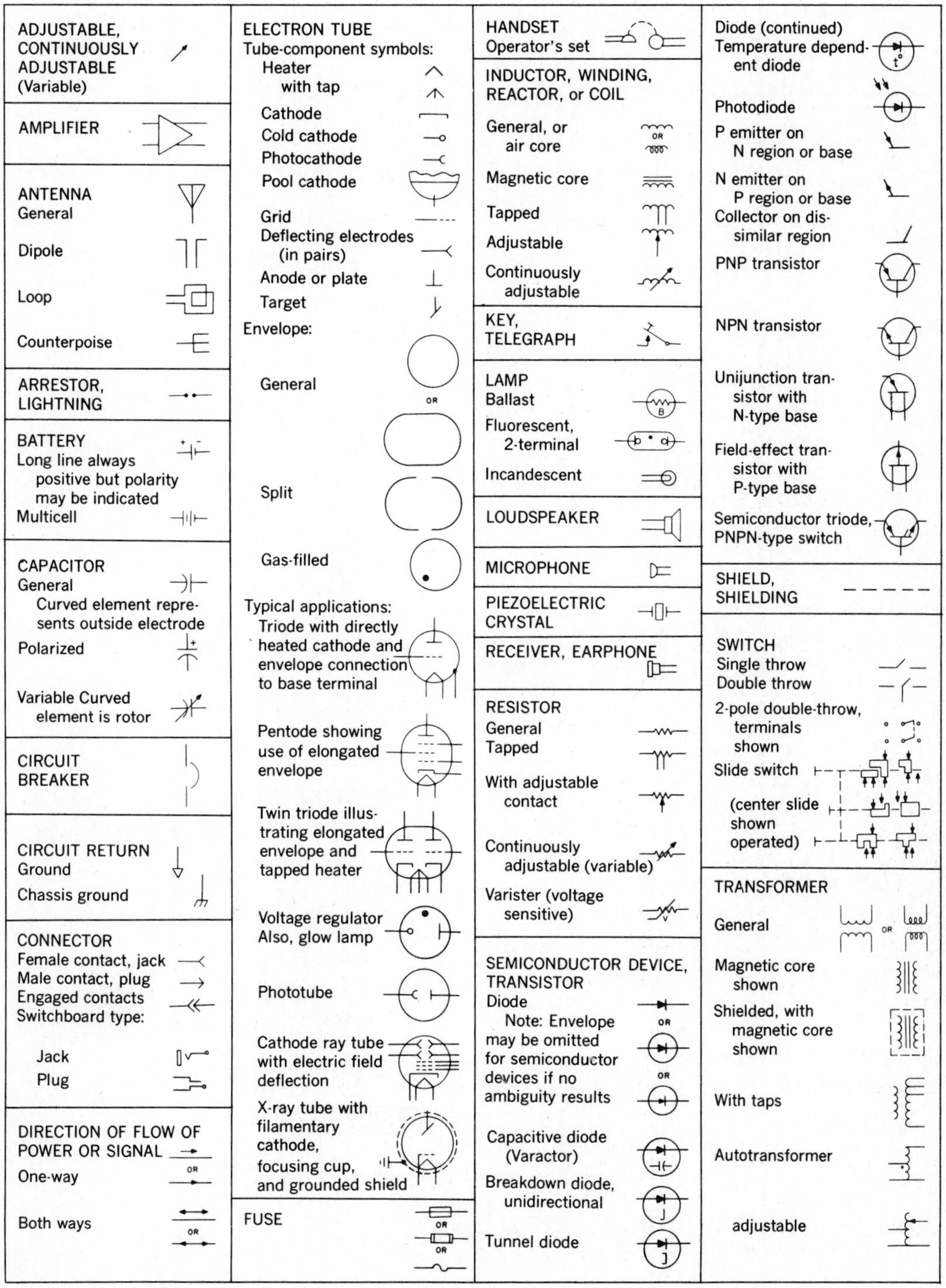

ADJUSTABLE, CONTINUOUSLY ADJUSTABLE (Variable)

AMPLIFIER

ANTENNA
General
Dipole
Loop
Counterpoise

ARRESTOR, LIGHTNING

BATTERY
Long line always positive but polarity may be indicated
Multicell

CAPACITOR
General
Curved element represents outside electrode
Polarized
Variable Curved element is rotor

CIRCUIT BREAKER

CIRCUIT RETURN
Ground
Chassis ground

CONNECTOR
Female contact, jack
Male contact, plug
Engaged contacts
Switchboard type:
Jack
Plug

DIRECTION OF FLOW OF POWER OR SIGNAL
One-way
OR
Both ways
OR

ELECTRON TUBE
Tube-component symbols:
Heater
with tap
Cathode
Cold cathode
Photocathode
Pool cathode
Grid
Deflecting electrodes (in pairs)
Anode or plate
Target
Envelope:
General
OR
Split
Gas-filled
Typical applications:
Triode with directly heated cathode and envelope connection to base terminal
Pentode showing use of elongated envelope
Twin triode illustrating elongated envelope and tapped heater
Voltage regulator
Also, glow lamp
Phototube
Cathode ray tube with electric field deflection
X-ray tube with filamentary cathode, focusing cup, and grounded shield

FUSE
OR
OR

HANDSET
Operator's set

INDUCTOR, WINDING, REACTOR, or COIL
General, or air core
OR
Magnetic core
Tapped
Adjustable
Continuously adjustable

KEY, TELEGRAPH

LAMP
Ballast
Fluorescent, 2-terminal
Incandescent

LOUDSPEAKER

MICROPHONE

PIEZOELECTRIC CRYSTAL

RECEIVER, EARPHONE

RESISTOR
General
Tapped
With adjustable contact
Continuously adjustable (variable)
Varister (voltage sensitive)

SEMICONDUCTOR DEVICE, TRANSISTOR
Diode
Note: Envelope may be omitted for semiconductor devices if no ambiguity results
OR
OR
Capacitive diode (Varactor)
Breakdown diode, unidirectional
Tunnel diode

Diode (continued)
Temperature dependent diode
Photodiode
P emitter on N region or base
N emitter on P region or base
Collector on dissimilar region
PNP transistor
NPN transistor
Unijunction transistor with N-type base
Field-effect transistor with P-type base
Semiconductor triode, PNPN-type switch

SHIELD, SHIELDING

SWITCH
Single throw
Double throw
2-pole double-throw, terminals shown
Slide switch
(center slide shown operated)

TRANSFORMER
General
OR
Magnetic core shown
Shielded, with magnetic core shown
With taps
Autotransformer
adjustable

[a] ANSI Y32.2–1967.

## 32 American National Standard Wrought Steel Pipe[a] and Taper Pipe Threads[b]

All dimensions are in inches except those in last two columns.

| Nominal Pipe Size | D Outside Diameter of Pipe | Threads per Inch | $L_1$[c] Normal Engagement by Hand Between External and Internal Threads | $L_2$[c] Length of Effective Thread | Nominal Wall Thickness | | | | | | | | | | Length of Pipe, Feet, per Square Foot External Surface[f] | Length of Standard-Weight Pipe, Feet, Containing 1 cu. ft.[f] |
|---|---|---|---|---|---|---|---|---|---|---|---|---|---|---|---|---|
| | | | | | Sched. 10 | Sched. 20[d] | Sched. 30[d] | Sched. 40[d] | Sched. 60[e] | Sched. 80[e] | Sched. 100 | Sched. 120 | Sched. 140 | Sched. 160 | | |
| ⅛ | 0.405 | 27 | 0.1615 | 0.2639 | .... | .... | .... | **0.068** | .... | **0.095** | .... | .... | .... | .... | 9.431 | 2,533.8 |
| ¼ | 0.540 | 18 | 0.2278 | 0.4018 | .... | .... | .... | **0.088** | .... | **0.119** | .... | .... | .... | .... | 7.073 | 1,383.8 |
| ⅜ | 0.675 | 18 | 0.240 | 0.4078 | .... | .... | .... | **0.091** | .... | **0.126** | .... | .... | .... | .... | 5.658 | 754.36 |
| ½ | 0.840 | 14 | 0.320 | 0.5337 | .... | .... | .... | **0.109** | .... | **0.147** | .... | .... | .... | 0.188 | 4.547 | 473.91 |
| ¾ | 1.050 | 14 | 0.339 | 0.5457 | .... | .... | .... | **0.113** | .... | **0.154** | .... | .... | .... | 0.219 | 3.637 | 270.03 |
| 1 | 1.315 | 11.5 | 0.400 | 0.6828 | .... | .... | .... | **0.133** | .... | **0.179** | .... | .... | .... | 0.250 | 2.904 | 166.62 |
| 1¼ | 1.660 | 11.5 | 0.420 | 0.7068 | .... | .... | .... | **0.140** | .... | **0.191** | .... | .... | .... | 0.250 | 2.301 | 96.275 |
| 1½ | 1.900 | 11.5 | 0.420 | 0.7235 | .... | .... | .... | **0.145** | .... | **0.200** | .... | .... | .... | 0.281 | 2.010 | 70.733 |
| 2 | 2.375 | 11.5 | 0.436 | 0.7565 | .... | .... | .... | **0.154** | .... | **0.218** | .... | .... | .... | 0.344 | 1.608 | 42.913 |
| 2½ | 2.875 | 8 | 0.682 | 1.1375 | .... | .... | .... | **0.203** | .... | **0.276** | .... | .... | .... | 0.375 | 1.328 | 30.077 |
| 3 | 3.500 | 8 | 0.766 | 1.2000 | .... | .... | .... | **0.216** | .... | **0.300** | .... | .... | .... | 0.438 | 1.091 | 19.479 |
| 3½ | 4.000 | 8 | 0.821 | 1.2500 | .... | .... | .... | **0.226** | .... | **0.318** | .... | .... | .... | .... | 0.954 | 14.565 |
| 4 | 4.500 | 8 | 0.844 | 1.3000 | .... | .... | .... | **0.237** | .... | **0.337** | .... | 0.438 | .... | 0.531 | 0.848 | 11.312 |
| 5 | 5.563 | 8 | 0.937 | 1.4063 | .... | .... | .... | **0.258** | .... | **0.375** | .... | 0.500 | .... | 0.625 | 0.686 | 7.199 |
| 6 | 6.625 | 8 | 0.958 | 1.5125 | .... | .... | .... | **0.280** | .... | **0.432** | .... | 0.562 | .... | 0.719 | 0.576 | 4.984 |
| 8 | 8.625 | 8 | 1.063 | 1.7125 | .... | 0.250 | 0.277 | **0.322** | 0.406 | 0.500 | 0.594 | 0.719 | 0.812 | 0.906 | 0.443 | 2.878 |
| 10 | 10.750 | 8 | 1.210 | 1.9250 | .... | 0.250 | 0.307 | **0.365** | **0.500** | 0.594 | 0.719 | 0.844 | 1.000 | 1.125 | 0.355 | 1.826 |
| 12 | 12.750 | 8 | 1.360 | 2.1250 | .... | 0.250 | 0.330 | 0.406 | **0.562** | 0.688 | 0.844 | 1.000 | 1.125 | 1.312 | 0.299 | 1.273 |
| 14 OD | 14.000 | 8 | 1.562 | 2.2500 | 0.250 | 0.312 | **0.375** | 0.438 | 0.594 | 0.750 | 0.938 | 1.094 | 1.250 | 1.406 | 0.273 | 1.065 |
| 16 OD | 16.000 | 8 | 1.812 | 2.4500 | 0.250 | 0.312 | **0.375** | **0.500** | 0.656 | 0.844 | 1.031 | 1.219 | 1.438 | 1.594 | 0.239 | 0.815 |
| 18 OD | 18.000 | 8 | 2.000 | 2.6500 | 0.250 | 0.312 | 0.438 | 0.562 | 0.750 | 0.938 | 1.156 | 1.375 | 1.562 | 1.781 | 0.212 | 0.644 |
| 20 OD | 20.000 | 8 | 2.125 | 2.8500 | 0.250 | **0.375** | 0.500 | 0.594 | 0.812 | 1.031 | 1.281 | 1.500 | 1.750 | 1.969 | 0.191 | 0.518 |
| 24 OD | 24.000 | 8 | 2.375 | 3.2500 | 0.250 | **0.375** | 0.562 | 0.688 | 0.969 | 1.219 | 1.531 | 1.812 | 2.062 | 2.344 | 0.159 | 0.358 |

[a] ANSI B36.10–1970.
[b] ANSI B2.1–1968.
[c] Refer to §13.22 and Fig. 13.21.
[d] Boldface figures correspond to "standard" pipe.
[e] Boldface figures correspond to "extra strong" pipe.
[f] Calculated values for Schedule 40 pipe.

## 33 American National Standard Thicknesses and Weights of Cast-Iron Pipe[a]

| Size, inches | Thickness, inches | Outside Diameter, inches | Weight (lb) Based on 16 ft Laying Length: Avg. per Foot[b] | Weight (lb) Based on 16 ft Laying Length: Per Length |
|---|---|---|---|---|
| **Class 50: 50 psi Pressure—115 ft Head** | | | | |
| 3 | 0.32 | 3.96 | 12.4 | 195 |
| 4 | 0.35 | 4.80 | 16.5 | 265 |
| 6 | 0.38 | 6.90 | 25.9 | 415 |
| 8 | 0.41 | 9.05 | 37.0 | 590 |
| 10 | 0.44 | 11.10 | 49.1 | 785 |
| 12 | 0.48 | 13.20 | 63.7 | 1,020 |
| 14 | 0.48 | 15.30 | 74.6 | 1,195 |
| 16 | 0.54 | 17.40 | 95.2 | 1,525 |
| 18 | 0.54 | 19.50 | 107.6 | 1,720 |
| 20 | 0.57 | 21.60 | 125.9 | 2,015 |
| 24 | 0.63 | 25.80 | 166.0 | 2,655 |
| 30 | 0.79 | 32.00 | 257.6 | 4,120 |
| 36 | 0.87 | 38.30 | 340.9 | 5,455 |
| 42 | 0.97 | 44.50 | 442.0 | 7,070 |
| 48 | 1.06 | 50.80 | 551.6 | 8,825 |
| **Class 100: 100 psi Pressure—231 ft Head** | | | | |
| 3 | 0.32 | 3.96 | 12.4 | 195 |
| 4 | 0.35 | 4.80 | 16.5 | 265 |
| 6 | 0.38 | 6.90 | 25.9 | 415 |
| 8 | 0.41 | 9.05 | 37.0 | 590 |
| 10 | 0.44 | 11.10 | 49.1 | 785 |
| 12 | 0.48 | 13.20 | 63.7 | 1,020 |
| 14 | 0.51 | 15.30 | 78.8 | 1,260 |
| 16 | 0.54 | 17.40 | 95.2 | 1,525 |
| 18 | 0.58 | 19.50 | 114.8 | 1,835 |
| 20 | 0.62 | 21.60 | 135.9 | 2,175 |
| 24 | 0.68 | 25.80 | 178.1 | 2,850 |
| 30 | 0.79 | 32.00 | 257.6 | 4,120 |
| 36 | 0.87 | 38.30 | 340.9 | 5,455 |
| 42 | 0.97 | 44.50 | 442.0 | 7,070 |
| 48 | 1.06 | 50.80 | 551.6 | 8,825 |
| **Class 150: 150 psi Pressure—346 ft Head** | | | | |
| 3 | 0.32 | 3.96 | 12.4 | 195 |
| 4 | 0.35 | 4.80 | 16.5 | 265 |
| 6 | 0.38 | 6.90 | 25.9 | 415 |
| 8 | 0.41 | 9.05 | 37.0 | 590 |
| 10 | 0.44 | 11.10 | 49.1 | 785 |
| 12 | 0.48 | 13.20 | 63.7 | 1,020 |
| 14 | 0.51 | 15.30 | 78.8 | 1,260 |
| 16 | 0.54 | 17.40 | 95.2 | 1,525 |
| 18 | 0.58 | 19.50 | 114.8 | 1,835 |
| 20 | 0.62 | 21.60 | 135.9 | 2,175 |
| 24 | 0.73 | 25.80 | 190.1 | 3,040 |
| 30 | 0.85 | 32.00 | 275.4 | 4,405 |
| 36 | 0.94 | 38.30 | 365.9 | 5,855 |
| 42 | 1.05 | 44.50 | 475.3 | 7,605 |
| 48 | 1.14 | 50.80 | 589.6 | 9,435 |
| **Class 200: 200 psi Pressure—462 ft Head** | | | | |
| 3 | 0.32 | 3.96 | 12.4 | 195 |
| 4 | 0.35 | 4.80 | 16.5 | 265 |
| 6 | 0.38 | 6.90 | 25.9 | 415 |
| **Class 200: 200 psi Pressure—462 ft Head (*cont'd*)** | | | | |
| 8 | 0.41 | 9.05 | 37.0 | 590 |
| 10 | 0.44 | 11.10 | 49.1 | 785 |
| 12 | 0.48 | 13.20 | 63.7 | 1,020 |
| 14 | 0.55 | 15.30 | 84.4 | 1,350 |
| 16 | 0.58 | 17.40 | 101.6 | 1,625 |
| 18 | 0.63 | 19.50 | 123.7 | 1,980 |
| 20 | 0.67 | 21.60 | 145.9 | 2,335 |
| 24 | 0.79 | 25.80 | 205.6 | 3,290 |
| 30 | 0.92 | 32.00 | 297.8 | 4,765 |
| 36 | 1.02 | 38.30 | 397.1 | 6,355 |
| 42 | 1.13 | 44.50 | 512.3 | 8,195 |
| 48 | 1.23 | 50.80 | 637.2 | 10,195 |
| **Class 250: 250 psi Pressure—577 ft Head** | | | | |
| 3 | 0.32 | 3.96 | 12.4 | 195 |
| 4 | 0.35 | 4.80 | 16.5 | 265 |
| 6 | 0.38 | 6.90 | 25.9 | 415 |
| 8 | 0.41 | 9.05 | 37.0 | 590 |
| 10 | 0.44 | 11.10 | 49.1 | 785 |
| 12 | 0.52 | 13.20 | 68.5 | 1,095 |
| 14 | 0.59 | 15.30 | 90.6 | 1,450 |
| 16 | 0.63 | 17.40 | 110.4 | 1,765 |
| 18 | 0.68 | 19.50 | 133.4 | 2,135 |
| 20 | 0.72 | 21.60 | 156.7 | 2,505 |
| 24 | 0.79 | 25.80 | 205.6 | 3,290 |
| 30 | 0.99 | 32.00 | 318.4 | 5,095 |
| 36 | 1.10 | 38.30 | 425.5 | 6,810 |
| 42 | 1.22 | 44.50 | 549.5 | 8,790 |
| 48 | 1.33 | 50.80 | 684.5 | 10,950 |
| **Class 300: 300 psi Pressure—693 ft Head** | | | | |
| 3 | 0.32 | 3.96 | 12.4 | 195 |
| 4 | 0.35 | 4.80 | 16.5 | 265 |
| 6 | 0.38 | 6.90 | 25.9 | 415 |
| 8 | 0.41 | 9.05 | 37.0 | 590 |
| 10 | 0.48 | 11.10 | 53.1 | 850 |
| 12 | 0.52 | 13.20 | 68.5 | 1,095 |
| 14 | 0.59 | 15.30 | 90.6 | 1,450 |
| 16 | 0.68 | 17.40 | 118.2 | 1,890 |
| 18 | 0.73 | 19.50 | 142.3 | 2,275 |
| 20 | 0.78 | 21.60 | 168.5 | 2,695 |
| 24 | 0.85 | 25.80 | 219.8 | 3,515 |
| **Class 350: 350 psi Pressure—808 ft Head** | | | | |
| 3 | 0.32 | 3.96 | 12.4 | 195 |
| 4 | 0.35 | 4.80 | 16.5 | 265 |
| 6 | 0.38 | 6.90 | 25.9 | 415 |
| 8 | 0.41 | 9.05 | 37.0 | 590 |
| 10 | 0.52 | 11.10 | 57.4 | 920 |
| 12 | 0.56 | 13.20 | 73.8 | 1,180 |
| 14 | 0.64 | 15.30 | 97.5 | 1,605 |
| 16 | 0.68 | 17.40 | 118.2 | 1,945 |
| 18 | 0.79 | 19.50 | 152.9 | 2,520 |
| 20 | 0.84 | 21.60 | 180.2 | 2,970 |
| 24 | 0.92 | 25.80 | 236.3 | 3,895 |

[a] ANSI A21.8-1962.
[b] Average weight per foot based on calculated weight of pipe before rounding.

# 34 American National Standard 125-lb Cast-Iron Screwed Fittings[a]

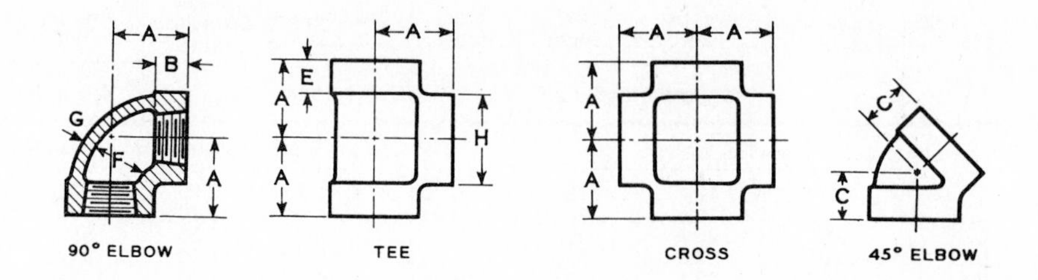


DIMENSIONS OF 90° AND 45° ELBOWS, TEES, AND CROSSES (STRAIGHT SIZES)

All dimensions given in inches.
Fittings having right- and left-hand threads shall have four or more ribs or the letter "L" cast on the band at end with left-hand thread.

| Nominal Pipe Size | Center to End, Elbows, Tees, and Crosses A | Center to End, 45° Elbows C | Length of Thread, Min. B | Width of Band, Min. E | Inside Diameter of Fitting F | | Metal Thick-ness G | Diameter of Band, Min. H |
|---|---|---|---|---|---|---|---|---|
| | | | | | Max. | Min. | | |
| ¼ | 0.81 | 0.73 | 0.32 | 0.38 | 0.584 | 0.540 | 0.110 | 0.93 |
| ⅜ | 0.95 | 0.80 | 0.36 | 0.44 | 0.719 | 0.675 | 0.120 | 1.12 |
| ½ | 1.12 | 0.88 | 0.43 | 0.50 | 0.897 | 0.840 | 0.130 | 1.34 |
| ¾ | 1.31 | 0.98 | 0.50 | 0.56 | 1.107 | 1.050 | 0.155 | 1.63 |
| 1 | 1.50 | 1.12 | 0.58 | 0.62 | 1.385 | 1.315 | 0.170 | 1.95 |
| 1¼ | 1.75 | 1.29 | 0.67 | 0.69 | 1.730 | 1.660 | 0.185 | 2.39 |
| 1½ | 1.94 | 1.43 | 0.70 | 0.75 | 1.970 | 1.900 | 0.200 | 2.68 |
| 2 | 2.25 | 1.68 | 0.75 | 0.84 | 2.445 | 2.375 | 0.220 | 3.28 |
| 2½ | 2.70 | 1.95 | 0.92 | 0.94 | 2.975 | 2.875 | 0.240 | 3.86 |
| 3 | 3.08 | 2.17 | 0.98 | 1.00 | 3.600 | 3.500 | 0.260 | 4.62 |
| 3½ | 3.42 | 2.39 | 1.03 | 1.06 | 4.100 | 4.000 | 0.280 | 5.20 |
| 4 | 3.79 | 2.61 | 1.08 | 1.12 | 4.600 | 4.500 | 0.310 | 5.79 |
| 5 | 4.50 | 3.05 | 1.18 | 1.18 | 5.663 | 5.563 | 0.380 | 7.05 |
| 6 | 5.13 | 3.46 | 1.28 | 1.28 | 6.725 | 6.625 | 0.430 | 8.28 |
| 8 | 6.56 | 4.28 | 1.47 | 1.47 | 8.725 | 8.625 | 0.550 | 10.63 |
| 10 | 8.08[b] | 5.16 | 1.68 | 1.68 | 10.850 | 10.750 | 0.690 | 13.12 |
| 12 | 9.50[b] | 5.97 | 1.88 | 1.88 | 12.850 | 12.750 | 0.800 | 15.47 |

[a] From ANSI B16.4-1963.
[b] This applies to elbows and tees only.

# 35 American National Standard 250-lb. Cast-Iron Screwed Fittings[a]

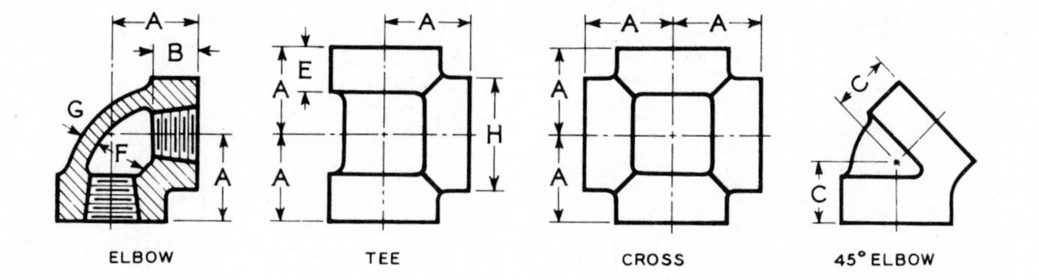


DIMENSIONS OF 90° AND 45° ELBOWS, TEES, AND CROSSES (STRAIGHT SIZES)

All dimensions given in inches.
The 250 lb standard for screwed fittings covers only the straight sizes of 90° and 45° elbows, tees, and crosses.

| Nominal Pipe Size | Center to End, Elbows, Tees, and Crosses A | Center to End, 45° Elbows C | Length of Thread, Min. B | Width of Band, Min. E | Inside Diameter of Fitting F | | Metal Thick-ness G | Outside Diameter of Band, Min. H |
|---|---|---|---|---|---|---|---|---|
| | | | | | Max. | Min. | | |
| ¼ | 0.94 | 0.81 | 0.43 | 0.49 | 0.584 | 0.540 | 0.18 | 1.17 |
| ⅜ | 1.06 | 0.88 | 0.47 | 0.55 | 0.719 | 0.675 | 0.18 | 1.36 |
| ½ | 1.25 | 1.00 | 0.57 | 0.60 | 0.897 | 0.840 | 0.20 | 1.59 |
| ¾ | 1.44 | 1.13 | 0.64 | 0.68 | 1.107 | 1.050 | 0.23 | 1.88 |
| 1 | 1.63 | 1.31 | 0.75 | 0.76 | 1.385 | 1.315 | 0.28 | 2.24 |
| 1¼ | 1.94 | 1.50 | 0.84 | 0.88 | 1.730 | 1.660 | 0.33 | 2.73 |
| 1½ | 2.13 | 1.69 | 0.87 | 0.97 | 1.970 | 1.900 | 0.35 | 3.07 |
| 2 | 2.50 | 2.00 | 1.00 | 1.12 | 2.445 | 2.375 | 0.39 | 3.74 |
| 2½ | 2.94 | 2.25 | 1.17 | 1.30 | 2.975 | 2.875 | 0.43 | 4.60 |
| 3 | 3.38 | 2.50 | 1.23 | 1.40 | 3.600 | 3.500 | 0.48 | 5.36 |
| 3½ | 3.75 | 2.63 | 1.28 | 1.49 | 4.100 | 4.000 | 0.52 | 5.98 |
| 4 | 4.13 | 2.81 | 1.33 | 1.57 | 4.600 | 4.500 | 0.56 | 6.61 |
| 5 | 4.88 | 3.19 | 1.43 | 1.74 | 5.663 | 5.563 | 0.66 | 7.92 |
| 6 | 5.63 | 3.50 | 1.53 | 1.91 | 6.725 | 6.625 | 0.74 | 9.24 |
| 8 | 7.00 | 4.31 | 1.72 | 2.24 | 8.725 | 8.625 | 0.90 | 11.73 |
| 10 | 8.63 | 5.19 | 1.93 | 2.58 | 10.850 | 10.750 | 1.08 | 14.37 |
| 12 | 10.00 | 6.00 | 2.13 | 2.91 | 12.850 | 12.750 | 1.24 | 16.84 |

[a] From ANSI B16.4–1963.

# 36 American National Standard Class 125 Cast-Iron Flanges and Fittings[a]

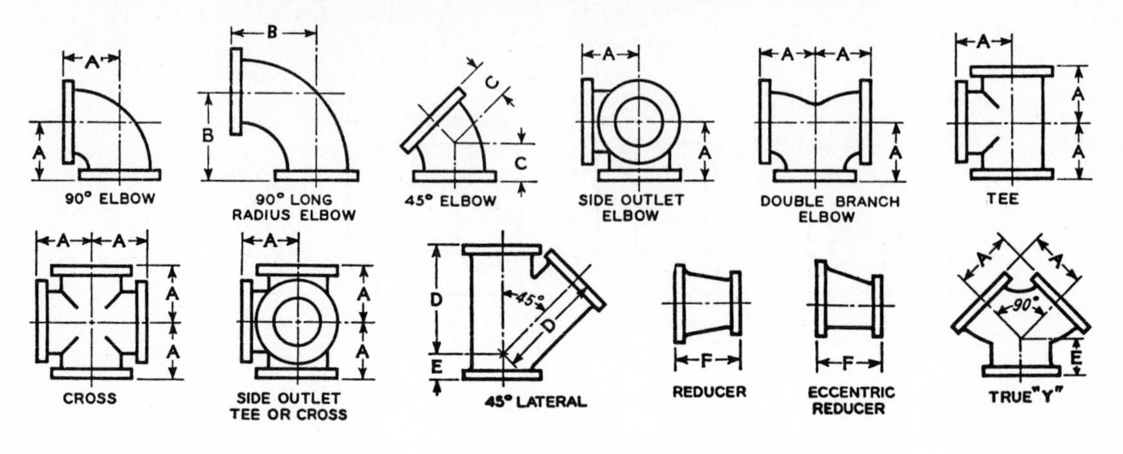


DIMENSIONS OF ELBOWS, DOUBLE BRANCH ELBOWS, TEES, CROSSES, LATERALS, TRUE Y's (STRAIGHT SIZES), AND REDUCERS

All dimensions in inches.

| Nominal Pipe Size | Inside Diameter of Fittings | Center to Face 90° Elbow, Tees, Crosses True "Y" and Double Branch Elbow A | Center to Face, 90° Long Radius Elbow B | Center to Face 45° Elbow C | Center to Face Lateral D | Short Center to Face True "Y" and Lateral E | Face to Face Reducer F | Diameter of Flange | Thickness of Flange, Min. | Wall Thickness |
|---|---|---|---|---|---|---|---|---|---|---|
| 1 | 1 | 3½ | 5 | 1¾ | 5¾ | 1¾ | .... | 4¼ | 7/16 | 5/16 |
| 1¼ | 1¼ | 3¾ | 5½ | 2 | 6¼ | 1¾ | .... | 4⅝ | ½ | 5/16 |
| 1½ | 1½ | 4 | 6 | 2¼ | 7 | 2 | .... | 5 | 9/16 | 5/16 |
| 2 | 2 | 4½ | 6½ | 2½ | 8 | 2½ | 5 | 6 | ⅝ | 5/16 |
| 2½ | 2½ | 5 | 7 | 3 | 9½ | 2½ | 5½ | 7 | 11/16 | 5/16 |
| 3 | 3 | 5½ | 7¾ | 3 | 10 | 3 | 6 | 7½ | ¾ | ⅜ |
| 3½ | 3½ | 6 | 8½ | 3½ | 11½ | 3 | 6½ | 8½ | 13/16 | 7/16 |
| 4 | 4 | 6½ | 9 | 4 | 12 | 3 | 7 | 9 | 15/16 | ½ |
| 5 | 5 | 7½ | 10¼ | 4½ | 13½ | 3½ | 8 | 10 | 15/16 | ½ |
| 6 | 6 | 8 | 11½ | 5 | 14½ | 3½ | 9 | 11 | 1 | 9/16 |
| 8 | 8 | 9 | 14 | 5½ | 17½ | 4½ | 11 | 13½ | 1⅛ | ⅝ |
| 10 | 10 | 11 | 16½ | 6½ | 20½ | 5 | 12 | 16 | 1 3/16 | ¾ |
| 12 | 12 | 12 | 19 | 7½ | 24½ | 5½ | 14 | 19 | 1¼ | 13/16 |
| 14 OD | 14 | 14 | 21½ | 7½ | 27 | 6 | 16 | 21 | 1⅜ | ⅞ |
| 16 OD | 16 | 15 | 24 | 8 | 30 | 6½ | 18 | 23½ | 1 7/16 | 1 |
| 18 OD | 18 | 16½ | 26½ | 8½ | 32 | 7 | 19 | 25 | 1 9/16 | 1 1/16 |
| 20 OD | 20 | 18 | 29 | 9½ | 35 | 8 | 20 | 27½ | 1 11/16 | 1⅛ |
| 24 OD | 24 | 22 | 34 | 11 | 40½ | 9 | 24 | 32 | 1⅞ | 1¼ |
| 30 OD | 30 | 25 | 41½ | 15 | 49 | 10 | 30 | 38¾ | 2⅛ | 1 7/16 |
| 36 OD | 36 | 28 | 49 | 18 | .... | .... | 36 | 46 | 2⅜ | 1⅝ |
| 42 OD | 42 | 31 | 56½ | 21 | .... | .... | 42 | 53 | 2⅝ | 1 13/16 |
| 48 OD | 48 | 34 | 64 | 24 | .... | .... | 48 | 59½ | 2¾ | 2 |

[a] ANSI B16.1-1967.

## 37 American National Standard Class 125 Cast-Iron Flanges, Drilling for Bolts and Their Lengths[a]

| Nominal Pipe Size | Diameter of Flange | Thickness of Flange, Min. | Diameter of Bolt Circle | Number of Bolts | Diameter of Bolts | Diameter of Bolt Holes | Length of Bolts |
|---|---|---|---|---|---|---|---|
| 1 | 4¼ | 7/16 | 3⅛ | 4 | ½ | ⅝ | 1¾ |
| 1¼ | 4⅝ | ½ | 3½ | 4 | ½ | ⅝ | 2 |
| 1½ | 5 | 9/16 | 3⅞ | 4 | ½ | ⅝ | 2 |
| 2 | 6 | ⅝ | 4¾ | 4 | ⅝ | ¾ | 2¼ |
| 2½ | 7 | 11/16 | 5½ | 4 | ⅝ | ¾ | 2½ |
| 3 | 7½ | ¾ | 6 | 4 | ⅝ | ¾ | 2½ |
| 3½ | 8½ | 13/16 | 7 | 8 | ⅝ | ¾ | 2¾ |
| 4 | 9 | 15/16 | 7½ | 8 | ⅝ | ¾ | 3 |
| 5 | 10 | 15/16 | 8½ | 8 | ¾ | ⅞ | 3 |
| 6 | 11 | 1 | 9½ | 8 | ¾ | ⅞ | 3¼ |
| 8 | 13½ | 1⅛ | 11¾ | 8 | ¾ | ⅞ | 3½ |
| 10 | 16 | 1 3/16 | 14¼ | 12 | ⅞ | 1 | 3¾ |
| 12 | 19 | 1¼ | 17 | 12 | ⅞ | 1 | 3¾ |
| 14 OD | 21 | 1⅜ | 18¾ | 12 | 1 | 1⅛ | 4¼ |
| 16 OD | 23½ | 1 7/16 | 21¼ | 16 | 1 | 1⅛ | 4½ |
| 18 OD | 25 | 1 9/16 | 22¾ | 16 | 1⅛ | 1¼ | 4¾ |
| 20 OD | 27½ | 1 11/16 | 25 | 20 | 1⅛ | 1¼ | 5 |
| 24 OD | 32 | 1⅞ | 29½ | 20 | 1¼ | 1⅜ | 5½ |
| 30 OD | 38¾ | 2⅛ | 36 | 28 | 1¼ | 1⅜ | 6¼ |
| 36 OD | 46 | 2⅜ | 42¾ | 32 | 1½ | 1⅝ | 7 |
| 42 OD | 53 | 2⅝ | 49½ | 36 | 1½ | 1⅝ | 7½ |
| 48 OD | 59½ | 2¾ | 56 | 44 | 1½ | 1⅝ | 7¾ |

[a] ANSI B16.1-1967.

## 38 Shaft Center Sizes

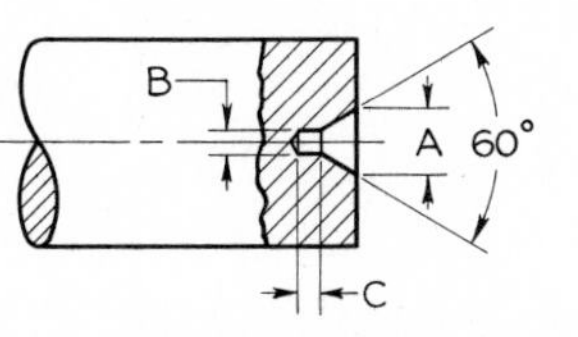


| Shaft Diameter D | A | B | C | Shaft Diameter D | A | B | C |
|---|---|---|---|---|---|---|---|
| 3/16 to 7/32 | 5/64 | 3/64 | 1/16 | 1⅛ to 1 15/32 | 5/16 | 5/32 | 5/32 |
| ¼ to 11/32 | 3/32 | 3/64 | 1/16 | 1½ to 1 31/32 | ⅜ | 3/32 | 5/32 |
| ⅜ to 17/32 | ⅛ | 1/16 | 5/64 | 2 to 2 31/32 | 7/16 | 7/32 | 3/16 |
| 9/16 to 25/32 | 3/16 | 5/64 | 3/32 | 3 to 3 31/32 | ½ | 7/32 | 7/32 |
| 13/16 to 1 3/32 | ¼ | 3/32 | 3/32 | 4 and over | 9/16 | 7/32 | 7/32 |

# 39 American National Standard Class 250 Cast-Iron Flanges and Fittings[a]

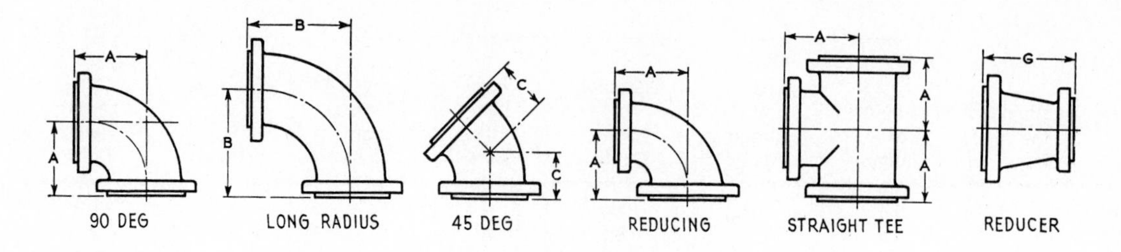


DIMENSIONS OF ELBOWS, TEES, AND REDUCERS

All dimensions are given in inches.

| Nominal Pipe Size | Inside Diameter of Fitting, Min. | Wall Thickness of Body | Diameter of Flange | Thickness of Flange, Min. | Diameter of Raised Face | Center-to-Face Elbow and Tee A | Center-to-Face Long Radius Elbow B | Center-to-Face 45° Elbow C | Face-to-Face Reducer G |
|---|---|---|---|---|---|---|---|---|---|
| 2 | 2 | 7/16 | 6½ | 7/8 | 4 3/16 | 5 | 6½ | 3 | 5 |
| 2½ | 2½ | ½ | 7½ | 1 | 4 15/16 | 5½ | 7 | 3½ | 5½ |
| 3 | 3 | 9/16 | 8¼ | 1 1/8 | 5 11/16 | 6 | 7¾ | 3½ | 6 |
| 3½ | 3½ | 9/16 | 9 | 1 3/16 | 6 5/16 | 6½ | 8½ | 4 | 6½ |
| 4 | 4 | 5/8 | 10 | 1¼ | 6 15/16 | 7 | 9 | 4½ | 7 |
| 5 | 5 | 11/16 | 11 | 1 3/8 | 8 5/16 | 8 | 10¼ | 5 | 8 |
| 6 | 6 | ¾ | 12½ | 1 7/16 | 9 11/16 | 8½ | 11½ | 5½ | 9 |
| 8 | 8 | 13/16 | 15 | 1 5/8 | 11 15/16 | 10 | 14 | 6 | 11 |
| 10 | 10 | 15/16 | 17½ | 1 7/8 | 14 1/16 | 11½ | 16½ | 7 | 12 |
| 12 | 12 | 1 | 20½ | 2 | 16 7/16 | 13 | 19 | 8 | 14 |
| 14 OD | 13¼ | 1 1/8 | 23 | 2 1/8 | 18 15/16 | 15 | 21½ | 8½ | 16 |
| 16 OD | 15¼ | 1¼ | 25½ | 2¼ | 21 1/16 | 16½ | 24 | 9½ | 18 |
| 18 OD | 17 | 1 3/8 | 28 | 2 3/8 | 23 5/16 | 18 | 26½ | 10 | 19 |
| 20 OD | 19 | 1½ | 30½ | 2½ | 25 9/16 | 19½ | 29 | 10½ | 20 |
| 24 OD | 23 | 1 5/8 | 36 | 2¾ | 30 5/16 | 22½ | 34 | 12 | 24 |

[a] ANSI B16.1-1967.

## 40 American National Standard Class 250 Cast-Iron Flanges, Drilling for Bolts and Their Lengths[a]

| Nominal Pipe Size | Diameter of Flange | Thickness of Flange, Min. | Diameter of Raised Face | Diameter of Bolt Circle | Diameter of Bolt Holes | Number of Bolts | Size of Bolts | Length of Bolts | Length of Bolt Studs with Two Nuts |
|---|---|---|---|---|---|---|---|---|---|
| 1 | 4⅞ | 11/16 | 2 11/16 | 3½ | ¾ | 4 | ⅝ | 2½ | ..... |
| 1¼ | 5¼ | ¾ | 3 1/16 | 3⅞ | ¾ | 4 | ⅝ | 2½ | ..... |
| 1½ | 6⅛ | 13/16 | 3 9/16 | 4½ | ⅞ | 4 | ¾ | 2¾ | ..... |
| 2 | 6½ | ⅞ | 4 3/16 | 5 | ¾ | 8 | ⅝ | 2¾ | ..... |
| 2½ | 7½ | 1 | 4 15/16 | 5⅞ | ⅞ | 8 | ¾ | 3¼ | ..... |
| 3 | 8¼ | 1⅛ | 5 11/16 | 6⅝ | ⅞ | 8 | ¾ | 3½ | ..... |
| 3½ | 9 | 1 3/16 | 6 5/16 | 7¼ | ⅞ | 8 | ¾ | 3½ | ..... |
| 4 | 10 | 1¼ | 6 15/16 | 7⅞ | ⅞ | 8 | ¾ | 3¾ | ..... |
| 5 | 11 | 1⅜ | 8 5/16 | 9¼ | ⅞ | 8 | ¾ | 4 | ..... |
| 6 | 12½ | 1 7/16 | 9 11/16 | 10⅝ | ⅞ | 12 | ¾ | 4 | ..... |
| 8 | 15 | 1⅝ | 11 15/16 | 13 | 1 | 12 | ⅞ | 4½ | ..... |
| 10 | 17½ | 1⅞ | 14 1/16 | 15¼ | 1⅛ | 16 | 1 | 5¼ | ..... |
| 12 | 20½ | 2 | 16 7/16 | 17¾ | 1¼ | 16 | 1⅛ | 5½ | ..... |
| 14 OD | 23 | 2⅛ | 18 15/16 | 20¼ | 1¼ | 20 | 1⅛ | 6 | ..... |
| 16 OD | 25½ | 2¼ | 21 1/16 | 22½ | 1⅜ | 20 | 1¼ | 6¼ | ..... |
| 18 OD | 28 | 2⅜ | 23 5/16 | 24¾ | 1⅜ | 24 | 1¼ | 6½ | ..... |
| 20 OD | 30½ | 2½ | 25 9/16 | 27 | 1⅜ | 24 | 1¼ | 6¾ | ..... |
| 24 OD | 36 | 2¾ | 30 5/16 | 32 | 1⅝ | 24 | 1½ | 7½ | 9½ |

[a] ANSI B16.1-1967.

# index

(NUMBERS LISTED ARE PAGE NUMBERS)